Second Edition

Handbook of Animal-Based Fermented Food and Beverage Technology

Second Edition

Handbook of
Animal-Based Fermented Food and Beverage Technology

Edited by Y. H. Hui

Administrative Associate Editor
E. Özgül Evranuz

Associate Editors
Ramesh C. Chandan
Luca Cocolin
Eleftherios H. Drosinos
Lisbeth Goddik
Ana Rodríguez
Fidel Toldrá

CRC Press
Taylor & Francis Group
Boca Raton London New York

CRC Press is an imprint of the
Taylor & Francis Group, an **informa** business

Previous edition was published with a different title: Handbook of Food and Beverage Fermentation Technology (ISBN 9780824747800).

CRC Press
Taylor & Francis Group
6000 Broken Sound Parkway NW, Suite 300
Boca Raton, FL 33487-2742

First issued in paperback 2019

ISBN-13: 978-1-4398-5022-0 (hbk)
ISBN-13: 978-1-138-37443-0 (pbk)

Library of Congress Cataloging-in-Publication Data

Handbook of animal-based fermented food and beverage technology / edited by Y. H. Hui, E. Özgül Evranuz.
    p. cm.
   Rev. ed. of: Handbook of food and beverage fermentation technology / edited by Y.H. Hui ... [et al.]. 2004.
   "Handbook of food and beverage fermentation technology" is the first edition of "Handbook of plant-based fermented food and beverage technology" and "Handbook of animal-based fermented food and beverage technology."
   Includes bibliographical references and index.
   ISBN 978-1-4398-5022-0 (hardback)
   1. Fermented foods--Handbooks, manuals, etc. 2. Beverages--Microbiology--Handbooks, manuals, etc. 3. Fermentation--Handbooks, manuals, etc. 4. Food of animal origin--Handbooks, manuals, etc. 5. Animal products--Handbooks, manuals, etc. I. Hui, Y. H. (Yiu H.) II. Evranuz, E. Özgül. III. Handbook of food and beverage fermentation technology.

TP371.44.H358 2012
664'.024--dc23
                                                2012000810

Visit the Taylor & Francis Web site at
http://www.taylorandfrancis.com

and the CRC Press Web site at
http://www.crcpress.com

Contents

Part I   Introduction

Part II   Fermented Milk and Semisolid Cheeses

Part III   Solid Cheeses

Part IV   Meat and Fish Products

Part V   Probiotics and Fermented Products

Preface

Fermented food is a very interesting category of food products. Not only can it be produced with inexpensive ingredients and simple techniques, it also makes a significant contribution to human diet, especially in rural households and village communities in many areas of the world. In every ethnic group in the world there are fermented foods produced from recipes handed down from generation to generation. Such food products play an important role in cultural identity, local economy, and gastronomical delight. With progresses in the biological and microbiological sciences, the manufacture of some of the more popular fermented food products has been commercialized while others are still produced at home or in small-scale industries using traditional methods.

Fermentation changes the initial characteristics of a food into a product that is significantly different but highly acceptable to consumers. Of course, consumer preference for fermented food varies within and between cultures. For example, in the United States, many consumers like sour milk, although some do not. The trend in North America is toward acceptance and preference of foreign fermented non-European food products. You can find Chinese fermented sausages and Philippine fermented fish sauces or pastes in most Asian markets and some major grocery chains in North America.

Although reference books on fermented foods have been in existence for at least 50 years, those with details on the science, technology, and engineering of food fermentation began to appear after 1980. Scientific literature in the past decade has been flooded with new applications of genetic engineering in the fermentation of food products, especially in the dairy field.

This book provides an up-to-date reference for fermented foods and beverages derived from animal products. Almost every book on food fermentation has something not found in others. The *Handbook of Animal-Based Fermented Food and Beverage Technology* provides a detailed background of history, microorganisms, quality assurance, and the manufacture of general fermented food products and discusses the production of six categories of fermented foods and beverages:

- Semisolid dairy products, for example, sour cream
- Solid dairy products, for example, cheese
- Meat products, for example, sausages
- Marine products, for example, fish sauce
- Probiotics involved in some fermented foods and beverages
- Food ingredients from fermentation

Traditional fermented products are discussed, including yogurt, cheese, sausage, and ham. We also present details of the manufacture and quality characteristics of some fermented foods that are less common in other English-language books on fermentation: fermented products from East Asia, for example, fermented fish, and fermented meat from Turkey. Although this book covers a variety of products, many topics are omitted for different reasons including space limitation, product selection, and the contributors' areas of expertise.

This book is unique in several aspects: it is an updated and comprehensive reference source, it contains topics not covered in similar books, and its contributors include experts from government, industry, and academia worldwide. The book has 44 chapters and is divided into five parts. It is the cooperative effort of more than 60 contributors from 15 countries (Argentina, Brazil, China, Finland, France, Greece, Himalaya, Hungary, India, Italy, Malaysia, Spain, Taiwan, Turkey, and the United States) with expertise in one or more fermented products, led by an editorial team of eight academicians from five countries (Greece, Italy, Spain, Turkey, and the United States). The associate editors of this team are all experts in their areas of specialization as indicated below:

- Dr. Ramesh C. Chandan is associated with Global Technologies of Coon Rapids, Minnesota. He is the author of *Manufacturing Yogurt and Fermented Milks*, among other works.
- Dr. Luca Cocolin is the editor-in-chief of *International Journal of Food Microbiology*, with expertise in food fermentation. He is with the University of Turin, Italy.
- Dr. Eleftherios H. Drosinos is the Head of Laboratory of Food Quality Control and Hygiene and Vice President of the Department of Food Science and Technology, Agricultural University of Athens, Greece. His expertise in meat and meat safety is documented in numerous professional publications.
- Dr. Lisbeth Goddik is an associate professor with the Department of Food Science and Technology at Oregon State University. She was one of the original associate editors for the first edition of this book.
- Dr. Ana Rodríguez is a staff scientist with the Department of Technology and Biotechnology of Dairy Products (Intituto de Productos Lácteos de Asturias, Consejo Superior de Investigaciones Científicas), Asturias, Spain. She is an expert in dairy starter cultures, as documented by her numerous professional publications in books and journals in science and technology.
- Dr. Fidel Toldrá is a staff scientist with the Intituto de Agroquimica y Tecnologïa de Alimentos, the Spanish National Research Council, Valencia, Spain. He has authored, coauthored, edited, and coedited many books in meat science and technology, including the *Handbook of Fermented Meat and Poultry*. He was one of the original associate editors for the first edition of this book.

In sum, the approach for this book makes it an essential reference on animal-based fermented foods and beverages. The editorial team thanks all the contributors for sharing their experience in their fields of expertise. They are the people who made this book possible. We hope you enjoy and benefit from the fruits of their labor.

We know how hard it is to develop the content of a book. However, we believe that the production of a professional book of this nature is even more difficult. We thank the production team at CRC Press. You are the best judge of the quality of this book.

Y. H. Hui
United States

E. Özgül Evranuz
Turkey

Editors

Y. H. Hui received his PhD from the University of California at Berkeley. He is currently a senior scientist with the consulting firm of Science Technology System at West Sacramento, California. He has authored, coauthored, edited, and coedited more than 30 books in food science, food technology, food engineering, and food laws, including *Handbook of Food Product Manufacturing* (2 volumes, Wiley, 2007) and *Handbook of Food Science, Technology, and Engineering* (4 volumes, CRC Press, 2006). He has guest-lectured in universities in the Americas (North, Central, and South), Europe, and Asia and is currently a consultant for the food industries, with an emphasis on food sanitation and food safety.

E. Özgül Evranuz is a professor (1994–present) of food processing at the Food Engineering Department of İstanbul Technical University (ITU), Turkey. After graduating from the Chemical Engineering Department of Middle East Technical University (METU; 1971), Ankara, Turkey, she has conveyed her academic career in the area of food technology. She holds MS degrees in Chemical Engineering (METU) and Food Science and Management (University of London, Queen Elizabeth College, London) and a PhD in Chemical Engineering (ITU). She is among the builders of the Food Engineering Department of ITU. Her main interests are preservation, storage, and packaging of food plant materials. She has served as an editor of a book and has several publications in the field.

Associate Editors

Ramesh C. Chandan a consultant in dairy food science and technology, has served for more than 50 years in various food companies including Unilever, Land O'Lakes, and General Mills. He has also served on the faculty of University of Nebraska–Lincoln and Michigan State University, where he taught dairy technology courses for several years. He has established expertise in the manufacture of fermented dairy foods, especially cheese and yogurt. His research publications center on microbiology of fermented dairy products and biochemistry of milk and lactic cultures. He has authored or edited seven books. More recently, he has coedited *Manufacturing Yogurt and Fermented Milks* (2006), *Dairy Processing & Quality Assurance* (2008), and *Dairy Ingredients for Food Processing* (2011).

Luca Cocolin is an associate professor at the University of Turin, Italy; an executive board member of the International Committee on Food Microbiology and Hygiene, part of the International Union of Microbiological Societies; and the editor-in-chief of *International Journal of Food Microbiology*. He is a member of the editorial board of *Applied and Environmental Microbiology*. He has coauthored more than 150 papers in both national and international journals. Cocolin is an expert in the field of food fermentations and more specifically in the application of culture-independent molecular methods for the study of the microbiota of fermented foods. He has published several papers on the microbiological aspects of fermented sausages considering specifically ecological aspects of lactic acid bacteria, coagulase-negative cocci, and yeasts.

Eleftherios H. Drosinos graduated from the Agricultural College of Athens and received his PhD in food microbiology from the School of Biology and Biochemistry, University of Bath, U.K. He is an associate professor and the head of the Laboratory of Food Quality Control and Hygiene, Department of

Food Science and Technology, Agricultural University of Athens. His research profile includes studies on biodiversity of autochthonous microbiota and safety of fermented sausages, application, and assessment of food safety management systems. He has participated in international projects on traditional fermented meat products. In his work, particular emphasis is placed on research with bacteriocinogenic strains and application of bacteriocins in meat products.

Lisbeth Goddik holds graduate degrees in food science from Cornell University and Oregon State University. She has held the position of extension dairy processing specialist at Oregon State University since 1999, where she teaches dairy processing classes, supervises graduate student research projects, and leads the OSU dairy foods extension program. Her research has focused on milk quality, cheese technology, and economics of artisan cheese processing. She holds two professorships endowed by the dairy industry: the ODI-Bodyfelt Professorship and the Paul and Sandra Arbuthnot Professorship. Prior to her current position, she worked in Denmark, Norway, France, New Zealand, and Canada.

Ana Rodríguez is a senior scientist at the Instituto de Productos Lácteos de Asturias (IPLA-CSIC), where she has been working on biotechnology of dairy products for 23 years. In 1984, she earned her PhD degree in biology (minor in microbiology) at the University of Oviedo, working in the research project *Cycle of Development of the Actinophage φC31 Infecting Streptomyces sp.* Her research activities at IPLA-CSIC focused on the potential of wild lactic acid strains to be used as cheese starter strains (physiology and metabolism of strains, bacteriophage resistance, exopolysaccharide production, or antimicrobial activity). At present, biopreservation has become the main topic of her research. Bacteriocin-producing strains have been used as protective starters in cheese production. Apart from bacteriocins, "bacteriophages infecting pathogens" and "enzybiotics" are being studied to be exploited as biopreservatives in dairy products. She is the author of over 70 scientific and technique articles and several book chapters. She has been the supervisor of PhD and master's projects.

Fidel Toldrá is a research professor at the Instituto de Agroquímica y Tecnología de Alimentos (CSIC). He has filed 11 patents, authored 1 book, edited/coedited more than 30 books for major U.S. publishers, and published over 200 manuscripts in SCI journals and 90 chapters of books. He received the International Prize for Meat Science and Technology (International Meat Secretariat, 2002) and the Distinguished Research Award from the American Meat Science Association (2010). Dr. Toldrá is the European editor of *Trends in Food Science and Technology* (Elsevier), the Editor-in-Chief of *Current Nutrition & Food Science* (2005–11, Bentham), and a member of the editorial board of 9 journals. He was elected as a fellow of the International Academy of Food Science and Technology (2008) and a fellow of the Institute of Food Technologists (2009). He is a member of the Scientific Panel on Food Additives (2003–2008) and on food flavorings, enzymes, processing aids, and contact materials (2008–2014) at the European Food Safety Authority.

Contributors

Shantanu Agarwal
Dairy Research Institute
Rosemont, Illinois

Levent Akkaya
Food Hygiene and Technology Department
Afyon Kocatepe University
Afyonkarahisar, Turkey

Diana Ansorena
Faculty of Pharmacy
University of Navarra
Pamplona, Navarra, Spain

Lucia Aquilanti
Dipartimento di Scienze Agrarie, Alimentari ed
   Ambientali
Università Politecnica delle Marche
Ancona, Italy

Emiliane Andrade Araújo
Instituto de Ciências Agrárias
Universidade Federal de Viçosa
Rio Paranaíba, Minas Gerais, Brazil

R. Arenas
Facultad de Veterinaria
Universidad de León
Leon, Spain

Iciar Astiasarán
Faculty of Pharmacy
University of Navarra
Pamplona, Navarra, Spain

Marta Ávila
Departamento de Tecnología de Alimentos
Instituto Nacional de Investigación y Tecnología
   Agraria y Alimentaria (INIA)
Madrid, Spain

Jose Maria Fresno Baro
Facultad de Veterinaria
Universidad de León
Leon, Spain

Catherine Béal
AgroParisTech, Institut National de la Recherche
   Agronomique (INRA)
UMR Génie et Microbiologie des Procédés
   Alimentaires
Thiverval-Grignon, France

K. Bülent Belibağlı
Department of Food Engineering
Faculty of Engineering
University of Gaziantep
Gaziantep, Turkey

Hüseyin Bozkurt
Department of Food Engineering
Faculty of Engineering
University of Gaziantep
Gaziantep, Turkey

Flávia Carolina Alonso Buriti
Embrapa Goats and Sheep
Sobral, Ceará, Brazil

Domenico Carminati
Fodder and Dairy Productions Research Center
Lodi, Italy

Antônio Fernandes de Carvalho
Departamento de Tecnologia de Alimentos,
   Centro de Ciências Exatas
Universidade Federal de Viçosa
Viçosa, Minas Gerais, Brazil

Gisèle Chammas
UMR Génie et Microbiologie des Procédés
   Alimentaires
Thiverval-Grignon, France

Ramesh C. Chandan
Global Technologies, Inc.
Coon Rapids, Minnesota

Ming-Ju Chen
Department of Animal Science and Technology
National Taiwan University
Taipei, Taiwan, Republic of China

Stephanie Clark
Department of Food Science and Human
   Nutrition
Iowa State University
Ames, Iowa

Francesca Clementi
Dipartimento di Scienze Agrarie,
   Alimentari ed Ambientali
Università Politecnica delle Marche
Ancona, Italy

Luca Cocolin
Di.Va.P.R.A. Faculty of Agriculture
University of Turin
Torino, Italy

Pietrino Deiana
Dipartimento di Scienze Ambientali Agrarie e
   Biotecnologie Agro-alimentari
Università degli Studi di Sassari
Sassari, Italy

Eleftherios H. Drosinos
Laboratory of Food Quality Control and Hygiene,
   Department of Food Science and Technology
Agricultural University of Athens
Athens, Greece

D. Fernández
Facultad de Veterinaria
Universidad de León
Leon, Spain

R. E. Ferrazza
Departamento de Agroindustria de Alimentos
Universidade Federal de Pelotas
Rio Grande do Sul, Brazil

Sonia Garde
Departamento de Tecnología de Alimentos
Instituto Nacional de Investigación y Tecnología
   Agraria y Alimentaria (INIA)
Madrid, Spain

Cristiana Garofalo
Dipartimento di Scienze Agrarie,
   Alimentari ed Ambientali
Università Politecnica delle Marche
Ancona, Italy

Giorgio Giraffa
Fodder and Dairy Productions Research Center
Lodi, Italy

Lisbeth Goddik
Department of Food Science and Technology
Oregon State University
Corvallis, Oregon

Veli Gök
Food Engineering Department
Afyon Kocatepe University
Afyonkarahisar, Turkey

Denis Hemme
Previously a member of Institut National de la
   Recherche Agronomique
INRA Research Center
Jouy-en-Josas, France

Claudio Hidalgo-Cantabrana
Instituto de Productos Lácteos de Asturias-
   Consejo Superior de Investigaciones
   Científicas
Asturias, Spain

Nurul Huda
School of Industrial Technology
Universiti Sains Malaysia
Minden, Malaysia

Y. H. Hui
Science Technology System
West Sacramento, California

Kálmán Incze
Hungarian Meat Research Institute
Budapest, Hungary

Vesa Joutsjoki
MTT Agrifood Research Finland
Biotechnology and Food Research
Jokioinen, Finland

Minna Kahala
MTT Agrifood Research Finland
Biotechnology and Food Research
Jokioinen, Finland

Andrew Laws
Department of Chemical and Biological Sciences
University of Huddersfield
Huddersfield, United Kingdom

Evanthia Litopoulou-Tzanetaki
Faculty of Agriculture
Aristotle University of Thessaloniki
Thessaloniki, Greece

Nicoletta Pasqualina Mangia
Dipartimento di Scienze Ambientali Agrarie e
    Biotecnologie Agro-alimentari
Università degli Studi di Sassari
Sassari, Italy

Abelardo Margolles
Instituto de Productos Lácteos de Asturias-Consejo
    Superior de Investigaciones Científicas
Asturias, Spain

Beatriz Martínez
Instituto de Productos Lácteos de Asturias
Consejo Superior de Investigaciones Cientificas
Asturias, Spain

K. R. Nauth
Nauth Consulting Inc.
Bartlett, Illinois

Manuel Nuñez
Departamento de Tecnología de Alimentos
Instituto Nacional de Investigación y Tecnología
    Agraria y Alimentaria (INIA)
Madrid, Spain

G.-J. E. Nychas
Laboratory of Food Microbiology and Biotechnology
Department of Food Science and Technology
Agricultural University of Athens
Athens, Greece

Ersel Obuz
Food Engineering Department
Celal Bayar University
Muradiye, Manisa, Turkey

Andrea Osimani
Dipartimento di Scienze Agrarie, Alimentari ed
    Ambientali
Università Politecnica delle Marche
Ancona, Italy

Maria Papagianni
School of Veterinary Medicine
Aristotle University of Thessaloniki
Thessaloniki, Greece

E. C. Pappa
National Agricultural Research Foundation
Dairy Research Institute
Katsikas, Ioannina, Greece

S. Paramithiotis
Laboratory of Food Quality Control and Hygiene
Department of Food Science and Technology
Agricultural University of Athens
Athens, Greece

Antonia Picon
Departamento de Technología de Alimentos
Instituto Nacional de Investigación y Tecnología
    Agraria y Alimentaria (INIA)
Madrid, Spain

Maximiliano Soares Pinto
Instituto de Ciências Agrárias
Universidade Federal de Minas Gerais
Montes Claros, Minas Gerais, Brazil

Anna Polychroniadou-Alichanidou
Aristotle University of Thessaloniki
Thessaloniki, Greece

Kalliopi Rantsiou
Di.Va.P.R.A. Faculty of Agriculture
University of Turin
Torino, Italy

Ana Rodríguez
Instituto de Productos Lácteos de Asturias
Asturias, Spain

Patricia Ruas-Madiedo
Instituto de Productos Lácteos de Asturias-
    Consejo Superior de Investigaciones
    Científicas
Asturias, Spain

Amelia C. Rubiolo
Instituto de Desarrollo Tecnológico para la
    Industria Química
Universidad Nacional del Litoral
Santa Fe, Argentina

Susana Marta Isay Saad
Faculty of Pharmaceutical Sciences
University of São Paulo
São Paulo, Brazil

Borja Sánchez
Instituto de Productos Lácteos de Asturias-
  Consejo Superior de Investigaciones
  Científicas
Asturias, Spain

Ana Clarissa dos Santos Pires
Departamento de Tecnologia de Alimentos,
  Centro de Ciências Exatas
Universidade Federal de Viçosa
Viçosa, Minas Gerais, Brazil

Robert Scott
Danville Area Community College
Danville, Illinois

Guillermo A. Sihufe
Instituto de Desarrollo Tecnológico para la
  Industria Química
Universidad Nacional del Litoral
Santa Fe, Argentina

Cínthia Hoch Batista de Souza
Faculty of Pharmaceutical Sciences
University of São Paulo
São Paulo, Brazil

Juan E. Suárez
Área de Microbiología
Unidad Asociada al CSIC
Universidad de Oviedo
Oviedo, Spain

Jyoti Prakash Tamang
Department of Microbiology
Sikkim University
Tadong, Sikkim, India

Chung-Miao Tien
Wan-Yu-Chiuan Food Co.
Taipei City, Taiwan, Republic of China

Fidel Toldrá
Instituto de Agroquímica y Tecnología de
  Alimentos (CSIC)
Valencia, Spain

M. E. Tornadijo
Facultad de Veterinaria
Universidad de León
Leon, Spain

Rung-Jen Tu
Livestock Research Institute
Council of Agriculture, Executive Yuan
Tainan, Taiwan, Republic of China

Hsiang-Yun Wu
Livestock Research Institute
Council of Agriculture, Executive Yuan
Tainan, Taiwan, Republic of China

G. K. Zerfiridis
School of Agriculture
Aristotle University of Thessaloniki
Thessaloniki, Greece

Heping Zhang
School of Food Science and Engineering
Inner Mongolia Agricultural University
Huhhot, China

Wenyi Zhang
School of Food Science and Engineering
Inner Mongolia Agricultural University
Huhhot, China

Susana E. Zorrilla
Instituto de Desarrollo Tecnológico para la
  Industria Química
Universidad Nacional del Litoral
Santa Fe, Argentina

Part I

Introduction

1

Fermented Animal Products and Their Manufacture

Y. H. Hui

CONTENTS

1.1 Introduction

The availability of fermented foods has a long history among different cultures. Acceptability of fermented foods also differs because of cultural habits. A product highly acceptable in one culture may not be so acceptable by consumers in another culture. The number of fermented food products is countless. Manufacturing processes of fermented products vary considerably owing to variables such as food groups, form and characteristics of final products, kind of ingredients used, and cultural diversity. It is beyond the scope of this chapter to address all the manufacturing processes used to produce animal-based fermented foods. Instead, this chapter is organized to address fermented food products based on food groups such as dairy, meat, and fish. Within each food group, manufacturing processes of typical products are addressed. This chapter is only an introduction of manufacturing processes of selected fermented food products. Readers should consult the references at the end of this chapter and other available literature for detailed information.

1.2 Fermented Dairy Products

1.2.1 Ingredients and Kinds of Products

Fermented dairy products are commonly produced in milk-producing countries and by nomadic people. These products are highly acceptable in these cultures. They have been gradually accepted by other cultures because of cultural exchange. It is generally accepted that most fermented dairy products were first discovered and developed by nomadic people. The production of a fermented dairy product nowadays can be a highly sophisticated process. However, the production of another fermented dairy product can still be conducted in a fairly primitive manner in another location. The quality of a fermented dairy product varies due to the milk, microorganisms, and other ingredients used in the manufacturing process. Many factors affect the gross composition of milk (Early 1998; Jenness 1988; Kosikowski and Mistry 1997; Spreer 1998; Walstra et al. 1999). The factors most significant to the processing of milk products are breed, feed, season, region, and herb health. Reviews of animals' milk are available in the literature. Table 1.1 lists the approximate composition of cow's milk (Early 1998; Jenness 1988; Kosikowski and Mistry 1997; Robinson 1986; Spreer 1998; Walstra et al. 1999). In industrial countries, milk composition is standardized to meet a country's requirements. However, it is understood that the requirements in one country may not be the same as those in another; thus, the composition may vary for the same product. International agreements to standardize some products are now available. However, products produced in different locations still can vary because of microorganisms and culturing practices used in their production.

Fermented dairy products can be grossly divided into three big categories—cheeses, yogurts, and fermented liquid milks. Within each of these categories, there are subcategories. Table 1.2 presents

TABLE 1.1

Approximate Composition of Milk

Components	Average Content in Milk (% w/w)	Range (% w/w)	Average Content in Dry Matter (% w/w)
Water	87.1	85.3–88.7	
Solids-not-fat	8.9	7.9–10.0	−69
Fat in dry matter	31	22–38	−31
Lactose	4.6	3.8–5.3	36
Fat	4	2.5–5.5	31
Protein	3.25	2.3–4.4	25
Casein	2.6	1.7–3.5	20
Mineral substances	0.7	0.57–0.83	5.4
Organic acids	0.17	0.12–0.21	1.3
Miscellaneous	0.15		1.2

Source: Early, R., editor, *The Technology of Dairy Products*, Blackie Academic & Professional, London, 1998. With permission. Jenness, R., in *Fundamentals of Dairy Chemistry*, 3rd edition, edited by N.P. Wang et al., Kluwer Academic Publishers, New York, pp. 1–38, 1988. With permission. Kosikowski, F.V. and Mistry, V.V., *Cheese and Fermented Milk Foods*, 3rd edition, Vols. I and II, F.V. Kosikowski, Westport, CT, 1997. With permission. Reprinted from *Modern Dairy Technology*, Vol. 2, Robinson, R.K., editor, Copyright 1986, with permission from Elsevier. Spreer, E. (A. Mixa translator), *Milk and Dairy Technology*, Marcel Dekker, Inc., New York, 1998. With permission. Walstra, P. et al., *Dairy Technology: Principles of Milk Properties and Processes*, Marcel Dekker, Inc., New York, 1999. With permission.

TABLE 1.2

Kinds of Fermented Dairy Products with Examples

Kinds	Examples
Fermented Liquid Milks	
Lactic fermentation	Buttermilk, Acidophilus
With alcohol and lactic acid	Kefir, Koumiss
With mold and lactic acid	Villi
Concentrated	Ymer, Skyr, Chakka
Yogurts	
Viscous/liquid	Yogurt
Semisolid	Strained yogurt
Solid	Soft/hard frozen yogurt
Powder	Dried yogurt
Cheeses	
Extra hard	Parmesan, Romano, Sbrinz
Hard with eyes	Emmental, Gruyere, Swiss
Hard without eyes	Cheddar, Chester, Provolone
Semihard	Gouda, Edam, Caerphilly
Semihard, internally mold ripened	Roquefort, Blue, Gorgonzola
Semisoft, surface ripened with bacteria	Limburger, Brick, Munster
Soft, surface mold ripened	Brie, Camembert, Neufchatel
Soft, unripened	Cream, Mozzarella, US-Cottage

Source: Early, R., editor, *The Technology of Dairy Products*, Blackie Academic & Professional, London, 1998. With permission. Jenness, R., in *Fundamentals of Dairy Chemistry*, 3rd edition, edited by N.P. Wang et al., Kluwer Academic Publishers, New York, pp. 1–38, 1988. With permission. Kosikowski, F.V. and Mistry, V.V., *Cheese and Fermented Milk Foods*, 3rd edition, Vols. I and II, F.V. Kosikowski, Westport, CT, 1997. With permission. Reprinted from *Modern Dairy Technology*, Vol. 2, Robinson, R.K., editor, Copyright 1986, with permission from Elsevier. Spreer, E. (A. Mixa translator), *Milk and Dairy Technology*, Marcel Dekker, Inc., New York, 1998. With permission. Walstra, P. et al., *Dairy Technology: Principles of Milk Properties and Processes*, Marcel Dekker, Inc., New York, 1999. With permission.

examples for each of these categories (Early 1998; Jenness 1988; Kosikowski and Mistry 1997; Robinson 1986; Spreer 1998; Walstra et al. 1999).

In the manufacturing of fermented dairy products, various ingredients such as the milk itself, microorganism(s), coagulants, salt, sugar, vitamins, buffering salts, bleaching (decolorizing) agents, dyes (coloring agents), flavoring compounds, stabilizers, and emulsifiers may be used. The use of these ingredients in fermented liquid milks, yogurts, and natural and processed cheeses is summarized in Table 1.3 (Early 1998; Jenness 1988; Kosikowski and Mistry 1997; Robinson 1986; Spreer 1998; Walstra et al. 1999).

Various microorganisms such as lactic acid bacteria, yeasts, and molds are used in the manufacturing of fermented dairy products to give the various characteristics in these products. Table 1.4 lists some of the more common dairy microorganisms and their uses in fermented liquid dairy products, yogurts, and cheeses (Davies and Law 1984; Jay 1996; Robinson 1990).

Cultures of the different microorganisms are available in various forms such as liquid, frozen, or freeze-dried. Examples of their usage in the manufacturing of fermented dairy products are listed in Table 1.5 (Early 1998; Jenness 1988; Kosikowski and Mistry 1997; Robinson 1986; Spreer 1998; Walstra et al. 1999).

Because the starter cultures are available in various forms, the preparation steps of these cultures before inoculation are different. Table 1.6 lists some of the preparation procedures used in the industry

TABLE 1.3

Ingredients for Fermented Dairy Food Production

Ingredients	Fermented Liquid Milk Products	Yogurt	Natural Cheese	Processed Cheese Products
Milk				
Raw	Optional	Optional	Optional	Optional
Standardized (fat and milk solids)	Preferred	Preferred	Preferred	Preferred
Milk powders	Optional	Optional	Optional	Optional
Microorganisms				
Starter bacteria	Required	Required	Required	Required
Mold	Optional	Optional	Optional	Optional
Yeast	Optional	Optional	Optional	Optional
Genetically modified microorganisms	Optional	Optional	Optional	Optional
Coagulant				
Rennet	Preferred	Preferred	Preferred	Preferred
Acid	Optional	Optional	Optional	Optional
Microbial protease(s)	Optional	Optional	Optional	Optional
Common Salt				
Sodium chloride	No	No	Required	Required
Sugar	Optional	Optional	No	No
Vitamins	Preferred	Preferred	Preferred	Preferred
Buffering Salts				
Calcium chloride, hydroxide phosphates, sodium, or potassium phosphates	Optional	Optional	Optional	Optional
Bleaching (Decolorizing) Agents	No	No	Optional	Optional
Antimicrobial Agents	Optional	Optional	No	Preferred
Dyes (Coloring Agents)	No	No	Optional	Optional
Flavoring Compounds				
Fruits, spices, spice oils, fruits, fruit flavors, artificial smoke	Optional	Optional	Optional	Optional
Stabilizers	No	Preferred	No	Preferred
Emulsifiers	Optional	Optional	No	Preferred

Source: Early, R., editor, *The Technology of Dairy Products*, Blackie Academic & Professional, London, 1998. With permission. Jenness, R., in *Fundamentals of Dairy Chemistry*, 3rd edition, edited by N.P. Wang et al., Kluwer Academic Publishers, New York, pp. 1–38, 1988. With permission. Kosikowski, F.V. and Mistry, V.V., *Cheese and Fermented Milk Foods*, 3rd edition, Vols. I and II, F.V. Kosikowski, Westport, CT, 1997. With permission. Reprinted from *Modern Dairy Technology*, Vol. 2, Robinson, R.K., editor, Copyright 1986, with permission from Elsevier. Spreer, E. (A. Mixa translator), *Milk and Dairy Technology*, Marcel Dekker, Inc., New York, 1998. With permission. Walstra, P. et al., *Dairy Technology: Principles of Milk Properties and Processes*, Marcel Dekker, Inc., New York, 1999. With permission.

for different forms of starter cultures (Early 1998; Jenness 1988; Kosikowski and Mistry 1997; Robinson 1986; Spreer 1998; Walstra et al. 1999).

Different microorganisms have different temperature requirements for their optimum growth and functioning. Some fermented dairy products such as mold-ripen cheeses may require more than one microorganism to complete the manufacturing process. These molds function best during the long ripening period and, therefore, have standard incubation temperatures in the refrigerated range. This is also true for some cheeses that require long ripening periods. Microorganisms requiring higher incubation

TABLE 1.4

Some Common Organisms Used in Fermented Milk Products

Microorganisms	Buttermilk	Cream	Fermented Milk	Yogurt	Kefir	Cheese
Bifidobacterium bifidum			X	X		X
Enterococcus durans						X
Enterococcus faecalis						X
Geotrichum candidum						X
Lactobacillus acidophilus			X			
Lactobacillus casei						X
Lactobacillus delbrueckii subsp. *bulgaricus*	X	X				X
Lactobacillus helveticus						X
Lactobacillus kefir					X	
Lactobacillus lactis						X
Lactobacillus lactis biovar. *diacetylactis*		X				X
Lactobacillus lactis subsp. *cremoris*	X	X				X
Lactobacillus lactis subsp. *lactis*						X
Lactobacillus lactis var. *hollandicus*						X
Leuconostoc mesenteroides subsp. *cremoris*						X
Leuconostoc mesenteroides subsp. *dextranicum*						X
Propionibacterium freudenreichii subsp. *shermanii*						X
Penicillium camemberti						X
Penicillium glaucum						X
Penicillium roqueforti						X
Streptococcus thermophilus				X		X

Source: Reprinted from *Advances in the Microbiology and Biochemistry of Cheese and Fermented Milk*, Davies, F.L. and Law, B.A., editors, Copyright 1984, with permission from Elsevier. Jay, J.M., *Modern Food Microbiology*, Chapman & Hall, New York, 1996. With permission. Robinson, R.K., editor, *Dairy Microbiology*, Vols. 1 and 2, Applied Science, London, 1990. With permission.

temperatures are used in the production of fermented liquid milks that require only a short incubation time. Table 1.7 lists some of the dairy microorganisms used in some products and their incubation temperatures (Davies and Law 1984; Emmons 2000; Jay 1996; Law 1997; Robinson 1990; Nath 1993; Scott et al. 1998; Specialist Cheesemakers Association 1997).

1.2.2 Cheeses

Cheeses can be classified into different categories based on their moisture, the way the milk is processed, and the types of microorganisms used for the ripening process (Table 1.8; Early 1998; Jenness 1988; Robinson 1986).

TABLE 1.5

Dairy Starter Cultures

Physical Form	Usage
Liquid cultures in skim milk or whole milk (antibiotic free)	For inoculation of intermediate cultures
Liquid culture (frozen)	For inoculation of intermediate cultures
	For inoculation into bulk cultures
Dried culture (from normal liquid culture)	For inoculation of intermediate culture
Spray-dried cultures	For inoculation into bulk cultures
	For direct-to-vat inoculation
Frozen cultures in special media (frozen at −40°C)	For inoculation into bulk cultures
	For direct-to-vat inoculation
Frozen concentrated culture (in sealed containers at −196°C)	For inoculation into bulk cultures
	For direct-to-vat inoculation
Single strain lyophilized cultures (in foil sachets with known activity)	For inoculation into bulk cultures
	For direct-to-vat inoculation

Source: Early, R., editor, *The Technology of Dairy Products*, Blackie Academic & Professional, London, 1998. With permission. Jenness, R., in *Fundamentals of Dairy Chemistry*, 3rd edition, edited by N.P. Wang et al., Kluwer Academic Publishers, New York, pp. 1–38, 1988. With permission. Kosikowski, F.V. and Mistry, V.V., *Cheese and Fermented Milk Foods*, 3rd edition, Vols. I and II, F.V. Kosikowski, Westport, CT, 1997. With permission. Reprinted from *Modern Dairy Technology*, Vol. 2, Robinson, R.K., editor, Copyright 1986, with permission from Elsevier. Spreer, E. (A. Mixa translator), *Milk and Dairy Technology*, Marcel Dekker, Inc., New York, 1998. With permission. Walstra, P. et al., *Dairy Technology: Principles of Milk Properties and Processes*, Marcel Dekker, Inc., New York, 1999. With permission.

TABLE 1.6

Types of Starter Cultures and Their Preparation Prior to Usage

Kinds	Preparation Steps	Timing
Regular starter culture	Preparation of starter culture blanks	8:00 a.m.
	Storing milk blanks	11:00 a.m.
	Activating lyophilized culture powder	3:00 p.m.
	Daily mother culture preparation	3:00 p.m.
	Semi-bulk and bulk starter preparation	3:00 p.m.
Frozen culture and bulk starter application	Store frozen culture at −40°C or less	
	Warm to 31°C and use directly	
Reconstituted milk or whey-based starter	Reconstitution	8:00 a.m.
	Heating and tempering	8:30 a.m.
	Inoculating and incubating	10:00 a.m.
Bulk starter from ultrafiltered milk	Ultrafiltration	1:00 p.m.
	Heating and tempering	3:30 p.m.
	Inoculating and incubating	5:00 p.m.

Source: Early, R., editor, *The Technology of Dairy Products*, Blackie Academic & Professional, London, 1998. With permission. Jenness, R., in *Fundamentals of Dairy Chemistry*, 3rd edition, edited by N.P. Wang et al., Kluwer Academic Publishers, New York, pp. 1–38, 1988. With permission. Kosikowski, F.V. and Mistry, V.V., *Cheese and Fermented Milk Foods*, 3rd edition, Vols. I and II, F.V. Kosikowski, Westport, CT, 1997. With permission. Reprinted from *Modern Dairy Technology*, Vol. 2, Robinson, R.K., editor, Copyright 1986, with permission from Elsevier. Spreer, E. (A. Mixa translator), *Milk and Dairy Technology*, Marcel Dekker, Inc., New York, 1998. With permission. Walstra, P. et al., *Dairy Technology: Principles of Milk Properties and Processes*, Marcel Dekker, Inc., New York, 1999. With permission.

TABLE 1.7

Temperature Requirements and Acid Production for Some Dairy Microbes

Microorganisms	Product Group	Standard Temperature for Incubation (°C)	General Maximum Titratable Acidity Produced in Milk (%)
Bacteria			
Bifidobacterium bifidum	1	36–38	0.9–1.0
Lactobacillus acidophilus	1	38–44	1.2–2.0
Lactobacillus delbrueckii subsp. bulgaricus	1	43–47	2.0–4.0
Lactobacillus lactis subsp. cremoris	2	22	0.9–1.0
Lactobacillus subsp. *lactis*	2	22	0.9–1.0
Leuconostoc mesenteroides subsp. *cremoris*	2	20	0.1–0.3
Streptococcus durans	2	31	0.9–1.1
Streptococcus thermophilus	2	38–44	0.9–1.1
Molds			
Penicillium roqueforti	3	11–16	NA
Penicillium camemberti	3	10–22	NA

Source: Reprinted from *Advances in the Microbiology and Biochemistry of Cheese and Fermented Milk*, Davies, F.L. and Law, B.A., editors, Copyright 1984, with permission from Elsevier. Emmons, D.B., *Practical Guide for Control of Cheese Yield*, International Dairy Federation, Brussels, Belgium, 2000. With permission. Scott, R. et al., *Cheese Making Practice*, Chapman & Hall, New York, 1998. With permission. Jay, J.M., *Modern Food Microbiology*, Chapman & Hall, New York, 1996. With permission. Law, B.A., editor, *Microbiology and Biochemistry of Cheese and Fermented Milk*, Blackie Academic & Professional, London, 1997. With permission. Nath, K.R., in *Dairy Science and Technology Handbook*, Vol. 2, edited by Y.H. Hui, VCH Publishers, Inc., New York, pp. 161–255, 1993. With permission. Robinson, R.K., editor, *Dairy Microbiology*, Vols. 1 and 2, Applied Science, London, 1990. With permission. Specialist Cheesemakers Association, *The Specialist Cheesemakers: Code of Best Practice*, Specialist Cheesemakers Association, Staffordshire, England, Great Britain, 1997. With permission.

Note: Product group—1: yogurt, 2: fermented liquid milk, 3: cheese.

TABLE 1.8

Classification of Cheese according to Moisture Content, Scald Temperature, and Method of Ripening

Hard Cheese (Moisture 20%–42%; Fat in Dry Matter, 32%–50%, Min.)			
Low scald, lactic starter	Medium scald, lactic starter	High scald, propionic eyes	Plastic curd, lactic starter, or propionic eyes
Gouda	Cheddar	Parmesan	Provolone
Cheshire	Svecia	Beaufort	Mozzarella

Semihard Cheese (Moisture 45%–55%; Fat in Dry Matter, 40%–50%, Min.)		
Lactic starter	Smear coat	Blue-veined mold
St. Paul	Limburg	Roquefort
Lancaster	Munster	Danablue

Soft Cheese (Moisture > 55%; Fat in Dry Matter, 4%–51%, Min.)				
Acid coagulated	Smear coat or surface mold	Surface mold	Normal lactic starter	Unripened fresh
Cottage cheese (USA)	Brie	Camembert	Quarg	Cottage (UK)
Queso Blanco	Bel Paese	Neufchatel	Petit Suisse	York

Source: Early, R., editor, *The Technology of Dairy Products*, Blackie Academic & Professional, London, 1998. With permission. Jenness, R., in *Fundamentals of Dairy Chemistry*, 3rd edition, edited by N.P. Wang et al., Kluwer Academic Publishers, New York, pp. 1–38, 1988. With permission. Reprinted from *Modern Dairy Technology*, Vol. 2, Robinson, R.K., editor, Copyright 1986, with permission from Elsevier.

In the processing of cheese, the amount of curd used for each block of cheese to be made differs considerably, thus producing different block weights (Table 1.9; Early 1998; Jenness 1988; Kosikowski and Mistry 1997; Law 1997; Nath 1993; Robinson 1986; Scott et al. 1998; Specialist Cheesemakers Association 1997). Harder cheeses have much larger blocks than the soft cheeses. This may be due to the ease of handling after ripening.

Cheeses are packaged in different forms to satisfy consumer consumption patterns and, to some extent, be compatible with the way the cheese is ripened and for marketing purposes. The various packaging materials are selected to protect the cheeses and preserve their sanitary condition, extend their shelf life, and delay the deterioration of the final products. Table 1.10 lists some of the requirements of cheese packaging materials (Early 1998; Emmons 2000; Kosikowski and Mistry 1997; Nath 1993; Robinson 1986; Scott et al. 1998; Specialist Cheesemakers Association 1997; Spreer 1998; Walstra et al. 1999).

All cheeses produced must be coagulated from acceptable milk to form the curd followed by removal of the whey. Most cheeses are made from standardized and pasteurized milk. Nonpasteurized milk is also used in some exceptional cases provided that they do not carry pathogens. The majority of cheeses

TABLE 1.9

Approximate Weight of Cheese Varieties

Cheese Variety	Approximate Weight (kg)
Hard to Semihard or Semisoft	
Wensleydale	3–5
Caerphilly	3–6
White Stilton	4–8
Single Gloucester	10–12
Leicester	13–18
Derby	14–16
Sage Derby	14–16
Cheddar	18–28
Cheshire	20–22
Dunlap	20–27
Double Gloucester	22–28
Lancashire	22
Internally Mold-Ripen (Blue-Veined) Cheese	
Blue Wensleydale	3–5
Blue Vinney	5–7
Blue Stilton	6–8
Blue Cheshire	10–20
Soft Cheese	
Colwich	0.25–0.50
Cambridge	0.25–1.0
Melbury	2.5

Source: Early, R., editor, *The Technology of Dairy Products*, Blackie Academic & Professional, London, 1998. With permission. Jenness, R., in *Fundamentals of Dairy Chemistry*, 3rd edition, edited by N.P. Wang et al., Kluwer Academic Publishers, New York, pp. 1–38, 1988. With permission. Kosikowski, F.V. and Mistry, V.V., *Cheese and Fermented Milk Foods*, 3rd edition, Vols. I and II, F.V. Kosikowski, Westport, CT, 1997. With permission. Law, B.A., editor, *Microbiology and Biochemistry of Cheese and Fermented Milk*, Blackie Academic & Professional, London, 1997. With permission. Reprinted from *Modern Dairy Technology*, Vol. 2, Robinson, R.K., editor, Copyright 1986, with permission from Elsevier. Scott, R. et al., *Cheese Making Practice*, Chapman & Hall, New York, 1998. With permission. Specialist Cheesemakers Association, *The Specialist Cheesemakers: Code of Best Practice*, Specialist Cheesemakers Association, Staffordshire, England, Great Britain, 1997. With permission.

TABLE 1.10

Requirements of Cheese Packaging Materials

1. Low permeability to oxygen, carbon dioxide, and water vapor
2. Strength and thickness of film
3. Stability under cold or warm conditions
4. Stability to fats and lactic acid
5. Resistance to light, especially ultraviolet
6. Ease of application, stiffness, and elasticity
7. Ability to seal and accept adhesives
8. Laminated films to retain lamination
9. Low shrinkage or aging unless shrinkage is a requisite
10. Ability to take printed matter
11. Should not impart odors to the cheese
12. Suitability for mechanization of packaging
13. Hygienic considerations in storage and use
14. Cost effectiveness as a protective wrapping

Source: Early, R., editor, *The Technology of Dairy Products*, Blackie Academic & Professional, London, 1998. With permission. Emmons, D.B., *Practical Guide for Control of Cheese Yield*, International Dairy Federation, Brussels, Belgium, 2000. With permission. Kosikowski, F.V. and Mistry, V.V., *Cheese and Fermented Milk Foods*, 3rd edition, Vols. I and II, F.V. Kosikowski, Westport, CT, 1997. With permission. Nath, K.R., in *Dairy Science and Technology Handbook*, Vol. 2, edited by Y.H. Hui, VCH Publishers, Inc., New York, pp. 161–255, 1993. With permission. Reprinted from *Modern Dairy Technology*, Vol. 2, Robinson, R.K., editor, Copyright 1986, with permission from Elsevier. Scott, R. et al., *Cheese Making Practice*, Chapman & Hall, New York, 1998. With permission. Specialist Cheesemakers Association, *The Specialist Cheesemakers: Code of Best Practice*, Specialist Cheesemakers Association, Staffordshire, England, Great Britain, 1997. With permission. Spreer, E. (A. Mixa translator), *Milk and Dairy Technology*, Marcel Dekker, Inc., New York, 1998. With permission. Walstra, P. et al., *Dairy Technology: Principles of Milk Properties and Processes*, Marcel Dekker, Inc., New York, 1999. With permission.

are made from cow's milk. Milks from other animals are also used for specialty products. The coagulation process is conducted through the addition of the coagulant (rennin or chymosin) and incubation of appropriate lactic acid bacteria in milk to produce enough acid and appropriate pH for curdling of the milk. After the casein is recovered, it is salted and subject to fermentation with or without the inoculation with other microorganisms to produce the desirable characteristics of the various cheeses. Variations in the different manufacturing steps thus produce a wide variety of cheeses with various characteristics. Table 1.11 summarizes the basic steps in the cheese manufacturing process (Early 1998; Davies and Law 1984; Jenness 1988; Robinson 1986, 1990; Nath 1993; Jay 1996; Kosikowski and Mistry 1997; Scott et al. 1998; Specialist Cheesemakers Association 1997; Spreer 1998; Walstra et al. 1999). Table 1.12 summarizes the ripening conditions for various cheeses. Selected examples are introduced to provide an overview of the complexity of cheese manufacturing (Early 1998; Davies and Law 1984; Jay 1996; Jenness 1988; Kosikowski and Mistry 1997; Nath 1993; Robinson 1986, 1990; Robinson and Tamime 1991; Scott et al. 1998; Specialist Cheesemakers Association 1997; Spreer 1998; Walstra et al. 1999).

1.2.2.1 Cottage Cheese Manufacturing

Cottage cheese is a product with very mild fermentation treatment. It is produced by incubating (fermenting) the standardized and pasteurized skim milk with the starter lactic acid bacteria to produce enough acid and appropriate pH for the curdling of milk. The curd is then recovered and washed, followed by optional salting and creaming. The product is then packed and ready for marketing. No further ripening is required for this product. This is different from most fermented cheeses that require a ripening process. Table 1.13 lists the various steps involved in the production of cottage cheese (Early 1998; Nath 1993; Kosikowski and Mistry 1997; Robinson 1986; Scott et al. 1998; Spreer 1998; Walstra et al. 1999).

TABLE 1.11

Basic Cheese Making Steps

---
1. Standardization of cheese milk
2. Homogenization of cheese milk
3. Heat treatment or pasteurization of cheese milk
4. Starter addition
5. Addition of color and additives
6. Coagulation/curdling
7. Cutting the coagulum/curd
8. Stirring and scalding
9. Washing of curd cheese
10. Salting of cheese
11. Pressing of cheese
12. Coating, bandaging, and wrapping of cheese
13. Ripening
14. Retail packaging
15. Storage

Source: Reprinted from *Advances in the Microbiology and Biochemistry of Cheese and Fermented Milk*, Davies, F.L. and Law, B.A., editors, Copyright 1984, with permission from Elsevier. Early, R., editor, *The Technology of Dairy Products*, Blackie Academic & Professional, London, 1998. With permission. Jay, J.M., *Modern Food Microbiology*, Chapman & Hall, New York, 1996. With permission. Jenness, R., in *Fundamentals of Dairy Chemistry*, 3rd edition, edited by N.P. Wang et al., Kluwer Academic Publishers, New York, pp. 1–38, 1988. With permission. Kosikowski, F.V. and Mistry, V.V., *Cheese and Fermented Milk Foods*, 3rd edition, Vols. I and II, F.V. Kosikowski, Westport, CT, 1997. With permission. Nath, K.R., in *Dairy Science and Technology Handbook*, Vol. 2, edited by Y.H. Hui, VCH Publishers, Inc., New York, pp. 161–255, 1993. With permission. Reprinted from *Modern Dairy Technology*, Vol. 2, Robinson, R.K., editor, Copyright 1986, with permission from Elsevier. Robinson, R.K., editor, *Dairy Microbiology*, Vols. 1 and 2, Applied Science, London, 1990. With permission. Scott, R. et al., *Cheese Making Practice*, Chapman & Hall, New York, 1998. With permission. Specialist Cheesemakers Association, *The Specialist Cheesemakers: Code of Best Practice*, Specialist Cheesemakers Association, Staffordshire, England, Great Britain, 1997. With permission. Spreer, E. (A. Mixa translator), *Milk and Dairy Technology*, Marcel Dekker, Inc., New York, 1998. With permission. Walstra, P. et al., *Dairy Technology: Principles of Milk Properties and Processes*, Marcel Dekker, Inc., New York, 1999. With permission.

1.2.2.2 Cheddar Cheese Manufacturing

Cheddar cheese is a common hard cheese without eyes used in the fast-food industry and in households. Its production process is characterized by a requirement for milling and cheddaring of the curd. This cheese can be ripened with a wax rind or rindless (sealed under vacuum in plastic bags). It is also categorized as either regular, mild, or sharp based on the aging period (45–360 days). The longer the aging period is, the sharper the flavor becomes, and sometimes more costly. It is packaged as a large block or in slices. Table 1.14 lists the basic steps in the manufacturing of cheddar cheese (Early 1998; Kosikowski and Mistry 1997; Nath 1993; Robinson 1986; Scott et al. 1998; Spreer 1998; Walstra et al. 1999).

1.2.2.3 Swiss Cheese Manufacturing

Swiss cheese is also a common cheese used in the fast-food industry and in households. It is characterized by having irregular eyes inside the cheese. These eyes are produced by *Propionibacterium freudenreichii* subsp. *shermanii* that produces gases trapped inside the block of cheese during fermentation and ripening. A cheese with eyes like Swiss cheese has become the icon for cheeses in graphics. Swiss cheese is also characterized by its propionic acid odor. The salting process for Swiss cheese is that it utilizes both the dry and brine salting processes. Like cheddar cheese, it can be categorized as either regular, mild, or sharp depending on the length of the curing process. Table 1.15 lists the basic steps in the manufacture of Swiss cheese (Early 1998; Kosikowski and Mistry 1997; Nath 1993; Robinson 1986; Scott et al. 1998; Spreer 1998; Walstra et al. 1999).

TABLE 1.12

Cheese Ripening Conditions

Types of Cheese	Storage Period (Days)	Temperature (°C)	Relative Humidity (%)
Soft	12–30	10–14	90–95
Mold ripened	15–60	4–12	85–95
Cooked, for example, Emmental			
Cold room	7–25	10–15	80–85
Warm room	25–60	18–25	80–85
Hard, for example, Cheddar	45–360	5–12	87–95

Source: Early, R., editor, *The Technology of Dairy Products*, Blackie Academic & Professional, London, 1998. With permission. Reprinted from *Advances in the Microbiology and Biochemistry of Cheese and Fermented Milk*, Davies, F.L. and Law, B.A., editors, Copyright 1984, with permission from Elsevier. Jay, J.M., *Modern Food Microbiology*, Chapman & Hall, New York, 1996. With permission. Jenness, R., in *Fundamentals of Dairy Chemistry*, 3rd edition, edited by N.P. Wang et al., Kluwer Academic Publishers, New York, pp. 1–38, 1988. With permission. Kosikowski, F.V. and Mistry, V.V., *Cheese and Fermented Milk Foods*, 3rd edition, Vols. I and II, F.V. Kosikowski, Westport, CT, 1997. With permission. Nath, K.R., in *Dairy Science and Technology Handbook*, Vol. 2, edited by Y.H. Hui, VCH Publishers, Inc., New York, pp. 161–255, 1993. With permission. Reprinted from *Modern Dairy Technology*, Vol. 2, Robinson, R.K., editor, Copyright 1986, with permission from Elsevier. Robinson, R.K., editor, *Dairy Microbiology*, Vols. 1 and 2, Applied Science, London, 1990. With permission. Robinson, R.K. and Tamime, A.Y., editors, *Feta and Related Cheeses*, Chapman & Hall (Ellis Horwood, Ltd.), New York, 1991. With permission. Scott, R. et al., *Cheese Making Practice*, Chapman & Hall, New York, 1998. With permission. Specialist Cheesemakers Association, *The Specialist Cheesemakers: Code of Best Practice*, Specialist Cheesemakers Association, Staffordshire, England, Great Britain, 1997. With permission. Spreer, E. (A. Mixa translator), *Milk and Dairy Technology*, Marcel Dekker, Inc., New York, 1998. With permission. Walstra, P. et al., *Dairy Technology: Principles of Milk Properties and Processes*, Marcel Dekker, Inc., New York, 1999. With permission.

1.2.2.4 Blue Cheese

Blue cheese is characterized by its strong flavor and by blue mold filaments from *Penicillium roqueforti* inside the cheese. It is commonly consumed as cheese or made into a salad dressing. In the manufacturing of blue cheese, as in that of Swiss cheese, salting is accomplished by the application of dry salting and brining processes. It is characterized by a cream-bleaching step to show off the blue mold filament with a lighter background and by needling of the block of curd so that the mold can spread the blue mold filaments inside the block. It also has a soft and crummy texture due to the needling process and the gravity draining procedure to drain the curd. The curing period of 2–4 months is shorter than for hard cheeses. Its shelf life of 2 months is also shorter than that of its harder counterparts. Table 1.16 lists the basic steps in the manufacture of blue cheese (Early 1998; Kosikowski and Mistry 1997; Nath 1993; Robinson 1986; Scott et al. 1998; Spreer 1998; Walstra et al. 1999).

1.2.2.5 American-Style Camembert Cheese

American-style Camembert cheese is categorized as a soft cheese. It is characterized by having a shell of mold filament on the surface produced by *Penicillium camemberti*. Brie cheese is a similar product. Addition of annatto color is optional. Like blue cheese, it is gravity drained. Therefore, it has a soft, smooth texture. This cheese is surface-salted and has a total curing period of 3 weeks before distribution. It is usually cut into wedges and wrapped individually for direct consumption. Table 1.17 lists the basic steps in the manufacture of American-style Camembert cheese (Early 1998; Kosikowski and Mistry 1997; Nath 1993; Robinson 1986; Scott et al. 1998; Spreer 1998; Walstra et al. 1999).

1.2.2.6 Feta Cheese Manufacturing

Feta cheese is a common cheese in the Mediterranean countries. It is a soft cheese and is characterized by its brine curing (maturation) process that is not common in cheese making. Instead, it has similarity to the manufacture of sufu (Chinese fermented tofu). Like the other soft cheeses, the curing period is only

TABLE 1.13

Basic Steps in Cottage Cheese Making

1. Standardization of skim milk
2. Pasteurization of milk with standard procedure and cooling to 32°C
3. Inoculation of active lactic starter, addition of rennet, and setting of curd

Rennet addition—at 2 ml single strength (prediluted, 1:40) per 1000 kg milk within 30 min of starter addition

Type of Activity	Short Set	Medium Set	Long Set
Starter concentration	5%	3%	0.5%
Temperature of milk set	32°C	27°C	22°C
Setting to cutting	5 h	8 h	14–16 h

Final pH and whey titratable acidity—4.6 and 0.52% whey titratable acidity, respectively.

4. Cutting of curd with 1.3, 1.6, or 1.9 cm wire cheese knife
5. Cooking of curd

Let curd cubes stand for 15–30 min and cook to 51°C–54°C with 1.7°C per 10 min

Roll the curds gently every 10 min after initial 15–30 min wait

Test curd firmness and hold 10–30 min longer to obtain proper firmness

6. Washing of curd

First, wash with 29°C water temperature

Second, wash with 16°C water temperature

Third, wash with 4°C water temperature

7. Gravitational draining of washed curd for about 2.5 h
8. Salting and creaming at 152 kg creaming mixture per 454 kg with final 0.5%–0.75% salt content and 4% fat content (varies with products and optional)
9. Packaging in containers
10. Storage at refrigerated temperature

Source: Early, R., editor, *The Technology of Dairy Products*, Blackie Academic & Professional, London, 1998. With permission. Kosikowski, F.V. and Mistry, V.V., *Cheese and Fermented Milk Foods*, 3rd edition, Vols. I and II, F.V. Kosikowski, Westport, CT, 1997. With permission. Nath, K.R., in *Dairy Science and Technology Handbook*, Vol. 2, edited by Y.H. Hui, VCH Publishers, Inc., New York, pp. 161–255, 1993. With permission. Reprinted from *Modern Dairy Technology*, Vol. 2, Robinson, R.K., editor, Copyright 1986, with permission from Elsevier. Scott, R. et al., *Cheese Making Practice*, Chapman & Hall, New York, 1998. With permission. Spreer, E. (A. Mixa translator), *Milk and Dairy Technology*, Marcel Dekker, Inc., New York, 1998. With permission. Walstra, P. et al., *Dairy Technology: Principles of Milk Properties and Processes*, Marcel Dekker, Inc., New York, 1999. With permission.

2–3 months. Table 1.18 lists the basic steps in the manufacture of Feta cheese (Robinson and Tamime 1991).

1.2.3 Yogurt

Yogurt can be considered as a curdled milk product. Plain yogurt is yogurt without added flavor, stabilizer, or coagulant. Its acceptance is limited to those who really enjoy eating it. With the development of technology, other forms of yogurt, such as flavored and sweetened yogurt, stirred yogurt, yogurt drinks, and frozen yogurt, are now available. Its popularity varies from location to location. It is considered as a health food when active or live cultures are added to the final product. Table 1.19 lists the basic steps involved in the manufacture of yogurt. Table 1.3 should also be consulted for reference to other ingredients (Chandan and Shahani 1993; Tamime and Robinson 1999).

Most commercially produced yogurt and its products contain sweeteners, stabilizers, or gums (Table 1.20); fruit pieces and natural and synthetic flavors (Table 1.21); and coloring compounds (Table 1.22; Chandan and Shahani 1993; Tamime and Robinson 1999).

Different countries also have different standards on the percentages of fat and solids-not-fat (SNF) contents in their yogurt products (Table 1.23; Chandan and Shahani 1993; Tamime and Robinson 1999).

The different variables described above make the situation complicated. The term "yogurt" in one country may not have the same meaning in another country. This also creates difficulties for international

TABLE 1.14

Basic Steps in the Making of Cheddar Cheese

1. Standardization of cheese milk.
2. Homogenization of milk.
3. Pasteurization and additional heating of milk.
4. Cooling of milk to 31°C.
5. Inoculation of milk with lactic starter (0.5%–2% active mesophilic lactic starter).
6. Addition of rennet or other protease(s)—198 ml single strength (1:15,000) rennet per 1000 kg milk. Dilute the measured rennet 1:40 before use. Agitate at medium speed.
7. Setting the milk to proper acidity—25 min.
8. Cutting the curd using 0.64 cm or wider wire knife. Stir for 5 min at slow speed.
9. Cooking the curd at 38°C for 30 min with 1°C every 5 min increment. Maintain temperature for another 4–5 min and agitate periodically at medium speed.
10. Draining the curd at 38°C.
11. Cheddaring the curd at pH 5.2–5.3.
12. Milling the curd slabs.
13. Salting the curd at 2.3–3.5 kg salt per 100 kg curd in three portions for 30 min.
14. Waxed cheddar cheese.

    Hooping and pressing at 172 kPa for 30–60 s then 172–344 kPa overnight.

    Drying the cheese at 13°C at 70% relative humidity (RH) for 2–3 days.

    Paraffin whole cheese at 118°C for 6 s.
15. Rindless cheddar cheese.

    Pressing at 276 kPa for 6–18 h.

    Prepress for 1 min followed by 45 min under 686 mm vacuum.

    Remove and press at 345 kPa for 60 min.

    Remove and vacuum seal in bags with hot water shrinkage at 93°C for 2 s.
16. Ripening at 85% RH at 4°C for 60 days or longer, up to 9–12 months, or at 3°C for 2 months then 10°C for 4–7 months, up to 6–9 months.

Source: Early, R., editor, *The Technology of Dairy Products*, Blackie Academic & Professional, London, 1998. With permission; Kosikowski, F.V. and Mistry, V.V., *Cheese and Fermented Milk Foods*, 3rd edition, Vols. I and II, F.V. Kosikowski, Westport, CT, 1997. With permission; Nath, K.R., in *Dairy Science and Technology Handbook*, Vol. 2, edited by Y.H. Hui, VCH Publishers, Inc., New York, pp. 161–255, 1993. With permission. Reprinted from *Modern Dairy Technology*, Vol. 2, Robinson, R.K., editor, Copyright 1986, with permission from Elsevier; Scott, R. et al., *Cheese Making Practice*, Chapman & Hall, New York, 1998. With permission; Spreer, E. (A. Mixa translator), *Milk and Dairy Technology*, Marcel Dekker, Inc., New York, 1998. With permission; Walstra, P. et al., *Dairy Technology: Principles of Milk Properties and Processes*, Marcel Dekker, Inc., New York, 1999. With permission.

trade. Consensus or agreement among countries and proper labeling are needed to identify the products properly.

1.2.4 Fermented Liquid Milks

In milk-producing countries, it is common to have fermented milk products. These products were first discovered or developed by accident. Later, the process is modified for commercial production. Fermented liquid milks are similar to plain yogurt drinks. It was basically milk that has gone through an acid and/or alcoholic fermentation. The final product is maintained in liquid form rather than in the usual soft-gel form of yogurt. There are different fermented liquid milks available, but only sour milk, kefir, and acidophilus milk are discussed below. Readers should refer to the references listed below and other available literature on related products.

1.2.4.1 Sour Milk Manufacturing

Table 1.24 presents the basic steps in the manufacturing of the most basic fermented liquid milk—sour milk. The milk is standardized, pasteurized, inoculated, incubated, homogenized, and packaged. It is

TABLE 1.15

Basic Steps in Swiss Cheese Making

1. Standardization of cheese milk to 3% milk fat—Treatment with H_2O_2–catalase optional.
2. Pasteurization of the milk.
3. Inoculation with starters.
   Streptococcus thermophilus, 330 ml per 1000 kg milk
   Lactobacillus delbrueckii subsp. *bulgaricus*, 330 ml per 1000 kg milk
   Propionibacterium freudenreichii subsp. *shermanii*, 55 ml per 1000 kg milk
4. Addition of rennet, 10–20 min after inoculation.
   Rennet extract (154 ml single strength; 1:15,000) per 1000 kg milk, prediluted 1:40 with tap water before addition
   Stir for 3 min
5. Setting (coagulation) of milk in 25–30 min.
6. Cutting the curd with 0.64 wire knife and let curd stand undisturbed for 5 min and stir at medium speed for 40 min.
7. Cooking the curd slowly to 50°C–53°C for about 30 min and stir at medium speed; and then turn off steam and continue stirring for 30–60 min with pH reaching 6.3–6.4.
8. Dripping the curd for 30 min.
9. Pressing the curd with preliminary pressing then at 69 kPa overnight.
10. First salting—In 23% salt brine for 2–3 days at 10°C.
11. Second salting—At 10°C–16°C, 90% RH for 10–14 days by wiping the cheese surface from brine soaking followed by sprinkling of salt over cheese surface daily.
12. Third salting—At 20°C–24°C, 80%–85% RH. Wash cheese surface with salt water and sprinkle with dry salt two to three times weekly for 2–3 weeks.
13. Rinded Block Swiss cheese.
    Curing—At 7°C or lower (US) or 10°C–25°C (Europe) for 4–12 months
    Packaging in container and stored at cool temperature
14. Rindless Block Swiss cheese.
    Wrap the block or vacuum pack the blocks
    Curing stacked cheese at 3°C–4°C for 3–6 weeks
    Stored at cool temperature

Source: Early, R., editor, *The Technology of Dairy Products*, Blackie Academic & Professional, London, 1998. With permission. Kosikowski, F.V. and Mistry, V.V., *Cheese and Fermented Milk Foods*, 3rd edition, Vols. I and II, F.V. Kosikowski, Westport, CT, 1997. With permission. Nath, K.R., in *Dairy Science and Technology Handbook*, Vol. 2, edited by Y.H. Hui, VCH Publishers, Inc., New York, pp. 161–255, 1993. With permission. Reprinted from *Modern Dairy Technology*, Vol. 2, Robinson, R.K., editor, Copyright 1986, with permission from Elsevier. Scott, R. et al., *Cheese Making Practice*, Chapman & Hall, New York, 1998. With permission. Spreer, E. (A. Mixa translator), *Milk and Dairy Technology*, Marcel Dekker, Inc., New York, 1998. With permission. Walstra, P. et al., *Dairy Technology: Principles of Milk Properties and Processes*, Marcel Dekker, Inc., New York, 1999. With permission.

a very straightforward procedure compared to those for the other two products—kefir and acidophilus milk (Early 1998; Davies and Law 1984; Jay 1996; Jenness 1988; Kosikowski and Mistry 1997; Robinson 1990; Spreer 1998; Walstra et al. 1999).

1.2.4.2 Kefir Manufacturing

Kefir is a fermented liquid milk product characterized by the small amount of alcohol it contains and its inoculant (the kefir grains). It is a common product in the Eastern European countries and is considered to have health benefits. Among all the fermented dairy products, only this and similar products contain a small amount of alcohol. Moreover, among all the fermented dairy products, pure cultures of bacteria, yeasts, and/or mold are used, but in kefir, the kefir grains are used and recycled. Kefir grains are masses of bacteria, yeasts, polysaccharides, and other products of bacterial metabolism, together with curds of milk protein. Production of kefir is a two-step process: (1) the production of mother kefir and (2) the production of the kefir drink. Table 1.25 lists the basic steps in kefir manufacturing (Early 1998; Davies

TABLE 1.16

Basic Steps in Blue Cheese Making

1. Milk preparation.

   Separation of cream and skim milk

   Pasteurize skim milk by high temperature–short time, cool to 30°C

   Bleach cream with benzoyl peroxide (optional) and heating to 63°C for 30 s

   Homogenize hot cream at 6–9 mPa and then 3.5 mPa, cool and mix with pasteurized skim milk

2. Inoculation with 0.5% active lactic starter to milk at 30°C. Let stand for 1 h.

3. Addition of rennet.

   Rennet (158 ml single strength; prediluted 1:40) per 1000 kg milk and mix well

4. Coagulation or setting in 30 min.

5. Cutting curd with 1.6 cm standard wire knife.

6. Cooking of curd at 30°C followed by 5 min standing and then agitation every 5 min for 1 h. Whey should have 0.11–0.14 titratable acidity.

7. Draining of whey by gravity for 15 min.

8. Inoculation with *Penicillium roqueforti* spores—2 kg coarse salt and 28 g *P. roqueforti* spore powder per 100 kg curd followed by thorough mixing. Addition of food grade lipase optional.

9. First salting by dipping the curd in 23% brine for 15 min followed by pressing or molding at 22°C with turning every 15 min for 2 h and every 90 min for rest of day.

10. Second salting on cheese surface everyday for 5 days at 16°C, 85% RH.

11. Final dry salting or brine salting in 23% brine for 24–48 h. Final salt concentration about 4%.

12. Incubation for 6 days at 16°C, 90% RH.

13. Waxing and needling air holes or vacuum pack and needling air holes.

14. Mold filament development in air holes at 16°C for 6–8 days.

15. Curing at 11°C, 95% RH for 60–120 days.

16. Cleaning and storing.

    Stripe off the wax or vacuum packaging bag.

    Clean cheese, dry, and repack in aluminum foil or vacuum packaging bags

    Store at 2°C

17. Product shelf life—2 months.

Source: Early, R., editor, *The Technology of Dairy Products*, Blackie Academic & Professional, London, 1998. With permission. Kosikowski, F.V. and Mistry, V.V., *Cheese and Fermented Milk Foods*, 3rd edition, Vols. I and II, F.V. Kosikowski, Westport, CT, 1997. With permission. Nath, K.R., in *Dairy Science and Technology Handbook*, Vol. 2, edited by Y.H. Hui, VCH Publishers, Inc., New York, pp. 161–255, 1993. With permission. Reprinted from *Modern Dairy Technology*, Vol. 2, Robinson, R.K., editor, Copyright 1986, with permission from Elsevier. Scott, R., et al., *Cheese Making Practice*, Chapman & Hall, New York, 1998. With permission. Spreer, E. (A. Mixa translator), *Milk and Dairy Technology*, Marcel Dekker, Inc., New York, 1998. With permission. Walstra, P. et al., *Dairy Technology: Principles of Milk Properties and Processes*, Marcel Dekker, Inc., New York, 1999. With permission.

and Law 1984; Farnworth 1999; Jay 1996; Jenness 1988; Kosikowski and Mistry 1997; Robinson 1986, 1990; Spreer 1998; Walstra et al. 1999).

1.2.4.3 Acidophilus Milk

Acidophilus milk is considered to have probiotic benefits. Like yogurt, it is advertised as having live cultures of *Lactobacillus acidophilus* and *Bifidobacterium bifidum* (optional). These live cultures are claimed to provide the benefit of maintaining a healthy intestinal microflora. Traditional acidophilus milk has considerable amount of lactic acid and is considered to be too sour for the regular consumers in some locations. Therefore, a small amount of sugar is added to the final product to make it more palatable. This later product is called sweet acidophilus milk. Table 1.26 lists the basic steps in the manufacture of acidophilus milk (Early 1998; Davies and Law 1984; Jay 1996; Jenness 1988; Robinson 1986, 1990; Kosikowski and Mistry 1997; Spreer 1998; Walstra et al. 1999).

TABLE 1.17

Basic Steps in American-Style Camembert Cheese

1. Standardization of milk
2. Homogenization of milk
3. Pasteurization of milk at 72°C for 6 s
4. Cool milk to 32°C
5. Inoculation with 2% active lactic starter followed by 15–30 min acid ripening to 0.22% titratable acidity
6. Addition of annatto color at 15.4 ml per 1000 kg milk (optional)
7. Addition of rennet—220 ml single strength (prediluted 1:40) per 1000 ml followed by mixing for 3 min and standing for 45 min
8. Cutting of curd with 1.6 cm standard wire knife
9. Cooking of curd at 32°C for 15 min with medium speed stirring
10. Draining of curd at 22°C for 6 h with occasional turning
11. Inoculation with *Penicillium camemberti* spores through spray gun on both sides of cheese once
12. Pressing and molding curd by pressing for 5–6 h at 22°C without any weight on surface
13. Surface salting of cheese and let stand for about 9 h
14. Curing at 10°C, 95% RH for 5 days undisturbed, then turned once and continue curing for 14 days
15. Packaging, storage, and distribution
    Wrap cheese and store at 10°C, 95%–98% RH for another 7 days
    Move to cold room at 4°C and cut into wedges, if required, and rewrap
    Distribute immediately

Source: Early, R., editor, *The Technology of Dairy Products*, Blackie Academic & Professional, London, 1998. With permission. Kosikowski, F.V. and Mistry, V.V., *Cheese and Fermented Milk Foods*, 3rd edition, Vols. I and II, F.V. Kosikowski, Westport, CT, 1997. With permission. Nath, K.R., Cheese, in *Dairy Science and Technology Handbook*, Vol. 2, edited by Y.H. Hui, VCH Publishers, Inc., New York, pp. 161–255, 1993. With permission. Reprinted from *Modern Dairy Technology*, Vol. 2, Robinson, R.K., editor, Copyright 1986, with permission from Elsevier. Scott, R. et al., *Cheese Making Practice*, Chapman & Hall, New York, 1998. With permission. Spreer, E. (A. Mixa translator), *Milk and Dairy Technology*, Marcel Dekker, Inc., New York, 1998. With permission. Walstra, P. et al., *Dairy Technology: Principles of Milk Properties and Processes*, Marcel Dekker, Inc., New York, 1999. With permission.

TABLE 1.18

Basic Steps in Feta Cheese Making

1. Standardization of milk with 5% fat, enzyme treated and decolorized
2. Homogenization of milk
3. Pasteurization by standard procedure and cooling to 32°C
4. Inoculation with 2% active lactic starter as cheddar cheese followed by 1 h ripening
5. Addition of rennet at 198 ml single strength (prediluted, 1:40) per 1000 kg milk followed by 30–40 min setting
6. Cutting of the curd with 1.6 cm standard wire knife followed by 15–20 min standing
7. Dripping of curd for 18–20 h at 12–18 kg on 2000 cm^2 with pH and titratable acidity developed to 4.6% and 0.55%, respectively
8. Preparation of cheese blocks of 13 × 13 × 10 cm each
9. Salting in 23% salt brine for 1 day at 10°C
10. Canning and boxing cheese blocks in 14% salt brine (sealed container)
11. Curing for 2–3 months at 10°C
12. Soak cured cheese in skim milk for 1–2 days before consumption to reduce salt
13. Yield—15 kg/100 kg of 5% fat milk

Source: Robinson, R.K. and Tamime, A.Y., editors, *Feta and Related Cheeses*, Chapman & Hall (Ellis Horwood, Ltd.), New York, 1991. With permission.

TABLE 1.19

Basic Steps in the Production of Yogurt

1. Standardization of liquid milk
2. Homogenization of liquid milk
3. Heat treatment or pasteurization of liquid milk at 90°C for 5 min or equivalent
4. Cooling of pasteurized milk to 1°C–2°C above inoculation temperature
5. Addition of starter (inoculation), 1%–3% operational culture
6. Addition of flavor, sweetener, gums, and/or color (optional)
7. Incubation at 40°C–45°C for 2.5–3.0 h for standard cultures
8. Breaking of curd (optional)
9. Cooling to 15°C–20°C in 1–1.5 h
10. Addition of live culture (optional)
11. Packaging
12. Storage at ≤10°C

Source: Chandan, R.C. and Shahani, K.M., in *Dairy Science and Technology Handbook*, Vol. 2, edited by Y.H. Hui, VCH Publishers, Inc., New York, pp. 1–56, 1993. With permission. Tamime, A.Y. and Robinson, R.K., *Yogurt: Science and Technology*, CRC Press, Boca Raton, FL, 1999. With permission.

1.3 Meat Products

1.3.1 Ingredients and Types

Fermented meat products such as ham and sausages have been available to different cultures for centuries. It is interesting to learn that the ways these products are produced are basically very similar in different cultures. Besides the meat, nitrite and salt, and sugar (optional), pure cultures are sometimes used, especially in fermented sausages. Microorganisms do not merely provide the characteristic flavor for the products; the lactic acid bacteria also produce lactic and other acids that can lower the pH of the products. Pure cultures are sometimes used in hams to lower the pH and thus inhibit the growth of *Clostridium botulinum*. The raw meat for ham manufacturing is basically a large chunk of meat, and it is difficult for microorganisms to penetrate into the center, unless they are injected into the interior. Microbial growth is mainly on the surface, and the microbial enzymes are gradually diffused into the center. By contrast, in sausages, the cultures, if used, are mixed with the ingredients (ground or chopped meats) at the beginning, and the fermentation is carried out without difficulty. Besides, sausages are much smaller than hams. Table 1.27 lists some of the ingredients used in the manufacture of hams and sausages (Cassens 1990; Hammes et al. 1990; Huang and Nip 2001; Incze 1998; Roca and Incze 1990; Skrokki 1998; Toldra et al. 2001; Townsend and Olsen 1987; Xiong et al. 1999).

1.3.2 Hams

Hams as indicated earlier are made from large chunks of meat. Western cultures manufacture ham using either a dry cure and/or a brine cure process, sometimes followed by a smoking process. Tables 1.28 and 1.29 list the basic steps involved in the dry cure and brine cure of hams, respectively. These two processes are similar except for the salting step (Cassens 1990; Townsend and Olsen 1987).

Chinese hams are basically manufactured using a dry curing process. Procedures differ slightly depending on the regions where the hams are made. The most famous Chinese ham is the Jinhua ham made in Central China. Yunan ham from Southern China also has a good reputation. In the old days, without refrigeration facilities during processing, transportation, and storage, it is believed that hams completed their aging process during the transportation and storage stages. Today, with controlled temperature and relative humidity rooms, the hams are produced under controlled conditions. Table 1.30 lists the current process used in China for Jinhua ham (Huang and Nip 2001; Xiong et al. 1999).

TABLE 1.20

Some Common Gums That Could Be Used in
Yogurt Manufacturing

Kind	Name of Gum
Natural	Agar
	Alginates
	Carrageenan
	Carob gum
	Corn starch
	Casein
	Furcellaran
	Gelatin
	Gum arabic
	Guar gum
	Karaya gum
	Pectins
	Soy protein
	Tragacanth gum
	Wheat starch
Modified gums	Cellulose derivatives
	Dextran
	Low-methoxy pectin
	Modified starches
	Pregelatinized starches
	Propylene glycol alginate
	Xanthin
Synthetic gums	Polyethylene derivatives
	Polyvinyl derivatives

Source: Chandan, R.C. and Shahani, K.M., in *Dairy Science and Technology Handbook*, Vol. 2, edited by Y.H. Hui, VCH Publishers, Inc., New York, pp. 1–56, 1993. With permission. Tamime, A.Y. and Robinson, R.K., *Yogurt: Science and Technology*, CRC Press, Boca Raton, FL, 1999. With permission.

1.3.3 Sausages

Many European-type sausages are manufactured using a fermentation process. These sausages have their own characteristic flavors due to the formulations and curing processes used. It is not the intent of this chapter to list the various formulations. Readers should consult the references in this chapter and other references available elsewhere. Commercial inocula are available. Bacteria and some yeasts grow inside the sausage during the ripening period, producing the characteristic flavor. Molds can grow on the surface during storage if the sausages are not properly packaged and stored in the refrigerator. Because these sausages are not sterilized, fermentation is an on-going process and aged sausages carry a stronger flavor. Table 1.31 lists the basic steps in the manufacture of dry fermented sausages (Hammes et al. 1990; Incze 1998; Roca and Incze 1990; Toldra et al. 2001).

1.4 Fish and Shrimp Products

Many Asian and Pacific Rim countries have traditional fermented fish and shrimp products, especially those mentioned in Table 1.32.

TABLE 1.21

Some Common Flavors for Yogurt

Retail Flavor	Natural Characteristic-Impact Compound	Synthetic Flavoring Compound Available
Apricot	NA	*g*-Undecalactone
Banana	3-Methylbutyl acetate	NA
Bilberry	NA	NA
Black currant	NA	*trans-* and *cis-p-* Methane-8-thiol-3-one
Grape, Concord	Methyl antranilate	NA
Lemon	Citral	15 compounds
Peach	*g*-Decalactone	*g*-Undecalactone
Pineapple	NA	Allyl hexanoate
Raspberry	1-*p*-Hydroxyphenyl-3-butanone	NA
Strawberry	NA	Ethyl-3-methyl-3-phenylglycidate

Source: Chandan, R.C. and Shahani, K.M., in *Dairy Science and Technology Handbook*, Vol. 2, edited by Y.H. Hui, VCH Publishers, Inc., New York, pp. 1–56, 1993. With permission. Tamime, A.Y. and Robinson, R.K., *Yogurt: Science and Technology*, CRC Press, Boca Raton, FL, 1999. With permission.

The fermented products are mainly sauce or paste and are used as condiments or ingredients in food preparation, such as fermented fish sauce and fermented shrimp paste. Although both products enjoy the same popularity in these countries, only fish sauce is well known among non-Asian consumers in the world, especially those countries with Asian grocery stores or supermarkets, such as Australia and countries in Western Europe and North America. The recognition of fish sauce among non-Asians is a result of the popularity of Thai dishes, though Vietnamese cooking is also gaining popularity. Shrimp paste is not as well known because it is not a common ingredient for dishes by non-Asian consumers. Moreover, shrimp paste has a pungent odor and flavor and is not preferred by most consumers in the West.

We will briefly discuss two such products—fish sauce and shrimp paste. Fish sauce and shrimp paste are often the most important flavoring ingredients of the diet. They can be added to soups, curries, noodles, rice, and other dishes.

TABLE 1.22

Permitted Yogurt Colorings

Name of Color	Maximum Level (mg/kg)
Indigotin	6
Brilliant black PN	12
Sunset yellow FCF	12
Tartrazine	18
Cochineal	20
Carminic acid	20
Erythrosine	27
Red 2G	30
Ponceau	48
Caramel	150
Brilliant blue FCF	200

Source: Chandan, R.C. and Shahani, K.M., in *Dairy Science and Technology Handbook*, Vol. 2, edited by Y.H. Hui, VCH Publishers, Inc., New York, pp. 1–56, 1993. With permission. Tamime, A.Y. and Robinson, R.K., *Yogurt: Science and Technology*, CRC Press, Boca Raton, FL, 1999. With permission.

TABLE 1.23

Existing or Proposed Standards for Commercial Yogurt Composition
[Percentages of Fat and Solids-Not-Fat (SNF)] in Selected Countries

| Country | Percentage of Fat | | | Percentage of SNF |
	Low	Medium	Normal	
Australia	NA	0.5–1.5	3	NA
France	0.5	NA	3	NA
Italy	1	NA	3	NA
Netherlands	1	NA	3	NA
New Zealand	0.3	NA	3.2	NA
UK	0.3	1.0–2.0	3.5	8.5
USA	0.5–1.0	2	3.25	8.5
West Germany	0.5	1.5–1.8	3.5	8.25–8.5
FAO/WHO	0.5	0.5–3.0	3	8.2
Range	0.3–1.0	0.5–3.0	3.3.5	8.2–8.5

Source: Chandan, R.C. and Shahani, K.M., in *Dairy Science and Technology
Handbook*, Vol. 2, edited by Y.H. Hui, VCH Publishers, Inc., New
York, pp. 1–56, 1993. With permission. Tamime, A.Y. and Robinson,
R.K., *Yogurt: Science and Technology*, CRC Press, Boca Raton, FL,
1999. With permission.

TABLE 1.24

Basic Steps in Sour Milk Processing

1. Standardization of milk
2. Heating of milk to 85°C–95°C followed by homogenization
3. Cooling of milk to 19°C–25°C and transfer of milk to fermentation tank
4. Addition of 1%–2% start culture (inoculation)
5. Shock-free fermentation to pH 4.65–4.55
6. Homogenization of gel
7. Cooling to 4°C–6°C
8. Filling into bottles, jars, or one-way packs or wholesale packs

Source: Early, R., editor, *The Technology of Dairy Products*, Blackie
Academic & Professional, London, 1998. With permission. Reprinted
from *Advances in the Microbiology and Biochemistry of Cheese and
Fermented Milk*, Davies, F.L. and Law, B.A., editors, Copyright
1984, with permission from Elsevier. Jay, J.M., *Modern Food
Microbiology*, Chapman & Hall, New York, 1996. With permission.
Jenness, R., in *Fundamentals of Dairy Chemistry*, 3rd edition, edited
by N.P. Wang et al., Kluwer Academic Publishers, New York, pp.
1–38, 1988. With permission. Kosikowski, F.V. and Mistry, V.V.,
Cheese and Fermented Milk Foods, 3rd edition, Vols. I and II, F.V.
Kosikowski, Westport, CT, 1997. With permission. Robinson, R.K.
editor, *Dairy Microbiology*, Vols. 1 and 2, Applied Science, London,
1990. With permission. Spreer, E. (A. Mixa translator), *Milk and
Dairy Technology*, Marcel Dekker, Inc., New York, 1998. With permission. Walstra, P. et al., *Dairy Technology: Principles of Milk
Properties and Processes*, Marcel Dekker, Inc., New York, 1999.
With permission.

TABLE 1.25

Basic Steps in Kefir Processing

Preparation of mother "kefir"

1. Standardization of milk for preparation of mother "kefir."
2. Pasteurize milk at 90°C–95°C for 15 min and cool to 18°C–22°C.
3. Spread kefir grains at the bottom of a container (5–10 cm thick) and add pasteurized milk (20–30 times the amount of kefir grains).
4. Ferment for 18–24 h with mixing two to three times. Kefir grains float to the surface.
5. Filter out the kefir grains with a fine sieve; wash the grains with water and save for the next fermentation.
6. Save the fermented milk for the next-step inoculation.

Preparation of drinkable kefir

1. Blend fermented milk from above with 8–10 times fresh, pasteurized, and untreated milk.
2. Fill into bottles, closed and fermented for 1–3 days at 18°C–22°C.
   [Another option is to mix the fermented milk with fresh milk at 1%–5%; ferment at 20°C–25°C for 12–15 h until pH 4.4–4.5 followed by ripening in storage tanks 1–3 days at 10°C. Product is not as traditional but acceptable.]
3. Cool to refrigerated temperature.
4. Store and distribute.

Source: Early, R., editor, *The Technology of Dairy Products*, Blackie Academic & Professional, London, 1998. With permission. Reprinted from *Advances in the Microbiology and Biochemistry of Cheese and Fermented Milk*, Davies, F.L. and Law, B.A., editors, Copyright 1984, with permission from Elsevier. Jay, J.M., *Modern Food Microbiology*, Chapman & Hall, New York, 1996. With permission. Jenness, R., in *Fundamentals of Dairy Chemistry*, 3rd edition, edited by N.P. Wang et al., Kluwer Academic Publishers, New York, pp. 1–38, 1988. With permission. Kosikowski, F.V. and Mistry, V.V., *Cheese and Fermented Milk Foods*, 3rd edition, Vols. I and II, F.V. Kosikowski, Westport, CT, 1997. With permission. Reprinted from *Modern Dairy Technology*, Vol. 2, Robinson, R.K., editor, Copyright 1986, with permission from Elsevier. Robinson, R.K., editor, *Dairy Microbiology*, Vols. 1 and 2, Applied Science, London, 1990. With permission. Spreer, E. (A. Mixa translator), *Milk and Dairy Technology*, Marcel Dekker, Inc., New York, 1998. With permission. Farnworth, E.R., *J. Nutra. Funct. Med. Foods* 1(4), 57–68, 1999. With permission. Walstra, P. et al., *Dairy Technology: Principles of Milk Properties and Processes*, Marcel Dekker, Inc., New York, 1999. With permission.

1.4.1 Fish Sauce

Fish sauce is made from species of fish with little commercial value as a food or dietary source. Traditionally, this means small fish such as anchovies or whatever small fish species available to local residents. The production of commercial fermented fish sauce (Ang et al. 1999; Hutkins 2006; Steinkraus 1996, 2004) depends on three factors, among others:

1. Fish species and freshness
2. Fermentation parameters
3. Curing duration and technique

The above factors will determine the flavor, aroma, and cooking properties of the final product. Some examples of better known fish sauces from different countries are shown in Table 1.33. Table 1.34 provides the preparation steps for the commercial production of fish sauce (Ang et al. 1999; Hutkins 2006; Steinkraus 1996, 2004).

1.4.1.1 Process

The process varies with different countries, though the fundamental approach is the same. Whole fish (fresh or previously frozen) are rinsed and drained and then mixed with salt (3:1). The salted fish are packed into tanks and allowed to ferment. Fermentation results in autolysis and production of volatile flavors and other substances responsible for the taste, aroma, and cooking properties. The tanks are sealed and left undisturbed for 6–18 months. The temperature during the fermentation period is kept high. In

TABLE 1.26

Basic Steps for Sweet Acidophilus Milk Processing

Procedure #1

1. Standardization of milk
2. Heating milk to 95°C for 60 min, cooled to 37°C and hold for 3–4 h, reheat to 95°C for 10–15 min, cool to 37°C
3. Inoculation with 2%–5% bulk starter
4. Incubation for up to 24 h or to 1% lactic acid
5. Cooling to 5°C
6. Packing and distribution

Procedure #2

1. Standardization of milk
2. Homogenization of milk at 14.5 MPa
3. Heating to 95°C for 60 min
4. Cooling to 37°C
5. Inoculation with direct vat inoculation starter
6. Incubation for 12–16 h or to about 0.65% lactic acid
7. Ultra high temperature of 140°C–145°C for 2–3 s to eliminate undesirable contaminants
8. Cooling to 10°C or lower
9. Packaging and distribution

Source: Early, R., editor, *The Technology of Dairy Products*, Blackie Academic & Professional, London, 1998. With permission. Reprinted from *Advances in the Microbiology and Biochemistry of Cheese and Fermented Milk*, Davies, F.L. and Law, B.A., editors, Copyright 1984, with permission from Elsevier. Jay, J.M., *Modern Food Microbiology*, Chapman & Hall, New York, 1996. With permission. Jenness, R., in *Fundamentals of Dairy Chemistry*, 3rd edition, edited by N.P. Wang et al., Kluwer Academic Publishers, New York, pp. 1–38, 1988. With permission. Kosikowski, F.V. and Mistry, V.V., *Cheese and Fermented Milk Foods*, 3rd edition, Vols. I and II, F.V. Kosikowski, Westport, CT, 1997. With permission. Reprinted from *Modern Dairy Technology*, Vol. 2, Robinson, R.K., editor, Copyright 1986, with permission from Elsevier. Robinson, R.K., editor, *Dairy Microbiology*, Vols. 1 and 2, Applied Science, London, 1990. With permission. Spreer, E. (A. Mixa translator), *Milk and Dairy Technology*, Marcel Dekker, Inc., New York, 1998. With permission. Walstra, P. et al., *Dairy Technology: Principles of Milk Properties and Processes*, Marcel Dekker, Inc., New York, 1999. With permission.

general, high salt content slows the rate of fermentation but allows the final product to have a longer shelf life (see Ang et al. 1999; Hutkins 2006; Steinkraus 1996, 2004).

During fermentation, the fish protein is hydrolyzed by natural or added autolytic enzymes and becomes liquefied. At the end of the fermentation process, we can harvest two products.

The solid or filtrate is arranged to be sun ripened for 1–3 months using various exposed containers that can best utilize the sun's heat.

The liquid is drained off and further filtered to remove any sediment. The filtrate is transferred to clean jars and ripened in the sun for 1–3 months; vaporization allows some of the strong fish odors to dissipate. The clear, dark brown liquid is the fish sauce, which is ready for bottling and distribution.

In many parts of the world, traditional production of fish sauce depends on a recipe passed on from generation to generation. Little attention is paid to all the variables that must be considered in the commercial manufacture of fish sauce. In addition to those mentioned earlier, some more specific parameters are discussed below.

The low content of carbohydrate in fish limits fermentation. Carbohydrates such as garlic, rice, and even sugar may be added to facilitate fermentation. This increases the rate of fermentation, and such techniques can make fish sauce available in a few weeks of fermentation. Large volumes of fish sauce can be produced in a short period of time.

The use of enzymes in hastening the fermentation process has been explored by large corporations specializing in such Asian products.

One big problem with the commercial production of fish sauce is the assurance of a safe product for the consumers. As a result, the participation of food safety experts in the production of fish sauce is

TABLE 1.27

Raw Ingredients for Fermented Meat Products

Ingredient	Ham	Sausage
Meat		
Pork	Yes	Yes
Beef	No	Optional
Casing	No	Yes
Salt	Yes	Yes
Sugar	Optional	Optional
Starter Microorganisms	Optional	Optional
Lactobacillus sakei, L. curvatus,		
L. plantarum, L. pentosus, L. pentoaceus		
Pediococcus pentosaceus, P. acidilactici		
Staphylococcus xylosus, S. carnosus		
Kocuria varians		
Debaryomyces hansenii		
Candida famata		
Penicillium nalgiovense, P. chrysogenum		
Spices	Optional	Optional
Other Flavoring Compounds	Optional	Optional
Moisture Retention Salts	Optional	Optional
Preservatives	No	No

Source: Cassens, R.G., *Nitrite-Cured Meat: A Food Safety Issue in Perspective*, Food & Nutrition Press, Trumbull, CT, 1990. With permission. Hammes, W.P. et al., *FEMS Microbiol. Lett.* 87(1/2), 165–174, 1990. With permission. Incze, K., *Meat Sci.* 49(Supplement), S169–S177, 1998. With permission. Roca, M. and Incze, K., *Food Rev. Int.* 6(1), 91–118, 1990. With permission. Huang, T.C. and Nip, W.K., in *Meat Science and Applications*, edited by Y.H. Hui et al., Marcel Dekker, Inc., New York, pp. 403–442, 2001. With permission. Incze, K., *Meat Sci.* 49(Supplement), S169–S177, 1998. With permission. Roca, M. and Incze, K., *Food Rev. Int.* 6(1), 91–118, 1990. With permission. Toldra, F. et al., in *Meat Science and Applications*, edited by Y.H. Hui et al., Marcel Dekker, Inc., New York, pp. 538–591, 2001. With permission. Townsend, W.E. and Olsen, D.G., in *The Science of Meat and Meat Products*, edited by J.F. Price and B.S. Scheweigert, Food & Nutrition Press, Westport, CT, pp. 431–456, 1987. With permission. Xiong, Y.L. et al., in *Asian Food Products: Science and Technology*, edited by C.Y.W. Ang et al., Technomic Publishing Co., Inc., Lancaster, PA, pp. 201–213, 1999. With permission.

strongly recommended in view of known food poisoning from fish sauce (see Ang et al. 1999; Hutkins 2006; Steinkraus 1996, 2004).

1.4.2 Shrimp Paste

Shrimp paste or shrimp sauce is a common ingredient used in Southeast Asian and Southern Chinese cuisines. Different Asian countries have a different name for it. Table 1.35 shows the shrimp pastes available in selected Asian countries.

TABLE 1.28

Basic Steps in Dry Cured Ham Processing

---

1. Preparation of pork for dry curing
2. Mixing of the proper ratio of ingredients [salt, sugar, nitrite, and inocula (optional)]
3. Rubbing of the curing mixture to the meat
4. Stacking of the green ham for initial dry curing at 36°C–40°C
5. Rerubbing of the green ham and stacking for additional curing at 36°C–40°C
   [The ham should be left in the cure for the equivalent of 3 days per pound of meat.]
6. Soaking of the cured ham for 2–3 h followed by thorough scrubbing
7. Place green ham in tight-fitting stockinette and hang in smokehouse to dry overnight
8. Smoking at about 60°C or 80°C with 60% RH for 12–36 h
9. Cooling
10. Vacuum packaging and cool storage

Source: Cassens, R.G., *Nitrite-Cured Meat: A Food Safety Issue in Perspective*, Food & Nutrition Press, Trumbull, CT, 1990. With permission. Townsend, W.E. and Olsen, D.G., in *The Science of Meat and Meat Products*, edited by J.F. Price and B.S. Scheweigert, Food & Nutrition Press, Westport, CT, pp. 431–456, 1987. With permission.

1.4.2.1 Preparation

Preparation techniques can vary greatly; however, the following procedure is most common in China and Southeast Asia. The stages are as follows:

1. Small shrimp rinsed and drained
2. Shrimp mixed with salt
3. Sun dried on ground, stilts, and stacks
4. After several days, mixture becomes dark and turns into a thick pulp
5. Fermentation time is shorter for small shrimp and longer for larger shrimp
6. Grinding of pulp to a smooth consistency
7. Fermentation and grinding are repeated a few times to make sure that the paste is mature
8. Shrimp paste dried
9. Dried shrimp paste cut into blocks

Such shrimp paste is available in grocery stores and supermarkets in many Western countries—Europe, North America, Australia, New Zealand, South America, Central America, and so on.

TABLE 1.29

Basic Steps in Brine Cured Ham Processing

---

1. Preparation of pork for brine curing
2. Mixing of the proper ratio of ingredients [salt, sugar, and nitrite with inocula (optional)]; (5 gal of brine for 100 lb meat)
3. Soaking of the meat in the prepared brine, or stitch pumping of the brine into the meat (10% of the original weight of the meat) followed by soaking in the brine for 3–7 days vacuum tumbling or massaging (optional)
4. Remove the meat from the cover brine and wash
5. Place green ham in tight-fitting stockinette and hang in smokehouse to dry overnight
6. Smoking at about 60°C or 80°C with 60% RH for 12–36 h
7. Cooling
8. Vacuum packaging and cool storage

Source: Cassens, R.G., *Nitrite-Cured Meat: A Food Safety Issue in Perspective*, Food & Nutrition Press, Trumbull, CT, 1990. With permission; Townsend, W.E. and Olsen, D.G., in *The Science of Meat and Meat Products*, edited by J.F. Price and B.S. Scheweigert, Food & Nutrition Press, Westport, CT, pp. 431–456, 1987. With permission.

TABLE 1.30

Basic Steps in Chinese Jinhua Ham Processing

---

1. Selection of pork hind leg, 5–7.5 kg
2. Trimming
3. Salting, 7–8 kg salt per 10 kg ham
4. Stacking and overhauling at 0°C–10°C for 33–40 days
5. Washing with cold water and brush
6. Drying in the sun for 5–6 days
7. Fermentation (curing) for 2–3 months at 0°C–10°C with harmless green mold developing on surface
8. Brushing off the mold and trimming
9. Aging for 3–4 months, maximum 9 months; alternate aging process in temperature programmable room with 60% for 1–2 months
10. Grading
11. (Yield: about 55–60)
12. Packaging and distribution

Source: Huang, T.C. and Nip, W.K., in *Meat Science and Applications*, edited by Y.H. Hui et al., Marcel Dekker, Inc., New York, pp. 403–442, 2001. With permission. Xiong, Y.L. et al., in *Asian Food Products: Science and Technology*, edited by C.Y.W. Ang et al., Technomic Publishing Co., Inc., Lancaster, PA, pp. 201–213, 1999. With permission.

TABLE 1.31

Basic Steps in Dry (Fermented) Sausage Processing

---

1. Selection of meat for processing
2. Chopping and mixing of chopped meat with spices, seasonings, and inocula at temperature of about 10°C
3. Stuffing the mixture in suitable casings
4. Linking
5. Curing or drying for 1–3 months in rooms with temperature, relative humidity, and air circulation regulated according to the type of sausages being produced
6. Packaging and cool storage

Source: Hammes, W.P. et al., *FEMS Microbiol. Lett.* 87(1/2), 165–174, 1990. With permission. Incze, K., *Meat Sci.* 49(Supplement), S169–S177, 1998. With permission. Roca, M. and Incze, K., *Food Rev. Int.* 6(1), 91–118, 1990. With permission. Toldra, F. et al., in *Meat Science and Applications*, edited by Y.H. Hui et al., Marcel Dekker, Inc., New York, pp. 538–591, 2001. With permission.

TABLE 1.32

Selected Asian Countries with Traditional Fermented Fish and Shrimp Products

---

Burma (Myanmar)
Cambodia
China
Hong Kong
India
Japan
Laos
Philippines
Singapore
Thai
Vietnam

TABLE 1.33

Selected Fish Sauces: Country of Origin

Country	Fish Sauce
Burma (Myanmar)	Ngapi
Cambodia	Nuoc-mam
Indonesia	Ketjap-ikan
Philippines	Patis
Thailand	Nampla
Vietnam	Nuoc-mam

TABLE 1.34

Steps in Preparing Fish Sauce

1. Fish (gutted or ungutted) is mixed with salt (3:1)
2. Salted fish transferred to a fermentation tank; temperature at 38°C–42°C
3. Tanks filled completely and sealed
4. Fermented for 6–18 months
5. Solids separated from liquid by filtration
6. Liquid and solid

Note: The clear, dark brown liquid is the fish sauce, which is ready for bottling and distribution. The solid or filtrate is arranged to be sun ripened for 1–3 months, using various exposed containers that can best utilize the sun's heat.

TABLE 1.35

Shrimp Paste in Selected Countries

Country	Shrimp Paste
Burma (Myanmar)	Ngapi
Cambodia	Ngapi
China, South (Hong Kong)	Haam ha
Indonesia	Terasi
Malaysia	Belacan
Philippines	Bagoong alamang
Thailand	Kapi
Vietnam	Ngapi, chao tom

1.4.2.2 Factors in Manufacture

Some of the factors involved in producing shrimp paste in small villages or individual houses include the following.

1. The raw ingredient is mainly a marine product such as shrimp, prawns, crawfish, fish (krill), or a mixture of such marine products, depending on availability and cost.
2. The raw ingredients may undergo preparation before storage and fermentation, for example, steaming, boiling, and grinding.
3. It is made from fermented ground shrimp or even fish, sun dried, and then cut into rectangular blocks of various sizes as indicated in Table 1.35.
4. The product exists as paste, sauce, liquid, or blocks; the most common is paste. For liquid shrimp paste, it is usually served as a dip condiment.
5. The ground shrimps may no longer resemble actual shrimps, though in some regions, the shrimps are still distinguishable in shape and size.

6. Most finished products are raw and have to be cooked in some preparation format, such as boiling, steaming, marinating, and so on. However, among some ethnic groups, the shrimp paste is eaten raw as a dipping sauce.
7. Shrimp paste serves multiple functions in the diet of many Asian groups:
   a. An essential ingredient in many dishes including meat, fish, vegetables, mixed dishes (curries, stews, etc.) and sauces
   b. A condiment for dipping, rice, potatoes, buns, fish, and vegetables

A shrimp paste is characterized as follows:

1. The aroma (smell) can be pungent to mild. Higher quality products have a mild aroma.
2. Shrimp pastes vary in appearance from pale (pinkish gray) liquid sauces to solid chocolate-colored or dark blocks. Some are in powder formats.
3. When paste is produced by families or small regional businesses, it has better attributes, though with potentially higher chances of food poisoning. Commercial preparations are less attractive but definitely safer. The major concern is of course the occurrence of pathogens. The magnitude of any poisoning is reflected by the number of consumers affected. If only a small number of victims are involved, it usually means that they have consumed a tainted family-produced paste product.
4. Smell, texture, taste, and saltiness vary among different produced paste product prepared under unique Asian cultures.
5. The use of additional ingredients is common whether the paste is produced by a small or large establishment. The manufacturers add spice, fragrance, and tasty materials (pepper, garlic, shallot, sugar). In many modern establishments, chemical food additives and colors are used extensively.
6. For some products, the raw ingredients may be pretreated (fried, boiled, steamed, ground, and so on).

REFERENCES

Ang, C.Y.W., Liu, K., and Huang, Y.W., editors. 1999. *Asian Foods: Science and Technology*. Boca Raton, FL: CRC Press.

Cassens, R.G. 1990. *Nitrite-Cured Meat: A Food Safety Issue in Perspective*. Trumbull, CT: Food & Nutrition Press.

Chandan, R.C. and Shahani, K.M. 1993. Yogurt. In: *Dairy Science and Technology Handbook*, Vol. 2, edited by Y.H. Hui, pp. 1–56. New York: VCH Publishers, Inc.

Davies, F.L. and Law, B.A., editors. 1984. *Advances in the Microbiology and Biochemistry of Cheese and Fermented Milk*. New York: Elsevier.

Early, R., editor. 1998. *The Technology of Dairy Products*. London: Blackie Academic & Professional.

Emmons, D.B. 2000. *Practical Guide for Control of Cheese Yield*. Brussels, Belgium: International Dairy Federation.

Farnworth, E.R. 1999. From folklore to regulatory approval. *J. Nutra. Funct. Med. Foods* 1(4):57–68.

Hammes, W.P., Bantleon, A., and Min, S. 1990. Lactic acid bacteria in meat fermentation. *FEMS Microbiol. Lett.* 87(1/2):165–174.

Huang, T.C. and Nip, W.K. 2001. Intermediate-moisture meat and dehydrated meat. In: *Meat Science and Applications*, edited by Y.H. Hui, W.K. Nip, R.W. Rogers, and O.A. Young, pp. 403–442. New York: Marcel Dekker, Inc.

Hutkins, R.W. 2006. *Microbiology and Technology of Fermented Foods*. Hoboken, NJ: Wiley.

Incze, K. 1998. Dry fermented sausages. *Meat Sci.* 49(Supplement):S169–S177.

Jay, J.M. 1996. *Modern Food Microbiology*. New York: Chapman & Hall.

Jenness, R. 1988. Composition of milk. In: *Fundamentals of Dairy Chemistry*, 3rd edition, edited by N.P. Wang, R. Jenness, R.M. Keeney, and E.H. Marth, pp. 1–38. New York: Kluwer Academic Publishers.

Kosikowski, F.V. and Mistry, V.V. 1997. *Cheese and Fermented Milk Foods*, 3rd edition, Vols. I and II. Westport, CT: F.V. Kosikowski.

Law, B.A., editor. 1997. *Microbiology and Biochemistry of Cheese and Fermented Milk*. London: Blackie Academic & Professional.

Nath, K.R. 1993. Cheese. In: *Dairy Science and Technology Handbook*, Vol. 2, edited by Y.H. Hui, pp. 161–255. New York: VCH Publishers, Inc.

Robinson, R.K., editor. 1986. *Modern Dairy Technology*, Vol. 2. New York: Elsevier Applied Science Publishers.

Robinson, R.K., editor. 1990. *Dairy Microbiology*, Vols. 1 & 2. London: Applied Science.

Robinson, R.K. and Tamime, A.Y., editors. 1991. *Feta and Related Cheeses*. New York: Chapman & Hall (Ellis Horwood, Ltd.).

Roca, M. and Incze, K. 1990. Fermented sausages. *Food Rev. Int.* 6(1):91–118.

Scott, R., Robinson, R.K., and Wilbey, R.A. 1998. *Cheese Making Practice*. New York: Chapman & Hall.

Skrokki, A. 1998. Additives in Finnish sausages and other meat products. *Meat Sci.* 39(2):311–315.

Specialist Cheesemakers Association. 1997. *The Specialist Cheesemakers: Code of Best Practice*. Staffordshire, England, Great Britain: Specialist Cheesemakers Association.

Spreer, E. (A. Mixa translator). 1998. *Milk and Dairy Technology*. New York: Marcel Dekker, Inc.

Steinkraus, K.H., editor. 1996. *Handbook of Indigenous Fermented Foods, Second Edition, Revised and Expanded*. Boca Raton, FL: CRC Press.

Steinkraus, K.H., editor. 2004. *Industrialization of Indigenous Fermented Foods, Revised and Expanded*. Boca Raton, FL: CRC Press.

Tamime, A.Y. and Robinson, R.K. 1999. *Yogurt: Science and Technology*. Boca Raton, FL: CRC Press.

Toldra, F., Sanz, Y., and Flores, M. 2001. Meat fermentation technology. In: *Meat Science and Applications*, edited by Y.H. Hui, W.K. Nip, R.W. Rogers, and O.A. Young, pp. 538–591. New York: Marcel Dekker, Inc.

Townsend, W.E. and Olsen, D.G. 1987. Cured meat and meat products processing. In: *The Science of Meat and Meat Products*, edited by J.F. Price and B.S. Scheweigert, pp. 431–456. Westport, CT: Food & Nutrition Press.

Walstra, P., Geurts, T.J., Noomen, A., Jellema, A., and van Boekel, M.A.J.S. 1999. *Dairy Technology: Principles of Milk Properties and Processes*. New York: Marcel Dekker, Inc.

Xiong, Y.L., Yang, F.Q., and Lou, X.Q. 1999. Chinese meat products. In: *Asian Food Products: Science and Technology*, edited by C.Y.W. Ang, K.S. Liu, and Y.W. Huang, pp. 201–213. Lancaster, PA: Technomic Publishing Co., Inc.

2

Dairy Starter Cultures

Ana Rodríguez, Beatriz Martínez, and Juan E. Suárez

CONTENTS

2.1 Introduction

Starter cultures have been used for centuries in the production and preservation of fermented dairy products. However, it was unknown until the nineteenth century that indigenous lactic acid bacteria (LAB) present in raw milk were responsible for the coagulation of milk and cream ripening. The use of small portions of the coagulated milk as inoculum (back-slopping) for cheese or fermented milk manufacture purposes became the origin of the current starters, as this practice resulted in fermented products with valued/enhanced organoleptic properties. In 1890, Conn in the United States, Storch in Denmark, and Weigmann in Germany observed that pure cultures of *Streptococcus lactis* and/or *Streptococcus cremoris* (acid producers) prompted the production of good flavored butter. However, the traditional butter flavor was not achieved until 1919 when mixed cultures of acid producers and aroma producers (*Streptococcus diacetilactis* and *Leuconostoc*) were used by Hammer and Bailey (United States), Storch (Denmark), and de Vries (Holland). Diacetyl was identified as the main butter aroma compound with citrate as its precursor (Michaelian et al. 1938).

Nowadays, most of the cheeses and fermented milks are manufactured with specific starters that provide controlled fermentation conditions, with the main role of LAB starters being the production of lactic acid from lactose, but they also contribute to the production of relevant aroma compounds. Besides starter performance, new attributes are currently sought to meet consumer's demands for safe, health-promoting fermented products as well as to improve starter production technologies.

2.2 Composition of Dairy Starters

A starter culture is defined as a microbial preparation of large numbers of cells of at least one microorganism to be added to raw or pasteurized milk to produce a dairy fermented product. The main components of dairy starter cultures are LAB, a phylogenetically and functionally related group of low G + C Gram-positive bacteria that produce lactic acid as unique or main product from lactose, the major fermentable sugar of milk. The homofermentative and heterofermentative pathways used by LAB to convert carbohydrates to lactate have been extensively studied (Kandler 1983; Cocaign-Bousquet et al. 1996; Figure 2.1 illustrates the pathways of lactose catabolism by LAB).

The bacterial species involved in starter cultures belong to the genera *Lactococcus, Leuconostoc, Lactobacillus,* and *Streptococcus.* Recently, *Enterococcus* and *Bifidobacterium,* a high G + C Gram-positive actinobacterium, have been incorporated into dairy starter cultures. Other bacteria, yeast, and molds are also involved in the manufacture of cheese and fermented milk, producing biochemical and organoleptic changes in the fermented products (Table 2.1 shows some starter strains used in fermented dairy products).

Rapid growth and acid production in milk require efficient systems for lactose fermentation, as well as milk protein degradation systems. Lactose uptake occurs by two different transport systems—lactose entering the cell either as free sugar (permease system) or as sugar phosphate [phosphoenol

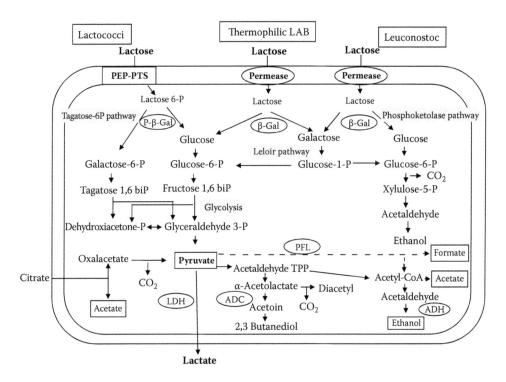

FIGURE 2.1 Pathways of lactose catabolism by LAB and some enzymes involved. P-β-gal, phospho-β-D-galactosidase; β-gal, β-galactosidase; LDH, lactate dehydrogenase; ADC, acetolactate decarboxylase; PLF, pyruvate-formate lyase; ADH, alcohol dehydrogenase.

TABLE 2.1

Some Starter Strains Used in Fermented Dairy Products

Starter Organism	Lactose Fermentation	Metabolic Products	Fermented Products
Lactic Acid Bacteria (Mesophilic)			
Lactococcus lactis subsp. *lactis,* *Lac. lactis* subsp. *cremoris*	Homofermentative	L(+)-lactate	Cheddar, Cottage, Feta, Edam, Gouda, Camembert cheeses
Lac. lactis subsp. *cremoris*	Homofermentative	L(+)-lactate	Viili
Lac. lactis subsp. *lactis* biovar. *diacetylactis*	Homofermentative	L(+)-lactate; diacetyl	Gouda, Edam, Cheddar, Buttermilk, Nordic milks
Leuconostoc mesenteroides subsp. *cremoris*	Heterofermentative	D(+)-lactate, diacetyl, ethanol, CO_2	Cheddar, Buttermilk, Sour cream, Viili
Lactobacillus fermentum, *L. kefiranofaciens* *L. casei, L. plantarum, L. curvatus*	Heterofermentative	D,L-lactate	Yogurt, Kefir NSLAB in long ripened cheeses
Lactic Acid Bacteria (Thermophilic)			
Streptococcus thermophilus	Homofermentative	L(+)-lactate, acetaldehyde, diacetyl	Yogurt, Gruyere, Emmental, Grana, Mozzarella
L. delbrueckii subsp. *bulgaricus*	Homofermentative	D(−)-lactate, acetaldehyde, diacetyl	
L. acidophilus, L. helveticus	Homofermentative	D,L-lactate	Acidophilus milk, Gruyere, Emmental, Grana cheeses
Probiotics			
L. rhamnosus GG	Heterofermentative	L(+)-lactate	Viili
L. casei Shirota	Heterofermentative	D,L-lactate	Yakult
Bifidobacterium adolescentis, *B. animalis, B. bifidum, B. breve,* *B. infantis, B. lactis, B. longum*	Heterofermentative	L(+)-lactate, acetate	Yogurt, Yakult, Buttermilk, Sour cream
Other Bacteria			
Propionibacterium freuderenreichii		Propionate, acetate, CO_2	Emmental, Gruyere
Brevibacterium linens		Volatile sulfur compounds	Munster, Limburger, Brick (surface-ripened cheeses)
Yeasts			
Kluyveromyces marxianus, *K. lactis, Saccharomyces* *cerevisiae, Debaryomyces* *hansenii, Pichia fermentans*		Ethanol, CO_2, Propanal	Kefir, Koumiss, Viili, surface-ripened cheeses, blue-veined cheeses
Molds			
Penicillium roqueforti (blue mold) *Pen. camemberti* (white mold)		Methyl ketones, secondary alcohols, esters, aldehydes, lactones	Danish Blue, Blue Stilton, Roquefort, Gorgonzola, Spanish blue cheeses Camembert, Brie
Geotrichum candidum		Dimethyl disulfide, methyl ketones, ammonia, free fatty acids	Viili, Kefir, Camembert, Saint Nectaire cheeses

pyruvate–phosphotransferase system (PEP–PTS)]. In *Lactococcus lactis* and *Lactobacillus casei* strains, lactose enters the cytoplasm as lactose phosphate by the PEP–PTS (phosphorylation occurs at the galactose moiety), and it is further hydrolyzed by phopho-β-D-galactosidase (P-β-gal) to glucose and galactose-6-phosphate (Gal6P), which are metabolized by the glycolytic and the tagatose-6-phosphate pathways, respectively, to pyruvate that is mainly converted into lactic acid by the enzyme lactate dehydrogenase. However, a shift from homolactic (lactate production) to acid-mixed metabolism (etanol, acetate, formate) occurs in anaerobiosis when a limited concentration of carbohydrates (lactose and glucose) is available, whereas acetate, acetaldehyde, etanol, acetoin, diacetyl, and 2,3-butanediol are produced in aerobiosis (Thomas et al. 1979; Axelsson 2004).

In other lactic acid species such as *Leuconostoc* subsp., *Streptococcus thermophilus*, *Lactobacillus delbrueckii* subsp. *bulgaricus*, *L. delbrueckii* subsp. *lactis*, *Lactobacillus acidophilus*, and *Lactobacillus helveticus*, lactose enters the cytoplasm as lactose by the permease system. A subsequent cleavage by β-galactosidase (β-gal) to glucose and galactose occurs (Crow and Thomas 1982). The energy required for this system is provided in the form of proton motive force or other transmembrane potential. Glucose is metabolized by the glycolytic pathway in *Lactobacillus* sp. and *S. thermophilus* and by the phosphoketolase pathway in *Leuconostoc*. Galactose is metabolized to glucose-1-phosphate via the Leloir pathway. However, galactose is excreted into the medium by *S. thermophilus*, *L. delbrueckii* subsp. *bulgaricus*, and *L. delbrueckii* subsp. *lactis* (Grossiord et al. 1998).

Besides lactose fermentation, LAB possess a proficient proteolytic system (cell-envelope proteinase; transport systems that translocate peptides across the cytoplasm membrane; peptidases) responsible for the enzymatic degradation of caseins into peptides and free amino acids, which are precursors of volatile compounds involved in flavor development. Amino acid catabolism (decarboxylation, deamination, transamination, desulfurization, and hydrolysis of side chains) produces a number of metabolites such as amines, alcohols, ketones, aldehydes, esters, sulfur-containing compounds, and carboxylic acids (Fernández and Zúñiga 2006; Fernández et al. 2008). The proteolytic activity is also important to decrease the presence of hydrophobic peptides involved in cheese bitterness (Sridhar et al. 2005; Figure 2.2 illustrates the proteolytic system in LAB).

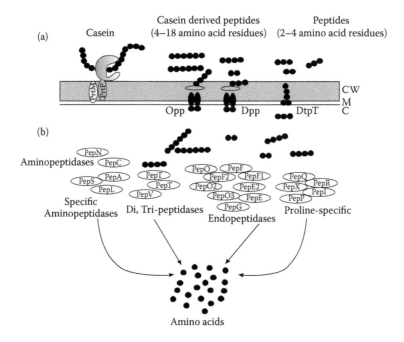

FIGURE 2.2   Diagram of the proteolytic system in LAB. (a) PrtP, cell-envelope proteinase; PrtM, membrane-bound lipoprotein (proteinase activator); Opp, oligopeptide permease; Dpp, the ABC transporter for peptides; DtpT, the ion-linked transporter for di- and tripeptides. (b) Intracellular peptidases. CW, cell wall; M, cell membrane; C, cytoplasm.

Some LAB species (*Lactococcus lactis* subsp. *lactis* var. *diacetylactis* and *Leuconostoc* species) produce a variety of flavor compounds (acetaldehyde, diacetyl, acetoin, and 2,3-butylene glycol) from the citrate present in milk. These compounds have aromatic properties and provide aroma to several dairy products.

2.3 Starter Function

The primary function of dairy starter bacteria is to produce lactic acid from lactose during the fermentation of milk, resulting in a pH decrease that is of great importance for milk coagulation and prevention of growth of spoilage and pathogenic microorganisms. They also contribute to flavor and aroma development in dairy fermented products, with the enzymatic activity of LAB having an important contribution to the rheological, organoleptic, and nutritional characteristics of dairy fermented products (Leroy and De Vuyst 2004).

The acidification produced by the dairy starter during the manufacture of cheese promotes the curd syneresis (whey expulsion) and improves rennet activity and its retention in curd, promoting the conversion of milk protein into peptides and amino acids by the proteolytic system of LAB and other protein hydrolyzing enzymes (rennet, milk plasmin). Consequently, acidification also influences cheese texture and flavor and the meltability of cheese by solubilizing the colloidal calcium phosphate (Cogan and Daly 1987). Moreover, CO_2 production from citrate (homofermentative and heterofermentative species) or lactose (heterofermentative species) results in eye formation in the cheese matrix. In the case of fresh cheese, butter, and some fermented milks, diacetyl, produced from citrate by *Lac. lactis* subsp. *lactis* var. *diacetylactis* strains, is the main aroma component (Hugenholtz et al. 1993). The fresh and distinctive acidic flavor of the fermented milks is complemented with other flavor components. In the case of yogurt, the degradation of threonine by *L. delbrueckii* subsp. *lactis* produces acetaldehyde, the major flavor component of yogurt.

2.4 Types of Starter Cultures

Starter cultures are classified according to their main function (primary and secondary starters), temperature of growth (mesophilic and thermophilic starters), and composition (natural whey starters, natural milk starters, mixed-strain starters, defined-strain starters). *Primary starters* are responsible for the production of lactic acid from lactose, whereas *secondary starters* are a diverse microbiota only involved in cheese ripening, comprising nonstarter lactic acid bacteria (NSLAB) (lactobacilli, enterococci), propionibacteria, corynebacteria, yeasts, and molds.

2.4.1 Primary Starters

2.4.1.1 Mesophilic and Thermophilic Starters

According to the optimum temperature of growth, primary starters are classified as *mesophilic* (about 30°C) and *thermophilic* (about 42°C).

Mesophilic starters contain acid-producing species *Lac. lactis* subsp. *lactis* and/or *Lac. lactis* subsp. *cremoris*, but aromatic species such as citrated-utilizing strains of *Lac. lactis* subsp. *lactis* and/or *Leuconostoc mesenteroides* subsp. *cremoris* and *Leu. lactis* are often included. The main function of the aromatic strains is the production of CO_2 and flavor compounds. They are used in the production of a great number of cheeses (Gouda, Edam, Cheddar, Camembert, etc.), fermented milks, and ripened cream butter (Parente and Cogan 2004).

Thermophilic starters contain species such as *S. thermophilus*, *L. delbrueckii* subsp. *bulgaricus*, *L. delbrueckii* subsp. *lactis*, *L. acidophilus*, and *L. helveticus*. These starter cultures are used in the production of yogurt and high-cooking temperature cheeses (Gruyere, Emmental, Grana, Pecorino, etc.). *S. thermophilus* and *L. delbrueckii* subsp. *bulgaricus* show a synergistic relationship as the former produces formic acid, promoting the growth of *L. bulgaricus*, whereas the latter stimulates *S. thermophilus* by releasing amino acids. Mixtures of mesophilic and thermophilic species are used in the manufacture of some cheeses (Parente et al. 1997).

2.4.1.2 Natural Whey and Natural Milk Starters

Current starter cultures are derived from *artisanal* or *natural* starters containing an undefined mixture of different species and/or strains that have to be reproduced and maintained daily in cheese factories under selective conditions (temperature, pH, etc.), aiming at selecting the best adapted strains. These starters are used widely in Europe. According to the composition, two different subtypes are distinguished.

Natural whey starters are prepared by incubating the whey from the previous milk fermentation under specific conditions, for example, overnight incubation at 45°C resulting in pH 3.3. *L. helveticus* is the dominant species but *L. delbrueckii* subsp. *lactis*, *L. fermentum*, and *S. thermophilus* are also present. These starters are used in the manufacture of extra hard (Grana Padano and Parmigiano Reggiano) and *pasta filata* Italian cheeses, hard cheese Argentinian varieties, and Comté cheese in France. Deproteinized whey starters are used for Pecorino cheese, and they are supplemented with rennet for the manufacture of Swiss-type cheeses (Parente and Cogan 2004).

Natural milk starter preparations involve heat treatment of raw milk (62°C–65°C, 10–15 min) followed by incubation (37°C–45°C) in order to select the desired microbiota. *S. thermophilus* is in this case the dominant species, but *Enterococcus faecalis*, *Enterococcus faecium*, *L. helveticus*, *L. delbrueckii* subsp. *bulgaricus/lactis*, and *L. casei* are usually present. These starters are used in Italy for the production of traditional cheeses (Limsowtin et al. 1997; Parente and Cogan 2004; Carminati et al. 2010).

2.4.1.3 Mixed- and Defined-Strain Starters

Nowadays, commercial *mixed-strain starters* derived from selected natural starters are often used in cheesemaking. They are a mixture of an unknown number of LAB types with a well-recognized phage resistance (Mäyrä-Makinen and Bigret 1998). They show a great stability in composition and performance because they are prepared under controlled conditions by specialized starter producers. Concentrated cultures for direct inoculation of the cheese milk are available, which avoid the need for transfers to build up the bulk starter and reduce the risk of variability in starter composition and performance. Mixed starter cultures are classified as mesophilic and thermophilic according to the optimum temperature of growth. The dominant species present in the mesophilic starters are *Lac. lactis* subsp. *lactis* and *Lac. lactis* subsp. *cremoris*, but citrate-fermenting species (*Lac. lactis* subsp. *lactis* var. *diacetylactis* and *Leuconostoc* species) are usually present (Hugenholtz 1986). Their performance is exemplified by the successful use in the production of Dutch (Stadhouders and Leenders 1984) and Swiss cheese varieties (Glättli 1990), respectively.

In contrast to mixed-strain starters, the species and strain composition and their ratio are known in *defined-strain starters*. Their technological performance is highly reproducible, but the risk of phage attack in these cultures is higher than in undefined mixed cultures where blends of two or more strains are present. To overcome this problem, phage-unrelated strains and selection of bacteriophage-insensitive mutants are used in rotation (Limsowtin et al. 1997; Heap 1998; Moineau 1999). Mesophilic defined-strain starters are used in New Zealand, Australia, the United States, and Ireland. On the other hand, thermophilic defined-strain starters are used for the production of Swiss- and Italian-type cheeses. In the manufacture of Mozzarella cheese, single or multiple strains of *S. thermophilus* or mixtures of *S. thermophilus* and *L. delbrueckii* subsp. *bulgaricus* are used depending on the manufacture of high- or low-moisture cheese, respectively.

Traditionally, "bulk starter" in liquid form was used to inoculate milk in the manufacture of cheese, yogurt, buttermilk, and other fermented products, but nowadays, concentrated freeze-dried or frozen starters designated as either direct vat set or direct vat inoculation (DVI) cultures are often used, particularly in small plants. The use of DVI cultures reduces the chance of contamination of the starter by phages present in the factory environment because handling of bulk starters is avoided.

2.4.1.4 Sources of New Primary Starters

The natural starters are a valuable source of strains with desired physiological and metabolic features (proper acidification rate, aroma compound production, antimicrobial production, phage resistance, etc.). However, the daily propagation can lead to unwanted composition fluctuations and, consequently, to an unpredictable performance. On the other hand, the extensive use of standardized starters has had a great positive effect on the quality of fermented dairy products, but the diversity of their organoleptic characteristics has clearly been diminished.

Therefore, efforts have been made to isolate new strains from raw milk and fermented dairy products produced without starter cultures to complement those that are currently available in industrial starters (Wouters et al. 2002). The new strains should have unique and diverse properties compared to commercial starters, such as higher acidifying activity (Parente and Cogan 2004), higher bacteriophage resistance (Madera et al. 2003), a distinct flavor profile (Ayad et al. 2001; Smit et al. 2005), and production of antimicrobials (De Vuyst and Leroy 2007). Traditional Spanish starter-free cheeses made with raw milk have been shown as a source of *Lac. lactis* bacteriocin-producing strains (Martínez et al. 1995; Alegría et al. 2010). Texture properties (exopolysaccharide production), postacidification, strain interaction, and resistance to specific manufacture conditions are particularly taken into account when screening new strains for the manufacture of fermented milk (Tamine et al. 2006).

The new strains will contribute to expand the range of fermented dairy products with differential characteristics.

2.4.2 Secondary Starters

Secondary starter cultures are defined as those cultures used for the manufacture of cheese, in which their main function is to develop and control the flavor, aroma, and texture of the cheese. Their development is preceded by the fermentation of lactose by the primary starter cultures.

2.4.2.1 Nonstarter Lactic Acid Bacteria

The secondary microbiota of several varieties of cheeses consists mostly of facultative heterofermentative (mesophilic) lactobacilli and is referred to as nonstarter lactic acid bacteria (NSLAB). Lactobacilli, which are present in raw milk and the dairy environment, are regarded as desirable contaminants of milk due to their contribution to cheese flavor by forming small peptides and amino acids, which are the precursors of flavor compounds (Wouters et al. 2002). Their heat resistance facilitates their presence in cheeses made with pasteurized milk.

NSLAB consist of a wide variety of strains that vary among cheeses and the ripening time. Indeed, this secondary microbiota also depends on the composition of the primary starter used for the manufacture of cheese (Martley and Crow 1993). They grow to considerable numbers in cheeses ripened for long periods. *L. casei*, *L. paracasei*, *L. plantarum*, *L. curvatus*, and *L. rhamnosus* are the facultative heterofermentative species found in cheeses, but obligatory heterofermentative lactobacilli (*L. brevis*, *L. buchneri*, and *L. fermentum*) are also found and have been considered as spoilage and pathogenic bacteria (Chamba and Irlinger 2004). For instance, strains of *L. buchneri* and *L. brevis* are responsible for the production of histamine in Swiss-type cheese and tyramine in Gouda cheese, respectively (Taylor et al. 1982; Joosten and Northolt 1989). NSLAB are also responsible for the formation of unappetizing calcium lactate crystals in Cheddar cheese due to production of D(-)-lactate (Chou et al. 2003).

2.4.2.2 Propionibacteria

The propionic acid bacteria (PAB) constitute the essential secondary microbiota in Swiss-type cheeses. They are Gram-positive bacteria, nonmotile, nonsporulating, small rods, and anaerobic to aerotolerant mesophiles. *Propionibacterium freudenreichii* is the most common species found in hard cooked cheese (up to 10^9 CFU/g in Emmental cheese). The main role of *P. freudenreichii* is the conversion of lactate produced by homofermentative LAB into propionate, acetate, and carbon dioxide. The latter is the one responsible for the presence of big eyes in the Swiss-type cheeses. Propionate and acetate contribute to the preservation and the taste of these cheeses. PAB have an important caseinolytic potential and synthesize a wide spectrum of intracellular aminopeptidases that hardly get into the cheese matrix because of the limited autolysis of these bacteria (Ostlie et al. 1999). PAB show a lipolytic activity up to 100-fold higher than LAB and produce aroma compounds through the release of free fatty acids and their further catabolism (Kerjean et al. 2000).

2.4.2.3 Corynebacteria

These aerobic, mesophilic, and salt-tolerant bacteria are present on the surface of surface-ripened cheeses (Munster, Limburger, Brick, etc.) and show a strong proteolytic activity, with lipolytic and esterolytic activities also detected (Wouters et al. 2002). They also produce red-orange pigments. The growth and metabolism of the smear population during ripening have an unambiguous impact on flavor development, the textural characteristics, and final appearance of the cheese. The dominant genera found in cheeses are *Brevibacterium*, *Arthrobacter*, *Corynebacterium*, and *Microbacterium*, depending on the type of cheese.

Brevibacterium linens is the main adjunct culture available for smear cheeses. Dipeptidyl peptidase and aminopeptidases that cleaved various N-terminal amino acids were detected in *Brevibacterium* and *Corynebacterium* strains. Some of them are able to form thiols from sulfur-containing amino acid precursors (Williams et al. 2004), producing volatile sulfur compounds involved in the distinct smear cheese aroma and flavor.

2.4.2.4 Yeasts

Yeast species are essential for the typical characteristics of some fermented milks (kefir, koumiss, viili) and surface-ripened and blue-veined cheese varieties. They grow well throughout manufacture and ripening of fermented dairy products because of their tolerance to low pH and low water activity. Some species are lactose-fermenting and others are able to metabolize lactate, which increases the pH and promotes the growth of pH-sensitive microbiota. In addition, yeasts show proteolytic and lipolytic activities, which can play substantial roles in the development of taste and aroma (Jakobsen and Narvhus 1996). Positive interactions between yeast and LAB include the production of carbon dioxide, pyruvate, propionate, and succinate (Leroy and Pidoux 1993). In addition, yeasts seem to promote the development of *Penicillium roqueforti* in blue cheeses and contribute to the open texture of these types of cheeses by formation of gas (van den Tempel and Jakobsen 2000). *Debaryomyces hansenii*, *Kluyveromyces marxianus*, *K. lactis*, *Yarrovia lipolitica*, *Saccharomyces cerevisiae*, *Zygosaccharomyces rouxi*, *Pichia fermentans*, and *Rhodotorula mucilaginosa* are the predominant yeasts in cheeses (Jakobsen and Narvhus 1996).

In the fermented milk kefir, the predominant yeasts species are *K. marxianus*, *Candida kefir*, and *Sac. cerevisiae*. *Kluyveromyces* species (lactose-fermenting) are permanently present and provide the typical yeast aroma in the fermented milk, whereas species of *Saccharomyces* are mainly detected in koumiss. Yeasts perform the alcoholic fermentation providing ethanol and carbon dioxide that account for the organoleptic properties of these fermented milks along with lactic acid and diacetyl produced by LAB (Wszolek et al. 2006).

2.4.2.5 Molds

Molds are used in the manufacture of semisoft cheese varieties in which they contribute to aroma and flavor enhancing, and also affect the body and the structure of cheese. *Penicillium camemberti* (white mold) and *Pen. roqueforti* (blue mold) are the species usually found in cheeses. The former grows on the surface of Camembert and Brie cheeses, whereas the latter grows inside of blue cheeses (Danish Blue, Blue Stilton, Roquefort, Gorgonzola; Wouters et al. 2002). *Pen. roqueforti* is also present in Spanish blue-veined cheeses (Fernández-Bodega et al. 2009). Both show high proteolysis and lipolysis that result in the typical volatile flavor compounds of cheeses such as methyl ketones, secondary alcohols, esters, aldehydes, and lactones (Gripon 1993).

Other highly proteolytic species such as *Mucor mucedo* and *M. racemosus* are involved in Gamalost cheese ripening and grow on the surface and inside the cheese (Rage 1993).

The yeast-like mold *Geotrichum candidum* has been detected on the surface of soft cheeses such as Camembert (Boutrou et al. 2006) and of semihard cheeses such as Saint Nectaire contributing to the ripening process by the breakdown of protein and fat, resulting in the production of important aroma compounds (Jollivet et al. 1994). *G. candidum* has been also isolated in the blue cheese Cabrales (Alvarez-Martín et al. 2008). This species is also a component of the starter used for the manufacture of viili, a Finnish fermented milk, on which it produces a white cream top layer (Fondén et al. 2006).

2.5 LAB Cultures with Technological and Functional Advantages

Dairy starter culture research has been conducted for the development of starter cultures with satisfactory performance and capacity to produce innovative dairy fermented products.

2.5.1 Phage-Resistant Cultures

One of the most critical problems in the fermentation processes conducted by LAB is the contamination of starters by bacteriophages that cause bacterial lysis and significant economic losses. Several strategies have been developed to control phage contamination in the dairy environment. These strategies have been based on strain selection and rotation, phage-insensitive mutants, design of phage-resistant starter strains, and antiphage systems. Nevertheless, *Lac. lactis* and *S. thermophilus* strains, involved in most of the industrial dairy fermentations, suffer the highest number of phage infections (Rodríguez et al. 2010). Phage resistance may arise by natural resistance mechanisms (inhibition of phage adsorption, DNA injection blocking, restriction–modification systems, and abortive infection; Moineau and Levésque 2005). Some of these mechanisms are encoded by plasmid genes that can be transferred by conjugation and seen as a natural strategy to generate bacteriophage-resistant starter strains, but the lack of suitable selective markers on plasmids makes this approach not always possible. The isolation of spontaneous bacteriophage-insensitive mutants (BIMs) is an alternative to the conjugation processes that do not require genetic manipulation. Indeed, many phage-resistant variants with high technological performance have been isolated and successfully used in industrial dairy fermentation (Quiberoni et al. 1999). The generation of BIMs was attributed to mutations in the phage receptor, but, recently, whole genome mining has revealed the presence of clustered regularly interspaced short palindromic repeats (CRISPRs) as an antiphage mechanism acquired by bacteria after phage infection. CRISPR loci have been found in the genome of *Lactobacillus*, *Streptococcus*, *Enterococcus*, and *Bifidobacterium*, but are absent in *Lactococcus* and *Leuconostoc* (Horvath et al. 2009). Three different CRISPR loci have been found in *S. thermophilus* (Deveau et al. 2008).

The construction of genetically engineered strains is nowadays possible, and it basically aims to exploit native phage defense mechanisms and some phage genetic elements. Phage-triggered suicide systems and antisense RNA technology have already been proposed (Djordejevic et al. 1997; Walker and Klaenhammer 2000). However, these genetically engineered starters have not made it to the market yet due to the restricted European legislation regarding genetically modified organisms (Renault 2010).

The increasing role of lactobacilli as probiotics in dairy fermentations has been accompanied by an increase in phage contaminations. These strains grow slowly, thus becoming more vulnerable to phages (Capra et al. 2006). In addition, a high incidence of lysogeny in *Lactobacillus* strains has been reported, and spontaneous prophage induction is thought to contribute to the generation of new virulent phages (Mercanti et al. 2011). Thus far, the development of phage-resistant probiotic starters is very limited (Carminati et al. 2010).

2.5.2 Bacteriocin-Producing Cultures

Preservation of fermented dairy products by LAB is primarily due to lactose conversion into lactic acid and other organic acids with the concomitant pH reduction (Stiles 1996). Besides, LAB synthesize other compounds with antimicrobial activity including hydrogen peroxide, diacetyl, acetaldehyde, carbon dioxide, reuterin, reutericyclin, and bacteriocins. The latter are small, heat-stable cationic peptides that display a wide spectrum of inhibition against Gram-positive foodborne and spoilage bacteria (Cotter et al. 2005).

LAB bacteriocins are a relatively heterogeneous group of peptides with regard to their primary structure, composition, and physicochemical properties. Different classifications have been proposed. A first classification comprised four major classes (Klaenhammer 1993), but Cotter et al. (2005) have simplified it and proposed two major classes—Class I (lantibiotics) and Class II (non-lantibiotic bacteriocins). Class I include single and two- post-translationally modified peptides containing the unusual amino acids lanthionine, dehydroalanine, and dehydrobutyrine. Up to 11 subclasses based on the sequence of the unmodified pro-peptides have been proposed. The peptide length ranges from 18 to 38 amino acids. Class II is a heterogeneous class of small peptides in which four subclasses are distinguished—IIa (pediocin-like bacteriocins), IIb (two-peptide bacteriocins), IIc (cyclic peptides), and IId (miscellaneous).

The lantibiotic nisin produced by *Lac. lactis* strains was the first described Gram-positive bacteriocin. It was initially applied to inhibit *Clostridium tyrobutyricum*, responsible for late blowing in cheese (Hirsch et al. 1951). Nowadays, it is the only bacteriocin authorized as a food preservative (E234) and has found widespread applications (Delves-Broughton et al. 1996). Lantibiotics (nisin, lacticin 481, lacticin 3147) with a wide spectrum of inhibition against foodborne pathogens and spoilage bacteria (*Staphylococcus aureus*, *Clostridium* species, *Bacillus* species, *Listeria monocytogenes*, etc.) and the pediocin-like bacteriocins, displaying a high specific activity against *Lis. monocytogenes*, are the two most important bacteriocin groups with applications in food preservation. On the other hand, enterocin AS-48, a representative of class IIc, has been proved very effective against a wide range of foodborne pathogens and spoilage bacteria in several food matrices.

Since LAB have been unintentionally consumed for centuries, having thus a long history of safe use, bacteriocinogenic strains and their bacteriocins can play an important role in dairy products' biopreservation. The relative tolerance of bacteriocins to technologically relevant conditions such as pH, NaCl, and heat treatment as well as the lack of toxicity toward eukaryotic cells support their use as food biopreservatives (Gálvez et al. 2007).

Bacteriocin producers can be used for *ex situ* production of bacteriocins, which are regarded as an ingredient that also contains other fermentation metabolites (ALTA2431 and Micrograd) or as an additive in a semipurified or purified preparation (Nisaplin; García et al. 2010). However, provided that the bacteriocin producer is able to grow in the food matrix (e.g., milk), *in situ* production is an interesting alternative because it is not affected by legal regulations and it is cost effective.

Several reports have shown the efficacy of bacteriocin producers as adjunct starters in combination with primary starters to improve the safety of fermented dairy products. For instance, nisin-producing strains of *Lac. lactis* have been successfully used for inhibiting *Lis. monocytogenes* in Camembert cheese (Maisnier-Patin et al. 1992), *Lis. innocua* in raw ewe's milk cheese (Rodríguez et al. 1998), and *Sta. aureus* in raw milk semihard cheese (Rodríguez et al. 2000) or in pasteurized milk acid-coagulated cheese (Rilla et al. 2004). Nisin-producing strains were also effective in preventing the late-blowing in cheese caused by *Clo. tyrobutyricum* (Hugenholtz et al. 1995; Rilla et al. 2003). *Lac. lactis* strains producing the two-component broad-spectrum bacteriocin lacticin 3147 reduced the level of *Lis. monocytogenes* on the surface of smear-ripened cheese (O'Sullivan et al. 2006) and in Cottage cheese (McAuliffe et al. 1999).

Besides food preservation, bacteriocin producers have also been applied to accelerate cheese ripening by increasing starter cell lysis, which results in higher concentrations of intracellular enzymes and the subsequent increase in free amino acids levels (Morgan et al. 2002). The effectiveness of bacteriocin producers to control the growth of the nonstarter lactic acid bacteria responsible for off-flavor defects in cheeses has been assessed. This strategy has been successfully tested in the manufacture of Cheddar cheese as the use of commercial starter strains to which the genetic determinants encoding lacticin 3147 have been transferred (Ryan et al. 1996) and lactococcal strains producing lacticin 481 (O'Sullivan et al. 2003) resulted in a significant reduction in the levels of nonstarter lactic acid bacteria.

Among LAB, enterococci produce a number of enterocins—bacteriocins able to inactivate pathogenic bacteria such as *Lis. monocytogenes*—suggesting their role as protective bacteria (Giraffa 1995). With regard to this, *E. faecalis* AS 48-32, enterocin AS 48 producer, has been used as an adjunct starter in combination with a commercial starter in the manufacture of both nonfat hard cheese and fresh cheese to control the growth of enterotoxigenic *Bacillus cereus* (Muñoz et al. 2004) and *Sta. aureus* (Muñoz et al. 2007), respectively. Enterococci have a long history of safe use because they are present in fermented dairy products manufactured in Mediterranean countries (Giraffa 2002). However, some strains harbor antibiotic resistance and virulence genes. Therefore, the safety of enterocin-producing strains must be assessed prior to application in food (Ogier and Serror 2008).

2.5.3 Exopolysaccharide-Producing Cultures

Some LAB species produce long-chain, high-molecular mass polymers, named exopolysaccharides (EPSs), which positively contribute to the texture, stability, mouth-feel, and taste perception of some dairy products (Jolly et al. 2002). Based on their sugar composition, EPSs are divided into homopolysaccharides (one sugar type) and heteropolysaccharides (repeating units of two or more sugars). The capacity of the latter group to improve the rheological characteristics of fermented milks has been reported and shown to be responsible for the "ropy" phenotype of these fermented products (De Vuyst and Degeest 1999). More recently, health benefits have also been attributed to these polymers (Ruas-Madiedo et al. 2002). Indeed, they are suggested as prebiotics (nondigestible food ingredients that stimulate the growth and/or activity of bacteria in the gastrointestinal tract), cholesterol-lowering, and immunomodulants.

EPS-producing strains show a clear technological advantage when used in the manufacture of dairy products. For instance, the use of EPS-producing strains of *S. thermophilus* and *L. delbrueckii* subsp. *bulgaricus* reduces syneresis and graininess in yogurt and enhances its texture and viscosity (Hassan et al. 2003). They also contribute to the consistency of stirred-type yogurts because they suffer less mechanical damage throughout the manufacture process (pumping, blending, filling; De Vuyst and Degeest 1999). However, an excessive production of EPS may negatively alter firmness and elasticity of the yogurt due to the formation of large voids within the protein matrix as a consequence of EPS interference with protein binding (Rawson and Marshall 1997).

The use of EPS-producing strains is widespread in countries where the addition of animal/plant-derived stabilizers is banned (Cerning 1995). Indeed, dairy starter cultures containing EPS-producing LAB strains are commercially available in Europe and the United States. The firm, thick, slimy consistency typical of Scandinavian fermented milk drinks (i.e., viili) is mostly due to dominant microbiota composed of EPS-producing mesophilic strains (*Lac. lactis* subsp. *lactis* and *Lac. lactis* subsp. *cremoris*), whereas in others, EPS-producing *S. thermophilus* strains are used to provide proper texture. These thermophilic strains have been isolated from kefir and acid and alcoholic fermented milk from Eastern Europe (Duboc and Mollet 2001).

EPS-producing strains have also been used to improve the texture and functionality of cheese. However, the presence of EPS in cheese whey increases viscosity, reducing the efficiency of membrane processing and limiting the interest on EPS producers for cheesemaking. Nevertheless, EPS-producing *S. thermophilus* has been successfully used in the manufacture of low-fat Mozzarella to counteract the undesirable effects of removing fat on the physical properties of cheese (rubbery, melt properties, etc.). In this way, cheese moisture level, yield, and melt properties were clearly enhanced (Petersen et al. 2000).

2.5.4 Probiotic Cultures

Probiotics are defined as strains of living microorganisms whose ingestion in certain doses confers health benefits beyond inherent basic nutrition (FAO/WHO 2002). For this reason, the consumption of dairy products supplemented with probiotic strains has shown a great increase in the last 15 years, turning the production of these dairy products into a multimillion Euro business. Indeed, a large number of health benefits have been attributed to products containing probiotic bacteria—alleviation of lactose intolerance, prevention and reduction of diarrhea symptoms, treatment and prevention of allergy, hypocholesterolemic effect, prevention of inflammatory bowel disease, inhibition of intestinal pathogens and *Helicobacter pylori*, and modulation of the immune system (Vasiljevic and Shah 2008).

The selection of probiotic strains should fulfill several requirements—human origin, nonpathogenicity, survival during gastric transit, tolerance to bile salts, and adhesion to gut epithelial tissue. Finally, probiotics should confer a health benefit to consumers. Nowadays, commercial probiotic cultures are available to be used as adjunct cultures. They include strains of *Lactobacillus* (*Lac. acidophilus*, *Lac. delbrueckii* subsp. *bulgaricus*, *Lac. paracasei*, *Lac. jonhsonii*, *Lac. rhamnosus*, *Lac. casei*, *Lac. reuteri*) and *Bifidobacterium* (*B. animalis* subsp. *lactis*, *B. lactis*, *B. longum*, and *B. breve*; Shah 2004).

To guarantee their functionality, probiotics need to be delivered in active and viable form, with the critical concentration being over 10^7 CFU/g or mL (Lourens-Hattingh and Viljeon 2001). Some probiotic strains, however, grow poorly in milk, and consequently, the final concentration in fermented products is very low (Masco et al. 2005). Therefore, probiotics have been traditionally incorporated in yogurt and fermented milks as adjunct cultures, the starter strains (i.e., *S. thermophilus* and *L. delbrueckii* subsp. *bulgaricus*) leading milk fermentation (Shah 2004). These species have been recently regarded as microorganisms with probiotic potential. Indeed, they can release enzymes (β-galactosidase) that improve nutrient digestion (Guarner et al. 2005).

Cheese can be an alternative to yogurt and fermented milks as a delivery system for probiotic strains. In fact, it has some advantages over fermented milks, such as higher pH, tighter matrix, and relatively high buffering capacity, all of them contributing to a better survival of probiotics in the product. In this regard, the probiotic strain *L. delbrueckii* subsp. *lactis* UO 004 was able to survive at high cell numbers (10^8–10^9 CFU/g) after 28 days of ripening in washed-curd cheese (Vidiago type) made with goat's milk. The incorporation of this probiotic strain did not negatively affect the sensory properties of cheese (Fernández et al. 2005). Other probiotic lactobacilli (*L. plantarum*, *L. fermentum*, and *L. rhamnosus*) used as adjunct starters in cheeses have also improved the intensity of cheese flavor (Briggiler-Marcó et al. 2007).

2.6 Trends in Starter Culture Research: Advent of "Omics"

The availability of whole genome sequences of the microorganisms involved in dairy fermentations offers new means to address the production of tastier and healthier dairy products. Indeed, analysis of genome sequences provides a first insight into the strains' metabolic potential, and metabolic models have been constructed in *Lac. lactis* subsp. *lactis*, (Oliveira et al. 2005), *L. plantarum* (Teusink et al. 2005), and *S. thermophilus* (Pastink et al. 2009).

At present, 25 LAB genomes have been sequenced and annotated, comprising strains of *Lactobacillus*, *Lactococcus*, *Streptococcus*, *Leuconostoc*, *Pediococcus*, and *Oenococcus* (Mayo et al. 2008). Particular attention has been paid to the genome sequences of the *Lac. lactis* strains published to date because they are one of the main components in both artisanal and industrial cultures. Comparative and functional genomics of the strains *Lac. lactis* subsp. *lactis* IL1403, *Lac. lactis* subsp. *cremoris* MG1363, and the "true" *Lac. lactis* subsp. *cremoris* SK11 have revealed that the genome size of the two *cremoris* strains (2.53 and 2.44 Mbp) is larger than that of the *lactis* strain (2.37 Mbp). The *cremoris* strains share a 97.7% identity in the coding domains, whereas only 85% identity is found between *Lac. lactis* IL1403 and *Lac. cremoris* SK11. The presence of 17 genes involved in carbohydrate metabolism and transport in MG1363 suggests a greater capacity to grow on different sugars such as those found in plant material. Its adaptation to the milk environment appears to have involved changes in its metabolic activity along with the acquisition of plasmids and mobile elements from other bacteria. Indeed, many lactococcal traits essential for dairy fermentation are encoded on plasmids (lactose fermentation, casein breakdown, bacteriophage resistance, exopolysaccharide production) and are responsible for the strain individuality (Hols et al. 2005).

S. thermophilus is also an important thermophilic starter traditionally used in combination with *L. delbrueckii* subsp. *bulgaricus* or *L. helveticus* for the manufacture of yogurt and hard cooked and *pasta filata* cheeses. The three genome sequences available (strains CNRZ 1066, LMD-9, LMG18311) are approximately 1.8 Mbp in size. Genome mining indicated that *S. thermophilus* strains have diverged from its pathogen relatives (*S. pneumoniae, S. pyogenes, S. agalactiae*) by loss of most important pathogenic determinants. In addition, genome analysis has established differences regarding the presence or absence of the CRISPR loci (see phage-resistant starters), which have been recently characterized. On the other hand, the genome-scale metabolic model of *S. thermophilus* LMG18311 has determined its minimal autotrophy and its broad range of volatiles compared to *Lac. lactis* and *L. plantarum*. Moreover, the pathway for acetaldehyde production involved in yogurt flavor could also be identified (Pastink et al. 2009). In contrast to lactococci, the presence of plasmids in *S. thermophilus* plays an insignificant role because few industrially useful phenotypic traits are plasmid encoded (Renye and Somkuti 2008).

The available genomes of thermophilic lactobacilli include the starter strains *L. delbrueckii* subsp. *bulgaricus* ATCC11842 and ATTCC BAA-365 (1.8 Mbp) and *L. helveticus* DPC4571 (2.0 Mbp; Mayo et al. 2008). The transition to a nutritionally rich environment must have been behind the metabolic simplification shown by these strains characterized by incomplete metabolic pathways and few regulatory functions (Callanan et al. 2008). Of note, the complex symbiotic relationship between *L. delbrueckii* subsp. *bulgaricus* and *S. thermophilus* established during the manufacture of yogurt has promoted horizontal transfer of several genes between the two species as observed by the bioinformatic analysis of both genomes (Liu et al. 2009).

Once genome information became available, a main step forward has been made to understand the complexity behind milk fermentation because a platform for *in situ* transcriptomic studies has now been established. Recent reports based on transcriptomics during milk fermentation have revealed, for example, that it is not only the presence or absence of genes that matters but also tuned regulation that clearly impacts starter performance (Tan-a-ram et al. 2011). Also, an in-depth analysis of the symbiosis between yogurt strains has been approached as well as the *in situ* stress responses of starter bacteria (Sieuwerts et al. 2010; Cretenet et al. 2011).

2.7 Concluding Remarks

The dairy starter industry has successfully solved bottlenecks encountered in bulk production and starter performance and robustness. However, new ones are envisaged to meet the increasing demand of distinct, traditionally tasting, safe, and healthy dairy products by todays' consumers. Major challenges are found in the field of probiotic cultures that are very demanding in terms of large-scale production and maintenance. Special attention is currently given to design proper delivery systems to keep them viable and functional, which is the basis of their health-promoting properties. Moreover, new means to face the development of dairy starters are foreseen with the advent of "omics" technologies and system approaches that will definitely help in understanding dairy fermentation from a more general perspective.

REFERENCES

Alegría, A., Delgado, S., Roces, C., López, B., and Mayo, B. 2010. Bacteriocins produced by wild *Lactococcus lactis* strains isolated from traditional, starter-free cheeses made of raw milk. *Int. J. Food Microbiol.* 143:61–66.

Alvarez-Martín, P., Flórez, A.B., Hernández-Barranco, A., and Mayo, B. 2008. Interaction between dairy yeasts and lactic acid bacteria strains during milk fermentation. *Food Control* 19:62–70.

Axelsson, L. 2004. Lactic acid bacteria: Classification and physiology. In: *Lactic Acid Bacteria. Microbiological and Functional Aspects*, 3rd edition, edited by S. Salminen, A. von Wright, and E. Ouwehand, pp. 1–68. New York: Marcel Dekker, Inc.

Ayad, E.H.E., Verheul, A., Bruinenberg, P., Wouters, J.T.M., and Smit, G. 2001. Enhanced flavour formation by combination of selected lactococci from industrial and artisanal origin with focus on completion of a metabolic pathway. *J. Appl. Microbiol.* 90:59–67.

Boutrou, R., Kerriou, L., and Gassi, J.Y. 2006. Contribution of *Geotrichum candidum* to the proteolysis of soft cheese. *Int. Dairy J.* 16:775–783.

Briggiler-Marcó, M., Capra, M.L., Quiberoni, A., Vinderola, G., Reinheimer, J.A., and Hynes, E. 2007. Nonstarter lactobacillus strains as adjunct cultures for cheese making: In vitro characterization and performance in two model cheeses. *J. Dairy Sci.* 90:4532–4542.

Callanan, M., Kaleta, P., O'Callaghan, J., O'Sullivan, O., Jordan, K., McAuliffe, O., Sangrador-Vegas, A., Slattery, L., Fitzgerald, G.F., Beresford, T., and Ross, R.P. 2008. Genome sequence of *Lactobacillus helveticus*, an organism distinguished by selective gene loss and insertion sequence element expansion. *J. Bacteriol.* 190:727–735.

Capra, M.L., Quiberoni, A., and Reinheimer, J.A. 2006. Phages of *Lactobacillus casei/paracasei*: Response to environmental factors and interaction with collection and commercial strain. *J. Appl. Microbiol.* 100:334–342.

Carminati, D., Giraffa, G., Quiberoni, A., Binetti, A., Suárez, V., and Reinheimer, J. 2010. Advances and trends in starter cultures for dairy fermentations. In: *Biotechnology of Lactic Acid Bacteria. Novel Applications*, edited by F. Mozzi, R.R. Raya, and G. Vignolo, pp. 177–192. Singapore: Wiley-Blackwell.

Cerning, J. 1995. Production of exopolysaccharides by lactic acid bacteria and dairy propionibacteria. *Lait* 75:463–472.

Chamba, J.F. and Irlinger, F. Secondary and adjunct cultures. 2004. In: *Cheese: Chemistry, Physics and Microbiology*, Vol. I, edited by P.F. Fox, P.J.H. McSweeney, T.M. Cogan, and T.P. Guinee, pp. 190–206. Amsterdam: Elsevier.

Chou, Y.E., Edwards, C.G., Luedecke, L.O., Bates, M.P., and Clark, S. 2003. Non starter lactic acid bacteria and aging temperature affect calcium lactate crystallization in Cheddar cheese. *J. Dairy Sci.* 86:2516–2524.

Cocaign-Bousquet, M., Garrigues, C., Loubiere, P., and Lindley, N.D. 1996. Physiology of pyruvate metabolism in *Lactococcus lactis*. *Antonie van Leeuwenhoek* 70:253–267.

Cogan, T.M. and Daly, C. 1987. Cheese starter cultures. In: *Cheese: Chemistry, Physics and Microbiology*, Vol. I, edited by P.F. Fox, pp. 179–250. London: Elsevier Applied Science Publishers Ltd.

Cotter, P.D., Hill, C., and Ross, R.P. 2005. Bacteriocins: Developing innate immunity for food. *Nat. Rev. Microbiol.* 3:777–788.

Cretenet, M., Laroute, V., Ulve, V., Jeanson, S., Noualille, S., Even, S., Piot, M., Girbal, L., Le Loir, Y., Loubiere, P., Lortal., S., and Cocaign-Bousquet, M. 2011. Dynamic analysis of the *Lactococcus lactis* transcriptome in cheeses made from milk concentrated by ultrafiltration reveals multiple strategies of adaptation to stresses. *Appl. Environ. Microbiol.* 77:247–257.

Crow, V.L. and Thomas, T.D. 1982. D-tagatose 1,6-diphosphate aldolase from lactic streptococci: Purification, properties, and use in measuring intracellular tagatose 1,6-diphosphate. *J. Bacteriol.* 151:600–608.

Delves-Broughton, J.P., Blackburn, P., Evans, R.E., and Hugenholtz, J. 1996. Applications of the bacteriocin nisin. *Antonie van Leeuwenhoek* 69:193–202.

De Vuyst, L. and Degeest, B. 1999. Heteropolysaccharides from lactic acid bacteria. *FEMS Microbiol. Rev.* 23:153–177.

De Vuyst, L. and Leroy, F. 2007. Bacteriocins from lactic acid bacteria: Production, purification, and food applications. *J. Mol. Microbiol. Biotechnol.* 13:194–199.

Deveau, H.R., Barrangou, J.E., Garneau, J., Labonté, C., Fremaux, P., Boyaral, D.A., Romero, P., Horvath, P., and Moineau, S. 2008. Phage response to CRISPR-encoded resistance in *Streptococcus thermophilus*. *J. Bacteriol.* 190:1390–1400.

Djordejevic, G.M., O'Sullivan, D.J., Walker, S.A., Conkling, M.A., and Klaenhammer, T.R. 1997. A triggered-suicide system designed as a defense against bacteriophages. *J. Bacteriol.* 179:6441–6448.

Duboc, P. and Mollet, B. 2001. Applications of exopolysaccharides in the dairy industry. *Int. Dairy J.* 11:759–768.

FAO/WHO. 2002. Guidelines for evaluation of probiotics in food. Report of a joint FAO/WHO working group report on drafting guidelines for the evaluation of probiotics in food. London, Ontario, Canada.

Fernández, F., Delgado, T., Boris, S., Rodríguez, A., and Barbés, C. 2005. A washed-curd goat's cheese as a vehicle for delivery of a potential probiotic bacterium: *Lactobacillus delbrueckii* subsp. *lactis* UO 004. *J. Food Prot.* 68:2665–2671.

Fernández M., Álvarez, M.A., and Zúñiga, M. 2008. Proteolysis and amino acid catabolism in lactic acid bacteria. In: *Molecular Aspects of Acid Lactic Bacteria for Traditional and New Applications*, edited by B. Mayo, P. López, and G. Pérez-Martínez, pp. 89–136. Kerala, India: Research Signpost.

Fernández, M. and Zúñiga, M. 2006. Amino acid catabolic pathways of lactic acid bacteria. *Crit. Rev. Microbiol.* 32:155–183.

Fernández-Bodega, M.A., Mauriz, E., Gómez, A., and Martín, J.F. 2009. Proteolytic activity, mycotoxins and andrastin A in *Penicillium roqueforti* strains isolated from Cabrales, Valdeón and Bejes-Tresviso local varieties of blue-veined cheeses. *Int. J. Food Microbiol.* 136:18–25.

Fondén, R., Leporanta, K., and Svensson, V. 2006. Nordic/Scandinavian fermented milk products. In: *Fermented Milks*, edited by A.Y. Tamine, pp. 156–173. Ames, IA: Blackwell Publishing.

Gálvez, A., Abriouel, H., Lápez, R.L., and Ben, O.N. 2007. Bacteriocin-based strategies for food biopreservation. *Int. J. Food Microbiol.* 120:51–70.

García, P., Rodríguez, L., Rodríguez, A., and Martínez, B. 2010. Food biopreservation: Promising strategies using bacteriocins, bacteriophages and endolysins. *Trends Food Sci. Technol.* 21:373–382.

Giraffa, G. 1995. Enterococcal bacteriocins: Their potential as anti-*Listeria* factors in dairy technology. *Food Microbiol.* 12:291–299.

Giraffa, G. 2002. Enterococci from foods. *FEMS Microbiol. Rev.* 26:163–171.

Glättli, H. 1990. Starter and starter production for Swiss type cheeses. Proc. 2nd Cheese Symposium, Teagasc, Dublin, pp. 23–30.

Gripon, J.C. 1993. Mould-ripened cheeses. In: *Cheese: Chemistry, Physics and Microbiology*, Vol. 2, 2nd edition, edited by P.F. Fox, pp. 111–136. London: Chapman & Hall.

Grossiord, B., Vaughan, E.E., Luesink, E., and de Vos, W.M. 1998. Genetics of galactose utilisation via the Leloir pathway in lactic acid bacteria. *Lait* 78:77–84.

Guarner, F., Perdigón, G., Corthier, G., Salminen, S., Koletzko, B., and Morelli, L. 2005. Should yogurt cultures be considered probiotic? *Br. J. Nutr.* 93:783–786.

Hassan, A.N., Ipsen, R., Janzen, T., and Qvist, K.B. 2003. Microstructure and rheology of yogurt made with cultures differing only in their ability to produce exopolysaccharides. *J. Dairy Sci.* 86:1632–1638.

Heap, H.A. 1998. Optimising starter culture performance in New Zealand cheese plants. *Aust. J. Dairy Technol.* 53:74–78.

Hirsch, A., Grinsted, E., Chapman, H.R., and Mattick, A. 1951. A note on the inhibition of an anaerobic spore former in Swiss-type cheese by a nisin-producing *Streptococcus. J. Dairy Res.* 18:205–206.

Hols, P., Hancy, F., Fontaine, L., Grossiord, B., Proxy, D., Leblond-Bourget, N., Bernard Decaris, B., Bolotin, A., Delorme, C., Ehrlich, S.D., Guedon, E., Monne, V., Renault, P., and Kleerebezem, M. 2005. New insights in the molecular biology and physiology of *Streptococcus thermophilus* revealed by comparative genomics. *FEMS Microbiol. Rev.* 29:435–463.

Horvath, P., Coûté-Monvoisin, A.C., Romero, D.A., Boyadal, P., Fremaux, C., and Barrangou, R. 2009. Comparative analysis of CRISPR loci in lactic acid bacteria genomes. *Int. J. Food Microbiol.* 131:62–70.

Hugenholtz, J. 1986. Population dynamics of mixed starter cultures. *Neth. Milk. Dairy J.* 40:129–140.

Hugenholtz, J., Perdon, L., and Abee, T. 1993. Growth and energy generation by *Lactococcus lactis* subsp. *lactis* biovar *diacetylactis* during citrate metabolism. *Appl. Environ. Microbiol.* 59:4216–4222.

Hugenholtz, J., Twigt, M., Slomp, M., Smith, M.R., and Knorr, D. 1995. Development of nisin-producing starters for Gouda cheese manufacture. International Dairy Lactic Acid Bacteria Conference, Palmerston North, New Zealand, Book of Abstracts, S. 2.4.

Jakobsen, M. and Narvhus, J. 1996. Yeasts and their possible beneficial and negative effects on the quality of dairy products. *Int. Dairy J.* 6:755–768.

Jollivet, N., Chataud, J., Vayssier, Y., Bensoussan, M., and Belin, J.M. 1994. Production of volatile compounds in model milk and cheese media by eight strains of *Geotrichum candidum* link. *J. Dairy Res.* 61: 241–248.

Jolly, L., Vincent, S.J.F., Duboc, P., and Neeser, J.R. 2002. Exploiting exopolysaccharides from lactic acid bacteria. *Antonie van Leeuwenhoek* 82:367–374.

Joosten, H.M. and Northolt, M.D. 1989. Detection, growth and amine-producing capacity of lactobacilli in cheese. *Appl. Environ. Microbiol.* 55:2356–2359.

Kandler, O. 1983. Carbohydrate metabolism in lactic acid bacteria. *Antonie van Leeuwenhoek* 49:209–224.

Kerjean, J.R., Condon, S., Lodi, R., Kalantzopoulos, G., Chamba, J.E., Suomailainen, T., Cogan, T., and Moreau, D. 2000. Improving the quality of European hard cheese by controlling interactions between lactic acid bacteria and propionibacteria. *Food Res. Int.* 33:284–287.

Klaenhammer, T.R. 1993. Genetics of bacteriocins produced by lactic acid bacteria. *FEMS Microbiol. Rev.* 12:39–85.

Leroy, F. and De Vuyst, L. 2004. Lactic acid bacteria as functional starter cultures for the food fermentation industry. *Trends Food Sci. Technol.* 15:67–78.

Leroy, F. and Pidoux, M. 1993. Detection of interactions between yeasts and lactic acid bacteria isolated from sugary kefir grains. *J. Appl. Bacteriol.* 74:48–53.

Limsowtin, G.K.Y., Bruinenberg, E.G., and Powell, I.B. 1997. A strategy for cheese starter culture management in Australia. *J. Microbiol. Biotechnol.* 7:1–7.

Liu, M., Siezen, R.J., and Nauta, A. 2009. In silico prediction of horizontal gene transfer events in *Lactobacillus delbrueckii* ssp. *bulgaricus* and *Streptococcus thermophilus* reveals proto-cooperation in yogurt manufacturing. *Appl. Environ. Microbiol.* 75:4120–4129.

Lourens-Hattingh, A. and Viljeon, C.B. 2001. Yogurt as probiotic carrier food. *Int. Dairy J.* 11:1–17.

Madera, C., García, P., Janzen, T., Rodríguez, A., and Suárez, J.E. 2003 Characterization of technologically proficient wild *Lactococcus lactis* strains resistant to phage infection. *Int. J. Food Microbiol.* 86:213–222.

Maisnier-Patin, S., Deschamps, N., Tatini, S.R., and Richard, J. 1992. Inhibition of *Listeria monocytogenes* in camembert cheese made with a nisin-producing starter. *Lait* 72:249–263.

Martínez, B., Suárez, J.E., and Rodríguez, A. 1995. Antimicrobials produced by wild lactococcal strains isolated from homemade cheeses. *J. Food Prot.* 58:1118–1123.

Martley, F.G. and Crow, V.L. 1993. Interaction between non-starter microorganisms during cheese manufacture and ripening. *Int. Dairy J.* 3:461–483.

Masco, L., Huys, G., De Brandt, E., Temmerman, R., and Swings, J. 2005. Culture dependent and culture-independent qualitative analysis of probiotic products claimed to contain bifidobacteria. *Int. J. Food Microbiol.* 102:221–230.

Mayo, B., van Sinderen, D., and Ventura, M. 2008. Genome analysis of food grade lactic acid-producing bacteria: From basics to applications. *Curr. Genomics* 9:169–183.

Mäyrä-Makinen, A. and Bigret, M. 1998. Industrial use and production of lactic acid bacteria. In: *Lactic Acid Bacteria. Microbiological and Functional Aspects*, edited by S. Salminen, and A. von Wright, pp. 73–102. New York: Marcel Dekker, Inc.

McAuliffe, O., Hill, C., and Ross, R.P. 1999. Inhibition of *Listeria monocytogenes* in Cottage cheese manufactured with a lacticin 3147-producing starter culture. *J. Appl. Microbiol.* 86:251–256.

Mercanti, D.J., Carminati, D., Reinheimer, J.A., and Quiberoni, A. 2011. Widely distributed lysogeny in probiotic lactobacilli represents a potentially high risk for the fermentative dairy industry. *Int. Food Microbiol.* 144:503–510.

Michaelian, M.B., Hoecker, W.H., and Hammer, B.W. 1938. Effect of pH on the production of acetylmethylcarbinol plus diacetyl in milk by the citric acid fermenting streptococci. *J. Dairy Sci.* 21:213–218.

Moineau, S. 1999. Applications of phage resistance in lactic acid bacteria. In: *Proceedings of the Sixth Symposium on Lactic Acid Bacteria: Genetics, Metabolism and Applications*, edited by W.N. Konings, O.P. Kuipers, and J.H.J. Huis in't Veld, pp. 377–382. Veldhoven, The Netherlands: Kluwer Academic Publishers.

Moineau, S. and Levésque, C. 2005. Control of bacteriophages in industrial fermentations. In: *Bacteriophages: Biology and Applications*, edited by E. Kutter and A. Sulakvelidze, pp. 285–296. Boca Raton, FL: CRC Press.

Morgan, S.M., O'Sullivan, L., Ross, R.P., and Hill, C. 2002. The design of a three strain starter system for Cheddar cheese manufacture exploiting bacteriocin-induced starter lysis. *Int. Dairy J.* 12:985–993.

Muñoz, A., Ananou. S., Gálvez, A., Martínez-Bueno, M., Rodríguez, A., Maqueda, M., and Valdivia, E. 2007. Inhibition of *Staphylococcus aureus* in dairy products by enterocin AS-48 produced *in situ* and *ex situ*: Bactericidal synergism through heat and AS-48. *Int. Dairy J.* 17:760–769.

Muñoz, A., Maqueda, M., Gálvez, A., Martínez-Bueno, M., Rodríguez, A., and Valdivia, E. 2004. Control of psychotrophic *B. cereus* in a non-fat hard type cheese by an enterococcal strain producing enterococcal strain producing enterococcin AS-48. *J. Food Prot.* 67:1517–1521.

Ogier, J.C. and Serror, P. 2008. Safety assessment of dairy microorganisms: The *Enterococcus* genus. *Int. J. Food Microbiol.* 126:291–301.

Oliveira, A.P., Nielsen, J., and Forster, J. 2005. Modeling *Lactococcus lactis* using a genome-scale flux model. *BMC Microbiol.* 5:39.

Ostlie, H.M., Vegarud, G., and Langsrud, T. 1999. Autolytic systems in propionic acid bacteria. *Lait* 79:105–112.

O'Sullivan, L., O'Connor, E.B., Ross, R.P., and Hill, C. 2006. Evaluation of live-culture-producing lacticin 3147 as a treatment for the control of *Listeria monocytogenes* on the surface of smear-ripened cheese. *J. Appl. Microbiol.* 100:135–143

O'Sullivan, L., Ross, R.P., and Hill, C. 2003. A lacticin 481-producing adjunct culture increases starter lysis while inhibiting non-starter lactic acid bacteria proliferation during Cheddar cheese ripening. *J. Appl. Microbiol.* 95:1235–1241.

Parente, E. and Cogan, T.M. 2004. Starter cultures: General aspects. In: *Cheese: Chemistry, Physics and Microbiology*, Vol. I, edited by P.F. Fox, P.J.H. McSweeney, T.M. Cogan, and T.P. Guinee, pp. 123–147. Amsterdam: Elsevier.

Parente, E., Rota, M.A., Ricciardi, A., and Clementi, E. 1997. Characterization of natural starter cultures used in the manufacture of *Pasta Filata* cheese in Basilicata (Southern Italy). *Int. Dairy J.* 7:775–783.

Pastink, M.I., Teusink, B., Hols, P., Visser, S., de Vos, W.M., and Hugenholtz, J. 2009. Genome-scale model of *Streptococcus thermophilus* LMG18311 for metabolic comparison of lactic acid bacteria. *Appl. Environ. Microbiol.* 75:3627–3633.

Petersen, B.L., Dave, R.I., McMahon, D.J., Oberg, C.J., and Broadbent, J.R. 2000. Influence of capsular and ropy exopolysaccharide-producing *Streptococcus thermophilus* on Mozzarella cheese and cheese whey. *J. Dairy Sci.* 83:1952–1956.

Quiberoni, A., Reinheimer, J., and Suárez, V.B. 1999. Performance of *Lactobacillus delbrueckii* bacteriophages by heat and biocides. *Int. J. Food Microbiol.* 84:51–62.

Rage, A. 1993. North European varieties of cheese. IV. Norwegian cheese varieties. In: *Cheese: Chemistry, Physics and Microbiology*, Vol. 2, 2nd edition, edited by P.F. Fox, pp. 257–260. London: Chapman & Hall.

Rawson, H.L. and Marshall, V.M. 1997. Effect of 'propy' strains of *Lactobacillus delbrueckii* ssp. *bulgaricus* and *Streptococcus thermophilus* on rheology of stirred yogurt. *Int. J. Food Sc. Technol.* 32:213–220.

Renault, P. 2010. Genetically modified lactic acid bacteria. In: *Biotechnology of Lactic Acid Bacteria. Novel Applications*, edited by F. Mozzi, R.R. Raya, and G. Vignolo, pp. 361–381. Singapore: Wiley-Blackwell.

Renye, J.A. and Somkuti, G.A. 2008. Cloning of milk-derived bioactive peptides in *Streptococcus thermophilus*. *Biotechnol. Lett.* 30:723–730.

Rilla, N., Martínez, B., Delgado, T., and Rodríguez, A. 2003. Inhibition of *Clostridium tyrobutyricum* in Vidiago cheese by *Lactococcus lactis* ssp. *lactis* IPLA 729, a nisin Z producer. *Int. J. Food Microbiol.* 85:22–33.

Rilla, N., Martínez, B., and Rodríguez, A. 2004. Inhibition of a methicillin-resistant *Staphylococcus aureus* strain in Afuega'l Pitu cheese by the nisin Z producing strain *Lactococcus lactis* subsp. *lactis* IPLA 729. *J. Food Prot.* 67:928–933.

Rodríguez, E., Arques, J.L., Gaya, P., Tomillo, J., Nuñez, M., and Medina, M. 2000. Behaviour of *Staphylococcus aureus* in semihard cheese made from raw milk with nisin-producing starter cultures. *Milchwissenschaft* 55:633–635.

Rodríguez, A., García, P., and Raya, R.R. 2010. Bacteriophages of lactic acid bacteria. In: *Biotechnology of Lactic Acid Bacteria. Novel Applications*, edited by F. Mozzi, R.R. Raya, and G. Vignolo, pp. 111–123. Singapore: Wiley-Blackwell.

Rodríguez, E., Gaya, P., Nuñez, M., and Medina, M. 1998. Inhibitory activity of a nisin-producing starter culture on *Listeria innocua* in raw ewes milk Manchego cheese. *Int. J. Food Microbiol.* 39:129–132.

Ruas-Madiedo, P., Hugenholtz, J., and Zoon, P. 2002. An overview of the functionality of exopolysaccharides produced by lactic acid bacteria. *Int. Dairy J.* 12:163–171.

Ryan, M.P., Rea, M.C., Hill, C., and Ross, R.P. 1996. An application in cheddar cheese manufacture for a strain of *Lactococcus lactis* producing a novel broad-spectrum bacteriocin, lacticin 3147. *Appl. Environ. Microbiol.* 1:612–619.

Shah, N.P. 2004. Probiotics and prebiotics. *Agro Food Ind. Hi-Tech.* 15:13–16.

Sieuwerts, S., Molenaar, D., van Hijum, S.A., Beerthuyzen, M., Stevens, M.J., Janssen, P.W., Ingham, C.J., de Bok, F.A., de Vos, W.M., and van Hylckama Vlieg, J.E. 2010. Mixed-culture transcriptome analysis reveals the molecular basis of mixed-culture growth in *Streptococcus thermophilus* and *Lactobacillus bulgaricus*. *Appl. Environ. Microbiol.* 76:7775–7784.

Smit, G., Smit, B.A., and Engels, W.J.M. 2005. Flavour formation by lactic acid bacteria and biochemical flavour profiling of cheese products. *FEMS Microbiol. Rev.* 29:591–610.

Sridhar, V.R., Hughes, J.E., Welker, D.L., Broadbent, J.R., and Steele, L. 2005. *Appl. Environ. Microbiol.* 71:3025–3032.

Stadhouders, J. and Leenders, G.J.M. 1984. Spontaneously developed mixed-strain cheese starters: Their behaviour toward phages and their use in the Dutch cheese industry. *Neth. Milk Dairy J.* 38:157–181.

Stiles, M.E. 1996. Biopreservation by lactic acid bacteria. *Antonie van Leeuwenhoek* 70:331–345.

Tamine, A.Y., Skriver, A., and Nilsson, L.E. 2006. Starter cultures. In: *Fermented Milks*, edited by A.Y. Tamine, pp. 11–52. Ames, IA: Blackwell Publishing.

Tan-A-ram, P., Cardoso, T., Daveran-Mingot, M.L., Kanchanatawee, S., Loubière, P., Girbal, L., and Cocaign-Bousquet, M. 2011. Assessment of the diversity of dairy *Lactococcus lactis* subsp. *lactis* isolates by an integrated approach combining phenotypic, genomic, and transcriptomic analyses. *Appl. Environ. Microbiol.* 77:739–748.

Taylor, L., Keefe, T., Windham, E.S., and Howell, J.F. 1982. Outbreak of histamine poisoning associated with consumption of Swiss cheese. *J. Food Prot.* 45:455–457.

Teusink, B., van Enckevort, F.H., Francke, C., Wiersma, A., Wegkamp, A., Smid, E.J., and Siezen, R.J. 2005. *In silico* reconstruction of the metabolic pathways of *Lactobacillus plantarum*: Comparing predictions of nutrient requirements with those from growth experiments. *Appl. Environ. Microbiol.* 71:7253–7262.

Thomas, T.D., Ellwood, D.C., and Longyear, C.M. 1979. Changes from homo- to heterolactic fermentation by *Streptococcus lactis* resulting from glucose limitation in anaerobic chemostat cultures. *J. Bacteriol.* 138:109–117.

van den Tempel, T. and Jakobsen, M. 2000. The technological characteristics of *D. hansenii* and *Y. lipolytica* and their potential as starter cultures for production of Danablu. *Int. Dairy J.* 10:263–270.

Vasiljevic, T. and Shah, N.P. 2008. Probiotics—From Metchnikoff to bioactives. *Int. Dairy J.* 18:714–728.

Walker, S.A. and Klaenhammer, T.R. 2000. An explosive antisense RNA strategy for inhibition of a lactococcal bacteriophage. *Appl. Environ. Microbiol.* 66:310–319.

Williams, A.G., Beattie, S.H., and Banks, J.M. 2004. Enzymes involved in flavour formation by bacteria isolated from the smear population of surface-ripened cheese. *Int. J. Dairy Technol.* 57:7–13.

Wouters, J.T.M., Ayad, E.H.E., Hugenholtz, J., and Smit, G. 2002. Microbes from raw milk for fermented dairy products. *Int. Dairy J.* 12:91–109.

Wszolek, M.B., Kupiec-Teahan, H.S., Gulgader, H.S., and Tamine, A.Y. 2006. Production of kefir, koumiss and other related products. In: *Fermented Milks*, edited by A.Y. Tamine, pp. 174–216. Ames, IA: Blackwell Publishing.

3

Microorganisms and Food Fermentation

Giorgio Giraffa and Domenico Carminati

CONTENTS

3.1 Introduction

Fermentation is one of the oldest forms of food preservation technologies in the world. It is responsible for many properties of fermented foods, such as flavor, shelf life, texture, and health benefits. Food fermentation covers a wide range of microbial and enzymatic processing of food and ingredients to achieve desirable characteristics such as prolonged shelf life, improved safety, attractive flavor, nutritional enrichment, and promotion of health (Giraffa 2004; Sieuwerts et al. 2008). Fermentation is a relatively efficient, low-energy preservation process, which increases the shelf life and decreases the need for refrigeration or other forms of food preservation technology. Fermentation processes enhance food safety by reducing toxic compounds such as aflatoxins and cyanogens and producing antimicrobial factors such as lactic acid, bacteriocins, carbon dioxide, hydrogen peroxide, and ethanol, which facilitate inhibition or elimination of foodborne pathogens. Fermentation also improves the nutritional value of foods through the biosynthesis of vitamins, essential amino acids, and proteins, by improving protein and fiber digestibility, by enhancing micronutrient bioavailability, and by degrading antinutritional factors. It also provides a source of calories when used in the conversion into human foods or substrates that are unsuitable for human consumption. Therapeutic properties of fermented foods have also been reported. In addition to its nutritive, safety, and preservative effects, fermentation enriches the diet through the production of a diversity of flavors, textures, and aromas. It improves the shelf life of foods while reducing energy consumption required for their preparation. The production of fermented foods is also important in adding value to agricultural raw materials, thus providing income and generating employment (Giraffa 2004).

The consumption of fermented foods has increased greatly since the 1970s, and, today, fermented foods are among the most popular types of consumed foods. In recent years, there has been massive product diversification, and many prebiotic and probiotic products with a high added value have emerged. Simultaneously, artisanal products have gained popularity due to their particular flavor and aroma characteristics (Sieuwerts et al. 2008). A variety of fermented foods can be found throughout the world. They include common foods like dairy products (yogurt, cheeses, buttermilk), fermented sausages, sourdoughs, fermented alcoholic beverages, vegetables, fruits, and sauces, as well as ethnic foods such as kefir and koumiss. Collectively, sales of fermented foods on a global basis exceeds $1 trillion, with an ever greater overall economic impact. Fermented dairy products represent 20% of the total economic value of fermented foods produced worldwide (O'Brien 2004; Hutkins 2006).

3.2 Microorganisms in Fermented Foods

Within the fermentation industry, microorganisms are used for the production of specific metabolites such as acids, alcohols, enzymes, antibiotics, and carbohydrates. Major fermentation microorganisms include lactic acid bacteria (LAB), molds, and yeasts. In particular, LAB are one of the most industrially important groups of bacteria and have a very long history of use in fermentation. These organisms are used in a variety of ways, including food production, health improvement, and production of macromolecules, enzymes, and metabolites. LAB are arguably second only to yeast in importance in their services to mankind. They have been used worldwide in the generation of safe, storable, and organoleptically pleasant foodstuffs for centuries. Today, LAB play a prominent role in the world food supply, performing the main bioconversions in fermented dairy products, meats, and vegetables. LAB are also essential for the production of wine, coffee, silage, cocoa, sourdough, and numerous indigenous food fermentations (Makarova et al. 2006).

Agricultural products of animal or vegetable origin are fermented by either the indigenous microflora or an added starter culture to improve or obtain shelf life, nutritional value, health benefit, flavor, or texture. Unlike the Western world, Asia has developed many foods based on vegetable proteins using fungi, often in a solid-state fermentation process. In Europe and the United States, the main focus of food fermentation has been on food preservation by means of acid fermentation, whereas properties such as taste, nutritional value, and health effect are more important in Asian (fungal) fermentation (Oyewole 1997; Caplice and Fitzgerald 1999; Wood 1998; Hansen 2002; Holzapfel 2002).

3.2.1 Starter Cultures

The basic role of starter cultures is to drive the fermentation process. Concomitantly, they contribute to the sensorial characteristics of the products and to their safety. In fermentation, the raw materials are converted by microorganisms (bacteria, yeasts, and molds) to products that have acceptable food qualities. Spontaneous (natural) fermentation, that is, a process initiated without the use of a starter inoculum, has been applied to food preservation for millennia. In a natural fermentation, the conditions are set so that the desirable microorganisms grow preferentially and produce metabolic by-products that give the unique characteristics of the product. The majority of small-scale fermentation in developing countries and even some industrial processes such as sauerkraut fermentation are still conducted as spontaneous processes. Various types of starter cultures are widely used in such fermentations, even in industrialized countries (Carminati et al. 2010).

However, spontaneous food fermentations are neither predictable nor controllable. The natural microflora of the raw material is either inefficient, uncontrollable, and unpredictable or is destroyed altogether by the heat treatments given to the food. Moreover, the industrialization of food production led to the optimization and upscaling of many fermentation processes. Similarly, industrially produced starter cultures have emerged, leading to improved and reproducible fermentation control and product quality. In a controlled fermentation, the fermentative microorganisms (usually LAB or yeasts) are isolated and characterized and then maintained for use. Because the starter industry relies on the use of selected strains of given species with known metabolic properties, the introduction of commercial starter cultures has undoubtedly improved product quality and process standardization (Giraffa 2004; Sieuwerts et al. 2008).

3.3 Microbial Estimation: Ecological and Biochemical Considerations

Searching for the presence, numbers, and types of microorganisms in fermented foods is of paramount importance for the starter and food industry. Whatever the primary objective of these microbial analyses (e.g., control of food quality, food preservation, efficiency of starter cultures, monitoring of particular species/strains), the taxonomic level of the microbial discrimination needed may depend upon the sensitivity of the technique (either phenotypic or genotypic) used and may range from genus (or species) to subspecies or strain level (subtyping). While the above methods may describe microbial identity and composition at given spatial/temporal combinations, it must be considered that populations in a food microbial system are continuously evolving. The dynamics of growth, survival, and biochemical activity of microorganisms in foods are the result of stress reactions in response to changing physical and chemical conditions that occur in food microenvironments (e.g., pH, salt, temperature), the ability of microorganisms to colonize the food matrix and to grow into spatial heterogeneity (e.g., microcolonies and biofilms), and the *in situ* cell-to-cell ecological interactions that often take place in a solid phase (Fleet 1999).

To effectively manage the growth and activities of microorganisms in fermented foods, the following points should be raised: (1) information on diversity, taxonomic identity, growth cycle, quantitative changes, and spatial distribution of microbial species that ferment into the food at every stage of production; (2) biochemical and physiological data on the food colonization process; (3) impact of the intrinsic, extrinsic, and processing factors influencing microbial growth, survival, and biochemical activity; and (4) relationship between growth and activity of individual microorganisms and product quality and safety. Therefore, to evaluate microbial diversity in fermented food is problematic because of the concomitant action of different ecological and biochemical factors. This task is further complicated because of the difficulty to cultivate a portion of the viable bacteria (Fleet 1999; Giraffa 2004; Giraffa and Carminati 2008).

3.4 Problems Related to Microbial Identification and Estimation in Fermented Foods

In many microbial systems, previously unrecovered species may not easily be isolated. This is the case of the complex microflora colonizing the human gut or the microbial communities characterizing many traditional fermented foods. This made it necessary to apply more than a single method to identify and/ or type bacteria (Giraffa and Neviani 2001; Ventura et al. 2004). Moreover, the viability of microorganisms can make it difficult to evaluate microbial composition in food systems. Determination of bacterial viability is a complex issue, as illustrated by the numerous scientific papers published on the topic (Kell et al. 1998; Colwell 2000; Nystrom 2001). Traditionally, plate counting has been the method of choice for viability assays. However, it is widely accepted that plate culturing techniques reveal little of the true microbial population in natural ecosystems. This phenomenon can be explained by the inability to detect novel microorganisms, which might not be cultivable using known media, and/or the inability to recover known microorganisms that either are stressed or enter a nonculturable state (Fleet 1999). The determination of bacterial viability in fermented foods, especially in probiotic products, is of economic, technological, and clinical significance (Lahtinen et al. 2005).

3.4.1 Survival Mechanisms of Microorganisms in Foods

Stress is any change in the genome, proteome, or environment that imposes either reduced growth or survival potential. Such changes lead to attempts by a cell to restore a pattern of metabolism that fits it either for survival or for faster growth. For any stress, the bacterial cell has a defined range within which the rate of increase of colony forming units is positive (growth), zero (survival), or negative (death). In the first two cases, that is, growth and survival, the cells are sublethally injured, whereas in the case of death (e.g., after a bacteriophage attack), the cells are lethally damaged and rapidly autolyze. The individual values at which the cell moves from one physiological state into the next are conditional on the degree of stress imposed by other environmental conditions (Booth 2002).

Stress can influence cell viability and culturability in different ways. Kell et al. (1998) have suggested four terms to describe different life stages of microorganisms: (1) viable (active and readily culturable), (2) dormant (inactive but ultimately culturable), (3) active but nonculturable (VBNC), and (4) dead (inactive and nonculturable). In microbial populations, viable cells are usually countable on both nonselective and selective media, whereas stressed cells are able to form colonies on nonselective media but are not countable on selective media. In foods, many adverse conditions such as nutrient depletion, low temperature, and other stresses can sublethally damage microorganisms. In a recent study, it has been shown that probiotic bacteria become dormant during storage (Lahtinen et al. 2005). The VBNC state is a kind of stress that induces healthy, cultivable cells to enter a phase in which they are still capable of metabolic activity but do not produce colonies on media (both nonselective and selective) that normally support their growth. Many food-associated bacteria, including a variety of important human pathogens, are known to respond to various environmental stresses by entry into a VBNC state (Giraffa and Carminati 2008; Oliver 2010).

3.4.2 Microbial Communication and Quorum Sensing

A wide range of communication mechanisms have been described so far within bacteria, such as production of bacteriocins, pheromones, and signaling molecules (e.g., acyl-L-homoserine lactones). In addition to releasing the signaling molecules, bacteria are also able to measure the number (concentration) of the molecules within a population. The term "quorum sensing" (QS) is applied to describe the phenomenon whereby the accumulation of signaling molecules allows a single cell to sense the number of bacteria (cell density), thus enabling microorganisms to coordinate their behavior (Konaklieva and Plotkin 2006). Mechanisms regulating the multitude of language signals that diffuse through different microbial (including food) communities are rapidly being elucidated (Di Cagno et al. 2011).

Little is still known on the role of QS in food ecosystems. It has been shown that this mechanism regulates the *in situ* phenotypic expression and population behavior of food spoilage bacteria (Gram et al. 2002). An implication of QS in the expression of genes that code for bacteriocins in LAB has recently been demonstrated (Diţu et al. 2009). QS signal molecules released by probiotics may also interact with human epithelial cells from the intestine, thus modulating several physiological functions (Gobbetti et al. 2010).

3.5 Microbial Dynamics in Food

3.5.1 Culture-Dependent Methods

Culture-dependent methods include traditional cultivation methods in combination with phenotypic (physiological and biochemical) and genotypic [species-specific and randomly amplified polymorphic DNA–polymerase chain reaction (RAPD-PCR)] identification and typing techniques of isolated strains. Generally, microbiota present during fermentation are enumerated by cultivating on media with the intention of selecting different groups of bacteria (e.g., total aerobic count on plate agar count medium, presumptive streptococcal and lactococcal counts on LM17 and LM17 plus cycloheximide, respectively, enterococci on Kanamycin Aesculin Azide (KAA) medium, leuconostocs on Mayeux, Sandine and Elliker (MSE) medium, mesophilic lactobacillus on Facultatively Heterofermentative (FH) medium, thermophilic lactobacillus on deMan, Rogosa and Sharpe (MRS) medium, etc.). Differences in oxygen tolerance, nutritional requirements, antibiotic susceptibility, and colony morphology and color constitute the bases of differentiation among these methods. Selectivity of enumeration, however, can be strongly improved by altering incubation temperatures, incubation times, and the pH of the medium. Media allowing the separate (or simultaneous) enumeration of probiotic bacterial species are available today for quality control of fermented milks (Giraffa 2004).

Classical plating methodologies, therefore, only allow a rough measure of microbial groups or genera and rarely is their sensitivity down to the species level. As stated above, the isolation, species identification, and (facultative) typing of microbial isolates from agar plates are needed to increase sensitivity. For the latter two purposes, either phenotyping or genotyping (or both) is applied (Olive and Bean 1999; Giraffa and Neviani 2000). Concerning phenotyping, identification down to species level can be performed by API galleries or other phenotypic methods such as the *Phene Plate* System (Iversen et al. 2002; Lund et al. 2002). These techniques allow dynamic trends of cultivable, dominant microbial populations or microbial groups to be highlighted.

Species-specific PCR has been introduced as a molecular tool for microbial identification down to the species level. The online availability of DNA sequences of ribosomal RNA (rRNA) genes and rRNA gene spacers of practically all the known microbial species and the presence of taxa-specific oligonucleotide stretches within the ribosomal locus have enabled these genes (or portions of them) to be routinely PCR-amplified and examined for differences indicative of species identity. Several PCR amplification protocols are presently available for practically all food-associated LAB species (Giraffa and Carminati 2008). Among PCR-based typing techniques, RAPD-PCR is the most popular applied technique in studying the microbial ecology of fermented foods. In recent years, hundreds of articles reported the application of RAPD-PCR to identify the presence, succession, and persistence of microorganisms (both useful and pathogens) in both fermented food and industrial environments (Cocolin and Ercolini 2008).

Classical phenotyping is laborious and time consuming, especially when both spatial and temporal distribution of microbial populations are needed. Most importantly, culture-dependent (both phenotypic and genotypic) identification and/or typing methods are not effective if, for any reason, microorganisms are not cultivable.

3.5.2 Culture-Independent Methods

In the last two decades, culture-independent, nucleic acid-based, molecular approaches have undergone considerable development in microbial ecology. Compared to traditional culturing, these methods aim to obtain a picture of a microbial population without the need to isolate and culture its single components. This is possible because these techniques are based on a "community DNA/RNA isolation approach." Although there are limitations to these methods, they can nevertheless be very useful once these limitations are taken into consideration (for a review, see Forney et al. 2004). Such limitations include technical problems, such as obtaining representative genomic DNA from food samples, to conceptual questions, such as using universally accepted and meaningful definitions of microbial species. Such techniques enable analyses of total microbial communities and greatly improve our understanding of their composition, dynamics, and activity (Wilmes and Bond 2009).

Culture-independent methods are being increasingly applied to detect, quantify, and study microbial populations in food or during food processes. Culture-independent techniques and their application to fermented food have been exhaustively reviewed in a recent book (Cocolin and Ercolini 2008). They can be generally grouped into *in situ* methods and PCR-based techniques. The common trait of the *in situ* methods is that morphologically intact cells (both cultivable and uncultivable) can be identified and counted directly, that is, in minimally disturbed samples. The fluorescence *in situ* hybridization (FISH) with rRNA targeted oligonucleotide probes is the most popular *in situ* method, and its applications in food microbiology have recently been reviewed (Bottari et al. 2009). FISH has been used to evaluate bacterial community structure and location in fermented foods, probiotics, and cheeses (Giraffa and Carminati 2008).

PCR-based, culture-independent techniques have been the subject of considerable focus in food microbiology. Although most of these methods are generally based on the amplification of only the variable regions or the totality of the 16S rRNA genes, amplified fragments can derive also from total RNA extracted from food and amplified by reverse transcriptase–PCR (RT-PCR). Because active bacteria have a higher number of ribosomes than dead cells, the use of RNA instead of DNA highlights the metabolically active populations present in the ecosystem. PCR-based methods such as PCR-denaturing gradient gel electrophoresis (PCR-DGGE) are routinely used to quantify either pathogens or beneficial populations such as starter microbes or probiotics in fermented foods (Le Dréan et al. 2010; Giraffa and Carminati 2008; Malorni et al. 2008). It should be mentioned that by combining different methods (e.g., PCR-DGGE, cloning and sequencing of rRNA gene amplicons, and classical cultivation techniques) in a "polyphasic ecology" approach, it is now possible to profile time-dependent specific shifts in the composition of complex food microflora, to evaluate and quantify noncultivable food populations, and, among these latter, to monitor the metabolically active microbial groups (Giraffa 2004). For example, DGGE and other techniques have successfully been applied in polyphasic studies to monitor the microbial dynamics of food ecosystems (Ercolini 2004; Ercolini et al. 2004).

3.5.3 Present Trends

3.5.3.1 Phenomics

In recent years, a breakthrough technology, the Phenotype MicroArrays (PM), made it possible to quantitatively measure thousands of cellular phenotypes all at once (Bochner 2009). This technology, following on to proteomics and genomics technologies, can be categorized as "phenomics." The advantage of phenomic research is the ability to conduct simultaneous testing of numerous bacterial phenotypes by using an automated instrument to measure growth in the presence of various substrates. This instrument determines cell growth every 5 min by monitoring absorbance of substrate color change of an appropriate indicator. For substrate metabolism, 700–2000 different substrates located in microtiter well plates could be tested at one time.

The application of PM technology to food microbiology for a global phenotypic characterization of bacteria is still in its early stages and concerns the phenotypic analysis of human and animal pathogens (Guard-Bouldin et al. 2007; Viti et al. 2008; Tang et al. 2010). The information provided may be complementary to, and often more easily interpretable than, information provided by global molecular analytical methods such as gene chips and proteomics. Phenomics could help in discovering the physiology and the variability in the phenotypic expression of starter or probiotic bacteria in many different growth conditions, thus helping to predict their behavior in fermented food products.

3.5.3.2 Quantitative PCR

It is well known that standard PCR reactions are not quantitative. A promising tool for the advancement of studies on food-associated microbial populations, either cultivable or not cultivable, is the application of quantitative PCR (qPCR) to food systems. Typical qPCR utilizes approaches originally developed in clinical microbiology, with the 5′-fluorogenic exonuclease (TaqMan) assay representing the widest applied development. By using an internal probe, which is labeled with fluorescent dyes, in addition to standard PCR amplification primers, TaqMan chemistry provides in-tube, real-time detection of PCR product accumulation (real-time PCR) during each amplification cycle and at very early stages in the amplification process. Using DNA as starting material, qPCR offers the possibility to quantify microbial populations through measurement of gene numbers. Using RNA as a template combined with reverse transcription (RT), qPCR can also estimate transcript amounts, therefore providing data on microbial activity (Giraffa and Carminati 2008).

In the last years, (RT-)qPCR applications in food microbiology have strikingly developed. Real-time PCR is increasingly applied for specific detection and quantification of pathogens and spoiling agents in food (McGuinness et al. 2009; Juvonen et al. 2008) and for the enumeration of beneficial microbial populations in fermented milk products (Bleve et al. 2003; Furet et al. 2004; Masco et al. 2007; Zago et al. 2009; Le Dréan et al. 2010), dairy starters (Friedrich and Lenke 2006), and wine (Neeley et al. 2005). Innovation is moving toward simultaneous identification of several microbial species by applying multiplex amplification. By TaqMan chemistry, several species-specific DNA probes can be applied using different fluorophores, thus enabling different targets to be co-amplified and quantified within a single reaction tube (Smith and Osborn 2009).

3.5.3.3 Chip, DNA Array-Based, Technology

The DNA chip microarray technology is a direct result of the availability of genome sequence information. The technique involves very large (approximately 100,000) cDNA sequences or synthetic DNA oligomers being attached onto a glass slide (the chip) in known locations on a grid. An RNA sample is then labeled and hybridized to the grid, and relative amounts of RNA bound to each square in the grid are measured. Such DNA chips can be used for simultaneous monitoring of levels of expression of all of the genes in a cell in order to study whole genome expression patterns in various matrices during development. Moreover, because parallel hybridizations to hundreds or thousands of genes in a single experiment can be performed by high-throughput DNA microarrays, direct profiling of microbial populations is achievable. Rudi et al. (2002) combined the specificity obtained by enzymatic labeling of species-specific, oligonucleotide probes with the possibility of detecting several targets simultaneously by DNA array hybridization with 16S rRNA gene from pure cultures. By hybridization of chip-bound probes with bulk DNA extracted from food, this is a promising tool for microbial community analyses in foods. In one development of this basic technique, Bae et al. (2005) described genome-probing microarrays (GPMs), which consist of depositing hundreds of microbial genomes as labeled probes on a glass slide and hybridizing with bulk community DNAs. GPM enabled quantitative, high-throughput monitoring of LAB community dynamics during fermentation of kimchi, a traditional Korean food. Compared to currently used oligonucleotide microarrays, the specificity and sensitivity of GPM were remarkably increased (Bae et al. 2005).

3.5.3.4 Functional Proteomics

Because the level of bacterial mRNA does not always correlate with the amount of expressed protein in the cells and because it has a short average life span, alternatives to DNA microarrays for gene expression have started to emerge. Functional proteomics, coupled to protein microarrays, is considered one of the leading technologies for bacterial and gene identification. Proteins and antibodies can be attached to chemically pretreated slides to build a protein microarray chip that is able to identify bacteria, antibodies, and viruses. Unlike DNA, proteins are difficult to attach to a glass slide and to synthesize and are easily denatured. Therefore, building a protein array can be more challenging than building the DNA chip (Al-Khaldi and Mossoba 2004). Because of the amount of data we may uncover by proteomics, protein microarray could be helpful for identification and functional characterization of food microorganisms.

3.5.3.5 Data Analysis

Driven by automated DNA sequencing technology and the different genome projects, analysis of DNA and protein sequence data has spurred the dramatic growth of a new scientific discipline—bioinformatics—in the 1990s. Bioinformatics can be defined as the use of computers for the acquisition, management, and analysis of biological information. Bioinformatics combines *in silico* biological techniques with the DNA sequencing analysis approach. *In silico* biology combines statistical and mathematical algorithms with the need to manage and elaborate huge numbers of biological data. The development of bioinformatics has enabled improvement of the interpretation and elaboration of microbiological data. The acquisition of specialized, commercially available software packages, which are expensive and demand a high level of technical skill for their efficient use, is necessary so that the most important international microbial collections can manage, compare, and implement databases holding information on nucleic acid (or protein) sequences, electrophoretic profiles, and phenotypic data.

One of the advantages of bioinformatics in relation to studying bacterial taxonomy and diversity concerns the possibility of sharing databases. As stated above, diagnostic tools based on PCR have been developed for rapid inexpensive genotype assay. In addition, several microbial genomes have now been sequenced (for a review, see Klaenhammer et al. 2005), while large numbers of DNA sequences have been compiled and are available via the World Wide Web. There is an urgent need to process this mass of information into a useful classification tool, which will require further automation and software development in order to effectively link different databases. In addition, bioinformatics could allow advances in functional genomics, that is, conversion of the mass of sequence data presently available in public data banks into knowledge, so that microbial diversity could be assessed not only at the molecular level but also at the functional level (Perego and Hoch 2001).

3.6 Final Remarks and Future Needs

Modern food microbiologists are fortunate in having a variety of tools that provide very advanced molecular differentiation of microorganisms and that can be tailored to fit the needs of both research laboratories and the food industry. Both cultivable and noncultivable bacteria can be analyzed, microbial populations can be quantified, and new microbial species can be isolated and characterized. Once efficiently integrated via advances in bioinformatics, molecular identification and fingerprinting techniques will provide more precise information on microbial taxonomy and functional diversity of a given food system at a particular time and space. Several of these molecular methods, once applied to the food industry, could enable the creation of large reference libraries of typed organisms to which new strains can be compared either within the same laboratory or across different laboratories. Aspects such as changes in microbial populations, identification of contamination sources, management of customer/supplier disputes, assessment of sanitation programs, authentication of starter cultures, and verification of laboratory culture integrity could then be efficiently monitored. The choice of a molecular typing method will depend upon the needs, level of skill, and resources of the laboratory concerned.

However, more specific problems arising in the analysis of microbial communities of food ecosystems need specific answers, for example, how we can increase our knowledge of cell physiology, cell-to-cell interactions, and *in situ* modification of the microbial metabolism in natural ecosystems, especially in response to adverse environmental conditions; how we can detect and possibly quantify nonculturable and/or nondominant species/strains; and whether there are nondestructive methods of sample preparation to better evaluate spatial distribution and colonization of microorganisms in heterogeneous food matrices.

Another key point to be addressed is that many of the methods reported above are PCR based. PCR is an excellent technique for examining mixed microbial communities. Amplification, however, often introduces bias that makes quantification of natural populations difficult. The ability to detect target rRNA sequences without amplification (i.e., direct profiling of rRNA sequences) would greatly improve our capacity in determining the relative number of sequences representing natural microbial populations. In this regard, high-throughput DNA microarrays, which provide for parallel hybridizations to hundreds or thousands of genes in a single experiment, may play a major role in the study of food systems. Furthermore, standard PCR reactions are not quantitative. Promising tools for the advancement of studies on food-associated microbial populations, either cultivable or not cultivable, include qPCR, which has the potential for accurate and highly sensitive enumeration of microorganisms.

Finally, more knowledge is needed on the physiological state (viability), the stress response, and the survival of microorganisms into foods, especially probiotic foods. Stressed, sublethally injured, or otherwise "viable but nonculturable" cells often go undetected when using traditional microbiological techniques. In this regard, flow cytometry (FCM) has been recently applied to analyze, in different proportions, subpopulations of variably stressed (or damaged) bacteria in probiotic products and dairy starters. The sensitivity of the FCM method was significantly higher than agar plate techniques (Budde and Rasch 2001; Bunthof and Abee 2002).

REFERENCES

Al-Khaldi, S.F. and Mossoba, M.M. 2004. Gene and bacterial identification using high-throughput technologies: Genomics, proteomics, and phenomics. *Nutrition* 20:32–38.

Bae, J.W., Rhee, S.K., Park, J.R., Chung, W.H., Nam, J.D., Lee, I., Kim, H., and Park, Y.H. 2005. Development and evaluation of genome-probing microarrays for monitoring lactic acid bacteria. *Appl. Environ. Microbiol.* 71:8825–8835.

Bleve, G., Rizzotti, L., Dellaglio, F., and Torriani, S. 2003. Development of reverse transcription (RT)-PCR and real-time RT-PCR assays for rapid detection and quantification of viable yeasts and molds contaminating yogurts and pasteurized food products. *Appl. Environ. Microbiol.* 69:4116–4122.

Bochner, B.B. 2009. Global phenotypic characterization of bacteria. *FEMS Microbiol. Rev.* 33:191–205.

Booth, I.R. 2002. Stress and the single cell: Intrapopulation diversity is a mechanism to ensure survival upon exposure to stress. *Int. J. Food Microbiol.* 78:19–30.

Bottari, B., Ercolini, D., Gatti, M., and Neviani, E. 2009. FISH in food microbiology. In: *Fluorescence In Situ Hybridization (FISH)—Application Guide*, edited by T. Liehr, pp. 395–410. Berlin, Heidelberg, Germany: Springer-Verlag.

Budde, B.J. and Rasch, M. 2001. A comparative study on the use of flow cytometry and colony forming units for assessment of the antibacterial effect of bacteriocins. *Int. J. Food Microbiol.* 63:65–72.

Bunthof, C.J. and Abee, T. 2002. Development of a flow cytometric method to analyze subpopulations of bacteria in probiotic products and dairy starters. *Appl. Environ. Microbiol.* 68:2934–2942.

Caplice, E. and Fitzgerald, G.F. 1999. Food fermentations: Role of microorganisms in food production and preservation. *Int. J. Food Microbiol.* 50:131–149.

Carminati, D., Giraffa, G., Quiberoni, A., Binetti, A., Suarez, V., and Reinheimer, J. 2010. Advances and trends in starter cultures for dairy fermentations. In: *Biotechnology of Lactic Acid Bacteria: Novel Applications*, edited by F. Mozzi, R.R. Raya, and G.M. Vignolo, pp. 177–192. Ames, IA: Wiley-Blackwell Publishing.

Cocolin, L. and Ercolini, D. 2008. *Molecular Techniques in the Microbial Ecology of Fermented Foods*. New York: Springer Science.

Colwell, R.R. 2000. Viable but nonculturable bacteria: A survival strategy. *J. Infect. Chemother.* 6:121–125.

Di Cagno, R., De Angelis, M., Calasso, M., and Gobbetti, M. 2011. Proteomics of the bacterial cross-talk by quorum sensing. *J. Proteomics* 74:19–34.

Diţu, L.M., Chifiriuc, C., Lazăr, V., and Mihăescu, G. 2009. Implication of quorum sensing phenomenon in the expression of genes that code for bacteriocins in lactic bacteria. *Bacteriol. Virusol. Parazitol. Epidemiol.* 54:147–166.

Ercolini, D. 2004. PCR-DGGE fingerprinting: Novel strategies for detection of microbes in food. *J. Microbiol. Methods* 56:297–314.

Ercolini, D., Mauriello, G., Blaiotta, G., Moschetti, G., and Coppola, S. 2004. PCR-DGGE fingerprints of microbial succession during a manufacture of traditional water buffalo mozzarella cheese. *Lett. Appl. Microbiol.* 96:263–270.

Fleet, G.H. 1999. Microorganisms in food ecosystems. *Int. J. Food Microbiol.* 50:101–117.

Forney, L.J., Zhou, X., and Brown, C.J. 2004. Molecular microbial ecology: Land of the one-eyed king. *Curr. Opin. Microbiol.* 7:210–220.

Friedrich, U. and Lenke, J. 2006. Improved enumeration of lactic acid bacteria in mesophilic dairy starter cultures by using multiplex quantitative real-time PCR and flow cytometry-fluorescence in situ hybridization. *Appl. Environ. Microbiol.* 72:4163–4171.

Furet, J.P., Quénée, P., and Tailliez, P. 2004. Molecular quantification of lactic acid bacteria in fermented milk products using real-time quantitative PCR. *Int. J. Food Microbiol.* 97:197–207.

Giraffa, G. 2004. Studying the dynamics of microbial populations during food fermentation. *FEMS Microbiol. Rev.* 28:251–260.

Giraffa, G. and Carminati, D. 2008. Molecular techniques in food fermentation: Principles and applications. In: *Molecular Techniques in the Microbial Ecology of Fermented Foods*, edited by L. Cocolin and D. Ercolini, pp. 1–30. New York: Springer Science.

Giraffa, G. and Neviani, E. 2000. Molecular identification and characterization of food-associated lactobacilli. *Ital. J. Food Sci.* 12:403–423.

Giraffa, G. and Neviani, E. 2001. DNA-based, culture-independent strategies for evaluating microbial communities in food-associated ecosystems. *Int. J. Food. Microbiol.* 67:19–34.

Gobbetti, M., Cagno, R.D., and De Angelis, M. 2010. Functional microorganisms for functional food quality. *Crit. Rev. Food Sci. Nutr.* 50:716–727.

Gram, L., Ravn, L., Rasch, M., Bruhn, J.B., Christensen, A.B., and Givskov, M. 2002. Food spoilage— Interactions between food spoilage bacteria. *Int. J. Food Microbiol.* 78:79–97.

Guard-Bouldin, J., Morales, C.A., Frye, J.G., Gast, R.K., and Musgrove, M. 2007. Detection of *Salmonella enterica* subpopulations by phenotype microarray antibiotic resistance patterns. *Appl. Environ. Microbiol.* 23:7753–7756.

Hansen, E.B. 2002. Commercial bacterial starter cultures for fermented foods of the future. *Int. J. Food Microbiol.* 78:119–131.

Holzapfel, W.H. 2002. Appropriate starter culture technologies for small-scale fermentation in developing countries. *Int. J. Food Microbiol.* 75:197–212.

Hutkins, R.W. 2006. *Microbiology and Technology of Fermented Foods*. Ames, IA: Blackwell Publishing Professional.

Iversen, A., Kühn, I., Franklin, A., and Möllby, R. 2002. High prevalence of vancomycin-resistant enterococci in Swedish sewage. *Appl. Environ. Microbiol.* 68:2838–2842.

Juvonen, R., Koivula, T., and Haikara, A. 2008. Group-specific PCR-RFLP and real-time PCR methods for detection and tentative discrimination of strictly anaerobic beer-spoilage bacteria of the class Clostridia. *Int. J. Food Microbiol.* 125:162–169.

Kell, D.B., Kaprelyants, A.S., Weichart, D.H., Harwood, C.R., and Barer, M.R. 1998. Viability and activity in readily culturable bacteria: A review and discussion of the practical issues. *Antonie van Leeuwenhoek* 73:169–187.

Klaenhammer, T.R., Barrangou, R., Buck, B.L., Azcarate-Peril, M.A., and Altermann, E. 2005. Genomic features of lactic acid bacteria effecting bioprocessing and health. *FEMS Microbiol. Rev.* 29:393–409.

Konaklieva, M.I. and Plotkin, M.J. 2006. Chemical communication—Do we have a quorum? *Mini-Rev. Med. Chem.* 6:817–825.

Lahtinen, S.J., Gueimonde, M., Ouwehand, A.C., Reinikainen, J.P., and Salminen, S. 2005. Probiotic bacteria may become dormant during storage. *Appl. Environ. Microbiol.* 71:1662–1663.

Le Dréan, G., Mounier, J., Vasseur, V., Arzur, D., Habrylo, O., and Barbier, G. 2010. Quantification of *Penicillium camemberti* and *P. roqueforti* mycelium by real-time PCR to assess their growth dynamics during ripening cheese. *Int. J. Food Microbiol.* 138:100–107.

Lund, B., Adamsson, A., and Edlund, C. 2002. Gastrointestinal transit survival of an *Enterococcus faecium* probiotic strain administered with or without vancomycin. *Int. J. Food Microbiol.* 77:109–115.

Makarova, K.A., Slesarev, Y., Wolf, A., Sorokin, B., Mirkin, E., Koonin, A., Pavlov, N., Pavlova, V., Karamychev, N., Polouchine, V., Shakhova, I., Grigoriev, Y., Lou, D., Rohksar, S., Lucas, K. et al. 2006. Comparative genomics of the lactic acid bacteria. *Proc. Natl. Acad. Sci. U. S. A.* 103:15611–15616.

Malorni, B., Lofstrom, C., Wagner, M., Kramer, N., and Hoorfar, J. 2008. Enumeration of *Salmonella* bacteria in food and feed samples by real-time PCR for quantitative microbial risk assessment. *Appl. Environ. Microbiol.* 74:1299–1304.

Masco, L., Vanhoutte, T., Temmerman, R., Swings, J., and Huys, G. 2007. Evaluation of real-time PCR targeting the 16S rRNA and recA genes for the enumeration of bifidobacteria in probiotic products. *Int. J. Food Microbiol.* 113:351–357.

McGuinness, S., McCabe, E., O'Regan, E., Dolan, A., Duffy, G., Burgess, C., Fanning, S., Barry, T., and O'Grady, J. 2009. Development and validation of a rapid real-time based method for the specific detection of *Salmonella* on fresh meat. *Meat Sci.* 83:555–562.

Neeley, E.T., Phister, T.G., and Mills, D.A. 2005. Differential real-time PCR assay for enumeration of lactic acid bacteria in wine. *Appl. Environ. Microbiol.* 71:8954–8957.

Nystrom, T. 2001. Not quite dead enough: On bacterial life, culturability, senescence, and death. *Arch. Microbiol.* 176:159–164.

O'Brien, J.W. 2004. Global dairy demand: Where do we go? *Eur. Dairy Mag.* 16:22–25.

Olive, D.M. and Bean, P. 1999. Principles and applications of methods for DNA-based typing of microbial organisms. *J. Clin. Microbiol.* 37:1661–1669.

Oliver, J.D. 2010. Recent findings on the viable but nonculturable state in pathogenic bacteria. *FEMS Microbiol. Rev.* 34:415–425.

Oyewole, O.B. 1997. Lactic fermented foods in Africa and their benefits. *Food Control* 8:289–297.

Perego, M. and Hoch, J.A. 2001. Functional genomics of Gram-positive microorganisms: Review of the meeting, San Diego, California, 24 to 28 June 2001. *J. Bacteriol.* 183:6973–6978.

Rudi, K., Nogva, H.K., Moen, B., Nissen, H., Bredholt, S., Møretrø, T., Naterstad, K., and Holck, A. 2002. Development and application of new nucleic acid-based technologies for microbial community analyses in foods. *Int. J. Food Microbiol.* 78:171–180.

Sieuwerts, S., de Bok, F.A.M., Hugenholtz, J., and van Hylckama Vlieg, J.E.T. 2008. Unravelling microbial interactions in food fermentations: From classical to genomics approaches. *Appl. Environ. Microbiol.* 74:4997–5007.

Smith, C.J. and Osborn, A.M. 2009. Advantages and limitations of quantitative PCR (Q-PCR)-based approaches in microbial ecology. *FEMS Microbiol. Ecol.* 67:6–20.

Tang, J.Y.H., Carlson, J., Mohamad Ghazali, F., Saleha, A.A., Nishibuchi, M., Nakaguchi, Y., and Radu, S. 2010. Phenotypic MicroArray (PM) profiles (carbon sources and sensitivity to osmolytes and pH) of *Campylobacter jejuni* ATCC 33560 in response to temperature. *Int. Food Res. J.* 17:837–844.

Ventura, M., van Sinderen, D., Fitzgerald, G.F., and Zink, R. 2004. Insights into the taxonomy, genetics and physiology of bifidobacteria. *Antonie van Leeuwenhoek* 86:205–223.

Viti, C., Bochner, B.B., and Giovannetti, L. 2008. Florence Conference on phenotype microarray analysis of microorganisms—The environment, agriculture, and human health. *Ann. Microbiol.* 58:347–349.

Wilmes, P. and Bond, P.L. 2009. Microbial community proteomics: Elucidating the catalysts and metabolic mechanisms that drive the Earth's biochemical cycles. *Curr. Opin. Microbiol.* 12:310–317.

Wood, B.J.B. 1998. *Microbiology of Fermented Foods*. London: Blackie Academic and Professional.

Zago, M., Bonvini, B., Carminati, D., and Giraffa, G. 2009. Detection and quantification of *Enterococcus gilvus* in cheese by real-time PCR. *Syst. Appl. Microbiol.* 32:514–521.

4

Animal-Based Fermented Foods of Asia

Jyoti Prakash Tamang

CONTENTS

4.1 Introduction

Consumption of animal-based foods in Asia is mostly influenced by religions, dietary laws such as taboos imposed on consumptions, availability of animal sources, geographical location, and environmental factors. There is no historical record of any ethnic fermented milk product in Japan, Korea, and many Southeast Asian countries; however, production of fermented milk products has been reported in ancient China. Traditionally, preserved fish products are largely confined to East and Southeast Asia. Culturally, fish is a main dish for people residing near sea coasts, rivers, streams, lakes, and ponds where there are plenty of fish available. Fermentation, salting, drying, and smoking are the principal methods of perishable fish preservation innovated by the ethnic people of Asia to enrich their diets. Fish fermentation technology is a home-based traditional technique where varieties of fermented fish products, mostly fish sauce, are prepared and used as staple foods, side dishes, and condiments in Asia. Chicken, duck, fish, pork, beef, mutton, and eggs are consumed by all communities in Asia except by the majority of vegetarian Hindu and a few Buddhist communities due to religion taboo. Unlike Europe, traditionally fermented meats, sausages, hams, and bacons are uncommon in Asia, except for a few in Thailand, China, India, and Bhutan. Table 4.1 shows a list of animal-based fermented meat products of Asia.

4.2 Fermented Milks

Ethnic fermented animal milk (cow, buffalo, yak, goat, camel, donkey, and horse) products are prepared from whole milk by microbial fermentation in North and West Asian countries including the Indian subcontinent. Historical evidences prove that the present-day fermented milk products were originally developed by the West Asian nomads who used to rear and breed cattle for milk and milk products (Tamang and Samuel 2010). Traditionally, consumption of animal milk and its fermented products is not a food culture of ethnic Chinese, Koreans, Japanese, and many races of Mongolian origin despite an abundance of cows in their possession (Tamang and Samuel 2010). Instead of animal milk and milk products, a soybean called "cow of China" (Hymowitz 1970) is processed to make soya milk and *tofu* and fermented into a number of ethnic fermented soybean products such as *miso, shoyu, natto, douchi,* and *sufu*. However, descriptions of some fermented milk drinks made from mare's milk such as *koumiss, airag,* and *chirag* were reported from Medieval China (Sabban 1988; Huang 2000). This indicates that dairy products were once a part of the Chinese food culture.

TABLE 4.1

Animal-Based Fermented Foods of Asia

Country	Animal Source	Fermented Food	Nature	Use
Fermented Milks				
All parts of Asia	Animal milk	Butter	Soft paste	Butter
Bangladesh	Cow/Buffalo milk	*Mishti doi*	Sweet, thick gel	Curd, savory
Mongolia	Mare or camel milk	*Airag*	Acidic, sour, mildly alcoholic	Drink
Middle East, Asia, India	Cow milk	Buttermilk	Acidic, sour	Drink
India, Nepal, Bhutan	Cow milk	*Chhurpi* (soft)	Less sour, soft, cheese-like	Curry, pickle
China, India, Nepal, Bhutan	Yak milk	*Chhurpi* (hard)	Hard mass	Masticator, gum-like
China, India, Bhutan	Cow or yak milk	*Chhu* or *sheden*	Soft, strong flavored	Curry, soup
China, India, Bhutan	Yak milk	*Chur yuupa*	Mildly acidic, soft, flavored	Curry, soup
India, Nepal, Pakistan, Sri Lanka, Bangladesh, Bhutan	Cow milk	*Dahi*	Acidic, viscous	Curd, savory
Bhutan	Cow or yak milk	*Dachi*	Soft, cheese-like, strong flavored	Hot curry
India, Nepal, Bhutan	Cow milk	*Dudh chhurpi*	Hard mass	Masticator, chewing gum-like
India, Nepal, Bhutan, Bangladesh, Pakistan	Cow milk	Gheu/ghee	Soft, oily mass, solid	Butter
Middle East, Iran	Sheep milk–wheat	*Kishk*	Mildly acidic, dried balls	Drink
Mongolia, China	Horse, donkey, or camel milk	*Koumiss* or *Kumiss*	Acidic, mildly alcoholic, liquid	Drink
India, Nepal, Bhutan, Bangladesh, Pakistan, Middle East	Cow milk	*Lassi*	Acidic, buttermilk	Refreshing beverage
China, India, Bhutan	Yak milk	*Maa*	Mildly acidic, viscous	Butter
Nepal, India, Bhutan	Cow milk	*Mohi*	Acidic, buttermilk	Refreshing beverage
India	Buffalo/Cow milk	*Misti dahi*	Mildly acidic, thick gel, sweet	Curd, savory
India, China	Yak milk	*Phrung*	Mildly acidic, hard mass	Masticator
India, China, Bhutan	Cow/Yak milk	*Philu*	Cream	Curry
India, China, Bhutan, Nepal	Tea–yak butter, salt	*Pheuja* or *suja*	Salty with buttery flavor, liquid	Refreshing tea
India, Nepal, Pakistan, Bangladesh, Middle East	Buffalo or cow milk	*Paneer*	Whey, soft, cheese-like product	Fried snacks
India	Cow, buffalo milk	*Shrikhand*	Acidic, sweet, viscous	Savory
India, Nepal	Cow or yak milk	*Somar*	Strong flavored, paste	Condiment
China, Bhutan, India	Yak milk	*Shyow*	Acidic, thick gel, viscous	Curd-like, savory
Mongolia	Cow, yak, goat milk	*Tarag*	Acidic, sour	Drink
Japan	Cow milk	Yakult	Mildly acidic, sweet, savory	Probiotic—yogurt

(continued)

TABLE 4.1 (Continued)

Animal-Based Fermented Foods of Asia

Country	Animal Source	Fermented Food	Nature	Use
Ethnic Fish Products				
India	Fish	*Ayaiba*	Smoked fish	Pickle, curry
Philippines	Fish, shrimp	*Bagoong*	Paste	Condiment
Philippines	Shrimp	*Balao balao*	Fermented shrimp	Condiment
Malaysia	Shrimp	*Belacan*	Paste	Condiment
India	Fish	*Bordia*	Dried, salted	Curry
Malaysia	Anchovies	*Budu*	Fish sauce	Condiment
Philippines	Fish–rice	*Burong isda*	Fermented rice–fish mixture	Sauce, staple
India	Hill river fish	*Gnuchi*	Smoked fish	Curry
Korea	Shellfish	*Gulbi*	Salted and dried	Side dish
India	Fish and petioles of aroid plants	*Hentak*	Fermented fish paste	Curry
Sri Lanka	Marine fish	*Jaadi*	Fermented fish paste	Curry, condiment
Korea	Fish	*Jeot kal*	High-salt fermented	Staple
India	Fish	*Karati*	Dried, salted	Curry
Thailand	Small fish	*Kapi*	Paste	Condiment
Indonesia	Fish	*Kecap ikan*	Liquid, sauce	Seasoning
Thailand	Shrimp, salt, sweetened rice	*Kung chao*	Paste	Side dish
India	Fish	*Lashim*	Dried, salted	Curry
Middle East, Asia	Marine fish	*Mehiawah*	Fermented paste	Side dish
India	Fish	*Mio*	Dried	Curry
India	Fish	*Naakangba*	Dried	Pickle, curry
Myanmar	Fish	*Ngapi*	Fermented paste	Condiment
Thailand	Anchovies	*Nam pla*	Fish sauce	Condiment
Japan	Sea fish, cooked millet	*Narezushi*	Fermented paste	Side dish
Myanmar	Fish	*Ngan pyaye*	Fish sauce	Condiment
India	Fish	*Ngari*	Fermented fish	Curry
Vietnam	Marine fish	*Nuoc mam*	Fish sauce	Condiment
Philippines	Marine fish	*Patis*	Fish sauce	Condiment
Thailand	Fish, rice	*Pla ra*	Fermented paste	Condiment, staple
Thailand	Fish, rice, garlic	*Plaa-som*	Fermented paste	Condiment, staple
Indonesia	Mackerel	*Pedah*	Partly dried and salted	Side dish
Malaysia	Freshwater fish–rice	*Pekasam*	Fermented fish	Side dish
Indonesia	Fish	*Pindang*	Dried, salted	Side dish
Japan	Squid	*Shiokara*	Fermented; side dish	Side dish
Japan	Marine fish	*Shottsuru*	Fermented fish	Condiment
Nepal, India	Fish	*Sidra*	Dried fish	Curry
Korea	Fish–cereals	*Sikhae*	Low-salt fermented	Sauce
Nepal, India	River fish	*Suka ko maacha*	Smoked, dried	Curry
Nepal, India	Fish	*Sukuti*	Dried fish	Curry
Indonesia	Shrimps/Fish	*Trassi*	Fermented paste	Side dish
India	Fish	*Tungtap*	Fermented fish, paste	Pickle
Thailand	Crabs	*Fermented crabs*	Flavored, solid	Side dish

(*continued*)

TABLE 4.1 (Continued)

Animal-Based Fermented Foods of Asia

Country	Animal Source	Fermented Food	Nature	Use
Ethnic Meat Products				
India, Nepal	Large intestine of chevon	*Arjia*	Sausage	Curry
India	Pork	*Bagjinam*	Fermented pork	Curry
India	Chevon	*Chartaysha*	Dried, smoked meat	Curry
India, Nepal, China, Bhutan	Pork	*Faak kargyong*	Sausage, soft or hard, brownish	Curry
India	Intestine of chevon, finger millet	*Jamma*	Sausage, soft	Curry
India, Nepal, China, Bhutan	Beef	*Lang kargyong*	Sausage, soft	Curry
India, Nepal, China, Bhutan	Beef	*Lang satchu*	Dried, smoked meat	Curry
India, China, Bhutan	Beef fat	*Lang chilu*	Hard, used as an edible oil	
India, China, Bhutan	Sheep fat	*Luk chilu*	Hard, solid, oily	Edible oil
India, Nepal, China, Bhutan	Beef	*Lang kheuri*	Chopped intestine of beef	Curry
Thailand	Pork	*Nham*	Dry, semi-dry	Sausage
Vietnam	Pork, salt cooked rice	*Nem chua*	Hard, salty	Sausage
Iraq	Chopped beef	*Pastirma*	Dry, semi-dry sausage	Sausage
Thailand	Pork, rice	*Sai-krok-prieo*	Sausage	Sausage
India, Nepal	Buffalo meat	*Suka ko masu*	Dried, smoked	Curry
Nepal	Buffalo	*Sukula*	Dried, smoked	Curry
India, China, Bhutan	Yak fat	*Yak chilu*	Hard, oily	Edible oil
India, Nepal, China, Bhutan	Yak	*Yak kargyong*	Sausage, soft	Curry
India, China, Bhutan	Yak	*Yak kheuri*	Chopped intestine of yak	Curry
India, China, Bhutan	Yak meat	*Yak satchu*	Dried, smoked meat	Curry

Traditionally, Indians, Semites, and the nomadic tribesmen of North Central Asia are animal milk drinkers. Historical evidence of milk processing was found at Teil Ubaid in the mid-East Asia in the form of sculptured relief around 2900–2460 BC (Prajapati and Nair 2003). Milk and fermented milk products constitute the important items in Indian diets and have been consumed for more than 5000 years (Tamang and Samuel 2010). For the Hindus, cow is regarded as a sacred animal, and its milk and milk products are used in every religious and cultural function besides consumption. *Rig Veda* and *Upanishads*, the oldest sacred books of the Hindus, mentioned the origin of *dahi* (a yogurt-like fermented milk product and is one of the oldest fermented milk products of the Hindus) and other fermented milk products in India during 6000–4000 BC (Yegna Narayan Aiyar 1953). It is well known in ancient Indian history that *dahi*, buttermilk, and *ghee* (butter) were widely consumed milk products during Lord Krishna's time about 3000 BC (Prajapati and Nair 2003). Some common ethnic fermented milk products of modern India are *dahi, lassi, misti dahi, ghee, shrikant, rabadi, kalari, chhurpi, somar, philu,* and *chhu.*

Dahi, a curd, is the most popular fermented milk product in India and Nepal and also in Pakistan, Bangladesh, and Sri Lanka, and is obtained by lactic acid fermentation of cow or buffalo milk. *Dahi* is well known for its palatability and nutritive value (Rathi et al. 1990). It resembles plain yogurt in appearance and consistency but differs in having less acidity. During the traditional method of preparation of *dahi*, fresh milk of cow or buffalo is boiled in a vessel, cooled to room temperature, and transferred to a hollow wooden vessel or container. A small quantity of previously prepared *dahi* (serves as source of inoculums; back-slopping technique) is added to boiled and cooled milk. This is left for natural fermentation for 1–2 days in summer and for 2–4 days in winter at room temperature. The duration of fermentation depends on the season as well as on the geographical location of the place. *Dahi* is consumed directly as a refreshing nonalcoholic beverage and savory. *Lactobacillus bulgaricus, L. acidophilus, L. helveticus, L. casei, L. brevis, L. bifermentans, L. alimentarius, L. paracasei* subsp. *pseudoplantarum, Lactococcus lactis* subsp. *lactis, Lac. lactis* subsp. *cremoris, Enterococcus faecalis, Streptococcus thermophilus* as lactic flora, and yeasts species of *Saccharomycopsis* and *Candida* were isolated from *dahi* (Ramakrishnan 1979; Mohanan et al. 1984; Dewan and Tamang 2007). *Dahi* is considered by both Hindus and Buddhists as a sacred item in many of their festivals and religious ceremonies. It is also used as an adhesive to make "tika" with rice and colored powder during the Hindu festival of Nepali called "dashain" and is applied to foreheads by the family elders (Tamang 2005). It is used to prepare many ethnic milk by-products such as *lassi* (buttermilk) in India, *mohi* in Nepal (a nonalcoholic refreshing beverage), and *ghee* (butter). *Misti dahi* (sweetened *dahi, mishti doi, lal dahi,* or *payodhi*) is a sweetened fermented milk product of India and Bangladesh. *Shrikhand* is an ethnic concentrated sweetened fermented milk product of west India. *Rabadi*, an ethnic fermented milk-based, thick slurry-like product, is prepared by fermenting cereals and pulses including wheat, barley, maize, and pearl-millet in north and west India and Pakistan (Tamang 2010b).

Natural fermented milks of the Himalayan regions of India, Nepal, Bhutan, and China (Tibet) prepared by the mountain people from cow and yak milk are *dahi, mohi, gheu, chhurpi, chhu, somar, maa, philu, maa,* and *shyow* (Tamang 2010a). Some of the milk products *dahi* (curd), *mohi* (buttermilk), and *gheu* (butter) are familiar in all regions of the Himalayas, whereas others (*chhurpi, chhu, philu,* etc.) are confined mostly to the Tibetans, Bhutia, Drukpa, and Ladakhi. Hard *chhurpi* is eaten as chewing gum and masticator, which gives an extra energy to the body by continuous movement of jaws and gum in high mountains. *Somar* is exclusively prepared and consumed by the Sherpa of Nepal and Sikkim living in high altitudes. The practice of using standard starter culture during milk fermentation is uncommon. Instead, a back-slopping technique to freshly boiled milk is more commonly used (Dewan and Tamang 2006). *L. plantarum, L. alimentarius, L. casei* subsp. *pseudoplantarum, L. casei* subsp. *casei, L. farciminis* and *L. salivarius, L. bifermentans, L. hilgardii, L. kefir, L. brevis, L. confusus, Lac. lactis* subsp. *lactis, Lac. lactis* subsp. *cremoris,* and *E. faecium* were isolated from ethnic yak and cow milk products of the Himalayas (Tamang et al. 2000; Dewan and Tamang 2006, 2007). High content of protein and fat is observed in all milk products. Ethnic fermented milk products are high-calorie content foods of which *maa* (butter) made from yak has a high calorie value of 876.3 kcal/100 g (Tamang 2005). Yak-made *chhurpi* give extra energy and are a nutritious source of protein and fat (Katiyar et al. 1989) to the people living in snow-bound areas. Probiotic property is observed in many ethnic Himalayan fermented milk products (Tamang 2007).

The Aryan–Hindu pastoral system has influenced the consumption of milk and dairy products in the early settlement in the Himalayas. Cattle rearing was one of the important pastoral systems in Nepal during the Gopala dynasty in 900–700 BC (Adhikari and Ghimirey 2000). The milk and milk products of the Himalayas might have originated from the main Indian Hindu culture. *Shyow* (curd) is served exclusively during the *shoton* festival of the Tibetans. For both Hindus and Buddhists, *gheu* or butter is a sacred item in all their religious ceremonies and is used in birth, marriage, and death, as well as in other prayer and sacred offerings. *Somar*, an ethnic fermented milk product from cow or yak, is generally consumed by the Sherpa highlanders in the Himalayas to increase the appetite and to cure digestive problems (Tamang 2005).

Airag or *chigee*, similar to *koumiss*, is ethnic fermented milk prepared from mare or camel milk beverage in Mongolia, Kazakhstan, Kyrgyzstan, and some Central Asian regions of Russia (Kosikowski and Mistry 1997; Watanabe et al. 2008). It is a mildly alcoholic, sour-tasting fermented drink that is made by native lactic acid bacterial fermentation; this is followed by alcoholic fermentation of the residual sugar by yeasts. *Airag* contains a small amount of carbon dioxide and as much as 2% alcohol, but it does not contain kefir grains. *Tarag* is an ethnic fermented milk product of Mongolia similar to yogurt and is made from cow, yak, goat, or camel milk (Watanabe et al. 2008). The traditional manufacture of *airag* involves storing of mare's milk in animal skin bags, where a natural or induced (inoculated) acidification process takes place. Dry *airag* is usually kept from season to season, and the distinctive starters are transferred from one generation to another within the families for back-slopping (Mayo et al. 2010). Microorganisms present in ethnic Mongolia fermented milk products are LAB [*L. casei, L. helveticus, L. plantarum, L. coryniformis, L. paracasei, L. kefiranofaciens, L. curvatus,* and *L. fermentum* (Ying et al. 2004; Watanabe et al. 2008; Wu et al. 2009)] and yeasts [*Saccharomyces unisporus, Kluyveromyces marxianus, Pichia membranaefaciens,* and *S. cerevisiae* (Ni et al. 2007; Watanabe et al. 2008)].

The ancient Turkish nomads in Asia are believed to be the first to make yogurt, which they called *yoghurut* (Rasic and Kurmann 1978). The inhabitants of Thrace used to make soured milks called *prokish* from sheep's milk, which later became yogurt (Chomakow 1973). The ancient Turks who were Buddhists used to offer yogurt to the angels and stars that protected them (Rasic and Kurmann 1978). *Kishk* is a fermented milk–wheat mixture stored in the form of dried balls in Egypt (Abd-el-Malek and Demerdash 1977) and is popular among the rural populations of some West Asian countries such as Syria, Lebanon, Jordan, and Iraq (Basson 1981). *Kishk* is a cultural as well as traditional functional food with excellent keeping quality, is richer in B vitamins than either wheat or milk, is well adapted to hot climates by its content of lactic acid, and has a therapeutic value (Morcos et al. 1973).

4.3 Fermented Fish

Fermented fish as sauce or pastes and dried and salted fish products are widely prepared and consumed in Southeast Asia as seasoning, condiments, and side dish. Fermented fish products are prepared from freshwater and marine finfish, shellfish, and crustaceans that are processed with salt to cause fermentation and, thereby, to prevent putrefaction (Ishige 1993). The Mekong basin (Mae Nam Khong), which covers the Yunnan province of China, Myanmar, Laos, Thailand, Cambodia, and Vietnam, is considered as the origin of fermented fish products in Southeast Asia (Ishige 1993). *Narezushi* (cooked rice added to the fish and salt mixture) remains common in the Mekong basin areas of Laos, Cambodia, and Thailand (Ruddle 1993). Fermentation of fish was historically associated with salt production, irrigated rice cultivation, and the seasonal behavior of fish stock (Lee et al. 1993). If no vegetables are added, the salt–fish mixture yields fish sauce, which is commonly used as a condiment, and if the product of fish and salt preserves the whole or partial shape of the original fish, it is called *shiokara*, which is then made into paste (Ishige and Ruddle 1987). *Shiokara* paste has synonyms in Southeast Asia: it is known as *ngapi* in Myanmar and as *pra-hoc* in Cambodia (Ruddle 1986). *Shiokara* is used mostly as a side dish by the ethnic people of Cambodia, Laos, Thailand, Myanmar, Luzon and the Visayas in the Philippines, and Korea. Squid *shiokara* is the most popular fermented seafood in Japan (Fujii et al. 1999).

Lactic fermented fish products are often associated with inland areas such as the Central Luzon region of the Philippines and the northeast of Thailand where the freshwater fish are the usual raw material (Adams 1998). In northeast Thailand, it was found that several different sources were used for the fermented fish—the flooded rice fields, paddy ponds beside rice fields used for collecting fish when the field has dried up, and a large local freshwater reservoir (Dhanmatitta et al. 1989). Culturally, in hot countries, fermented fish products continue to play a vital role in adding protein, flavor, and variety to rice-based diets (Campbell-Platt 1987). In Asia, among the fermented fish products, the most widely used are fish sauces and pastes. The fish sauce of Asia is a nutritious condiment made from traditionally fermented fish and salt mixture (Thongthai and Gildberg 2005). Some of the ethnic fish products of Asia are *patis* of the Philippines; *nam pla, pla ra,* and *plaa-som* of Thailand; *shottsuru* and *shiokara* of Japan; *jeot kal* of Korea; *pindang* of Indonesia; *budu* of Malaysia; *ngapi* of Myanmar; and *sukuti, sidra, ngari, hentak,* and *tungtak* of India and Nepal (Adams 1998; Salampessy et al. 2010). Species of *Lactobacillus, Pediococcus, Micrococcus, Bacillus,* and yeast including species of *Candida* and *Saccharomyces* are reported from *nam pla* and *kapi*, which are fermented fish products of Thailand (Watanaputi et al. 1983). *L. plantarum* and *L. reuteri* were isolated from *plaa-som*, a Thai fermented fish prepared from freshwater fish (Saithong et al. 2010). *Micrococcus* and *Staphylococcus* are dominant microorganisms during ripening of *shiokara* (Tanasupawat et al. 1991; Wu et al. 2000). *Haloanaerobium fermentans* (Kobayashi et al. 2000a,b), *Tetragenococcus muriaticus*, and *T. halophilus* are isolated from the Japanese fermented puffer fish ovaries (Kobayashi et al. 2000c). *Bacillus stearothermophilus, B. shaercus, B. circulans*, etc. are predominant microflora in *ngapi*, which is the fermented fish paste of Myanmar (Tyn 1993). Species of *Halobacterium* and *Halococcus* were isolated from *nam pla* (Thongthai and Suntinanalert 1991).

Fish preservation is common in foothill and low attitudes in the Himalayan regions of India, Nepal, and Bhutan, mostly in nearby rivers and their tributaries, lakes, and ponds. Ethnic fermented fish products are *ngari* and *hentak* of Manipur and *tungtap* of Meghalaya in India; the rest of the fish products are dried or smoked, which include *sidra, sukuti*, and *gnuchi* of Nepal, Darjeeling hills and Sikkim in India and Bhutan, and *karati, lashim* and *bordia* of Assam in India (Thapa et al. 2004, 2006, 2007). Nonconsumption of fish products by the Tibetans may be due to the religious taboo, as fish is worshipped for longevity and Buddhists strongly believe that releasing captured fish into the rivers may prolong their longevity through prayers (Tamang 2010a). Moreover, lakes are regarded as sacred by the Buddhists in the Himalayas, barring them from angling. Another reason for not consuming fish in the local diet may be due to the taste of animal meats and dairy products and difficulty in minute bones and scales of fish that hinder the Tibetans from eating easily, as compared to animal meats with larger bones. Fermented fish foods are deeply associated with food culture of the Meitei in Manipur, which are prepared and eaten in every festival and religious occasions. Consumption of fish products in the Himalayas, though, is important in the diet and is comparatively less than other fermented products such as vegetable and dairy products. This may be due to adoption of a pastoral system of agriculture and the consumption of dairy products. *Lac. lactis* subsp. *cremoris, Lac. plantarum, Lac. lactis* subsp. *lactis, Leuconostoc mesenteroides, E. faecium, E. faecalis, Pediococcus pentosaceus, Lac. confusus, L. fructosus, L. amylophilus, L. coryniformis* subsp. *torquens, L. plantarum, B. subtilis, B. pumilus, Micrococcus*, and yeasts *Candida chiropterorum, C. bombicola*, and *Saccharomycopsis* spp. were isolated from *ngari, hentak, tungtap, sidra, sukuti, gnuchi, karati, lashim*, and *bordia* (Thapa et al. 2004, 2006, 2007).

4.4 Fermented Meat

In Asia, fermentation, smoking, and drying of available animals are restricted to a few countries. Sausage was prepared and consumed by ancient Babylonians as far as 1500 BC and by the ancient Chinese (Pederson 1979). Some common fermented meat products of Thailand are *nham* (fermented beef or pork sausage), *naang* (fermented pork or beef), *nang-khem* (fermented buffalo skin), and *sai-krork-prieo* (fermented sausage; Phithakpol et al. 1995; Visessanguan et al. 2006; Chokesajjawatee et al. 2009), whereas China has *lup cheong* (Leistner 1995). *Nham* is the ethnic semi-dry, uncooked, fermented pork or beef sausage of Thailand (Yanasugondha 1977). It is one of the popular meat products of the country prepared from ground pork, shredded cooked pork rind, sucrose, garlic, salt, cooked rice, sodium erythrobate, trisodium polyphosphate, monosodium glutamate, whole bird chili, and potassium nitrite, thoroughly mixed, stuffed into a plastic casing, and sealed tightly (Visessanguan et al. 2004). LAB, notably lactobacilli and staphylococci/micrococci, were dominant with the lower counts of yeasts and molds (Visessanguan et al. 2006). The predominant microorganisms in *nham* are *L. plantarum, L. pentosus, L. sakei, P. acidilactici*, and *P. pentosaceus* (Valyasevi and Rolle 2002). The fermentation of *nham* takes 3–5 days and relies mainly on adventitious microorganisms, which are normally found in raw materials (Khieokhachee et al. 1997). The occurrence of pathogenic bacteria, namely, *Salmonella* sp., *Staphylococcus aureus*, and *Listeria monocytogenes*, were reported in *nham* (Paukatong and Kunawasen 2001). However, Chokesajjawatee et al. (2009) found nonproduction of enterotoxins and low risk of staphylococcal food poisoning in properly fermented *nham*. Application of *L. curvatus* as a starter culture in *nham* provides a tool for preventing the outgrowth of pathogenic bacteria (Visessanguan et al. 2006).

The Himalayan people have a variety of traditionally processed smoked, sun-dried, air-dried, or fermented meat products including ethnic sausages from yak, beef, pork, sheep, and goats (Tamang 2010a). Some common as well as some less well-known traditionally processed ethnic meat products of the Himalayas are *kargyong, satchu, suka ko masu, kheuri, chilu, chartayshya, jamma*, and *arjia* (Rai et al. 2009). These are naturally cured without starter cultures or the addition of nitrites/nitrates. *Kargyong* is an ethnic sausage-like product prepared from yak, beef, and pork meat in the Himalayan regions of India, Nepal, Bhutan, and China (Rai et al. 2009). Three varieties of *kargyong* are prepared and consumed: *yak kargyong* (prepared from yak meat), *lang kargyong* (prepared from beef), and *faak kargyong* (prepared from pork). *Yak kargyong* is a popular fermented sausage in Sikkim, Ladak, Tibet, Arunachal Pradesh, and Bhutan in the Himalayas. During the preparation of *kargyong*, meat and its fat are finely chopped and combined with crushed garlic, ginger, and salt and mixed with water. The mixture is stuffed into the segment of the gastrointestinal tract called *gyuma*, which is used as a natural casing 3–4 cm in diameter and 40–60 cm in length. One end of the casing is tied up with rope, and the other end is sealed after stuffing; the sausages are then boiled for 20–30 min. Cooked sausages are hung in bamboo strips above the kitchen oven for smoking and drying for 10–15 days or more to make *kargyong*. It is sliced and fried in edible oil, adding onion, tomato, powdered or ground chilies, and salt, and is made into curry.

Satchu is an air-dried or smoked meat product of the Indian Himalayas, Bhutan, and China, mostly prepared from yak and beef meat (Tamang 2010a). Red meat is cut into several strands of about 60–90 cm with thickness of 2–5 cm and is mixed thoroughly with turmeric powder, edible oil or butter, and salt. The meat strands are hung in bamboo strips and are kept in open air in the corridor of the house, or are smoked above the kitchen oven for 10–15 days as per the convenience of the consumers. This is a natural type of preservation of perishable fresh raw meat in the absence of refrigeration. Deep-fried *satchu* is a popular side dish of the ethnic people, which is eaten with traditional alcoholic beverages. Bacteria (*L. sake*, *L. curvatus*, *L. divergens*, *L. carnis*, *L. sanfrancisco*, *E. faecium*, *E. cecorum*, *Leu. mesenteroides*, *L. plantarum*, *L. casei*, *L. brevis*, *P. pentosaceus*, *Staphylococcus aureus*, *Micrococcus* spp.) and yeasts (*Debaryomyces hansenii*, *D. polymorphus*, *D. pseudopolymorphus*, *P. burtonii*, *P. anomala*, *Candida famata*, *C. albicans*, and *C. humicola*) were isolated from ethnic meat products of the Himalayas (Rai et al. 2010).

During festivals, goats are ritually sacrificed after the ceremony and the fresh meat is cooked and eaten during a family feast; the remaining meat is smoked above the earthen oven to make *suka ko masu* for future consumption. Ethnic people in the high mountain areas of the Himalayas slaughter yaks occasionally and consume the fresh meat, and the remaining flesh of the meat is smoked or preserved in open air, which is called *satchu*. The making of *kargyong*, which are ethnic fermented sausages made from yak minced meats using intestine as natural casing, might have started a long time before Buddhism was embraced by the Tibetans probably 3000 years ago (Tamang 2010a). The ethnic people of the Western Himalayas prepare many chevon (goat)-based ethnic meat products such as *chartaysha*, *jamma*, and *arjia* (Rai et al. 2009). By 16S rRNA and phenylalanyl–tRNA synthase (*pheS*) genes sequencing, LAB isolates from *chartaysha*, *jamma*, and *arjia* were identified as *E. durans*, *E. faecalis*, *E. faecium*, *E. hirae*, *Leu. citreum*, *Leu. mesenteroides*, *P. pentosaceus*, and *Weissella cibaria* (Oki, K., Rai, A., Sato, S., Watanabe, K., and Tamang, J.P., unpublished). The native skills of the Himalayan people in preservation methods of locally available raw meat can be justified by making sausage-like products using unappealing animal parts such as scraps, organ meats, fat, and blood. A traditional sausage-like meat product is also made with the leftover parts of the animal. Hence, the making of *kargyong* (ethnic sausage-like products) may have been first influenced by the high-altitude dwellers in their search for the uses of leftover meats (Rai et al. 2010).

REFERENCES

Abd-el-Malek, Y. and Demerdash, M. 1977. Microbiology of kishk. *Proceedings of the Symposium on Indigenous Fermented Foods*, GIAM-V, Bangkok, November 21–27, 1977.
Adams, M.R. 1998. Fermented fish. In: *Microbiology Handbook Fish and Seafood*, Vol. 3, edited by R.A. Lawley and P. Gibbs, pp. 157–177. Leatherhead: Leatherhead Food RA.
Adhikari, R.R. and Ghimirey, H. 2000. *Nepalese Society and Culture*. Kathmandu, Nepal: Vidyarthi Pustak Bhandar.
Basson, P. 1981. Women and traditional food technologies. Changes in rural Jordan. *Ecol. Food Nutr.* 11:17–23.
Campbell-Platt, G. 1987. *Fermented Foods of the World: A Dictionary and Guide*. London: Butterworth.
Chokesajjawatee, N., Pornaem, S., Zo, Y.-G., Kamdee, S., Luxananil, P., Wanasen, S., and Valyasevi, R. 2009. Incidence of *Staphylococcus aureus* and associated risk factors in Nham, a Thai fermented pork product. *Food Microbiol.* 26:547–551.

Chomakow, H. 1973. *The Dairy Industry in the People's Republic of Bulgaria*, booklet. Bulgaria: Agriculture Academy.

Dewan, S. and Tamang, J.P. 2006. Microbial and analytical characterization of *Chhu*, a traditional fermented milk product of the Sikkim Himalayas. *J. Sci. Ind. Res.* 65:747–752.

Dewan, S. and Tamang, J.P. 2007. Dominant lactic acid bacteria and their technological properties isolated from the Himalayan ethnic fermented milk products. *Antonie van Leeuwenhoek* 92(3):343–352.

Dhanmatitta, S., Puwastien, P., and Yhoung-Aree, J. 1989. *Fermented Fish Products Study (Thailand) Phase 1. Baseline Study (1987–1988)*. Bangkok: Institute of Nutrition, Mahidol University.

Fujii, T., Wu, Y.C., Suzuki, T., and Kimura, B. 1999. Production of organic acids by bacteria during the fermentation of squid shiokara. *Fish. Sci.* 65(4):671–672.

Huang, H.T. 2000. *Science and Civilisation in China, Vol. 6, Biology and Biological Technology, Part 5, Fermentations and Food Science*. Cambridge: University Press.

Hymowitz, T. 1970. On the domestication of the soybean. *Econ. Bot.* 24(4):408–421.

Ishige, N. 1993. Cultural aspects of fermented fish products in Asia. In: *Fish Fermentation Technology*, edited by C.-H. Lee, K.H. Steinkraus, and P.J. Alan Reilly, pp. 13–32. Tokyo: United Nations University Press.

Ishige, N. and Ruddle, K. 1987. Gyosho in Northeast Asia—A study of fermented aquatic products (3). *Bull. Natl. Mus. Ethnol.* 12(2):235–314 (in Japanese).

Katiyar, S.K., Kumar, N., and Bhatia, A.K. 1989. Traditional milk products of Ladakh tribes. *Arogya J. Health Sci.* XV:49–52.

Khieokhachee, T., Praphailong, W., Chowvalitnitithum, C., Kunawasen, S., Kumphati, S., Chavasith, V., Bhumiratana, S., and Valyasevi, E. 1997. Microbial interaction in the fermentation of Thai pork sausage. *Proceedings of the Sixth ASEAN Food Conference*, December 24–27, 1997, Singapore, pp. 312–318.

Kobayashi, T., Kimura, B. and Fujii, T. 2000a. Strictly anaerobic halophiles isolated from canned Swedish fermented herrings (Suströmming). *Int. J. Food Microbiol.* 54:81–89.

Kobayashi, T., Kimura, B. and Fujii, T. 2000b. *Haloanaerobium fermentans* sp. nov., a strictly anaerobic, fermentative halophile isolated from fermented puffer fish ovaries. *Int. J. Syst. Evol. Microbiol.* 50:1621–1627.

Kobayashi, T., Kimura, B. and Fujii, T. 2000c. Differentiation of *Tetragenococcus* populations occurring in products and manufacturing processes of puffer fish ovaries fermented with rice-bran. *Int. J. Food Microbiol.* 56:211–218.

Kosikowski, F.V. and Mistry, V.V. 1997. *Cheese and Fermented Milk Foods*, 3rd edition. Westport, CT: F.V. Kosikowski LLC.

Lee, C.-H., Steinkraus, K.H., and Alan Reilly, P.J. 1993. *Fish Fermentation Technology*. Tokyo: United Nations University Press.

Leistner, L. 1995. Stable and safe fermented sausages world-wide. In: *Fermented Meats*, edited by G. Campbell-Platt and P.E. Cook, pp. 160–175. London: Blackie Academic and Professional.

Mayo, B., Ammor, M.S., Delgado, S., and Alegría, A. 2010. Fermented milk products. In: *Fermented Foods and Beverages of the World*, edited by J.P. Tamang and K. Kailasapathy, pp. 263–288. New York: CRC Press, Taylor & Francis Group.

Mohanan, K.R., Shankar, P.A., and Laxminarayana, H. 1984. Microflora of *dahi* prepared under household conditions of Bangalore. *J. Food Sci. Technol.* 21:45–46.

Morcos, S.R., Hegazi, S.M., and El-Damhougy, S.T. 1973. Fermented foods in common use in Egypt I the nutritive value of kishk. *J. Sci. Food Agric.* 24:1153–1156.

Ni, H.J., Bao, Q.H., Sun, T.S., Chen, X., and Zhang, H.P. 2007. Identification and biodiversity of yeasts isolated from Koumiss in Xianjiang of China. *Weishengwu Xuebao* 47:578–582.

Paukatong, K.V. and Kunawasen, S. 2001. The hazard analysis and critical control points (HACCP) generic model for the production of Thai fermented pork sausage (Nham). *Berl. Munch. Tierarztl. Wochenschr.* 114:327–330.

Pederson, C.S. 1979. *Microbiology of Food Fermentations*, 2nd edition. Westport, CT: AVI Publishing Company.

Phithakpol, B., Varanyanond, W., Reungmaneepaitoon, S., and Wood, H. 1995. *The Traditional Fermented Foods of Thailand*. Bangkok: Institute of Food Research and Product Development, Kasetsart University.

Prajapati, J.B. and Nair, B.M. 2003. The history of fermented foods. In: *Handbook of Fermented Functional Foods*, edited by R. Farnworth, pp. 1–25. New York: CRC Press.

Rai, A.K., Palni, U., and Tamang, J.P. 2009. Traditional knowledge of the Himalayan people on production of indigenous meat products. *Indian J. Tradit. Knowl.* 8(1):104–109.

Rai, A.K., Tamang, J.P., and Palni, U. 2010. Microbiological studies of ethnic meat products of the Eastern Himalayas. *Meat Sci.* 85:560–567.

Ramakrishnan, C.V. 1979. Studies on Indian fermented foods. *Baroda J. Nutr.* 6:1–57.

Rasic, J.L. and Kurmann, J.A. 1978. *Yoghurt—Scientific Grounds, Technology, Manufacture and Preparations.* Copenhagen: Technical Dairy Publishing House.

Rathi, S.D., Deshmukh, D.K., Ingle, U.M., and Syed, H.M. 1990. Studies on the physico-chemical properties of freeze-dried dahi. *Indian J. Dairy Sci.* 43:249–251.

Ruddle, K. 1986. The supply of marine fish species for fermentation in Southeast Asia. *Bull. Natl. Mus. Ethnol.* 11(4):997–1036.

Ruddle, K. 1993. The availability and supply of fish for fermentation in Southeast Asia. In: *Fish Fermentation Technology*, edited by C.-H. Lee, K.H. Steinkraus, and P.J. Alan Reilly, pp. 45–84. Tokyo: United Nations University Press.

Sabban, F. 1988. Insights into the problem of preservation by fermentation in 6th century China. In: *Food Conservation*, edited by: A. Riddervold and A. Ropeid, pp. 45–55. London: Prospect Books.

Saithong, P., Panthavee, W., Boonyaratanakornkit, M., and Sikkhamondhol, C. 2010. Use of a starter culture of lactic acid bacteria in *plaa-som*, a Thai fermented fish. *J Biosci Bioeng* 110 (5): 553–557.

Salampessy, J., Kailasapathy, K., and Thapa, N. 2010. Fermented fish products. In: *Fermented Foods and Beverages of the World*, edited by J.P. Tamang and K. Kailasapathy, pp. 289–307. New York: CRC Press, Taylor & Francis Group.

Tamang, J.P. 2005. *Food Culture of Sikkim.* Sikkim Study Series, Vol. 4. Gangtok: Information and Public Relations Department, Government of Sikkim.

Tamang, J.P. 2007. Fermented foods for human life. In: *Microbes for Human Life*, edited by A.K. Chauhan, A. Verma, and H. Kharakwal, pp. 73–87. New Delhi: I.K. International Publishing House Pvt. Limited.

Tamang, J.P. 2010a. *Himalayan Fermented Foods: Microbiology, Nutrition, and Ethnic Values.* New York: CRC Press, Taylor & Francis Group.

Tamang, J.P. 2010b. Diversity of fermented foods. In: *Fermented Foods and Beverages of the World*, edited by J.P. Tamang and K. Kailasapathy, pp. 41–84. New York: CRC Press, Taylor & Francis Group.

Tamang, J.P. and Samuel, D. 2010. Dietary culture and antiquity of fermented foods and beverages. In: *Fermented Foods and Beverages of the World*, edited by J.P. Tamang and K. Kailasapathy, pp. 1–40. New York: CRC Press, Taylor & Francis Group.

Tamang, J.P., Dewan, S., Thapa, S., Olasupo, N.A., Schillinger, U., and Holzapfel, W.H. 2000. Identification and enzymatic profiles of predominant lactic acid bacteria isolated from soft-variety *chhurpi*, a traditional cheese typical of the Sikkim Himalayas. *Food Biotechnol.* 14(1 and 2):99–112.

Tanasupawat, S., Hashimoto, Y., Ezaki, T., Kozaki, M. and Komagata, K. 1991. Identification of *Staphylococcus carnosus* strains from fermented fish and soy sauce mash. *J. Gen. Appl. Microbiol.* 37:479–494.

Thapa, N., Pal, J., and Tamang, J.P. 2004. Microbial diversity in ngari, hentak and tungtap, fermented fish products of Northeast India. *World J. Microbiol. Biotechnol.* 20(6):599–607.

Thapa, N., Pal, J., and Tamang, J.P. 2006. Phenotypic identification and technological properties of lactic acid bacteria isolated from traditionally processed fish products of the Eastern Himalayas. *Int. J. Food Microbiol.* 107(1):33–38.

Thapa, N., Pal, J., and Tamang, J.P. 2007. Microbiological profile of dried fish products of Assam. *Indian J. Fish.* 54(1):121–125.

Thongthai, C. and Gildberg, A. 2005. Asian fish sauce as a source of nutrition. In: *Asian Functional Foods*, edited by J. Shi, C.-T. Ho, and F. Shahidi, pp. 215–265. New York: Taylor & Francis.

Thongthai, C. and Suntinanalert, P. 1991. Halophiles in Thai fish sauce (*Nam pla*). In: *General and Applied Aspects of Halophilic Microorganisms*, edited by F. Rodriguez-Valera, pp. 381–388. New York: Plenum Press.

Tyn, M.T. 1993. Trends of fermented fish technology in Burma. In: *Fish Fermentation Technology*, edited by C.H. Lee, K.H. Steinkraus, and P.J. Alan Reilly, pp. 129–153. Tokyo: United Nations University Press.

Valyasevi, R. and Rolle, R.S. 2002. An overview of small-scale food fermentation technologies in developing countries with special reference to Thailand: Scope for their improvement. *Int. J. Food Microbiol.* 75:231–239.

Visessanguan, W., Benjakul, S., Riebroy, S., and Thepkasikul, P. 2004. Changes in composition and functional properties of proteins and their contributions to Nham characteristics. *Meat Sci.* 66:579–588.

Visessanguan, W., Benjakul, S., Smitinont, T., Kittikun, C., Thepkasikul, P., and Panya, A. 2006. Changes in microbiological, biochemical and physico-chemical properties of nham inoculated with different inoculum levels of *Lactobacillus curvatus*. *LWT* 39:814–826.

Watanabe, K., Fujimoto, K., Sasamoto, M., Dugersuren, J., Tumursuh, T., and Demberel, D. 2008. Diversity of lactic acid bacteria and yeasts in airag and tarag, traditional fermented milk products from Mongolia. *World J. Microbiol. Biotechnol.* 24:1313–1325.

Watanaputi, S.P., Chanyavongse, R., Tubplean, S., Tanasuphavatana, S., and Srimahasongkhraam, S. 1983. Microbiological analysis of Thai fermented foods. Journal of the Graduate School, Chulalongkorn University 4:11–24.

Wu, R., Wang, L., Wang, J., Li, H., Menghe, B., Wu, J., Guo, M., and Zhang, H. 2009. Isolation and preliminary probiotic selection of lactobacilli from Koumiss in Inner Mongolia. *J. Basic Microbiol.* 49:318–326.

Wu, Y.C., Kimura, B., and Fujii, T. 2000. Comparison of three culture methods for the identification of *Micrococcus* and *Staphylococcus* in fermented squid shiokara. *Fish. Sci.* 66:142–146.

Yanasugondha, D. 1977. Thai nham, fermented pork and other fermented Thai foods. Symposium on Indigenous Fermented Foods, Bangkok, Thailand.

Yegna Narayan Aiyar, A.K. 1953. Dairying in ancient India. *Indian Dairyman* 5:77–83.

Ying, A., Yoshikazu, A., and Yasuki, O. 2004. Classification of lactic acid bacteria isolated from chigee and mare milk collected in Inner Mongolia. *Anim. Sci. J.* 75: 245–252.

5

Leuconostoc and Its Use in Dairy Technology*

Denis Hemme

CONTENTS

---

* This review is dedicated to the memory of my old "Leuconostoc friends" Jean-Pierre Accolas (Institut National de la Recherche Agronomique, INRA) at Jouy en Josas, France and Francis Martley (from his time at the New Zealand Dairy Research Institute at Palmerston) who both passed away.

5.1 Introduction

The genus *Leuconostoc* is a member of the lactic acid bacteria (LAB) that has been less studied than others such as *Lactococcus* and *Lactobacillus*, although specific properties are of great importance in numerous foods and feed fermentation processes. *Leuconostoc* was defined by van Tieghem in 1878 for dextran-producing bacteria that develop in the sugar industry. In the dairy field, Storch initiated studies on dairy starters 140 years ago in Denmark. His main aim was to obtain cultures that could be used to ensure a cream fermentation resulting in aroma butter. It took him 30 years to finally isolate in 1911 the aroma bacterium "X," which was for a long time called *Betacoccus cremoris* and later identified as *Leuconostoc*. He was never pleased with his defined starters because artisan cultures that were exchanged among creameries generally proved to be superior aroma producers (see the detailed history in Knudsen 1931). This example shows that the importance of *Leuconostoc* sp. in dairy products was recognized quite early. Furthermore, the isolation and culture of the dairy species need good basic knowledge of these bacteria.

Presently, there is still a great potential for the exploitation of this genus by selection of efficient and adapted strains, obtention of specific mutants, construction of tailor-made strains with combined properties, or utilization of *Leuconostoc* enzymes transferred in other LAB or non-LAB genus, giving them new abilities. This chapter is an update of a previous exhaustive review (Hemme and Foucaud-Scheunemann 2004, cited in the present chapter as "Rev2004"). The reader could refer to it for some aspects that have been summarized in this chapter. "Rev2004" also replaces references (often oldest ones) cited in 2004 that are not given in the present chapter to shorten the bibliography.

5.2 Habitat, Characteristics, and Taxonomy

5.2.1 Natural Habitat and Presence in Fermented Foods

Leuconostoc strains are present in many environments, but their natural ecological niche is green vegetation and their roots (Rev2004), where they often represent less than 1% of the microbial flora. They are present on various plants of natural and man-made sources (malt, maize, rice, soybeans, cabbage, tef, millet, sorghum, bamboo, coffee, beans, olives, beetroot, carrot, cucumber, sweet pepper, taro, cassava, durian fruit, banana, etc.).

In fermented foods of plant or animal origin, *Leuconostoc* strains are actors in the transformation process (Rev2004) where they originate from four sources:

1. The raw material—depending on the process, they could remain without growth or develop in foods prepared from raw, nonheated, vegetal matters
2. The processing environment, for example, vats and other utensils utilized
3. A fraction of the previous production used as an inoculum for the next one ("old–young technique" or back-shuffling)
4. A commercial *Leuconostoc* starter, in single-strain, multi-strain, or multi-strain mixed culture preparation (Mäyrä-Mäkinen and Bigret 1998)

Leuconostoc species are involved in the production of numerous fermented foods such as cheeses, fermented milks, sourdough, gari (manioc), sauce foods, juices and beverages, kimchi, sauerkraut, pickles, olives, cocoa, bamboo, mais, coffee, tempoyak, and kocho. Furthermore, they are also implicated in the fermentation of silage used as animal feed and may play a role at the surface of meat, fish or meat slices or fillets, shrimp, etc. Dairy products will be considered further in this review.

Alone or during a phase of the fermentation process, *Leuconostoc* strains contribute in making the final product more edible, safe, and healthy (digestibility, vitamins, and free amino acid and peptide enrichment) and to be of pleasant texture and aroma.

5.2.2 Spoilage and Health Infections

Leuconostoc could be detected in spoiling products (in particular, in vacuum-packaged minced herring or lamprey, poultry) but commonly as a fraction of the spoiling association. Finally, *Leuconostoc* species have also been isolated from the microflora of cattle, fish, insects, and soils.

Sometimes isolated from human sources, they are not usually part of the autochthonous human flora (Rev2004) but rather have been associated with human infections. In clinical testing, *Leuconostoc* shared characteristics with α-hemophylic *Streptococcus viridans* or *Enterococcus* (Rev2004). A *Leuconostoc* infection should be suspected in any case of vancomycin-resistant Gram-positive bacteria detected (Dhodapkar and Henry 1996). A wide range of infections has been reported (Rev2004; Florescu et al. 2008; Yossuck et al. 2009; Ishihyama et al. 2010).

As *Leuconostoc*-caused infections are often linked to the central venous catheter, they could be considered as nosocomial infections (Yossuck et al. 2009). These catheter-linked infections may be linked to the potential of *Leuconostoc* to rapidly establish biofilms. With healthy patients, no case of infections linked to the consumption of foods including dairy products containing *Leuconostoc* has been described. Contaminated infant powder, utensils used for preparation, or parenteral catheters have been involved, but these infections were probably linked to an underlying disease such as disrupted bowel mucosa. Successful treatment is achieved using high dosages of penicillin with or without gentamycin.

Considering the rarity of *Leuconostoc* infections and the large presence of *Leuconostoc* in foods and the environment, their generally recognized as safe (GRAS) or qualified presumption of safety (QPS) status is always recognized. *Leuconostoc* could not be considered as an opportunistic pathogen, although it has natural intrinsic vancomycin resistance.

Some health-related metabolic activities of LAB have been described (Rev2004). Biogenic amines are organic bases implicated in food poisoning. Their production by *Leuconostoc* has never been reported. Bile salt hydrolysis that could be involved in various illnesses is absent in *Leuconostoc*. D-lactate is slowly metabolized by humans so that the intake (WHO norm) should be null for babies and lower than 100 mg/day/kg body weight, a higher level than what could be reached by a normal consumption of dairy products.

5.2.3 Phenotypical Characterization

Leuconostoc belong to the low G + C branch of Gram-positive bacteria. They are usually present as ovoid nonmotile and non-spore-forming cocci in pair or short chains. They are facultative anaerobe mesophilic bacteria (growth at 8°C, no growth at 45°C), are catalase-negative, and do not hydrolyze Arg. Their glucose catabolism is heterofermentative via the pentose pathway, producing CO_2, D-lactate, and ethanol or acetate. Fructose is also metabolized with concomitant production of mannitol. Most strains produce slime in the presence of sucrose (Table 5.1).

5.2.4 Isolation and Identification

Selective media are used for the isolation and enumeration of *Leuconostoc*, in particular, media containing sodium azide and sucrose (MSE medium) and vancomycin (Rev2004). Sucrose allows the identification of slimy colonies formed by slime (dextrans)-producing strains. Vancomycin alone is not sufficiently selective as some *Pediococcus* and *Lactobacillus* strains are also resistant.

Identification tools were for a long time based on biochemical characters or specific properties, some of them being time consuming, fastidious, and not precise, leading to some misidentification (Rev2004; Ogier et al. 2008). Thus, the accurate differentiation of *Ln. paramesenteroides* and *Ln. pseudomesenteroides* (now both *Weissella*) from *Ln. mesenteroides* was not possible. Nevertheless, some phenotypical characters are useful, such as the absence of slime formation by *Ln. lactis* and *Ln. mesenteroides* subsp. *cremoris*, the absence of esculin hydrolysis by *Ln. lactis*, the absence of malate decarboxylation by *Ln. fallax*, and the absence of acid production from fructose and maltose by *Ln. mesenteroides* subsp. *cremoris* or from raffinose and mellibiose by *Ln. citreum* that is also the only *Leuconostoc* to produce a yellow pigment (Table 5.1).

Apart from the phenotypical characters cited above, all *Leuconostoc* species possess a similar and specific peptidoglycan with interpeptide bridges containing Ser and/or Ala, which is never found in other

TABLE 5.1

Presumptive Phenotypical Identification of *Leuconostoc*

General Characteristics

Gram-positive
Cocci (ovoid shaped), nonmotile, non-spore-forming
Facultative anaerobic, catalase-negative
Vancomycin resistant
Production of gas from glucose
No arginine hydrolysis
Production of D-lactate from glucose

Additional Characteristics

Growth at 8°C, no growth at 45°C
No growth at pH 4.8
Growth with NaCl 7%
No H_2S formation
Acid production from glucose (all strains); arabinose, arbutin, cellulose, cellobiose, fructose, galactose, lactose, maltose, mannitol, mannose, melibiose, raffinose, ribose, salicin, sucrose, trehalose, and xylose (variable within the species or subspecies)

Features of Some Species or Subspecies

No production of slime from sucrose by *Ln. mesenteroides* subsp. *cremoris*, *Ln. lactis*
No acid production from fructose by *Ln. mesenteroides* subsp. *cremoris* (and some *Ln. lactis* strains), from maltose by *Ln. mesenteroides* subsp. *cremoris* (and some *Ln. gelidum* strains)
No malate decarboxylation by *Ln. fallax*
Yellow pigment produced by *Ln. citreum*

LAB. The peptidoglycan also contains a pentadepsipeptide with a C-terminal D-lactate that is less discriminant as this structure is also present in *Weissella* and lactobacilli resistant to vancomycin (Delcour et al. 1999).

Protein patterns have also been used for identification of *Leuconostoc* (Piraino et al. 2006). They are now replaced by genotypic DNA-based methods, comparing sequences or polymerase chain reaction (PCR) products obtained (Rev2004; review by Randazzo et al. 2007). These molecular techniques are often based on variable regions of 16S or 23S ribosomal genes. They allow differentiation of *Leuconostoc* from *Lactococcus* and *Lactobacillus*, differentiation of *Leuconostoc* species and subspecies, definition of new *Leuconostoc* species, for instance, *Ln. fallax*, and differentiation of strains by randomly amplified polymorphic DNA (RAPD) fingerprints.

These new PCR-based molecular techniques could be coupled to phenotypical methods for differentiating with accuracy and rapidity (Nieminen et al. 2011). The efficiency is linked to the choice of primers that allows discrimination at the genus or at the species level, sometimes at the strain level with highly specific primers. Pulse field gel electrophoresis (PFGE) is a good tool for the estimation of the diversity and assessment of genomic organization of *Leuconostoc* and closely related bacteria (Chelo et al. 2010). RAPD fingerprints were used to describe the diversity of 221 *Leuconostoc* dairy strains, with the classification being confirmed by 16S rDNA sequence analysis (Cibik et al. 2000). Strains formerly described as *Ln.* (now *Weissella*) *paramesenteroides* were in fact non-slime-producing variants of *Ln. mesenteroides*, and these authors proposed that the *Ln. mesenteroides* subsp. *mesenteroides* is only a biovar in the species. Denaturing gradient gel electrophoresis (DGGE) and temperature gradient gel electrophoresis allow for *Leuconostoc* detection in dairy products without the cultivation step with a semiquantitative approach (Ogier et al. 2002). Restriction fragment length polymorphism (RFLP) and terminal RFLP (T-RFLP), based on the use of a unique universal primer, have a higher potential than DGGE, and avoiding the gel variability, T-RFLP is rapid and reproducible and gives a proportion of each of the detected species (Randazzo et al. 2007; Nieminen et al. 2011). Repetitive sequence-based PCR method, based on the (GTG)5 primer, is a powerful tool to classify LAB (Gevers et al. 2001). Multiplex PCR, amplified ribosomal DNA restriction analysis, or restriction of internal spacer region-amplified fragments was also used to follow the whole bacterial communities in different products (Cho et al. 2009) or for identification of *Leuconostoc* and *Weissella* (Rev2004). From the 16S rRNA gene sequence, a primer specific for *Leuconostoc* and another for *Weissella* have been recently defined as the basis of a reliable method that allows an accurate unequivocal allocation of all *Leuconostoc*, including *Ln. palmae* and *Weissella* species as well as 106 new isolates (Schillinger et al. 2008; Schillinger 2011, personal communication). No or very few PCR products were obtained for *Fructobacillus* and for all *Lactobacillus* tested. PFGE confirms this particularity (Chelo et al. 2010).

The relationship between the observed phenotypes (and their variations with external conditions) and genotypes remains a big challenge. No correlation between amplified fragment length polymorphism and biochemical data was observed for 83 *Oenococcus* isolates (Cappello et al. 2010). One of the factors affecting this absence of correlation is the type of DNA fragment utilized. The knowledge of the entire sequence will give more information, although some detected genes or operons are silenced.

At present, it remains somewhat more difficult to discriminate at the strain level, and, therefore, the genotypic differences observed are not reliable with regard to phenotypic or biochemical characteristics. Specific *Leuconostoc* biovars are described by one or several additional specific properties, often linked to selection of *Leuconostoc* strains with a predominant role in food technology and industrial applications (see Section 5.6.3). Among these characters, the most important are related to *Leuconostoc* utilization in the dairy industry:

1. Intensity of gas and slime production
2. Fermentation of citrate, fructose, and lactose
3. Growth in mixed cultures with lactococci as well as at low temperatures

4. Time of survival during a process or in the product
5. Type of enzymatic activities liberated after cell lysis
6. Viability and metabolism as starter or non-starter
7. Resistance to deep freezing or freeze-drying
8. The (pro)-phage content
9. Facility of recovering the biomass after culture

5.2.5 Classification, Taxonomy, and Phylogeny

As for all bacteria, *Leuconostoc* was classified on the basis of phenotypical characters. The genotypic characterization, mainly based only on the type strains, has led to some changes in the classification and taxonomy. Phylogenic trees of *Leuconostoc* were obtained by comparing the sequence of the 16S rRNA gene or analyzing concatenated gene sequences of different genes (multilocus sequence analysis or typing). Doubts subsist concerning the good choice as tree discrepancies, but a good correlation exists between the phylogenetic tree based on the 16S rRNA gene and trees derived from intergenic spacer region or individual genes such as *rpo*C, *rec*A, *dna*A, *dna*K, or *gyr*B. The resulting phylogeny of *Leuconostoc*, their evolution, and their place in the bacterial tree are now more precise (Chelo et al. 2007; Endo and Okada 2008). The *Leuconostoc* group is tachytelic, showing a faster evolution of a factor of 1.7–1.9, compared to the *Lactobacillus–Pediococcus* group, with *Oenococcus* being the most divergent genus, evolving 1.6 times more than *Leuconostoc* (Makarova and Koonin 2007).

Nevertheless, both genotypic analysis results and biochemical and physiological characteristics are considered to differentiate between species and for the description of new species (Endo and Okada 2008). *Ln. fructosus* and related *Leuconostoc* are now classified as the new genus *Fructobacillus* and thus are no longer in the *Leuconostoc* group *sensu stricto*. The status of *Ln. fallax* is not clear. Showing in particular large gaps (approximately 16 bp) in the V1 region of the 16S rRNA and waiting for new data, *Ln. fallax* constitutes the most peripheral line in the phylogenic tree probably due to a slower evolutionary rate than other *Leuconostoc* species. Separation of *Ln. citreum* and *Ln. gelidum* appeared clearly; that of *Ln. mesenteroides* and *Ln. pseudomesenteroides* was less supported.

The genus *Leuconostoc* was first described in 1878 by van Tieghem. Four species were included in the *Bergey's Manual of Systematic Bacteriology* (Garvie 1986). For historical background, see previous reviews (Thunell 1995; Björkroth and Holzapfel 2006; Rev2004). This last review listed 13 species; this number results from the addition of new species and the reclassification of *Ln. oenos* as *Oenococcus oeni* and of *Ln. paramesenteroides*, with some heterofermentative lactobacilli, as *Weissella paramesenteroides* in the new genus *Weissella*. The number of species cited in Table 5.2 results from the addition of six new species: *Ln. garlicum, Ln. durionis, Ln. pseudoficulneum, Ln. holzapfelii, Ln. palmae, Ln. miyukkimchii and the new* subsp. *Ln. mesenteroides* subsp. *suionicum*; the reclassification of *Ln. argentinum* (later heterotypic synonym of *Ln. lactis*), and the reclassification of the four species *Ln. durionis, ficulneum, pseudoficulneum, and fructosum* (previously *Lactobacillus*) in the new genus *Fructobacillus* (Endo and Okada 2008). Strains isolated from the microflora of the wasp *Vespula germanica* presented a high homology with the known species but have not been proposed as a new species up to now (Rev2004).

5.2.6 Genomics

The sequence of *Oenococcus* was first described (Mills et al. 2005); that of the *Ln. mesenteroides* type strain ATCC 8293 was done at the U.S. Department of Energy Joint Genome Institute. Recently, many draft sequences of *Leuconostoc* have been published by Korean institutes: *Ln. citreum* (Kim et al. 2008b), *Ln. kimchii* (Oh et al. 2010), *Ln. gelidum* (Kim et al. 2011a), *Ln. lactis* (*argentinum*) (Nam et al. 2011a), *Ln. fallax* (Nam et al. 2011a), and *Ln. inhae* (Kim et al. 2011b), *Ln. gasicomitatum* (Johansson et al. 2011), *Ln. pseudomesenteroides* (Kim et al. 2011c), and *Ln. carnosum* (Nam et al. 2011b). The genome comprises between 1639 and 2102 kbp, the contigs > 100 bp varied from 30 to 98, and the coding sequences from 1774 to 2205. Single or multiple copies of the 5, 16, and 23S rRNA are found and GC varied from 36% to 43%.

TABLE 5.2

Leuconostoc Species

Present Species	Date of Description	Observations
Ln. mesenteroides	1878 van Tieghem[c]	
subsp. *mesenteroides*[a]	1960 Garvie[c]	
subsp. *dextranicum*	1960 Garvie[c]	
subsp. *cremoris*[b]	1960 Garvie[c]	
subsp. *suionicum*	2011 Gu et al.	
Ln. lactis[a]	1986 Garvie[c]	*Ln. argentinum* is a later synonym of *Ln. lactis* (Vancanneyt et al. 2006)
Ln. pseudomesenteroides[a]	1989 Farrow et al.[c]	
Ln. carnosum[a]	1989 Shaw and Harding[c]	
Ln. gelidum[a]	1989 Shaw and Harding[c]	
Ln. fallax[a]	1991 Martinez-Murcia and Collins	
Ln. citreum[a]	1992 Takahashi et al.[c]	Formerly *Ln. amelibiosum* (Schillinger et al. 1989[c])
Ln. gasicomitatum[a]	2000 Björkroth et al.	
Ln. kimchi[a]	2000 Kim et al.	
Ln. garlicum	2002 Kim and Kyung	
*Ln. inhae**	2003 Kim et al.	
Ln. holzapfelii	2007 De Bruyne et al.	
Ln. palmae	2009 Ehrmann et al.	
Ln. miyukkimchii	2011 Lee et al.	
Reclassified	**Date of description**	**Now classified as**
Ln. paramesenteroides	1967 Garvie[c]	*Weissella paramesenteroides* (Collins et al. 1993)
Ln. fructosum	2002 Antunes et al.	*Fructobacillus fructosum*[d] first described as *Lactobacillus fructosus*
Ln. ficulneum	2002 Antunes et al.	*Fructobacillus ficulneum*[d]
Ln. durionis	2005 Leisner et al.	*Fructobacillus durionis*[d]
Ln. pseudoficulneum	2006 Chambel et al.	*Fructobacillus pseudoficulneum*[d]

Note: In chronological order of description.

[a] Indicates at least a draft sequence published.

[b] *Ln. mesenteroides* subsp. *cremoris* has the more complex history[c]—first included in the genus *Betacoccus* (Orla-Jensen 1919[c]), then successively called *Betacoccus cremoris* (Knudsen and Sorensen 1929[c]), *Ln. citrovorus* (Hucker and Pederson 1931[c]), *Ln. citrovorum* (Garvie 1960[c]), and *Ln. mesenteroides* subsp. *cremoris* (Garvie 1983[c]).

[c] See Thunell 1995, Hemme and Foucaud-Scheunemann (Rev2004), Holzapfel, Björkroth and Dicks 2006 (Bergey's Manual of Sytematic Bacteriology).

[d] Endo and Okada 2008.

* Indicates new draft sequences published during the reviewing process.

5.3 Growth and Properties

5.3.1 Enumeration

Classical methods for enumeration require growth of the strains. Dilution of the samples should be done in chilled 0.1% peptone water or 0.9% sodium chloride, but not in phosphate buffer that lowers the results. Diluted samples must not be frozen. As for *Lactococcus*, a disrupting mechanical treatment should be done to ensure a reliable cell count and not a chain count (see Rev2004).

For monocultures of *Leuconostoc* with high recovery rates, various classical media can be used, such as all-purpose medium with Tween, Briggs, MRS, La, and brain heart infusion supplemented with yeast extract (Rev2004). An absolute selective medium to enumerate *Leuconostoc* from source containing mixed populations is not yet available. The media based on citrate utilization are unsatisfactory when other citrate-positive strains of *Lactobacillus* or *Lc. lactis* subsp. *lactis* biovar. *diacetylactis* are present because some *Leuconostoc* strains are citrate negative. Some other media consider the preference of microaerophilic to anaerobic conditions or the aciduric nature of *Leuconostoc*. Diverse inhibitory compounds tend to give good selectivity, that is, potassium sorbate (for MRSS pH 5.7), thallous acetate (MRST pH 6.5), sodium azide at 0.075% (MSE), tomato juice, and antibiotics such as vancomycin (30 μg/mL) or tetracycline (Rev2004), some of them leading to a slower growth (MSE, due to azide). Sodium chloride (3%) delayed growth (Garvie 1986); sucrose (5%) could be added to facilitate the distinction of slime-producing *Leuconostoc* from other non-slime-forming bacteria, in particular, *Lactococcus* and *Lactobacillus*. Using 5% glucose instead of 2% allows obtaining good-size colonies after 24 h for *Lactococcus* and only 48 h for *Leuconostoc* (Rev2004). Spreading X-gal on the plates gives a blue color resulting from the *Leuconostoc* β-galactosidase (β-gal) activity, not present in *Lactococcus* (P-β-gal activity). It could also be used for discriminating β-gal-positive and β-gal-negative strains (Rev2004; Demirci and Hemme 1995). Fluorescence *in situ* hybridization method, based on the use of fluorescent-labeled oligonucleotide probes targeted to 16S rRNA, allows both identification and quantification of the *Leuconostoc* population (Olsen et al. 2007).

5.3.2 Preservation of Cultures

For up to 2 weeks, *Leuconostoc* cultures (broth or plates) could be stored at 4°C without growth, except for some psychrotrophic strains (Rev2004; Eom et al. 2007; Matamoros et al. 2009a,b). Longer storage periods require freezing of a young culture with cells in exponential phase (often litmus milk + 5% yeast extract and 10% glycerol) at −20°C (up to 1 year) or −80°C (longer). Freeze-drying of cultures (in milk with 4% lactose) also allows long survival.

5.3.3 Optimal Culture Media

The classical MRS medium, modified or not by the omission of citrate and meat extract, could be used for the cultivation of pure strains. The presence of Tween-80 gives a better growth (and dextran production). Some modifications are introduced for specific studies (Rev2004). A chemically defined medium containing a great number of amino acids and vitamins allows the same growth as in a complex medium (Foucaud et al. 1997). All 14 strains tested were auxotroph for Gln and branched amino acids, but auxotrophy was strain dependent for His (11 about 14), Met (8), Trp (5), Arg (2), and Cys (1). With large variations, the growth of *Leuconostoc* strains in milk was limited and stimulated by the addition of yeast extract, tryptone, protease peptone, or casamino acids (Demirci and Hemme, unpublished). The effect of Mn^{2+} in milk is season dependent, and low content could be elevated by Mn^{2+} addition to milk for an adequate growth.

5.3.4 Industrial Production

Complex ready-to-use laboratory media are too expensive for industrial production of *Leuconostoc* starters. Papaïn-hydrolyzed casein or milk- or whey-based medium containing yeast extract, tryptone, lactose, and the obligatory ingredients (Tween-80, Mg- and Mn-cations) may be used. Regulation of the batch pH at 6.5 is recommended for high cellular yield (Rev2004). A cheap, simplified semisynthetic medium allows both good growth and heteropolysaccharide (HePS) production and is adequate in obtaining purity of the HePS (Vijayendra and Babu 2008). Carob syrup is also a cheap base that is used after supplementation with peptone, beef and yeast extract, Tween-80, Mg^{2+}, and Mn^{2+} for the growth of *Leuconostoc* (Carvalheiro et al. 2011). The use of lactose instead of glucose avoids development of contaminant bacteria because most of them do not ferment lactose. Compared to *Lactobacillus* or *Lactococcus*, the yield for *Leuconostoc* is lower, not exceeding 5×10^9 cells/mL in the final culture. To predict the best growth parameters, response surface methodology and artificial surface networks have been used (Garcia-Gimeno et al. 2005).

The recovery of the cells from the culture medium is classically achieved by centrifugation. The bacterial concentrates can be frozen or freeze-dried in the presence of cryoprotectants such as glycerol 2% and maltodextrin 5%. These compounds allow obtaining higher viability during the process, although that of *Leuconostoc* is lower than that of *Lactococcus* and that of *Lb. plantarum* (Coulibaly et al. 2009). These authors also showed that a good survival during the storage is linked to conditions that did not affect the stability of linoleic and linolenic acids [low temperature, absence of oxygen avoiding the formation of reactive oxygen species (ROS), low moisture].

5.3.5 Growth in Mixed Cultures

The growth of *Leuconostoc* in mixed cultures with *Lactococcus* is complex and strain dependent (Rev2004). To sum up, it has been described as a synergistic functional relationship. In particular, *Leuconostoc* is able to produce aroma compounds only after *Lactococcus* have acidified the milk. Growth rate and final biomass of *Leuconostoc* in these mixed cultures are limited due to competition with *Lactococcus* for peptides or amino acids in milk, and this effect is suppressed by supplementation with these compounds. Stimulation of *Ln. lactis* growth in the presence of *Lc. cremoris* AM2 and co-aggregation of *Leuconostoc* with *Lactococcus* have been described. The interactions are also temperature dependent, with *Lactococcus* being favored at 25°C, and an imbalance of the *Leuconostoc/Lactococcus* ratio could rapidly occur. When acetaldehyde enhances growth of pure *Leuconostoc* culture, it is no longer the case in mixed cultures. *Enterococci* and yeasts are able to stimulate *Leuconostoc* growth through degradation of casein and subsequent production of peptides and amino acids.

The industrial production of mixed cultures of *Leuconostoc* and at least one other LAB has been tested, but mixed starters were more regular when issued from a blend of pure strain starters. To follow the implantation of specific strains or to estimate or at least to know the proportion of specific strains, PCR-multiplex methods are now developed. Some are based on the intergenic region between the 16S and 23S genes and/or specific genes.

5.3.6 Inhibition of Other Microorganisms by *Leuconostoc*

The biopreservative role of LAB, including *Leuconostoc*, is linked to the production of compounds that inhibit the growth of pathogenic and spoilage competitors. Active molecules are mostly small degradation products of the carbohydrate or citrate metabolism, in particular, hydrogen peroxide, organic acids like lactic acid, acetic acid, diacetyl (lethal for Gram-negative bacteria, yeasts, and molds; bacteriostatic for Gram-positive bacteria), and ethanol that affect the membrane and the cell integrity at the level of membrane potential, lipids, and transport systems (Rev2004). Inhibition of *Penicillium* sp. by ethanol produced by *Leuconostoc* was described (Bauquis et al. 2002), as well as inhibition of pathogens (Uraz et al. 2004). The antifungal activity of *Ln. citreum* could replace the addition of calcium propionate (0.3%) in some foods (Valerio et al. 2009). A recent review has summarized the potential of LAB as inhibitors of molds (Dallié et al. 2010). The dextrans produced by *Ln. gelidum*, *Ln. mesenteroides* subsp. *cremoris*, or *Ln. mesenteroides* subsp. *mesenteroides* have a profound inhibitory effect on the formation of *S. mutans* biofilm and on the proliferation of *S. mutans* (Kang et al. 2007).

5.3.7 Production of Bacteriocins

A biopreservation-related remarkable character of *Leuconostoc* is the potential of production of bacteriocin (Rev2004). These are 5–10 kDa peptides, with 84% of them containing 20–60 amino acids, showing a bactericidal or a bacteriostatic effect. The first *Leuconostoc* bacteriocin described belongs to class II that are pediocin-like heat-stable peptides containing the YGNGV consensus sequence and do not contain lanthionin. Leucocin A-UAL 187 from *Ln. gelidum* is a 37-amino acid peptide resulting after a maturation before its synthesis, with the gene being plasmid encoded. Leucocins B and C were also described on the basis of differing spectra. The best studied bacteriocins are the class IIa type mesenterocin Mes52A (identical to MesY105) and the class IIc type mesenterocin Mes52B (=MesB105), which is a 32-amino acid peptide (Rev2004; Limonet et al. 2004). The gene sequences are nevertheless more complex than those of *Pediococcus* (Makarova and Koonin 2007). A new PRC-based method will allow a better knowledge of genes (Macwana and Muriana 2011). Sensitive *Leuconostoc* strains showed a negatively charged surface and a fluid cell membrane that allows penetration of the cationic Mes52A in the membrane, probably at the site of the mannose transporter. The pore formed is the source of a K^+

efflux that leads to a decrease in Δ-PSI, thus explaining the observed bacteriostatic effect (Jasnievski et al. 2008). Natural insensitive or selected resistant strains exhibit a less negative surface and a more rigid membrane that prevent action of the mesenterocin. The immunity proteins produced accumulate in the cytoplasm and are not membrane proteins such as colicins. Some *Leuconostoc* strains produce multiple (two or three) bacteriocins, with similar or closed spectra. Four novel bacteriocins with similar spectra but different intensities have been described by *Ln. pseudomesenteroides* QU15. Two of them, named leucocins Q and N and encoded by adjacent open reading frame (ORF) in the same operon, constitute the novel class IId (Sawa et al. 2010). The production of bacteriocins is growth associated, displaying primary metabolite kinetics, and is influenced both by amount of carbon and peptide source (Mataragas et al. 2004). The influence of oxygen concentration appeared somewhat complex and should not be too high (Vázquez et al. 2005).

5.3.8 Resistance to Antibiotics

Intrinsic resistance of *Leuconostoc* to the glycopeptide vancomycin is explained by the presence of a pentadepsipeptide with a C-terminal D-lactate instead of a D-Ala in the peptidoglycan (Delcour et al. 1999). Vancomycin cannot bind D-Ala–D-lactate but binds D-Ala–D-Ala by sensitive bacteria, inhibiting the cross-linking by the transpeptidase. The D-Ala–D-lactate ligase gene from *Ln. mesenteroides* has been cloned (Rev2004). Some *Pediococcus* and *Lactobacillus* strains are also vancomycin resistant. *Leuconostoc* and viridans streptococci (Enterocci) have the same clinical phenotype, but streptococci are sensitive to vancomycin.

Few reports are available on other antibiotics, and the studies were either partial or the antimicrobial agents tested and susceptibility methods used (type of broth, broth dilution, disk diffusion, or Etest) differed, so that few concordant results are available, often being controversial (Rev2004). These difficulties with resistance testing (in particular, determination of microbiological breakpoint values) were confirmed by Hummel et al. (2007) who concluded the necessity of determining both the phenotypic and genotypic resistance profiles, although some resistances detected could not be explained. Minimum inhibitory concentration values of the two *Leuconostoc* strains are low for ampicillin, penicillin G, erythromycin, chloramphenicol, tetracycline, and gentamycin, higher (and higher than values that can be obtainable in blood of treated patients) for the quinolone ciprofloxacin (>32 and 16 μg/mL), and variable for streptomycin (8 and 192). In *Leuconostoc* as in other LAB, it is the first time that chloramphenicol acetyltransferase genes are detected but were found in an inactive status. The resistances present great variations, as shown by a recent clinical study (Ishihyama et al. 2010). The starter strain 195 from Boll purchaser was resistant to a large variety of antibiotics. With such difficulties in determining resistances to antibiotics, evaluation of safety is complicated. Nevertheless, as most of *Leuconostoc* resistances are intrinsic, they are usually considered as nontransferable. Thus, according to the European Food Safety Authority, the use of *Leuconostoc* commercial starter culture bacteria in the production of dairy (and other) products represents a low potential for the spread of gene encoding resistance.

Leuconostoc strains (one *Ln. citreum*, one *Ln. mesenteroides* subsp. *mesenteroides*, and two *Ln. pseudomesenteroides*) resistant to mesentericin 52A or 52B were also resistant to lysozyme 2.5 mg/mL, even when mesenterocin was present (Limonet et al. 2004). A more cationic membrane is probably the cause of the interaction of both compounds. A bacteriocin-producing strain of *Ln. mesenteroides* subsp. *dextranicum* was also found to be lysozyme-resistant for a long time (8 h) and a high concentration (1000 ppm; Fantuzzi et al. 2007).

5.3.9 Biofilm Production

Biofilms produced by *Leuconostoc* are principally involved in default caused in the sugar industry. By a new and elegant method using magnetic particles and specific hydrolases, Badel et al. (2008) showed that dextrans, proteins, and nucleic acids have a role in the biofilm formation. Model systems have been recently reviewed (Coenye and Nelis 2010). A recent study has indicated that the biofilm-forming capacity is not linked to a type of polysaccharide produced and is strain-specific both for *Ln. mesenteroides* and *Ln. citreum*, with wide variation (Leathers and Bischoff 2010).

5.3.10 Lysis and Survival of *Leuconostoc*

Leuconostoc species are bacteria that are able to survive for a long time in unfavorable conditions, including dryness, for instance, on cheesemaking molds (Rev2004). This resistance is probably linked to the production of slime or glycocalix, even at a very low concentration, which may form a biofilm-like structure. The cell lysis is less for lactose-grown cells having produced exopolysaccharides (EPSs) than glucose-grown ones, and factors found in cheese (Ca, Na, pH value) affect lysis. The internal pH of the *Leuconostoc* cells decreases when the culture pH falls, which explains that growth stops when internal values of 5.4–5.7 are reached. Citrate that contributes to a better growth also acts as a buffer to prevent lysis induced by acidic conditions. The autolysis of *Leuconostoc* is similar to *Lactococcus*, both being less than that of *Lactobacillus*. Lysis potential is also very strain-dependent in buffer, varying from 10% to 50% after 1 day at pH 6.5 (Hemme, personal data), or in cheese (Ayad et al. 2004). Two major peptidoglycan hydrolases (PGHs) have been found in *Leuconostoc* and two additional PGHs only in *Ln. lactis*. These properties constitute a specific differential tool (Cibik et al. 2001). One PGH (*Ln*Mur) is a 209-amino acid protein with a putative 31-amino acid signal peptide, and its gene was cloned and is active in *Lc. lactis*.

Like *Halobacteria*, *Ln. lactis* and *Ln. mesenteroides* cells synthesize γ-Glu–Cys as the major intracellular low molecular weight (MW) thiol and not bacillithiol present in many Gram-positive bacteria (Kim et al. 2008a). The γ-Glu–Cys concentration is higher in response to H_2O_2 treatment. Glutathione was also found in two *Ln. mesenteroides* subsp. *cremoris* strains (Fernandes and Steele 1993). Stress proteins with N-terminal sequence close to DnaK and GroEL are expressed after heat shock, and regulation by Ctsr might exist. Pasteurization has the same effect on *Leuconostoc* and other bacteria with a major inactivation when using 67°C for 30 s, with *Ln. lactis* strains being more resistant (Thunell 1995; Samelis et al. 2009). The survival of freeze-dried *Leuconostoc* cells is related to a decrease in unsaturated long chain fatty acids (Coulibaly et al. 2009). *Leuconostoc* was inhibited at the same level as other bacteria by UV-C treatment (Allende et al. 2006). High hydrostatic pressure (HHP) was effective on *Leuconostoc* and enhanced when conjugated with the action of pediocin (Rev2004). The general effect of HHP has been recently reviewed (Rendueles et al. 2010).

5.4 Metabolism

The reader can report our general review (Rev2004) and the specific review of Zaunmüller et al. (2006). The main traits of *Leuconostoc* metabolism are: (1) the heterofermentation of hexoses with production of D-lactate, CO_2, and ethanol; (2) the utilization of the acetyl-P pathway branch with production of acetate at the expense of ethanol and a better growth in the presence of citrate, pyruvate, fructose, or oxygen that allows more adenosine triphosphate (ATP) formation and favors the coenzyme reoxidation; and (3) the production of exopolysaccharides (at least for some of them) mostly from saccharose.

Genomics have brought a new insight concerning metabolism by indicating the presence of the genetic background (Makarova et al. 2006; Makarova and Koonin 2007). This DNA analysis also revealed putative (inactive, pseudo) genes, but the functional annotation of genomes is sometimes inconsistent, often restricted to known gene families and did not consider broad substrate specificities. Improvement of functional annotation has been proposed in the case of flavor-forming pathways of LAB, including *Leuconostoc* (Liu et al. 2008). With the restriction that only the sequenced-type strain is considered, some specific traits are the presence of metabolic pathway genes, including two for D-hydroxyacid dehydrogenase (DH; related to D-lactate production), six genes for aromatic aminotransferase (ArAT), two for AspAT, two for Ser acetyltransferase, and two for cystathionine β-lyase/cystathionine γ-lyase (common with *Oenococcus*, only one or absent in other LAB). A comparative global transcriptional analysis, based on the types of mRNA produced for enzyme synthesis and their level in *Leuconostoc*, will be more accurate in describing the solutes' transport and their metabolisms (Barrangou et al. 2006 for *Lb. acidophilus*).

5.4.1 Carbohydrate Metabolism

As with other LAB, *Leuconostoc* lack catalase, functional cytochromes, and some enzymes of the Krebs cycle. Some pseudogenes are also detected. *Leuconostoc* ferment a large variety of monosaccharides and disaccharides, and the fermentation spectrum is used for discriminating species, subspecies, or sometimes strains or selecting them for application processes. Of importance are, in particular, arabinose, fructose, and saccharose, as well as ribose and xylose (see Sections 5.2.2 and 5.6.3). *Leuconostoc* does not metabolize starch, except *Lb. amylovorus* whose α-amylase has been expressed in *Ln. citreum* (Eom et al. 2009).

5.4.1.1 Fermentation of Sugars as Sole Energy Source

Pentose catabolism occurs via the pentose-P, cleaved by phosphoketolase to glyceraldehyde-3-P and acetyl-P, further to D-lactate and acetate with only one ATP formed and, thus, without acetaldehyde or ethanol production.

Hexoses such as glucose or fructose are also catabolized by the same pathway via two preceding steps, with production of CO_2 and no acetate. In this case, only one ATP is produced from one molecule of hexose, diminishing the growth yield by a factor of 2, and three nicotinamide adenine dinucleotide phosphate [NAD(P)H] instead of one are obtained, explaining the drop of the growth rate by a factor of 3, which is linked to the formation of ethanol being the limiting step. Alternative reoxidation pathways could be used to bypass this limiting step, such as erythritol or glycerol formation (Zaunmüller et al. 2006). This catabolic pathway leading to ethanol via acetyl-CoA is also limited by availability of D-panthothenate, a precursor of HSCoA, explaining the need for supplementation of culture medium with this growth factor. The glucose 6-phosphate DH of *Leuconostoc* is rare, able to function with nicotinamide adenine dinucleotide (NADH) and also with NADPH, which is usually utilized for biosynthetic pathways (Naylor et al. 2001).

Mannitol is produced when *Leuconostoc*, as all heterofermentative LAB, catabolizes fructose (Mills et al. 2005). The metabolism, studied by Nuclear Magnetic Resonance (NMR), is identical by *Ln. ficulneum*, *Ln. mesenteroides*, and *Lb. fructosus* (Antunes et al. 2002). As a sole sugar source and at a nonlimiting concentration, fructose has two roles—one as energy source via the heterolactic fermentation and the other as an electron acceptor to form mannitol via a specific and probably adapted mannitol DH, which differs from those of other bacteria and is used in industry (Aarnikunnas et al. 2002; see Section 5.6.3). The mannitol produced maintains the cell turgor, stabilizes membranes, and has a role in scavenging the ROS. No sorbitol production is observed, and acetate production is favored at the expense of ethanol because of the limited reoxidation capacities of the ethanol pathway. Detection of mannitol is a method used for appreciating the sugar cane deterioration.

D-lactate is produced from pyruvate by *Leuconostoc*, as by some heterofermentative *Lactobacillus*, a part being incorporated in the peptidoglycan. The D-lactate dehydrogenase (LDH) is not activated by fructose-1-6-di-P, and several properties vary within the strains: the affinity for pyruvate (K_m values from 200 to 700 µM), those for NADH (14–150 µM), the optimal pH (5.7–7), and optimal temperature (30°C–37°C). Some *Ln. mesenteroides* strains show a D-LDH activity that is very specific for pyruvate; others are also active on α-keto acids (Letort and Hemme, unpublished). It could reflect the presence of a D-2-hydroxy acid DH (HicDH) as described by *Ln. lactis* (Rev2004). This was confirmed by *Ln. citreum*, where, after replacement of the D-LDH gene by the L-LDH gene of *Lb. plantarum* D-lactate is always produced via a HicDH or by a second D-LDH (Jin et al. 2009).

Lactose metabolism is important for strains used in dairy technology. By *Leuconostoc*, it enters the cells by permeases without phosphorylation. The lactose permease gene *lac*S complements *Escherichia coli*–deficient mutants and presents a high homology with the genes of the *E. coli* Gln and His transporters. The intracellular lactose is typically cleaved to galactose and glucose with the two overlapping genes *lac*L and *lac*M being homologous to *Lb. casei*. However, two strains also presented a minor P-β-gal activity (Smart et al. 1993). Galactose, as mannose, is then metabolized via the Leloir pathway. Some *Leuconostoc* strains required high concentration of lactose to grow, sometimes more than 50 g/L, which is the normal milk concentration (D. Hemme, unpublished). Occurrence of lac-negative strains in culture is frequent perhaps due to plasmid loss (Demirci and D. Hemme 1995). Part of the lactose-positive wild strains isolated from various French (D. Hemme, unpublished) or Spanish (Sánchez et al. 2005) cheeses are also galactose-positive, thus of interest for dairy applications (see Section 5.6.3). The β-gal activity is constitutive and reveals two subgroups of *Leuconostoc*: one with low specific activities (between 0.4 and 8 μmol s^{-1} g^{-1}) and the other with high activities (between 17 and 258) and could not be related to subspecies or growth rate (Letort and Hemme, unpublished).

Sucrose (saccharose) metabolism with production of EPSs is typical of *Leuconostoc*. This property of the cells or of their purified enzymes is widely used in the industry (see Section 5.6.3). The EPSs produced out of the cell are homopolysaccharides formed from glucose units of two types: dextrans α-1,6 linkage and variable between the units and alternans arranged only with both α-1,6 and α-1,3 linkages. Their formation implies secreted specific glycosyltransferase and dextran or levan sucrase that have been well studied, with the gene being expressed by other industrial bacteria (Bozonnet et al. 2002; Ruas-Madiedo et al. 2010; see Section 5.6.3). The fructose moiety enters the cell as fructose-6P that is converted to glucose-6P. Apart from these industrial EPSs, other minor EPSs, less or not described but very active, and although produced in small concentrations in dairy products, have roles in the texture (see Section 5.6.3).

Leuconostoc could also use some unusual fermentation reactions (Zaunmüller et al. 2006). The trisaccharide raffinose metabolism involves α-gal and/or β-fructosidase. Most of the *Leuconostoc* strains present an inducible α-gal activity with different levels within the strains (Rev2004). Thus, *Leuconostoc* strains are good candidates as probiotics for digestibility of raffinose derived from some food polyosides.

The catabolism of some polysaccharides, for instance, cellulose or amylose, has also been reported (Rev2004). An extracellular enolase from *Ln. mesenteroides* 512FMCM has been cloned in *E. coli* (Lee et al. 2006). Some enzymes of the catabolism have been characterized and their gene cloned.

5.4.1.2 Fermentation of Sugars in the Presence of Exogenous Electron Acceptors

Citrate, fructose, oxygen, and pyruvate could favor a more efficient fermentation of sugars by their role in a more rapid NAD(P)H reoxidation and a doubled ATP formation (Zaunmüller et al. 2006). In the co-metabolism of fructose plus glucose, fructose is only used for mannitol production. On the contrary, fructose follows the pentose pathway when the powerful electron acceptor pyruvate is present. Fructose is particularly utilized by the *Fructobacillus* ex-*Leuconostoc* species (Table 5.2).

The glucose metabolism in the presence of oxygen is similar to that of fructose, leading also to acetate production and formation of $2H_2O_2$, instead of mannitol, by activation of the NADH oxidase. Some strains convert a part of the acetate in acetaldehyde via the bifunctional alcohol/acetaldehyde DH (Koo et al. 2005). The aerobic growth is largely higher than in anaerobiosis, resulting from the activation of the acetyl-P branch that leads to the doubling of ATP formation via a higher acetate kinase activity. The endogenous O^{2-} deleterious action in the *Leuconostoc* cells is counteracted by the high level of Mn^{2+} (6–10 mM) as a defense mechanism that replaces the usual superoxydismutase or manganicatalase systems, probably reflecting the usual high Mn^{2+} original niche of *Leuconostoc*. When hemin (10 ppm), which is not synthesized by *Leuconostoc*, supplements the medium, *Leuconostoc* growth is enhanced in aerobic conditions, resulting from the presence of functional cytochromes a2 and b1 (Rev2004).

5.4.1.3 Fermentation of Organic Acids

Pyruvate is also utilized by *Leuconostoc* (and *Oenococcus*) as a sole carbon source; otherwise, this property is found in a *Eubacterium* from the rumen. This is the fourth way of pyruvate utilization, which is uncommon by anaerobic bacteria, via a pyruvate DH decarboxylase that leads to acetyl-CoA and then acetate. Lactate is formed by consumption of a second pyruvate to regenerate NAD(P).

As a large variety of LAB, *Leuconostoc* is able to utilize citrate alone under fermentative conditions at low pH values (Turner 1988). At this level, *Leuconostoc* differs from citrate-positive *Lc. lactis* strains that require the presence of a fermentable carbohydrate to metabolize citrate, leading to CO_2, acetate, diacetyl, acetoin, and 2,3-butanediol (Rev2004). In milk at high pH values, a co-metabolism of citrate (8 mM) and lactose is effective on *Leuconostoc* (Jordan and Cogan 1995). Citrate allows better growth rate and, to some extent, also greater yield and resistance to acidity. Nevertheless, the effect is species or strain dependent. In milk, *Ln. lactis* can grow in the presence of citrate to pH 5.4, better than *Ln.*

mesenteroides subsp. *cremoris* and *Ln. mesenteroides* subsp. *mesenteroides* in which growth stopped at pH 5.9. The metabolism of citrate by *Leuconostoc* is a typical citrolactic fermentation that allows a better maintenance of a neutral-pH homeostasis, particularly true for *Ln. mesenteroides* subsp. *mesenteroides* (review by Bourel et al. 2001). The coupling of citrate with lactate is thus at the level of the redox state of the cell, whereas it is at the level of the end product of glycolysis by *Lactococcus* (Konings 2002). The citrate permease (CitP) acts as an exchanger that imports two dianionic citrate molecules, or other dicarboxylates or tricarboxylates such as malate, from the medium and exports two lactates. As a consequence, the membrane potential is enhanced, and the cell has more energy for growth. The CitP gene is homologous to MleP that exchanges malate and lactate by *Lactococcus*, but CitP is inducible by *Leuconostoc* and constitutive by *Lactococcus*. The affinity for lactate is low, whereas citrate shows high affinity for CitP linked to the presence of the positive group of Arg425 (Rev2004). The regulation of citrate metabolism involves the transcription factor CitI, whose gene is located in the citrate cluster, located upstream of the citrate operon with divergent polarity. CitI is the key of citrate sensing, enhancing its affinity after binding citrate, both for its own promoter and for the operon promoter containing all the genes implicated in citrate metabolism (Martin et al. 2005). The genes of citrate metabolism are chromosomal in *Ln. lactis*, whereas CitP is plasmid-encoded in *Ln. mesenteroides*.

In this co-metabolism, more acetate is produced, which comes from the splitting of citrate by citrate lyase. The by-product oxaloacetate is then decarboxylated, enhancing the pyruvate pool that is essentially reduced to D-lactate, in contrast to *Lc. lactis*, where substantial amounts of diacetyl, acetoin, 2-3-butanediol, and acetaldehyde are produced. By *Leuconostoc*, only a minor part of pyruvate is transformed to diacetyl and acetoin (Rev2004). After condensation of pyruvate with acetaldehyde by acetolactate synthase, acetolactate is spontaneously decarboxylated to diacetyl and then enzymatically to acetoin under acidic conditions (pH of the medium <5). Compared to *Lactococcus*, *Leuconostoc* has a high and irreversible diacetyl reductase activity that leads to a higher proportion of acetoin in the resulting aroma profile.

The co-metabolism of citrate and xylose stimulates the growth of *Ln. mesenteroides* with an increase in D-lactate and acetate production correlated with a larger acetyl-P pool (Schmitt et al. 1997). Most of the *Leuconostoc* strains (and *Oenococcus* but not *Weissella*) are able to exert the malo-lactic fermentation, with the genes being upstream the citrate utilization locus (Konings 2002).

Leuconostoc (except *Ln. fallax*), like *Oenococcus*, is able to ferment malate in L-lactate and CO_2 (Zaunmüller et al. 2006). As for citrate, the exchange of lactate with malate results in a deacidification of the medium. Malate is not present in dairy products, and thus, this does not concern dairy *Leuconostoc*.

5.4.2 Metabolism of Nitrogenous Compounds

5.4.2.1 Amino Acids

The requirement of *Leuconostoc* for amino acids was determined in chemically defined medium and was observed to be generally high but with variations within species and strains (see Section 5.3.5). The amino acids enter the cell by a proton motive force-driven transport system (Winters et al. 1991). Some multiple amino acid strain-dependent transport systems are also present, and cross-inhibition takes place, for instance, inhibition of Glu transport by Asp (Gendrot et al. 2002). The fact that Ser stimulates growth is in accordance with the absence of Ser (and Gly) biosynthesis genes revealed in genomics studies (Makarova et al. 2006). Ala is never required: Asp is synthesized from oxaloacetate and is converted to Asn.

The enzymatic equipment of *Leuconostoc* indicates a potential to catabolize amino acids, thus contributing to the production of amino acid–derived aroma flavor. Genes analysis revealed the presence of aminotransferases and keto acids DH genes, whereas keto acid decarboxylase, D-hydroxyacids DH, and glutamate dehydrogenase (GluDH) genes are absent (Liu et al. 2008). This theoretical potential derived from *in silico* analysis of only one sequence strain has to be confirmed by strain screening. Thus, GluDH activity, which leads to α-keto-glutarate that is the key of all amino acid transamination, is detected for three of six strains investigated (Fernández de Palencia et al. 2006; Hanniffy et al. 2009). Aminotransferase activity, catalyzing another important degradation step, is present in *Leuconostoc* at least for aromatic and branched-chain amino acid, as compared with *Lactococcus* at a reduced level (Fernández de Palencia et al. 2006; Hanniffy et al. 2009). Lyase activity for C–S-containing compounds is also present at a low level, revealing a lower potential of enzymatic equipment for Met and Cys catabolism than found in other LAB and, furthermore, the absence of Met γ-lyase (Hanniffy et al. 2009). The His decarboxylase gene was detected in 16 of the 24 dairy strains tested (A. Lonvaud and D. Hemme, unpublished data), but none of 17 other grape must and wine strains tested produced histamine, whereas three produced tyramine and four produced putrescine (Moreno-Arribas et al. 2003).

5.4.2.2 Peptides

Dipeptides, tripeptides, and oligopeptides up to seven amino acids are utilized by *Leuconostoc* for growth (Foucaud et al. 2001) or by nonproliferating cells (D. Hemme and A.-S. Lepeuple, unpublished data), with variation among the strains. Dipeptide transport systems differ from the amino acid systems, and a specific system stimulated by Mg^{2+} or Ca^{2+} exists for oligopeptides containing more than four amino acids (Germain-Alpettaz and Foucaud-Scheunemann 2002). Carboxypeptidase activities are absent in all strains tested. Aminopeptidase activities are located only in the cytoplasm, and are numerous, varying from 14 to 39, comparable to those of *Lc. lactis*, including PepN, PepT, and PepX activities that are considered to be "technological peptidases." Some strains show a γ-glutamyl transferase, which is absent in *Lactococcus*.

5.4.2.3 Proteolytic Activity

Leuconostoc poorly utilizes proteins for growth, thus resulting in a limited growth in milk (10^8 cells/mL), which reflects the utilization of amino acids and peptides present and is similar to the growth of other protease-negative bacteria in milk (Demirci and Hemme 1994). Nevertheless, incubation of some (4 above 14) *Leuconostoc* strains in milk indicates some casein proteolysis. It is probably linked to some lysis of the cells that liberated internal proteolytic enzymes.

5.4.3 Other Metabolisms

Some other metabolic activities are also found in *Leuconostoc* (Rev2004). Esterase activities are detected in *Leuconostoc* for C2 to C4 compounds, but not for longer length of chains. Hydrolysis of tributyrin was described for one strain of *Ln. mesenteroides*. *Ln. lactis* synthesizes butyl and ethyl esters from tributyrin and ethanol by a transferase reaction. Cells of *Leuconostoc* incubated with methane thiol produce S–Me–thioacetate. Menaquinones-9 and -10 are produced by some *Leuconostoc* strains.

Transport of vitamins by a novel class of transporters has been described. They comprise an integral membrane protein (S component) forming an active transporter in the presence of an energy AT module (Rodionov et al. 2009). *Leuconostoc* probably possesses universal transporters. The transport system of Met precursor presents its specific AT module, with the mtsTUV operon being regulated by the Met-specific T-box attenuator. The transport of folate by *Ln. mesenteroides* is also mediated by an energy coupling factor transporter but requires the universal energy coupling AT module (EcfAAT) that is encoded by a separate gene cassette.

5.5 Genetics

In the last years, new insights in *Leuconostoc* genetics mainly concern the utilization of some genes transferred in industrial bacterial strains of other species and also construction of modified *Leuconostoc* strains (mutants or engineered; see Section 5.6.3).

5.5.1 Phages

Up to now, *Leuconostoc* phage attacks have never been described in the dairy field, although *Leuconostoc* phages were isolated. The low growth and poor acidification due to *Leuconostoc* in mixed cultures could explain the absence of detection. Nevertheless phages have been isolated (Dessart and Steenson 1995). Six *Ln. fallax* phages genetically distinct were isolated from industrial sauerkraut (Rev2004) and one from *Ln. gelidum* (Greer et al. 2007). A phage from *Ln. pseudomesenteroides* was obtained after UV induction (Jang et al. 2010). The first *Leuconostoc* phage sequence was Phi1-A4 from *Ln. mesenteroides* (Lu et al. 2010). It is a linear double-stranded DNA phage of 29 kbp, a 36% GC, 50 putative ORF, and a 30% inversion (including the lysis module), clustering with *Lactococcus* phages and not with other *Leuconostoc* phages. All these phages have a long and noncontractile tail typical of the *Siphoviridae* family. The thermal resistance of phages have been tested (Atamer et al. 2012).

5.5.2 Plasmid Biology

As with other LAB, *Leuconostoc* strains often harbor plasmids. Their number could vary from 1 to 6 in commercial starters (Johansen and Kibenich 1992). One or two (most strains of the INRA-CNRZ collection (CNRZ was the previous name, Centre de Recherches Zootechniques, of the INRA Research Center of Jouy en Josas, France), three (strains CAF1 and CH), or four plasmids (MW 2.3, 5.4, 35, and 80 for the commercial strain 195) were detected in another study (A. Arias 1985, unpublished). Some of them are cryptic or encode phage resistance, lactose utilization (explaining rapid detection of lactose-negative variants), bacteriocin production, and diacetyl reductase (Rev2004). Mutants cured from the plasmid showed loss of the diacetyl reductase activity or lactose. Some plasmids exhibit a rolling circle replication mechanism such as pIHO1 (Park et al. 2005), whereas others such as pFMBL1 replicate via a theta mechanism (Jeong et al. 2007).

5.5.3 Horizontal Gene Transfer Ability

Exchange of genetic material with other bacteria is indicated by the presence of insertion elements, with similarity to genes of other LAB. Amino acid sequences of the *Ln. mesenteroides* bacteriocins of *Leuconostoc* are quite similar to those of *Ln. gelidum*. The *Ln. mesenteroides* cluster citR-mae-citCDEF for citrate utilization is closely related to the *Lb. plantarum* cluster, whereas the genes for lactose transport and hydrolysis are related to the genes of *Ln. lactis* (Rev2004). A large homology (99.2%) was described between citP of *Ln. lactis* and *Lc. lactis*. The genomics studies indicated that if *Leuconostoc*, as all other LAB, has lost a part of the ancestor genome during the evolution leading to a metabolic simplification, some 100 genes have been acquired after the divergence from the common ancestor (Makarova and Koonin 2007).

5.5.4 Cloning Vectors

Some difficulties appeared when trying to transform *Leuconostoc* because they do not present the natural competence that gives the capacity to accept and integrate DNA. The low transformation efficiency of some strains may also be due to the presence of a plasmid. Some large broad vectors (pAMB1, pIP501) and transposons were first used to transform *Leuconostoc* strains by conjugal transfer or transformation and then by electroporation (Rev2004). Finally, specific tools were developed often based on plasmids. The plasmid pCI411 from *Ln. lactis* was able to transform *Ln. mesenteroides* and *Leuconostoc/Weissella paramesenteroides* strains. The plasmid pFR18 of *Ln. mesenteroides* subsp. *mesenteroides* was used to transform strains of *Ln. mesenteroides* subsp. *cremoris*, *Ln. mesenteroides* subsp. *dextranicum*, and also *Lb. sakei*. A food-grade vector able to provide mesenterocin production ability was developed (Biet et al. 2002). The first shuttle vector, derived from the *Ln. citreum* plasmid pIH01, is able to transform by electroporation other *Leuconostoc* strains and also *Lb. plantarum* and *Lactococcus* strains (Park et al. 2005). Other shuttle vectors able to transform other *Leuconostoc*, *Lactobacillus* or *E.coli* strains, have been constructed from *Leuconostoc* plasmids (see Eom et al. 2012). In the same way, a new vector, stable for 100 generations in the absence of the erythromycin marker, is based on pCC3, a cryptic plasmid of *Ln. citreum* (Chang and Chang 2010).

5.5.5 Gene Cloning and Heterologous Expression

The first studies described the cloning of glucose-6-P DH and P-glucose isomerase from the pNZ63 plasmid of *Ln. lactis* or the *rec*A gene of *Ln. mesenteroides* in *E. coli*. Numerous other *Leuconostoc* genes have now been cloned (Rev2004) such as the genes of D-Ala–D-lactate ligase of *Ln. mesenteroides*, mannitol DH, diacetyl/acetoin reductase of *Ln. pseudomesenteroides*, and, recently, HPr kinase/phosphorylase from *Ln. mesenteroides* (Park et al. 2008).

Nowadays, cloning results concern mostly heterologous transfer of enzymes with a technological interest (see Section 5.6.3). Genes of *Leuconostoc* such as D-LDH and carbohydrate metabolism enzymes, polyoside synthesis enzymes, and peptidoglycan hydrolase are expressed in other bacteria. In the same way, genes from other bacteria could modify the potential of *Leuconostoc*; for example, L-LDH from *Lb. plantarum* (Jin et al. 2009) or alpha-amylase from *Lb. amylovorus* (Eom et al. 2009) are expressed in *Ln. citreum*, and the lacG gene of *Lc. lactis* is expressed in *W.* (ex-*Ln.*) *paramesenteroides* (Dessart and Steenson 1995).

5.6 *Leuconostoc* in Dairy Technology

Leuconostoc species are important agents in a wide range of dairy products made under controlled conditions and probably also in many other less studied artisanal cheeses where *Leuconostoc* strains have been detected. As for the other dairy LAB *Lactococcus* and *Lactobacillus*, the roles of *Leuconostoc* (the "third L dairy species") are mostly strain-dependent, and thus, application in the dairy field is somewhat away from academic studies, which often only consider the type strains. The development of new more accurate tools will help in describing the strain particularities and their potential uses in technology. Genetic tools adapted to *Leuconostoc* will also allow a fine understanding of the *Leuconostoc* metabolism and its regulations, as formerly done for *Lactococcus* and *Lactobacillus*, where genome reduction or inactivation by insertion sequence is better described (Douglas and Klaenhammer 2010). It appears achievable to analyze differences between strains that are either useful or not, in particular, when the strain variations are dependent from regulation genes. Only one mutation in a regulatory gene could lead to a different expression of half of the total genome. As shown for *Lb. acidophilus* (Azcarate-Peril et al. 2005), the strain diversity could be better explained and utilized for the strain selection, for instance, coupled levels of gas or aroma production adapted to the diverse cheese technologies.

5.6.1 Presence in Dairy Products

Besides technologies where *Leuconostoc* species are deliberately added as starters, these species are also present in a major part of dairy products, in particular, those utilizing raw milks, the so-called old–young technique, and/or processing vessels (vats or molds) that inoculate the milk.

Leuconostoc represented 10% of the extensive study of 4379 isolates from 35 products, including 24 artisanal cheeses (Cogan et al. 1997). An extensive collection of wild *Leuconostoc* strains from French representative cheese varieties is present in the INRA-CNRZ collection (Devoyod and Poullain 1988). The *Leuconostoc* population comprised between 10^4 and 10^7 per gram depending on the cheese. The testing of more than 200 strains of this collection by RAPD and 16S RNA sequencing indicates some correlation at the subspecies or species level between phenotype and cheese technologies, suggesting a better adaptation of certain strains to a particular technological environment (Cibik et al. 2000). Thus, Reblochon (mountain cow milk soft cheese) contains mostly *Ln. mesenteroides* subsp. *dextranicum* strains, whereas *Ln. citreum* is the major species found in Crottin de Chavignol (soft to dry goat milk cheese). *Ln. mesenteroides* subsp. *mesenteroides* strains are mostly isolated from Roquefort (blue-veined ewe-milk soft cheese). This confirms the selection of strains belonging to these subspecies for the starter production in Roquefort technology (Devoyod and Poullain 1988; G. Pradel, personal communication).

The *Leuconostoc* population in milk, as for other genera, reflects the general milk quality resulting from diverse contaminations. It has to be considered in all cheeses made from raw milk, mostly used in the high-quality cheeses with a European Registered Denomination of Origin (RDO) or a French Indication Géographique de Provenance (IGP). The absence of an effect or a reduced effect after deliberate addition of *Leuconostoc* to raw milk in certain French regions was in fact due to the anticipated high *Leuconostoc* contamination of the milk (D. Hemme, personal observation). Such a presence of *Leuconostoc* in the dominant flora was also observed for the raw milk used for the Sicilian pasta filata-type cheese Ragusano

(Randazzo et al. 2002). Interestingly, the authors indicated that *Leuconostoc* cells revealed by PCR-based methods show an inactive state. This is an important character from a technological point of view because classical methods detect only cultivable cells. The deduced "inactivity" revealed by the presence of weak DGGE bands of rRNA does not mean a true inactivity, as even without lysis, the permeability of the resting cells increases, thus facilitating exchange with the cheese curd [see Section 5.6.3.2 and refer to the old work of Reiter et al. (1967) and the description of the different bacterial stages by Niven and Mulholland (1998)].

For the analysis of cheese flora, the reader may refer to our previous review (Rev2004). After that review, works dealing with the detection of *Leuconostoc* (and general flora) in cheeses have continued to be published, in particular, for RDO European cheeses and also for other dairy products around the world. Mostly a combination of classical and molecular approaches was used. They concern, for instance, the description of the *Leuconostoc* flora, which is the dominant flora of the Tibetan Qula yak cheese (Tan et al. 2010), or part of the natural flora for the Italian Pecorino Crotonese cheese (Randazzo et al. 2010) and for the Norwegian Präst cheese (Ostlie et al. 2005). The microflora of fermented milks is also studied, for instance, the complex flora of kefir (Witthuhn et al. 2005) and koumiss (Hao et al. 2010). As has been done for a long time, the goal of these studies is often to select new starter strains (i.e., Sánchez et al. 2005; Nieto-Arribas et al. 2010 for the Spanish Afuega'l Pitu and Manchego cheeses).

5.6.2 Judgment of Cheese Quality

Sensory analysis and physical parameters are usually used for judging the quality of the cheese. For some cheeses, a good correlation exists between measures in the young cheese (curd) and the final quality of the long-term ripened cheese (i.e., the value of the pH after 24 h in the French hard cheese Comté). For other cheeses, it is rather more difficult to do. New data such as the DGGE pattern, which is similar in cheeses of the same quality, may help in monitoring the quality as shown for Ragusano cheese (Randazzo et al. 2002).

5.6.3 Roles in Dairy Technology

The roles of *Leuconostoc* in the dairy field differ largely from the classical role in vegetable fermentations (sauerkraut and silage). Among the different *Leuconostoc* species, the subspecies *Ln. mesenteroides* subsp. *cremoris* is probably most used in the dairy field. This is due to its large utilization as an aroma producer in the dairy industry in starters for fermented milk, cream, butter, and cheese where only a limited opening is required (Rev2004). For a long time, these *Ln. mesenteroides* subsp. *cremoris* strains were the sole starters available on the starter market. This explains why most trials for their use in cheesemaking were unsuccessful (Turner 1988; D. Hemme, personal data; see Section 5.7.2). In Cheddar cheese, *Leuconostoc* does not contribute to flavor production as it is not part of the non-starter LAB (NSLAB) flora (mostly lactobacilli; Rev2004), except in cheeses with texture defect (Turner 1988).

Empirical observations and more precise ecological studies on ewe milk and cheese flora both indicate that the *Leuconostoc* found (or useful in cheesemaking) belong mostly to the two subspecies *Ln. mesenteroides* subsp. *mesenteroides* and *Ln. mesenteroides* subsp. *dextranicum*, and sometimes to *Ln. citreum* (Devoyod and Poullain 1988; Cogan et al. 1997; Cibik et al. 2001; Nieto-Arribas et al. 2010). In some specific cheeses, such as some soft or blue-veined cheeses, achieving openness is the major role of *Leuconostoc*. In artisanal raw milk cheeses, *Leuconostoc* are present as part of the natural flora in particular because of their persistence on vats, tools, and molds, in relation with the production of slime that favors adhesion on various supports. For controlled industrial production, in particular, with high hygienic conditions and pasteurized milk, *Leuconostoc* starters are used. In this case, artisanal starter cultures are prepared for some local productions and freeze-dried cultures by starter providers for industrial products (blue-veined cheeses such as the French Saint Agur and Bresse Bleu). Besides openness, *Leuconostoc* produced aroma compounds or precursors in the cheese and have probably other more complex, not well-known roles, for instance, linked to production of slimy products that play a role in the texture quality, the production of D-lactate or bacteriocins, and some specific activities (see Section 5.4). It is undoubtedly true for cheeses where *Leuconostoc* is the dominant and stable flora, for instance, in artisanal goat milk Spanish cheeses (Rev2004). It was also demonstrated for semi-hard cheeses with deliberate adjunct of *Leuconostoc* (D. Hemme et al. 1996; Turner 1988; D. Hemme unpublished).

5.6.3.1 Use as Starters in Dairy Technologies

The classical nomenclatures for starters were D for the citrate-positive *Lc. lactis* var. *diacetylactis*-containing ones, B (or L) for those containing *Leuconostoc*, and BD (or DL) for mixed starters, that is, containing both *Leuconostoc* and *Lactococcus*. Inoculation of *Leuconostoc* should be at least 10^6 cells/mL of milk to obtain a significant effect (Martley and Crow 1993, 1996). For instance, it should be 10^7 log representing 10% of the *Lactococcus* dose for the fabrication of the French Pyrénées cheese to obtain eyes formation. All cheeses made with *Leuconostoc* contain D-lactate. The total D- and L-lactate represent somewhat 75% of the organic acids present (acetic, pyruvic, formic, uric, propionic, orotic) depending on the cheese type (Rev2004).

5.6.3.1.1 Use for Cheese Openness

Leuconostoc is a decisive contributor to the openness by its CO_2 production. In some cheese varieties, this production results from the action of other bacteria, in particular, propionibacteria in hard cheeses where H_2 is additionally produced. Yeasts are also able to produce gas, but they also give a too strong flavor. The formation of eyes or slits results from CO_2 that is no longer soluble in the cheese curd. The solubilities of CO_2 in water, fat, and proteins are similar, and variations in the eye formation after saturation are linked to pH (solubility of 5% at pH 5.2 and 2% at pH 4.8) or temperature (approximately 50% more at 10°C than at 20°C). In Gouda cheese, the saturation concentration is around 35 mmol/kg, but the formation of eyes begins already at 20 mmol/kg due to technological conditions (Martley and Crow 1993). To obtain this concentration, the *Leuconostoc* inoculum should be at least 10^6 cells/mL of vat milk. The selected strain(s) must grow during the early cheesemaking stages at around 30°C, although this commonly applied temperature is unfavorable to *Leuconostoc* (as above the optimum; see Section 5.3.5). The forms of openings are dependent on the curd texture, with rough slits being more present in acid cheeses with short texture, regular or irregular round-formed shiny holes, or eyes in flexible past. Openness is sometimes related to the process but mostly to gas produced by heterofermentative bacteria (Martley and Crow 1996; McSweeney and Sousa 2000). For some productions, the choice of strains is adapted to the production of a curd with fewer openings, with the maximum concentration being only 16 mM.

In pressed ripened Dutch cheeses such as Gouda, Edam, and other brine salted varieties that are pressed under the whey to avoid air entrapment, no mechanical openness occurred. The small and shiny openings are only due to CO_2 produced by the traditional adjunct of *Leuconostoc* together with the *Lactococcus* starter (mixed starter). Their formation is initiated at the level of microscopic undissolved nitrogen bubbles present in the curd. In some other cheeses such as Pyrénées cheese, the openings are numerous and more irregular due to differences in technology. In Havarti and blue cheeses, openness results both from air trapped between the curd grains resulting from stirring of the curd and from the formation of gas by *Leuconostoc* strains. The form and repartition of openings are quite different in appearance in these cases.

For blue cheeses, the opening is fundamental because the galleries formed are colonized by the *Penicillium roqueforti*. The curd is pierced or pricked to admit air in the case of industrial blue-veined cheeses, but some more artisanal productions of top quality avoid this step in their rules or RDO definition. The numerous (different) blue-veined cheese types produced in France result from the choice of different *Leuconostoc* strains producing various gas quantities and also those of the *P. roqueforti* strain (i.e., chosen from >70 strains available), giving a different coloration and shape of the characteristic blue veins. It is also the case for Roquefort, a ewe's milk blue-veined cheese produced in the Massif Central (Center of France), exported around the world and considered as the top blue cheese. The historical process does not use the mechanical step, and only the natural *Leuconostoc* flora participates in the formation of gas (Devoyod and Poullain 1988; Martley and Crow 1993, 1996; D. Hemme, unpublished). As soon as the 70th, the use of concentrated suspensions of *Leuconostoc* was proposed by Devoyod et al. (1974) for milk containing an insufficient number of wild *Leuconostoc*. The initial inoculation should be around $10^{6.5}$ cells/mL of milk, representing 10% of the *Lactococcus* starter concentration. The Roquefort cheeses of high quality grew up from 70% to 95%. Such a *Leuconostoc* addition is nowadays the rule considering the increasing bacteriological quality of raw milk or its use after pasteurization. Moreover, it is well known from producers that strains of the subspecies *Ln. mesenteroides* subsp. *mesenteroides*, and not *Ln. mesenteroides* subsp. *dextranicum* strains, allow the optimal high-quality Roquefort cheeses (Devoyod and Poullain 1988; G. Pradel, personal communication).

Although empirical, the choice of strains is also probably related to their survival in the curd where they are able to continue to metabolize as resting cells in the absence of sugars. This is the case for the catabolism of citrate, which could persist without the presence of sugar and at low pH. Some other activities are pH-dependent and will differ with the cheese type. It was shown that *Leuconostoc* strains that are very active in low pH curds are inactive in washed-curd Dutch-type cheeses having a high pH (Turner 1988; Martley and Crow 1993). Although considered good gas producers, some specific (and rare) strains are not adapted for cheese usage as they undergo rapid lysis in young cheese curd (D. Hemme, unpublished). This could be related to the specific lysis equipment of *Leuconostoc* (Rev2004; Nieto-Arribas et al. 2010).

Leuconostoc has been involved in early and late blowing of some cheeses such as Saint Nectaire containing $10^{7.5}$ cells/g (Devoyod and Poullain 1988). This blowing, also caused by other bacterial types, is linked to the fermentation of abnormal residual lactose in the curd. An atypical eye formation has been detected in a Norwegian Gouda-type cheese mostly (9/13) related to the presence of *Leuconostoc* or heterofermentative *Lactobacillus* at 10^7 cells/g (Narvhus et al. 1992). *Leuconostoc* were present in Cheddar cheese with slit defect (Turner 1988; Martley and Crow 1993). When Cheddar presents openness (considered as a defect) due to *Leuconostoc*, its population was at least 10^6 per gram.

5.6.3.1.2 Use for Aroma Production

The primary *Lactococcus* starter alone contributes to a major part of the flavor of the cheese, as shown by results obtained with products made under sterile conditions. However, secondary flora is needed to obtain the full flavor of mature cheeses. It is evident in two main types: (1) soft cheeses with an aerobic surface white flora, with yeasts and molds or more complex for the red/pink smear flora, and (2) soft blue-veined cheeses with internal blue *P. roqueforti*. It is also true for a large variety of other cheeses with specific aroma. This secondary flora could originate from the adventitious milk flora and factory environment or could be a deliberately added adjunct (mixed *Leuconostoc*/*Lactococcus* starter). When present at a sufficient level, *Leuconostoc* has a role in aroma production that could be direct by production of compounds or precursors or indirect by influencing the development of undesirable bacteria being part of the so-called NSLAB (Martley and Crow 1993).

The major compounds produced by *Leuconostoc* are diacetyl, ethanol, and acetate in particular by citrate–glucose co-metabolism (see Section 5.4). Some other minor compounds are also produced. When using *Lactococcus* starter strains that do not ferment citrate, the *Leuconostoc* strain(s) should be present at a sufficiently high population to metabolize citrate, with the advantage that *Leuconostoc* are able to express this metabolism in both the presence and absence of sugar. The activity will be higher in cheeses having a low pH than in cheeses where acidification is limited by replacing a part of the whey with water (lactose removal technology). Due to its low flavor threshold, diacetyl produced at concentrations between 1.5 and 5 ppm is sufficient to give the typical buttery perception. When allowed by legislation, addition of citrate to milk (i.e., 0.15% in the United States) or to wash water allows an enhanced diacetyl production. The incorporation of oxygen in products such as fermented milks also favors the diacetyl production.

The diacetyl produced is further transformed into acetoin and 2,3-butanediol, which do not contribute to aroma. This unfavorable transformation could be lowered when products such as fermented milks are cooled after the aroma production. Some worldwide known industrial St. Paulin–type cheeses are also ripened at 4°C to avoid too high and too rapid ripening. However, most of the cheeses are ripened at higher temperatures (10°C–13°C) where the process could not be stopped. As it is true for mutant genetically modified *Lactococcus* starters with inactivation of a specific gene, it could be of great importance to use strains having lost *Leuconostoc* activities that transform the diacetyl, either wild or engineered strains. The level of diacetyl and acetoin production varies greatly within the strain (Bellengier et al. 1994; Nieto-Arribas et al. 2010).

Ethanol is the most common alcohol found in cheese and an obligatory precursor of a lot of ethyl esters, in particular, light chain fatty acid ethyl esters (mostly butanoic and hexanoic acids, as well as isovalerates). Ester biosynthesis probably occurred more by alcoholysis (hydrolysis by ethanol) than by esterification due to esterase activities (Liu et al. 2004). These esters are known to give a fruity flavor to the products, which could be a defect if present at a too high concentration in some cheeses (Cheddar) or a part of the normal aroma of numerous soft and semi-hard cheeses (Liu et al. 2004). These esters have very low perception thresholds (expressed in parts per billion). However, ester formation could be low due to the availability of ethanol, which is often the limiting factor, in particular, when only homofermentative *Lactococcus* strains are used. This is not the case if *Leuconostoc* are present because they are ethanol producers as a result of their heterofermentative sugar metabolism and their ability to reduce the acetaldehyde produced. The second activity counteracts the "green" defects, which are caused by too high acetaldehyde concentration produced by the *Lactococcus* starter in butter and fermented milk or by *S. thermophilus* now widely used in some soft cheeses. In cheese, the impact of some other off-flavors such as malty or brothy aroma or bitterness is attenuated by using *Ln. lactis* that presents a maximal activity at the usual low ripening temperatures. The low pH, increased salt level, and lower water activity are factors that lower the acetaldehyde reduction (Liu et al. 2004). The use of *Leuconostoc* both with *Lactococcus* as a primary starter in Camembert cheese leads to an enhanced aroma, which is perhaps related to the higher ethanol concentration detected that also has a role by lowering the *P. camemberti* covering of the cheese (Bauquis et al. 2002).

Acetate is also a precursor of some aroma products; in particular, thioesters become formed when thiols, mostly methane thiol, are produced by the smear bacterial flora present at the surface of smear-coated cheeses, essentially by *Brevibacterium linens* and *Arthrobacter* sp. (Hemme and Desmazeaud 1999). As for ethanol, the availability of acetate could be the limiting factor of thioester production. Thus, the formation of acetate is important in smeared soft cheeses such as Livarot, Munster, Epoisses, Pont l'Evêque, Reblochon, Limburger, or Tilsitt and probably some other cheeses having a yeast flora that is also able to produce methanethiol. Methyl thioacetate is the major compound of the typical aroma of Pont l'Evêque. Most of these cheeses are RDO cheeses made with raw milk that contains the adventitious *Leuconostoc* flora. For such soft cheeses presenting a large exchange surface with the atmosphere, a sufficient O_2 concentration may allow the formation of acetate by *Leuconostoc*. The thioester formation is a pure chemical reaction due to the high reactivity of thiols in the cheese conditions, which also react with each other or with aldehydes.

The α-keto-glutarate availability could limit the transamination reactions that are crucial steps in the production of amino acid–derived aroma products in cheese ripening (Yvon 2006). This compound is produced from Glu, which is present at high concentrations during cheese ripening, by GluDH activity that is present and is important for some *Leuconostoc* strains (Fernández de Palencia et al. 2006; Hanniffy et al. 2009), and also remains active after cell lysis. As for ethanol, *Leuconostoc* is thus a good candidate to obtain this precursor.

5.6.3.2 Deliberated Adjunct of Leuconostoc

A lot of investigations concerning adjunct NSLAB cells or extracts have been reported (Shakeel-Ur-Rehman et al. 2000). They were performed mostly with lactobacilli, recently for Gouda production with a mixed commercial starter including *Lb. paracasei* (Van Hoorde et al. 2010) and a minor part with *Leuconostoc*. As proposed by Martley and Crow (1993), the deliberate addition of selected adjunct flora may be important for controlling the cheese quality. *Leuconostoc* strains are potential members of this added secondary flora, perhaps more in some countries such as France, the "country with 1000 cheeses," where the proportion of soft and semi-hard cheeses with complex flora is higher than in countries where Cheddar types predominate.

Ln. mesenteroides subsp. *cremoris* strains usually used for production of butter, cream, or fresh cheeses are not recovered when used in other cheeses and have only a minor effect, if any. *Ln. mesenteroides* subsp. *mesenteroides* or *Ln. mesenteroides* subsp. *dextranicum* strains are preferred; the cell numbers in the products need to be at least 10^7 cells/mL of milk and in the curd, and this is not favored by the presence of *Lactococcus* (see Section 5.3.5). Milk can also be matured with about 10^6 cells/mL for 15 h at 13°C to reach this desired level (Hemme and Boulanger, unpublished).

Objective results deduced from adjunct studies first imply good adventitious microflora knowledge and/or the use of pasteurized milk or microfiltration, which is the best way to avoid temperature-linked modifications. In Gouda-type cheese manufactured under aseptic industrial conditions, the effect of added *Leuconostoc* was not clear despite the fact that the citrate content had been decreased (Martley, personal communication). Some of our first experiences in the industry using milk that contained at least a fraction of raw milk indicated only a little effect of *Leuconostoc* adjunct; this was explained by a high level of the endogenous *Leuconostoc* flora (Hemme and Ferchichi, unpublished). Further trials were conducted with pasteurized or microfiltrated milk.

Adjuncts of *Leuconostoc* cells at a high density close to that used for Roquefort (10^7 cells/mL of milk to obtain 10^8 per gram of curd) were tested at a pilot scale in the INRA Experimental Atelier from Jouy-en-Josas for pressed semi-hard St. Paulin–type cheeses (Hemme et al. 1996; Hemme, Boulanger, Münchner, and Ogier, unpublished). Sensory analysis indicated that the time of ripening at 12°C of such experimental cheeses was reduced by 50%, being the same after 2 weeks as the control cheeses after 4 weeks. An accelerated consumption of lactose was observed, although the final pH remained 0.2 unit higher, corresponding to a reduced (10%) total acid production detected, with D-lactate representing 25%. This higher pH probably favors proteolysis that was confirmed by higher content of hydrophilic soluble peptides on the high-performance liquid chromatography profiles and the higher amino acid (>10%) concentration found. The level of gas production and acceleration of ripening was directly linked to the initial concentration of cells, with an optimum of 10^7–10^8 per gram for obtaining the best compromise between faster ripening and adequate openness. When *Leuconostoc* was present at 10^9 in the curd, a detectable flavor of silage, acetate, and bitterness occurred. At the adequate level, *Leuconostoc* selected strains are thus good candidates to counteract acidifying starters and accelerate the cheese ripening.

These results were confirmed by trials under aseptic conditions using the micromodel developed for pressed-type washed-curd cheeses (Hynes et al. 2000). Aroma was modified when cell concentrations were greater than $10^{6.5}$ per milliliter in milk, whereas opening appeared only between $10^{6.5}$ and $10^{7.5}$, depending on the *Leuconostoc* strain, and originating possibly from its autolytic properties (Hemme, Bienvenu, and Ogier, unpublished).

Adjuncts of *Leuconostoc* cell extracts or heat-shocked cells have also been tested in some cheese productions. In Ras cheese, addition of such cells or extracts accelerated the ripening and decreased the bitterness of the product. Manchego cheeses made from pasteurized milk using defined starters comprising *Ln. mesenteroides* obtained higher scores for flavor quality and intensity and overall impression than cheeses made with commercial starters (Rev2004).

5.6.3.3 Other Roles in Cheeses

5.6.3.3.1 Inhibition of Specific Flora

Some *Leuconostoc* strains exert a positive influence on cheese quality by avoiding some negative activities of NSLAB, in particular, lactobacilli that produce malty or brothy off-flavor or bitterness. It could be due to the production of bacteriocins active against NSLAB, the reduction of some growth factors and substrates, or other indirect actions. A *Ln. mesenteroides* strain inhibited NSLAB growth by 2 logs during ripening of Cheddar cheese (Martley and Crow 1993). This atypical strain was thermoresistant and showed a high gas production potential from lactose, causing a defective open texture in the cheese from which it was isolated. Possible explanations could be the utilization of growth nutrients and the inhibitory role of ethanol that was demonstrated for Camembert cheeses (Bauquis et al. 2002). In this case, where the obtained *Leuconostoc* level was 10^8 cells/g, the produced ethanol limited the development of *P. camemberti*, thus obtaining the traditional surface aspect of RDO/IGP raw milk French Camembert and not the "carton-like" white surface often present for industrial Camembert or Brie produced from pasteurized milk often found. This inhibition was confirmed on laboratory agar medium plates. The limited mold growth leads to a decrease in some disadvantageous events during the ripening, in particular, formation of bitter peptides or too high ammonia production in soft cheeses with moldy surface flora. It is also of great importance that *Leuconostoc* present exhibit protease/peptidase activities. At the opposite of *Lactobacillus*, which are often present at high numbers and cause aroma defects, *Leuconostoc*, even at high concentrations, are not involved in such defects linked to excessive proteolysis and resulting in peptide-linked bitterness aroma. It is the reason they were called "gentle" and "strictly associative" bacteria by my INRA colleague J.-J. Devoyod, a pioneer in Roquefort research. In some cheeses, the obtained slimy texture is linked to the production of slime by *Leuconostoc*. This sliminess property is major in the case of some slimy fermented milks.

5.6.3.3.2 *Bacteriocin Utilization as Preservative*

Compared to chemical preservatives, bacteriocins are of interest in food technology. As biopreservatives, they are harmless for eukaryotic cells, pH and heat tolerant, and, as being peptides, easily hydrolyzed in the digestive tract. As enterocins that are now proposed as preservatives against spoilage or pathogenic microorganisms (Khan et al. 2010), *Leuconostoc* bacteriocins could be utilized. The bacteriocin-producing character of *Leuconostoc* is scarce, and effort is made for isolating positive strains (Voulgan et al. 2009). None of the 210 strains, isolated from a large variety of cheeses, from the INRA-CNRZ collection was found to inhibit *Listeria* (N. Deschamps, personal communication). Bacteriocinogenic *Leuconostoc* were isolated from the Bulgarian drink boza (Todorov 2010). *Leuconostoc* strains with antimicrobial bacteriocin-related properties have been selected for their protease and peptidase activities for their use in the traditional Spanish Genostoso cheese (González et al. 2010). This character was also considered for various seafood products, revealing 132 colonies (above 5575) with inhibitory properties, including 52 LAB and three *Leuconostoc* (Matamoros et al. 2009a). The psychrotrophic strains belong to *Ln. gelidum*, and when sprayed at the surface of sliced food kept under refrigeration, the inoculum doubled the shelf life of the products by improving the safety and quality by inhibiting both spoilage and pathogenic bacteria (Matamoros et al. 2009b). Similarly, the use of the bacteriocin-producing strain *Ln. citreum* GJ7 as a starter is the first example of application that allows an extension of the shelf life of kimchi (Chang and Chang 2010). The production of kimchi is enhanced by the presence of the sensitive strain *Lb. plantarum* KFRI 464 that liberated an inducing factor, which is a 6500-Da peptide homologous to the 30S ribosomal protein S16 (Chang et al. 2010).

The addition of purified bacteriocin in fermented foods or at the surface (by spray or contained in films) of nonfermented products could in some countries require prior approval by the regulatory authorities, thus presently limiting their potential use.

5.6.4 Roles in Functional Foods

5.6.4.1 *Role as Potential Probiotics*

As with some other microorganisms currently proposed to the consumers, *Leuconostoc* strains do not colonize the intestinal tract but may survive during the limited time of transit. Their effect on the host through bacterial actions is thus expected to be small except when ingested at high cell concentrations. Feeding children with the Indian fermented milk Dahi containing 10^8 cells/g of *Lc. lactis* and *Ln. mesenteroides* was described as a tool for reducing the duration of diarrhea by 0.3 day. The supplementation of such *Leuconostoc* fermented milk beverages with tryptone (100 mg/L) enhances their therapeutic value linked to a higher iron, zinc, magnesium, and fatty acid bioavailability (Shobharani and Agrawal 2009). *Leuconostoc* strains (and *S. thermophilus*) were shown to be more potent inducers of Th1-type cytokines IL-12 and IFN-γ than the probiotic *Lactobacillus* strains (Kekkonen et al. 2008), thus modulating the Th1/Th2 imbalance (Kang et al. 2009). Thus, *Leuconostoc* belongs to the NSLAB, which are selected on the basis of their health benefits (review by Settanni and Moschetti 2010).

5.6.4.2 *Hydrolysis of α-Galactosides*

α-Galactosides such as stachyose and raffinose present in some foods result from the hydrolysis of polyosides commonly found in plants. They are not metabolized by humans or animals owing to the lack of α-galactosides in intestinal mucosa, hence causing flatulence. To overcome these drawbacks and to boost the consumption of otherwise highly nutritional food products, trials have been made to eliminate α-galactosides using physical methods or α-gal (Hugenholtz et al. 2002). *Leuconostoc* strains could be good candidates as probiotics for their ability to hydrolyze such α-galactosides. Presently, attempts have not come out, and perhaps other bacteria able to develop in the digestive tract and having acquired the *Leuconostoc* α-galactosidase would be better adapted.

5.6.4.3 Production of Prebiotic Oligosaccharides

Numerous oligosaccharides are produced by *Leuconostoc* strains (review by Seibel and Buchholz 2010). Thus, they are the object of numerous recent studies as most known prebiotics are oligosaccharides. The α-gluco-oligosaccharides produced by *Ln. mesenteroides* NRRL-B-18242 are highly resistant against the attack by digestive enzymes. They have been proposed to have a potential prebiotic effect in human neonates as they stimulate the growth of beneficial bacteria of the intestinal flora. Such oligosaccharides were catabolized by bifidobacteria and lactobacilli but not by *Salmonella* or *E. coli*, pointing toward their effect on intestinal microflora modification (Chung and Day 2002). In fermented milk, addition of *Ln. citreum* affects only slightly the growth of the *Lactococcus* and *Lactobacillus* starters, but allows the production of oligosaccharides in the presence of lactose or other sugars. It constitutes a means for the production of a variety of oligosaccharides in fermented milks, thus contributing to the development of synbiotic dairy foods (Seo et al. 2007). The choice of the adequate strain is fundamental for the production of a specific type of oligosaccharide (Côté and Leathers 2005), which is also the case for fermented vegetables (Eom et al. 2007). In the presence of maltose, more branched oligosaccharides are produced from sucrose by constitutive mutants of *Ln. mesenteroides* Lm28 having a higher glucosyltransferase activity (Iliev et al. 2008).

5.6.4.4 Production of Exopolysaccharides

Leuconostoc strains are the more efficient exopolysaccharide (EPS) producers, and the cells or the enzymes of specific, often mutants, are currently used for their production in the industry. These polysaccharides also have roles when produced directly in the food: for the food itself, due to a specific texture, and for humans ingesting the food due to positive action in the intestinal microbiota. The EPSs produced by *Leuconostoc* are homopolysaccharides with a variable strain-related degree of polymerization. Dextrans are mainly composed of α-1,6 linked residues with variable (strain-specific) degrees of α-1,2 branching, whereas alternans present α-1,3 and α-1,6 linkages. The biosynthesis process is extracellular, requires sucrose, and is achieved by excreted specific glycosyltransferases and dextran or levan sucrase enzymes that are classified according to the strain-dependent acceptor product patterns (Côté and Leathers 2005). Eight glucansucrase-encoding genes were cloned (Bozonnet et al. 2002), and some could be expressed in *Lactococcus* to achieve dextran synthesis (Neubauer et al. 2003). The presence of additional sugars to sucrose promotes the reduction of the degree of polymerization, therefore allowing us to obtain EPSs with different properties (Santos et al. 2005). A non-gelling, non-film–forming, and water-soluble heteropolysaccharides has been described (Vijayendra and Babu 2008). These properties are of great interest in the food industry, permitting rapid and low-cost production conditions nether found before.

Dextrans were first utilized as a blood plasma expander and as blood flow improvers. EPSs are now food additives used as texturizers by increasing viscosity and as stabilizers through strengthening the rigidity of the casein network by binding hydration water and interacting with milk constituents. As a consequence, EPSs decrease syneresis and improve product stability. They play a recognized role in the manufacture of fermented milks, cultured cream, milk-based dessert, flavored milks, and probably also some soft cheeses.

Certain EPSs are also claimed to have beneficial physiological effects on the consumer. It is assumed that the increased viscosity of EPS-containing foods may increase the residence time of ingested fermented milks in the gastrointestinal tract and therefore be beneficial to a transient colonization by probiotic bacteria.

5.6.4.5 Production of Mannitol

Mannitol is claimed to possess health-promoting properties because it can be converted in short-chain fatty acids such as butyrate that acts against cancer (Neves et al. 2005). It is a low-calorie sugar that could replace sucrose, lactose, glucose, or fructose in food products. It is metabolized independently of insulin and is also applicable in diabetic food products. *Ln. pseudomesenteroides* and *Ln. mesenteroides* are known for their ability to produce mannitol in the fermentation of fructose (see Section 5.4). Numerous studies described improved industrial production by *Leuconostoc* strains (Carvalheiro et al. 2011), including by organisms expressing *Leuconostoc* enzymes, that is, the *Leuconostoc* mannitol DH in *Bacillus megaterium* (Bäumchen et al. 2007) or the *Leuconostoc* mannitol permease by *E. coli* (Heuser et al. 2009).

5.6.5 Industrial Use out of the Dairy Field

Leuconostoc and their enzymes are used for the production of specific compounds:

1. New well-defined oligosaccharides (Berensmeier et al. 2004)
2. Salicin analogs with high blood anticoagulant activity (Seo et al. 2005)
3. New dextrans with novel properties (Richard et al. 2005)
4. The antioxidant epigallocatechin gallate 7-*O*-α-D-glucopyranoside (Hyun et al. 2007)
5. A new dextran with only α-1-6-glycosidic linkage having significant industrial perspectives (Sarwat et al. 2008)
6. A hydroquinone fructoside synthesized by *Ln. mesenteroides* levan sucrase that could replace hydroquinone as a skin whitening agent (Kang et al. 2009a)
7. Alkyl glucosides with emulsification activities close to those of Triton X-100 (Kim et al. 2010)
8. L-DOPA α-glycosides for use in Parkinson's disease obtained by the action of glucansucrases from four *Ln. mesenteroides* strains (Yoon et al. 2010)
9. L-Ascorbic-2-glucosides having a potential as antioxidant in industrial applications, obtained by a novel glucansucrase of *Ln. lactis* EG001 (Kim et al. 2010)

Industrial productions of some compounds could be improved by *Leuconostoc* enzymes such as D-lactate (utilized in plastic manufacturing) produced by *Saccharomyces cerevisiae* expressing the D-lactate DH of *Ln. mesenteroides* subsp. *mesenteroides* (Ishida et al. 2006), α-D-glucose-1-P by *E. coli* with the *Ln. mesenteroides* sucrose phosphorylase (Goedl et al. 2007), and ethanol synthesis from glycerol that is enhanced by *E. coli* mutants expressing adhE from *Ln. mesenteroides* (Nikel et al. 2010). Dextrans are currently used for gel filtration products and in the textile industry.

5.7 Conclusions and Prospects

5.7.1 Academic Studies

The potential of the different *Leuconostoc* strains and not only that of type strains, which is also true for other LAB strains, could be better described in the near future due to the progress in rapid and efficient molecular methods. As previously done for *Lactococcus*, new genetic tools will be available for *Leuconostoc* that can allow easier gene transfers, thus obtaining modified strains with new potential. Nevertheless, the metabolic engineering of metabolic flux is not easy as shown in studies done in the last 10 years with *Lactococcus* modified strains. The limit of such genetically modified organisms will be the authorization of their use by the food industry.

5.7.2 Strain Selection

The selection of *Leuconostoc* strain(s) adapted to the specific process and product(s) will continue to be of prime importance. For this purpose, studies are continuously conducted for isolating new strains with various potentials, in particular, in artisanal products that have not yet been investigated. Perhaps some strains empirically chosen as the best ones for a specific technology will always be used in the future, although no explanations of this adaptation occurred.

Examples of selection have been given in this review. Considering the INRA-CNRZ collection, a strain selection has been conducted considering useful parameters with meaning in the dairy field (Server-Busson et al. 1999). The use of defined starters containing *Lactococcus* and *Leuconostoc* gave the Manchego cheeses with the highest score for flavor quality and intensity (Poveda et al. 2003). Using new criteria, only three strains above 27 (one *Ln. lactis*, two *Ln. mesenteroides* subsp. *dextranicum*, no *Ln. mesenteroides* subsp. *mesenteroides*) were selected (Nieto-Arribas et al. 2010). A rational selection of *Ln. citreum* strains (2/11) was also published for Afuega'l Pitu cheese (Sánchez et al. 2005). It is interesting to note that these two strains were both lactose- and galactose-positive. This last character could be an additional potential role for *Leuconostoc*, as the dairy industry asks for galactose-positive safe bacteria in certain productions (D. Hemme, personal communication).

As suggested by Crow et al. (2002), multiple adjunct NSLAB strains could be of great interest, with some being selected for obtaining a typical flavor and others for limiting a specific fraction of the cheese flora. Growth conditions of the adjunct strain could also influence its potentiality as resting or lysed cells in the cheese and should be studied. We have clearly shown the effects of *Leuconostoc* at high concentrations (what was called biomass) for acceleration of cheese ripening. Because of the cost, it could be a great advantage to find or construct *Leuconostoc* strains that are able to develop better in the cheese curd.

Strain selection is of great importance for its use as an adjunct. Thus, *Ln. mesenteroides* subsp. *cremoris* LC60 was unable to grow in Gouda cheese, whereas LC83 grew up to more than 10^7 colony forming units per gram, utilizing the citrate (Turner 1988; Martley and Crow 1993). Depending on the cheese, the selected *Leuconostoc* strains may develop in the cheese milk and/or curd or at least remain at a level sufficient to be active. For instance, one of the strains we have tested in St. Paulin cheese disappeared as viable cells recovered, although added at a high level to the cheese milk (Hemme, unpublished). It will be interesting to know if the cells remained viable but not culturable or if lysis occurred; this property could vary within the strains or subspecies as shown for *Lc. lactis* and *Lc. cremoris* (Martley and Crow 1993; Niven and Mulholland 1998). The effects of these strains on cheese ripening were lower than for strains remaining culturable, suggesting that at least some of the acceleration process modifications described were due to living resting cells, in particular, the production of D-lactate and gas.

5.7.3 Dairy Product Prospects

In dairy products where ethanol and/or acetate are obligatory aroma precursors, in particular, soft smeared cheeses, *Leuconostoc* are good actors. For plant- or mixed dairy–plant products, the fermentation of xylose or other pentoses will decrease the ethanol content of the product and enhance that of diacetyl, thus answering some demand for alcohol-free foods. As obtained for *Lactococcus* strains, *Leuconostoc* mutants that do not transform the diacetyl formed could be of great interest by improving the flavor stability. For low products with reduced D-lactate content, in particular, dairy preparations for babies, *Leuconostoc* strains containing a heterologous L-LDH could be used, as such strains produce L-lactate and less D-lactate (Jin et al. 2009). *Leuconostoc* strains are mannitol producers and thus could be utilized when mannitol is required. The same is also true for EPSs, which could give the adequate (or modify the) texture of the products, in particular, fermented milks. The rather high survival rate of *Leuconostoc* in the cheese curd could probably explain a continuous metabolism as resting cells. It will be interesting to use strains that are able to develop in the cheese curd by acquiring enzymes of some other safe LAB. This is no longer a free hypothesis with the availability of *Leuconostoc* genetic tools that were introduced in the last years. *Leuconostoc* enzymes could also be transferred to classical starters to adapt their activities as has been proposed recently for *Lactococcus* strains having *Leuconostoc* β-gal (Schwab et al. 2010).

5.7.4 Health, Safety, and Benefits

Considering the wide distribution of *Leuconostoc* in the environment, the large number that are ingested daily in the vast range of fermented foods such as dairy and plant products, and the absence of infection due to food consumption, the reasonable conclusion is that *Leuconostoc* are safe (GRAS/ QPS status). Products containing *Leuconostoc* (as *Lactobacillus*) may also be a non-negligible source of Mn^{2+} as cells contain millimolar concentrations. Variants with increased concentration could be of potential interest. Other properties of *Leuconostoc* are of interest. The ability to hydrolyze α-gal such as raffinose could be exploited for food containing such sugars. The α-gal could interestingly be transferred in bacteria that are normally resident in the digestive tract and not only transient as *Leuconostoc*. Another way is the utilization of the enzyme for treatment of the food constituents before their incorporation in the products, as is done for lactose hydrolysis of milk by β-gal. Similarly, *Leuconostoc* is able to express the amylase of *Lb. amylovorus* (Eom et al. 2009) and in this way could acquire such an activity to obtain some starch hydrolysis in the product or after its ingestion. *Leuconostoc* are able to synthesize acarbose analogs, which may improve the glycemic profile and insulin sensitivity in patients with type 2 diabetes.

The formation of a wide range of oligosides, for instance, inulin-like polymers, having a probiotic potential is also a major concern linked to *Leuconostoc*. There is no doubt that *Leuconostoc* remains to be the preferred genus for such production in the near future. Moreover, its enzymes are able to exert glycosylation of a wide variety of molecules as shown with some examples in this review. A good example has recently been reported for sourdoughs (Galle et al. 2010). *Leuconostoc* cells are better potent inducers of Th1-type cytokines IL-12 and IFN-γ than *Lactobacillus* strains. This constitutes also a challenge to develop in the future to explain and utilize this ability. For these different and interesting properties, it will be interesting to obtain *Leuconostoc* strains that are able to colonize the digestive tract.

5.7.5 Other Utilization of *Leuconostoc* or Their Enzymes

The glycosyltransferase of *Leuconostoc* will be utilized for industrial production of a wide range of glycosylated compounds for application in various domains. Production of D-lactate and mannitol by *Leuconostoc* or *Leuconostoc* enzyme could certainly be improved. EPS production is also a large field of research and applications, in and out of the dairy field, such as in bioremediation processes (Singh et al. 2006) or coating of steel for corrosion protection (Finkenstadt et al. 2011), which will be satisfied by new polyosides produced by novel wild strains or known mutants.

ACKNOWLEDGMENTS

The author greatly appreciates the expert help of Pr. Dr. Walter P. Hammes for his advice and rechecking of the manuscript. INRA colleagues Agnès Delacroix-Buchet and Philippe Gaudu are also thanked for their help in accession to recent references.

REFERENCES

Aarnikunnas, J., Ronnholm, K., and Palva, A. 2002. The mannitol dehydrogenase gene (*mdh*) from *Leuconostoc mesenteroides* is distinct from other known bacterial *mdh* genes. *Appl. Microbiol. Biotechnol.* 59:665–671.

Allende, A., McEvoy, J.L., Luo, Y., Artes, F., and Wang, C.Y. 2006. Effectiveness of two-sided UV-C treatments in inhibiting natural microflora and extending the shelf-life of minimally processed 'red oakleaf' lettuce. *Food Microbiol.* 23:241–249.

Antunes, A., Rainey, F.A., Nobre, M.F., Schumann, P., Ferreira, A.M., Ramos, A., Santos, H., and de Costa M.S. 2002. *Leuconostoc ficulneum* sp. nov., a novel lactic acid bacterium isolated from a ripe fig, and reclassification of *Lactobacillus fructosus* as *Leuconostoc fructosum* comb. nov. *Int. J. Syst. Evol. Microbiol.* 52:647–655.

Atamer, Z., Ali, Z., Neve, H., Heller, K., and Hinrichs, J. 2011. Thermal resistance of bacteriophages attacking flavor-producing dairy *Leuconostoc* starter cultures. *Int. Dairy J.* 21:327–334.

Ayad, E.H.E., Nashat, S., El-Sadek, N., Metwaly, H., and El-Soda, M. 2004. Selection of wild lactic acid bacteria isolated from traditional Egyptian dairy products according to production and technological criteria. *Food Microbiol.* 21:715–725.

Azcarate-Peril, A., McAuliffe, O., Altermann, E., Lick, S., Russell, W.M., and Klaenhammer, T.R. 2005. Microarray analysis of a two-component regulatory system involved in acid resistance and proteolytic activity in *Lactobacillus acidophilus*. *Appl. Environ. Microbiol.* 71:5794–5804.

Badel, S., Laroche, C., Gardarin, C., Bernardi, T., and Michaud, P. 2008. New method showing the influence of matrix components in *Leuconostoc mesenteroides*. *Appl. Biochem. Biotechnol.* 151:364–370.

Barrangou, R., Azcarate-Peril, A., Duong, T., Conners, S.B., Kelly, R.M., and Klaenhammer, T.R. 2006. Global analysis of carbohydrate utilization by *Lactobacillus acidophilus* using cDNA microarrays. *Proc. Natl. Acad. Sci. U. S. A.* 103:3816–3821.

Bäumchen, C., Roth, A.H., Biedendieck, R., Malten, M., Follmann, M., Sahm, H., Bringer-Meyer, S., and Jahn, D. 2007. D-mannitol production by resting state whole cell biotransformation of D-fructose by heterologous mannitol and formate dehydrogenase gene expression in *Bacillus megaterium*. *Biotechnol. J.* 2:1408–1416.

Bauquis, A.C., Raynaud, V., le Tual, A.G., Eppert, I., Bercetche, J.C., and Roustan, G. 2002. [Appearance and flavour of traditional mould ripened soft cheese influenced by selected combinations of primary and secondary cultures] in French. *Sci. Aliments* 22:169–175.

Bellengier, P., Hemme, D., and Foucaud, C. 1994. Citrate metabolism in sixteen *Leuconostoc mesenteroides* subsp. *mesenteroides* and subsp. *dextranicum* strains. *J. Appl. Bacteriol.* 77:54–60.

Berensmeier, S., Ergezinger, M., Bohnet, M., and Buchholz, K. 2004. Design of immobilized dextransucrase for fluidized bed application. *J. Biotechnol.* 114:255–267.

Biet, F., Cenatiempo, Y., and Frémaux, C. 2002. Identification of a replicon from pTXL1, a small cryptic plasmid from *Leuconostoc mesenteroides* subsp. *mesenteroides* Y110, and development of a food-grade vector. *Appl. Environ. Microbiol.* 68:6451–6456.

Björkroth, K.J., Geisen, R., Schillinger, U., Weiss, N., de Vos, P., Holzapfel, W.H., Korkeala, H.J., and Vandamme, P. 2000. Characterization of *Leuconostoc gasicomitatum* sp. nov., associated with spoiled raw tomato-marinated broiler meat strips packaged under modified-atmosphere conditions. *Appl. Environ. Microbiol.* 66:3764–3772.

Bourel, G., Henini, S., Krantar, K., Oraby, M., Diviès, C., and Garmyn, D. 2001. Métabolisme sucre-citrate chez *Leuconostoc mesenteroides*. *Lait* 81:75–82.

Bozonnet, S., Dols-Laffargue, M., Fabre, E., Pizzut, S., Remaud-Simeon, M., Monsan, P., and Willemot, R.M. 2002. Molecular characterization of DSR-E, an alpha-1,2 linkage-synthesizing dextransucrase with two catalytic domains. *J. Bacteriol.* 184:57–61.

Cappello, M.S., Zapparoli, G., Stefani, D., Logrieco, A. 2010. Molecular and biochemical diversity of *Oenococcus oeni* strains isolated during spontaneous malolactic fermentation of Malvasia Nera wine. *Syst. Appl. Microbiol.* 33:461–467.

Carvalheiro, F., Moniz, P., Duarte, L.C., Esteves, M.P., and Girio, F.M. 2011. Mannitol production by lactic acid bacteria grown in supplemented carob syrup. *J. Ind. Microbiol. Biotechnol.* 38:221–227.

Chambel, L., Chelo, I.M., Zé-Zé, L., Pedro, L.G., Santos, M.A., and Tenreiro, R. 2006. *Leuconostoc pseudoficulneum* sp. nov., isolated from a ripe fig. *Int. J. Syst. Evol. Microbiol.* 56:1375–1381.

Chang, J.Y. and Chang, H.C. 2010. Improvements in the quality and shelf life of kimchi by fermentation with the induced bacteriocin-producing strain *Leuconostoc citreum* GJ7 as a starter. *J. Food Sci.* 75M:103–110.

Chang, J.Y., Lee, H.J., and Chang, H.C. 2010. Identification if the agent from *Lactobacillus plantarum* KFRI 464 that enhances bacteriocin production by *Leuconostoc citreum*. *J. Appl. Microbiol.* 103: 2504–2515.

Chelo, I.M., Zé-Zé, L., and Tenreiro, R. 2007. Congruence of evolutionary relationships inside the *Leuconostoc–Oenococcus–Weissella* clade assessed by phylogenetic analysis of the 16S RNA. *Int. J. Syst. Evol. Microbiol.* 57:276–286.

Chelo, I.M., Zé-Zé, L., and Tenreiro, R. 2010. Genome diversity in the genera *Fructobacillus*, *Leuconostoc* and *Weissella* determined by physical and genetic mapping. *Microbiology* 156:420–430.

Cho, K.M., Math, R.K., Islam, S.M., Lim, W.J., Hong, S.Y., Kim, J.M., Yun, M.G., Cho, J.J., and Yun, H.D. 2009. Novel multiplex PCR for the detection of lactic acid bacteria during kimchi fermentation. *Mol. Cell. Probes* 23:90–94.

Chung, C.H. and Day, D.F. 2002. Glucooligosaccharides from *Leuconostoc mesenteroides* B-742 (ATCC 13146): A potential prebiotic. *J. Indus Microbiol. Biotechnol.* 29:196–199.

Cibik, R., Lepage, E., and Tailliez, P. 2000. Molecular diversity of *Leuconostoc mesenteroides* and *Leuconostoc citreum* isolated from traditional French cheeses as revealed by RAPD fingerprinting, 16S rDNA sequencing and 16S rDNA fragment amplification. *Syst. Appl. Microbiol.* 23:267–278.

Cibik, R., Tailliez, P., Langella, P., and Chapot-Chartier, M.-P. 2001. Identification of Mur, an atypical peptidoglycan hydrolase derived from *Leuconostoc citreum*. *Appl. Environ. Microbiol.* 67:858–864.

Coenye, T. and Nelis, H.J. 2010. *In vitro* and *in vivo* model systems to study microbial biofilm formation. *J. Microbiol. Methods* 83:89–105.

Cogan, T.M., Barbosa, M., Beuvier, E., Bianchi-Salvadori, B., Cocconcelli, P.S., Fernandes, I., Gomez, J., Gomez, R., Kalantzopoulos, G., Ledda, A., Medina, M., Rea, M.C., and Rodriguez, E. 1997. Characterization of the lactic acid bacteria in artisanal dairy products. *J. Dairy Res.* 64:409–421.

Collins, M.D., Samelis, J., Metaxopoulos, J., and Wallbanks, S. 1993. Taxonomic studies on some leuconostoc-like organisms from fermented sausages: Description of a new genus *Weissella* for the *Leuconostoc paramesenteroides* group of species. *J. Appl. Bacteriol.* 75:595–603.

Côté, G.L. and Leathers, T.D. 2005. A method for surveying and classifying *Leuconostoc* spp. glucansucrases according to strain-dependent acceptor product patterns. *J. Ind. Microbiol. Biotechnol.* 32:53–60.

Coulibaly, I., Amenan, A.Y., Lognay, G., Fauconnier, M.L., and Thonart, P. 2009. Survival of freeze-dried *Leuconostoc mesenteroides* and *Lactobacillus plantarum* related to their cellular fatty acids composition during storage. *Appl. Biochem. Biotechnol.* 157:70–84.

Crow, V.L., Curry, B., Christison, M., Hellier, K., Holland, R., and Liu, S.Q. 2002. Raw milk flora and NSLAB as adjuncts. *Aust. J. Dairy Technol.* 57:99–105.

Dallié, D.K.D., Deschamps, A.M., and Richard-Forget, F. 2010. Lactic acid bacteria—Potential for control of mould growth and mycotoxins: A review. *Food Control* 21:370–380.

De Bruyne, K., Schillinger, U., Caroline, L., Boehringer, B., Cleenwerck, I., Vancanneyt, M., De Vuyst, L., Franz, C.M., and Vandamme, P. 2007. *Leuconostoc holzapfelii* sp. nov., isolated from Ethiopian coffee fermentation and assessment of sequence analysis of housekeeping genes for delineation of *Leuconostoc* species. *Int. J. Syst. Evol. Microbiol.* 57:2952–2959.

Delcour, J., Ferain, T., Deghorain, M., Palumbo, E., and Hols, P. 1999. The biosynthesis and functionality of the cell-wall of lactic acid bacteria. *Antonie van Leeuwenhoek* 76:159–184.

Demirci, Y. and Hemme, D. 1994. Growth of *Leuconostoc mesenteroides* strains isolated from French raw milk cheeses in a reference milk. *Milchwissenschaft* 43:483–485.

Demirci, Y. and Hemme, D. 1995. Acidification test based on the use of biomass for screening of *Leuconostoc*. Application to *Ln. mesenteroides* strains isolated from French raw milk cheeses. *J. Dairy Res.* 62:521–527.

Dessart, S.R. and Steenson, L.R. 1995. Biotechnology of dairy *Leuconostoc*. In: *Food Biotechnology: Microorganisms*, edited by Y.H. Hui and G.G. Khatchatourians, pp. 665–702. New York: VCH Publishers.

Devoyod, J.-J. and Poullain, F. 1988. Les leuconostocs. Propriétés: leur rôle en technologie laitière. *Lait* 68:249–280.

Devoyod, J.-J., Labbe, M., and Auclair, J. 1974. Concentrated frozen-suspensions of *Leuconostoc* for Roquefort cheese-making. 19th IDF International Dairy Congress, New Delhi, Vol. 1E, p. 419.

Dhodapkar, K.M. and Henry, N.K. 1996. *Leuconostoc* bacteremia in an infant with short-gut syndrome: Case report and literature review. *Mayo Clin. Proc.* 71:1171–1174.

Douglas, G.L. and Klaenhammer, T.R. 2010. Genomic evolution of domesticated microorganisms. *Annu. Rev. Food Sci. Technol.* 1:397–414.

Ehrmann, M.A., Freiding, S., and Vogel, R.F. 2009. *Leuconostoc palmae* sp. nov., a novel lactic acid bacterium isolated from palm wine. *Int. J. Syst. Evol. Microbiol.* 59:943–947.

Endo, A. and Okada, S. 2008. Reclassification of the genus *Leuconostoc* and proposals of *Fructobacillus fructosus* gen. nov., comb. nov., *Fructobacillus durionis* comb. nov., *Fructobacillus ficulneus* comb. nov. and *Fructobacillus pseudoficulneus* comb. nov. *Int. J. Syst. Evol. Microbiol.* 58:2195–2205.

Eom, H.J., Moon, J.S., Cho, S.K., Kim, J.H., and Han, N.S. 2012. Construction of theta-type shuttle vector for *Leuconostoc* and other lactic acid bacteria using pCB42 isolated from kimchi. *Plasmid* 67:35–43.

Eom, H.J., Moon, J.S., Seo, E.Y., and Han, N.S. 2009. Heterologous expression and secretion of *Lactobacillus amylovorus* alpha-amylase in *Leuconostoc citreum*. *Biotechnol. Lett.* 31:1783–1788.

Eom, H.J., Seo, D.M., and Han, N.S. 2007. Selection of psychrotrophic *Leuconostoc* spp. producing highly active dextransucrase from lactate fermented vegetables. *Int. J. Food Microbiol.* 117:61–67.

Fantuzzi, L., Raffellini, S., and Gonzalez, F.C. 2007. Effect of lysozyme on two autochthonous lactic acid strains treated under different conditions. *Sci. Tec. Latt.-Casearia* 58:227–242.

Fernandes, L. and Steele, J.L. 1993. Glutathione content of lactic acid bacteria. *J. Dairy Sci.* 76:1233–1242.

Fernández de Palencia, P., de la Plaza, M., Amarita, F., Requena, T., and Pelaez, C. 2006. Diversity of amino acid converting enzymes in wild lactic acid bacteria. *Enzyme Microbiol. Technol.* 38:88–93.

Finkenstadt, V.L., Côté, G.L., and Willett, J.L. 2011. Corrosion protection of low-carbon steel using exopolysaccharide coatings from *Leuconostoc mesenteroides*. *Biotechnol. Lett.*, Online, 3 February 2011.

Florescu, D., Hill, L., Sudan, D., and Iwen, P.C. 2008. *Leuconostoc* bacteremia in pediatric patients with short bowel syndrome: Case series and review. *Pediatr. Infect. Dis. J.* 27:1013–1019.

Foucaud, C., François, A., and Richard, J. 1997. Development of a chemically defined medium for the growth of *Leuconostoc mesenteroides*. *Appl. Environ. Microbiol.* 63:301–304.

Foucaud, C., Hemme, D., and Desmazeaud, M. 2001. Peptide utilization by *Lactococcus lactis* and *Leuconostoc mesenteroides*. *Lett. Appl. Microbiol.* 31:20–25.

Galle, S., Schwab, C., Arendt, E., and Gänzle, M. 2010. Exopolysaccharide-forming *Weissella* strains as starter cultures for sorghum and wheat sourdoughs. *J. Agric. Food Chem.* 58:5834–5841.

Garcia-Gimeno, R.M., Hervas-Martinez, C., Rodriguez-Pérez, R., and Zurera-Cosano, G. 2005. Modelling the growth of *Leuconostoc mesenteroides* by artificial neural networks. *Int. J. Food Microbiol.* 105:317–332.

Garvie, E.I. 1986. Genus *Leuconostoc* van Tieghem 1878. In: *Bergey's Manual of Systematic Bacteriology*, edited by P.H.A. Sneath, N.S. Mair, M.E. Sharpe, and J.G. Holt, pp. 1071–1075. Baltimore: Williams and Wilkins.

Gendrot, F., Foucaud-Scheunemann, C., Ferchichi, M., and Hemme, D. 2002. Characterization of amino acid transport in the dairy strain *Leuconostoc mesenteroides* subsp. *mesenteroides* CNRZ 1273. *Lett. Appl. Microbiol.* 35:291–295.

Germain-Alpettaz, V. and Foucaud-Scheunemann, C. 2002. Identification and characterization of an oligo-peptide transport system in *Leuconostoc mesenteroides* subsp. *mesenteroides* CNRZ 1463. *Lett. Appl. Microbiol.* 35:68–73.

Gevers, D., Huys, G., and Swings, J. 2001. Applicability of rep-PCR fingerprinting for identification of *Lactobacillus* species. *FEMS Microbiol. Lett.* 205:31–36.

Goedl, C., Schwarz, A., Minani, A., and Nidetzky, B. 2007. Recombinant sucrose phosphorylase from *Leuconostoc mesenteroides*: Characterization, kinetic studies of transglucosylation, and application of immobilised enzyme for production of alpha-D-glucose-1-phosphate. *J. Biotechnol.* 129:77–86.

González, L., Sacristán, N., Arenas, R., Fresno, J.M., and Eugenia Tornadijo, M. 2010. Enzymatic activity of lactic acid bacteria (with antimicrobial properties) isolated from a traditional Spanish cheese. *Food Microbiol.* 27:592–597.

Greer, G.G., Dilts, B.D., and Ackermann, H.W. 2007. Characterization of a *Leuconostoc gelidum* bacterio-phage from pork. *Int. J. Food Microbiol.* 114:370–375.

Gu, C.T., Wang, F., Li, C.Y., Liu, F., and Huo, G.C. *Leuconostoc mesenteroides* subsp. *suionicum* subsp. nov., a novel subspecies within *Leuconostoc mesenteroides*. *Int. J. Syst. Evol. Microbiol.* published ahead of print August 19.

Hanniffy, S.B., Peláez, C., Martínez-Bartolomé, M.A., Requena, T., and Martínez-Cuesta, M.C. 2009. Key enzymes involved in methionine catabolism by cheese lactic acid bacteria. *Int. J. Food Microbiol.* 135:223–230.

Hao, Y., Zhao, L., Zhang, H., Zhai, Y., Liu, X., and Zhang, L. 2010. Identification of the bacterial diversity in koumiss by denaturing gradient gel electrophoresis and species-specific polymerase chain reaction. *J. Dairy Sci.* 93:1926–1933.

Hemme, D., and Desmazeaud, M.J. 1999. Lactic acid bacteria starters and other starters cultures: present and future use. In: *Biology of Lactation*, edited by J. Martinet, L.-M. Houdebine, and H.H. Head, Chap. 24: pp. 619–640. Versailles, France: INRA Editions.

Hemme, D. and Foucaud-Scheunemann, C. 2004. *Leuconostoc*, characteristics, use in dairy technology and prospects in functional foods. *Int. Dairy J.* 14:467–494.

Hemme, D., Giraudon, C., and Vassal, L. 1996. Addition of *Leuconostoc* to the milk in cheesemaking. 5th Symposium on Lactic Acid Bacteria, September 8–12, 1996, Veldhoven, The Netherlands.

Heuser, F., Marin, K., Bringer, S., and Sahm, H. 2009. Improving D-mannitol productivity of *Escherichia coli*: Impact of NAD, CO_2 and expression of a putative sugar permease from *Leuconostoc mesenteroides*. *Metab. Eng.* 11:173–183.

Holzapfel, W.H., Björkroth, J., and Dicks, L.M.T. 2006. Genus I. *Leuconostoc* van Tieghem 1878, 198AL emend mut. Char. Hucker and Pederson 1930, 66AL. In: *Bergey's Manual of Systematic Bacteriology* 2nd ed., vol. 3: The low G + C Gram-positive bacteria, edited by G.M. Garrity. New York: Springer.

Hugenholtz, J., Sybesma, W., Groot, M.N., Wisselink, W., Ladero, V., Burgess, K., van Sinderen, D., Piard, J.C., Eggink, G., Smid, E.J., Savoy, G., Sesma, F., Jansen, T., Hols, P., and Kleerebezem, M. 2002. Metabolic engineering of lactic acid bacteria for the production of nutraceuticals. *Antonie van Leeuwenhoek* 82:217–235.

Hummel, A.S., Hertel, C., Holzapfel, W.H., and Franz, C.M. 2007. Antibiotic resistances of starter and probiotic strains of lactic acid bacteria. *Appl. Environ. Microbiol.* 73:730–739.

Hynes, E., Ogier, J.-C., and Delacroix-Buchet, A. 2000. Protocol for the manufacture of miniature washed-curd cheeses under controlled microbiological conditions. *Int. Dairy J.* 10:733–737.

Hyun, E.K., Park, H.Y., Kim, H.J., Lee, J.K., Kim, D., and Oh, D.K. 2007. Production of epigallocatechin gallate 7-*O*-alpha-D-glucopyranoside (EGCG-G1) using the glucosyltransferase from *Leuconostoc mesenteroides. Biotechnol. Progr.* 23:1082–1086.

Iliev, I., Vassileva, T., Ignatova, C., Ivanova, I., Haertlé, T., Monsan, P., and Chobert, J.M. 2008. Gluco-oligosaccharides synthesized by glucosyltransferases from constitutive mutants of *Leuconostoc mesenteroides* strain Lm 28. *J. Appl. Microbiol.* 104:243–250.

Ishida, N., Suzuki, T., Tokuhiro, K., Nagamori, E., Onishi, T., Saitoh, S., Kitamoto, K., and Takahashi, H. 2006. D-lactic acid production by metabolically engineered *Saccharomyces cerevisiae. J. Biosci. Bioeng.* 101:172–177.

Ishihyama, K., Yamazaki, H., Senda, Y., Yamauchi, H., and Nakao, S. 2010. *Leuconostoc* bacteremia in three patients with malignacies. *J. Infect. Chemother.*, Online, 14 September 2010.

Jang, S.H., Hwang, M.H., and Chang, H.I. 2010. Complete genome sequence of ΦMH1, a *Leuconostoc* temperate phage. *Arch. Virol.* 155:1883–1885.

Jasnievski, J., Cailliez-Grimal, C., Younsi, M., Millière, J.-B., and Revol-Junelles, A.-M. 2008. Functional differences in *Leuconostoc* sensitive and resistant strains to mesenterocin 52A, a class IIa bacteriocin. *FEMS Microbiol. Lett.* 289:193–201.

Jin, Q., Jung, J.Y., Kim, Y.J., Eom, H.J., Kim, S.Y., Kim, T.J., and Han, N.S. 2009. Production of L-lactate in *Leuconostoc citreum* via heterologous expression of L-lactate dehydrogenase gene. *J. Biotechnol.* 144:160–164.

Johansson, P., Paulin, L., Säde, E., Salovuori, N., Alatalo, E.R., Björkroth, K.J., and Auvinen, P. 2011. Genome sequence of a food spoilage lactic acid bacterium, *Leuconostoc gasicomitatum* LMG 18811^T, in association with specific spoilage reactions. *Appl. Environ. Microbiol.* 77:4344–4351.

Jordan, K.N. and Cogan, T.M. 1995. Growth and metabolite production by mixed-strain starter cultures. *Ir. J. Agric. Food Res.* 34:39–47.

Kang, H., Myung, E.J., Ahn, K.S., Eom, H.J., Han, N.S., Kim, Y.B., Kiml, Y.J., and Sohn, N.W. 2009b. Induction of Th1 cytokines by *Leuconostoc mesenteroides* subsp. *mesenteroides* (KCTC 3100) under Th2-type conditions and the requirement of NF-kappaB and p38/JNK. *Cytokine* 46:283–289.

Kang, J., Kim, Y.M., Kim, N., Kim, D.W., Nam, S.H., and Kim, D. 2009a. Synthesis and characterization of hydroquinone fructoside using *Leuconostoc mesenteroides* levansucrase. *Appl. Microbiol. Biotechnol.* 83:1009–1016.

Kang, M.S., Kang, I.C., Kim, S.M., Lee, H.C., and Oh, J.S. 2007. Effect of *Leuconostoc* spp. on the formation of *Streptococcus mutans* biofilm. *J. Microbiol.* 45:291–296.

Kekkonen, R.A., Kajasto, E., Miettinen, M., Veckman, V., Korpela, R., and Julkunen, I. 2008. Probiotic *Leuconostoc mesenteroides* ssp. *cremoris* and *Streptococcus thermophilus* induce IL-12 and IFN-gamma production. *World J. Gastroenterol.* 14:1192–1203.

Khan, H., Flint, S., and Yu, P.L. 2010. Enterocins in food preservation. *Int. J. Food Microbiol.* 141:1–10.

Kim, B., Lee, J., Jang, J., Kim, J., and Han, H. 2003. *Leuconostoc inhae* sp. nov, a lactic acid bacterium isolated from kimchi. *Int. J. Syst. Evol. Microbiol.* 53:1123–1126.

Kim, D.S., Choi, S.H., Kim, D.W., Kim, R.N., Nam, S.H., Kang, A., Kim, A., and Park, H.S. 2011a. Genome sequence of *Leuconostoc gelidum* KCTC 3527 isolated from kimchi. *J. Bacteriol.* 193:799–800.

Kim, D.S., Choi, S.H., Kim, D.W., Kim, R.N., Nam, S.H., Kang, A., Kim, A., and Park, H.S. 2011b. Genome sequence of *Leuconostoc inhae* KCTC 3774 isolated from kimchi. *J. Bacteriol.* 193:1278–1279.

Kim, D.W., Choi, S.H., Nam, S.H., Kim, R.N., Kim, A., Kim, D.-S., and Park, H.S. 2011c. Genome sequence of *Leuconostoc pseudomesenteroides* KCTC 3652. *J. Bacteriol.* 193:4299.

Kim, E.K., Cha, C.J., Cho, Y.J., Cho, Y.B., and Roe, J.H. 2008a. Synthesis of gamma-glutamylcysteine as a major low-molecular-weight thiol in lactic acid bacteria *Leuconostoc* spp. *Biochem. Biophys. Res. Commun.* 369:1047–1051.

Kim, J., Chun, J., and Han, H.U. 2000. *Leuconostoc kimchii* sp. nov., a new species from kimchi. *Int. J. Syst. Evol. Microbiol.* 50:1915–1919.

Kim, J.F., Jeong, H., Lee, J.S., Choi, S.H., Ha, M., Hur, C.G., Kim, J.S., Lee, S., Park, H.S., Park, Y.H., and Oh, T.K. 2008b. Complete genome sequence of *Leuconostoc citreum* KM20. *J. Bacteriol.* 190:3093–3094.

Kim, Y.M., Kim, B.H., Ahn, J.S., Kim, G.E., Jin, S.D., Nguyen, T.H., and Kim, D. 2009. Enzymatic synthesis of alkyl glucosides using *Leuconostoc mesenteroides* dextransucrase. *Biotechnol. Lett.* 31:1433–1438.

Kim, Y.M., Yeon, M.J., Choi, N.S., Chang, Y.H., Jung, M.Y., Song, J.J., and Kim, J.S. 2010. Purification and characterization of a novel glucansucrase from *Leuconostoc lactis* EG001. *Microbiol. Res.* 165:384–391.

Knudsen, S. 1931. Starters. *J. Dairy Res.* 2:137–163.

Konings, W. 2002. The cell membrane and the struggle for life of lactic acid bacteria. *Antonie van Leeuwenhoek* 82:3–27.

Koo, O.K., Jeong, D.W., Lee, J.M., Kim, M.J., Lee, J.H., Chang, H.C., Kim, J.H., and Lee, H.J. 2005. Cloning and characterization of the bifunctional alcohol/acetaldehyde dehydrogenase gene (adhE) in *Leuconostoc mesenteroides* isolated from kimchi. *Biotechnol. Lett.* 27:505–510.

Leathers, T.D. and Bischoff, T.D. 2010. Biofilm formation by strains of *Leuconostoc citreum* and *L. mesenteroides*. *Biotechnol. Lett.*, Online, 3 November 2010.

Lee, J.H., Kang, H.K., Moon, Y.H., Cho, D.L., Kim, D., Choe, J.Y., Honzatko, R., and Robyt, J.F. 2006. Cloning, expression and characterization of an extracellular enolase from *Leuconostoc mesenteroides*. *FEMS Microbiol. Lett.* 259:240–248.

Lee, S.H., Park, M.S., Jung, J.Y., and Jeon, C.O. 2011. *Leuconostoc miyukkimchii* sp. nov., isolated from brown algae (*Undaria pinnatifida L.*) kimchi in Korea *Int. J. Syst. Evol. Microbiol.* published ahead of print June 24.

Leisner, J.J., Vancanneyt, M., van der Meulen, R., Lefebvre, K., Engelbeen, K., Hoste, B., Laursen, B.G., Bay, L., Rusul, G., de Vuyst, L., and Swings, J. 2005. *Leuconostoc durionis* sp. nov., a heterofermenter with no detectable gas production from glucose. *Int. J. Syst. Evol. Microbiol.* 55:1267–1270.

Limonet, M., Cailliez-Grimal, C., Linder, M., Revol-Junelles, A.-M., and Millière, J.-B. 2004. Cell envelope analysis of insensitive, susceptible or resistant strains of *Leuconostoc* and *Weissella* genus to *Leuconostoc mesenteroides* FR52 bacteriocins. *FEMS Microbiol. Rev.* 241:49–55.

Liu, M., Nauta, A., Francke, C., and Siezen, R.J. 2008. Comparative genomics of enzymes in flavor-forming pathways from amino acids in lactic acid bacteria. *Appl. Environ. Microbiol.* 74:4590–4600.

Liu, S.Q., Holland, R., and Crow, V.L. 2004. Ester and their biosynthesis in fermented dairy products. *Int. Dairy J.* 14:923–945.

Lu, Z., Altermann, E., Breidt, F., and Kozyavkin, S. 2010. Sequence analysis of *Leuconostoc mesenteroides* bacteriophage Φ-A4 isolated from an industrial vegetable fermentation. *Appl. Environ. Microbiol.* 76:1955–1966.

Macwana, S.J., and Muriana, P.M. 2011. A "bacteriocin PCR array" for identification of bacteriocin-related structural genes in lactic acid bacteria. *J. Microbiol. Methods*: online 18 November 2011.

Makarova, K., Slesarev, A., Wolf, Y., Sorokin, A., Mirkin, B., Koonin, E.V. et al. 2006. Comparative genomics of the lactic acid bacteria. *Proc. Natl. Acad. Sci. U. S. A.* 103:15611–15616.

Makarova, K.S. and Koonin, E.V. 2007. Evolutionary genomics of lactic acid bacteria. *J. Bacteriol.* 189:1199–1208.

Martin, M.G., Magni, C., de Mendoza, D., and López, P. 2005. CitI, a transcription factor involved in regulation of citrate metabolism in lactic acid bacteria. *J. Bacteriol.* 187:5146–5155.

Martinez-Murcia, A.J. and Collins, M.D. 1991. A phylogenetic analysis of an atypical *Leuconostoc*: Description of *Leuconostoc fallax* sp. nov. *FEMS Microbiol. Lett.* 82:55–60.

Martley, F.G. and Crow, V.L. 1993. Interactions between non-starter microorganisms during cheese manufacture and ripening. *Int. Dairy J.* 3:461–483.

Martley, F.G. and Crow, V.L. 1996. Open texture in cheese: The contributions of gas production by microorganisms and cheese manufacturing practices. *J. Dairy Res.* 63:489–507.

Matamoros, S., Leroi, F., Cardinal, M., Gigout, F., Kasbi Chadli, F., Cornet, J., Prévost, H., and Pilet, M.F. 2009b. Psychrotrophic lactic acid bacteria used to improve the safety and quality of vacuum-packaged cooked and peeled tropical shrimp and cold-smoked salmon. *J. Food Prot.* 72:365–374.

Matamoros, S., Pilet, M.F., Gigout, F., Prévost, H., and Leroi, F. 2009a. Selection and evaluation of seafood-borne psychrotrophic lactic acid bacteria as inhibitors of pathogenic and spoilage bacteria. *Food Microbiol.* 26:638–644.

Mataragas, M., Drosinos, E.H., Tsakalidou, E., and Metaxopoulos, J. 2004. Influence of nutrients on growth and bacteriocin production by *Leuconostoc mesenteroides* L124 and *Lactobacillus curvatus* L442. *Antonie van Leeuwenhoek* 85:191–198.

Mäyrä-Mäkinen, A. and Bigret, M. 1998. Industrial use and production of lactic acid bacteria. In: *Lactic Acid Bacteria: Microbiology and Functional Aspects*, edited by S. Salminen and A. von Wright. New York: Marcel Dekker Inc., pp. 73–102.

McSweeney, P.L.H. and Sousa, M.J. 2000. Biochemical pathways for the production of flavour compounds in cheeses during ripening. A review. *Lait* 80:293–324.

Mills, D.A., Rawsthorne, H., Parker, C., Tamir, D., and Makarova, K. 2005. Genomic analysis of *Oenococcus oeni* PSU-1 and its relevance to winemaking. *FEMS Microbiol. Rev.* 29:465–475.

Moreno-Arribas, V., Polo, C., Jorganes, F., and Munoz, R. 2003. Screening of biogenic amine production by lactic acid bacteria isolated from grape must and wine. *Int. J. Food Microbiol.* 84:117–123.

Nam, S.H., Choi, S.H., Kang, A., Kim, D.W., Kim, R.N., Kim, A., and Park, H.S. 2010. Genome sequence of *Leuconostoc argentinum* KCTC 3773. *J. Bacteriol.* 192:6490–6491.

Nam, S.H., Choi, S.H., Kang, A., Kim, D.W., Kim, D.S., Kim, R.N., and Park, H.S. 2011a. Genome sequence of *Leuconostoc fallax* KCTC 3537. *J. Bacteriol.* 193:588–589.

Nam, S.H., Kim, A., Choi, S.H., Kang, A., Kim, D.W., Kim, R.N., Kim, D.S., and Park, H.S. 2011b. Genome sequence of *Leuconostoc carnosum* KCTC 3525. *J. Bacteriol.* 193:6100–6101.

Narvhus, J.A., Hulbaekdal, A., and Abrahamsen, R.K. 1992. Occurrence of lactobacilli and leuconostoc in Norwegian cheese of Gouda type and their possible influence on cheese ripening and quality. Symposium FIL/IDF on Cheese Ripening, Lund, Sweden.

Naylor, C.E., Gover, S., Basak, A.K., Cosgrove, M.S., Levy, H.R., and Adams, M.J. 2001. $NADP^+$ and NAD^+ binding to the dual coenzyme specific enzyme *Leuconostoc mesenteroides* glucose 6-phosphate dehydrogenase: Different interdomain binary and ternary complexes. *Acta Crystallogr., Sect. D* 57:635–648.

Neubauer, H., Bauche, A., and Mollet, B. 2003. Molecular characterization and expression analysis of the dextransucrase DsrD of *Leuconostoc mesenteroides* Lcc4 in homologous and heterologous *Lactococcus lactis* cultures. *Microbiology* 149:973–982.

Neves, A.R., Pool, W.A., Kok, J., Kuipers, O.P., and Santos, H. 2005. Overview on sugar metabolism and its control in *Lactococcus lactis*—The input from *in vivo* NMR. *FEMS Microbiol. Rev.* 29:531–554.

Nieminen, T.T., Vihavainen, E., Paloranta, A., Lehto, J., Paulin, L., Auvinen, P., Solismaa, M., and Björkroth, K.J. 2011. Characterization of psychrotrophic bacterial communities in modified atmosphere-packed meat with terminal restriction fragment length polymorphism. *Int. J. Food Microbiol.* 144:360–366.

Nieto-Arribas, P., Seseña, S., Poveda, J.M., Palop, L., and Cabezas, L. 2010. Genotypic and technological characterization of *Leuconostoc* isolates to be used as adjunct starters in Manchego cheese manufacture. *Food Microbiol.* 27:85–93.

Nikel, P.I., Ramirez, M.C., Pettinari, M.J., Mendez, B.S., and Galvagno, M.A. 2010. Ethanol synthesis from glycerol by *Escherichia coli* redoc mutants expressing adhE from *Leuconostoc mesenteroides*. *J. Appl. Microbiol.* 109:492–504.

Niven, G.W. and Mulholland, F. 1998. Cell membrane integrity and lysis in *Lactococcus lactis*: The detection of a population of permeable cells in post-logarithmic phase cultures. *J. Appl. Microbiol.* 84:90–96.

Ogier, J.C., Casalta, E., Farrokh, C., and Saïhi, A. 2008. Safety assessment of dairy microorganisms: The *Leuconostoc* genus. *Int. J. Food Microbiol.* 126:286–290.

Ogier, J.-C., Son, O., Gruss, A., Tailliez, P., and Delacroix-Buchet, A. 2002. Identification of the bacterial microflora in dairy products by temporal temperature gradient gel electrophoresis. *Appl. Environ. Microbiol.* 68:3691–3701.

Oh, H.M., Cho, Y.J., Kim, B.K., Roe, J.H., Kang, S.O., Nahm, B.H., Jeong, G., Han, H.U., and Chun, J. 2010. Complete genome sequence analysis of *Leuconostoc kimchii* IMSNU 11154. *J. Bacteriol.* 192:3844–3845.

Olsen, K.N., Brockmann, E., and Molin, S. 2007. Quantification of *Leuconostoc* populations in mixed dairy starter cultures using fluorescence in situ hybridization. *J. Appl. Microbiol.* 103:855–863.

Ostlie, H.M., Eliassen, L., Florvaag, A., and Skeie, S. 2005. Phenotypic and PCR-based characterization of the microflora in Präst cheese during ripening. *Int. Dairy J.* 15:911–920.

Park, J., Lee, M., Jung, J., and Kim, J. 2005. pIHO1, a small cryptic plasmid from *Leuconostoc citreum* IH3. *Plasmid* 54:184–189.

Park, J.Y., Lee, K.W., Lee, A.R., Jeong, W.J., Chun, J., Lee, J.H., and Kim, J.H. 2008. Characterization of a bifunctional HPr kinase/phosphorylase from *Leuconostoc mesenteroides* SY1. *J. Microbiol. Biotechnol.* 18:746–753.

Piraino, P., Ricciardi, A., Salzano, G., Zotta, T., and Parente, E. 2006. Use of unsupervised and supervised artificial neural networks for the identification of lactic acid bacteria on the basis of SDS-PAGE patterns of whole cell proteins. *J. Microbiol. Methods* 66:336–346.

Poveda, J.M., Sousa, M.J., Cabezas, L., and McSweeney, P.L.H. 2003. Preliminary observations on proteolysis in Manchego cheese made with a defined-strain starter culture and adjunct starter (*Lactobacillus plantarum*) or a commercial starter. *Int. Dairy J.* 13:169–178.

Randazzo, C.L., Pitino, I., Ribbera, A., and Caggia, C. 2010. Pecorino Crotonese cheese: Study of bacterial population and flavour compounds. *Food Microbiol.* 27:363–374.

Randazzo, C.L., Torriani, S., Akkermans, A.D.L., de Vos, W.M., and Vaughan, E.E. 2002. Diversity, dynamics, and activity of bacterial communities during production of an artisanal Sicilian cheese as evaluated by 16S rRNA analysis. *Appl. Environ. Microbiol.* 68:1882–1892.

Reiter, B., Fryer, T.F., Pickering, A., Chapman, H.R., Lawrence, R.C., and Sharpe, M.E. 1967. The effect of microbial flora on the flavour and free fatty acid composition of Cheddar cheese. *J. Dairy Res.* 34:257–272.

Rendueles, E., Omer, M.K., Alvseike, O., Alonso-Calleja, C., Capita, R., and Prieto, M. 2010. Microbiological food safety assessment of high hydrostatic pressure processing: A review. *LWT—Food Sci. Technol.*, Online, 5 November 2010.

Richard, G., Yu, S., Monsan, P., Remaud-Simeon, M., and Morel, S. 2005. A novel family of glucosyl 1,5-anhydro-D-fructose derivatives synthesised by transglycosylation with dextransucrase from *Leuconostoc mesenteroides* NRRL B-512F. *Carbohydr. Res.* 340:395–401.

Rodionov, D.A., Hebbeln, P., Eudes, A., ter Beek, J., Rodionova, I.A., Erkens, G.B., Slotboom, D.J., Gelfand, M.S., Osterman, A.L., Hanson, A.D., and Eitinger, T. 2009. A novel class of modular transporters for vitamins in prokaryotes. *J. Bacteriol.* 191:42–51.

Ruas-Madiedo, P., Salazar, N., and de los Reyes-Gavilan, C.G. 2010. Exopolysaccharides produced by lactic acid bacteria in foods and probiotic applications. In *Microbial Glycobiology: structure, relevance and applications*, edited by Eds A.P. Moran, O. Holst, P.J. Brennan, and M. von Izstein, pp. 887–902. London: Elsevier.

Samelis, J., Lianou, A., Kakouri, A., Delbès, C., Rogelj, I., Bogovic-Matijašić, B., and Montel, M.C. 2009. Changes in the microbial composition of raw milk induced by thermization treatments applied prior to traditional Greek hard cheese processing. *J. Food Prot.* 72:783–790.

Sánchez, J.I., Martinez, B., and Rodriguez, A. 2005. Rational selection of *Leuconostoc* strains for mixed starters based on the physiological biodiversity found in raw milk fermentations. *Int. J. Food. Microbiol.* 105:377–387.

Santos, M., Rodrigues, A., and Teixeira, J.A. 2005. Production of dextran and fructose from carob pod extract and cheese whey by *Leuconostoc mesenteroides* NRRL B512(f). *Biochem. Eng. J.* 25:1–6.

Sarwat, F., Ul Qader, S.A., Aman, A., and Ahmed, N. 2008. Production and characterization of a unique dextran from an indigenous *Leuconostoc mesenteroides* CMG713. *Int. J. Biol. Sci.* 4:379–386.

Sawa, N., Okamura, K., Zendo, T., Himeno, K., Nakayama, J., and Sonomoto, K. 2010. Identification and characterization of novel multiple bacteriocins produced by *Leuconostoc pseudomesenteroides* QU 15. *J. Appl. Microbiol.* 109:282–291.

Schillinger, U., Boehringer, B., Wallbaum, S., Caroline, L., Gonfa, A., Huch Née Kostinek, M., Holzapfel, W.H., and Franz, C.M. 2008. A genus-specific PCR method for differentiation between *Leuconostoc* and *Weissella* and its application in identification of heterofermentative lactic acid bacteria from coffee fermentation. *FEMS Microbiol. Lett.* 286:222–226.

Schmitt, P., Vasseur, C., Phalip, V., Huang, D.Q., Diviès, C., and Prévost, H. 1997. Diacetyl and acetoin production from the co-metabolism of citrate and xylose by *Leuconostoc mesenteroides* subsp. *mesenteroides*. *Appl. Microbiol. Biotechnol.* 47:715–718.

Schwab, C., Sørensen, K.I., and Gänzle, M.G. 2010. Heterologous expression of glycoside hydrolase family 2 and 42 β-galactosidases of lactic acid bacteria in *Lactococcus lactis*. *Syst. Appl. Microbiol.* 33:300–307.

Seibel, J. and Buchholz, K. 2010. Tools in oligosaccharide synthesis: Current research and application. *Adv. Carbohydr. Chem. Biochem.* 63:101–138.

Seo, D.M., Kim, S.Y., Eom, H.J., and Han, N.S. 2007. Synbiotic synthesis of oligosaccharides during milk fermentation by addition of *Leuconostoc* starter and sugars. *J. Microbiol. Biotechnol.* 11:1758–1764.

Seo, E.S., Lee, J.H., Park, J.Y., Kim, D., Han, H.J., and Robyt, J.F. 2005. Enzymatic synthesis and anti-coagulant effect of salicin analogs by using the *Leuconostoc mesenteroides* glucansucrase acceptor reaction. *J. Biotechnol.* 117:31–38.

Server-Busson, C., Foucaud, C., and Leveau, J.Y. 1999. Selection of dairy *Leuconostoc* isolates for important technological properties. *J. Dairy Res.* 66:245–256.

Settanni, L. and Moschetti, G. 2010. Non-starter lactic acid bacteria used to improve cheese quality and provide health benefits. *Food Microbiol.* 27:691–697.

Shakeel-Ur-Rehman, Fox, P.F., and McSweeney, P.L.H. 2000. Methods used to study non-starter microorganisms in cheese: A review. *Int. J. Dairy Technol.* 53:113–119.

Shobharani, P. and Agrawal, R. 2009. Supplementation of adjuvants for increasing the nutritive value and cell viability of probiotic fermented milk beverage. *Int. J. Food Sci. Nutr.* 25:1–14.

Singh, R., Paul, D., and Jain, R.K. 2006. Biofilms: Implication in bioremediation. *Trends Microbiol.* 14:389–397.

Smart, J.B., Pillidge, C.J., and Garman, J.H. 1993. Growth of lactic acid bacteria and bifidobacteria on lactose and lactose-related mono-, di- and trisaccharides and correlation with distribution of β-galactosidase and phosphor-β-galactosidase. *J. Dairy Res.* 60:557–568.

Tan, Z., Pang, H., Duan, Y., Qin, G., and Cai, Y. 2010. 16S ribosomal DNA analysis and characterization of lactic acid bacteria associated with traditional Tibetan Qula cheese made from yak milk. *Anim. Sci.* 81:706–713.

Thunell, R.K. 1995. Taxonomy of the leuconostocs. *J. Dairy Sci.* 78:2514–2522.

Todorov, S.D. 2010. Diversity of bacteriocinogenic lactic acid bacteria isolated from boza, a cereal-based fermented beverage from Bulgaria. *Food Control* 21:1011–1021.

Turner, K.W. 1988. Some aspects of the microbiology of cheese ripening investigated using aseptic manufacturing techniques. PhD Thesis, Massey University, Palmerston North, New Zealand.

Uraz, G., Adigüzel, S., and Oncül, O. 2004. Inhibitory effects of *Leuconostoc* and *Pediococcus* strains isolated from raw milk on some pathogenic bacteria. *Egypt. J. Dairy Sci.* 32:31–37.

Valerio, F., Favilla, M., de Bellis, P., Sisto, A., de Candia, S., and Lavermicocca, P. 2009. Antifungal activity of strains of lactic acid bacteria isolated from a semolina ecosystem against *Penicillium roqueforti*, *Aspergillus niger* and *Endomyces fibuliger* contaminating bakery products. *Syst. Appl. Microbiol.* 32:438–448.

Van Hoorde, K., van Leuven, I., Dirinck, P., Heyndrickx, M., Coudijzer, K., Vandamme, P., and Huys, G. 2010. Selection, application and monitoring of *Lactobacillus paracasei* strains as adjunct cultures in the production of Gouda-type cheeses. *Int. J. Food Microbiol.* 144:226–235.

Vancanneyt, M., Zamfir, M., de Wachter, M., Cleenwerk, I., Hoste, B., Rossi, F., Dellaglio, F., de Vuyst, L., and Swings, J. 2006. Reclassification of *Leuconostoc argentinum* as a later synonym od *Leuconostoc lactis*. *Int. J. Sys. Evol. Microbiol.* 56:213–216.

Vázquez, J.A., Miron, J., Gonzalez, M.P., and Murado, M.A. 2005. Effects of aeration on growth and on production of bacteriocins and other metabolites incultures of eight strains of lactic acid bacteria. *Appl. Biochem. Biotechnol.* 127:111–124.

Vijayendra, S.V.N. and Babu, R.S. 2008. Optimization of a new heteropolysaccharide production by a native isolate of *Leuconostoc* sp. CFR-2181. *Lett. Appl. Microbiol.* 46:643–648.

Voulgan, K., Hatzikamari, E., Delepoglou, A., Georgapoulos, P., Litopoulou-Tzanetaki, E., and Tzanetakis, N. 2009. Antifungal activity of non-starter lactic acid bacteria isolates from dairy products. *Food Control* 21:136–142.

Winters, D.A., Poolman, B., Hemme, D., and Konings, W.N. 1991. Branched-chain amino acid transport in cytoplasmic membranes of *Leuconostoc mesenteroides* subsp. *dextranicum* CNRZ 1273. *Appl. Environ. Microbiol.* 57:3350–3354.

Witthuhn, R.C., Schoeman, T., and Britz, T.J. 2005. Characterisation of the microbial population at different stages of Kefir production and Kefir grain mass cultivation. *Int. Dairy J.* 15:383–389.

Yoon, S.H., Fulton, D.B., and Robyt, J.F. 2010. Enzymatic synthesis of L-DOPA alpha-glycosides by reaction with sucrose catalysed by four different glucansucrases from four strains of *Leuconostoc mesenteroides*. *Carbohydr. Res.* 345:1730–1735.

Yossuck, P., Miller-Canfield, P., Moffett, K., and Graeber, J. 2009. *Leuconostoc* spp sepsis in an extremely low birth weight infant: A case report and review of the literature. *W. V. Med.* 105:24–27.

Yvon, M. 2006. Key enzymes for flavour formation by lactic acid bacteria. *Aust. J. Dairy Technol.* 61:88–95.

Zaunmüller, T., Eichert, M., Richter, H., and Unden, G. 2006. Variations in the energy metabolism of biotechnologically relevant heterofermentative lactic acid bacteria during growth on sugars and organic acids. *Appl. Microbiol. Biotechnol.* 72:421–429.

6

Food Fermentation and Production of Biopreservatives

Maria Papagianni

CONTENTS

6.1 Introduction

The modern food industry is undergoing dramatic changes due to increasing consumer demands for "natural," "fresh," and nonprocessed or minimally processed foods with reduced use of chemical preservatives, yet remaining healthy and safe. Minimizing processing and reducing the use of preservatives, however, could lead to increased safety risks. In addition, the large number of outbreaks caused by the pathogen *Listeria monocytogenes* indicates that current preservation practices need improvement. The use of antibiotics used in human therapy as an auxiliary treatment to control contaminants is a highly criticized practice because of the possible interference in the control of human diseases. During recent years, these facts have led to the growing interest in naturally produced antimicrobial agents and, therefore, biopreservation.

Biopreservation is the extension of storage life and safety of foods using natural or controlled microflora and/or their antimicrobial products (Montville and Chen 1998; Stiles 1996). Preservatives of biological origin have a long history of use in extending shelf life and improving safety of foods. Nisin is an antimicrobial peptide naturally produced by selected strains of the lactic acid bacterium *Lactococcus lactis*. It has a 'generally regarded as safe' status for certain foods' preservation in the U.S.A., and it is commercially used worldwide. Lysozyme, weak organic acids, and lactoferrin are only some—among many other molecules—naturally produced by eukaryotic and prokaryotic organisms known to be safe for consumption for a very long time. Many of these antimicrobials act synergistically (Franchi et al. 2003; Gill and Holley 2000; Razavi-Rohani and Griffiths 1996), and their combination can be applied in hurdle technology for food preservation.

Biopreservation of food is achieved through microbial antagonism or microbial interference. Microbial antagonism, however, is not only due to the production of antimicrobial metabolites but is also caused by the competition of microorganisms for nutrients and substrates. Selected lactic acid bacteria (LAB), in pure cultures, have been used routinely since the beginning of the twentieth century as starter cultures for the production of fermented food products. Growth and metabolism of these cultures contribute in a number of different ways to the control of undesired microorganisms, including pathogens, and the extension of shelf life in addition to the modification of the organoleptic properties of foods. Antagonism between two or more different species or genera of microorganisms takes place when they compete for a common niche, when an antagonistic extracellular metabolite is produced by a microorganism, or even when a modification of the microenvironment is caused by a microorganism that is not friendly for the others to grow (Brashears et al. 2006). This fact has prompted interest in recent years in using these bacteria or their fermentation products as biopreservatives for other foods. Utilization of cells refers to "protective" cultures (rather than starter cultures) and has gained significant importance during the last few years. Among the substances produced by LAB in fermentation that act as antimicrobial agents are lactic and acetic acids, formic acid, propionic acid, bacteriocins, hydrogen peroxide, diacetyl, reuterin, and various inhibitory substances—mostly unidentified—with antifungal activities (Lindgren and Dobrogosz 1990; Schnürer and Magnusson 2005).

6.2 Lactic Acid Bacteria

The group consists of Gram-positive, non-spore-forming cocci and rods, which ferment carbohydrate-forming lactic acid. The term "lactic acid bacteria" has become commonly associated with the genera *Lactobacillus*, *Lactococcus*, *Streptococcus*, *Leuconostoc*, *Pediococcus*, and *Enterococcus* (Wood and Holzapfel 1995). However, newer taxonomic studies as well as descriptions of new genera have included the group of LAB the genera *Aerococcus*, *Alloiococcus*, *Vagococcus*, *Tetragenococcus*, *Oenococcus*, *Carnobacterium*, *Weissella*, and others (Axelsson 1998; Stiles and Holzapfel 1997). Taxonomic criteria include (1) morphological characteristics, (2) the metabolic pathways used to ferment carbohydrates, (3) acid, alkaline, and salt tolerance, and (4) temperature requirements for growth. Currently, however, intensive ribosomal ribonucleic acid sequence studies are used extensively in taxonomy research.

LAB are described as facultative anaerobes or anaerobes, are generally catalase-negative and non-motile, and do not reduce nitrate (Brashears et al. 2006). Their growth requirements for carbohydrates, amino acids, peptides, fatty acids, nucleic acid derivatives, vitamins, and salts are complex and therefore dictate the type of microenvironment in which they occur in nature. Consequently, they are common in milk and dairy products, fermented foods, vegetables, rotting vegetable material, silage, and the intestinal tracks and mucous membranes of humans and animals.

LAB have evolved and been selected for exploitation in a variety of food fermentation processes. As fermentative bacteria, they impart flavor and texture to fermented foods while playing a major role in preservation. Currently, selective species of LAB make valuable industrial microorganisms that are cultivated to produce (1) biomass for starter cultures, with extensive use by the food industry, (2) lactic acid, with widespread use in food technology and the chemical industry, (3) antimicrobial metabolites for food applications, (4) various metabolites and enzymes that are used as flavor enhancers, (5) biomass that is used in probiotic preparations, which have been increasingly popular as health-promoting factors for humans and animals and constitute a rapidly expanding market, and finally, (6) extracellular polysaccharides that as food additives represent functional ingredients, known as prebiotics, with again health-promoting properties.

6.2.1 Microbial Antagonism and Lactic Acid Bacteria in Fermented Foods

Selected LAB are used as starter cultures for the production of sausages, fermented dairy products, and fermented vegetables. Fermentation offers development of particular substances for each type of fermented food organoleptic properties and preservation for later consumption. The latter fact has led to increasing interest over the years in utilizing these bacteria or their metabolites as biopreservatives for other foods. This might involve the addition of either a culture (starter and/or protective culture) or an antimicrobial metabolite (e.g., a bacteriocin).

Lactic and acetic acids, bacteriocins, and hydrogen peroxide are the most important antimicrobial metabolites produced by LAB. Lactic acid is the main end product of fermentation of these bacteria, whereas the more inhibitory acetic acid is produced under the heterofermentative mode in smaller quantities. Because both weak organic acids require a low pH for maximum inhibitory activity, they could be of limited use in most nonfermented foods. Members of LAB produce peptides or proteins with antimicrobial properties, the bacteriocins, which attract the strong interest of the academia and the food industry for use as biopreservatives (Papagianni 2003). At the moment, only the lantibiotic nisin has been approved for safe use in foods, while several others have been found to exhibit technologically important properties like thermostability or stability to wide pH ranges, as well as to various proteases (Anastasiadou et al. 2008a,b; Bauer et al. 2005; Bhunia et al. 1991), which along with antimicrobial activity make them important candidates for use as biopreservatives. The rather narrow spectrum of activity of most bacteriocins and their interactions with food components are limiting factors for their use in foods. Some LAB produce inhibitory levels of hydrogen peroxide at low temperatures. Hydrogen peroxide is an effective antimicrobial due to its strong oxidizing effect on the bacterial cells, while it can react with other components to form inhibitory substances (Brashears et al. 2006). The use of hydrogen peroxide–producing LAB at refrigeration temperature may control microbial growth in certain types of nonfermented refrigerated foods (Gilliland 2003).

6.2.1.1 Antimicrobials Produced by Lactic Acid Bacteria in Fermentation

6.2.1.1.1 Weak Organic Acids

LAB generate adenosine triphosphate (ATP) by the fermentation of carbohydrates coupled to substrate-level phosphorylation. The two major pathways for the metabolism of hexoses are the homofermentative glycolytic pathway (the Embden–Meyerhof pathway), with lactic acid as the main end product, and the heterofermentative (phosphoketolase) pathway, in which other end products such as acetic acid, propionic acid, CO_2, and ethanol are produced in addition to lactic acid (Kandler 1983).

Weak organic acids are known to have strong antimicrobial activity and have been used for centuries to preserve foods. In modern times, weak acids such as acetic, propionic, lactic, citric, and sorbic acids have been widely used as food preservatives as well as in the preharvest and postharvest phases of agriculture (Ricke 2003). However, despite their widespread use, the antimicrobial mode of action of weak organic acids is still not fully understood. It is generally agreed that the ability of weak acids to inhibit microbial growth is related to their lipid permeability (Brul and Coote 1999; Cherrington et al. 1991). Weak acids, such as acetic acid, can be either charged or uncharged, depending on the protonation state of their acidic group, which is in turn determined by the dissociation constant pK_a of the acidic group and the pH of the environment surrounding it. Low pH values favor the uncharged, undissociated state of a molecule that is lipid permeable and can freely diffuse into the cytoplasm of microbial cells. Subsequently, upon encountering a higher pH inside the cell (as what happens with most neutralophilic microorganisms), the molecule will dissociate, resulting in the release and accumulation of charged anions and protons, which cannot cross the cell membrane (Brashears et al. 2006; Hirshfield et al. 2003).

A number of explanations have been offered on the antimicrobial activity of weak acids. Early studies were concentrated on the effects on intracellular pH (pH_i) and the toxicity of a drop in pH_i that resulted from the accumulation of weak acid anions and protons in the cytoplasm (Cole and Keenan 1986; Salmond et al. 1984). However, Roe et al. (1998) showed that *Escherichia coli* cultures grown in the presence of different weak acids grew at the same rate even when their pH_i values were significantly different. In addition, Young and Foegeding (1993) showed in their work with *L. monocytogenes* that different weak acids inhibited the growth of *Listeria* to different extents even when the pH_i values were the same. Obviously, it is not the lowering of pH_i alone that is responsible for the antimicrobial activity of weak acids, but as the available evidence indicates, it can also exert effects on the membrane function. Undissociated weak acids are known to partition into membrane lipid bilayers at concentrations, depending on their lipophilicity. A good correlation was found to exist between the growth inhibition of yeasts by weak acids and their lipophilicity (Stratford and Anslow 1998). Furthermore, the increased levels of weak acid anions that accumulate in the cytoplasm can have osmotic effects on the cell (Roe et al. 1998) and on metabolic pathways by influencing the activity of enzymes in reactions that take place within the cytoplasm (Roe et al. 2003).

Growing concern, however, over the development of resistance to weak organic acids is a major issue. Although weak acids have been used as preservatives for centuries, it is only in recent years that their mode of action and the relationship between adaptation to acid stress and resistance to organic acids have been investigated. The ability of foodborne pathogens to adapt to weak acids may allow longer survival in various commodities and the gastric acid environment of the stomach (Hirshfield et al. 2003). Experiments with *E. coli* and *Salmonella* have demonstrated that resistance to organic acids can be induced by inorganic acids or by short-chain organic acids used as food preservatives and that preexposure to organic acids can induce resistance in both organic and inorganic acids (Bearson et al. 1997; Goodson and Rowbury 1989; Guilfoyle and Hirshfield 1996; Foster 2000). Genomic and proteomic approaches [gene fusion technology, two-dimensional gel electrophoresis, and deoxyribonucleic acid (DNA) microarray technology] have been applied in further investigations into the genes and proteins that enable bacteria to respond to and resist the detrimental effects of organic acids (Hirshfield et al. 2003).

Organic acids produced by LAB also act as fungal inhibitory metabolites. Lactic, acetic, and propionic acids are known to reduce fungal growth, the latter two especially at lower pH values (Eklund 1989; Woolford 1984a). Propionic acid affects fungal membranes at pH values below 4.5 (Hunter and Segel 1973). Salts of propionic acid were also shown to have similar effects against yeasts and filamentous fungi at low pH values (Woolford 1984b). Propionic and acetic acids were found to inhibit the amino acid uptake in fungi (Eklund 1989). Mixtures of the three weak acids inhibited yeasts that normally grow well at concentrations of 100 mM of the individual acids, except for propionic acid.

Apart from the control of foodborne bacterial pathogens, the food industry is equally concerned about the control of fungal spoilage of foods and beverages, and therefore, resistance to the major organic acid food preservatives poses a serious problem, because it is often necessary to use these preservatives at millimolar rather than micromolar concentrations to prevent yeast spoilage in low-pH foods and beverages (Piper et al. 2001). Weak acid stress response has been studied extensively for yeast species that are of major importance in this regard, the "spoilage yeasts," for example, *Zygosaccharomyces rouxii*, *Z. bailii*, and *Z. lentus* (see the work of Piper et al. 2001 and the references cited therein). Considerably more is known, however, for *Saccharomyces cerevisiae*, which seems to achieve adaptation to weak acid stress by a different strategy from other yeasts (Mollapour et al. 2006; Piper et al. 2001; Simões et al. 2006). *S. cerevisiae* possesses an active process for acid removal from a cell in which high levels of a specific ATP-binding cassette (ABC) transporter (Pdr12) is induced in its plasma membrane in order to catalyze the active efflux. *Z. bailii*, in contrast, does not show any major changes to the protein composition of its plasma membrane during weak acid stress, but instead, it actively limits the initial diffusional entry of the acid to the cells.

In recent years, in the context of hurdle technology in food preservation, the combined efficacy of weak organic acids or their salts with various bacteriocins has been evaluated in many cases, mostly against *L. monocytogenes*. Organic acids and their salts can increase the activity of bacteriocins, while low pH levels enhance the activity of both weak organic acids and bacteriocins (Stiles 1996). Many reports confirm the increased antibacterial activity of nisin, pediocin, lacticin, or enterocin when combined with lactate, acetate, or other weak acids in several food systems (Bari et al. 2005; Gálvez et al. 2007; Grande et al. 2006; Ukuku and Fett 2004). Organic acids have also been incorporated along with nisin in edible films with antimicrobial activity against *L. monocytogenes* (Pintado et al. 2009).

6.2.1.1.2 Bacteriocins

Bacterial ribosomally synthesized antimicrobial peptides are generally referred to as bacteriocins. The biosynthesis, mode of action, characterization, classification, and genetics of bacteriocins have been extensively reviewed and published (Abee et al. 1995; Papagianni 2003; Papagianni and Anastasiadou 2009; Skaugen et al. 2003; Venema et al. 1995). Bacteriocins produced by LAB are divided into three main groups—the modified bacteriocins, known as lantibiotics (class I), the heat-stable unmodified bacteriocins (class II), and the larger, heat-labile bacteriocins (class III). A fourth group (class IV) with complex bacteriocins carrying lipid or carbohydrate moieties in their molecules is often included in bacteriocins' classifications.

The lantibiotics have been subdivided into two major groups based on their structure (Jung 1991; Sahl et al. 1995; Sahl and Bierbaum 1998). Type A lantibiotics include elongated, screw-shaped, flexible molecules that have a positive charge and act via membrane depolarization. They have a molecular weight in the range of 2–5 kDa. Nisin belongs to this group of lantibiotics and is perhaps the most thoroughly characterized lantibiotic. Type B lantibiotics are globular in shape, and they have a molecular weight of about 2 kDa. They act by interfering with cellular enzymatic reactions; for example, the type B lantibiotics mersacidin and actagardine interfere with the cell wall synthesis in Gram-positive bacteria (Sahl et al. 1995).

Nisin is the best known and characterized lantibiotic and the only one to have realized widespread commercial use. Nisin has been the subject of a wide variety of fundamental studies on its structure and genetics (De Vuist and Vandamme 1994; Dutton et al. 2002). It is composed of 34 amino acids and has a pentacyclic structure (Shiba et al. 1991) with one lanthionine residue (ring A) and four β-methyllanthionine residues (rings B, C, D, and E). Nisin can be effective at nanomolar concentrations, depending on the target strain. An interesting feature of this lantibiotic is its unusually high specificity compared to eukaryotic-derived adenosine monophosphates. Earlier studies demonstrated that it inhibits peptidoglycan biosynthesis (Linnet and Strominger 1973) and that it interacts with either lipid I or II structures (Reisinger et al. 1980). Later, it was found that it causes pore formation in the membranes of sensitive bacteria (Benz et al. 1991; Ruhr and Sahl 1985; Sahl et al. 1987). More recently, it has been shown that nisin interacts with a docking molecule, lipid II, which is a membrane-bound precursor involved in cell wall biosynthesis (Breukink et al. 1999; Wiedemann et al. 2001). Mutations in N-terminal rings indicated that these are involved in lipid II binding, whereas mutations in the flexible hinge region severely affected the ability of the bacteriocin to form membrane pores. Such experiments revealed the dual functionality of the nisin molecule involving initial binding to lipid II followed by pore formation, resulting in rapid killing of the target cell.

Bacteriocins of class II are generally small unmodified peptides of <5-kDa molecular weight (MW), which are subdivided into the groups of pediocin-like and two-peptide bacteriocins (Papagianni and Anastasiadou 2009). Pediocin-like bacteriocins (36–48 residues) are produced by a variety of LAB and have similar amino acid sequences (40%–60% sequence similarity). The peptides of this group are also known as *"Listeria*-active" peptides, and they are characterized by an N-terminus that contains the sequence –Y–G–N–G–V–. The hydrophilic N-terminal region is well conserved. The N-terminal region of all pediocin-like bacteriocins presently identified contains two cysteines joined by a disulfide bond found in the "pediocin box" motif: $–Y–G–N–G–V–X_1–C–X_2–K/N–X_3–X_4–C–$, where X_{1-4} represent polar uncharged or charged residues.

Several pediocins have been isolated and characterized. Among the first were pediocin PA1 (Henderson et al. 1992; Nieto Lozano et al. 1992), sakacin A (Holck et al. 1992), sakasin P (Tichaczek et al. 1992), leucocin A (Hastings et al. 1991), curvacin A (Holck et al. 1992; Tichaczek et al. 1992), mesentericin Y105 (Héchard et al. 1992), and carnobacteriocins BM1 and B2 (Quadri et al. 1994). More recently isolated and characterized are pediocin PD-1 (Green et al. 1997), pediocins SA-1 and SM-1 (Anastasiadou et al. 2008a,b), and a pediocin isolated from *Weissella paramesenteroides* culture broths (Papagianni and Papamichael 2010; see Table 6.1). Despite the structural similarities, their molecular weight varies, as well as their spectrum of antimicrobial activity. By constructing hybrid bacteriocins, it has been shown for pediocin-like bacteriocins that the C-terminal region is an important determinant of target cell specificity (Fimland et al. 1996). This region also appears to interact with the cell membrane, thereby causing membrane leakage, because the region, in contrast to the N-terminal region, is either hydrophobic or amphiphilic (Chikindas et al. 1993; Papagianni and Anastasiadou 2009). Pediocins exhibit important technological properties, for example, thermostability and maintenance of activity at a wide pH range, which along with the bactericidal action against Gram-positive food spoilage and pathogenic bacteria, make them an important class of biopreservatives. Much new information regarding pediocins has emerged during the last few years (Table 6.1).

Class III bacteriocins are the least well-characterized group and consist of heat-labile proteins that are generally of >30-kDa MW. The group includes Helvetin J produced by *Lb. helveticus* (Joerger and Klaenhammer 1986) and enterolysin produced by *Enterococcus faecium* (Nilson 1999).

So far, only nisin and pediocin PA-1/AcH are available in commercial preparations for use as food biopreservatives, while several others offer potential applications in food preservation. The use of bacteriocins in the food industry can help reduce the use of chemical preservatives and the intensity of heat treatments, resulting in more naturally preserved foods and with better preserved organoleptic and nutritional properties (Gálvez et al. 2007). In recent years, the application of bacteriocins in the context of hurdle technology has gained much attention. Several bacteriocins show synergistic effects when used in combination with chemical preservatives (e.g., organic acids and their salts), natural phenolic compounds, or other antimicrobial proteins (e.g., lysozyme and lactoferrin). This approach may help reduce the result of the reduced efficacy of several limitations that bacteriocins face in food environments

TABLE 6.1

Properties of a Newly Isolated Class IIa Bacteriocin from the LAB *Weissella paramesenteroides* in Comparison with Properties of Known Pediocins Reported in the Literature

Bacteriocin	Temperature		pH		Pepsin	Proteolytic Enzymes				Reference
	100°C/60 min	121°C/60 min	2–10	4–7		Papain	Trypsin	α-Chymotrypsin	Proteinase K	
Bacteriocin from *W. paramesenteroides*	+	+	+	+	+	+	+	+	−	Papagianni and Papamichael 2010
p SA-1	+	+	+	+	+	+	+	+	−	Anastasiadou et al. 2008a,b
p ACCEL	+	±	+	+	+	−	−	−	−	Wu et al. 2004
p PD-1	+	+	+	+	+	+	+	+	−	Green et al. 1997
p SJ-1	+	+	−	−	+	ND	−	−	−	Schved et al. 1993
p N5p	+	+	−	−	+	ND	ND	ND	ND	Strässer de Saad et al. 1995
p AcH	+	+	+	+	+	ND	−	−	−	Bhunia et al. 1988
p PA-1	+	±	+	+	+	−	−	ND	ND	Gonzalez and Kunka 1987

Note: +, Activity; −, absence of any activity; ND, not determined; and p, pediocin.

(from food-related factors) and even extend the antimicrobial spectrum to Gram-negative bacteria. For example, the spectrum of nisin–lysozyme combinations could be extended to Gram-negative bacteria by the addition of chelating agents (Gill and Holley 2000). The combination of bacteriocins and physical treatments like high-pressure processing or pulsed electric fields also offers opportunities for more effective preservation of foods (Arqués et al. 2005; Dutreux et al. 2000; Gálvez et al. 2007). The combined application of many other technologies such as irradiation, microwave, and ohmic heating or ultrasonication still remains to be studied.

Combinations of bacteriocins have also been tested in order to increase their antimicrobial efficacy. The combined use, for example, of nisin and pediocin AcH provides greater activity than each one separately (Hanlin et al. 1993). Several combinations have been tested so far, for example, lactacin/nisin, lactacin/pediocin, nisin/leucosin, and nisin/curvaticin, and it was shown that combined use not only has better antimicrobial results but may also help avoid regrowth of adapted to bacteriocin cells. However, in such cases of combined use, cross-resistance development is not impossible, especially when both bacteriocins belong to the same class (Gravesen et al. 2002a,b).

Bacteriocins are added to foods (nisin and pediocin) in the form of concentrated crude preparations as preservatives, shelf life extenders, additives, or ingredients (Gálvez et al. 2007). In addition, they can be produced *in situ* by bacteriocinogenic starter, adjunct, or protective cultures. Immobilization techniques may offer further potential applications for bacteriocins in the form, for instance, of microcapsules for improved stability and availability in the food matrix (Bennech et al. 2002) or in active packaging systems (Pintado et al. 2009).

6.2.1.1.3 Hydrogen Peroxide

Most LAB have flavoprotein oxidases that enable them to produce hydrogen peroxide (H_2O_2) in the presence of oxygen as a means of protecting themselves against oxygen toxicity. This mechanism along with the absence of the heme protein catalase (Condon 1987) results in the formation of H_2O_2 in amounts that are in excess of the capacity of these microorganisms to degrade it. Thus, H_2O_2 accumulates in the environment of LAB and may inhibit growth or kill other members of the present microflora. The antimicrobial effect of H_2O_2 is well documented and is attributed to its strong oxidizing effect; sulfhydril groups of cell proteins and membrane lipids can be oxidized (Davidson 1997; Juven and Pierson 1996; Lindgren and Dobrogosz 1990).

Hydrogen peroxide can react with food components to form inhibitory substances. In raw milk, for example, H_2O_2 generated by LAB can react with endogenous thiocyanate, the reaction catalyzed by the enzyme lactoperoxidase, to form hypothiocyanate and other intermediate products with inhibitory activities. This way, the antimicrobial effect of H_2O_2 at noninhibitory concentrations is potentiated in milk by lactoperoxidase and thiocyanate (Condon 1987). The mechanism, known as the lactoperoxidase antibacterial system (LP system or LP-s), has been well documented (Brashears et al. 2006; Schnürer and Magnusson 2005). Lactoperoxidase is an enzyme that is naturally present in milk. Hydrogen peroxide and thiocyanate are also naturally occurring in milk, in concentrations of around 5 ppm. The method of activating the LP-s in milk is to add about 10 ppm of thiocyanate to the raw milk to increase the overall level to 15 ppm. The solution is mixed for 30 s, and then, an equimolar amount (8.5 ppm) of hydrogen peroxide is added (Björck 1978; Björck et al. 1975; FAO/WHO 2005). The activation of the lactoperoxidase has a bacteriostatic effect on the raw milk and effectively extends its shelf life for 7–8 h at 30°C or longer at lower temperatures. This way, there is adequate time for the milk to be transported from the collection point to a processing point without refrigeration.

The LP-s has received much attention as a system for raw milk preservation. The antimicrobial activity of the LP-s against a wide variety of milk spoilage and pathogenic microorganisms, including bacteria, yeasts, molds, mycoplasma, viruses, and protozoa, has been well documented in both laboratory and practical settings (FAO/WHO 2005). The overall activity of the LP-s is primarily bacteriostatic, while its extent depends on the initial bacterial load, the species and strains of contaminating bacteria, and the time/temperature combinations. The LP-s can be used without any expensive infrastructure or equipment, and it represents a viable alternative preservation option, especially for small rural milk producers. Furthermore, as recent strategies for controlling spoilage and pathogenic bacteria tend to apply hurdle technology, the combination of the LP-s and nisin has attracted interest, and the effects of combined applications have been evaluated in raw and reconstituted milk. Although the mechanisms of action are different for the LP-s and nisin, both cause damage to the cytoplasmic membrane, a fact that according to Zapico et al. (1998) could explain their synergistic action. Reports, however, on the synergistic action of the LP-s and nisin have been contradictory (Brashears et al. 2006; Rodríguez et al. 1997; Sobrino-López and Martín-Belloso 2008; Zapico et al. 1998), and therefore, further research is required to elucidate the field.

6.2.1.1.4 Proteinaceus Compounds with Antifungal Activities

Recently, a number of antifungal metabolites—cyclic dipeptides, proteinaceous compounds, phenyl-lactic acid, and 3-hydroxylated fatty acids—have been isolated from LAB and have attracted interest for their potential applications in preservation of foods and feeds (Schnürer and Magnusson 2005; Sjogren 2005; Rouse et al. 2008). Unlike the vast literature on bacteriocins, only few reports exist on antifungal peptides and proteins of LAB. Okkers et al. (1999) reported on the peptide TV35b from *Lb. pentosus* with activity against *Candida albicans*. Another case is a 3-kDa heat-stable peptide produced by a *Lb. coryniformis* strain that exhibits antifungal activities against several molds and yeasts, as reported by Magnusson and Schnürer (2001). However, studies in this field are complicated by the general antimicrobial effects of LAB fermentation products such as lactic and acetic acids, which act synergistically with more targeted antifungal compounds.

6.2.1.1.5 Reuterin

Reuterin [3-hydroxypropionaldehyde (3-HPA)] is a broad-spectrum antimicrobial substance first described as a product of the anaerobic growth of *Lb. reuteri* in the presence of glycerol. Production of reuterin has also been reported from *Lb. brevis* and *Lb. buchneri* (Schutz and Radler 1984), *Lb. collinoides* (Claisse and Lonvaud-Funel 2000), and *Lb. coryniformis* (Magnusson et al. 2003). Reuterin is an intermediate product of glycerol conversion to 1,3-propanediol, a pathway proposed to generate NAD^+ from NADH and to contribute to increased growth yields (Luthi-Peng et al. 2002). Reuterin is one of the most intensively studied low-molecular-mass inhibitory compounds of LAB.

Reuterin is active against Gram-positive and Gram-negative bacteria, yeasts, molds, and viruses (Chung et al. 1989; Cleusix et al. 2007). Despite 20 years of investigation, the mechanism of action by which reuterin exerts its effects still remains unclear. Active reuterin is an equilibrium mixture of monomeric, hydrated monomeric, and cyclic dimeric forms of hydroxypropionaldehyde (Talarico et al. 1988), and because it can be converted into these compounds, it has been difficult to determine its exact mechanism of action. The inhibitory effect of reuterin has been associated with its action on DNA synthesis by acting as an inhibitor of the substrate-binding subunit of the enzyme ribonucleotide reductase (decreases ribonucleotide reductase activity; Talarico and Dobrogosz 1989). Most recent research, however, has provided evidence that reuterin induces oxidative stress in cells most likely by modifying thiol groups in proteins and small molecules (Schaefer et al. 2010). Schaefer et al. (2010) have also demonstrated that reuterin is produced solely by the enzyme glycerol dehydratase and that direct contact with other bacteria can stimulate its production and secretion, while *in vivo* and *in vitro* evidence indicate that the aldehyde form of reuterin is the bioactive form that causes an oxidative stress response by modifying thiol groups inside cells.

Reuterin has been proposed to partly mediate the probiotic health benefits ascribed to its producer microorganism. Strains of *Lb. reuteri* have been shown to be effective against a variety of ailments, including diarrhea and colic (Savino et al. 2007; Weizman et al. 2005). Reuterin has also been studied as a possible additive to prevent food spoilage and pathogen growth in food. As a chemical compound, it is water soluble, active at a wide range of pH values, and resistant to proteolytic and lipolytic enzymes (El-Ziney et al. 1999). El-Ziney et al. (1999) reported the antimicrobial activity of reuterin against *E. coli* O157:H7 and *L. monocytogenes* on the surface of cooked pork at 7°C. Its bactericidal activity against the same microorganisms in milk and cottage cheese has also been shown (El-Ziney and Debevere 1998). Effects of reuterin alone or in combinations with nisin and the LP-s on *L. monocytogenes* and *Staphylococcus aureus* have been studied by Arqués et al. (2008). The most pronounced decrease in pathogen counts was achieved by the triple combination of nisin–reuterin–LP-s, which acted synergistically on the inactivation of *L. monocytogenes* and *S. aureus* in cuajada, a semisolid Spanish dairy product, over 12 days at 10°C. The treatment, combining these three natural biopreservatives at low concentrations, within the hurdle concept of food preservation, might be a useful tool for controlling the growth of pathogenic microorganisms in nonacidified dairy products.

6.2.1.1.6 Diacetyl

Diacetyl (2,3-butanedione) is formed from the metabolism of citrate via pyruvate as a fermentation by-product by many microorganisms. In dairy fermentations, it is produced by the utilization of the minor component of milk citric acid by species of the genera *Lactobacillus, Leuconostoc, Pediococcus,* and *Lactococcus* (Jay 1982). Diacetyl has a strong, buttery flavor and is an essential compound at low concentrations in many dairy products such as butter, buttermilk, and fresh cheese. In addition to dairy products, diacetyl is found in wines, brandy, ensilage, coffee, and many other fermented foods. The antibacterial activities of diacetyl have been identified before the 1930s (Lemoigne 1927). In 1982, Jay reported on the parameters that affect the antimicrobial activity of diacetyl in an extensive study that included 40 different cultures, of which 10 were LAB, 14 were Gram-negative, 12 were Gram-positive bacteria, and 4 were yeasts. He found that diacetyl is more effective at pH levels lower than 7.0 and against Gram-negative bacteria, yeasts, and molds than against Gram-positive bacteria.

Diacetyl is known to react with the amino acid arginine in proteins (Riordan 1979) and to inhibit enzymes that are important for protecting cells from oxidative damage, such as superoxide dismutase (Borders et al. 1985), glutathione reductase (Boggaram and Mannervik 1982), and glyoxalase I (Lupidi et al. 2001). Studies have shown that diacetyl can be used as a food ingredient during meat preservation to control *E. coli* O157:H7 and *Salmonella* in the presence of the starter *P. acidilactici* (Kang and Fung 1999). The amounts of diacetyl, however, needed to exert antimicrobial activity in foods may affect their flavor and aroma (Piard and Desmazeaud 1991).

6.2.1.2 Use of Lactic Acid Bacteria Cultures as Protective Cultures for Biopreservation of Food

Protective cultures (PCs), which are antagonistic toward targeted food spoilage and pathogenic microorganisms, are used in the "controlled microflora applications" approach for biopreservation (Stiles 1996). This approach has been of growing interest in processes in which the freshness of products is important and should be preserved until they reach the consumer. Extended shelf life cook–chill meals are becoming increasingly popular, and the use of traditional chemical preservatives would be damaging for their fresh image. Extended shelf life cook–chill meals can be produced by a variety of methods such as sous-vide, hot-fill, hygienic, aseptic, modified atmosphere, or vacuum packaged in pouches or stores in specially designed containers. These technologies are used widely today in the U.S., Europe, and Australia, and their products could be soups, sauces, casseroles, roasts, entrees, desserts, fresh pasta, salads (including pasta, seafood, and meat), and complete meals for "in-house" use or for supply of external customers, which can be another food service or a retail outlet. There have been a number of studies that demonstrate the effectiveness of PCs in cook–chill foods. In such applications, however, the protective effect of the cultures depends on many parameters such as the size of the inoculum and the state of the culture (fresh, frozen, or dried), the temperature at which inhibition should take place, and the ability of the culture to produce antimicrobials at the particular temperature or the population of spoilage and pathogenic organisms in the food matrix. Thus, the candidates for PCs must grow and produce antimicrobial metabolites at refrigeration temperatures if they are to be used as biocontrol agents (Vescovo et al. 1996). Crandall et al. (1994) demonstrated the failure of *P. pentosaceus* ATCC 43200 to inhibit *Clostridium botulinum*, as it is not able to grow at 10°C. On the other hand, *Lc. lactis* grows and produces nisin at 10°C (Wessels and Huss 1996).

Hugas et al. (1998) utilized a strain of *Lb. sake* (CTC 494) to inhibit the growth of *L. inocua* in meat products (poultry breasts, cooked pork, and raw minced pork) under different conditions of packaging (vacuum package, oxygen permeable films, and modified atmosphere using 20% CO_2 and 80% O_2) at 7°C for 8 days. *L. inocua* was chosen instead of *L. monocytogenes,* because the former had shown to be more resistant to the sakacin produced by the particular strain of *Lb. sake* and also because the two microorganisms show similar behavior in meat packaged under modified atmospheres. In all studied samples, inoculation with *Lb. sake* or its bacteriocin sakacin K led to a bacteriostatic effect over *L. inocua* after 8 days at 7°C. The more pronounced inhibitory effect, however, was observed in the vacuum packaged samples of poultry breasts and cooked pork and in the modified atmosphere packaged samples of raw minced pork, suggesting that the type of meat and the levels of indigenous flora are important parameters for the growth of *Lb. sake* and the production of sakacin K.

Amezquita and Brashears (2000) isolated species of LAB from refrigerated deli meats and hot dogs and screened the isolates for their ability to inhibit *L. monocytogenes* during refrigeration without growing. From the obtained isolates, only one species eliminated completely the growth of *L. monocytogenes* during refrigerated storage. An important observation in the study was that there was no increase in the number of LAB during refrigerated storage.

While much of the research in this area has focused on the inhibition of *L. monocytogenes*, other foodborne pathogens can be inhibited by PCs in meat and other products during refrigerated storage. Holzapfel et al. (1995) suggested psychrotrophic *Clostridium* and *Bacillus* spp. as targets in sous-vide products along with *Listeria*. *Salmonella* species and *E. coli* O175:H7 in ground beef, raw chicken breasts, and other products were either completely eliminated or severely reduced in numbers by *L. lactis* alone or in combinations with other LAB (Brashears et al. 2003a,b; Brashears and Durre 1999).

Despite the obvious advantages of using live microorganisms as protective cultures for biopreservation, their commercial use is hindered by factors such as their sensitivity to heat (which requires replacing of hot-filling with aseptic packaging and protection during processing); the parameters mentioned above, such as the population and the state of the culture, the temperature at which inhibition takes place, and the ability of the culture to produce antimicrobials at that temperature, as well as the population of spoilage and pathogenic organisms in the food product; and their effect on the sensory properties of foods (which may require reformulation of certain foods to mask the presence of a PC; Rodgers et al. 2002).

6.3 Concluding Remarks

Biopreservation of food has a long history of applications. In recent years, however, revolutionary changes in the food industry have led to a growing demand for large-scale biopreservation. This fact has led to intensive research, a large number of publications, and the adoption of new approaches to long known practices. The application of microbial cultures as protective cultures is a novel concept for the food industry that is expected to grow as a solution for the preservation and safety of minimally processed foods. Antimicrobial metabolites of LAB such as weak organic acids, hydrogen peroxide, and the bacteriocin nisin have widespread use. However, newer approaches adopt their use in combinations, because many of these act synergistically and are therefore good candidates for hurdle technology applications in food preservation. Antifungal compounds produced by LAB could also be employed as novel biopreservatives during storage of food, although research in the particular area is far behind the respective on bacteriocins.

So far, the commercial use of new biopreservatives is hindered by many factors. If using cultures, these need to be specifically targeted to ensure an extended storage life, and their survival will depend on the qualities of the food matrix, temperature, storage time, and many other issues. Additional concerns relate to altering product composition to mask flavor problems. For some bacteriocins, interaction with food components can be a limiting factor for use in foods. Economic issues include extra costs of production and premarket studies by regulatory agencies. Finally, there is always concern on how to deal with label declarations and whether consumers will buy products that contain large numbers of live bacteria as something beneficial to them.

REFERENCES

Abee, T., Krockel, L., and Hill, C. 1995. Bacteriocins: Modes of action and potentials in food preservation and control of food poisoning. *Int. J. Food Microbiol.* 28:169–185.

Amezquita, A. and Brashears, M. M. 2000. Competitive inhibition of *Listeria monocytogenes* by lactic acid bacteria in ready-to-eat cooked products. *J. Food Prot.* 65:316–325.

Anastasiadou, S., Papagianni, M., Filiousis, G., Ambrosiadis, I., and Koidis, P. 2008a. Pediocin SA-1, an antimicrobial peptide from *Pediococcus acidilactici* NRRL B5627: Production conditions, purification and characterization. *Bioresour. Technol.* 99:5384–5390.

Anastasiadou, S., Papagianni, M., Filiousis, G., Ambrosiadis, I., and Koidis, P. 2008b. Growth and metabolism of a meat isolated strain of *Pediococcus pentosaceus* in submerged fermentation: Purification, characterization and properties of the produced pediocin SM-1. *Enzyme Microb. Technol.* 43:448–454.

Arqués, J. L., Rodríguez, E., Gaya, P., Medina, M., and Nuñez, M. 2005. Effect of combinations of high-pressure treatment and bacteriocin-producing lactic acid bacteria on the survival of *Listeria monocytogenes* in raw milk cheese. *Int. Dairy J.* 15:893–900.

Arqués, J. L., Rodríguez, E., Nuñez, M., and Medina, M. 2008. Antimicrobial activity of nisin, reuterin, and the lactoperoxidase system on *Listeria monocytogenes* and *Staphylococcus aureus* in cuajada, a semisolid dairy product manufactured in Spain. *J. Dairy Sci.* 91:70–75.

Axelsson, L. 1998. Lactic acid bacteria: Classification and physiology. In: *Lactic Acid Bacteria: Microbiology and Functional Aspects*, 2nd edition, edited by S. Salminen and A. von Wright, pp. 1–72. New York: Marcel Dekker Inc.

Bari, M. L., Ukuku, D. O., Kawasaki, T., Inatsu, Y., Isshiki, K., and Kawamoto, S. 2005. Combined efficacy of nisin and pediocin with sodium lactate, citric acid, phytic acid, and potassium sorbate and EDTA in reducing the *Listeria monocytogenes* population of inoculated fresh-cut produce. *J. Food Prot.* 68:1381–1387.

Bauer, R., Chikindas, M. L., and Dicks, L. M. T. 2005. Purification, partial amino acid sequence and mode of action of pediocin PD-1, a bacteriocin produced by *Pediococcus damnosus* NCFB1832. *Int. J. Food Microbiol.* 101:17–27.

Bearson, S., Bearson, B., and Foster, J. 1997. Acid stress responses in enterobacteria. *FEMS Microbiol. Lett.* 147:173–180.

Bennech, R. O., Kheadr, E. E., Lacroix, C., and Fliss, I. 2002. Antibacterial activities of nisin Z encapsulated in liposomes or produced in situ by mixed culture during cheddar cheese ripening. *Appl. Environ. Microbiol.* 68:5607–5619.

Benz, R., Jung, G., and Sahl, H.-G. 1991. Mechanism of channel-forming lantibiotics in black lipid membranes. In: *Nisin and Novel Lantibiotics*, edited by G. Jung and H.-G. Sahl, pp. 359–372. Leiden: ESCOM.

Bhunia, A. K., Johnson, M. C., and Ray, B. 1988. Purification, characterization and antimicrobial spectrum of a bacteriocin produced by Pediococcus acidilactici. *J. Appl. Bacteriol.* 65:261–268.

Bhunia, A. K., Johnson, M. C., Ray, B., and Kalchayanad, N. 1991. Mode of action of pediocin AcH from *Pediococcus acidilactici* H on sensitive bacterial strains. *J. Appl. Bacteriol.* 70:23–25.

Björck, L. 1978. Antibacterial effect of the lactoperoxidase system on phsychrotrophic bacteria in milk. *J. Dairy Res.* 45:109–118.

Björck, L., Rosen, C., Marshall, V., and Reiter, B. 1975. Antibacterial activity of the lactoperoxidase system in milk against pseudomonads and other Gram-negative bacteria. *Appl. Microbiol.* 30:199–204.

Boggaram, V. and Mannervik, B. 1982. Essential arginine residues in the pyridine nucleotide binding sites of glutathione reductase. *Biochim. Biophys. Acta* 701:119–126.

Borders, C. L. Jr., Saunders, J. E., Blech, D. M., and Frovich, I. 1985. Essentiality of the active-site arginine residue for the normal catalytic activity of Cu,Zn superoxide dismutase. *Biochem. J.* 230:771–776.

Brashears, M. M., Amezquita, A., and Jaroni, D. 2006. Control of foodborne bacterial pathogens in animals and animal products through microbial antagonism. In: *Food Biotechnology*, 2nd edition, edited by K. Shetty, G. Paliyath, A. Pometto, and R. E. Levin, pp. 1359–1389. Boca Raton, FL: CRC Press—Taylor & Francis.

Brashears, M. M. and Durre, W. A. 1999. Inhibitory action of *Lactococcus lactis* towards *Salmonella* spp. and *E. coli* O157:H7 during growth and refrigerated storage. *J. Food Prot.* 62:1336–1340.

Brashears, M. M., Galyean, M. L., Mann, J. E., Killinger-Mann, K., and Loneragan, G. 2003a. Reduction of *Escherichia coli* O157 and improvement in performance in beef feedlot cattle with *Lactobacillus* direct fed microbial. *J. Food Prot.* 66:748–754.

Brashears, M. M., Jaroni, D., and Trimble, J. 2003b. Isolation, selection and characterization of lactic acid bacteria for competitive exclusion product to reduce shedding of *E. coli* O157:H7 in cattle. *J. Food Prot.* 66:355.

Breukink, E., Wiedemann, I., van Kraaij, C., Kuipers, O. P., Sahl, H.-G., and de Kruijff, B. 1999. Use of the cell wall precursor lipid II by a pore-forming peptide antibiotic. *Science* 286:2361–2364.

Brul, S. and Coote, P. 1999. Preservative agents in foods: Mode of action and microbial resistance mechanisms. *Int. J. Food Microbiol.* 50:1–17.

Cherrington, C. A., Hinton, M., Mead, G. C., and Chopra, I. 1991. Organic acids: Chemistry, antibacterial activity and practical applications. *Adv. Microb. Physiol.* 32:87–108.

Chikindas, M. L., Garcia-Garcera, M. J., Driessen, A. M., Lederboer, A. M., Nissen-Meyer, J., Nes, I. F., Abee, T., Konings, W. N., and Venema, G. 1993. Pediocin PA-1, a bacteriocin from *Pediococcus acidilactici* PAC1.0, forms hydrophilic pores in the cytoplasmic membrane of target cells. *Appl. Environ. Microbiol.* 59:3577–3584.

Chung, T. C., Axelsson, L., Lindgren, S. E., and Dobrogosz, W. J. 1989. *In vitro* studies on reuterin synthesis by *Lactobacillus reuteri*. *Microb. Ecol. Health Dis.* 2:137–144.

Claisse, O. and Lonvaud-Funel, A. 2000. Assimilation of glycerol by a strain of *Lactobacillus collinoides* isolated from cider. *Food Microbiol.* 17:513–519.

Cleusix, V., Lacroix, C., Vollenweider, S., Dudoux, M., and Le Blay, G. 2007. Inhibitory activity spectrum of reuterin produced by *Lactobacillus reuteri* against intestinal bacteria. *BMC Microbiol.* 7:101.

Cole, M. B. and Keenan, M. H. 1986. Synergistic effects of weak-acid preservatives and pH on the growth of *Zygosaccharomyces bailii*. *Yeast* 2:93–100.

Condon, S. 1987. Responses of lactic acid bacteria to oxygen. *FEMS Microbiol. Rev.* 46:269–280.

Crandall, A. D., Winkowski, K., and Montville, T. J. 1994. Inability of *Pediococcus pentosaceus* to inhibit *Clostridium botulinum* in sous-vide beef with gravy at 4 and 10°C. *J. Food Prot.* 57:104–107.

Davidson, P. M. 1997. Chemical preservatives and natural antimicrobial compounds. In: *Food Microbiology: Fundamentals and Frontiers*, edited by M.P. Doyle, L.R. Beuchat, and T.J. Montville, pp. 520–556. Washington, DC: ASM Press.

De Vuist, L. and Vandamme, E. J. 1994. *Nisin, a Lantibiotic Produced by Lactococcus lactis subsp. lactis: Properties, Biosynthesis, Fermentation and Application*, edited by L. De Vuist and E.J. Vandamme, pp. 151–222. Glasgow: Blackie Academic Press.

Dutreux, N., Notermans, S., Góngora-Nieto, M. M., Barbosa-Cánovas, G. V., and Swanson, B. G. 2000. Effects of combined exposure of *Micrococcus luteus* to nisin and pulsed field electric fields. *Int. J. Food Microbiol.* 60:147–152.

Dutton, C. J., Haxell, M. A., McArthur, H. A. I., and Wax, R. G. 2002. *Peptide Antibiotics, Discovery, Modes of Action and Applications*. New York: Marcel Dekker.

Eklund, T. 1989. Organic acids and esters. In: *Mechanisms of Action of Food Preservation Procedures*, edited by G. W. Gould, pp. 161–200. New York: Elsevier, pp. 161–200.

El-Ziney, M. G. and Debevere, J. M. 1998. The effect of reuterin on *Listeria monocytogenes* and *Escherichia coli* O157:H7 in milk and cottage cheese. *J. Food Prot.* 61:1275–1280.

El-Ziney, M. G., van den Tempel, T., Debevere, J., and Jakobsen, M. 1999. Application of reuterin produced by *Lactobacillus reuteri* 12002 for meat decontamination and preservation. *J. Food Prot.* 62:257–261.

Fimland, G., Blingsmo, O., Sletten, K., Jung, G., Nes, I. F., and Nissen-Meyer, J. 1996. New biologically active hybrid bacteriocins constructed by combining regions from various pediocin-like bacteriocins: The C-terminal region is important for determining specificity. *Appl. Environ. Microbiol.* 62: 3313–3318.

Food and Agricultural Organization of the United Nations/World Health Organization (FAO/WHO). 2005. Benefits and potential risks of the lactoperoxidase system of raw milk preservation. FAO/WHO Technical Meeting, Rome, November 28–December 2, 2005.

Foster, J. W. 2000. Microbial responses to acid stress. In: *Bacterial Stress Responses*, 1st edition, edited by G. Storge and R. Hengge-Aronis, pp. 99–115. Washington, DC: ASM Press.

Franchi, M. A., Serra, G. E., and Cristiani, M. 2003. The use of biopreservatives in the control of bacterial contaminants of sugarcane alcohol fermentation. *J. Food Sci.* 68:2310–2315.

Gálvez, A., Abriouel, H., López, R. L., and Omar, N. B. 2007. Bacteriocin-based strategies for food preservation. *Int. J. Food Microbiol.* 120:51–70.

Gill, A. O. and Holley, R. A. 2000. Inhibition of bacterial growth on ham and bologna by lysozyme, nisin, and EDTA. *Food Res. Int.* 33:83–90.

Gilliland, S. E. 2003. Lactic acid bacteria as biopreservatives in the food industry. IFT Annual Meeting, July 12–16, 2003, Chicago, IL, 63-2.

Gonzalez, C. F. and Kunka, B. S. 1987. Plasmid-associated bacteriocin production and sucrose fermentation in *Pediococcus acidilactici. Appl. Environ. Microbiol.* 53:2534–2538.

Goodson, M. and Rowbury, R. J. 1989. Habituation to normal lethal acidity by prior growth of *Escherichia coli* at a sublethal acid pH value. *Lett. Appl. Microbiol.* 8:77–79.

Grande, M. J., Lucas, R., Abriouel, H., Valdivia, E., Ben Omar, N., Maqueda, M., Martínez-Bueno, M., Martínez-Cañamero, M., and Gálvez, A. 2006. Inhibition of toxicogenic *Bacillus cereus* in rice-based foods by enterocin AS-48. *Int. J. Food Microbiol.* 106:185–194.

Gravesen, A., Jydegaard Axelsen, A. M., Mendes da Silva, J., Hansen, T. B., and Knøchel, S. 2002a. Frequency of bacteriocin resistance development and associated fitness costs in *Listeria monocytogenes. Appl. Environ. Microbiol.* 68:756–764.

Gravesen, A., Ramnath, M., Rechinger, K. B., Andersen, N., Jansch, L., Hechard, Y., Hastings, J. W., and Knøchel, S. 2002b. High-level resistance to class IIa bacteriocins is associated with one general mechanism in *Listeria monocytogenes. Microbiology* 148:2361–2369.

Green, G., Dicks, L. M. T., Bruggeman, G., Vandamme, E. J., and Chikindas, M. L. 1997. Pediocin PD-1, a bactericidal antimicrobial peptide from *Pediococcus damnosus* NCFB 1832. *J. Appl. Microbiol.* 83:127–132.

Guilfoyle, D. E. and Hirshfield, I. N. 1996. The survival benefit of short-chain organic acids and the inducible arginine and lysine decarboxylase genes for *Escherichia coli. Lett. Appl. Microbiol.* 22:393–396.

Hanlin, M. B., Kalchayanand, N., Ray, P., and Ray, B. 1993. Bacteriocins of lactic acid bacteria in combination have greater antibacterial activity. *J. Food Prot.* 56:252–255.

Hastings, J. W., Sailer, M., Johnson, K., Roy, K. L., Vederas, J. C., and Stiles, M. E. 1991. Characterization of leucocin A-UAL 187 and cloning of the bacteriocin gene from *Leuconostoc gelidurn. J, Bacteriol.* 173:7491–7500.

Héchard, Y., Derijard, D. B., Latellier, F., and Cenatiampo, Y. 1992. Characterization and purification of mesentericin Y105, an anti-*Listeria* bacteriocin from *Leuconostoc mesenteroides. J. Gen. Microbiol.* 138:2725–2731.

Henderson, J. T., Chopko, A. L., and van Wassenaar, P. D. 1992. Purification and primary structure of pediocin PA-1 produced by *Pediococcus acidilactici* PAC-1.0. *Arch. Biochem. Biophys.* 295:5–12.

Hirshfield, I. N., Terzulli, S., and O'Byrne, C. 2003. Weak organic acids: A panoply of effects on bacteria. *Sci. Prog.* 86:245–296.

Holck, A., Axelsson, L., Birkeland, S.-E., Aukrust, T., and Blom, H. 1992. Purification and amino acid sequence of sakacin A, a bacteriocin from *Lactobacillus sake* Lb706. *J. Gen. Microbiol.* 138: 2715–2720.

Holzapfel, W. H., Deisen, R., and Schillinger, U. 1995. Biological preservation of foods with reference to protective cultures, bacteriocins and food-grade enzymes. *Int. J. Food Microbiol.* 24:343–362.

Hugas, M., Pages, F., and Garriga, J. M. 1998. Application of the bacteriocinogenic *Lactobacillus sakei* CTC494 to prevent growth of *Listeria* in fresh and cooked meat products packed with different atmospheres. *Food Microbiol.* 15:639–650.

Hunter, D. R. and Segel, I. H. 1973. Effect of weak acids on amino acid transport by *Penicillium chrysogenum*: Evidence for a proton or charge gradient as the driving force. *J. Bacteriol.* 113:1184–1192.

Jay, J. M. 1982. Antimicrobial properties of diacetyl. *Appl. Environ. Microbiol.* 44:525–532.

Joerger, M. C. and Klaenhammer, T. R. 1986. Characterization and purification of helveticin J and evidence for a chromosomally determined bacteriocin produced by *Lactobacillus helveticus* 481. *J. Bacteriol.* 167:439–446.

Jung, G. 1991. *Lantibiotics: A Survey, Nisin and Novel Lantibiotics*, edited by G. Jung and H.-G. Sahl, pp. 1–34. Leiden: ESCOM Science Publishers.

Juven, B. J. and Pierson, M. D. 1996. Antibacterial effects of hydrogen peroxide and methods for its detection and quantification. *J. Food Prot.* 59:1233–1241.

Kandler, O. 1983. Carbohydrate metabolism in lactic acid bacteria. *Antonie van Leeuwenhoek* 49:202–224.

Kang, D. H. and Fung, D. Y. C. 1999. Effect of diacetyl on controlling *Escherichia coli* O157:H7 and *Salmonella typhimurium* in the presence of starter culture in a laboratory medium and during meat fermentation. *J. Food Prot.* 62:975–979.

Lemoigne, M. 1927. Sur le metabolisme du diacetyle. *C. R. Soc. Biol.* 97:1749–1781.

Lindgren, S. E. and Dobrogosz, W. J. 1990. Antagonistic activities of lactic acid bacteria in food and feed fermentations. *FEMS Microbiol. Rev.* 87:149–164.

Linnet, P. E. and Strominger, J. L. 1973. Additional antibiotic inhibitors of peptidoglycan biosynthesis. *Antimicrob. Agents Chemother.* 4:231–236.

Lupidi, G., Bollettini, M., Venardi, G., Marmocchi, F., and Rotilio, G. 2001. Functional residues on the enzyme active site of glyoxalase I from bovine brain. *Prep. Biochem. Biotechnol.* 31:317–329.

Luthi-Peng, Q., Dileme, F. B., and Puhan, Z. 2002. Effect of glucose on glycerol bioconversion by *Lactobacillus reuteri. Appl. Microbiol. Biotechnol.* 59:289–296.

Magnusson, J. and Schnürer, J. 2001. *Lactobacillus coryniformis* subsp. *coryniformis* strain Si3 produces a broad spectrum proteinaceus antifungal compound. *Appl. Environ. Microbiol.* 67:1–5.

Magnusson, J., Ström, K., Roos, S., Sjøgren, J., and Schnürer, J. 2003. Broad and complex antifungal activity among environmental isolates of lactic acid bacteria. *FEMS Microbiol. Lett.* 219:129–135.

Mollapour, M., Phelan, J. P., Millson, S. H., Piper, P. W., and Cooke, F. T. 2006. Weak acid and alkali stress regulate phosphatidylinositol bisphosphate synthesis in *Saccharomyces cerevisiae. Biochem. J.* 395:73–80.

Montville, T. J. and Chen, Y. 1998. Mechanistic action of pediocin and nisin: Recent progress and unresolved questions. *Appl. Microbiol. Biotechnol.* 50:511–519.

Nieto Lozano, J. C., Nissen-Meyer, J., Sletten, K., Pelaz, C., and Nes, I. F. 1992. Purification and amino acid sequence of a bacteriocin produced by *Pediococcus acidilactici. J. Gen. Microbiol.* 138:1985–1990.

Nilson, T. 1999. Novel enterococcal bacteriocins: Optimization of production, purification, biochemical and genetic characterization. PhD Thesis, Agricultural University of Norway, As, Norway.

Okkers, D. J., Dicks, L. M. T., Silvester, M., Joubert, J. J., and Odendaal, H. J. 1999. Characterization of pentocin TV35b, a bacteriocin-like peptide isolated from *Lactobacillus pentosus* with fungistatic effect on *Candida albicans. J. Appl. Microbiol.* 87:726–734.

Papagianni, M. 2003. Ribosomally synthesized peptides with antimicrobial properties: Biosynthesis, structure, function, and applications. *Biotechnol. Adv.* 21:465–499.

Papagianni, M. and Anastasiadou, S. 2009. Pediocins: The bacteriocins of pediococci. *Microb. Cell Fact.* 8:3.

Papagianni, M. and Papamichael, E. 2010. A class IIa bacteriocin produced by an atypical sausage-isolated strain of *Weissella paramesenteroides* in fermentation systems. Proceedings of the 1st International Conference on Advances in Industrial Microbial Biotechnology, November 3–5, Thessaloniki, Greece, pp. 16–17.

Piard, J. C. and Desmazeaud, M. 1991. Inhibiting factors produced by lactic acid bacteria—Part 1: Oxygen metabolites and catabolism end products. *Lait* 71:525–541.

Pintado, C. M., Ferreira, M. A., and Sousa, I. 2009. Properties of whey protein-based films containing organic acids and nisin to control *Listeria monocytogenes. J. Food Prot.* 72:1891–1896.

Piper, P., Ortiz Calderon, C., Hatzixanthis, K., and Mollapour, M. 2001. Weak acid adaptation: The stress response that confers yeasts with resistance to organic acid food preservatives. *Microbiology* 147:2635–2642.

Quadri, L. E. N., Sailer, M., Roy, K. L., Vederas, J. C., and Stiles, M. E. 1994. Chemical and genetic characterization of bacteriocins produced by *Carnobacterium piscicola* LV17B. *J. Biol. Chem.* 269:12204–12221.

Razavi-Rohani, S. M. and Griffiths, M. W. 1996. Inhibition of spoilage and pathogenic bacteria associated with foods by combinations of antimicrobial agents. *J. Food Saf.* (USA) 16:87–104.

R. E. *Food Biotechnology*, 2nd edition, pp. 1359–1389. Boca Raton, FL: CRC Press—Taylor & Francis.

Reisinger, P., Seidel, H., Tschesche, H., and Hammes, W. P. 1980. The effect of nisin on murein synthesis. *Arch. Microbiol.* 127:187–193.

Ricke, S. C. 2003. Perspectives on the use of organic acids and short chain fatty acids as antimicrobials. *Poult. Sci.* 82:632–639.

Riordan, J. F. 1979. Arginyl residues and anion binding sites in proteins. *Mol. Cell Biochem.* 26:71–92.

Rodgers, S., Kailasapathy, K., Cox, J., and Peiris, P. 2002. Bacteriocin production by protective cultures. *Food Serv. Technol.* 2:59–68.

Rodríguez, E., Tomillo, J., Nuñez, M., and Medina, M. 1997. Combined effect of bacteriocin producing lactic acid bacteria and lactoperoxidase system activation on *Listeria monocytogenes* in refrigerated raw milk. *J. Appl. Microbiol.* 83:389–395.

Roe, A. J., McLaggan, D., Davidson, I., O'Byrne, C., and Booth, I. R. 1998. Perturbation of anion balance during inhibition of growth of *Escherichia coli* by weak acids. *J. Bacteriol.* 180:767–772.

Roe, A. J., O'Byrne, C., McLaggan, D., and Booth, I. R. 2003. Inhibition of *Escherichia coli* growth by acetic acid: A problem with methionine biosynthesis and homocysteine toxicity. *Microbiology* 148:2215–2222.

Rouse, S., Harnett, D., Vaughan, A., and van Sinderen, D. 2008. Lactic acid bacteria with potential to eliminate fungal spoilage in foods. *J. Appl. Microbiol.* 104:915–923.

Ruhr, E. and Sahl, H.-G. 1985. Mode of action of the peptide antibiotic nisin and influence on the membrane potential of whole cells and on artificial membrane vesicles. *Antimicrob. Agents Chemother.* 27:841–845.

Sahl, H.-G. and Bierbaum, G. 1998. Lantibiotics: Biosynthesis and biological activities of uniquely modified peptides from gram-positive bacteria. *Annu. Rev. Microbiol.* 52:41–79.

Sahl, H.-G., Jack, R. W., and Bierbaum, G. 1995. Biosynthesis and biological activities of lantibiotics with unique posttranslational modifications. *Eur. J. Biochem.* 230:827–853.

Sahl, H.-G., Kordel, M., and Benz, R. 1987. Voltage-dependent depolarization of bacterial membranes and artificial lipid bilayers by the peptide antibiotic nisin. *Arch. Microbiol.* 149:120–124.

Salmond, C. V., Kroll, R. G., and Booth, I. R. 1984. The effect of food preservatives on pH homeostasis in *Escherichia coli*. *J. Gen. Microbiol.* 130:2845–2850.

Savino, F., Pelle, E., Palumeri, E., Oggero, R., and Miniero, R. 2007. *Lactobacillus reuteri* (American Type Culture Collection Strain 55730) versus simethicone in the treatment of infant colic: A prospective randomized study. *Pediatrics* 119:e124–e130.

Schaefer, L., Auchtung, T. A., Hermans, K. E., Whitehead, D., Borhan, B., and Britton, R. A. 2010. The antimicrobial compound reuterin (3-hydroxypropionaldehyde) induces oxidative stress via interaction with thiol groups. *Microbiology* 156:1589–1599.

Schnürer, J. and Magnusson, J. 2005. Antifungal lactic acid bacteria as biopreservatives. *Trends Food Sci. Technol.* 16:70–78.

Schutz, H. and Radler, F. 1984. Anaerobic reduction of glycerol to propanediol-1,3, by *L. brevis* and *L. buchneri*. *Syst. Appl. Microbiol.* 5:169–178.

Schved, F., Lalazar, Y., Henis, Y., and Juven, B. J. 1993. Purification, partial characterization and plasmid-linkage of pediocin SJ-1, a bacteriocin produced by *Pediococcus acidilactici*. *J. Appl. Bacteriol.* 74:67–77.

Shiba, T., Wakamiya, T., Fukase, K., Ueki, Y., Teshima, T., and Nishikawa, M. 1991. *Nisin and Novel Lantibiotics. Structure of the Lanthionine Peptide Nisin, Ancovenin and Lanthiopeptin*, edited by G. Jung and H.-G. Sahl, pp. 113–122. Leiden: ESCOM Science Publishers.

Simões, T., Mira, N. P., Fernandes, A. R., and Sá-Correia, I. 2006. The *SPI1* gene, encoding a glycosylphosphatidylinositol-anchored cell wall protein, plays a prominent role in the development of yeast resistance to lipophilic weak-acid food preservatives. *Appl. Environ. Microbiol.* 72:7168–7175.

Sjogren, J. 2005. Bioassay-guided isolation and characterization of antifungal metabolites—Studies of lactic acid bacteria and propionic acid bacteria. PhD Thesis, University of Uppsala, Sweden.

Skaugen, M., Cintas, L. M., and Nes, I. F. 2003. Genetics of bacteriocin production in lactic acid bacteria. In: *Genetics of Lactic Acid Bacteria*, edited by B. J. Wood and P. J. Warner. New York: Kluwer Academic/Plenum Publishers.

Sobrino-López, A. and Martín-Belloso, O. 2008. Use of nisin and other bacteriocins for preservation of dairy products. *Int. Dairy J.* 18:329–343.

Stiles, M. E. 1996. Biopreservation by lactic acid bacteria. *Antonie van Leeuwenhoek* 70:331–345.

Stiles, M. E. and Holzapfel, W. H. 1997. Lactic acid bacteria of foods and their current taxonomy. *Int. J. Food Microbiol.* 36:1–29.

Strässer de Saad, A. M., Pasteris, S. E., and Manca de Nadra, M. C. 1995. Production and stability of pediocin N5p in grape juice medium. *J. Appl. Bacteriol.* 78:473–476.

Stratford, M. and Anslow, P. A. 1998. Evidence that sorbic acid does not inhibit yeast as a classic "weak acid preservative." *Lett. Appl. Microbiol.* 27:203–206.

Talarico, T. L., Casas, I. A., Chung, T. C., and Dobrogosz, W. J. 1988. Production and isolation of reuterin, a growth inhibitor produced by *Lactobacillus reuteri*. *Antimicrob. Agents Chemother.* 32:1854–1858.

Talarico, T. L. and Dobrogosz, W. J. 1989. Chemical characterization of a chemical substance produced by *Lactobacillus reuteri*. *Antimicrob. Agents Chemother.* 33:674–679.

Tichaczek, P. S., Nissen-Meyer, J., Nes, I. F., Vogel, R. F., and Hammes, W. P. 1992. Characterization of the bacteriocin curvacin A from *Lactobacillus curvatus* LTH1174 and sacacin from *L. sake* LTH673. *Syst. Appl. Microbiol.* 15:460–468.

Ukuku, D. O. and Fett, W. F. 2004. Effect of nisin in combination with EDTA, sodium lactate, and potassium sorbate for reducing *Salmonella* on whole and fresh-cut cantaloupe. *J. Food Prot.* 67:2143–2150.

Venema, K., Venema, G., and Kok, J. 1995. Lactococcal bacteriocins: Mode of action and immunity. *Trends Microbiol.* 3:299–304.

Vescovo, M., Torriani, S., Orsi, C., Macchiarolo, F., and Scolari, G. 1996. Application of anti-microbial-producing lactic acid bacteria to control pathogens in ready-to-use vegetables. *J. Appl. Bacteriol.* 81: 113–119.

Weizman, Z., Asli, G., and Alsheikh, A. 2005. Effect of a probiotic infant formula on infections in child care centers: Comparison of two probiotic agents. *Pediatrics* 115:5–9.

Wessels, S. and Huss, H. H. 1996. Suitability of *Lactococcus lactis* subsp. *lactis* ATCC 11454 as a protective culture for lightly preserved fish products. *Food Microbiol.* 13:323–332.

Wiedemann, I., Breukink, E., van Kraaij, C., Kuipers, O., Bierbaum, G., de Kruijff, B., and Sahl, H.-G. 2001. Specific binding of nisin to the peptidoglycan precursor lipid II combines pore formation and the inhibition of cell wall biosynthesis for potent antibiotic activity. *J. Biol. Chem.* 276:1772–1779.

Wood, B. J. B. and Holzapfel, W. H. 1995. *The Genera of Lactic Acid Bacteria*. Glasgow: Blackie Academic and Professional.

Woolford, M. K. 1984a. The antimicrobial spectra of organic compounds with respect to their potential as hay preservatives. *Grass Forage Sci.* 39:75–79.

Woolford, M. K. 1984b. The antimicrobial spectra of some salts of organic acids and glutaraldehyde in respect to their potential as silage additives. *Grass Forage Sci.* 39:53–57.

Wu, C. W., Yin, L. J., and Jiang, S. T. 2004. Purification and characterization of bacteriocin from *Pediococcus pentosaceus* ACCEL. *J. Agric. Food Chem.* 52:1146–1151.

Young, K. M. and Foegeding, P. M. 1993. Acetic, lactic and citric acids and pH inhibition of *Listeria monocytogenes* Scott A and the effect on intracellular pH. *J. Appl. Bacteriol.* 74:515–520.

Zapico, P., Medina, M., Gaya, P., and Nunez, M. 1998. Synergistic effect of nisin and the lactoperoxidase system on *Listeria monocytogenes* in skim milk. *Int. J. Food Microbiol.* 40:35–42.

7

Exopolysaccharides from Lactic Acid Bacteria and Bifidobacteria

Patricia Ruas-Madiedo, Borja Sánchez, Claudio Hidalgo-Cantabrana,
Abelardo Margolles, and Andrew Laws

CONTENTS

7.1 Introduction

Most bacterial cells are covered by an envelope of similar architecture consisting of a cytoplasmic membrane, a cell wall, and, if present, other external structures such as the outer membrane in Gram-negative bacteria or layers of polysaccharides or proteins, which could be present either in Gram-negative or Gram-positive bacteria. Particularly, the cell wall of a Gram-positive envelope consists of a phospholipid bilayer membrane, in which some proteins are embedded, surrounded by a thick layer of peptidoglycan (murein), which is a structural polysaccharide consisting of alternating β-$(1\rightarrow4)$-linked N-acetyl-D-glucosamine and N-acetyl-muramic acid residues cross-linked by peptide side chains (Figure 7.1). Several secreted proteins associated with peptidoglycan are present, as well as other carbohydrate structures, namely, lipoteichoic acids, teichoic acids, and polysaccharides. The latter components comprise the capsular polysaccharides (CPS), which remain attached to the peptidoglycan forming a capsule like in Gram-negative bacteria, and the slime exopolysaccharides (EPS), which are secreted into the environment (Holts et al. 2009). In addition to the structural function, these outer cellular molecules act as a barrier to protect bacteria against adverse environmental conditions and could also be sensors to communicate with other biotic and abiotic components of the ecosystems they inhabit.

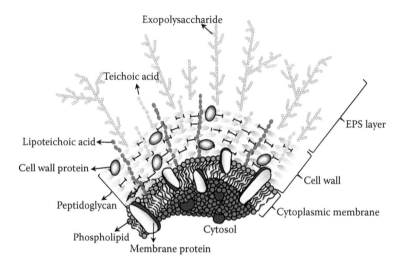

FIGURE 7.1 Schematic representation of the Gram-positive cell envelope. The cytoplasmic membrane is formed by a phospholipid bilayer, with proteins embedded, which is covered by a cell wall built by a structural polysaccharide (peptidoglycan). Several proteins and carbohydrates (teichoic and lipoteichoic acids) are associated with peptidoglycan, and surrounding this wall, layers of exopolysaccharide are present.

Foods constitute complex and rich niches supporting many microorganisms, and in contrast, some of the microorganisms are themselves of special relevance for food manufacture such as the case of two Gram-positive bacteria groups, namely, lactic acid bacteria (LAB) and bifidobacteria. LAB were empirically used during ancient times for preservation of raw food materials in spontaneous fermentations. Currently, the development of food technologies has provided the concept of functional starters, which are defined as those able to "contribute to food safety and/or offer one or more organoleptic, technological, nutritional, or health advantages" (Leroy and De Vuyst 2004). Thus, a great variety of fermented foods with distinguishable sensorial and flavor attributes derived from the activity of specific LAB starters are currently available. Nowadays, there is consumer demand for foods that improve health—the so-called functional foods. One of these foods, with the highest market value, is that containing probiotics, which are defined by the World Health Organization/Food and Agriculture Organization of the United Nations (WHO/FAO) in 2002 as "live microorganisms which when administered in adequate amounts confer a health benefit on the host" (WHO/FAO 2006). The genera most commonly used as probiotics for foods in human consumption are *Lactobacillus*, belonging to the LAB group, and *Bifidobacterium* (Margolles et al. 2009). However, only a few specific strains have health-promoting characteristics, and in some cases, the positive effect could be attributable to some of their surface components as is the case of EPS. This book chapter collects background information about the EPS produced by LAB and bifidobacteria and their role in food production, with special focus on dairy products, as well as the current methodology used for the detection of EPS-producing strains and polymer characterization.

7.1.1 Lactic Acid Bacteria and Bifidobacteria

As previously stated, LAB and bifidobacteria are two bacterial groups currently used in food processing and are included in the formulation of a variety of products. Additionally, some species of these bacteria also share common natural habitats, such as mucosa of animals, being members of the intestinal microbiota in humans. However, they belong to different Gram-positive taxonomic phyla. LAB are non-spore-forming, cocci, and rod-shaped bacteria, having a G + C content lower than 50%, and in general, they are catalase-negative, nonmotile, nonpigmented, and oxygen-tolerant anaerobes. The main end products from glucose catabolism are lactic acid (homofermentative species) or lactic and acetic acid, ethanol, and CO_2 (heterofermentative species). The natural habitats are animal-related mucosa and secretions (e.g., milk) and nutritive-rich plants; thus, they could be naturally present in many foods. Taxonomy classifies LAB as belonging to phylum *Fimicutes*, class *Bacilli*, enclosing 13 genera—*Aerococcus, Alloiococcus, Carnobacterium, Enterococcus, Lactobacillus, Lactococcus, Leuconostoc, Oenococcus, Pediococcus, Streptococcus, Tetragenococcus, Vagococcus,* and *Weissella* (Axelsson 2004; Pot 2008).

The genus *Bifidobacterium* is constituted by non-spore-forming and nonmotile rods displaying different shapes, with the most typical ones being bifurcated or with "bifid" morphology. They are strict anaerobes and ferment hexoses by an exclusive pathway whose key enzyme is the fructose-6-phosphate phosphoketolase, with acetic and lactic acid being the main end products. They naturally inhabit the oral cavity and gastrointestinal tract of animals, and some species are isolated from human milk, but bifidobacteria can also be present in sewage, indicating fecal contamination. They are also included in functional (probiotic) foods (Margolles et al. 2009). Members of the genus *Bifidobacterium*, currently with more than 30 species, have high G + C content (>50%) and, thus, are phylogenetically separated from LAB. They are included in the phylum *Actinobacteria*, class *Actinobacteria*, belonging to the family *Bifidobacteriaceae* together with the genera *Aeriscardovia, Alloscardovia, Gardnerella, Metascardovia, Parascardovia,* and *Scardovia* (Felis and Dellaglio 2007; Ribosomal Database Project, http://rdp.cme.msu.edu).

Finally, for the application in the food industry, it is worth mentioning that some species of these bacterial groups are 'generally recognized as safe' microorganisms according to the Food and Drug Administration (FDA 2009). Similarly, due to the long history of safe use in human consumption, they also have the 'qualified presumption of safety' (QPS) status given by the European Food Safety Authority (EFSA). In the case of the last organization, five *Bifidobacterium* species and several LAB species (33 *Lactobacillus*, 3 *Leuconostoc*, 3 *Pediococcus*, *Lactococcus lactis*, and *Streptococcus thermophilus*) are included in the QPS list (EFSA 2007), meaning that strains unambiguously assigned to a QPS group would be free from further safety assessment to be included in foods and feeds, unless additional information is required by EFSA.

7.1.2 Exopolysaccharides from Lactic Acid Bacteria and Bifidobacteria

EPS from LAB and bifidobacteria are exocellular carbohydrate polymers that, according to their chemical composition and mode of synthesis, could be classified into two categories—homopolysaccharides (HoPS) and heteropolysaccharides (HePS).

As their name suggests, HoPS are polymers built by a single monosaccharide type, either D-glucose (glucans) or D-fructose (fructans), in pyranose ring conformation, with residues varying in glycosidic linkage and branching degree. To date, production of HoPS has been described in different genera of LAB, but as far as we know, no HoPS-producing bifidobacteria have been reported (Ruas-Madiedo et al. 2009a,b). There are two glucan types, α- and β-glucan, and the second one is formed by β-D-Glc*p*-(1→3) repetitions (Dueñas-Chasco et al. 1998). Generalizing, α-glucans are subdivided into dextran [α-D-Glc*p*-(1→6)], mutan [α-D-Glc*p*-(1→3)], alternan [α-D-Glc*p*-(1→6)/α-D-Glc*p*-(1→3)], and reuteran [α-D-Glc*p*-(1→4)], with Glc*p* denoting glucose pyranose ring conformation. Regarding fructans, two types have been described: levan, linked by β-D-Fru*p*-(2→6), and inulin-like, linked by β-D-Fru*p*-(2→1) osidic bonds, with Fru*p* denoting fructose pyranose ring conformation (Korakli and Vogel 2006; Monsan et al. 2001; van Hijum et al. 2006).

Most HoPS are synthesized from sucrose through the action of extracellular enzymes called glycansucrases, which are members of the glycoside hydrolase (GH) family ("Carbohydrate Active Enzymes" classification available at http://www.cazy.org). Glucansucrases (GS; family GH70) and fructansucrases (FS; family GH68) are involved in the synthesis of α-glucans and β-fructans, respectively (van Hijum et al. 2006). On the other hand, β-glucans are synthesized by LAB strains isolated from fermented beverages (with primary alcoholic fermentation, such as wine and cider) by means of glucosyltransferases bound to the cell membrane; the mechanism involved in this process still remains unclear (Werning et al. 2006; Dols-Lafargue et al. 2008).

HePS polymers are formed by several repeated units, those described to date varying from two to eight monomers (De Vuyst et al. 2001; Laws and Marshall 2001; Ruas-Madiedo et al. 2002a, 2009a). These units are composed of monosaccharides (the most abundant being D-glucose, D-galactose, and L-rhamnose), derivative monosaccharides (currently N-acetyl-D-glucosamine and N-acetyl-D-galactosamine, uronic acids), and organic and inorganic substituents (mainly acetyl, glycerol, glycerol-phosphate, and phosphate). Additionally, HePS are synthesized in a different way from HoPS, because a number of genes, which are organized in *eps* clusters, are involved in their anabolic pathway. The structural organization of *eps* clusters has been described for several LAB strains, mainly members of the genera *Lactococcus*, *Lactobacillus*, and *Streptococcus*, most of them isolated from fermented foods. The location of *eps* clusters could be either in the chromosome or in plasmids, the last case being the reason for the instability of EPS production detected in some strains used for food (dairy) manufacture (Ruas-Madiedo et al. 2009a). The first description of genes directing the synthesis of HePS in LAB was achieved in the strain *S. thermophilus* Sfi6 (Stingele et al. 1996), although almost simultaneously, the molecular characterization of the *eps* cluster in *Lc. lactis* subsp. *cremoris* NIZO B40 was available (van Kranenburg et al. 1997). Nowadays, the structural and functional organization of several *eps* clusters from LAB strains has been published (see the review of Ruas-Madiedo et al. 2009a). Based on homology analysis and also on some functional studies, it seems that *eps* clusters in LAB are organized as an operon, which could be transcribed in a single messenger ribonucleic acid, and they share common structural features. Four functional regions could be identified in LAB *eps* clusters: (1) genes coding for enzymes involved in regulation (currently located at the 5′ end); (2) genes involved in chain length determination–polymerization (located at the 5′ end); (3) genes related to the biosynthesis and assembly of the repeated units, which code for glycosyltransferases (GTF; located in the core region); and (4) genes related to export–polymerization (located at the 3′ end). The best characterized genes are those coding GTF, enzymes that catalyze the transfer of sugar moieties from activated donor molecules (several nucleotide-sugars) to specific acceptor molecules, thereby forming a glycosidic bond (Boels et al. 2001). Special attention has been paid to the priming-GTF (generally annotated as undecaprenyl-phosphate-glycosyl-1-phosphate transferase), the enzyme responsible for the initiation of the synthesis of the HePS repeated unit, which transfers a sugar-1-phosphate residue to the membrane-bound lipid carrier C55. In fact, priming-GTF shows highly conserved functional homology in Gram-negative and Gram-positive bacteria, especially in the C-terminal region. The amino acid sequence of this region in several LAB defines a conserved region highly similar to families Bactransf (bacterial sugar transferases of Pfam, accession number PF02397) and WcaJ (sugar transferases involved in the lipopolysaccharide synthesis, accession number COG2148; Ruas-Madiedo et al. 2009a). Regarding *eps* clusters present in bifidobacteria, its description is more recent and is mainly based on the available genomic sequences (Schell et al. 2002; Ventura et al. 2009; Lee and O'Sullivan, 2010). However, up until now, the functional characterization of these bifidobacterial clusters has not been undertaken yet.

The pathway of HePS synthesis in LAB is not clearly understood, but a mechanism has been proposed based on functional studies and homology comparison with HePS-producing pathogenic Gram-positives, such as *Streptococcus pneumoniae*, and the Gram-negative *Xanthomonas campestris*. The working model proposes that the assembly of the repeated unit would occur in the cytoplasm; afterward, this unit would be translocated through the membrane and polymerization of the repeated units would take place outside the cell (Boels et al. 2001; Broadbent et al. 2003; van Kranenburg et al. 1999). This mechanism is schematically depicted in Figure 7.2 for the polymer produced by *Lc. lactis* subsp. *cremoris* NIZO B40, which is built from a pentasaccharide repeated unit (van Casteren et al. 1998). B40 HePS synthesis is initiated by the activity of the priming-GTF (codified by *epsD*), which transfers the UDP-glucose to the lipid carrier C55. Furthermore, glucose, galactose, rhamnose, and galactose-phosphate moieties (coming from their nucleotide-sugar precursors) are sequentially linked by means of GTF activities codified by *epsE*/*epsF*, *epsG*, *epsH*, and *epsJ* genes, respectively. Then, this pentasaccharide unit is predicted to be translocated across the membrane by the activity of *epsK* gene coding for a "flippase-like" protein. Extracellular polymerization of repeated units will be carried out by a polymerase (codified by *epsI*), and several enzymes (*epsA*, *epsB*) will determine the chain length of EPS B40. Finally, the lipid carrier will be retranslocated and intracellularly dephosphorylated, ready for a new cycle (Boels et al. 2001). This complex mechanism, which shares intermediates used in the central carbon metabolism and the lipid carrier C55 involved in the cell wall peptidoglycan synthesis, is still far from being totally understood in LAB (Ruas-Madiedo et al. 2009a).

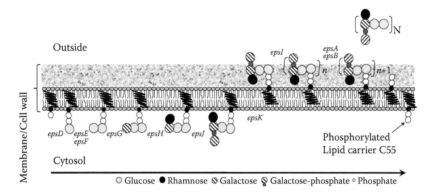

FIGURE 7.2   Mechanism proposed for the synthesis of the HePS produced by *Lc. lactis* subsp. *cremoris* B40 (modified from Boels, I. C., van Kranenburg, R., Hugenholtz, J., Kleerebezem, M., and de Vos, W. M. 2001. *Int. Dairy J.* 11:723–732). Assembly of the repeated unit begins with the attachment of the first nucleotide diphosphorylated-sugar (UDP-glucose in this example) to the undecaprenyl-phosphate lipid carrier C55, reaction catalyzed by the priming-glycosyltransferase (p-GTF) codified by *epsD*. Afterward, several GTF (*epsE/epsF*, *epsG*, *epsH*, and *epsJ*) catalyze the sequential addition of glucose, galactose, rhamnose, and galactose-phosphate to build the pentasaccharide unit of EPS B40. Then, this repeated unit is translocated across the membrane by means of a "flippase-like" protein (*epsK*), and extracellularly, the addition of the new unit to the polymer under construction is catalyzed by a polymerase (*epsI*). Several enzymes (*epsA* and *epsB*) are involved in the chain length determination. Finally, the lipid carrier C55 is retranslocated through the membrane and intracellularly dephosphorylated.

7.1.3 Functional Properties of EPS: Special Focus on Dairy Products

In the past decades, research on EPS has been focused on the ability of EPS-producing strains to improve the physicochemical properties of fermented foods, mainly dairy products. Nowadays, EPS are receiving special interest due to their potential as health-promoting effectors and also because it is becoming more evident of the involvement of these molecules in the protection and colonization of different ecosystems by producing strains (Figure 7.3). Recent reviews about the ecological and physiological functionality of LAB and bifidobacteria can be found in the work of Ruas-Madiedo et al. (2008, 2009b).

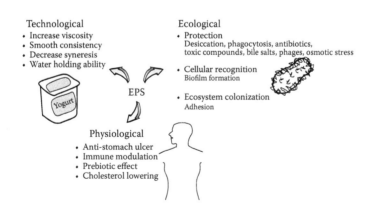

FIGURE 7.3   Functionality of exopolysaccharides (EPS) and EPS-producing strains. EPS produced by LAB are used in the dairy industry due to their ability to act as natural biothickeners and stabilizers. The role that EPS play for the producing strain is not clear, but they seem to be involved in protection against adverse factors and environmental colonization. Additionally, several health-promoting properties have been recently attributed to EPS synthesized by LAB and bifidobacteria.

Regarding their technological application, EPS-producing strains are used in the dairy industry because of their ability to improve the rheological, textural, and mouthfeel properties of fermented products (Welman and Maddox 2003). Several dairy fermented products include EPS-producing LAB in their formulation, such as yogurt, kefir, Scandinavian fermented milk (e.g., viili), and cheese (e.g., Mozzarella and Feta; Ruas-Madiedo et al. 2009b). However, the presence of EPS-producing strains is considered negative in the manufacture of fermented beverages (beer, wine, or cider) and in "ready to eat" processed meat, because in both cases, they are responsible for their spoilage. In the dairy industry, EPS synthesized *in situ* during milk fermentations act as natural biothickeners, emulsifiers, stabilizers, or gelling agents (Laws and Marshall 2001). However, their use as additives is limited due to the low yield reported for LAB, which varies accordingly to strain, polymer type, and culture conditions. In general, EPS are produced in concentrations ranging from 25 to 600 mg/L of LAB cultures, and HoPS are synthesized in higher amounts than HePS. However, this EPS yield is not comparable to that of polymers used for application in the food industry (more than 10 g/L) such as xanthan, gellan, or dextran (Ruas-Madiedo et al. 2009b). Only a few species of lactobacilli are reported to produce high amounts of polymer under certain optimized conditions. This is the case for *Lb. reuteri* Lb121, which can be synthesized around 10 g/L of two glucan and fructan HoPS types (van Geel-Schutten et al. 1999), or the HePS produced at a concentration of 2 g/L by *Lb. rhamnosus* RW-9595M (Bergmaier et al. 2005). Data of bifidobacterial EPS yield are scarce; those obtained in our group indicate that a small amount of HePS is formed (from 0.2 to 5.2 mg/L of milk–whey; unpublished data) by intestinal bifidobacteria strains growing in pasteurized milk (supplemented with 0.2% yeast extract), and these were not able to significantly increase the viscosity of the fermented milk (Salazar et al. 2009).

The amount of EPS synthesized in milk is directly related to their ability to increase the viscosity of the fermented product, but other parameters play a pivotal role, because at the same EPS concentration, the viscosifying capability is variable according to the EPS type. These viscosity-intensifying parameters are, among others, the size [or molar mass (MM)] and the primary repeated-unit structure of the EPS (Laws and Marshall 2001; Ruas-Madiedo et al. 2002a). The MM reported for LAB ranged from 10^4 up to 10^6 Da, that of HoPS being higher, in general, than HePS ones (Ruas-Madiedo et al. 2009a). The MM of both fructan and glucan HoPS is around 10^6 Da, although lower MM have been reported as well (van Hijum et al. 2006). The MM values reported for HePS in the literature vary from 4×10^4 to 9×10^6 Da for *S. thermophilus*, from 1×10^5 to 2×10^6 Da for *Lc. lactis* subsp. *cremoris*, from 2×10^4 to 1.4×10^6 Da for *Lactobacillus*, and from 9.4×10^3 to 1.5×10^6 Da for *Bifidobacterium* strains (Ruas-Madiedo et al. 2009a). An interesting feature about the MM of HePS is the simultaneous presence of more than one MM-size fraction in the same polymer, being a common characteristic in LAB and bifidobacteria (Mozzi et al. 2006; Salazar et al. 2009). This multisize MM distribution could be due to the production of two EPS fractions with different MMs but with the same composition, as is the case in *S. thermophilus* LY03 (Degeest and De Vuyst 1999), or due to the simultaneous production of two HePS varying in size and composition, as, for example, the EPS synthesized by *Lactobacillus pentosus* LPS26 (Sánchez et al. 2006).

The relationship of the physical EPS parameters with the intrinsic viscosity (η_0) of the polymer in aqueous solution is described in Section 7.3. However, it is clear that these parameters have a direct influence on the viscosity of the fermented products. In general, EPS having high MM are those able to enhance the fermented milk viscosity. This correlation has been found for EPS produced by *S. thermophilus* (Faber et al. 1998), *Lc. lactis* subsp. *cremoris* (Ruas-Madiedo et al. 2002b; Tuinier et al. 2001), and *Lb. delbrueckii* subsp. *bulgaricus* (Petry et al. 2003), among other works. Summarizing, the effect of EPS on the rheological and textural properties of dairy products depends not only on the amount of EPS synthesized during fermentation but also on the physicochemical properties of the polymers. Interactions between EPS and other components of the food matrix, such as the casein network in fermented milks, are also important factors (Hassan 2008).

7.2 Methods for Screening and Detection of EPS-Producing Strains

In recent years, a renewed interest in research about EPS-producing LAB and bifidobacteria has arisen; several reviews on this topic are thereby available and have been reported in the work of Ruas-Madiedo and de los Reyes-Gavilán (2005). This review is a compilation of phenotypic methods for EPS working, in which several protocols for the screening, identification, and characterization of these polymers were collected. This book chapter summarizes these phenotypic methods and introduces those related with molecular approaches, which become more relevant with the advanced knowledge on bacterial genomes and the implementation of new powerful bioinformatic tools.

7.2.1 Phenotypic Tools

Several terms have been used to describe the EPS-producing phenotype in bacteria; most common are the "mucoid" and "ropy" descriptors (Hassan 2008). Both have been indistinctly used but do not mean exactly the same, because not all mucoid EPS are ropy (Ruas-Madiedo and de los Reyes-Gavilán 2005). Mucoid and ropy characters can easily be distinguished by means of macroscopic observation of bacterial cultures (Figure 7.4a). In the surface of agar plates, both mucoid and ropy strains have a glistening and smooth aspect, but ropy strains are the only ones forming a long filament when a loop is introduced into the colony (Vescovo et al. 1989). Similarly, the ropy strand can also be visualized in a liquid culture (Vedamuthu and Neville 1986). Additionally, mucoid and ropy EPS-producing bacteria can be detected in liquid cultures in which cells are not easily harvested at low centrifugal forces for short periods (typically 2000 × *g*, 5 min, Figure 7.4a). The inability to obtain a compact pellet is due to the presence of the polymer near the bacterial surface, and this can be visualized under a microscope (Figure 7.4b). Using optical microscopy, the EPS are detected as a transparent halo surrounding bacteria; under higher magnification, using different electron microscopy techniques, the presence of EPS is denoted as a matrix in which bacteria are embedded (Figure 7.4b) or that can be separated from bacteria (Hassan et al. 2003; Ruas-Madiedo et al. 2009a,c). Finally, it is worth indicating that ropy strains are able to confer, in general, better technological properties to fermented products (Rawson and Marshall 1997) and that a given strain could express both mucoid and ropy phenotypes depending on culturing conditions (Dierksen et al. 1997).

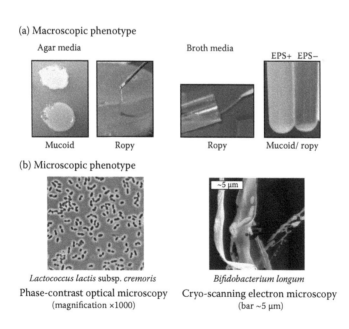

(a) Macroscopic phenotype

Agar media
Broth media

EPS+ EPS−

Mucoid   Ropy   Ropy   Mucoid/ ropy

(b) Microscopic phenotype

~5 µm

Lactococcus lactis subsp. *cremoris*
Phase-contrast optical microscopy
(magnification ×1000)

Bifidobacterium longum
Cryo-scanning electron microscopy
(bar ~5 µm)

FIGURE 7.4 Detection of the EPS-producing phenotype by means of macroscopic (a) and microscopic (b) tools. EPS-producing strains growing in agar media present two macroscopic phenotypes, mucoid and ropy, with the last one characterized by the formation of a long filament that could also be detected in cultures grown in broth media. Additionally, EPS-producing strains are detected in liquid cultures due to the difficulty of harvesting cells under low centrifugal forces (example showed 2000 × *g*, 5 min). EPS production can be visualized as well under an optical or electron microscope; in the first case, EPS form a shining halo surrounding the bacteria (contrasted in black), and in the second, EPS are detected as a matrix (gray area) in which bacteria (white rods) are embedded.

7.2.2 Molecular Tools

Several genes coding for enzymes involved in bacterial EPS biosynthesis, which have been described in previous sections, were used as the targets of different polymerase chain reaction (PCR) protocols aimed at detecting potential EPS-producing strains (Table 7.1). Regarding EPS production in LAB, Dols-Lafargue et al. (2008) demonstrated that a single glucosyltransferase controls the synthesis of a ropy polysaccharide (β-glucan) in *Oenococcus oeni* and the corresponding gene was found to be present on the chromosomes of several *O. oeni* strains isolated from wines. Furthermore, a PCR assay has been developed for the detection of β-glucan-producing LAB based on primers designed to amplify the *glucosyltransferase* gene of a *Pediococcus parvulus* strain isolated from ropy cider. The method was valid for the specific detection of the gene in β-glucan-producing strains of *Pediococcus*, *Lactobacillus*, and *Oenococcus* (Werning et al. 2006). On the other hand, by the use of PCR primers designed to detect FS and GS genes, Tieking et al. were able to detect several β-fructan- and α-glucan-producing lactobacilli (Tieking et al. 2003, 2005). However, due to the wide biodiversity among lactobacilli species and strains, as well as the different kinds of HoPS that they are able to produce, the use of a unique PCR method to target *eps* genes in this genus has not been established. With regard to the synthesis of HePS, as previously stated, the C-terminus domain of the priming-GTF (the enzyme that initiates the synthesis of the repeated unit) is highly conserved among Gram-positive and Gram-negative bacteria. This *priming-gtf* constitutes a promising target for PCR techniques aimed at correlating the EPS production phenotype with the presence of specific gene markers. In this regard, a primer design strategy to analyze the presence of the *priming-gtf* gene in LAB strains was able to detect this gene in *S. thermophilus*, *Lb. casei*, and *Lb. rhamnosus* strains, either in EPS+ or EPS− producers (Provencher et al. 2003). Regarding molecular detection of *eps* genes in bifidobacteria, at the time being, specific primers based on bifidobacterial genes have not been designed, although *eps* clusters related to HePS synthesis have been described in the available genomes (see Section 7.2.4). Additionally, some HePS polymers have been characterized (see Section 7.3.3); however, no glucan and fructan HoPS types have been described among *Bifidobacterium* to date.

The currently available information about *eps* genes is enormous and shows a high variability even among strains belonging to the same species, thus making the design of universal primers for PCR detection techniques of each polymer type difficult, if not impossible. In addition, no correlation between the presence of *eps* genes and EPS production was always established. A combination of phenotypic and genetic tools is thereby the desirable approach in order to identify and select potential EPS-producing strains.

7.2.3 Bioinformatic Tools

During the last few years, an increasing number of bacterial genomic sequences have been released and became freely available for researchers. This has facilitated, to a great extent, the identification of genes or whole operons potentially involved in EPS metabolism, by simply comparing the target sequence against public databases. In order to perform this process, different developers have created a vast array of computer programs, either as free distributions or implemented within web servers. In this section, we will give an overview of several of these bioinformatic tools, as well as the rationale behind their use in the search for genes involved in specific metabolic pathways, for example, in EPS biosynthesis (Table 7.2).

The National Center for Biotechnology Information (NCBI; http://www.ncbi.nlm.nih.gov) is a resource for molecular biology information, funded by the U.S. government. The NCBI provides access to many public databases and other references, including thousands of complete/incomplete microbial genomes and millions of nucleotide/amino acid sequences. For instance, we can use the e-PCR application (http://www.ncbi.nlm.nih.gov/sutils/e-pcr) for designing appropriate primers, allowing the amplification of a known gene, for instance, a GTF-encoding gene. Among the other bioinformatic tools offered at NCBI (a complete list can be obtained at http://www.ncbi.nlm.nih.gov/guide/all), a basic local alignment search tool (BLAST) is noteworthy. BLAST is the basic system for identifying homologs to a known gene or protein. BLAST can be used to query the whole NCBI database with a single protein or gene sequence, returning

TABLE 7.1

Primer Pairs Used to Detect *eps* Genes in Lactic Acid Bacteria and Bifidobacteria

Primer Pairs	Sequence (5′–3′)	EPS Type	Target Gene (Species Codon Usage)	Reference
FQ1	CTGTGTCTGGCGTTTCTGTAGG	β-Glucan	*Glucosyltransferase (Oenococcus oeni)*	Dols-Lafargue et al. 2008
FQ2	GCCACCGCATAGGGTATTTGTC			
PF1	GATTGTAATAAAATAAAAAGACCC	β-Glucan	*Glucosyltransferase (O. oeni)*	Dols-Lafargue et al. 2008
PF8	CATATGATAACACGCAGGGC			
GTFF	CGGTAATGAAGCGTTTCCTG	β-Glucan	*Glucosyltransferase (Pediococcus parvulus)*	Werning et al. 2006
GTFR	GCTAGTACGGTAGACTTG			
dexwobV	GAYGCIGTIGAYAAYGTI	α-Glucan	*Glucansucrase* (Lactobacilli)	Tieking et al. 2005
dexwobR	YTTGRAARTTISWRAAICC			
DexreuV	GTGAAGGTAACTATGTTG	α-Glucan	*Glucansucrase* (Lactobacilli)	Tieking et al. 2005
DexreuR	ATCCGCATTAAAGAATGG			
LevV	GA(CT)GTITGGGA(CT)(AT)(GC)ITGGC	β-Fructan	*Fructansucrase* (Lactobacilli)	Tieking et al. 2003
LevR	TCIT(CT)(CT)TC(AG)TCI(GC)(AT)I(AG)(AC)CAT			
G-Lr-Bact-b-F-20	TTGCCAAATATTGGAGGGGT	HePS	*Priming-glycosyltransferase (Lb. rhamnosus)*	Provencher et al. 2003
G-Lr-Bact-b-R-20	TTTAATAGGCTCCAGTTGGA			
G-*-Bact-a-F-36	TCATTTTATTCGTAAAACCTCAATTGAYGARYTNCC	HePS	*Priming-glycosyltransferase* (hybrid, *Lb. rhamnosus*)	Provencher et al. 2003
G-*-Bact-a-R-27	AATATTATTACGACCTSWNAYYTGCCA			

TABLE 7.2

List of Some Common Web Servers Used in Bioinformatic Analysis

Web Server	Address	Bioinformatic Tools Implemented[a]
National Center for Biotechnology Information	http://www.ncbi.nlm.nih.gov	e-PCR, BLAST, CD search
European Bioinformatics Institute	http://www.ebi.ac.uk	Sequence similarity and analysis, pattern and motif searches, structure analysis, Ensembl Genomes, ClustalW2
The Sanger Institute	http://www.sanger.ac.uk	BLAST, Gene DB, Pfam, Artemis
Swiss Institute of Bioinformatics	http://www.isb-sib.ch	UniProtKB/Swiss-Prot, PROSITE, PdbViewer
Expasy Server	http://www.expasy.org	Tools for the analysis of a protein sequence regarding its primary, secondary, tertiary, and quaternary structure (signal peptide detection, subcellular localization, hydrophobicity, ...)
SoftBerry	http://www.softberry.com	Free online applications for automatic genome annotation, alignment of whole genomes, sequence comparison, alignment of sequences, analysis of gene regulation, search for functional motifs or protein 3D structure analysis and modeling, among others
Center for Biological Sequence Analysis	http://www.cbs.dtu.dk	Generation of protein subsets, FeatureExtract, FeatureMap3D
The Sequence Manipulation Suite	http://www.bioinformatics.org/sms2	Several applications for generating, formatting, and analyzing short DNA and protein sequences
Kyoto Encyclopedia of Genes and Genomes	http://www.genome.jp/kegg	Gene annotation and pathway association
Enzyme Database—BRENDA	http://www.brenda-enzymes.org	Enzyme information

[a] See text for details.

similar sequences and calculating the statistical significance of matches. Another helpful application is the Conserved Domain Search Service (CD Search; http://www.ncbi.nlm.nih.gov/Structure/cdd/wrpsb .cgi), which allows for searching specific domains in the protein (or translated gene) of interest, such as sugar-binding domains and carrier lipid-binding domains, and it could be of special interest in the analysis of bacterial EPS biosynthesis.

The European Bioinformatics Institute (EBI) constitutes Europe's primary nucleotide sequence resource and is part of the European Molecular Biology Laboratory (EMBL). Single deoxyribonucleic acid (DNA) and ribonucleic acid sequences, genome projects, and environmental metagenomes are submitted to EBI by individual researchers or genome-sequencing projects. One of the tools implemented at EBI and other molecular biology resources is the ClustalW2 (http://www.ebi.ac.uk/Tools/msa/clustal w2/). This software is a multiple-sequence alignment program for DNA or proteins, and it can be either accessed online or downloaded as stand-alone software (ftp://ftp.ebi.ac.uk/pub/software/clustalw2). The ClustalW2 output can be used either for comparing the similarity of the same gene/protein domain in different strains or for inferring phylogenetic relationships among different bacterial taxa. As with NCBI, the EBI site allows sequence similarity and analysis, pattern and motif searches, and structure analysis against its databases. Furthermore, the Ensembl Genomes database is also implemented. This database is a joint project between EMBL–EBI and the Sanger Institute and aims at producing automatic annotation on selected prokaryotic genomes.

The Sanger Institute (http://www.sanger.ac.uk) is a genome research institute primarily funded by the Wellcome Trust. In this server, several bacterial genomes and metagenomes of complex communities can be queried. Some of the bioinformatic tools implemented in this server are BLAST, Gene DB, which is a database containing annotation information of a growing number of organisms, and Pfam, which provides a classification of proteins into families (helpful for homolog search among different bacterial species). Concerning the different applications that can be downloaded at the Sanger Institute site,

Artemis (http://www.sanger.ac.uk/resources/software/artemis), a free, widely used, genome viewer and annotation tool, is worth mentioning.

The Swiss Institute of Bioinformatics (SIB; http://www.isb-sib.ch) is an academic, nonprofit institute that coordinates research and education in bioinformatics throughout Switzerland, providing bioinformatic services to the international research community. Several databases are maintained in this site (http://www.isb-sib.ch/services/databases.html), among which are UniProtKB/Swiss-Prot (a non-redundant protein database that provides high level of annotation) and PROSITE (a database of protein families and domains). Concerning the free software that can be downloaded at the site, DeepView (Swiss-PdbViewer) can be highlighted. This software is a widely used application for tridimensional protein visualization, modeling, and analysis of several proteins at the same time. This program is linked to SWISS-MODEL and the Protein Databank (PDB), an automated homology modeling server at the SIB that allows the generation of protein models from a given amino acid sequence (the creation of a free account at the server is mandatory). With this program and this server, the creation of a three-dimensional (3-D) template from an amino acid sequence is greatly facilitated.

The Expert Protein Analysis System (ExPASy; http://www.expasy.org) proteomics server is part of the SIB and is dedicated, among others, to the analysis of protein sequences. In this server, several tools are implemented for the analysis of the primary, secondary, tertiary, and quaternary structure of proteins (from computing molecular masses and isoelectric points or signal peptide detection to defining protein structures based on crystallography data or predicting the subcellular localization of a given protein). Interestingly, the Expasy Server contains several free websites for PDF file generation that can be used as an input for DeepView (Swiss-PdbViewer).

The SoftBerry (http://www.softberry.com) server contains a huge number of free online applications that can be used mainly for automatic genome annotation, visualization of annotations, visualization of bacterial genome comparisons and annotations, alignment of whole genomes, sequence comparison, alignment of sequences, analysis of gene regulation, search for functional motifs, conserved motifs and regulatory motifs, protein 3-D structure analysis and modeling, etc.

The Center for Biological Sequence (CBS) analysis (http://www.cbs.dtu.uk) is an advanced site for bioinformatics and conducts research on cellular information processing, relations between sequence composition, and the content and order of macromolecular structure, among others. In addition, several genomes can be queried, and interesting information such as secreted proteins, membrane proteins, or surface-associated protein subsets can be retrieved. Several free bioinformatic tools are available for direct downloading, such as FeatureExtract [extraction of sequences and annotation from GenBank (NCBI) format files] and FeatureMap3D (combines protein sequence–based information with structural data from the PDB).

The Sequence Manipulation Suite 2 (SMS2; http://bioinformatics.org/sms2) comprises a collection of JavaScript programs for the manipulation of short DNA and protein sequences. Among others, the tools implemented in SMS allow DNA translation, reverse complement generation, primer design, *in silico* PCR, and generation of restriction maps.

The Kyoto Encyclopedia of Genes and Genomes (KEGG, http://www.genome.jp/kegg) is a bioinformatic resource for linking genomes to life and the environment. KEGG databases mainly integrate the known molecular pathways and metabolic routes to the available information concerning genome annotation. Databases can be queried by enzyme name, product, metabolic pathway, and, noteworthy, genes annotated for a given route, and the selected microorganism are highlighted. The latter may be very valuable for identifying genes organized in an operon, such as the example of *eps* clusters reviewed in this chapter.

To end this general compilation of freely available bioinformatic tools, BRENDA (http://www.brenda-enzymes.org) is another enzyme database that provides information on enzyme activity, nomenclature, reactions, and its presence and experimental evidence in different organisms. It is linked to other public databases, such as KEGG and PubMed (NCBI).

7.2.4 Methods' Application

In this section, we will describe the application of methods described above to the detection of EPS-producing lactobacilli and bifidobacteria isolated from a complex microbiological ecosystem, such as the human intestinal microbiota, and a comparative genome analysis of *eps* clusters from *Bifidobacterium* species.

7.2.4.1 Screening, Detection, and Identification of EPS-Producing Strains

Intestinal lactobacilli and bifidobacteria were isolated by Delgado et al. (2006a) as follows—fecal and mucosal samples of healthy adults were collected and homogenized in a reducing medium under anaerobic conditions; afterward, serial dilutions of homogenized samples were plated on the surface of specific media (MRS agar with 0.25% L-cysteine = MRSC), and plates were incubated in selective conditions (37°C, anaerobiosis) to favor the growth of cultivable lactobacilli and bifidobacteria. Single colonies were picked up, and their aspect is visualized under an optical microscope to select those with desirable morphology. Stocks of these selected isolates were kept at −80°C in MRSC + 20% glycerol, and they were identified at the species level by using PCR techniques, with primers targeting different regions of the ribosomal ribonucleic acid gene (Delgado et al. 2006b). A collection of 362 *Lactobacillus* and *Bifidobacterium* isolates from human intestinal microbiota were obtained, which was screened for the detection of EPS-producing strains. Isolates were grown for 72 h in anaerobic conditions (80% N_2, 10% CO_2, 10% H_2) in the surface of MRSC agar supplemented with 2% glucose, fructose, lactose, or sucrose, added separately, and the phenotypic detection of EPS+ strains was achieved by searching for colonies with mucoid or ropy phenotype (Figure 7.4). A number of 60 putative EPS+ strains (35 bifidobacteria and 25 lactobacilli) were detected, most of them having a mucoid aspect (92%) and only 8% presenting the ropy strand (Ruas-Madiedo et al. 2007). Among the 60 putative EPS+ strains, a genetic screening was carried out to detect genes involved in the synthesis of HoPS and HePS. Primers (Table 7.1) employed for the detection of GS, FS, and glucosyltransferase genes (Tieking et al. 2003, 2005; Werning et al. 2006) failed in the PCR amplification; this indicates that, under conditions employed in this study, no HoPS-producing intestinal lactobacilli and bifidobacteria strains were found. For the detection of HePS, the *priming-gtf* was used as a target gene, and the hybrid primers (Table 7.1) designed by Provencher et al. (2003) were used. Positive amplification was obtained for 25 human isolates (11 bifidobacteria and 14 lactobacilli) and 3 reference strains of *Bifidobacterium longum*, whereas none of the human isolates or reference *Bifidobacterium animalis* strains amplified *priming-gtf* using this pair of primers (Ruas-Madiedo et al. 2007). In order to demonstrate correlation among EPS production and the presence of *eps* genes, 21 out of 60 putative EPS+ and 2 reference strains (*B. longum* NB667 and *B. animalis* IPLA-R1) were selected to isolate their polymers. EPS-like fractions were isolated from the 23 strains grown in the surface of MRSC agar plates (see procedure in Section 7.3.1); however, only 12 of them presented positive amplification for the *priming-gtf* gene (Ruas-Madiedo et al. 2007); thus, half of the EPS-producing strains pass unnoticed in the genetic detection. This result could be due to the sequence heterogeneity of the priming-GTF among bacterial species, because the primers used were designed based on the codon usage of *Lb. rhamnosus*. However, it also suggests that most intestinal lactobacilli and bifidobacteria were HePS producers. This was further proved by analyzing the chemical composition of the 21 purified polymers from intestinal origin (Salazar et al. 2009). This work underlines the relevance of using both phenotypic and genetic tools as complementary techniques for the detection of EPS-producing strains.

7.2.4.2 Comparative Analysis of Putative EPS Clusters in Bifidobacteria

As indicated in previous sections, nowadays, the functionality of any *eps* cluster in the genus *Bifidobacterium* has not been described. However, the genomes available in the GenBank database make it possible and interesting to achieve a comparative study between the putative *eps* clusters of different bifidobacteria species (Lee and O'Sullivan 2010). Fifteen bifidobacterial genomes were publicly available at the NCBI web page at the end of 2010, from strains belonging to the species *B. longum* subsp. *infantis*, *B. longum* subsp. *longum*, *B. animalis* subsp. *lactis*, *B. adolescentis*, *B. Bifidum*, and *B. dentium*, and 13 of them appear to possess several genes involved in HePS biosynthesis, with only two *B. bifidum* strains lacking them. This suggests that potential EPS-producing *Bifidobacterium* strains are very ubiquitous.

In this book section, a comparative *in silico* analysis was done for searching the putative *eps* clusters of six strains belonging to different species—*B. adolescentis* ATCC15703 (GenBank accession number AP009256), *B. dentium* Bd1 (CP001750; Ventura et al. 2009), *B. longum* subsp. *longum* NCC2705 (AE014295; Schell et al. 2002), *B. longum* subsp. *infantis* ATCC15697 (CP001095; Sela et al. 2008), *B. animalis* subsp. *lactis* DSM10140 (CP001606; Barrangou et al. 2009), and *B. bifidum* S17 (CP002220; Zhurina et al. 2011). The main criterion used for initially defining the *eps* clusters was the location of the gene coding for the priming-GTF. In *B. longum* NCC2705, whose genome was the first publicly available, there are two putative priming-GTF annotated as "undecaprenyl-phosphate sugar phosphotransferase" (RfbP, GenBank accession number NP_695455) and "galactosyl-transferase" (CpsD, NP_695447). Because the *eps* clusters have an operon-like structure in most LAB, this study in bifidobacteria continued searching for closely placed gene-coding proteins putatively implicated in the synthesis of the HePS (GTF, export, polymerization, and chain length determination). Most bifidobacteria strains analyzed showed a putative *eps* cluster structure. However, in *B. bifidum* S17, only a protein having homology with the RfbP priming-GTF was detected. Physical maps of the *eps* clusters from five out of six strains under study, showing the putative functions of the proteins coded by these *eps* clusters, are thereby collected in Figure 7.5.

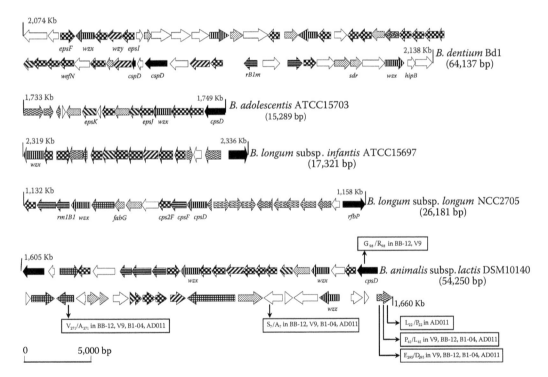

FIGURE 7.5 Physical maps of the *eps* clusters detected in five species of *Bifidobacterium* genus, whose genomes are publicly available (see text). The size of each *eps* cluster is indicated between brackets. Putative functions of protein coded by these genes are presented in different textures: ■▶, priming-GTF (glycosyltransferase); ▨, GTF; ▨, polymerization–chain length determination; ▥, export–polymerization; ▧, polysaccharide biosynthesis; ⇉, rhamnose biosynthesis; ▨, involved in DNA mobility (transposase/IS and integrase); ▦, membrane protein; ▨, other known functions (acyl-synthetase, acyl-transferase, acetyl-transferase, pyruvyl-transferase, dehydrogenase/phosphorylase/reductase and pyridoxine biosynthesis); ▭, unknown function. An additional comparison among *eps* clusters of five strains from species *B. animalis* subsp. *lactis* was carried out. Modified amino acids are included in a frame—the first corresponds with that present in strain DSM10140, used as template; the second is that changing in the other strains (AD011, BB-12, B1-04, and V9); the subscript indicates the position of modified amino acid. The genome locations of the genes clusters are indicated as kilobases.

The size of the *eps* clusters was highly variable and strain dependent; the biggest was that of *B. dentium* Bd1 (64,137 bp, 53 genes), and the smallest was that of *B. adolescentis* (15,289 bp, 15 genes). The size variability could be related to the different role of EPS in the ecological niche where these strains inhabit; *B. dentium* is localized in the oral cavity, and *B. adolescentis* is localized in the colon, or it may be due to the lack of complete definition of the *eps* clusters. Additionally, from the inspection of these bifidobacterial *eps* clusters, it was not possible to describe a common functional organization, as previously indicated for LAB *eps* clusters. In general, these bifidobacterial *eps* clusters lack genes involved in regulation, except for the transcription regulator *hipB* found at the 3' end in *B. dentium* Bd1. This function could be coded by proteins with an unknown function in the defined clusters, or these regulator genes may remain still unallocated upstream or downstream these putative *eps* clusters. Similarly, proteins implicated in polymerization–chain length determination were not localized in *B. adolescentis* ATCC15703 and *B. longum* subsp. *longum* NCC2705 clusters.

It is remarkable to locate the presence of more than one priming-GTF in several of these bifidobacterial *eps* clusters. This is the case of *B. animalis* subsp. *lactis* DSM10140, *B. longum* subsp. *longum* NCC2705, and *B. dentium* Bd1. The amino acid sequences of priming-GTF CpsD have, on the average, a 45% identity among them when compared using the ClustalW2 server (Table 7.2), and the second priming-GTF (RfbP in *B. longum* NCC2705) also showed high interspecies homology. These results showed a high degree of conservation for this protein among bifidobacteria, as well as among bifidobacteria and LAB, although priming-GTF from bifidobacteria were clearly separated from LAB in the phylogenetic tree constructed by Ruas-Madiedo et al. (2007) using the amino acid sequences of 56 priming-GTF from LAB and bifidobacteria. Genes coding for different GTF were detected in the five genomes under study, and they were the most abundant genes in all *eps* cluster structures. Comparison of these GTF, using the Blastp tool (Table 7.2), showed a variable percentage of homology among GTF and strains. GTF from *B. dentium* showed, in general, the lowest homology percentage with respect to the other species reflecting, probably, its different ecological habitat adaptation in which EPS production could play a relevant role (e.g., in dental plaque biofilm formation). In addition, all genomes harbored at least one gene coding for proteins involved in EPS export–polymerization. Interestingly, the five *eps* clusters presented one or more genes with high homology to *wzx*, the gene coding for a "flippase" in *Escherichia coli*. This protein seems to translocate the undecaprenyl-linked O-antigen subunit across the membrane to show the carbohydrate moiety in the periplasmic side of this Gram-negative bacterium (Whitfield and Roberts 1999). This protein has been found as well in LAB *eps* clusters (Ruas-Madiedo et al. 2009a), thus suggesting a similar function in Gram-positive and Gram-negative bacteria. Another common feature between bifidobacterial and LAB *eps* clusters was the presence of insertion sequences and transposases with the exception, again, of *B. dentium* Bd1. It has been proposed that these mobile elements could be responsible for the horizontal transfer of *eps* genes among different genera and could also be the reason for the instability of the EPS production phenotype in some strains with *eps* clusters located into their chromosomes (Ruas-Madiedo et al. 2009a). Although no mechanism of HePS synthesis has been proposed for bifidobacteria, the high homology of their priming-GTF with those of LAB, as well as the presence of most of the *eps* genes described in LAB clusters, makes it tempting to hypothesize that HePS synthesis in bifidobacteria could be similar to that described for LAB in Section 7.1.2.

However, apart from the above-described common characteristics shared by bifidobacterial and LAB *eps* clusters, there are some genes that are not usually present in LAB EPS biosynthesis loci. Into this category fit genes coding for proteins involved in the synthesis of rhamnose; as far as we know, only *Lb. rhamnosus, Lb. gasseri,* and *Lb. johnsonii* strains harbor these genes (Péant et al. 2005; Berger et al. 2007). However, rhamnose-precursor biosynthesis genes were present in three out of five bifidobacterial *eps* clusters analyzed, showing a high (80%) degree of protein homology (Blastp alignment) among them. In this context, it is worth mentioning that high rhamnose content was detected in EPS polymers isolated from intestinal strains with respect to those from food origin (52% versus 28%, respectively). This suggests a putative role of rhamnose in the interaction of EPS or the EPS-producing strain with the intestinal environment (Salazar et al. 2009). Finally, it is remarkable that in three out of the five bifidobacterial *eps* clusters, genes coding for proteins annotated as "membrane proteins" with unknown function were located.

The limited number of bifidobacteria strains used in this comparative analysis does not allow defining a functional *eps* cluster structure similar to that described for LAB. In addition, we do not know

TABLE 7.3

Structures of Heteropolysaccharide Repeat Units Produced by Several Strains of the Genus *Bifidobacterium*

Strain	Structure	Reference
B. longum JBL05[a]	(see structure 1 below)	Kohno et al. 2009
B. bifidum BIM B-465[b]	(see structure 2 below)	Zdorovenko et al. 2009
B. infantis ATCC15697 (PS2)[b]	(see structure 3 below)	Tone-Shimokawa et al. 1996
B. catenulatum YIT4016	(see structure 4 below)	Nagaoka et al. 1996
B. longum YIT4028[b]	(see structure 5 below)	Nagaoka et al. 1995
B. breve YIT4010[b]	(see structure 6 below)	Habu et al. 1987

Structure 1 (*B. longum* JBL05):

```
β-D-Glcp
1
↓
6
→4)-α-D-Galp-(1→4)-α-D-Galp-(1→4)-β-D-Glcp-(1→3)-α-D-Galp-(1→3)-β-L-Rhap-(1→
                                                                  4
                                                                  ↑
                                                                  1
                                                              β-D-Galp
```

Structure 2 (*B. bifidum* BIM B-465):

```
→6)-α-D-Glcp-(1→3)-β-D-Galf-(1→3)-α-D-Glcp-(1→2)-β-D-Galf-(1→3)-α-D-Galp-(1→3)-α-D-Galp-(1→
                  6
                  ↑
                  1
              α-D-Glcp
```

Structure 3 (*B. infantis* ATCC15697 (PS2)):

```
→3)-β-D-Galf-(1→3)-α-D-Galp-(1→
               6              6
               ↑              ↑
               1              1
           β-D-Glcp       β-D-Glcp
        (Absent in 10%)  (Absent in 70%)
```

Structure 4 (*B. catenulatum* YIT4016):

```
→6)-β-D-Galf-(1→5)-β-D-Galf-(1→
                6
                ↑
                1
            α-D-Galp
```

Structure 5 (*B. longum* YIT4028):

```
→2)-α-L-Rhap-(1→3)-α-D-Galp-(1→2)-α-L-Rhap-(1→3)-α-D-Galp-(1→
                6
                ↑
                1
            α-D-Galp
```

Structure 6 (*B. breve* YIT4010):

```
→3)-β-D-Glcp-(1→3)-β-D-Galp-(1→4)-α-D-Galp-(1→2)-α-D-Glcp-(1→
                6
                ↑
                1
            α-D-Glcp
```

[a] Extracellular polysaccharide.
[b] Cell wall polysaccharide.

whether *eps* clusters may be characteristic of a strain or are representative of species. To determine the intraspecies variability of bifidobacterial *eps* clusters, an additional *in silico* analysis was done using the five genomes available from species *B. animalis* subsp. *lactis*—DSM10140 (CP001606; Barrangou et al. 2009), AD011 (CP001213; Kim et al. 2009), BB-12 (CP001853; Garrigues et al. 2010), B1-04 (CP001515; Barrangou et al. 2009), and V9 (CP001892; Sun et al. 2010). Multiple alignment was done with the nucleotide sequence of these *eps* clusters using ClusalW2 (Table 7.2). When differences in nucleotides were found, sequences were translated to amino acids and were submitted to a new alignment. The variations detected among the five strains were represented in the *eps* cluster map of *B. animalis* subsp. *lactis* DSM10140 (Figure 7.5), indicating the amino acid changes with respect to DSM10140, its position, and the strain in which it has changed. The degree of homology between the five *eps* clusters of this species was very high, and only few amino acid changes were found in four genes. One of the most modified genes codes for a transposase protein is IS204/IS1001/IS1096/IS1165 (GenBank accession number YP_002970391), in which three amino acids varied among the five strains; this is not surprising in this mobile element, which could easily gain or lose nucleotides in each transposition. Remarkably, the priming-GTF CpsD (YP_002970370), a protein of 511 amino acids, presented a glycine (G) residue in strains DSM10140, AD011, and B1-04, whereas it was changed by arginine (R) in strains BB-12 and V9. However, we do not know whether this amino acid modification alters the functionality of this protein and then affects EPS production. The high genetic homology of *eps* clusters detected among the five strains of *B. animalis* subsp. *lactis* is not surprising, given that this species shows a scarce interstrain genetic variability (Briczinski et al. 2009). It is necessary to extend this study by analyzing strains of other bifidobacterial species before making any conclusion about intraspecies *eps* cluster's homology. In LAB, a genetic divergence in *eps* clusters is found among species closely related as, for example, those of lactobacilli species from *Lactobacillus acidophilus* group (Berger et al. 2007). A similar situation is also often found among *eps* clusters of strains from the same species or even subspecies, as is the case of *S. thermophilus* (Broadbent et al. 2003) and *Lc. lactis* subsp. *cremoris* (Dabour and LaPointe 2005; Knoshaug et al. 2007). However, in the case of four *Lb. rhamnosus* strains, the nucleotide sequence homology of their *eps* clusters is highly similar (Péant et al. 2005), as has occurred with the five *B. animalis* subsp. *lactis* strains used in the *in silico* analysis carried out in this review. The genetic variability that could be found in LAB and bifidobacterial *eps* clusters reflects the vast array of chemical structures reported mainly for EPS synthesized by LAB (Ruas-Madiedo et al. 2009a).

To end, it is necessary to remark that from the 10 strains used in these *in silico* studies, only the isolation and structural characterization of the EPS polymer produced by strain *B. longum* subsp. *infantis* ATCC15697 was achieved (Tone-Shimokawa et al. 1996; see Table 7.3). The EPS production in the other strains still remains to be detected; thus, it is not known if the bifidobacterial clusters described here are functional or what the conditions that trigger the expression of EPS production genes are. This way, we have shown that two strains of *B. animalis* subsp. *lactis* (IPLA4549 and 4549dOx), which under standard laboratory conditions do not produce EPS, are able to induce EPS synthesis in the presence of bile salts (Ruas-Madiedo et al. 2009c). Given that bile salts are present in the gastrointestinal tract, the production of EPS in the natural niche that bifidobacteria inhabit still remains to be elucidated.

7.3 Methods for Physicochemical Characterization of EPS

In order to understand the relationship between the EPS produced by LAB and bifidobacteria and their technological properties for food applications as well as their biological activity contributing to human health, it is necessary to determine the physicochemical characteristics of the polymers. The physicochemical properties, especially their solution behavior, will be influenced by their repeated unit structure and by their polysaccharide chain length. This part of the review will cover technical aspects of the EPS isolation and purification procedures that impact on repeated unit structure and polysaccharide chain length [expressed as the weight-averaged molecular mass (MM)], will discuss the analytical techniques used for the measurement of weight-averaged MM and for determining the structure of the repeated unit, and will conclude by covering work that links chain length and repeated unit structure to EPS solution properties.

7.3.1 Isolation and Quantification

There is a very large volume of work that describes the influence of culture media and fermentation conditions on the production and the purity of EPS isolated from LAB; this work has been the subject of a number of reviews (Laws et al. 2001; Ricciardi and Clementi 2000; Ruas-Madiedo and de los Reyes-Gavilán 2005; De Vuyst et al. 2001). There is a much smaller volume of literature describing the isolation of polysaccharides from bifidobacteria. Seeing a number of similarities between the structures reported for cell wall components and extracellular material, it should be noted that the procedures used to isolate the different classes of polysaccharides are similar, and sometimes, it is not absolutely clear if cell wall material, EPS, or a combination of the two has been isolated. It is also important to ensure that the isolation procedure itself does not influence the repeated unit structure and chain length. For example, a number of EPS contain sugar substituents, including pyruvate, phosphate, and acetyl groups that are moderately labile and are easily lost during purification. There is also the possibility that specific sugar linkages may be hydrolyzed in the acidic media used to isolate EPS, for example, glycosidic bonds from the anomeric carbon of furanose sugars; small amounts of midchain fission will markedly influence weight-averaged MM.

Methods used to isolate LAB-EPS are well established and have been developed from a protocol first introduced by Cerning et al. (1986). In summary, the isolation procedure involves four main steps: (1) the removal of cells by centrifugation; (2) the removal of proteins by either treatment with proteases or by the addition of trichloroacetic acid (TCA); (3) the precipitation of carbohydrates by the addition of chilled ethanol; and (4) the removal of small sugars/molecules by dialysis. When measuring the physicochemical characteristics of EPS samples, the isolation method is frequently supplemented with a final purification by chromatography. A comprehensive review of the methods used for the isolation of LAB–EPS has been published (Ruas-Madiedo and de los Reyes-Gavilán 2005). The literature describing the isolation of polysaccharides from bifidobacteria includes procedures for isolation of cell wall polysaccharides (Habu et al. 1987; Nagaoka et al. 1988, 1995, 1996; Tone-Shimokawa et al. 1996; Zdorovenko et al. 2009), procedures for isolating extracellular polysaccharides (Kohno et al. 2009), and procedures for isolating EPS (Alp and Aslim 2010; Amrouche et al. 2006; Audy et al. 2010; Chen et al. 2009; Liu et al. 2009; Roberts et al. 1995; Ruas-Madiedo et al. 2006; Wu et al. 2010). Authors specifically reporting EPS isolation from bifidobacteria (Bifido-EPS) use one of the two procedures to recover the EPS depending on whether the material is being recovered from broth cultures (Alp and Aslim 2010) or from plate cultures (Roberts et al. 1995; Ruas-Madiedo et al. 2006). The methods reported for the isolation of Bifido-EPS from broth cultures are adapted versions of those used for the isolation of LAB-EPS and involve various combinations of steps to remove cell material, to remove proteins, to precipitate the crude EPS, and then to purify the EPS. A typical procedure that is used when EPS is being recovered from cultures grown on plates (Ruas-Madiedo et al. 2006) involves harvesting the cell biomass from the plate with water, and the EPS is released by gently stirring overnight at room temperature with an equal volume of sodium hydroxide. Cell material is then removed by centrifugation, carbohydrates are precipitated by adding two volumes of chilled ethanol, and the EPS is isolated by centrifugation. The crude EPS is then resuspended in ultrapure water, and small molecule impurities are removed by dialysis. This method of EPS isolation from agar plates is used to avoid the coisolation of polysaccharides (currently glucomannans; Vaningelgem et al. 2004) present in the rich media needed to grow bifidobacteria.

As previously indicated, a number of LAB and bifidobacteria generate more than one EPS; frequently, these have the same repeat unit structure but have different MMs, although occasionally, EPS with different repeat unit structures have been isolated. When more than one EPS is produced, it is necessary to separate them before their physicochemical properties can be determined, and this has been accomplished using anion exchange chromatography (Kohno et al. 2009; Roberts et al. 1995), gel permeation chromatography (Gorret et al. 2003), or a combination of the two techniques.

The presence of impurities in EPS samples will influence the accurate recording of colligative properties that are, by definition, concentration dependent. In reporting the purity of EPS samples, most authors include the determination of the carbohydrate content, but in more detailed work, a measure of the levels of nucleic acid and protein impurities (Higashimura et al. 2000) and, occasionally, the moisture content is reported (Chadha 2010). The most popular method for the quantification of the carbohydrate content of both LAB– and Bifido–EPS is the phenol–sulphuric method described by Dubois et al. (1956); however, this method quantifies the total carbohydrate and does not differentiate between the EPS and small sugar impurities. Alternative methods for the estimation of the total carbohydrate content of the EPS involve either the measurement of reducing sugar content and the subtraction of this value from the total sugar measurement (Miller 1959) or the application of size exclusion chromatography (SEC) and the measurement of peak areas using refractive index (RI) detection (Salazar et al. 2009; Tuinier et al. 1999a). Protein content has been determined in LAB– and Bifido–EPS samples using either the Bradford method (Bradford 1976) or using the BCA/Pierce protein assay (Smith et al. 1985). The nucleic acid content of LAB–EPS has been determined using analytical SEC with UV detection (Chadha 2010), and the same author has reported the determination of water content by Karl–Fisher titration.

7.3.2 Molar Mass Determination

Before an EPS structure can be considered to be fully characterized, it is necessary to determine information about the MM of the material, to identify the composition and absolute configuration of the monomers, to determine the identity of any substituents and their point of attachment to monomers, and, finally, to determine the linkage pattern of the monomers.

The chain length (or weight-averaged MM) of LAB– and Bifido–EPS is most frequently determined using analytical SEC, eluting with aqueous sodium nitrate (Tuinier et al. 1999a; Higashimura et al. 2000) and using RI detection to monitor the retention time of EPS fractions. The weight-averaged MM of a given EPS can then be determined by the comparison of its retention time with either those of dextran standards (De Vuyst et al. 2003) or, less frequently, those of pullulan standards (Amrouche et al. 2006; Nakajima et al. 1992). Weight- and number-averaged MMs can also be obtained using a combination of SEC with a concentration-dependent differential refractometer and a multiangle laser light scattering detector (SEC-MALLS; Wyatt et al. 1988). Churns has reviewed the application of this technique to carbohydrate analysis (Churms 1996). Calculation of the weight- and number-averaged MMs using SEC-MALLS requires knowledge of the RI increment for the EPS samples, and these have been reported as 0.145 mL/g (Theisen et al. 1999) and 0.135 mL/g for *Lc. lactis* subsp. *cremoris* B40 (Tuinier et al. 1999a). A higher value of 0.41 mL/g has been reported for a neutral polysaccharide from *L. delbrueckii* (Goh et al. 2005) and a value of 0.2 mL/g for the EPS from *L. acidophilus* 5e2 (Chadha 2010). The measurement of the MM of a LAB-EPS has been recently reported using nuclear magnetic resonance (NMR) spectroscopy and translational diffusion experiments (Nordmark et al. 2005). Static and dynamic MALLS can also be used to determine information about the conformation adopted by EPS samples in solution (Goh et al. 2005; Higashimura et al. 2000; Ruas-Madiedo et al. 2002b; Tuinier et al. 1999a, 2001). Static MALLS has been used to determine the polydispersity and the number-averaged radius of gyration (R_g) for the native EPS from *Lc. lactis* subsp. *cremoris* strain NIZO B891 ($R_g = 70 \pm 2$ nm) and the deacetylated EPS from this strain ($R_g = 54 \pm 2$ nm) and from the same EPS from which, additional, side-chain galactose groups have been removed ($R_g = 43 \pm 1$ nm; Tuinier et al. 2001). Based on the exponent of the relationship between the number-averaged R_g and the weight-averaged MM of fractions eluting form a SEC column ($R_g = M^\upsilon$, where $\upsilon = 0.57$), the same authors were able to show that the EPS B40, synthesized by *Lc. lactis* subsp. *cremoris* NIZO B40 strain, adopts a random coil conformation in solution (Tuinier et al. 1999a). The relationship between R_g and MM has been used to estimate the Kuhn length parameter, which can be used as an estimation of the EPS chain flexibility (Tuinier et al. 2001; Ruas-Madiedo et al. 2002b). The same authors have used dynamic MALLS to monitor the diffusion of EPS B40 samples from which the hydrodynamic radius ($R_H = 86 \pm 4$ nm) was calculated; the relationship between R_g and R_H confirmed the random coil conformation (Tuinier et al. 1999a).

7.3.3 Chemical Composition and Primary Structure

A number of different methods are available for determining the monomer composition of EPS samples; their application to the analysis of the monomer composition of LAB-EPS has been reviewed (Ruas-Madiedo and de los Reyes-Gavilán 2005). Very similar methods have been applied for the determination of the monomer composition of Bifido–EPS. The first step in monomer analysis is the acid-catalyzed hydrolysis of the glycosidic links, and for LAB– and Bifido–EPS, this has most frequently been achieved by heating with 2M trifluoroacetic acid (TFA; Salazar et al. 2009) for between 1 and 12 h (Gorret et al. 2003) at 120°C. Other acids have been used, including mixtures of formic acid and TFA (Kohno et al. 2009), 2M HCl, and 1M/2M H_2SO_4 (Roberts et al. 1995). For the determination of neutral monosaccharides, the released monomers are converted into their alditol acetates (Sawardeker et al. 1965) or their aldononitrile acetates (Roberts et al. 1995), which are then analyzed by GC–MS. Another simpler method for the analysis of EPS monomers uses high-pressure anion exchange chromatography with pulsed amperometric detection, and this has been applied to the analysis of the monomer composition of cell wall polysaccharides of bifidobacteria (Hosono et al. 1997) and the EPS from *B. animalis* strains (Leivers et al. 2011).

A number of methods are also available for the determination of the absolute configurations of the monomers of LAB– and Bifido–EPS (Gerwig et al. 1978), and the most popular rely on the preparation of mixtures of anomers of acetylated alkyl-glycosides prepared from the reaction of the monomers with a single enantiomer of a chiral alcohol, (*S*)-(+)-2-butanol (Gerwig et al. 1978), or (*S*)-(+)-2-octanol (Leontein et al. 1978) and analysis by gas liquid chromatography. A number of authors have reported the use of enzyme assays using D-glucose and D-galactose oxidase in combination with peroxidase to determine the absolute configuration of these two sugars (Tone-Shimokawa et al. 1996). A more recent method relies on the direct treatment of monomers (2 mg) with (*S*)-(+)-2-methylbutyric anhydride (100 μL) and pyridine (100 μL) for 4 h at 120°C and the analysis of the ^{1}H NMR of the (*S*)-(+)-2-methylbutyrate derivatives (Säwén et al. 2010).

The linkage pattern of the monomers and the identification of the location of substituents in both LAB– and Bifido–EPS are determined using a combination of 'methylation' analysis (Stellner et al. 1973; Sweet et al. 1975) and NMR spectroscopy. NMR spectroscopy has been routinely used to characterize EPS structures; Duus et al. (2000) have reviewed methods in the field and list previous reviews published between 1992 and 1999. Many of these have collected reference spectroscopic data, chemical shifts, and coupling constants and are extremely useful for assigning spectra of complex carbohydrates. A review describing the application of two-dimensional (2-D) NMR in determining the primary structure of LAB–EPS has been published (Leeflang et al. 2000), and the same methods have been applied to the characterization of cell wall and EPS recovered from *Bifidobacterium*. In summary, one-dimensional ^{1}H and ^{13}C spectra are recorded for solutions of EPS samples in deuterium oxide, and spectra are recorded at temperatures of 70°C or above; the high sample temperature reduces the viscosity of the medium and shifts the residual-proton solvent signal upfield and away from important resonances. Vliegenthart et al. (1983) have suggested that spectra should be viewed as having 'structural reporter signals' and a 'bulk-region' (3–4 ppm). In proton spectra, the low-field reporter region includes resonances from the anomeric and ring atoms shifted out of the bulk region as a consequence of glycosylation. There is also a high-field reporter region including resonances from ring substituents such as acyl, alkyl, and acetal and ring substitutions such as *N*-acetylamino and H6 of 6-deoxy sugars. Homonuclear 2-D spectra [correlation spectroscopy (COSY) and total correlation spectroscopy (TOCSY)] are used to assign proton resonances to individual rings, and heteronuclear 2-D spectra, ^{13}C–^{1}H (HMQC, HMBC, or H2BC) are used to assign carbons and to obtain linkage information (HMBC). Further information about linkage, using nonscalar coupling, is available from rotating-frame Overhauser effect spectroscopy spectra.

Applying most of the methods mentioned above, a significant number of bacterial EPS structures have been analyzed. The structures for LAB–EPS that have been published before 2003 have been collated in several reviews (Broadbent et al. 2003; De Vuyst et al. 2001; Laws et al. 2001; Ruas-Madiedo et al. 2002b) and an additional update until 2008 has recently been compiled (Ruas-Madiedo et al. 2009a). Since this last review, new LAB–EPS structures have been reported; these include HePS derived from *Lb. johnsonii* 142 (Gorska et al. 2010), *Lb. fermentum* TDS030603 (Fukuda et al. 2010), *Lactobacillus* sp. CFR-2182 (Vijayendra et al. 2009), and *S. thermophilus* ST1 (Säwén et al. 2010). The structures of a very limited number of polysaccharides isolated from *Bifidobacterium* have been published to date (Table 7.3). These include cell wall polysaccharides (Habu et al. 1987; Nagaoka et al. 1995, 1996; Tone-Shimokawa et al. 1996; Zdorovenko et al. 2009) and a structure described as an extracellular polysaccharide from *B. longum* JBL05 (Kohno et al. 2009). Hosono et al. have also reported incomplete structural data for a water-soluble polysaccharide prepared by the sonication of *B. adolescentis* M101-4 (Hosono et al. 1997). It is clear that while a growing number of EPS structures are available from LAB, this is not the case for Bifido–EPS and this fact will need to be addressed if we are to understand the biological activity of these bifidobacterial EPS.

7.3.4 Rheological Properties of EPS Solutions

One of the few physical properties of EPS solutions that have been investigated in any detail is their rheological behavior, including the measurement of η_0 and the EPS response to shear.

The η_0 can be considered a measure of the hydrodynamic volume of the EPS in solution, and this parameter has been determined for a number of LAB– and Bifido–EPS by capillary viscometry (Canquil et al. 2007; Goh et al. 2005; Higashimura et al. 2000; Yang et al. 1999). Measurement of the reduced viscosity (η_{red}) at low concentrations of EPS and extrapolation of the data to infinite dilution using the Huggins and Kraemer equations gives the intrinsic viscosity. Intrinsic viscosities for aqueous solutions of EPS have been reported for *L. acidophilus* 5e2 (2.1–0.640 dL/g for MM_w varying between 4.8E + 6 g/mol and 0.16E + 6 g/mol; Chadha 2010), *L. delbrueckii* subsp. *bulgaricus* NCFB 2483 (20.13 dL/g; Goh et al. 2005), *Lc. lactis* subsp. *cremoris* ARH53 (19.62 dL/g; Yang et al. 1999), *Lc. lactis* subsp. *cremoris* SBT 0495 (22.60–4.70 dL/g for MM_w varying between 2.64E + 6 g/mol and 0.37E + 6 g/mol recorded at an ionic strength of 0.1M NaCl; Higashimura et al. 2000), *B. longum* subsp. *infantis* ATCC15697 (0.19 dL/g recorded at an ionic strength of 0.1M NaCl; Canquil et al. 2007), *L. delbrueckii* (0.75 dL/g recorded at an ionic strength of 0.1M NaCl; Canquil et al. 2007), and *S. thermophilus* Th4 (4.23 dL/g recorded at an ionic strength of 0.1M NaCl; Canquil et al. 2007). Intrinsic viscosities have also been calculated from the SEC-MALLS measurement of the weight-averaged MM and radius R_g (Ruas-Madiedo et al. 2002b) and from monitoring the zero-shear viscosity of EPS solutions (Tuinier et al. 1999b). Many of the EPS analyzed are charged and demonstrate polyelectrolyte behavior; for such systems, the variation of the η_0 with ionic strength has been used to determine the Smidsrød *B* values (Smidsrød and Haug 1971), which give a measure of the EPS chain stiffness—the higher the Smidsrød *B* value, the more flexible the polymer chain. The EPS from *Lc. lactis* subsp. *cremoris* SBT0495 was determined to behave as an immediately stiff rigid/flexible coil polyelectrolyte (*B* = 0.03; Higashimura et al. 2000). Canquil et al. (2007) reported *B* values of 0.224 for *S. thermophilus* Th4, 0.022 for *B. infantis* ATCC15697, and 0.015 for *L. delbrueckii*, with strain Th4 having the most flexible chain and the *L. delbrueckii* strain being the most rigid. Based on the limited data that have been published, it is clear that η_0 is determined by a number of factors, including the repeated unit structure and the polymer chain length. With regard to the influence of repeated unit structure, it has been suggested that the stiffer the backbone of the repeat unit, the higher the hydrodynamic volume of the EPS in solution (Tuinier et al. 2001). It is also clear that for a specific EPS structure, there is a direct relationship between weight-averaged MM (chain length) and the hydrodynamic volume of the EPS (Chadha 2010; Yang et al. 1999).

The response of EPS solutions to shear gives information about the solution conformation of the EPS and also provides details of the extent to which EPS chains interact. The response to shear has been reported for *L. delbrueckii* subsp. *bulgaricus* NCFB 2483 (Goh et al. 2005), *Lc. lactis* subsp. *cremoris* SBT 0495 (Oba et al. 1999), *Lc. lactis* subsp. *cremoris* B40 (Tuinier et al. 1999b, 2000), *B. infantis* ATCC15697, *L. delbrueckii* 0.82 and *S. thermophilus* Th4 (Cinquin et al. 2006), and *Lactobacillus sake* 0-1 (Vandenberg et al. 1995). The majority of EPS are subject to shear thinning, and frequently, they demonstrate a concentration dependence on the critical shear rate (shear rate at which the zero shear viscosity starts to decrease), with the critical shear rate reducing with increased concentration of the polymer, indicating entangled polymer chains at high concentrations.

In summary, while most of EPS rheological studies were originally conducted to determine the influence that polymers have on the rheological behavior of fermented dairy products, the information will also be valuable in determining the interaction of EPS with other biological macromolecules. An understanding of the rheological properties of EPS solutions will help provide an insight into the role that EPS play in assisting bacteria to colonize and adhere to specific environments and how they exert their biological activity.

7.4 Concluding Remarks

EPS-producing LAB have been used in dairy manufacture from a long time ago, and more recently, the growing knowledge about the putative implication of these polymers on the health benefits of some producing strains has rejuvenated the interest in EPS research. It is clear that the use of purified polymers as additives is not applicable in the food industry due to their scarce yield. Moreover, several strategies conducted to engineer EPS production in LAB have not been successful to date due to the complex biosynthesis mechanism of HePS and the direct competition of this anabolic pathway with that of peptiglycan or with the central carbohydrate catabolism. Additionally, under the current legal framework of many European countries, genetically modified (micro)organisms cannot be used for human food manufacture. Thus, the exploration of traditional (food) or new (animal) ecological niches has special relevance to increase the biodiversity of EPS-producing strains that could be used for the manufacture of food with desired properties. Therefore, the update of methodology used for the detection and characterization of EPS-producing strains and their polymers, which has been collected in this book chapter, will be a valuable tool for the scientific community interested in this research field.

ACKNOWLEDGMENTS

The authors thank the Spanish Ministry of Science and Innovation (MICINN) for the financial support of this research work through Project AGL2009-09445. B. Sánchez and C. Hidalgo-Cantabrana are grateful for their postdoctoral "Juan de la Cierva" contract and predoctoral FPI fellowship, respectively.

REFERENCES

Alp, G. and Aslim, B. 2010. Relationship between the resistance to bile salts and low pH with exopolysaccharide (EPS) production of *Bifidobacterium* spp. isolated from infants feces and breast milk. *Anaerobe* 16:101–105.

Amrouche, T., Boutin, Y., Prioult, G., and Fliss, I. 2006. Effects of bifidobacterial cytoplasm, cell wall and exopolysaccharide on mouse lymphocyte proliferation and cytokine production. *Int. Dairy J.* 16:70–80.

Audy, J., Labrie, S., Roy, D., and LaPointe, G. 2010. Sugar source modulates exopolysaccharide biosynthesis in *Bifidobacterium longum* subsp. *longum* CRC002. *Microbiology* 156:653–664.

Axelsson, L. 2004. Lactic acid bacteria: Classification and physiology. In: *Lactic Acid Bacteria. Microbial and Functional Aspects*, edited by S. Salminen, A. von Wright, and A. Ouwenhand, pp.1–66. New York: Marcel Dekker Inc.

Barrangou, R., Briczinski, E. P., Traeger, L. L., Loquasto, J. R., Richards, M., Hovath, P., Coûte-Monvoisin, A. C., Leyer, G., Rendulic, S., Steele, J. L., Broadbent, J. R., Oberg, T., Dudley, E. G., Schuster, S., Romero, D. A., and Roberts, R. F. 2009. Comparison of the complete genome of *Bifidobacterium animalis* subsp. *lactis* DSM10140 and Bi-04. *J. Bacteriol.* 191:4144–4151.

Berger, B., Pridmore, R. D., Barretto, C., Delmas-Julien, F., Schreiber, K., Arigoni, F., and Brüssow, H. 2007. Similarity and differences in the *Lactobacillus acidophilus* group identified by polyphasic analysis and comparative genomics. *J. Bacteriol.* 189:1311–1321.

Bergmaier, D., Champagne, C. P., and Lacroix, C. 2005. Growth and exopolysaccharide production during free and immobilized cell chemostat culture of *Lactobacillus rhamnosus* RW-9595M. *J. Appl. Microbiol.* 98:272–284.

Boels, I. C., van Kranenburg, R., Hugenholtz, J., Kleerebezem, M., and de Vos, W. M. 2001. Sugar catabolism and its impact on the biosynthesis and engineering of exopolysaccharide production in lactic acid bacteria. *Int. Dairy J.* 11:723–732.

Bradford, M. 1976. A rapid and sensitive method for quantification of microgram quantities of proteins utilizing the principles of protein dye binding. *Anal. Biochem.* 72:248–254.

Briczinski, E. P., Loquasto, J. R., Barrangou, R., Dudley, E. G., Roberts, A. M., and Roberts, R. F. 2009. Strain-specific genotyping of *Bifidobacterium animalis* subsp. *lactis* by using single-nucleotide polymorphisms, insertions, and deletions. *Appl. Environ. Microbiol.* 75:7501–7508.

Broadbent, J. R., McMahon, D. J., Welker, D. L., Oberg, C. J., and Moineau, S. 2003. Biochemistry, genetics, and applications of exopolysaccharide production in *Streptococcus thermophilus*: A review. *J. Dairy Sci.* 86:407–423.

Canquil, N., Villarroel, M., Bravo, S., Rubilar, M., and Shene, C. 2007. Behavior of the rheological parameters of exopolysaccharides synthesized by three lactic acid bacteria. *Carbohydr. Polym.* 68:270–279.

Cerning, J., Bouillanne, C., Desmazeaud, M. J., and Landon, M. 1986. Isolation and characterization of exocellular polysaccharide produced by *Lactobacillus bulgaricus*. *Biotechnol. Lett.* 8:625–628.

Chadha, M. 2010. Novel techniques for the characterization of exopolysaccharides secreted by lactic acid bacteria. PhD Thesis, Department of Chemical and Biological Sciences, University of Huddersfield, Huddersfield, p. 251.

Chen, X., Jiang, H., Yang, Y., and Liu, N. 2009. Effect of exopolysaccharide from *Bifidobacterium bifidum* on cell of gastric cancer and human telomerase reverse transcriptase. *Weishengwu Xuebao* (in Chinese) 49:117–122.

Churms, S. C. 1996. Recent progress in carbohydrate separation by high-performance liquid chromatography based on size exclusion. *J. Chromatogr., A* 720:151–166.

Cinquin, C., Le Blay, G., Fliss, I., and Lacroix, C. 2006. Comparative effects of exopolysaccharides from lactic acid bacteria and fructo-oligosaccharides on infant gut microbiota tested in an in vitro colonic model with immobilized cells. *FEMS Microbiol. Ecol.* 57:226–238.

Dabour, N. and LaPointe, G. 2005. Identification and molecular characterization of the chromosomal exopolysaccharide biosynthesis gene cluster from *Lactococcus lactis* subsp. *cremoris* SMQ-461. *Appl. Environ. Microbiol.* 71:7414–7425.

De Vuyst, L., de Vin, F., Vaningelgem, F., and Degeest, B. 2001. Recent development in the biosynthesis and applications of heteropolysaccharides from lactic acid bacteria. *Int. Dairy J.* 11:678–707.

De Vuyst, L., Zamfir, M., Mozzi, F., Adriany, T., Marshall, V., Degeest, B., and Vaningelgem, F. 2003. Exopolysaccharide-producing *Streptococcus thermophilus* strains as functional starter cultures in the production of fermented milks. *Int. Dairy J.* 13:707–717.

Degeest, B. and De Vuyst, L. 1999. Indication that the nitrogen source influences both amount and size of exopolysaccharides produced by *Streptococcus thermophilus* LY03 and modelling of the bacterial growth and exopolysaccharide production in a complex medium. *Appl. Environ. Microbiol.* 65:2863–2870.

Delgado, S., Ruas-Madiedo, P., Suárez, A., and Mayo, B. 2006a. Interindividual differences in microbial counts and biochemical-associated variables in the feces of healthy Spanish adults. *Dig. Dis. Sci.* 51:737–743.

Delgado, S., Suárez, A., and Mayo, B. 2006b. Identification of dominant bacteria in feces and colonic mucosa from healthy Spanish adults by culturing and 16S rDNA sequence analysis. *Dig. Dis. Sci.* 51:744–751.

Dierksen, K. P., Sandine, W. E., and Trempy, J. E. 1997. Expression of ropy and mucoid phenotypes in *Lactococcus lactis*. *J. Dairy Sci.* 80:1528–1536.

Dols-Lafargue, M., Lee, H. Y., Le Marrec, C., Heyraud, A., Chambat, G., and Lonvaud-Funel, A. 2008. Characterization of *gtf*, a glucosyltransferase gene in the genomes of *Pediococcus parvulus* and *Oenococcus oeni*, two bacterial species commonly found in wine. *Appl. Environ. Microbiol.* 74:4079–4090.

Dubois, M., Gillies, K., Hamilton, J., Reberes, P., and Smith, F. 1956. A colorimetric method for the determination of sugars. *Anal. Chem.* 28:349–356.

Dueñas-Chasco, M. T., Rodríguez-Carvajal, M. A., Tejero-Mateo, P., Espartero, J. L., Irastorza-Iribas, A., and Gil-Serrano, A. M. 1998. Structural analysis of the exopolysaccharides produced by *Lactobacillus* sp. G-77. *Carbohydr. Res.* 307:125–133.

Duus, J. O., Gotfredsen, C. H., and Bock, K. 2000. Carbohydrate structural determination by NMR spectroscopy: Modern methods and limitations. *Chem. Rev.* 100:4589–4614.

European Food Safety Authority (EFSA) 2007. Introduction of a qualified presumption of safety (QPS) approach for assessment of selected microorganism referred to EFSA. *EFSA J.* 587:1–16.

Faber, E. J., Zoon, P., Kamerling, J. P., and Vliegenthart, J. F. G. 1998. The exopolysaccharides produced by *Streptococcus thermophilus* Rs and Sts have the same repeating unit but differ in viscosity of their milk cultures. *Carbohydr. Res.* 310:269–276.

Felis, G. E. and Dellaglio, F. 2007. Taxonomy of lactobacilli and bifidobacteria. *Curr. Issues Intest. Microbiol.* 8:44–61.

Food and Drug Administration (FDA). 2009. Generally Recognized as Safe. Available at http://www.fda.gov.

Fukuda, K., Shi, T., Nagami, K., Leo, F., Nakamura, T., Yasuda, K., Senda, A., Motoshima, H., and Urashima, T. 2010. Effects of carbohydrate source on physicochemical properties of the exopolysaccharide produced by *Lactobacillus fermentum* TDS030603 in a chemically defined medium. *Carbohydr. Polym.* 79:1040–1045.

Garrigues, C., Johansen, E., and Pedersen, M. B. 2010. Complete genome sequence of *Bifidobacterium animalis* subsp. *lactis* BB-12, a widely consumed probiotic strain. *J. Bacteriol.* 192:2467–2468.

Gerwig, G. J., Kamerling, J. P., and Vliegenthart, J. F. G. 1978. Determination of the D- and L- configuration of neutral monosaccharides by high-resolution capillary GLC. *Carbohydr. Res.* 62:349–357.

Goh, K. K. T., Hemar, Y., and Singh, H. 2005. Viscometric and static light scattering studies on an exopolysaccharide produced by *Lactobacillus delbrueckii* subspecies *bulgaricus* NCFB 2483. *Biopolymers* 77:98–106.

Gorret, N., Renard, C., Famelart, M. H., Maubois, J. L., and Doublier, J. L. 2003. Rheological characterization of the EPS produced by *Propionibacterium acidipropionici* on milk microfiltrate. *Carbohydr. Polym.* 51:149–158.

Gorska, S., Jachyrnek, W., Rybka, J., Strus, M., Heczko, P. B., and Gamian, A. 2010. Structural and immunochemical studies of neutral exopolysaccharide produced by *Lactobacillus johnsonii* 142. *Carbohydr. Res.* 345:108–114.

Habu, Y., Nagaoka, M., Yokokura, T., and Azuma, I. 1987. Structural studies of cell-wall polysaccharides from *Bifidobacterium breve* YIT4010 and related *Bifidobacterium* species. *J. Biochem.* 102:1423–1432.

Hassan, A. N. 2008. Possibilities and challenges of exopolysaccharide-producing lactic cultures in dairy foods. *J. Dairy Sci.* 91:1282–1298.

Hassan, A. N., Frank, J. F., and Elsoda, M. 2003. Observation of bacterial exopolysaccharides in dairy products using cryo-scanning electron microscopy. *Int. Dairy J.* 13:755–762.

Higashimura, M., Mulder-Bosman, B. W., Reich, R., Iwasaki, T., and Robijn, G. W. 2000. Solution properties of viilian, the exopolysaccharide from *Lactococcus lactis* subsp. *cremoris* SBT 0495. *Biopolymers* 54:143–158.

Holts O., Moran A., and Brennan P. J. 2009. Overview of the glycosylated components of the bacterial cell envelope. In: *Microbial Glycobiology. Structures, Relevance and Applications*, 1st edition, edited by A. P. Moran, O. Holst, P. J. Brennan, and M. von Itzstein, pp. 3–13. London: Academic Press.

Hosono, A., Lee, J.W., Ametani, A., Natsume, M., Hirayama, M., Adachi, T., and Kaminogawa, S. 1997. Characterization of a water-soluble polysaccharide fraction with immunopotentiating activity from *Bifidobacterium adolescentis* M101-4. *Biosci. Biotechnol. Biochem.* 61:312–316.

Kim, J. F., Jeong, H., Yu, D. S., Choi, S. H., Hur, C. G., Park, M. S., Yoon, S. H., Kim, D. W., Ji, G. E., Park, H. S., and Oh, T. K. 2009. Genome sequence of the probiotic bacterium *Bifidobacterium animalis* subsp. *lactis* AD011. *J. Bacteriol.* 191:678–679.

Knoshaug, E. P., Ahlgren, J. A., and Trempy, J. E. 2007. Exopolysaccharide expression in *Lactococcus lactis* subsp. *cremoris* Ropy352: Evidence for novel gene organization. *Appl. Environ. Microbiol.* 73:897–905.

Kohno, M., Suzuki, S., Kanaya, T., Yoshino, T., Matsuura, Y., Asada, M., and Kitamura, S. 2009. Structural characterization of the extracellular polysaccharide produced by *Bifidobacterium longum* JBL05. *Carbohydr. Polym.* 77:351–357.

Korakli, M. and Vogel, R. F. 2006. Structure–function relationship of homopolysaccharides producing glycansucrases and therapeutic potential for their synthesised glycans. *Appl. Microbiol. Biotechnol.* 71:790–803.

Laws, A., Gu Y. C., and Marshall, V. 2001. Biosynthesis, characterization, and design of bacterial exopolysaccharides from lactic acid bacteria. *Biotechnol. Adv.* 19:597–625.

Laws, A. P. and Marshall, V. M. 2001. The relevance of exopolysaccharides to the rheological properties in milk fermented with ropy strains of lactic acid bacteria. *Int. Dairy J.* 11:709–721.

Lee, J. H. and O'Sullivan, J. 2010. Genomic insights into bifidobacteria. *Microbiol. Mol. Biol. Rev.* 74:378–416.

Leeflang, B. R., Faber, E. J., Erbel, P., and Vliegenthart, J. F. G. 2000. Structure elucidation of glycoprotein glycans and of polysaccharides by NMR spectroscopy. *J. Biotechnol.* 77:115–122.

Leivers, S., Hidalgo-Cantabrana, C., Robinson, G., Margolles, A., Ruas-Madiedo, P., and Laws, A. P. 2011. Structure of the high molecular weight exopolysaccharide produced by *Bifidobacterium animalis* subsp. *lactis* IPLA-R1 and sequence analysis of its putative eps cluster. *Carbohydr. Res.* 346:2710–2717.

Leontein, K., Lindberg, B., and Lonngren, J. 1978. Assignment of absolute-configuration of sugars by G.I.C. of their acetylated glycosides formed from chiral alcohols. *Carbohydr. Res.* 62:359–362.

Leroy, F. and De Vuyst, L. 2004. Lactic acid bacteria as functional starter cultures for the food fermentation industry. *Trends Food Sci. Technol.* 15:67–68.

Liu, C. T., Hsu, I. T., Chou, C. C., Lo, P. R., and Yu, R. C. 2009. Exopolysaccharide production of *Lactobacillus salivarius* BCRC 14759 and *Bifidobacterium bifidum* BCRC 14615. *World J. Microbiol. Biotechnol.* 25:883–890.

Margolles A., Mayo B., and Ruas-Madiedo P. 2009. Screening, identification, and characterization of *Lactobacillus* and *Bifidobacterium* strains. In: *Handbook of Probiotics and Prebiotics*, 2nd edition, edited by K. Nomoto, S. Salminen, and Y. K. Lee, pp. 4–43. New Jersey: John Wiley & Sons Inc.

Miller, G. 1959. Use of dinitrosalicylic acid reagent for determination of reducing sugars. *Anal. Chem.* 31:426–428.

Monsan, P., Bozonnet, S., Albenne, C., Joucla, G., Willemont, R. M., and Remaud-Siméon, M. 2001. Homopolysaccharides from lactic acid bacteria. *Int. Dairy J.* 11:675–685.

Mozzi, F., Vaningelgem, F., Hébert, E. M., van der Meulen, R., Foulquié-Moreno, M. R., Font de Valdez, G., and De Vuyst, L. 2006. Diversity of heteropolysaccharides producing lactic acid bacterium strains and their biopolymers. *Appl. Environ. Microbiol.* 72:4431–4435.

Nagaoka, M., Hashimoto, S., Shibata, H., Kimura, I., Kimura, K., Sawada, H., and Yokokura, T. 1996. Structure of a galactan from cell walls of *Bifidobacterium catenulatum* YIT4016. *Carbohydr. Res.* 281:285–291.

Nagaoka, M., Muto, M., Yokokura, T., and Mutai, M. 1988. Structure of 6-deoxytalose-containing polysaccharide from the cell-wall of *Bifidobacterium adolescentis*. *J. Biochem.* 103:618–621.

Nagaoka, M., Shibata, H., Kimura, I., Hashimoto, S., Kimura, K., Sawada, H., and Yokokura, T. 1995. Structural studies on a cell-wall polysaccharide from *Bifidobacterium longum* YIT4028. *Carbohydr Res.* 274:245–249.

Nakajima, H., Hirota, T., Toba, T., Itoh, T., and Adachi, S. 1992. Structure of the extracellular polysaccharide from slime-forming *Lactococcus lactis* subsp. *cremoris* SBT-0495. *Carbohydr. Res.* 224:245–253.

Nordmark, E. L., Yang, Z. N., Huttunen, E., and Widmalm, G. 2005. Structural studies of an exopolysaccharide produced by *Streptococcus thermophilus* THS. *Biomacromolecules* 6:105–108.

Oba, T., Higashimura, M., Iwasaki, T., Matser, A. M., Steeneken, P. A. M., Robijn, G. W., and Sikkema, J. 1999. Viscoelastic properties of aqueous solutions of the phosphopolysaccharide "viilian" from *Lactococcus lactis* subsp *cremoris* SBT 0495. *Carbohydr. Polym.* 39:275–281.

Péant, B., LaPointe, G., Gilbert, C., Atlan, D., Ward, P., and Roy, D. 2005. Comparative analysis of the exopolysaccharide biosynthesis gene clusters from four strains of *Lactobacillus rhamnosus*. *Microbiology* 151:1839–1851.

Petry, S., Furlan, S., Waghorne, E., Saulnier, L., Cerning, J., and Maguin, E. 2003. Comparison of the thickening properties of four *Lactobacillus delbrueckii* subsp. *bulgaricus* strains and physicochemical characterization of their exopolysaccharides. *FEMS Microbiol. Lett.* 221:285–291.

Pot, B. 2008. The taxonomy of lactic acid bacteria. In: *Bactéries Lactiques. De la génétique aux ferments*, edited by G. Corrieu and F.M. Luquet, pp. 1–152. Paris: Lavoisier.

Provencher, C., LaPointe, G., Sirois, S., Van Calsteren, M. R., and Roy, D. 2003. Consensus-degenerate hybrid oligonucleotide primers for amplification of priming glycosyltransferase genes of the exopolysaccharide locus in strains of the *Lactobacillus casei* group. *Appl. Environ. Microbiol.* 69:3299–3307.

Rawson, H. L. and Marshall, V. 1997. Effect of "ropy" strains of *Lactobacillus delbrueckii* ssp. *bulgaricus* and *Streptococcus thermophilus* on rheology of stirred yoghurt. *Int. J. Food Sci. Technol.* 32:213–220.

Ricciardi, A. and Clementi, F. 2000. Exopolysaccharides from lactic acid bacteria: Structure, production and technological applications. *Ital. J. Food Sci.* 12:23–45.

Roberts, C. M., Fett, W. F., Osman, S. F., Wijey, C., O'Connor, J.V., and Hoover, D. G. 1995. Exopolysaccharide production by *Bifidobacterium longum* BB-79. *J. Appl. Bacteriol.* 78:463–468.

Ruas-Madiedo, P., Abraham, A., Mozzi, F., and de los Reyes-Gavilán, C. G. 2008. Functionality of exopolysaccharides produced by lactic acid bacteria. In: *Molecular Aspects of Lactic Acid Bacteria for Traditional and New Applications*, edited by B. Mayo, P. López, and G. Pérez-Martínez, pp. 137–166. Kerala, India: Research Signpost.

Ruas-Madiedo, P. and de los Reyes-Gavilán, C. G. 2005. Invited review: Methods for the screening, isolation, and characterization of exopolysaccharides produced by lactic acid bacteria. *J. Dairy Sci.* 88:842–856.

Ruas-Madiedo, P., Gueimonde, M., Arigoni, F., de los Reyes-Gavilán, C. G., and Margolles, A. 2009c. Bile affects the synthesis of exopolysaccharides by *Bifidobacterium animalis*. *Appl. Environ. Microbiol.* 75:1204–1207.

Ruas-Madiedo, P., Gueimonde, M., Margolles, A., de los Reyes-Gavilán, C. G., and Salminen, S. 2006. Exopolysaccharides produced by probiotic strains modify the adhesion of probiotics and enteropathogens to human intestinal mucus. *J. Food Prot.* 69:2011–2015.

Ruas-Madiedo, P., Hugenholtz, J., and Zoon, P. 2002a. An overview of the functionality of exopolysaccharides produced by lactic acid bacteria. *Int. Dairy J.* 12:163–171.

Ruas-Madiedo, P., Moreno, J. A., Salazar, N., Delgado, S., Mayo, M., Margolles, A., and de los Reyes-Gavilán, C. G. 2007. Screening of exopolysaccharide-producing *Lactobacillus* and *Bifidobacterium* strains isolated from the human intestinal microbiota. *Appl. Environ. Microbiol.* 73:4385–4388.

Ruas-Madiedo, P., Salazar, N., and de los Reyes-Gavilán, C. G. 2009a. Biosynthesis and chemical composition of exopolysaccharides produced by lactic acid bacteria. In: *Bacterial Polysaccharides. Current Innovations and Future Trends*, edited by M. Ullrich, pp. 279–310. Norfolk: Caister Academic Press.

Ruas-Madiedo, P., Salazar, N., and de los Reyes-Gavilán, C. G. 2009b. Exopolysaccharides produced by lactic acid bacteria in food and probiotic applications. In: *Microbial Glycobiology. Structures, Relevance and Applications*, edited by A.P. Moran, O. Holst, P.J. Brennan, and M. von Itzstein, pp. 887–902. London: Elsevier Inc.

Ruas-Madiedo, P., Tuinier, R., Kanning, M., and Zoon, P. 2002b. Role of exopolysaccharides produced by *Lactococcus lactis* subsp. *cremoris* on the viscosity of fermented milks. *Int. Dairy J.* 12:689–695.

Salazar, N., Prieto, A., Leal, J. A., Mayo, B., Bada-Gancedo, J. C., de los Reyes-Gavilán, C. G., and Ruas-Madiedo, P. 2009. Production of exopolysaccharides by *Lactobacillus* and *Bifidobacterium* strains of human origin, and metabolic activity of the producing bacteria in milk. *J. Dairy Sci.* 92:4158–4168.

Sánchez, J. I., Martínez, B., Guillén, R., Jiménez, R., and Rodríguez, A. 2006. Culture conditions determine the balance between two different exopolysaccharides produced by *Lactobacillus pentosus* LPS26. *Appl. Environ. Microbiol.* 72:7495–7502.

Sawardeker, J., Sloneker, J., and Jeanes, A. 1965. Quantitative determination of monosaccharides as their alditol acetates by gas liquid chromatography. *Anal. Chem.* 37:1602–1604.

Säwén, E., Huttunen, E., Zhang, X., Yang, Z. N., and Widmalm, G. 2010. Structural analysis of the exopolysaccharide produced by *Streptococcus thermophilus* ST1 solely by NMR spectroscopy. *J. Biomol. NMR* 47:125–134.

Schell, M. A., Karmirantzou, M., Snell, B., Vilanova, D., Berger, B., Pessi, G., Zwahlen, M. C., Desiere, F., Bork, P., Delley, M., Pridmore, R. D., and Arigoni, F. 2002. The genome sequence of *Bifidobacterium longum* reflects its adaptation to the human gastrointestinal tract. *Proc. Natl. Acad. Sci. U. S. A.* 99:14422–14427.

Sela, D., Chapman, J., Adeuya, A., Kim, J. H., Chen, F., Whitehead, T. R., Lapidus, A., Rokhsar, D. S., Lebrilla, C. B., German, J. B., Price, N. P., Richardson, P. M., and Mills, D. A. 2008. The genome sequence of *Bifidobacterium longum* subsp. *infantis* reveals adaptations for milk utilization within the infant microbiome. *Proc. Natl. Acad. Sci. U. S. A.* 105:18964–18969.

Smidsrød, O. and Haug, A. 1971. Estimation of relative stiffness of molecular chain in polyelectrolytes from measurements of viscosity at different ionic strengths. *Biopolymers* 10:1213–12127.

Smith, P., Krohn, R., Hermanson, G., Mallia, A., Gartner, F., Provenzano, M., Fujimoto, E., Goeke, N., Olson, B., and Klenk, D. 1985. Measurement of protein using bicinchoninic acid. *Anal. Biochem.* 150:76–85.

Stellner, K., Saito, H., and Hakomori, S. I. 1973. Determination of aminosugar linkages in glycolipids by methylation: Aminosugar linkages of ceramide pentasaccharides of rabbit erythrocytes and Forssman antigen. *Arch. Biochem. Biophys.* 155:464–472.

Stingele, F., Neeser, J. R., and Mollet, B. 1996. Identification of the *eps* (exopolysaccharide) gene cluster from *Streptococcus thermophilus* Sfi6. *J. Bacteriol.* 178:1680–1690.

Sun, Z., Chen, X., Wang, J., Gao, P., Zhou, Z., Ren, Y., Sun, T., Wang, L., Meng, H., Chen, W., and Zhang, H. 2010. Complete genome sequence of probiotic *Bifidobacterium animalis* subsp. *lactis* strain V9. *J. Bacteriol.* 192:4080–4081.

Sweet, D. P., Shapiro, R. H., and Albersheim, P. 1975. Quantitative analysis by various glc response factor theories for partially methylated and partially ethylated alditol acetates. *Carbohydr. Res.* 40:217–225.

Theisen, A., Deacon, M. P., Johann, C., and Harding, S. 1999. *Refractive Increment Data-Book: For Polymer and Biomolecular Scientists*, p. 64. Nottingham: Nottingham University Press.

Tieking, M., Kaditzky, S., Valcheva, R., Korakli, M., Vogel, R. F., and Gänzle, M. G. 2005. Extracellular homopolysaccharides and oligosaccharides from intestinal lactobacilli. *J. Appl. Microbiol.* 99:692–702.

Tieking, M., Korakli, M., Ehrmann, M. A., Gänzle, M. G., and Vogel, R. F. 2003. *In situ* production of exopolysaccharides during sourdough fermentation by cereal and intestinal isolates of lactic acid bacteria. *Appl. Environ. Microbiol.* 69:945–952.

Tone-Shimokawa, Y., Toida, T., and Kawashima, T. 1996. Isolation and structural analysis of polysaccharide containing galactofuranose from the cell walls of *Bifidobacterium infantis*. *J. Bacteriol.* 178:317–320.

Tuinier, R., Oomen, C. J., Zoon, P., Stuart, M. A. C., and de Kruif, C. G. 2000. Viscoelastic properties of an exocellular polysaccharide produced by a *Lactococcus lactis*. *Biomacromolecules* 1:219–223.

Tuinier, R., van Casteren, W. H. M., Looijesteinj, P. J., Schols, H. A., Voragen, A. G. J., and Zoon, P. 2001. Effects of structural modifications on some physical characteristics of exopolysaccharides from *Lactococcus lactis*. *Biopolymers* 59:160–166.

Tuinier, R., Zoon, P., Olieman, C., Stuart, M. A. C., Fleer, G. J., and de Kruif, C. G. 1999a. Isolation and physical characterization of an exocellular polysaccharide. *Biopolymers* 49:1–9.

Tuinier, R., Zoon, P., Stuart, M. A. C., Fleer, G. J., and de Kruif, C. G. 1999b. Concentration and shear-rate dependence of the viscosity of an exocellular polysaccharide. *Biopolymers* 50:641–646.

van Casteren, W. H. M., Dijkema, C., Schols, H. A., Beldman, G., and Voragen, A. G. J. 1998. Characterization and modification of the exopolysaccharide produced by *Lactococcus lactis* subsp. *cremoris* B40. *Carbohydr. Polym.* 37:123–130.

Vandenberg, D. J. C., Robijn, G. W., Janssen, A. C., Giuseppin, M. L. F., Vreeker, R., Kamerling, J. P., Vliegenthart, J. F. G., Ledeboer, A. M., and Verrips, C. T. 1995. Production of a novel extracellular polysaccharide by *Lactobacillus sake* 0-1 and characterization of the polysaccharide. *Appl. Environ. Microbiol.* 61:2840–2844.

van Geel-Schutten, G. H., Faber, E. J., Smit, E., Monting, K., Smith, M. R., ten Brink, B., Kamerling, J. P., Vliegenthart, J. F. G., and Dijkhuizen, L. 1999. Biochemical and structural characterization of the glucan and fructan exopolysaccharides synthesised by the *Lactobacillus reuteri* wild-type strains and by mutant strains. *Appl. Environ. Microbiol.* 65:3008–3014.

van Hijum, S. A. F. T., Kralj, S., Ozimek, L. K., Dijkhuizen, L., and van Geel-Schutten, I. G. H. 2006. Structure–function relationship of glucansucrase and fructansucrase enzymes from lactic acid bacteria. *Microbiol. Mol. Biol. Rev.* 70:157–176.

Vaningelgem, F., Zamfir, M., Mozzi, F., Adriany, T., Vancanneyt, M., Swings, J., and De Vuyst, L. 2004. Biodiversity of exopolysaccharides produced by *Streptotoccus thermophilus* strains is reflected in their production and their molecular and functional characteristics. *Appl. Environ. Microbiol.* 70:900–912.

van Kranenburg, R., Marugg, J. D., van Swan, I. I., Willem, N. J., and de Vos, W. M. 1997. Molecular characterization of the plasmid-encoded *eps* gene cluster essential for exopolysaccharide biosynthesis in *Lactococcus lactis*. *Mol. Microbiol.* 24:387–397.

van Kranenburg, R., Vos, R. G., van Swan, I. I., Kleerebezem, M., and de Vos, W. M. 1999. Functional analysis of glycosyltransferase genes from *Lactococcus lactis* and other gram-positive cocci: Complementation, expression and diversity. *J. Bacteriol.* 181:6347–6353.

Vedamuthu, E. R. and Neville, J. M. 1986. Involvement of a plasmid in production of ropiness (mucoidness) in milk cultures by *Streptococcus cremoris* MS. *Appl. Environ. Microbiol.* 51:677–682.

Ventura, M., Turroni, F., Zomer, A., Foroli, E., Giubellini, V., Bottacini, F., Canchaya, C., Clasesson, M. J., He, F., Mantzourani, M., Mulas, L., Ferrarini, A., Gao, B., Delledonne, M., Henrissat, B., Coutinho, P., Oggioni, M., Gupta, R.S., Zhang, Z., Bieghton, D., Fitzgerald, G.F., O'Toole, P. W., and van Sinderen, D. 2009. The *Bifidobacterium* dentium Bd1 genome sequence reflects its genetic adaptation to the human oral cavity. *PLoS Genet.* 5(12):e1000785.

Vescovo, M., Scolari, G. L., and Botazzi, V. 1989. Plasmid-encoded ropiness production in *Lactobacillus casei* ssp. *casei*. *Biotechnol. Lett.* 10:709–712.

Vijayendra, S., Palanivel, G., Mahadevamma, S., and Tharanathan, R. 2009. Physico-chemical characterization of a new heteropolysaccharide produced by a native isolate of heterofermentative *Lactobacillus* sp CFR-2182. *Arch. Microbiol.* 191:303–310.

Vliegenthart, J. F. G, Dorland, L., and Vanhalbeek, H. 1983. High-resolution, H-1-nuclear magnetic resonance spectroscopy as a tool in the structural analysis of carbohydrates related to glycoproteins. *Adv. Carbohydr. Chem. Biochem.* 41:209–374.

Welman, A. D. and Maddox, I. S. 2003. Exopolysaccharides from lactic acid bacteria: Perspectives and challenges. *Trends Biotechnol.* 21:269–274.

Werning, M. L., Ibarburu, I., Dueñas, M. T., Irastorza, A., Navas, J., and López, P. 2006. *Pediococcus parvulus* *gtf* gene encoding the GTF glycosyltransferase and its application for specific PCR detection of β-D-glucan producing bacteria in foods and beverages. *J. Food Prot.* 69:161–169.

Whitfield, C. and Roberts, I. S. 1999. Structure, assembly and regulation of expression of capsules in *Escherichia coli*. *Mol. Microbiol.* 31:1307–1319.

World Health Organization/Food and Agriculture Organization of the United Nations (WHO/FAO) 2006. Probiotics in foods: Health and nutritional properties and guidelines for evaluation. FAO and Nutrition Paper 85 (ISBN-92-5-105513-0).

Wu, M. H., Pan, T. M., Wu, Y. J., Chang, S. J., Chang, M. S., and Hu, C. Y. 2010. Exopolysaccharide activities from probiotic *Bifidobacterium*: Immunomodulatory effects (on J774A.1 macrophages) and antimicrobial properties. *Int. J. Food Microbiol.* 144:104–110.

Wyatt, P. J., Jackson, C., and Wyatt, G. K. 1988. Absolute GPC determinations of molecular-weights and sizes from light scattering. *Am. Lab.* 20:86–89.

Yang, Z. N., Huttunen, E., Staaf, M., Widmalm, G., and Tenhu, H. 1999. Separation, purification and characterization of extracellular polysaccharides produced by slime-forming *Lactococcus lactis* ssp. *cremoris* strains. *Int. Dairy J.* 9:631–638.

Zdorovenko, E. L., Kachala, V. V., Sidarenka, A. V., Izhik, A. V., Kisileva, E. P., Shashkov, A. S., Novik, G. I., and Knirel, Y. A. 2009. Structure of the cell wall polysaccharides of probiotic bifidobacteria *Bifidobacterimum bifidum* BIM B-465. *Carbohydr. Res.* 344:2417–2420.

Zhurina, D., Zomer, A., Gleinser, M., Brancaccio, V. F., Auchter, M., Waidmann, M. S., Westermann, C., van Sinderen, D., and Riedel, C. U. 2011. Complete genome sequence of *Bifidobacterium bifidum* S17. *J. Bacteriol.* 193:301–302.

8

Fermentation Ecosystems

Robert Scott

CONTENTS

8.1 Fermentation as an Ecological Process

In everyday usage, the concepts of environment and ecology are generally associated with large, natural systems such as the global ecosystem. This leaves microscopic ecological processes in the blind spot of the quotidian observer. Yet, humans encounter fermentation ecosystems everyday in their food. In this paper, I explore the potential for looking at fermentation processes ecologically and culturally. I would like to highlight the importance of the indoor, microbial ecology of fermented foods, such as beer, cheese, bread, yogurt, and other foods that constitute microbial ecosystems. Beyond describing fermentation as a human ecological process, I suggest that fermentation is a suitable vehicle for teaching ecological concepts in science classrooms.

Food fermentation is practiced by human cultures all over the world. It is a major component of human survival in places where preserved food is a necessity. Production of fermented food does not require knowledge of the biologically mediated nature of fermentation, because the biota that carry out fermentation are present worldwide. Fermentation organisms are ambient in the environment, whether humans make use of them or not. In recent years, there has been a commercial effort to improve the reputation of "probiotic" food products (Sanders and Marco 2010). This paper goes a step further to formulate the relationships embodied in preserved foods as microbial ecosystems, sitting on countertops and stored in food cellars around the world.

The main components of the fermentation ecosystem include microbes (yeast and bacteria), organic material to be fermented, a medium in which the fermentation takes place, a vessel with a controlled gate, and various tools that may be used to develop and monitor the fermentation (e.g., thermometer, hydrometer, and siphoning tubes). This is an ecosystem in that it is a complex of living and nonliving components that are viewed in terms of their interactions in a specific place. Beer biochemistry has recently been used to teach food chemistry (Pelter 2006); as would be expected, the study emphasized the understanding of chemical pathways and metabolism. In this chapter, we will treat the chemical pathways of microbial metabolism as a black box in order to shift attention toward the interactions within the microbial *environment* (see Figure 8.1).

Reformulating a food as a fermentation ecosystem begins with apprehending the vocabulary of the food as a cultural form. Development of fermented foods was driven by necessity and culture preference,

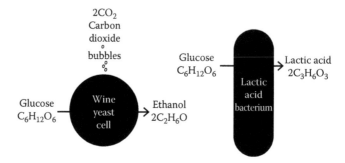

FIGURE 8.1 In order to focus attention to interspatial relationships and relationships to the fermentation substrate, microbial metabolism is treated as a black box in this chapter. Alcoholic fermentation carried on by yeast (left) and lactic acid fermentation carried on by bacteria (right) are central to the fermentation of food and drink. Note that yeast and bacteria cells come in different sizes and shapes; this figure is not an illustration of scale.

rather than awareness or concern for the microbes involved. Generally speaking, the organisms that ferment human foods are ambient in the environment. The most basic fermentation organisms such as *Lactobacillus* and *Saccharomyces* are found on all continents. Thus, fermentation ecosystems were originally cultivated by people who were not aware of the underlying microbiology. Before Pasteur (1858), the vocabulary of microbial fermentation simply was not available.

Archeological sites around the world have yielded evidence of controlled fermentations dating back to the earliest agricultural settlements. Chinese fermentation vessels containing traces of rice, honey, and fruit have been dated to early Neolithic villages (McGovern et al. 2004). Egyptian hieroglyphics depict viticultural practices during the Old Kingdom, and evidence of white wine has been found in Tutankhamun's tomb (Guasch-Jané et al. 2005). Chicha, an alcoholic beverage produced from corn, was produced in South America before the Incan Empire (Moseley et al. 2005). None of this should be too surprising, for food fermentation occurs naturally when sugars are exposed to fermentation organisms. All human cultures develop and rely on fermented food products and the microbes that produce them. This chapter describes the underlying fermentation ecology, with a focus on pickled foods, alcoholic beverages, microbial evolution, and domestication.

8.2 Fermented Foods

For better or for worse, fermentation affects the human food supply and thus impacts on human life. Fermentation is the process by which alcohol or lactic acid is produced by living cells in solutions that contain sugars (Ribéreau-Gayon et al. 2000). Wild fermentation bacteria and yeast cover the continents and permeate ecosystems, in the air, soil, water, and guts of animals. They may be viewed as a natural resource or an agent of spoilage—under the right conditions, bread dough naturally becomes sourdough, whereas a fruit juice may become a disgusting solution of vinegar and alcohol. The human health implications of fermentation are not trivial. Fermented foods often introduce microflora that inhabit the human body (Ross et al. 2002; Reid et al. 2003; Picard et al. 2005). Many common foods contain fermentation microorganisms and their by-products (see Table 8.1).

There are two kingdoms of life in fermentation ecosystems, namely, fungi and bacteria. Fermentation fungi include yeasts (such as *Saccharomyces*, which produce alcoholic beverages) and molds (such as those found in blue cheese). Fermentation bacteria are responsible for pickles, cheese, and cured sausages. Generally, yeasts are used to produce ethanol, and bacteria are used to produce lactic acid. There are exceptions: bread yeast is employed to produce carbon dioxide bubbles in leavened dough, and *Gluconobacter* is a genus of bacteria that can produce acetic acid (vinegar).

Fermented foods are generally produced using plant or animal ingredients in combination with fungi or bacteria that are either present in the environment or carefully conserved in cultures maintained by humans. Just as living organisms cover the surface of the earth, fermentation microbes cover the surface

TABLE 8.1

Some Common Fermented Foods

Food	Raw Material	Fermentors (Selected Species)
Beer	Grain malt	*Saccharomyces cerevisiae*
Bread	Grain flour	*Saccharomyces cerevisiae*
Butter	Milk	*Streptococcus* spp.
		Leuconostoc spp.
Cheese	Milk	*Lactobacillus* spp. (primary)
		Lactococcus spp. (primary)
		Pediococcus spp. (primary)
		Streptococcus spp. (primary)
		Leuconostoc spp. (secondary)
		Propionibacter spp. (secondary)
Chocolate	Cacao bean	*Saccharomyces cerevisiae*
		Candida rugosa
		Kluyveromyces marxianus
Coffee	Coffee bean	*Erwinia dissolvens*
Kefir	Milk	*Saccharomyces kefir*
		Torula kefir
Kimchi	Napa, Daikon	*Lactobacillus plantarum*
		Lactobacillus brevis
		Streptococcus faecalis
		Leuconostoc mesenteroides
		Pediococcus pentosaceus
Miso	Soybean	*Aspergillus oryzae*
Olives	Olives	*Candida* spp.
		Cryptococcus spp.
		Debaryomyces hanseii
		Lactobacillus spp.
		Saccharomyces spp.
Pickles	Cucumber	*Leuconostoc mesenteroides*
		Lactobacillus spp.
Sauerkraut	Cabbage	*Coliform* spp.
		Leuconostoc plantarum
		Lactobacillus spp.
Sausages	Meat and spices	*Pediococcus* spp.
		Lactobacillus spp.
Soy sauce	Soybean	*Aspergillus oryzae*
Tempeh	Soybean	*Rhyzopus oligosporus*
Vinegar	Fruit juice	*Saccharomyces cerevisiae*
		Gluconobacter spp.
		Acetobacter spp.
Wine	Grapes	*Saccharomyces cerevisiae*
		Saccharomyces bayanus
Yogurt	Milk	*Lactobacillus bulgaricus*
		Streptococcus thermophilus

Note: The fermentation organisms listed comprise only some of the most well-known members of the fermentation ecosystem. There are many more.

of the organisms. Wild yeasts are found living on grapes (Chamberlain et al. 1997), and bacteria line the human digestive tract. Kimchi, a spicy pickled food that Koreans eat at almost every meal, has scores, perhaps hundreds of species living in it (Lee et al. 2005), and previously unknown species of bacteria are being discovered in kimchi microflora (Kim et al. 2000, 2003; Lee et al. 2002; Yoon et al. 2000). The ecology of fermented foods is still a frontier of discovery.

8.3 Ecosystem Succession in Pickled Foods

One can observe ecological succession in the fermentation of sauerkraut. Ecological succession involves a community of organisms changing their environment such that the environment triggers changes in the species populating that community over time (Odum 1971). The word *succession* conjures an image of a reasonably directional and, therefore, predictable process. Ecological succession in sauerkraut involves populations of bacteria creating conditions for subsequent bacteria to thrive. Making sauerkraut is simple: chop the cabbage, mix with salt, and then press down the cabbage/salt mixture in a container (see Figure 8.2). Call this the "disturbance phase" of the sauerkraut ecosystem. Three tablespoons of salt will extract the juice from a gallon of chopped cabbage via osmosis. The salt must not contain iodine or other anticaking agents. The salty solution or brine inhibits the growth of putrefying organisms while creating ideal conditions for the growth of *Coliform* bacteria, which are ubiquitous in the air. Call *Coliform* the "pioneer species" of the sauerkraut ecosystem. *Coliform* bacteria produce acids, lower the pH of the kraut, and set the stage for *Leuconostoc* bacteria to colonize the medium. *Leuconostoc* bacteria lower the pH of the sauerkraut further still, thus creating the conditions for *Lactobacillus* to grow (Katz 2003). *Lactobacillus* adds the characteristic taste of lactic acid to the food, a smooth note in the otherwise tangy kraut. Lactic acid can create a temporary equilibrium state—call that moment the "climax" of the sauerkraut ecosystem. Refrigeration halts the fermentation process. Of course, unrefrigerated kraut will continue to ferment, dry, decompose, and attract insects. It becomes compost. The phenomenon is similar to the aging of mineral soils, which lose organic matter and become more acidic as they age (Brady and Weil 2002).

Ecological succession is present in cheese fermentation as well. Dairy fermentations involve the conversion of lactose to lactic acid through several successional phases. Raw milk, after it is collected, contains multiple bacterial populations. Fresh milk has a neutral pH and a warm temperature—perfect conditions for the rapid multiplication of *Lactococcus* bacteria. *Lactococcus* immediately begins producing lactic acid, lowering pH, and thus creating favorable conditions for the growth of *Lactobacillus* bacteria, which convert lactose to lactic acid and lower the pH further still (Flórez and Mayo 2006). By using up nutrients, lowering pH, and producing metabolic wastes, each microbe creates conditions for the next species. Many cheeses are also surface ripened by molds, yeast, or other bacteria that significantly alter the flavor. Some yeasts, coming after *Lactobacillus*, can consume lactic acid. This is the end stage for some live-rind cheeses, where the fermentation is halted (Marcellino and Benson 1992). The site-specific adaptation of wild microbes can be quite strong: in one study, cheeses inoculated with human-selected bacterial strains were instead ripened via a sequence of local microflora that crowded out the introduced bacteria and rendered the inoculum superfluous (Brennan et al. 2002). Cheeses, like sauerkraut, ultimately end up being

FIGURE 8.2   Components of a lactic acid fermentation (sauerkraut)—vegetable (cabbage), noniodized salt, weight, and time.

decomposed and returned to organic matter in soils or waterways; within the larger ecosystem, cheese fermentation is a microcosm of ecological processes that result from human choices.

Cheese recipes vary in temperature, water/salt proportions, and storing location during aging. In dairy fermentation ecosystems, these are variations of microclimate, moisture regime, and geography. For centuries, cheesemakers have refined their methods of guiding the succession of bacteria and yeast that produce the best cheeses by creating controlled conditions in which selection pressures determine microbe survival (Marcellino et al. 2001). Although cheesemakers may have simply thought of themselves as altering temperature, water, and salt (and, indeed, they were), they were also altering microbial population growth dynamics, ecosystem species composition, and microbial survival such that the desired flavor compounds were produced in a stable manner. The geographic location of each fermentation ecosystem affected species composition and, thus, the ecological stopping point at which point a cheese's ecological succession tends toward the equilibrium. Depending on the region and cheese variety, a stable cheese can last for weeks, months, or even years. The cultural value comes in the variety of nutritional compounds of the dairy product preserved in cheese for longer than it would have lasted without the controlled conditions. In this sense, human fermentation can be seen in terms of slowing down the ecological process of decomposition.

The same equation roughly applies for all pickled foods and hard sausages—lactic acid fermentation generally involves reducing the amount of water in the food, increasing the amount of salt, and controlling the temperature such that the desired microorganisms lower the pH and ultimately produce lactic acid, effectively preserving the food while producing otherwise unavailable flavors and nutritional compounds. This is the archetypical lactic acid fermentation ecosystem. Most lactic acid fermentations also produce a salty liquid solution (e.g., brine, whey, and tamari) that also may be considered a biologically active fermented food. This is in contrast to alcoholic fermentations, in which the final product is usually dead and sterile.

8.4 Alcoholic Fermentation as a Model Ecosystem

Alcoholic fermentation ecosystems tend to be dominated by *Saccharomyces*. Beer is produced by fermenting maltose from malted grains. Wines are produced by fermenting fruit juices. Species of *Saccharomyces* are ambient in the environment and are present on the skins of fruits such as ripe grapes (Holloway et al. 1990). Traditional European grape wines are products of environmental conditions in wine-growing areas of Europe, which have distinct soils, weather patterns, and even yeasts. Research suggests that there is a succession of wine-fermenting microorganisms, that yeast strain populations differ between batches and years, and that wine quality is therefore vineyard and vintage dependent (Schütz and Gafner 1994). Wine yeasts can outcompete the pathogenic onslaught introduced during barefoot stomping without pasteurization and produce a 12% alcohol (and, thus, pathogen-free) wine for the table. Domesticated *Saccharomyces* (store-bought wine yeast or brewer's yeast) combined with different ingredients and environmental conditions can be manipulated using simple equipment (see Figure 8.3) to model environmental chemistry, population dynamics, nutrient limitation, and the competition for finite resources in ecosystems.

FIGURE 8.3 Components of wine fermentation—fruit (elderberry), pot, bucket, carboy, siphon tube, and airlocks to allow carbon dioxide to escape from the containers.

Controlled alcoholic fermentation involves pasteurizing a sugary solution called a must and then later adding selected strains of *Saccharomyces* to the cooled must to convert the sugar to alcohol. The components of a finished wine or beer include flavor compounds that also function as nonliving chemical components of the fermentation environment. The different tastes of wines, ports, meads, beers, and distilled liquors come from the different relative proportions of alcohol and the other compounds remaining in the final product as a consequence of must preparation, yeast-strain selection, finishing technique, and storage conditions. Distinguishable flavor compounds include (but are not limited to) organic acids (acetic acid, citric acid, tartaric acid, malic acid, and lactic acid), esters, carbonyl compounds (diacetyl and aldehydes), sulfur compounds (dimethyl sulfide and hydrogen sulfide), and residual sugars (Berry and Slaughter 2003). This is the environmental chemistry of the must; there are several books that focus primarily on the chemistry aspects of alcoholic fermentation in relation to the activity of yeast and malolactic bacteria (Janson 1996; Ribéreau-Gayon et al. 2000).

One can observe a clear application of Liebig's law (also known as "the law of the minimum") in the production of wine. Liebig's law states that the growth of an organism is not based on the combined resources available to it, but rather on the availability of the scarcest resource (van der Ploeg et al. 1999). The scarcest resource is often called the "limiting factor." A common example is nitrogen availability in terrestrial plant growth—so long as water, sunlight, oxygen, and other necessities are available, a plant's growth is typically limited by the scarcity of nitrogen in the soil medium. In nature, *Saccharomyces* are limited by the unavailability of sugars. Home-scale alcoholic fermentations are often started by producing a "starter culture" in a plastic bottle consisting of sugar, fruit juice, and yeast. Adding a packet of wine yeast, beer yeast, or even bread yeast should work, because they are all the same species (although bread yeast strains will not produce delicious beverages). The yeast multiplies rapidly and begins releasing carbon dioxide (CO_2) and ethanol within a few hours. The CO_2 must be released periodically or else the pressure will build up and burst the bottle. The goal of making a starter culture is to produce a large colony of live *Saccharomyces* to inoculate a must. This is achieved by supplying all of the necessary elements for yeast metabolism in abundance.

Fermented food ecosystems are marked by a lack of primary production (i.e. photosynthesis), and the result is a competition among microbes for scarce carbon. Whatever carbon is added at the beginning of fermentation will be the maximum carbon available for the duration of the fermentation (assuming no new ingredients are added). Competition for carbon is probably what led to the evolution of "killer yeasts." These yeasts secrete proteins that kill other yeast strains (Woods and Bevan 1968). Killer *Saccharomyces* strains are sometimes used in unpasteurized musts to ensure fast fermentation, for the killer yeasts also tend to work very quickly in scavenging the finite carbon available. The killer character has also been studied for its potential to control undesirable yeast growth in nonfermented food products (Palpacelli et al. 1991); it may also help diagnose and treat yeast infections in the human body (Polonelli and Conti 2009; Polonelli et al. 1994).

In the absence of primary producers, the fermentation ecosystem is thus an ecosystem of consumers. Nutrient cycling, habitat/niche, and competition for resources may be demonstrated by manipulating the availability of nutrients provided at the outset by the humans who set up the system. There is evident cycling of nutrients from dead yeast and sediments in fermentation musts. There are different fermentation "climates": ale yeasts prefer to ferment at 70°F, whereas lager yeasts do better around 55°F. *Rhyzopus oligosporus* (a mold fungus) cannot ferment soybeans to tempeh unless it is very warm (>80°F). The pods of coffee, chocolate, and vanilla beans are fermented outdoors in the sun in tropical parts of the world and likely provide a unique habitat to the microbes that fulfill that role. In the absence of oxygen, yeasts (*Saccharomyces*) ferment grape juice to wine, but in the presence of oxygen, acetic acid bacteria (*Acetobacter*) can oxidize the ethanol to acetic acid (vinegar; Fleet 1992). These are the climatic zones of the global fermentation ecosystem.

8.5 Fermentation Ecosystems around the Globe

Ecological studies of fermentations are outnumbered by reports on production practices. In spite of their ubiquity in terrestrial and aquatic systems, the ecological functionality of yeasts remains relatively unexplored in comparison with other microorganisms (Herrera and Pozo 2010). There may be practical applications to the mapping of ecosystem science onto fermentations. This will be complicated by the current trend away from the study of discrete ecosystems. Today, the concept of a *whole ecosystem* is usually applied to the global ecosystem, as many have abandoned the idea that natural ecosystems have boundaries. However, anthropogenic ecosystems such as agroecosystems do have discrete boundaries, and fermentations illustrate natural phenomena bounded by human choices/constraints.

The repeated fermentation of a single food product in a given place over long periods of time may be regarded as a process of domestication. Registered appellations (the official wine-growing regions of a country) are a product of local yeasts, local soils, and local cave conditions that produce distinct regional wines, even when different regions are using the same grape cultivar. The same may be said for cheeses. For centuries, cheesemakers have domesticated the yeast strains that produce the best cheeses by creating controlled conditions in which human domestic selection pressures determine yeast survival (Marcellino et al. 2001). By altering salt, water, and temperature, cheesemakers were altering microbial environments, affecting microbial survival based on adaptation to the cheesemaking processes over centuries. The influence of human choice over fermentation processes reveals different ecosystem responses to these perturbations.

Indeed, some fermented food cultures are themselves biodiverse and are comprised of symbiotically associated bacteria and fungi. Kefir is a fermented milk product in which lactose has been converted to alcohol and lactic acid. Kefir is produced by kefir "grains," which are tapioca-like colonies of bacteria and yeast (see Figure 8.4). Recent studies have found kefir grains to include populations of *Zygosaccharomyces*, *Candida*, *Leuconostoc*, *Lactococcus*, *Lactobacillus*, and *Cryptococcus*, which grow differentially in successional phases (Witthuhn et al. 2005). Adding live kefir grains to milk produces a sour yogurt-like beverage within about 48 h, and if the solution is sealed during fermentation, it will develop effervescence (Katz 2003). Complete kefir fermentation produces a lactose-free food that lactose-intolerant people can digest. Kefir has been investigated as a leavener for bread (Plessas et al. 2005), as a starter culture for cheeses (Goncu and Alpkent 2005), and as a source of kefiran, an insoluble polysaccharide with antibacterial and cicatrizing properties that show potential for use in medicine (Rodrigues et al. 2005). Kefir is a stable probiotic food that may be kept for months without spoiling, and the kefir colonies provide a tangible, visible sense of the biodiversity that inhabits this fermented food. Kefir points to a means of solving environmental problems by using diversity instead of eliminating it.

The various fermentations of the world intersect with the study of human cultures and histories (Steinkraus 1995). Miso, a fairly fundamental part of the Japanese diet, dates back to the introduction of Buddhism in Japan when there was a movement away from fish ferments toward the savory soybean miso (Shurtleff and Aoyagi 2001). In harsh habitats such as the Darfur region of Sudan, the extremely small window of ecological primary production necessitates the fermentation of almost everything, down to the bones of cattle, in order to make food available year-round (Dirar 1993). In Europe, fermented meats (e.g., bologna and salami), bread, cheese, and wine are indispensible and all date back to before the Roman Empire. The archeology of prehistoric fermented foods has its own literature and distinct applications for ecological information (McGovern et al. 1996).

FIGURE 8.4 Kefir "grains" and kefir beverage. Kefir combines many of the properties of various fermented foods in a single product.

8.6　Fermentation Ecosystem as an Educational Model

Fermented foods provide an excellent classroom activity for ecological instruction, because they provide simple models and the potential complex experimentation. Additionally, it is often the case that ecological education will not make its way into high school classrooms, unless it is hybridized with other units such as biology and chemistry (DiEnno and Hilton 2005). One challenge facing the environmental educator is the difficulty of modeling a discrete ecosystem at the classroom scale. While the project method of learning has long since been established as a method of inspiring student interest (Kilpatrick 1918), the synthetic power of classroom experiments has largely been applied in the arts (Knoll 1997). Fermented foods can be used to construct ecosystem demonstrations that are small scale and short term. In the case of lactic acid fermentations, they may also provide a classroom snack.

A loaf of sourdough bread is easier to explain than a large ecosystem such as an arctic tundra, and it is more tangible for students as well. Sourdough could be used to address the lack of autonomy of open biotic communities mentioned above in reference to the decline of the concept of discrete ecosystems. San Francisco sourdough cultures taken to other parts of the world lose their San Francisco flavor qualities after about 6 months. Thus, a baker who produces San Francisco sourdough bread in New Jersey will have to constantly import new starter cultures to maintain the integrity of a San Francisco sourdough bread operation. Tasting the fermented foods at different points in time allows for a primary experience of the link between flavor and an evolving biotic community of yeast populations. It is otherwise difficult for humans to have a tangible experience of the consequences of a biotic community, changing over millions of generations in a short period of time.

Alcohol fermentation could provide an excellent mechanism for demonstrating nutrient limitations in biology classrooms. *Saccharomyces* cannot live on sugar alone; thus, alcoholic fermentations can be limited by the absence of other nutrients. To illustrate, prepare a 1-gal must of sugar dissolved in water and add wine yeast to ferment the sugar. After a short flurry of activity, the sugar water fermentation will become stagnant or "stuck," because the microbial ecosystem is nutrient limited. It should be possible to overcome stuck sugar water ferments by adding a source of nutrients—a cup of fruit juice, for instance, provided it does not contain potassium sorbate (yeast killer). Thus, one can demonstrate that fermentation is not purely the conversion of sugar to alcohol any more than plant respiration is purely the conversion of H_2O and CO_2 to sugar. This exercise would fit nicely into a unit on the 16 chemical elements that are essential for life on earth; in some cases, students could experiment with delimiting chemical elements one by one.

Fermenting foods can be as easy as following a recipe (see Table 8.2), and concepts can be illustrated by modifying one variable at a time. It is difficult to capture the essence of continental ecosystems, but it is easy to distinguish the boundary of a fermentation container and to observe the processes taking place within those boundaries on a microbiological scale. These processes include but are not limited to (1) the effect of temperature/climate on species composition, (2) the effect of species composition on environmental chemistry, ecological succession, and population dynamics such as competition, climax,

TABLE 8.2

Some *How-To* Books for First-Time Fermentors

Title	Author	Publisher	Year
And That's How You Make Cheese!	Sokol, S.	iUniverse	2001
Classic Sourdoughs: A Home Baker's Handbook	Wood, E.	Ten Speed Press	2001
Home Cheese Making	Carroll, R.	Storey Books	2002
The Compleat Meadmaker	Schramm, K.	Brewers Publications	2003
The Book of Miso	Shurtleff, W.	Ten Speed Press	2001
The Complete Joy of Home Brewing, 3rd Edition	Papazian, C.	Collins	2003
The Joy of Home Winemaking	Garey, T.	Collins	1996
The Permaculture Book of Ferment and Human Nutrition	Mollison, B.	Tagari Publications	1993
Wild Fermentation	Katz, S.	Chelsea Green Publications	2003

and equilibrium, and (3) the concepts of nutrient limitation, selection pressures, domestication, and bio-diversity. The intentional use of fermented foods shows how all life processes are inextricably linked with ecological processes; the most enduring patterns of culture may be those that mimic the dynamics of ecosystems.

ACKNOWLEDGMENTS

The author is grateful to William Sullivan for his work on an earlier version of this text.

REFERENCES

Berry, D. R. and Slaughter, J. C. 2003. Alcoholic beverage fermentations. In: *Fermented Beverage Production*, 2nd edition, edited by A. G. H. Lea and J. R. Piggott, pp. 25–39. New York: Kluwer Academic/Plenum Publishers.

Brady, N. C. and Weil, R. R. 2002. *The Nature and Properties of Soils*, 13th edition, Upper Saddle River, NJ: Prentice Hall.

Brennan, N. M., Ward, A. C., Beresford, T. P., Fox, P. F., Goodfellow, M., and Cogan, T. M. 2002. Biodiversity of the bacterial flora on the surface of a smear cheese. *Appl. Environ. Microbiol.* 68:820–830.

Chamberlain, G., Husnik, J., and Subden, R. E. 1997. Freeze–desiccation survival in wild yeasts in the bloom of icewine grapes. *Can. Inst. Food Sci. Technol. J.* 30:435–439.

DiEnno, C. M. and Hilton, S. C. 2005. High school students' knowledge, attitudes, and levels of enjoyment of an environmental education unit on nonnative plants. *J. Environ. Educ.* 37(1):13–25.

Dirar, H. 1993. *The Indigenous Fermented Foods of the Sudan: A Study in African Food and Nutrition*. London: CAB International.

Fleet, G. H. 1992. The microorganisms of winemaking—Isolation, enumeration, and identification. In: *Wine: Microbiology and Biotechnology*, edited by G. H. Fleet, pp. 1–26. Newark, NJ: Harwood Academic Publishers.

Flórez, A. B. and Mayo, B. 2006. Microbial diversity and succession during the manufacture and ripening of traditional, Spanish, blue-veined Cabrales cheese, as determined by PCR-DGGE. *Int. J. Food Microbiol.* 110:165–171.

Goncu, A. and Alpkent, Z. 2005. Sensory and chemical properties of white pickled cheese produced using kefir, yoghurt or a commercial cheese culture as a starter. *Int. Dairy J.* 15:771–776.

Guasch-Jané, M. R., Andrés-Lacueva, C., Jáuregui, O., and Lamuela-Raventós, R. M. 2005. First evidence of white wine in ancient Egypt from Tutankhamun's tomb. *J. Archaeol. Sci.* 33:1075–1080.

Herrera, C. M. and Pozo, M. I. 2010. Nectar yeasts warm the flowers of a winter-blooming plant. *Proc. R. Soc. B* 277:1827–1834.

Holloway, P., Subden, R. E., and LaChance, M. A. 1990. The yeasts in a Riesling must from the Niagara grape-growing region of Ontario. *Can. Inst. Food Sci. Technol. J.* 23:212–216.

Janson, L. W. 1996. *Brew Chem 101: The Basics of Homebrewing Chemistry*. North Adams, MA: Storey Books.

Katz, S. E. 2003. *Wild Fermentation: The Flavor, Nutrition, and Craft of Live-Culture Foods*. White River Junction, VT: Chelsea Green Publishing Company.

Kilpatrick, W. H. 1918. The project method. *Teachers College Record* 19:319–335.

Kim, B., Lee, J., Jang, J., Kim, J., and Han, H. G. 2003. *Leuconostoc inhae* sp *nov.*, a lactic acid bacterium isolated from kimchi. *Int. J. Syst. Evol. Microbiol.* 53:1123–1126.

Kim, J., Chun, J., and Han, H. U. 2000. *Leuconostoc kimchii* sp *nov.*, a new species from kimchi. *Int. J. Syst. Evol. Microbiol.* 50:1915–1919.

Knoll, M. 1997. The project method: Its vocational educational origin and international development. *J. Ind. Educ.* 34(3):59–80.

Lee, J. S., Heo, G. Y., Lee, J. W., Oh, Y. J., Park, J. A., Park, Y. H., Pyun, Y. R., and Ahn, J. S. 2005. Analysis of kimchi microflora using denaturing gradient gel electrophoresis. *Int. J. Food Microbiol.* 102:143–150.

Lee, J. S., Lee, K. C., Ahn, J. S., Mheen, T. I., Pyun, Y. R., and Park, Y. H. 2002. *Weissella koreensis* sp *nov.*, isolated from kimchi. *Int. J. Syst. Evol. Microbiol.* 52:1257–1261.

Marcellino, N. and Benson, D. R. 1992. Scanning electron and light microscopic study of microbial succession on Bethlehem St-Nectaire cheese. *Appl. Environ. Microbiol.* 58:3448–3454.

Marcellino, N., Beuvier, E., Grappin, R., Guéguen, M., and Benson, D. R. 2001. Diversity of *Geotrichum candidum* strains isolated from traditional cheesemaking fabrications in France. *Appl. Environ. Microbiol.* 67:4752–4759.

McGovern, P. E., Fleming, S. J., and Katz, S. H. Eds. 1996. *The Origins and Ancient History of Wine.* Amsterdam: Overseas Publishers Association.

McGovern, P. E., Zhang, J. H., Tang, J. G., Zhang, Z. Q., Hall, G. R., Moreau, R. A., Nuñez, A., Butrym, E. D., Richards, M. P., Wang, C., Cheng, G., Zhao, Z., and Wang, C. 2004. Fermented beverages of pre- and proto-historic China. *Proc. Natl. Acad. Sci. U. S. A.* 101:17593–17598.

Mollison, B. 1993. *The Permaculture Book of Ferment and Human Nutrition.* Tyalgum, Australia: Tagari Publications.

Moseley, M. E., Nash, D. J., Williams, P. R., deFrance, S. D., Miranda, A., and Ruales, M. 2005. Burning down the brewery: Establishing and evacuating an ancient imperial colony at Cerro Baúl, Peru. *Proc. Natl. Acad. Sci. U. S. A.* 102:17264–17271.

Odum, E. P. 1971. *Fundamentals of Ecology.* Philadelphia: W.B. Saunders Company.

Palpacelli, C., Ciani, M., and Rosini, G. 1991. Activity of different 'killer' yeasts on strains of yeast species undesirable in the food industry. *FEMS Microbiol. Lett.* 84:75–78.

Pasteur, L. 1858. *Mémoiressur La Fermentation Appeléelactique. Annales De Chimieet De Physique*, 3rd series 52:404–418.

Pelter, M. 2006. Brewing science. *J. Coll. Sci. Teach.* 35:48–52.

Picard, C., Fioramonti, J., Francois, A., Robinson, T., Neant, F., and Matuchansky, C. 2005. Review article: Bifidobacteria as probiotic agents—Physiological effects and clinical benefits. *Aliment. Pharmacol. Ther.* 22:495–512.

Plessas, S., Pherson, L., Bekatorou, A., Nigam, P., and Koutinas, A. A. 2005. Bread making using kefir grains as baker's yeast. *Food Chem.* 93:585–589.

Polonelli, L., Bernardis, F. D., Conti, S., Boccanera, M., Gerloni, M., Morace, G., Magliani, W., Chezzi, C., and Cassone, A. 1994. Idiotypic intravaginal vaccination to protect against candidal vaginitis by secretory, yeast killer toxin-like anti-idiotypic antibodies. *J. Immunol.* 152(6):3175–3182.

Polonelli, L. and Conti, S. 2009. Biotyping of *Candida albicans* and other fungi by yeast killer toxin sensitivity. *Methods Mol. Biol.* 499:97–115.

Reid, G., Jass, J., Sebulsky, M. T., and McCormick, J. K. 2003. Potential uses of probiotics in clinical practice. *Clin. Microbiol. Rev.* 16:658–672.

Ribéreau-Gayon, P., Dubourdieu, D., Douèche, B., and Londvand, A. 2000. *Handbook of Enology*—Volume 1: *The Microbiology of Wine.* New York: John Wiley & Sons.

Rodrigues, K. L., Caputo, L. R. G., Carvalho, J. C. T., Evangelista, J., and Schneedorf, J. M. 2005. Antimicrobial and healing activity of kefir and kefiran extract. *Int. J. Antimicrob. Agents* 25:404–408.

Ross, R. P., Fitzgerald, G., Collins, K., and Stanton, C. 2002. Cheese delivering biocultures—Probiotic cheese. *Aust. J. Dairy Technol.* 57:71–78.

Sanders, M. E. and Marco, M. L. 2010. Food formats for effective delivery of probiotics. *Annu. Rev. Food Sci. Technol.* 1:65–85.

Schütz, M. and Gafner, J. 1994. Dynamics of the yeast-strain population during spontaneous alcoholic fermentation determined by chef gel-electrophoresis. *Lett. Appl. Microbiol.* 19:253–257.

Shurtleff, W. and Aoyagi, A. 2001. *The Book of Miso: Savory, High-Protein Seasoning*, Vol. 1, 2nd edition. Berkeley, CA: Ten Speed Press.

Steinkraus, K. H., editor. 1995. *Handbook of Indigenous Fermented Foods.* New York: Marcel Dekker, Inc.

van der Ploeg, R. R., Böhm, W., and Kirkham, M. B. 1999. On the origin of the theory of mineral nutrition of plants and the Law of the Minimum. *Soil Sci. Soc. Am. J.* 63:1055–1062.

Witthuhn, R. C., Schoeman, T., and Britz, T. J. 2005. Characterization of the microbial population at different stages of kefir production and kefir grain mass cultivation. *Int. Dairy J.* 15:383–389.

Woods, R. A. and Bevan, E. A. 1968. Studies on the nature of the killer factor produced by *Saccharomyces cerevisiae. J. Gen. Microbiol.* 51:115–126.

Yoon, J. H., Kang, S. S., Mheen, T. I., Ahn, J. S., Lee, H. J., Kim, T. K., Park, C. S., Kho, Y. H., Kang, K. H., and Park, Y. H. 2000. *Lactobacillus kimchii* sp *nov.*, a new species from kimchi. *Int. J. Syst. Evol. Microbiol.* 50:1789–1795.

Part II

Fermented Milk
and Semisolid Cheeses

9

Fermentation and Koumiss

Wenyi Zhang and Heping Zhang

CONTENTS

9.1 Introduction

Fermentation is one of the oldest methods exploited for the preservation of food and beverages (Ross et al. 2002). Fermentation depends on the biological activities of edible microorganisms and their enzymes, which could convert food substrates into nontoxic products with desirable aromas, flavors, and textures, as well as nutritional properties (Gadaga et al. 1999). Applications of fermentation could be traced back to the dawn of history on earth. Advances in the knowledge of fermentation are recognized to have played key roles in the development of human culture and technology (Steinkraus 1997). Throughout the history and around the world, there are several traditional fermented products that have been documented, such as cheeses, pancakes, and porridges. Some of these foods resulting from the fermentation were used by local people to mark major life events including victories, auspicious events, and harvests (McGovern et al. 2004).

Koumiss, also called chige, chigo, arrag, or airag (in Mongolian language), is a fermented low alcohol–containing beverage. The name koumiss is probably derived from a tribe, the Kumanes, who live in the area along the Kuma River in the Asiatic Steppes (Robinson et al. 2002). For centuries, koumiss has been a popular drink and considered a complete nutriment with medicinal properties to assist in the treatment of ailments (Ishii and Konagaya 2002). In addition, it is served in special celebratory activities, particularly among populations in Mongolia and Inner Mongolia of China. People in those regions of China imbibed koumiss during grand festivities and sacrificial offerings (Hasisurong et al. 2003).

Typical koumiss contains 0.6%–3% alcohol and a small amount of carbon dioxide (Park et al. 2006). It is traditionally made from unpasteurized fresh mare's milk. There are two types of koumiss, namely, sweet koumiss (0.6%–0.8% acidity and 0.7%–1.0% ethanol) and strong koumiss (1%–1.2% acidity and 1.8%–2.3% ethanol). During its production, lactic acid bacteria (LAB) acidify the milk, followed by the alcoholic fermentation of the residual sugar by yeasts (Wu et al. 2009). This chapter aims at providing an overview of the current scientific knowledge on koumiss regarding its origin, historical background, and microbiological and beneficial properties, as well as technological improvement that has been achieved on its production.

9.2 Origin of Koumiss

Koumiss originated in nomadic tribes of Central Asia, who lived between the Caucasian mountains and Mongolia. The emergence of koumiss is closely linked to the nomadic ways of life and the domestication of certain mammals such as goat, sheep, cow, camel, yak, and mare.

Initially, the reason for making koumiss could have been simply to preserve the valuable nutrients in milk. The prototype koumiss was therefore assumed to have been made accidentally by the spontaneous fermentation of mare's milk during nomadic migration. Mare's milk, which was carried in a back pocket on horseback and turned over on occasion or strapped to the saddle and joggled around over the course of a day's riding, could have resulted in the formation of a koumiss-like product. For nomadic people, semisolid, highly nutritious foods probably were representative of the foods that they consumed. In response to variations in climatic and environmental conditions in different grasslands, nomads together with their livestock had to travel from place to place. Consequently, they adhered to a simple lifestyle. Hence, it is not surprising that conveniently transportable drinks such as koumiss were favored.

9.3 Historical Background of Koumiss

Koumiss is widely known to be an ancient beverage. The production of koumiss could be traced back to the ancient Greeks and Romans. The Greek historian Herodotos (485–425 BC) reported that a refreshing drink produced from mare's milk was popular among the Ghets tribes (Wszolek et al. 2006). By the 13th century, similar accounts of the koumiss-making process were recorded by the traveler William of Rubruck. Of course, the preparation of koumiss developed as an artisanal practice without any knowledge of the role of microorganisms involved. In China, koumiss also has a long history of use as a popular drink among traditionally nomadic people. The earliest record of koumiss appeared in the Han dynasty of China (202 BC–202 AD) and attained widespread popularity during the Yuan dynasty (1271 AD–1368 AD). Nowadays, koumiss is popular in Central Asia, Eastern Europe, Russia, Mongolia, and some areas of China such as Inner Mongolia and Xinjiang with different product names (Park et al. 2006). In some of those areas, it is manufactured on an industrial scale (Park et al. 2006).

9.4 Production of Koumiss

Because a mare can only be milked as long as she has an unweaned foal, koumiss is normally made during the short milking period, usually from July to September through the summer. During this period, a lactating mare should be milked at 2-h intervals about five to six times per day. Around 3–5 kg of milk can be obtained from one lactating animal. Depending on the breed, nutrition, environment, health, and management conditions, the exact quantity of milk production varies among individual mares (Doreau and Boulot 1989). Around the world, horses are still milked by hand in many places (Levin 1997), as shown in Figure 9.1.

Traditionally, fermentation of koumiss took place in a wooden cask or animal skins (Figure 9.2a and b). At present, urns may be used. To make koumiss, two urns are left outside the yurt and placed side by side, with the bottom of the urn buried to a depth of about 30 cm into the ground. Then, raw, filtrated mare's milk is poured into these containers and churned (Figure 9.2c), beaten with a wooden stick, and held at

FIGURE 9.1   Mare milking. (Photo by Heping Zhang.)

ambient temperature (approximately 20°C) for 1–3 days for the propagation of microflora. Every day, after removing the first portion for consumption (Figure 9.2d), a small aliquot of koumiss from one of the urns is retained for use as starters for the next day, and more milk is added to provide an ongoing fermentation. Agitation of the fermentation liquor (using a wooden paddle) at frequent intervals is believed to be a key factor in developing desirable flavors in the product and in avoiding abnormal fermentation. Sometimes, leaves of certain flowering plants or a small portion of horse meat are added for accelerating the fermentation process.

In a modern production system, pasteurized mare's milk and starter composed of pure cultures of thermophilic LAB and yeasts are used. Before use, the LAB and yeast cultures are mixed together to produce a composite bulk starter culture. The bulk starter culture is added at the rate of 30% to processed mare's milk. Fermentation is carried out at a temperature of around 25°C for 2 h with frequent stirring. After packaging, fermentation is continued in bottles at 18°C–20°C for 2–3 h, then cooled to 4°C–6°C, and stored until consumption (Park et al. 2006).

FIGURE 9.2   (a) Koumiss fermented in wooden cask. (b) Koumiss fermented in a leather bag. (c) Starter and churning rod for koumiss fermentation. (d) Koumiss for drinking. (Photos by Heping Zhang.)

9.5 Raw Materials for Making Koumiss

Koumiss is produced mostly from mare's milk. As an example of the volume of mare's milk used for koumiss production, Steinkraus (1996) reported one document that described that 230,000 horses were kept in the Union of Soviet Socialist Republics specifically for producing koumiss.

Compared with milk from other species, the major composition of mare's milk is quite similar to that of human milk but is significantly different from that of cow's milk. Both mare's milk and human milk contain comparable levels of whole salt and protein content (Malacarne et al. 2002). It has been shown that the richness in whey protein in mare milk makes it more favorable for human consumption because of the higher content of essential amino acids (Malacarne et al. 2002). As does human milk, mare's milk has a lower proportion of saturated fatty acids with a low and high number of carbon atoms (C4:0, C6:0, C16:0, and C18:0; Malacarne et al. 2002). Furthermore, mare's milk is rich in polyunsaturated fatty acids, especially linoleic acid, and it has high vitamin C and lactose content (Pagliarini et al. 1993).

Because the availability of mare's milk falls short of the demand for industrial-scale manufacture of koumiss, there has been a growing interest in the use of cow's milk for koumiss production. Much research was carried out to determine the physical, chemical, microbiological, and organoleptic properties of koumiss made using cow's milk. Research showed that cow's milk should be suitably modified to be close to the composition of mare's milk. A simple method of modifying cow's milk was firstly reported by Lutskova (1957). Proper amount of sucrose was added to cow's milk that was diluted with water. Similar attempts to modify cow's milk by the addition of water, lactose, whey, and ascorbic acid were also reported (Davidov and Sokolovskii 1963; Khrisanfova 1965; Seleznev and Artykova 1970; Gallmann and Puhan 1978; Kielwein and Daun 1978). Kucukecetin et al. (2003) developed a successful method of modifying cow's milk using membrane technologies, which yielded an end product very similar to koumiss made from mare's milk in terms of chemical and sensory properties.

9.6 Appearance and Composition of Koumiss

Koumiss prepared from mare's milk does not coagulate. It is ivory or slightly yellow in color and has a sharp acidic and alcohol flavor. Koumiss generally contains 0.7%–1.8% lactic acid, 0.6%–2.5% alcohol, 0.5%–0.88% carbon dioxide, 2%–4% milk sugar, and 2% fat (Oberman and Libudzisz 1985). The chemical composition of koumiss originating in Mongolia consists approximately of 3.9%–8.2% dry mass, 1.7%–2.1% protein, 1.2%–1.9% lipid, 0.3% ash, 1.0%–1.3% lactic acid, and 1.8%–2.9% alcohol (Ishii et al. 1997). Similar values can be found in koumiss from the Xinjiang province in China (Zhang et al. 1990). The fermentation process of koumiss in Mongolia is dependent on the indigenous microflora of the available milk, which produces lactic acid and alcohol as end products (Wang et al. 2008). Therefore, it has been postulated that the fluctuation in the composition of koumiss is mainly contributed by variations in temperature and time of incubation, which influences the growth and physiology of the microorganisms.

9.7 Microbiology of Koumiss

The microorganisms in koumiss mainly consist of LAB and yeast. Sometimes, acetic acid bacteria are also found. The combined metabolic activities of these microorganisms impart the unique product characteristics and quality to koumiss. During the natural fermentation process, LAB tend to grow faster than yeasts and produce lactate from lactose, giving a more suitable environment for yeasts to grow, while yeasts may produce vitamins and other stimulative factors to enhance the growth of LAB (Viljoen 2001). More complicated interactions between LAB and yeasts were postulated, resulting in a complementary metabolism, where a compound produced by one organism may be utilized by another (Narvhus and Gadaga 2003).

9.7.1 Lactic Acid Bacteria

The LAB composition in koumiss appears somewhat controversial. An earlier document suggested that common LAB in koumiss mainly consisted of *Lactobacillus* and *Lactococcus* species (Oberman and Libudzisz 1985). Thermophilic lactobacilli such as *L. delbrueckii* subsp. *bulgaricus* and *L. helveticus* play a major role in the process of lactic acid development. Recent investigations revealed that the koumiss made by traditional methods contains an abundance of LAB species, which varied considerably.

Sun et al. (2010a,b) systematically investigated the lactobacilli involved in homemade koumiss in Xinjiang, Inner Mongolia, and Qinghai in China by the combined use of conventional and molecular methods. A total of 46 koumiss samples were aseptically collected from nomadic families in these three areas, and 171 bacterial isolates were considered presumptive LAB. The predominance of *L. helveticus*, *L. casei*, and *L. plantarum* was observed in samples from Xinjiang, Inner Mongolia, and Qinghai, respectively, with a lower isolation frequency for *L. fermentum*, *L. diolivorans*, *L. acidophilus*, *L. pontis*, *L. kefiri*, *L. delbrueckii*, and *L. reuteri*. The overall distribution pattern showed that the *Lactobacillus* groups from the three geographically distant regions varied a lot. The number of identified species ranged from 3 to 7.

As might be expected, similar findings were reported in several other studies. However, the ratio between the frequently isolated species was different from each sampling site. For example, An et al. (2004) found that the isolates from koumiss in Inner Mongolia consisted of mainly *Lactobacillus* sp., *L. plantarum*, *L. pentosus*, and *Lactococcus lactis* ssp. *cremoris*. They were isolated at the rates of 48%, 33%, and 19%, respectively; *L. helveticus* and *L. kefiri* were major microflora in koumiss collected from the Donto-Govi prefecture in Mongolia (Uchida et al. 2007); and *L. helveticus* and *L. kefiranofaciens* were isolated from koumiss as the predominant LAB strains in samples from the Mongolian provinces of Arhangai, Bulgan, Dundgobi, Tov, Uburhangai, and Umnugobi (Watanabe et al. 2008). Overall, a few cocci strains have been isolated from koumiss partly due to the low pH values of the koumiss samples.

9.7.2 Yeast

As mentioned in some publications, yeasts are the secondary major component in koumiss. Taxonomically, strains from genera *Saccharomyces*, *Torula*, *Torulopsis*, and *Candida* could be commonly detected. Yeasts occurring in koumiss consist of three main types, namely, lactose fermenting, lactose nonfermenting, and carbohydrate nonfermenting yeasts. Lactose-fermenting yeasts that have been identified include *S. lactis*, *Kluyveromyces marxianus* subsp. *marxianus*, and *Candida kefyr*, whereas the other two types include *S. cartilaginosus* and *Mycoderma*.

Ni et al. (2007) studied 87 strains isolated from 28 homemade koumiss samples in the Xinjiang province in China. All the samples contained at least three species of yeasts, and *S. unisporus* were the most common. However, when considering the sampling site and time, differences among actual populations of the isolates were not apparent. In the same study, six strains of *K. marxianus*, five strains of *S. unisporus*, and one strain of *S. cerevisiae* were characterized in four samples from Wusu of the Yili Hasaka Autonomous Prefecture, and eight strains of *K. marxianus*, six strains of *S. unisporus*, and *Pichia membranaefaciens* were identified in seven samples from Xinyuan of the Yili Hasaka Autonomous Prefecture. It is obvious from the data that *K. marxianus* isolates are the predominant microorganisms in samples from these two areas.

Several investigations of yeasts in koumiss produced in Mongolia also reported similar results. Watanabe et al. (2008) isolated 68 yeast strains from 22 samples, which were grouped into eight species: *C. pararugosa*, *Dekkera anomala*, *Issatchenkia orientalis*, *Kazachstania unispora*, *K. marxianus*, *P. manshurica*, *S. cerevisiae*, and *T. delbrueckii*. Significant difference was observed in the cell count of *K. marxinsu* and *S. cerevisiae*, and *K. marxinsu* isolated as the predominant species from all koumiss samples. Using the denaturing gradient gel electrophoresis analysis, strains belonging to *D. bruxellensis* and *Kluyveromyces* were identified from six koumiss samples by Miyamoto et al. (2010). In a study performed with three koumiss samples, *S. dairensis* was found to be the most prevalent among yeast species *S. cerevisiae*, *Issatchenkia orientailis*, and *K. wickerhamii* (Uchida et al. 2007).

9.7.3 Acetic Acid Bacteria

Very little information about acetic acid bacteria in koumiss is available from the literature, probably because these bacteria could not be isolated using the agar plate technique. According to Park et al. (2006), the presence of *Acetobacter aceti* in koumiss links it to its typical vinegar-like acetic acid flavor development.

9.8 Therapeutic Properties of Koumiss

Koumiss is considered a wholesome beverage in areas where it is widely consumed. However, most documentary evidence in ancient books is based on traditional medicine, not on modern nutritional knowledge and analytical data. Overall, it is believed to be beneficial to a wide range of diseases, especially for postoperative care.

Using koumiss for the cure of chronic diseases, particularly for the treatment of tuberculosis patients, has been practiced for a long time. In the 19th century, Cossacks introduced koumiss for military rations to prevent tuberculosis (Ishii and Samejima 2001). Around that time, koumiss was often used as medicine for treating early-stage tuberculosis by people from Russia, Mongolia, and Inner Mongolia of China (Kosikowski 1977). This beneficial effect was finally confirmed by the long history of the therapeutic use of koumiss sanitaria built in local areas (Park et al. 2006). In the sanitaria, patients who suffered from tuberculosis were routinely dispensed koumiss for regular intake. A building that served as a sanitarium is depicted in Figure 9.3.

It is reported that koumiss could stimulate the immune system and promote antibacterial activities. These beneficial effects may be attributed to some of the ingredients added during fermentation and to certain microbial metabolic by-products such as peptides, bactericidal substances, synthesized vitamins, and fatty acids of the $n-3$ series (Park et al. 2006). Koumiss also has a hypocholesterolemic effect. Ishii and Samejima (2001) found that feeding rats with koumiss powder suppressed the serum cholesterol and triglyceride levels, but the potential mechanism has not been determined. In this study, the koumiss used was made using microorganisms isolated from koumiss itself.

Very recently, Chen et al. (2010) have indicated that koumiss is rich in angiotensin I-converting enzyme (ACE) inhibitory peptides. The peptides have an antihypertensive effect. Furthermore, they are suggested to be beneficial for cardiovascular health. They analyzed the variations of the ACE inhibitor in koumiss samples, which were digested by ACE, pepsin, tyrosinase, and chymotrypsin. From the digests, four novel ACE inhibitory peptides were isolated. Consistent evidence regarding the peptide attributes was subsequently reported by this research team. Sun et al. (2010a,b) determined that 16 out of the 81 strains of *Lactobacillus* isolated from koumiss exhibited ACE inhibitory activity.

Other favorable influences on the alimentary canal's activity, the circulatory and nervous systems, blood-forming organs, kidney functions, and endocrine glands have been reported by Fedechko et al. (1995), Stoianova et al. (1988), and Sukhov et al. (1986).

FIGURE 9.3   Koumiss sanitaria in Mongolia. (Photo by Heping Zhang.)

9.9 Conclusion

Koumiss is currently the most widely consumed mare's milk product where traditional horse breeding for dairy production has been developed. Although research on koumiss has made some progress in the past decades, the information obtained from controlled experimentations appears limited. Accordingly, further scientific input on koumiss is still necessary. Some important research areas, including the bioactive substances, interaction with microorganism metabolites, and the selection and production of pure cultures, will lead to a more successful industrialized production, and the functionality and consistency between koumiss made by industrial and traditional methods should be addressed.

ACKNOWLEDGMENTS

The authors wish to thank Dr. Ebenezer R. Vedamuthu and Dr. Sarn Settachaimongkon for their critical reading of the manuscript and English editing.

REFERENCES

An, Y., Adachi, Y., and Ogawa, Y. 2004. Classification of lactic acid bacteria isolated from chigee and mare milk collected in Inner Mongolia. *Anim. Sci. J.* 75:245–252.

Chen, Y., Wang, Z., Chen, X., Liu, Y., Zhang, H., and Sun, T. 2010. Identification of angiotensin I-converting enzyme inhibitory peptides from koumiss, a traditional fermented mare's milk. *J. Dairy Sci.* 93(3):884–892.

Davidov, R. B. and Sokolovskii, V. P. 1963. Koumiss from cows' milk. *Med. Prom.* 18:30–31.

Doreau, M. and Boulot, S. 1989. Recent knowledge on mare milk production: A review. *Livest. Prod. Sci.* 22:213–235.

Fedechko, I. M., Hrytsko, R., and Herasun, B. A. 1995. The anti-immunodepressive action of kumiss made from cow's milk. *Lik. Sprava* (9–12):104–106.

Gadaga, T. H., Mutukumira, A. N., Narvhus, J. A., and Feresu, S. B. 1999. A review of traditional fermented foods and beverages of Zimbabwe. *Int. J. Food Microbiol.* 53:1–11.

Gallmann, P. and Puhan, Z. 1978. Anwendung der ultrafiltration zur herstellung von kumys aus kuhmilch. *Schweiz. Milchwirtsch. Forsch.* 7:23–32.

Hasisurong, Amuguleng, Manglai. 2003. Koumiss and its value in medicine. *Zhongguo Zhongyao Zazhi* 28(1):11–14.

Ishii, S., Kikuchi, M., and Takao, S. 1997. Isolation and identification of lactic acid bacteria and yeasts from "chigo" in Inner Mongolia, China. *Anim. Feed Sci. Technol.* 68:325–329.

Ishii, S. and Konagaya, Y. 2002. Beneficial role of koumiss intake of Mongolian nomads. *J. Jpn. Soc. Nutr. Food Sci.* 55:281–285.

Ishii, S. and Samejima, K. 2001. Feeding rats with koumiss suppresses the serum cholesterol and triglyceride levels. *Milk Sci.* 50:113–116.

Khrisanfova, L. P. 1965. Manufacture and microflora of koumiss made from cow's skim milk. *Med. Prom.* 26:38–40.

Kielwein, G. and Daun, U. 1978. Ein neues getraenk nach nomadenart auf der basis von kuhmilcheiwei. *Dtsch. Apoth. Ztg.* 99:724–726.

Kosikowski, F. V. 1977. *Cheese and Fermented Milk Foods*, 2nd edition, pp. 40–46, Michigan: Edwards Bros.

Kucukecetin, A., Yaygin, H., Hinrichs, J., and Kulozik, U. 2003. Adaptation of bovine milk towards mare's milk composition by means of membrane technology for koumiss manufacture. *Int. Dairy J.* 13:945–951.

Levin, M. A. 1997. Eating horses: The evolutionary significance of hippophagy. *Antiquity* 72:90–100.

Lutskova, M. 1957. Simplified method for the preparation of koumiss from cow's milk. *Med. Prom.* 18(12):30–31.

Malacarne, M., Martuzzi, F., Summer, A., and Mariani, P. 2002. Protein and fat composition of mare's milk: Some nutritional remarks with reference to human and cow's milk. *Int. Dairy J.* 12:869–877.

McGovern, P. E., Zhang, J., Tang, J., Zhang, Z., Hall, G. R., Moreau, R. A., Nuñez, A., Butrym, E. D., Richards, M. P., Wang, C. S., Cheng, G., Zhao, Z., and Wang, C. 2004. Fermented beverages of pre- and proto-historic China. *Proc. Natl. Acad. Sci. U. S. A.* 101:17593–17598.

Miyamoto, M., Seto, Y., Nakajima, H., Burenjargal, S., Gombojav, A., Demberel, S., and Miyamoto, T. 2010. Denaturing gradient gel electrophoresis analysis of lactic acid bacteria and yeasts in traditional Mongolian fermented milk. *Food Sci. Technol. Res.* 16:319–326.

Narvhus, J. A. and Gadaga, T. H. 2003. The role of interaction between yeasts and lactic acid bacteria in African fermented milks: A review. *Int. J. Food Microbiol.* 86:51–60.

Oberman, H. and Libudzisz, Z. 1985. Fermented milks. In: *Microbiology of Fermented Foods*, 2nd edition, edited by B. J. B. Wood, pp. 308–345. London: Elsevier Applied Science Publishers.

Pagliarini, E., Solaroli, G., and Peri, C. 1993. Chemical and physical characteristics of mare's milk. *Ital. J. Food Sci.* 4:323–332.

Park, W., Zhang, H., Zhang, B., and Zhang, L. 2006. Mare milk. In: *Handbook of Milk of Non-Bovine Mammals*, 1st edition, edited by Y. W. Park and G. F. W. Haenlein, pp. 275–296. U.K.: Blackwell Publishing.

Robinson, R. K., Tamime, A. Y., and Wszolek, M. 2002. Microbiology of fermented milks. In: *Dairy Microbiology Handbook*, 3rd edition, edited by R. K. Robinson, pp. 367–421. United States: Wiley-Interscience.

Ross, R. P., Morgan, S., and Hill, C. 2002. Preservation and fermentation: Past, present and future. *Int. J. Food Microbiol.* 79:3–16.

Seleznev, V. I. and Artykova, L. A. 1970. Koumiss from cows' milk. *Med. Prom.* 27:86–91.

Steinkraus, K. H. 1996. *Handbook of Indigenous Fermented Foods*, 2nd edition, pp. 303–304. New York: Marcel Dekker.

Steinkraus, K. H. 1997. Classification of fermented foods: Worldwide review of household fermentation techniques. *Food Control* 8:311–317.

Stoianova, L. G., Abramova, L. A., and Ladodo, K. S. 1988. Sublimation-dried mare's milk and the possibility of its use in creating infant and dietary food products. *Vopr. Pitan.* 3:64–67.

Sukhov, S. V., Kalamkarova, L. I., Ilchenko, L. A., and Zhangabylov, A. K. 1986. Microfloral changes in the small and large intestines of chronic enteritis patients on diet therapy including sour milk products. *Vopr. Pitan.* 4:14–17.

Sun, T. S., Zhao, S. P., Wang, H. K., Cai, C. K., Chen, Y. F., and Zhang, H. P. 2010a. ACE-inhibitory activity and gamma-aminobutyric acid content of fermented skim milk by *Lactobacillus helveticus* isolated from Xinjiang koumiss in China. *Eur. Food Res. Technol.* 228:607–612.

Sun, Z., Liu, W., Zhang, J., Yu, J., Zhang, W., Cai, C., Menghe, B., Sun, T., and Zhang, H. 2010b. Identification and characterization of the dominant lactobacilli isolated from koumiss in China. *J. Gen. Appl. Microbiol.* 56:257–265.

Uchida, K., Hirata, M., Motoshima, H., Urashima, T., and Arai, K. 2007. Microbiota of 'airag,' 'tarag' and other kinds of fermented dairy products from nomads in Mongolia. *Anim. Sci. J.* 78:650–658.

Viljoen, B. C. 2001. The interaction between yeasts and bacteria in dairy environments. *Int. J. Food Microbiol.* 69:37–44.

Wang, J., Chen, X., Liu, W., Yang, M., and Zhang, H. 2008. Identification of *Lactobacillus* from koumiss by conventional and molecular methods. *Eur. Food Res. Technol.* 227:1555–1561.

Watanabe, K., Fujimoto, J., Sasamoto, M., Dugersuren, J., Tumursuh, T., and Demberel, S. 2008. Diversity of lactic acid bacteria and yeasts in airag and tarag, traditional fermented milk products of Mongolia. *World J. Microbiol. Biotechnol.* 24:1313–1325.

Wszolek, M., Kupiec-Teahan, B., Guldager, H. S., and Tamime, A. Y. 2006. Production of kefir, koumiss and other related products. In: *Fermented Milks*, 1st edition, edited by A. Y. Tamime, pp. 174–175. U.K.: Blackwell Publishing.

Wu, R., Wang, L., Wang, J., Li, H., Menghe, B., Wu, J., Guo, M., and Zhang, H. 2009. Isolation and preliminary probiotic selection of lactobacilli from koumiss in Inner Mongolia. *J. Basic Microbiol.* 49:318–326.

Zhang, L. B., Liu, J. S., and Li, F. G. 1990. Studies on orthogonal test of blend fermenting of three phases of intermediate culture of mare's milk and flavor evaluation of koumiss. *China Dairy Ind.* 18:246–251.

10

Traditional Finnish Fermented Milk "Viili"

Minna Kahala and Vesa Joutsjoki

CONTENTS

10.1 Traditional Nordic Fermented Milk

The preparation of sour milk products has been known in all parts of the world as a means of preserving milk against spoilage. Traditional sour milk foods were obtained by spontaneous acidification with lactic acid bacteria (LAB) naturally present in milk, which during their growth reduce the pH and denature milk proteins, causing appearance and texture changes in the product. In certain cases, yeasts and molds, which occur along with LAB, give a special feature to the sour milk product. Due to the specific characteristics of milk and LAB and the conditions of fermentation, various types of regionally specific fermented milk have been developed throughout the world.

The dominating microflora in traditional fermented milk typical of Nordic countries consists mainly of mesophilic lactococci, which are often accompanied by leuconostocs. Of these mesophilic cultures, *Lactococcus lactis* subsp. *lactis*, and *L. lactis* subsp. *cremoris* produce mainly lactic acid and are usually referred to as acid producers. *L. lactis* subsp. *lactis* biovar. *diacetylactis* and *Leuconostoc* species, which also ferment citric acid and produce metabolites like CO_2, acetaldehyde, and diacetyl, are referred to as flavor producers. A wide spectrum of metabolic activity of strains belonging to the same species has given rise to a variety of different fermented, regionally specific milk products. Fermented milk products typical of the Scandinavian region or more restricted local specialties are listed in Table 10.1.

Ropy or slimy variants of *L. lactis* subsp. *lactis* and *L. lactis* subsp. *cremoris*, which produce a glycoprotein-like substance in milk, are a prerequisite to generate the high viscosity typical of many traditional Nordic fermented milk products. The ropy texture has been shown to be plasmid mediated (Vedamuthu and Neville 1986; Neve et al. 1988), which explains why this trait is occasionally lost in ropy strains.

As mesophilic lactococci and leuconostocs are known to produce a wide variety of inhibitory substances acting against spoilage microorganisms and, on occasion, also against other starter strains in mixed-strain starters (Hugenholtz 1986), the antagonistic effect against other starter strains is a risk also in traditional fermentation processes. Yet, it seems that traditional mixed-strain starters, used in domestic conditions, can evolve to harmonious coexistence. This coexistence should be one of the main criteria for choosing strains to design starters for the commercial manufacture of dairy products based on traditional fermented milk.

TABLE 10.1

Traditional Nordic Fermented Milk

Product	Microbes	Technology	Region
Cultured buttermilk	*Lactococcus lactis* subsp. *cremoris* *Lactococcus lactis* subsp. *lactis*	Stirred fermented milk	Scandinavia
Långfil	*Lactococcus lactis* biovar. *longi* *Leuconostoc mesenteroides* subsp. *cremoris* (EPS-producing strains)	Set-type product, fermented milk with very high viscosity and ropy texture	Sweden
Skyr	*Lactobacillus delbrueckii* subsp. *bulgaricus* *Lactobacillus helveticus* Yeasts (Thermophilic starter culture)	Concentrated product, concentration of milk after fermentation	Iceland
Tjukkmjölk	Not identified (EPS-producing strains)	Set-type product, fermented milk with very high viscosity and ropy texture	Norway
Viili	*Lactococcus lactis* subsp. *cremoris* *Lactococcus lactis* subsp. *lactis* *Lactococcus lactis* subsp. *lactis* biovar. *diacetylactis* *Leuconostoc mesenteroides* subsp. *cremoris* *Geotrichum candidum* (EPS-producing strains)	Set-type product, fermented milk with very high viscosity and ropy texture Mold *Geotrichum candidum* used together with unhomogenized milk (velvet-like creamy layer on the surface)	Finland
Ymer	*Lactococcus lactis* subsp. *lactis* biovar. *diacetylactis* *Leuconostoc mesenteroides* subsp. *cremoris*	Concentrated product, concentration of milk before or after fermentation	Denmark

Source: Leporanta, K. 2010. Traditional Nordic Fermented Milk Products. Oral presentation in the IDF Symposium on Science and Technology of Fermented Milk, Tromsö, Norway, 7–9 June 2010.

10.2 Overview of Viili

Already in the 1800s, viili manufacturing was known in all regions of Finland. Originally, manufacturing of viili started in Sweden and was later imported to Finland. However, currently, viili is mainly produced in Finland. As with most traditional sour milk products, the home manufacture of natural viili was initiated with a small quantity of a previously prepared product. The industrial manufacture of viili began in the 1950s (Leporanta 2003), and to date, the production of viili has evolved to highly automated industrial-scale manufacture. Yet, there are still people making viili at home for household use by inoculating milk with a portion of commercial product and incubating it overnight at room temperature. Currently, yearly consumption of viili in Finland is 5.1 kg per capita, of which flavored viili accounts for over 40%. The traditional Finnish fermented milk viili has a pleasant, mild acid taste, a good diacetyl flavor, and a thick, stringy, or ropy consistency, which can be cut with a spoon (Figure 10.1). Even though

FIGURE 10.1 Structure of viili. (Photograph by Lucia Blasco, 2010.)

the production of viili has nowadays evolved to industrial-scale manufacture, the starter cultures in use are still originating in traditional mixed populations of lactococci and leuconostocs. The starter is composed of mesophilic acid–forming, polysaccharide-producing, and citrate-utilizing strains of LAB and the mold *Geotrichum candidum*. The polysaccharides that are excreted into milk during the exponential growth phase of starter LAB prevent syneresis and graininess and cause the thickness characteristic and ropy consistency of the product (Macura and Townsley 1984). The mold generates a special characteristic taste, aroma, and appearance to the product. For the manufacture of products with uniform and high quality, the technologically desirable features of mixed-strain starter cultures are extremely important. As the composition and specific characteristics of starters have a key role in successful manufacturing processes, there is a constant demand for rapid and reliable methods of screening the microbial populations of mixed-strain starter cultures.

10.3 Industrial Manufacturing of Viili

Commercially, natural viili is produced from unhomogenized bovine milk. The milk is separated, standardized, and heat treated before use for the manufacture of viili. The manufacture of viili is presented in Figure 10.2. During separation, the cream is skimmed from milk, and in a standardization step, fat content is adjusted (Thure 2000). The fat content of industrial viili products ranges from 1.0% to 3.5% (www.valio.fi; www.arlaingman.fi). After standardization, the milk is heat treated (Thure 2000). The

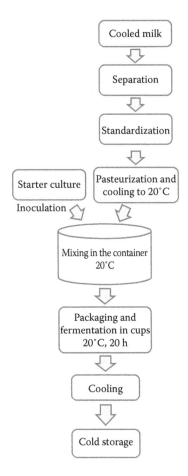

FIGURE 10.2 Industrial manufacture of viili.

starter mix inoculate is produced by incubating the microbes in milk at 20°C. Pasteurized milk cooled to 20°C is inoculated with 3%–6% starter mix in a container, after which the mixture is packed in cups. The cups are placed on trays and transferred to a ripening storage, where mesophilic fermentation is carried out at 20°C for about 20 h (Thure 2000; Leporanta 2003). After fermentation, the pH and structure of the product are checked, and the cups are cooled below 6°C. The shelf life of the product is about 3 weeks (Moilanen 2008). During the manufacturing procedure, the cream rises up to form an upper layer, and the mold grows, forming a velvet-like surface layer on the cream of viili (Meriläinen 1984; Leporanta 2003). No addition of stabilizers or milk solids is needed to achieve the final structure of the product (Leporanta 2003). In addition to natural viili, commercial fruit-flavored versions of viili with a slightly modified manufacturing process have been developed. In addition, the need for the development and production of low-lactose and lactose-free viili has grown in recent years because of consumers' awareness of lactose intolerance.

10.4 Starter Composition

Viili is produced with a complex mesophilic starter mix consisting of acid-forming, capsule-producing strains of *L. lactis* subsp. *lactis* and *L. lactis* subsp. *cremoris*, also of citrate-utilizing strains *L. lactis* subsp. *lactis* biovar. *diacetylactis* and *Leuconostoc mesenteroides* subsp. *cremoris*, and the mold *G. candidum* (Saxelin et al. 1986). The starter composition affects the final flavor and texture of the fermented product by producing typical flavor compounds, lactic acid, and exopolysaccharides (EPS), and by the coagulation of proteins due to acidification (Pastink et al. 2008). Mesophilic LAB have an optimum growth temperature between 20°C and 30°C (Wouters et al. 2002).

Currently, almost all industrial- and small-scale manufacture processes of fermented dairy products rely on industrially prepared starters (Wouters et al. 2002). Because viili is manufactured almost exclusively in Finland, the microbes used in viili manufacture are usually stored in dairies, because commercial starters for viili are not available. The starter mixes for industrial viili production currently in use are composed of the old traditional starter strains. Because the composition of mixed-strain starters may change during the course of time, dairy companies must have developed preservation methods, such as freezing or lyophilization, and cultivation methods for viili starter strains. The stable composition of starter cultures is important for the dairy industry to manufacture products of uniform quality.

Traditionally, identification of the bacteria has been based on physiological and biochemical properties, such as carbohydrate fermentation and certain enzyme activities. However, closely related species of LAB often have only minor differences in phenotypic traits, the characterization of strains by phenotypic methods can sometimes produce ambiguous results, and misclassification at the species level may occur (Wouters et al. 2002). In recent years, the application of genomic methods, such as polymerase chain reaction methods, has enabled rapid and reliable identification of closely related species. Studies on the identification of strains isolated from viili have revealed some differences in starter composition. Originally, three lactococci (*L. lactis* subsp. *cremoris*, *L. lactis* subsp. *lactis*, and *L. lactis* subsp. *lactis* biovar. *diacetylactis*) have been isolated from Finnish viili (Forsén 1966; Kahala et al. 2008), whereas Saxelin et al. (1986) found only *L. lactis* subsp. *cremoris* and *L. lactis* subsp. *lactis* biovar. *diacetylactis*. It has been assumed that the traditional viili starters contain *L. lactis* subsp. *lactis* only as a minority. However, recent results have shown that, instead of *L. lactis* subsp. *cremoris*, *L. lactis* subsp. *lactis* may exist as a main acid-forming lactococcal strain (Kahala et al. 2008). In addition to lactococci, *L. mesenteroides* subsp. *cremoris* has been identified in viili starters. Strain-level differentiation of the isolates of viili starter by pulsed-field gel electrophoresis has revealed different genotypes of strains belonging to the same species, indicating the presence of a wide variety of individual strains within starter cultures (Kahala et al. 2008). Diversity of the characteristics of strains, like flavor-forming abilities, occurs not only at the species level but also within species (Pastink et al. 2008). The change in composition of a starter culture or phenotypic pattern of single strains may occur by time in starter cultures. In particular, plasmid-encoded characteristics may cause instability of particular traits. Viili strains displaying unusual metabolic characteristics have been found, and by combining phenotypic tests with genomic differentiation methods, a reliable identification of the strains could be ensured (Kahala et al. 2005).

10.4.1 Acid-Forming and Exopolysaccharide-Producing Strains of Viili

Lactococci are mesophilic organisms with an optimum growth temperature of 25°C–30°C. Their role in fermentation is primarily to acidify the product by producing lactic acid, which also initiates curd and texture formation, aroma, and flavor production (Mäyrä-Mäkinen and Bigret 1998).

The ropy consistency of a fermented milk product is caused by the EPS excreted into milk during fermentation. Lactococcal strains producing EPS at the fermentation temperature of viili are responsible for the strong ropy character of viili and are essential for the proper consistency of the product (Forsén 1966; Kontusaari and Forsén 1988). EPS can either be attached to the producing cell as capsules or released into the growth medium as dispersed slime (Tamime et al. 2007). These polysaccharides contribute to the texture, viscosity, stability, and mouthfeel of fermented milk (Duboc and Mollet 2001). The fermentation of milk by the ropy *L. lactis* subsp. *cremoris* strain isolated from viili was also shown to decrease the susceptibility to syneresis (Toba et al. 1990). EPS from viili are heteropolysaccharides, have been found to consist of galactose, glucose, and rhamnose, and have an anionic nature due to the presence of a phosphate group (Nakajima et al. 1990; Yang et al. 1999; Ruas-Madiedo et al. 2006; Purohit et al. 2009). A cell surface–associated protein of a lactococcal viili strain has been shown to be associated with slime production (Kontusaari and Forsén 1988). Toba et al. (1990) found out that the ropy polysaccharides produced by *L. lactis* subsp. *cremoris* strains isolated from Finnish viili form a network together with the bacterial cells and the cell surface-associated protein matrix.

The ability to produce the ropy polymer is associated with plasmid-encoded genes (Vedamuthu and Neville 1986; Neve et al. 1988), which may cause instability of the mucoid characteristic in lactococci (Macura and Townsley 1984; Oberman and Libudzisz 1998). Several plasmids were detected in the mucoid *L. lactis* subsp. *cremoris* isolates derived from Finnish viili. A 30-MDa plasmid was found to determine the ropy phenotype, while the lactose-fermenting ability was linked to other plasmids (von Wright and Tynkkynen 1987; Neve et al. 1988). The slime-forming capacity of viili strains has been demonstrated to be higher at low-temperature incubation around 18°C. Increase in temperature resulted in considerable reduction or loss of the desirable high viscosity and mucoidness (Forsén et al. 1973; Macura and Townsley 1984). The EPS yield measured in viili was found to be around 71 mg/kg (Ruas-Madiedo et al. 2006), which was similar to that obtained in skimmed milk fermented with lactococcal strains of viili (40–80 mg/L; Ruas-Madiedo et al. 2002).

10.4.2 Flavor-Forming Strains of Viili

The flavor of the product is a combination of lactic acid and various aroma compounds. The main flavor producers in viili are the citrate-utilizing *L. lactis* subsp. *lactis* biovar. *diacetylactis* and *Leuconostoc mesenteroides* subsp. *cremoris*.

Citrate is present in milk in low concentrations and is metabolized by lactococci and leuconostocs, resulting in the production of carbon dioxide and the flavor compound diacetyl, which is the primary source of aroma and flavor in cultured dairy products and essential in dairy products such as butter, buttermilk, some young cheeses (Starrenburg and Hugenholtz 1991; Cogan and Jordan 1994), and also in viili. Citrate transport in lactococci and leuconostocs is encoded on plasmids (Cogan and Jordan 1994).

In the dairy industry, leuconostocs are normally grown in association with lactococci, described as a synergistic functional relationship (Vedamuthu 1994). Lactococci produce acid from lactose, and the role of leuconostocs is to ferment the citric acid of milk to aroma compounds (Vedamuthu 1994). The balance in the number of cells between leuconostocs and lactococci in starter culture is important for fermentation. It has been shown that the utilization of citrate is strongly dependent on the medium pH. Citrate can be completely converted at pH values below 6.0, indicating that sufficient acid production by lactococci is needed for the aroma-producing bacteria to convert citrate to aromatic compounds (Starrenburg and Hugenholtz 1991; Vedamuthu 1994). Furthermore, lactococci can produce stimulatory substances necessary for the growth of leuconostocs (Vedamuthu 1994).

10.4.3 Mold *Geotrichum candidum*

The mold *G. candidum* has been isolated from milk and is also utilized in the manufacture of some dairy products. The major role of this mold in dairy products is to enhance the flavor and modify the texture. Due to the proteolytic, peptidolytic, and lipolytic activity of *G. candidum*, important flavor compounds are also produced (Medveďová et al. 2008). Álvarez-Martín et al. (2008) found that *G. candidum* strains produce noticeable amounts of lactic acid, which is comparable to that produced by *L. lactis* strains. The pH level was also lowered to around 4.0 (Álvarez-Martín et al. 2008).

In viili production, *G. candidum* is added to the starter mix, and the growth of the mold results in a velvet-like surface on the cream layer. It also consumes oxygen from the cup of viili and produces carbon dioxide, which partly dissolves in the milk and forms carbonic acid and carbonates (Meriläinen 1984). The growth of the mold is limited due to the restricted amount of oxygen present in the sealed cup (Leporanta 2003). The mold generates a characteristic taste, aroma, and appearance to the product compared with other types of fermented milk products manufactured without *G. candidum*. *G. candidum* is also used as a starter in the production of Camembert cheese, having a key role in the ripening of this cheese (Wouters et al. 2002).

10.4.4 Bacteriophages of Viili Starters

The manufacture of viili is susceptible to disturbances, which impact on productivity and product quality. Disturbances in the production of viili may be caused by the loss of plasmid of a starter strain or contamination by bacteriophages, which produce changes in microbial relations. Dairy starters are susceptible to phage infection—pasteurized milk may contain phages, and phages are easily dispersed in milk (Courtney 2000). Moreover, repeated use of pure cultures provides permanent hosts for phage proliferation (Courtney 2000). Broad strain composition and the special viscous starter strains used in viili manufacture may be the reasons for appearance of several distinct types of phages in viili, including some rare types of phages (Saxelin et al. 1986). Traditional starter cultures contain a wide variety of strains with different phage sensitivities, and so, disturbances in a starter do not always cause a lack of fermentation (Saxelin et al. 1986). To prevent starter disturbances, the conditions for storing and cultivating have to be strictly controlled.

10.5 Potential Physiological and Health Effects

Microorganisms supplied to the intestines with fermented milk produce different enzymes capable of hydrolyzing various compounds in the diet, which allows a more thorough use of food nutrients. Humans lacking lactose-hydrolyzing enzymes due to a diseased state benefit from the enzymes of starter microbes, which are able to split lactose and predigest milk proteins. Moreover, allergic milk proteins may be hydrolyzed by fermentation, which might be advantageous for allergic individuals (Rasic and Kurmann 1978). Generally, the health value of fermented milk is considered to result from the increased availability of milk nutrients, reduced content of lactose and increased content of some vitamins of the B group, increased absorption of calcium and iron, control of the composition of intestinal microflora (inhibition of the growth of pathogenic microorganisms), inhibitory action on some types of cancer, and decrease of cholesterol level in the blood (data adapted from the work of Oberman and Libudzisz 1998).

The EPS produced by certain LAB strains can be used in the food industry as viscosifiers, stabilizers, emulsifiers, or gelling agents to alter the rheological properties and consistency of products. In fermented milk products, EPS-producing LAB strains serve as natural food biothickeners (Ruas-Madiedo and de los Reyes-Gavilán 2005). In addition to the technological advantages in food manufacture, some EPS produced by LAB may have beneficial health effects, such as the ability to lower cholesterol (Nakajima et al. 1992; Pigeon et al. 2002), immunomodulating and antitumoral activities (Kitazawa et al. 1992, 1998; Chabot et al. 2001), and prebiotic effects (Dal Bello et al. 2001; Korakli et al. 2002).

Due to its strong ropy character, viili is a potential source of health benefits mediated by the EPS fraction. The high extracellular EPS concentration may have an effect on the adhesion of microbes to the human intestinal mucus. For probiotic bacteria, adhesion is important for survival in the gastrointestinal tract. Adhesion is usually mediated by bacterial cell surface adhesions, which may be proteins, polysaccharides, or other cell wall–associated components. Ruas-Madiedo et al. (2006) reported that the EPS fraction of viili competitively inhibited the adhesion of two commercial probiotics, *Lactobacillus rhamnosus* and *Bifidobacterium lactis*, to the human intestinal mucus. The inhibition of adhesion was dependent on the dose of EPS and the probiotic strain used. Thus, the potential effect of EPS present in fermented milk on the intestinal adhesion of probiotic bacteria should be considered in the design of new probiotic products (Ruas-Madiedo et al. 2006).

REFERENCES

Álvarez-Martín, P., Flórez, A. B., Hernández-Barranco, A., and Mayo, B. 2008. Interaction between dairy yeasts and lactic acid bacteria strains during milk fermentation. *Food Control* 19:62–70.

Chabot, S., Yu, H. L., De Léséleuc, L., Cloutier, D., van Calsteren, M. R., Lessard, M., Roy, D., Lacroix, M., and Oth, D. 2001. Exopolysaccharide from *Lactobacillus rhanmnosus* RW-959M stimulate TNF, IL-6 and IL-12 in human and mouse cultured immunocompetent cells, and IFN-g in mouse splenocytes. *Lait* 81:683–697.

Cogan, T. M. and Jordan, K. N. 1994. Metabolism of *Leuconostoc* bacteria. *J. Dairy Sci.* 77:2704–2717.

Courtney, P. D. 2000. *Lactococcus lactis* subspecies *lactis* and *cremoris*. In: *Encyclopedia of Food Microbiology*, Vol. 2, edited by R. K. Robinson, C. A. Batt, and P. D. Patel, pp. 1164–1171. San Diego, CA: Academic Press.

Dal Bello, F., Walter, J., Hertel, C., and Hammes, W. P. 2001. *In vitro* study of prebiotic properties of levantype exopolysaccharides from lactobacilli and nondigestible carbohydrates using denaturing gradient gel electrophoresis. *Syst. Appl. Microbiol.* 24:232–237.

Duboc, P. and Mollet, B. 2001. Applications of exopolysaccharides in the dairy industry. *Int. Dairy J.* 11:759–768.

Forsén, R. 1966. Die Langmilch (Pitkäpiimä). *Finn. J. Dairy Sci.* 26:1–76.

Forsén, R., Raunio, V., and Myllymaa, R. 1973. Studies on slime forming group N *Streptococcus* strains—Part I: Differentiation between some lactic streptococcus strains by polyacrylamide gel electrophoresis of soluble cell proteins. Acta Univ. Ouluensis Ser. A12, *Biochem.* 3:1–19.

Hugenholtz, J. 1986. Population dynamics of mixed-strain cultures. *Neth. Milk Dairy J.* 40:129–140.

Kahala, M., Mäki, M., Lehtovaara, A., Tapanainen, J.-M., and Joutsjoki, V. 2005. *Leuconostoc* strains unable to split a lactose analogue revealed by characterization of mesophilic dairy starters. *Food Technol. Biotechnol.* 43:207–209.

Kahala, M., Mäki, M., Lehtovaara, A., Tapanainen, J.-M., Katiska, R., Juuruskorpi, M., Juhola, J., and Joutsjoki, V. 2008. Characterization of starter lactic acid bacteria from the Finnish fermented milk product viili. *J. Appl. Microbiol.* 105:1929–1938.

Kitazawa, H., Harata, T., Uemura, J., Saito, T., Kaneko, T., and Itoh, T. 1998. Phosphate group requirement for mitogenic activation of lymphocytes by an extracellular phosphopolysaccharide from *Lactobacillus delbrueckii* ssp. *bulgaricus*. *Int. J. Food Microbiol.* 40:169–175.

Kitazawa, H., Yamaguchi, T., and Itoh, T. 1992. B-cell mitogenic activity of slime products produced from slime-forming, encapsulated *Lactococcus lactis* ssp. *cremoris*. *J. Dairy Sci.* 75:2946–2951.

Kontusaari, S. and Forsén, R. 1988. Finnish fermented milk "viili": Involvement of two cell surface proteins in production of slime by *Streptococcus lactis* ssp. *cremoris*. *J. Dairy Sci.* 71:3197–3202.

Korakli, M., Gänzle, M. G., and Vogel, R. F. 2002. Metabolism by bifidobacteria and lactic acid bacteria of polysaccharides from wheat and rye, and exopolysaccharides produced by *Lactobacillus sanfranciscensis*. *J. Appl. Microbiol.* 92:958–965.

Leporanta, K. 2003. Viili and Långfil—Exotic fermented products from Scandinavia. *Valio—Foods & Functionals* 2:3–5.

Leporanta, K. 2010. Traditional Nordic Fermented Milk Products. Oral presentation in the IDF Symposium on Science and Technology of Fermented Milk, Tromsö, Norway, 7–9 June 2010.

Macura, D. and Townsley, P. M. 1984. Scandinavian ropy milk—Identification and characterization of endogenous ropy lactic streptococci and their extracellular excretion. *J. Dairy Sci.* 67:735–744.

Mäyrä-Mäkinen, A. and Bigret, M. 1998. Industrial use and production of lactic acid bacteria. In: *Lactic Acid Bacteria, Microbiology and Functional Aspects*, edited by S. Salminen and A. von Wright, pp. 73–103. New York: Marcel Dekker Inc.

Medveďová, A., Liptáková, D., Hudecová, A., and Valík, L. 2008. Quantification of the growth competition of lactic acid bacteria: A case of coculture with *Geotrichum candidum* and *Staphylococcus aureus*. *Acta Chimica Slovaca* 1:192–207.

Meriläinen, V.T. 1984. Microorganisms in fermented milks: Other microorganisms. *Bull. Int. Dairy Fed.* 179:89–93.

Moilanen, J. 2008. Viiliä Oulusta. *Maitotalous* 4:8–9.

Nakajima, H., Suzuki, Y., Kaizu, H., and Hirota, T. 1992. Cholesterol lowering activity of ropy fermented milk. *J. Food Sci.* 57:1327–1329.

Nakajima, H., Toyoda, S., Toba, T., Itoh, T., Mukai, T., Kitazawa, H., and Adachi, S. 1990. A novel phospho-polysaccharide from slime-forming *Lactococcus lactis* subspecies *cremoris* SBT 0495. *J. Dairy Sci.* 73:1472–1477.

Neve, H., Geis, A., and Teuber, M. 1988. Plasmid-encoded functions of ropy lactic acid streptococcal strains from Scandinavian fermented milk. *Biochimie* 70:437–442.

Oberman, H. and Libudzisz, Z. 1998. Fermented milks. In: *Microbiology of Fermented Foods*, Vol. 1, 2nd edition, edited by B.J.B. Wood, pp. 308–350. London: Blackie Academic & Professional.

Pastink, M. I., Sieuwerts, S., de Bok, F. A. M., Janssen, P. W. M., Teusink, B., van Hylckama Vlieg J. E. T., and Hugenholz, J. 2008. Genomics and high-throughput screening approaches for optimal flavor production in dairy fermentation. *Int. Dairy J.* 18:781–789.

Pigeon, R. M., Cuesta, E. P. and Gilliland, S. E. 2002. Binding of free bile acids by cells of yogurt starter culture bacteria. *J. Dairy Sci.* 85:2705–2710.

Purohit, D. H., Hassan, A. N., Bhatia, E., Zhang, X., and Dwivedi, C. 2009. Rheological, sensorial, and che-mopreventive properties of milk fermented with exopolysaccharide-producing lactic cultures. *J. Dairy Sci.* 92:847–856.

Rasic, J. L. and Kurmann, J. A. 1978. *Yoghurt: Scientific Grounds, Technology, Manufacture and Preparations*. Copenhagen, Denmark: Technical Dairy Publishing House.

Ruas-Madiedo, P. and de los Reyes-Gavilán, C. G. 2005. Invited review: Methods for the screening, isolation and characterization of exopolysaccharides produced by lactic acid bacteria. *J. Dairy Sci.* 88:843–856.

Ruas-Madiedo, P., Guiemonde, M., de los Reyes-Gavilán, C.G., and Salminen, S. 2006. Effect of exopolysac-charide isolated from "viili" on the adhesion of probiotics and pathogens to intestinal mucus. *J. Dairy Sci.* 89:2355–2358.

Ruas-Madiedo, P., Tuinier, R., Kanning, M., and Zoon, P. 2002. Role of exopolysaccharides produced by *Lactococcus lactis* subsp. *cremoris* on the viscosity of fermented milks. *Int. Dairy J.* 12:689–695.

Saxelin, M.-L., Nurmiaho-Lassila, E.-L., Meriläinen, V. T., and Forsen, R. I. 1986. Ultrastructure and host specificity of bacteriophages of *Streptococcus cremoris, Streptococcus lactis* subsp. *diacetylactis*, and *Leuconostoc cremoris* from Finnish fermented milk "viili." *Appl. Environ. Microbiol.* 52:771–777.

Starrenburg, M. J. C. and Hugenholtz, J. 1991. Citrate fermentation by *Lactococcus* and *Leuconostoc* spp. *Appl. Environ. Microbiol.* 57:3535–3540.

Tamime, A. Y., Hassan, A., Farnworth, E., and Toba, T. 2007. Structure of fermented milks. In: *Structure of Dairy Products, Society of Dairy Technology Series*, edited by A. Y. Tamime, pp. 134–167. Wiley-Blackwell. Oxford, UK: Blackwell Publishing Ltd.

Thure, T. 2000. Identification of bacteria from the Finnish sour-milk product "viili" and composition of a new starter. MSc Thesis, University of Helsinki, 65 pp.

Toba, T., Nakajima, H., Tobitani, A., and Adachi, S. 1990. Scanning electron microscopic and texture studies on characteristic consistency of Nordic ropy sour milk. *Int. J. Food Microbiol.* 11:313–320.

Vedamuthu, E. R. 1994. The dairy Leuconostocs: Use in dairy products. *J. Dairy Sci.* 77:2725–2737.

Vedamuthu, E. R. and Neville, J. M. 1986. Involvement of a plasmid in production of ropiness (mucoidness) in milk cultures by *Streptococcus cremoris* MS. *Appl. Environ. Microbiol.* 51:677–682.

von Wright, A. and Tynkkynen, S. 1987. Construction of *Streptococcus lactis* subsp. *lactis* strains with a single plasmid associated with mucoid phenotype. *Appl. Environ. Microbiol.* 53:1385–1386.

Wouters, J. T. M., Ayad, E. H. E., Hugenholtz, J., and Smit, G. 2002. Microbes from raw milk for fermented dairy products. *Int. Dairy J.* 12:91–109.

Yang, Z., Huttunen, E., Staaf, M., Widmalm, G., and Tenhu, H. 1999. Separation, purification and characteriza-tion of extracellular polysaccharides produced by slime-forming *Lactococcus lactis* ssp. *cremoris* strains. *Int. Dairy J.* 9:631–638.

11

Production of Laban

Catherine Béal and Gisèle Chammas

CONTENTS

11.1 Introduction

Laban is a fermented milk produced in Lebanon and some Arab countries. It is obtained from lactic acid fermentation, thus leading to acidification and coagulation of milk. Historically, these fermented dairy products appeared as a means to preserve milk, but they now gain a high interest due to their pleasant sensory properties (freshness, acidity, and mouth coating). They represent an interesting alternative to milk and cheese consumption. The historical and geographical origin of laban has never been precisely established, but the first known products appeared in the Middle East 10,000 or 15,000 years ago. They are now commercialized within a large number of countries, as they are essential in the Arab diet. Lebanon and Arab countries are the largest consumers, together with countries in North Africa.

Laban is obtained through the lactic acid fermentation of heat-treated cow milk by thermophilic starters such as *Streptococcus thermophilus*, *Lactobacillus acidophilus*, and *Lb. delbrueckii* subsp. *bulgaricus*, at 40°C–45°C (Baroudi and Collins 1976; Chammas et al. 2006b). Depending on the product, the milk used may be full-fat, partially skimmed, or fully skimmed, thus leading to the commercialization of whole-fat, low-fat, and fat-free laban (Libnor 1999).

In Lebanon, laban is manufactured by both industrial and small traditional producers, each of them using different manufacturing practices (Surono and Hosono 2003). Traditional producers mainly use artisanal starters, consisting of a mixed culture of unknown number of undefined and indigenous thermophilic strains. This procedure generates diversified and typically flavored products that are generally preferred by consumers (Wemekamp-Kamphuis et al. 2002). The industrial production of laban is carried out by associating selected starters showing well-characterized properties and by using a standardized process. This practice allows achieving the desired metabolic activity and technological properties of the strains during their growth in milk. It also leads to a constant quality of the products and a high level of reproducibility of the processes, but to the detriment of the diversity of flavors, which is lower compared to traditional products.

The composition and characteristics of laban differ according to the production scale. As a general rule, laban has a titratable acidity of about 1% and a pH of 4.0 (Baroudi and Collins 1976). Table 11.1 shows the gross composition of laban, which has been determined recently by Musaiger et al. (1998) and Guizani et al. (2001). The fat content depends on the kind of laban—a whole milk laban contains 3.1% of fat, whereas a low-fat product comprised only 1.3% of fat (Musaiger et al. 1998).

11.2 Laban Market in Lebanon and Other Countries

In the last few decades, the consumption of dairy products increased at extremely high rates in the Middle East due to their nutritional value—they are considered as cheap sources of animal protein and a

TABLE 11.1

Physicochemical Composition of Laban

Kind of Product Parameter	Experimental Laban	Whole-Fat Laban	Low-Fat Laban	Homemade Laban	Commercial Laban	Artisanal Laban	Tunisian Leben
pH	4.25	4.2–4.9	4.3–4.6	3.98 ± 0.13	4.52 ± 0.03	Nd	4.29–4.45
Titratable acidity (%)	0.9–1.2	0.6–1.1	0.8	1.12 ± 0.12	0.77 ± 0.04	0.9–1.75	Nd
Total solids (%)	Nd	11.1–13.1	9.7–11.5	6.29 ± 0.21	10.87 ± 0.32	Nd	7.05–7.40
Protein (%)	Nd	2.5–4.6	2.9–3.5	2.11 ± 0.17	3.15 ± 0.15	Nd	1.86–2.56
Fat (%)	Nd	2.5–4.3	1.2–1.4	1.12 ± 0.35	3.50 ± 0.17	3.3–4.2	1.48–3.50
Lactose content (%)	Nd	3.3–4.2	3.7	Nd	Nd	2.65–3.99	1.90–2.59
Reference	Baroudi and Collins (1976)	Musaiger et al. (1998)	Musaiger et al. (1998)	Guizani et al. (2001)	Guizani et al. (2001)	Chammas et al. (2006a)	Samet-Bali et al. (2009)

Note: Nd, not determined.

well-known source of calcium in the human diet. Currently, the per capita consumption of dairy products is 152 kg/year (equivalent to milk quantity) in Lebanon, 117 kg/year in Syria, 78 kg/year in Jordan, and 54 kg/year in Saudi Arabia (Al Ammouri 2006; Lebanese Ministry of Agriculture 2007; Alqaisi et al. 2009; BMI Business Monitor International 2010). In Lebanon, laban is consumed as such or used in the preparation of a wide variety of dishes. Therefore, the annual consumption of this fermented milk is the highest among all the countries of the region.

Laban is the main fermented dairy product manufactured in the Middle East and the most consumed dairy product in Lebanon (Nasreddine et al. 2006). In Lebanon, the laban consumption reaches 25 kg/year/person, followed by cheese (12.8 kg/year/person), and then by labneh or strained laban (10 kg/year/person). According to Lemoine (2002), 40 tons of laban is produced annually by about 135 factories in Lebanon. Other 165 small-scale dairy processing units are scattered all over the country and deliver their production to neighborhood markets. The biggest dairy companies in Lebanon are Karoun Dairy and Liban Lait. Almarai is a well-known Saudi Arabian dairy company. It is currently the largest producer and exporter of milk and dairy products in the Middle East. The company production capacity is 1.8 million liters of milk per day. According to Sadi and Henderson (2007), about 60% of Almarai raw milk is fermented to make laban. Probiotic culture laban is also manufactured.

The level of export of dairy products is still very low in some Arab countries compared to Saudi Arabia where dairy exports are increasing at a high rate. Although Syria is the biggest producer among these countries, small quantities are processed by the dairy factories with restrained infrastructure and technology, as compared to on-farm and artisanal production. On the other hand, in Saudi Arabia, the processing units are more developed, and sophisticated technology is used with a higher percentage of milk delivered to dairy (Alqaisi et al. 2010). Laban is one of the most exported dairy products, followed by unsweetened milk. The main destinations of these exports are the neighboring gulf countries, Iraq, Jordan, Lebanon, Yemen, and selected African markets (Alqaisi et al. 2010).

Due to the long tradition of milk and fermented milk product consumption in the Middle East, the potential for further development in the dairy sector of this region is greatly possible. Dairy cow population is estimated at 77,000, 773,000, 50,000, and 111,600 cows for Lebanon, Syria, Jordan, and Saudi Arabia, respectively (Lebanese Ministry of Agriculture 2007; Anonymous 2008; Alqaisi et al. 2009, 2010). Total milk produced in 2007 was estimated at 0.24, 2.63, 0.28, and 1.34 million tons in Lebanon, Syria, Jordan, and Saudi Arabia, respectively (Lebanese Ministry of Agriculture 2007; Alqaisi et al. 2010).

In all Arab countries, increasing feed prices is a major risk factor for dairy industry, mainly because these countries rely basically on imported feedstuff in feeding animals. Additionally, the availability of water resources, particularly in Saudi Arabia and Jordan, is another limiting factor in fodder production, and thus for dairy industry.

11.3 Lactic Acid Fermentation of Laban

As a fermented dairy product, laban is obtained from lactic acid fermentation that corresponds to the transformation of carbohydrates into lactic acid as a major metabolic end product, with the aid of specific microorganisms called lactic acid bacteria. This bioreaction leads to important microbiological, biochemical, physicochemical, and sensory changes in milk.

The first outcome of the lactic acid fermentation is to increase the shelf life of the products by inhibiting the growth of microbial spoilage and the occurrence of enzymatic reactions. It allows obtaining safe products, free of pathogenic or undesirable microorganisms, as a result of product acidity. It also confers the products some specific nutritional and sensory properties such as texture and flavor.

11.3.1 Microbiology of Laban Fermentation

11.3.1.1 Microorganisms

The microbial composition of laban differs according to the authors and to the countries. Baroudi and Collins (1976) identified five microorganisms in artisanal products, corresponding to three bacterial species, *S. thermophilus*, *Lb. acidophilus*, and *Leuconostoc lactis*, and two kinds of yeasts, *Kluyveromyces fragilis* and *Saccharomyces cerevisiae*. Guizani et al. (2001) characterized some artisanal and commercial laban from the Sultanate of Oman. They indicated that *Lactococcus lactis* subsp. *lactis*, *Lc. lactis* subsp. *lactis* biovar. *diacetylactis*, *Lc. lactis* subsp. *cremoris*, and *Lb. plantarum* were responsible for the fermentation of traditional laban, whereas *Lb. acidophilus*, bifidobacteria, and *S. thermophilus* were responsible for the production of commercial laban. Chammas et al. (2006b) identified 96 strains isolated from 18 laban products that were collected from small-scale farms located in 15 areas distributed throughout the Lebanese territory and one commercial product manufactured by using a mixture of European commercial starters. The herds corresponded to Friesland cattle and to Baladi, a draft breed in Lebanon. In all of these products, two bacterial species were identified, belonging to *S. thermophilus* and *Lb. delbrueckii* subsp. *bulgaricus* species. As an illustration, Figure 11.1 allows visualizing some cells of *S. thermophilus* and *Lb. delbrueckii* subsp. *bulgaricus* enclosed in a milk coagulum. More recently, Samet-Bali et al. (2009) demonstrated that the production of industrial laban in Tunisia was triggered by the action of *Lc. lactis* subsp. *lactis*, *Lc. lactis* subsp. *diacetylactis*, and *Lc. lactis* subsp. *cremoris*. From this information, different bacterial species are mentioned as triggering lactic acid fermentation in laban.

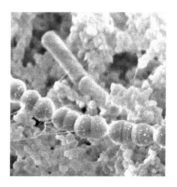

FIGURE 11.1 Photograph of laban comprising *S. thermophilus* and *Lb. bulgaricus* subsp. *bulgaricus* cells.

The differences may be ascribed to the different countries from which the samples have been collected and to the level of industrialization of the products.

Whatever the considered study, the microorganisms are always used in mixed cultures, combining at least two species and often many strains of each species. In most of the studies, the strains were not selected by the authors as they were isolated from artisanal products. On the contrary, in industrial plants, only the two species *S. thermophilus* and *Lb. bulgaricus* are encountered, thus indicating that microbiological composition of industrial laban is close to yogurt. By analogy with yogurt, the criterion of choice of the strains relies on technological considerations (acidification activity and bacteriophage resistance) and sensory properties (exopolysaccharide production, aroma compounds synthesis, and post-acidification).

11.3.1.2 General Characteristics of Lactic Acid Bacteria Used for Laban Fermentation

Lactic acid bacteria are Gram-positive bacteria, nonsporulating, and nonrespiring rods or cocci. Their low guanine + cytosine (G + C) content is between 33% and 54%. They are known to synthesize large amounts of lactic acid from lactose, either as the main end product (homofermentative) or in combination with carbon dioxide and ethanol (heterofermentative). L(+), D(–), or racemic lactic acid is synthesized according to the species. For example, *S. thermophilus* produces only L(+) lactic acid, whereas *Lb. delbrueckii* subsp. *bulgaricus* synthesizes D(–) lactic acid, with both of these species being homofermentative. According to the genus, they grow at temperatures of 25°C–30°C (mesophilic bacteria) or 37°C–45°C (thermophilic bacteria), but not at 15°C. These characteristics have to be taken into account when associating two species in the same product.

11.3.1.3 Associative Growth

Simultaneous growth of *S. thermophilus* and *Lb. delbrueckii* subsp. *bulgaricus* leads to a positive indirect interaction between these two species that is used to be termed as symbiosis (Rajagopal and Sandine 1990). This interaction results in an increase in acidification rate and bacterial concentrations in mixed culture, as compared to pure cultures (Béal and Corrieu 1998). The production of aroma compounds (mainly acetaldehyde) and physical stability of the product (low syneresis) are also improved in mixed cultures. The stimulation of *S. thermophilus* by *Lb. delbrueckii* subsp. *bulgaricus* operates through the proteolytic activity of the lactobacilli, which liberates peptides and amino acids that stimulate the growth of the streptococci (Courtin et al. 2002). In return, *S. thermophilus* synthesize formic acid and CO_2 that promote the growth of *Lb. delbrueckii* subsp. *bulgaricus* (Perez et al. 1991; Ascon-Reyes et al. 1995).

When other bacteria are associated with *S. thermophilus* and *Lb. delbrueckii* subsp. *bulgaricus*, other kinds of interaction take place. *Lb. delbrueckii* subsp. *bulgaricus* limits the development of *Lb. acidophilus* (competition and inhibition phenomena), whereas associative growth is observed between *S. thermophilus* and *Lb. acidophilus* (Vinderola et al. 2002).

Finally, specific inhibition phenomena may occur as a result of bacteriocin production. Bacteriocins are small and thermostable proteins produced by some lactic acid bacteria to inhibit the growth of closely related strains. By considering *S. thermophilus* and *Lb. delbrueckii* subsp. *bulgaricus*, less than 10 bacteriocins have been identified (Tamime and Robinson 1999b). Nevertheless, this phenomenon makes it necessary to verify the biological compatibility of strains before associating them for laban production.

11.3.2 Environmental Factors That Affect Lactic Acid Bacteria Metabolism

Growth and acidification of lactic acid bacteria are strongly influenced by physical, chemical, and microbiological environmental factors.

11.3.2.1 Physicochemical Factors

Temperature is the first environmental factor to be taken into account during laban manufacture. It acts directly on chemical and biochemical reaction rates. As at least two species of microorganisms are associated during laban fermentation, they first have to be compatible by considering their optimal temperature for growth. By considering the species involved in laban fermentation, the optimal temperature is between 37°C and 42°C (Béal et al. 1989; Adamberg et al. 2003; Chammas et al. 2006b). Nevertheless, in artisanal productions, the temperature is generally lower, thus favoring the growth of other microorganisms (Baroudi and Collins 1976). The fermentation temperature also affects the physical properties of the gels, with a lower temperature leading to a stronger texture (Béal et al. 1999).

Water activity (aw) in milk products is mainly related to NaCl or sugar concentration (Labuza 1980). By reducing aw, the fraction of free water decreases and the availability of the nutriments is affected (Fajardo-Lira et al. 1997). In fermented dairy products, NaCl has been shown to limit post-acidification together with an increase in bacterial mortality (Lacroix and Lachance 1988). Nevertheless, as in laban, no saccharose nor NaCl is introduced in milk before or during fermentation, water activity is generally not affected.

Milk composition has a strong influence on the growth and acidification by lactic acid bacteria. If lactose and mineral concentrations are high enough in milk (Desmazeaud 1990), the nitrogenous fraction (amino acids and oligopeptides) is insufficient, and starvation may occur at the end of fermentation. Moreover, when present in milk, some specific components have a negative impact on bacterial growth—lactenins, lactoperoxidase/thiocyanate/hydrogen peroxide, H_2O_2, antibiotic residues, detergent and disinfectant residues, insecticide residues, or somatic cells (Reiter 1978).

Thermal treatment of milk before fermentation acts positively on bacterial metabolism, thus reducing the fermentation time. This stimulatory effect is explained by several factors—elimination of undesirable and pathogenic microorganisms, thus reducing competition phenomena during growth, restriction of some antibacterial substance levels that are naturally present in milk (agglutinins, lactoperoxidase, and toxic sulfides), release of cysteine and glutathione that act as antioxidants, and production of small quantities of formic acid from lactose (Tamime and Robinson 2007b). Finally, it contributes in slightly increasing the amounts of amino acids and small peptides in milk.

As a last environmental factor, pH is very important for growth of lactic acid bacteria (Béal and Corrieu 1991). It affects the nutrient's availability, the cellular membrane permeability, and the enzymatic reaction rates (Rault et al. 2009). During laban production, pH is not controlled and thus decreases during fermentation. This acidification is a major factor that slows down bacterial metabolism and confers the product some important technological characteristics. Moreover, undissociated lactic acid concentration increases in milk as a result of fermentation and pH decrease. As bacterial membrane is permeable to nondissociated lactic acid, the inhibitory effect of acidification is strong (Amrane and Prigent 1999), thus leading to low bacterial concentration at the end of the laban fermentation.

11.3.2.2 Microbiological Factors

The choice of bacterial strains depends on several factors referring to their technological properties and performances. The main technological properties of lactic acid bacteria concern their acidification activity (Corrieu et al. 1988), their capacity to metabolize carbohydrates, their proteolytic and lipolytic activities, and their ability to produce aroma compounds, exopolysaccharides, or bacteriocins (Béal et al. 2008). The selection of a lactic acid starter also takes into account the bacterial growth and metabolic rates, the interactions of the strains with other species, and their sensitivity to bacteriophages. The ability of the bacteria to be frozen or freeze-dried is also important in order to permit their stabilization before use (Béal et al. 2008).

The inoculum level affects the fermentation rate, which is more rapid with a high inoculum rate. As a general rule, inoculation is done at a concentration of 10^6 CFU/mL in order to simultaneously shorten the fermentation and limit the costs. When using a direct inoculation, the inoculum rate in milk varies between 2.5 and 70 g/100 L, whereas 1 L/100 L is generally necessary for an indirect inoculation.

The balance between the bacterial species influences the fermentation kinetic. When streptococci are involved in the fermentation, they always predominate at the end of acidification, even if the initial ratio is generally well balanced (1:1) between the species *S. thermophilus* and *Lb. delbrueckii* subsp. *bulgaricus* (Béal and Corrieu 1991). The existence of a positive mutual interaction between these two species partly explains this phenomenon (Rajagopal and Sandine 1990).

Finally, when bacteriophages arise in the production plant, they strongly affect the fermentation, thus leading to serious economic losses in laban industry. By considering *S. thermophilus* and *Lb. delbrueckii* subsp. *bulgaricus*, two kinds of bacteriophages may be encountered, both of them belonging to the *Siphoviridae* family (Krusch et al. 1987). The virulant or lytic bacteriophages lyse the bacterial host during infection, whereas the temperate, prophage, or lysogenic bacteriophages, which insert their genome in the host chromosome, do not lyse the host cells (Neve 1996; Quiberoni et al. 2004). In order to avoid phage attack, the combined use of aseptic techniques for the propagation of starter cultures, proper heat treatment of the milk, and daily rotation of bacteriophage unrelated strains or phage-resistant strains is strongly recommended.

11.3.3 Biochemistry of Laban Fermentation

Lactic acid bacteria commonly use milk nutrients to permit their growth as well as metabolite production. Most of these reactions are answerable to lactose catabolism that corresponds to the main functionality of lactic acid bacteria and is essential in obtaining a high-grade fermented product (flavor and stability). Some anabolism reactions are also important as they act on the synthesis of exopolysaccharides, aroma compounds, or preservative compounds.

11.3.3.1 Lactose Catabolism

Lactic acid fermentation is defined by the following simplified biochemical reaction:

$$\text{Lactose} + \text{Water} \rightarrow \text{Lactic acid}$$

$$C_{12}H_{22}O_{11} + H_2O \quad 4C_3H_6O_3$$

The catabolism of lactose into lactic acid occurs according to four steps that are summarized in Figure 11.2. Because the lactose catabolism takes place inside the cells, the entry of lactose represents the first step of this pathway. In *S. thermophilus* and *Lb. delbrueckii* subsp. *bulgaricus*, it involves cytoplasmic proteins such as ATP-dependent permease that translocates lactose without chemical modification. This LacS permease acts according to two ways: a symport lactose/H^+ and an antiport lactose/galactose, with the last one being preferred when galactose is present (Poolman et al. 1996). Lactose is then hydrolyzed into glucose and galactose by β-galactosidase (Monnet et al. 2008). Glucose is catabolized to pyruvate according to the Embden–Meyerhof pathway, whereas galactose is excreted from the cells in *S. thermophilus* and *Lb. delbrueckii* subsp. *bulgaricus*. In mesophilic lactic acid bacteria, galactose is catabolized into pyruvate through Leloir pathway. These reactions drive the synthesis of energy (2 ATP/glucose). The third step corresponds to the reduction of pyruvate into lactate using lactate dehydrogenase as catalyst (Monnet et al. 2008). This reaction allows the cofactor NAD^+ to be oxidized. According to the species, L(+), D(–), or racemic lactic acid is synthesized. Finally, lactate is secreted from the cells due to a symport with H^+. The increase in lactate concentration in milk undergoes acidification, together with an inhibition of bacterial growth that stops before lactose starvation. This inhibition combines the effects of lactic acid accumulation with resulting pH decrease.

Figure 11.3 shows the typical acidification kinetics obtained during manufacture of two laban samples. Acidification kinetics are different among the two samples, with the first one being highly acidifying as

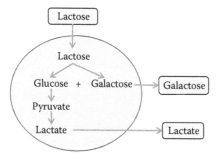

FIGURE 11.2   Main oxydo-reduction reactions involved in lactose catabolism during laban fermentation.

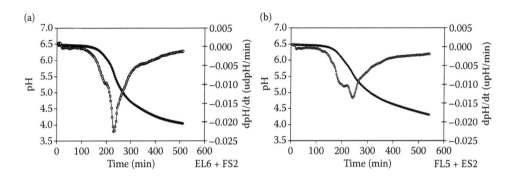

FIGURE 11.3 Acidification kinetics of two laban samples. (a) Strains of *S. thermophilus* FS2 and *Lb. bulgaricus* subsp. *bulgaricus* EL6. (b) Strains of *S. thermophilus* ES2 and *Lb. bulgaricus* subsp. *bulgaricus* FL5. (From Chammas, G.I., Caractérisation et utilisation de souches de bactéries lactiques thermophiles locales dans le développement de produits laitiers fermentés au Liban, Institut National Agronomique Paris-Grignon, Paris, 2006. With permission.)

pH 4.5 was obtained after 334 min, whereas the second one needed 485 min to reach pH 4.5, thus indicating that these last strains were less active.

11.3.3.2 Nitrogen Catabolism

Because they are unable to catabolize mineral nitrogen, lactic acid bacteria use the milk proteins as a nitrogen source during laban fermentation. Milk contains about 32 g/L nitrogen compounds that are shared into a soluble fraction (free amino acids, small peptides, whey proteins, nitrogenous bases, urea, and group B vitamins) and an insoluble fraction composed of the caseins (Desmazeaud 1990). Lactic acid bacteria may use the free amino acids, the small peptides, and the whey proteins contained in milk as a consequence of their proteolytic activity, although it is considered as weak and variable according to the species (Stefanitsi and Garel 1997; Fernandez-Espla et al. 2000). Nevertheless, as they cannot synthesize all amino acids, an exogenous supply of at least leucine and valine is necessary to fulfill the bacterial needs (Letort and Juillard 2001). These components take part in many cellular functions such as protein and enzyme synthesis that are used for the biosynthesis of cell constituents. The liberation of some amino acids into milk is important as they stimulate the growth of some species such as *S. thermophilus* (Courtin et al. 2002). Some peptides that are liberated may also act as precursors for the synthesis of flavor compounds. Among them, threonine is converted, thanks to a threonine aldolase, into acetaldehyde that is a major compound of laban aroma (Ott et al. 2000a).

11.3.3.3 Lipid Metabolism

Milk lipids represent between 33 and 47 g/L of solid fraction of cow milk, mainly in the form of triacyl glycerols (96%–98%). They act as a source of essential fatty acids, which cannot be synthesized by animals, and of fat-soluble vitamins (A, D, E, and K). Lactic acid bacteria display limited lipolysis, as a consequence of lipase and esterase activities, which are species- and strain-dependent (Kilcawley et al. 1998). The lipase activity may also contribute toward the flavor and rheological properties of the dairy products.

11.3.3.4 Other Metabolisms

Bacterial metabolism also concerns compounds that are present in lower quantities such as nitrogenous bases, vitamins, and minerals. By considering the vitamins, lactic acid bacteria used for laban fermentation need biotin, niacin, riboflavin, vitamin B_{12}, and pantothenic acid, whose concentrations decrease in milk as a result of microbial catabolism (Béal et al. 2008). Conversely, depending on the strain, niacin, folic acid, and to a lesser extent riboflavin, thiamin, vitamin B_6, and vitamin B_{12} are actively synthesized by these lactic acid bacteria during milk fermentation (Rao et al. 1984).

Some nitrogenous bases that are essential for nucleic acid synthesis are brought within milk (Zink et al. 2000). Nevertheless, a high variability among the needs exists within species and strains, thus explaining some differences observed for bacterial growth in milk.

Finally, some cationic compounds are involved in metabolic pathways. Among them, manganese, magnesium, and potassium are often concerned (Boyaval 1989).

11.4 Influence of Lactic Acid Fermentation on Laban Properties

During fermentation, lactic acid bacteria induce strong changes in milk composition and properties. These changes concern the physical properties of the gel, the flavor of the product, the shelf life of the laban, and its nutritional characteristics.

11.4.1 Impact of Lactic Acid Fermentation on Physical Properties of Laban

11.4.1.1 Microstructure of Laban

The microstructure of laban consists of a protein matrix composed of aggregated casein micelle chains and clusters, surrounding fat globules. Among these proteins, caseins αS1, αS2, β, and κ (relative proportion: 4/1/3.7/1.4) account for around 80% of total nitrogen fraction of cow milk.

During fermentation, a destabilization of the casein complex occurs as a result of acidification, thus leading to aggregation and gel formation. The lactic acid that is excreted by lactic acid bacteria progressively converts the colloidal calcium/phosphate complex in the micelle to a soluble calcium phosphate fraction. The casein micelles are then gradually depleted of calcium, which leads to coagulation of the proteins at pH 4.6–4.8. The reactions of dissociation and aggregation of casein micelles can be summarized as follows:

1. Between 6.7 and 5.8, calcium phosphate is partly solubilized, thus inducing a dissociation of the casein micelles.
2. Between pH 5.8 and 5.3, there is a complete solubilization of calcium phosphate. Protein interactions decrease inside the micelles and gelification starts, thanks to appearance of hydrophobic links between serum proteins.
3. Between pH 5.3 and 4.8, dissociated caseins are integrated into acido-micelles whose hydration is reduced as compared to native micelles.
4. Below pH 4.8, no electrostatic repulsion occurs between acido-micelles because of their isoelectric pH (pH 4.6). As a consequence, hydrophobic interactions increase, thus enhancing aggregation of casein and strengthening the protein network.

This phenomenon is displayed in Figure 11.4 that shows the time course of coagulation as a function of acidification.

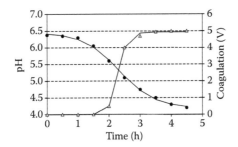

FIGURE 11.4 Time course of acidification and coagulation during laban manufacture. (●) pH; (△) coagulation.

Depending on the acidification rate and the strains used, the coagulation is accompanied by syneresis. This phenomenon that is commonly observed in laban, corresponds to the shrinkage of the gel that induces whey formation on the surface of the product. In laban, it can be reduced either by increasing the nonfat solid content of milk prior to fermentation, or by heating the milk, for example, at 90°C for 10 min or at 120°C for 2 min. Heating causes the denaturation of whey proteins, especially β-lactoglobulin, and weakens their interaction with the casein micelles. The whey protein-coated micelles form a finer gel than that formed from unheated or high-temperature short time pasteurized milk, with less tendency to syneresis.

11.4.1.2 Physical Properties of Laban

The physical structure of laban has been extensively studied by Chammas et al. (2006b). It is mainly the consequence of milk coagulation, as exopolysaccharide production has never been demonstrated in this fermented product (Duboc and Mollet 2001). Six main descriptors of textural properties are used to characterize laban (Chammas et al. 2006b).

Quantification of laban texture is first obtained by measuring the firmness, the cohesiveness, and the fracturability of the product. These characteristics can be determined by means of a texture analyzer, which allows measuring the force of penetration of a probe in the laban sample at a well-defined speed and at a given temperature, thus leading to the determination of the texture profile of the sample (Chammas et al. 2006b). For each force–time curve, obtained by two successive compressions of each sample, the force (in newtons) required to disrupt the sample is used as a measure of fracturability, the maximum force ($F1$, in newtons) recorded during the first compression is used as a measure of firmness, and the ratio (dimensionless) of the area under the force curve measured during the second compression ($A2$) to the one measured during the first compression ($A1$) is used as a measure of cohesiveness. By measuring these parameters for 96 samples of milk fermented with *S. thermophilus* and *Lb. bulgaricus* isolates taken from laban samples, Chammas et al. (2006b) obtained high firmness values between 0.32 and 0.52 N. Fracturability of these samples ranged between 0.21 and 0.4 N and cohesiveness varied from 0.5 and 1.1, thus showing a high diversity among strains.

Laban also exhibits thixotropic rheological properties, that is, the viscosity of the product decreases as the rate of shear increases. A typical relationship between viscosity and shear rate is shown in Figure 11.5 (Guizani et al. 2001). These properties are quantified by the apparent viscosity (pascal-second), which represents the flow behavior (where the sample structure is destroyed) and the complex viscosity (pascal-second) that represents the viscoelastic behavior (where the sample structure is not affected) of the product. These rheological measurements are generally performed by means of a rheometer,

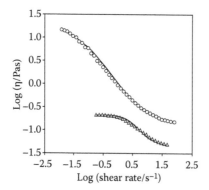

FIGURE 11.5 Viscosity variation as a function of shear rate for commercial (O) and home-made (Δ) laban. (From Guizani, N. et al., *Int. J. Food Sci. Technol.* 36(2), 199–205, 2001. With permission.)

equipped with a cone-and-plate system (Chammas et al. 2006b). These authors measured the rheological properties of 96 fermented milks made with strains isolated from laban samples. The apparent viscosity slightly varied between 0.11 and 0.32 N, whereas the complex viscosity strongly differed, from 0.24 to 12.63 N. These values were lower than those measured with yogurt samples, thus indicating that strains used for laban manufacture can be considered as nonropy strains, which is in agreement with the low cohesiveness and the high firmness of the products obtained with these bacteria (Chammas et al. 2006b) and with the results obtained by Duboc and Mollet (2001).

Finally, as syneresis may occur, it has been determined by Chammas et al. (2006b) for different laban products. Syneresis can be measured by gently collecting the whey on the surface of the samples at 4°C. The amount of collected whey is expressed in percentage (v/v) of fermented milk sample. It varied between 1.9% and 13.7% by considering the previous 96 different fermented milks (Chammas et al. 2006b).

These rheological and textural properties are major parameters of the quality of laban. They can be controlled by varying not only the strains used but also the total solids content of the milk, heat treatment, and homogenization of the milk.

Besides laban production, the manufacture process of labneh and other Middle Eastern fermented milks differs as these fermented products are concentrated by removing part of the serum (whey). This is done traditionally by stirring and then straining the laban in cloth or animal-skin bags. More recent technologies are now available using membrane techniques (ultrafiltration and reverse osmosis) or centrifugation (Nsabimana et al. 2005). The main differences with laban concern the rheological properties of labneh, which are the consequence of its higher total solids level (230–250 g/kg; Ozer and Robinson 1999).

11.4.2 Impact of Lactic Acid Fermentation on Laban Flavor

Lactic acid bacteria strongly determine the flavor of laban for two main reasons: acidification and aroma compound synthesis.

11.4.2.1 Taste of Laban

Lactic acid, which is the main fermentation metabolite, confers the laban a typical sharp and acidic taste. Depending on the product, the acid level may differ, with artisanal products being more variable than industrial products. From sensory analysis or sourness, Chammas et al. (2006a) demonstrated that the acidity of five labans differs, as the ratings are comprised between 1 and 11, on a 0–12 scale (Figure 11.6). This sourness was related to titratable acidity that varies between 9.2 and 17.5 g/kg. Among the tested samples, traditional laban samples show lower and more variable acidity (9.2–17.4 g/kg) than laban elaborated with commercial strains (17.5 g/kg). These values are higher than those observed for yogurts, where lactic acid content varies between 9 and 10 g/kg (Tamime and Robinson 1999b).

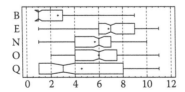

FIGURE 11.6 Boxplot showing the distribution of ratings by panelists for five samples of laban for sourness. (+) represents the mean value ratings given by the panelists for each sample. (From Chammas, G.I., Caractérisation et utilisation de souches de bactéries lactiques thermophiles locales dans le développement de produits laitiers fermentés au Liban, Institut National Agronomique Paris-Grignon, Paris, 2006. With permission.)

11.4.2.2 Aroma Compounds of Laban

Lactic acid bacteria are responsible for the synthesis of aroma compounds, which contribute to the aroma of laban. These compounds include nonvolatile acids (lactic and pyruvic acids), volatile acids (formic, acetic, and butyric acids), carbonyl compounds (acetaldehyde, acetone, acetoin, or diacetyl), and some miscellaneous compounds (certain amino acids, compounds formed by thermal degradation of proteins, fat, or lactose).

The identification of volatile compounds of fermented milks is generally done by gas chromatography. From Chammas et al. (2006b), aroma molecules have to be first extracted by using solid-phase microextraction technique before being analyzed by gas chromatography coupled to mass spectrometry.

These authors demonstrated that 12 main aroma compounds are found in laban samples fermented by *S. thermophilus* and *Lb. bulgaricus*. Two compounds, furancarboxaldehyde and furanmethanol, were brought into the cultures by the heat-treated milk, where their relative content was equal to 18% and 28%, respectively. Four compounds (acetone, acetic, butanoic, and hexanoic acids) were synthesized by both *S. thermophilus* and *Lb. bulgaricus*, but in different amounts (Table 11.2). They represented around 50% of the total flavor profile of *Lb. bulgaricus* cultures, but only 5% in *S. thermophilus* cultures. Milks fermented with *S. thermophilus* allow obtaining fermented milks with high contents of 2,3-butanedione, acetoin, and 2,3-pentanedione (87%), whereas the flavor profile of *Lb. delbrueckii* subsp. *bulgaricus* fermented milks includes acetic, butanoic and hexanoic acids (48%), and acetaldehyde and acetone (16%). Besides these species-related differences, some diversity exists within a bacterial species (Chammas et al. 2006b).

These sensory properties are the major parameters of the flavor of laban. They can be controlled by varying the kind and the number of strains used, as well as the balance between strains, in order to achieve fermented products with various flavor characteristics. From Table 11.2, strong differences exist between pure and mixed cultures, after milk fermentation, whereas differences between mixed cultures are more subtle. All labans obtained from mixed cultures display a complex aroma composition, including all kinds of aroma compounds produced by the two species *S. thermophilus* and *Lb. delbrueckii* subsp. *bulgaricus*.

11.4.3 Impact of Lactic Acid Fermentation on Shelf Life of Laban

The shelf life of laban is defined as the time during which the product can be stored and consumed, under specific conditions, without damaging the quality of the product. It is defined according to hygienic and sensory features. By considering hygienic properties, the low pH of the laban allows increasing the shelf life of the product as compared to that of pasteurized milk. The growth of bacterial pathogens and psychrotrophic bacteria is definitively restrained, whereas yeasts and molds that are acid-tolerant can spoil retail products. This resistance limits the shelf life of the laban to 30 days.

By considering sensory characteristics, post-acidification may occur, depending on the strains used (Kneifel et al. 1993; Sofu and Ekinci 2007). As a general rule, strains exhibiting high acidification activity are not recommended for mild product manufacture because they may favor post-acidification during storage at 4°C. *Lb. delbrueckii* subsp. *bulgaricus* is generally involved in post-acidification as mainly D(–) lactic acid increases in the product during storage (Imhof and Bosset 1994). This species is acid-tolerant and has the ability to produce lactic acid to levels of 1.7 g/100 mL, or even above, depending on the strain (Robinson et al. 2006).

The post-acidification is often associated with an increase in viscosity of the product (Martin et al. 1999), due to the increase in lactic acid and to exopolysaccharide production (Saint-Eve et al. 2008), and with other defects such as syneresis (Kneifel et al. 1993). Thus, after fermentation, cooling must be done as rapidly as possible, so that post-acidification in the product remains less than 0.3 pH units (Spreer 1998). Besides this, starter cultures with reduced post-acidification behavior have been developed and are offered by many culture manufacturers (Kneifel et al. 1993).

From these considerations, in Lebanon, the usual shelf life of laban is between 14 and 30 days.

TABLE 11.2

Main Aroma Compounds Identified in Laban Produced with Two Species of *S. thermophilus* (ES2 and FS2) and Two Species of *Lb. delbrueckii* subsp. *bulgaricus* (EL6 and FL5)

Strains	Acetaldehyde	Acetone	2,3-Butanedione	2,3-Pentanedione	Acetoin	Acetic Acid	Butanoic Acid	Hexanoic Acid
ES2	0	2.53	59.84	4.83	24.13	1.16	2.76	1.70
FS2	0	4.67	54.24	5.41	24.97	1.03	3.03	1.81
EL6	5.00	8.40	0.00	0.00	0.00	18.05	16.47	10.47
FL5	6.43	12.93	0.00	0.00	0.00	24.55	15.92	11.01
EL6 + ES2	1.44^b	3.53^b	25.18^a	23.61^c	24.91^f	5.88^b	4.97^b	3.73^b
EL6 + FS2	1.14^{a,b}	1.94^a	47.29^{c,d}	22.18^c	13.80^{b,c}	5.01^{a,b}	3.46^{a,b}	2.46^{a,b}
EL6 + ES2 + FS2	1.11^{a,b}	2.10^a	37.11^b	29.21^d	16.80^{d,c}	4.20^{a,b}	3.47^{a,b}	2.46^{a,b}
FL5 + ES2	0.85^a	1.62^a	52.83^{d,e}	18.34^b	15.36^{c,d}	3.37^{a,b}	3.09^{a,b}	2.12^a
FL5 + FS2	0.85^a	1.47^a	58.94^{e,f}	16.52^b	10.61^a	2.83^a	3.33^{a,b}	2.38^{a,b}
FL5 + ES2 + FS2	0.60^a	1.24^a	55.01^f	17.86^b	14.52^c	2.60^a	2.77^a	1.90^a
EL6 + FL5 + ES2 + FS2	0.94^{a,b}	1.67^a	45.15^c	24.39^c	15.83^{c,d}	3.00^a	3.64^{a,b}	2.37^{a,b}
EL6 + FL5 + ES2	0.69^a	1.29^a	36.60^b	31.34^d	18.42^e	4.24^{a,b}	2.85^a	1.84^a
EL6 + FL5 + FS2	0.86^{a,b}	1.27^a	62.71^e	11.66^a	12.39^{a,b}	3.15^{a,b}	2.60^a	2.20^a

Source: Chammas, G.I., Caractérisation et utilisation de souches de bactéries lactiques thermophiles locales dans le développement de produits laitiers fermentés au Liban, Institut National Agronomique Paris-Grignon, Paris, 2006. With permission.

Note: Different letters account for statistically different values, according to multiple comparison test of Newman–Keuls at 99.9% for acetaldehyde, acetone and acetic, hexanoic, and butanoic acids, 99% for 2,3-pentanedione, and 95% for 2,3-butanedione and acetoin.

11.4.4 Modification of Nutritional Value of Milk

Fermented milks have been recognized by nutritionists as being beneficial to human health. By comparing with liquid milk, consumption of these products improves lactose digestion in lactose-intolerant individuals (Parra and Martinez 2007) and protein digestibility (Beshkova et al. 1998; Serra et al. 2009), increases the level of some B-complex vitamins (Crittenden et al. 2003; Fabian et al. 2008), enhances calcium assimilation (Parra et al. 2007), and may assure health-promoting effects related to probiotic bacteria (Uyeno et al. 2008).

11.4.4.1 Decreasing Lactose Malabsorption

Lactose intolerance is the inability of some human populations to digest significant amounts of lactose, the predominant sugar of milk. This inability results from a shortage of the enzyme lactase, which is normally produced by the cells that line the distal ileum (McBean and Miller 1998). Lactase breaks down milk sugar into simpler forms (glucose and galactose) that can be absorbed into the bloodstream. Lactose maldigestion is a relevant factor influencing milk and dairy product consumption because lactase deficiency often produces gastrointestinal symptoms after lactose intake (Parra et al. 2007). As undigested lactose reaches the colon, it is fermented by gas-producing bacteria and may cause bloating, flatulence, abdominal pain, and diarrhea (Jarvis and Miller 2002).

Laban and other fermented milks are better tolerated than milk by lactose-intolerant individuals (Varela-Moreiras et al. 1992). During the fermentation of milk, lactic acid bacteria hydrolyze the lactose and produce lactic acid as their main product, thus causing the lactose content of milk to drop from around 5% to 3%. In addition, after ingestion, some viable starter cultures reach the small intestine and participate in lactose digestion as they contain β-galactosidase activity (Pochart et al. 1989). These effects are observed only when living bacteria reach the intestine, as heat treatment of the fermented milk diminishes the effect on lactose digestibility due to enzymatic inactivation (Parra et al. 2007).

11.4.4.2 Increasing Protein Digestibility

Biochemical changes of milk during laban fermentation include not only carbohydrate metabolism and production of flavor components but also a slight but significant degree of proteolysis. Many authors studied the properties of the proteolytic systems of lactic acid bacteria, which revealed that proteolytic activity presents important variations not only among the species but also among the strains belonging to the same species (Law and Haandrikman 1997; Shihata and Shah 2000). By considering casein hydrolysis, these variations are related to cell wall proteases, to transport systems of peptides from extracellular to intracellular media, and to cytoplasmic peptidases that produce amino acids (Savijoki et al. 2006; Tzvetkova et al. 2007). Conversely, the cleavage of whey proteins during the fermentation by lactic acid bacteria involved in laban production remains undetectable (Bertrand-Harb et al. 2003).

Many studies show the effect of lactic acid fermentation on protein digestibility and human health. According to Béal and Sodini (2003), the digestibility of the protein fraction of milk increases after yogurt fermentation due to a modification of protein structure that facilitates the action of the proteolytic enzymes during the intestinal transit. In addition, fresh yogurt intake results in higher acute leucine assimilation than during intake of pasteurized product (Parra and Martinez 2007). By considering some specific strains, proteolysis leads to the formation of bioactive peptides, which are encrypted within the primary structure of milk proteins (Hayes et al. 2007) and are of special interest for their potential biological activities (Serra et al. 2009). These bioactive health-beneficial peptides are called β-casomorphins (Schieber and Brückner 2000). They act as histamine releaser and are believed to be produced from milk proteins by *Lactobacillus* strains as reported by Savijoki et al. (2006).

Casein hydrolysis continues during laban storage, thus leading to an increase in the total concentration of free amino acids that becomes four times higher than in milk after 2 days of storage at 4°C (Beshkova et al. 1998). An increase in hydrophobic bioactive peptides also occurs (Serra et al. 2009). Taking into account these considerations, the maintenance of living microorganisms in these products would be crucial for the health benefits claimed for fermented milks.

11.4.4.3 Effect on Vitamins Content

During milk fermentation, some vitamins are utilized by lactic acid bacteria, while others are actively synthesized. The extent of synthesis and metabolism of these vitamins by bacterial cultures depends on the strain of bacteria used, the size of the inoculum, and the conditions of fermentation (Kneifel et al. 1989; Fabian et al. 2008). Among the B-complex vitamins, an increase in the vitamin content of yogurt was reported for folic acid, pyridoxine, and biotin (Béal and Sodini 2003; Crittenden et al. 2003). However, the folic acid level remains relatively low in terms of recommended daily allowance, even though inoculum species were judiciously selected (Crittenden et al. 2003). The results of the study conducted by Fabian et al. (2008) indicate that daily consumption of 200 g of yogurt for 2 weeks can contribute to the total recommended intake of thiamin (vitamin B1) and riboflavin (vitamin B2).

Based on their ability to synthesize vitamins, it is supposed that lactic acid bacteria ingested with laban, which are the same as that in yogurt, could be a supplementary source of vitamins in human nutrition.

11.4.4.4 Effect on Calcium Assimilation

Dairy products and specifically laban, are important sources of calcium (Gambelli et al. 1999). However, calcium absorbability, or the availability of calcium for absorption by the intestines, is the first step toward bioavailability of calcium from these sources. According to Takano and Yamamoto (2003), calcium assimilation after consumption of both milk and fermented milk is comparable, whereas Béal and Sodini (2003) reported higher calcium uptake from fermented milk compared with milk. In fact, by containing live starters, laban is considered as an important calcium source as it leads to optimized calcium assimilation (Parra et al. 2007).

Lower acute calcium assimilation is obtained from pasteurized milk as compared to that obtained from fresh fermented milk products. The negative effect of pasteurization is due, on the one hand, to a decrease in lactose absorption and consequently to a decrease in calcium assimilation, and, on the other hand, to changes in the structure of the yogurt related to the pasteurization process that could also modify the calcium disposal (Parra et al. 2007). In fact, some proteins are denatured and aggregated during the heating process, and a nonspecific binding of proteins to calcium is described, decreasing the availability of calcium to be absorbed by the human body (Halpern 1993).

11.4.4.5 Health Effects from Probiotic Bacteria

Fermented milks are considered as good probiotic-carrier food with health-promoting effects. Probiotics are defined as "live microorganisms which, when administered in adequate amounts, confer a health benefit on the host" (Sanders 2003). The application of the term "probiotic" to the starter bacteria *Lb. delbrueckii* subsp. *bulgaricus* and *S. thermophilus* is still debated. Moreover, no specific study has been conducted by considering the bacteria involved in laban production.

Survival during passage through the gastrointestinal tract is generally considered a key feature for probiotics to preserve their expected health-promoting effects (Holzapfel et al. 2001). There have been conflicting studies concerning the recovery of *Lb. delbrueckii* subsp. *bulgaricus* and *S. thermophilus* from fecal samples after daily ingestion of fermented milk. Some authors confirmed that yogurt bacteria can be retrieved from feces (Guarner and Malagelada 2003; Mater et al. 2005; Elli et al. 2006), whereas others obtained negative results (Pedrosa et al. 1995; Del Campo et al. 2005). As an illustration, the results obtained by Marteau et al. (1997) show that survival of *Lb. delbrueckii* subsp. *bulgaricus* and *S. thermophilus* is two times lower in a gastric juice than that of *Lb. acidophilus* and *Bifidobacterium bifidum* (Figure 11.7). By sampling intestinal liquid by intubation, Pochart et al. (1989) observed that after ingestion of 10^{11} CFU/mL, only 1% survived in the duodenum (10^9 CFU/mL). These bacteria were characterized by a high lactase activity. These probiotic traits are strictly strain specific, and, therefore, the diversity of these results is explained by the use of different strains in the different studies.

The consumption of probiotic products is helpful in maintaining good health, as demonstrated by the following beneficial effects of probiotic bacteria in humans:

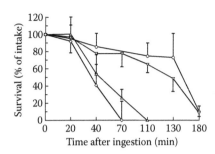

FIGURE 11.7 Comparative survival of ingested *B. bifidum* (○), *Lb. acidophilus* (▽), *Lb. bulgaricus* (△), and *S. thermophilus* (◇) in a gastric juice. Values are expressed as mean percentages (±SE) of live bacteria relative to the ingested numbers (*n* = 6 for each strain). (From Marteau, P. et al., *J. Dairy Sci.* 80(6), 1031–1037, 1997. With permission.)

1. Control of intestinal infections by producing inhibitory/antimicrobial substances such as organic acids and bacteriocins and by stimulating the immune system (Lourens-Hattingh and Viljoen 2001; Fioramonti et al. 2003; Ayar et al. 2005; Uyeno et al. 2008)

2. Reducing lactose intolerance by producing β-galactosidase (Lourens-Hattingh and Viljoen 2001)

3. Control of hypercholesterolemia (Lourens-Hattingh and Viljoen 2001; Ayar et al. 2005)

4. Anticarcinogenic activity (Lourens-Hattingh and Viljoen 2001)

From this information, interest for the consumption of products containing probiotic microorganisms is increasing, and the market is continuously expanding, mainly in European countries (Uyeno et al. 2008). In Lebanon and other Arab countries, the market of these fermented products remains very small but is going to increase.

11.5 General Process of Laban Manufacture

The general flow diagrams of laban manufacture are shown in Figure 11.8 for small-scale and large-scale production plants. They differ significantly according to artisanal and industrial units, the last one being more complicated and controlled than the first one.

11.5.1 Milk Reception and Analysis

After being collected, fresh milk is stored in refrigerated tanks and transported into refrigerated road tankers to the production plant. At arrival, milk is first controlled, then pumped and filtrated to discard any solid residues (straw, leaves), and stored at least at 5°C in silos. These tanks, which are made of stainless steel, may reach 100,000 L at industrial scale. Temperature control is achieved by using a jacket in which cold water is injected and circulates.

11.5.2 Standardization of Milk Fat

Cow milk is constituted of water, lactose, fat, proteinaceous compounds (caseins and native serum proteins), other nitrogen compounds, and minerals. The average composition of milk varies according to the breed, the lactation stage, the cattle feeding of the animals, and the season. For a better regularity of the quality of the industrial products, it is standardized by considering both fat and protein contents.

Fat content of cow milk fluctuates between 3.8% and 4.2%. Standardization allows this fat content to achieve the desired value in the product, which varies between 0.5% and more than 3%, depending on the laban product (fat-free, low-fat, or whole-fat laban). This is done by first skimming the fat off the milk and then by mixing the skimmed milk with cream in convenient proportions.

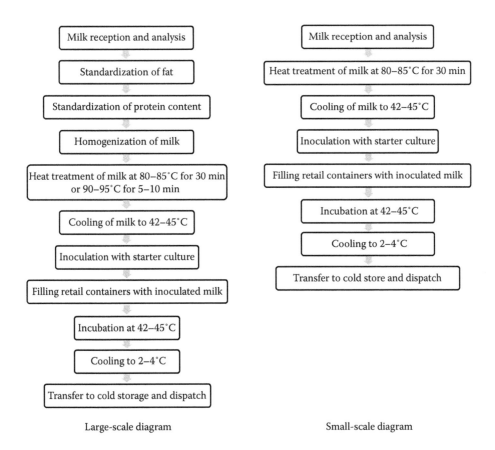

Large-scale diagram   Small-scale diagram

FIGURE 11.8   General flow diagrams of laban manufacture for large-scale and small-scale production plants.

As an example, one shall use the following calculation to prepare a mix of 10,000 kg of milk at 1.6% of fat. If m is the quantity of skimmed milk containing 0.1% of fat and c is the quantity of cream composed of 50% of fat, the following number system has to be used:

$$m + c = 10,000$$

$$0.1m + 50c = 1.6 \times 10,000$$

Then $m = 9700$ and $c = 300$.

Finally, one shall mix 300 kg of cream with 9700 kg of skimmed milk to obtain 10,000 kg of milk with 1.6% of fat.

Fat separation is done by using centrifugal separators at a temperature of 70°C. The skimming efficiency of these separators allows obtaining a residual fat in skimmed milk lower than 0.07%. The final fat content in milk is achieved by proportional mixing of skimmed milk and cream. At industrial scale, this operation is carried out automatically by using flow meters, densimeters, and pressure gauges to allow online calculation and monitoring of the fat content in the final product. Flow rates range between 7000 and 45,000 L/h, and accuracy of the final fat content is higher than 0.03%.

11.5.3 Standardization of Protein Content

The total nitrogen content of cow milk varies between 2.9% and 3.7% all year long. In order to reduce this variability and to allow the protein content to reach a value between 3.2% and 5% in laban, milk is supplemented with dairy proteins. This standardization allows the laban to become more firm and to reduce syneresis in the product. It also increases the viscosity of the fermented milks.

Various practices are used to complete protein fortification. In traditional small production plants, milk is heated before fermentation to favor partial evaporation and to reduce the volume, in order to achieve a protein content of around 5%. In larger production plants, milk supplementation is done by adding skimmed milk powder, concentrated milk, buttermilk powder, whey powder, whey protein concentrates, or casein hydrolysates (Tamime and Robinson 2007a). Concentration by vacuum evaporation is also employed by incorporating a single effect plate evaporator into a processing line.

As an example, one shall use the following calculation to prepare a mix of 10,000 kg of milk at 4.5% of protein. If m is the quantity of skimmed milk containing 3.3% of protein and p is the quantity of skimmed milk composed of 34% of protein, the following number system has to be used:

$$m + p = 10,000$$

$$3.3m + 34p = 4.5 \times 10,000$$

Then $m = 9610$ and $p = 390$.

Finally, one shall mix 390 kg of skimmed milk powder with 9610 kg of skimmed milk to obtain 10,000 kg of milk with 4.5% of protein content.

During incorporation of dry ingredients such as milk powder or derived milk products into the aqueous phase (milk standardized in fat content), a complete dispersion and hydration is required in order to avoid formation of lumps. It is also important to circumvent the incorporation of air into the product in order to limit foam formation during mixing. Different kinds of equipment exist, either batch or continuous, all of them allowing high-quality cleaning in place. The mixing funnel involving a Venturi valve is the simplest system. As a general procedure, a tank is first filled with the fat standardized milk at 40°C–45°C that is pumped at a flow rate of about 25 m³/h. The dry ingredients are then added into the funnel where they are progressively incorporated into the milk, thanks to the Venturi effect and the circulation of the milk. As an alternative, online systems that allow continuous incorporation of the milk powder also exist. Mixing rates as high as 45 kg/min are obtained with these systems.

11.5.4 Homogenization

As it contains fat, milk is homogenized to provide a stable fat-in-water emulsion in order to prevent creaming during fermentation at industrial scale. By reducing the size of fat particles, homogenization also helps in increasing the viscosity of laban, reducing syneresis during storage of the product, and enhancing the whiteness of the milk. It is generally conducted before heat treatment or during heating of milk. Nevertheless, the study of McKenna (1987) demonstrates that downstream homogenization (post-pasteurization) allows obtaining a more stable laban by considering viscosity and syneresis.

Homogenization is a mechanical treatment that decreases the average diameter and increases the number and surface area of the fat globules in milk. Homogenizers used in laban production are composed of one or two stages. With high fat content products, two stages are needed. Homogenization is done by passing standardized milk under high pressure through a tiny orifice (0.1 mm diameter). The velocity is generally between 100 and 400 m/s between the valve and the valve seat. The residence time ranges between 10 and 15 µs, and the flow rate is between 4000 and 20,000 L/h. During homogenization, pressure and temperature are controlled at 10–20 MPa and 65°C–70°C, respectively. After homogenization, the mean diameter of fat globules is reduced from 3 µm (range 1–10 µm) to 0.5 µm (range 0.2–2 µm).

11.5.5 Heat Treatment and Cooling

After homogenization, the mix is heat-treated in order to reduce the microbial contaminations and to improve the physical properties of the laban (viscosity and water holding). It aims to degrade 80% of the two main serum proteins: α-lactalbumin and β-lactoglobulin. When being denaturated, these two proteins attach themselves to casein surface through disulfur bonds. This phenomenon allows avoiding the casein micelles to link in large aggregates during further acidification, thus reducing syneresis and permitting a good consistency of the products.

Manufacturers provide equipment for batch or continuous heat treatment. Batch processes are conducted in a jacketed vat by injecting hot water or steam in the double jacket or by means of heating coils surrounding the inner jacket. During heating and holding periods, the mix shall be agitated. Barem used for batch heat treatment of laban is generally 85°C–90°C for 15–30 min.

Continuous processes are generally preferred for large-scale production, as bulk milk is processed in a short span of time. It also allows reducing the vat volume, thus increasing productivity and saving costs. For continuous heat treatment, either scraped surface heat exchangers or plate heat exchangers are used, with the last one being more popular. The main advantage of plate heat exchangers is that they offer a large transfer surface that is readily accessible for cleaning. Overall heat transfer coefficients are in the range of 2400–6000 $J/m^2 \cdot s \cdot °C$. Barems for continuous heat treatment of laban vary according to the industrial plant: 30 min at 85°C, 5 min at 90°C–95°C, or 3 s at 115°C. The most used conditions are 92°C for 5–7 min, with a flow rate between 4000 and 20,000 L/h.

After heat treatment, milk is cooled down to fermentation temperature. Sometimes, cooling the milk at 4°C is necessary before or after inoculation in order to delay incubation. It is then stored in vats under overpressure for a few hours.

11.5.6 Inoculation and Fermentation

Fermentation of heat-treated milk starts by inoculating the selected bacteria (mainly lactobacilli and streptococci) into the milk. After inoculation, the liquid product is packed, and the fermentation takes place in the retail containers. The process involves seven stages:

1. Stabilization of milk temperature at fermentation temperature (generally 42°C during laban production)
2. Inoculation by addition of the starter culture to the prepared milk
3. Mixing in order to obtain a well-homogenized mix
4. Aseptic dispatching of the product into the retail containers
5. Sealing of the retail containers
6. Incubation at 42°C for a few hours in controlled temperature cabinets or tunnels
7. Cooling of the product in order to stop the fermentation

11.5.6.1 Inoculation Modes

Inoculation mode depends on the manufacturing scale. During traditional laban production, inoculation is done by using an artisanal starter consisting of an unknown number of undefined strains. These starter cultures are composed of fermented samples taken from a previous laban production. This procedure implies that artisanal starter composition is unknown and strongly variable, thus leading to a variable quality of the product (Chammas et al. 2006b).

Industrial starters are characterized by well-defined strains, which are combined in well-defined balance. This procedure allows providing products with the desired characteristics that result from the metabolic activity and the technological properties of the strains during their growth in milk. This leads to a standard quality of the laban and a high level of reproducibility of the processes, but lower sensory properties (Chammas et al. 2006a). By considering large-scale processes, direct vat inoculation is commonly employed. The starters are bought from industrial starter manufacturers that sell them either in frozen form (storage below −40°C) or in freeze-dried form (storage at or below 4°C). The inoculation rate is generally between 5×10^5 and 5×10^6 CFU/mL. Semi-direct inoculation is also employed, thus involving a preculture step before laban inoculation. The cost of the starters is then reduced.

The balance between the bacterial species strongly influences the fermentation kinetic and the quality of the final product. During laban production, a ratio of 1:1 between *S. thermophilus* and *Lb. delbrueckii* subsp. *bulgaricus* is generally employed. The choice of the strain also affects the acidification rates and the sensory characteristics of the products, as shown by Chammas et al. (2006b).

11.5.6.2 Incubation of Laban

Incubation of inoculated milk takes place directly in the cups. Depending on the size of the production plant, it is carried out either in warm rooms or tunnels at a temperature of 42°C. By considering small-scale manufacturers, fermentation generally takes place in insulated chambers, in which forced hot air is circulated inside. When incubation is finished, it is replaced by chilled air to cool down the product to 4°C.

In larger production plants, tunnel systems are preferred as they allow a continuous operation and using them saves energy as compared to warm rooms. Two different tunnels are generally combined (one for incubation and one for cooling), but single tunnels with an incubation section and a cooling section are also available. The pallets with the laban cups move inside the tunnel with the help of conveyors. The length of the tunnel and the moving speed are calculated as a function of fermentation duration.

11.5.6.3 Kinetic of Laban Fermentation

During laban production, fermentation occurs according to Figure 11.9, which shows the time course of bacterial growth, substrate, and lactic acid concentration as well as pH of milk. This kinetic was obtained by using a mixed culture (1:1) of *S. thermophilus* and *Lb. delbrueckii* subsp. *bulgaricus*. From this figure, the growth of *S. thermophilus* starts earlier than that of *Lb. delbrueckii* subsp. *bulgaricus*. This is explained by the requirements of the lactobacilli in growth factors, formic acid, and CO_2 that are synthesized by the streptococci (Ascon-Reyes et al. 1995). The development of *S. thermophilus* stops after 3.5 h at pH 4.5 because of the sensitivity of this species to low pH. Figure 11.9 also shows that *Lb. delbrueckii* subsp. *bulgaricus* starts to grow after 2 h of fermentation. As lactobacilli are more resistant to acidity, their growth continues until 6 h. The final ratio between these two species is generally unbalanced in favor of the streptococci.

The lactose concentration, which is the main carbon substrate for bacterial growth, decreases from 49 to 31 g/L. A significant lactose concentration remains in the final product. At the same time, the galactose and lactic acid concentrations increase in the fermented milk, thus reaching 7 and 8 g/L, respectively, in laban. The molar yield of lactic acid (calculated as the ratio between formed lactic acid and consumed lactose) is equal to 84% of the theoretical yield, thus corresponding to a homofermentative metabolism.

In the same time, the pH of the milk decreases from pH 6.4 to pH 4.2. This acidification is inversely correlated to the increase in the lactic acid concentration. As explained previously, this pH decrease induces the gel formation. This is confirmed by the measurement of the electrical signal obtained from a coagulation sensor, which varies as a function of pH as previously observed in Figure 11.4.

Finally, the laban composition is characterized by a bacterial concentration that is higher than 10^9 CFU/g, a level of acidity that depends on the time at which the fermentation is stopped, a lactose concen-

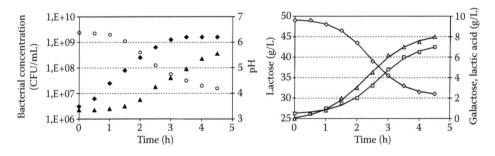

FIGURE 11.9 Time course of bacterial growth, substrate consumption, metabolite production, and pH decrease during laban manufacture. (◆) *S. thermophilus*; (■) *Lb. delbrueckii* subsp. *bulgaricus*; (○) pH; (◇) lactose; (▲) galactose; (△) lactic acid.

tration that is lower than that found in milk, and the presence of specific compounds that may influence the sensory properties of the final product.

11.5.6.4 Controls during Laban Fermentation

The controls during laban production are quite basic. They mainly concern the temperature and sometimes the pH or the titratable acidity that are checked in the samples taken, both factors determining the fermentation duration.

During incubation, the first operating condition to be controlled is the temperature. For laban fermentation, it is maintained between 40°C and 45°C, with the precise value depending on the strains used. In some artisanal production units, the temperature may be lower (30°C–37°C), thus leading to a longer fermentation time.

The length of the incubation is the second important parameter to be controlled. It depends on the final pH and acidity that are required. By considering laban, the length is between 3 h (competitive industrial plants) and 12 h (artisanal production units). These values correspond to a final pH of 4.8 to 4.2 in the product and a final titratable acidity level of 9.2 to 17.5 g/kg.

11.5.7 Cooling

A rapid cooling of the product is necessary to stop the fermentation, thus allowing most of the metabolic and enzymatic activities to cease. A quick cooling is also useful to achieve a more uniform quality of the product and to prevent post-acidification. Cooling starts when the acidity of the product reaches the required value and lasts between 30 min and 1 h. For laban production, the cooling begins when the lactic acid level reaches 0.8%–1% (pH 4.8–4.7) in order to obtain a final acidity in the product between 1.2% and 1.4%.

The cooling is done by directly decreasing the temperature from incubation temperature to 5°C either in a cold chamber, by circulating cooled air, or in a chill tunnel. This last practice allows a more rapid cooling and then a lower post-acidification than the first one. Nevertheless, very rapid cooling may lead to whey separation, and the investment is much more expensive.

After cooling, laban cups are maintained at low temperature, that is, between 4°C and 8°C during storage, transport, and distribution.

11.6 Packaging and Storage of Laban

Packaging is crucial for safe delivery of laban to consumers. It constitutes also the most effective means of communication between a dairy product manufacturer and eventual consumers.

Packaging is defined by Brody (2008) as "the totality of all elements required to confine the product within an envelope that functions as a barrier between the product and the environment, which is invariably hostile to the contained product unless the protection afforded by packaging is present." It should meet specific requirements such as providing protection, easy handling, and offering a vehicle for a message. Environmental factors including temperature, moisture, oxygen, shock, compression, and human disturbances have to be taken into account.

11.6.1 Packaging of Laban

11.6.1.1 Packaging Materials

Among the primary packages used for fermented milks, plastic tubs (capacity 500 g and 1, 2, and 5 kg) and cups (capacity 125 and 500 g) are the most common. The term "plastic" describes a family of materials derived from petrochemical sources, capable of being shaped or molded. Plastic package materials are characterized by their light weight, relative ease of fabrication, low cost, and malleability. The most commonly used plastic packaging materials for storage of fermented milks such as laban are high-density polyethylene (HDPE), polypropylene (PP), and polystyrene (PS) (Cutter 2002; Cooper 2007). HDPE is used for the manufacture of tubs and their lids, whereas PP and PS are used for the manufacture of cups.

1. HDPE is a semi-rigid translucent plastic. This polymer has good moisture and water resistance, but has very poor barrier properties to gases. It has good heat resistance (up to about 100°C–120°C), making it suitable for "hot fill" and pasteurization. Usually, HDPE is used to form bottles for milk, drinkable yogurt, and laban.

2. PP is stiffer than HDPE and has good clarity and high moisture resistance, but has low gas barrier properties. PP is converted with the addition of thermoforming to make injection-molded cups for yogurt and laban.

3. PS is a hard polymer, with excellent transparency and good structural properties, but with poor oxygen and water vapor barrier properties. High-impact polystyrene (HIPS), made with incorporation of a blowing agent, is a material used commonly for packaging of laban and related products worldwide. However, oxygen diffuses into the product through the HIPS packaging during storage (Talwalkar et al. 2004). Packaging alternatives to HIPS have been developed: PS-based gas barrier that is effective in preventing diffusion of oxygen into the product during storage (Talwalkar et al. 2004) and PS-based gas barrier with an active packaging film that can actively scavenge oxygen from the product (Cutter 2002; Talwalkar et al. 2004).

Among these materials, PS packaging is preferred for limiting aroma compound losses from the product and for avoiding the development of odor and aroma defects (Saint-Eve et al. 2008).

11.6.1.2 Packaging Systems

As laban is a set-style plain fermented milk, fermentation takes place in the final cups. In Lebanon and other Arab countries, traditional laban is produced by an old-age practice that entails a manual filling (for tubs capacity 1, 2, and 5 kg) and container uncovering during incubation. This process is slow, cumbersome, labor intensive, and unhygienic. The use of modern processing lines, which minimize manual handling during production and contamination during incubation, potentially yields a product with superior microbiological quality as compared to the traditional method. After inoculation, milk is pumped to the filling machine and containers are sealed immediately. Filled and sealed cups and tubs are placed on pallets and transferred to the incubation room. After fermentation, they are transported into the cooling room.

Filling machines are designed to fill laban in two types of packaging systems (Robinson et al. 2006):

1. Preformed containers are filled with laban and then covered with an aluminum foil lid that is heat-sealed to the container (cups capacity 125 g) or covered with a snap-on lid for tub capacity 500 g, 1 kg, and 2 kg.

2. Form-fill-seal containers are produced during the filling operation. Filling machine is fed by a roll of film, thermoforms the cups, fills them, and seals them with a foil lid. This system is applicable to 125-g-capacity cups and is suitable for large-scale operations.

The choice of the packaging machine and the type of container are influenced, among other considerations, by the marketing concepts and consumer acceptability (Tamime et al. 2001).

11.6.1.3 Storage of the Products

Laban must be stored and transported in such a way that the product is not negatively influenced by the environment. For this purpose, to safeguard the quality of the product, temperatures below 4°C are chosen by most manufacturers.

Laban containers are stored in cold refrigerated rooms (temperature between 2°C and 4°C) for delivery to grocery stores or warehouses for distribution. The vehicles used for the distribution are mechanically refrigerated at the same temperatures as in the storage facilities in the factory (between 2°C and 4°C).

11.6.2 Shelf Life of Laban

The shelf life of a product is the recommended time during which the defined quality remains acceptable under specified conditions of distribution, storage, and display. The shelf life of foods is determined according to the proportion of product failure tolerated by the health risks that might result from consumption of failed/expired products (Labuza 2000). For products that, practically, do not pose health hazards to consumers, shelf life is determined at probability of failure of 50% (Hough et al. 1999).

Due to its inherent inability to support growth of pathogens, due to its low pH (Muir 1996), laban has an excellent safety record. *S. thermophilus* and *Lb. delbrueckii* subsp. *bulgaricus* prevent the survival of *Escherichia coli* (Kasimoğlu and Akgün 2004), *Listeria monocytogenes* (Gohil et al. 1995), and *Enterobacter* (Shaker et al. 2008) during the processing and storage stages.

However, as laban is sold with live bacteria, sensory properties of the product will change until they reach a limit beyond which the consumer will reject the product. According to Hough et al. (1999), the shelf life of food products depends on the interaction of the food with the consumer—hence, a laban with a long storage period may be accepted by a consumer who likes high-acid flavor, but rejected by another consumer who does not like high-acid products (Curia et al. 2005; Salvador et al. 2005). Practically, changes in physical, chemical, and microbiological structure of laban determine the storage and shelf life of this product.

A study conducted by Saint-Eve et al. (2008) shows that a rapid evolution of low-fat yogurts stored at 4°C occurred during the first 14 days of storage, at the sensory and physicochemical levels. However, defects of the sensorial quality of whole-fat yogurts take longer time to appear than that of fat-free yogurts (Curia et al. 2005; Salvador et al. 2005; Saint-Eve et al. 2008).

Although low pH of laban inhibits the growth of many bacterial pathogens, other microorganisms such as psychrotrophic bacteria, yeasts, and molds are acid-tolerant and can spoil retail products within an anticipated shelf life of 21 days (Tamime et al. 2001). Thus, in addition to deterioration of physicochemical properties, microbiological counts are used as indices for the end of shelf life (Muir 1996; Hough et al. 2003). Psychrotrophic bacteria have been reported as being the major determinant of shelf life of yogurt and related products (Lewis and Dale 2000). The detrimental level to flavor quality of fermented milks is higher than 10^7 CFU/g (Bishop and White 1986). Flavor defects have been reported when counts of yeasts and molds reach levels of 10^5 CFU/g (Kadamany et al. 2003; Tamime and Robinson 2007b). Al-Tahiri (2005) reported a significant production of gas and the presence of unpleasant flavors in highly contaminated samples by yeasts and molds. According to Sofu and Ekinci (2007), there is an increase in the percentage of the total area having colors of pale-greenish-yellow, grayish-yellow, light grayish-green, and yellowish-green, 14 days after the storage of yogurt samples. The presence of these colors is associated with microbial spoilage of this product.

Consequently, the shelf life of laban depends on the level of contamination and the production procedure (artisanal or industrial scale). Yogurt may be sold up to 21 days (Robinson et al. 2006), 28 days (Salvador et al. 2005), or 35 days (Curia et al. 2005; Robinson et al. 2006) after manufacture. In Lebanon, the shelf life of laban varies between 14 and 30 days.

11.7 Evaluation of Laban Quality

Before commercialization, laban is evaluated in terms of microbiological, physicochemical, and sensory aspects. These controls are done systematically for industrial products, whereas they remain insufficient for artisanal products.

11.7.1 Microbiological Characteristics

The microbiological examination of the finished product includes checks for the survival of the starter organisms, as well as for the presence of undesirable spoilage species. According to the Lebanese legislation (Libnor 1999), *Lb. delbrueckii* subsp. *bulgaricus* and *S. thermophilus* dominate in the product, whereas other species of lactic acid bacteria may also be present in artisanal products. According to Libnor (1999), live starter bacteria must be abundant in laban until the end of the shelf life of the product, but the required counts are not specified. As a comparison, in yogurt, the total bacterial population of starter origin should be at least 10^6 CFU/mL from French legislation (Loones 1994), but it generally exceeds 10^9 CFU/mL (Tamime et al. 2001).

Laban has a pH below 4.3 and, therefore, should be considered as a safe product. However, Robinson et al. (2006) reported that occasional checks for specific pathogens such as *Salmonella* spp. and *L. monocytogenes* are necessary. As fixed by the Lebanese standards, the total count of *Staphylococcus aureus* must be lower than 10 CFU/g and *Salmonella* should be absent in 25 g of sample (Libnor 2007).

Microbiological check covers also yeast and molds, coliform bacteria, and eventually foreign flora. Acid-tolerant yeasts that may grow in laban and affect its quality involve some species of *Saccharomyces*, *K. marxianus* var. *lactis*, and *K. marxianus* var. *marxianus* (Robinson and Itsaranuwat 2006; Robinson et al. 2006). Molds such as *Mucor* spp., *Rhizopus* spp., *Penicillium* spp., and *Aspergillus* spp. can grow at the product–air interface (Robinson et al. 2006). In yogurt, total colony counts for non-starter bacteria should be lower than 10^3 CFU/g, and yeast and mold counts should be inferior to 10 CFU/mL (Tamime et al. 2001). However, the Lebanese standards tolerate higher values. Total and fecal coliform counts must be lower than 5×10^2 CFU/g, and yeast and mold counts should remain lower than 10^3 CFU/g (Libnor 2007).

11.7.2 Physicochemical Characteristics

The physicochemical tests of laban products cover fat content, solids-non-fat (SNF) content, pH value, titratable acidity, foreign water content, and gel firmness.

Retail products are designed as fat-free, low-fat, and whole-fat laban with a fat content set by the legislation as follows (Libnor 1999): less than 0.5% for fat-free products, between 0.5% and 3% for low-fat products, and not less than 3% for whole-fat products. In Lebanon, the mean value of fat content in commercial whole-fat laban is 3.5%, whereas in traditional laban, the fat content varies between 3.3% and 4.2% (Chammas et al. 2006a). According to Musaiger et al. (1998), the fat content of low-fat laban is between 1.2% and 1.4%. The production of diversified laban according to its fat content is carried out to satisfy diet-conscious consumers, but it has an effect on the acceptability of the products. Fat content affects the creamy smell, the flavor, and the thickness of laban (Chammas et al. 2006a).

SNF content indicates protein, lactose, and mineral contents of the product. It depends on the degree of fortification of milk solids. The minimum SNF required in laban is 8.5% (Libnor 1999). This level is essential in obtaining a firm coagulum, as stabilizers are not permitted under local regulations. In the products analyzed by Musaiger et al. (1998), the SNF content was between 8.8% and 9.3%, thus satisfying these rules. The mineral content was mainly characterized by potassium (132–146 mg/100 g), calcium (120–128 mg/100 g), phosphorus (101–104 mg/100 g), sodium (56–81 mg/100 g), and magnesium (11–12 mg/100 g) (Musaiger et al. 1998).

Acidity of laban is important with respect to public safety and product quality. Mild products with pH values higher than 4.5 can allow the survival of *Salmonella* for up to 10 days (Al-Haddad et al. 2003) or *E. coli* for up to 7 days (Massa et al. 1977). The pH of laban has been determined by many authors. According to Musaiger et al. (1998), it is between 4.4 and 4.5; for Baroudi and Collins (1976), it is lower (pH 4.25); and for Guizani et al. (2001), it depends on the manufacture scale (pH 4.0 in artisanal products and pH 4.5 in commercial products).

The acidity of the final product is monitored according to consumer preference. Lactic acid content should not exceed 1.5% in laban (Libnor 1999). However, products available in the market show variable acidity values, ranging between 0.8% and 1.75% (Musaiger et al. 1998; Chammas et al. 2006a). As a consequence, laban is more acidic than other fermented milk products, where the lactic acid content is 0.9%–1% in yogurt

(Musaiger et al. 1998; Tamime and Robinson 1999b), 0.9%–1.2% in zabady produced in Egypt (Abd El-Salam 2003), and 0.77% in laban from Sultanate of Oman (Guizani et al. 2001), respectively.

Although firmness and viscosity of laban are requested by the consumers, there is no legislation set to evaluate these two parameters. The study carried out by Chammas et al. (2006b) shows that there are big differences between laban samples in terms of gel firmness and apparent viscosity due to variations in starter culture strains. From Chammas et al. (2006b), apparent viscosity of laban is between 0.18 and 0.28 Pa·s, which is consistent with that of yogurt prepared with 3% fat in milk (apparent viscosity of 0.23 Pa·s; Shaker et al. 2000). Finally, even though no specific study concerns laban, processing parameters may also affect the physical characteristics of this product, as demonstrated with yogurt (Hassan et al. 1996; Lucey and Singh 1998; Béal et al. 1999).

11.7.3 Sensory Evaluation

Consumers' perception of fermented milk products is mainly related to health, nutrition, sensory characteristics, and pleasure (Ares et al. 2008). Manufacturers should, therefore, ensure that their product meets the expectations of the consumers concerning natural laban. In Lebanon, labans with different acidity, texture, and aroma profiles are present in the market to comply to consumers' preference.

From a sensory point of view, laban is characterized by its acidity, texture, and aroma. As compared to other fermented milks, laban is more acidic (Musaiger et al. 1998; Chammas et al. 2006a). Increase in fat and total solids content decreases the perception of acidity in yogurt (Robinson et al. 2006) and leads to the appearance of granulated texture (Trachoo and Mistry 1998).

Laban is composed of a firm gel as a result of the increase in lactic acid level (Saint-Eve et al. 2008). The composition of the starter strongly influences the gel firmness (Chammas et al. 2006b). Laban is also characterized by its aroma, the intensity of which reflects the presence of aromatic compounds such as acetaldehyde, organic acids, 2,3-butanedione, and acetoin (Ott et al. 2000b; Chammas et al. 2006b).

Sensory analysis is considered as an important technique in determining product quality. Murray et al. (2001) reported that descriptive sensory tests profiled products on all their perceived sensory characteristics by training a group of panelists to reliably identify and score product attributes. Pre-established lists of sensory attributes related to fermented milk products can be found in the literature (Bodyfelt et al. 1988; Civille and Lyon 1996; Martin et al. 1999). Chammas et al. (2006a) used a set of 23 attributes to differentiate 19 samples of laban in terms of appearance, color, texture, odor, taste, and aftertaste. Eight sensory attributes were related to appearance, color, and texture. Five attributes were used to describe odor (yogurt, creamy, butter, burnt, and yeast), five characteristics to represent flavor (sourness, bitterness, astringency, sweetness, and saltiness), and five descriptors to characterize aftertaste (sourness, bitterness, astringency, sweetness, and saltiness). These attributes, as well as their definitions, are given in Table 11.3. From the study of Chammas et al. (2006a), these properties of laban strongly vary according to the product under consideration.

TABLE 11.3

Attributes Used in the Sensory Evaluation of Physical Properties of Laban

Perception Category	Attributes	Definition
Appearance	Gel firmness	Absence of syneresis
	Color	Scale yields from white to yellow
Texture assessed with the spoon	Gel-like	Product's ability to wiggle similar to a gelatin dessert
	Smooth	Presence of grain-size particles in the gel quantified by visual inspection of the spoon's back
	Thick in spoon	Product's ability to flow from the spoon
	Slimy	Product's ability to flow in a continuous way from the spoon
Texture assessed in the mouth	Thick in mouth	Product's flowing resistance assessed by pressing one spoonful of the product between the tongue and the palate
	Mouth coating	Product's ability to form a film lining the mouth

Source: Chammas, G.I. et al., *J. Food Sci.* 71(2), 5156–5162, 2006.

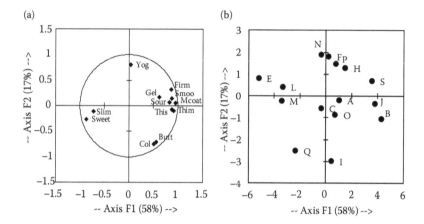

FIGURE 11.10   Graphical representation of the principal component analysis of the sensory ratings for 15 labans show-ing PC1 versus PC2: (a) factor loadings; (b) factor scores. Abbreviations: gel firmness (Firm), color (Col), gel-like (Gel), smooth (Smoo), thickness assessed by the spoon (This), slimy (Slim), thickness assessed by the mouth (Thim), mouth coat-ing (Mcoat), odor of yogurt (Yog), odor of butter (Butt), sourness (Sour), and sweetness (Sweet). Letters A to S refer to 15 different laban samples. (From Chammas, G.I. et al., *J. Food Sci.* 71(2), S156–S162, 2006. With permission.)

Figure 11.10 shows the principal component analysis of the sensory ratings obtained from 15 laban sam-ples, collected all around Lebanese territory. This figure confirms that laban is a product with various sensory characteristics, thus encountering the diverse expectations of a wide range of consumers.

11.8 Conclusion

Laban is a major dairy product in the diet of many Middle Eastern countries. Production of laban is the result of the use of lactic acid bacteria, which leads to milk acidification, aroma compound synthesis, and texture development. Strong differences exist between artisanal and commercial products, owing to the strains used and also to the manufacture process. These differences generate a great diversity in sensory properties, with artisanal products being more typically flavored than commercial labans. On the other hand, microbiological traits are less controlled in artisanal products, thus inducing flavor defects, leading to insufficient process regularity and decreasing hygienic safety.

In the future, some progress has to be done in order to develop specific strains for laban, thus reach-ing consumers' preferences together with guaranteeing high safety and regularity of the products. This progress may also concern the definition of well-defined starters by combining these specific strains at well-defined balances. Moreover, based on their health benefits for consumers, the development of pro-biotic strains for laban manufacture will permit an increase in laban consumption.

REFERENCES

Abd El-Salam, M.H. 2003. Fermented milks: Middle East. In: *Encyclopedia of Dairy Sciences*, edited by H. Roginski, J.W. Fuquay, and P.F. Fox, pp. 1041–1045. London, England: Academic Press.

Adamberg, K., Kask, S., Laht, T.M., and Paalme, T. 2003. The effect of temperature and pH on the growth of lactic acid bacteria: A pH-auxostat study. *Int. J. Food Microbiol.* 85(1–2):171–183.

Al-Ammouri, N. 2006. *Comparative Advantage of Cow Milk in Syria*, Vol. 25, p. 20. Syria: National Agricultural Policy Center.

Al-Haddad, K.S.H. and Robinson, R.K. 2003. Survival of salmonellae in bio-yoghurts. *Dairy Ind. Int.* 69(7):16–18.

Alqaisi, O., Ndambi, O.A., and Hemme, T. 2009. Development of milk production and the dairy industry in Jordan. *Livest. Res. Rural Dev.* 21(7), online edition.

Alqaisi, O., Ndambi, O.A., Uddin, M.M., and Hemme, T. 2010. Current situation and the development of the dairy industry in Jordan, Saudi Arabia, and Syria. *Trop. Anim. Health Prod.* 42:1063–1071.

Al-Tahiri, R. 2005. A comparison on microbial conditions between traditional dairy products sold in Karak and same products produced by modern dairies. *Pak. J. Nutr.* 4(5):345–348.

Amrane, A. and Prigent, Y. 1999. Differentiation of pH and free lactic acid effects on the various growth and production phases of *Lactobacillus helveticus. J. Chem. Technol. Biotechnol.* 74:33–40.

Anonymous. 2008. Syrian Agricultural Database, NAPC National Agricultural Policy Center: http://www.napcsyr.org/sadb.htm.

Ares, G., Giménez, A., and Gámbaro, A. 2008. Understanding consumers' perception of conventional and functional yogurts using word association and hard laddering. *Food Qual. Prefer.* 19(7):636–653.

Ascon-Reyes, D.B., Ascon-Cabrera, M.A., Cochet, N., and Lebeault, J.M. 1995. Indirect conductance for measurements of carbon dioxide produced by *Streptococcus salivarius* spp. *thermophilus* TJ 160 in pure and mixed cultures. *J. Dairy Sci.* 78(1):8–16.

Ayar, A., Elgün, A., and Yazici, F. 2005. Production of a high nutritional value, aromatised yogurt with the addition of non-fat wheat germ. *Aust. J. Dairy Technol.* 60(1):14–18.

Baroudi, A.A.G. and Collins, E.B. 1976. Microorganisms and characteristics of laban. *J. Dairy Sci.* 59(2):200–202.

Béal, C. and Corrieu, G. 1991. Influence of pH, temperature and inoculum composition on mixed cultures of *Streptococcus thermophilus* 404 and *Lactobacillus bulgaricus* 398. *Biotechnol. Bioeng.* 38(1):90–98.

Béal, C. and Corrieu, G. 1998. Production of thermophilic lactic acid starters in mixed cultures. *Lait* 78:99–105.

Béal, C., Louvet, P., and Corrieu, G. 1989. Optimal controlled pH and temperature for growth and acidification of *Streptococcus thermophilus* 404 and *Lactobacillus bulgaricus* 398. EEC Sectorial Meeting, Cultures Collections, Genova, Italy.

Béal, C., Marin, M., Fontaine, E., Fonseca, F., and Obert, J.P. 2008. Production et conservation des ferments lactiques et probiotiques. In: *Bactéries lactiques: De la génétique aux ferments*, edited by G. Corrieu and F.M. Luquet, pp. 661–785. Paris, France: Tec&Doc Lavoisier.

Béal, C., Skokanova, J., Latrille, E., Martin, N., and Corrieu, G. 1999. Combined effects of culture conditions and storage time on acidification and viscosity of stirred yoghurt. *J. Dairy Sci.* 82:673–681.

Béal, C. and Sodini, I. 2003. Fabrication des yaourts et des laits fermentés. *Techniques de l'Ingénieur* F6315:1–16. Paris, France.

Bertrand-Harb, C., Ivanova, I.V., Dalgalarrondo, M., and Haertle, T. 2003. Evolution of β-lactoglobulin and α-lactalbumin content during yoghurt fermentation. *Int. Dairy J.* 13(1):39–45.

Beshkova, D.M., Simova, E.D., Frengova, G.I., Simov, Z.I., and Adilov, E.F. 1998. Production of amino acids by yogurt bacteria. *Biotechnol. Prog.* 14:963–965.

Bishop, J.R. and White, C.H. 1986. Assessment of dairy product quality and potential shelf life. *J. Food Prot.* 49:739–753.

BMI Business Monitor International. 2010. Saudi Arabia Agribusiness Report Q3.

Bodyfelt, F.W., Tobias, J., and Trout, G.M. 1988. *The Sensory Evaluation of Dairy Products.* New York: Van Nostrand Reinhold.

Boyaval, P. 1989. Lactic acid bacteria and metal ions. *Lait* 69(2):87–113.

Brody, A.L. 2008. Packaging milk and milk products. In: *Dairy Processing and Quality Control*, edited by R.C. Chandan, A. Kilara, and N.P. Shah, pp. 443–464. Ames, IA: Wiley-Blackwell.

Chammas, G.I. 2006. Caractérisation et utilisation de souches de bactéries lactiques thermophiles locales dans le développement de produits laitiers fermentés au Liban. Thesis, Paris: Institut National Agronomique Paris-Grignon.

Chammas, G.I., Saliba, R., and Béal, C. 2006a. Characterization of the fermented milk "laban" with sensory analysis and instrumental measurements. *J. Food Sci.* 71(2):S156–S162.

Chammas, G.I., Saliba, R., Corrieu, G., and Béal, C. 2006b. Characterization of lactic acid bacteria isolated from fermented milk "laban." *Int. J. Food Microbiol.* 110(1):52–61.

Civille, G.V. and Lyon, B.G. 1996. *Aroma and Flavor Lexicon for Sensory Evaluation: Terms, Definitions, References, and Examples.* West Conshohocken, PA: American Society for Testing and Materials.

Cooper, I. 2007. Plastics and chemical migration into food. In: *Chemical Migration and Food Contact Materials*, edited by K.A. Barnes, R. Sinclair, and D.H. Watson, pp. 228–250. Cambridge, England: Woodhead Publishing Ltd.

Corrieu, G., Spinnler, H.E., Jomier, Y., and Picque, D. 1988. *Automated System to Follow Up and Control the Acidification Activity of Lactic Acid Starters.* Patent n° FR 2629612. France: INRA.

Courtin, P., Monnet, V., and Rul, F. 2002. Cell-wall proteinases PrtS and PrtB have a different role in *Strepto-coccus thermophilus/Lactobacillus bulgaricus* mixed cultures in milk. *Microbiology* 148:3413–3421.

Crittenden, R.G., Martinez, N.R., and Playne, M.J. 2003. Synthesis and utilisation of folate by yoghurt starter cultures and probiotic bacteria. *Int. J. Food Microbiol.* 80:217–222.

Curia, A., Aguerrido, M., Langohr, K., and Hough, G. 2005. Survival analysis applied to sensory shelf life of yogurts—I: Argentine formulations. *J. Food Sci.* 70(7):S442–S445.

Cutter, C.N. 2002. Microbial control by packaging: A review. *Crit. Rev. Food Sci. Nutr.* 42(2):151–161.

Del Campo, R., Bravo, D., Cantón, R., Ruiz-Garbajosa, P., García-Albiach, R., Montesi-Libois, A., Yuste, F.-J., Abraira, V., and Baquero, F. 2005. Scarce evidence of yogurt lactic acid bacteria in human feces after daily yogurt consumption by healthy volunteers. *Appl. Environ. Microbiol.* 71(1):547–549.

Desmazeaud, M. 1990. Le lait milieu de culture. *Microbiol., Aliments, Nutr.* 8:313–325.

Duboc, P. and Mollet, B. 2001. Applications of exopolysaccharides in the dairy industry. *Int. Dairy J.* 11(9):759–768.

Elli, M., Callegari, M.L., Ferrari, S., Bessi, E., Cattivelli, D., Soldi, S., Morelli, L., Feuillerat, N.G., and Antoine, J.-M. 2006. Survival of yogurt bacteria in the human gut. *Appl. Environ. Microbiol.* 72(7):5113–5117.

Fabian, E., Majchrzak, D., Dieminger, B., Meyer, E., and Elmadfa, I. 2008. Influence of probiotic and con-ventional yoghurt on the status of vitamins B1, B2 and B6 in young healthy women. *Ann. Nutr. Metab.* 52:29–36.

Fajardo-Lira, C., Garcia-Garibay, M., Wacher-Rodarte, C., Farrés, A., and Marshall, V.M. 1997. Influence of water activity on the fermentation of yogurt made with extracellular polysaccharide-producing or non-producing starters. *Int. Dairy J.* 7(4):279–281.

Fernandez-Espla, M.D., Garault, P., Monnet, V., and Rul, F. 2000. *Streptococcus thermophilus* cell wall-anchored proteinase: Release, purification, and biochemical and genetic characterization. *Appl. Environ. Microbiol.* 66(11):4772–4778.

Fioramonti, J., Theodorou, V., and Bueno, L. 2003. Probiotics: What are they? What are their effects on gut physiology? *Best Pract. Res., Clin. Gastroenterol.* 17(5):711–724.

Gambelli, L., Manzi, P., Panfili, G., Vivanti, V., and Pizzoferrato, L. 1999. Constituents of nutritional relevance in fermented milk products commercialised in Italy. *Food Chem.* 66:353–358.

Gohil, V.S., Ahmed, M.A., Davies, R., and. Robinson, R.K. 1995. The incidence of *Listeria* in foods in the United Arab Emirates. *J. Food Prot.* 58:102–104.

Guarner, F. and Malagelada, J.R. 2003. Gut flora in the health and disease. *Lancet* 360:512–519.

Guizani, N., Kasapis, S., and Al-Ruzeiki, M. 2001. Microbial, chemical and rheological properties of laban (cultured milk). *Int. J. Food Sci. Technol.* 36(2):199–205.

Halpern, G.M. 1993. Benefits of yogurt. *Int. J. Immunother.* 9:65–68.

Hassan, A.N., Frank, J.F., Schmidt, K.A., and Shalabi, S.I. 1996. Textural properties of yogurt made with encapsulated nonropy lactic cultures. *J. Dairy Sci.* 79(12):2098–2103.

Hayes, M., Stanton, C., Fitzgerald, G., and Ross, R.P. 2007. Putting microbes to work: Dairy fermentation, cell factories and bioactive peptides. Part II: Bioactive peptides functions. *Biotechnol. J.* 2:435–449.

Holzapfel, W.H., Haberer, P., Geisen, R., Bjorkroth, J., and Schillinger, U. 2001. Taxonomy and important fea-tures of probiotic microorganisms in food and nutrition. *Am. J. Clin. Nutr.* 73(2):365S–373S.

Hough, G., Langohr, K., Gómez, G., and Curia, A. 2003. Survival analysis applied to sensory shelf life of foods. *J. Food Sci.* 68:359–362.

Hough, L., Puglieso, M.L., Sanchez, R., and Da Silva, O.M. 1999. Sensory and microbiological shelf-life of commercial ricotta cheese. *J. Dairy Sci.* 82:454–459.

Imhof, R. and Bosset, J.O. 1994. Relationships between microorganisms and formation of aroma compounds in fermented dairy products. *Z. Lebensm.-Unters. Forsch.* 198:267–276.

Jarvis, J.K. and Miller, G.D. 2002. Overcoming the barrier of lactose maldigestion to reduce health disparities. *J. Natl. Med. Assoc.* 94(2):55–66.

Kadamany, E.A., Khattar, M., Haddad, T., and Toufeili, I. 2003. Estimation of shelf-life of concentrated yogurt by monitoring selected microbiological and physicochemical changes during storage. *Lebensm.-Wiss. Technol.* 36:407–414.

Kasimoğlu, A. and Akgün, S. 2004. Survival of *Escherichia coli* O157:H7 in the processing and post-process-ing stages of acidophilus yogurt. *Int. J. Food Sci. Technol.* 39:563–568.

Kilcawley, K.N., Wilkinson, M.G., and Fox, P.F. 1998. Review: Enzyme modified cheese. *Int. Dairy J.* 8:1–10.

Kneifel, W., Jaros, D., and Erhard, F. 1993. Microflora and acidification properties of yogurt and yogurt-related products fermented with commercially available starter cultures. *Int. J. Food Microbiol.* 18:179–189.

Kneifel, W., Holub, S., and Wirthmann, M. 1989. Monitoring of B-complex vitamins in yogurt during fermentation. *J. Dairy Res.* 56:651–656.

Krusch, U., Neve, H., Luschei, B., and Teuber, M. 1987. Characterization of virulent bacteriophages of *Streptococcus salivarius* subsp. *thermophilus* by host specificity and electron microscopy. *Kiel. Milchwirtsch. Forschungsber.* 39(3):155–167.

Labuza, T.P. 1980. Influence of water activity on food product stability. *Food Technol.* 34(4):36–41.

Labuza, T.P. 2000. The search for shelf life. *Food Test. Anal.* 6:26–36.

Lacroix, C. and Lachance, O. 1988. Effect de l'Aw sur la survie de *Lactobacillus bulgaricus* et *Streptococcus thermophilus* et le développement d'acidité dans le yogourt conservé au froid. *Can. Inst. Food Sci. Technol. J.* 21(5):501–510.

Law, J. and Haandrikman, A. 1997. Review article: Proteolytic enzymes of lactic acid bacteria. *Int. Dairy J.* 7:1–11.

Lebanese Ministry of Agriculture. 2007. *State of Agriculture in Lebanon 2006–2007.* Lebanon: Lebanese Ministry of Agriculture.

Lemoine, R. 2002. La filière laitière libanaise: Un potentiel d'investissements dans l'élevage et l'industrie. *Rev. Lait. Fr.* 619:12–15.

Letort, C. and Juillard, V. 2001. Development of a minimal chemically-defined medium for the exponential growth of *Streptococcus thermophilus*. *J. Appl. Microbiol.* 91(6):1023–1029.

Lewis, M. and Dale, R.H. 2000. Chilled yogurt and other dairy desserts. In: *Shelf Life Evaluation of Foods*, edited by D. Man and A. Jones, pp. 89–109. Frederick, MD: Aspen Publishers.

Libnor. 1999. Fermented Milks, standard no. 33. Beirut, Lebanon: Lebanese Standards Institution, Ministry of Industry.

Libnor. 2007. Lebanese Standards of Dairy Products: Microbiological Limits, standard no. 510. Beirut, Lebanon: Lebanese Standards Institution, Ministry of Industry.

Loones, A. 1994. Laits fermentés par les bactéries lactiques. In: *Bactéries lactiques*, edited by H. de Roissart and F.M. Luquet, pp. 135–154. Uriage, France: Lorica.

Lourens-Hattingh, A. and Viljoen, B.C. 2001. Yogurt as probiotic carrier food. *Int. Dairy J.* 11(1–2):1–17.

Lucey, J.A. and Singh, H. 1998. Formation and physical properties of acid milk gels: A review. *Food Res. Int.* 30(7):529–542.

Marteau, P., Minekus, M., Havenaar, R., and Huis In't Veld, J.H.J. 1997. Survival of lactic acid bacteria in a dynamic model of the stomach and small intestine: Validation and the effects of bile. *J. Dairy Sci.* 80(6):1031–1037.

Martin, N., Skokanova, J., Latrille, E., Béal, C., and Corrieu, G. 1999. Influence of fermentation and storage conditions on the organoleptic properties of plain low fat stirred yoghurts. *J. Sens. Stud.* 14(2):139–160.

Massa, S., Altieri, V., and Quarante de Pace, R. 1977. Survival of *Escherichia coli* 0157:H7 in yoghurt during preparation and storage at 4°C. *Lett. Appl. Microbiol.* 24:347–350.

Mater, D.D., Bretigny, L., Firmesse, O., Flores, M.-J., Mogenet, A., Bresson, J.-L., and Corthier, G. 2005. *Streptococcus thermophilus* and *Lactobacillus delbrueckii* subsp. *bulgaricus* survive gastrointestinal transit of healthy volunteers consuming yoghurt. *FEMS Microbiol. Lett.* 250:185–187.

McBean, L.D. and Miller, G.D. 1998. Allaying fears and fallacies about lactose maldigestion. *J. Am. Diet. Assoc.* 98(6):671–676.

McKenna, A.B. 1987. Effects of homogenization pressure and stabilizer concentration on the physical stability of longlife laban. *N. Z. J. Dairy Sci. Technol.* 22(2):167–174.

Monnet, V., Atlan, D., Béal, C., Champomier-Vergès, M.C., Chapot-Chartier, M.P., Chouayekh, H., Cocaign-Bousquet, M., Deghorain, M., Gaudu, P., Gilbert, C., Guedon, E., Guillouard, I., Goffin, P., Guzzo, J., Hols, P., Juillard, V., Ladero, V., Lindley, N., Lortal, S., Loubiere, P., Maguin, E., Monnet, C., Rul, F., Tourdot-Maréchal, R., and Yvon, M. 2008. Métabolisme et ingénierie métabolique. In: *Bactéries lactiques: De la génétique aux ferments*, edited by G. Corrieu and F.M. Luquet, pp. 271–509. Paris, France: Tec&Doc Lavoisier.

Muir, D.D. 1996. The shelf life of dairy products: 2. Raw milk and fresh products. *J. Soc. Dairy Technol.* 49:44–48.

Murray, J.M., Delahunty, C.M., and Baxter, I.A. 2001. Descriptive sensory analysis: Past, present and future. *Food Res. Int.* 34:461–471.

Musaiger, A.O., Al-Saad, J.A., Al-Hooti, D.S., and Khunji, Z.A. 1998. Chemical composition of fermented dairy products consumed in Bahrain. *Food Chem.* 61(1–2):49–52.

Nasreddine, L., Hwalla, N., Sibai, A., Hamzé, M., and Parent-Massin, D. 2006. Food consumption patterns in an adult urban population in Beirut, Lebanon. *Public Health Nutr.* 9(2):194–203.

Neve, H. 1996. Bacteriophage. In: *Dairy Starter Cultures*, edited by T.M. Cogan and J.-P. Accolas, pp. 157–189. New York: VCH Publishers.

Nsabimana, C., Jiang, B., and Kossah, R. 2005. Manufacturing, properties and shelf life of labneh: A review. *Int. J. Dairy Technol.* 58(3):129–137.

Ott, A., Germond, J.E., and Chaintreau, A. 2000a. Origin of acetaldehyde during milk fermentation using C-13-labeled precursors. *J. Agric. Food Chem.* 48(5):1512–1517.

Ott, A., Hugi, A., Baumgartner, M., and Chaintreau, A. 2000b. Sensory investigation of yoghurt flavour perception: Mutual influence of volatiles and acidity. *J. Agric. Food Chem.* 48:441–450.

Ozer, B.H. and Robinson, R.K. 1999. The behaviour of starter cultures in concentrated yoghurt (Labneh) produced by different techniques. *Lebensm.-Wiss. Technol.* 32(7):391–395.

Parra, D.M. and Martínez, J.A. 2007. Amino acid uptake in lactose intolerance from a probiotic milk. *Br. J. Nutr.* 98(Suppl. 1):S101–S104.

Parra, D.M., Martinez de Morentin, B.E., Cobo, J.M., Lenoir-Wijnkoop, I., and Martinez, J.A. 2007. Acute calcium assimilation from fresh or pasteurized yogurt depending on the lactose digestibility status. *J. Am. Coll. Nutr.* 26(3):288–294.

Pedrosa, M.C., Golner, B.B., Goldin, B.R., Barakat, S., Dallal, G.E., and Russell, R.M. 1995. Survival of yogurt-containing organisms and *Lactobacillus gasseri* (ADH) and their effect on bacterial enzyme activity in the gastrointestinal tract of healthy and hypochlorhydric elderly subjects. *Am. J. Clin. Nutr.* 61:353–359.

Perez, P.F., De Antoni, G.L., and Anon, C. 1991. Formate production by *Streptococcus thermophilus* cultures. *J. Dairy Sci.* 74(9):2850–2854.

Pochart, P., Dewit, O. Desjeux, J.-F., and Bourlioux, P. 1989. Viable starter culture, b-galactosidase activity, and lactose in duodenum after yogurt ingestion in lactase-deficient humans. *Am. J. Clin. Nutr.* 49:828–831.

Poolman, B., Knol, J., Henderson, P.J., Liang, W.J., Le Blanc, G., Pourcher, T., and Mus-Veteau, I. 1996. Cation and sugar selectivity determinants in a novel family of transport proteins. *Mol. Microbiol.* 19:911–922.

Quiberoni, A., Guglielmotti, D., Binetti, A., and Reinheimer, J. 2004. Characterization of three *Lactobacillus delbrueckii* subsp. *bulgaricus* phages and the physicochemical analysis of phage adsorption. *J. Appl. Microbiol.* 96(2):340–351.

Rajagopal, S.N. and Sandine, W.E. 1990. Associative growth and proteolysis of *Streptococcus thermophilus* and *Lactobacillus bulgaricus* in skim milk. *J. Dairy Sci.* 73(4):894–899.

Rao, D.R., Reddy, A.V., Pulusani, S.R., and Cornwell, P.E. 1984. Biosynthesis and utilization of folic acid and vitamin B12 by lactic acid cultures in skim milk. *J. Dairy Sci.* 67:1169–1174.

Rault, A., Bouix, M., and Béal, C. 2009. Fermentation pH influences the physiological-state dynamics of *Lactobacillus bulgaricus* CFL1 pH-controlled culture. *Appl. Environ. Microbiol.* 75(13):4374–4381.

Reiter, B. 1978. Antimicrobial systems in milk. *J. Dairy Res.* 45(1):131–147.

Robinson, R.K. and Itsaranuwat, P. 2006. Properties of yoghurt and their appraisal. In: *Fermented Milks*, edited by A.Y. Tamime, pp. 76–94. Oxford, England: Blackwell Science Ltd.

Robinson, R.K., Lucey, J.A., and Tamime, A.Y. 2006. Manufacture of yoghurt. In: *Fermented Milks*, edited by A.Y. Tamime, pp. 53–75. Oxford, England: Blackwell Science Ltd.

Sadi, M.A. and Henderson, J.C. 2007. In search of greener pastures, Al-Marai and dairy food business in Saudi Arabia. *Br. Food J.* 109(8):637–647.

Saint-Eve, A., Lévy, C., Le Moigne, M., Ducruet, V., and Souchon, I. 2008. Quality changes in yogurt during storage in different packaging materials. *Food Chem.* 110:285–293.

Salvador, A., Fiszman, S.M., Curia, A., and Hough, G. 2005. Survival analysis applied to sensory shelf life of yogurts—II: Spanish formulations. *Sens. Nutr. Qual. Food Sci.* 70(7):S446–S449.

Samet-Bali, O., Bellila, A., Ayadi, M.A., Marzouk, B., and Attia, H. 2009. A comparison of the physicochemical, microbiological and aromatic composition of traditional and industrial Leben in Tunisia. *Int. J. Dairy Technol.* 63(1):98–104.

Sanders, M.E. 2003. Probiotics: Considerations for human health. *Nutr. Rev.* 61:91–99.

Savijoki, K., Ingmer, H., and Varmanen, P. 2006. Proteolytic systems of lactic acid bacteria. *Appl. Microbiol. Biotechnol.* 71(4):394–406.

Schieber, A. and Brückner, H. 2000. Characterization of oligo- and polypeptides isolated from yoghurt. *Eur. Food Res. Technol.* 210:310–313.

Serra, M., Trujillo, A.J., Guamis, B., and Ferragut, V. 2009. Proteolysis of yogurts made from ultra-high-pressure homogenized milk during cold storage. *J. Dairy Sci.* 92:71–78.

Shaker, R.R., Jumah, R.Y., and Abu-Jdayil, B. 2000. Rheological properties of plain yogurt during coagulation process: Impact of fat content and preheat treatment of milk. *J. Food Eng.* 44(3):175–180.

Shaker, R.R., Osaili, T.M., and Ayyash, M. 2008. Effect of thermophilic lactic acid bacteria on the fate of *Enterobacter zakazakii* during processing and storage of plain yogurt. *J. Food Saf.* 28:170–182.

Shihata, A. and Shah, N.P. 2000. Proteolytic profiles of yogurt and probiotic bacteria. *Int. Dairy J.* 10(5–6):401–408.

Sofu, A. and Ekinci, F.Y. 2007. Estimation of storage time of yogurt with artificial neural network modeling. *J. Dairy Sci.* 90(7):3118–3125.

Spreer, E. 1998. *Milk and Dairy Product Technology*. New York: Marcel Dekker Inc.

Stefanitsi, D. and Garel, J. 1997. A zinc dependent proteinase from the cell wall of *Lactobacillus delbrueckii* subsp. *bulgaricus. Lett. Appl. Microbiol.* 24:180–184.

Surono, I.S. and Hosono, A. 2003. Fermented milks: Types and standards of identity. In: *Encyclopedia of Dairy Sciences*, Vol. 2, edited by H. Roginski, J.W. Fuquay, and P.F. Fox, pp. 1018–1023. London, England: Academic Press.

Takano, T. and Yamamoto, N. 2003. Fermented milks: Health effects of fermented milks. In: *Encyclopedia of Dairy Sciences*, Vol. 2, edited by H. Roginski, J.W. Fuquay, and P.F. Fox, pp. 1063–1069. London, England: Academic Press.

Talwalkar, A., Miller, C.W., Kailasapathy, K., and Nguyen, M.H. 2004. Effect of materials and dissolved oxygen on the survival of probiotic bacteria in yoghurt. *Int. J. Food Sci. Technol.* 39:605–611.

Tamime, A.Y. and Robinson, R.K. 1999a. Traditional and recent developments in yoghurt production and related products. *Yoghurt: Science and Technology*, Chapter 5, pp. 306–388. Cambridge, England: Woodhead Publishing Limited and CRC Press.

Tamime, A.Y. and Robinson, R.K. 1999b. *Yoghurt: Science and Technology*. Cambridge, England: Woodhead Publishing Limited.

Tamime, A.Y. and Robinson, R.K. 2007a. Background to manufacturing practice. In: *Yoghurt: Science and Technology*, edited by A.Y. Tamime and R.K. Robinson, pp. 11–128. Boca Raton, FL: CRC Press.

Tamime, A.Y. and Robinson, R.K. 2007b. *Yoghurt: Science and Technology*. Boca Raton, FL: CRC Press.

Tamime, A.Y., Robinson, R.K., and Latrille, E. 2001. Yogurt and other fermented milks. In: *Mechanisation and Automation in Dairy Technology*, edited by A.Y. Tamime and B.A. Law, pp. 152–203. Sheffield, England: Sheffield Academic Press.

Trachoo, N. and Mistry, V.V. 1998. Application of ultrafiltered sweet buttermilk and buttermilk powder in the manufacture of nonfat and low fat yogurts. *J. Dairy Sci.* 81:133–138.

Tzvetkova, I., Dalgalarrondo, M., Danova, S., Iliev, I., Ivanova, I., Chobert, J.M., and Haertle, T. 2007. Hydrolysis of major dairy proteins by lactic acid bacteria from Bulgarian yogurts. *J. Food Biochem.* 31(5):680–702.

Uyeno, Y., Sekiguchi, Y., and Kamagata, Y. 2008. Impact of consumption of probiotic lactobacilli-containing yogurt on microbial composition in human feces. *Int. J. Food Microbiol.* 122:16–22.

Varela-Moreiras, G., Antoine, J.M., Ruiz-Roso, B., and Varela, G. 1992. Effects of yogurt and fermented-then-pasteurized milk on lactose absorption in an institutionalized elderly group. *J. Am. Coll. Nutr.* 11(2):168–171.

Vinderola, C.G., Mocchiutti, P., and Reinheimer, J.A. 2002. Interactions between lactic acid starter and probiotic bacteria used for fermented milk products. *J. Dairy Sci.* 85(4):721–729.

Wemekamp-Kamphuis, H.H., Karatzas, A.K., Wouters, J.A., and Abee, T. 2002. Enhanced levels of cold shock proteins in *Listeria monocytogenes* LO28 upon exposure to low temperature and high hydrostatic pressure. *Appl. Environ. Microbiol.* 68(2):456–463.

Zink, R., Elli, M., Reniero, R., and Morelli, L. 2000. *Growth Medium for Lactobacilli Containing Amino Acids, Nucleosides and Iron*. Patent n° US 6521443. USA: Nestec S.A.

12

Yogurt

Ramesh C. Chandan and K. R. Nauth

CONTENTS

12.1 Introduction to Fermented/Cultured Dairy Products and Yogurt

Fermented dairy foods have constituted a vital part of the human diet in many regions of the world, having been consumed ever since the domestication of animals (Chandan 2006). Archeological findings (Chandan 2004, 2006; Chandan and Shahani 1993, 1995; Hutkins 2006; Tamime and Robinson 2007) associated with the Sumerians and Babylonians of Mesopotamia, the Pharaohs of northeast Africa, and the Indo-Aryans of the Indian subcontinent provide convincing evidence for their use dating back to thousands of years. Furthermore, the ancient Indian Ayurvedic system of medicine cites *dadhi* (modern *dahi*) for its health-giving and disease-fighting properties (Aneja et al. 2002; Vedamuthu 2006). In general, cultured milk products resulted in the conservation of valuable nutrients and permitted their consumption over a period significantly longer than milk. Besides, the conversion of milk to fermented milk generated distinct consistency, smooth texture, and unmistakable flavor coupled with food safety, portability, and novelty for consumers. Modern research data have revealed that fermentation modifies certain milk constituents to enhance the nutritional status of the product. In addition, it is established that live and active cultures in significant numbers in cultured milk confer distinct heath benefits beyond conventional nutrition. Fermented milk products may be termed as "functional foods" that have health benefits beyond conventional nutrition. [For more information on functional foods, the reader is referred to the publications of Shah (2001), Chandan and Shah (2006), and Chandan (2007).]

12.1.1 Diversity

The variety of fermented milks in the world may be ascribed to various factors (Chandan 2002; Hirahara 2002; Ahmed and Wangsai 2007).

12.1.1.1 Milk of Various Species

The milk of various mammals exhibits significant differences in total solids, fat, mineral, and protein content. The viscosity and texture characteristics of yogurt are primarily related to its moisture content and protein level. Apart from quantitative levels, protein fractions and their ratios play a significant role in gel formation and strength. Milk proteins further consist of caseins and whey proteins that have distinct functional properties. In turn, caseins are comprise of α_{s1}-, β-, and κ-caseins. The ratio of casein fractions and the ratio of casein to whey protein differ widely in milk of various milch animals. Furthermore, the pretreatment of milk of different species prior to fermentation produces varying magnitudes of protein denaturation. These factors have a profound effect on the rheological characteristics of yogurt, leading to bodies and textures ranging from drinkable fluid to firm curd. The fermentation of the milk of buffalo, sheep, and yak produces a well-defined custard-like body and firm curd, whereas the milk of other animals tends to generate a soft-gel consistency.

Cow milk is used for the production of yogurt in a majority of countries around the world. In the Indian subcontinent, buffalo milk and blends of buffalo and cow milk are used widely for dahi making using mixed mesophilic cultures (Aneja et al. 2002). In certain countries, buffalo milk is the base for making yogurt using thermophilic cultures. Sheep, goat, or camel milk is the starting material of choice for fermented milk in several Middle Eastern countries.

12.1.1.2 Cultures for the Production of Fermented Milk

Various microorganisms characterize the diversity of fermented milk around the world. In general, lactic fermentation by bacteria transforms milk into the majority of products. Combinations of lactic starters and yeasts are used for some products, and in a few cases, lactic fermentation combined with molds make up the flora.

12.1.2 Forms of Fermented Milk

Yogurt may be mixed with water to make a refreshing beverage (e.g., liquid yogurt or smoothies). Salt, sugar, spices, or fruits may be added to enhance the taste. Spoonable yogurt has significant commercial importance all over the world. It comes in cups and tubes. To enhance its health appeal, the trend now is to deliver prebiotics as well as probiotic organisms through conventional yogurt. In many countries, probiotic yogurt made with defined cultures is available. These products claim scientifically documented health benefits.

Yogurt/buttermilk may be concentrated by removing whey by straining through cloth or by mechanical centrifugation to generate a cheese-like product. The concentrate may be mixed with herbs, fruits, sugar, or flavorings to yield shrikhand in India, Greek yogurt in North America and Europe, quarg/tvorog/topfen/taho/kwarg in central Europe, and fromage frais in France.

For the extension of shelf life, fermented milk and yogurt may be sun or spray dried to get a powder form. Leben zeer of Egypt and than/tan of Armenia are examples of concentrated yogurt without whey removal. In Lebanon, the concentrated yogurt is salted, compressed into balls, sun dried, and preserved in oil. Another way of preserving yogurt is by smoking and dipping in oil. Labneh anbaris and shanklish are partially dried yogurt products preserved in oil. Spices are added to shanklish, and the balls made therefrom are kept in oil. In Iran, Iraq, Lebanon, Syria, and Turkey, the concentrated yogurt is mixed with wheat products and sun dried to get kishk.

Yogurt may constitute a meal or may be consumed as an accompaniment to a meal. Yogurt is commonly used as a snack, drink, dessert, condiment, or spread. It may be used as an ingredient of cooked dishes.

In the past 30 years, the annual per capita yogurt consumption in the United States (Table 12.1) has grown from 1.13 kg in 1980 to 5.67 kg in 2010 (International Dairy Foods Association 2010).

With the introduction of new flavors, varieties, and packaging innovations, the yogurt market continues to grow at a rate of 3%–4% per year. The U.S. yogurt category is segmented into various parts (see Figure 12.1).

TABLE 12.1

Production and Consumption of Yogurt in the United States

Year	Total Production		Per Capita Consumption	
	(lb × 10⁶) or million pounds	(kg × 10⁶)	(lb)	(kg)
1980	570	258.5	2.5	1.13
1990	1055	478.5	4.2	1.90
1991	1109	503.0	4.4	1.99
1992	1154	523.4	4.5	2.04
1993	1286	583.3	4.9	2.22
1994	1392	631.4	5.3	2.40
1995	1646	746.6	6.2	2.81
1996	1588	720.3	5.9	2.68
1997	1574	713.9	5.8	2.63
1998	1639	743.4	5.9	2.68
1999	1717	778.8	6.2	2.81
2000	1837	833.2	6.5	2.94
2001	2003	908.5	7.0	3.18
2002	2311	1048.2	7.4	3.36
2003	2507	1137.1	8.2	3.72
2004	2707	1227.8	9.2	4.17
2005	3058	1387.1	10.3	4.67
2006	3301	1497.3	11.1	5.03
2007	3476	1576.7	11.5	5.21
2008	3570	1619.0	11.8	5.35
2009	3832	1738.2	12.5	5.67

Source: International Dairy Foods Association, *Dairy Facts*, Washington, DC, 2010.

FIGURE 12.1 Segmentation of yogurt market. (Adapted from Chandan, R. C., Editor, C. H. White, A. Kilara, and Y. H. Hui, Associate Editors, *Manufacturing Yogurt and Fermented Milks*, pp. 3–15, Blackwell Publishing, Ames, IA, 2006.)

Yogurt is available in full-fat, low-fat, and nonfat varieties. In addition, packaging formats [single cup of 4–6 oz in size or large-size (16–32 oz), multipack cups, tubes, and bottles] segment the yogurt market further (Pannell and Schoenfuss 2007).

Worldwide, yogurt is produced from the milk of cow, buffalo, goat, sheep, yak, and other mammals. In the industrial production of yogurt, cow milk is the predominant starting material. To get a custard-like consistency, cow milk is generally fortified with nonfat dry milk, milk protein concentrate, or condensed skimmed milk.

This brief chapter on yogurt is written to bring a clear understanding of this simple product that is now becoming a vehicle for introducing probiotics and prebiotics, believed to impart good health by maintaining a balanced and healthy gut. For more extensive discussion on yogurt technology and health benefits, see the books of Tamime and Robinson (2007) and Chandan et al. (2006).

12.2 Definition of Yogurt, Low-Fat Yogurt, and Nonfat Yogurt

12.2.1 Yogurt

Yogurt is defined by the U.S. Food and Drug Administration (FDA; CFR 2011) as the food produced by culturing standardized yogurt mix (defined below) with a characterizing bacterial culture that contains the lactic acid–producing bacteria *Lactobacillus delbrueckii* subsp. *bulgaricus* and *Streptococcus thermophilus*.

The dairy ingredients permitted in yogurt mix are cream, milk, partially skimmed milk, or skimmed milk, used alone or in combination. One or more of the other optional ingredients are also allowed to increase the nonfat solids of the food, provided that the ratio of protein to nonfat solids and the protein efficiency ratio of all the protein present in the mix will not be decreased as a result of adding such ingredients. These optional ingredients include concentrated skimmed milk, nonfat dry milk, buttermilk, whey, lactose, lactalbumins, lactoglobulins, or whey modified by the partial or complete removal of lactose or minerals to increase the non-fat solids of the food. Yogurt, before the addition of bulky flavors, contains not less than 3.25% milk fat and not less than 8.25% milk solids-not-fat (SNF) and has a titratable acidity of not less than 0.9%, expressed as lactic acid. The food may be homogenized and shall be pasteurized or ultrapasteurized prior to the addition of the bacterial culture. Flavoring ingredients may be added after pasteurization or ultrapasteurization. To extend the shelf life of the food, yogurt may be heat treated after culturing is completed to destroy viable microorganisms. However, the parenthetical phrase "heat treated after culturing" shall follow the name of the food (yogurt) on the label.

Optional ingredients are also defined in the regulations.

1. Vitamins A and D [If added, vitamin A shall be present in such quantity that each 946 mL (quart) of the food contains not less than 2000 International Units (IU) thereof, within limits of the current good manufacturing practice. If added, vitamin D shall be present in such quantity that each 946 mL (quart) of the food contains 400 IU thereof, within limits of the current good manufacturing practice.]
2. Nutritive carbohydrate sweeteners [sugar (sucrose), beet, or cane; invert sugar (in paste or syrup form); brown sugar; refiner's syrup; molasses (other than blackstrap); high-fructose corn syrup; fructose; fructose syrup; maltose; maltose syrup, dried maltose syrup; malt extract, dried malt extract; malt syrup, dried malt syrup; honey; and maple sugar, except for table syrup]
3. Flavoring ingredients
4. Color additives
5. Stabilizers

12.2.2 Low-Fat Yogurt

Low-fat yogurt has a similar description as given earlier for yogurt, except that the milk fat of the product before the addition of bulky flavors is not less than 0.5% and not more than 2%. The minimum SNF content is the same (8.25%) as in yogurt.

12.2.3 Nonfat Yogurt

Nonfat yogurt is the product as per the previous description of yogurt, except that the milk fat content before the addition of bulky flavors is less than 0.5%. The minimum SNF content is the same (8.25%) as in yogurt.

12.2.4 National Yogurt Association Criteria for Live and Active Culture Yogurt

According to the National Yogurt Association (1996), live and active culture yogurt is the food produced by culturing permitted dairy ingredients with a characterizing culture in accordance with the FDA standards of identity for yogurt. In addition to the use of *S. thermophilus* and *Lb. delbrueckii* subsp. *bulgaricus*, live and active culture yogurt may contain other safe and suitable food-grade bacterial cultures.

Heat treatment of live and active cultures in yogurt with the intent to kill the culture is not consistent with the maintenance of live and active cultures in the product. Producers of live and active culture yogurt should ensure proper practices of distribution, code dates, and product handling conducive to the maintenance and activity of the culture in the product.

Live and active culture yogurt must satisfy the following requirements. The product must be fermented with both *Lb. delbrueckii* subsp. *bulgaricus* and *S. thermophilus*. The cultures must be active at the end of the stated shelf life. Compliance with this requirement shall be determined by conducting an activity test on a representative sample of yogurt that has been stored at a temperature between 0°C and 7°C (32°F and 45°F) for refrigerated cup yogurt. The activity test is carried out by pasteurizing a 12% nonfat dry milk (NFDM) at 92°C (198°F) for 7 min, cooling to 43°C (110°F), adding 3% inoculum of the material under test, and fermenting at 43°C (110°F) for 4 h. The total organisms are to be enumerated in the test material both before and after fermentation by official standard methodology. The activity test is met if there is an increase of 1 log or more during fermentation. In the case of refrigerated cup yogurt, the total population of organisms in live and active culture yogurt must be at least 10^8 cfu/g at the time of manufacture. It is anticipated that, if proper distribution practices and handling instructions are followed, the total organisms in refrigerated cup live and active culture yogurt at the time of consumption will be at least 10^7 cfu/g.

12.2.5 Frozen Yogurt

Frozen yogurt resembles ice cream in its physical state. Both soft-serve and hard-frozen yogurt products are popular. These are available in nonfat and low-fat varieties. These products are not very acidic. The industry standards require a minimum titratable acidity of 0.30%, with a minimum contribution of 0.15% as a consequence of fermentation by yogurt bacteria (Chandan and Shahani 1993; Chandan et al. 2006). Technology for the production of frozen yogurt involves limited fermentation in a single mix and arresting further acid development by rapid cooling or a standardization of titratable acidity to a desirable level by blending plain yogurt with ice cream mix containing fruit/syrup base, stabilizers, and sugar and then freezing the mix in a conventional ice cream freezer. The mix is frozen at −6°C (21°F) and hardened at −40°C (−40°F). The finished product's pH level may vary between 5.5 and 6.0, depending on the consumer acceptance. Typical composition of nonfat hard-pack frozen yogurt may contain 0% fat, 13% milk solids-non-fat, 13% sucrose, 6% corn syrup solids 36 Dextrose Equivalent (DE), 2% maltodextrin 10 Dextrose Equivalent (DE), and 1.2% stabilizer (Chandan and Shahani 1993).

12.3 Yogurt Starter Organisms

For the manufacture of legal yogurt in the United States, *S. thermophilus* and *Lb. delbrueckii* subsp. *bulgaricus* must be added for the basic fermentation of yogurt mix. Additional, optional organisms, preferably of human intestinal origin, may be incorporated in the yogurt either through the starter culture or blended in after fermentation is complete. The optional organisms (Table 12.2) may be selected from a long list of candidates (Chandan et al. 2006; Rasic and Kurmann 1978; Tamime and Robinson 2007).

TABLE 12.2

Optional Cultures for Addition to Yogurt

Lactobacillus acidophilus
Lactobacillus rhamnosus
Lactobacillus reuteri
Lactobacillus casei
Lactobacillus gasseri
Lactobacillus johnsonii
Lactobacillus plantarum
Bifidobacterium bifidum
Bifidobacterium breve
Bifidobacterium adolescentis
Bifidobacterium infantis
Bifidobacterium lactis
Bifidobacterium longum
Enterococcus faecalis
Enterococcus faecium
Saccharomyces boulardi
Pediococcus acidilactici
Propionibacterium freudenreichii

Source: Chandan, R. C., Editor, C. H. White, A. Kilara, and Y. H. Hui, Associate Editors, *Manufacturing Yogurt and Fermented Milks*, pp. 3–15, Blackwell Publishing, Ames, IA, 2006.

12.3.1 *S. thermophilus*

This organism is a Gram-positive, catalase–negative, anaerobic cocci largely used in the manufacture of hard-cheese varieties, mozzarella, and yogurt. It does not grow at 10°C (50°F) but grows well at 45°C (113°F). Most strains can survive at 60°C (140°F) for 30 min (Rasic and Kurmann 1978). It is very sensitive to antibiotics. Penicillin (0.005 IU/mL) can interfere with milk acidification (Chandan and Shahani 1993). It grows well in milk and ferments lactose and sucrose. Sodium chloride (2%) may prevent the growth of many strains. These streptococci possess a weak proteolytic system. It is often combined with the more proteolytic lactobacilli in starter cultures. Most *S. thermophilus* cells grow more readily in milk than lactococci and produce acid faster. These streptococci strains possess β-galactosidase (β-gal) and utilize only the glucose moiety of lactose and leave galactose in the medium (Hutkins and Morris 1987).

The proteolytic activities of nine strains of *S. thermophilus* and nine strains of *Lb. delbrueckii* subsp. *bulgaricus* cultures incubated in pasteurized reconstituted nonfat dry milk at 42°C (108°F) as single and mixed cultures were studied (Rajagopal and Sandine 1990). Lactobacilli were highly proteolytic (61.0–144.6 µg of tyrosine/mL of milk), and *S. thermophilus* was less proteolytic (2.4–14.8 µg of tyrosine/mL of milk). Mixed cultures, with the exception of one combination, liberated more tyrosine (92.6–419.9 µg/mL) than the sum of the individual cultures. Mixed cultures also produced more acid (lower pH). Of the 81 combinations of *Lb. delbrueckii* subsp. *bulgaricus* and *S. thermophilus* cultures, only one combination was less proteolytic (92.6 µg of tyrosine/mL) than the corresponding *Lb. delbrueckii* subsp. *bulgaricus* strain in pure culture (125 µg of tyrosine/mL). *S. thermophilus* requires fewer amino acids than lactococci and lactobacilli. Only glutamine and glutamic acid, along with sulfur amino acids, are essential for all of the strains that have been tested (Neviani et al. 1995; Garault et al. 2000). It has been shown that *S. thermophilus* possesses branched-chain amino acid (leucine, isoleucine, and valine) biosynthesis pathway as an essential pathway for optimal growth in milk. This pathway is thought to play a role in maintaining the internal pH of the organism by converting acetolactate to amino acids (Garault et al. 2000). These organisms also have urease that produces ammonia from urea in milk to counteract the acid effects. Dairy lactococci do not have this pathway.

Factors influencing the lipase activity of *S. thermophilus* have been studied by DeMoraes and Chandan (1982). Of the 32 strains studied, all but two showed lipase activity. The optimum temperature for activity was 44°C (111°F), and the optimum pH was 9.0. The enzyme was partially inactivated by pasteurization treatment. The production of lipase activity was inhibited by the addition of butter oil (29%), milk (41%), and casein (12%) but was stimulated by soybean oil (39%), cream (27%), and corn oil (21%) to the medium.

12.3.2 *Lactobacillus delbrueckii* subsp. *bulgaricus*

These organisms are Gram-positive, catalase-negative, anaerobic/aerotolerant homofermentative and produce D(–) lactate (1.8%) and hydrogen peroxide (Nauth 1992). These cultures have β-galactosidase activity; only the glucose moiety of lactose is utilized, and galactose is released in the medium (Premi et al. 1972). *Lb. delbrueckii* subsp. *bulgaricus* has a high level of protease activity in milk that reaches its maximum during the log phase; *S. thermophilus* produces highly active peptidase instead of protease (Oberg and Broadbent 1993). In another study (Kawai et al. 1999), it was demonstrated that *Lb. delbrueckii* subsp. *bulgaricus* reached maximum protease activity between 4 and 8 h after incubation and then declined rapidly. When grown singly in pasteurized reconstituted NFDM at 42°C (108°F), *S. thermophilus* and *Lb. delbrueckii* subsp. *bulgaricus* were less proteolytic than the mixed culture growth (Rajagopal and Sandine 1990).

Production of cell-bound proteinase by *Lb. delbrueckii* subsp. *bulgaricus* NCDO 1489 was studied (Argyle et al. 1976a). The strain produced a single, cell-bound proteinase during its growth. The cell-bound enzyme could be liberated under conditions favoring autolysis or by treatment with lysozyme. The enzyme activity was optimal at 45°C–50°C (113°F–122°F). It was more active on dissolved caseinate than on native micellar casein (Chandan et al. 1982). β-Casein was a more susceptible fraction, and the activity against whey proteins was low.

Physical properties of the acid-generated coagulum of milk were attributed to the slow, irreversible aggregation of whey proteins (Argyle et al. 1976b). The aggregation was enhanced by casein and was not altered by the presence of milk fat or the pasteurization treatment of milk.

12.3.3 Associative Growth of *S. thermophilus* and *Lb. delbrueckii* subsp. *bulgaricus*

A symbiotic relationship exists between *S. thermophilus* and *Lb. delbrueckii* subsp. *bulgaricus* in mixed cultures (Radke-Mitchell and Sandine 1984); carbon dioxide, formate, peptides, and numerous amino acids liberated from casein are involved. The associative growth of rods and cocci results in greater acid production and flavor development than the single culture growth (Labropoulos et al. 1982; Moon and Reinbold 1974). It has been established that numerous amino acids liberated from casein by the proteases of *Lb. delbrueckii* subsp. *bulgaricus* stimulate the growth of *S. thermophilus* (Bautista et al. 1966; Pettie and Lolkema 1950). In turn, *S. thermophilus* produces CO_2 and formate, which stimulate *Lb. delbrueckii* subsp. *bulgaricus* (Driessen et al. 1982; Galesloot et al. 1968). During the early part of the incubation, *S. thermophilus* grows faster, removes excess oxygen, and produces the stimulants noted above. After the growth of *S. thermophilus* has slowed because of the increasing concentrations of lactic acid, the more acid-tolerant *Lb. delbrueckii* subsp. *bulgaricus* increases in number (Oberg and Broadbent 1993; Pulsani and Rao 1984). For a discreet ratio of *Lb. delbrueckii* subsp. *bulgaricus* and *S. thermophilus*, the inoculum level, time, and temperature of incubation must be controlled, and the bulk starter should be cooled promptly. However, contemporary commercial yogurt products generally contain *S. thermophilus* and *Lb. delbrueckii* subsp. *bulgaricus* in ratios of 3:1–9:1.

Yogurt and other cultured milk contain up to 50 ppm of benzoic acid attributed to the metabolic activity of the culture (Chandan et al. 1977).

12.3.4 Bifidobacteria

This Y-shape organism was isolated from infant stool in 1899 at the Pasteur Institute by Tissier (Sgorbati et al. 1995; Rasic and Kurmann 1978). There are 24 species in this group. Nine of these are of human origin, and the other 15 come from animals (Ballongue 1998). These Gram-positive organisms are strictly anaerobic. The degree of tolerance to oxygen depends on the species and the growth medium. It appears that the strains of *Bifidobacterium bifidus* are relatively aerotolerant. Optimum growth temperature for species of human origin is 36°C–38°C (97°F–100°F); there is no growth at 20°C (68°F). Bifidobacteria do not tolerate heat; *B. bifidus* is inactivated at 60°C (140°F). Two moles of glucose are fermented by the fructose-6-phosphate phosphoketolase pathway to 2 moles of L(+) lactate and 3 moles of acetate. Some formic acid and ethanol may also be produced. Bifidobacteria of human origin synthesize several vitamins, thiamine (B1), riboflavin (B2), pyridoxine (B6), folic acid (B9), cyanocobalamin (B12), and nicotinic acid (P; Ballongue 1998).

It has been observed that the number of bifidobacteria is reduced significantly in the stool of adults and the elderly. The proportions of various species of bifidobacteria vary with the age of humans. It appears that *B. infantis* and *B. breve* are favored in breast-fed infants and *B. adolescentis* predominates in bottle-fed infants and in adults (Ballongue 1998). Allergic infants were reported to be colonized mainly by *B. adolescentis* species with a lower mucous-binding capacity than bifidobacteria from healthy infants (Vaughan et al. 2002). Bifidobacteria population may be reduced after the administration of Western diet. All strains of bifidobacteria do not behave similarly. The Morinaga Milk Industry Co. Ltd. (Morinaga) claims that their *B. bifidum* (BB536) is hardier and superior to other strains.

12.3.5 Starter Culture Propagation

These days, the yogurt manufacturing plants are large and highly automated. Trouble-free functioning of these operations require predictable and dependable starter culture performance in the context of the type of products and their sensory attributes. The following are some helpful hints.

1. Milk should be of good microbiological quality, free of antibiotics and inhibitors of bacterial origin.

2. Fresh milk heat treated at 90°C–95°C (194°F–203°F) for 5 min or 85°C (185°F) for 30 min tend to give a balanced growth, with a 1:1 or 2:1 ratio for *S. thermophilus:Lb. delbrueckii* subsp. *bulgaricus*. When cultures are propagated in reconstituted NFDM, *S. thermophilus* tends to show abnormally large cells within a chain. More severe heat treatment, such as autoclaving of milk, is somewhat inhibitory to *S. thermophilus*, favors the growth of *Lb. delbrueckii* subsp. *bulgaricus*, and can cause culture imbalance in favor of *Lb. delbrueckii* subsp. *bulgaricus*. It should be noted that *S. thermophilus* first initiates growth, followed by *Lb. delbrueckii* subsp. *bulgaricus* (Rasic and Kurmann 1978). A low population of *S. thermophilus* will delay the completion of starter or yogurt fermentation. Heat treatment at 90°C (194°F) for a few minutes generally inactivates bacteriophages present in the milk.

3. Many yogurt producers prefer to use skimmed milk and condensed skimmed milk over nonfat dry milk to raise solids in the growth medium. Handling of nonfat dry milk is labor intensive and invariably results in the dusting of plant equipment and overhead fixtures. Under these conditions, keeping the plant clean is difficult and can result in higher incidence of yeast and mold. If nonfat dry milk is used, the area should be enclosed and isolated. The exhaust air should be filtered.

4. It is important to cool the starter when appropriate pH/% titratable acidity (TA) is attained. Higher acidity tends to reduce the *S. thermophilus* count (Rasic and Kurmann 1978).

5. At any given time, the availability of phage-unrelated cultures suitable for specific yogurt attributes is rather small. These cultures should be handled carefully to grant them long life in the plant.

6. The phage in the starter and in the environment on a regular basis should be monitored.

7. Many of the defined cultures may contain up to six strains of lactobacilli. The culturing conditions should be carefully verified and controlled for uniform culture activity. If the starter contains probiotic cultures or other adjuncts, their numbers should be verified in the starter and product.

Plants should work closely with culture suppliers for phage monitoring and culture performance.

12.4 Yogurt Manufacture

12.4.1 Bulk Starter Preparation

This is one of the key operations and should be attended to carefully by trained personnel dedicated to this duty. The starter tank valves, pipes, and hatch with gasket should be assembled and sterilized with live steam at low pressure (3–5 lb). Keep the bottom valve open for the condensate to drain. Continue to steam the tank for 30 min after the surface temperature in the tank has reached ~99°C (210°F). At this point, the steam is turned off, and the bottom valve is closed.

Skimmed milk with total solids raised to 10%–12% is either pumped cold and heated to 90°C (194°F) and held for 60 min, or the starter mix is heated to 90°C (194°F) in a plate heat exchanger and then held at that temperature in the tank for 60 min. Some plants prefer to use reconstituted nonfat dry milk at 10%–12% total solids (TS) for their bulk starter. In certain operations where a large amount of starter is used for yogurt inoculation, yogurt base is used for starter culture preparation. This brings ease of operation and eliminates yogurt composition variation. The mix is cooled to 43°C (109°F). The chilled water valve is closed early enough such that the temperature does not fall below 43°C (194°F). For 500 gal, thaw a can (350 mL) of frozen culture concentrate (10^{10} cfu/g) in 5 gal of tepid water containing 100 ppm of chlorine. Inoculate the tank and stir it for 5 min. Turn off the agitator and let it incubate quiescently for 6–8 h to reach 0.9% titratable acidity. Cool the starter to 5°C (41°F) using slow agitation. Then, turn off agitation. Turn on agitation for a few minutes before the starter is to be pumped. This rate of inoculation yields ~10^6 cfu/g of starter mix. For a healthy culture, it may take 8–10 h to reach 0.9% acidity. The starter has 2–5×10^8 cfu/mL. The time to reach 0.9% titratable acidity also depends on the strain composition of the frozen culture concentrate.

It is advisable to make a Gram strain of the fresh starter and run an activity test as described earlier using 1% inoculum. Such data are invaluable in tracking the performance of culture(s) and in preventing failed yogurt fermentation.

12.4.2 Yogurt Mix Preparation

Fermented milk products have delicate flavor and aroma and require milk of good microbiological quality. A variety of yogurt mix can be formulated and standardized from whole milk, partially skimmed milk, condensed skimmed milk, nonfat dry milk, whey compositions, and cream. These mixes should be formulated to comply with regulations and meet consumer expectations. Clarified, fat-adjusted milk at 50°C (122°F) should be blended with appropriate dry ingredients using a powder funnel. It should be allowed to circulate for a few minutes. The solids content of separated or whole milk can also be raised to 12% and 15%, respectively, by evaporation. The increased solids content prevents whey separation and improves the texture (Chandan and O'Rell 2006a,b; Lucey 2002).

12.4.3 Sweeteners

Sweeteners may be added to yogurt as part of the mix before fermentation and/or through fruit preserved with sweeteners. Sucrose (sugar) is widely used in yogurt production. It provides a clean sweet taste that has no other taste or odor. It complements flavors and contributes to desirable flavor blends. It can be used as a dry, granulated, free-flowing, and crystalline form or as liquid syrup (67% sucrose). Inclusion of more than 5% sucrose in a yogurt mix of 16%–20% total solids may cause culture inhibition and lack of flavor development (Tamime and Robinson 2007). Several corn syrup preparations and other sweeteners are also available. Nonnutritive sweeteners such as Aspartame are used in light products. These sweeteners have a lingering aftertaste, but it can be moderated by combining them (Chandan and O'Rell 2006a,b). The choice of sweetener(s) is determined by availability, cost, and legal status for use in yogurt.

12.4.4 Stabilizers

The set yogurt gel structure results from an acid–casein interaction where casein micelles at or near their isoelectric point flocculate and the colloidal calcium phosphate partially solubilize as acidity increases. During the fermentation of milk, the pH gradually declines to around 4.5, and destabilized micelles aggregate into a three-dimensional (3-D) network in which whey is entrapped (Chandan et al. 2006; Jaras et al. 2002). The appearance of whey on the surface of yogurt gel is due to syneresis, that is, the separation of serum from curd. In yogurt, this defect is called wheying-off (Chandan and O'Rell 2006c).

In stirred-style yogurt, the 3-D network is disturbed when fruits and flavors are mixed into the plain yogurt. The texture and physical properties of the yogurt depend on the fruit, stabilizer, and the rate of cooling (Lucey 2002; Chandan and O'Rell 2006c).

Stabilizers are added to prevent surface appearance of whey and to improve and maintain body, texture, viscosity, and mouthfeel. Yogurt with lower milk solids has a greater tendency to synerese. Numerous stabilizers are available on the market. Generally, a combination of several stabilizers is included in the formulation to avoid defects that may result from the use of a single stabilizer. A partial list of stabilizers used in yogurt include the following:

1. Gelatin is a protein of animal origin. It is derived from the hydrolysis of collagen. Only high-bloom gelatin should be used in yogurt making due to improved gelatin/casein interactions, its higher melting point, and its higher stabilizing ability (Chandan et al. 2006). The term *bloom* refers to the gel strength. It disperses in cold but requires heat for activation. It is used at 0.3%–0.5% level. The microstructure of yogurt made with 0.5% gelatin under scanning electron microscopy did not show gelatin, and the structure did not differ from that of a plain, unfortified yogurt. This yogurt was rated smooth in sensory evaluation (Modler et al. 1983).

2. Whey protein concentrates (WPCs) are used at 1%–2% level. In a study with skimmed milk yogurt fortified with dairy-based proteins, yogurt made with casein-based products was coarser than and inferior to those made with WPCs at 1%–1.5%. It was recommended that WPCs should be used along with other stabilizers (Modler et al. 1983; Chandan and O'Rell 2006b).

3. Gums are water-soluble or dispersible polysaccharides and their derivatives. In general, they thicken or turn into gel aqueous systems when used at low concentrations. Gums are used to stabilize emulsion and prevent wheying-off. Food gums are tasteless, odorless, colorless, and nontoxic (BeMiller 1991). All are essentially noncaloric and are classified as soluble fiber. These are used at 0.2%–1.5%, depending on the application (Chandan and O'Rell 2006a).

Locust bean gum is a seed gum. It has low cold water solubility. It is generally used where delayed viscosity development is needed. Dispersion of this gum when heated to about 185°F and then allowed to cool is high in viscosity. It works synergistically with carrageenan in some applications.

Guar gum is very similar to locust bean gum but is more soluble in cold water. It hydrates readily at pH 6–9. Its solubility is not affected by pH in the 4.8–5.0 range. It does not cross link well with carrageenan.

Carrageenan is derived from red seaweed. It is a mixture of various types—kappa, iota, and lambda. It may contain 60% of kappa form and 40% lambda. The kappa type forms a gel, whereas lambda does not. The polymer is stable at pH above 7.0, has a tendency to degrade slightly at pH 5–7, and degrades rapidly below pH 5.0. The potassium salt of this gum is the best gel former, but the gels are brittle and prone to syneresis. This defect is prevented by the addition of a small amount of locust bean gum. It interacts with casein in milk and promotes the stabilization of the yogurt gel (Chandan et al. 2006).

Xanthan is produced by microbial fermentation. It is readily soluble in cold and hot water. It is not affected by pH changes. A synergistic increase in viscosity results from the interaction of xanthan with κ-carrageenan and locust bean gum. These gels are prone to shear thinning. It also gives sheen to products, which may not be desirable in yogurt (Chandan et al. 2006).

Protein, starch, modified starch, and tapioca-based starches can be used without affecting the flavor of yogurt. The stabilizer used in yogurt is generally a blend of stabilizers incorporated at 0.5%–0.7% or less. The amount used also depends on milk solids level (Chandan and O'Rell 2006b).

12.4.5 Fruits and Fruit Flavorings for Yogurt

The growth and popularity of yogurt is largely due to fruits and sugar. For fruit-containing yogurt, the primary component of yogurt taste is the perceived degree of sweetness. This attribute of yogurt is believed to be responsible for its spectacular growth (O'Rell and Chandan 2006).

Many fruit flavors, single or blended, are popular and may vary with the season. Fruit preparations are added to 10%–20% in the final product. A fruit preserve consists of 55% sugar and 45% fruit (O'Rell and Chandan 2006). These are cooked until the final solids reach 65%–68%. The pH of these preparations is adjusted to 3.0–3.5 with citric acid or other food-grade acid. The processed fruit in most cases is filled aseptically in totes and shipped to yogurt plants. The transfer of fruit to yogurt should be done through sterile equipment to avoid yeast and mold contamination. The blending and the filler areas are very crucial to the microbiological quality of the yogurt with respect to yeast and mold. These areas should have air that has been filtered using high-efficiency particulate air filters to keep out airborne yeast, mold, and other contaminants. Cardboard boxes should not be brought in these areas. In addition, high-pressure water hoses should be avoided while fruit blending and packaging is ongoing.

12.4.6 Heat Treatment

The high-solids mix is given higher heat treatment than the conventional pasteurization (Chandan and O'Rell 2006a–c). Generally, milk is heat treated at 85°C–95°C (185°F–203°F) and then held for 10–40 min. At this temperature range, bacteriophages and vegetative bacterial cells are inactivated, and the growth of starter bacteria improves. Up to 60°C (140°F), there is no effect on whey protein. At

60°C–100°C (140°F–232°F), the whey proteins interact with each other and κ-casein. This interaction decreases the dissociation of $α_s$- and β-casein and increases κ-casein dissociation (Anema and Li 2000). When skimmed milk was pre–heat treated at 85°C (185°F) for 30 min and 90°C (194°F) for 2 min, whey protein denaturation was 76.5% and 55.0%, respectively. This heat treatment and the resulting interactions increase the water-binding capacity of the protein system.

12.4.7 Homogenization

Homogenization of mix is carried out in two stages: the first stage is at 10–20 MPa, and the second stage is at 3.5 MPa. It reduces the fat globule size to less than 3 μm, which gives a rich mouthfeel. This prevents creaming of the mix upon storage. Homogenization also improves gel strength upon fermentation due to greater protein–protein interaction.

12.4.8 Yogurt Mix Inoculation and Incubation

The inoculation rate may vary from 0.5% to 6%, depending on the type of yogurt and system setup. For yogurt fermented in a cup (set style), small surges of mix (20–100 gal) may be inoculated at 5%, packaged, and then incubated at 43°C–45°C (109°F–113°F). Yogurt may reach pH levels of 4.7–4.8 in less than 2 h and is then sent out of the hot room. If more time is available, the mix can be inoculated at 0.5%–1.5%, which may take 6–10 h to reach pH 4.4–4.5. The inoculation rate may vary from 0.5% to 6%, depending on the plant layout and equipment available.

12.5 Styles of Yogurt

Yogurt is marketed in various styles (Chandan and O'Rell 2006a–c). The main categories are given below.

1. *Plain yogurt.* This contains no sugar and is made by cup or vat incubation.
2. *Fruit-flavored yogurt.* This is popular in the world.
   – *Stirred- or Swiss-style yogurt.* In this type of product, fully fermented, plain, cooled yogurt at pH 4.3–4.4 is cooled to ~20°C (68°F), blended with the fruit preparation, and filled in cups. The palletized product is placed in the cooler. The texture and physical properties of the yogurt depend on the fruit, stabilizer, and rate of cooling.
   – *Set-style fruit-on-the-bottom.* In this kind of yogurt, cups receive fruit preparation (15–20%, by weight), followed by inoculated mix at ~44°C (111°F) in the filler room. The cups are placed in cases, and pallets are moved to the hot room maintained at ~48°C (118°F). The pH of the product reaches 4.7–4.85 in ~100 min, and then, it is moved to the cooling tunnel. At the end of the tunnel, the temperature of the yogurt is 16°C–18°C (61°F–64°F). It is then moved to the cooler.
   – *Light yogurt.* It is made without added sugar. High-intensity sweeteners are added either in the fruit preparation or directly to the yogurt base.
   – *Custard-style yogurt.* This fruit-flavored yogurt contains enough starch to create a custard-like consistency.
3. *Yogurt whips/mousse.* This fluffy and light-textured yogurt is produced as foam. It contains more sugar and stabilizers. Gelatin is an essential ingredient. A gas is injected to create foam.
4. *Greek-style/strained, concentrated yogurt.* This style is obtained by straining or centrifuging plain yogurt. It has a cream cheese–like smooth texture and contains much higher protein content.
5. *Frozen yogurt.* Typically, it contains 10%–15% plain yogurt blended with ice milk mix. It tastes more like ice milk and comes in soft- or hard-frozen varieties. It is either a low-fat or a nonfat product.

6. *Yogurt drinks/smoothies.* These are consumed as a drink or shake. The product is specially formulated to give it a beverage-like consistency.

12.6 Yogurt Fermentation

Yogurt fermentation is a homolactic fermentation. Glucose metabolism by *S. thermophilus, Lb. delbrueckii* subsp. *bulgaricus,* and *Lb. acidophilus* proceeds by the Embden–Meyerhof pathway. Lactose utilization in *S. thermophilus, Lb. delbrueckii* subsp. *bulgaricus,* and bifidobacteria involves lactose transport into the cells via cytoplasmic proteins (permeases). This translocation of lactose takes place without its chemical modification. This unphosphorylated lactose is hydrolyzed by β-galactosidase to glucose and galactose. Glucose is catabolized, and galactose is secreted from the cells (Vedamuthu 2006). The lactose permease is an active transport system, and the energy is provided in the form of a proton motive force developed by the expulsion of protons. The excreted amount of galactose is proportional to the amount of lactose taken up. The current model for lactose transport in these bacteria is that a single transmembrane antiport permease simultaneously translocates lactose molecules into the cytoplasm and galactose out of the cell. The energy generated through galactose efflux thus supports lactose uptake into these cells. The average lactose content of yogurt mix of 13% milk SNF was about 8.5% (Goodenough and Kleyn 1976a). During fermentation, it was reduced to about 5.75%. The initial galactose content of the mix was a trace but increased to 1.20% during fermentation. Only a trace amount of glucose was noted. In commercial yogurt, lactose ranged from 3.31% to 4.74%; galactose varied from 1.48% to 2.50%.

There are a number of inhibitors for the yogurt culture that can impede or slow down lactose fermentation. Some of these are listed here; for further details, refer to the work of Chandan and Shahani (1993), Monnet et al. (1995), and Nauth and Wagner (1973). These are heat-sensitive lactenins, lactoperoxidase/thiocyanate/hydrogen peroxide, agglutinins, mastitic milk, antibiotic residues, hydrogen peroxide, detergents and sanitizer residues, and bacteriophages. Many of the inhibitors mentioned here may be seasonal and sporadic or accidental. Bacteriophages, on the other hand, are pernicious and can be devastating if not managed properly.

12.6.1 Bacteriophages

Bacteriophages (phages) are viruses that can infect bacteria and destroy one or more components of the yogurt culture. Phages are differentiated into virulent (lytic) and temperate phages, which reflect different growth responses in the bacterial host. Phages that infect and lyse the host cell are called virulent phages, whereas phages that do not necessarily lyse their bacterial hosts but instead insert their genome into the host cell chromosome are called temperate phages. The propagation of virulent phages in the bacterial cell is called the lytic or vegetative cycle of phage multiplication and results in the release of new infectious phage progeny (Chandan and O'Rell 2006a,b). Bacteriophages multiply much faster than the bacteria. A bacteriophage with a burst size of 100 can destroy a culture within a couple of generations. This can cause huge economic losses and result in inferior yogurt. Due to the explosive growth of yogurt and mozzarella cheese production, a greater incidence of phage against *S. thermophilus* has been reported (Oberg and Broadbent 1993; Chandan and Shahani 1993, 1995). It is also noted that the phage for lactobacilli appears less frequently (Chandan and O'Rell 2006a–c). Research work has shown that *S. thermophilus* phages are closely related at both the genetic and morphological levels, making differentiation difficult. Electron microscopy studies revealed that both temperate and lytic phages were nearly identical, having small isometric heads and long noncontractile tails (Sturino and Klaenhammer 2002). Some lysogenic strains were autolytic at 45°C. New strategies to develop phage-resistant strains include antisense ribonucleic acid technology and origin-conferred phage-encoded resistance.

Both lytic and temperate bacteriophages have been found in *Lb. delbrueckii* subsp. *bulgaricus* and subsp. *lactis* (Cluzel et al. 1987). Strains of *Lb. acidophilus* isolated from dairy products harbored temperate phages and some produced bacteriocins. One induced phage lysed nine other dairy lactobacilli, including *Lb. delbrueckii* subsp. *bulgaricus* (Kilic et al. 1996). Some of the *Lb. delbrueckii* subsp. *bulgaricus* cells were also sensitive to the bacteriocin produced by *Lb. acidophilus.*

Primary yogurt cultures, *S. thermophilus* and *Lb. delbrueckii* subsp. *bulgaricus*, should be carefully chosen and evaluated with respect to phage and compatibility with other adjuncts. Because many of the organisms used in yogurt may harbor temperate phages, the propagated starter culture, yogurt, and environment should be monitored for phages. The use of aseptic techniques for propagation and production by properly trained personnel along with proper mix heat treatment sufficient to kill phages are essential to keep phages under control.

12.6.2 Flavor of Yogurt

The flavor of yogurt depends on the milk, its heat treatment, the starter strains used, the incubation temperature, and the balance of the organisms in the yogurt. Biochemically, flavor compounds of yogurt include but are not limited to lactic acid, acetic acid, formic acid, propionic acid, butyric acid, acetaldehyde, acetone, diacetyl, acetoin, and several other compounds. In milk, *S. thermophilus* produces formic, acetic, propionic, butyric, isovaleric, and caproic acid; diacetyl; acetone; and some acetaldehyde. Lactobacilli, on the other hand, produce large quantities of lactic acid, acetaldehyde, diacetyl, peptides, and amino acids. Many of the compounds are derived from lactose and some from other components in milk (Chandan and O'Rell 2006b). The flavor of yogurt can turn acidic and bitter during storage through shelf life. Protein degradation can continue during cold storage of yogurt, and some of the peptides released may be bitter (Robinson et al. 2002). The incubation temperature of yogurt below 30°C may also cause bitterness. It is again emphasized that the cultures used should be carefully selected to deliver quality attributes of yogurt through shelf life.

12.7 Nutritional and Health Aspects of Yogurt

The nutritional value of yogurt is derived from milk. The value of milk and its products was recognized long ago. Yogurt has higher nutrient density at 13%–18% milk solids than milk at 12.3%. Yogurt is a good source of calcium. One 8-oz serving will provide about 400 mg of calcium. In the literature, a number of healthful benefits of fermented milk have been assigned against various disease states (Shahani and Chandan 1979; Fuller 1989; Chandan et al. 2006). These disease states are colitis, constipation, various kinds of diarrhea, gastric acidity, gastroenteritis, indigestion, intoxication (bacterial toxins), diabetes, hypercholesterolemia, kidney and bladder disorders, lactose intolerance, liver and bile disorders, obesity, skin disorders, tuberculosis, vaginitis and urinary tract infections, cancer prevention, prevention and treatment of *Helicobacter pylori* gastritis, and irritable bowel syndrome.

In 1908, in explanation to his longevity-without-aging theory, Metchinkoff stated that lactic acid bacteria in sour milk displaced toxin-producing bacteria, thus promoting health. In 1899, Tisser isolated bifidobacteria from the stool of infants and recommended the administration of the same to infants suffering from diarrhea (Robinson et al. 2002). Since then, numerous studies on the use of lactic cultures in foods have continued. These studies have yielded promising results with regard to the health benefits of probiotics (Fernandes et al. 1992; Chandan 1999; Fonden et al. 2000; Brandt 2001; Ouwehand et al. 2002; Playne 2002; Salminen and Ouwehand 2003; Tamime 2005; U.S. Probiotics 2010; Chandan and Kilara 2008; Miller et al. 2007). The reader is referred to these references for more information on probiotics.

It is generally agreed that *Lb. delbrueckii* subsp. *bulgaricus* and *S. thermophilus*, the yogurt bacteria, do not adhere to the mucosal surfaces of the intestinal tract during their transit through the gut. Feeding trials with Gottingen minipigs appear to indicate that these yogurt organisms do survive the passage to the terminal ileum (Brandt 2001). The numbers detected (10^6–10^7 cfu/g of chyme) are regarded to be high enough to be considered for potential probiotic. It is believed that the gut is home to 400–500 species of organisms. Recent studies have indicated that the total microbiota of each adult individual had a unique pattern reflecting their differences in composition, which is partly dependent on the host genotype (Vaughan et al. 2002). What happens to the balance of these in different human beings with different dietary habits and age is not known. How the organisms are established in the gut in different segments is also not precisely known.

During the past few years, the use of probiotics in yogurt and fermented products has exploded. They are available in the form of supplements (powders, capsules, and tablets) and may be added to various foods. Still, yogurt is a preferred carrier of probiotics.

12.7.1 Definition of Probiotics

There are several ways of defining probiotics (Sanders 2007). A commonly used definition is given by the Food and Agricultural Organization of the United Nations (FAo) as: "Probiotics are live microorganisms that, when administered in adequate amounts, confer a health benefit on the host." Other definitions include the following:

1. Preparation of viable microorganisms that are consumed by humans or other animals, with the aim of inducing beneficial effects by qualitatively or quantitatively influencing their gut microflora and/or modifying their immune status (Fuller 1989, 2004)
2. Specific live or inactivated microbial cultures that have documented targets in reducing the risk of human disease in their nutritional management (Isolauri et al. 2002)
3. A live microbial culture of cultured dairy product that beneficially influences the health of the host (Salminen 1996)
4. A live microbial feed supplement that beneficially affects the host animal by improving its intestinal microbial balance (Fuller 2004)
5. Organisms and substances that have a beneficial effect on the host animal by contributing to its intestinal microbial balance (Parker 1974)
6. A preparation or a product containing viable, defined microorganisms in sufficient numbers that alter the microflora (by implantation or colonization) in a compartment of the host and therefore exert beneficial health benefits in this host (Screzenmeir and de Verse 2001)

12.7.1.1 Benefits of Probiotics

Health benefits imparted by probiotic bacteria are strain specific and not species or genus specific (Sanders 2007; Khurana and Kanawjia 2007; Maity and Misra 2009).

Table 12.3 lists various proprietary strains of probiotics that have been shown to confer specific health benefits.

Food formats for carrying probiotics are discussed (Sanders and Marco 2010). Probiotics stabilize gut mucosal barrier (Salminen et al. 1996). Yogurt and some probiotics alleviate symptoms of lactose intolerance (Goodenough and Kleyn 1976b; Kilara and Shahani 1976; Kolars et al. 1984; Garvie et al. 1984).

An expert meeting on the current level of consensus was held in London on 23 November 2009. The experts agreed that the survival of probiotic bacteria in the gut is not essential for efficacy. Furthermore, a combination of probiotic strains in a product does not necessarily increase the benefits of each strain. The panel agreed that the benefits of some probiotics include the following:

1. Prevention of diarrhea caused by antibiotics
2. Reduction of the risk of *Clostridium difficile* infection
3. Effective treatment of infectious diarrhea in children
4. Modulation of the immune system and regulation of intestinal immunity
5. Improvement of global symptoms of irritable bowel syndrome
6. Relief symptoms of functional gastrointestinal disorders (bloating and abdominal pain)
7. Reduction of the frequency and severity of attacks of necrotizing enterocolitis in premature infants

Promising areas include the regulation of systemic immunity, relief from constipation by enhancing transit time, prevention of pouchitis, prevention of vaginal tract infections, prevention of atopic dermatitis in infants and children, lowering the risk of colon cancer, and effective treatment of diarrhea in adults.

Areas currently unsupported are the prevention of human immunodeficiency virus infections, management of dyslipidemia, autism, obesity, and diabetes, prevention of the recurrence of Crohn's disease, and prevention of the infections of the urinary tract.

TABLE 12.3

Specific Strains of Probiotics and the Documented Benefits

Strain	Source	Commercial Product	Benefit Reported in Human Trials
Lactobacillus acidophilus NCFM	Danisco	Ingredient	Lactose digestion; reduces bacterial overgrowth in the small intestines
Lb. acidophilus LA5	Chr. Hansen	Ingredient	Lactose digestion
Lb. acidophilus LB	Lacteol Laboratory	Ingredient	Lactose digestion; reduces bacterial overgrowth in the small intestines
Lb. casei DN114-001	Danone	DanActive fermented milk	Enhances immune function
Lb. casei Shirota YIT9029	Yakult	Yakult	Enhances immune function; balances intestinal microbiota; combats the recurrence of superficial bladder cancer
Lb. fermentum VR1003	Probiomics	Ingredient	Reduces fecal toxin of *Clostridium difficile*
Lb. johnsonii La1/Lj1	Nestlé	LC1	Enhances immune function; eradicates *Helicobacter pylori* infection
Lb. paracasei CRL 431	Chr. Hansen	Ingredient	Improves digestion and immune function
Lb. paracasei F19	Medipharm	Ingredient	Improves digestion and immune function
Lb. plantarum 299V	Probi AB/ NextFoods	Ingredient/ GoodBelly juice	Relieves irritable bowel syndrome; postsurgical gut nutrition
Lb. plantarum OM	Bio-Energy Systems, Inc.	Ingredient	Relieves irritable bowel syndrome; postsurgical gut nutrition
Lb. reuteri RC14 + *Lb. rhamnosus* GR-1	Chr. Hansen, Urex Biotech, Jarrow Formulas	Fem-Dophilus	Controls children's diarrhea; eradicates *H. pylori* infection
Lb. rhamnosus GG	Valio, Dannon	Culturelle, Dannon Danimals	Prevents children's infectious diarrhea and atopic dermatitis; enhances immune function
Lb. rhamnosus 271	Probi AB	Ingredient	Prevents diarrhea; enhances immune function
Lb. rhamnosus HN001 (DR20)	Danisco	Ingredient	Controls diarrhea
Lb. rhamnosus PBO1 + *Lb. gasserri* EB01	Bifodan	Ecovag	Alleviates diarrhea and gastrointestinal distress
Lb. rhamnosus R0011 *Lb. acidophilus* R0052	Institut Rosell	Ingredient	Prevents diarrhea; enhances immune function
Lb. rhamnosus LB21 + *Lc. lactis* L1A	Essum AB	Ingredient	Prevents diarrhea; enhances immune function
Lb. salivarius UCC118	University College, Cork	Not available	Alleviates inflammatory bowel disease
Bifidobacterium animalis DN173-010	Danone	Activia yogurt	Normalizes/regulates intestine transit time
B. infantis 35624	Proctor & Gamble	Align	Alleviates irritable bowel syndrome
B. lactis BB-12	Chr. Hansen	Ingredient	Enhances immune function; alleviates diarrhea in children
B. lactis HN019 (DR10)	Danisco	Ingredient	Enhances immune function in the elderly
B. longum BB536	Morinaga	Ingredient	Alleviates allergy symptoms; balances microbial ecology
Saccharomyces cerevisiae (*boulardii*)	Biocodex	Florastor	Alleviates antibiotic-induced diarrhea and *C. difficile* infections
Bacillus coagulans BC30	Ganeden Biotech, Inc.	Sustenex, Digestive Advantage, Ingredient	Improves digestion

Source: Chandan, R. C., Editor, C. H. White, A. Kilara, and Y. H. Hui, Associate Editors, *Manufacturing Yogurt and Fermented Milks*, pp. 3–15, Blackwell Publishing, Ames, IA, 2006; Sanders, M. E., 2007. *Functional Foods & Nutraceuticals*, June 2007 (www.ffnmag.com); Maity, T. K. and Misra, A. K. 2009. *Afr. J. Food, Agric., Nutr. Dev.* 9(8), 1778–1796, online, November 2009; U.S. Probiotics.org, available at: http://www.usprobiotics.org/basics.asp, 2010.

12.7.1.2 Survival of Probiotics in Yogurt

It is known that bifidobacteria and lactobacilli are members of the human gut flora. In addition, a large number of yogurt brands claim the addition of probiotics. For the probiotics to be effective, the organisms have to be alive (Shah and Lankaputhra 1997; Chandan et al. 2006). In market-bought samples, the fresh product had 10^6–10^7 cfu/g. Their number of probiotics declined to $\leq 10^3$ cfu/g in some products. Some strategies to improve the number of probiotics in yogurt include the reduction of regular *S. thermophilus* and *Lb. delbrueckii* subsp. *bulgaricus* in the product and the addition of cultured probiotics to yogurt.

The safety aspects of probiotics have been addressed (Sanders et al. 2010). They are generally considered safe for consumption as food.

12.7.2 Prebiotics

Prebiotics are defined as a nondigestible food ingredient that beneficially affects the host by selectively stimulating the growth and/or activity of one or a limited number of bacteria in the colon that can improve the host's health (Holzapfel and Schillinger 2002). The beneficial effects of the presence of bifidobacteria in the gut are dependent on their viability and metabolic activity. Their growth is stimulated by the presence of complex carbohydrates known as oligosaccharides. Some of these are considered prebiotics (Gopal et al. 2001; Rastall 2010). Fructo-oligosaccharides (FOS) are well-known prebiotics that are found in 36,000 plants. FOS may contain two to eight units in a chain. Inulin, a type of FOS extracted from chicory root, has a degree of polymerization up to 60. FOS and inulin occur naturally in a variety of fruits, vegetables, and grains, especially chicory, Jerusalem artichoke, bananas, mangoes, onion, garlic, asparagus, barley, wheat, and tomatoes.

12.7.2.1 Benefits of Prebiotics

Prebiotic fermentation leads to health benefits such as increased fecal biomass and increased stool weight and/or frequency. Prebiotics are fermented by bifidobacteria with the production of short chain fatty acids (SCFAs), mainly acetate, propionate, and butyrate, hydrogen, and carbon dioxide. Production of SCFAs leads to lower pH in the colon, which facilitates the absorption of calcium, magnesium, and zinc (Rastall 2010). Lower pH also restricts pathogenic and other harmful bacteria, thus reducing or eliminating precarcinogenic activity.

Several materials are considered prebiotics. These are oligosaccharide (FOS Rafitilose P95), inulin, pyrodextrine, galacto-oligosaccharides, soy-oligosaccharides, xylo-oligosaccharides, isomalto-oligosaccharides, lactulose, and transoligosaccharides.

12.7.3 Synbiotics

Synbiotics refer to a product in which probiotics and prebiotics are combined (Bielecka et al. 2002). The synbiotic effect may be directed toward two different regions, both large and small intestines. The combination of prebiotics and probiotics in one product has been shown to confer benefits beyond those of either on their own.

We have come a long way in the development of yogurt into a respectable dairy product with noted health benefits. Producers and marketers of this product are making every effort to keep the yogurt category growing through product development and packaging innovations while delivering a good-for-you flavorful product suited for all occasions of gastronomic indulgence.

REFERENCES

Ahmed, S. I. and Wangsai, J. 2007. Fermented milk in Asia—Chapter 68. In: *Handbook of Food Products Manufacturing*, Vol. 2, edited by Y. H. Hui (Editor), R. C. Chandan, S. Clak, N. Cross, J. Dobbs, W. J. Hurst, L. M. l. Nollet, E. Shimoni, N. Sinha, E. B. Smith, S. Suripat, A. Tilchenal, and F. Toldra (Associate Editors), pp. 431–448. New York: John Wiley-Interscience Publishers.

Aneja, R. P., Mathur, B. N., Chandan, R. C., and Banerjee, A. K. 2002. *Technology of Indian Milk Products*, pp. 183–196. New Delhi, India: Dairy India Yearbook.

Anema, S. G. and Li, Y. 2000. Further studies on the heat-induced, pH-dependent dissociation of casein from the micelles in reconstituted skim milk. *Lebensm.-Wiss. Technol.* 30:335–343.

Argyle, P. J., Mathison, G. E., and Chandan, R. C. 1976a. Production of cell-bound proteinase by *Lactobacillus bulgaricus* and its location in the bacterial cell. *J. Appl. Bacteriol.* 41:175–184.

Argyle, P. J., Jones, N., Chandan, R. C., and Gordon, J. F. 1976b. Aggregation of whey proteins during storage of acidified milk. *J. Dairy Res.* 43:45–51.

Ballongue, J. 1998. Bifidobacteria and probiotic action. In: *Lactic Acid Bacteria Microbiology and Functional Aspects*, edited by S. Salminen and A. V. Wright, pp. 519–587. New York: Marcel Dekker.

Bautista, E. S., Dahiya, R. S., and Speck, M. L. 1966. Identification of compounds causing symbiotic growth of *Streptococcus thermophilus* and *Lactobaccillus bulgaricus* in milk. *J. Dairy Res.* 33:209–307.

BeMiller, J. N. 1991. Gums. In: *Encyclopedia of Food Science and Technology*, Vol. 2, edited by Y. H. Hui, pp. 1338–1344. New York: John Wiley & Sons.

Bielecka, M., Biedrzycka, E., and Majkowska, A. 2002. Selection of probiotics and prebiotics for synbiotics and confirmation of their *in vivo* effectiveness. *Food Res. Int.* 35:125–131.

Brandt, L. A. 2001. Prebiotics enhance gut health. *Prepared Foods* 170:NS7–NS10.

Chandan, R. C. 1999. Enhancing market value of milk by adding cultures. *J. Dairy Sci.* 82:2245–2256.

Chandan, R. C. 2002. Benefits of live fermented milks: Present diversity of products. *Proc. Int. Dairy Congr.* (in CD-ROM), Paris, France.

Chandan, R. C. 2004. Dairy—Chapter 16: Yogurt. In: *Food Processing—Principles and Applications*, edited by J. Scott Smith and Y. H. Hui, pp. 297–318. Ames, IA: Blackwell Publishing.

Chandan, R. C. 2006. Chapter 1: History and consumption trends. In: *Manufacturing Yogurt and Fermented Milks*, edited by R. C. Chandan (Editor), C. H. White, A. Kilara, and Y. H. Hui (Associate Editors), pp. 3–15. Ames, IA: Blackwell Publishing.

Chandan, R. C. 2007. Chapter 43: Functional properties of milk constituents. In: *Handbook of Food Products Manufacturing*, Vol. 2, edited by Y. H. Hui (Editor), R. C. Chandan, S. Clak, N. Cross, J. Dobbs, W. J. Hurst, L. M. l. Nollet, E. Shimoni, N. Sinha, E. B. Smith, S. Suripat, A. Tilchenal, and F. Toldra (Associate Editors), pp. 971–987. New York: Wiley-Interscience Publishers.

Chandan, R. C., Argyle, P. J., and Mathison, G. E. 1982. Action of *Lactobacillus bulgaricus* protease preparations on milk proteins. *J. Dairy Sci.* 65:1408–1413.

Chandan, R. C., Gordon, J. F., and Morrison, A. 1977. Natural benzoate content of dairy products. *Milchwissenschaft* 32(9):534–537.

Chandan, R. C. and Kilara, A. 2008. Chapter 18: Role of milk and dairy foods in nutrition and health. In: *Dairy Processing and Quality Assurance*, edited by R. C. Chandan, A. Kilara, and N. P. Shah, pp. 411–428. Ames, IA: Wiley-Blackwell Publishers.

Chandan, R. C. and O'Rell, K. R. 2006a. Chapter 11: Ingredients for yogurt manufacture. In: *Manufacturing Yogurt and Fermented Milks*, edited by R. C. Chandan (Editor), C. H. White, A. Kilara, and Y. H. Hui (Associate Editors), pp. 179–193. Ames, IA: Blackwell Publishing.

Chandan, R. C. and O'Rell, K. R. 2006b. Chapter 12: Principles of yogurt processing. In: *Manufacturing Yogurt and Fermented Milks*, edited by R. C. Chandan (Editor), C. H. White, A. Kilara, and Y. H. Hui (Associate Editors), pp. 195–209. Ames, IA: Blackwell Publishing.

Chandan, R. C. and O'Rell, K. R. 2006c. Chapter 15: Yogurt plant—Quality assurance. In: *Manufacturing Yogurt and Fermented Milks*, edited by R. C. Chandan (Editor), C. H. White, A. Kilara, and Y. H. Hui (Associate Editors), pp. 247–264. Ames, IA: Blackwell Publishing.

Chandan, R. C. and O'Rell, K. R. 2006d. Chapter 13: Manufacture of various types of yogurt. In: *Manufacturing Yogurt and Fermented Milks*, edited by R. C. Chandan (Editor), C. H. White, A. Kilara, and Y. H. Hui (Associate Editors), pp. 211–236. Ames, IA: Blackwell Publishing.

Chandan, R. C. and Shah, N. P. 2006. Chapter 20: Functional foods and disease prevention. In: *Manufacturing Yogurt and Fermented Milks*, edited by R. C. Chandan (Editor), C. H. White, A. Kilara, and Y. H. Hui (Associate Editors), pp. 311–325. Ames, IA: Blackwell Publishing.

Chandan, R. C. and Shahani, K. M. 1993. Yogurt. In: *Dairy Science and Technology Handbook*, Vol. 2, pp. 1–56. New York: VCH Publications.

Chandan, R. C. and Shahani, K. M. 1995. Other fermented dairy products. In: *Biotechnology*, 2nd edition, Vol. 9, edited by G. Reed and T. W. Nagodawithana, pp. 386–418. Weinheim, Germany: VCH Publications.

Chandan, R. C. (Editor), White, C. H., Kilara, A., and Hui, Y. H. (Associate Editors). 2006. *Manufacturing Yogurt and Fermented Milks*. Ames, IA: Blackwell Publishing.

Cluzel, P. J., Serio, J., and Accolas, J. P. 1987. Interaction of *Lactobacillus bulgaricus* temperate bacteriophage 0448 with host strain. *Appl. Environ. Microbiol.* 53:1850–1854.

DeMoraes, J. and Chandan, R. C. 1982. Factors influencing the production and activity of *Streptococcus thermophilus* lipase. *J. Food Sci.* 47:1579–1583.

Driessen, R. M., Ubbels, J., and Stadhouders, J. 1982. Evidence that *Lactobacillus bulgaricus* is stimulated by carbon dioxide produced by *Streptococcus thermophilus*. *Neth. Milk Dairy J.* 36:135–144.

Fernandes, C. F., Chandan, R. C., and Shahani, K. M. 1992. Fermented dairy products and health. In: *The Lactic Acid Bacteria*, Vol. 1, edited by B. J. B. Wood, pp. 279–339. New York: Elsevier Applied Science.

Fonden, R., Mogensen, G., Tanaka, R., and Salminen, S. 2000. Effect of culture-containing dairy products on intestinal microflora, human nutrition and health—Current knowledge and future perspectives. IDF Bull. No. 352/2000, Brussels, Belgium.

Food and Agriculture Organization of the United Nations. Health and nutritional properties of probiotics in food including powder milk with live lactic acid bacteria. Available at: http://www.who.int/foodsafety/publications/fs_management/en/probiotics.pdf.

Food and Drug Administration, Code of Federal Regulations (CFR), 21 CFR, Parts 131, p. 31–39. GMP Publications, Washington, D.C. Revised as of April 1, 2011.

Fuller, R. 1989. Probiotics in man and animals. *J. Appl. Bacteriol.* 66:365–378.

Fuller, R. 2004. What is a probiotic? *Biologist* 51:232.

Galesloot, T. E., Hassing, F., and Veringa, H. A. 1968. Symbiosis in yogurt.—Part 1: Stimulation of *Lactobacillus bulgaricus* by a factor produced by *Streptococcus thermophilus*. *Neth. Milk Dairy J.* 22:50–63.

Garault, P., Letort, C., Juillard, V., and Monnet, V. 2000. Branched-chain amino acid biosynthesis is essential for optimal growth of *Streptococcus thermophilus* in milk. *Appl. Environ. Microbiol.* 66:5128–5133.

Garvie, E. I., Cole, C., Fuller, B. R., and Hewitt, D. 1984. The effect of yogurt on some components of gut microflora and on the metabolism of lactose in the rat. *J. Appl. Bacteriol.* 56:237–245.

Goodenough, E. R. and Kleyn, D. H. 1976a. Qualitative and quantitative changes in carbohydrates during the manufacture of yogurt. *J. Dairy Sci.* 59:45–57.

Goodenough, E. R. and Kleyn, D. H. 1976b. Influence of viable yogurt microflora on digestion of lactose by the rat. *J. Dairy Sci.* 59:601–606.

Gopal, P. K., Sullivan, P. A., and Smart, J. B. 2001. Utilization of galacto-oligosaccharides as selective substrates for growth by lactic acid bacteria including *Bifidobacterium lactis* DR10 and *Lactobacillus rhamnosus* DR20. *Int. Dairy J.* 11:19–25.

Hirahara, T. 2002. Trend and evolution of fermented milk. *Proc. Int. Dairy Congr.* CD-ROM, Paris, France.

Holzapfel, W. H. and Schillinger, U. 2002. Introduction to pre- and pro-biotics. *Food Res. Int.* 35:109–116.

Hutkins, R. W. 2006. *Microbiology and Technology of Fermented Foods*, pp. 1–14. Ames, IA: Blackwell Publishing.

Hutkins, R. W. and Morris H. A. 1987. Carbohydrate metabolism by *Streptococcus thermophilus*. A review. *J. Food Prot.* 50:876–884.

Isolauri, E., Rautava, S., and Kalliomaki, M. 2002. Role of probiotics in food hypersensitivity. *Curr. Opin. Allergy Clin. Immunol.* 2:263–271.

International Dairy Foods Association. 2010 Edition. *Dairy Facts*. Washington, DC: International Dairy Foods Association.

Jaras, D., Rohm, H., Hague, A., Bonaparte, C., and Kneifel, W. 2002. Influence of the starter culture on the relationship between dry matter content and physical properties of set-style yogurt. *Milchwissenschaft* 57:325–328.

Kawai, Y., Tadokoro, K., Konomi, R., Itoh, K., Saito, T., Kitazawa, H., and Itoh, T. 1999. A novel method for the detection of protease and the development of extracellular protease in early growth stages of *Lactobacillus delbrueckii* ssp. *bulgaricus*. *J. Dairy Sci.* 82:481–485.

Khurana, H. K. and Kanawjia, S. K. 2007. Recent trends in development of fermented milks. *Curr. Nutr. Food Sci.* 3:91–108.

Kilara, A. and Shahani K. M. 1976. Lactase activity of cultured and acidified dairy products. *J. Dairy Sci.* 59:2031–2035.

Kilic, A. O., Pavlova, S. I., Ma, W. G., and Tao, L. 1996. Analysis of *Lactobacillus* phages and bacteriocins in American dairy products and characterization of a phage isolated from yogurt. *Appl. Environ. Microbiol.* 62:2111–2116.

Kolars, J. C., Levitt, M. D., Aouji, M., and Savaiano, D. A. 1984. Yogurt—An autodigesting source of lactose. *N. Engl. J. Med.* 310:1–3.

Labropoulos, A. E., Collins, W., and Stone, W. K. 1982. Starter culture effects on yogurt fermentation. *Cult. Dairy Prod. J.* 17:15–17.

Lucey, J. A. 2002. Formation and physical properties of milk protein gels. *J. Dairy Sci.* 85:281–294.

Maity, T. K. and Misra, A. K. 2009. Probiotics and human health: Synoptic review. *Afr. J. Food, Agric., Nutr. Dev.* 9(8):1778–1796, online, November.

Miller, G. D., Jarvis, J. K., and McBean, L. D. 2007. *Handbook of Dairy Foods and Nutrition*, 3rd edition. Boca Raton, FL: CRC Press.

Modler, H. W., Larmond, M. E., Lin, C. S., Froehlic, D., and Emmons, D. B. 1983. Physical and sensory properties of yogurt stabilized with milk proteins. *J. Dairy Sci.* 66:422–429.

Monnet, V., Condon, S., Cogan, T. M., and Gripon, J. C. 1995. Metabolism of starter cultures. In: *Dairy Starter Cultures*, edited by T.M. Cogan and J.P. Accolas, pp. 53–55. New York: VCH Publishers Inc.

Moon, N. J. and Reinbold, G. W. 1974. Selection of active and compatible starters for yogurts. *Cult. Dairy Prod. J.* 9:10–12.

Morinaga Milk Industry Co. Ltd., 33-1 Shiba 5-chome, Minato-Ku, Tokyo 108-8384, Japan.

National Yogurt Association. 1996. Live and Active Culture Program Procedure. January 1996, McLean, Virginia.

Nauth, K. R. 1992. Cheese. In: *Dairy Science and Technology Handbook*, Vol. 2, edited by Y. H. Hui, pp. 174–179. New York: VCH Publishers Inc.

Nauth, K. R. and Wagner, B. J. 1973. Stimulation of lactic acid bacteria by a *Micrococcus* isolate: Evidence for multiple effects. *Appl. Microbiol.* 26:49–55.

Neviani, E., Giraffa, G., Brizzi, A., and Carminati, D. 1995. Amino acid requirements and peptidase activities of *Streptococcus salivarious* subsp. *thermophilus. J. Appl. Bacteriol.* 79:302–307.

Oberg, C. J. and Broadbent, J. R. 1993. Thermophillic starter culture: Another set of problems. *J. Dairy Sci.* 76:2392–2406.

O'Rell, K. R. and Chandan, R. C. 2006. Yogurt—Chapter 9: Fruit preparations and flavoring materials. In: *Manufacturing Yogurt and Fermented Milks*, edited by R. C. Chandan (Editor), C. H. White, A. Kilara, and Y. H. Hui (Associate Editors), pp. 151–166. Ames, IA: Blackwell Publishing.

Ouwehand, A. C., Salminen, S., and Isolauri, E. 2002. Probiotics: An overview of beneficial effects. *Antonie van Leeuwenhoek* 82:279–289.

Pannell, L. and Schoenfuss, T. C. 2007. Chapter 76: Yogurt. In: *Handbook of Food Products Manufacturing*, Vol. 2, edited by Y. H. Hui (Editor), R. C. Chandan, S. Clak, N. Cross, J. Dobbs, W. J. Hurst, L. M. l. Nollet, E. Shimoni, N. Sinha, E. B. Smith, S. Suripat, A.Tilchenal, and F. Toldra (Associate Editors), pp. 647–676. New York: Wiley-Interscience Publishers.

Parker, R. B. 1974. Probiotics—The other half of the antibiotic story. *Anim. Nutr. Health* 29:4–8.

Pettie, J. W. and Lolkema, H. 1950. Yogurt—Part II: Growth stimulating factors for *S. thermophilus. Neth. Milk Dairy J.* 4:209–224.

Playne, M.J. 2002. The health benefits of probiotics. *Food Aust.* 54:71–74.

Premi, L., Sandine, W. E., and Elliker, P. R. 1972. Lactose hydrolyzing enzymes of *Lactobacillus* species. *Appl. Microbiol.* 24:51–57.

Pulsani, S. R. and Rao, D. R. 1984. Stimulation by formate of antimicrobial activity of *Lactobacillus bulgaricus* in milk. *J. Food Sci.* 49:652–653.

Radke-Mitchell, L. and Sandine, W. E. 1984. Associative growth and differential enumeration of *Streptococcus thermophilus* and *Lactobacillus bulgaricus*: A review. *J. Food Prot.* 47:245–248.

Rajagopal, S. N. and Sandine, W. E. 1990. Associative growth and proteolysis of *Streptococcus thermophilus* and *Lactobacillus bulgaricus* in skim milk. *J. Dairy Sci.* 73:894–899.

Rasic, J. L. and Kurmann, J. A. 1978. *Yogurt—Scientific Grounds, Technology, Manufacturing and Preparations*, pp. 26–99, 297–301. Copenhagen, Denmark: Technical Publishing Dairy House.

Rastall, R. A. 2010. Functional oligosaccharides: Application and manufacture. *Annu. Rev. Food Sci. Technol.* 1:305–339.

Robinson, R. K., Tamime, A. Y., and Wszolek, M. 2002. Microbiology of fermented milks. In: *Dairy Microbiology Handbook*, edited by K.K. Robinson, pp. 367–430. New York: John Wiley & Sons Inc.

Salminen, S. 1996. Uniqueness of probiotic strains. *Int. Dairy Fed. Nutr. Newsl.* 5:16–18.

Salminen, S., Isolauri, E., and Salminen, E. 1996. Clinical uses of probiotics for stabilizing the gut mucosal barrier: Successful strains and future challenges. *Antonie van Leeuwenhoek* 70:347–358.

Salminen, S. and Ouwehand, A. C. 2003. Probiotics, applications in dairy products. *Encyclopedia of Dairy Sciences*, Vol. 4, pp. 2315–2322. London: Academic Press.

Sanders, M. E. 2007. Probiotics: Strains matter. *Functional Foods & Nutraceuticals*. June, (pages 34–41).

Sanders, M. E. and Marco, M. L. 2010. Food formats for effective delivery of probiotics. *Annu. Rev. Food Sci. Technol.* 1:65–85.

Sanders, M. E., Akkermans, L. M. A., Haller, D., Hammerman, C., Heimbach, J., Hormannsperger, G., Huys, G., Levy, D. D., Lutgendorff, F., Mack, D., Photirath, P., Solsno-Aguilar, G., and Vaughan. E. 2010. Safety assessment of probiotics for human use. *Gut Microbes*: 1(3): 164–185.

Screzenmeir, J. and de Verse, M. 2001. Probiotics, prebiotics and synbiotics—Approaching a definition. *Am. J. Clin. Nutr.* 73(2):361S–364S.

Sgorbati, B., Biavati, B., and Palenzona, D. 1995. The genus *Bifidobacterium*. In: *The Lactic Acid Bacteria—Vol. 2: The Genera of Lactic Acid Bacteria*, edited by B. J. B. Wood and W. H. Holzapfel, pp. 279–306. New York: Chapman and Hall.

Shah, N. P. 2001. Functional foods from probiotics and prebiotics. *Food Technol.* 55:46–53.

Shah, N. P. and Lankaputhra, W. E. V. 1997. Improving viability of *Lactobacillus acidophilus* and *Bifidobacterium* spp. in yogurt. *Int. Dairy J.* 7:349–359.

Shahani, K. M. and Chandan, R. C. 1979. Nutritional and healthful aspects of cultured and culture-containing dairy foods. *J. Dairy Sci.* 62:1685–1694.

Sturino, J. M. and Klaenhammer, T. R. 2002. Expression of antisense RNA targeted against *Streptococcus thermophilus* bacteriophages. *Appl. Environ. Microbiol.* 68:588–596.

Tamime, A. Y. 2005. *Probiotic Dairy Products*. Oxford, UK: Blackwell Publishing Limited.

Tamime, A. Y. and Robinson, R. K. 2007. *Yogurt Science and Technology*, 3rd edition. Boca Raton, FL: CRC Press.

U.S. Probiotics.org. 2010. Available at: http://www.usprobiotics.org/basics.asp (accessed on December 18, 2011).

Vaughan, E. E., deVries, M. C., Zoetendal, E. G., Ben-Amor, K., Akkermans, A. D. L., and deVos, W. M. 2002. The intestinal LABs. *Antonie van Leeuwenhoek* 82:341–352.

Vedamuthu, E. R. 2006. Chapter 19: Other fermented and culture-containing milks. In: *Manufacturing Yogurt and Fermented Milks*, edited by R. C. Chandan (Editor), C. H. White, A. Kilara, and Y. H. Hui (Associate Editors), pp. 295–310. Ames, IA: Blackwell Publishing.

13

Sour Cream and Crème Fraîche

Lisbeth Goddik

CONTENTS

13.1 Introduction

Sour cream is a relatively heavy, viscous product with a glossy sheen. It has a delicate, lactic acid taste with a balanced, pleasant, buttery-like (diacetyl) aroma (Bodyfelt 1981). Various types of sour cream are found in many regions of the world. The products vary with regard to fat content and by the presence or absence of nondairy ingredients. Furthermore, both cultured and direct acidifications are utilized to lower pH. This chapter covers sour cream as it is produced in the U.S. and its French counterpart—crème fraîche.

13.2 Sour Cream

13.2.1 Definition

The U.S. Food and Drug Administration (FDA) [21 Code of Federal Regulations (CFR) 131.160] defines sour cream as follows: "Sour cream results from the souring, by lactic acid producing bacteria, of pasteurized cream. Sour cream contains not less than 18 percent milkfat;.... Sour cream has a titratable acidity of not less than 0.5 percent, calculated as lactic acid." If stabilizers are used, the fat content of the dairy fraction must be at least 18% fat and above 14.4% of the entire product. Optional ingredients permitted in sour cream are "(1) safe and suitable ingredients that improve texture, prevent syneresis, or extend the shelf life of the product, (2) sodium citrate in an amount not more than 0.1 percent that may be added prior to culturing as a flavor precursor, (3) rennet, (4) safe and suitable nutritive sweeteners, (5) salt, and (6) flavoring ingredients, with or without safe and suitable coloring, as follows: (i) fruit and fruit juice (including concentrated fruit and fruit juice) and (ii) safe and suitable natural and artificial food flavoring."

Consumers' desire for decreasing dietary fat content has created a market for low-fat sour cream. Among these products, reduced-fat, light (at least 25% or 50% fat reduction), and nonfat are common, in part due to FDA's labeling requirements for reduced-fat products (21 CFR 101). Sales data over the past 25 years for the U.S. market (USDA 2010) are illustrated in Figure 13.1. The trend clearly shows increased sales. In 2009, nearly 400 kg × 10⁶ of sour cream was sold. Per capita sales of sour cream and dips was 1.88 kg. In comparison, per capita sales for yogurt was 5.661 kg (USDA 2010).

13.2.2 Sensory Characteristics

Traditionally, the flavor of sour cream was well characterized by "sour." However, the trend for cultured dairy products is toward a milder flavor (Barnes et al. 1991) in part due to consumers' dislike of "too sour" fermented products (Thompson et al. 2007). Reduced acidity permits the sensation of aromatic compounds produced by lactic acid cultures. Lindsay et al. (1967) found that important flavor compounds in sour cream include diacetyl, acetic acid, acetaldehyde, and dimethyl sulfide. All aroma compounds are associated with mesophilic heterofermentative starter culture metabolism (Vasiljevic and Shah 2008). Sour cream is highly viscous and should be smooth and free of particulate matter. As for appearance, a homogeneous, glossy surface is preferred, and no whey separation should be visible in the container (Costello 2008).

13.2.3 Utilization

Sour cream is predominantly utilized as an accompaniment with warm entrees such as baked potatoes and burritos. This usage imposes certain demands on the sensory characteristics of the product, especially with regard to texture when in contact with warm surfaces. Sour cream must remain viscous without whey separation when placed on warm food. In addition, flavor characteristics become less significant when mixed with high-intensity savory flavor notes such as those encountered in Mexican cuisine. In fact, for some usages, the absence of off-flavors may be considered the primary flavor attribute. This general shift in emphasis away from flavor toward texture has led to a renewed interest in a "back to basics" sour cream such as crème fraîche, which is described later in this chapter.

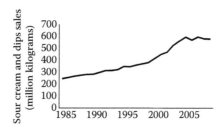

FIGURE 13.1   Sale, in million kilograms, of sour cream and dips in the U.S. between 1975 and 2008. (From U.S. Department of Agriculture (USDA). 2010. Economic Research Service Report. Agricultural Marketing Service, Washington, DC. Available at www.ers.usda.gov/publications/LDP/xlstables/FLUIDSALES.xls (accessed 13 October 2010). With permission.)

Sour cream is frequently used as a base in dips and sauces. Traditionally, flavors such as onion and spinach have been popular. Currently, the trend is more diverse, with flavors ranging from apricot ginger to chipotle (Jones 2010).

13.3 Fermentation

As with all fermented dairy products, the choice of starter culture is crucial for the production of high-quality sour cream (Folkenberg and Skriver 2001). Currently, the focus for sour cream starters is placed on accelerating the activity to complete the fermentation within 12 h. Mixed strains of mesophilic lactic acid bacteria are used for sour cream. In general, both acid and aroma producers are utilized. Acid producers include *Lactococcus lactis* ssp. *lactis* and *Lc. lactis* ssp. *cremoris*. *Lc. lactis* ssp. *lactis* biovar. *diacetylactis* (or *Cit+ Lactococci*) and *Leuconostoc mesenteroides* ssp. *cremoris* are commonly used aroma producers.

Acid producers convert lactose into L-lactate through a homofermentative pathway. They can produce up to 0.8% lactic acid in milk (Cogan 1995) and are responsible for lowering pH in the fermented product.

In contrast, aroma producers are heterofermentative and can convert lactose into D-lactate, ethanol, acetate, and CO_2. In addition, these strains convert citrate into diacetyl, which is one of the major flavor compounds responsible for typical sour cream flavor. Diacetyl is subsequently partially converted into acetoin, which is a flavorless compound (Monnet et al. 1995). Extensive research at starter culture companies have led to the development of *Leuconostoc* strains that show less of a tendency to convert diacetyl into acetoin, thus retaining high levels of diacetyl. Diacetyl levels may be elevated by the addition of citrate (Levata-Jovanovic and Sandine 1997). The use of such strains can extend the shelf life of sour cream, as it takes longer for the product to turn stale. Leuconostocs also reduce acetaldehyde to ethanol (Dessart and Steenson 1995; Keenan et al. 1966). In fact, acetaldehyde has been shown to promote the growth of *L. mesenteroides* ssp. *cremoris* (Lindsay et al. 1965; Collins and Speckman 1974). Acetaldehyde is typically associated with yogurt flavor (green apple) but is considered an off-flavor in sour cream.

The choice of starter cultures will affect product texture as well. Strains of acid producers that increase viscosity through the production of exopolysaccharides have been developed (Folkenberg et al. 2005). These polysaccharide chains contain galactose, glucose, fructose, mannose, and other sugars. Quantity and type depend on the bacteria strain and growth conditions (Duboc 2001; Ruas-Madiedo 2005). The exopolysaccharides interact with the protein matrix, creating a firmer network and increasing water-binding capacity. The importance of this behavior was confirmed by Adapa and Schmidt (1998), who found that low-fat sour cream, fermented by exopolysaccharide-producing lactic acid bacteria, was less susceptible to syneresis and had higher viscosity.

Production of high-quality sour cream requires a fine balance of acid, viscosity, and flavor-producing bacteria. While this balance varies among commercially available strains, a typical combination would be 60% acid producers, 25% acid and viscosity producers, and 15% flavor producers. However, fermentation conditions such as temperature and pH end point will impact the ratio during fermentation (Savoie et al. 2007).

13.4 Gel Formation

Fermentation leads to a significant increase in viscosity. Two physicochemical changes cause this behavior (Fox et al. 2000; Walstra and van Vliet 1986). The casein submicelles disaggregate because of the solubilization of colloidal calcium phosphate. In addition, the negative surface charge on the casein micelles decrease as the pH level approaches the isoelectric point. This creates the opportunity for casein micelles to enter into a more ordered system. Temperature during fermentation impacts the rate of fermentation and the viscoelastic properties of the gels (Lee and Lucey 2004). Besides the protein network, the cream gains viscosity from the formation of homogenization clusters (Mulder and Walstra 1974). Following single-stage homogenization at room temperature, milk fat globules will cluster, and these clusters may contain up to about 10^5 globules (Walstra et al. 1999). Casein molecules adsorb onto newly formed fat globule membranes and, in the case of high fat content, form bridges between fat globules. Clustering increases viscosity because (1) serum is entrapped between the globules and because of (2) the formation of irregularly shaped clusters.

13.5 Stabilizers

The gel structure may not be sufficiently firm to withstand abuse during transportation, handling, and storage. This could result in a weak-bodied sour cream and whey syneresis in the container. These defects are especially noticeable for low-fat products. To ensure consistent firm texture, dairy processors can either build up milk solids by adding dairy proteins or, more often, choose to add nondairy stabilizers (Hunt and Maynes 1997). Stabilizers commonly found in sour cream include different polysaccharides.

Stabilizers must be food grade and approved. The type and quantity used vary widely, depending on the fat content, starter culture, and required sensory characteristics of the final product. Types and quantities of potential stabilizer mixtures used in sour cream are outlined in Table 13.1. In particular, the nonfat formulation contains other ingredients such as emulsifiers, color, and protein.

Polysaccharides bind water and increase viscosity. Commonly used plant polysaccharides include carrageenans, guar gums, and cellulose derivatives. Modified starches are frequently utilized as well. It is necessary to fully hydrate these polysaccharides to optimize their functionality. Depending on the ingredient, this may require efficient blending systems for the incorporation of the ingredient into the cream, although care should be taken to avoid churning the cream. Complete hydration can sometimes only be accomplished following heating and cooling steps, which are conveniently done by the pasteurization process. Time may also be a factor for hydration to occur. Besides binding with water molecules, polysaccharides may also interact with milk proteins and form a network, which limits the movement of water and increases viscosity. A short description of the stabilizers is provided below.

Carrageenans: Extract of seaweed. Three types of carrageenans are commercially available—lambda, iota, and kappa—which differ based on the amount of sulfate. They have low viscosity at high temperature, but viscosity increases during cooling. Lambda has the highest sulfate content, is soluble in cold milk, and forms weak gels. Iota is soluble in hot milk (55°C) and prevents syneresis. Kappa only dissolves in hot milk (<70°C) and forms brittle gels (Marshall and Arbuckle 1996).

Guar gum: Endosperm of seed from the plant *Cyamopsis tetragonolobus*. Different types of guar gum are available to fit processing conditions. Maximum viscosity develops over time. All are soluble in cold milk. The main component is mannose with attached galactose units.

Methylcellulose: It is a cellulose that improves freeze–thaw stability and prevents melt upon heating (Hunt and Maynes 1997).

Lately, the trend is toward natural sour cream. No stabilizers are added to this product, and instead, body is improved by the addition of milk solids.

TABLE 13.1

Example of Stabilizer Used in Sour Cream

Product	Stabilizers	Usage Level
Sour cream	Modified food starch, grade A whey, sodium phosphate, guar gum, sodium citrate, calcium sulfate, carrageenan, locust bean gum	1.5%–3.0%
Light sour cream	Same as above	1.75%–3.5%
Nonfat sour cream	Modified food starch, microcrystalline cellulose, propylene glycol monoester, gum arabic, artificial color, cellulose gum	3.5%–6.5%

Source: Continental Custom Ingredients, Inc. 2002. Technical Bulletin Regarding Sour Cream Formulation and Processing. Continental Custom Ingredients, Inc., West Chicago, IL 60185. With permission.

13.6 Processing

Throughout the processing of sour cream, extra care should be taken to protect the cream. Prior to pasteurization, rough cream treatment could lead to rancid off-flavors due to lipolysis. Following fermentation, it is important to treat the coagulum gently to retain body and texture. This includes the use of positive displacement pumps instead of centrifugal pumps, round pipe elbows instead of 90° angles, and the use of gravity feed wherever possible. In addition, special cream pasteurizers may be used (Figure 13.2).

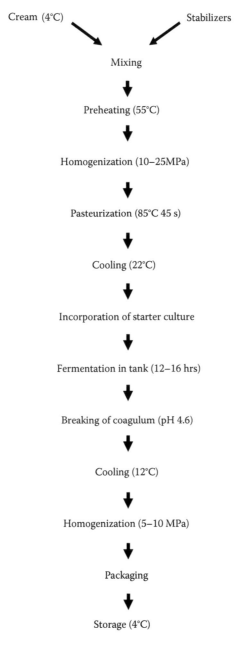

Cream (4°C)   Stabilizers

Mixing

↓

Preheating (55°C)

↓

Homogenization (10–25MPa)

↓

Pasteurization (85°C 45 s)

↓

Cooling (22°C)

↓

Incorporation of starter culture

↓

Fermentation in tank (12–16 hrs)

↓

Breaking of coagulum (pH 4.6)

↓

Cooling (12°C)

↓

Homogenization (5–10 MPa)

↓

Packaging

↓

Storage (4°C)

FIGURE 13.2 Process flow chart of a typical sour cream process.

Ingredients can be incorporated directly into standardized cream by mixing equipment such as a triblender. Another option is to incorporate the dry ingredients into the milk portion before standardizing the cream. The mix is preheated and homogenized (~65°C, 10–25 MPa; Kosikowski and Mistry 1999; Okuyama et al. 1994) immediately prior to pasteurization. Dairy homogenizers are normally double stage to prevent homogenization clusters. However, in sour cream production, single-stage homogenization is preferred to build up the body of the product. Homogenizing twice can build up texture but is inefficient. Instead, homogenization followed by passage through a shearing pump may be utilized. Additional viscosity is obtained if the cream is homogenized downstream from the pasteurizer, although such a process increases the potential for postpasteurization contamination. Pasteurization is done at relatively high temperatures (85°C–90°C for 10–45 s), well above what is required for the destruction of pathogens. The more severe heat treatment lowers the potential for oxidative and rancid off-flavors during storage, and it may help improve product viscosity due to the partial denaturation of serum proteins. The cream is cooled to 22°C–25°C and pumped into the fermentation tank, and the starter culture is added. Gentle mixing should continue until the culture and the cream are properly mixed (maximum of 30 min). At this point, mixing is stopped until fermentation is complete. The fermentation tank may be double jacketed to allow for better temperature control. However, in reality, this is not essential if the temperature of the processing room remains relatively constant at around 22°C. Fermentation temperature may vary slightly from plant to plant. Higher temperatures lead to faster fermentation and, potentially, a more acidic product, whereas lower fermentation temperatures may give a more flavorful product. The fermentation is slowed down/stopped by cooling when the desired acidity (~pH 4.5 or titratable acidity around 0.7%–0.8%) is achieved. Typically, this takes 14–18 h. The coagulum is broken by gentle stirring, and the product is cooled either by pumping cooling water into the double-jacketed area of the tank or by pumping the cream through a special plate cooler. The cream should be cooled to around 8°C–12°C, which slows starter culture activity before packaging. Prior to packaging, it can also be passed through a homogenizer screen (smoothing plug) or a similar type of flow restrictor to smoothen and improve texture (Continental Custom Ingredients 2002). The final cooling to around 4°C must occur slowly in the package in the cooler in order to allow the cream to obtain the appropriate viscosity. It is essential that the cream not be moved during this cooling step.

The above process assumes a large-scale production. However, numerous process variations exist.

13.6.1 "Shortcuts"

Throughout the process described above, special attention is focused on the gentle treatment of the product to assure proper body and texture. In reality, the stabilizers currently used permit more flexibility in the process. A certain amount of product abuse can be tolerated without lowering the product quality, because the stabilizers, when properly used, create a firm texture and prevent whey separation.

13.6.2 Artisan Production

It is possible to significantly simplify the process when producing small quantities of a product. Sour cream can be made in a double-jacketed pasteurization tank which doubles as fermentation tank with gravity feed to the filler. The absence of a final in-line cooling step would require an efficient cooling procedure for the packaged product.

13.6.3 Chymosin Addition

Low quantities of chymosin may be added at the same time as the starter culture. This creates a more "spoonable" sour cream. Lee and White (1993) found that chymosin addition (e.g., 0.066 mL/L) to low-fat sour cream resulted in increased viscosity and whey separation. Sensory scores were lower for the chymosin-containing sour cream with regard to flavor, body/texture, and appearance. This indicates that it may be preferable to modify the stabilizer mixture rather than to add chymosin when trying to increase product viscosity.

13.6.4 Set Sour Cream

The standardized, pasteurized cream can be mixed with starter culture and immediately filled into the package. The cream is then fermented within the final package, which leaves the coagulum undisturbed. When the appropriate acidity is obtained, the products are cooled either by passing through a blast cooler or by placement in a cooler. The advantage of this method is the possibility of lowering or eliminating stabilizers and yet obtaining excellent body and texture. The disadvantages are the large space requirement for fermenting the packaged product and the relatively slow cooling.

13.6.5 Direct Acidification

A product somewhat similar to sour cream can be obtained by direct acidification by organic acids such as lactic acid instead of fermentation. However, Kwan et al. (1982) and Hempenius et al. (1969) found that sensory panelists preferred cultured sour cream instead of chemically acidified cream. Product temperature at the time of acidification is critical and should be around 20°C–25°C. Higher temperatures increase the likelihood that graininess occurs, and lower temperatures increase the time required for gel formation (Continental Custom Ingredients 2002).

13.6.6 Low-Fat and Nonfat Sour Cream

Vitamin A fortification is required in these products. The processes are often similar to traditional sour cream, although nonfat sour cream mix should be homogenized at a much lower pressure. The main difference is observed in the stabilizer mix as described in Section 13.5.

13.7 Shelf Life

Sour cream should have a shelf life of around 25–45 days. One study documents that, when properly stored undisturbed at 4°C, sour cream has an acceptable shelf life of up to 6 weeks (Warren 1987). In another study, Folkenberg and Skriver (2001) evaluated the change of sensory properties of sour cream during storage time. As storage time approached 28 days, the intensity of prickling mouthfeel, sour odor, and bitter taste increased. The samples were stored under ideal conditions, which suggest that real-life distribution and storage temperature abuse would likely decrease the shelf life of this product below 28 days.

The single most important factor determining shelf-life is cream quality. Unless the cream is of excellent quality, the sour cream quickly develops off-flavors. Two parameters that impact cream quality are (1) raw milk quality and (2) pretreatment of milk. Good-quality raw milk has low bacterial content (low standard plate count) and comes from healthy cows (low somatic cell count). Even good-quality raw milk spoils unless quickly cooled and kept at low temperatures until pasteurization. Furthermore, the time interval between milking and pasteurization should be as short as possible to limit the growth of psychotropic microorganisms. Other factors to consider are proper cleaning and sanitation of all milk contact surfaces, well-installed and sized pumps, and no unnecessary milk handling.

Assuming that high-quality cream is utilized, the parameters that limit shelf life tend to be associated with either flavor defects or surface growth of yeasts and molds. When using appropriate stabilizers, body and texture should remain adequate throughout the shelf life. A guide on how to prevent flavor defects is included below. Yeasts and molds are controlled by improving sanitation throughout the process. As with many other dairy products, sanitation trouble spots are often associated with the filler machines, which are difficult to clean properly.

13.8 Sensory Defects in Sour Cream

13.8.1 Flavor

The high lipid content makes sour cream extremely vulnerable to lipid-associated off-flavors such as rancidity and oxidation. Other flavor defects include flatness, lack of cultured flavor, and high acidity.

Rancid. Hydrolytic rancidity or lipolysis is caused by the release of free fatty acids from the glycerol backbone of triglycerides. The reaction is catalyzed by the lipase enzyme, which can be a native milk lipoprotein lipase or can originate from bacterial sources. Triglycerides are generally protected from lipase activity, as long as the milk fat globule remains intact. However, damage to the globule will lead to rapid lipolysis, because lipase, which is situated on the surface of the globule, gains access to the triglycerides. Therefore, precautions must be taken to prevent damage to the milk fat globule until pasteurization, which denatures most types of lipase. This means that raw milk/cream must be pasteurized before or immediately after homogenization to assure denaturation of lipase. Likewise, it is strongly recommended never to recycle pasteurized milk/cream back into raw milk/cream storage, which is essentially an issue of rework handling. Cream from poor-quality raw milk can also develop rancid off-flavors during storage, as some bacterial lipases may be quite heat stable and not denature during pasteurization.

Oxidized. Autoxidation of milk fat is a reaction with oxygen that proceeds through a free radical mechanism. Unsaturated fatty acids and phospholipids are the prime substrates that are broken down into smaller molecular weight compounds such as aldehydes and ketones. Oxidized cream exhibits off-flavors and aromas that have been characterized as cardboardy, metallic, oily, painty, fishy, and tallowy (Bodyfelt et al. 1988). Oxidation is catalyzed by divalent cations such as iron or copper. Thus, the best prevention is to avoid contact of milk/cream with these metals. This requires attention to details, as a single fitting or pipe made of these metals can cause significant autoxidation. Milk and cream can be more susceptible to autoxidation following changes in the cows' diet. For example, flax seed can increase the content of polyunsaturated fatty acids in the milk. The feed and water can also be a source of catalysts such as copper (Timmons et al. 2001). The problem is likely more evident toward the end of winter, as the levels of antioxidants (e.g., vitamin E) in the feed tends to be lower.

Light-induced oxidation can also influence flavor characteristics of sour cream. Light-induced oxidation is a defect associated with milk proteins and involves the degradation of milk fat, proteins, and vitamins (Webster et al. 2009; Intawiwat et al. 2010). It has been demonstrated that the light barrier properties of the packaging materials impact sour cream flavor over the shelf life, with high-light barrier properties minimizing the defect and low-light barrier properties leading to off-flavors within hours of exposure to fluorescent light sources typically associated with retail conditions (Larsen et al. 2009).

Lacks fine or cultured flavor. Both flavor defects tend to be associated with the choice of starter culture. It may be possible to improve flavor by switching to culture systems with more aroma-producing capacity or to strains that retard the transfer of diacetyl into acetoin. It is also possible to add low concentrations of citric acid (below 0.1%), which is then converted to diacetyl by the aroma-producing starter cultures. A slight change (usually decrease) in fermentation temperature has also been found to improve the concentration of aroma compounds. The defect can also result from flavors imparted by stabilizers. Lowering the stabilizer dose or changing to another stabilizer system may be required.

Has high acidity. If the final product pH is very low (e.g., around pH 4.0), the product gets an unpleasant sour flavor. While it is possible to stop the fermentation at a higher pH level, this does not necessarily solve the problem, because slow fermentation continues in the cooled and packaged product. Therefore, it is often preferable to change the starter culture mixture to lower the ratio of acid-producing bacteria.

Bitter. Bitter off-flavors are often indicators of excess proteolytic activity. Poor-quality raw milk may contain heat-stable proteases that remain active throughout storage. The defect is especially noticeable at the end of the shelf life. Improving raw milk quality, increasing pasteurization temperature, or shortening code dates are possible solutions (Folkenberg and Skriver 2001).

13.8.2 Body and Texture

As described above, texture is an essential quality parameter. Sour cream must remain highly viscous when in contact with warm food surfaces such as baked potatoes.

Too firm or weak. Improper choice of stabilizers can cause overstabilized sour cream that clings to the spoon. Alternatively, the sour cream can be weak bodied and "melt" on the hot food surface.

Grainy. Graininess is primarily a mouthfeel problem, although it can be visually distracting as well in extreme cases. Grainy sour cream can be an indication of poor blending or incomplete hydration of ingredients. A different choice of stabilizers or a modification of incorporation procedure may improve the product. Another solution is to pass the product through a single-stage homogenizer valve prior to packaging. Grains can also indicate that the fermentation was stopped at too high a pH level and the caseins are at their isoelectric point around pH 4.6.

Free whey. Whey syneresis on top of the sour cream in the package is considered a significant quality defect. There are three solutions available to solving the problem: (1) change or increase the concentration of stabilizer, (2) increase fat content (higher fat sour creams have a better water-binding capacity) or (3) reevaluate the entire process and eliminate points of product abuse. This would primarily include all steps following fermentation.

13.9 Functional Properties

Although conjugated linoleic acid (CLA) is a minor component of milk fat, dairy is one of the major sources of CLA in our diet. The CLA content in milk and dairy products varies greatly due to the cows' diet and can range from 0.1% up to 2% of milk lipids (Khanal and Olson 2004). High intake of CLA has been linked to decreased risk of several cancers and enhanced function of immune cells (Ip et al. 1994; Krichevsky 2004). Moreover, data suggest that the consumption of dairy products high in fat such as sour cream may reduce the risk of colorectal cancer (Larsson et al. 2005). Sphingolipids in dairy fats have also been linked to anticarcinogenic effects. In addition, clearly higher fat dairy products such as sour cream contain more of these beneficial lipids (Ribar et al. 2007).

13.10 Crème Fraîche

Crème fraîche or, more correctly, crème fraîche épaisse fermentée is the European counterpart of the U.S. sour cream product. Crème fraîche has fat content around 30%–45% and has a mild, aromatic cream flavor. The differences between the two products originate in the manner of usage. The usage of sour cream is described above. Crème fraîche is used cold on desserts such as fruits or cakes, or warm as a foundation in cream sauces that are commonly used in French cuisine. This double usage creates a unique demand for specific product attributes. The dessert utilization requires a clean, not too sour (Barnes et al. 1991), cultured flavor that does not overpower flavors from other dessert components. The cultured flavor should be refreshing so that it covers the impression of fat in the product. This emphasis on flavor has led to significant research at starter culture and dairy processing companies to develop starter cultures that cause optimum flavor development. Body and texture should be smooth and less firm than sour cream. Crème fraîche should be "spoonable," not "pourable," and should spread slightly on the dessert without being a sauce.

The incorporation of crème fraîche into a warm sauce requires thermostability; otherwise, the protein would precipitate and flocculate in the sauce. For regular crème fraîche (>30% fat), flocculation is rarely a problem. In contrast, low-fat crème fraîche (~15% fat) is less stable when heated. The addition of stabilizers such as xanthan gum can stabilize low-fat crème fraîche. However, based on European labeling legislation, crème fraîche cannot contain stabilizers, and a stabilized product would therefore need to be marketed under another name.

Crème fraîche is produced by a process similar to that of sour cream, with the exception that no ingredients are added. Without stabilizers, it becomes a challenge to obtain good body and texture.

Each processing step requires attention to producing and maintaining high viscosity. In this case, the homogenizer becomes an essential tool for building viscosity. Only single-stage homogenization is utilized. The product is sometimes homogenized twice, either in subsequent runs before pasteurization or, more commonly, both before and after pasteurization. Homogenization after pasteurization promotes better viscosity and, equally important, better thermostability. An additional homogenization following fermentation gives a homogeneous product with a smooth mouthfeel. Homogenization downstream from the pasteurizer (i.e., after pasteurization) should raise concerns with regard to postpasteurization contamination. Ideally, an aseptic homogenizer should be used. However, the high price of such homogenizers makes this an unsuitable alternative. Instead, great emphasis must be placed on the proper cleaning and sanitizing of the downstream homogenizer. In addition, food safety issues are normally controlled because of the high content of lactic acid bacteria and the low pH level.

There is some discussion as to the final pH level of crème fraîche fermentée. Kosikowski et al. (1999) and Kurmann et al. (1992) state that the cream is fermented to pH 6.2–6.3. However, commercially, it is commonly fermented to an end pH around 4.5. The mild flavor is not obtained by a higher pH but, rather, through selection of aroma-producing starter cultures. It is the combination of aroma compounds and the high fat content that mask the sour flavor in crème fraîche.

Crème fraîche is a new product in the U.S. market. The high fat content and small-scale processing contribute to a retail price that is at least twice as expensive as the traditional sour cream. The product is frequently made by artisan dairy processors and is sold in outlets such as farmers' markets and high-end restaurants. Its increasing popularity is an indication of changing culinary habits promoted by growing population diversity and exposure to other culinary culture.

REFERENCES

Adapa, A. and Schmidt, K. A. 1998. Physical properties of low-fat sour cream containing exopolysaccharide-producing lactic acid. *J. Food Sci.* 63:901–903.

Barnes, D., Harper, S. J., Bodyfelt, F. W., and McDaniel, M. R. 1991. Correlation of descriptive and consumer panel flavor ratings for commercial prestirred strawberry and lemon yogurts. *J. Dairy Sci.* 74:2089–2099.

Bodyfelt, F. W. 1981 Cultured sour cream: Always good, always consistent. *Dairy Rec.* 82(4):84–87.

Bodyfelt, F. W., Tobias, J., and Trout, G. M. 1988. *The Sensory Evaluation of Dairy Products*, pp. 247–251. New York: Van Nostrand Reinhold.

Code of Federal Regulations 2010, Title 21. Section 131. Washington, DC: U.S. Government Printing Office.

Cogan, T. M. 1995. History and taxonomy of starter cultures. In: *Dairy Starter Cultures*, edited by T. M. Cogan and J. P. Accolas, pp. 1–23. New York: VCH Publishers.

Collins, E. B. and Speckman, R. A. 1974. Influence of acetaldehyde on growth and acetoin production by *Leuconostoc citrovorum. J. Dairy Sci.* 57:1428–1431.

Continental Custom Ingredients, Inc. 2002. Technical bulletin regarding sour cream formulation and processing. Continental Custom Ingredients, Inc., 245 West Roosevelt Road, West Chicago, IL 60185.

Costello, M. 2008. Sour cream and related products. In: *The Sensory Evaluation of Dairy Products*, 2nd edition, edited by S. Clark, M. Costello, M. Drake, and F. Bodyfelt, pp. 403–426. New York: Springer.

Dessart, S. R. and Steenson, L. R. 1995. Biotechnology of dairy *Leuconostoc*. In: *Food Biotechnology*, edited by Y. H. Hui and G. G. Khachatourians, pp. 665–702. New York: VCH Publishers.

Duboc, P. 2001. Application of exopolysaccharides in the dairy industry. *Int. Dairy J.* 11:759–768.

Folkenberg, D. M. and Skriver, A. 2001. Sensory properties of sour cream as affected by fermentation culture and storage time. *Milchwissenschaft* 56:261–264.

Folkenberg, D. M., Dejmek, P., Skriver, A., and Ipsen, R. 2005. Relation between sensory texture properties and exopolysaccharide distribution in set and in stirred yogurts produced with different starter cultures. *J. Texture Stud.* 36:174–189.

Fox, P. F., Guinee, T. P., Cogan, T. M., and McSweeney, P. L. H. 2000. *Fundamentals of Cheese Science*, pp. 363–387. Gaithersburg, MD: Aspen.

Hempenius, W. L., Liska, B. J., and Harrington, R. B. 1969. Selected factors affecting consumer detection and preference of flavor levels in sour cream. *J. Dairy Sci.* 52:588–593.

Hunt, C. C. and Maynes, J. R. 1997. Current issues in the stabilization of cultured dairy products. *J. Dairy Sci.* 80:2639–2643.

Intawiwat, N., Petterson, M. K., Rukke, E. O., Meier, M. A., Vogt, G., Dahl, A. V., Skaret, J., Keller, D., and Vold, J. P. 2010. Effect of different colored filters on photooxidation in pasteurized milk. *J. Dairy Sci.* 93:1372–1382.

Ip, C., Singh, M., Thompson, H. J., and Scimeca, J. A. 1994. Conjugated linoleic acid suppresses mammary carcinogenesis and proliferative activity of the mammary gland in the rat. *Cancer Res.* 54:1212–1215.

Jones, W. 2010. Dips and sauces defy traditional trends. Prepared Foods Network. July 2010. Available at: http://www.preparedfoods.com/Articles (accessed 9 October 2010).

Keenan, T. W., Lindsay, R. C., and Day, E. A. 1966. Acetaldehyde utilization by *Leuconostoc* species. *Appl. Environ. Microbiol.* 14:802–806.

Khanal, R. C. and Olson, K. C. 2004. Factors affecting conjugated linoleic acid (CLA) content in milk, meat, and egg: A review. *Pak. J. Nutr.* 3:82–98.

Kosikowski, F. and Mistry, V. V. 1999. *Cheese and Fermented Milk Foods*, 3rd edition, pp. 6–14. Great Falls, VA: F.V. Kosikowski, L.L.C.

Krichevsky, D. 2004. Conjugated linoleic acid. In: *Handbook of Functional Dairy Products*, edited by C. Shortt and J. O'Brien, pp. 155–168. New York: CRC Press.

Kurmann, J. A., Rasic, J. L., and Kroger, M. 1992. *Encyclopedia of Fermented Fresh Milk Products*, pp. 94–95. New York: Van Nostrand Reinhold.

Kwan, A. J., Kilara, A., Friend, B. A., and Shahani, K. M. 1982. Comparative B-vitamin content and organoleptic qualities of cultured and acidified sour cream. *J. Dairy Sci.* 65:697–701.

Larsen, H., Veberg, A., and Geiner, S. B. 2009. Quality of sour cream packaged in cups with different light barrier properties measured by fluorescence spectroscopy and sensory analysis. *J. Food Sci.* 74:345–350.

Larsson, S. C., Bergkvist, L., and Wolk, A. 2005. High-fat dairy food and conjugated linoleic acid intakes in relation to colorectal cancer incidence in the Swedish mammography Cohort. *Am. J. Clin. Nutr.* 82:894–900.

Lee, F. Y. and White, C. H. 1993. Effect of rennin on stabilized lowfat sour cream. *Cult. Dairy Prod. J.* 28:4–13.

Lee, W. J. and Lucey, J. A. 2004. Structure and physical properties of yogurt gels: Effect of inoculation rate and incubation temperature. *J. Dairy Sci.* 87:3153–3164.

Levata-Jovanovic, M. and Sandine, W. E. 1997. A method to use *Leuconostoc mesenteroides* ssp. *cremoris* 91404 to improve milk fermentations. *J. Dairy Sci.* 80:11–18.

Lindsay, R. C., Day, E. A., and Sandine, W. E. 1965. Green flavor defect in lactic starter cultures. *J. Dairy Sci.* 48:863–869.

Lindsay, R. C., Day, E. A., and Sather, L. A. 1967. Preparation and evaluation of butter flavor concentrates. *J. Dairy Sci.* 50:25–31.

Marshall, R. T. and Arbuckle, W. S. 1996. *Ice Cream*, 5th edition. New York: Chapman & Hall.

Monnet, V., Condon, S., Cogan, T. M., and Gripon, K. C. 1995. Metabolism of starter cultures. In: *Dairy Starter Cultures*, edited by T. M. Cogan and J. P. Accolas, pp. 47–100. New York: VCH Publishers.

Mulder, H. and Walstra, P. 1974. The milk fat globule. In: *Emulsion Science as Applied to Milk Products and Comparable Foods*, pp. 163–192. Wageningen: Pudoc.

Okuyama, S., Uozumi, M., and Tomita, M. 1994. Effect of homogenization pressure on physical properties of sour cream. *Nippon Shokuhin Gakkaishi* 41:407–412.

Ribar, S., Karmelic, I., and Mesaric, M. 2007. Sphingoid bases in dairy products. *Food Res. Int.* 40:848–854.

Ruas-Madiedo, P. 2005. Effect of exopolysaccharides and proteolytic activity of *Lactococcus lactis* subsp. *cremoris* strains on the viscosity and structure of fermented milks. *Int. Dairy J.* 15:155–164.

Savoie, S., Audet, P., Chiasson, S., and Champagne, C. P. 2007. Media and process parameters affecting the growth, strain ratios and specific acidifying activities of a mixed lactic starter containing aroma-producing and probiotic strains. *J. Appl. Microbiol.* 103:163–174.

Thompson, J., Lopetcharat, K., and Drake, M. A. 2007. Preferences for commercial strawberry drinkable yogurts among African American, Caucasian, and Hispanic consumers in the United States. *J. Dairy Sci.* 90:4974–4987.

Timmons, J. S., Weiss, W. P., Palmquist, D. L., and Harper, W. J. 2001. Relationships among dietary roasted soybeans, milk components, and spontaneous oxidized flavor of milk. *J. Dairy Sci.* 84:2440–2449.

U.S. Department of Agriculture (USDA). 2010. Economic Research Service Report, Agricultural Marketing Service, Washington, DC. Available at: www.ers.usda.gov/publications/LDP/xlstables/FLUIDSALES .xls (accessed 13 October 13 2010).

Vasiljevic, T. and Shah, N. P. 2008. Cultured milk and yogurt. In: *Dairy Processing & Quality Assurance*, edited by R. C. Chandan, A. Kilara, and N. P. Shah. Ames, IA: Wiley Blackwell.

Walstra, P. and van Vliet, P. 1986. The physical chemistry of curd-making. *Neth. Milk Dairy J.* 40:241–259.

Walstra, P., Geurts, T. J., Noomen, A., Jellema, A., and van Boekel, M. A. J. S. 1999. Dairy technology. In: *Principles of Milk Properties and Processes*, pp. 245–264. New York: Marcel Dekker.

Warren, S. 1987. Influence of storage conditions on quality characteristics of sour cream. *Cult. Dairy Prod. J.* 8:13–14, 16.

Webster, J. B., O'Keefe, S. F., Marcy, J. F., and Duncan, S. F. 2009. Controlling light oxidation flavor in milk by blocking riboflavin excitation wavelengths by interference. *J. Food Sci.* 74:390–398.

14

Fresh Cheese

Lisbeth Goddik

CONTENTS

14.1 Introduction

14.1.1 Cheese Classification

Hundreds of different cheeses are produced around the world. Officially, approximately 500 cheese types have been recognized (Burkhalter 1981). Several classification schemes have been developed to group these cheeses, and a good description of these classifications can be found in *Fundamentals of Cheese Science* (Fox et al. 2000). One system classifies cheeses based on hardness and spans the spectra from very hard, hard, semisoft, and soft. No category is developed for semisolid cheeses, and as such, cheeses should not be described in this segment of the handbook, which covers semisolid dairy products. However, the authors felt that several fresh high-moisture cheeses do belong in the semisolid dairy products group along with yogurt and sour cream. Thus, this chapter describes one of these semisolid or "spoonable" cheeses, that is, fromage frais from France.

14.1.2 Characteristics of Fromage Frais

Fromage frais is a fresh acid-curd cheese that can be consumed immediately after production. Its name "fromage frais" (translated as fresh cheese) clearly communicates to the consumer that it is a nonripened cheese and freshness is a major attribute of this cheese. Fromage frais is very similar to the German

TABLE 14.1

Examples of Composition (% w/w) of Fromage Frais with Different Fat Contents

Composition	Low Fat	20% Fat in Dry Matter	40% Fat in Dry Matter
Total solids (%)	17	21	26
Fat (%)	0.05	4.5	10.4
Protein (%)	13	13	12
Lactose (%)	3	3	3
Salt (%)	0.5–1	0.5–1	0.8–1

Source: Adapted from Lehmann et al. 1991. *Processing Lines for the Production of Soft Cheese.* 3rd edition, pp. 26–47, Westfalia Separator, Oelde.

cheese quark, which has slightly higher solid content. In appearance, it resembles yogurt—a resemblance that is amplified by the use of packaging similar to yogurt cups. The flavor is subdued, even bland, with a refreshing, mildly sour characteristic. Yet, the product has a clear cultured flavor, which is not to be mistaken with acetaldehyde in yogurt. The mouthfeel should be smooth, rich, and homogeneous.

Compositions of fromage frais with different fat contents are outlined in Table 14.1. Declaration of fat content must be made on the label as percentage (grams of fat/100 g of product; Dellaglio 1992). Some fromage frais is sold plain, that is, no flavor/fruit is added. However, many versions exist with added fruit, herbs, flavors, and sweeteners. French legislation requires the cheese to be made from pasteurized milk. The total solid content should be at least 15% when fat in dry matter is above 20% and at least 10% when fat in dry matter is below 20%. Permissible ingredients include sucrose, honey, natural flavors, fruit, pulp, juice, jam, fibers, and colorants. Nonnatural flavors to reinforce natural flavors may be added in low quantities. Sorbate or potassium sorbate is permitted in fruit mixtures, as long as the quantity remains below 2% (the current dosage is generally less than 0.2%).

14.1.3 Utilization

Fromage frais is frequently utilized as a dessert and may be served with fruit and sugar. The preflavored products are often sweetened (10%–12%), packed in small cups (30 g/cup), and marketed to children. In addition, it can be served plain or with herbs, although the latter option is not currently used in France. In particular, the low-fat products provide an excellent source of proteins low in fat and calories.

Although fromage frais is traditionally a French cheese, it is now sold in many European countries (Hilliam 1997). The sale of fromage frais is increasing in part due to the product attributes listed below. The popularity with children is reflected in the sale of flavored products that are generally targeted toward children (Figure 14.1).

1. Excellent nutritional quality, that is, high protein and low fat (for low-fat varieties).
2. Mild taste well adapted for children.
3. Higher viscosity than European yogurt and, thus, very convenient for young children learning to eat with a spoon.

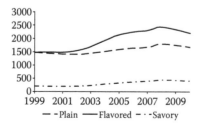

FIGURE 14.1   Retail value from 1999 to 2010 (in U.S. dollars) for fromage frais and quark sold in Western Europe. (Courtesy of Sabrina Kinckle, Euromonitor International, Chicago, IL, 2011.)

4. Excellent safety record. The heat treatment, continuous postfermentation processing procedures, and low pH level minimize the potential for postpasteurization contamination.

5. Low production costs due to highly automated processing.

6. High yield as several methods that lower protein content of the whey exist.

7. Ease of product diversification. Different fat contents, flavors, sweeteners, and fruits can be incorporated into the same cheese base prior to packaging.

8. Ease of packaging. The product is packaged on equipment similar to yogurt fillers and can be packaged into a wide variety of cup shapes and sizes.

14.2 Fermentation

Mixed strains of mesophilic acid producers are utilized, principally *Lactococcus lactis* ssp. *lactis* and *Lc. lactis* ssp. *cremoris*. In addition, aroma producers such as *Lc. lactis* ssp. *lactis* biovar. *diacetylactis* (*Cit⁺ Lactococci*) and *Leuconostoc mesenteroides* ssp. *cremoris* are included.

The acid producers are homofermentative and convert lactose into lactic acid. Lactic acid is the basis for the clean sour flavor of the product and is also responsible for coagulating the cheesemilk. The conversion of lactose into lactic acid has been extensively studied and is well defined (Fox et al. 1990; Sanders 1995). Lactose is transported into the bacterial cell by the phosphoenolpyruvate–phosphotransferase lactose transport system and by lactose permease. During the transport, galactose is phosphorylated and transformed into glyceraldehyde-3-phosphate through the D-tagatose-6-phosphate pathway and into glycolysis. The glucose moiety is phosphorylated and metabolized through the glycolytic (Embden–Meyerhof) pathway (Sanders 1995; McKay 1981). The overall homofermentative reaction can be described in simple terms, although the reaction continues beyond the conversion into lactic acid:

$$\text{Lactose} + 4\text{ADP} + 4\text{H}_3\text{PO}_4 \rightarrow 4\text{Lactic acid} + 4\text{ATP} + 3\text{H}_2\text{O}. \tag{14.1}$$

Aroma producers do not follow the Embden–Meyerhof pathway, because they lack aldolases. Instead, they convert lactose into a variety of end products such as lactic acid, ethanol, acetic acid, and CO_2. More important is their ability to metabolize citrate (Cachon and Divies 1993; Hache et al. 1999). Milk contains approximately 1600 mg of citrate per 1 kg of milk. Citrate permease transports citrate into the bacterial cell. Citrate is converted to oxaloacetate, pyruvate, and, eventually, diacetyl (Lee 1996), which is an important flavor compound in fromage frais. Continued conversion into the flavorless compound acetoin is possible (Monnet et al. 1995) and depends on residual citrate levels, pH, and enzyme activity, which are strain specific.

14.3 Formation of Acid Coagulum

The mesophilic fermentation slowly converts lactose to lactic acid, which leads to increased acidity of the cheesemilk. The acid development causes several physicochemical changes in the milk (Lucey 2002) primarily based on the conversion of casein from micellar form to a casein matrix. It is generally recognized that colloidal calcium phosphate connects and bridges caseins with the caseins micelle. Most or all of the colloidal calcium phosphate is solubilized by the time that acidification is complete around pH 4.5 (Van Hooydonk et al. 1986; Dalgleish and Law 1989). This solubilization initially leads to the dissociation of individual caseins from the casein micelles (maximum around pH 5.2–5.4; Dalgleish and Law 1988; Gastaldi et al. 1997). However, as the acidification continues, the proportion of serum casein decreases and is practically absent at the isoelectric point. This reversal is thought to be due to the reduction of surface charge on the proteins as they approach their pI coupled with the increased concentration of calcium ions, which neutralizes charged regions. In this environment, soluble casein reattaches onto the surface of casein micelles.

Gelation is thought to progress through several stages, which have been classified as pregel and stages 1–4 (Tranchant et al. 2001), and acidification and gel formation are considered concurrent processes.

Increases in viscosity are observed starting around pH 5.2–5.3 as casein aggregates become evident (Gastaldi et al. 1996). These initial aggregates consist of altered casein micelles, small dissociated sub-micelles, and micelles with reattached dissociated casein. As pH reduction continues, the aggregates touch and form a three-dimensional network (Lucey 2002). Hydrophobic, hydrogen, and electrostatic interactions are all involved in the structure of the final gel (Lefebvre-Cases et al. 1998).

Rennet may be added once fermentation has commenced. Rennet cuts κ-casein and releases the hydro-philic glycomacro peptide, which leads to lower surface charges on the casein micelles. In addition, less casein dissociates from the micelles, causing less changes in the shape and size of the original micelles. These factors initiate casein aggregation at a higher pH level, which provides additional time for a firmer network to form.

Obtaining the correct gel structure is essential for product characteristics such as mouthfeel, consistency, and water-binding capacity (Lucey 2002). For example, a slowly formed gel consists of a highly ordered casein network with high water-holding capacity, whereas a quickly formed gel (fast fermentation) results in a coarse network, with large serum pockets, which may lead to syneresis (Green 1989). Coarse casein network, accompanied by syneresis, has also been attributed to higher gelation temperature (Lagoueyte et al. 1994). In contrast, protein standardization leads to a stronger, firmer gel (Gastaldi et al. 1997) with smaller serum pockets that lower the tendency for syneresis (Modler and Kalab 1983; Modler et al. 1983). Severe heat treatment (above the required pasteurization temperature) increases product viscosity, because heat-denatured whey proteins (α-lactalbumin and β-lactoglobulin) will associate with κ-casein on the micelle surfaces and self-aggregate as well, thereby becoming part of the protein network instead of being dispersed in the serum phase (Hinrichs 2001).

14.4 Processing

A typical process for fromage frais is outlined in Figure 14.2 (Lehmann et al. 1991). This process is highly automated and is used in large commercial processes. Each processing step is described in further detail below. It should be kept in mind that many alternative and more traditional processes are utilized in smaller plants.

14.4.1 Centrifugation and Thermization

Thermization or heat shock treatment (65°C, 15–30 s) generally follows the initial centrifugation, which separates raw milk into skim milk and cream. This heat treatment is not a legal pasteurization but suffices to destroy most spoilage bacteria. In general, this heating increases process flexibility by allowing more time for protein standardization without the deterioration of skim milk quality.

14.4.2 Protein Standardization

While cheeses traditionally require casein for fat standardization, fromage frais is prepared from skim milk, and only protein standardization occurs. The reasons for standardizing protein content are three-fold: (1) to ensure constant protein content throughout the year, (2) to increase the cheesemaking capacity of the processing line, and (3) to promote a firm coagulum (see Section 14.3). The standardization can be done by adding protein or by concentrating the skim milk. Typically, proteins are added as nonfat milk powder or milk protein concentrates. It is important that time is available for the added powders to become properly hydrated. Hydration rates are increased if proper mixing equipment such as a triblender is utilized. The subsequent pasteurization and cooling also promote hydration.

Concentration of skim milk can be achieved through membrane technology or by evaporation. The choice of method affects the composition of the standardized skim milk. Ultrafiltration (UF) retentates contain less low-molecular-weight compounds such as minerals and lactose than evaporated concentrates, because low-molecular-weight compounds are removed in the UF permeate. Compositional differences can have profound influence on cheese characteristics and must be controlled by changing processing parameters such as rennet addition and fermentation rate and extent (Pouliot 2008).

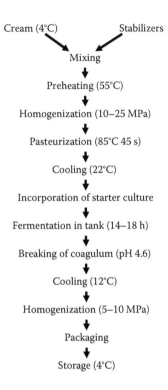

Cream (4°C)   Stabilizers
↓   ↓
Mixing
↓
Preheating (55°C)
↓
Homogenization (10–25 MPa)
↓
Pasteurization (85°C 45 s)
↓
Cooling (22°C)
↓
Incorporation of starter culture
↓
Fermentation in tank (14–18 h)
↓
Breaking of coagulum (pH 4.6)
↓
Cooling (12°C)
↓
Homogenization (5–10 MPa)
↓
Packaging
↓
Storage (4°C)

FIGURE 14.2   Processing of fromage frais using curd centrifuge.

14.4.3 Pasteurization

It is important to pasteurize as quickly as possible to prevent the growth of psychrotrophic spoilage bacteria. Pasteurization conditions are well above the minimum requirements for destroying potential raw milk pathogens. The severe heat treatment creates a higher viscosity of the coagulum, in part because whey proteins are heat sensitive and denature at severe heat treatment. Denatured whey proteins are retained in the casein network, thus leading to increased cheese yield. Westfalia (1976) obtained a patent in 1977 for the thermoquark process, which is frequently used in fromage frais processing. The patent covers the process of pasteurizing cheesemilk to 90°C–95°C for 2–3 min, followed by fermentation and then a second heat treatment of the coagulum at 60°C for 1–2 min. The advantage of this process is a decrease in protein content of the whey by up to 50%.

Following the holding tube, the milk is cooled to the appropriate fermentation temperature in the regeneration section of the pasteurizer. Typical fermentation temperatures are around 25°C–27°C.

14.4.4 Fermentation

The quantity of starter cultures added depends on the activity of the culture, incubation temperature, time available for fermentation, and sensory characteristics expected of the finished product. In general, 0.5%–3% culture is added. Approximately 90 min after the addition of starter culture, the milk/coagulum reaches a pH of 6.3, and rennet is added (0.5–1 mL single strength/100 L of cheesemilk). Fromage frais is a fresh acid-curd cheese and can be made without rennet. However, rennet firms the coagulum and subsequently minimizes the loss of shattered casein curd into the whey. If severe heat treatment has been utilized to denature whey proteins, it is beneficial to increase the rennet quantity, because κ-casein is protected by the denatured whey protein. However, compared to traditional rennet cheeses such as cheddar, the rennet addition to fromage frais is low.

The fermentation takes 14–16 h and is complete when the pH reaches 4.5, whereupon the coagulum is well mixed to assure a smooth texture. Fermentation time above 18 h indicates slow cultures and could pose a safety risk. In contrast, overly fast fermentation may contribute to syneresis as described above (Section 14.3). Fermentation time is a function of starter activity as well as fermentation temperature.

The coagulum is heat treated at around 62°C for 120 s to improve whey separation in the subsequent step. In addition, this heat treatment destroys some bacteria and enzymes, thus increasing the shelf life of the final product. The disadvantage of this heat treatment is that lactic acid bacteria from the starter culture are destroyed as well. It is estimated that only 10% of lactic acid bacteria survive the treatment (Lehmann et al. 1991).

14.4.5 Separation of the Whey

Syneresis is required to increase the solid content of the cheese. Following fermentation, the coagulum physically retains moisture within serum pockets. To permit moisture release, it is necessary to break the aggregated protein strands around serum pockets. Breaking the strands release moisture, and the broken strands will reaggregate into a more compact structure, thus encapsulating less moisture (Lucey 2002).

Traditionally, whey separation in cheesemaking is obtained by cutting the curd, cooking, and draining the whey. The automated and continuous whey separation step in fromage frais processing utilizes a centrifuge to mechanically remove the whey. The coagulum is cooled to around 40°C–44°C before entering the centrifuge. Within the centrifuge, curd is fed into the center of the bowl and through the distributor into the rising channels of the disc stack (Lehmann et al. 1991). The centrifuge separates the coagulum into curd ($\delta = 1.05$) and whey ($\delta = 1.02$). Curd centrifuges are specifically developed for viscous products and have capacities from 1000 to 10,000 L/h. The centrifuge is cooled by circulating cold water through the hood. The solid content of the curd is determined by the feed rate as well as the nozzle discharge capacity. It is possible to install different diameter nozzles when changes in solid content are required. Solid nonfat contents range from 13% to 24%, depending on the equipment and the cheese produced. Constant solid content is obtained by having a homogeneous coagulum composition and constant feed rate to the centrifuge. A well-installed centrifuge can keep variations in total solids within ±0.05%. Besides variations in composition, failing to optimize the operation of the centrifuge can lead to defects such as sandy mouthfeel. The whey should be clear and free of curd particles. The whey can be tested by placing a sample in a lab centrifuge and centrifuging at 3500 rpm for 9 min (Lehmann et al. 1991). If suspended particles are present, it is necessary to optimize variables such as temperature, feed rate, pretreatment, and solid content of the coagulum. For nonfat coagula, the whey constitutes the light phase, which moves toward the interior while the curd is pushed outward. High-fat coagula require different centrifuges, because in this case, the light phase is the fat-containing curd, and the heavier whey is pushed outward.

To optimize yields, a UF system can be connected to the whey stream to collect whey proteins, which are then reintroduced into the curd (Darrington 1995). However, the thermoquark/thermosoft method, which denatures whey proteins, is often considered a more efficient process for yield improvement. Another approach for increasing yield is to minimize nonprotein nitrogen by selecting a starter culture with low proteolytic activity.

UF can also be used for concentrating milk solids in place of the curd centrifuge. The UF system can be installed either before or after fermentation. In the former scenario, only retentate is fermented as the concentration step occurs before fermentation. Concentration of cheesemilk up to final total solid content is utilized in the production of other cheeses as well such as feta, camembert, and brie. This process was developed by Maubois et al. in 1969. To obtain the correct calcium balance and, ultimately, a comparable coagulum, it is necessary to slightly preacidify the milk prior to UF. Traditionally, whey contains high calcium content, because the solubilization of the colloidal calcium phosphate occurring during fermentation shifts calcium from the casein fraction to the serum fraction. UF retentate retains the colloidal calcium phosphate, unless the milk is acidified prior to filtration. Furthermore, UF retentate contains all whey proteins, which lead to a weak coagulum unless the proteins are denatured by severe heat treatment (Hinrichs 2001).

As stated above, UF can also be performed following fermentation. In this case, the calcium balance will likely be correct, because the fermentation has caused the solubilization of the colloidal calcium phosphate. When installed after fermentation, the UF system basically replaces the curd centrifuge. It has been stated that this process gives higher yield than the thermoquark method (Nieuwoundt 1997) and provides higher process flexibility, because higher fat coagula can be processed (Ottosen 1996).

14.4.6 Cooling

The incorporation of refrigerated cream has a cooling effect; however, additional cooling is required. Tubular heat exchangers are frequently utilized for cooling because of the gentle effect on product viscosity. Likewise, positive displacement pumps must be utilized at all times following curd separation to assure minimal product abuse.

14.4.7 Fruit Ingredients

The low pH level makes fromage frais sensitive to yeast and mold growth (Rohm et al. 1992). Traditionally, these contaminants were often introduced with the fruit. Dairy processors responded to this challenge by demanding high microbiological standards from fruit suppliers. The risk can be controlled through the addition of sorbate (permitted by the European legislation), which is effective against a wide range of yeast flora (Mihyar et al. 1997). However, processing plants that practice high levels of sanitation, process control, ultraclean packaging rooms, filtered laminar air flow, etc., can produce fromage frais without utilizing sorbate or any other preservative.

14.4.8 Packaging

Fromage frais is packaged in cups ranging in size from 0.03 to 1 kg. The fillers are similar to yogurt fillers and work at high capacity. Packaging materials are often polystyrene or polypropylene, and the cups are either preformed by injection molding or thermoformed prior to filling. As with other dairy products, the current trend is to focus on minimizing packaging materials to promote sustainable practices.

14.5 Skim Milk Requirement

Lehmann et al. (1991) provide a formula for calculating the skim milk requirement for production of 1-kg low-fat fromage frais:

$$S_m = \frac{DM_Q - DM_w}{DM_{SM} - DM_w},$$
(14.2)

where S_m is measured in kilograms of skim milk per 1 kg of low-fat fromage frais, DM_Q is the percentage of dry matter in the fromage frais, DM_w is the percentage of dry matter in the whey, and DM_{SM} is the percentage of dry matter in the skim milk.

14.6 Shelf Life

Typical shelf life is around 24 days [maximum of 30 days to retain the label "frais" (fresh)]. Limiting factors are microbiological deterioration (yeast, mold, and spoilage bacteria), development of bitter off-flavors, whey separation, and grains. Defects and potential remedies are summarized in Table 14.2.

TABLE 14.2

Product Defects and Potential Remedies

Defect	Cause	Remedial Action
Unclean flavors	High bacterial content of raw milk Postpasteurization contamination Excessive levels of potassium sorbate	Improve the quality of raw milk; decrease raw milk storage temperature and storage time Improve cleaning and sanitation Decrease product storage and distribution temperatures Decrease the concentration of potassium sorbate
Yeast and mold contamination	Addition of contaminated fruit products	Improve the quality of fruit products added Add sorbate
Bitter off-flavors	Proteolysis	Improve the quality of raw milk to lower the concentration of bacterial heat-stable proteases Lower the concentration of rennet or change to a different rennet product
Grainy texture	Product pH too close to isoelectric point	Lower the pH of the product
Syneresis	Incorrect gel formation Product abuse during storage and distribution	Decrease the rate of fermentation, increase the protein content, and increase the pasteurization temperature Increase cold storage time before distribution Decrease storage and distribution temperature Decrease the physical movement (shaking or bumping) of products during transportation

14.7 Summary

Fromage frais is a popular cheese because of its nutritional qualities, multiple diversification opportunities, and excellent safety record. Its make-procedure lends itself to automated, high-yield processes. These advantages translate into increased market shares in the European market. It is certainly possible that fromage frais eventually will become a standard item in the U.S. portfolio of dairy products.

REFERENCES

Burkhalter, G. 1981. *Catalogue of Cheeses*, Bulletin International Dairy Federation, Document 141, pp. 3–40. Brussels: International Dairy Federation.

Cachon, R. and Divies, C. 1993. A descriptive model for citrate utilization by *Lactococcus lactis* ssp. *lactis* bv *diacetylactis*. *Biotechnol. Lett.* 15:837–842.

Dalgleish, D. G. and Law, A. J. R. 1988. pH-induced dissociation of bovine casein micelles—Part I: Analysis of liberated caseins. *J. Dairy Res.* 55:529–538.

Dalgleish, D. G. and Law, A. J. R. 1989. pH-induced dissociation of bovine casein micelles—Part II: Mineral solubilization and its relation to casein release. *J. Dairy Res.* 56:727–735.

Darrington, H. 1995. Ten years of food technology. *Int. Food Ingredients* 4:59–60, 62.

Dellaglio, F. 1992. *General Standard of Identity for Fermented Milks*, International IDF Standard 163, pp. 1–4. Brussels: International Dairy Federation.

Fox, P. F., Guinee, T. P., Cogan, T. M., and McSweeney, P. L. 2000. *Fundamentals of Cheese Science*, pp. 363–428. Gaithersburg, MD: Aspen Publishers, Inc.

Fox, P. F., Lucey, J. A., and Cogan, T. M. 1990. Glycolysis and related reactions during cheese manufacture and ripening. CRC—*Crit. Rev. Food Sci. Nutr.* 29:237–253.

Gastaldi, E., Lagaude, A., Marchesseau, S., and Tarodo de la Fuente, B. 1997. Acid milk gel formation as affected by total solids content. *J. Food Sci.* 4:671–675, 687.

Gastaldi, E., Lagaude, A., and Tarodo de la Fuente, B. 1996. Micellar transition state in casein between pH 5.5 and 5.0. *J. Food Sci.* 1:59–64, 68.

Green, M. L. 1989. The formation and structure of milk protein gels. *J. Food Chem.* 6:41–49.

Hache, C., Cachon, R., Wache, Y., Belguendouz, T., Riondet, C., Deraedt, A., and Divies, C. 1999. Influence of lactose–citrate cometabolism on the differences of growth and energetics in *Leuconostoc lactis*, *Leuconostoc mesenteroides* ssp. *mesenteroides* and *Leuconostoc mesenteroides* ssp. *cremoris*. *Syst. Appl. Microbiol.* 22(4):507–513.

Hilliam, M. 1997. Dairy Desserts. *World of Ingredients* 12:10–12.

Hinrichs, J. 2001. Incorporation of whey proteins in cheese. *Int. Dairy J.* 11:495–503.

Lagoueyte, N., Lablee, J., Lagaude, A., and Tarodo de la Fuente, B. 1994. Temperature affects microstructure of renneted milk gel. *J. Food Sci.* 5:956–959.

Lee, B. H. 1996. Bacterial-based processes and products. In: *Fundamentals of Food Biotechnology*, edited by Y.H. Hui, pp. 219–242. New York: VCH.

Lefebvre-Cases, E., Gastaldi, E., Vidal, V., Marchesseau, S., Lagaude, A., Cuq, J. L., and Tarodo de la Fuente, B. 1988. Identification of interactions among casein gels using dissociating chemical agents. *J. Dairy Sci.* 81:932–938.

Lehmann, H. R., Dolle, E., and Bucker, H. 1991. *Processing Lines for the Production of Soft Cheese*, 3rd edition, pp. 26–47. Oelde: Westfalia Separator.

Lucey, J. A. 2002. Formation and physical properties of milk protein gels. *J. Dairy Sci.* 85:281–294.

Maubois, J. L., Mocquot, G., and Vassal, L. 1969. A method for processing milk and dairy products. French Patent No. 2,052,121.

McKay, L. L. 1981. Regulation of lactose metabolism in dairy streptococci. In: *Developments in Food Microbiology*, edited by R. Davies, pp. 153–182. Essex: Applied Science Publishers.

Mihyar, G. F., Yamani, M. I., and Al-sa'ed, A. K. 1997. Resistance of yeast flora of Labaneh to potassium sorbate and sodium benzoate. *J. Dairy Sci.* 80:2304–2309.

Modler, H. W. and Kalab, M. 1983. Microstructure of yogurt stabilized with milk proteins. *J. Dairy Sci.* 66:430–437.

Modler, H. W., Larmond, M. E., Lin, C. S., Froehlich, D., and Emmons, D. B. 1983. Physical and sensory properties of yogurt stabilized with milk proteins. *J. Dairy Sci.* 66:422–429.

Monnet, V., Condon, S., Cogan, T. M., and Gripon, K. C. 1995. Metabolism of starter cultures. In: *Dairy Starter Cultures*, edited by T. M. Cogan and J. P. Accolas, pp. 47–100. New York: VCH Publishers.

Nieuwoundt, J. 1997. Membrane technology. *Food Rev.* 24:19–20, 23.

Ottosen, N. 1996. The use of membranes for the production of fermented cheese. *Bull. Int. Dairy Fed.* 311:18–20.

Pouliot, Y. 2008. Membrane processes in dairy technology—From a simple idea to worldwide panacea. *Int. Dairy J.* 18:735–740.

Rohm, H., Eliskases-Lechner, F., and Bräuer, M. 1992. Diversity of yeasts in selected dairy products. *J. Appl. Microbiol.* 72:370–376.

Sanders, M. E. 1995. Lactococci. In: *Food Biotechnology*, edited by Y. H. Hui and G. G. Khachatourians, pp. 645–664. New York: VCH.

Tranchant, C. C., Dalgleish, D. G., and Hill, A. R. 2001. Different coagulation behavior of bacteriologically acidified and renneted milk: The importance of fine-tuning acid production and rennet action. *Int. Dairy J.* 11:483–494.

Van Hooydonk, A. C. M., Hagedoorn, H. G., and Boerrigter, J. J. 1986. pH-induced physico-chemical changes of casein micelles in milk and their effect on renneting—Part I: Effect of acidification on physico-chemical properties. *Neth. Milk Dairy J.* 40:281–296.

Westfalia Separator, Patent for thermoquark processing. German patent # P 26 36 882.1. Date: 17 August 1976.

15

Cottage Cheese: Fundamentals and Technology

Emiliane Andrade Araújo, Ana Clarissa dos Santos Pires,
Maximiliano Soares Pinto, and Antônio Fernandes de Carvalho

CONTENTS

15.1 Introduction: Colloidal Aspects of Milk and Cheese

Milk is defined as the secretion of the mammary glands of mammals, its primary natural function being the nutrition of the young. The milk of some animals, especially cows, buffaloes, goats, and sheep, is also used for human consumption, either as such or in the form of a range of dairy products (Wasltra et al. 2006).

The most important compounds of milk are water (85.3%–88.7% w/w), lactose (3.8%–5.3% w/w), fat (2.5%–5.5% w/w), protein (2.3%–4.4% w/w), and salts (0.57%–0.83% w/w; Wasltra et al. 2006).

Physicochemically, milk can be defined as a colloidal system where fat globules, proteins, and salts are dispersed in a continuous phase formed by an aqueous solution of lactose and soluble salts, as can be observed in Figure 15.1.

Between the continuous and the disperse phases, there is a region with an excess of Gibbs free energy, called interface. This excess energy occurs as a consequence of the unbalanced intermolecular interaction field between molecules of phases α and β. Excess Gibbs free energy gives rise to various interfacial phenomena such as interfacial tension (γ), wetting, adsorption (Γ), and adhesion. The resulting interfacial properties govern the interactions between colloidal particles and, therewith, the macroscopic behavior and characteristics of a colloidal system, such as its rheological and optical properties and its stability against aggregation (Pires et al. 2009).

Interfacial tension γ can be defined thermodynamically as the increment of Gibbs free energy when reversibly extending the interfacial area by one unit, at constant temperature, pressure, and composition of the system (Norde 2003; Equation 15.1):

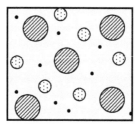

FIGURE 15.1   Schematic representation of the colloidal structure of milk. (◯): Aqueous solution of lactose and salts, (●): fat globules, (◎): casein micelles, and (●): whey proteins.

$$\gamma = \frac{dG}{dA}, \tag{15.1}$$

where γ is the interfacial tension, G is the Gibbs free energy, and A is the area.

In cheese production, proteins play a fundamental role. Milk proteins are divided into two main groups—casein (80%) and whey proteins (20%). Casein micelles consist of water, protein, and salts. Caseins are divided into micelles that are named α_{s1}, α_{s2}, and β-caseins, which precipitate in the presence of calcium and κ-casein, which is soluble in the presence of calcium.

The colloidal aspect of milk is fundamental in cheese science, particularly when related to stability during the protein phase. The stability of casein micelles is kept by hydrophobic interactions between casein fragments (calcium phosphate bound and arranged in nanoclusters) and electrostatic forces mainly due to the hairy layer of κ-casein. In products such as fluid milk, the stability of the colloidal system is essential for product quality. However, in cheese manufacture, it is necessary to disturb the protein phase, by enzyme action or acid addition, for curd production.

Colloidal destabilization of casein micelles by enzyme action occurs, because chymosin hydrolyzes specifically at the Phe_{105}–Met_{106} bond of κ-casein. Therefore, the hydrophilic portion of the protein is released to the whey, and the hydrophobic part, being insoluble in the presence of calcium, forms the initial curd together with the other casein fragments (α_{s1}, α_{s2}, and β; Fox et al. 2000).

Many cheeses such as mozzarella, cheddar, emmental, and parmesan are produced by rennet coagulation. Other cheeses such as cottage cheese are produced by acid coagulation, where there is a tendency toward the disaggregation of the casein micelles.

The loss of casein stability results from the solubilization of the colloidal calcium phosphate, which occurs at pH 5.2–5.3, dissociation of the β-casein from micelles, reduction of the negative surface charge on the casein micelles, and a decrease in casein hydration in the pH range 5.3–4.6. Moreover, there is an increase in the ionic strength of the whey milk due to the increased concentrations of calcium and phosphate ions, which have a shrinking effect on the casein micelles (Fox et al. 2000).

All these conditions promote the precipitation of the casein and, consequently, the curd formation. The process of manufacturing cottage cheese includes the addition of acid or lactic acid produced by bacteria. In the presence of H^+ in the solution, the attractive forces become higher than the repulsive forces, and when the pH reaches the isoelectric point (pH = 4.6), the casein micelles aggregate.

During acidification, the most important change in salt equilibrium is that colloidal calcium and phosphate are released from micelles into the continuous phase, as can be observed by the equilibrium (Equation 15.2):

$$Ca^{2+} + H_2PO_4^- \leftrightarrow CaHPO_4 + H^+. \tag{15.2}$$

The presence of the H^+ ion changes the equilibrium toward the formation of $H_2PO_4^-$, because the interaction between hydrogen and phosphate is more favorable than the interaction between calcium and phosphate.

As caseins aggregate, the interfacial area of such a protein phase reduces, and therefore, the colloidal system becomes more stable through the decrease in Gibbs free energy, as can be seen in Equation 15.1.

In this chapter, the technological steps involved in the cottage cheese manufacturing process are discussed, as well as some perspectives of its application, because it has an enormous potential on the market due to its technological flexibility and health claims.

15.2 Cottage Cheese: Manufacturing Process

The specific origin of cottage cheese is unknown, although many authors state that this cheese originated in Britain, because it is widespread in all Anglo-Saxon countries. Its industrial production began in the United States around 1916, and it became a very popular type of cheese, representing about 8% of the total cheese production in the United States. The name "cottage" is related to its origins as a cheese typical of towns and villages (Farkye 2004). Nowadays, new ways of cottage cheese production have been developed. As a consequence of market competition, the production and manufacture of cheese are currently at the stage of innovative dynamics (Jelinski et al. 2007).

Cottage cheese, quark, cream cheese, fromage frais, and ricotta are commercially the most important types of fresh acid-curd cheese whose consumption has increased because they are perceived as healthy by diet-conscious consumers. In general, the fat content of these cheeses is lower than that of rennet curd cheeses (Fox et al. 2000).

Currently, there are several varieties of cottage cheese that have different consistencies and different fat and moisture contents. The average composition for cottage cheese is about 80% moisture content, fat content in the range of 4%, and 25% fat in dry matter and are, therefore, classified as low-fat cheese (Table 15.1; Araújo 2007). According to U.S. law, the maximum fat is 0.5% for dry curd grains and 4% for cottage cheese with dressing (Farkye 2004).

Cottage cheese is a tasty, nutritious, easily digested, and surprisingly low-calorie food. Moreover, it is a fresh cheese, creamy, slightly acidic, and not ripened. The body has a near-white color and a granular texture consisting of discrete individual soft curd granules of relatively uniform size, from approximately 3 to 12 mm, depending on whether a small or a large type of curd is desired, and possibly covered with a creamy mixture (Souza et al. 2002; Codex Alimentarius Commission 1984).

The manufacture of most cheese varieties involves combining four ingredients—milk, rennet, microorganisms, and salt—which are processed through a number of common steps such as gel formation, whey expulsion, acid production, and salt addition. Variations in ingredient blends and subsequent processing have led to the evolution of all these cheese varieties (Beresford et al. 2001). Figure 15.2 shows a basic flow chart of the manufacturing process of cottage cheese.

15.2.1 Skim and Pasteurization

Milk, the main raw material used in cheese, should be of good quality, that is, obtained from healthy animals. Good manufacturing practices have to be followed to obtain a safe product for consumers. Milk with acidity between 15°Dornic (°D) and 18°D must be pasteurized, and its fat must be reduced up to 0.5% for the production of cottage cheese.

For use in the production of cottage cheese, milk is usually pasteurized to inactivate pathogenic and spoilage bacteria that may be present. During pasteurization, milk is heated from its storage temperature

TABLE 15.1

Physicochemical Composition of Cottage Cheese

Components	Percentage
Moisture	80.0%
Fat	4.0%
Protein	11.5%
Salt	1.0%

Source: Araújo, E. A. et al., *J. Funct. Foods* 2, 85–89, 2010. With permission.

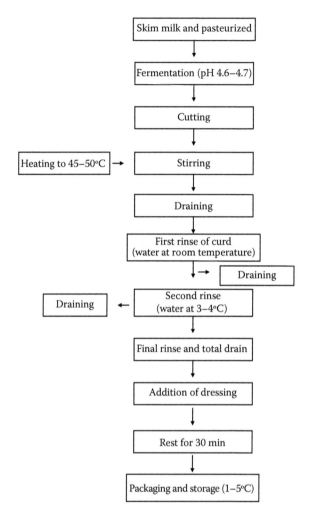

FIGURE 15.2 Processing for cottage cheese manufacture.

of less than 10°C up to 72°C, held at this temperature for 15 s, and then cooled to a temperature for use in cheese production, which, in this case, is about 30°C (Knight et al. 2004).

15.2.2 Fermentation

After the pasteurization process comes milk acidification. Cottage cheese curd can be manufactured by the acid coagulation of skim milk, typically via starter culture activity. Milk may also be directly acidified. Several years ago, it was common for cottage cheese dry curd to be manufactured by direct acidification with food-grade acidulants. Nowadays, starter culture fermentation and direct acidification are both used (Drake et al. 2009).

According to Walstra et al. (2001), three widely different processing methods—short, medium, and long set—are used industrially to produce cottage cheese. The main differences between these involve coagulation temperature and starter concentration. The incubation times are approximately 5, 8, and 12–16 h for short, medium, and long set, respectively. In the short-set method, milk is generally inoculated with ~5% of non-gas-producing mesophilic culture and incubated at 32°C. For the medium- and long-set methods, the usual incubation temperature and starter concentration are 27°C and ~2.75%, and 22°C and ~0.5%, respectively.

Castillo et al. (2006a,c) showed that cottage cheese curds were influenced by the parameters of incubation temperature and inoculum concentration, because higher values of both parameters resulted in more flexible gels.

Coagulation is an important step in cheese manufacture, because it aims at concentrating the milk protein and also at retaining the fat. In cottage cheese, as previously mentioned, there is an acid coagulation step that leads to a loss of casein stability. The bacteria produce lactic acid, which reduces the pH, reaching the isoelectric point of casein. Thus, the electrical charges of colloidal particles are neutralized. The repulsive forces cease, allowing the colloidal particles to come together to form a gel. In addition, the increase in the ionic strength of the milk whey due to the increased concentrations of calcium and phosphate ions also contributes to the aggregation of the casein micelles.

It is a common practice during cottage cheese manufacturing, after the starter addition, to also add a small quantity of rennet to the milk, which does not influence the coagulation process greatly. When rennet is used, the gel obtained is predominantly acid, but the gel's characteristics are modified to some degree by renneting, improving curd firmness and whey separation, and reducing curd fines.

Typical rennet addition levels are 30–60 rennet units (or 0.5–1.0 mL of single-strength rennet) per 100 L. The rennet hydrolyzes some κ-casein, and there are decreases in the negative charge on the micelles and casein dissociation from them. These changes contribute to the enhanced aggregation of the micelles (Fox et al. 2000). The rennet changes curd texture, increases the ability of each curd cub to remain intact, and permits a curd bed to be cut at a slightly higher pH, thus providing a sweeter cheese.

It is important to stress that acidification may also be directly obtained by the addition of acids, resulting in a faster production of curd. Some acids such as phosphoric and hydrochloric acids have become popular for direct use in the milk for cottage cheese containing small amounts of rennet and $CaCl_2$.

15.2.2.1 Starter Microorganisms

The main bacteria used in milk fermentation for cottage cheese manufacture are *Lactococcus lactis* subsp. *lactis* and *Lc. lactis* subsp. *cremoris*, which give cottage cheese its acidic taste. In the manufacturing process, it is also common to add *Leuconostoc mesenteroides* subsp. *cremoris* as a producer of dominant flavor. The metabolism of citrate by the latter results in the production of the flavor compounds diacetyl and acetate, as well as CO_2, which vaporizes upon subsequent heating. Other acids present include formic and acetic acids, whereas the concentrations of propionic and butyric acids are <1 mg/kg (Brocklehurst and Lund 1985).

Certain dairy products such as butter, cultured buttermilk, cottage cheese, and cultured sour cream are fermented with lactic culture and, therefore, usually contain some aromatic compounds such as diacetyl. Flavor is an important feature for foods, mainly for low-fat products, because it is a determinant of food acceptance by consumers (Gonzáles-Tomás et al. 2007; Guichard 2006). The cottage cheese flavor results mainly from the production of volatile compounds during fermentation by starter cultures and nonstarter lactic acid bacteria.

Leuconostoc is a minor component of mesophilic starter cultures in complex L (with *Leuconostoc* strains as flavor cultures) and DL cultures (with *Lc. lactis* subsp. *lactis* biovar. *diacetylactis* strains as flavor cultures). Among the three species present in dairies, *Leu. lactis* plays a minor role, and *Leu. mesenteroides* and *Leu. pseudomesenteroides* constitute the major population (Atamer et al. 2011; Zamfir et al. 2006). These bacteria have significant importance in the fermentation of milk and the production of cottage cheese. The presence of *Leuconostoc* phages in the milk for cottage cheese can be the cause of a decrease in compounds responsible for flavor.

The influence of lactic acid bacteria on cheese flavor has been the subject of studies (Awad et al. 2007; Holland et al. 2005; Gummalla and Broadbent 1999; Kebary et al. 1997). Lactic acid bacteria can influence cheese flavor in different ways, including (1) controlling the growth and composition of contaminant or endogenous microbiota, (2) reducing the pH, (3) contributing to casein hydrolysis, and (4) synthesis of flavor compounds.

For many varieties of cheese, there is hydrolysis of casein by proteolytic enzymes to yield free amino acids and peptides. The liberated amino acids are thought to be flavor precursors and contribute to the background flavor of cheese (Kok 1993). However, cottage cheese is a fresh cheese, and the use of culture does not provide free amino acids that can contribute significantly to cheese flavor the way they do in ripened cheese. In many fermented dairy products, such as yogurt and cottage cheese, the end point of the fermentation process is a decisive factor in product quality.

The use of culture in the cottage cheese manufacturing process is important in order to achieve low curd pH (4.8–4.6). In cheese manufacture, the pH level is an important attribute that affects not only the formation of the curd but also draining and, consequently, moisture and texture of cheese (McSweeney 2007).

15.2.2.2 Probiotic Cottage Cheese

Cottage cheese consumption has been decreasing. The possible reasons for such a reduction are mainly related to the image of cottage cheese, in association with diet and consumption behavior (Reiter 1993). Other issues that have been associated with cottage cheese are inconsistency in product quality, particularly in relation to flavor and texture, and reduced curd-to-dressing ratio (Bodyfelt and Potter 2009).

Much effort has been undertaken to recover cottage cheese consumption, and one of the most successful ones is the addition of prebiotic and probiotic compounds conferring value-added benefits to the product (Ramirez 2007).

Throughout the world, the consumption of probiotic foods has greatly increased in recent years. Probiotics are live microorganisms administered in quantities sufficient to confer health benefits (FAO/WHO 2001). Some effects have been attributed to probiotic bacteria, and they are clinically proven, such as anticarcinogenic and antimutagenic activities, combating infection by *Helicobacter pylori*, treatment of inflammatory bowel disease, prevention and treatment of gastrointestinal disorders, increase in the activity of the immunological system, antimicrobial action, reduction of lactose intolerance, and reduction of blood cholesterol levels (Agrawal 2009; Shah 2007). Other potential benefits of probiotics have been reported as well, such as the benefits for the human skin (Krutman 2009) and protection against colds and flu (Leyer et al. 2009).

Dairy products are the most widely used food carriers for delivering probiotics. A wide range of probiotic dairy products is available in different markets; examples include pasteurized milk, ice cream, fermented milk, cheeses, and infant milk powder. Due to its limited acidity, low oxygen level, and low storage temperature, cheese appears to be a suitable carrier for delivering live probiotic bacteria (Grattepanche et al. 2008). The matrix of the cheese, its buffering capacity, and its fat content may offer protection to cells during their passage through the digestive tract (Stanton et al. 1998). Furthermore, probiotic cheeses often retain their technological properties.

The addition of probiotic cells was tested in some cheeses. Most of these tests were successful in maintaining the viability of microorganisms and the adequate technological properties of the final product (Buriti et al. 2005). Cheeses of different origin have commonly been studied as carriers for probiotic strains—cheddar, gouda, festive, fresh cheese, and cottage cheese (Gomes et al. 1998; Stanton et al. 1998; Blanchette et al. 1996; Buriti et al. 2005).

Cottage cheese shows an adequate profile for incorporating either probiotic cells and/or prebiotic substances. In addition, it is a healthy alternative by virtue of its low-fat content (Araújo et al. 2010). The probiotics *Lactobacillus paracasei* and *Lb. rhamnosus* GG have been used in cottage cheese manufacture. Usually, probiotic bacteria are introduced into cheese as adjunct cultures along with the lactic starter cultures (Tamime et al. 2005). Araújo et al. (2010) produced symbiotic cottage cheese and found that, in addition to exhibiting higher levels than those recommended for beneficial effects, probiotic bacteria in the product showed good survival rates in the acidic conditions of the gastrointestinal tract, suggesting that cheese characteristics protected the cells.

Probiotic bacteria used in food products, such as *Lactobacillus* spp. and *Bifidobacterium* spp., present microaerophile or anaerobic metabolism. Hence, the presence of oxygen may represent a threat for their survival. In general, *Bifidobacterium* spp. is more sensitive to oxygen than *Lb. acidophilus* due to its strict anaerobe nature (Talwalkar and Kailasapathy 2004). On the other hand, cottage cheese shows potential for the use of *Bifidobacterium* spp.; for example, Blanchette et al. (1996) developed a cottage cheese containing *B. infantis*. Cheeses presented populations as high as 1×10^6 CFU g^{-1} during 10 days of refrigerated storage.

Probiotic cells could be added to the dressing, the creamy liquid that surrounds the granules of cheese. Moreover, prebiotic substance can be added to the dressing, making the product a symbiotic food. Production of symbiotic cottage cheese offers great opportunities for the market because of its claim as a functional food, that is, a healthier cheese.

15.2.2.3 Cottage Cheese Shelf Life

Lactic acid bacteria are often used in fermented foods, and fermentation acts to retain and optimize the microbial viability and productivity. During fermentation, several metabolic products appear in foods, including lactic acid, acetic acid, and bacteriocins, and the pH of the product decreases.

Fresh cheese usually exhibits high moisture content, which renders these products susceptible to microbial spoilage and, consequently, short shelf life. Neugebauer and Gilliland (2005) inoculated cells of *Pseudomonas fluorescens* to evaluate the ability of *Lb. delbrueckii* ssp. *lactis* RM2-5 to inhibit the growth of bacteria. Gram-negative psychrotrophic bacteria are the most important spoilage organisms encountered in milk and on equipment used in the manufacture of cottage cheese. The growth of lactic acid bacteria in the cheese is a favorable factor, because they present the ability to produce antimicrobial substances such as hydrogen peroxide at low temperatures. The authors observed that the number of *P. fluorescens* in the control samples increased 2 log cycles, whereas in the sample containing the highest level of *Lb. delbrueckii* ssp. *lactis* RM2-5, their numbers did not change over the 21-day period. Other experiments were carried out, which proved that the lactic acid bacterium studied was effective at controlling Gram-negative bacteria in cottage cheese, retarding its spoilage.

As mentioned above, cottage cheese is highly perishable and a good growth medium for many spoilage organisms. Thus, the U.S. Food and Drug Administration approved the use of microgard in cheese in order to extend the shelf life of the product. Microgard is a bacteriocin-like inhibitory product obtained by the fermentation of skim milk or dextrose by the action of *Propionibacterium shermanii* or specific *Lactococci*, and their antimicrobial compounds are diacetyl, lactic, propionic, and acetic acids (Lehrke et al. 2011).

15.2.3 Cutting and Draining

The curd point is reached when it shows a homogeneous gelatinous consistency and smooth expulsion of whey on the surface. The acidity is about 65°D–70°D, and the pH is in the range of 4.6–4.7.

During cottage cheese manufacture, it is important to correctly determine the optimum cutting time in order to optimize cheese texture, homogeneity, and yield. Cutting time is usually determined by an objective pH measurement. Cutting the curd at the right pH is the most important factor in producing high-quality cottage cheese (Emmons and Beckett 1984).

The cut aims at transforming the curd into uniformly sized grains, allowing an important release of whey. This step must be performed carefully to avoid excessive breakage of the grains. First, the lyre is used horizontally lengthwise, and then, the cut starts with the lyre vertically in both directions. After 15 min of letting the curd rest, its heating slowly begins to expel whey from the interior of the grains, thereby preventing the formation of a film that keeps the whey in the grain and can cause loss of consistency. Heating must increase the temperature in increments of 1°C every 5 min until the temperature reaches 45°C–50°C. The indirect heating step occurs while the curd is being stirred; the stirring speed is very slow at first and gradually increases to avoid excessive crowding of the mass.

Collision of the pieces of gel during concentration forces them into close proximity and thus contributes to further casein aggregation. The moisture content of the curd is inversely related to the degree of aggregation. Factors that enhance casein aggregation reduce the moisture content.

The initial pH of the curd and the intensity of cooking determine the curd retention capacity for cream dressing. Cutting the curd at a high pH, near 4.8, and cooking it to high temperatures for long periods reduce the absorbing properties of curd for cream. Curds cut at pH 4.6–4.5 are softer, are more fragile, and retain more moisture, but they absorb the cream better than curds cut at a higher pH.

Draining is considered one of the most important steps in cottage cheese production due to its influence on the moisture, acidity, and texture of the product. Rearrangement of casein micelles during the syneresis process is responsible for the shrinkage of the casein matrix and the subsequent expulsion of whey from curd pieces. Better control of the syneresis process would result in an improvement of the final cheese product homogeneity, texture, color, and quality (Castillo et al. 2006b). The rate and extent of syneresis depend on the equilibrium between the pressure gradient within the gel network and the resistance to whey expulsion (Walstra et al. 1985).

In cottage cheese production, the cutting time is currently determined on the subjective evaluation of curd texture and on objective pH measurement. Factors such as protein, calcium and enzyme concentration, pH, and temperature showed an important influence on the cutting time. The development of optical sensor technology helps control coagulation and allows the prediction of the cutting time, significantly contributing to cottage cheese quality. In the work of Payne et al. (1998), a fiber-optic probe was used to check the changes in light backscatter during the fermentation stage of cottage cheese.

Once the right consistency of the grains is reached, as determined by a firm and compact grain texture, all stirring is stopped, and the mass is allowed to rest until the syneresis of whey begins. Then, the mass is washed three times in succession—the first washing takes place at room temperature, the other two at 3°C. The water used for washing the curd has to be of good physicochemical and microbiological quality and should be added in an amount corresponding to the volume of drained whey. After the water's addition, the curd is stirred and proceeds to new syneresis.

15.2.4 Dressing

The dressing is composed of salt, cream, and skim milk. The fat content of the mixture should be 16%–18%, and the salt is about 3%. The mixture is pasteurized, cooled to 5°C, and maintained at this temperature until the next day. On the average, about 7% of dressing over the initial volume of milk is added to the cheese mass, and the cheese is left to rest for 30 min before packaging begins. Table 15.2 shows some formulations of cream dressings for cottage cheese.

15.2.5 Packaging

Packaging usually occurs in polyethylene pots with lids. It is important to ensure that the containers were previously sanitized. The product is stored in cold chambers at a temperature of 5°C.

15.2.6 Yield

Cheese yield is high due to the high moisture of the cheese. Cottage cheese uses an average of 4.5 L of milk per 1 kg of product. Yield is influenced by the concentration and condition of casein in milk solids. A total solid of 9.5% in skim milk gives better yield than 9.0% because of the higher casein content. Fragility of the curd before cutting and excessive agitation of the curd reduces cheese yield and results in inferior-quality cheese.

TABLE 15.2

Formulation of Cream Dressings for Cottage Cheese

Simplified Sweet Cream	Fortified-Stabilized Sweet Cream
Whole milk (3.7% fat)	Whole milk (3.7% fat)
Heavy sweet cream (40% fat)	Heavy sweet cream (40% fat)
Salt (NaCl)	Salt (NaCl)
	Skim milk power and stabilizer (<0.5%)
Composition: fat (18%); total solids (25.7%); and water (74.3%)	Composition: fat (17.5%); total solids (24.5%); and water (75%)

15.2.7 Cottage Cheese Defects

Cottage cheese can suffer some defects during production, which may reduce its shelf life. General defects arise from spoilage microorganisms such as coliforms, yeasts, molds, and bacteria psychrotrophs. Gonçalves et al. (2009) developed an antimicrobial sachet containing allyl isothiocyanate used in an active packing system to preserve cottage cheese. They observed that the sachets were very effective against yeast and molds and therefore enhanced the shelf life of cottage cheese to 35 days almost without preservatives in the cheese mass.

High acidity is the most prevalent defect of chemical nature in cottage cheese, with too much acid developing before and during the cooking of the curd or too much whey being retained in the curd. Others reasons that contribute to this defect include high temperature during the incubation of the starter culture, long time of fermentation, lactic culture possessing high activity, or contaminated milk. An acidic flavor can also be due to the fact that the curd has not been sufficiently washed and drained. It is important to emphasize that the water used in all procedures should be of good microbiological and physicochemical quality. Other defects of the cheese can be observed, such as the presence of whey in the product, the presence of gas in the packaging, as well as a bitter flavor, normally caused by the action of spoilage microorganisms.

15.3 Concluding Remarks

Cottage cheese is a soft, unripened, mild acid cheese produced by the acid coagulation of pasteurized skim milk or reconstituted extra low heat skim milk powder. During the twentieth century, cottage cheese developed along with the evolution of dairy science and technology. Changes in its production occurred during the 1970s with the introduction of milk ultrafiltration and in the 1990s when nanofiltration was introduced. These processes have a positive impact on the quality of the curd and the final product by increasing its protein concentration.

A major change in the production process was made by developing direct acidification, where the lactic acid bacteria fermentation is replaced by the addition of lactic acid, glucono-γ-lactone, or other food-grade phosphoric, citric, and hydrochloric acids. Currently, thousands of tons of cottage cheese is produced using this technology.

Although this traditional dairy product is well adapted to the health necessity of the modern population, its consumption has been decreasing over the past years. By developing new production processes, cottage cheese, apart from carrying the nutritional qualities of milk, may also transmit lactic acid bacteria, as well as probiotic microorganisms and prebiotics. The lactic acid bacteria perform more critical functions in cottage cheese than merely producing lactic acid, as they also aid the fabrication process and increase the final rheological and sensorial quality of the cheese. The control of the fermentation process with lactic acid bacteria allows for the enhancement of the sensorial quality of the cheese and could hence play a crucial role in raising the consumption of cottage cheese.

REFERENCES

Agrawal, R. 2009. Probiotics: An emerging food supplement with health benefits. *Food Biotechnol.* 19:227–246.

Araújo, E. A. 2007. Desenvolvimento de queijo tipo cottage simbiótico e análise de sobrevivência do *Lactobacillus delbrueckii* UFV H2b20 em condições de simulação do trato gastro digestório. MS Thesis, Departamento de Tecnologia de Alimentos, Universidade Federal de Viçosa, Viçosa, Brasil.

Araújo, E. A., Carvalho, A. F., Leandro, E. S., Furtado, M. M., and Moraes, C. A. 2010. Development of a symbiotic cottage cheese added with *Lactobacillus delbrueckii* UFV H2b20 and inulin. *J. Funct. Foods* 2:85–89.

Atamer, Z., Ali, Y., Neve, H., Heller, K. J., and Hinrichs, J. 2011. Thermal resistance of bacteriophages attacking flavor-producing dairy *Leuconostoc* starter cultures. *Int. Dairy J.* 21(5):327–334.

Awad, S., Ahmed, N., and El Soda, M. 2007. Evaluation of isolated starter lactic acid bacteria in ras cheese ripening and flavor development. *Food Chem.* 104:1192–1199.

Beresford, T. P., Fitzsimons, N. A., Brennan, N. L., and Cogan, T. M. 2001. Recent advances in cheese microbiology. *Int. Dairy J.* 11:259–274.

Blanchette, L., Roy, D., Belanger, G., and Gauthier, S. F. 1996. Production of cottage cheese using dressing fermented by *Bifidobacteria. J. Dairy Sci.* 79:8–15.

Bodyfelt, F. W. and Potter, D. 2009. Sensory evaluation of creamed cottage cheese. In: *The Sensory Evaluation of Dairy Products*, edited by S. Clark, M. Costello, M. A. Drake, and F. Bodyfelt, pp. 167–190. New York: Van Nostrand Reinhold.

Brocklehurst, T. E. and Lund, B. M. 1985. Microbiological changes in cottage cheese varieties during storage at 7°C. *Food Microbiol.* 2:207–233.

Buriti, F. C. A., Rocha, J. S., Assis, E. G., and Saad, S. M. I. 2005. Probiotic potential of Minas Fresh cheese prepared with the addition of *Lactobacillus paracasei. LWT* 38:173–180.

Castillo, M., Lucey, J. A., and Payne, F. A. 2006a. The effect of temperature and inoculum concentration on rheological and light scatter properties of milk coagulated by a combination of bacterial fermentation and chymosin. Cottage cheese-type gels. *Int. Dairy J.* 16:131–146.

Castillo, M., Lucey, J. A., Wanga, T., and Payne, F. A. 2006b. Effect of temperature and inoculum concentration on gel microstructure, permeability and syneresis kinetics: Cottage cheese–type gels. *Int. Dairy J.* 16:153–163.

Castillo, M., Payne, F. A., Wang, T., and Lucey, J. A. 2006c. Effect of temperature and inoculum concentration on prediction of both gelation time and cutting time: Cottage cheese–type gels. *Int. Dairy J.* 16:147–152.

Codex Alimentarius Commission. 1984. International individual standard for cottage cheese, including creamed cottage cheese, Standard C-16 (1968). *Codex Alimentarius*, Volume XVI. Rome: FAO/WHO.

Drake, S. L., Lopetcharat, K., and Drake, M. A. 2009. Comparison of two methods to explore consumer preferences for cottage cheese. *J. Dairy Sci.* 92:5883–5897.

Emmons, D. B. and Beckett, D. C. 1984. Effect of pH at cutting and during cooking on cottage cheese. *J. Dairy Sci.* 67:2200–2209.

Farkye, N. Y. 2004. Cheese technology. *Int. J. Dairy Technol.* 57:91–98.

Food and Agriculture Organization of the United Nations/World Health Organization (FAO/WHO). 2001. Available from Evaluation of health and nutritional properties of probiotics in food including powder milk with live lactic acid bacteria. Report of a joint FAO/WHO expert consultation, Córdoba, Argentina. Available at: ftp://ftp.fao.org/es/esn/food/probioreport_en.pdf. Accessed Dec. 10, 2009.

Fox, P. F., Guinee, T. P., Cogan, T. M., and McSweeney, P. L. H. 2000. *Fundamentals of Cheese Science*. Gaithersburg, MD: Aspen Publication, Inc.

Gomes, A. M. P., Vieira, M. M., and Malcata, F. X. 1998. Survival the probiotic microbial strains in a cheese matrix during ripening: Simulation of rates of salt diffusion and microorganism survival. *J. Food Eng.* 36:281–301.

Gonçalves, M. P. J. C., Pires, A. C. S., Soares, N. F. F., and Araújo, E. A. 2009. Use of allyl isothiocyanate sachet to preserve cottage cheese. *J. Foodservice* 20(6):275–279.

Gonzáles-Tomás, L., Bayarri, S., Taylor, A. J., and Costell, E. 2007. Flavor release and perception from model dairy custards. *Food Res. Int.* 40:520–528.

Grattepanche, F., Miescher-Schwenninger, S., Meile, L., and Lacroix, C. 2008. Recent developments in cheese cultures with protective and probiotic functionalities. *Dairy Sci. Technol.* 88:421–444.

Guichard, E. 2006. Flavor retention and release from protein solutions. *Biotechnol. Adv.* 24:226–229.

Gummalla, S., and Broadbent, J.R. 1999. Tryptophan catabolism by Lactobacillus casei and Lactobacillus helveticus cheese flavor adjunct. *J. Dairy Sci.* 82(10):2070–2077.

Holland, R., Liu, S.-Q., Crow, V. L., Delabre, M.-L., Lubbers, M., Bennett, M., and Norris, G. 2005. Esterases of lactic acid bacteria and cheese flavor: Milk fat hydrolysis, alcoholysis and esterification. *Int. Dairy J.* 15:711–718.

Jelinski, T., Du, C. J., Sun, D. W., and Fornal, J. 2007. Inspection of the distribution and amount of ingredients in pasteurized cheese by computer vision. *J. Food Eng.* 83:3–9.

Kebary, K. M. K., Salem, O. M., Hamed, A. I., and El-Sisi, A. S. 1997. Flavor enhancement of direct acidified kareish cheese using attenuated lactic acid bacteria. *Food Res. Int.* 30(3–4):265–272.

Knight, G. C., Nicol, R. S., and McMeekin, T. A. 2004. Temperature step changes: A novel approach to control biofilms of *Streptococcus thermophilus* in a pilot plant-scale cheese-milk pasteurization plant. *Int. J. Food Microbiol.* 93:305–318.

Kok, J. 1993. Microbiology of cheese flavor development: Genetics of proteolytic enzymes of Lactococci and their role in cheese flavor development. *J. Dairy Sci.* 76:2056–2064.

Krutman, J. 2009. Pre- and pro-biotics for human skin. *J. Dermatol. Sci.* 54(1):1–5.

Lehrke, G., Hernaez, L., Mugliaroli, S. L., von Staszewski, M., and Jagus, R. L. 2011. Sensitization of *Listeria innocua* to inorganic and organic acids by natural antimicrobials. *LWT–Food Sci. Technol.* 44:984–991.

Leyer, G. J., Li, S., Mubshaer, M. E., Reifer, C., and Ouwehan, A. C. 2009. Probiotics effects on cold and influenza-like incidence and duration in children. *Pediatrics* 124:172–178.

McSweeney, P. L. H. 2007. *Cheese Problems Solved*, 402 pp. Abington, Cambridge, England: Woodhead Publishing Limited.

Neugebauer, K. A. and Gilliland, S. E. 2005. Antagonistic action of *Lactobacillus delbrueckii* ssp. *lactis* RM2-5 toward spoilage organisms in cottage cheese. *J. Dairy Sci.* 88:1335–1341.

Norde, W. 2003. *Colloids and Interfaces in Life Sciences*, p. 430. New York: Marcel Dekker, Inc.

Payne, F. A., Freels, R. C., Nokes, S. E., and Gates, R. S. 1998. Diffuse reflectance changes during the culture of cottage cheese. *Trans. ASAE* 41:709–713.

Pires, A. C. S., Silva, M. C. H., and Silva, L. H. M. 2009. Physical chemistry of colloidal systems applied to food engineering. In: *Engineering Aspects of Milk and Dairy Products*, 1st edition, edited by J. S. R. Coimbra, pp. 1–26. New York: CRC Press, Taylor & Francis Group.

Ramirez, J. C. 2007. Cottage on the cusp. Available at: http://www.dairyfoods.com/Archives_Davinci? article=1169. Accessed on 15 December 2010. *Sciences* 54:1–5.

Reiter, J. 1993. Saving a category—Cottage cheese. http://findarticles.com/p/articles/mi_m3301/is_n3_v94/ ai_14080734. Accessed Dec.15, 2009.

Shah, N. P. 2007. Functional cultures and health benefits. *Int. Dairy J.* 17:1262–1277.

Souza, R. M. B., Rangel, F. F., Rapini, L. S., Penna, C. F. A., Cerqueira, M. M. O. P., and Souza, M. R. 2002. Avaliação de caracterísiticas físico-químicas de queijo cottage e Ricota comercializados em Belo Horizonte (MG). Anais do XIX Congresso Nacional de Laticínios, Vol. 327, pp. 291–294.

Stanton, C., Gardiner, G., Lynch, P. B., Collins, J. K., Fitzgerald, G., and Ross, R. P. 1998. Probiotic cheese. *Int. Dairy J.* 8:491–496.

Talwalkar, A., and Kailasapathy, K. 2004. The role of oxygen in the viability of probiotic bacteria with reference to *L. acidophilus and Bifidobacterium* spp. *Curr. Issues Intest. Microbiol.* 5(1):1–8.

Tamime, A. Y., Saarela, M., Korslund Søndergaard, A., Mistry, V. V., and Shah, N. P. 2005. Production and maintenance of viability of probiotic microorganisms in dairy products. In: *Probiotic Dairy Products*, edited by A. Y. Tamime, pp. 39–72. Oxford: Blackwell Publishing Ltd.

Walstra, P., Geurts, T. J., Noomen, A., Jellema, A., and van Boekel, M. A. J. S. 2001. Ciencia de la leche y tecnología de los productos lácteos. Zaragoza: Acribia S. A.

Walstra, P., Van Dijk, H. J. M., and Geurts, T. J. 1985. The syneresis of curd—Part 1: General considerations and literature review. *Neth. Milk Dairy J.* 39:209–246.

Walstra, P., Wouters, J.T.M., and Geurts, T.J. 2006. *Dairy Science and Technology*, 763 pp. Boca Raton, Florida: CRC Press, Taylor & Francis Group.

Zamfir, M., Vancanneyt, M., Makras, L., Vaningelgem, F., Lefebvre, K., Pot, B., Swings, J., and De Vuyst, L. 2006. Biodiversity of lactic acid bacteria in Romanian dairy products. *Syst. Appl. Microbiol.* 29(6):487–495.

16

Teleme Cheese

E. C. Pappa and G. K. Zerfiridis

CONTENTS

16.1 Introduction

Teleme cheese belongs to the group of white brined cheeses (WBCs), often called pickled cheeses. Common features of this group are that the cheeses are white in color and they are ripened and kept in brine. Brine storage has a determinable effect on the biochemical, textural, and structural changes that occur in the cheeses and leads to the development of their characteristic flavor. Traditionally, their manufacture was limited to the Mediterranean basin and the Balkan countries. However, their production has been extended to several parts of the world as a result of their popularity and increased demand in the international market (Mann 1999). According to Scott (1986), WBC is the oldest known group of cheeses widely spread and produced in many countries and in many varieties. The well-known ones are feta and telemes (Greece); telemea/branza de Braila (Romania); bjalo salamureno sirene/bjalo sirene (Bulgaria); bieno sirenje (FYROM); mohant (Slovenia); sjenicki, homoljski, zlatarski, and svrljiški (Serbia); pljevaljski, polimsko-vasojevaski, and ulcinjski (Montenegro); travnicki/vlasicki (Bosnia-Herzegovina); beyaz peynir and edirne peyniri (Turkey); liqvan and Iranian white (Iran); brinza (Israel); akawi (Lebanon); and domiati and mish (Egypt; Alichanidis and Polychroniadou 2008). There are also WBCs in different countries known with local names (e.g., the "cheese of Chalkidiki" and Gidotyri in Greece, which are WBCs made from goat milk).

Originally, WBCs were made from sheep, goat, or buffalo milk, but nowadays, cow and mixed milk is also used, and the cheeses are produced in modern or automated plants. Teleme cheese is a soft, rind-less cheese salted, matured, and kept in brine; it is acidic and slightly salty. It originates in Romania and named telemea or branza de Braila because of the town of Braila on the Danube (Eekhof-Stork 1976). Teleme was spread to other Balkan countries (Greece, Turkey, and Bulgaria). In Greece, it was introduced around 1906 by the Greek refugees of East Romylia (Zygouris 1952). This cheese is now consumed in Greece as well as in other countries. It is made from more than one type of milk, with different manufacturing technologies. Therefore, its manufacture has to meet different national standards and different consumer habits, affecting remarkably its organoleptic characteristics. For that reason, teleme cheese lacks standard production technology and composition in contrast to feta cheese, which is a standard product with respect to composition and other qualities, and it has been recognized as a Protected Designation of Origin cheese by the European Commission.

It is worth noticing that WBCs are produced in countries where either the climate is warm or the conditions of milk production permit severe contamination by microorganisms and cheese deteriorates before it ripens. Under these conditions, teleme cheese is a very good product in exploiting milk, as the cheese is always in brine, after a short drainage. Other good reasons for using milk in teleme cheesemaking are the acidity and saltiness of the cheese, properties that make up for its good keeping quality. In addition, its organoleptic characteristics fit well to human nutrition in warm climates (Abd El-Salam and Alichanidis 2004; Zerfiridis 2001). This explains the fact that, among the WBCs, feta is the principal cheese consumed in Greece, where someone saying "cheese" always has in mind feta just like an Anglo-Saxon saying "cheese" would mean cheddar. However, feta is originally produced from sheep milk, and this creates a problem for the industry, as it is practically impossible to duplicate the feta flavor obtained from sheep milk when using cow (Zerfiridis 1968) or any other kind of milk. Therefore, the majority of sheep milk production in Greece is used in feta making, and hence, teleme, the main replacement of feta cheese, has to be made primarily from other kinds of milk.

Data mentioned in this chapter refer to the traditional teleme cheese, unless otherwise stated.

16.2 Manufacture of Teleme Cheese

16.2.1 Cheese Milk and Starters

To obtain a good-quality teleme cheese, a good quality of sheep, goat, or cow milk alone or any mixture of them can be used in its production. While feta cheese is made only from pure sheep milk or a mixture of sheep and up to 30% goat milk into the mixture (Greek Codex for Foodstuff and Beverages 2009), there is no such regulation for teleme cheese.

Traditionally, teleme cheese was manufactured as artisanal from nonpasteurized milk in small family premises with rough equipment. However, the use of raw milk leads to unpredictable chemical and biological changes or the possible survival of various pathogens during manufacture and ripening. Nowadays, pasteurized milk is used instead of raw milk in industrial scale, and rigorous control of the production and maturation process is essential, making the use of lactic acid bacteria as starter cultures necessary. Oftentimes, a 24-h-old yogurt of pH 4.2 is used as a starter culture in small cheese enterprises. However, the use of yogurt does not give satisfactory results regarding the rate of acid production during the first hours of cheese manufacture, which is essential for the syneresis and drainage of the curd as well as the prevention of undesirable microorganism development (Zerfiridis 2001). A blend of mesophilic starter culture (e.g., *Lactococcus lactis* subsp. *lactis* and *Lc. lactis* subsp. *cremoris*) or a mixture of thermophilic and mesophilic starter cultures (e.g., *Lc. lactis* subsp. *cremoris*, *Lc. lactis* subsp. *lactis*, *Lactobacillus delbrueckii* subsp. *bulgaricus*, and *Streptococcus thermophilus*) is widely used (Anifantakis 1991, 2004). The use of mesophilic or mixed culture (thermophilic and mesophilic) results in teleme cheese showing a faster rate of curd acidification, whereas the use of thermophilic culture results in cheeses of slower acidification rate (Pappa et al. 2006a). Figure 16.1 shows the acid development in curd when using different combinations of microorganisms. This figure shows that thermophilic culture causes a rate of pH decrease in curd slower than both the mixed culture and the mesophilic culture, regardless of the kind of milk used, whereas the pH in the two latter cultures did not differ significantly. One can choose the starter from a wide range of lactic acid bacteria that fit better the circumstances prevailing in the cheese plant, as long as none of the starter bacteria is producing gas and the rate of pH drop is fast at the early stages of draining. Usually, a quantity of 0.5% is added and left for ripening into cheese milk for 20–30 min before renneting (Anifantakis 1991; Veinoglou et al. 1980; Zerfiridis 2001).

The milk is usually filtered and standardized. The casein-to-fat ratio has a significant effect on cheese yield. Usually, the ratio of 0.75:1 for ewe milk, 0.64:1 for goat milk, and 0.73–0.80:1 for cow milk is used (Mallatou et al. 2003; Pappa et al. 2006a; Zerfiridis 2001). The milk is pasteurized at 72°C/15 s or

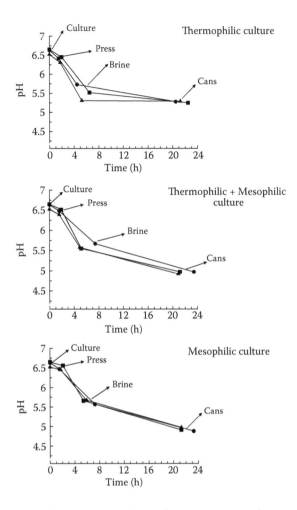

FIGURE 16.1 Effect of the kind of culture on the pH of curd of teleme cheese made from ewe milk (–●–), goat milk (–▲–), or cow milk (–■–) during the first 24 h. (Reprinted from *Food Control*, 17, Pappa, E. C. et al., 570–581, Copyright 2006, with permission from Elsevier.)

thermized usually at 63°C/20 min and then cooled at 32°C before the addition of starter culture. A 40% solution of CaCl₂ is added at 200–300 mL/ton of milk to ensure better coagulation and quality of the curd. Chlorophyll solution can be added according to the manufacturer's instructions, making the cheese white, compensating for the carotene content of cow milk, especially when cows are fed with abundant green grass. Other whitening treatments may also be employed. In addition, lipases can be added to the milk to enhance the taste of the cheese (Anifantakis 1991, 2004; Zerfiridis 2001).

16.2.2 Coagulation, Cutting, and Draining

Usually, rennet is added 20–30 min after inoculating the milk with the starter cultures. The temperature of coagulation is 32°C for fresh milk; however, when the cheese milk has developed some acidity or the weather is hot, coagulation may take place at as low as 25°C (Anifantakis 1991; Zerfiridis 2001; Zygouris 1952). The amount of rennet added should be sufficient to clot the milk and provide good curd firmness ready for cutting after 50–60 min. However, when the weather is hot or the milk is high in acid, the coagulation should take place in a shorter time. The rennet used can be in liquid or powder form. When powder rennet is used, it should be dissolved in cold water, a small quantity of salt may be added, and the solution should be used immediately. When liquid rennet is used, it is added to water at a ratio of 1:8 and then slowly added to milk under gentle stirring (Zerfiridis 2001).

The curd, after coagulation, is cut crosswise into cubes of ~2 cm, allowed to rest for 10 min, and then transferred to rectangular bottomless molds that are dressed with cheese cloths. The molds can be wooden or stainless steel (side: 40–60 cm; height: ~20 cm). A weight is placed on top of the curd in the mold to assist draining, usually equal to the curd weight or even more (Zygouris 1952). Pressure is applied at room temperature (19°C–20°C). When no more whey is expelled from the curd and the height of the curd in the mold is about 9–11 cm, normally in the same afternoon (Anifantakis 1991; Zerfiridis 2001; Zerfiridis et al. 1989), the cheese is taken out of the mold, the cloth is removed, and the cheese mass is cut into equal blocks of dimensions 11 × 11 × 8 cm (~1-kg weight each) or 7 × 7 × 6 cm (~0.25-kg weight each), depending on the dimensions of the final container for packaging and storage.

16.2.3 Salting, Packing, and Storage

The cheese blocks are immersed in brine solution (density 16–18°Be) side by side to retain their shape; coarse salt is sprinkled on the surface to allow for slow penetration of the salt into the cheese and left in ripening rooms at ~18°C. As whey is continuously expelled from the cheese, the surface of the cheese should be resalted after a few hours (Alichanidis et al. 1981; Anifantakis 1991; Veinoglou et al. 1980; Zerfiridis 2001; Zygouris 1952).

Curd acidification at the appropriate rate and time is the key step in the manufacture of good-quality cheese. For that reason, the pH of the cheese the following morning should be below 5, the salt-in-moisture content should be above 2.5%, and the moisture content should be 62%–65% (Zerfiridis 2001). At the end of brine salting by the next morning, the blocks are transferred into open cans in three layers, four cheese blocks per layer, and are again salted *in situ* with grain-size salt. Some salt is spread on the bottom of the can, and the rest is equally spread between the layers and the top of the cheese. In a short time, whey is expelled, and the cheese blocks are completely submerged in whey brine, thus avoiding mold growth on the surface. The cans are then loosely covered by their lids, and the cheese may stay there for up to 3 weeks (prematuring time) at 18°C. After this period, cheese blocks are transferred to their final cans, filled with brine (7%–8% NaCl), closed by their lids, or sealed and transferred to cold storage (Anifantakis 1991; Zerfiridis 2001). However, in industries, after 2–4 days from the manufacture, cheeses are packed into the final tin-plated cans, and the lids are sealed. This practice may save cost, but the cheese develops defects and lacks flavor. Figure 16.2 shows blocks of teleme cheese, in dimensions of 11 × 11 × 8 cm, kept in tin cans, four in each of the four rows to fill up the can. Several other modern ways of packaging are acceptable, as long as the cheese is in brine.

When both the pH and moisture of the cheese decrease to ≤4.6 and ≤56%, respectively, and the salt-in-moisture content is over 5% (Zerfiridis 2001), the tin cans are transferred to the cold room (3°C–5°C) for further ripening and storage. Cheese is ready for the market 2 months after making (Greek Codex for Foodstuff and Beverages 2009), but the flavor is enhanced when it is kept longer.

FIGURE 16.2 Teleme cheese. (Reprinted with permission from Anifantakis, E. M., *Greek Cheeses—A Tradition of Centuries*, pp. 27–42, National Dairy Committee of Greece, Athens, 1991.)

Teleme cheese, being always in brine, has limited chances of spoilage, as it is not exposed to surface bacterial and mold off-flavor development. From this point of view, teleme cheese is an easy and secure WBC to be made even by amateur cheesemakers.

WBCs have been manufactured using the teleme cheese technology from either ultrafiltrated (Antoniou et al. 1995; Raphaelides and Antoniou 1996; Raphaelides et al. 1995; Veinoglou and Boyazoglu 1982; Veinoglou et al. 1977) or low-fat (Romeih et al. 2002) milk. The same is also true when deep-frozen curd (Alichanidis et al. 1981) and refrigerated stored milk (Kalogridou-Vassiliadou and Alichanidis 1984) are used. The WBCs obtained in these cases were different with respect to many features of the traditional teleme cheese, and therefore, comments on them are omitted.

16.3 Ripening

16.3.1 Compositional Characteristics

According to the Greek Codex for Foodstuff and Beverages (2009), the first quality of teleme cheese has a maximum moisture content of 56% and a minimum fat-in-dry matter of 43%. The physicochemical composition (pH, moisture content, ash, protein, salt-in-moisture, fat-in-dry matter, and the yield on a basis of 56% moisture) of mature (60–180 days) teleme cheese is shown in Tables 16.1 and 16.2. More specifically, the moisture of the market cheese varies widely. In most cases, it is 55%–56%; however, it is sometimes over the mandatory limit of 56%. The use of different kinds of milk also has been found to affect the physicochemical characteristics of teleme cheese; the highest and the lowest moisture contents were obtained when cow milk and goat milk were used in cheesemaking, respectively (Mallatou and Pappa 2005; Pappa et al. 2006a). Fat-in-dry matter content is 46%–56%, which fulfills well the mandatory limit of 43%. These differences may be attributed to the cheese milk composition and the technology used in cheesemaking. The protein content has a wide variation between 12.9% and 18.6% due to the moisture variation and loss of soluble protein in the brine. The salt-in-moisture content is 4.3%–6.3%; however, 5% is the low limit to keep the cheese from being spoiled (Davis 1965). The ash content, being affected by the salt content, is in the range of 3.4%–4.6%. Generally, the highest ash content is observed in cheeses made from ewe milk (Mallatou and Pappa 2005). The yield, on a basis of 56% moisture, differs significantly with respect to the kind of milk used. In mature cheese made from ewe, goat, and cow milk, the yield is 25.25%–27.04%, 15.44%–17.95%, and 13.30%–15.11%, respectively (Mallatou and Pappa 2005; Pappa et al. 2006a; Veinoglou et al. 1980). This is due to the large differences in the composition of the kind of milk used in cheesemaking, particularly in protein, and to the standardization of milk with respect to the casein-to-fat ratio. The optimum pH for teleme cheese is 4.5–4.6, but a variation of ±0.2 that occurs is acceptable.

Stef et al. (2009) studied the quality of Telemea cheese from different producers and the kinds of milk within the Banat region in Romania. They concluded that, according to the Romanian standards, the quality was rather poor (Table 16.1). Specifically, they found that the range of moisture content was 52.4%–54.9% (with a maximum value of 55%), whereas that of protein and fat-in-dry matter content was 15.8%–17.2% (minimum 16%) and 46.6%–53.4% (minimum 50%), respectively.

Other data on the chemical composition of mature teleme cheese concern lactose content (about 1%; Manolkidis et al. 1970a) and cholesterol content. The mean values of cholesterol/fat ratio (in milligrams per gram) in 60-day-old teleme cheese made from different types of milk ranged from 2.46 to 3.31 (Mallatou and Pappa 2005). Similar results were reported by Andrikopoulos et al. (2003) for commercial teleme cheese samples from the Greek market. Minerals that have been reported per 100 g of 60-day-old teleme cheese are K (53.50–77.35 mg), Mg (6.25–20.30 mg), Zn (0.84–1.87 mg), Mn (2.50–18.15 µg), Cu (57.40–72.50 µg), and Fe (175–334 µg), depending on the kind of milk used (Mallatou and Pappa 2005). Research on market teleme cheese for Ca, P, K, and Na (mg/100 g) gives an average content of 528.27, 396.02, 90.58, and 1064.70, respectively (Vafopoulou-Mastroyannaki 1977).

The composition of teleme cheese varies because of the kind of milk used for its manufacture, the lack of fixed standardization of the casein-to-fat ratio of the cheese milk (Abd El-Salam and Alichanidis 2004; Anifantakis 1991), and the lack of standard procedures for the manufacture of this cheese. Several factors have studied to improve the manufacture and quality of teleme cheese. It is now well understood that

TABLE 16.1

Reported Values of the Main Components of Mature Teleme Cheese

Fat-in-Dry Matter (%)		Moisture (%)		Ash (%)		Proteins (%; $N \times 6.38$)		Kind of Milk	Reference
60 days	180 days	60 days	180 days	60 days	180 days	60 days	180 days		
52.3						14.74		C	(1)
51.1		52.47		3.59		18.63		S	(1)
45.6–48.4	48.0–50.8	53.62				16.97		90% S, 10% G	(2)
		55.00				16.21		C	(3)
49.5	50.9	50.30	50.50	3.69			18.42	90% S, 10% G	(4)
43.4		53.79				15.25		C	(5)
			50.50–52.77		3.61			C	(6)
45.3		57.25				16.56		C	(7)
								C	(8)
54.6 (47.3–63.8)		56.20 (54.10–60.20)		3.4 (3.2–3.7)		15.50 (14.60–16.20)		a	(9)
46.1	48.2	56.63	56.40	4.42	4.56	16.14	15.26	S	(10)
56.2	55.1	53.42	54.01	3.65	3.84	16.82	15.25	G	(10)
51.3	51.8	55.77	56.52	4.26	4.38	16.65	15.26	50% S, 50% G	(10)
54.0	55.8	54.34	54.89	3.90	4.09	16.74	15.03	C	(10)
54.4–55.4	52.5–54.9	54.69–56.33	55.69–56.44			14.43–15.58	14.20–15.43	S	(11)
47.0–52.0	48.9–49.9	55.77–56.75	56.34–56.64			15.70–16.31	15.12–15.65	G	(11)
48.2–50.2	49.4–52.9	57.73–57.76	56.74–57.69			14.61–15.06	12.91–14.65	C	(11)
46.6–53.4		52.4–54.9				15.8–17.2		b	(12)

Note: S, sheep; G, goat; and C, cow. (1) Veinoglou et al. 1977; (2) Polychroniadou and Vlachos 1979; (3) Veinoglou et al. 1980; (4) Alichanidis et al. 1981; (5) Veinoglou and Boyazoglu 1982; (6) Kalogridou-Vassiliadou and Alichanidis 1984; (7) Zerfiridis et al. 1989; (8) Raphaelides et al. 1995; (9) Andrikopoulos et al. 2003; (10) Mallatou and Pappa 2005; (11) Pappa et al. 2006a; and (12) Stef et al. 2009.

a　Commercial cheese samples. The cheese milk used was cow milk or a mixture of a little sheep or goat milk. The values in parentheses show the range of minimum–maximum values.

b　Commercial cheese samples from different regions of manufacture. The type of milk used was cow or sheep milk.

TABLE 16.2

Reported Values of the Physicochemical Characteristics of Mature Teleme Cheese

pH		Salt-in-Moisture (%)		Yield in 56% Moisture (%)		Kind of	
60 days	180 days	60 days	180 days	60 days	180 days	Milk	Reference
4.60				14.90		C	(1)
4.65		5.60		26.39		S	(1)
4.81						90% S, 10% G	(2)
4.52–4.84	4.45–4.73				13.30–15.11	C	(3)
4.80						90% S, 10% G	(4)
4.20	4.18	4.79	5.42			C	(5)
4.81						C	(6)
	4.40–4.60					C	(7)
4.58		5.98				C	(8)
		5.19	5.68			S	(9)
		4.27 (3.88–4.49)				a	(10)
4.45	4.40	6.28	6.25	27.04	26.80	S	(11)
4.39	4.44	6.17	6.26	17.95	17.41	G	(11)
4.52	4.52	6.14	6.34	23.02	22.20	50% S, 50% G	(11)
4.51	4.49	6.11	6.13	13.69	13.32	C	(11)
4.23–4.25	4.16–4.25	4.75–5.34	5.01–5.33	26.41–26.86	25.25–25.71	S	(12)
4.32–4.38	4.30–4.35	5.03–5.08	5.09–5.39	15.44–16.77	15.81–16.22	G	(12)
4.20–4.33	4.22–4.28	4.49–4.81	5.00–5.15	13.60–14.30	13.51–14.71	C	(12)

Note: S, sheep; G, goat; and C, cow. (1) Veinoglou et al. 1977; (2) Polychroniadou and Vlachos 1979; (3) Veinoglou et al. 1980; (4) Alichanidis et al. 1981; (5) Veinoglou and Boyazoglu 1982; (6) Kalogridou-Vassiliadou and Alichanidis 1984; (7) Zerfiridis et al. 1989; (8) Raphaelides et al. 1995; (9) Tzanetakis and Litopoulou-Tzanetaki 1992; (10) Andrikopoulos et al. 2003; (11) Mallatou and Pappa 2005; and (12) Pappa et al. 2006a.

[a] Commercial cheese samples. The cheese milk used was cow milk or a mixture of a little sheep or goat milk. The values in parentheses show the range of minimum–maximum values.

the pH, which is crucial for teleme cheese, is significantly affected by the kind of starter culture used—a fact that has been extensively studied (Pappa et al. 2006a; Tzanetakis et al. 1991; Zerfiridis et al. 1989).

16.3.2 Microbiological Characteristics

Teleme cheese is nowadays made from pasteurized milk with added starter cultures. Yogurt (at a level of 0.5%) was the traditional starter culture for the production of teleme cheese. Currently, thermophilic, mesophilic, or a mixture of thermophilic–mesophilic starter cultures are widely used. Therefore, lactic acid bacteria constitute the main flora of teleme cheese at the early stages of ripening, attaining the maxima counts of about 8 log cfu/mL in 60 days (Tzanetakis and Litopoulou-Tzanetaki 1992). Generally, when cheeses are in the warm room ripening, lactic acid bacteria reach their highest counts and then decline during cold room ripening. During the warm room ripening, nonstarter lactic acid bacteria increase and remain constant, as they have the ability to grow at low pH values (Manolopoulou et al. 2003; Tzanetakis and Litopoulou-Tzanetaki 1992). Therefore, lactococci (*Lc. lactis* ssp. *lactis*) are present only in the early stages of teleme cheese, probably because of the inhibitory effect of low pH and high salt-in-moisture content in the mature cheese. However, the low pH of the cheese, as a result of acid production by cheese microorganisms and the high salt content, favors the growth of lactobacilli, which predominate over the other lactic acid bacteria (Litopoulou-Tzanetaki et al. 1992; Tzanetakis and Litopoulou-Tzanetaki 1992). The lactobacilli belong to nonstarter lactic acid bacteria and play an important role in cheese ripening, contribute to an increase in peptidase activity, enhance the production of peptides and amino acids, and affect the characteristic flavor of the cheese. Nonstarter lactic acid bacteria can originate in indigenous milk flora and the environment of the cheese plant (Peterson et al. 1990). *Lb. plantarum*, which is resistant to NaCl, predominates throughout the maturation period of teleme cheese from ewe milk, representing about 65% of the isolates. *Lb. paracasei*, *Lb. casei*, and *Lb. brevis* have also been identified in teleme cheese made from ewe milk (Tzanetakis and Litopoulou-Tzanetaki 1992).

Leuconostocs and enterococci are present in lower numbers. Leuconostocs are more frequently encountered at the beginning of the cheese ripening, with the NaCl-resistant species *Leuconostoc paramesenteroides* being the most predominant. Isolates of enterococci were identified as *Enterococcus faecium* (Litopoulou-Tzanetaki et al. 1992; Tzanetakis and Litopoulou-Tzanetaki 1992). The presence of staphylococci, micrococci, yeasts, phychrotrophs, and coliforms in teleme cheese is also reported in the literature (Alichanidis et al. 1981; Manolkidis et al. 1975; Tzanetakis et al. 1991; Veinoglou et al. 1980; Zerfiridis et al. 1989).

16.3.3 Lipolysis and Formation of Volatile Compounds

Lipolysis in teleme cheese is not very extensive. Free fatty acids (FFAs) are released usually by the action of lipases (from different sources) on fat during ripening and contribute directly to the cheese flavor, particularly when they are properly balanced with products of proteolysis and other reactions (Fox et al. 1995). The level of total FFAs ranges from 100 to 250 mg/100 g of cheese (Abd El-Salam and Alichanidis 2004; Alichanidis 1981; Anifantakis and Moatsou 2006). Nevertheless, acetic acid, the principal volatile FFA in teleme cheese, is not produced by lipolysis but originates mainly in the fermentation of lactate during the early stages of ripening and, to a limited extent, in amino acids and contributes greatly to the final flavor of teleme cheese (Abd El-Salam et al. 1993). The level of acetic acid is a characteristic feature of WBCs giving a harsh but not rancid flavor to the cheese and ranges from 55% to 86% of the total volatile acids (Alichanidis et al. 1981; Buruiana and El-Senaity 1986; Efthymiou 1967; Mallatou et al. 2003). Mallatou et al. (2003) found the percentage of acetic acid and short-chain FFAs (C_2–C_8) of mature (60–120 days) teleme cheese to be from 40.1 to 50.7 mg/100 g. The percentage of short-chain carboxylic acids has a significant impact on the development of characteristic aroma and flavor of teleme cheese. Other major FFAs without any special impact on flavor are palmitic and oleic acids (Alichanidis 1981; Efthymiou 1967; Mallatou et al. 2003).

Generally, the concentration of volatile fatty acids is higher in teleme cheese made from cow milk than that made from ewe or goat milk, the lowest concentration being in the cheese made from goat milk (Mallatou et al. 2003; Massouras et al. 2006). However, the results of the organoleptic examination of teleme cheeses made from the three kinds of milk showed no statistical difference in cheese flavor ($P > 0.05$; Mallatou et al. 2003).

Data on other volatile compounds are limited. Headspace analysis of teleme cheese reveals that the main groups of compounds found in the volatile fraction of teleme cheese are aldehydes, ketones, alcohols, and fatty acids, including large quantities of acetaldehyde, ethanol, 2-butanol, and 2-butanone (Massouras et al. 2006).

16.3.4 Proteolysis

Proteolysis is an important biochemical process in the ripening of most cheese varieties. The total nitrogen (TN) values of teleme cheese show a slight decrease during ripening due to the diffusion of the hydrolytic products of casein and the whey proteins from the cheese into the brine, whereas soluble nitrogen (SN) values [as estimated by the fractionation of nitrogen substances with various solvents such as water, trichloroacetic acid (TCA), or phosphotungstic acid (PTA)] increase throughout aging (Alichanidis et al. 1981; Mallatou et al. 2004; Mallatou and Pappa 2005; Manolkidis et al. 1970b; Pappa et al. 2006b; Raphaelides et al. 1995). Obviously, protein breakdown is high enough to cover up for the increase in soluble protein and some loss of it in the brine throughout ripening. This fact is not closely controllable, and therefore, the water-soluble fraction in teleme cheese as well as in other WBCs is not always a reliable maturation index of the cheese. The greater part of nitrogen fraction changes occur during the warm room ripening.

Some reported values of proteolysis in mature (over 60 days) teleme cheese are shown in Table 16.3. Thus, in a 2-month-old teleme cheese, 9.1%–16.1% of TN is in the form of water-soluble nitrogen (WSN),

TABLE 16.3

Reported Values of Proteolysis in Mature Teleme Cheese

WSN %TN			TCA-SN %TN			PTA-SN %TN			Kind of Milk	Reference
60 days	120 days	180 days	60 days	120 days	180 days	60 days	120 days	180 days		
		13.80–25.01	9.00	13.50					90% S, 10% G	(1)
			8.88	8.97					C	(2)
									C	(3)
				16.51			6.03		C	(4)
14.03	18.40		10.60	15.50					C	(5)
14.07		14.87	9.13		11.30	1.91		2.23	S	(6)
8.99		10.29	6.86		7.60	1.59		2.28	G	(6)
12.14		14.47	9.40		10.61	1.87		2.30	50% S, 50% G	(6)
10.19–16.08	11.28–17.19	11.74–19.02	6.88–8.18	8.07–8.74	8.51–11.92	1.76–2.14	1.82–2.04	2.14–2.69	S	(7)
7.41–10.19	7.76–11.71	9.00–13.00	5.81–8.43	5.90–10.05	6.35–10.32	1.62–2.11	1.57–1.99	1.95–2.86	G	(7)
9.06–10.11	9.04–11.75	10.15–17.10	6.73–7.04	6.05–7.89	7.31–10.31	1.81–2.00	1.87–2.20	2.28–2.94	C	(7)

Note: WSN %TN, water-soluble nitrogen, expressed as a percentage of the total nitrogen. TCA-SN %TN, nitrogen soluble in 12% trichloroacetic acid (medium- and small-size peptides and amino acids), expressed as a percentage of the total nitrogen; and PTA-SN %TN, nitrogen soluble in 5% phosphotungstic acid (small-size peptides and amino acids), expressed as a percentage of the total nitrogen. S, sheep; G, goat; and C, cow. (1) Alichanidis et al. 1981; (2) Kalogridou-Vassiliadou and Alichanidis 1984; (3) Zerfiridis et al. 1989; (4) Tzanetakis et al. 1991; (5) Raphaelides et al. 1995; (6) Mallatou et al. 2004; and (7) Pappa et al. 2006b.

5.8%–10.6% is soluble in 12% TCA (TCA-SN, i.e., peptide and amino acid nitrogen), and 1.6%–2.14% is soluble in 5% PTA (PTA-SN, i.e., amino acid nitrogen). The nitrogen fractions are affected greatly by the type of starter culture (Pappa et al. 2006b; Zerfiridis et al. 1989) and by the kind of milk used in cheesemaking (Mallatou et al. 2004; Pappa et al. 2006b).

Proteolysis in teleme cheese follows the general rules of cheese proteolysis, more or less, with some quantitative and qualitative differences. The degradation of the caseins of teleme cheese made from cow milk during ripening is shown in Figure 16.3. Both α_{s1}- and β-caseins decrease over the course of ripening.

Electrophoretic studies show that 11%–51% of α_{s1}-casein and 5%–19% of β-casein were degraded in a 60-day-old teleme cheese, 15%–74% of α_{s1}-casein and 9%–25% of β-casein in a 120-day-old teleme cheese, and 21%–66% of α_{s1}-casein and 15%–30% of β-casein in a 180-day-old cheese, depending on the type of milk used in cheesemaking (Alichanidis et al. 1981; Mallatou et al. 2004; Pappa et al. 2006b). Degradation of caseins was the lowest in goat teleme cheese and the highest in cow teleme cheese but ranged in intermediate levels in sheep teleme cheese (Mallatou et al. 2004; Pappa et al. 2006b). Polyacrylamide gel electrophoresis (PAGE) reveals that some of the peptides produced show electrophoretic mobility higher than that of α_{s1}-caseins and some are of lower mobility than that of β-casein or of intermediate mobility between those of α_{s1}- and β-caseins (Figure 16.3). The hydrolysis products showing mobility higher than that of α_{s1}-CN are produced from the action of the coagulant enzyme chymosin on α_{s1}-casein, whereas the products showing mobility lower than that of β-casein are produced from the action of indigenous milk enzymes (plasmin) on β-casein (γ-caseins; Abd El-Salam and Alichanidis 2004; McSweeney et al. 1993). However, the activity of plasmin is relatively low due to low pH and high NaCl content (Abd El-Salam and Alichanidis 2004). Generally, factors that increase proteolysis in teleme cheese include the pH value (around 4.5) of the cheese that favors the proteolytic activity of chymosin, the residual rennet that is retained at higher levels because of the low pH during draining and the high moisture content of the curd (De Roos et al. 1998; Samal et al. 1993; Van den Berg and Exterkate 1993), and the lack of curd cooking in the teleme cheese manufacture (Alichanidis and Polychroniadou 2008).

Proteinases and peptidases released from the starter and nonstarter lactic acid bacteria are responsible for the production of medium- and small-size peptides and free amino acids during ripening. Figure 16.4 shows the chromatograms [reversed phase high-performance liquid chromatography (RP-HPLC)] of the water-soluble extract of teleme cheese made from goat milk, at different ripening days. Studies by Mallatou et al. (2004) and Pappa et al. (2006b) show that a large number of peaks exist, indicating a heterogeneous mixture of proteolysis products in the water-soluble extracts. In addition, as the age of teleme cheese increases, new peaks appear, whereas peaks that existed at the initial stage of ripening increase or decrease in size (Figure 16.4). Generally, peptides eluted in the front region of RP-HPLC profiles are mainly hydrophilic, with a molecular weight of <3000 Da (Belitz and Kaiser 1993; Kaiser et al. 1992). Hydrophobic peptides are eluted mainly in the rear region of chromatograms, and large-size peptides are generally eluted later than those with low molecular weight. Nevertheless, differences may exist in

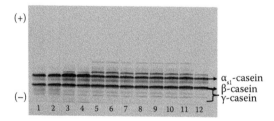

FIGURE 16.3 Electrophoretogram on urea–PAGE of teleme cheese made from cow milk during ripening. Lanes 1 and 2: total cow casein; lanes 3 and 4: 1-day-old cheese; lanes 5 and 6: 15-day-old cheese (before transferring to the cold room); lanes 7 and 8: 60-day-old cheese; lanes 9 and 10: 120-day-old cheese; and lanes 11 and 12: 180-day-old cheese. (From Pappa, E. C., Study of the physicochemical characteristics of Teleme cheese made from different types of milk and culture. PhD Thesis. Agricultural University of Athens, Greece (in Greek), 2003.)

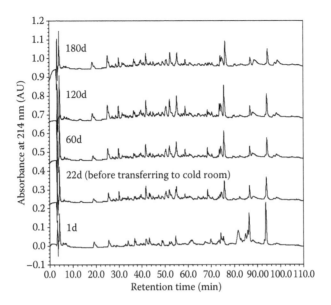

FIGURE 16.4 RP-HPLC chromatograms of the water-soluble extract of teleme cheese made from goat milk, at different ripening days. (From Pappa, E. C., Study of the physicochemical characteristics of Teleme cheese made from different types of milk and culture. PhD Thesis. Agricultural University of Athens, Greece (in Greek), 2003.)

the retention time of medium- and large-size peptides with the same amino acid composition but slightly different sequences due to conformation effects (Lee and Warthesen 1996; Polo et al. 1992). During the ripening of teleme cheese, the ratio of hydrophobic to hydrophilic peptides decreases (Mallatou et al. 2004; Pappa et al. 2006b). This decrease can be attributed to the degradation of water-soluble hydrophobic peptides and the formation of hydrophilic peptides (Cliffe et al. 1989; Engels and Visser 1994) as well as to highly hydrophobic peptides that are no longer water soluble (Lau et al. 1991). Another reason contributing to the decrease of hydrophobic area is the diffusion of whey proteins into the brine as these proteins are eluted at the rear region of the chromatogram (Michaelidou et al. 1998, 2005).

The range of the total free amino acid content is 424–590 (mg/100 g dry matter of cheese) in 60-day-old teleme cheese and 492–1030 in 120-day-old teleme cheese, that is, the total free amino acid content of teleme cheese increases during ripening. The principal free amino acids are lysine, leucine, valine, glutamic acid, and phenylalanine at all stages of ripening (Alichanidis et al. 1981; Mallatou et al. 2004; Pappa and Sotirakoglou 2008; Polychroniadou and Vlachos 1979).

In a recent study (Pappa et al. 2008), the water-soluble extracts of teleme cheese prepared from ewe, goat, or cow milk and matured for 120 days have been analyzed for constituent peptides and proteins using proteomics. The water-soluble extract of the cheese made from ewe milk gave a broader spectrum of peptides than those made from goat and cow milk, obviously due to differences in protein and peptide hydrolysis during ripening. The results show that two-dimensional electrophoresis with matrix-assisted laser desorption ionization/mass spectrometry is suitable to separate and identify proteins that are not hydrolyzed during ripening such as the whey proteins (α-lactalbumin, β-lactoglobulin, and serum albumin) but the resolution power of this method is limited as whey proteins predominate in the water-soluble extracts. The use of HPLC in conjunction with mass spectrometry and Edman degradation shows that each peak contains at least one constituent that could either be α-lactalbumin or β-lactoglobulin or could be peptides originating in α_{s1}- or β-casein. Tandem mass spectrometry leads to the identification of small peptides (with mass range from 3000 to 1500 Da) from the water-soluble extract, most of them originating in β-casein and only one found from α_{s1}-casein.

Keeping the main manufacturing practices constant, the degree of proteolysis of teleme cheese is generally affected by the kind of milk and culture used (Buruiana and Farag 1980; Pappa et al. 2006b) and the pasteurization of cheese milk (Yetismeayen et al. 2003).

In conclusion, biochemical changes of teleme cheese are not extensive during ripening because of the low pH, the high salt content, and the relatively short ripening period. However, the rate of proteolysis is constant throughout the maturation period (Alichanidis et al. 1981; Mallatou et al. 2004; Manolkidis et al. 1970b; Pappa et al. 2006b).

16.3.5 Rheological Characteristics and Organoleptic Evaluation

Teleme cheese is a WBC without rind. The body and texture is smooth and nongranular, without gas holes, but small mechanical openings are desirable. These features and other specifications of teleme cheese are evaluated by Manolkidis et al. (1974). Traditionally, teleme cheese was made from ewe milk, but now, it is made from cow, goat, and buffalo milk or a mixture of them. Therefore, the color of the cheese depends on the kind of milk used. When ewe, goat, or buffalo milk is used, the color is white and described as porcelain white, marble white, or snow white. However, when cow milk is used, the color ranges from off white to yellowish, and in some countries, "decolorants" (e.g., chlorophylls and titanium dioxide) or other treatments are used to eliminate or cover the yellowish color (Alichanidis and Polychroniadou 2008). Teleme cheese is soft but firm enough to be sliced by a knife. The flavor of teleme is described as acidic and slightly salty. The flavor is best when ewe milk is used. The use of goat milk gives a piquant flavor to the cheese, especially when it is very mature (Abd El-Salam and Alichanidis 2004; Zerfiridis 2001). However, Stef et al. (2009) concluded that cow Romanian Telemea cheese was better (sensorial estimation) than ewe Romanian Telemea cheese. This confirms that the concept of flavor differs greatly from one country to another. Generally, teleme cheese can become brittle when it is very ripe and the pH is very low (Abd El-Salam and Alichanidis 2004; Alichanidis and Polychroniadou 2008). Body and texture depends on the kind of milk used. When ewe milk is used for the manufacture of the cheese, the texture is smooth and creamy, but when goat milk is used, the sensory feeling of texture is dry and hard, whereas it is crumbly when cow milk is used. Body and texture as well as flavor also depend on the way of manufacturing teleme, such as the pasteurization temperature, which according to Yetismeayen et al. (2003) causes the decrease of curd firmness as the temperature is increased and the kind and level of starter cultures used in cheese-making (Kalogridou-Vassiliadou and Alichanidis 1984; Pappa et al. 2006a; Tzanetakis et al. 1991; Veinoglou et al. 1980).

The rheological characteristics of teleme cheese are usually studied with the method of uniaxial force compression using the Instron universal testing machine. Rheological measurements of mature teleme cheese can be summarized as follows: the fracture of the sample (fracturability), that is, the force needed to fracture the cheese sample, and hardness, that is, the maximum force necessary to cause a certain deformation, usually 70%–80%. Other parameters such as the percentage compression at the point of fracture, cohesiveness, springiness, gumminess, and chewiness can be measured too. The reported values for fracturability range between 0.6 and 4.3 kg for 60-day-old teleme cheese and between 0.9 and 3.5 kg for 120-day-old teleme cheese. The reported values for hardness are 2.5–8.9 kg for 60-day-old and 2.4–7.3 kg for 120-day-old teleme cheese (Pappa et al. 2007; Raphaelides and Antoniou 1996; Raphaelides et al. 1995). Generally, hardness and fracturability decreased during ripening, obviously due to cheese proteolysis, which results in the weak structure and softening of the cheese. Teleme cheese made in a traditional way is firmer and more rigid than the ultrafiltrated type (Raphaelides and Antoniou 1996; Raphaelides et al. 1995). In addition, the use of different types of milk and starter culture affects the rheological characteristics of teleme cheese. The use of goat milk or mesophilic culture results in a harder teleme cheese with a higher force and a higher compression at the point of fracture than cheeses made from sheep or cow milk or than cheeses made with a thermophilic or a mixed thermophilic–mesophilic culture, respectively (Pappa et al. 2007). This difference is easily noticeable even by organoleptic evaluation.

The dynamic rheometry by employing a textural analyzer has also been used to evaluate teleme textural properties. Among the techniques of dynamic rheometry, texture profile analysis was the most commonly used method for the assessment of teleme cheese texture (Romeih et al. 2002; Volikakis et al. 2004). Although results are expressed in a different way and they are more detailed than those received by the Instron instrument, generally, they end up with the similar conclusions.

ACKNOWLEDGMENTS

The authors would like to thank professors emeriti E. Alichanidis and A. Polychroniadou, School of Agriculture, Aristotle University of Thessaloniki, Thessaloniki, Greece, for their valuable suggestions in reviewing this chapter.

REFERENCES

Abd El-Salam, M. H. and Alichanidis, E. 2004. Cheese varieties ripened in brine. In: *Cheese: Chemistry, Physics and Microbiology—Vol. 2: Major Cheese Groups*, edited by P. F. Fox, P. L. H. McSweeney, T. M. Cogan, and T. P. Guinee, pp. 227–249. Amsterdam: Elsevier Applied Science.

Abd El-Salam, M. H., Alichanidis, E., and Zerfiridis, G. K. 1993. Domiati and feta type cheeses. In: *Cheese: Chemistry, Physics and Microbiology*, edited by P. F. Fox, pp. 301–335. London: Elsevier Applied Science.

Alichanidis, E. 1981. Manufacture of Teleme cheese from ewes' milk treated with H_2O_2. Thesis, University of Thessaloniki, Greece (in Greek).

Alichanidis, E. and Polychroniadou, A. 2008. Characteristics of major traditional regional cheese varieties of East-Mediterranean countries: A review. *Dairy Sci. Technol.* 88:405–510.

Alichanidis, E., Polychroniadou, A., Tzanetakis, N., and Vafopoulou, A. 1981. Teleme cheese from deep-frozen curd. *J. Dairy Sci.* 64:732–739.

Andrikopoulos, N. K., Kalogeropoulos, N., Zerva, A., Zerva, U., Hassapidou, M., and Kapoulas, V. M. 2003. Evaluation of cholesterol and other nutrient parameters of Greek cheese varieties. *J. Food Compos. Anal.* 16:155–167.

Anifantakis, E. M. 1991. *Greek Cheeses—A Tradition of Centuries*, pp. 27–42. Athens: National Dairy Committee of Greece.

Anifantakis, E. M. 2004. Microorganisms used in cheesemaking. In: *Cheesemaking. Chemistry–Physicochemistry–Microbiology*, 2nd edition, pp. 185–221. Athens, Greece: Stamoulis (in Greek).

Anifantakis, E. M. and Moatsou, G. 2006. Feta and other Balcan cheeses. In: *Brined Cheeses*, edited by A. Y. Tamine, pp. 43–76. Oxford: Blackwell Publishing.

Antoniou, K. D., Kioulafli, P., and Sakellaropoulos, G. 1995. Studies on the application of ultrafiltration for the manufacture of Teleme cheese. *Milchwissenschaft* 50:560–565.

Belitz, H.-D. and Kaiser, K.-P. 1993. Monitoring cheddar cheese ripening by chemical indices of proteolysis—Part III: Identification of several high molecular mass peptides. *Z. Lebensm.-Unters. Forsch.* 197:118–122.

Buruiana, L. M. and El-Senaity, M. H. 1986. Volatile fatty acids in Teleme cheese during ripening. *Egypt. J. Dairy Sci.* 14:201–205.

Buruiana, L. M. and Farag, S. I. 1980. Determination and significance of the proteolytic activity of Romanian Telemea cheese made from cow's and sheep's milk. *Egypt. J. Dairy Sci.* 8:57–65.

Cliffe, A. J., Revell, D., and Law, B. A. 1989. A method for the reverse phase HPLC of peptides from cheddar cheese. *Food Chem.* 34:147–160.

Davis, J. G. 1965. *Cheese Basic Technology*. London: Churchill.

De Roos, A. L., Geurts, T. G., and Walstra, P. 1998. On the mechanism of rennet retention in cheese. Bulletin No. 332, pp. 15–19. Brussels: I.D.F.

Eekhof-Stork, N. 1976. *The World Atlas of Cheese*, 1st edition, p. 142. New York: Paddington.

Efthymiou, C. 1967. Major fatty acids of feta cheese. *J. Dairy Sci.* 50:20–24.

Engels, W. J. M. and Visser, S. 1994. Isolation and comparative characterization of components that contribute to the flavour of different types of cheese. *Neth. Milk Dairy J.* 48:127–140.

Fox, P. F., Singh, T. K., and McSweeney, P. L. H. 1995. Biogenesis of flavor compounds in cheese. In: *Chemistry of Structure–Function Relationships in Cheese*, edited by E. L. Malin and M. H. Tunick, pp. 59–98. New York: Plenum Press.

Greek Codex for Foodstuff and Beverages. 2009. *Greek Codex Alimentarius*. Official Journal of the Hellenic Republic. Vol. B, Article 83, Section A, No. 899 Article 83 Paragraph 1.10. Athens: National Printing Office (in Greek).

Kaiser, K.-P., Belitz, H.-D., and Fritsch, R. J. 1992. Monitoring cheddar cheese ripening by chemical indices of proteolysis—Part II: Peptide mapping of casein fragments by reverse-phase high-performance liquid chromatography. *Z. Lebensm.-Unters. Forsch.* 195:8–14.

Kalogridou-Vassiliadou, D. and Alichanidis, E. 1984. Effect of refrigerated storage of milk on the manufacture and quality of teleme cheese. *J. Dairy Res.* 51:629–636.

Lau, K. Y., Barbano, D. M., and Rasmussen, R. R. 1991. Influence of pasteurization of milk on protein breakdown in cheddar cheese during aging. *J. Dairy Sci.* 74:727–740.

Lee, K. D. and Warthesen, J. J. 1996. Mobile phases in reverse-phase HPLC for the determination of bitter peptides in cheese. *J. Food Sci.* 61:291–294.

Litopoulou-Tzanetaki, E., Kalogridou-Vassiliadou, D., and Tzanetakis, N. 1992. Evolution de la flore microbienne au cours de la fabrication et de l'affinage du fromage Telemes. *Microbiol. Aliments. Nutr.* 10:283–288.

Mallatou, H. and Pappa, E. C. 2005. Comparison of the characteristics of Teleme cheese made from ewe's, goat's and cow's milk or a mixture of ewe's and goat's milk. *Int. J. Dairy Technol.* 58:158–163.

Mallatou, H., Pappa, E., and Boumba, V. A. 2004. Proteolysis in Teleme cheese made from ewes', goats' or mixture of ewes' and goats' milk. *Int. Dairy J.* 14:977–987.

Mallatou, E., Pappa, E. C., and Massouras, T. 2003. Changes in free fatty acids during ripening of Teleme cheese made with ewe's, goat's, cow's or a mixture of ewe's and goat's milk. *Int. Dairy J.* 13:211–219.

Mann, E. 1999. Feta cheese. *Dairy Ind. Int.* 64:11–12.

Manolkidis, C., Litopoulou-Tzanetaki, E., Kalogridou-Vassiliadou, D., and Tzanetakis, N. 1975. Sur les contamination par des bacteries coliformes au cours des premieres stages de la fabrication du fromage "Teleme." Extrait des "Archives de l' Institute Pasteur Hellenique," Athenes, XXI:53–61.

Manolkidis, C., Polychroniadou, A., and Alichanidis, E. 1970b. Observations suivies sur la proteolyse pendant la maturation du fromage "Teleme." *Lait* 50:128–136.

Manolkidis, K., Polychroniadou, A., and Alichanidis, E. 1970a. Variations dans la composition du fromage "Teleme" au cours de sa maturation. *Lait* 50:38–48.

Manolkidis, K. S., Zerfiridis, G. K., and Karazanos, G. P. 1974. Organoleptic examination and specifications of Teleme cheese. *Geoponika* 21:295–300 (in Greek).

Manolopoulou, E., Sarantinopoulos, P., Zoidou, E., Aktypis, A., Moschopoulou, E., Kandarakis, I. G., and Anifantakis, E. M. 2003. Evolution of microbial populations during traditional feta cheese manufacture and ripening. *Int. J. Food Microbiol.* 82(2):153–161.

Massouras, T., Pappa, E. C., and Mallatou, H. 2006. Headspace analysis of volatile flavor compounds of Teleme cheese made from sheep and goat milk. *Int. J. Dairy Technol.* 59:250–256.

McSweeney, P. L. H., Olson, N. F., Fox, P. F., Healy, A., and Hojrup, P. 1993. Proteolytic specificity of chymosin on bovine α_{s1}-casein. *J. Dairy Res.* 60:401–412.

Michaelidou, A., Alichanidis, E., Polychroniadou, A., and Zerfiridis, G. 2005. Migration of water-soluble nitrogenous compounds of Feta cheese from the cheese blocks into the brine. *Int. Dairy J.* 15:179–189.

Michaelidou, A., Alichanidis, E., Urlaub, H., Polychroniadou, A., and Zerfiridis, G. K. 1998. Isolation and identification of some major water-soluble peptides in feta cheese. *J. Dairy Sci.* 81:3109–3116.

Pappa, E. C. 2003. Study of the physicochemical characteristics of Teleme cheese made from different types of milk and culture. PhD Thesis, Agricultural University of Athens, Greece (in Greek).

Pappa, E. C., Kandarakis, I., Anifantakis, E. M., and Zerfiridis, G. K. 2006a. Influence of types of milk and culture on the manufacturing practices, composition and sensory characteristics of Teleme cheese during ripening. *Food Control* 17:570–581.

Pappa, E. C., Kandarakis, I., and Mallatou, H. 2007. Effect of different types of milks and cultures on the rheological characteristics of Teleme cheese. *J. Food Eng.* 79:143–149.

Pappa, E. C., Kandarakis, I. K., Zerfiridis, G. K., Anifantakis, E. M., and Sotirakoglou, K. 2006b. Influence of starter cultures on the proteolysis of Teleme cheese made from different types of milk. *Lait* 86:273–290.

Pappa, E. C., Robertson, J. A., Rigby, N. M., Mellon, F., Kandarakis, I., and Mills, E. N. C. 2008. Application of proteomic techniques to protein and peptide profiling of Teleme cheese made from different types of milk. *Int. Dairy J.* 18:605–614.

Pappa, E. C. and Sotirakoglou, K. 2008. Changes of free amino acid content of Teleme cheese made with different types of milk and culture. *Food Chem.* 111:606–615.

Peterson, S. D., Marshall, R. T., and Heyman, H. 1990. Peptidase profiling of lactobacilli associated with cheddar cheese and its application to identification and selection of strains for cheese ripening studies. *J. Dairy Sci.* 73:1454–1464.

Polo, M. C., de Gonzalez Llano, D., and Ramos, M. 1992. Determination and liquid chromatographic separation of peptides. In: *Food Analysis by HPLC*, edited by L. Nollet, pp. 123–125. New York: Marcel Dekker.

Polychroniadou, A. and Vlachos, J. 1979. Les acides amines du fromage Teleme. *Lait* 585–586:234–243.

Raphaelides, S. and Antoniou, K. D. 1996. The effect of ripening and mechanical properties of traditional and ultrafiltrated Teleme cheese. *Milchwissenschaft* 51:82–85.

Raphaelides, S., Antoniou, K. D., and Petridis, D. 1995. Texture evaluation of ultrafiltrated Teleme cheese. *J. Food Sci.* 60:1211–1215.

Romeih, E. A., Michaelidou, A., Biliaderis, C. G., and Zerfiridis, G. K. 2002. Low-fat white-brined cheese made from bovine milk and two commercial fat mimmetics: Chemical, physical and sensory attributes. *Int. Dairy J.* 12:525–540.

Samal, P. K., Pearce, K. N., Bennet, R. J., and Dunlop, F. R. 1993. Influence of residual rennet and proteolysis on the exudation of whey from feta cheese during ripening. *Int. Dairy J.* 3:729–745.

Scott, R. 1986. *Cheese Making Practice*, 2nd edition. London: Elsevier Applied Science.

Stef, D. S., Stef, L., Druga, M., Heghedus-Mandru, G., and Biron, R. 2009. Telemea cheese quality from Banat region. *J. Food. Agric. Environ.* 7:24–26.

Tzanetakis, N. and Litopoulou-Tzanetaki, E. 1992. Changes in numbers and kinds of lactic acid bacteria in Feta and Teleme, two Greek cheeses from ewe's milk. *J. Dairy Sci.* 75(6):1389–1393.

Tzanetakis, N., Litopoulou-Tzanetaki, E., and Vafopoulou-Mastrojisnnski, A. 1991. Effect of *Pediococcus pentosaceus* on microbiology and biochemistry of Teleme cheese. *Lebensm.-Wiss. Technol.* 24:173–176.

Vafopoulou-Mastroyannaki, A. 1977. Retention of Ca, P, K, and Na in Teleme cheese industrially processed from ewes' milk. *Milchwissenschaft* 32:475–476.

Van den Berg, G. and Exterkate, F. A. 1993. Technological parameters involved in cheese ripening. *Int. Dairy J.* 3:487–505.

Veinoglou, B. C. and Boyazoglu, E. S. 1982. Improvement in the quality of Teleme cheese produced from ultrafiltrated cow's milk. *Int. J. Dairy Technol.* 35:54–56.

Veinoglou, B. C., Boyazoglu, E., and Anifantakis, E. 1977. Production of Feta and Teleme cheese from ultrafiltrated cow's and sheep's milk. *Sodobna Proizvodnja in Predelava Mleka* 5–7:512–521.

Veinoglou, B. C., Boyazoglu, E. S., and Kotouza, E. D. 1980. The effect of starters on the production of Teleme cheese. *Dairy Ind. Int.* 45:11–19.

Volikakis, P., Biliaderis, C. G., Vamvakas, C., and Zerfiridis, G. K. 2004. Effects of a commercial oat β-glucan concentrate on the chemical physico-chemical and sensory attributes of a low-fat white-brined cheese product. *Food Res. Int.* 37:83–94.

Yetismeayen, A., Gencer, N., Gürsoy, A., Deveci, O., Karademir, E., Senel, E., and Öztekin, S. 2003. The effect of some technological processes on the properties of Teleme from goat cheese. *Milchwissenschaft* 58:286–289.

Zerfiridis, G. K. 1968. Feta cheese from pasteurized cow's milk, pp. 1–2. Thesis, Ohio State University, Columbus, OH, USA.

Zerfiridis, G. K. 2001. Teleme cheese. In: *Technology of Dairy Products*, 2nd edition, pp. 176–192. Thessaloniki, Greece: Giachoudis-Giapoulis (in Greek).

Zerfiridis, G. K., Alichanidis, E., and Tzanetakis, N. M. 1989. Effect of processing parameters on the ripening of Teleme cheese. *Lebensm.-Wiss. Technol.* 22:169–174.

Zygouris, N. 1952. *The Milk Industry*, 2nd edition, pp. 391–427. Athens: Ministry of Agriculture (in Greek).

17

Goat Milk Cheeses

Evanthia Litopoulou-Tzanetaki

CONTENTS

17.1 Introduction

Goats are of great importance in developed countries for the production of high-quality cheeses and the sustainable development of rural areas and even in developing countries to exploit marginal agricultural resources. A good dairy goat gives about 900–1800 kg milk in a 305-day lactation period (Haenlein 1993). Goat milk production has been concentrated mainly in the Mediterranean and Middle East countries, as well as some East and West European countries. There is now archeological evidence (Luikart et al. 2006; Zeder 2008) that probably the Neolithic farmers migrated out of the Near East and across Europe following two main routes—through the continental heartland up to the Danube valley or along the Mediterranean coast, and findings suggest their initial settlement in the Balkans and southern Italy (Vigne 1999). In post-Neolithic times, civilizations from the Mediterranean area, such as Greek, Roman, Phoenician, and Berber, probably introduced new species of animals and new breeds of livestock in southwest Europe, while some colonists may have imported stock from overseas and improved local livestock, explaining the high diversity in breeds of domestic goats (Pereira et al. 2006). The migration history of the species played a pivotal role in the present-day structure of the breeds, and it seems that coastal routes were easy for migrating. A westward coastal route to Italy through Greece could have led to gene flow along the Northern Mediterranean (Pariset et al. 2009).

More than 95% of the goat population is found in developing countries, with the European Union (EU) countries having 1.6% of the population but producing 13.2% of goat milk and 2.0% of goat meat generated in the world annually. The highest rankings for goat production are Greece with 5.4 million heads, Spain with 2.8 million, followed by France and Italy with 1.2 and 0.9 million heads, respectively (FAO 2006). Annual goat milk production in France, Greece, and Spain is as high as 583, 511, and 423 million L, respectively (83% of the total goat milk produced in the EU), but only France and Spain have increased their production in the last decade due to the high productivity of their goats. Milk and, above all, meat obtained from goat herds meet the nutritional needs of the rural population in developing countries, whereas in developed countries, goat breeders have developed a wide range of products with excellent nutritional qualities linked to natural and sustainable systems (Boyazoglu et al. 2005). There is an increasing demand of the growing population of people for goat milk consumption, for goat milk products, cheeses, and yogurt, related especially to people with cow milk allergies and other gastrointestinal ailments (Haenlein 2004).

Numerous varieties of goat milk cheeses are produced worldwide, depending on diversity in the locality, milk composition, and manufacturing techniques used. The Mediterranean region is known for its ancient tradition of goat cheese production, and the quality of the "typical" products manufactured is strongly influenced by the local production area and its traditions. Safeguarding these products means safeguarding the uniqueness of their historical and cultural environment (Boyazoglu and Morand-Fehr 2001). Consumers in southern Europe (France, Italy, Spain, Portugal, and Greece) are giving much more importance to their "typicality," a recent way of understanding the quality of a product, and for this reason, the EU has produced a law ruling the products and the factors determining the typicality.

17.2 Gross Compositional and Physicochemical Parameters of Goat Milk

The gross composition of goat milk is similar to that of cow milk and varies with breed, animals within breed, parity, diet, feeding and management conditions, locality, season, and stage of lactation (Park 2006).

Goat milk is white in color and has a stronger flavor than sheep milk; it is also alkaline in nature due to higher protein content and a different arrangement of phosphates. This alkalinity is very useful for people with acidity problems (Jandal 1996). The compositional characteristics of goat milk are presented in Table 17.1.

Goat milk contains lower quantities of fat, solids-not-fat, proteins, caseins, whey proteins, and ash compared with sheep milk but has more fat, protein, and ash and less lactose than cow milk (Alichanidis and Polychroniadou 2008; Jandal 1996; Park 2006). Significant variations occur in milk composition and yield during different seasons and stages of lactation of goats, with high contents in fat, total solids, and protein, higher in early lactation and then decreasing rapidly toward the end (Park 2006). Gross chemical composition may also vary in milk from different breeds with respect to total solids, fat, solids-not-fat (SNF), and protein, whereas during the lactation period, fat, protein, chlorides, ash, SNF, and total solids (TS) increase, and the lactose content declines (Agnihotri and Prasad 1993; Alichanidis and Polychroniadou 2008; Slacanac et al. 2010; Voutsinas et al. 2009). It also seems that in the regions of Mediterranean countries, where goats are traditionally raised for milk production, as well as in tropical countries, the local, well-adapted breeds give milk richer in fat and protein than imported goats of improved European breeds. Variation due to breed in certain physicochemical properties of goat milk is also observed. Although the pH of the milk does not vary much, titratable acidity (TA) may differ (Agnihotri and Prasad 1993). In addition, TA increases with the progress of lactation in all breeds, whereas pH shows a decreasing trend. This parameter is highly affected by the ambient temperature, the raising conditions of the herd, and the hygienic conditions (Alichanidis and Polychroniadou 2008). Morning milk was also found to be richer in fat than evening milk from Greek native goats in Epirus, Greece. Goat milk is sensitive to heating; temperatures above 60°C also cause a destabilization of whey proteins. The density of caprine milk is lower than ovine and similar

TABLE 17.1

Comparative Composition of Milk from Different Sources

Component	Values per 100 g of Milk (Fat)[a]			
	Goat	**Cow**	**Sheep**	**Human**
Gross Composition[b]				
Fat	3.80	3.67	7.62	3.67–4.70
Solids-not-fat	8.68	9.02	10.33	8.9
Lactose	4.08	4.78	3.7	6.92
Protein	2.90	3.23	6.21	1.10
Casein	2.47	2.63	5.16	0.40
Whey proteins	0.43	0.60	0.81	0.70
Ash	0.79	0.73	0.90	0.31
Minerals[c]				
Ca (mg)	170.0	120.0	–[d]	28.0
P (mg)	120.0	90.0	–	11.0
Fe (mg)	0.3	0.2	–	–
Na (mg)	11.0	16.0	–	–
K (mg)	110.0	140.0	–	–
Vitamins[e]				
Vitamin A (IU)	185.0	126.0	–	190.0
Vitamin C (mg)	1.29	0.94	–	5.0
Thiamine (mg)	0.068	0.045	–	0.017
Riboflavin (mg)	0.21	0.16	–	0.02
Niacin (mg)	0.27	0.08	–	0.17
Folate (μg)	1.0	5.0	–	5.5
Vitamin B12 (μg)	0.065	0.357	–	0.03
Fatty Acids[b,e]				
C4:0	2.6	3.3	4.0	–
C6:0	2.9	1.6	2.6	Tr[f]
C8:0	2.7	1.3	2.5	Tr
C10:0	8.4	3.0	7.5	1.3
C12:0	3.3	3.1	3.7	3.1
C14:0	10.3	9.5	11.9	5.1
C16:0	24.6	26.5	25.2	20.2
C16:1	2.2	2.3	2.2	5.7
C18:0	12.5	14.6	12.6	6.0
C18:1	28.5	29.8	20.0	46.4
C18:2	2.2	2.5	2.1	13.0
C18:3	Tr	1.8	–	1.4

[a] For fatty acids.
[b] Jandal, J. M. 1996. *Small Ruminant Res.* 22:177–185.
[c] Agnihotri, M. K. and Prasad, V. S. S. 1993. *Small Ruminant Res.* 12:151–170.
[d] No data.
[e] Park, Y. W. 2006. *Handbook of Milk of Nonbovine Mammals*, edited by Y. W. Park and G. F. W. Haenlein, pp. 34–58. Oxford, U.K.: Blackwell Publishing Professional.
[f] Trace amounts.

to bovine milk, and its freezing point values are similar to the low point values reported for cow milk (Slacanac et al. 2010).

Different structures of caprine and bovine milk have been significantly appreciable in cheeses and fermented milk as differences in consistency, flavor, odor, color, stability during storage, and syneresis degree (Slacanac et al. 2010).

17.3 Goat Milk in Human Nutrition

There are several reasons that goat milk is considered to have significant nutritional values in human nutrition as well as high nutrient bioavailability. In general, goat milk is very rich in nutrients, but usually, it is not used for drinking as an alternative to cow milk (Alichanidis and Polychroniadou 2008).

The nutritional advantages of goat milk over cow milk come not from its protein, mineral, or vitamin differences but from another component in goat milk, the lipids, and more specifically, the fatty acids within the lipids (Jandal 1996; Park 2006). Goat milk exceeds cow milk in monounsaturated and poly-unsaturated fatty acids, as well as medium-chain triglycerides, which are all known to be beneficial to human health. The fat globules of goat milk are smaller, and almost 20% of the fatty acids fall into the short-chain (C4:0–C12:0) fatty acid category. These properties make the way easy for lipases in the gut to attack both, lipids and fatty acids, rapidly, which means a more rapid digestion of goat milk (Jandal 1996; Park 2006).

Goat milk proteins differ in genetic polymorphisms from cow milk due to amino acid substitutions in the protein chains, which in turn are responsible for the differences in digestibility, cheesemaking properties, and flavor of goat milk products (Haenlein 2004). Goat milk with low or no α_{s1}-casein has less curd yield, longer rennet coagulation time, more heat liability, and weaker curd, which could explain the benefits in digestibility in the human digestive tract (Haenlein 2004). In addition, goat milk is richer in iron than human, cow, and sheep milk but poorer in copper and folate (Alichanidis and Polychroniadou 2008); it also has selenium content similar to human milk but higher than bovine milk, and this makes it important for the prevention of cancer and cardiovascular diseases.

Experiments with goat milk as a substitute for cow milk with undernourished children showed that substitution caused higher rates of intestinal fat absorption, weight gain, height, skeletal mineralization, and blood serum contents of vitamin A, calcium, and other vitamins (Haenlein 2004). It was also found that the consumption of goat milk reduces total cholesterol levels and low-density lipoprotein due to higher levels of medium-chain triglycerides in goat milk, resulting in a decrease in the synthesis of endogenous cholesterol (Alférez et al. 2001).

17.4 Therapeutic and Hypoallergenic Value of Goat Milk

The most significant health benefit of goat milk compared to cow milk is its hypoallergenic value (Haenlein 2004; Park 1994), because a high percentage of people suffering bovine milk allergy tolerated caprine milk.

Cow milk protein is probably the most common allergen during infancy and early childhood, with a prevalence of 2.5% in children during the first 3 years of life (Buscino and Bellanti 1993). The reason for the hypoallergenic value of goat milk in comparison to cow milk is the difference between their protein structures. Beta-lactoglobulin is mostly important for milk allergy, and because both cow and goat milk contain this fragment, people allergic to this fragment would be unable to tolerate either cow or goat milk (Robinson 2001). It was also shown (Bellioni-Businco et al. 1999) that goat milk is not an appropriate substitute for cow milk in children with IgE-mediated cow milk allergy. Because the α_{s1}-casein content of goat milk is low or absent, it could be assumed that people with sensitivity to α_{s1}-casein of cow milk may show no clinical symptoms when goat milk is consumed (Haenlein 2004; Robinson 2001). *In vitro* studies (Almaas et al. 2006) showed that goat milk proteins were digested faster than cow milk proteins by human gastric and duodenal enzymes.

Short- and medium-chain fatty acids of goat milk may be beneficial in the treatment of many physiological disorders in humans, such as coronary, pulmonary, and intestinal disorders, cystic fibrosis, malabsorption, and cholesterol level regulation (Haenlein 2004; Jandal 1996). Caprine milk consumption may prevent atherosclerosis (Kullisaar et al. 2003), improves iron bioavailability in rats (Alférez et al. 2006), and affects beneficially the nutritive utilization of protein, magnesium, calcium, phosphorus, zinc, and selenium of rats (Slacanac et al. 2010).

17.5 Typicality and Biodiversity of Goat Milk Products

Goats are distributed over all types of ecology, with more concentration in the tropic dry zones and developing countries (FAOSTAT 2001) so that they can now be found in every environment of the continent. Their widespread cultural acceptance forms a sound basis for development. With such a wide distribution and a capacity to adapt to quite different environments, the goat is expected to show a great deal of diversity within the 570 recognized breeds. An index on the number of breeds per million of population was used to calculate the degree of biodiversity (Galal 2005). This index for the goat was 0.8 compared to 1.2 for sheep, 0.9 for cattle, and 0.5 for buffalo. Europe has the heaviest goat breed with the largest milk production, whereas Latin America and the Caribbean have the lowest. Among breed, variability was lowest in Europe and highest in Africa.

The notion of "typical product" combines the characteristics of the products with its localization and, above all, with its origin—geographic and historic (Rubino et al. 1999). The definition "typical" includes the sensory characteristics of the product, the geographical origin of both, the raw material and the transformation process, and its relationship with the social and cultural tradition of the area of production (Scintu and Piredda 2007). Typicality is, in fact, a recent way of understanding the quality of the product and is given an appreciable importance by the consumers in southern Europe. EU established rules for the protection of agricultural products intended for human consumption by Council Regulation (EEC) No. 2081/92 and distinguished between two categories of protected designations: Protected Geographical Indications (PGI) and Protected Designation of Origin (PDO). The PDO/PGI labels guarantee specific sensory properties and imply that the product is qualified by particular characteristics. Characteristics related to milk composition obtained from a specific animal, with respect to species and/or breed nourished under specific climatic conditions and feeding and management systems, are linked to certain natural and human factors. Due to the rich diversity of the above, traditional cheeses exhibit diversity in their characteristics, closely related to the ecosystem of the production zone. This diversity is highlighted by the producers and supporters of PDO cheeses.

Production factors affecting typicality and biodiversity are (Rubino et al. 1999; Scintu and Piredda 2007) the following:

1. *Breed.* Milk and cheese characteristics are influenced by the existence of a polymorphism in the casein fractions. The polymorphism varies either among animals or among populations. The technological properties of milk change according to the casein variants (A, B, C—strong alleles; high level of casein; D, F—low alleles; null type, lacking α_{s1}-casein, with low casein content). The presence of strong alleles α_{s1}-casein, A, B, and C, is related to higher fat content and cheese yield and a lesser dimension of casein micelle. On the other hand, the "goaty flavor" is less strong in cheese prepared with the casein A milk type. The allelic frequencies vary among breeds. Consequently, the breed determines specific organoleptic characteristics of a cheese because of its casein polymorphism. "Strong" alleles have a high frequency in breeds from the Mediterranean area. On the other hand, the poor clotting ability of goat milk in French breeds was attributed to its low casein content. For these reasons, several cheese varieties are made from the milk of a specific breed. For example, Serena cheese in Spain and Pecorino di Filiano in Italy are produced from milk provided by the Merino sheep breed.

2. *Feeding system.* The feeding system affects milk composition and, consequently, cheese flavor. Aromatic compounds deriving from plants (e.g., Thymus, Mentha, Origanum, and Salvia)

and other secondary metabolites may be transferred into the animal's blood and from there to the milk. Flavor compounds such as carbonyl compounds and corresponding alcohol come from unsaturated fatty acids in the grass lipids. Desirable flavors are also associated with high levels of polyisoprene from the pasture. The milk of pasture-grazed animals is characterized by the presence of certain terpenoids in the milk fat, and cheeses produced in Alpine pastures were found to contain more terpene (α- and β-pinene) than those produced in the low lands. Concentrate supplements given to the animals, especially during early lactation and late pregnancy, modify the acetic/propionic acids ratio in the rumen and may contribute to the different characteristic flavors in milk and cheese.

3. *Fat composition, which reflects the effect of the environment.* Many factors, such as season, stage of lactation, feeding practices, and breed, change the fat content and the fatty acid profile. The development of goat milk flavor is related with not only the occurrence of volatile branched chain fatty acids but also the structure of triglycerides and the lipolytic action of the milk lipoprotein lipase, which in goat milk is largely bound to cream. Apart from the genetic factors, dietary factors also affect the levels of rumeric and vaccenic acids in goat milk, and the fatty acid profile varies according to different forage (alfalfa, sula, clover, ryegrass, and *Chrysanthemum coronarum*; Addis et al. 2005). Considering the beneficial effects of conjugated linoleic acid on human health and the putative negative effects of trans-isomers, one may envisage the choice of milk with the best characteristics for the production of typical cheeses.

4. *Rennets.* Animal or vegetable rennets used as coagulants reflect the specificity and biodiversity of areas where the goats are raised, due to the different enzymes, according to the rennet type. The lamb or kid paste rennet contains chymosine, pepsin, and lipolytic enzymes, and PDO cheeses, such as feta in Greece, Pecorino Romano in Italy, and idiazabal in Spain, obtain a characteristic piquant taste due to the production of free fatty acids (FFAs) from triglycerides by the activity of lipolytic enzymes from rennet. Some PDO cheeses from Portugal and Spain are made by another kind of rennet linked to the production area, coming from the vegetables *Cynara cardunculus* and *C. humilis*. Cardesins a- and b-, enzymes of the cardoon, are similar to chymosin and pepsin, respectively, in terms of specificity and kinetic parameters. Different rennets contribute to differential ripening characteristics, as well as the flavor of the cheeses, and determine a strong relationship between the area where the local cardoon flowers grow and/or animals (from where the stomachs are obtained) are bred and the cheese characteristics. For these reasons, the use of paste rennet to make Pecorino Romano and Pecorino di Filiano is obligatory by regulations in Italy.

5. *Microbial composition of the milk.* The milk used to make cheese is either raw or thermized (60°C for 15 s) and/or pasteurized. The consumers recognize typicality factors in cheeses made from raw milk and cheeses produced at farms. The microflora of the raw milk is composed of various microbes deriving from the local ecosystem, which have a marked impact on the characteristics of the final product. Studies on the biodiversity of the autochthonous lactic microflora in goat milk cheeses showed a high biodiversity in the numbers and kinds of the "wild" lactic acid bacteria (LAB) present in the milk and cheeses throughout the whole lactation season (Hatzikamari et al. 1999; Psoni et al. 2003), as well as in their genetic and technological properties (Psoni et al. 2006, 2007). This biodiversity contributes to the peculiar sensory properties of raw milk traditional cheeses and can be considered the basis of the typicality and the uniqueness of the products. Therefore, the protection of the complexity of microbial microflora in typical cheeses is important for the maintenance of the characteristics of each cheese in its traditional area of production.

6. *Ripening conditions.* Natural cheese cellars confer optimal environmental conditions, which affect greatly the cheese typicality. For example, the quality of Roquefort cheese is determined by the ideal conditions of the Roquefort area, and the typicality of Formaggio di Fossa is attributed to its ripening in pits dug in the tufa at the borders of the Romagna and Marche regions.

17.6 European Traditional Goat Milk Cheeses

17.6.1 Goat Milk Cheeses in Southern EU Countries

17.6.1.1 France

According to the information presented in the *Atlas of Goat Products* (Rubino et al. 2004), this excellent effort "to describe goat cheeses, where they are processed and by whom, what about their consumption and the facilities of their commercialization," in France, 17 main cheese varieties are made in various areas. Medina and Nuñez (2004) presented the information that more than 50 goat cheese varieties are produced in France, of which nine are under a PDO status.

Bastelicaccia cheese is a soft cheese made in Corsica from raw ewe and/or goat milk. The traditional method involves both, coagulation by traditional rennet and acid production by the indigenous microflora. In cheese from goat milk made at two farms, lactococci and leuconostoc dominated (10^8 cfu/g) in the cheese, whereas enterococci were less numerous (10^5 cfu/g). Cheese made in winter had significantly lower dry matter, level of lipolysis, and firmness but significantly higher degree of proteolysis from cheeses made in spring. The cheeses also differed significantly with respect to physicochemical and rheological properties due to variabilities in cheesemaking conditions, depending on the farmhouse producer (Casalta et al. 2001).

Boxe goat cheese from Poitou (PDO) is made from fresh, whole goat milk. The milk is renneted, and the curd is poured into molds, salted, and then aged for 8 days. At this time, the cheese is covered with a white to mottled rind. *Brocciu*, also called Broccio (PDO), is a whey cheese made from either goat or ewe serum at goat farms of dairy plants of Corsica. For the production of *Cabecou d'Autan*, the milk is coagulated directly in small molds, salted, and ripened for 10 days. *Cabecou de Perigord* is made in Perigord from local goat milk. It has a creamy, fine texture and a smooth, light yellow rind. The lactic curd ages for at least 7 days. *Calenzana cheese* is made from either sheep or goat milk from local breeds. The milk is renneted, the cheese is lightly salted, and then, it is either consumed fresh or ripened for 4 (spring cheese) or 10–12 months. *Chabichou du Poitou* (PDO) is made from whole milk. The milk is coagulated, and after draining, the cheese is salted by dry salt and ripened for at least 10 days. *Charolais* is a farmhouse cheese with a soft paste. The milk is curdled with lactic culture for 24 h, and after draining (2–3 days), it is dry salted and ripened for 15 days. During ripening, the cheese is covered with a natural moldy rind. *Crottin de Chavignol* (PDO) is made from raw full cream goat milk that is slightly renneted, and the lactic coagulation lasts about 2 days. After draining, the cheese is salted and then aged for 10 days. *Feuille du Limusine* has a fine, white, and soft paste. The milk is coagulated by lactic fermentation. After removal from leaf-shaped molds and salting, the cheese ripens for different times. *Mâconnais* is also called Chevroton de Macon and is made by farmers and artisan cheesemakers with flocks of goats fed in the fields between the vineyards of the Maconnais area. After curdling and salting, the cheese ripens for 12 days or longer. During ripening, the cheese must be at least covered by *Geotrichum* mold. Its body is white, smooth, and firm. *Mothais sur Feille* cheese is made at farms by slow curdling. After removal from the molds, the cheese is salted and then set on a chestnut leaf to dry and obtain its characteristic taste. The cheese develops a fine, white, soft paste and a yellow-ochre thick rind with small, gray–blue spots. *Pelardon cheese* (PDO) is mainly made at farms by a slow lactic fermentation for 24 h. The cheese ripens for a maximum of 11 days. Its body is fine and white and has a smooth rind covered in a lightly yellow, blond, or blue-colored mold. *Picodon cheese* (PDO) is made from milk curdled over 24 h at 20°C by the addition of rennet and whey. It is then drained for 24 h. Aging lasts for at least 8 days. The paste of the cheese is spicy and dry, and its texture is smooth and fine. The young cheese has a soft white rind and a fresh taste. *Pouligny Saint-Pierre* (PDO) cheese is made from goat milk from Alpine, Saanen, and Poitevine breeds fed with a specific diet. The milk is slowly curdled by animal rennet, molded, and then surface salted. The cheese ripens for at least 7 days. *Rocamadour* (PDO) cheese is manufactured from milk obtained from Alpine or Saanen breeds. The milk is coagulated for 24 h. The cheese is molded and ripened in cellars for at least 6 days, where it develops a soft paste and buttery taste. Its rind is grooved skin, slightly soft, and velvet. *Saint-Maure de Touraine* (PDO) cheese is made by a slow curdling. After removal from the molds, the cheese is transferred onto pyroengraved rye straw. The cheese is then salted and displayed on a wicker, where the draining will end. The product ripens for at least 11 days. The cheese has a white, soft paste under a grayish moldy rind (Rubino et al. 2004).

In Saint-Maure cheese made from raw milk sprayed with *Geotrichum* mold spores and then ripened for 31 days, the cheese pH was low (~4.3) in the fresh cheese and remained low in the cheese center during ripening; however, a large pH increase was noticed on the cheese surface (from 4.4 in the fresh cheese to 5.1 on the 31-day-old cheese) due to the mold surface flora (Le Quere et al. 1998). The casein level was decreased only by 5% during ripening, and this increased the nonprotein nitrogen (NPN) fraction by 3.5%. None of the casein fragments (α_{s1}-, α_{s2}-, and β-CN) was preferentially hydrolyzed. This pattern of proteolysis indicates activity by *Geotrichum candidum*. A high lipolysis level at the end of ripening was noticed due to the activity of the milk lipase initially, which gave rise to the saturated C6–C10 FFA, and then to the *Geotrichum* lipase, which caused a high increase in the C18:1 and a lower increase in the C16:1 and C18:2 amounts at the end of ripening. Among the 38 compounds identified in the volatile fraction, seven were characterized as having a specific cheese or goat cheese aroma—hexanoic, octanoic, nonanoic, decanoic, 3-methyl-butanoic, 4-methyl-octanoic, and 4-ethyl-octanoic. The last two have been reported as bearing a nuttony/goaty flavor, and they were present at levels in the range of their threshold values along with C8 and C10 in the fresh cheese. On the whole, eight main aromatic compounds contributing to the goat cheese flavor were identified in the 31-day-old cheese. In addition to the above-mentioned four found in the fresh cheese, butanoic, 3-methyl-butanoic, hexanoic, and heptanoic also reached their threshold values at the end of ripening.

Tomme des Pyrenees is a pressed, uncooked cheese made from cow, sheep, and/or goat milk. The milk used for its manufacture is kept at 15°C for 12 h for ripening and then fermented before renneting. After draining for 24 h, the cheese is salted and aged for 30 days for the small size and 60 days for the larger sizes. The color of the cheese paste is white (Rubino et al. 2004).

17.6.1.2 Italy

Italy is very rich with respect to cheese varieties manufactured in each region. In general, northern Italian areas have a terrain better suited to cows, and southern Italian regions have long been the domain of shepherds, so cow and sheep milk cheeses prevail in the respective regions. Goat farming is located mainly in the Alps and the Center-Southern and major islands (Sardinia and Sicily) of Italy (Di Cagno et al. 2007). Almost all milk from goats coming from many different goat breeds, alone or mixed with either cow or ewe milk, is transformed into cheese. A PDO status is unusual for the Italian goat milk cheese, and the scientific reports on their characteristics are very few. However, there are some PDO cheeses in Italy (Bitto, Bra, Castermango, Raschera, Valle d'Aosta, and Fromadzo) made from cow milk to which small amounts of goat milk are added (Medina and Nuñez 2004).

Rubino et al. (2004) described 29 Italian goat milk cheese types with respect to their origin, appearance, and technique of processing.

Agrino delle Orobie is a mountain cheese made from the Orobica goat breed. Fresh raw goat milk is renneted, and a glass of whey from the previous cheesemaking is then added. The curd is drained, salted, and either consumed fresh or after ripening for some months. The aged cheese is often covered by a natural mold. *Blu Grater* cheese is produced from milk from local goats. The milk is curdled, the curd is salted, and after a week, it is perforated to ease the growth of mold (*Penicillium roqueforti*). The cheese (soft and creamy) taste is slightly spicy. *Cachet* cheese is a combination of goat milk and goat cheese. Leftover cheese and fresh goat milk are mixed, and this way, the milk is fermented quickly. After remixing, the cheese is seasoned with a distillate of juniper or an infusion based on leeks. The cheese ripens for a few days or several weeks. After maturing, the cheese has a soft, creamy, spreadable consistency, white or creamy color, and intense aroma and taste. *Caccioricotta* cheese is made from the bulk milk of Mediterranean goat breeds. The cheese is made from half cheese protein and half whey protein. The milk is heated to 85°C–90°C and cooled to 48°C. At this point, a paste of kidskin or lamp is added. The curd is put in wicker baskets, and it is dry salted.

Albenzio et al. (2006) studied the characteristics of caccioricotta cheese made from Garganica goat milk. The cheeses were made from farm milk and four different flocks. Cheeses showed mean values of pH and water activity of 6.38 and 0.95, respectively. The fat content of the cheese differed significantly from 11.2% to 22.7%, possibly due to differences in fat recovery during cheesemaking as well as to different enzymatic compositions of kid rennet produced on each farm. The protein content ranged from 15.4% to 19.4%, and the nitrogen fractions deriving from protein hydrolysis differed between the cheese batches; the same was observed for moisture content and plasmin activity. Cheeses with low moisture content were characterized by a more accelerated ripening process.

Taurine is the most abundant amino acid in goat milk and was found to be particularly high in milk from Mediterranean goat breeds; its importance in the growth of newborns and young children has been well established (Pasqualone et al. 2000). Caccioricotta cheese appears to be a good source of taurine, especially fresh cheese, because a significant decrease in taurine content during caccioricotta ripening is observed.

Caciotta caprina del Matese is produced from whole raw milk renneted (with kid rennet), cut, drained, and salted by dry salt. The cheese ripens for 1–2 months. The mature cheese tastes sour and strong, and its aroma is delicate and pleasant. The cheese paste is white, and its structure is compact. *Canestrato d'Aspromonte cheese* is produced from pure goat milk, pure sheep milk, or a mixture of both. The milk is coagulated by liquid rennet, and the curd is sliced with a wooden knife from the wild pear tree. The curd is transferred in rush baskets and pressed, dry salted, and ripened for 2–3 months. The mature cheese has a light covering of white mold (Rubino et al. 2004).

The cheese is also known as Caprino dell' Aspromonte. Micari et al. (2007) studied several physico-chemical parameters and hygienic characteristics of Caprino dell' Aspromonte cheese made in winter and spring. The mean pH of the cheese made in winter was 5.67, whereas the spring cheeses exhibited a mean pH value of 5.42. In addition, a mean dry matter value of 56.68% was recorded for the cheeses made in winter and a higher mean value (62.71%) for cheeses made in spring. The mean value for fat content (% dry matter) of the cheeses made in winter was 47.43%, and the NaCl content had a mean value of 2.09%. The mean fat content of spring cheeses was 46.76% and that of NaCl% was 2.58. Crude protein content (% dry matter) in the cheeses produced in winter had a mean value of 36.22 and was higher than mean values (26.49%) exhibited by spring samples. A variability in the numbers of coliforms (0–5.04 $\log_{10}$ cfu/g), fecal coliforms (0–4.96 $\log_{10}$ cfu/g), and yeasts (2.60–4.95 $\log_{10}$ cfu/g) was recorded.

Under the name "caprino" (from "capra," the Italian name for goat), several cheese varieties are made in Italy, soft, semihard, and hard, which usually are named after the area of production—caprino a pasta cruda, caprino della carnia, caprino della val vigezza, caprino semicotto, caprino valsesiano, caprino di rimella, caprino ossolano, caprino di cavalese, caprino della valbrevenna, caprino dell' aspromonte, caprino della limina, and caprino di montefalcone del sannio (Rubino et al. 2004).

Di Cagno et al. (2007) studied the compositional, microbiological, and biochemical characteristics of four Italian semihard cheeses in order to identify the most appropriate characters suitable for obtaining a "denomination of origin." Moisture content at the end of ripening was measured in the range of 28.2% (caprino di valsassina)—34.0% (flor di capra). The lowest value of fat content was 30.3% (flor di capra), and the highest was 33.1% (caprino di cavalese). The NaCl content was <1.1% (w/w) in all cheeses. The lowest and highest contents were recorded for flor di capra (23.0%) and caprino di valsassina (27.5%), respectively. The pH values varied by 0.3 between the four cheeses and were >5.0 for all. The concentration of FFA, on the other hand, was highest for caprino di valsassina (17.94 mg/g) and caprino di cavalese (15.76 mg/g), intermediate for flor di capra (10.39 mg/g), and lowest for capritilla (5.39 mg/g). The four cheese types had been produced from milk obtained from the following respective goat breeds—Orobica, Saanen, Sarda, and Alpina. The determination of free amino acid (FAA) profile suggested a clear differentiation of caprino di valsassina and caprino di cavalese from the other two cheeses. In addition, only the concentration of five FAAs (Glu, Cys, Tyr, His, and Trp) significantly differentiated the profiles of caprino di valsassina and caprino di cavalese. However, FFAs and esters qualitatively and quantitatively differentiated these two cheeses. The concentrations of butanoic, nonanoic, decanoic, dodecanoic, tridecanoic, tetradecanoic, pentadecanoic, *n*-hexadecanoic, and 9-octadecanoic acids differentiated caprino di valsassina from the other cheeses. Almost all the above, along with 3-methyl-butanoic, hexanoic, heptanoic, and octanoic acids differentiated caprino di cavalese cheese. In addition, the alcohols 11-dodecanol, 2-methyl-2-buten-1-ol, and 9-octadecen-1-ol characterized caprino di valsassina, whereas 2-methyl-2-buten-1-ol, 4-methyl-phenol, and phenylethyl alcohol statistically differentiated caprino di cavalese cheese. Among ketones, 2-heptanone and acetophenone distinguished caprino di valsassina and caprino di cavalese from the other cheeses, respectively. The lactones γ-nonalactone, δ-decenolactone, and δ-undecalactone were the most abundant in all the four cheese types and are well known as cheese flavor components, giving pleasant, buttery, and fruity sensory attributes. The highest concentrations of alcohols, esters, ketones, and lactones were recorded for caprino di valsassina cheese. At the end of ripening, the populations of nonstarter LAB in the four cheese types ranged between 7.98 and 8.51 $\log_{10}$ cfu/g, with *Lactobacillus paracasei*, *Lb. casei*, and *Lb. plantarum* being dominant in almost all cheeses.

Enterococci constituted a significant part of the microbial population of *Semicotto Caprino* cheese, with *Enterococcus faecalis* and *Ent. faecium* being the most frequently isolated species, followed by *Ent. durans*. The isolates exhibited proteolytic activity and antagonistic activity toward *Listeria innocua*. The highest acidifying potential in skim milk was shown in *Ent. faecalis* isolates (Suzzi et al. 2000).

Caprino dei Nebrodi is a raw goat (from the breed Capra Messinese) cheese produced in Sicily. Raw goat milk from morning and evening milking is coagulated at 35°C with kid rennet within 30–40 min. The curd is cut into grain-size pieces and then drained in molds for 24 h. The cheese is surface salted and ripened on wooden boards for 30 days at 18°C. Microbiological studies on milk and cheese showed that LAB represented the most prominent microbial group in the milk, and the number of microorganisms indicative of the bacteriological quality (*Enterobacteriaceae*, *Escherichia coli*, staphylococci, and micrococci) was also high. Lactobacilli and lactococci predominated and induced a notable decrease in pH (5.32 at 30 days). Lactobacilli increased significantly during maturation, and enterococci were also found in abundance and tended to increase their numbers during maturation. Ripening time had an influence on the survival of *Enterobacteriaceae* and *E. coli*, because they underwent a significant decline during ripening. *Micrococcaceae*, on the other hand, were counted at significant numbers in the ripening cheeses (Zumbo et al. 2009).

Casieddu cheese is made from pure goat milk. The milk is initially filtered through ferns, seasoned with a little bag of Nepota leaves, and then heated to 90°C and renneted (kid rennet). After curdling, the coagulum is shaped into orange-size balls. Each of them is enveloped with fern leaves and covered with hay. *Caso conzato* is produced from raw goat and sheep milk. After curdling (kid rennet), the coagulum is cut, drained, and then salted. The cheeses are washed with cooking water of a traditional paste (pettola), dried, and rubbed with an emulsion of olive oil, white vinegar, white thyme, and local red chili peppers. They are then put in clay containers for 6–24 months. *Caso peruto* is produced from raw goat milk coagulated with vegetable rennet (*C. cardunculus*). The coagulum is then cut to drain and dry salted. After 1–2 weeks, the cheeses are seasoned with vinegar and olive oil and covered with aromatic wild thyme leaves. Aging lasts for at least 1 year. *Cevrin di Coazze cheese* is made from goat milk or a mixture of goat and cow milk (at 60%–70%). The milk is coagulated, and the curd is cut and drained. After dry salting, the product is ripened for at least 20 days but frequently longer than 120 days. *Fatuli cheese* is produced from goat milk obtained from a local breed called Biond dell' Adamello. Animal rennet is used to curdle the milk. The curd is cut and then mildly scalded, pressed to drain, dry salted, and smoked in a juniper wood fire. The cheese matures for 30–180 days. *Felciata cheese* is produced from goat milk fed exclusively from pastures. The milk is filtered with fresh fern leaves and coagulated with rennet paste. The cheesemaker then lines a mulberry or walnut mold with green fern leaves and transfers the curd with a perforated maple wood ladle into it in layers separated by fern leaves. *Formagella del Luinese* is made from raw, whole goat milk obtained mainly from three goat breeds (Camosciata delle Alpi, Saanen, and Nera di Verzasca). It is curdled by natural rennet after adding a natural selected starter. The cheese is aged for at least 20 days. *Formaggio a pasta molle della Valle d' Aosta* is produced from the milk of animals fed with pastures. The milk is coagulated with liquid or powdered rennet. The cheese is usually dry salted and ripened for 20–25 days. *Jama cheese* is a soft goat cheese made from raw or thermized goat milk with selected cultures. The milk is curdled, and then, its paste is kneaded and salted. Sometimes, its paste is mixed with aromatic herbs. *Juncata Calabrese cheese* is produced from either goat or sheep milk or a mixture of the two. The milk is curdled with kid rennet, and the coagulum is cut and put into rush baskets to drain. The cheese is consumed either fresh or after smoking. *Musulupu cheese* is made, mainly, from goat and sheep milk coming from animals that graze principally on the pasture. The milk is coagulated by liquid or paste rennet, and the coagulum is cut and pressed to drain.

The curd is then scorched in whey and put in "Musulupara," a specific mold made of mulberry wood (Rubino et al. 2004).

Goat ricotta cheese is produced from whey from goat cheesemaking and some quantity of milk (up to 20%). Pizzillo et al. (2005) studied the effect of goat breed on the sensory, chemical, and nutritional characteristics of ricotta cheese and found that properties such as softness, granulocity, and greasiness were affected significantly according to the breed. They also observed that goat odor was more pronounced in cheese made from the whey from Syriana and Maltese goats and the same was observed for the fatty acid profile. Higher percentages of butyric acid as well as saturated caprinic, lauric, and palmitic acids were detected in ricotta cheese made from the whey of local breed compared to the other breeds. The cheese made from the whey coming from Girgentana goats had greater softness, lower granulocity and "goat flavor," and higher levels of monosaturated and polyunsaturated fatty acids. Therefore, this product satisfies the consumers' demand with respect to sensory and nutritional properties.

Robiola del bec cheese is made from partially skimmed goat milk, coagulated by a few drops of calf rennet. The curd is drained, dry salted, and usually aged. The cheese is made in September and October, when the goats go into the heat, and was named after "bec," the male goat (Rubino et al. 2004).

Robiola di roccaverano cheese is the only Italian goat cheese with a PDO. The artisanal cheese is made from milk coming from two milkings, which is curdled by calf rennet and dry salted after draining. The cheese normally ripens for 3 days, but some producers keep it in glass jars for months. Bonetta et al. (2008) studied the microbiological characteristics of this cheese. Pathogenic bacteria such as *Staphylococcus aureus*, *Salmonella* spp., and *Listeria* spp. were not detected in the cheese. Cheese ripening for 20 days did not affect significantly the growth of the various microbial populations, although an increase of some microbial parameters was recorded. Similarly, no significant difference in microbial counts was observed according to the season, but in general, higher counts were detected in the summer. Bacterial counts in the industrial cheese samples were lower than the artisanal products. *Lc. lactis* subsp. *lactis* and *Lc. lactis* subsp. *cremoris* were the most widespread bacteria, but no lactobacilli were detected. Some species, such as *Lc. garvieae*, *Streptococcus parauberis*, and *Str. macedonicus*, were detected only in the artisanal cheese. With respect to the bacterial ecology, a comparison of the denaturing gradient gel electrophoresis (DGGE) profiles corresponding to the winter and spring seasons did not show any differences among cheese products, whereas the industrial products and the starters exhibited similar profiles in winter. On the contrary, in the summer season, the cluster analysis of DGGE profiles discriminated the cheeses according to the producers. Therefore, different manufacturing and climatic conditions allowed the differentiation of the ecology of each product. The DGGE profiles of yeasts were affected by ripening and the season.

Scuete cheese is a whey cheese and exists in two types—fresh and smoked. *Tomino di talucco cheese* is a fresh cheese made from goat or a mixture of goat and cow milk. Whole milk is heated to 90°C and then cooled and curdled by liquid rennet. The curd is broken and then molded and salted. *Vecjo di cjavre cheese* is a hard/semihard cheese made from raw or mildly heated milk by the addition of milk from the previous production or selected bacteria in warm milk and liquid rennet or kid rennet paste. After curdling, the coagulum is cut and then cooked (42°C–43°C), molded, and dry salted (Rubino et al. 2004).

17.6.1.3 Spain

Among Mediterranean countries, Spain occupies the third position after France and Greece in milk production (FAO 2006). However, only around 12% of the amount of milk is transformed into cheese (MAPA 2002). Twenty-eight varieties are made exclusively from goat milk, and four are PDO cheeses (Medina and Nuñez 2004), whereas for the production of around 20 non-PDO traditional cheeses, goat milk mixed with cow and/or ewe milk is used.

Queso Alhama de Granada is an uncooked, pressed cheese made from fresh, whole, raw goat milk with animal rennet. The curd is cut with a fig stick, and it is drained and salted and is then ripened on straw mats for 45–70 days. *Queso Aracena* is made from fresh raw goat milk. The milk is renneted with animal rennet. After curdling, the curd is crumbled and scooped into traditional molds from tin or fig bark. After pressing, the cheese is dry salted and then aged on pine and chestnut slats for 60–90 days (Rubino et al. 2004).

Armada is a farm cheese made from raw milk by the addition of whey from the previous cheesemaking and animal rennet. After coagulation, the curd is cut and drained in cheese cloths for 48 h, kneaded, and continued draining for 72 h, when salt is added. After a new kneading, the curds are molded, wrapped in a cloth, and hung to ripen (at 10°C–15°C) for 2–4 months. Lactose content disappears almost completely within the first 7 days of ripening (Fresno et al. 1996), when the pH declines to levels 4.31–4.68, increasing again later to 4.89–5.25 in day 120 (Tornadijo et al. 1993). Water activity (a_w) levels decline progressively during ripening, reaching very low values (0.895; Fresno et al. 1996), whereas the DM increases to 75%–82% on day 120 (Tornadijo et al. 1993). The NPN increases progressively from 5% on day 7 to 7.3% after 120 days of ripening, when residual α_s- and β-CN are found at significant levels (93% and 98%, respectively). Total FAAs increase from 2.1 to 3.6g/kg from day 7 to day 120, respectively, whereas the respective amounts of FFAs at these times were 5.9 and 44.5g/kg (Fresno et al. 1997). Magnesium, calcium, phosphorus, and zinc were the only minerals that underwent variation in content during ripening and reduced significantly in the first days of ripening, coincidently with the decrease in pH values (Fresno et al. 1996).

The highest counts in all microbial groups were recorded after 1 week of ripening (Tornadijo et al. 1995). They then dropped gradually until ripening for 120 days, when the numbers in cheeses made in summer were significantly lower than those determined in autumn. LAB predominated over the other microbial groups during manufacture and ripening, with lactococci (*Lc. lactis* subsp. *lactis* and *Lc. lactis* subsp. *cremoris*) dominating in the curd (70% of the isolates) and 1-week-old cheese (70%). Lactobacilli (mainly *Lb. casei* subsp. *casei* and *Lb. plantarum*) prevailed during the last stages of ripening.

Babia-Laciana cheese is made from raw goat milk from indigenous breeds. The milk is curdled, and the curd is slowly pressed. The product is acidic in taste and buttery in the mouth. The TS content in the cheese increased progressively on ripening and reached mean values 78% after 60 days. In addition, protein and fat contents (%TS) changed very little, reaching final mean values of 61.1% and 32%, respectively. The final respective mean levels of NaCl and ash were 1.1% and 2.8% (%TS), respectively. Lactose was degraded gradually with ripening to a mean value of 1.6% (%TS). The cheese pH decreased from a mean value of 6.28 in the curd to 4.60 after 3 days and remained more or less constant thereafter. Water activity underwent a gradual decrease from 0- (mean 0.993) to 60-day-old cheese (mean 0.934), and among the mineral elements, calcium and phosphorus were the most abundant. Babia-Laciana cheese undergoes very little proteolysis during ripening. Proline was the main amino acid in the curd and gamma-aminobutyric acid after 60 days. The fat acidity increased approximately 4.5 times during cheese ripening, and the total FFA content showed a similar evolution (Franco et al. 2003).

Queso Cadiz is produced from fresh, raw, whole goat milk, curdled with animal rennet. The curd is cut with a thorn wooden spatula and drained in esparto molds. The cheese is dry salted. *Fromatge de Cabra de Pirineu* is made from raw goat milk by lactic coagulation and slow draining (Rubino et al. 2004).

Queso Garrotxa is made from pasteurized milk by enzymatic curdling. Curds are pressed and salted in brine. The cheese is ripened (14°C) until the development of a proper mold (6–8 weeks). Increases in the proteolysis of Garrotxa cheese were found to be of small magnitude after treatment at 50 MPa, but treatment at 400 MPa at the beginning of cheese ripening has been revealed as a convenient alternative to accelerate Garrotxa cheese ripening (Saldo et al. 2000). Pressure treatment at 400 MPa for 5 min had as a result a lower amount of FFAs and, in general, less volatile compounds than in the untreated cheese (Saldo et al. 2003). On the contrary, noncasein and NPN as well as humidity were higher with pressurized cheese. Bacterial counts were significantly reduced compared to the untreated cheese, and the two cheese types differed substantially.

For the production of *Queso Gomero*, the milk is curdled by kid rennet, after being salted first. The curd is broken, pressed by hand in the mold, and dry salted. After a day, the cheese is smoked, and it is consumed either fresh or after aging from 3 to 6 months (Rubino et al. 2004).

Queso Gredos is a farm-made cheese. Milk is renneted, and the curd is cut and then scooped into molds made out of *Celtis australis* bark and pressed by hand. It is dry salted, ripened for 15 days, and then immersed in olive oil for 45 days at 8°C–10°C (Medina et al. 1992). The microbial population in 1-day Gredos cheese ranged between 2.62 (coagulase-positive staphylococci) and 6.95 $\log_{10}$ cfu/g (total viable counts). *Enterobacteriaceae*, coliforms, and fecal coliforms reached their maximum levels in the

4-day-old cheese and lactobacilli in the 15-day-old cheese. After the immersion of the cheeses in olive oil at 15 days, a small reduction in the levels of most microbial groups was observed, but after 2 months, *Enterobacteriaceae*, coliforms, and fecal coliforms were still present in the cheese (10^3 and 10^2 cfu/g, respectively). Coagulase-positive staphylococci were not detected in the cheese from 1 month onward. The cheese pH fell from 6.79 in the curd to 4.64 in the 45-day-old cheese, and moisture was 58.0% on day 15 and then stabilized. Residual α_s- and β-CN in the 60-day-old cheese were 22% and 40%, respectively, and although a decrease in N fractions was recorded in the 1-day-old cheese, due to prolonged whey drainage, a gradual increase in all soluble N fractions was observed from this time until the end of ripening.

Queso Ibores is produced from milk obtained from the Serrana, Verata, and Retinta goat breeds. When pasteurized milk is used for its production, specific starters prepared with autochthonous microorganisms should be used (Mas et al. 2002). It is a semihard cheese made with lactic and enzymatic techniques. The aging period of Ibores cheese is 2 months.

Gonzalez et al. (2003) studied the effect of *Lc. lactis* subsp. *lactis* autochthonous strains on the microbiological, physicochemical, and sensorial characteristics of Ibores cheese, along with cheeses made without starter and/or commercial starter. The cheeses were differentiated, with respect to pH only at the first day of ripening, with the pH of cheese made with any type of starter lower than in cheese without starter. The levels of nitrogen fractions SN/TN and TCA SN/TN were significantly higher in cheeses without starter after 7 and 30 days of ripening. The evolution of LAB counts during ripening was similar for all cheese types, but enterococci in cheeses made with the commercial starter were significantly lower throughout ripening. Moreover, in cheeses made with the autochthonous starters, the levels of coliforms and coagulase-positive staphylococci declined faster than the other cheese types. The cheeses made with two of the autochthonous starters received the better scores. Mas et al. (2002) also investigated the microbiological and physicochemical changes throughout the ripening of Ibores cheese from raw goat milk. The cheese pH decreased sharply to 4.98 on day 30 and then increased to 5.18 at day 60. The NaCl content increased gradually to 2.5% by day 60, when the mean fat content was 52.6%. The products of proteolysis (SN/TN and TCA SN/TN) increased with the progress of ripening. During ripening, the counts of total LAB and lactococci reached their peak growth on day 15, whereas those of enterococci on day 30. Coliforms reached maximum counts after 3 days and decreased thereafter by about 3 log units, and *S. aureus* declined gradually and significantly with ripening time. Lactococci (*Lc. lactis* subsp. *lactis*) formed the prevalent group throughout ripening, followed by leuconostocs (*Ln. mesenteroides* ssp. *dextranicum* and ssp. *mesenteroides* and *Weisella paramesenteroides*) and enterococci (*Ent. faecium* and *Ent. faecalis*) found at similar levels, whereas lactobacilli (*Lb. casei* and *Lb. plantarum*) were counted at low levels.

Lactic curd cheese is made from pasteurized goat milk, curdled by a slow lactic fermentation. The curd is drained and accepted two dry saltings. The cheese is then ripened for 2–3 weeks on shelves at 10°C–12°C (Rubino et al. 2004).

Queso Majorero is made from goat milk from the Majorera breed. It is a PDO cheese, made from raw or pasteurized milk. The milk is curdled at 32°C with animal rennet, and the coagulum is broken, molded, and hand pressed. The cheese is dry salted and ripened for 8–20, 20–60 or over 60 days, depending on the cheese type. The fresh cheeses have a sweet, delicate flavor, and the aged are slightly acid and spicy. In the fresh cheese (2 days), coliforms and staphylococci reached counts of 10^6–10^7 and 10^4–10^5 cfu/g, respectively, and declined to $<10^1$ cfu/g after ripening for 90 days. *Lc. lactis* subsp. *lactis* and *Lc. lactis* subsp. *cremoris* predominated during ripening. *Lb. casei* and *Lb. plantarum* were also present, as well as *Ln. mesenteroides* and *Ln. paramesenteroides*, *Ent. faecalis* var. *faecalis*, and *Ent. faecalis* var. *liquefasciens*. At the end of ripening (90 days), the NPN was 19% of the total N, residual α_s- and β-caseins were 27% and 76%, respectively, and the cheese pH was 5.44 (Fontecha et al. 1990).

An investigation on the evolution of microbial flora in Majorero cheese made from pasteurized milk and lactococci (*Lc. lactis* subsp. *lactis* and *Lc. lactis* subsp. *cremoris*) as starter revealed that, initially, the total counts increased rapidly and later stabilized or decreased. Lactobacilli increased their counts strongly during the first 30–60 days of ripening and were the main components after 90 days. *Lb. casei* var. *casei* was found to be the most representative species (36% of the lactobacilli). The enterococci never exceeded 5×10^3 cfu/g and disappeared by the end of the second month. Micrococci and staphylococci reached their maximum levels after 10–30 days and then decreased to 10^2 cfu/g. Coliforms disappeared early, and yeasts disappeared after 2–3 months (Gomez et al. 1989).

High levels of neutral protease activities in Majorero cheese is reflected in the high soluble nitrogen/TN level. The increased protease activity was due not only to the enzymes of the residual rennet but also to endocellular and exocellular enzymes deriving from the cheese microflora. Aminopeptidase activity was also detected, with that of leu-aminopeptidase being increased with ripening. No carboxypeptidase activity was detected (Gomez et al. 1988).

Martin-Hernandez and Juarez (1992) studied three different types of goat cheese—fresh, semihard washed curd, and Majorero. The biochemical characteristics of the fresh cheese did not alter on storage for 15 days. The water-soluble N and the NPN fractions in the semihard, washed curd, and Majorero cheeses increased during ripening. At the end of ripening (2 and 3 months, respectively), the percentages of water-soluble N in the two cheeses were 41.1% and 28.1%, and those of the NPN fraction were 55.7% and 51.6%, respectively. The percentages of degradation of the α_s- and β-caseins were similar (35%) at 2 months for the semihard and washed curd cheeses. The α_s- and β-caseins of Majorero had been degraded by 54% and 19%, respectively, after ripening for 2 months. The FFA contents increased slightly in all three goat milk cheeses, indicating a low level of lipolysis in these cheeses.

Queso Malaga is made from raw whole milk curdled by animal rennet. The curd is cut with an oak stick and drained in esparto molds, where it is hand pressed. The cheese is dry salted, and it is consumed either fresh or after ripening for 30–60 days. *Queso Mató* is produced with pasteurized milk heated to 80°C and then cooled to 35°C–36°C for curdling, usually with animal rennet. It is poured into containers and is ready for consumption. *Queso Murcia al vino* has a pressed paste, washed and uncooked, with creamy and elastic texture (Rubino et al. 2004). The cheese curd is washed, and its taste is pleasantly acidic with a light aroma. Its rind is rubbed twice with red wine (Tejada et al. 2008a).

Tejada et al. (2008b) studied the compositional characteristics of Murcia al vino cheese curdled with either animal rennet or powdered vegetable coagulant. The pH and TA were affected by the type of coagulants, with TA being significantly higher until up to 30 days of ripening in cheese made with vegetable rennet due to a higher concentration of FAAs in this cheese type. The total solids of the cheese were not affected significantly by the kind of coagulant but by the type of ripening. On the contrary, fat, protein, and ash content did not change significantly during ripening. However, while fat and protein content were unaffected by the kind of coagulant, the ash content was affected significantly, and after 15 days of ripening, it was higher in cheese made with animal rennet. A significant increase in NaCl content with the time of ripening was also observed. There was an increase in the Na, Ca, P, Zn, Cu, and Fe components with ripening. The esterified fatty acids were not affected significantly by either the type of coagulant or the ripening time, with the most abundant being the C16 and C18:1.

In the cheeses produced with plant coagulant, the proteolysis was more intense compared to cheese made with animal rennet, and after 60 days of ripening, values of 34.79%, 11.16%, 0.59%, and 0.88% for water-soluble N, nonprotein N, ammonia N, and amino acid N were recorded. Similarly, casein proteolysis was more intense in cheese manufactured with plant coagulant, and the same was observed with hydrophilic and hydrophobic peptides (Tejada et al. 2008a).

Queso Palmero is an artisanal smoked PDO cheese produced from the Palmera breed. The milk is curdled with kid paste rennet, and the curd is broken and then molded to drain with strong pressure. The cheese is salted by sea salt and then ripened for 7–15 (fresh), 15–60 (semiripened), or >60 (aged) days. Before ripening, the cheese may be smoked (Rubino et al. 2004). The sensory properties of this product are influenced by the local breed and the climatic, geographic, and cheesemaking practices (Fresno et al. 2005). In cheeses smoked with almond shells (*Prunus dulcis*), segmented cactus (*Opuntia ficus indica*), and needles and wood of Canary Pine (*Pinus canariensis*), authorized smoking materials were tested by experts, and a good distinction between the products was found (Picon et al. 2007).

Guillen et al. (2007) studied the safety of smoked cheeses by almond shells and/or dry prickly pear with respect to the occurrence of polycyclic aromatic hydrocarbons. It seems, that the nature of the material used for smoking influences the type of polycyclic aromatic hydrocarbons accumulated, especially the alkyl derivatives and some light polycyclic aromatic hydrocarbons. The two cheese types had a very similar profile with respect to these substances, in amounts that guarantee the cheese safety.

Queso de cabra de l'Alt Urgell is made by the acid coagulation of milk. In the milk, rennet is also added. Without breaking, the curd is slightly drained and put into molds to age for about 2 months. It is rubbed with olive oil and aromatic herbs. *Queso de Suero* is made from the leftover whey of various goat cheese types, warmed at 90°C–95°C. *Queso Sierra Morena* is made from fresh, whole goat milk curdled by animal rennet. The curd is cut by a big stick and then molded (esparto or other molds), squeezed and lightly hand pressed, and then dry salted. Fresh or ripened (45–60 days) cheese is made (Rubino et al. 2004).

Tenerife is a traditional farmhouse cheese variety made from the raw goat milk of local goats. The milk is coagulated by animal rennet, and the curd is molded, pressed by hands, and salted by coarse salt. The cheese is consumed either fresh or after ripening for 30–60 days at 15°C–25°C (Zarate et al. 1997). *Enterobacteriaceae*, coliforms, and *S. aureus* decreased rapidly with the ripening of this cheese. Lactococci, leuconostocs, and lactobacilli predominated during ripening, with *Lc. lactis* subsp. *lactis* being the most abundant *Lactococcus* (78.9% of lactococci). *Lb. plantarum* predominated over the other lactobacilli initially, but *Lb. paracasei* subsp. *paracasei* increased its population as ripening progressed and dominated in the 60-day-old cheese. *Ln. mesenteroides* subsp. *dextranicum* was the main species among leuconostocs, and its numbers decreased, especially during the last month of ripening, whereas the enterococci *Ent. faecalis* and *Ent. faecium*, micrococci, yeasts, and molds were found in significant populations (Zarate et al. 1997).

For the production of *Trochón*, goat milk is curdled with animal rennet, and the curd is cut, mildly heated, molded, hand pressed, and dry salted. The cheese is consumed fresh or after ripening for 2 months (Rubino et al. 2004).

Valdeteja cheese is a homemade hard cheese type produced from raw, whole goat milk, and enzymatic coagulation. The cheese ripens for 3–4 weeks and has a yellowish, dry rind, its paste is white, with little holes, and it has a strong flavor. Lactococci, lactobacilli, and leuconostocs were the dominant microflora throughout cheese ripening, with lactococci predominating over lactobacilli until up to 27 days of ripening. Leuconostocs stabilized their population from day 2 onward, and final log counts of enterococci were 3.62 cfu/g. *Enterobacteriaceae*, *Micrococcaceae*, and aerobic mesophilic bacteria reached their peak growth in the curd and decreased significantly during ripening. On the contrary, yeasts and molds increased significantly to 10^6 cfu/g (Alonso-Calleja et al. 2002).

The cheese has high total solids content (72.53%) and fat (59.5%) and a low pH value (~4.5). Studies on cheese samples showed average values of TA of 1.39 g/100 g TS, L-lactate 0.91 g LA/100 g TS, and D-lactate 0.47 g/100 g TS; a_w 0.906. In ripening cheese, lactose disappeared completely after day 17, in parallel with a continuous increase of D-lactic acid. The D-lactic acid increased rapidly until the fifth day and then decreased, with an appreciable fall in pH and an increase in TA until 5–10 days. Total solids content increased almost linearly with time (from 45% in the curd to 72.4%) at 27 days, and a_w progressively and constantly decreased. The ash and NaCl percentage of TS increased quickly in the fresh cheese as a result of salt penetration. Short-chain fatty acids, acetic and butyric, increased significantly after 5 days from the amounts in the curd, and hexanoic acid rose after day 2. Nitrogen compounds were not degraded much (Carballo et al. 1994).

17.6.1.4 Portugal

There are two Portuguese cheeses produced from plain goat milk—Algavre and Cabra Transmontano (Rubino et al. 2004).

Algavre cheese is produced from the milk of pasture-fed breed called Algavre. The milk is curdled by thistle rennet and is then broken and shaped. The cheese is consumed either fresh or aged. *Cabra Transmontano* is a cheese under a PDO status and is made from the milk from Serrana goat. The milk is coagulated with animal rennet and thoroughly cut. It is then molded, pressed, salted, and left to dry. The cheese ripens (9°C) for over 50 days.

17.6.1.5 Greece

Although goat milk is allowed to be added to ewe milk at a limited proportion (usually up to 30%) for the production of Greek traditional cheeses, only five of the cheeses are made exclusively from goat milk— anevato, batzos, manouri, white-brined cheese from goat milk (katsikisio tyri), and xinotyri.

Anevato cheese was traditionally produced in western Macedonia by shepherds with large flocks of goats and sheep. They renneted milk obtained in the morning just before taking the cattle out for feeding. During the day, the curd was 'raised' ("anevato" means cheese that is raised) and was ready to be drained on their return late in the afternoon. Whole milk from local herds is left at 18°C–20°C to sour, liquid rennet is added to the milk at a pH of about 6.2, and coagulation takes place in 12 h. The curd is cut, put in cheese cloths, and then drained for 24 h. After draining, salt is added, and the curd is thoroughly mixed and put in containers (about 5 kg). The cheese is then transferred to cold rooms (4°C) until sold. Nine batches of anevato cheese were examined throughout a 60-day storage time at three different periods within the lactation season of the goat. High mean log counts per gram of cheese for aerobic bacteria (7.92–9.56), LAB (7.78–9.32), Gram-negative organisms (5.64–9.67), psychrotrophs (7.90–11.79), and proteolytic bacteria (7.57–9.36) were found. *Enterobacteriaceae*, coliforms, and yeasts were considerably lower. *Enterobacteriaceae* and coliforms in the curd of cheese made in May were lower than counts in curd made in January and March. This coincided with lower pH and higher counts of LAB in cheese made in March and May. Yeast populations were affected by the season and were higher in May than March and/or January. Lactococci dominated in the cheese until 15 days, but lactobacilli became predominant after 30 days. Citrate fermenting *Lc. lactis* subsp. *lactis* was abundant in anevato cheese (Hatzikamari et al. 1999).

Batzos cheese was traditionally made in the same area as above from either raw goat milk or raw ewe milk as a by-product during the production of manouri, a whey cheese, or whey butter, respectively. The real intention was to obtain fat-rich, high-quality manouri or a large quantity of butter from the whey. For this reason, the milk was "hitted" with a thick wooden stick about 500 times during coagulation, and thus, a large proportion of fat was transferred to the whey (Zygouris 1956). The changes in microbial flora in batzos during ripening were studied throughout the whole lactation season in nine cheese batches manufactured, three in each, winter, spring and summer. High counts of *Enterobacteriaceae* and coliforms were recorded early in ripening (10^6 cfu/g), but their levels decreased significantly on ripening and storage. LAB predominated over the other microbial groups throughout ripening (10^6–10^8 cfu/g) during the whole lactation season. The season affected the composition of the lactic microflora. Thus, in winter, enterococci were abundant, with *Ent. durans* being predominant, whereas in spring and summer, lactobacilli isolates were found more frequently (mainly *Lb. plantarum* and *Lb. paraplantarum*). *Lc. lactis* subsp. *lactis* was the most frequently isolated species in the cheese (Psoni et al. 2003).

Manouri is a whey cheese, and it has long been made from the whey obtained during the production of batzos. Six batches of manouri cheese were manufactured, and selected microbial groups were counted throughout storage for 20 days at 4°C. The counts of all the microbial groups increased throughout storage and reached higher levels in cheeses made in summer than in those made in spring. Moreover, the microorganisms developed better on the cheese surfaces than in the interiors, especially in summer. The pH (6.78–7.33) and salt-in-moisture content (2.53–3.72) of the cheeses did not seem to affect the growth of the present bacteria and yeasts. The isolates of *Enterobacteriaceae* were mainly *Hafnia* (68.75%), whereas the isolates from the Baird–Parker medium were mainly staphylococci. There was a great diversity of yeast species, but *Debaryomyces hansenii* and *Pichia membranefasciens* predominated (Lioliou et al. 2001).

Xinotyri is a farm cheese variety manufactured from raw goat milk from indigenous breeds in the island of Naxos. The milk is renneted and curdled in about 24 h at room temperature. Sometimes, a small amount of cheese whey from the previous day is used as starter. The curd is drained, dry salted, and then kneaded for the uniform dispersion of salt. The cheese ripens for 30–45 days. The cheese microflora is composed of high populations of mesophilic LAB and lactococci declined during ripening. The same was observed for enterococci and yeasts. *Enterobacteriaceae* and *Micrococcaceae* were counted at high levels. The cheese was free of *Salmonella* and *Listeria* (Bontinis et al. 2008).

17.6.2 Goat Milk Cheeses from Other European Countries

Cheeses from plain goat milk are made in several other countries in Europe, such as Hungary, Norway, Poland, and Switzerland (Rubino et al. 2004). Pasteurized goat milk is used, in general, for their manufacture; the curd is mildly cooked for some of them, and the majority is salted in brine (Table 17.2). It also seems that goat milk cheeses are also made in Sweden.

In a study on the bacteriological quality of on-farm manufactured goat cheeses (Tham et al. 1990), samples of soft or semisoft goat cheeses obtained from dairy farms and retail trade in Sweden were examined. In the cheeses made with starter, the pH level was <6.0. Higher levels of total aerobic counts were measured in cheeses from raw milk and, thus, were accompanied by quite high levels of coliforms and fecal coliform bacteria. In general, cheeses made from raw milk without starters were unsatisfactory from a hygienic point of view, with respect to the presence of undesirable microorganisms.

17.6.3 Goat Milk Cheeses from Africa

In South Africa, goat milk is produced by many small-scale milk producers and processed into various types of cheeses. Within the counties of southern and eastern Africa, there does not appear to be any community that has traditionally made cheese from milk by precipitating the casein by coagulating enzymes. However, in Zimbabwe, the Shona people have the tradition of heating colostral milk, and the precipitated curd was eaten by the children. In Tanzania, some families in the community of Chagga filter the whey from soured milk in a piece of cloth, which is then hung over the fireplace in the kitchen for about 1 week before being consumed (FAO 1990). Milk production from goats seems to have been increased over the last few years, as well as projects aimed at promoting the production of goat milk by both householders and small-scale farmers (Donkin 1998). Gouda cheese is a cheese variety manufactured by these producers.

To contribute to the bacteriological quality of this cheese, goat milk was activated by the lactoperoxidase system (LP) and then used to make cheese (Seifu et al. 2004). The system contributed significantly to lower coliforms and coagulase-positive staphylococci counts and to lower lipolysis of the cheese than those in the control cheese. In addition, the cheese made from LP-activated goat milk had a milder flavor. The proteolysis of the two Gouda cheese types was comparable.

In Sudan, not only goat milk but also cow or sheep milk or mixtures are used to make *gibna bayda*, which was introduced by Greek immigrants and resembles feta with respect to appearance, texture, and flavor (Osman 1987).

In countries in North Africa, such as Algeria, Egypt, and Morocco, there are cheeses made from goat milk, even though other milk types are also used for their production (Rubino et al. 2004). *Djben cheese* of Algeria is a fresh cheese made with either lamp's rennet or rennet obtained from the flower of wild thistle, produced in rural areas, often for family consumption. *Dhani cheese* is made in Egypt on a small scale as a cottage industry. *Karish cheese* is also produced in the rural areas of Egypt by raw goat (or camel) milk left for fermentation for 3 days in a warm place. The fat raised on the surface is removed, and the curd is spread on a wooden dish for another 3 days for the exclusion of the whey. From the curd, small cubes (7 cm) are formed and salted in brine.

J'ben in Morocco is made from goat (or cow) milk in leather goat bags or in terracotta bags. The milk is fermented by spontaneous fermentation for 24–48 h, and the curd is transferred in thin cloth and hung to drain. Surface salting facilitates whey drainage. The cheese is consumed either fresh or after salting (dry salt or brine) or sun drying in a more ripened form (Belaiche 2007; Rubino et al. 2004). There are several studies on the physicochemical characteristics of this cheese. The microbial flora of this cheese is rich and is dominated by LAB (lactococci, 5.1×10^8 cfu/g; lactobacilli, 3.2×10^8 cfu/g; leuconostocs 2.6×10^8 cfu/g). *Lc. lactis* subsp. *lactis* and the biovar. *diacetylactis*, *Lb. casei* subsp. *casei*, and *Ln. lactis* are the dominant species. The cheese has a low pH (4.05–4.22), and its mean dry matter ranges between 29.4 and 45.6, depending on the draining time.

17.6.4 America

17.6.4.1 North American Goat Milk Cheeses

Queso fresco is a fresh Mexican cheese made mostly by artisans from fresh goat milk and calf rennet. After drainage, the cheese is shaped spherical. *Queso fresco de Aro* is a soft Mexican cheese made

TABLE 17.2

Manufacture of Goat Milk Cheeses Made in Some European Countries

Country	Product	Milk	Curdling	Cooking	Ripening
Hungary	Berettyo	Pasteurized	Culture + rennet, 32°C–35°C	No	Brine salting, 2 months ripening
	Keeske Sajt Borsodi	Pasteurized	Butter culture + rennet, 32°C–35°C	Yes	Brine salting, not ripened
	"Csongradi" Bio Kecske Sajt	Pasteurized	Culture + rennet, 38°C	No	Curd salting + seasoning; not ripened
	"Gida" Ecsedar Kecske Sajt	Pasteurized	Culture + rennet, 35°C	Cheddaring + smoking	2 months, 4°C–8°C
	"Gida" Fanfar Kecske Sajt	Pasteurized	Culture + rennet, 32°C	No	Brine salting; 2 months ripening
	"Gida" Felkene Kecske Sajt	Pasteurized	Culture + rennet, 32°C	No	Brine salting; 1.5 months ripening 4°C–8°C
	Kecske Gomolya	Pasteurized	Culture + rennet, ~30°C	Yes	Brine salting, no ripening
	Laci Kescke Trappista Sajt	Pasteurized	Culture + rennet, ~30°C	Yes	Brine salting; ripening 20 days
	Laci Lagy Kescke Sajt	Pasteurized	Culture + rennet, ~30°C	Yes	Curd salting + seasoning; 1 week ripening
	Menfoi Kescke sajt	Pasteurized	Culture + rennet	Yes	Curd seasoning; brine salting; 2 months, 4°C–8°C
	Soma's Kescke Sajt	Pasteurized	Butter culture + rennet, 34°C	No	Curd salting + seasoning; 4–5 days, 18°C
Norway	Ekte Geitost	Goat whey + cream + milk			
	Kvit Geitost Snofrisk	Goat milk + cow's cream			
Poland	Blekitna Kraina	Pasteurized	Culture + rennet, 32°C	Yes	Brine salting + *Penicillium roqueforti*; 3 weeks ripening
	Camembert zmlekakoziego	Pasteurized	Starters + rennet, 32°C	Yes	Brine salting + *P. candidum*
	Czarnuszka	Pasteurized	Culture + rennet, 30°C	No	Brine salting; 4 weeks; rind covering with herbs
	Ser kozi pelnotlustry	Pasteurized	Culture + rennet, 30°C		Dry salting; 2 weeks 12°C–14°C
	Ser kozi Swiezy	Pasteurized	Culture + rennet, 26°C		Seasoning
	Ser podpuszczkowy dojrzewajacy	Raw	Rennet, 38°C	No	Brine salting; 6 weeks
	Ser typu Feta	Pasteurized	Starters + rennet, 38°C	Yes	Brine salting; ripening in pickling oil
	Ser zloty	Pasteurized	Culture + rennet, 30°C	No	Brine salting; 3 months ripening
	Sockewka	Pasteurized	Culture + rennet, 30°C	No	Ripening in brine, 4 weeks
	Twarozek zmleka koziego	Pasteurized	Culture + rennet	No	
Switzerland	Dallenwiler Geisschase	Pasteurized	Starters + rennet, 28°C–32°C	No	Brine salting; 2 weeks ripening, mold development

Source: Rubino, R. et al., *Atlas of Goat Products*, La Biblioteca Di CASEUS, Potenza, Italy, 2004.

from whole goat milk that is heated to 60°C–65°C and coagulated (kid rennet). The curd is broken, immediately molded, pressed, and then dry salted. *Ranchero de Cabra de Quere taro* is a soft goat cheese made by farmers from the raw goat milk of native goats. The milk is renneted (by kid rennet paste), and the curd is broken and drained in a wooden form. It is then salted and aged for up to 6 months. *Ranchero molido* is produced from goat (cow or a mixture) milk. The curd is cut and drained in cloth sacks. It is then crumbled, dry salted, and ground in a mill. After 2–3 h, the sacks are refrigerated, and the next day, the cheese is ground again. The curd is then put in 0.5-kg forms and pressed for 1 h. It is afterward ready for consumption. *Chevre logs* is a fresh cheese produced from pasteurized goat milk. After cutting and draining, flavorings are sometimes added. The cheese is coated by herbs or pepper. *Gray's Chapel* is a semihard cheese made from raw goat milk. Starters and rennet are added, and the coagulum is cut, stirred, and then heated mildly and washed with warm (38°C) water. The cheese is molded, pressed, and then brined for 8 h. Its ripening lasts 100 days. *Silk Hope cheese* is made from pasteurized goat milk with starter, rennet, and white penicillium. The curd is drained, dry salted, and then ripened for 45–60 days at 3°C–4°C. *The Smokey Mountain Round cheese* is a fresh cheese made from pasteurized goat milk, fermented by starter. After draining in a bag, the curd is molded, dry salted, dried for 10 days, and then smoked around 30 min over soaked pecan wood logs (Rubino et al. 2004).

17.6.4.2 South American Goat Milk Cheeses

Quesillo is a fresh cheese made from the milk of Griolla goats. Fresh milk and rennet are added to the whey heated to boiling point. The curd is cut and pressed. Under the name quesillo, a soft cheese is also made by women when the milk production exceeds the needs of kids. *Queijo fundido* is a processed cheese. *Queijo Minas Frescal* is a fresh soft cheese made from goat milk since the late 1970s, when dairy goat breeding started. *Queijo temperado* is a creamy curd cheese seasoned with herbs. *Quesillo* or *Requesón* is a cheese type made by soft enzymatic clotting (from kid's stomach) of fresh goat milk without salting. *Queso fresco* is also made from raw milk and kid rennet. *Queso blanco* is a typical white cheese made from raw goat milk.

Farm goat cheese is a very typical cheese made in Chile by goat keepers using primitive methods and very simple facilities. The curd is pressed for the exclusion of the whey, and the cheese ripens during the transportation of the goat keepers and their families from the Andes Mountains to the valleys (FAO 1990; Rubino et al. 2004).

17.6.5 Goat Milk Cheeses in the Middle East and Other Asian Countries

In Turkey, the tradition in cheesemaking has been influenced by various civilizations (Durlu-Ozkaya and Gun 2007). Çerkez, akçakatic, and vrfa cheeses may be produced from goat milk. *Çerkez cheese* has a Circassian origin. The milk is curdled, and the clot is boiled, then salted, and put into baskets. *Akçakatic cheese* is a kind of cheese dried in cloths or goat stomach. In the cheese, carnation salt and/ or *Nigella sativa* are added. The cheese is either sold fresh or after keeping for 2–3 months. *Vrfa cheese* is also called "kiz memesi" or "kuzu başi," and it ripens for 4–6 months and can be kept in brine for a long time.

In Syria, goats produce 5.4% of the total milk. The mountain goats constitute 96% of the goats, and the Shami goats constitute 4% of the total goat population (El Mayda 2007). *Halloumi cheese* is made from goat (or sheep or mixtures) milk. After clotting, the curd is cut and sometimes mildly cooked, drained in cloths by light pressing, and cut into small blocks, which are then put in heated whey for 30–60 min (90°C–92°C) after the removal of the proteins. The cheese is afterward transferred onto a draining table, each block into a "U-shape," and after cooling, dry salt is sprinkled over the surface of the cheese. *Nabelsi cheese* is sometimes made from goat milk. After processing a fresh cheese, the cheese is boiled and preserved. Before coagulation, gum arabic and ground seeds of *P. mahaleb* are added to the milk to give a special flavor. The salt is homogenized in the curd. *Baladi or green cheese* is made from raw goat (or sheep) milk, to which yogurt and rennet are added. The drained curd is immersed in brine for 1 h and then is stored.

Darfyieh cheese has been produced for years in Lebanon from raw milk from the local Baladi breed of goats. The milk is left to sour for 24–48 h, rennet is added, and the curd is broken, put into baskets, and salted. The whey is heated to produce *arichi*, and layers of cheese and arichi are put in the pouch of goat skin (darrif). The cheese ripens in natural caves for 1–6 months. *Djamid* or *jameed* is a hard cheese made in some Middle East countries (commonly in Jordan and Syria), principally from sheep milk and, to a lesser extent, from goat milk. The milk is curdled by natural acidification. The clot is churned for the extraction of butter, and the liquid is used to make djamid by heating at 40°C–60°C, coagulating, and then cooking. Natural colors (Safran) and officinal herbs are put in the curd. The curd is drained in warm cheese cloth and pressed. Balls of cheese are dried in the sun. *Labaneh cheese* is a traditional Middle Eastern cheese sometimes made from goat milk. Whole yogurt is drained in cloth bags over 24–30 h, and the mass is salted (1%–2%) and then packed in tins that are refrigerated in ice water. Bedouins dry heavily salted labaneh in the sun (Rubino et al. 2004).

17.6.6 Australia

In Australia, private breeders have imported several breeds of goats, and goat milk products are gaining a reputation for quality, particularly goat cheeses, some of which have won major awards. Soft fresh cheeses, mold varieties, typical of French styles and Mediterranean technologies such as that for feta, are made especially since the 1970s, when interest in goat dairying has started to increase (Rubino et al. 2004).

REFERENCES

Addis, M., Cabiddu, A., Pinna, G., Decandia, M., Piredda, G., Pirisi, A., and Molle, G. 2005. Milk and cheese fatty acid composition in sheep fed Mediterranean forages with reference to conjugated linoleic acid *cis*-9, *trans*-11. *J. Dairy Sci.* 88:3443–3454.

Agnihotri, M. K. and Prasad, V. S. S. 1993. Biochemistry and processing of goat milk and milk products. *Small Ruminant Res.* 12:151–170.

Albenzio, M., Caroprese, M., Marino, R., Muscio, A., Santillo, A., and Sevi, A. 2006. Characteristics of Garganica goat milk and Cacioricotta cheese. *Small Ruminant Res.* 64:35–44.

Alférez, M. J. M., Barrionuevo, M., Lopez-Aliaga, I., Sanz Sampelayo, M. R., Lisbona, F., Robles, J. C., and Campos, M. S. 2001. Digestive utilization of goat and cow milk fat in malabsorption syndrome. *J. Dairy Res.* 68:451–461.

Alférez, M. J. M., López-Aliaga, I., Nestares, T., Díaz-Castro, J., Barrionuevo, M., Ros, P. B., and Campos, M. S. 2006. Dietary goat milk improves iron bioavailability in rats with induced ferropenic anaemia in comparison with cow milk. *Int. Dairy J.* 16:813–821.

Alichanidis, E. and Polychroniadou, A. 2008. Characteristics of major traditional regional cheese varieties of East-Mediterranean countries: A review. *Dairy Sci. Technol.* 88:495–510.

Almaas, H., Cases, A. L., Devold, T. G., Holm, H., Langsrud, T., Aabakken, L., Aadnoey, T., and Vegarud, G. E. 2006. *In vitro* digestion of bovine and caprine milk by human gastric and duodenal enzymes. *Int. Dairy J.* 16:961–968.

Alonso-Calleja, C., Carballo, J., Capita, R., Bernardo, A., and Garcia-Lopez, M. L. 2002. Changes in the microflora of Valdeteja raw goat's milk cheese throughout manufacturing and ripening. *Lebensm.-Wiss. Technol.* 35:222–232.

Belaiche, T. 2007. Etude de deux produits laitiers traditionnels du Maroc: Le Jben et le Smen. In: Historical Cheeses of Countries around the Archipelago Mediterraneo. International Symposium, December 6–8, Thessaloniki, Greece, edited by E. Litopoulou-Tzanetaki, N. Tzanetakis, and S. Kourletaki-Belibasaki, pp. 45–54.

Bellioni-Businco, B., Paganelli, R., Lucenti, P., Giampietro, P. G., Perborn, H., and Businco, L. 1999. Allergenicity of goat's milk in children with cow's milk allergy. *J. Allergy Clin. Immunol.* 103:1191–1194.

Bonetta, S., Carraro, E., Rantziou, K., and Cocolin, L. 2008. Microbiological characterization of Robiola di Roccaverano cheese using PCR–DGGE. *Food Microbiol.* 25:786–792.

Bontinis, T. G., Mallatou, H., Alichanidis, E., Kakouri, A., and Samelis, J. 2008. Physicochemical, microbiological and sensory changes during ripening and storage of Xinotyri, a traditional Greek cheese from raw goat's milk. *Int. J. Dairy Technol.* 61:229–236.

Boyazoglu, J., Hatziminaoglou, I., and Morand-Fehr, P. 2005. The role of the goat in society: Past, present and perspectives for the future. *Small Ruminant Res.* 60:13–23.

Boyazoglu, J. and Morand-Fehr, P. 2001. Mediterranean dairy sheep and goat products and their quality: A critical review. *Small Ruminant Res.* 40:1–11.

Buscino, L. and Bellanti, J. 1993. Food allergy in childhood: Hypersensitivity to cow's milk allergens. *Clin. Exp. Allergy* 23:481–483.

Carballo, J., Fresno, J. M., Tuero, J. R., Prieto, J. G., Bernardo, A., and Martin Sarmiento, R. 1994. Characterization and biochemical changes during the ripening of a Spanish hard goat cheese. *Food Chem.* 49:77–82.

Casalta, E., Noel, Y., Le Bars, D., Carre, C., Achilleos, C., and Maroselli, M.-X. 2001. Characterisation du fromage Bastelicaccia. *Lait* 81:529–546.

Di Cagno, R., Miracle, R. E., De Angelis, M., Minervini, F., Rizzello, C. G., Drake, M. A., Fox, P. F., and Gobbetti, M. 2007. Compositional, microbiological, biochemical, volatile profile and sensory characterization of four Italian semihard goats' cheeses. *J. Dairy Res.* 74:468–477.

Donkin, E. F. 1998. Milk production from goats for households and small-scale farmers in South Africa. In: *Proceedings of a Workshop Research and Training Strategies for Goat Production Systems in South Africa*, edited by E. C. Webb, P. B. Cronje, and E. F. Donkin, pp. 27–32. Hogsback, South Africa.

Durlu-Ozkaya, F. and Gun, I. 2007. Traditional Turkish cheeses. In: Historical Cheeses of Countries around the Archipelago Mediterraneo. International Symposium, December 6-8, Thessaloniki, Greece, edited by E. Litopoulou-Tzanetaki, N. Tzanetakis, and S. Kourletaki-Belibasaki, pp. 65–88.

El Mayda, E. 2007. Manufacture of local cheese from raw milk in Syria. In: Historical Cheeses of Countries around the Archipelago Mediterraneo. International Symposium, December 6-8, Thessaloniki, Greece, edited by E. Litopoulou-Tzanetaki, N. Tzanetakis, and S. Kourletaki-Belibasaki, pp. 55–64.

FAOSTAT. 2001. Food and Agriculture Organization of the United Nations. Rome, Italy. Ed. CD-ROM.

Fontecha, J. C., Pelaez, C., Juarez, M., Requena, T., Gomez, C., and Ramos, M. 1990. Biochemical and microbiological characteristics of artisanal hard goat's cheese. *J. Dairy Sci.* 73:1150.

Food and Agriculture Organization of the United Nations (FAO). 1990. *Animal Production and Health.* FAO: Rome, Italy.

Food and Agriculture Organization of the United Nations (FAO). 2006. *Food and Agriculture Organization of the United Nations (FAO) Official Statistics.* FAO: Rome, Italy.

Franco, I., Prieto, B., Bernardo, A., Gonzales-Prieto, M. J., and Carballo, J. 2003. Biochemical changes throughout the ripening of a traditional Spanish goat cheese variety (Babia-Laciana). *Int. Dairy J.* 13:221–230.

Fresno, J. M., Tornadijo, E., Carballo, J., Bernardo, A., and Gonzalez-Prieto, J. 1997. Proteolytic and lipolytic changes during the ripening of a Spanish craft goat cheese (Armada variety). *J. Sci. Food Agric.* 75:148–154.

Fresno, J. M., Tornadijo, M. E., Carballo, J., Gonzales-Prieto, M. J., and Bernardo, A. 1996. Characterization and biochemical changes during the ripening of a Spanish goats' milk cheese (Armada variety). *Food Chem.* 55:225–230.

Fresno, M., Pino, V., Álvarez, S., Darmanin, N., Fernández, M., and Guillén, M. D. 2005. The effects of the smoking materials used in the sensory characterization of the Palmero (PDO) cheeses. *Options Mediterraneennes, Series A* 67:195–199.

Galal, S. 2005. Biodiversity in goats. *Small Ruminant Res.* 60:75–81.

Gomez, C., Pelaez, C., and De La Torre, E. 1989. Microbiological study of semihard goat's milk cheese (Majorero). *Int. J. Food Sci. Technol.* 24:147–151.

Gomez, C., Pelaez, C., and Martin-Hernandez, M. C. 1988. Enzyme activity in Spanish goat's cheeses. *Food Chem.* 28:159–165.

Gonzalez, J., Mas, M., Tabla, R., Moriche, J., Roa, I., Rebollo, J. E., and Caceres, P. 2003. Autochthonous starter effect on the microbiological, physicochemical and sensorial characteristics of Ibores goat's milk cheeses. *Lait* 83:193–202.

Guillen, M. D., Palencia, G., Sopelana, P., and Ibargoritia, M. L. 2007. Occurrence of polycyclic aromatic hydrocarbons in artisanal Palmero cheese smoked with two types of vegetable matter. *J. Dairy Sci.* 90:2717–2725.

Haenlein, G. F. W. 1993. Producing quality goat milk. *Int. J. Anim. Sci.* 8:79–84.

Haenlein, G. F. W. 2004. Goat milk in human nutrition. *Small Ruminant Res.* 51:155–163.

Hatzikamari, M., Litopoulou-Tzanetaki, E., and Tzanetakis, N. 1999. Microbiological characteristics of Anevato: A traditional Greek cheese. *J. Appl. Microbiol.* 87:595–601.

Jandal, J. M. 1996. Comparative aspects of goat and sheep milk. *Small Ruminant Res.* 22:177–185.

Kullisaar, T., Songisepp, E., Mikelsaar, M., Zilmer, K., Vihalemm, T., and Zilmer, M.. 2003. Antioxidative probiotic fermented goats' milk decreases oxidative stress-mediated atherogenicity in human subjects. *Br. J. Nutr.* 90:449–456.

Le Quere, J.-L., Pierre, A., Riaublanc, A., and Demaizieres, D. 1998. Characterization of aroma compounds in the volatile fraction of soft goat cheese during ripening. *Lait* 78:279–290.

Lioliou, K., Litopoulou-Tzanetaki, E., Tzanetakis, N., and Robinson, R. K. 2001. Changes in the microflora of manouri, a traditional Greek whey cheese, during storage. *Int. J. Dairy Technol.* 54:100–106.

Luikart, G., Fernandez, H., Mashkour, M., England, P., Taberlet, P., Zeder, A. M., Emshwiller, E., Smith, B. D., and Bradley, D. G. 2006. Origins and diffusion of domestic goats inferred from DNA markers. In: *Documenting Domestication*, edited by A.M. Zeder and B.D. Smith, pp. 294–305. Berkeley: University of California Press.

Martin-Hernandez, M. C. and Juarez, M. 1992. Biochemical characteristics of three types of goat cheese. *J. Dairy Sci.* 75:1747–1752.

Mas, M., Tabla, R., Moriche, J., Roa, I., Gonzalez, J., Rebollo, J. E., and Caceres, P. 2002. Ibores goat's milk cheese: Microbiological and physicochemical changes throughout ripening. *Lait* 82:579–587.

Medina, M., Gaya, P., and Nuñez, M. 1992. Gredos goats' milk cheese: Microbiological and chemical changes throughout ripening. *J. Dairy Res.* 59:563–566.

Medina, M. and Nuñez, M. 2004. Cheese made from ewe's and goat's milk. In: *Cheese: Chemistry, Physics and Microbiology*, edited by P. F. Fox, P. L. H. McSweeney, T. M. Cogan, and T. P. Guinee, pp. 279–299. London: Elsevier Ltd.

Micari, P., Sarullo, V., Sidari, R., and Caridi, A. 2007. Physicochemical and hygienic characteristics of the Calabrian raw milk cheese, Caprino d' Aspromonte Turkish. *J. Anim. Sci.* 31:55–60.

Ministerio de Agricultura, Pesca y Alimentacion (MAPA). 2002. Anuario de Estadística Agroalimentaria. Madrid, España.

Osman, A. O. 1987. The technology of the Sudanese white cheese "Gibna bayda." In: *Bulletin of the International Dairy Federation*. pp. 113–115. Brussels, Belgium: FIL-IDF, Secreteriat General.

Pariset, L., Cuteri, A., Ligda, C., Ajmone-Marsan, P., Valentini, A., and Consortium, E. 2009. Geographical pattering of sixteen goat breeds from Italy, Albania and Greece assessed by single nucleotide polymorphisms. *BMC Ecol.* 9:1–9.

Park, Y. W. 1994. Hypoallergenic and therapeutic significance of goat milk. *Small Ruminant Res.* 14:151–159.

Park, Y. W. 2006. Goat milk—Chemistry and nutrition. In: *Handbook of Milk of Nonbovine Mammals*, edited by Y. W. Park and G. F. W. Haenlein, pp. 34–58. Oxford, U.K.: Blackwell Publishing Professional.

Pasqualone, A., Caponio, F., Allogio, V., and Gomes, T. 2000. Content of taurine in Apulian Cacioricotta goat's cheese. *Eur. Food Res. Technol.* 211:158–160.

Pereira, F., Davis, S. J. M., Pereira, L., McEvoy, B., Bradley, D. G., and Amorim, A. 2006. Genetic signatures of a Mediterranean influence in Iberian Peninsula sheep husbandry. *Mol. Biol. Evol.* 23:1420–1426.

Picon, A., Medina, M., and Nunez, M. 2007. Spanish cheeses made from raw ewe's or goat milk with protected designation of origin. In: Historical Cheeses of Countries around the Archipelago Mediterraneo. International Symposium, December 6–8, Thessaloniki, Greece, edited by E. Litopoulou-Tzanetaki, N. Tzanetakis, and S. Kourletaki-Belibasaki, pp. 123–142.

Pizzillo, M., Claps, S., Cifuni, G. F., Fedele, V., and Rubino, R. 2005. Effect of goat breed on the sensory, chemical and nutritional characteristics of ricotta cheese. *Livest. Prod. Sci.* 94:33–40.

Psoni, L., Kotzamanidis, C., Yiangou, M., Tzanetakis, N., and Litopoulou-Tzanetaki, E. 2007. Genotypic and phenotypic diversity of *Lactococcus lactis* isolates from Batzos, a Greek PDO raw goat milk cheese. *Int. J. Food Microbiol.* 114:211–220.

Psoni, L., Kotzamanidis, C., Andrighetto, C., Lombardi, A., Tzanetakis, N., and Litopoulou-Tzanetaki, E. 2006. Genotypic and phenotypic heterogeneity in *Enterococcus* isolates from Batzos, a raw goat milk cheese. *Int. J. Food Microbiol.* 109:109–120.

Psoni, L., Tzanetakis, N., and Litopoulou-Tzanetaki, E. 2003. Microbiological characteristics of Batzos, a traditional Greek cheese from raw goat's milk. *Food Microbiol.* 20:575–582.

Robinson, F. 2001. Goats milk—A suitable hypoallergenic alternative? *Br. Food J.* 103:198–208.

Rubino, R., Morand-Fehr, P., Renieri, C., Peraza, C., and Sarti, F. M. 1999. Typical products of the small ruminant sector and the factors affecting their quality. *Small Ruminant Res.* 34:289–302.

Rubino, R., Morand-Fehr, P., and Sepe, L. 2004. *Atlas of Goat Products*. Potenza, Italy: La Biblioteca Di CASEUS.

Saldo, J., Fernandez, A., Sendra, E., Butz, P., Tauscher, B., and Guamis, B. 2003. High-pressure treatment decelerated the lipolysis in a caprine cheese. *Food Res. Int.* 36:1061–1068.

Saldo, J., Sendra, E., and Guamis, B. 2000. High hydrostatic pressure for accelerating ripening of goat's milk cheese: Proteolysis and texture. *J. Food Sci.* 65:636–640.

Scintu, M. F. and Piredda, G. 2007. Typicity and biodiversity of goat and sheep milk products. *Small Ruminant Res.* 68:221–231.

Seifu, E., Buys, E. M., and Donkin, E. F. 2004. Quality aspects of Gouda cheese made from goat milk preserved by the lactoperoxidase system. *Int. Dairy J.* 14:581–589.

Slacanac, V., Bozanic, R., Hardi, J., Rezessyneszabo, J., Lucan, M., and Krstanovic, V. 2010 Nutritional and therapeutic value of fermented caprine milk. *Int. J. Dairy Technol.* 63:171–189.

Suzzi, G., Caruso, M., Gardini, F., Lombardi, A., Vannini, L., Guerzoni, M. E., Andrighetto, C., and Lanorte, M. T. 2000. A survey of the enterococci isolated from an artisanal Italian goat's cheese (semicotto caprino). *J. Appl. Microbiol.* 89:267–274.

Tejada, L., Abellan, A., Cayuela, J.M., Martinez-Cacha, A., and Fernandez-Salguero, J. 2008a. Proteolysis in goats' milk cheese made with calf rennet and plant coagulant. *Int. Dairy J.* 18:139–146.

Tejada, L., Abellan, A., Prados, F., and Cayuela, J. M. 2008b. Compositional characteristics of Murcia al Vino goat's cheese made with calf rennet and plant coagulant. *Int. J. Dairy Technol.* 61:119–125.

Tham, W. A., Hajdu, L. T., and Danielsson-Tham, M.-L.V. 1990. Bacteriological quality of on-farm manufactured goat cheese. *Epidemiol. Infect.* 104:87–100.

Tornadijo, E., Fresno, J. M., Carballo, J., and Martin Sarmiento, R. 1993. Study of Enterobacteriaceae throughout the manufacturing and ripening of hard goats' cheese. *J. Appl. Microbiol.* 75:240–246.

Tornadijo, M. E., Fresno, J. M., Bernardo, A., Martin Sarmiento, R., and Carballo, J. 1995. Microbiological changes throughout the manufacturing and ripening of a Spanish goat's milk cheese (Armada variety). *Lait* 75:551–570.

Vigne, J. D. 1999. The large "true" Mediterranean islands as a model for the Holocene human impact on the European vertebrate fauna? Recent data and new reflections. In: *The Holocene History of the European Vertebrate Fauna Modern Aspects of Research*, edited by Benneke, N. (Deutsches Archäologisches Institut, Eurasien-Abteilung), pp. 295–322. Verlag Marie Leidorf GmbH.

Voutsinas, L., Pappas, C., and Katsiari, M. 2009. The composition of Alpine goats' milk during lactation in Greece. *J. Dairy Res.* 57:41–51.

Zarate, V., Belda, C., Perez, C., and Cardell, E. 1997. Changes in the microbial flora of Tenerife goats' milk cheese during ripening. *Int. Dairy J.* 7:635–641.

Zeder, A. M. 2008. Domestication and early agriculture in the Mediterranean Basin: Origins, diffusion, and impact. *Proc. Natl. Acad. Sci. U. S. A.* 105:11597–11604.

Zumbo, A., Di Rosa, A. R., Billone, B., Carminati, D., Girgenti, P., and Di Marco, V. 2009. Ripening-induced changes in microbial groups of artisanal Sicilian goat's milk cheese. *Ital. J. Anim. Sci.* 8:450–452.

Zygouris, N. P. 1956. *Milk Industry*. Athens, Greece: Ministry of Agriculture.

18

Acidified Milk, Sour Cream, and Cream Cheese: Standards, Grades, and Specifications*

Y. H. Hui

CONTENTS

---

* The information in this chapter has been modified from the *Food Safety Manual* published by the Science Technology System (STS), West Sacramento, CA. Copyrighted 2012©. With permission.

18.1 Introduction

Food manufacturers in the United States must produce a product that consumers like in order to sell the product nationally. To achieve this goal, they depend on developments in science, technology, and engineering in order to produce a quality product. However, they must also pay attention to two other important considerations:

1. Are their products safe for public consumption?
2. Have they assured the economic integrity of their products? This means many things. The most important is the assurance that their products contain the proper quality and quantity of ingredients, that is, "No cheating."

To protect the consumers, U.S. government agencies have established many checks and balances to prevent "injuries" to the consumers in terms of health and economic fraud. The two major responsible agencies are

1. U.S. Food and Drug Administration (FDA; www.FDA.gov)
2. U.S. Department of Agriculture (USDA; www.USDA.gov)

The FDA establishes standards of identities for a number of cheeses for interstate commerce. This ensures that a particular type of cheese contains the minimal quality and quantity of the ingredients used to make this cheese. The FDA also issues safety requirements and recommendations in the manufacturing of cheeses. It makes periodic inspections of establishments producing cheeses to ensure full compliance. In effect, the FDA is monitoring the safety production of cheeses and their economic integrities.

The USDA plays the same important role in the commerce of cheeses in the United States. For example, a national restaurant chain that makes and sells pizzas will not buy cheeses from any manufacturer if the cheeses do not bear the USDA official identification or seal of approval. The USDA has issued official requirements and processing specifications that cheese manufacturers must follow if they want to obtain the USDA seal of approval for the products. To achieve this goal, the USDA sets up voluntary services of inspection under a standard fee arrangement. The inspection focuses on the safety and economic integrity of manufacturing cheeses. When the inspection demonstrates that the manufacturer has complied with safety and economic requirements, the USDA issues its official identification or seal of approval.

In essence, the FDA and USDA are the federal agencies that ensure that the product is safe and will not pose any economic fraud. Using the public information distributed at their web sites, this chapter discusses FDA standards of identities and USDA specifications for establishments in relation to the production of selected cheeses.

The information in this chapter is important, because it facilitates the commerce of cheeses and protects consumers from injuries of eating cheeses in the market. It also protects the consumers from economic fraud. For example, without FDA and USDA intervention, one may purchase a cheese that contains less than the normal amount of milk used. If so, the consumers lose money in the transaction.

All FDA standards of identities are described in

1. 21 CFR 131, Milk and cream
2. 21 CFR 133, Cheeses

All USDA requirements and standards for grade are described in 7 CFR 58.

For explanation, CFR refers to the United States Code of Federal Regulations. This chapter focuses on two areas:

1. Part 133: standards of identities issued by the FDA.
2. Part 58 (selected sections): USDA specifications and grade standards.

18.2 Acidified Milk

Acidified milk is the food produced by souring one or more of the optional dairy ingredients with one or more of the acidifying ingredients, with or without the addition of characterizing microbial organisms. One or more of the other optional ingredients may also be added. When one or more of the ingredients are used, they should be included in the souring process. All ingredients used are safe and suitable. Acidified milk contains not less than 3.25% milkfat and not less than 8.25% milk solids-not-fat and has a titratable acidity of not less than 0.5%, expressed as lactic acid. The food may be homogenized and should be pasteurized or ultrapasteurized prior to the addition of the microbial culture and, when applicable, the addition of flakes or granules of butterfat or milkfat.

18.2.1 Vitamin Addition (Optional)

If added, vitamin A should be present in such quantity that each 946 mL (quart) of the food contains not less than 2000 IU, within limits of the good manufacturing practice (GMP).

If added, vitamin D should be present in such quantity that each 946 mL (quart) of the food contains 400 IU, within limits of the GMP.

Optional dairy ingredients, namely, cream, milk, partially skimmed milk, or skimmed milk, may be used alone or in combination. Optional acidifying ingredients are acetic acid, adipic acid, citric acid, fumaric acid, glucono-*delta*-lactone, hydrochloric acid, lactic acid, malic acid, phosphoric acid, succinic acid, and tartaric acid.

18.2.2 Other Optional Ingredients

1. Concentrated skimmed milk, nonfat dry milk, buttermilk, whey, lactose, lactalbumins, lacto-globulins, or whey modified by the partial or complete removal of lactose and/or minerals to increase the nonfat solids content of the food, provided that the ratio of protein to total nonfat solids of the food and the protein efficiency ratio of all protein present should not be decreased as a result of adding such ingredients
2. Nutritive carbohydrate sweeteners—Sugar (sucrose), beet, or cane; invert sugar (in paste or syrup form); brown sugar; refiner's syrup; molasses (other than blackstrap); high-fructose corn syrup; fructose; fructose syrup; maltose; maltose syrup, dried maltose syrup; malt extract, dried malt extract; malt syrup, dried malt syrup; honey; maple sugar; or any other approved sweeteners.
3. Flavoring ingredients
4. Color additives that do not impart a color simulating that of milkfat or butterfat
5. Stabilizers
6. Butterfat or milkfat, which may or may not contain color additives, in the form of flakes or granules
7. Aroma- and flavor-producing microbial culture
8. Salt
9. Citric acid, in a maximum amount of 0.15% by the weight of the milk used, or an equivalent amount of sodium citrate as a flavor precursor

The name of the food is "acidified milk." The name of the food should be accompanied by a declaration indicating the presence of any characterizing flavoring and may be accompanied by a declaration such as a traditional name of the food or the generic name of the organisms used, thereby indicating the presence of the characterizing microbial organisms or ingredients when used, for example, "acidified kefir milk" or "acidified acidophilus milk," or when characterizing ingredients, the food may be named "acidified buttermilk."

18.3 Cultured Milk

Cultured milk is the food produced by culturing one or more of the optional dairy ingredients specified with characterizing microbial organisms. One or more of the other optional ingredients specified may also be added. When one or more of the ingredients specified are used, they should be included in the culturing process. All ingredients used are safe and suitable. Cultured milk contains not less than 3.25% milkfat and not less than 8.25% milk solids-not-fat and has a titratable acidity of not less than 0.5%, expressed as lactic acid. The food may be homogenized and should be pasteurized or ultrapasteurized prior to the addition to the microbial culture and, when applicable, the addition of flakes or granules of butterfat or milkfat.

18.3.1 Vitamin Addition (Optional)

If added, vitamin A should be present in such quantity that each 946 mL (quart) of the food contains not less than 2000 IU thereof, within limits of the GMP.

If added, vitamin D should be present in such quantity that each 946 mL (quart) of the food contains 400 IU, within limits of GMP.

Optional dairy ingredients, namely, cream, milk, partially skimmed milk, or skimmed milk, may be used alone or in combination.

18.3.2 Other Optional Ingredients

1. Concentrated skim milk, nonfat dry milk, buttermilk, whey, lactose, lactalbumins, lactoglobulins, or whey modified by the partial or complete removal of lactose and/or minerals to increase the nonfat solids content of the food, provided that the ratio of protein to total nonfat solids of the food and the protein efficiency ratio of all proteins present should not be decreased as a result of adding such ingredients

2. Nutritive carbohydrate sweeteners—Sugar (sucrose), beet, or cane; invert sugar (in paste or syrup form); brown sugar; refiner's syrup; molasses (other than blackstrap); high-fructose corn syrup; fructose; fructose syrup; maltose; maltose syrup, dried maltose syrup; malt extract, dried malt extract; malt syrup, dried malt syrup; honey; maple sugar; or other permitted sweeteners

3. Flavoring ingredients

4. Color additives that do not impart a color simulating that of milkfat or butterfat

5. Stabilizers

6. Butterfat or milkfat, which may or may not contain color additives, in the form of flakes or granules

7. Aroma- and flavor-producing microbial culture

8. Salt

9. Citric acid, in a maximum amount of 0.15% by the weight of the milk used, or an equivalent amount of sodium citrate as a flavor precursor

The name of the food is "cultured milk." The name of the food should be accompanied by a declaration indicating the presence of any characterizing flavoring and may be accompanied by a declaration such as a traditional name of the food or the generic name of the organisms used, thereby indicating the presence of the characterizing microbial organisms or ingredients, for example, "kefir cultured milk" or "acidophilus cultured milk," or when lactic acid-producing organisms are used, the food may be named "cultured buttermilk."

18.4 Sour Cream

Sour cream results from the souring, by lactic acid-producing bacteria, of pasteurized cream. Sour cream contains not less than 18% milkfat, except that when the food is characterized by the addition of nutritive sweeteners or bulky flavoring ingredients, the weight of the milkfat is not less than 18% of the remainder obtained by subtracting the weight of such optional ingredients from the weight of the food, but in no case does the food contain less than 14.4% milkfat. Sour cream has a titratable acidity of not less than 0.5%, calculated as lactic acid.

18.4.1 Optional Ingredients

These are safe and suitable ingredients that improve texture, prevent syneresis, or extend the shelf life of the product.

1. Sodium citrate in an amount not more than 0.1% may be added prior to culturing as a flavor precursor
2. Rennet
3. Safe and suitable nutritive sweeteners
4. Salt
5. Flavoring ingredients, with or without safe and suitable coloring, as follows:
   a. Fruit and fruit juice (including concentrated fruit and fruit juice)
   b. Safe and suitable natural and artificial food flavoring

The name of the food is "sour cream" or, alternatively, "cultured sour cream." The name of the food should be accompanied by a declaration indicating the presence of any flavoring that characterizes the product. If nutritive sweetener in an amount sufficient to characterize the food is added without the addition of characterizing flavoring, the name of the food should be preceded by the word "sweetened."

18.4.2 Acidified Sour Cream

Acidified sour cream results from the souring of pasteurized cream with safe and suitable acidifiers, with or without the addition of lactic acid-producing bacteria. Acidified sour cream contains not less than 18% milkfat, except that when the food is characterized by the addition of nutritive sweeteners or bulky flavoring ingredients, the weight of milkfat is not less than 18% of the remainder obtained by subtracting the weight of such optional ingredients from the weight of the food, but in no case does the food contain <14.4% milkfat. Acidified sour cream has a titratable acidity of not less than 0.5%, calculated as lactic acid.

18.4.3 Optional Ingredients

These are safe and suitable ingredients that improve texture, prevent syneresis, or extend the shelf life of the product.

1. Rennet
2. Safe and suitable nutritive sweeteners
3. Salt
4. Flavoring ingredients, with or without safe and suitable coloring, as follows:
   a. Fruit and fruit juice, including concentrated fruit and fruit juice
   b. Safe and suitable natural and artificial food flavoring

The name of the food is "acidified sour cream." The name of the food should be accompanied by a declaration indicating the presence of any flavoring that characterizes the product. If a nutritive sweetener in an amount sufficient to characterize the food is added without the addition of characterizing flavoring, the name of the food should be preceded by the word "sweetened."

18.5 USDA Specifications for Sour Cream and Acidified Sour Cream

Sour cream and acidified sour cream (sour creams) production must comply with the applicable FDA/ USDA requirements in:

1. Manufacture and packaging operations and procedures
2. Product grading and plant specifications and inspection

Sour creams should be maintained at 45°F (7.2°C) or less for at least 12 h prior to inspection.

18.5.1 Regulatory Requirements

Sour creams should comply with all applicable federal regulations, including those contained in the FDA Standards of Identity for Sour Cream or Acidified Sour Cream. Reduced-fat, "light" or "lite," low-fat, and fat-free sour creams should comply with all applicable FDA regulations, including those for sour cream, acidified sour cream, nutrient content claims for "light" or "lite" fat, and foods named by the use of a nutrient content claim and standardized term.

18.5.2 Composition Requirements

	Fat	Titratable Acidity
Sour cream and acidified sour cream contain	Not less than 18.0%	Not less than 0.5%, calculated as lactic acid
Sour cream and acidified sour cream with optional nutritive sweeteners and bulky flavorings should contain	Not less than 18% of the remainder obtained by subtracting the weight of the optional ingredients from the weight of the food. In no case should the product contain less than 14.4% total fat.	Not less than 0.5%, calculated as lactic acid
Reduced fat sour cream and reduced fat acidified sour cream should contain	Minimum of 25% reduction in total fat; contains 13.5% or less of total fat compared to sour cream meeting minimum compositional requirements for fat.	Not less than 0.5%, calculated as lactic acid
Light or lite sour cream and light or lite acidified sour cream should contain	Minimum of 50% reduction in total fat; contains 9.0% or less of total fat compared to sour cream meeting minimum compositional requirements for fat.	Not less than 0.5%, calculated as lactic acid
Low-fat sour cream and low-fat acidified sour cream should contain	Contains 3 g or less of fat per 50 g of product and 6.0% or less of total fat.	Not less than 0.5%, calculated as lactic acid
Nonfat sour cream and nonfat acidified sour cream should contain	Less than 0.5 g of fat per 50 g of product and less than 1.0% total fat.	Not less than 0.5%, calculated as lactic acid

18.5.3 Dairy and Nondairy Ingredient Requirements

All milk products used in the manufacture of sour creams and the plant in which the sour creams are processed should comply with all applicable FDA/USDA requirements. All optional nondairy ingredients should be clean and wholesome and should be approved by the FDA/USDA. Flavoring ingredients should be consistent in size and color to produce the desired appearance and appeal of the finished product, and they should be added at a level sufficient to impart a desirable characteristic flavor to the finished product.

18.5.4 Quality Requirements

The product should possess a pleasant, mild, aromatic acid and/or cultured flavor and be free from undesirable flavors such as rancid, oxidized, stale, yeasty, and unclean. Flavoring ingredients should be added at a level sufficient to impart a desirable characteristic flavor to the finished product. The product should be thick and smooth, uniform, free from lumps or graininess, and spoonable to a soft mound. Flavoring ingredients should be consistent in size and distribution in the finished product.

18.5.5 Color and Appearance

The product should present a clean natural color, with a smooth velvety appearance. Natural color may range from a bright white to a light cream color. The surface should appear smooth and dry without excessive whey separation. It should be free from visible sediment, mold, and surface discoloration. Flavoring ingredients should be consistent in size and color to produce the desired appearance and appeal of the finished product.

18.5.6 Analytical Testing and Microbial Requirements

Analytical and microbial analyses should be made in accordance with established procedures. Samples should be taken as often as necessary to determine that the product meets microbial, composition, and quality requirements. Analytical requirements include fat, titratable acidity, coliform not more than 10 per gram, and yeast and mold not more than 10 per gram.

18.5.7 Official Identification

Products officially inspected and found to meet these requirements may be identified with the official USDA Quality Approved Inspection Shield.

18.6 Cream Cheese

Cream cheese is the soft, uncured cheese prepared by FDA approved procedure. The minimum milkfat content is 33% by the weight of the finished food, and the maximum moisture content is 55% by weight. The dairy ingredients used are pasteurized.

One or more of the dairy ingredients specified may be homogenized and is subjected to the action of lactic acid–producing bacterial culture. One or more of the clotting enzymes specified is added to coagulate the dairy ingredients. The coagulated mass may be warmed and stirred, and it is drained. The moisture content may be adjusted with one or more of the optional ingredients specified. The curd may be pressed, chilled, and worked, and it may be heated until it becomes fluid. It may then be homogenized or otherwise mixed. One or more of the optional dairy ingredients and the other optional ingredients specified may be added during the procedure.

The following safe and suitable ingredients may be used:

1. Milk, nonfat milk, or cream, used alone or in combination
2. Rennet and/or other clotting enzymes of animal, plant, or microbial origin
3. Salt
4. Cheese whey, concentrated cheese whey, dried cheese whey, or reconstituted cheese whey prepared by the addition of water to concentrated cheese whey or dried cheese whey
5. Stabilizers, in a total amount not to exceed 0.5% of the weight of the finished food, with or without the addition of dioctyl sodium sulfosuccinate in a maximum amount of 0.5% of the weight of the stabilizer(s) used

18.7 Cream Cheese with Other Foods

"Cream cheese with other foods" is the class of foods :

1. Prepared by mixing the cream cheese with one or a mixture of two or more types of foods.
2. The foods added are in an amount sufficient to differentiate the mixture from cream cheese.
3. The mixing may be done with or without the aid of heat.

The following safe and suitable optional ingredients may be used

1. Properly prepared fresh, cooked, canned, or dried fruits or vegetables; cooked or canned meat; or relishes, pickles, or other suitable foods
2. Stabilizers, in a total amount not to exceed 0.8%, with or without the addition of dioctyl sodium sulfosuccinate in a maximum amount of 0.5% of the weight of the stabilizer(s) used
3. Coloring

The name of the food is "cream cheese with ___" or, alternatively, "cream cheese and ___," the blank being filled in with the name of the foods used in order of predominance by weight.

18.8 USDA Specifications for Cream Cheese, Cream Cheese with Other Foods, and Related Products

18.8.1 Plant Requirements

Cream cheese and related products should be manufactured and packaged in accordance with USDA requirements. The cheese should be cooled to 45°F prior to inspection and then tempered to 45°F–55°F for product evaluation. Cream cheese should comply with the FDA Standards of Identity for Cream Cheese.

Neufchatel cheese should comply with the FDA Standards of Identity for Neufchatel. Reduced-fat and light cream cheese should comply with all applicable FDA regulations, including those for cream cheese, for nutrient content claims for fat and for foods named by the use of a nutrient content claim and standardized term.

Cream cheese with other foods such as strawberries, chives, and salmon should comply with all applicable FDA regulations. Neufchatel cheese with other foods should comply with all applicable FDA Regulations for Neufchatel. Reduced-fat and light cream cheese with other foods should comply with all applicable FDA regulations for cream cheese, for cream cheese with other foods, for nutrient content claims for fat, and for foods named by the use of a nutrient content claim and a standardized term.

18.8.2 Composition Requirements

	Moisture	Milkfat	pH	Salt
Cream Cheese	Not more than 55%	Not less than 33% total fat (as marketed)	Range of 4.4–4.9	Not more than 1.4%
Neufchatel Cheese	Not more than 65%	Not less than 20% but <33% total fat (as marketed)	Range of 4.4–5.0	Not more than 1.4%
Reduced-Fat Cream Cheese	Not more than 70%	Not less than 16.5% but <20% total fat (as marketed)	Range of 4.4–5.1	Not more than 1.4%

	Moisture	Milkfat	pH	Salt
Light or Lite Cream Cheese	Not more than 70%	Not more than 16.5% total fat (as marketed)	Range of 4.4–5.2	Not more than 1.4%
Cream Cheese with Other Foods	Not more than 60%	Not less than 27% total fat (as marketed)		Not more than 1.4%
Neufchatel Cheese with Other Foods	Not more than 70%	Not less than 20% but <33% total fat (as marketed)		Not more than 1.4%
Reduced-Fat Cream Cheese with Other Foods	Not more than 70%	Not less than 16.5% but <20% total fat (as marketed)		Not more than 1.4%
Light or Lite Cream Cheese with Other Foods	Not more than 70%	Not more than 16.5% total fat (as marketed)		Not more than 1.4%

18.8.3 Dairy and Nondairy Ingredient Requirements

The quality of the cream used in the manufacture of cream cheese and related products should meet the USDA requirements of cream acceptable for the manufacture of butter that is U.S. Grade A or better. Dairy products used as ingredients for which there are U.S. grades established (nonfat dry milk, dry whole milk, and dry whey) should meet the criteria of U.S. Extra Grade. All stabilizers and emulsifiers should be clean and wholesome, and should be approved by the FDA.

Food colors should be those certified by the FDA as safe for human consumption. Salt should be free-flowing, white refined sodium chloride and should meet standard requirements. When other foods are added, they should be clean, wholesome, and of uniform good quality, free from visible mold, rancid flavor, or decomposed particles. Such ingredients should be consistent in size and color to produce the desired appearance and appeal of the finished product.

18.8.4 Quality Requirements

Cream cheese and related products should possess a slight lactic acid and cultured diacetyl flavor and aroma; no off-flavors or odors such as bitter, flat, sulfide, and yeasty should be present. When another food is added, it should be at a level sufficient to impart a desirable characteristic flavor to the finished product. The characterizing flavor should not be at an intensity that results in a harsh or unnatural flavor.

18.8.5 Body and Texture

Cream cheese should be smooth and free from lumps or grittiness. Reduced-fat cream cheese may be slightly weak or pasty. Light cream cheese may be weak or pasty to a pronounced degree. Droplets or beads of moisture on the surface of the cheese are permissible. Moisture droplets may not run together or pool. The cheese should be medium firm when cold (45°F) and be spreadable at room temperature (68°F). When labeled as "soft," the cream cheese should be spreadable at refrigeration temperature (45°F).

Cream cheese should have a uniform white to light cream color. When another food is added, it should be uniformly distributed and impart the desirable characteristic color to the finished product. The cheese should be free from visible mold or other surface discolorations.

18.8.6 Microbial Requirements

Microbial determinations should be made in accordance with methods prescribed. Samples should be taken as often as necessary to ensure microbial control.

1. *Coliform*—Not more than 10 per gram
2. *Escherichia coli*—Negative

3. *Yeast and mold*—Not more than 10 per gram
4. *Standard plate count*—Not more than 25,000 per gram

18.8.7 Official Identification

Products officially inspected and found to meet these requirements may be identified with the official USDA Quality Approved Inspection Shield.

19

Cottage Cheese and Yogurt: Standards, Grades, and Specifications*

Y. H. Hui

CONTENTS

---

* The information in this chapter has been modified from the *Food Safety Manual* published by the Science Technology System (STS) of West Sacramento, CA. Copyrighted 2012©. With permission.

19.1 Introduction

Food manufacturers in the United States must produce a product that consumers like in order to sell the product nationally. To achieve this goal, they depend on developments in science, technology, and engineering in order to produce a quality product. However, they must also pay attention to two other important considerations:

1. Are their products safe for public consumption?
2. Have they assured the economic integrity of their products? This means many things. The most important is the assurance that their products contain the proper quality and quantity of ingredients, that is, "No cheating."

To protect the consumers, U.S. government agencies have established many checks and balances to prevent "injuries" to the consumers in terms of health and economic fraud. The two major responsible agencies are

1. U.S. Food and Drug Administration (FDA; www.FDA.gov)
2. U.S. Department of Agriculture (USDA; www.USDA.gov)

The FDA establishes standards of identities for a number of cheeses for interstate commerce. This ensures that a particular type of cheese contains the minimal quality and quantity of the ingredients used to make this cheese. The FDA also issues safety requirements and recommendations in the manufacturing of cheeses. It makes periodic inspections of establishments producing cheeses to ensure full compliance. In effect, the FDA is monitoring the safety production of cheeses and their economic integrities.

The USDA plays the same important role in the commerce of cheeses in the United States. For example, a national restaurant chain that makes and sells pizzas will not buy cheeses from any manufacturer if the cheeses do not bear the USDA official identification or seal of approval. The USDA has issued official requirements and processing specifications that cheese manufacturers must follow if they want to obtain the USDA seal of approval for the products. To achieve this goal, the USDA sets up voluntary services of inspection under a standard fee arrangement. The inspection focuses on the safety and economic integrity of manufacturing cheeses. When the inspection demonstrates that the manufacturer has complied with safety and economic requirements, the USDA issues its official identification or seal of approval.

In essence, the FDA and USDA are the federal agencies that ensure that the product is safe and will not pose any economic fraud. Using the public information distributed at their web sites, this chapter discusses FDA standards of identities and USDA specifications for establishments in relation to the production of selected cheeses.

The information in this chapter is important, because it facilitates the commerce of cheeses and protects consumers from injuries of eating cheeses in the market. It also protects the consumers from economic fraud. For example, without FDA and USDA intervention, one may purchase cheese that contains less than the normal amount of milk used. If so, the consumers lose money in the transaction.

All FDA standards of identities are described in

1. 21 CFR 131, Milk and cream
2. 21 CFR 133, Cheeses

All USDA requirements and standards for grade are described in 7 CFR 58.

For explanation, CFR refers to United States Code of Federal Regulations. This chapter focuses on two areas:

1. Part 133: standards of identities issued by the FDA.
2. Part 58 (selected sections): USDA specifications and grade standards.

19.2 Cottage Cheese and Related Products

19.2.1 Cottage Cheese

Cottage cheese is a soft uncured cheese prepared by mixing cottage cheese dry curd with a creaming mixture. The milkfat content is not less than 4% by the weight of the finished food, within limits of good manufacturing practice (GMP). The finished food contains not more than 80% of moisture.

The creaming mixture is prepared from safe and suitable ingredients, including but not limited to milk or substances derived from milk. Any ingredients used that are not derived from milk should serve a useful function other than building the total solids content of the finished food and should be used in a quantity not greater than is reasonably required to accomplish their intended effect. The creaming mixture should be pasteurized; however, heat labile ingredients, such as bacterial starters, may be added following pasteurization.

When the optional process described is used to make the cottage cheese dry curd used in cottage cheese, the label should bear the statement "Directly set" or "Curd set by direct acidification." Wherever the name of the food appears on the label so conspicuously as to be seen under customary conditions of purchase, the statement specified in this paragraph, showing the optional process used, should immediately and conspicuously precede or follow such name without intervening written, printed, or graphic matter.

19.2.2 Dry Curd Cottage Cheese

Cottage cheese dry curd is a soft uncured cheese prepared by prescribed procedure. The finished food contains less than 0.5% milkfat. It contains not more than 80% of moisture.

One or more of the dairy ingredients specified is pasteurized; calcium chloride may be added in a quantity of not more than 0.02% (calculated as anhydrous calcium chloride) of the weight of the mix; thereafter, one of the following methods is employed:

1. Harmless lactic acid–producing bacteria, with or without rennet and/or other safe and suitable milk-clotting enzyme that produces equivalent curd formation, are added, and the mixture is held until it becomes coagulated. The coagulated mass may be cut, it may be warmed, it may be stirred, and it is then drained. The curd may be washed with water and further drained; it may be pressed, chilled, worked, and seasoned with salt.
2. Food-grade phosphoric acid, lactic acid, citric acid, or hydrochloric acid, with or without rennet and/or other safe and suitable milk-clotting enzyme that produces equivalent curd formation, is added in such amount as to reach a pH level between 4.5 and 4.7; coagulation to a firm curd is achieved while heating to a maximum of 120°F without agitation during a continuous process. The coagulated mass may be cut, it may be warmed, it may be stirred, and it is then drained.

The curd is washed with water, stirred, and further drained. It may be pressed, chilled, worked, and seasoned with salt.

3. Food-grade acids, D-glucono-delta-lactone with or without rennet, and/or other safe and suitable milk-clotting enzyme that produces equivalent curd formation are added in such amounts as to reach a final pH value in the range of 4.5–4.8, and the mixture is held until it becomes coagulated. The coagulated mass may be cut, it may be warmed, it may be stirred, and it is then drained. The curd is then washed with water and further drained. It may be pressed, chilled, worked, and seasoned with salt.

The dairy ingredients are sweet skim milk, concentrated skim milk, and nonfat dry milk. If concentrated skim milk or nonfat dry milk is used, water may be added in a quantity not in excess of that removed when the skim milk was concentrated or dried. "Skim milk" means the milk of cows from which the milk fat has been separated, and "concentrated skim milk" means skim milk from which a portion of the water has been removed by evaporation.

The name of the food consists of the following two phrases that should appear together:

1. The phrase "cottage cheese dry curd" or, alternatively, "dry curd cottage cheese" should all appear in type of the same size and style.
2. The phrase "less than 1/2% milkfat" should all appear in letters not less than one-half of the height of the letters but in no case less than one-eighth of an inch in height.
3. When either of the optional processes described is used to make cottage cheese dry curd, the label should bear the statement "Directly set" or "Curd set by direct acidification."
4. Wherever the name of the food appears on the label so conspicuously as to be seen under customary conditions of purchase, the statement specified, showing the optional process used, should immediately and conspicuously precede or follow such name without intervening written, printed, or graphic matter.
5. Each of the ingredients used in the food should be declared on the label as required by FDA regulations, except that milk-clotting enzymes may be declared by the word "enzymes."

19.3 USDA Specifications for Cottage Cheese and Dry Curd Cottage Cheese

19.3.1 General Requirements

Cottage cheese should be manufactured and packaged in accordance with USDA requirements. Cottage cheese should be cooled after packaging and maintained at a temperature of 45°F (7.2°C) or lower. Cottage cheese and dry curd cottage cheese should comply with all applicable federal regulations, including those contained in the FDA Standard of Identity for Cottage Cheese and Dry Curd Cottage Cheese discussed above. The same applies to reduced-fat, light, and fat-free cottage cheese or dry curd cottage cheese.

19.3.2 Composition Requirements

	Moisture	Milkfat	pH
Dry curd cottage cheese	Not more than 80% by the weight of the finished food	Not more than 0.5% by weight of the finished food	Not more than 5.2
Creamed cottage cheese	Not more than 80% by the weight of the finished food	Not less than 4% by the weight of the finished food	Not more than 5.2

19.3.3 Dairy and Nondairy Ingredients

Dry dairy products (such as nonfat dry milk) used as ingredients, for which there are U.S. grades established, should meet the criteria of U.S. Extra Grade. Dairy products for which there are no U.S. grades established should meet the applicable USDA requirements. Plants that produce dairy products for use in cottage cheese should be approved by the USDA. All optional nondairy ingredients should be clean and wholesome and should be approved by the FDA requirements.

19.3.4 Quality Requirements

The product should possess a pleasing and desirable flavor similar to fresh whole milk or cream (if creamed) and may possess the delicate flavor and aroma of lactic acid and diacetyl. The product may possess, to a slight degree, a feed, acid, flat, or salty flavor but should be free from chalky, utensil, fruity, yeasty, or other objectionable flavors. Flavoring ingredients should be uniformly distributed throughout the product. The flavor should be pleasing and characteristic of the flavoring ingredient used. The flavor should not be harsh or unnatural.

19.3.5 Body and Texture

The product should have a meaty texture, but if creamed, it should be sufficiently tender to permit proper absorption of cream or creaming mixture. The texture should be smooth and velvety and should not be mealy, crumbly, pasty, sticky, mushy, watery, or slimy or possess any other objectionable characteristics of body and texture. Small curd style (cut with 1/4-inch knives) should have curd particles approximately 1/4 inch or less in size. Large curd style (cut with knives over 1/4 inch) should have curd particles approximately 3/8 inch in size.

19.3.6 Color and Appearance

The product should present a clean, natural creamy white color. Cottage cheese should have uniform-sized particles (regardless of the style or cut of the curd). Creamed cottage cheese should have a uniform layer of cream around the curd particles with a minimum of free cream. Any excess cream should be of thick consistency (not whey-like or watery). Flavoring ingredients should be consistent in size, distribution, and color to produce a pleasing, natural appearance and appeal of the finished product.

19.3.7 Analytical and Microbial Requirements

Analytical requirements for fat, moisture, and pH should comply with the discussion earlier. For microbial requirements, note

1. Coliform—Not more than 10 per gram
2. Psychrotrophic—Not more than 100 per gram
3. Yeast and mold—Not more than 10 per gram

19.3.8 Official Identification

Products officially inspected and found to meet USDA requirements may be identified with the official USDA Quality Approved Inspection Shield.

19.4 Establishment Processing Cottage Cheese

The USDA has issued the following specifications for plants manufacturing and packaging cottage cheese.

19.4.1 Definitions

Condensed skim—Skim milk that has been condensed to approximately 1/3 the original volume in accordance with the standard commercial practice.

Cottage cheese—The soft uncured cheese meeting the requirements of the FDA for dry curd cottage cheese.

Cottage cheese—The soft uncured cheese meeting the requirements of the FDA for cottage cheese.

Reduced-fat, light, and fat-free cottage cheese—The products conforming to all applicable federal regulations, including regulations for cottage cheese, for dry curd cottage cheese, and for nutrient content claims for fat, fatty acid, and cholesterol content of foods and requirements for foods named by the use of a nutrient content claim and a standardized term.

Direct acidification—The production of cottage cheese, without the use of bacterial starter cultures, through the use of approved food-grade acids. This product should be labeled according to the requirements of the FDA.

Cottage cheese with fruits, nuts, chives, or other vegetables—Should consist of cottage cheese to which fruits, nuts, chives, and other vegetables have been added. The finished cheese should comply with the requirements of the FDA.

Cream—The milkfat portion of milk that rises to the surface of milk on standing or is separated from it by centrifugal force and contains not less than 18.0% of milkfat.

Creaming mixture—The creaming mixture consists of cream or a mixture of cream with milk or skim milk or both. To adjust the solids content, nonfat dry milk or concentrated skim milk may be added but not to exceed 3.0% by the weight of the creaming mixture. It may or may not contain a culture of harmless lactic acid and flavor-producing bacteria, food-grade acid, salt, and stabilizers with or without carriers. The creaming mixture in its final form may or may not be homogenized and should conform to the requirements of the FDA.

19.4.2 Rooms and Compartments

Processing operations with open cheese vats should be separated from other rooms or areas. Excessive personnel traffic or other possible contaminating conditions should be avoided. Rooms, compartments, coolers, and dry storage space in which any raw material, packaging, or ingredients, supplies, or finished products are handled, processed, packaged, or stored should be designed and constructed to ensure clean and orderly operations.

Ventilation. Processing and packaging rooms or compartments should be ventilated to maintain sanitary conditions, preclude the growth of mold and airborne bacterial contaminants, prevent undue condensation of water vapor, and minimize or eliminate objectionable odors. To minimize airborne contamination in processing and packaging rooms, a filtered air supply meeting the USDA requirements should be provided. The incoming air should exert an outward pressure so that the movement of air will be outward and prevent the movement of unfiltered air inward.

Starter facility. A separate starter room or properly designed starter tanks and satisfactory air movement techniques should be provided for the propagation and handling of starter cultures. All necessary precautions should be taken to prevent contamination of the room, equipment, and air therein. A filtered air supply with a minimum average efficiency of 90% should be provided so as to obtain an outward movement of air from the room to minimize contamination.

Coolers. Coolers should be equipped with facilities for maintaining proper temperature and humidity conditions, consistent with good commercial practices for the applicable product, to protect the quality and condition of the products. Coolers should be kept clean, orderly, and free from mold and maintained in good repair. They should be adequately lighted, and proper circulation of air should be maintained at all times. The floors, walls, and ceilings should be of such construction as to permit thorough cleaning.

19.4.3 Equipment and Utensils

The equipment and utensils used for the manufacture and handling of cottage cheese should be as specified. In addition, for certain other equipment, the following requirements should be met.

Cheese vats or tanks. Cheese vats or tanks should meet USDA requirements. When direct steam injection is used for heating the milk, the vat or tank may be of single shell construction. The steam should be culinary steam. Vats should be equipped with valves to control the heating and cooling medium and a suitable sanitary outlet valve. Vats used for creaming curd should be equipped with a refrigerated cooling medium. A circulating pump for the heating and cooling medium is recommended.

Agitators. Mechanical agitators should meet USDA requirements.

Container fillers. Should comply with the 3-A Sanitary Standards for Equipment for Packaging Frozen Desserts and Cottage Cheese.

Mixers. Only mixers that will mix the cheese carefully and keep shattering of the curd particles to a minimum should be used. They should be constructed in such a manner as to be readily cleanable. If shafts extend through the wall of the tank below the level of the product, they should be equipped with proper seals that are readily removable for cleaning and sanitizing. The mixer should be enclosed or equipped with tight-fitting covers.

Starter vats. Bulk starter vats should meet USDA requirements.

19.4.4 Quality Specifications for Raw Materials

Raw materials used for manufacturing cottage cheese should meet the following quality specifications.

Milk. The selection of raw milk for cottage cheese should be in accordance with USDA requirements. Dairy products include the following.

1. *Raw skim milk.* All raw skim milk obtained from a secondary source should be separated from milk meeting the USDA quality requirements for milk. Skim milk after being pasteurized and separated should be cooled to 45°F or lower, unless the skim milk is to be set for cheese within 2 h after pasteurizing. The skim milk should not be more than 48 h old from the time the milk was received at the plant and the skim milk is set for cheese.

2. *Nonfat dry milk.* Nonfat dry milk, when used, should be obtained from milk approved by the USDA and processed according to USDA requirements.

3. *Condensed skim milk.* Condensed skim milk, if used, should be prepared from raw milk or skim milk that meets the same quality requirements outlined above for raw milk or skim milk. It should be cooled promptly after drawing from the vacuum pan or evaporator and should have been pasteurized before concentrating or during the manufacture. The standard plate count of the concentrated milk should not exceed 30,000/mL at the time of use.

Cream. Any cream used for preparing the dressing for creamed cottage cheese should be separated from milk, meeting at least the same quality requirements as the skim milk used for making the curd. The flavor of the cream should be fresh and sweet. Cream obtained from a secondary source should meet the same requirements. The creaming mixture prepared from this cream, after pasteurization, should have a standard plate count of no more than 30,000/mL.

Nondairy ingredients include the following.

Calcium chloride. Calcium chloride, when used, should be of food-grade quality and free from extraneous materials.

Salt. Salt should be free-flowing, white refined sodium chloride and should meet USDA requirements.

Other ingredients. Other ingredients such as fruits, nuts, chives, or other vegetables used or blended with cottage cheese should be reasonably free of bacteria so as not to appreciably increase the bacterial count of the finished product. The various ingredients in kind should be consistent in size and color so as to produce the desired appearance and appeal of the finished product. The flavor of the ingredients used should be natural and represent the intended flavor and intensity desired in the finished product. Such ingredients should be clean, wholesome, of uniformly good quality, free from mold, rancid, or decomposed particles. Vegetables used in cottage cheese may first be soaked for 15–20 min in a cold

25–50 ppm chlorine solution to appreciably reduce the bacterial population. After soaking, the vegetables should be drained and used soon thereafter.

19.4.5 Operations and Opera

19.4.5.1 Pasteurization and Product Flow

The skim milk used for the manufacture of cottage cheese should be pasteurized not more than 24 h prior to the time of setting by heating every particle of skim milk to a temperature of 161°F for not less than 15 s or by any other combination of temperature and time, giving equivalent results. All skim milk must be cooled promptly to setting temperature. If held for more than 2 h between pasteurization and the time of setting, the skim milk should be cooled and held at 45°F or lower until set.

Cream or cheese dressing should be pasteurized at not less than 150°F for not less than 30 min, at not less than 166°F for not less than 15 s, or by any other combination of temperature and time treatment giving equivalent results. Cream and cheese dressing should be cooled promptly to 40°F or lower after pasteurization to aid in the further cooling of cottage cheese curd for improved keeping quality.

19.4.5.2 Reconstituted Nonfat Dry Milk

Reconstituted nonfat dry milk for cottage cheese manufacture need not be repasteurized, provided that it is reconstituted within 2 h prior to the time of setting using water that is free from viable pathogenic or otherwise harmful microorganisms as well as microorganisms that may cause spoilage of cottage cheese. Skim milk separated from pasteurized whole milk need not be repasteurized, provided that it is separated in equipment from which all traces of raw milk from previous operations have been removed by proper cleaning and sanitizing. Reconstituting nonfat dry milk should be reconstituted in a sanitary manner.

19.4.5.3 Laboratory and Quality Control Tests

Quality control tests should be made on samples as often as necessary to determine the shelf life and stability of the finished product. Routine analyses should be made on raw materials and finished product to assure satisfactory composition, shelf life, and stability.

19.4.5.4 Packaging and General Identification

Containers. Containers used for packaging cottage cheese should be any commercially acceptable multiple-use or single-service container or packaging material that will satisfactorily protect the contents through the regular channels of trade without significant impairment of quality with respect to flavor or contamination under normal conditions of handling. Caps or covers that extend over the lip of the container should be used on all cups or tubs containing 2 lbs or less to protect the product from contamination during subsequent handling.

Packaging. The cheese should be packaged in a sanitary manner, and automatic filling and capping equipment should be used on all small sizes. The containers should be checked and weighed during the filling operation to ensure that they are filled uniformly to not less than the stated net weight on the container. In addition, care should be taken such that the cottage cheese is of uniform consistency at the time of packaging to ensure legal composition in all packages.

General identification. Bulk packages containing cottage cheese should be adequately and legibly marked with the name of the product, net weight, name and address of the manufacturer, lot number, code or date of packaging, and any other identification as may be required. Consumer-size packaged products should meet the applicable regulations of the FDA.

19.4.6 Storage of Finished Product

Cottage cheese after packaging should be promptly stored at a temperature of 45°F or lower to maintain quality and condition until loaded for distribution. During distribution and storage prior to sale, the product should be maintained at a temperature of 45°F or lower. The product should not be exposed to foreign odors or conditions that might cause package or product damage, such as drippage or condensation. Packaged cottage cheese should not be placed directly on floors.

19.4.7 Official Identification and Requirements

Only cottage cheese manufactured and packaged in accordance with USDA requirements may be identified with the official USDA Quality Approved Inspection Shield. The following are some requirements.

Nonfat dry milk. Nonfat dry milk, when used in cottage cheese bearing official identification, should meet the requirements for U.S. Extra Grade (Spray Process) at the time of use and should be of U.S. Low Heat Classification (not less than 6.0-mg undenatured whey protein nitrogen per gram of nonfat dry milk). In addition, the nonfat dry milk should have a direct microscopic count not exceeding 75 million per gram. The age of the nonfat dry milk should be covered by a USDA grading certificate, evidencing compliance with quality requirements, dated not more than 6 months prior to use of the dry milk. In the interim between manufacture and use, the nonfat dry milk should be stored in a clean, dry, and vermin-free space. In any case, if the nonfat dry milk is more than 120 days old at the time of use, it should be examined for flavor to ascertain that it meets the requirements for U.S. Extra Grade.

Flavor. The cottage cheese should possess a mild pleasing flavor, similar to fresh whole milk or light cream, and may possess the delicate flavor and aroma of a good lactic starter. The product may possess, to a slight degree, a feed, acid, or salty flavor but should be free from chalky, bitter, utensil, fruity, yeasty, or other objectionable flavors.

Body and texture. The curd particles should have a meaty texture but sufficiently tender to permit the proper absorption of cream or cheese dressing. The texture should be smooth and velvety and should not be mealy, crumbly, pasty, sticky, mushy, watery, rubbery, or slimy or possess any other objectionable characteristics of body and texture. Small curd style (cut with 1/4-inch knives) should have curd particles approximately 1/4 inch or less in size. Large curd style (cut with knives over 1/4 inch) should have curd particles approximately 3/8 inch or more in size.

Color and appearance. The finished cottage cheese, creamed or plain curd, should have an attractive natural color and appearance with curd particles of reasonably uniform size. The creamed cottage cheese should be uniformly mixed with the cream or dressing properly absorbed or adhering to the curd so as to prevent excessive drainage.

Microbiological requirements. Compliance should be based on three out of five consecutive samples taken at the time of packaging.

1. Coliform—Not more than 10 per gram
2. Psychrotrophic—Not more than 100 per gram
3. Yeasts and molds—Not more than 10 per gram

Chemical requirements. Moisture and milkfat must comply with specific USDA requirements. The pH level should not be higher than 5.2. Phosphatase should not be more than 4 μg of phenol equivalent per gram of cheese.

Keeping quality requirements. Keeping quality samples taken from the packaging line should be held at 45°F for 10 days. At the end of the 10-day period, the samples should possess a satisfactory flavor and appearance and should be free from bitter, sour, fruity, or other objectionable tastes and odors. The surface should not be discolored, translucent, slimy, or show any other objectionable condition.

19.5 Yogurt and Related Products

19.5.1 Yogurt

Yogurt is the food produced by culturing one or more of the optional dairy ingredients specified in Section 19.5.1.2 with a characterizing bacterial culture that contains the lactic acid–producing bacteria,

Lactobacillus bulgaricus and *Streptococcus thermophilus*. One or more of the other optional ingredients may also be added. When one or more of the ingredients specified are used, they should be included in the culturing process. All ingredients used are safe and suitable. Yogurt, before the addition of bulky flavors, contains not less than 3.25% milkfat and not less than 8.25% milk solids-not-fat and has a titratable acidity of not less than 0.9%, expressed as lactic acid. The food may be homogenized and should be pasteurized or ultrapasteurized prior to the addition of the bacterial culture. Flavoring ingredients may be added after pasteurization or ultrapasteurization. To extend the shelf life of the food, yogurt may be heat treated after culturing is completed to destroy viable microorganisms.

19.5.1.1 Vitamin Addition (Optional)

1. If added, vitamin A should be present in such quantity that each 946 ml (quart) of the food contains not less than 2000 IU thereof, within limits of the current GMP.
2. If added, vitamin D should be present in such quantity that each 946 ml (quart) of the food contains 400 IU thereof, within limits of the current GMP.

19.5.1.2 Optional Dairy Ingredients

The following may be added, used alone or in combination:

1. Cream
2. Milk
3. Partially skimmed milk
4. Skim milk

19.5.1.3 Other Optional Ingredients

1. Concentrated skim milk, nonfat dry milk, buttermilk, whey, lactose, lactalbumins, lactoglobulins, or whey modified by the partial or complete removal of lactose and/or minerals to increase the nonfat solids content of the food, provided that the ratio of protein to total nonfat solids of the food and the protein efficiency ratio of all protein present should not be decreased as a result of adding such ingredients
2. Nutritive carbohydrate sweeteners—Sugar (sucrose), beet or cane; invert sugar (in paste or syrup form); brown sugar; refiner's syrup; molasses (other than blackstrap); high-fructose corn syrup; fructose; fructose syrup; maltose; maltose syrup, dried maltose syrup; malt extract, dried malt extract; malt syrup, dried malt syrup; honey; maple sugar; or any permitted sweeteners
3. Flavoring ingredients
4. Color additives
5. Stabilizers

The name of the food is "yogurt." The name of the food should be accompanied by a declaration indicating the presence of any characterizing flavoring. The word "sweetened" if nutritive carbohydrate sweetener is added, without the addition of characterizing flavor. The parenthetical phrase "(heat treated after culturing)" should follow the name of the food if the dairy ingredients have been heat treated after culturing.

19.5.2 Low-Fat Yogurt

Low-fat yogurt is the food produced by culturing one or more of the optional dairy ingredients with a characterizing bacterial culture that contains the lactic acid–producing bacteria, *Lb. bulgaricus* and *S. thermophilus*. One or more of the other optional ingredients specified may also be added. When one or more of the ingredients specified are used, they should be included in the culturing process. All ingredients used are safe and suitable. Low-fat yogurt, before the addition of bulky flavors, contains neither less than 0.5% nor more than 2% milkfat and not less than 8.25% milk solids-not-fat and has a titratable acidity of not less than 0.9%, expressed as lactic acid. The food may be homogenized and should be pasteurized or ultrapasteurized prior to the addition of the bacterial culture. Flavoring ingredients may be added after pasteurization or ultrapasteurization. To extend the shelf life of the food, low-fat yogurt may be heat treated after culturing is completed to destroy viable microorganisms.

19.5.2.1 Vitamin Addition (Optional)

1. If added, vitamin A should be present in such quantity that each 946 mL (quart) of the food contains not less than 2000 IU thereof, within limits of the current GMP.
2. If added, vitamin D should be present in such quantity that each 946 mL (quart) of the food contains 400 IU thereof, within limits of the current GMP.

19.5.2.2 Optional Dairy Ingredients

The following may be added, used alone or in combination:

1. Cream
2. Milk
3. Partially skimmed milk
4. Skim milk

19.5.2.3 Other Optional Ingredients

1. Concentrated skim milk, nonfat dry milk, buttermilk, whey, lactose, lactalbumins, lactoglobulins, or whey modified by the partial or complete removal of lactose and/or minerals to increase the nonfat solids content of the food, provided that the ratio of protein to total nonfat solids of the food and the protein efficiency ratio of all protein present should not be decreased as a result of adding such ingredients
2. Nutritive carbohydrate sweeteners—Sugar (sucrose), beet or cane; invert sugar (in paste or syrup form); brown sugar; refiner's syrup; molasses (other than blackstrap); high-fructose corn syrup; fructose; fructose syrup; maltose, maltose syrup, dried maltose syrup; malt extract, dried malt extract; malt syrup, dried malt syrup; honey; maple sugar; or any other permitted sweeteners
3. Flavoring ingredients
4. Color additives
5. Stabilizers

The name of the food is "low-fat yogurt." The name of the food should be accompanied by a declaration indicating the presence of any characterizing flavoring. The blank in the phrase "__% milkfat" is to be filled in with the fraction 1/2 or multiples thereof closest to the actual fat content of the food. The word "sweetened" is used if nutritive carbohydrate sweetener is added without the addition of characterizing flavoring. The parenthetical phrase "(heat treated after culturing)" should follow the name of the food if the dairy ingredients have been heat treated after culturing.

19.5.3 Nonfat Yogurt

Nonfat yogurt is the food produced by culturing one or more of the optional dairy ingredients with a characterizing bacterial culture that contains the lactic acid–producing bacteria, *Lb. bulgaricus* and *S. thermophilus*. One or more of the other optional ingredients specified may also be added. When one or more of the ingredients specified are used, they should be included in the culturing process. All ingredients used are safe and suitable. Nonfat yogurt, before the addition of bulky flavors, contains <0.5% milkfat and not less than 8.25%

milk solids-not-fat and has a titratable acidity of not less than 0.9%, expressed as lactic acid. The food may be homogenized and should be pasteurized or ultrapasteurized prior to the addition of the bacterial culture. Flavoring ingredients may be added after pasteurization or ultrapasteurization. To extend the shelf life of the food, nonfat yogurt may be heat treated after culturing is completed to destroy viable microorganisms.

19.5.3.1 Vitamin Addition (Optional)

1. If added, vitamin A should be present in such quantity that each 946 mL (quart) of the food contains not less than 2000 IU thereof, within limits of GMP.
2. If added, vitamin D should be present in such quantity that each 946 mL (quart) of the food contains 400 IU thereof, within limits of GMP.

19.5.3.2 Optional Dairy Ingredients

The following may be added, used alone or in combination:

1. Cream
2. Milk
3. Partially skimmed milk
4. Skim milk

19.5.3.3 Other Optional Ingredients

1. Concentrated skim milk, nonfat dry milk, buttermilk, whey, lactose, lactalbumins, lactoglobulins, or whey modified by the partial or complete removal of lactose and/or minerals to increase the nonfat solids content of the food, provided that the ratio of protein to total nonfat solids of the food and the protein efficiency ratio of all protein present should not be decreased as a result of adding such ingredients
2. Nutritive carbohydrate sweeteners—Sugar (sucrose), beet or cane; invert sugar (in paste or syrup form); brown sugar; refiner's syrup; molasses (other than blackstrap); high fructose corn syrup; fructose; fructose syrup; maltose; maltose syrup, dried maltose syrup; malt extract, dried malt extract; malt syrup, dried malt syrup; honey; maple sugar; or any of other permitted sweeteners
3. Flavoring ingredients
4. Color additives
5. Stabilizers

The name of the food is "nonfat yogurt." The name of the food should be accompanied by a declaration indicating the presence of any characterizing flavoring. The following terms should accompany the name of the food wherever it appears on the principal display panel:

1. The word "sweetened" is used if nutritive carbohydrate sweetener is added without the addition of characterizing flavoring.
2. The parenthetical phrase "(heat treated after culturing)" should follow the name of the food if the dairy ingredients have been heat treated after culturing.

19.6 USDA Specifications for Yogurt, Nonfat Yogurt, and Low-Fat Yogurt

19.6.1 General Requirements

Yogurt should be manufactured and packaged in accordance with the applicable USDA and FDA requirements. After the final steps in manufacturing and/or packaging, the yogurt should be cooled and maintained at

45°F (7.2°C) or less prior to inspection. Yogurt products should comply with all applicable federal regulations, including those contained in the FDA Standard of Identity for Yogurt, Low-Fat Yogurt, or Nonfat Yogurt.

Calorie-modified yogurt (such as "light" yogurt) should comply with all applicable FDA regulations, including nutrient content claims for "light" and "lite."

19.6.2 Composition Requirements

Composition requirements apply to the yogurt prior to the addition of bulky flavoring ingredients.

1. *Fat*—Not less than 3.25%
2. *Milk solids-not-fat*—Not less than 8.25%

19.6.3 Low-Fat Yogurt

1. *Fat*—Not less than 0.5% or more than 2%
2. *Milk solids-not-fat*—Not less than 8.25%

19.6.4 Nonfat Yogurt

1. *Fat*—Not more than 0.5%
2. *Milk solids-not-fat*—Not less than 8.25%

19.6.5 Dairy and Nondairy Ingredients

All milk products used in the manufacture of yogurt and the plant in which the yogurt is processed should comply with all applicable FDA and USDA requirements. All optional nondairy ingredients should be clean and wholesome and should be approved by the FDA.

19.6.6 Quality Requirements

The flavor of the yogurt should possess a pleasant, clean acid flavor. It should be free from undesirable flavors such as bitter, rancid, oxidized, stale, yeasty, and unclean. Bulk flavoring ingredients should be uniformly distributed throughout the product ("sundae style" yogurts will require mixing to evaluate uniformity). The flavor should be pleasing and characteristic of the flavoring ingredient used. The flavor should not be harsh or unnatural.

19.6.6.1 Body and Texture

Yogurt should possess a firm, custard-like body with a smooth, homogeneous texture. A spoonful of yogurt should maintain its form without displaying sharp edges. Bulk flavoring ingredients should be uniformly distributed throughout the product.

19.6.6.2 Color and Appearance

Yogurt should present a clean, natural color, with a smooth, velvety appearance. Natural color in unflavored yogurt may range from a bright white to an off-white color. The surface should appear smooth and not exhibit excessive whey separation, mold, or surface discoloration. Flavoring ingredients should be uniform in size, distribution, and color to produce a pleasing, natural appearance in the finished product.

19.6.7 Analytical Testing and Microbial Requirements

Refer to analytical requirements for fat and solids-not-fat in 7 CFR 58.

With regard to microbial requirements, coliform should not be more than 10 per gram, and yeast and mold should not be more than 50 per gram.

19.6.8 Official Identification

Products officially inspected and found to meet these requirements may be identified with the official USDA Quality Approved Inspection Shield.

Part III

Solid Cheeses

20

Cheddar and Related Hard Cheeses

Stephanie Clark and Shantanu Agarwal

CONTENTS

20.1 Introduction

The name cheddar stems from the fact that cheddar cheese originated in the village of Cheddar, in Somerset, England in the nineteenth century (Banks and Williams 2004). The term "cheddaring" specifies the process of piling and repiling of blocks of warm curd in cheese vats. During the cheddaring period of about 2 h, lactic acid increases rapidly, and the proteins stretch and align, which results in body and texture characteristics of cheddar cheese. The first cheddar cheese factory in the U.S., other than farmhouse cheesemaking, was established in New York in 1861 (Lawrence and Gilles 1987a). The procedures for cheddar cheese manufacture were popularized in the U.S. in 1876 by Robert McCadam, leading to the evolution of the American cheddar cheese industry (Kosikowski and Mistry 1997a).

The U.S. is the largest producer of cheddar cheese in the world. In 2009, the production of cheddar cheese in the U.S. exceeded 1.46 billion kg, and supermarket sales surpassed 270 million kg (IDFA 2010). In supermarket sales in 2009, cheddar had the highest sales (270 million kg), followed by American (195 million kg) and mozzarella (139 million kg) cheese (IDFA 2010). Supermarket sales of all other cheeses pale in comparison to these three favorites. In 2009, it was reported that consumers ate approximately 4.6 kg of cheddar cheese per capita (IDFA 2010), up from 4.5 kg per capita in 2008 (IDFA 2010). The greatest centers of American cheese production (including cheddar) in 2009 were Wisconsin (1.2 billion kg) with about 26% of the total U.S. production, followed by California (0.95 billion kg). Wisconsin, California, Idaho, New York, and Minnesota account for a total of 69% of the total U.S. cheese production (IDFA 2010).

20.2 Definitions

Cheddar cheese (FDA 2006a) is classified as a hard cheese, ranging in color from nearly white (particularly if made from goat or sheep milk) to yellow or orange (USDA 1978). Standards of identity in the U.S. Code of Federal Regulations (CFR Title 21.133.113) require that cheddar does not exceed 39% moisture and fat is at least 50% on dry basis (FDA 2006a). The term "reduced fat" means at least a 25% reduction in the total fat per reference amount in comparison to a reference food. The claim cannot be made if the reference food meets the definition for "low fat." Low-fat cheese contains a maximum of 3-g total fat per reference amount (when the reference amount is greater than 30 g or 2 tbsp). If the reference amount is less than 30 g or 2 tbsp or less, then to be called low fat, the cheese must contain a maximum of 3 g of total fat per 50 g of the food. Reduced-sodium cheese must have at least a 25% reduction in sodium per reference amount compared with an appropriate reference food. Low-sodium cheddar cheese contains not more than 140 mg of sodium per 50 g of finished food (CFR Title 21.133.116; FDA 2006c). Cheddar cheese for manufacturing (Title 21.133.114; FDA 2006b) conforms to the definition and standard of identity for cheddar cheese, except that the milk is not pasteurized, curing is not required, and the provisions for the use of antimycotic agents on slices or cuts do not apply (CFR Title 21.133.114; FDA 2006b).

Cheeses closely related to cheddar include colby (FDA 2006d) and monterey jack (FDA 2006e). Longhorn is not separately defined in the Code of Federal Regulations, but it may be found commercially. Longhorn cheddar is essentially a name that describes the round shape derived from Longhorn hoop usage during the pressing step. While cheddar is traditionally about 36.83 cm in diameter, is 30.48 cm thick, and weighs between 31.75 and 35.38 kg, Longhorn cheddar is 15.24 cm in diameter (round), is 33.02 cm long, and weighs 5.5–6 kg (USDA 1978).

Colby manufacture resembles cheddar, except that the curds are "washed" and stirred instead of matted and milled (Kosikowski and Mistry 1997a). Colby is aged for 2–3 months at 3°C–4°C, and the resulting cheese is more moist, softer, and more open in texture than cheddar (Fox et al. 2000d). Moisture must not exceed 40%, and fat in the solids must be at least 50% (FDA 2006d). Colby contains between 1.4% and 1.8% salt (USDA 1978).

Monterey, jack, or monterey jack cheese was first made in Monterey County, California in 1892 (USDA 1978). Monterey jack is made in a similar fashion to colby, but the make procedure requires less time, including a ripening period of only 5–7 weeks (Fox et al. 2000d). Monterey jack contains more moisture, commonly exhibits mechanical openings, and is softer than cheddar and colby (Fox et al. 2000d; USDA 1978). Standards of identity state that monterey jack must contain not more than 44% moisture and at least 50% fat in the solids (FDA 2006e). High-moisture jack cheese conforms to the definition and standard of identity and is subject to the requirement for the label statement of ingredients prescribed for monterey cheese, except that its moisture content is more than 44% but less than 50% (FDA 2006e). Monterey jack typically contains 1.5% salt (USDA 1978).

Colby jack cheese is a mixture of colby and jack curds prior to pressing and ripening. Approximately 38 million kg of colby jack was sold in supermarkets in 2006, more than both colby (11 million kg) and monterey jack (30 million kg) cheeses on their own (IDFA 2007).

20.3 Production of Cheddar and Related Hard Cheeses

Good cheese requires high-quality milk and carefully selected starter cultures. However, additional ingredients are often utilized to enhance visual appeal (annatto), coagulation properties (calcium chloride or enzymes), and flavor development (adjunct cultures or enzymes) to make great cheeses. How well the additional steps are employed determines whether one makes a great cheese. The following section elaborates on individual ingredients and their function in cheesemaking.

20.3.1 Ingredients

20.3.1.1 Milk

Cheese quality will never be better than the materials used to make it. Cheddar and most cheddar-like cheeses can be made from raw, heat-treated, heat-shocked, or pasteurized milk, milk of varying fat levels, or cream, alone or in combination (FDA 2006a). Legally, there is no distinction between raw, heat-treated, and heat-shocked milk. Because of the potential for pathogens to survive in cheese for up to 60 days, cheeses made from raw or heat-treated or heat-shocked milk must be aged for at least 60 days at >1.7°C prior to sale (FDA 2006a). Cheeses must be aged at >1.7°C to ensure microbial metabolic activity and progression through the life and death cycles. Cheeses made from pasteurized milk need not be aged prior to sale. Monterey jack cheese milk must be pasteurized because of its limited aging (FDA 2006e).

Regardless of heat treatment, low raw milk bacteria counts are essential for high-quality cheese, as high bacterial enzymes can lead to flavor and body defects. Psychrotrophic bacteria, including *Pseudomonas*, *Aeromonas*, *Flavobacterium*, *Acinetobacter*, *Bacillus*, *Micrococcus*, and other genera, can grow relatively rapidly in milk maintained at 7°C or lower (Banks and Williams 2004; Frank and Marth 1988; Richard and Desmazeaud 2000), so extended storage of milk (beyond 48 h) prior to pasteurization or cheesemaking is highly discouraged. Enzymes produced by psychrotrophs, including heat-stable lipases and proteinases, can act directly on milk proteins and lipids, reducing yield and contributing to quality defects development in the resultant cheese (Johnson 1988; Richard and Desmazeaud 2000). Raw milk with high spores can lead to cheese with splits and cracks due to secondary fermentation during the aging process (Martley and Crow 1996).

Because approximately 90% of both fat and protein from cheese milk are captured in the cheese (Table 20.1) and these components make up 91% of the solids in cheese (Johnson 1988), detrimental effects on either component will be realized in the cheese yield and quality. For consistency in cheese yield and product composition, milk for cheddar cheese is commonly standardized to a casein-to-fat ratio between 0.67 and 0.72 (Banks and Williams 2004; Lawrence and Gilles 1987a).

Somatic cells also have a negative impact on cheese yield. Barbano et al. (1991) demonstrated that milk casein as a percentage of true protein (C%TP) and cheese yield efficiency were low when the milk somatic cell count (SCC) was high. Cheese moisture, as well as fat and protein losses in whey, increased with increased SCC (Barbano et al. 1991). It was concluded that any increase in milk SCC above 100,000 cells/mL negatively affects cheese yield for milk from groups of cows with similar milk SCC (Barbano et al. 1991). The work of Marino et al. (2005) demonstrated that the proteolytic activity of somatic cells recovered from mastitic milk contributed directly to proteolysis in buffer, milk, and cheddar cheese but such activity was reduced by batch pasteurization.

Raw milk naturally contains low levels of endogenous enzymes, including alkaline phosphatase (ALP), plasmin, and lipoprotein lipase (Whitney 1988). ALP is slightly more heat stable than the most heat-resistant pathogenic microorganism in milk. Thus, ALP is a convenient indicator of pasteurization. Indeed, the *Pasteurized Milk Ordinance* defines legal limits for ALP for Grade A pasteurized milk and bulk shipped heat-treated milk products (U.S. Department of Health and Human Services et al. 1999). ALP assays should also measure negative in cheeses made from pasteurized milk. Plasmin, which is stable at pasteurization temperature, hydrolyzes both β- and α_{s1}-casein in milk and in cheese during maturation, thus contributing to cheese maturation (Banks and Williams 2004; Johnson 1988).

The lipoprotein lipase found in milk is identical to the lipoprotein lipase in blood and represents a spillover from the mammary tissues (Weihrauch 1988). Lipoprotein lipase, if activated by severe agitation, temperature fluctuations, or other means, leads to the hydrolysis of fatty acids from triacylglycerides and rancid off-flavors in milk and the subsequent cheese. Hydrolytic rancidity in milk may be detectable at an acid degree value exceeding 1.2 meq/L (Bodyfelt et al. 1988; Weihrauch 1988). Raw milk is somewhat resistant to lipase because of the protective nature of the milkfat globule membrane (Weihrauch 1988). Unlike contaminating bacterial lipases, natural milk lipase is heat labile. Heating of milk to 80°C for 20 s destroys all lipases in milk (Weihrauch 1988). However, any damage done to milkfat by lipase prior to pasteurization cannot be reversed by the pasteurization process; poor-quality milk will yield poor-quality cheese.

Extended storage of milk prior to pasteurization and cheesemaking is also discouraged, because it encourages the solubilization of colloidal calcium phosphate (CCP) and a shift in caseins from the micellar to soluble state (Johnson 1988). While soluble caseins constitute less than 15% of the total casein in normal milk directly from the udder, the proportion increases to up to 42% of the total casein during storage at 4°C (Johnson 1988). Soluble calcium phosphate and casein are lost during whey drainage, which reduces cheese yield (Johnson 1988).

TABLE 20.1

Recovery of Milk Components in Cheddar Cheese

Constituent	Percentage in Cheese Milk	Percentage Recovered in Cheese
Water	87.0–88.0	4.5
Fat	3.0–4.5	92.5
Casein	3.0–4.0	96.0
Lactose	4.5–5.0	4.0
Whey protein/salts	1.2–1.8	29.0

20.3.1.2 Calcium Chloride

Calcium chloride ($CaCl_2$) may be added to cheese milk as a coagulation aid, in an amount not more than 0.02% (calculated as anhydrous calcium chloride; CFR133.113; FDA 2006a) of the weight of the dairy ingredients. The addition of calcium chloride reduces the coagulation time and increases curd firmness, in part due to a decrease in pH and an increase in calcium ions (Lenoir et al. 2000b; Fox et al. 2000c). Calcium added to milk is solubilized during acidification and is thus lost in whey, so it does not contribute to the total calcium in the final cheese, yet the use of $CaCl_2$ may have positive effects on the recovery of fat and protein, and cheese yield (Fox et al. 2000c).

20.3.1.3 Starter Cultures

Starter bacteria can be defined as isolates that produce sufficient acid to reduce the pH of milk to below 5.3 in 6 h at 30°C–37°C and aid in curd digestion and flavor development (Lewis 1987). Cultures selected for cheddar cheese made throughout the world are typically "O" cultures, which are designated as cultures that produce lactic acid from lactose. The "O"-type cultures are composed of *Lactococcus lactis* subsp. *lactis* and *Lc. lactis* subsp. *cremoris* (Harrits 1997; Strauss 1997). *Lc. lactis* subsp. *lactis* and *Lc. lactis* subsp. *cremoris* can be differentiated by their ability to grow at 40°C and in the presence of salt. While *Lc. lactis* subsp. *lactis* will grow at 40°C and in the presence of 4% salt, *Lc. lactis* subsp. *cremoris* will not grow at 40°C and will grow in the presence of salt up to a 2% concentration (Harrits 1997). The "L"-type cultures consist of "O"-type cultures plus the citrate-fermenting culture *Leuconostoc mesenteroides* subsp. *cremoris*, which produces flavor compounds (e.g., diacetyl) plus small amounts of carbon dioxide (Harrits 1997). "L"-type cultures may also be used in the production of colby, because an open structure is allowed in colby (Harrits 1997). Depending on the type of culture preparation, usage rates vary between 0.75% and 1.25% traditional bulk starter and 0.5%–0.6% pH-controlled starter or direct vat set (DVS) starter (Strauss 1997).

Starter culture selection and uniform application are very important for flavor development of cheddar cheese (Shakeel Ur 2007). Ideally, acid should be formed quickly and at a predictable steady rate during curd formation. Bacteriophage (phage) resistance, salt sensitivity, and protease activity (desirable flavor development) are additional selection criteria for starter cultures (Strauss 1997). These homofermentative starter cultures are added deliberately to initiate cheddar cheese manufacture. The starter bacteria produce L(+)-lactate from lactose, and they grow, typically attaining cell densities of 10^8 cfu/g within hours of the beginning of manufacture. L(+) refers to an optically active substance that rotates the plane of polarized light counterclockwise; also called levorotatory. The optical isomer of L(+)-lactate is D(−)-lactate. The optical isomers are mirror images of each other and result from the tetrahedral geometry around the chiral carbon center.

The production of homogeneous, high-quality cheddar cheese requires uniform lactose fermentation, lipolysis, and proteolysis, each of which varies among bacterial strains. A relationship exists between the extent of starter cell autolysis and the level of lipolysis during cheddar cheese ripening (Collins et al. 2003). The rate and extent of both fermentation and proteolysis depend on the temperature and salt concentration (Thomas and Pearce 1981). One of the main roles of starter bacteria is to provide a suitable environment for enzyme activity from rennet/chymosin (an acid protease) and a favorable growth of secondary microflora with respect to redox potential, pH, and moisture content in cheese. Redox potential is a measure of the tendency of a system to donate or accept electrons and indicates aerobic or anaerobic conditions. Typically, the environment inside the cheese is anaerobic and reducing.

Extensive modifications in cheddar cheese manufacture have taken place with the introduction of continuous cheddar cheese manufacturing systems in large establishments. However, regardless of advances in automation, starter cultures are always used in the manufacturing of cheddar cheese, and considerable attention is given to culture selection. The primary function of starter cultures is to produce acid during the fermentation process, but starters also contribute to cheese ripening, as their enzymes contribute to proteolysis and the formation of flavor compounds (Wallace and Fox 1997).

20.3.1.4 Adjunct Cultures

An adjunct culture is one that is added, along with starter culture, for the desirable characteristics that it may impart on the cheese other than acid. Adjunct cultures are select nonstarter lactic acid bacteria (NSLAB), because not all NSLAB are desirable. Although the specific ripening mechanisms of NSLAB that contribute positively to cheddar flavor have not been fully determined, to be successful, NSLAB adjuncts require two important features. First, strains must provide a balance of beneficial ripening reactions in cheese (Crow et al. 2001). Second, strains need to be competitive against adventitious (not intentionally added) NSLAB and remain the dominant NSLAB during the ripening period (Crow et al. 2001). Fox et al. (2000a) reported that mesophilic lactobacilli are superior to thermophilic lactobacilli for cheddar cheese adjuncts. Additionally, they reported that the principal contribution of adjunct lactobacilli is the formation of amino acids.

Lactobacillus helveticus is one species that may be added to cheddar cheese milk for desirable cheese flavor development. *Lb. helveticus* species tend to be thermophilic and proteolytic and have the ability to utilize galactose after other sugars are fermented (Harrits 1997). Because of their proteolytic capabilities, *Lb. helveticus* adjunct cultures have shown to effectively improve the sensory quality of reduced-fat cheddar cheese (Drake et al. 1997).

20.3.1.5 Color

The recognizable yellow to orange color of cheddar and colby cheeses is derived from annatto, an extract from seeds of the "Lipstick tree," *Bixa orellana* (Walstra et al. 1999). Annatto was first used to make cheeses appear more fat rich when made during seasons of the year when the milk of cows produced less colorful cheese because the cows were fed diets lower in beta-carotene. Sheep and goat milk cheeses naturally are whiter than cow milk cheeses, because the beta-carotene is efficiently hydrolyzed to vitamin A in the digestive tracts of these species.

One disadvantage of adding annatto in cheese milk is that some of the annatto is transferred to the cheese whey, giving it an undesirable yellow color. Cheddar cheese processors who manufacture whey ingredients often bleach their whey in order to decolor the whey ingredients, as consumers and end users prefer pale whey ingredients. Recent research has shown that bleaching along with other processing steps leads to lipid oxidation products and creation of various volatile compounds that are associated with cardboard-like off-flavors (Croissant et al 2009; Kang et al 2010).

20.3.1.6 Enzymes

The Code of Federal Regulations permits the use of rennet, chymosin, and/or other clotting enzymes of animal, plant, or microbial origin as well as enzymes of animal, plant, or microbial origin, used in curing or flavor development for cheddar cheese (CFR 133.113; FDA 2006j). Chymosin, the primary enzyme added to milk for cheesemaking, has been traditionally obtained from the lining of calf stomach. Advancement in genetics has led to the production of chymosin using recombinant deoxyribonucleic acid technology. This advancement has led to the production of enzymes with consistent quality and has led to the large-scale production of cheese. Enzymes of starter, adjunct, and adventitious NSLAB naturally contribute to flavor and body/texture development in cheeses. Kheadr et al. (2003) were able to accelerate cheddar cheese proteolysis and lipolysis using various liposome-encapsulated enzymatic cocktails. Neutral bacterial protease, acid fungal protease, and lipase were individually entrapped or mixed as cocktails, entrapped in liposomes, and then added to cheese milk prior to renneting (Kheadr et al. 2003). Certain enzyme treatments resulted in cheeses with more mature texture and higher flavor intensity or cheddar flavor in a shorter time compared with control cheeses (Kheadr et al. 2003).

Milk naturally contains numerous enzymes, which may positively or negatively impact cheese quality. Plasmin is the predominant and most completely studied endogenous proteinase in cow milk (Ismail and Nielsen 2010). Although it has been implicated as the causative agent in a number of defects in dairy products, including reduced curd strength and curd yield and bitter off-flavors, plasmin is also important in cheese ripening (Crudden et al. 2005). Plasmin is part of a complex system composed of its inactive form plasminogen, plasminogen activators, and inhibitors, which interact with each other and other milk components to promote or inhibit proteolysis (Ismail and Nielsen 2010).

Other enzymes naturally found in milk include but are not limited to acid protease, ALP, α- and β-amylase, γ-glutamyltransferase, lactate dehydrogenase, lactoperoxidase, lysozyme, and xanthine oxidase (Whitney 1988). Native milk enzymes are susceptible to heat denaturation at varying degrees; while ALP is quite heat labile, lactoperoxidase is quite heat stable (Lorenzen et al. 2009). Enzymes may also be present in milk as a result of the lysis of psychrotrophic microorganisms. Lipolytic enzymes, which may enter milk from bacterial sources, cause the atypical flavor "rancid" in milk and cheddar cheese by hydrolyzing volatile short-chain fatty acids from milkfat triglycerides. Because lipase is heat labile, the pasteurization of milk soon after collection is recommended as one step to minimize the potential development of rancid off-flavors in milk and cheese.

20.3.1.7 Salt

Cheddar cheese typically contains 1.6%–1.8% salt (Williams et al. 2000) in the form of NaCl. Food-grade salt is essential to the production of safe, high-quality cheddar cheese, and the consistent salt grain size contributes to the uniformity of salt concentration throughout the cheese matrix. Salt is important, because it enhances flavor, encourages syneresis, slows or stops the growth of salt-sensitive bacteria, influences the activity of enzymes, and is used to maintain the expected body, texture, flavor, and shelf life of cheeses (Johnson et al. 2009). The level of salting in cheese and resulting salt in moisture (S/M) influence the amount of residual lactose, cheese pH, and rate and extent of protein hydrolysis in the cheese (Kapoor et al. 2007; Pastorino et al. 2003).

Investigators (Reddy and Marth 1993, 1995a,b) demonstrated that reduced-salt cheddar cheese can be made by replacing sodium chloride (NaCl) with potassium chloride (KCl) or a mixture of the two salts. The use of KCl to replace some of the NaCl for salting cheese had no detectable effect on the kinds of lactic acid bacteria, aerobic microorganisms, aerobic spores, coliforms, and yeasts and molds in cheeses made with KCl or NaCl/KCl compared to control cheeses (Reddy and Marth 1993). Authors concluded that low-sodium cheddar cheese can readily be produced without affecting its composition when 1/3 or more of the NaCl added to the cheese curd is replaced with KCl (Reddy and Marth 1993). However, because consumers like the taste of NaCl, sodium replacement does not always yield acceptable cheese. Cheddar cheeses produced with potassium chloride (KCl), magnesium chloride (MgCl$_2$), or calcium chloride (CaCl$_2$) at equivalent ionic strength to the control NaCl level (1.6% residual NaCl) were extremely bitter, rancid, metallic, crumbly, and unacceptable, likely due to low S/M (Fitzgerald and Buckley 1985). A 1:1 mixture of NaCl and KCl resulted in a cheese comparable to the control. Karagozlu et al. (2008) demonstrated the promise of partial NaCl substitution with KCl in Turkish white pickled cheese; a mixture of 75% NaCl and 25% KCl had acceptable physical, chemical, and sensory properties.

20.3.1.8 Other Optional Ingredients

The Code of Federal Regulations allows the use of antimycotic agents, applied to the surface of slices or cuts in consumer-sized packages (CRF133.113 Title 21.181.23; FDA 2006f). Some of the antimycotic substances allowed by the FDA are calcium propionate (FDA 2006h), methylparaben (methyl *p*-hydroxybenzoate; FDA 2006l), propylparaben (propyl *p*-hydroxybenzoate; FDA 2006i), sodium benzoate (FDA 2006m), calcium propionate (FDA 2006h), and sorbic acid (FDA 2006g).

Hydrogen peroxide is allowed if followed by a sufficient quantity of catalase preparation sufficient to eliminate the hydrogen peroxide. The weight of the hydrogen peroxide shall not exceed 0.05% of the weight of the milk (U.S. FDA 2006k), and the weight of the catalase shall not exceed 20 ppm of the weight of the milk treated (U.S. FDA 2006a).

20.3.2 Preparations for Cheesemaking

20.3.2.1 Culture Selection and/or Propagation

Culture quality is of the utmost importance in the production of high-quality cheese. Culture manufacturers commonly work closely with processing facility operators to effectively meet specific needs. Culture manufacturers continuously develop unique culture combinations for a given product. Several types of culture forms are available, including (1) liquid (for the propagation of the mother culture; rarely used today), (2) deep-frozen concentrated cultures (for the propagation of bulk starter), (3) freeze-dried concentrated cultures in powder form (for the propagation of bulk starter, for the preparation of mother culture), and (4) deep-frozen or freeze-dried, superconcentrated cultures in readily soluble form for direct inoculation of the product (DVS).

The availability of frozen or freeze-dried cultures eliminates the need for small dairy plants to make cultures or operate a culture room (Lewis 1987). The culture room is a separate room in the dairy plant reserved for the preparation and propagation of starters and an important element in the production of quality cheese, because it limits opportunities for contamination by airborne yeast, mold, and bacteriophage (Lewis 1987). Bacteriophages (phages) are viruses that infect specific strains of bacteria, which stress the importance of utilizing mixed-multiple strains and starter culture rotation in cheesemaking. The proliferation of phages will result in a failure of lactic acid production, termed "stuck vat," necessitating strict sanitation and whey handling practices to keep phage numbers to a minimum. Larger plants are typically supplied with frozen or freeze-dried cultures for the manufacture of bulk starters in aseptic bulk culture rooms (Lewis 1987). The cooling and storage conditions and the shelf life of cultures vary. Generally, deep-frozen and freeze-dried cultures can be stored for 9–12 months at −18°C and −45°C, respectively (Lewis 1987). The maintenance of consistency across cheese lots requires the constancy of culture handling.

20.3.2.2 Cheese Milk Pretreatments

Centrifugal clarifier–separators are used to separate the cream and skim fractions and to remove solid impurities from milk prior to standardization. Cheese milkfat and protein content are commonly standardized for the consistency of yield and composition. Fat may be increased with the addition of cream, while protein, particularly casein, may be increased with nonfat dry milk, skim milk, or condensed skim milk (Johnson 1988). A typical casein-to-fat ratio of between 0.67 and 0.72 may be used (Banks and Williams 2004; Lawrence and Gilles 1987b). Although cheese moisture is influenced by numerous factors during cheesemaking, higher fat levels in cheese milk are typically associated with lower moisture cheeses (Lawrence and Gilles 1987b). As a general rule, an increase of 0.05 in the casein-to-fat ratio in milk generally results in a decrease of about 1.4% in the fat on a dry basis and an increase of about 0.8% in moisture in cheddar cheese (Lawrence and Gilles 1980).

Most commonly in the U.S., whole milk is preheated to 55°C–65°C in the regeneration section of the high-temperature, short-time (HTST) pasteurizer prior to separation. Following separation, the cream is standardized to a preset fat level, and the fraction intended for standardization of milk is routed and remixed with the proper amount of skim milk to attain the desired fat and protein content. The surplus cream is directed to a separate cream pasteurizer, while the standardized milk flows through the pasteurizer.

El-Gazzar and Marth (1991) recommended the use of ultrafiltered milk for conversion into cheeses such as cheddar, cottage, havarti, feta, brick, colby, and domiati because of an increase in yield of product. Ultrafiltration results in the concentration of milk proteins, with reduction in lactose and monovalent and divalent cations. Additional benefits claimed for the use of ultrafiltered milk in cheesemaking include reduction in costs of energy, equipment, and labor, improved consistency of cheese flavor, and the potential production of new by-products (El-Gazzar and Marth 1991). A recent advancement in the area of membrane filtration has provided cheesemakers an option of using microfiltration to concentrate the casein fraction and remove whey proteins, lactose, and monovalent and divalent cations. Nelson and Barbano (2005) showed that the cheese made with microfiltered milk had increased yield due to increased fat retention and higher proteolysis.

20.3.2.3 Homogenization

Although the homogenization of cheese milk is typically not employed in the production of cheddar cheese, research has shown that the homogenization of cream may have applications to cheddar cheese. Homogenization is most efficient when fat globules are in the liquid state, so milk is preheated in the plate heat exchanger in the regeneration section of the HTST pasteurizer, where the temperature is raised to at least 60°C prior to homogenization (Morr and Richter 1988). In a two-stage homogenizer, pressures typically range from 10 to 25 MPa in the first stage and 5 MPa in the second stage.

In a study with cheddar cheese standardized to a casein-to-fat ratio of 0.70, Nair et al. (2000) demonstrated that cheese hardness was not influenced by homogenization and cheeses with homogenized cream had improved body, texture, and flavor over control cheeses. Cream homogenized at 6.9 and 3.5 MPa (first and second stages, respectively) was optimal for enhancing cheddar cheese yield and functionality (Nair et al. 2000). Metzger and Mistry (1994) demonstrated that cheese moisture and yield were higher in reduced-fat cheddar cheeses made with homogenized cream than controls. The body and texture of reduced-fat cheddar cheeses made from homogenized cream were improved over those for the control cheeses, which were hard, rubbery, and curdy (Metzger and Mistry 1994).

20.3.2.4 Pasteurization

The only step in the dairy processing system that guarantees the killing of pathogenic microorganisms is pasteurization. Thus, pasteurization may be considered the most critical segment of the cheese processing line. An added side benefit of pasteurization is that it also kills many spoilage microorganisms and inactivates enzymes that may contribute to quality defects in cheese. Pasteurization contributes to consistency in product quality. Of course, strict sanitation is critical up to and beyond pasteurization to assure the safety and quality of dairy products. In HTST pasteurization, milk must be held at a temperature of at least 72°C for a minimum of 15 s to be legally pasteurized (U.S. Department of Health and Human Services et al. 1999). In batch or low-temperature long-time systems (uncommon in large-scale operations), milk is continuously agitated in a single tank at a set temperature (legally at least 62.8°C) for a given time (legally at least 30 min if at 62.8°C) to guarantee the inactivation of pathogens (U.S. Department of Health and Human Services et al. 1999). Any lower temperature or shorter time than legal pasteurization means that the cheese must be treated as if made from raw milk, which means that products must be aged for at least 60 days at 1.66°C or higher. Regardless of the pasteurization method, cheese milk is then cooled, either to incubation temperature for selected starter cultures or to refrigeration temperature for future application.

While all pathogens and most spoilage microorganisms are killed by pasteurization, potentially beneficial or flavor-producing microorganisms are also killed. Thus, although cheeses made from pasteurized milk are safe, they also have more predictable but less overall flavor than raw milk cheeses. Buchin et al. (1998) studied the effects of pasteurization and fat makeup of experimental semihard cheeses with two different fat compositions. Raw milk cheeses had more intense flavor and volatile compounds than the pasteurized milk cheeses. Raw milk cheeses were characterized by higher amounts of numerous alcohols, fatty acids, and sulfur compounds, while pasteurized milk cheeses were characterized by higher amounts of ketones (Buchin et al. 1998). The differences were attributed to the high level of indigenous microflora in raw milk cheeses (Buchin et al. 1998).

In addition to modifying milk microflora, pasteurization acts on milk protein chemistry to influence cheese quality. Specifically, pasteurizing cheese milk influences the extent and characteristics of proteolysis during cheddar cheese aging (Lau et al. 1991). Pasteurization causes heat-induced precipitation of whey proteins on casein micelles, which results in the retention of additional whey protein in the cheese beyond that which is soluble in the aqueous phase of raw milk cheese. The presence of heat-denatured whey protein in cheese may influence the accessibility of caseins to proteases during cheese aging, a consequence of which would be differences in proteolysis during aging. These differences may be another factor that contributes to differences in flavor development in cheddar cheese made from pasteurized and raw milk (Lau et al. 1991). Temperatures higher than legal pasteurization (80°C) may be used to increase yield; however, gelling takes longer, the firming rate of the gel and its maximum firmness are reduced, and gel draining is more difficult and is incomplete (Lenoir et al. 2000b). These factors are a consequence of denatured whey proteins, particularly β-lactoglobulin, which bind with caseins, particularly κ-casein. Indeed, a complex between β-lactoglobulin and κ-casein leads to modification in the conformation of the κ-casein chain at the chymosin cleavage site, detrimentally affecting coagulation properties (Lenoir et al. 2000b). Denaturation of whey proteins is negligible at pasteurization temperatures but reaches 10% after a treatment of 75°C for 15 s and 20% after 85°C for 30 s (Lenoir et al. 2000b). Heat also decreases soluble calcium, ionized calcium, and soluble inorganic phosphorus (Lenoir et al. 2000b). In order to make a good aged cheddar, it is critical to avoid the denaturation of whey proteins by maintaining a consistent pasteurization temperature. Cheese milk pasteurized above 75°C tends to have pasty texture, whey taint defects, bitter taste, and lack of cheddar flavor.

20.3.3 Cheesemaking

A schematic diagram outlines the steps of cheesemaking, from fresh milk through aging cheese (Figure 20.1). In large automated plants, cheesemaking is typically held to a well-timed schedule. Culture, $CaCl_2$,

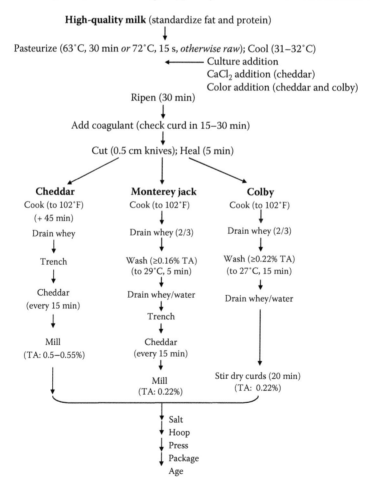

FIGURE 20.1 Production of cheddar and related cheeses.

and color are typically added as cheese milk enters the cheese vat after pasteurization and cooling. Chymosin is commonly added after the vat is completely filled with pasteurized milk. In pilot- or small-scale operations, culture, $CaCl_2$, and color are commonly added when an entire vat of cheese milk reaches target temperature. Chymosin is added after a ripening period of 15–30 min.

20.3.3.1 Calcium Chloride Addition

If $CaCl_2$ is to be added to the cheese milk, typically 0.2% of the cheese milk weight is adequate to improve coagulation properties (Kosikowski and Mistry 1997a; Lenoir et al. 2000a).

20.3.3.2 Color Addition

Annatto may be added to cheese milk at a rate of approximately 66 mL per 1000 kg of milk, adjusted to the desired product color (Kosikowski and Mistry 1997b). Annatto binds with protein to form straw to orange color in the final cheese, on concentration that occurs with whey expulsion (syneresis).

20.3.3.3 Culture Addition

Prior to culture addition, raw or pasteurized milk must be tempered to the appropriate temperature for starter culture multiplication, approximately 26°C–30°C for cheddar and related cheeses (Banks and Williams 2004). The inoculum level is defined by culture manufacturers, based on whether the culture is DVS or bulk culture, typically from 0.5% to 5%. Cheese manufacturers may increase or decrease the amount of culture based on seasonal variation in milk composition (Lawrence et al. 1999). Optional addition of adjunct culture typically varies from 0.1% to 1%.

20.3.3.4 Ripening

The titratable acidity (TA) of fresh milk is approximately 0.14–0.18, depending on the composition, and the pH level is about 6.6–6.8. When starter culture is added, cultures need time to equilibrate to their environment (lag phase), so only a small rise in TA is noted during the 30-min ripening period. Little lactose is converted to lactic acid during the lag phase of the cultures, but TA rises steadily during the log phase of growth, during which time culture numbers increase exponentially. Even after lactic acid formation begins, little change in pH is noted because of the milk's high buffering capacity due to the presence of proteins, citrate, and phosphate.

20.3.3.5 Enzyme Addition

In fresh fluid milk, charges on the κ-casein "hairs" are negative (–), so casein micelles repel each other. With the production of acid, the charges on some κ-casein hairs begin to change to positive (+). When the pH declines to near 5.2, calcium and phosphorus are solubilized, and the micelle structure changes (Brulé et al. 2000). At pH close to 4.6, coagulation occurs, as repulsive charges are neutralized and micelles come into contact with one another and coalesce (Brulé et al. 2000). Some cheeses are made exclusively with acid coagulation (i.e., cottage cheese). Because acid development is slow, cheesemaking procedures that rely entirely on acid coagulation are on the order of 10–18 h in length. The cheesemaking process is accelerated by the use of coagulating enzymes.

Chymosin, originally derived from the abomasum of milk-fed calves but now microbially or fungally derived, is the most common coagulating enzyme used in the manufacture of cheddar and related varieties (Ramet 2000b). Chymosin is an acid protease, which means that it is more active at an acid pH than neutral or basic pH. The highest activity is observed at pH 5.5 and 42°C (Ramet 2000b). Specifically, chymosin cleaves the peptide bond Phe105–Met106, which leads to the formation of κ-*para*-casein (1–105) and glycomacropeptide or caseinomacropeptide (CMP, 106–169; Brulé et al. 2000). CMP is soluble in whey. When chymosin is added to milk, coagulation occurs in three steps: (1) κ-casein hydrolysis, (2) aggregation of destabilized micelles, and (3) reorganization of calcium phosphate, or reticulation

(Brulé et al. 2000). The coagulation process is shortened, because rennet/chymosin cleaves the negatively charged κ-casein hairs off of the micelles, enabling the approach and coagulation of micelles. During the coagulation process, calcium phosphate bridges form between micelles and tighten as whey is expelled, forming a tight network of casein, which entraps some fat, water, and water-soluble components. As fermentation proceeds, Ca^{2+} is replaced by H^+, and the casein network continues to tighten.

Approximately 5–50 mL of single-strength liquid chymosin should be used to coagulate 100 L of milk (Brulé et al. 2000). Chymosin should always be diluted (approximately 1 part to 40 parts of water) prior to addition to the cheese vat to prevent localized coagulation. Dilution should always be done with cool (or room temperature) water immediately before adding to milk. Chymosin begins to lose its strength and activity immediately on dilution, which is why dilution should not be done in advance. In addition, chymosin is degraded by high temperatures and chlorine. Chymosin must not be overmixed into the milk, because the cleavage of κ-casein from casein micelle proteins begins immediately. In small operations, chymosin should only be mixed into milk for about 1 min to maximize the yield. During the incubation period, the cheese vat must not be agitated or disturbed in any way, or a soft or weak curd will result, and the yield will be affected.

Chymosin is allowed to set the cheese for 20–30 min prior to curd testing. In large automated plants, the curd is typically not checked, and cutting begins at a set time. To check the curd, a spatula or knife may be used. A spatula works best for checking curd set because of its rounded shape. The blade is cut through the curd in a 5-cm vertical orientation and is removed. The blade is then inserted at the bottom of the vertical cut, in a horizontal orientation, to form a T. The blade is pushed forward and lifted to encourage the curd to split open. The curd is ready for cutting when the curd is firm, breaks cleanly, and fills with clear yellow (not cloudy) whey.

20.3.3.6 Cutting

Cutting of the curd is an extremely important step in the cheesemaking process, because it influences whey drainage and cheese yield (Ramet 2000a). Cutting to a consistent size, with sharp knives, is critical to minimize small curd particles (fines), which may be lost during whey drainage (Ramet 2000a). Manually, the coagulated mass of cheese is cut with harps—knives constructed of stainless steel hardware and wire spaced at regular intervals. The wires on one harp are horizontally oriented, and the other harp wires are vertically oriented. The cutting progresses in such a way that the first horizontal and vertical sheets of curd are cut with the harp knives. The sheets are then cut into cubes by perpendicular cuts with the vertical harp knives (Figure 20.2).

Large dairy plants have automated cheese vats that vary in size from 2000- to 25,000-L capacity. These cheese vats are equipped with a shaft to which agitators are attached. These agitators are designed in such a way that they cut the cheese when the shaft rotates in one direction and agitate the cubes gently when the shaft rotates in the other direction. Cheese vats are automated and allow the cheese curds to heal and cook before pumping the curds and whey onto a perforated conveyor belt, where the cheese curd is separated from the whey. A schematic diagram of cheesemaking from cutting through cooking steps is included in Figure 20.3. Cutting the coagulum increases the surface area of the curd and enhances syneresis. On cutting, curd particles immediately begin to expel whey and shrink, and the TA that had been rising in the cheese milk immediately drops in the whey, because the whey has lower apparent acidity due to lower protein, citrate, and phosphate. The TA of whey will gradually increase as lactic acid is formed in curds and released with whey during syneresis.

20.3.3.7 Healing

Freshly cut curd is fragile and shatters easily, so curds are allowed to "heal" for 5–10 min prior to agitating and cooking. A healing period (10–15 min) is particularly important when goat cheddar cheeses are made, because the curds are naturally more fragile than the curd obtained from cow milk. During healing, a tender skin is formed around each freshly cut curd. As the skin firms, the curd becomes more resistant to shattering and yield losses.

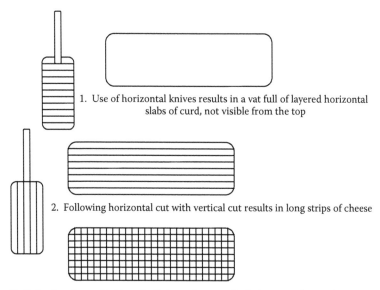

1. Use of horizontal knives results in a vat full of layered horizontal slabs of curd, not visible from the top

2. Following horizontal cut with vertical cut results in long strips of cheese

3. Only after the vertical harp knife is used perpendicularly across the previous cuts will cubes be made out of the curds, enabling optimal syneresis

FIGURE 20.2 Schematic diagram of steps involved in cheese coagulum cutting with harp knives.

20.3.3.8 Cooking

The cooking process is essentially a controlled increase in curd–whey temperature. Heating allows individual curd cubes to shrink, release whey, and firm. Cooking also increases reaction rates, specifically bacteria growth and metabolism, and enzyme activity. Temperature-sensitive bacteria strains are slowed down as temperature is raised.

Prior to raising the temperature of the curd–whey mixture, curds should be gently eased from the edges of the cheese vat, where they have matted. Curd cooking should begin slowly, with continual stirring of the curds. Hot water or steam may be used to increase the jacket temperature. The curd–whey

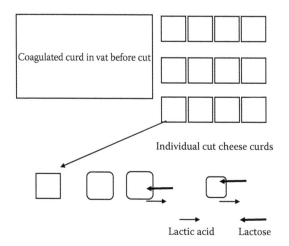

FIGURE 20.3 Equilibrium in curd–whey mixtures.

mixture temperature should be raised slowly, about 2°C every 5 min until 38°C is reached (about 35–45 min). The stirring speed may be increased as the curds firm, but stirring too fast will shatter curds and reduce yield. For a drier cheese, the temperature should be held at 38°C for an additional 45 min, with stirring. In small or start-up facilities, whey TA should be recorded every 15 min. Regardless of the plant size, good records of the entire cheesemaking procedure and final cheese quality should be kept. Failure to keep such records will reduce consistency.

For a short period after the curd is cut, lactose and lactic acid concentrations are at equilibrium in curd and whey. With time, the concentration of lactose drops faster in the curd than in the whey, because starter bacteria concentrated in the curd deplete lactose in the curd (Figure 20.3). As the lactose is fermented within the curd, replacement lactose diffuses into the curd from the whey (Figure 20.3, broad arrows; Lawrence and Gilles 1987a). As a neutral molecule, lactose diffuses easily through the matrix. Positively charged hydrogen ions exit the negatively charged curd much more slowly than lactose (Figure 20.3, narrow arrows). As fermentation progresses, hydrogen ions are neutralized by the negatively charged proteins and phosphates. As the buffering capacity of the caseins and CCP becomes saturated, the pH of the curd steadily drops. As the pH drops, CCP is solubilized and lost into the whey.

Once cheesemaking starts, managing acid development during the cooking stage of cheese manufacture is the most important factor in the control of cheese quality (Lawrence et al. 1984; Lawrence and Gilles 1987a). Acid development determines the basic structure, moisture, final pH, and flavor of cheese (Lawrence et al. 1984; Lawrence and Gilles 1987a). As the pH drops, the body of the cheese changes from rubbery (pH 5.4) to plastic (pH 5.3–5.2) to cheddar (pH 5.1–5.0) body and texture. When the vat TA rises too quickly, the curd will suffer an excessive loss of calcium but will not retain phosphate. The result is an increase in curd buffering capacity (Lawrence and Gilles 1987a). However, should a high TA result from an extended time between cutting and cheddaring, both calcium and phosphorus are lost in the whey (Lawrence and Gilles 1987a). The resulting cheese will have a low pH, an acid flavor, and a weak, pasty texture (Lawrence and Gilles 1987a). An objective for cheesemakers is to develop acid slowly during ripening and cooking and more quickly during cheddaring so that calcium phosphate is retained in the curd, as the loss of CCP alters the body and texture of the cheese (Lawrence and Gilles 1987a). Curd shrinks and tightens as syneresis proceeds during the cooking process, prior to whey drainage. Approximately 75% of the whey in the curd is expelled in the time from cutting to the end of stir-out. Longer stir-out times can significantly reduce moisture in cheese.

20.3.3.9 Draining

Whey may be drained entirely, as in the case of cheddar, or partially, with washing, as in the case of colby or monterey. In small plants, a cleaned and sanitized finishing table/vat should be aligned with the exit port of the cheese vat. The drain of the vat is opened, and curds and whey are allowed to flow onto the finishing table. Alternatively, in small plants, the cheese vat may double as a finishing table. A screen is installed ahead of the finishing table drain port to prevent curd loss as the whey is drained. Whey is commonly collected in a separate reservoir. Cheese curds should be allowed to settle into the vat or finishing table, at an even depth throughout the length, and permitted to mat for 15 min. Whey TA will rise more quickly during this interval and should be recorded every 15 min from this point forward, throughout the cheddaring process. In the largest plants, curds are delivered to a perforated conveyor belt for drainage and cheddaring, which allows the formation of a sheet of curd and continuous whey drainage. The conveyor is enclosed in a tunnel. On drainage, in the absence of the whey bath, the curd pH will drop at a faster rate, and the curds will continue to shrink and tighten. Much of the calcium is lost at drainage, particularly at low pH.

20.3.3.10 Washing

Washing is essential in colby and monterey cheese production. Washing or curd rinsing removes lactic acid and residual lactose and lactic acid from the curd, and the result is a higher pH in the final cheese.

Washing is rarely included during cheddar cheesemaking, but when it is, the duration of such rinsing is so limited that only the whey on the surface of the curds is removed (FDA 2006a).

In colby and monterey cheese production, whey is drained off until the curd on the bottom of the vat is visible, and then, sufficient cold water is introduced to reduce the temperature of the curd–whey mixture to 27°C (Lawrence and Gilles 1987a). The rate of syneresis is slowed if cool water is used, resulting in higher cheese moisture content. Long wash time removes more lactose, resulting in higher final pH of cheese. Temperature-sensitive strains may be revived if cool water is used.

20.3.3.11 Cheddaring

The step known as "cheddaring" was standardized into commercial practice by Joseph Harding in 1857 (Kosikowski and Mistry 1997a). During manual cheddaring, curds are flipped and stacked at regular intervals, naturally pressed under their own weight, which enhances syneresis yet still maintains a controllable level of moisture retention. The main purpose of cheddaring is to allow time for the acidity to increase and whey to be released (Lawrence and Gilles 1987a). Curd particles fuse into a solid mass, syneresis continues as acid builds, rennet/chymosin continues to act, and these forces cooperate to tighten the casein network. As lactic acid continues to build, curds begin to flow or stretch under the weight of piled slabs. Cheddar gains its characteristic body through the process of knitting and stretching and the orientation of the casein network during cheddaring, which requires pH below 5.8 (Lawrence and Gilles 1987a).

In manual operations, after the cooking step is completed and curds are allowed to settle for 15 min, matted curd should be trenched and then cut into equal-sized slabs. A wide (20–30 cm) trench is made at the center of the vat to facilitate syneresis and curd stretching. Slabs should be separated as they are cut to enable syneresis and stretching. Extra or broken curd should be placed on top of each slab to minimize the loss of fines. Slabs should be allowed to settle 15 min before the next step.

Cheddaring begins with the flipping of slabs, one by one. The bottom shall become the top, and the end toward the trench shall become the end toward the vat wall. Slabs should be allowed to settle 15 min between each subsequent step. The next step of cheddaring involves the flipping of one slab, followed by the placing of an adjacent slab on top of the flipped slab (without flipping). This step is called "flip–stay." The process continues for every pair of slabs. After 15 min, the top slab is placed (not flipped) into an empty spot in the vat. The previous bottom slab is then flipped and placed atop the new bottom slab. This step is called "stay–flip." Cheddaring continues with the flipping and stacking of slabs, alternating between "flip–stay" and "stay–flip" steps, until a whey TA of 0.35% as lactic acid is measured in a fresh sample of whey. In large plants, the process of cheddaring is automated. As the perforated conveyor mentioned previously transects a number of parallel planes during the approximately 90-min cheddaring process, the matted curds are flipped and stretched continuously in a tumbling motion.

20.3.3.12 Milling

Milling is the process of cutting the slabs into cubes about 5 cm in size, which enables more uniform salt distribution, encourages syneresis, and makes hooping more convenient (Lawrence and Gilles 1987a). When curds are milled, more whey is expelled, because milling greatly increases curd surface area and opens pores for syneresis. Salt distribution will be most uniform in cheese if curds are milled to a uniform size (Lawrence and Gilles 1987a). In large plants, as the mat of curd arrives at the discharge point of the perforated conveyor, it is cut to the desired size in a reciprocal dice-type mill or rotary curd mill.

An alternative to the milled-curd method is the stirred-curd method of manufacturing cheddar cheese, which is used by many large-scale commercial manufacturers (Shakeel Ur et al. 2007). The stirred-curd method differs in that curds are continuously stirred after whey drainage, hence eliminating the cheddaring and milling stages. Shakeel Ur et al. (2007) evaluated the quality of cheddar cheeses made by the milled- and stirred-curd methods of processing. Cheese diacetyl flavor and salty taste were higher in stirred-curd cheeses, while sour flavor was higher in milled-curd cheeses. Differences in proteolysis in cheese (percentage of water-soluble nitrogen) due to the manufacturing method were apparent by 270 days of aging. The investigators demonstrated that cultures and temperature had more impact on cheese flavor than the manufacturing method.

20.3.3.13 Salting

Milled curds of cheddar and related hard cheeses are dry salted rather than brine salted. Cheese is salted, because it (1) encourages further syneresis, (2) inhibits further growth and metabolism of most microorganisms (thus arresting lactic acid production), and (3) provides flavor. Approximately 2.5 kg of salt for every 100 kg of cheese curd is used. The salt is added in three equal applications and mixed for 5 min between applications. Adding salt too quickly will cause a "skin" to form on the curds, inhibiting salt absorption and syneresis. In large plants, the milled cheese quantity is determined continuously by weight prior to entering the salting machine. The salter automatically calculates the salt and sifts it over the milled cheese. The pH and TA will only change slightly beyond the point of salt addition.

Salt, more specifically salt in moisture (S/M), directly influences the final pH of cheese, the growth of microorganisms, and the overall flavor, body, and texture of cheese (Lawrence and Gilles 1987a). At S/M levels greater than 5.0, bitter flavors rarely occur (Lawrence and Gilles 1987a). Curd salted at low TA retains more salt (higher S/M) and is more plastic than curd salted at high TA (Lawrence and Gilles 1987a). Mistry and Kasperson (1998) reported that optimum S/M is approximately 4.5% and may be as low as 3.5%, depending on the fat reduction and moisture content of cheese. Lawrence et al. (1984) suggested a S/M of 4%–6% for premium-grade New Zealand cheddar cheese. Guinee and O'Kennedy (2007) stated that good-quality Irish cheddar cheese contains 4.5%–5.6% S/M (674–832 mg of sodium per 100 g).

20.3.3.14 Pressing and Packaging

Pressing gives cheese its final shape, reduces openings between curd particles, promotes fusion, and releases more free whey. In small plants, cheddar curds are pressed overnight using a batch method (Lawrence and Gilles 1987a). Pressure, approximately 1.4 atm, is applied to molds for 8–12 h at room temperature. After 1 or 2 h of pressing, cheeses may be flipped in the molds and lined with cheesecloth, which provides an attractive surface pattern. Large plants have a continuous "block-former" system (Lawrence and Gilles 1987a). Curds are fed into a tower under a partial vacuum, whey is siphoned off, and for a short period, mechanical pressure is applied at the base of the tower prior to packaging (Banks and Williams 2004; Lawrence and Gilles 1987a). A block former cuts 20-kg blocks from the stack at regular intervals, and the blocks are transferred to a vacuum packaging system prior to aging (Banks and Williams 2004).

20.3.3.15 Aging

Aging enables flavor and texture development of hard cheeses. Nearly all residual lactose should be fermented within about 48 h. The aging process begins with a cooling process. Differences in the cooling rate of cheddar cheeses from pressing (35°C) to aging temperature (3.5°C–12°C) influence flavor variation within and between production lots due to NSLAB activity (Grazer et al. 2007). With cold storage, between 5°C and 12°C, acid production slows down but continues until limiting conditions occur (Banks and Williams 2004). Starter bacteria lyse (burst) and release proteolytic enzymes into the matrix. Residual plasmin and coagulant also contribute to proteolysis during aging. Caseins are broken down into peptides and amino acids, which yield flavor and modify cheese body/texture. Secondary fermentations can occur if NSLAB are still active, which results in further changes in flavor and body/texture. Cheeses with low S/M have a higher rate of proteolysis, resulting in a softer texture, than cheeses with high S/M (Lawrence and Gilles 1987a).

Cheddar cheese is typically aged for 3–18 months at 7°C–13°C, but it is not unheard of to age cheddar for years in the case of specialty varieties (Banks and Williams 2004). Colby and monterey are aged for shorter periods of time due to their higher moisture content and milder expected flavor. As a result of the curd washing step and shorter aging periods, colby (550 mg/kg) and monterey jack (736 mg/kg) have lower typical concentrations of free fatty acids than cheddar (1028 mg/kg; Fox et al. 2000b).

20.4 Quality Control

20.4.1 Shelf Life

The shelf life of cheddar and related cheeses is limited by quality, not safety. The quality of good cheddar cheese improves with storage. Cheddar cheese may be removed from shelves due to flavor, body, or appearance defects. The most common flavor defects are high acid, bitter, unclean, and fermented/fruity. Common body defects are weak or crumbly body, gas holes, slits, surface discoloration, and the appearance of crystals on surfaces.

20.4.2 Evaluation

High-quality cheddar cheese has a full, balanced nutty, and sharp but not bitter flavor. The ideal texture should be closed (no gas holes or mechanical openings), and the body should be firm, smooth, and waxy (responds to moderate pressure). Colby and monterey jack cheeses are similar to cheddar but are milder in flavor and possess a softer body. Colby and monterey jack are prone to the same defects as cheddar. However, due to higher moisture content, lower acid and salt, and higher microbial and enzymatic activity, some sensory defects may reach greater intensity and frequency in colby and monterey jack cheeses than cheddar, particularly with extended aging.

Gas liquid chromatography analysis of cheddar cheese has shown that there are as many as 200 different compounds that may contribute to cheese flavor. However, flavor chemists believe that as few as 20 volatile compounds are pertinent to the determination of the eventual flavor of cheddar cheese. Cow diet, milk handling and sanitation practices, milk composition, and cheese manufacturing conditions all affect cheese chemistry (Buchin et al. 1998). What appears to be critical is the relative proportions of the key chemical flavor compounds in providing a "balanced cheddar flavor." Ammonia- and sulfur-like odors and bitter taste typically occur in aged cheeses, a consequence of amino acid breakdown. Common flavor attributes encountered in cheddar and related cheeses are included in Table 20.2. While some attributes, such as sulfide, may be considered desirable in an aged cheese, mild cheeses are discredited for pronounced attributes.

In addition to flavor attributes, consumers look for certain functional properties in cheese (melting, grinding, and slicing). Cheeses continually change during ripening, not only in flavor but also in body/texture. Common body/texture attributes encountered in cheddar and related cheeses are included in Table 20.3.

For a more extensive discussion of cheddar cheese evaluation, the reader is encouraged to review the chapter Cheddar Cheese in *The Sensory Evaluation of Dairy Products* (Partridge 2008).

20.4.3 Safety

Cheddar cheese and other semihard cheeses are generally considered safe and have rarely been associated with foodborne illness outbreaks (El-Gazzar and Marth 1992; Johnson et al. 1990; Leyer and Johnson 1992; Wood et al. 1984). However, some pathogens can survive the cheesemaking process and during ripening. Hargrove et al. (1969) demonstrated that the cheese pH, the rate and amount of acid produced during cheesemaking, and the type and amount of starter inoculum all influence the growth and survival of *Salmonella* in colby and cheddar cheeses. Raw milk cheddar cheese was linked to major

TABLE 20.2

Common Flavor Attributes in Hard Cheeses, Identification, and Their Probable Causes

Flavor	Identification	Probable Cause
Bitter	Sometimes perceived as throbbing/piercing; sensation perceived at the back of tongue; very common defect in aged cheddar	Excessive moisture; low salt; excessive acidity; proteolytic starter culture strains; microbial contaminants; poor-quality milk; plant sanitation issues
Feed	Grassy or green flavor	Feeding of strong-flavored feeds; feeding of cattle too close to milking
Fruity/Fermented	Pineapple-like	Psychrotrophic *Pseudomonas fragi*, which may produce ethylbutyrate and ethylhexanoate (esters); low acidity; excessive moisture; low salt level; poor milk quality
Flat/Lacks flavor	Lacks nutty flavor components; lacks typical cheddar flavor	Lack of acid production; use of milk low in fat; excessively high cooking temperature; use of low curing temperature; too short curing period
Heated	Sweet-like cooked flavor; reminiscent of Velveeta	High pasteurization or cooking temperature
High acid	Excessive acid taste; unbalanced acid taste	Development of excessive lactic acid; excessive moisture; use of too much starter culture; use of high acid milk; improper whey expulsion from curd; low salt level
Oxidized	Paperboard/cardboard; sometimes discoloration also; burnt hair aroma/flavor	Use of oxidized milk in cheesemaking; excessive exposure to ultraviolet light during aging
Rancid	Butyric, caproic, caprylic, capric acids; soapy; Romano cheese aroma/flavor; baby vomit aroma	Milk lipase activity; microbial lipase activity (from contaminants); accidental homogenization of raw milk; late lactation or mastitic milk
Sulfide	Eggy (cooked egg yolk) aroma, sharp flavor	Excessive breakdown of amino acids; only a defect when excessive for age of cheese
Unclean	Unpleasant off-flavor lingers	Microbial contamination; poor-quality off-flavored or old milk; allowing off-flavored cheese to be aged; improper techniques of cheddaring
Whey taint	Combination of acid, bitter, fermented; aftertaste does not linger like unclean	Poor whey expulsion from curd; improper cheddaring techniques; failure to drain whey from piles of curd slabs
Yeasty	Ethanol aroma; yeast (bread, beer) aroma	Development of ethanol flavors by contaminants; poor packaging procedures

salmonellosis outbreaks in Canada in 1982, 1984, and 1998 (D'Aoust et al. 1985; El-Gazzar and Marth 1992; Ratnam et al. 1999; Wood et al. 1984). *Salmonella* can survive in ripening cheddar cheese for 7–10 months (El-Gazzar and Marth 1992; Wood et al. 1984). Additionally, Ryser and Marth (1987) showed that *Listeria monocytogenes* can survive for as long as 434 days in cheddar cheese ripened at temperatures above 1.66°C. These facts highlight the importance of using the highest quality milk and good manufacturing practices, including strict sanitation and handling procedures, to prevent contamination and ensure cheese quality and safety.

With the focus on sodium in the diet by public health groups, manufacturers and brands are pursuing various reduced- or low-sodium cheddar cheese options. Until the publication of this book, little research has been done to look at the impact of lowering salt on the quality and safety of cheddar cheese.

TABLE 20.3

Common Body and Texture Attributes in Hard Cheeses, Identification, and Their Probable Causes

Body/Texture	Identification	Probable Cause
Corky	Dry, noncompressible; often crumbly as well	Lack of acid development; low fat
Crumbly	Falls apart while working	Excessive acid production; low moisture retention in cheese
Crystals	White crystals observed by visual examination	Tyrosine (only in aged cheese), calcium lactate, calcium citrate, calcium phosphate
Curdy	Resistant to compression	Inadequate aging conditions
Gassy	Smooth round gas holes	Contamination of cheese with CO_2-forming microorganisms
Mealy	Grainy (like corn meal)	Excessive acid production; formation of salt complexes
Open	Openings along curd lines	Improper mechanical pressing, lack of fusion between curds
Pasty	Sticky when working between fingers	High moisture retained by curd; excessive acid production
Short	Plug breaks quickly (snaps)	Excessive acid production
Weak	Plug is resistant to breaking (bends)	High moisture in cheese; excessive proteolysis

20.5 Troubleshooting

Although poor-quality milk will always result in poor-quality cheese, even the use of high-quality milk does not guarantee high-quality cheese. This section will summarize factors that influence cheese curd and crystal formation in cheddar cheese.

20.5.1 No Curd or Weak Curd Formation

A weak curd can result from at least one of the two main factors, namely, low starter numbers or poor chymosin activity. The following factors influence the starter and chymosin activity: the presence of natural inhibitors or antibiotics in the milk, residual cleaners/sanitizers on equipment, or the presence of bacteriophage in cultures or in the environment. Each factor will be discussed separately. Operator error is another reason for no curd formation. The bottom line is that personnel must be adequately trained to (1) measure appropriate levels of culture and coagulating enzyme to be added to cheese milk and (2) actually add the ingredients to the cheese milk.

20.5.1.1 Natural Inhibitors

Natural inhibitors in milk include lactenin L_1, L_2, and L_3 and the enzymes lysozyme and lactoperoxidase. Lactenin varies with individual animals and is inactivated by heat treatment (Desmazeaud 2000). Lysozyme attacks the glycosidic bonds found in Gram-positive bacterial cell walls, but it is unlikely to cause the inhibition of lactic acid starter bacteria (Desmazeaud 2000; Jensen 1995). When supplied with hydrogen peroxide (from lactic acid bacteria in the presence of oxygen) and thiocyanate (arises from the catalysis of thiosulfate or glucosides in liver), lactoperoxidase will catalyze the formation of lactococci bacteriocides (Ruden 1997), including bacteriocides that attack lactococci. Therefore, this method of preserving milk in areas with limited access to refrigeration is inappropriate for handling cheese milk. Mastitic milk naturally contains higher levels of leukocytes, which will also engulf and destroy lactic cultures (U.S. Department of Health and Human Services et al. 1999).

20.5.1.2 Antibiotics

Every tanker load of milk must test negative for the presence of beta-lactam antibiotics (Desmazeaud 2000); however, cultures can be inhibited by the presence of antibiotics at levels even lower than detectable by standard dairy lab testing methods. Lactic cultures can be inhibited by as much as 50% by 1.91 µg of cloxacillin, 0.13 µg of tetracycline, and 0.59 µg of streptomycin and as little as 0.12 µg of

penicillin per milliliter of milk (Desmazeaud 2000). Thermophilic bacteria are more resistant to streptomycin and more sensitive to penicillin than mesophilic starters (Ruden 1997).

20.5.1.3 Residual Cleansers/Sanitizers

Residual cleansers/sanitizers can slow a cheese vat precisely, because they are intended to kill microorganisms. Quaternary ammonium compounds or "quats" are not appropriate for use in a cheese plant
because they leave a residue on equipment that is effective against lactic acid bacteria. Quats will inhibit
many starter culture strains at concentrations as low as 10–20 μg/mL (Ruden 1997). Other effective bacteriocides include organic and inorganic chlorine compounds, chlorine dioxide, iodine compounds, acid
anionic sanitizers, and peroxyacetic acid, but they are unstable in the presence of organic matter such as
milk (Leach 1997). Regardless of the sanitizer, pipelines, vats, and other equipment must be allowed to
drain after bactericidal treatment to prevent contamination of the cheese milk supply.

20.5.1.4 Bacteriophage

Bacteriophage literally means "eaters of bacteria." Bacteriophages/phages are obligate intracellular parasites that attack and replicate within specific strains of bacterial cells (Leach 1997). Bacteriophages are
naturally present in the cheesemaking environment and can spread throughout a plant with poor sanitation practices. Bacteriophages are the largest single cause of slowed or failed vats of cheese (Leach 1997;
Tamime and Deeth 1980). Each strain of culture has a different level of sensitivity to bacteriophage.
Because it is impossible to entirely eliminate bacteriophages from a dairy plant operation, control measures must be employed. Careful selection of starter cultures, aseptic techniques for starter culture propagation, air filtration, equipment sterilization, plant sanitation, culture rotation of phage-unrelated strains,
or the use of phage-resistant strains are necessary techniques to control phages (Leach 1997). Starter
culture rotation essentially ensures that the bacteriophage population is diluted (through the process of
repeated sanitation efforts) to the point that their numbers are low enough to allow the starter culture to
function normally (Desmazeaud 2000).

20.5.1.5 Milk Pretreatment

Excessive heating, agitation, or aeration can reduce growth and acid development by lactic acid bacteria.
Heating of milk to pasteurization only slightly modifies the characteristics of milk for cultures. However,
temperatures above 80°C for 20 s induce chemical reactions that can either inhibit (by the destruction of
certain vitamins) or stimulate (by the destruction of lactoperoxidase, the production of formic acid from
lactose, and the release of nonprotein nitrogen) bacterial growth (Ruden 1997). Excessive heating of
milk should be avoided, not only because of subsequent effects on culture but also due to the detrimental
effects on curd formation, curd moisture retention, and cheese quality. Aeration or excessive agitation
will slow the vat by introducing dissolved oxygen. The presence of oxygen inhibits starters, because they
prefer a microaerophilic environment (McDowall and McDowell 1939).

20.5.2 Crystal Formation

The occurrence of undesirable crystals in cheddar cheese has been documented since the 1930s
(Conochie et al. 1960; Harper et al. 1953; McDowall and McDowell 1939; Pearce et al. 1973; Severn et
al. 1986; Tuckey et al. 1938; Washam et al. 1985), yet the problem still represents a challenge and expense
to cheese manufacturers (Chou et al. 2003; Agarwal et al. 2006a). Cheese crystals have been identified
as calcium lactate (Severn et al. 1986), a racemic mixture of L(+)- and D(–)-calcium lactate (Conochie
and Sutherland 1965), calcium phosphate (Conochie et al. 1960; Dorn and Dahlberg 1942; Harper et al.
1953), tyrosine (Bianchi et al. 1974; Harper et al. 1953), or mixtures of amino acids (Dybing et al. 1988;
Johnson et al. 1990a,b; Severn et al. 1986). However, most frequently, crystals in young cheese have been
identified as calcium lactate (Pearce et al. 1973).

The development of calcium lactate crystals (CLCs) may result from a number of causes, including milk composition (Dybing et al. 1988; Agarwal et al. 2008), cheesemaking procedure (Chou et al. 2003; Dybing et al. 1988; Johnson et al. 1990b; Pearce et al. 1973; Agarwal et al. 2006a), aging temperature (Chou et al. 2003; Johnson et al. 1990b; Somers et al. 2001), and the growth of NSLAB in cheese during aging (Dybing et al. 1988).

Dybing et al. (1988) concluded that casein-bound calcium is the major source of calcium in CLCs. The investigators showed that fast acid production and high milling acidities are associated with reduced CLC formation due to reduced concentrations of casein-bound calcium. Seasonal changes affecting milk casein and calcium affect CLC formation. A low casein to calcium ratio leads to increased amount of bound calcium in milk, contributing to increased calcium in cheese and greater predisposition to CLC formation (Agarwal et al. 2006a,b; Khalid and Marth 1990; Williams et al. 2000). Although starter bacteria make up the majority of cheese microflora initially, NSLAB dominate the viable population in cheese for much of the ripening period (Somers et al. 2001). Secondary NSLAB introduced at the cheese plant (Blake et al. 2005) or during cut and wrap may proliferate on stimulation by warm tempering temperatures (Dybing et al. 1988; Williams et al. 2000). Heterofermentative NSLAB utilize a variety of substrates for growth and produce an assortment of metabolites, including both L(+)- and D(–)-lactate (Johnson et al. 1990b). NSLAB that are capable of racemizing L(+)- to D(–)-lactate can contribute to CLCs, because D(–)-lactate is less soluble than L(+)-lactate, particularly at aging temperatures, in maturing cheese (1990b). In the early 1990s, Johnson et al. (Shakeel Ur et al. 2000; Turner and Thomas 1980) established correlations among CLCs, D(–)-lactic acid enantiomer, and numbers of NSLAB. Cheeses with racemase-positive *Lactobacillus* developed crystals compared to cheeses without *Lactobacillus*, which did not. Cheeses aged at lower temperatures developed crystals faster compared to cheeses aged at higher temperature. CLCs were never observed on cheeses with less than 20% of the lactic acid in the D(–) form. It follows, then, that because high temperatures favor the growth of NSLAB (Johnson et al. 1990b), aging cheese at high temperatures, as may be used to accelerate ripening, can result in elevated D(–)-lactate by NSLAB and induction of CLCs. Additionally, the aging of cheese at low temperatures may also increase CLCs due to the decreased solubility of calcium lactate at a low temperature (Chou et al. 2003). Chou et al. (2003) demonstrated the earliest and most extensive CLCs that occurred on cheeses aged at higher temperatures and then stored at lower temperatures. They also showed that specific NSLAB, the production of D(–)-lactate, and aging temperature affect CLCs in maturing cheddar cheese (Chou et al. 2003).

The elimination of racemizing NSLAB and control of storage temperature is critical to the prevention of CLCs. Agarwal (2006a,b) showed that, irrespective of the lactose to protein ratio, the contamination of cheese milk with racemizing NSLAB *Lb. curvatus* can lead to CLCs, particularly in combination with elevated storage temperatures. Additionally, Agarwal et al. (2005) demonstrated that, regardless of the presence of racemizing NSLAB, CLCs are more likely to form in cheeses flushed with gas than cheeses that are vacuum packaged.

Cheese milk composition and cheesemaking techniques have great impact on the occurrence of CLCs, particularly calcium L(+)-lactate crystals. Increased casein concentration in cheese milk is linked to increased colloidal calcium, which solubilizes as the pH of the cheese decreases (Upreti et al. 2006). In the work conducted by Agarwal et al. (2006a), CLCs were observed in cheeses made from skim milk with 3.14% protein at and below pH 5.1, whereas CLCs were observed in cheeses from skim milk supplemented with ultrafiltered milk to 6.80% protein and skim milk–supplemented nonfat dry milk to 6.80% protein at and below pH 5.3. The increased presence of soluble calcium enables CLCs to occur in cheese manufactured with increased concentrations of milk solids, particularly at and below pH 5.1 (Agarwal et al. 2006a). The control of pH and whey removal prior to packaging influence the occurrence of CLCs.

In a follow-up work by Agarwal et al. (2008), a commercial starter was selected based on its sensitivity to salt. Cheddar cheese was made by using either whole milk (3.25% protein and 3.85% fat) or whole milk supplemented with cream and ultrafiltered milk (4.50% protein and 5.30% fat). Calculated amounts of salt were added at milling (pH 5.40) to obtain cheeses with less than 3.6% and greater than 4.5% salt in moisture (S/M). All cheeses were vacuum packaged and gas flushed with nitrogen gas and aged at 7.2°C for 15 weeks. Concentration of total lactic acid in high S/M cheeses ranged from 0.73 to 0.80 g/100 g of cheese, whereas that in low S/M cheeses ranged from 1.86 to 1.97 g/100 g of cheese at the end of 15 weeks of aging because of the salt sensitivity of the starter culture. Concentrated milk cheeses with low and high S/M exhibited approximately 30% increases in total calcium compared with whole milk cheeses with low and high S/M throughout aging. Soluble calcium was approximately 40% greater in low S/M cheeses compared with high S/M cheeses. Because of the lower pH of the low S/M cheeses, CLCs were observed in low S/M cheeses. However, the greatest intensity of CLC was observed in gas-flushed cheeses made with milk containing increased protein concentration because of the increased content of calcium available for CLC formation. These results show that the occurrence of CLCs is dependent on the cheese milk concentration and pH of the cheese, which can be influenced by S/M and cheese microflora.

In summary, cleaning, sanitizing, prevention of contamination of cheese milk with lactate-racemizing NSLAB, controlled acidification and whey removal, consistent storage temperature, and vacuum packaging are encouraged to minimize CLCs.

20.6 Summary/Conclusion

Production of good-quality cheddar cheese is an art and poses unique challenges to novice cheesemakers. Understanding the science behind each cheesemaking step will help cheesemakers consistently produce high-quality cheddar cheese. Ongoing research in cheesemaking, cheese microflora, and sensory science will continue to help produce safe, high-quality cheddar cheese consistently. Although significant progress has been made in low-fat cheddar cheese that has similar taste and functionality as its medium-aged full-fat counterpart, much more research needs to be done to make high-quality, low-fat, and reduced-sodium aged cheddar cheeses.

REFERENCES

Agarwal, S., Costello, M., and Clark, S. 2005. Gas-flushed packaging contributes to calcium lactate crystals in cheddar cheese. *J. Dairy Sci.* 88(11):3773–3783.

Agarwal, S., Powers, J. R., Swanson, B. G., Chen, S., and Clark, S. 2006a. Cheese pH, protein concentration and formation of calcium lactate crystals. *J. Dairy Sci.* 89:4144–4155.

Agarwal, S., Powers, J. R., Swanson, B. G., Chen, S., and Clark, S. 2008. Influence of salt to moisture ratio on starter culture and calcium lactate crystal formation. *J. Dairy Sci.* 91:1–14.

Agarwal, S., Sharma, K., Swanson, B. G., Yüksel, G. Ü., and Clark, S. 2006b. Nonstarter lactic acid bacteria biofilms and calcium lactate crystals in cheddar cheese. *J. Dairy Sci.* 89:1452–1466.

Banks, J. M. and Williams, A. G. 2004. *Handbook of Food and Beverage Fermentation Technology.* New York: Marcel Dekker.

Barbano, D. M., Rasmussen, R. R., and Lynch, J. M. 1991. Influence of milk somatic cell count and milk age on cheese yield. *J. Dairy Sci.* 74(2):369–388.

Bianchi, A., Beretta, G., Caserio, G., and Giolitti, G. 1974. Amino acid composition of granules and spots in Grana Padano cheeses. *J. Dairy Sci.* 57:1504.

Blake, A. J., Powers, J. R., Luedecke, L. O., and Clark, S. 2005. Enhanced lactose cheese milk does not guarantee calcium lactate crystals in finished cheddar cheese. *J. Dairy Sci.* 88(7):2302–2311.

Bodyfelt, F. W., Tobias, J., and Trout, G. M. 1988. *The Sensory Evaluation of Dairy Products*, pp. 277–299. New York: Van Nostrand Reinhold Co.

Brulé, G., Lenour, J., and Remeuf, F. 2000. The casein micelle and milk coagulation. In: *Cheesemaking—From Science to Quality Assurance*, 2nd edition, edited by A. Eck and J. C. Gillis, pp. 6-40. Paris: Intercept/Lavoisier Publishing.

Buchin, S., Delague, V., Duboz, G., Berdague, J. L., Beuvier, E., Pochet, S., and Grappin, R. 1998. Influence of pasteurization and fat composition of milk on the volatile compounds and flavor characteristics of a semihard cheese. *J. Dairy Sci.* 81(12):3097–3108.

Campbell, R. E., Miracle, R. E., and Drake, M. A. 2010. The impact of starter culture and annatto on the flavor and functionality of whey protein concentrate. Abstract No. 128. National ADSA Dairy Foods Oral: Dairy Foods Oral Student Competition.

Chou, Y. E., Edwards, C. G., Luedecke, L. O., Bates, M. P., and Clark, S. 2003. Nonstarter lactic acid bacteria and aging temperature affect calcium lactate crystallization in cheddar cheese. *J. Dairy Sci.* 86:2516–2524.

Collins, Y. F., McSweeney, P. L. H., and Wilkinson, M. G. 2003. Evidence of a relationship between autolysis of starter bacteria and lipolysis in cheddar cheese during ripening. *J. Dairy Res.* 70(1):105–113.

Conochie, J., Czulak, J., Lawrence, A. J., and Cole, W. F. 1960. Tyrosine and calcium lactate crystals on rindless cheese. *Aust. J. Dairy Technol.* 15:120.

Conochie, J. and Sutherland, B. J. 1965. The nature and cause of seaminess in cheddar cheese. *J. Dairy Res.* 32:35.

Croissant, A. E., Kang, E. J., Campbell, R. E., Bastian, E., and Drake, M. A. 2009. The effect of bleaching agent on the flavor of liquid whey and whey protein concentrate. *J. Dairy Sci.* 92(12):5917–5927.

Crow, V., Curry, B., and Hayes, M. 2001. The ecology of nonstarter lactic acid bacteria (NSLAB) and their use as adjuncts in New Zealand cheddar. *Int. Dairy J.* 11(4/7):275–283.

Crudden, A., Fox, P. F., and Kelly, A. L. 2005. Factors affecting the hydrolytic action of plasmin in milk. *Int. Dairy J.* 15:305–313.

D'Aoust, J. Y., Warburton, D. W., and Sewell, A. M. 1985. *Salmonella typhimurium* phage type 10 from cheddar cheese implicated in a major Canadian foodborne outbreak. *J. Food Prot.* 48(12):1062–1066.

Desmazeaud, M. 2000. Suitability of the milk for the development of the lactic acid flora. In: *Cheesemaking— From Science to Quality Assurance*, 2nd edition, edited by A. Eck and J. C. Gillis, pp. 199–212. Paris: Intercept/Lavoisier Publishing.

Dorn, F. L. and Dahlberg, A. C. 1942. Identification of the white particles found on ripened cheddar cheese. *J. Dairy Sci.* 25:31–36.

Drake, M. A., Boylston, T. D., Spence, K. D., and Swanson, B. G. 1997. Improvement of sensory quality of reduced fat cheddar cheese by a *Lactobacillus* adjunct. *Food Res. Int.* 30(1):35–40.

Dybing, S. T., Wiegand, J. A., Brudvig, S. A., Huang, E. A., and Chandan, R. C. 1988. Effect of processing variables on the formation of calcium lactate crystals on cheddar cheese. *J. Dairy Sci.* 71(7):1701–1710.

El-Gazzar, F. E. and Marth, E. H. 1991. Ultrafiltration and reverse osmosis in dairy technology: A review. *J. Food Prot.* 54(10):801–809.

El-Gazzar, F. E. and Marth, E. H. 1992. Salmonellae, salmonellosis, and dairy foods: A review. *J. Dairy Sci.* 75:2327–2343.

Fitzgerald, E. and Buckley, J. 1985. Effect of total and partial substitution of sodium chloride on the quality of cheddar cheese. *J. Dairy Sci.* 68:3127–3134.

Fox, P. F., Guinee, T. P., Cogan, T. M., and McSweeney, P. L. H. 2000a. Acceleration of cheese ripening. In: *Fundamentals of Cheese Science*, pp. 349–362. Gaithersburg, MD: Aspen.

Fox, P. F., Guinee, T. P., Cogan, T. M., and McSweeney, P. L. H. 2000b. Biochemistry of cheese ripening. In: *Fundamentals of Cheese Science*, pp. 236–281. Gaithersburg, MD: Aspen.

Fox, P. F., Guinee, T. P., Cogan, T. M., and McSweeney, P. L. H. 2000c. Cheese yield. In: *Fundamentals of Cheese Science*, pp. 169–205. Gaithersburg, MD: Aspen.

Fox, P. F., Guinee, T. P., Cogan, T. M., and McSweeney, P. L. H. 2000d. Principal families of cheese. In: *Fundamentals of Cheese Science*, pp. 388–428. Gaithersburg, MD: Aspen.

Frank, J. F. and Marth, E. H. 1988. Fermentations. In: *Fundamentals of Dairy Chemistry*, 3rd edition, edited by N. P. Wong, R. Jenness, M. Keeney, and E. H. Marth, pp. 655–738. New York: Van Nostrand Reinhold Co.

Grazer, C. L., Simpson, R., Roncagliolo, S., Bodyfelt, F. W., and Torres, J. A. 2007. Modeling of time–temperature effects on bacteria populations during cooling of cheddar cheese blocks. *J. Food Process Eng.* 16(3):173–190.

Guinee, T. P. and O'Kennedy, B. T. 2007. Reducing salt in cheese and dairy spreads. In: *Reducing Salt in Foods: Practical Strategies*, Ch. 16, pp. 316–357. Cambridge, UK: Woodhead Publishing Limited.

Hargrove, R. E., McDonough, F. E., and Mattingly, W. A. 1969. Factors affecting survival of *Salmonella* in Cheddar and Colby cheese. *J. Milk Food Technol.* 32:480.

Harper, W. J., Swanson, A. M., and Sommer, H. H. 1953. Observations on the chemical composition of white particles in several lots of cheddar cheese. *J. Dairy Sci.* 36:368–372.

Harrits, J. 1997. Culture nomenclature. In: *Cultures for the Manufacture of Dairy Products*, pp. 117–122. Milwaukee, WI: Chr. Hansen, Inc.

International Dairy Foods Association (IDFA). 2007. *Cheese Facts*, 2007 edition. Washington, DC: International Dairy Foods Association.

International Dairy Foods Association (IDFA). 2010. *Dairy Facts*, 2010 edition. Washington, DC: International Dairy Foods Association.

Ismail, B. and Nielsen, S. S. 2010. Invited review—Plasmin protease in milk: Current knowledge and relevance to dairy industry. *J. Dairy Sci.* 93:4999–5009.

Jensen, R. G. 1995. Defense agents in bovine milk. In: *Handbook of Milk Composition*, edited by R. G. Jensen, pp. 746–748. San Diego: Academic Press.

Johnson, E. A., Nelson, J. H., and Johnson, M. 1990. Microbiological safety of cheese made from heat-treated milk—Part II: Microbiology. *J. Food Prot.* 53:519–540.

Johnson, M. E. 1988. Part II: Cheese chemistry. In: *Fundamentals of Dairy Chemistry*, 3rd edition, edited by N. P. Wong, R. Jenness, M. Keeney, and E. H. Marth, pp. 634–654. New York: Van Nostrand Reinhold Co.

Johnson, M. E., Kapoor, R., McMahon, D. J., McCoy, D. R., and Narasimmon, R. G. 2009. Reduction of sodium and fat levels in natural and processed cheeses: Scientific and technological aspects. *Compr. Rev. Food Sci. Food Saf.* 8:252–268.

Johnson, M. E., Riesterer, B. A., Chen, C., Tricomi, B., and Olson, N. F. 1990a. Effect of packaging and storage conditions on calcium lactate crystallization on the surface of cheddar cheese. *J. Dairy Sci.* 73(11):3033–3041 ill.

Johnson, M. E., Riesterer, B. A., and Olson, N. F. 1990b. Influence on nonstarter bacteria on calcium lactate crystallization on the surface of cheddar cheese. *J. Dairy Sci.* 73(5):1145–1149.

Kang, E. J., Croissant, A. E., Campbell, R. E., Bastian, E., and Drake, M. A. 2010. Invited review: Annatto usage and bleaching in dairy foods. *J. Dairy Sci.* 93(9):3891–3901.

Kapoor, R., Metzger, L. E., Biswas, A. C., and Muthukummarappan, K. 2007. Effect of natural cheese characteristics on process cheese properties. *J. Dairy Sci.* 90:1625–1634.

Karagozlu, C., Kinik, O., and Akbulut, N. 2008. Effects of fully and partial substitution of NaCl by KCl on physico-chemical and sensory properties of white pickled cheese. *Int. J. Food Sci. Nutr.* 59:181–191.

Khalid, N. M. and Marth, E. H. 1990. Lactobacilli—Their enzymes and role in ripening and spoilage of cheese: A review. *J. Dairy Sci.* 73(10):2669–2684.

Kheadr, E. E., Vuillemard, J. C., and El-Deeb, S. A. 2003. Impact of liposome encapsulated enzyme cocktails on cheddar cheese ripening. *Food Res. Int.* 36:241–252.

Kosikowski, F. V. and Mistry, V. V. 1997a. *Cheese and Fermented Milk Foods*, Vol. 1, 3rd edition. Westport, CT: F. V. Kosikowski.

Kosikowski, F. V. and Mistry, V. V. 1997b. *Cheese and Fermented Milk Foods*, Vol. 2, 3rd edition. Westport, CT: F. V. Kosikowski.

Lau, K. Y., Barbano, D. M., and Rasmussen, R. R. 1991. Influence of pasteurization of milk on protein breakdown in cheddar cheese during aging. *J. Dairy Sci.* 74(3):727–740 ill.

Lawrence, R. C. and Gilles, J. 1980. The assessment of the potential quality of young cheddar cheese. *N. Z. J. Dairy Sci. Technol.* 15(1):1–12 ill.

Lawrence, R. C. and Gilles, J. 1987a. Cheddar cheese and related dry-salted cheese varieties. In: *Cheese: Chemistry, Physics, and Microbiology*, edited by P. F. Fox, pp. 1–44. London: Elsevier Applied Science.

Lawrence, R. C. and Gilles, J. 1987b. Cheese composition and quality. In: *Milk—The Vital Force: Proceedings of the XXII International Dairy Congress*, The Hague, 29 September—3 October 1986, edited by [the] Organizing Committee of the XXII International Dairy Congress, pp. 111–121. Dordrecht: D. Reidel.

Lawrence, R. C., Gilles, J., and Creamer, L. K. 1999. Cheddar cheese and related dry-salted cheese varieties. In: *Cheese: Chemistry, Physics and Microbiology*, Vol. 2, 2nd edition, edited by P. F. Fox, pp. 1–38. London: Chapman & Hall.

Lawrence, R. C., Heap, H. A., and Gilles, J. 1984. A controlled approach to cheese technology. *J. Dairy Sci.* 67(8):1632–1645.

Leach, R. D. 1997. Causes for inhibition of starter cultures. In: *Cultures for the Manufacture of Dairy Products*, pp. 61–68. Milwaukee, WI: Chr. Hansen Inc.

Lenoir, J., Remeuf, F., and Schneid, N. 2000a. Correction of cooled and heated milks. In: *Cheesemaking Milk*. In: *Cheesemaking: From Science to Quality Assurance*, 2nd edition, edited by A. Eck and J. C. Gillis, pp. 213–238. Paris: Intercept/Lavoisier Publishing.

Lenoir, J., Remeuf, F., and Schneid, N. 2000b. Rennetability of milk. In: *Cheesemaking: From Science to Quality Assurance*, 2nd edition, edited by A. Eck and J. C. Gillis. Paris: Intercept/Lavoisier Publishing.

Lewis, J. E. 1987. *Cheese Starters: Development and Application of the Lewis System*, pp. 1–221. London: Elsevier.

Leyer, G. J. and Johnson, E. A. 1992. Acid adaptation promotes survival of *Salmonella* sp. in cheese. *Appl. Environ. Microbiol.* 58:2075–2080.

Lorenzen, P. Chr., Martin, D., Clawin-Radecker, I., Barth, K., and Knappstein, K. 2009. Activities of alkaline phosphatase, g-glutamyltransferase and lactoperoxidase in cow, sheep and goat's milk in relation to heat treatment. *Small Ruminant Res.* 89(1):18–23.

Marino, R., Considine, T., Sevi, A., McSweeney, P. L. H., and Kelly, A. L. 2005. Contribution of proteolytic activity associated with somatic cells in milk to cheese ripening. *Int. Dairy J.* 15(10):1026–1033.

Martley, F. G. and Crow, V. L. 1996. Open texture in cheese: The contributions of gas production by micro-organisms and cheese manufacturing practices. *J. Dairy Res.* 63:489–507.

McDowall, F. H. and McDowell, A. K. R. 1939. The white particles in mature cheddar cheese. *J. Dairy Res.* 10:118–119.

Metzger, L. E. and Mistry, V. V. 1994. A new approach using homogenization of cream in the manufacture of reduced fat cheddar cheese—Part 1: Manufacture, composition, and yield. *J. Dairy Sci.* 77(12):3506–3515.

Mistry, V. V. and Kasperson, K. M. 1998. Influence of salt on the quality of reduced-fat cheddar cheese. *J. Dairy Sci.* 81:1214–1221.

Morr, C. V. and Richter, R. L. 1988. Chemistry of processing. In: *Fundamentals of Dairy Chemistry*, edited by P. W. Noble, R. Jenness, M. Keeney, and E. H. Marth, pp. 739–766 ill. New York: Van Nostrand Reinhold Co.

Nair, M. G., Mistry, V. V., and Oommen, B. S. 2000. Yield and functionality of cheddar cheese as influenced by homogenization of cream. *Int. Dairy J.* 10(9):647–657.

Nelson, B. K. and Barbano, D. M. 2005. Yield and aging of cheddar cheeses manufactured from milks with different milk serum protein contents. *J. Dairy Sci.* 88(12):4183–4194.

Partridge, J. 2008. Cheddar cheese. In: *The Sensory Evaluation of Dairy Products*. pp. 225–270. New York: Springer.

Pastorino, A. J., Hansen, C. L., and McMahon, D. J. 2003. Effect of salt on structure-function relationships in cheese. *J. Dairy Sci.* 86:60–69.

Pearce, K. N., Creamer, L. K., and Gilles, J. 1973. Calcium lactate deposits on rindless cheddar cheese. *N. Z. J. Dairy Sci. Technol.* 8(1):3–7.

Ramet, J. P. 2000a. Milk transforming agents. In: *Cheesemaking: From Science to Quality Assurance*, 2nd edition, edited by A. Eck and J. C. Gillis, pp. 155–163. Paris: Intercept/Lavoisier Publishing.

Ramet, J. P. 2000b. The drainage of coagulum. In: *Cheesemaking: From Science to Quality Assurance*, 2nd edition, edited by A. Eck and J. C. Gillis, pp. 41–59. Paris: Intercept/Lavoisier Publishing.

Ratnam, S., Stratton, F., O'Keefe, C., Roberts, A., Coats, R., Yetman, M., Khakhria, R., and Hockin, J. 1999. *Salmonella enteritidis* outbreak due to contaminated cheese—Newfoundland. Can. Commun. Dis. Rep. 25:1–4.

Reddy, K. A. and Marth, E. H. 1993. Composition of cheddar cheese made with sodium chloride and potassium chloride either singly or as mixtures. *J. Food Compos. Anal.* 6(4):354–363.

Reddy, K. A. and Marth, E. H. 1995a. Lactic acid bacteria in cheddar cheese made with sodium chloride, potassium chloride or mixtures of the two salts. *J. Food Prot.* 58(1):62–69.

Reddy, K. A. and Marth, E. H. 1995b. Microflora of cheddar cheese made with sodium chloride, potassium chloride, or mixtures of sodium and potassium chloride. *J. Food Prot.* 58(1):54–61.

Richard, J. and Desmazeaud, M. 2000. Cheesemaking milk. In: *Cheesemaking: From Science to Quality Assurance*, 2nd edition, edited by A. Eck and J. C. Gillis, pp. 189–196. Paris: Intercept/Lavoisier Publishing.

Ruden, K. V. 1997. Non-phage-related causes of culture inhibition. In: *Cultures for the Manufacture of Dairy Products*, pp. 143–151. Milwaukee, WI: Chr. Hansen, Inc.

Ryser, E. T. and E. H. Marth. 1987. Behavior of Listeria monocytogenes during the manufacture and ripening of Cheddar cheese. *J. Food Protect.* 50:7–13.

Severn, D. J., Johnson, M. E., and Olson, N. F. 1986. Determination of lactic acid in cheddar cheese and calcium lactate crystals. *J. Dairy Sci.* 69(8):2027–2030.

Shakeel Ur, R., Banks, J. M., McSweeney, P. L. H., and Fox, P. F. 2000. Effect of ripening temperature on the growth and significance of nonstarter lactic acid bacteria in cheddar cheese made from raw of pasteurized milk. *Int. Dairy J.* 10(1/2):45–53.

Shakeel Ur, R., Drake, M. A., and Farkye, N. Y. 2007. Differences between cheddar cheese manufactured by the milled-curd and stirred-curd methods using different commercial strains. *J. Dairy Sci.* (91):76–84.

Somers, E. B., Johnson, M. E., and Wong, A. C. L. 2001. Biofilm formation and contamination of cheese by nonstarter lactic acid bacteria in the dairy environment. *J. Dairy Sci.* 84(9):1926–1936.

Strauss, K. 1997. American cheese types. In: *Cultures for the Manufacture of Dairy Products*, pp. 117–122. Milwaukee, WI: Chr. Hansen, Inc.

Tamime, A. Y. and Deeth, H. C. 1980. Yogurt: Technology and biochemistry. *J. Food Prot.* 43(12):939–977.

Thomas, T. D. and Pearce, K. N. 1981. Influence of salt on lactose fermentation and proteolysis in Cheddar cheese. *N. Z. J. Dairy Sci. Technol.* 16(3):253–259.

Tuckey, S. L., Rueke, H. A., and Clark, G. L. 1938. X-ray diffraction analysis of white specks in Cheddar cheese. *J. Dairy Sci.* 21:161.

Turner, K. W. and Thomas, T. D. 1980. Lactose fermentation in cheddar cheese and the effect of salt. *N. Z. J. Dairy Sci. Technol.* 15(3):265–276.

Upreti, P., McKay, L. L., and Metzger, L. E. 2006. Influence of calcium and phosphorus, lactose, and salt-to-moisture ratio on cheddar cheese quality: Changes in residual sugars and water-soluble organic acids during ripening. *J. Dairy Sci.* 89(2):429–443.

U.S. Department of Agriculture (USDA). 1978. Cheese varieties and descriptions. In: *Agriculture Handbook Number 54*, p. 151. Alexandria, VA: National Cheese Institute, U.S. Department of Agriculture.

U.S. Department of Health and Human Services, Department of Public Health Service, and Food and Drug Administration. 1999 Revision. Grade "A" Pasteurized Milk Ordinance. 1–323. Available online at: http://www.cfsan.fda.gov/~ear/pmo01toc.html (accessed 23 November 2010).

U.S. Food and Drug Administration (FDA). 2006a. Cheddar cheese. Code of Federal Regulations Title 21, Section 133.113. Pittsburgh, PA: U.S. Government Printing Office. Available online at: http://www.access.gpo.gov/nara/cfr/waisidx_06/21cfr133_06.html. Date accessed: 11/24/10.

U.S. Food and Drug Administration (FDA). 2006b. Cheddar cheese for manufacturing. Code of Federal Regulations Title 21, Section 133.114. Pittsburgh, PA: U.S. Government Printing Office. Available online at: http://www.accessdata.fda.gov/scripts/cdrh/cfdocs/cfcfr/CFRSearch.cfm?CFRPart=133 (accessed 24 November 2010).

U.S. Food and Drug Administration (FDA). 2006c. Low-sodium cheddar cheese. Code of Federal Regulations Title 21, Section 133.116. Pittsburgh, PA: U.S. Government Printing Office. Available online at: http://www.accessdata.fda.gov/scripts/cdrh/cfdocs/cfcfr/CFRSearch.cfm?CFRPart=133 (accessed 24 November 2010).

U.S. Food and Drug Administration (FDA). 2006d. Colby cheese. Code of Federal Regulations Title 21, Section 133.118. Pittsburgh, PA: U.S. Government Printing Office. Available online at: http://www.accessdata.fda.gov/scripts/cdrh/cfdocs/cfcfr/CFRSearch.cfm?CFRPart=133 (accessed 24 November 2010).

U.S. Food and Drug Administration (FDA). 2006e. Monterey cheese and monterey jack cheese. Code of Federal Regulations Title 21, Section 133.153. Pittsburgh, PA: U.S. Government Printing Office. Available online at: http://www.accessdata.fda.gov/scripts/cdrh/cfdocs/cfcfr/CFRSearch.cfm?CFRPart=133 (accessed 24 November 2010).

U.S. Food and Drug Administration (FDA). 2006f. Antimicotics. Code of Federal Regulations Title 21, Section 181.23. Pittsburgh, PA: U.S. Government Printing Office. Available online at: http://law.justia.com/us/cfr/title21/21-3.0.1.1.12.html (accessed 24 November 2010).

U.S. Food and Drug Administration (FDA). 2006g. Sorbic acid. Code of Federal Regulations Title 21, CFR 182.3089. Pittsburgh, PA: U.S. Government Printing Office. Available online at: http://www.access.gpo.gov/nara/cfr/waisidx_03/21cfr182_03.html (accessed 24 November 2010).

U.S. Food and Drug Administration (FDA). 2006h. Calcium propionate. Code of Federal Regulations Title 21, Section 184.1221. Pittsburgh, PA: U.S. Government Printing Office. Available online at: http://www.access.gpo.gov/nara/cfr/waisidx_04/21cfr184_04.html (accessed 24 November 2010).

U.S. Food and Drug Administration (FDA). 2006i. Propylparaben. Code of Federal Regulations Title 21, CFR 184.1670. Pittsburgh, PA: U.S. Government Printing Office. Available online at: http://www.access.gpo.gov/nara/cfr/waisidx_04/21cfr184_04.html (accessed 24 November 2010).

U.S. Food and Drug Administration (FDA). 2006j. Rennet (animal-derived) and chymosin preparation (fermentation-derived). Code of Federal Regulations Title 21, Section 184.1685. Pittsburgh, PA: U.S. Government Printing Office. Available online at: http://www.access.gpo.gov/nara/cfr/waisidx_04/21cfr184_04.html (accessed 24 November 2010).

U.S. Food and Drug Administration (FDA). 2006k. Hydrogen peroxide. Code of Federal Regulations Title 21, Section 184.1366. Pittsburgh, PA: U.S. Government Printing Office. Available online at: http://www .access.gpo.gov/nara/cfr/waisidx_04/21cfr184_04.html (accessed 24 November 2010).

U.S. Food and Drug Administration (FDA). 2006l. Methylparaben. Code of Federal Regulations Title 21, Section 184.1490. Pittsburgh, PA: U.S. Government Printing Office. Available online at: http://www .access.gpo.gov/nara/cfr/waisidx_04/21cfr184_04.html (accessed 24 November 2010).

U.S. Food and Drug Administration (FDA). 2006m. Sodium benzoate. Code of Federal Regulations Title 21, Section 184.1733. Pittsburgh, PA: U.S. Government Printing Office. Available online at: http://www .access.gpo.gov/nara/cfr/waisidx_04/21cfr184_04.html (accessed 24 November 2010).

Wallace, J. M. and Fox, P. F. 1997. Effect of adding free amino acids to cheddar cheese curd on proteolysis, flavor and texture development. *Int. Dairy J.* 7(2/3):157–167.

Walstra, P., Noomen, A., and Geurts, T. J. 1999. Dutch-type varieties. In: *Cheese: Chemistry, Physics and Microbiology*, Vol. 2., edited by P. F. Fox, pp. 39–82. London: Chapman & Hall.

Washam, C. J., Kerr, T. J., Hurst, V. J., and Rigsby, W. E. 1985. A scanning electron microscopy study of crystalline structures on commercial cheese. *Dev. Ind. Microbiol.* 26:749–761.

Weihrauch, J. L. 1988. Lipids of milk: Deterioration. In: *Fundamentals of Dairy Chemistry*, edited by N. P. Wong, R. Jenness, M. Keeney, and E. H. Marth, pp. 215–278. New York: Van Nostrand Reinhold Co.

Whitney, R. McL. 1988. Proteins of milk. In: *Fundamentals of Dairy Chemistry*, 3rd edition, edited by N. P. Wong, R. Jenness, M. Keeney, and E. H. Marth, pp. 81–169. New York: Kluwer Academic Publishers.

Williams, A. G., Withers, S. E., and Banks, J. M. 2000. Energy sources of nonstarter lactic acid bacteria isolated from cheddar cheese. *Int. Dairy J.* 10(1/2):17–23.

Wood, D. S., Collins-Thompson, D. L., Irvine, D. M., and Myhr, A. N. 1984. Source and persistence of *Salmonella muenster* in naturally contaminated cheddar cheese. *J. Food Prot.* 47:20–22.

21

Traditional Greek Feta

Anna Polychroniadou-Alichanidou

CONTENTS

21.1 Introduction

Cheese has always been an important component of the Greek diet. Greece has the world's highest cheese consumption—31.25 kg of cheese per capita per year (IDF 2009).

References to cheese production go back to the eight century BC. The cheesemaking technology described in Homer's *Odyssey* is similar to the technology used until recently by Greek shepherds to make feta cheese from the milk of their sheep and goats. Other ancient Greek authors also mention cheese made from goat or sheep milk as common Greek food. Aristotle (384–322 BC) goes a little further and says that cheese is composed of water, fat, and tyrine, a term used in Greece up to the early twentieth century to designate casein (in Greek, *tyri* means cheese).

Feta cheese is a white, semihard cheese made from sheep milk (or a mixture of sheep milk and up to 30% goat milk), which is ripened and stored in brine. Traditionally, the cheese was packed in blocks in wooden barrels; the shape of the cheese blocks looked like watermelon slices. The name feta, which means 'slice,' has probably come from the fact that, after elaboration, the curd is sliced into appropriate dimensions for salting and packaging. Besides feta, many named varieties of other cheeses in brine are traditionally produced in southeast Europe and the Middle East. The most well known among them are bjalo salamureno sirene (Bulgaria), beyaz peynir (Turkey), homoljski and zlatarski (Serbia), teleme or telemea (Greece, Romania), travnicki/vlasicki (Bosnia-Herzegovina), liqvan (Iran), akawi (Lebanon), and domiati (Egypt). The majority are made from sheep milk—sheep farming is very important in countries where climatic conditions are not favorable for cattle raising.

Feta, the most well-known variety of this group of cheeses, is very much appreciated in Greece and, in most cases, is synonymous to cheese. According to the data from the Hellenic Ministry of Agriculture, the annual production of feta in Greece is more than 130,000 metric tons, most of which is consumed within the country. However, because of its unique sensory properties and the fact that large Greek ethnic groups live in many countries around the world, small quantities of feta are exported, mainly to the United States and the European Union. Over time, the name feta has acquired an important trade value and is now being used to designate all cheeses in brine having similarities to feta but made from cow milk, even if made using a completely different technology. These cheeses cannot duplicate the typical flavor of traditional feta cheese, however, and do not fulfill consumers' expectations. In 1996, the European Commission recognized that the designation "feta" qualifies for a Protected Designation of Origin according to Regulation 2081/92 (Anonymous 1992). The designation was awarded only to the particular cheese in brine produced in the mainland of Greece and on the island of Lesvos from sheep milk or a mixture of sheep milk and up to 30% goat milk.

Traditionally, feta was made in small cheese plants using the milk produced in the region. The sensory characteristics of the product differed from place to place. During the past decades, large dairy industries, which were able to collect large quantities of sheep and goat milk, have started to produce feta cheese; now, the milk is pasteurized and the process is mechanized. This has led to the stabilization of high-quality and sensory properties throughout the year. Nevertheless, the special sensory characteristics of the cheeses made in farmhouses or in small plants on the mountains using the local milk are still appreciated by connoisseurs.

21.2 Manufacture of Feta Cheese

21.2.1 Milk

The most suitable milk for feta cheese is sheep milk, but a mixture of sheep and goat milk may be used. However, the percentage of goat milk in the mixture is not allowed to exceed 30% (Greek Food Code 1998). Goat milk alone can also be used for the manufacture of cheese in brine, but the cheese, although appreciated by consumers preferring a stronger taste, is not allowed to be called feta. The production of sheep and goat milk is mostly seasonal (December/January to June/July); therefore, the activities of feta cheese plants are attenuated during summer and autumn.

There is a significant variation in fat content during lactation, with lower values, at least for sheep milk, corresponding to the beginning of the warm period. Protein in sheep milk increases regularly as lactation advances (Voutsinas et al. 1988). Thus, the fat content of artisanal cheeses usually depends on the production time. In contrast, in modern cheese plants, the milk is standardized to a casein-to-fat ratio of 0.7–0.8; the fat content of the cheese is therefore more stable. For public health reasons, the milk is pasteurized and cooled to 34°C before the addition of starter culture. However, in small cheese plants or farms, the milk usually receives a thermal treatment lower than necessary for pasteurization conditions. In this case, part of the indigenous flora of the milk may survive, influencing the course of ripening and the release of flavor compounds.

21.2.2 Starters and Calcium Chloride

Starter culture is usually a combination of lactic acid bacteria at a ratio of lactococci to lactobacilli of 1:3 (e.g., *Lactococcus lactis* subsp. *lactis* and *Lactobacillus delbrueckii* subsp. *bulgaricus*). In small cheese plants, yogurt is sometimes used instead of pure cultures; in this case, acid development is slow (Abd El-Salam et al. 1993). Culture is added to cheese milk to a level of 0.5%–1% (v/v) and incubated for about 30 min. Calcium chloride up to 20 g per 100 kg milk may also be added, although this addition is not necessary when unpasteurized sheep milk is used. Indeed, sheep milk is rich in calcium—a fact related to the high casein content (Richardson et al. 1974; Polychroniadou and Vafopoulou 1985). Concentrations as high as 172–209 mg of Ca per 100 g are reported (Alichanidis and Polychroniadou 1996).

21.2.3 Coagulation

Coagulation by rennet is performed at 32°C–34°C for 45–50 min. Sheep milk is very sensitive to rennet, and because of the higher β/α_s-casein ratio, coagulation proceeds faster than with cow milk (Storry et al. 1983; Muir et al. 1993). Alternatively, less rennet is required to obtain the same coagulation time as with cow milk (Kalantzopoulos 1993). The use of artisanal (homemade) rennet was very common when the majority of feta cheese was manufactured at small plants in mountainous and semimountainous regions; it is still in use mainly by cheesemakers in Southern Greece. Artisanal rennet is prepared by cheese manufacturers from mixed dried lambs' and kids' whole abomasa extracted using NaCl solution. It is kept in liquid form at 4°C–5°C for 1 week at the maximum (Moschopoulou et al. 2007). Nowadays, modern large- and medium-sized cheese plants use commercial calf rennet.

21.2.4 Draining

After coagulation, the curd is cut crossways in cubes of 2–3 cm, left for about 10 min for partial exudation of the whey, and transferred into perforated molds. Molds are cylindrical when the cheese is to be packed in barrels and rectangular when intended for packing in tin cans. As layers of curd are piled in the mold, small mechanical openings are formed. These openings, full of whey, are typical of feta cheese structure.

The curd is left to drain at room temperature without pressing until it is firm enough to remove the molds. Ovine curd drains less than bovine curd drains because of the higher total solids, casein, and fat content and higher firmness. In contrast, caprine curd is usually drier than bovine curd, because goat milk contains less casein, especially α_{s1}-casein- and calcium, than sheep milk (Storry et al. 1983; Alichanidis and Polychroniadou 1996). Feta cheese producers profit from this behavioral difference and usually mix 10%–30% goat milk with sheep cheese milk, because the curd remains firm and drains better (Alichanidis and Polychroniadou 1996). The time needed for draining depends on the temperature. Environmental temperatures of 14°C–16°C are common, but higher temperatures are not rare during late spring. The composition of the starter culture also affects draining, because the rate of acidity development is crucial for the quality of feta cheese. The gradual cooling of the curd from 30°C to 16°C allows the starter microorganisms the time and the environmental conditions to develop acidity and enhance the draining process. Mesophilic starters are found to lower the pH faster than thermophilic ones and to shorten the draining time; however, a combination of mesophilic and thermophilic strains are more suitable (Kandarakis et al. 2001). A pH drop of 5.0–5.2 in 8 h and 4.8–5.0 in 18–24 h is considered necessary for proper drainage and ripening (Abd El-Salam et al. 1993). Table 21.1 shows the typical pH evolution during the manufacture of feta cheese.

21.2.5 Salting

Usually, the molds are inverted after 2–3 h, and the curd is usually left for another 2–3 h to complete draining. The molds are then removed, and the curd is cut into pieces, which are transferred on a salting table spread with granular salt. The pieces are placed side by side to retain their shape, and their exposed surfaces are sprinkled with salt the size of rice grains. The salt, dissolved in the exuded whey, slowly penetrates into the cheese without causing the formation of a dry rind. After about 12 h, the cheese pieces are turned, and two other surfaces are salted. This process is repeated four more times at 12-h intervals

TABLE 21.1

Typical pH Changes during Feta Cheese Manufacture

	Time from Start (Hours)	pH (Mean Values)	
		A	**B**
Milk	0	6.59	–
Milk incubated with starters[a] (before renneting)	0.5	6.39	–
Curd during dry salting	2	–	6.09
	4		5.72
Curd at the end of draining	8	5.12	5.16
Curd at the next morning	20–21	4.70	4.91

Source: Michaelidou, A. 1997. Investigation of protein hydrolysis during Feta cheese ripening, PhD Dissertation (in Greek), Aristotle University, Thessaloniki, Greece, p. 166; Kandarakis, I. et al. 2001. *Food Chem.* 72, 369–378.

[a] A, *Lactococcus lactis* and *Lactobacillus delbrueckii* var. *bulgaricus* (1:3). (Michaelidou 1997); B: *Lactococcus lactis* (2 strains) and *Lactobacillus delbrueckii* var. *bulgaricus* (2:2) (Kandarakis et al. 2001)

until each surface is salted twice and the cheese contains about 3% salt. During draining and salting, the pH of the curd continues decreasing to about 4.8 or less.

21.2.6 Ripening

Following salting, the cheeses remain on the table for 2 more days, depending on the ambient temperature. They are inverted at regular time intervals. During this period, a slimy layer is formed on their surfaces due to the growth of bacteria and yeasts (Anifantakis 1991); the composition of this surface flora depends on the environment of the cheese plant. Gradual dry salting, low pH, and slime formation are essential for the development of the typical flavor of feta cheese; it is found that the proteolytic and lipolytic activity of the surface microflora is significant during the first 15–20 days of ripening and contributes to the release of many peptides, amino acids, and fatty acids, which are precursors of most volatile flavor compounds (Abd El-Salam et al. 1993).

In modern dairy industries, molding, drainage, salting, and preripening are mechanically performed. The curd is transferred into the molds by gravity. Salting takes place in open tin cans, which can be automatically inverted when needed. In addition, salt is of high purity, and therefore, subsequent washing of the cheese pieces is avoided.

21.2.7 Packaging and Storage

Feta cheese is packed in wooden barrels or in tin cans. Barrels were the traditional containers for feta, and when stored in barrels, the cheese develops a stronger and spicier flavor, much appreciated by consumers. However, handling a filled barrel (weighing about 50 kg) is difficult, so currently, feta cheese is packaged mostly in tin cans, which weigh 16–17 kg, making their transportation easier.

Before packaging, the surface of the cheese pieces is carefully cleaned with water or brine. In the container, each layer is usually covered by a piece of parchment paper before the next layer is added. Finally, 1.0–1.5 kg of 6%–7% brine is added in order to cover the cheese. Cheese pieces must be tightly packed, allowing little space for the brine. If the volume of brine is larger than necessary, more low-molecular-weight compounds diffuse from cheese into the brine (see Section 5.1.1). From time to time, it is advisable to let the fermentation gases escape and to fill the container with brine if the level fails to cover the cheese. Salt concentration profiles of cheese in brine are temperature dependent; salt diffusion varies with temperature according to the Arrhenius equation (Turhan and Kaletunç 1992).

The barrels or tins are kept at 16°C–18°C until the pH of the cheese reaches 4.4–4.6 and the moisture drops to less than 56%. They are then transferred to cold rooms (4°C–5°C) to complete ripening.

Marketing of feta cheese is permitted only after 2 months postmanufacture (Greek Food Code 1998). Until sold to the consumer, the cheese must be kept in brine. If uncovered by brine, the surface becomes dry, changes color, and permits the growth of yeasts and molds. A good-quality feta cheese may be stored in brine for up to 12 months at 2°C–4°C.

After ripening, feta cheese slices may also be individually packed in plastic bags under vacuum for retail marketing. Once this packaging is opened, however, the cheese has to be consumed within a few days. An alternative consumer packaging is the placing of the cheese slice in a small plastic container filled with brine, permitting longer storage at home (at 4°C–5°C).

21.2.8 Yield

Because the average dry matter (DM) of sheep milk is 18%–20% (Alichanidis and Polychroniadou 1996), a yield of about 25% is expected (Alichanidis et al. 1984; Anifantakis 1991; Mallatou et al. 1994). This means that about 4 kg of milk is needed to produce 1 kg of cheese. The yield varies with the season (a little higher in winter than in spring) and with the percentage of goat milk mixed with sheep milk.

21.3 Gross Composition and Sensory Properties

21.3.1 Gross Composition

The gross composition of feta cheese is shown in Table 21.2. The compositional provisions of the Greek Food Code for feta are as follows: maximum moisture, 56% and minimum fat in DM, 43%.

Few rheological analyses have been conducted on feta cheese, as feta is a difficult material that easily breaks into pieces when compressed. Based on the data available, hardness seems to vary from 4 to 7 kg, fracture stress varies from 1.8 to 2.4 kg, and compression to fraction varies from 14% to 22% (Katsiari et al. 1997; Kandarakis et al. 2001). Increasing the proportion of goat milk in the cheese milk reduces the porosity of the casein network and increases the hardness of the cheese (Mallatou et al. 1994).

21.3.2 Sensory Properties

Feta has a short, firm, and smooth texture, has a moist surface without rind, and is sliceable. Mechanical openings distributed over the cheese are normal, but the presence of small round holes is regarded as a defect, indicating anomalous fermentations. The color of feta is bright white, because sheep and goats transfer very few carotenoids to their milk.

TABLE 21.2

Gross Composition of Feta Cheese

	Average Value
Moisture (%)	54.6
Fat (%DM)	49.1
Protein [%(N × 6.38)]	17.1
Salt (% of the aqueous phase)	5.3
pH	4.5

Source: Vafopoulou, A. et al. 1989. *J. Dairy Res.* 56, 285–296; Anifantakis, E. M. 1991. *Greek Cheeses: A Tradition of Centuries.* National Dairy Committee of Greece, Athens, pp. 27–42; Abd El-Salam, M. H. et al. 1993. In: *Cheese: Chemistry, Physics and Microbiology,* Vol. 2, 2nd edition, edited by P. F. Fox, pp. 301–350. London: Chapman and Hall; Katsiari, M. C. et al. 1997. *Int. Dairy J.* 7, 465–472; Valsamaki, K. et al. 2000. *Food Chem.* 71, 259–266.

The taste of feta cheese is slightly acidic and salty, and its flavor is rich and mildly rancid due to the relatively high level of low- and medium-chain free fatty acids. Feta made with mixtures of sheep and goat milk has, in general, a stronger but still pleasant flavor appreciated by some consumers. The sensory characteristics of traditionally produced feta in small dairies vary from place to place because of variations in the technology parameters applied, including the starter culture used and the indigenous microflora of the milk and of the environment in each plant. The use of artisanal rennet was found to contribute to higher ($p < 0.05$) scores than with commercial rennet, a fact attributed to the free fatty acid profile of the cheese (Moatsou et al. 2004). Additionally, the vegetation in local pastures is different in places having a different microclimate, and this imparts different flavors and aromas from the milk to the cheese. This variability is expected by the consumer. However, large cheese plants, with milk collected from a larger region, pasteurize the milk, use starter cultures and, because of a better control of the technological parameters, produce a more standardized taste and flavor.

21.4 Microbiology of Feta Cheese

Mesophilic lactic acid bacteria constitute the dominant flora of feta cheese (Tzanetakis and Litopoulou-Tzanetaki 1992; Sarantinopoulos et al. 2002; Manolopoulou et al. 2003). Their number significantly increases during the first 15-day period in the warm room (at about 16°C) and remains high throughout the ripening time (>8 log cfu/mL). This explains the high proteolytic activity observed at the beginning of the ripening period. Viable populations of *Streptococcus thermophilus* and *Lactococcus* spp. were also found in commercial feta using fluorescence *in situ* hybridization (Rantsiou et al. 2008).

Research has shown that nonstarter lactic acid bacteria (NSLAB) constitute a large part of this flora; they originate in milk (indigenous milk flora) and the environment of the cheese plant (Tzanetakis and Litopoulou-Tzanetaki 1992). Most species of NSLAB isolated from feta are lactobacilli. Presumptive *Leuconostoc* are also present at high numbers (Manolopoulou et al. 2003). Both groups have an important role in the ripening process and contribute to the development of the characteristic flavor of the product due to the environment of this type of cheese, which favors their development and activity (Tzanetakis and Litopoulou-Tzanetaki 1992; Rantsiou et al. 2008). *Lb. plantarum* is the dominant species, representing 40%–50% of the isolates, followed by *Lb. paracasei* subsp. *paracasei* (Bintsis et al. 2003). Many different strains of these microorganisms, isolated from feta cheese, have been characterized as to their enzyme and plasmid profiles, acidifying and proteolytic abilities, fermentation profiles, and other characteristics; a marked genotypic variability is observed, but phenotypic differences are small (Xanthopoulos et al. 2000; Manolopoulou et al. 2003). NSLAB are also the dominant group of bacteria in the brine; *Lb. paracasei* subsp. *paracasei* and *Lb. plantarum* are the principal species identified (Bintsis et al. 2000).

Pediococci and enterococci are also present, but in lower numbers (Tzanetakis and Litopoulou-Tzanetaki 1992; Manolopoulou et al. 2003). However, investigations showed that some strains of *Enterococcus durans* (Tzanetakis and Litopoulou-Tzanetaki 1992) or *E. faecium* (Litopoulou-Tzanetaki et al. 1993) could improve the sensory properties of feta cheese if used as adjunct cultures in combination with mesophilic starters (e.g., *Lc. lactis* subsp. *lactis*, *Lb. casei*, and *Leuconostoc mesenteroides* subsp. *cremoris*). Although enterococci decrease during cold room ripening, they continue to be found at high counts (5–6 log cfu/g) in cheese up to 4 months (Manolopoulou et al. 2003). Micrococci are also found, mainly as part of the surface flora (2–3 log cfu/g); they survive throughout ripening (Manolopoulou et al. 2003).

Salt-resistant yeasts grow at high numbers (6–8 log cfu/mL) on the surface of feta during dry salting, but their number decreases with ripening time (Tzanetakis et al. 1998; Manolopoulou et al. 2003). Among the species that have been isolated, *Saccharomyces cerevisiae* was predominant (47.9% of the isolates), followed by *Debaryomyces hansenii* and *Pichia farinosa* (30.9% and 11.2%, respectively). *Kluyveromyces lactis* and other *Piscia* species were also identified in commercial feta (Rantsiou et al. 2008). *D. hansenii* was found to be the dominant species in the brine (Bintsis et al. 2003). Although yeasts seem to be of minor significance for the ripening process (Vafopoulou-Mastrojiannaki et al. 1990; Litopoulou-Tzanetaki et al. 1993), they do exhibit some aminopeptidase and esterase activity (Tzanetakis et al. 1998).

Coliforms and *Escherichia coli* may be present in curd and reach their maximum values at 4 days, but they are eliminated in mature cheese (Manolopoulou et al. 2003).

21.5 Biochemistry of Feta Cheese Ripening

Ripening is a delicately balanced process of controlled enzymatic breakdown (Visser 1998). Proteolysis, lipolysis, and glycolysis lead to flavor compounds responsible for the sensory properties of the cheese. There are little data for glycolysis in feta cheese in the literature, but lipolysis and, especially, proteolysis have been the object of many studies.

21.5.1 Proteolysis

21.5.1.1 General Observations

It is generally accepted that proteolysis is a major event during cheese ripening. Casein breakdown involves the action of many enzymes from various sources, including indigenous milk proteinase (plasmin), chymosin and other milk-clotting enzymes, proteinases and peptidases from the starter microorganisms and NSLAB, and enzymes from adjunct cultures, yeasts, and molds.

The extent of proteolysis in feta cheese is, in general, lower than in most hard and semihard cheese varieties. During the 2-month minimum ripening time for feta [according to the Greek Food Code (1998)], water-soluble nitrogen, expressed as a percentage of the total N (WSN%TN), ranges from 12 to 20 (Alichanidis et al. 1984; Polychroniadou 1994; Valsamaki et al. 2000; Kandarakis et al. 2001; Moatsou et al. 2004); this depends somewhat on the residual rennet, the starter and nonstarter flora, and the time–temperature interaction during ripening. However, WSN%TN, a parameter generally accepted as a "ripening index," does not precisely reflect the extent of casein hydrolysis.

The water extract of a 3-day-old feta also includes milk serum proteins, mainly β-lactoglobulin (β-Lg) and α-lactalbumin (α-La). These proteins elute late from reversed-phase high-performance liquid chromatography (RP-HPLC) columns and form large peaks (about 25% of the total area) at the end of the chromatogram (Michaelidou et al. 1998; Katsiari et al. 2000a). When the cheese is placed in brine, another process that also influences the N content of the water extract starts—the diffusion of cheese constituents into the brine (and, eventually, of salt into the cheese) until equilibrium is reached. Along the storage period, small- and medium-sized nitrogenous compounds move into the brine. Milk serum proteins, being also water soluble, follow in the same way. It was found that in chromatograms produced by RP-HPLC, the area corresponding to α-La and β-Lg decreases by about 40% (Katsiari et al. 2000a). The nature of the compounds diffusing into the brine was investigated by Michaelidou et al. (2005). Analysis by RP-HPLC of the water extract of feta and of the brine showed that both contained serum proteins and various peptides but chromatographic profiles were substantially different. Furthermore, the distribution of the above compounds between cheese and brine changed during ripening. Size and hydrophobicity seem to be the parameters driving the solubility of peptides in brine. The same seems to be true also for the amino acids—hydrophobicity and the shape of the side chain of each particular amino acid may govern its ability to migrate into the brine. The large amount of serum proteins retained in the cheese and the migration of many nitrogenous compounds into the brine make WSN%TN a not very useful criterion of casein breakdown in feta cheese. A colorimetric procedure based on the reaction of 2,4,6-trinitrobenzenesulfonic acid with free amino groups was found to give useful information on the degree of ripening of feta cheese (Polychroniadou 1988).

21.5.1.2 Casein and Large Peptides

Most of the qualitative and quantitative changes occur during the first 15-day period, when the cheese is kept at a relatively high temperature (about 16°C). All soluble nitrogenous fractions show a significant increase at this time. Later, when the cheese is transferred to a refrigerator, the rate of casein and peptide degradation decreases (Table 21.3).

TABLE 21.3

Proteolysis Parameters during Feta Cheese Ripening

	Age (Days)						
	1	**3**	**15**	**30**	**60**	**90**	**120**
Residual α_{s1}-CN	100	92.5	53.6	49.4	41.2	39.2	35.4
Residual β-CN	100	96.1	93.2	88.6	84.7	79.8	73.6
WSN%TN	6.2	11.2	14.9	15.2	16.1	18.1	19.0
TCA-SN%TN	3.0	5.2	10.8	12.1	12.5	14.1	15.8
FAA (mg/kg)	546	1016	2430	2498	2922	3073	3978

Source: Alichanidis, E. et al. 1984. *J. Dairy Res.* 51, 141–147; Michaelidou, A. 1997. Investigation of protein hydrolysis during Feta cheese ripening, PhD Dissertation (in Greek), Aristotle University, Thessaloniki, Greece, p. 166; Valsamaki, K. et al. 2000. *Food Chem.* 71, 259–266; Katsiari, M.C. et al. 2000a. *Int. Dairy J.* 10, 635–646.

Note: CN, casein; WSN%TN, water-soluble N% total N; TCA-SN%TN, 12% (w/v) trichloroacetic acid-soluble N% total N; FAA, free amino acids.

Electrophoretograms of feta on urea polyacrylamide gels (Figure 21.1) reveal the degradation of casein during ripening. The breakdown of α_{s1}-casein starts very early. Many zones with an electrophoretic mobility higher than α_{s1}-casein are present in the 4-day sample; of these, the zones corresponding to the peptides produced by the action of chymosin, for example, α_{s1}-casein (f 24–199), are the most intense. Hydrolysis of β-casein starts later, generating peptides with lower electrophoretic mobility than the parent protein (γ-caseins region). It is worth noticing that the intensity of all zones decreases during ripening, as proteins and large peptides are hydrolyzed into smaller fragments not able to be fixed on the gel. Only the production of the peptides with slow (γ-caseins region) and intermediate mobility (zone X of unknown identity) increases with time.

The results of many investigations (Alichanidis et al. 1984; Michaelidou 1997; Katsiari et al. 2000a; Valsamaki et al. 2000) have shown that rennet plays an important role in feta cheese ripening. Based on investigations (Visser and de Groot-Moster 1977; van der Berg and Extercate 1993), it appears that the extended α_{s1}-casein hydrolysis during early ripening is due to the residual rennet; because of the low pH and high moisture of the feta curd, residual rennet is higher than for other cheese varieties (Samal

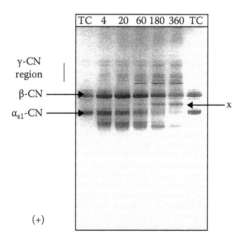

FIGURE 21.1 Electrophoretogram of feta cheese at various ages (4, 20, 60, 180, and 360 days). TC, total casein. (From Michaelidou, A. 1997. Investigation of protein hydrolysis during Feta cheese ripening, PhD Dissertation (in Greek), Aristotle University, Thessaloniki, Greece, p. 166. With permission.)

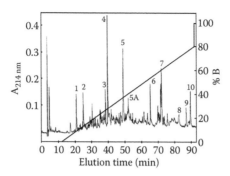

FIGURE 21.2 RP-HPLC profile of the water-soluble fraction of 6-month-old feta showing the peaks collected and identified. Eluent A was 1 mL of trifluoroacetic acid (TFA)/L of deionized water. Eluent B was 0.9 mL of TFA, 399.1 mL of deionized water, and 600 mL of acetonitrile/L. Gradient: 0–10 min, eluent A; 10–90 min, 0%–80% eluent B; 90–100 min, 100% eluent B. The flow rate was 0.8 mL/min. The absorbance of the eluate was monitored at 214 nm. Peak numbers correspond to the following compounds: 1, Tyr; 2, Phe; 3, α_{s1}-CN (f 4–14) and α_{s1}-CN (f 40–49); 4, α_{s1}-CN (f 1–14); 5, β-CN (f 164–180), α_{s1}-CN (f 102–109), and α_{s1}-CN (f 24–30); 5A, κ-CN (f 96–105) and α_{s1}-CN (f 91–98); 6, α_{s1}-CN (f 24–32); 7, β-CN (f 191–205); 8, α-LA; 9, β-LG (f 16–?); 10, β-LG. (Reprinted from Michaelidou, A. et al. 1998. *J. Dairy Sci.* 81, 3109–3116. With permission.)

et al. 1993; Fox and McSweeney 1998; de Roos et al. 1998). Additionally, the environmental conditions favor chymosin activity. Although chymosin is active on both α_{s1}- and β-casein, β-casein degradation is strongly retarded by the presence of salt (Fox and Walley 1971). On the other hand, the activity of plasmin is relatively low due to the low pH and the high NaCl content. The role of plasmin is mainly important in the ripening of cheeses in which rennet enzymes have been destroyed by high cooking temperature. This is not the case, however, with feta. As a result, α_{s1}-casein is the main substrate of the proteolytic activity; 40%–50% is hydrolyzed in 15–20 days (Michaelidou 1997; Valsamaki et al. 2000), whereas the percentage of hydrolyzed β-casein is less than 10%. The later breakdown of α_{s1}-casein could be attributed to the synergistic action of the residual rennet and the cell-bound proteinases of starter bacteria.

The RP-HPLC profile of the water-soluble fraction confirms the significant role of residual rennet in feta cheese ripening (Figure 21.2). Most of the identified peptides of that fraction originate in the N-terminal half of the α_{s1}-casein. The cleavage of the Phe[23]–Val[24], Phe[32]–Arg[33], Leu[98]–Leu[99], Leu[101]–Lys[102], and Leu[109]–Glu[110] bonds could be attributed to chymosin action. β-Casein is the main substrate for plasmin activity, which is reflected in the production of γ-caseins from its C-terminal part. However, the two identified peptides originating in β-casein in feta cheese extract apparently result from the cleavage of the Leu[190]–Tyr[191] and Ile[205]–Leu[206] bonds by chymosin (Michaelidou et al. 1998). Although the isolation of κ-casein peptides from cheese has not been reported, a peptide corresponding to κ-casein (f 96–105) was also detected in the water-soluble fraction of feta cheese. It presumably originated in *para*-κ-casein, and its formation could be the result of the action of lactococcal proteinase at Met[95]–Ala[96], which exhibits the characteristics of a susceptible cleavage site for such an enzyme (Reid et al. 1994). The total area of the RP-HPLC chromatograms is significantly ($p < 0.01$) correlated with cheese age, WSN%TN, TCA-SN%TN, and residual α_{s1}-casein content (Moatsou et al. 2002).

21.5.1.3 Small Peptides and Free Amino Acids

Important amounts of low-molecular-weight protein fragments (small peptides and free amino acids, FAAs) are produced within 15–20 days. At the end of that period, more than 65% of the water-soluble nitrogen is soluble in 12% trichloroacetic acid (TCA; Anifantakis 1991; Michaelidou 1997; Katsiari et al. 2000a), indicating a significant accumulation of FAAs and small peptides with 2–20 amino acid residues (Yvon et al. 1989). FAA content increases by about five times during that period (Valsamaki et al.

2000). It should be noted here that the FAA content of 1-day-old feta is significantly different from zero (Valsamaki et al. 2000). This is attributed to an accelerated amino acid release at the day of manufacture, when starter cultures are incubated for more than 30 min at temperatures favorable for their development and activity (Bütikofer and Fuchs 1997).

Because FAA release is attributed to the action of microbial peptidases (Visser 1993), the starter culture seems responsible for the massive production of FAAs during the first 15 days of ripening. However, the role of the NSLAB could not be overlooked. Sometimes, an insufficient thermal treatment of the cheese milk, combined with the technology of feta cheese (dry salting and further manipulations for 3–4 days), permits the curd surface to be contaminated with microorganisms of the cheese plant environment, and these can be the sources of NSLAB. The high temperature of the ripening room favors the development and proteolytic activity of both starter bacteria and NSLAB. Lactococcal peptidases are intracellular, and their action indicates cell lysis (Wilkinson et al. 1994). The high rate of FAA production during the first 2 weeks of ripening may be attributed to an early cell lysis, probably due to the low pH and high salt content of the curd.

After 20 days, the rate of proteolysis slows. Table 21.3 shows that the increase of the water-soluble and 12% TCA-soluble nitrogenous fraction (expressed as a percentage of the total N content of the cheese) is quite slow (Alichanidis et al. 1984; Michaelidou 1997; Valsamaki et al. 2000; Katsiari et al. 2000a). This is probably due to the low-temperature storage of feta cheese and to the already-mentioned diffusion of these compounds into the brine. Nevertheless, the hydrolysis of large peptides continues; at 60 days of ripening, the fraction soluble in 12% TCA constitutes 73%–80% of the water-soluble nitrogen (Moatsou et al. 2002). Consequently, the composition of the 12% TCA-soluble fraction continuously changes. This extract is enriched with peptides of a molecular weight lower than 600 Da and FAAs (Jarrett et al. 1982); the content in peptides and FAAs significantly increases with the ripening time. The FAA content of 2-month-old feta cheese varies from 2000 to 7000 mg/kg cheese. Leucine is the major FAA, ranging from 14% to 24% of the total, but levels of valine, lysine, and phenylalanine are also important (Alichanidis et al. 1984; Valsamaki et al. 2000; Katsiari et al. 2000a). Another significant increase in FAA content occurs after 3 months of ripening, probably due to massive starter flora and NSLAB lysis and the availability of substrate (small peptides) for the action of the peptidases released (Valsamaki et al. 2000).

21.5.1.4 Biogenic Amines

Biogenic amines in cheese are mainly generated from the decarboxylation of free amino acids by adventitious microorganisms rather than by starter bacteria (Joosten and Stadhouders 1987). The sheep and goat milk used for feta cheesemaking sometimes receives a thermal treatment below pasteurization conditions. In addition, the manipulation of the curd during dry salting facilitates a contamination by various bacterial species, eventually possessing decarboxylating properties. This could therefore cause the production of high amounts of biogenic amines. However, investigations have shown that the average total amine content of feta (390 mg/kg) is much lower than that reported for cheeses suspected in outbreaks of food poisoning. It seems, therefore, that the characteristic features of feta (low pH, high salt content, ripening, and storage in brine, and not extended proteolysis) create an environment unfavorable for amine accumulation (Valsamaki et al. 2000). Tyramine is the main biogenic amine in mature samples (about 42% of the total).

21.5.2 Lipolysis

Lipolysis is important for the development of the characteristic flavor of feta cheese. The free fatty acid (FFA) content in mature cheese ranges from 1000 to 6000 mg/kg (Alichanidis et al. 1984; Vafopoulou et al. 1989; Katsiari et al. 2000b; Moatsou et al. 2004). As with proteolysis, fat hydrolysis is more intense during the preripening period. The short-chain FFA (C2:0–C10:0) content increases more than two times from days 3 to 18 of ripening (Georgala et al. 2005).

The extent of lipolysis varies with the kind of rennet. Cheeses made with artisanal rennets, which are rich in lipases and pregastric esterase, have significantly ($p < 0.05$) higher FFA content, especially butanoic and decanoic acids, than cheeses made with commercial calf rennet (Moatsou et al. 2004). Other factors also affect FFA content such as the kinds of starter bacteria and NSLAB, and the temperature in draining and dry salting. High temperatures during draining enhance the lipolytic activity (Kandarakis et al. 2001). Short-chain volatile acids (C_2–C_8) constitute 30%–50% of the total FFA in mature cheese; among them, acetic acid is dominant at 29%–47% of the total FFA (Alichanidis et al. 1984; Vafopoulou et al. 1989; Georgala et al. 1999; Katsiari et al. 2000b; Kandarakis et al. 2001; Kondyli et al. 2002).

Dietary important FFAs, such as conjugated linoleic acid and monounsaturated fatty acids, were also determined in feta cheese in relatively high quantities (Zlatanos et al. 2002).

21.5.3 Other Flavor Compounds

Due to the lack of pressing, the curd of feta cheese retains a high moisture level and provides to the starter microorganisms and NSLAB enough lactose for fermentation.

Quick lactose metabolism to lactate and a subsequent drop of the pH to about 4.6 is vital for the quality of the cheese (Fox et al. 1993). Later, the high acidity and salt content of the cheese create an environment unfavorable for the growth and metabolism of starter bacteria. However, NSLAB survive and may continue to metabolize lactose, citrate, and other cheese constituents to flavor compounds (Abd El-Salam et al. 1993).

The main volatile compound of feta is ethanol, followed by acetic acid, acetaldehyde, acetoin, and other short-chain alcohols, aldehydes, and ketones (Horwood et al. 1981; Sarantinopoulos et al. 2002; Kondyli et al. 2002). High concentrations of ethanol (over 1000 mg/kg) in mature cheese were determined by dynamic headspace analysis. Ranges for acetaldehyde are 4–15 mg/kg; higher values are found in cheeses for which yogurt microorganisms were used as starter culture (Vafopoulou et al. 1989; Litopoulou-Tzanetaki et al. 1993; Sarantinopoulos et al. 2002). Diacetyl is not found in mature feta (Vafopoulou et al. 1989; Sarantinopoulos et al. 2002).

21.6 Defects

Early gas blowing is, by far, the most common defect for feta cheese. It occurs very rapidly, usually 22–48 h after curdling, and the visible outcome is the presence of numerous, generally small, holes within the cheese mass. It is much more frequent in cheeses made with raw milk of poor bacteriological quality and in cases of high environment temperature of the draining and salting rooms. Mainly, coliform bacteria are responsible for this defect, usually *Enterobacter aerogenes*, *E. coli*, and *Klebsiella aerogenes*, which ferment lactose into lactic acid, carbon dioxide (CO_2), and hydrogen (H_2; Alichanidis 2007).

Another type of gas blowing is the swelling of the cheese containers, not of the cheese. The problem is related to postpasteurization contaminations by several groups of microorganisms, usually heterofermentative lactic acid bacteria, producing gases that cause the inflation of tins or the ballooning of the plastic bags used for cheese packaging (Alichanidis 2007).

Moldiness of the surface of feta cheese is a rare defect and will never appear if the cheeses are continuously and completely submerged in the brine of the package. The most important consequence of moldiness is the formation of mycotoxins (e.g., aflatoxins), but no aflatoxin M_1 was detected in mature feta cheese (Karaioannoglou et al. 1989; Kaniou-Grigoriadou et al. 2005).

In some cases, the body of the cheese becomes soft by taking up water from the brine in the package, and the volume of the cheese blocks expands. The cheese blocks may stick together, sometimes making the effort to take them out of the container one at a time without damaging them impossible. The defect becomes serious when cheeses with insufficient acidity development and with high moisture content are prematurely transferred from the warm room (16°C–18°C) to the cold room (4°C–5°C) for slow ripening and storage (Alichanidis 2007).

Finally, ropiness is a defect of the brine in the package and is not usually associated with undesirable organoleptic properties, although it affects the appearance of the cheese and thus predisposes the consumer negatively. The rising of the brine viscosity is due to exopolysaccharides, which are compounds produced by some strains of mesophilic or thermophilic lactic acid bacteria (Alichanidis 2007).

REFERENCES

Abd El-Salam, M. H., Alichanidis, E., and Zerfiridis, G. K. 1993. Domiati and Feta type cheeses. In: *Cheese: Chemistry, Physics and Microbiology*, Vol. 2, 2nd edition, edited by P. F. Fox, pp. 301–350. London: Chapman and Hall.

Alichanidis, E. 2007. Cheeses ripened in brine. In: *Cheese Problems Solved*, edited by P. L. H. McSweeney, pp. 330–342. Cambridge, England: Woodhead.

Alichanidis, E., Anifantakis, E. M., Polychroniadou, A., and Nanou, M. 1984. Suitability of some microbial coagulants for Feta cheese manufacture. *J. Dairy Res.* 51:141–147.

Alichanidis, E. and Polychroniadou, A. 1996. Special features of dairy products from ewe and goat milk from the physicochemical and organoleptic point of view. In: *Production and Utilization of Ewe and Goat Milk*, edited by R. K. Robinson, P. Kastanas, F. Vallerand, J. C. Le Jaouen, and A. Tamime, pp. 21–43. Brussels: International Dairy Federation.

Anifantakis, E. M. 1991. *Greek Cheeses: A Tradition of Centuries*, pp. 27–42. Athens: National Dairy Committee of Greece.

Anonymous. 1992. Regulation 2081/92/EC on the protection of geographical indications and designations of origin for agricultural products and foodstuffs. *Official Journal of the European Commission* L 208, pp. 1–8.

Bintsis, T., Litopoulou-Tzanetaki, E., Davies, R., and Robinson, R. K. 2000. Microbiology of brines used to mature Feta cheese. *Int. J. Dairy Technol.* 53:106–112.

Bintsis, T., Vafopoulou-Mastrojiannaki, A., Litopoulou-Tzanetaki, E., and Robinson, R. K. 2003. Protease, peptidase and esterase activities by lactobacilli and yeast isolates from Feta cheese brine. *J. Appl. Microbiol.* 95:68–77.

Bütikofer, U. and Fuchs, D. 1997. Development of free amino acids in Appenzeller, Emmentaler, Gruyère, Racklette, Sbrinz and Tilsiter cheese. *Lait* 77:91–100.

de Roos, A. L., Geurts, T. J., and Walstra, P. 1998. On the mechanism of rennet retention in cheese. Bulletin No. 332:15–19. Brussels: International Dairy Federation.

Fox, P. F., Law, J., McSweeney, P. L. H., and Wallace, J. 1993. Biochemistry of cheese ripening. In: *Cheese: Chemistry, Physics and Microbiology*, Vol. 1, 2nd edition, edited by P. F. Fox, pp. 389–438. London: Chapman and Hall.

Fox, P. F. and McSweeney, P. L. H. 1998. *Dairy Chemistry and Biochemistry*, pp. 403–419. London: Blackie Academic & Professional.

Fox, P. F. and Walley, B. F. 1971. Influence of sodium chloride on the proteolysis of casein by rennet and by pepsin. *J. Dairy Res.* 38:165–170.

Georgala, A., Kandarakis, I. G., Kaminarides, S. E., and Anifantakis, E. M. 1999. Volatile free fatty acid content of feta and white-brined cheeses. *Aust. J. Dairy Technol.* 54:5–8.

Georgala, A., Moschopoulou, E., Aktypis, A., Massouras, T., Zoidou, E., Kandarakis, I., and Anifantakis, E. 2005. Evolution of lipolysis during the ripening of traditional Feta cheese. *Food Chem.* 93:73–80.

Horwood, J. F., Lloyd, G. T., and Stark, W. 1981. Some flavour components of feta cheese. *Aust. J. Dairy Technol.* 36:34–37.

International Dairy Federation (IDF). 2009. World Dairy Situation. Bulletin no. 438:39–40. Brussels: International Dairy Federation, p. 107.

Jarrett, W. D., Aston, J. W., and Dulley, J. R. 1982. A simple procedure for estimating free amino acids in cheddar cheese. *Aust. J. Dairy Technol.* 37:55–58.

Joosten, H. M. L. J. and Stadhouders, J. 1987. Conditions allowing the formation of biogenic amines in cheese—Part 1: Decarboxylative properties of starter bacteria. *Neth. Milk Dairy J.* 41:247–258.

Kalantzopoulos, G. C. 1993. Cheeses from ewes' and goats' milk. In: *Cheese: Chemistry, Physics and Microbiology*, Vol. 2, 2nd edition, edited by P. F. Fox, pp. 507–543. London: Chapman and Hall.

Kandarakis, I., Moatsou, G., Georgala, A. I. K., Kaminarides, S., and Anifantakis, E. 2001. Effect of draining temperature on the biochemical characteristics of Feta cheese. *Food Chem.* 72:369–378.

Kaniou-Grigoriadou, I., Eleftheriadou, A., Mouratidou, T., and Katikou, P. 2005. Determination of aflatoxin M_1 in ewe's milk samples and the produced curd and Feta cheese. *Food Control* 16:257–261.

Karaioannoglou, P. G., Mantis, A., Koufidis, D., Koidis, P., and Triantafillou, J. 1989. Occurrence of aflatoxin M_1 in raw and pasteurized milk and in Feta and Teleme cheese samples. *Milchwissenschaft* 44:746–748.

Katsiari, M. C., Alichanidis, E., Voutsinas, L. P., and Roussis, I. G. 2000a. Proteolysis in reduced sodium Feta cheese made by partial substitution of NaCl by KCl. *Int. Dairy J.* 10:635–646.

Katsiari, M. C., Voutsinas, L. P., Alichanidis, E., and Roussis, I. G. 2000b. Lipolysis in reduced sodium Feta cheese made by partial substitution of NaCl by KCl. *Int. Dairy J.* 10:369–373.

Katsiari, M. C., Voutsinas, L. P., Alichanidis, E., and Roussis, I. G. 1997. Reduction of sodium content in Feta cheese by partial substitution of NaCl by KCl. *Int. Dairy J.* 7:465–472.

Kondyli, E., Katsiari, M. C., Masouras, T., and Voutsinas, L. P. 2002. Free fatty acids and volatile compounds of low-fat Feta-type cheese made with commercial adjunct culture. *Food Chem.* 79:199–205.

Litopoulou-Tzanetaki, E., Tzanetakis, N., and Vafopoulou-Mastrojiannaki, A. 1993. Effect of the type of lactic starter on microbiological, chemical and sensory characteristics of Feta cheese. *Food Microbiol.* 10:31–41.

Mallatou, H., Pappas, C. P., and Voutsinas, L. P. 1994. Manufacture of Feta cheese from sheep's milk, goat's milk or mixtures of these milks. *Int. Dairy J.* 4:641–664.

Manolopoulou, E., Sarantinopoulos, P., Zoidou, E., Aktypis, A., Moschopoulou, E., Kandarakis, I. G., and Anifantakis, E. M. 2003. Evolution of microbial populations during traditional Feta cheese manufacture and ripening. *Int. J. Food Microbiol.* 82:153–161.

Michaelidou, A. 1997. Investigation of protein hydrolysis during Feta cheese ripening. PhD Dissertation (in Greek). Aristotle University, Thessaloniki, Greece, p. 166.

Michaelidou, A., Alichanidis, E., Urlaub, H., Polychroniadou, A., and Zerfiridis, G. 1998. Isolation and identification of some major water-soluble peptides in Feta cheese. *J. Dairy Sci.* 81:3109–3116.

Michaelidou, A., Alichanidis, E., Polychroniadou, A., and Zerfiridis, G. 2005. Migration of water-soluble nitrogenous compounds of Feta cheese from the cheese blocks into the brine. *Int. Dairy J.* 15:663–668.

Ministry of Economics. 1998. Chapter IX, Article 83. Greek Food Code. pp. 615–618. Athens, Greece: Ministry of Economics.

Moschopoulou, E., Kandarakis, I., and Anifantakis, E. 2007. Characteristics of lamb and kid artisanal rennet used for traditional Feta cheese manufacture. *Small Ruminant Res.* 72:237–241.

Moatsou, G., Massouras, T., Kandarakis, I., and Anifantakis, E. 2002. Evolution of proteolysis during ripening of traditional Feta cheese. *Lait* 82:601–611.

Moatsou, G., Moschopoulou, E., Georgala, Aik., Zoidou, E., Kandarakis, I., Kaminarides, S., and Anifantakis, E. 2004. Effect of artisanal liquid rennet from kids and lambs abomasa on the characteristics of Feta cheese. *Food Chem.* 88:517–525.

Muir, D. D., Horn, D. S., Law, A. J. R., and Sweetsur, A. W. M. 1993. Ovine milk—Part 2: Seasonal changes in indices of stability. *Milchwissenschaft* 48:442–445.

Polychroniadou, A. 1988. A simple procedure using trinitrobenzenesulphonic acid for monitoring proteolysis in cheese. *J. Dairy Res.* 55:585–596.

Polychroniadou, A. 1994. Objective indices of maturity of Feta and Teleme cheese. *Milchwissenschaft* 49:376–379.

Polychroniadou, A. and Vafopoulou, A. 1985. Variations of major mineral constituents of ewe milk during lactation. *J. Dairy Sci.* 68:147–150.

Rantsiou, K., Urso, R., Dolci, P., Comi, G., and Cocolin, L. 2008. Microflora of Feta cheese from four Greek manufacturers. *Int. J. Food Microbiol.* 126:36–42.

Reid, J. R., Coolbear, T., Pillidge, C. J., and Pritchard, G. G. 1994. Specificity of hydrolysis of bovine *n*-casein by cell envelope-associated proteinases from *Lactococcus lactis* strains. *Appl. Environ. Microbiol.* 60:801–806.

Richardson, B. C., Creamer, L. K., Pearce, K. N., and Munford, R. E. 1974. Comparative micelle structure—Part II: Structure and composition of casein micelles in ovine and caprine milk as compared with those in bovine milk. *J. Dairy Res.* 41:239–247.

Samal, P. K., Pearce, K. N., Bennett, R. J., and Dunlop, F. P. 1993. Influence of residual rennet and proteolysis on the exudation of whey from Feta cheese during ripening. *Int. Dairy J.* 3:729–745.

Sarantinopoulos, P., Kalantzopoulos, G., and Tsakalidou, E. 2002. Effect of *Enterococcus faecium* on microbiological, physicochemical and sensory characteristics of Greek Feta cheese. *Int. J. Food Microbiol.* 76:93–105.

Storry, J. E., Grandison, A. S., Millard, D., Owen, A.-J., and Ford, G. D. 1983. Chemical composition and coagulation properties of renneted milks from different breeds and species of ruminant. *J. Dairy Res.* 50:215–229.

Turhan, M. and Kaletunç, G. 1992. Modeling of salt diffusion in white cheese during long-term brining. *J. Food Sci.* 57:1082–1085.

Tzanetakis, N. and Litopoulou-Tzanetaki, E. 1992. Changes in numbers and kinds of lactic acid bacteria in Feta and Teleme, two Greek cheeses from ewes' milk. *J. Dairy Sci.* 75:1389–1393.

Tzanetakis, N., Hatzikamari, M., and Litopoulou-Tzanetaki, E. 1998. Yeasts of the surface microflora of Feta cheese. In: *Yeasts in the Dairy Industry: Positive and Negative Aspects*, edited by M. Jacobsen, J. Narvhus, and B. C. Viljoen, pp. 34–43. Brussels: International Dairy Federation.

Vafopoulou, A., Alichanidis, E., and Zerfiridis, G. 1989. Accelerated ripening of Feta cheese, with heatshocked cultures or microbial proteinases. *J. Dairy Res.* 56:285–296.

Vafopoulou-Mastrojiannaki, A., Litopoulou-Tzanetaki, E., and Tzanetakis, N. 1990. Effect of *Pediococcus pentosaceus* on ripening changes of Feta cheese. *Microbiol. Aliments. Nutr.* 8:53–62.

Valsamaki, K., Michaelidou, A., and Polychroniadou, A. 2000. Biogenic amine production in Feta cheese. *Food Chem.* 71:259–266.

van der Berg, G. and Extercate, F. A. 1993. Technological parameters involved in cheese ripening. *Int. Dairy J.* 3:485–507.

Visser, S. 1993. Proteolytic enzymes and their relation to cheese ripening and flavour: An overview. *J. Dairy Sci.* 76:329–350.

Visser, S. 1998. Enzymatic breakdown of milk proteins during cheese ripening. Bulletin No. 332:20–24. Brussels: International Dairy Federation.

Visser, F. M. W. and de Groot-Moster, A. E. A. 1977. Contribution of enzymes from rennet, starter bacteria and milk to proteolysis and flavour development in Gouda cheese—Part 4: Protein breakdown—A gel electrophoretical study. *Neth. Milk Dairy J.* 31:247–264.

Voutsinas, L. P., Delegiannis, C., Katsiari, M. C., and Pappas, C. 1988. Chemical composition of Boutsico ewe milk during lactation. *Milchwissenschaft* 43:766–771.

Wilkinson, A., Guinee, T. P., O'Callaghan, D. M., and Fox, P. F. 1994. Autolysis and proteolysis in different strains of starter bacteria during cheddar cheese ripening. *J. Dairy Res.* 61:249–262.

Xanthopoulos, V., Hatzikamari, M., Adamidis, T., Tsakalidou, E., Tzanetakis, N., and Litopoulou-Tzanetaki, E. 2000. Heterogeneity of *Lactobacillus plantarum* isolates from Feta cheese throughout ripening. *J. Appl. Microbiol.* 88:1056–1064.

Yvon, M., Chabanet, C., and Pelissier, J. P. 1989. Solubility of peptides in trichloroacetic acid (TCA) solution. *Int. J. Pept. Protein Res.* 34:166–176.

Zlatanos, S., Laskaridis, K., Feist, C., and Sagredos, A. 2002. CLA content and fatty acid composition of Greek Feta and hard cheeses. *Food Chem.* 78:471–477.

22

Reggianito Cheese: Hard Cheese Produced in Argentina

Guillermo A. Sihufe, Amelia C. Rubiolo, and Susana E. Zorrilla

CONTENTS

22.1 Introduction

Argentina has a strong tradition in the production and consumption of dairy products, with their annual cheese consumption per capita being similar to several European countries. In 2009, 500,000 tons of cheese was produced in Argentina with an annual consumption per capita of 11.75 kg (M.A.G. y P. 2010). In the same year, Argentina was the 13th cow milk–producing country and the 10th producing cheese country in the world rankings (FAO 2010).

Resulting from the large numbers of Italian immigrants in Argentina in late nineteenth and early twentieth centuries, hard cheeses similar in characteristics and manufacturing procedures to Italian hard cheeses are produced (Zalazar et al. 1999). Reggianito cheese is the most important hard cheese variety manufactured in Argentina. It has a higher moisture content, smaller size, and shorter ripening period than Italian hard cheeses. It is one of the well-known Argentinean cheeses because it is exported to different countries. It is worth mentioning that Argentina exported 7138 tons of hard cheeses in 2009, with the United States, Brazil, Chile, and Russia being the main export destinations (M.A.G. y P. 2010).

22.2 Manufacture of Reggianito Cheese

22.2.1 Milk

Reggianito cheese is made from cow milk, raw or pasteurized, with the milk being mainly obtained from pasture-fed animals with dietary supplements (Serrano 2008). The fat content of the cheese milk must be adjusted to satisfy the minimum fat-in-dry matter of Reggianito cheese according to the Food Code of Argentina. Most manufacturers normally made Reggianito cheese from pasteurized milk. After cooling to 31°C–34°C, $CaCl_2$ is added to a final concentration of 0.02% (w/v) before the addition of the starter (Gallino 1994). The initial acidity of milk must be 14°D–18°D, whereas the initial pH must be 6.6–6.75 (Serrano 2008).

22.2.2 Starter

The starter medium is whey-based. Dairy plants propagate starters from a previous batch of fermented product to inoculate a new batch. The whey is incubated during 18–22 h, reaching an acidity of 130°D–140°D and a pH of 3.4–3.5. The volume of the starter media inoculated into the milk depends on the milk–starter acidity, and it may vary between 3% and 4% v/v. The starter addition increases the initial milk acidity by 4°D (Gallino 1994; Serrano 2008). This natural whey starter is mainly composed of *Lactobacillus helveticus* and *L. delbrueckii* subsp. *lactis* (Reinheimer et al. 1996). Selected lyophilized or frozen cultures can also be used (Serrano 2008).

22.2.3 Coagulation

The bovine pepsin is the coagulant generally used, although rennets derived from fermentation are also used (Gallino 1994). The coagulation time is typically about 12–25 min. When the curd is considered ready for cutting, it is cut into grains with a size of 1–3 mm, with the whey acidity being 6°D–7°D lower than the acidity of the used milk (Serrano 2008).

22.2.4 Cooking

Following coagulation and cutting of curd, the curd is cooked and stirred to expel whey through syneresis in two steps (Serrano 2008):

1. The temperature is increased from 31°C–34°C to 45°C at 0.5°C min^{-1}.
2. If lyophilized starter was used, the temperature is increased from 45°C to 49°C at 1°C min^{-1}. If natural whey starter was used, the temperature is increased from 45°C to 51°C at 1°C min^{-1}.

22.2.5 Pressing

The curd grains are covered with whey to create a pre-pressing step that can be held for 15–20 min. After having been molded, the curd is subjected to final pressing for 12 h using pneumatic pressing systems as follows (Serrano 2008):

1. 2.0 kg cm^{-2} for 1 h
2. 2.5 kg cm^{-2} for 1 h
3. 3.0 kg cm^{-2} for 1 h
4. 3.5 kg cm^{-2} for 9 h

At the end of the pressing step, the cheese pH must be 4.95–5.1.

22.2.6 Salting

Cheeses are salted by immersion in brine solutions (23°Bé, < 30°D, pH 4.95–5.1) during 8–9 days at 12°C–14°C. After salting, cheeses are kept at 10°C–15°C and 85% relative humidity for 24 h (Serrano 2008).

22.2.7 Ripening

The Food Code of Argentina (CAA 2006) indicates a minimum ripening period of 6 months. It is ripened at 10°C–15°C and 82%–86% relative humidity (Serrano 2008). Cheeses are generally ripened with rinds; they can be covered with a paraffin coating or plastic emulsion mainly to avoid external mold contamination. Rindless cheeses can be covered with plastic film or a shrinkable plastic bag; during ripening, they are turned at regular intervals.

22.3 Gross Composition of Reggianito Cheese

CAA (2006) specifies that Reggianito cheese must have a low moisture content (<35.9 g/100 g cheese) and a minimum fat-in-dry matter of 32 g/100 g dry matter. A typical gross composition of Reggianito cheese at the end of the salting stage is shown in Table 22.1 (Sihufe et al. 2010a). During ripening, moisture content shows a decreasing trend with ripening time, which is typical for cheeses ripened without wrapping. On the other hand, for cheeses salted by brine immersion, salt is redistributed until a uniform concentration is reached throughout the cheese. Thus, after 6 months of ripening, Sihufe et al. (2010a) observed a cheese composition of approximately 35 g moisture/100 g cheese and 5 g NaCl/100 g moisture. Wolf et al. (2010) reported characteristic values of 18 samples of Reggianito cheese manufactured at six different dairy industries (Table 22.2).

TABLE 22.1

Gross Composition and pH of Reggianito Cheese at the Beginning of Ripening

Parameter	Average Value
Moisture (g/100 g cheese)	39.6 ± 0.5
Protein (g/100 g dry matter)	52.2 ± 2.0
Fat (g/100 g dry matter)	39.2 ± 0.4
NaCl (g/100 g moisture)	0.25 ± 0.08
pH	5.39 ± 0.02

Source: Sihufe G.A. et al., *J Sens Stud*, 25, 94–107, 2010a.

TABLE 22.2

Gross Composition and pH of 18 Samples of Reggianito Cheese from 6 Commercial Brands

Parameter	Average Value
Moisture (g/100 g cheese)	32.9 ± 1.6
Protein (g/100 g dry matter)	48.1 ± 3.6
Fat (g/100 g dry matter)	40.9 ± 3.5
NaCl (g/100 g moisture)	6.5 ± 1.9
pH	5.4 ± 0.1

Source: Wolf I.V. et al., *Food Res Int*, 43, 1204–1211, 2010. With permission.

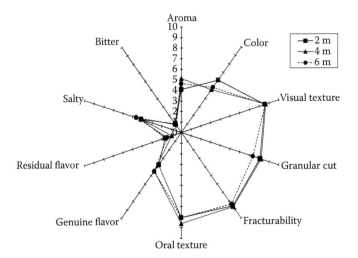

FIGURE 22.1 Average sensory attributes for Reggianito cheese ripened at different times (2, 4, and 6 months), 12°C, and 85% relative humidity. (From Sihufe G.A. et al., *J Sens Stud*, 25, 94–107, 2010. With permission.)

22.4 Sensory Properties of Reggianito Cheese

Reggianito cheese has a cylindrical shape (approximately 25 cm diameter × 15 cm height) and weighs about 7 kg (Serrano 2008). The rind is smooth, firm, and closed, and its characteristic color is pale yellow. The cheese has a compact, crumbly, and grainy texture (CAA 2006). It has no cracks or openness; only two openness (smaller than 3 mm) per cheese are allowed (Serrano 2008). The characteristic cheese color is light yellow, and the cheese flavor is slightly piquant and salty (CAA 2006).

There are a few studies related to sensorial characterization of Reggianito cheese. Hough et al. (1994) developed a sensory profile of 52 descriptors covering the following attributes: external appearance; color; visual, manual, and oral texture; aroma; and flavor. Hough et al. (1996) correlated parameters obtained from an instrumental compression test with visual, manual, and oral texture descriptors for Reggianito grating cheese. Sihufe et al. (2010a) analyzed the sensory characteristics of Reggianito cheese through a quantitative descriptive analysis carried out by a trained sensory panel. In Figure 22.1, the average values of the sensory attributes studied during the Reggianito cheese ripening at 12°C and 85% relative humidity are shown.

22.5 Biochemistry of Reggianito Cheese Ripening

22.5.1 Lipolysis

Lipolysis is important for the development of the characteristic flavor of Reggianito cheese. Free fatty acid (FFA) profiles in Reggianito cheese were determined by different authors (Perotti et al. 2005; Sihufe et al. 2007). The concentration values of nine FFAs determined in Reggianito cheese are given in Table 22.3 (Sihufe et al. 2007). Myristic ($C_{14:0}$), palmitic ($C_{16:0}$), stearic ($C_{18:0}$), and oleic ($C_{18:1}$) acids were reported as the FFAs present in higher concentration, which is expected because these four FFAs are the main components of cow milk triacylglycerides. Taking into account that Reggianito cheese is manufactured with pasteurized milk, the lipolytic agents that can be considered are principally the enzymes of starter and nonstarter lactic acid bacteria because indigenous lipoprotein lipase from milk is extensively (but not completely) denatured during pasteurization, and the commercial rennet used for milk coagulation does not have lipolytic activity. At the end of the ripening period, the level of total FFAs is approximately 2000 mg/kg of cheese (Sihufe et al. 2007; Wolf et al. 2010), which is

TABLE 22.3

Average Concentrations (mg/kg Cheese) of FFAs during Ripening of Reggianito Cheese at 12°C and 85% Relative Humidity

	Ripening Time (Months)		
Fatty Acid	**2**	**4**	**6**
$C_{6:0}$	29	38	40
$C_{8:0}$	26	31	33
$C_{10:0}$	55	64	69
$C_{12:0}$	62	77	83
$C_{14:0}$	203	309	272
$C_{16:0}$	528	752	686
$C_{18:0}$	228	275	256
$C_{18:1}$	539	787	719
$C_{18:2}$	84	112	102

Source: From Sihufe G.A. et al., *Food Res Int*, 40, 1220–1226, 2007. With permission.

considered as a moderate lipolysis degree and similar to cheeses as Cheddar or Emmental (Collins et al. 2004).

Short- and intermediate-chain FFAs ($C_{4:0}$–$C_{12:0}$) have considerably lower perception thresholds than long-chain FFAs (>12 carbon atoms), and each one gives a characteristic flavor note (Collins et al. 2004). Thus, it is usually considered that short- and intermediate-chain FFAs have a significant impact on the development of the characteristic cheese flavor (Georgala et al. 2005; Lopez et al. 2006). In the case of Reggianito cheese, the percentage of short- and intermediate-chain FFAs ($C_{4:0}$ and $C_{12:0}$) related to total FFAs is approximately 12% (Wolf et al. 2010).

22.5.2 Proteolysis

Proteolysis is the most complex and important event that occurs during ripening of a great number of cheese varieties (Fox and McSweeney 1996). A part of the casein is converted by proteolysis into water-soluble nitrogenous compounds such as peptides and amino acids, which contribute to the flavor and texture. Enzymes from coagulant, milk, starter bacteria, and nonstarter microflora catalyze proteolysis in the cheese (Fox 1987).

Proteolysis during cheese ripening can be followed by the maturation index (MI), which usually is calculated as a percentage of water-soluble nitrogen at pH 4.6 of the cheese total nitrogen. In the case of Reggianito cheese, the values for MI during ripening varied from 14% to 20% (Sihufe et al. 2007). Similar values were found by Candioti et al. (2002), Hynes et al. (2003), and Perotti et al. (2004). To obtain more information about Reggianito cheese proteolysis, specific techniques such as urea-polyacrylamide gel electrophoresis (urea-PAGE), peptide analysis by reversed-phase high-performance liquid chromatography (RP-HPLC), and amino acid analysis by RP-HPLC were used (Sihufe et al. 2010b).

22.5.2.1 Primary Proteolysis

A typical urea-PAGE electrophoretogram for Reggianito cheese is shown in Figure 22.2. The fractions F1 and F2 may be products of β-casein degradation, and F3 may be associated with products of α_{s1}- or α_{s1}-I-casein degradation. The α_{s1}- and β-casein fractions decreased significantly during the first 2 months of ripening, with the effect more pronounced being for α_{s1}-casein. During that period, α_{s1}-casein was hydrolyzed more than 30%, whereas β-casein was degraded less than 10% (Sihufe et al. 2010b). Hynes et al. (2004a,b) found that the residual rennet may be one of the main proteolytic agents responsible for the production of peptides derived from α_{s1}-casein in Reggianito cheese. The authors studied the

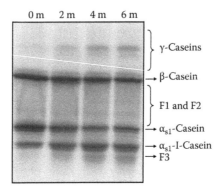

FIGURE 22.2   Urea-PAGE electrophoretogram for Reggianito cheese ripened at different times (0, 2, 4, and 6 months), 12°C, and 85% relative humidity. (Adapted from Sihufe G.A. et al., *LWT-Food Sci Tech*, 43, 247–253, 2010.)

influence of cooking temperature (45°C, 52°C, and 60°C) on milk-clotting enzyme residual activity and α_{s1}-casein hydrolysis during ripening. Rennet activity was inversely proportional to cooking temperature. The activity was quantifiable even in cheeses cooked at high temperature, suggesting renaturation or incomplete denaturation of the enzyme.

Following the degradation of α_{s1}-casein, α_{s1}-I-casein showed an important increase during the first 2 months of ripening with no changes during the rest of the ripening period (Sihufe et al. 2010b). Similar results were reported by Hynes et al. (2003).

22.5.2.2 Peptides

The information obtained from the MI is generally complemented by the RP-HPLC of the water-soluble fraction at pH 4.6. The chromatograms obtained are often referred to as "fingerprints" of the proteolytic process (Pripp et al. 1999). In Figure 22.3, a typical chromatogram of RP-HPLC of the water-soluble fraction at pH 4.6 of Reggianito cheese at 6 months of ripening is shown.

Multivariate analysis is an objective approach to analyze the data generated from chromatograms. Particularly, principal component analysis (PCA) of the peak data has been used to demonstrate difference in proteolysis in cheeses that were made using different manufacturing process (Coker et al. 2005). In the case of Reggianito cheese, Hynes et al. (2003) used this approach to compare the behavior of single and mixed cultures of *L. helveticus* selected strains with a whey-based starter culture for Reggianito cheese manufacturing and their influence on ripening. On the other hand, Sihufe et al. (2010b) used PCA to evaluate the influence of elevated ripening temperature and sampling site on proteolysis in Reggianito cheese.

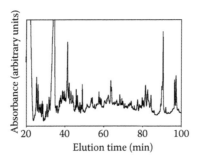

FIGURE 22.3   RP-HPLC chromatogram of the water-soluble fraction at pH 4.6 of Reggianito cheese ripened 6 months at 12°C and 85% relative humidity. Chromatographic conditions are described by Sihufe et al. (2010b). (From Sihufe G.A. et al., *LWT-Food Sci Tech*, 43, 247–253, 2010. With permission.)

22.5.2.3 Free Amino Acids

Phosphotungstic acid (PTA) leaves only amino acids and very small peptides in solution, and its use provides a good index of the progress of the late stages of proteolysis in cheese (Law 1987). In the case of Reggianito cheese, Wolf et al. (2010) determined the percentage of nitrogen soluble in 2.5% PTA of the cheese total nitrogen, ranging from 7.5% to 12.8%.

Amino acid degradation is one of the major processes for aroma formation in cheese, with aromatic amino acids (Phe, Tyr, Trp), branched-chain amino acids (Leu, Ile, Val), and Met being the major precursors of aroma compounds (Yvon and Rijnen 2001). Levels of free amino acids (FAAs) in Reggianito cheese are given in Table 22.4 (Sihufe et al. 2010b).

The principal FAAs were Glu, His, Val, Leu, and Lys, with amino acid profiles being similar to those previously reported for different cheese varieties (McSweeney and Sousa 2000; Frau et al. 1997). In cheeses ripened for 6 months and at 12°C, the individual concentration of those amino acids was higher than 1200 mg/kg cheese (Table 22.4). Taking into account its flavor-enhancing properties, Glu might contribute to the development of Reggianito cheese flavor, whereas Leu might be an important precursor of branched-chain volatile flavor compounds (Fallico et al. 2004; Sihufe et al. 2010b).

22.5.3 Volatile Compounds

In hard cheeses, flavor is the result of many enzymatic and nonenzymatic reactions such as decarboxylation, deamination, transamination, desulfurization, and cleavage of side chains that convert amino acids to aldehydes, alcohols, and acids. Also there are compounds that can be derived from other routes (e.g., lipolysis and catabolism of fatty acids; Gobbetti 2004). The volatile compounds that comprise Italian hard cheeses have been studied extensively (Qian and Burbank 2007). In the case of Reggianito cheese, Wolf et al. (2010) identified 53 compounds by solid phase microextraction, gas chromatography, and mass spectrometry/flame ionization detector. Compounds commonly reported in Italian grana-type cheeses (i.e., ketones, alcohols, acids, esters, and aldehydes) were identified in Reggianito cheese (Table 22.5).

TABLE 22.4

Average Concentrations (mg/kg Cheese) of FAAs during Ripening of Reggianito Cheese at 12°C and 85% Relative Humidity

Amino Acid	Ripening Time (months)		
	2	4	6
Asp	206	276	344
Glu	1769	2344	2867
Asn	695	831	996
Ser	213	234	341
His	908	1449	1455
Gln	786	902	905
Gly	144	211	340
Ala	320	353	379
Tyr	670	788	944
Met	341	464	543
Val	759	1044	1225
Phe	526	766	844
Ile	455	729	908
Leu	1248	1547	1627
Lys	1806	2225	2352

Source: From Sihufe G.A. et al., *LWT-Food Sci Tech*, 43, 247–253, 2010. With permission.

TABLE 22.5

Volatile Compounds Found in 18 Samples of 6 Commercial Brands of Reggianito Cheese

Chemical Family	Volatiles Common to Majority of Samples	Most Abundant Volatiles
Ketones	Methylketones from C_3 to C_9	Propanone, 2-pentanone + diacetyl, 2-heptanone, acetoin
Alcohols	Ethanol, 2-propanol, 2-pentanol, 2-heptanol, 1-butanol, 1-pentanol, 3-methyl 1-butanol	Ethanol
Esters	Ethyl acetate, ethyl butanoate, isopropyl hexanoate	Ethyl butanoate
Acids	Saturated fatty acids of even number of carbon atoms from C_2 to C_{10}	Acetic acid, butanoic acid
Aldehydes		Acetaldehyde, 2-methyl butanal, 3-methyl butanal

Source: Wolf I.V. et al., *Food Res Int*, 43, 1204–1211, 2010. With permission.

22.6 Acceleration of Reggianito Cheese Ripening

Generally, cheese ripening is a slow process. The capital cost associated with the storage facilities and the operating cost associated with the control of temperature and relative humidity of ripening rooms turn the ripening process into an expensive one. Therefore, acceleration of cheese ripening has received considerable attention. Various alternatives have been used to accelerate the ripening process of cheese including use of elevated ripening temperatures, addition of exogenous enzymes or attenuated starters, use of adjunct cultures, genetic modification of starter bacteria, and high-pressure treatment (Upadhyay and McSweeney 2003).

Undoubtedly, elevated ripening temperatures offer the most effective, and certainly the simplest and cheapest, method for the accelerating ripening (Fox et al. 1996). As expected, acceleration of cheese ripening is very attractive in the case of hard cheeses. However, it is worth mentioning that successful ripening at elevated temperatures requires careful control of cheese composition and microflora to ensure a typical flavor development (Upadhyay and McSweeney 2003).

In the case of Reggianito cheese, Sihufe et al. (2007, 2010a,b,c) studied the effect of elevated temperature on physicochemical, biochemical, and sensory characteristics, comparing results obtained from cheeses ripened at 12°C (control) with those obtained from cheeses ripened at 18°C. Elevated temperature accelerated the salt and moisture redistribution, and it increased the α_{s1}- and β-casein degradation and the peptide and amino acid formation (Sihufe et al. 2010b). Sihufe et al. (2007) reported that the total FFA content increases until 4 months of ripening in cheeses stored at 12°C, remaining almost constant the last 2 months of ripening. At 18°C, the total FFA content increases during the 6 months evaluated. However, the differences observed between cheeses ripened at different temperatures were significant only for the last 2 months of ripening. It is worth mentioning that the acceleration due to the elevated temperature in the lipolysis changes was slower than for the proteolysis changes (Sihufe et al. 2010c). In relation to the sensory analysis, although the attribute scores of cheeses ripened at 18°C were different from those ripened at 12°C, they did not result in atypical Reggianito sensory characteristics. Moreover, a Reggianito Argentino cheese ripened at 18°C during 2 months had similar sensory characteristics to a cheese ripened at 12°C during 6 months (Sihufe et al. 2010a). Finally, after studying the physicochemical, proteolysis, lipolysis, and sensory analyses as combined aspects, the authors determined that the optimal time for ripening Reggianito cheese at 18°C ranged between 2 and 3 months (Sihufe et al. 2010c).

REFERENCES

CAA. 2006. Código Alimentario Argentino. [Accessed 2010 Nov 23]. Available from: http://www.anmat.gov .ar/alimentos/normativas_alimentos_caa.asp.

Candioti MC, Hynes E, Quiberoni A, Palma SB, Sabbag N, Zalazar CA. 2002. Reggianito Argentino cheese: influence of *Lactobacillus helveticus* strains from natural whey cultures on cheese making and ripening processes. *Int Dairy J* 12:923–931.

Coker CJ, Crawford RA, Johnston KA, Singh H, Creamer LK. 2005. Towards the classification of cheese variety and maturity on the basis of statistical analysis of proteolysis data—a review. *Int Dairy J* 15:631–643.

Collins YF, McSweeney PLH, Wilkinson MG. 2004. Lipolysis and catabolism of fatty acids in cheese. In: Fox PF, McSweeney PLH, Cogan TM, Guinee TP, editors. *Cheese: Chemistry, Physics and Microbiology*. 3rd ed. Volume 1. Amsterdam: Elsevier Academic Press. pp. 373–389.

Fallico V, McSweeney PLH, Siebert KJ, Horne J, Carpino S, Licitra G. 2004. Chemometric analysis of proteolysis during ripening of Ragusano cheese. *J Dairy Sci* 87:3138–3152.

FAO. 2010. Food and Agriculture Organization of the United Nations. [Accessed 2010 Nov 23]. Available from: http://faostat.fao.org/.

Fox PF. 1987. Cheese: an overview. In: Fox PF, editor. *Cheese: Chemistry, Physics and Microbiology*. Volume 1. London: Elsevier Applied Science. pp. 1–32.

Fox PF, McSweeney PLH. 1996. Proteolysis in cheese during ripening. *Food Rev Int* 12(4):457–509.

Fox PF, Wallace JM, Morgan S, Lynch CM, Niland EJ, Tobin J. 1996. Acceleration of cheese ripening. *Antonie Van Leeuwenhoek* 70:271–297.

Frau M, Massanet J, Rosselló C, Simal S, Cañellas J. 1997. Evolution of free amino acid content during ripening of Mahon cheese. *Food Chem* 60(4):651–657.

Gallino RR. 1994. Queso Reggianito Argentino: tecnología de fabricación. In: *Ciencia y tecnología de los productos lácteos*. Diagramma S.A. Santa Fe: Medios Audiovisuales y Gráficos—CERIDE. pp. 244–287.

Georgala A, Moschopoulou E, Aktypis A, Massouras T, Zoidou E, Kandarakis I, Anifantakis E. 2005. Evolution of lipolysis during the ripening of traditional Feta cheese. *Food Chem* 93:73–80.

Gobbetti M. 2004. Extra-hard varieties. In: Fox PF, McSweeney PLH, Cogan TM, Guinee TP, editors. *Cheese: Chemistry, Physics and Microbiology*. 3rd ed. Volume 2. Amsterdam: Elsevier Academic Press. pp. 51–70.

Hough G, Califano AN, Bertola NC, Bevilacqua AE, Martinez E, Vega MJ, Zaritzky NE. 1996. Partial least squares correlations between sensory and instrumental measurements of flavor and texture for Reggianito grating cheese. *Food Qual Prefer* 7(1):47–53.

Hough G, Martinez E, Barbieri T, Contarini A, Vega MJ. 1994. Sensory profiling during ripening of Reggianito grating cheese, using both traditional ripening and in plastic wrapping. *Food Qual Prefer* 5:271–280.

Hynes E, Candioti MC, Zalazar CA, McSweeney PLH. 2004a. Rennet activity and proteolysis in Reggianito Argentino hard cooked cheese. *Aust J Dairy Technol* 59(3):209–213.

Hynes ER, Aparo L, Candioti MC. 2004b. Influence of residual milk-clotting enzyme on α_{s1} casein hydrolysis during ripening of Reggianito Argentino cheese. *J Dairy Sci* 87:565–573.

Hynes ER, Bergamini CV, Suárez VB, Zalazar CA. 2003. Proteolysis on Reggianito Argentino cheeses manufactured with natural whey cultures and selected strains of *Lactobacillus helveticus*. *J Dairy Sci* 86:3831–3840.

Law BA. 1987. Proteolysis in relation to normal and accelerated cheese ripening. In: Fox PF, editor. *Cheese: Chemistry, Physics and Microbiology*. Volume 1. London: Elsevier Applied Science. pp. 365–392.

Lopez C, Maillard MB, Briard-Bion V, Camier B, Hannon JA. 2006. Lipolysis during ripening of Emmental cheese considering organization of fat and preferential localization of bacteria. *J Agr Food Chem* 54:5855–5867.

M.A.G. y P. 2010. Ministerio de Agricultura, Ganadería y Pesca de la República Argentina. [Accessed 2010 Nov 23]. Available from: http://www.minagri.gob.ar/site/index.php.

McSweeney PLH, Sousa MJ. 2000. Biochemical pathways for the production of flavour compounds in cheeses during ripening: a review. *Lait* 80:293–324.

Perotti MC, Bernal SM, Meinardi CA, Candioti MC, Zalazar CA. 2004. Substitution of natural whey starter by mixed strains of *Lactobacillus helveticus* in the production of Reggianito Argentino cheese. *Int J Dairy Technol* 57(1):45–51.

Perotti MC, Bernal SM, Meinardi CA, Zalazar CA. 2005. Free fatty acid profiles of Reggianito Argentino cheese produced with different starters. *Int Dairy J* 15:1150–1155.

Pripp AH, Shakeel-Ur-Rehman, McSweeney PLH, Fox PF. 1999. Multivariate statistical analysis of peptide profiles and free amino acids to evaluate effects of single-strain starters on proteolysis in miniature Cheddar-type cheeses. *Int Dairy J* 9:473–479.

Qian MC, Burbank HM. 2007. Hard Italian cheeses: Parmiggiano-Reggiano and Grana Padano. In: Weimer BC, editor. *Improving the Flavour of Cheese*. Boca Raton: CRC Press. pp. 421–443.

Reinheimer JA, Quiberoni A, Tailliez P, Binetti AG, Suárez VB. 1996. The lactic acid microflora of natural whey starters used in Argentina for hard cheese production. *Int Dairy J* 6:869–879.

Serrano P. 2008. Protocolo de calidad para queso Reggianito. Resolución SAGPyA N°: 16/2008. [Accessed 2010 Nov 23]. Available from: http://www.inti.gov.ar/lacteos/pdf/queso_reggianito.pdf.

Sihufe GA, Zorrilla SE, Mercanti DJ, Perotti MC, Zalazar CA, Rubiolo AC. 2007. The influence of ripening temperature and sampling site on the lipolysis in Reggianito Argentino cheese. *Food Res Int* 40:1220–1226.

Sihufe GA, Zorrilla SE, Sabbag NG, Costa SC, Rubiolo AC. 2010a. The influence of ripening temperature on the sensory characteristics of Reggianito Argentino cheese. *J Sens Stud* 25:94–107.

Sihufe GA, Zorrilla SE, Rubiolo AC. 2010b. The influence of ripening temperature and sampling site on the proteolysis in Reggianito Argentino cheese. *LWT-Food Sci Tech* 43:247–253.

Sihufe GA, Zorrilla SE, Perotti MC, Wolf IV, Zalazar CA, Sabbag NG, Costa SC, Rubiolo AC. 2010c. Acceleration of cheese ripening at elevated temperature. An estimation of the optimal ripening time of a traditional Argentinean hard cheese. *Food Chem* 119:101–107.

Upadhyay VK, McSweeney PLH. 2003. Acceleration of cheese ripening. In: Smit G, editor. *Dairy Processing: Improving Quality*. Boca Raton: CRC Press. pp. 419–447.

Wolf IV, Perotti MC, Bernal SM, Zalazar CA. 2010. Study of the chemical composition, proteolysis, lipolysis and volatile compounds profile of commercial Reggianito Argentino cheese: characterization of Reggianito Argentino cheese. *Food Res Int* 43:1204–1211.

Yvon M, Rijnen L. 2001. Cheese flavour formation by amino acid catabolism. *Int Dairy J* 11:185–201.

Zalazar C, Meinardi C, Hynes E. 1999. *Quesos típicos argentinos. Una revisión general sobre producción y características*. Santa Fe: Centro de Publicaciones de la Universidad Nacional del Litoral. 59 pp.

23

Fiore Sardo Cheese

Pietrino Deiana and Nicoletta Pasqualina Mangia

CONTENTS

23.1 General Features

"Fiore Sardo" is a hard paste cheese manufactured exclusively from whole, raw ovine milk coagulated with lamb or kid rennet paste. Since long time ago, Fiore Sardo has been recognized as a typical food product (D.P.R. 30.10.1955 N 1279) by the Italian Government and in 1996 obtained the Protected Denomination of Origin (PDO) from the EU Commission (EC Regulation 1236/1996). Based on this latter EC Regulation, Fiore Sardo cheese can be manufactured exclusively within the Sardinian region (Italy). Currently, the consortium of producers of Fiore Sardo cheese ("Consorzio per la Tutela del Formaggio Fiore Sardo"; http://www.fioresardo.it/) monitors and guarantees the adoption, by the single producers, of strict cheese-making procedures and, more generally, is committed to the valorization of the cheese. Fiore Sardo is the typical cheese of the Sardinian agro-pastoral tradition and is produced seasonally at the farm level (approximately 500 tons/year) using ovine milk from flocks mainly raised at pasture. Fiore Sardo cheese plays a significant role for the economy of rural areas of Sardinia.

23.2 Origins

The origin of Fiore Sardo cheese, together with the culture that produced it, is lost in prehistory. It has been hypothesized that the name "Fiore Sardo" originated from the plant rennet historically used for its manufacturing. In particular, it seems that the flowers of wild artichokes (*Cynara cardunculus* L.), soaked with vinegar and salt, were used in the ancient production of Fiore Sardo: artichoke contains the *cinarase* enzyme, which is able to coagulate milk. Moreover, the flower that in the past was sculpted on the bottom of the wooden molds and the flower present nowadays in the Fiore Sardo logo adopted by the Consortium of producers seem to support this hypothesis. By the way, it should be mentioned that there is not a general agreement on this, and the origin of the name "Fiore Sardo" is currently still uncertain.

The first information on the production of what can be considered as the ancient Fiore Sardo cheese is provided by different authors such as Gemelli (1776) and Manca dell'Arca (1780). Later on, Cossu (1787) described the common manufacturing procedures adopted in the traditional production of Fiore Sardo, which will remain the most produced cheese in Sardinia until 1897 when the production of Pecorino Romano, by some Roman entrepreneurs, started in the island. The introduction of the Pecorino Romano

FIGURE 23.1   Traditional manufacturing of Fiore Sardo cheese.

FIGURE 23.2   *Pinnetu*: the typical ancient shepherd's house and his first "dairy-farm."

cheese-making technology favored the adoption of some changes in the manufacturing procedures of Fiore Sardo such as the milk filtering and heating and the addition of suitable amount of rennet able to coagulate the milk within 25–30 minutes.

In the ancient times, Fiore Sardo cheese was commonly manufactured inside the so called *pinnetu* (Figure 23.1), a circular hut made up by stones and covered with tree branches and Mediterranean scrubs, which constituted at the same time the shepherd's house and his "dairy farm" (Figure 23.2).

23.3 Cheese-Making

Currently, Fiore Sardo cheese manufacturing is still performed by the sheep farmers and is exclusively carried out in farms equipped with the compulsory structural requisites foreseen by the Italian law (EU Council, 1992). Fiore Sardo cheese is manufactured using the milk from the evening milking mixed with the fresh one milked in the morning. Traditionally, starter cultures of lactic acid bacteria (LAB) were not used. The rennet employed is constituted by the abomasa, suitably processed, of lactating lambs or kids. In particular, lambs or kids are slaughtered shortly after suckling, and the abomasa is collected, dried, cleaned, grounded, and stored under salt in a dry and cool place. Such a processing preserves the proteolytic and lipolytic enzymes present in the rennet.

A detailed flowchart illustrating the entire cheese-making procedure employed for Fiore Sardo production is shown in Figure 23.3.

The milk used for Fiore Sardo cheese manufacturing is heated to 35°C–37°C, and starter cultures, made up by autochthonous LAB, can be possibly added. The milk is coagulated by the addition of 30–35 g of rennet per 100 L of milk. The coagulation generally occurs in 15 minutes, and the coagulum is usually left to set for an additional 15 minutes. Afterward, the coagulum is cut into rice-sized pieces using a wood or a steel curd cutter called "Chiova" or with the fast rotation of the forearm until complete fragmentation. After 3 to 5 minutes, which is necessary to settle the small curd pieces, the latter are carefully compacted with hands (dipped into the whey) in a single curd mass; the curd is then cut in slices and distributed into plastic molds with a truncated-cone shape. The curd paste is added and

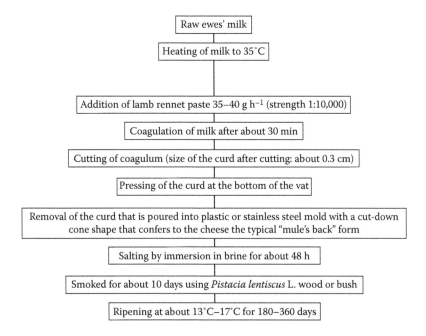

Raw ewes' milk

Heating of milk to 35°C

Addition of lamb rennet paste 35–40 g h⁻¹ (strength 1:10,000)

Coagulation of milk after about 30 min

Cutting of coagulum (size of the curd after cutting: about 0.3 cm)

Pressing of the curd at the bottom of the vat

Removal of the curd that is poured into plastic or stainless steel mold with a cut-down cone shape that confers to the cheese the typical "mule's back" form

Salting by immersion in brine for about 48 h

Smoked for about 10 days using *Pistacia lentiscus* L. wood or bush

Ripening at about 13°C–17°C for 180–360 days

FIGURE 23.3 Flowchart of the cheese-making procedure employed for Fiore Sardo production. (Adapted from Mangia N.P. et al., *Food Microbiol*, 25, 366–377, 2008.)

stratified to a height that is double with respect to that of the mold. During this step, the curd paste is pressed with hands to eliminate the most part of the whey. Finally, on top of the first mold, which is filled up with the pressed curd paste, a second mold is laid down, and the curd paste within molds is turned over several times in a way that the outer edges of the cheese acquire a characteristic shape (Figure 23.4). Once shaped, the cheeses are arranged in piles to facilitate the discharge of the whey, left to settle for approximately 1 hour, and then taken out from the molds and dipped into a whey/water mixture at 90°C for approximately 1 minute. This latter step is only carried out during the winter and not by all the producers. The cheese is salted in brine (25% NaCl) for approximately 36–48 hours and then laid on wood shelves where it undergoes a first ripening and a slight smoking (approximately 2 hours per day) using Mediterranean maquis scrub species. This phase can last for up to 15 days, and it is usually carried out at 10°C–18°C. After this, the ripening proceeds in different rooms at 15°C for about 6–8 months. During the first month, the cheese is turned upside down every day; afterward this operation is carried out only every 2 or 3 days. When the cheese surface is covered by molds (fungi), it is washed with vinegar and salt, left to dry, and covered with olive oil by brushing. This latter treatment is repeated four to five times during the ripening.

(a)   (b)

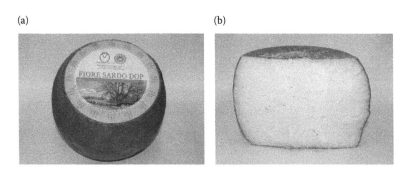

FIGURE 23.4 (a, b) Fiore Sardo cheese.

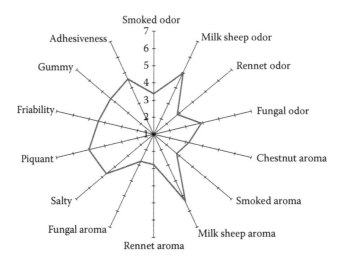

FIGURE 23.5   Sensory profile of ripened Fiore Sardo cheese.

Fiore Sardo cheese has a shape made up by two flattened truncated cones with horizontal bases joined by the main base conferring to the cheese its typical shape (Figure 23.4). At the end of the ripening, the cheese weight is between 3.6 and 4.6 kg. The rind is dark yellow or brown, the paste is white or pale yellow, and the cheese has a sheep milk odor and aroma, and a piquant and slightly salty taste (Figure 23.5). Recently, the majority of the Fiore Sardo cheese producers have improved and optimized several technological phases and, according to the new EU Hygiene Regulation 852 and 853/2004, also the overall hygienic quality of milk. According to these EU regulations, Fiore Sardo cheese should be produced from ewe's raw milk containing less than 500,000 bacterial cfu/ml. However, in several cases, this poor microbial content is not sufficient to carry out a suitable acidification and a proper ripening process. For this reason, the use of autochthonous starter cultures, allowed by the manufacturing procedures specified under the PDO Regulation 1236/1996, could be helpful to achieve a better management of the process and maintain the Fiore Sardo "typicality" as recently shown by our research group (Pisano et al. 2007; Mangia et al. 2008).

23.4 Microbiological Characteristics

The lactic microflora of raw ewe's milk is typically mesophilic and mainly composed by *Lactococcus lactis* subsp. *lactis* and *Enterococcus faecium*, among cocci, and by *Lactobacillus plantarum*, *Lb. casei*, *Lb. pentosus*, *Lb. paracasei*, and *Lb. brevis*, among rods (Mangia et al. 2008; Mannu et al. 2000; Di Cagno et al. 2003). Similarly, the majority of microorganisms usually recovered from Fiore Sardo cheese are mesophilic as are all cheeses manufactured from raw milk and without any heat treatment of the curd (Table 23.1).

TABLE 23.1

Most Representative Bacterial Species Isolated from Fiore Sardo Cheese

	Bacteria Groups Isolated from Fiore Sardo Cheese		
	Lactococci	**Lactobacilli**	**Enterococci**
Pisano et al. 2006	*L. lactis*	*Lb. plantarum*, *Lb. casei*, *Lb. brevis*	*E. faecium*, *E. durans*
Mangia et al. 2008	*L. lactis*	*Lb. plantarum*, *Lb. casei*, *Lb. pentosus*, *Lb. brevis*	*E. faecium*, *E. faecalis*
Comunian et al. 2010	*L. lactis*	*Lb. plantarum*, *Lb. casei*, *Lb. brevis*	Enterococci spp.

During the first ripening period of Fiore Sardo cheese (e.g., 30 days), lactic cocci predominate together with *E. faecium*, whereas, in the subsequent phases, facultative heterofermentative lactobacilli usually prevail (Mannu et al. 2000; Pisano et al. 2006; Mangia et al. 2008). In particular, *L. lactis* subsp. *lactis* was identified as the main responsible for the fermentative phase of Fiore Sardo cheese (Bottazzi et al. 1978; Ledda et al. 1994; Mangia et al. 2008). Mangia et al. (2008) showed that the population of this lactic species rapidly increases in the curd (about 7 $\log_{10}$ cfu/g), often reaching very high numbers around 30 days of ripening (about 10 $\log_{10}$ cfu/g). They also showed that the majority of *L. lactis* subsp. *lactis* strains isolated from traditional Fiore Sardo cheese possess a significantly fast acidifying capacity. This finding supports well the key role of *L. lactis* subsp. *lactis* during the fermentative phase of Fiore Sardo. The acidifying activity of *L. lactis* subsp. *lactis* limits the development of the acid-sensitive spoilage microflora and favors the activity of the rennet enzymes. All this has a direct positive influence on the microbiological and technological quality of cheese (e.g., its structural features). Moreover, *L. lactis* subsp. *lactis* can produce different bacteriocins, which are peptide compounds active against many potentially pathogenic microorganisms.

Different species of facultative heterofermentative lactobacilli dominate the later ripening phases of Fiore Sardo cheese. More in depth, *Lb. plantarum*, *Lb. casei.* and *Lb. pentosus* have been identified by Mangia et al. (2008) as the only facultative heterofermentative species in ripened Fiore Sardo. However, *Lb. curvatus*, *Lb. buchneri*, and *Lb. fermentum*, other than the above-mentioned lactobacilli species, were recovered by other research groups (Mannu et al. 2000; Pisano et al. 2006). These differences in the species composition of facultative heterofermentative lactobacilli of Fiore Sardo can be due to a different microbial composition of the raw milk employed for cheese manufacturing and/or to the specific microbial environment characterizing the distinct dairy farms in which Fiore Sardo was produced (Mannu et al. 2000). Among lactobacilli, *Lb. plantarum* was found by Mangia et al. (2008) as the predominant species during the first 150 days, while subsequently, its number declined in favor of *Lb. brevis* and *Lb. casei* that were dominating the last part of the ripening. Interestingly, these latter authors reported about a synergistic effect between mesophilic lactococci (e.g., *L. lactis* subsp. *lactis*) and *Lb. casei/Lb. plantarum*, which determines a significant increase in the growth rates of both species when grown together. Nevertheless, this synergistic effect seems to be more relevant for rods and for *Lb. plantarum* in particular, suggesting a good adaptability of this species to complex substrates such as raw ewe's milk. The abundant presence of mesophilic lactobacilli during the last part of the ripening is significant since these microorganisms are able to ferment citrate producing pyruvate, acetate, and acetoin, which contribute to the development of Fiore Sardo's typical taste and flavor (Mannu et al. 2000). In particular, it has been shown that *Lb. plantarum* and *Lb. casei* are important in secondary proteolysis phenomena, for example, in the hydrolysis of bitter peptides and flavor formation (Milesi et al. 2008; Poveda et al. 2004). Figure 23.6 illustrates the trend of the mesophilic microflora of Fiore Sardo during 210 days of ripening.

The recovery of different species of *Enterococcus* in Fiore Sardo cheese, and in cheeses made from raw ovine milk, is not rare (Ledda et al. 1994; Mangia et al. 2008). However, in Fiore Sardo cheese made up with the addition of mesophilic starter cultures, this microbial group progressively disappears during the ripening (Mangia et al. 2008). This can be considered as a positive result as enterococci can represent a spoilage problem (Giraffa et al. 1997).

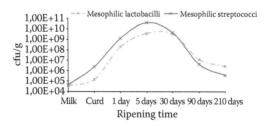

FIGURE 23.6 Evolution of Fiore Sardo cheese mesophilic microflora during 240 days of ripening.

TABLE 23.2

Chemical Composition of Fiore Sardo Cheese (Mean Values ± S.D.)

Total solid	73.20 ± 1.32	
Ashes	0.21 ± 0.00	
Fat	29.21 ± 0.04	28.70 ± 0.00
Proteins	30.60 ± 0.80	28.50 ± 1.80
Lactose	0.01 ± 0.00	
Acid lactic	2.23 ± 0.02	
pH	5.22 ± 0.02	5.31 ± 0.01
NaCl	2.50 ± 0.20	1.70 ± 0.01

Note: Except for pH, all data were expressed as g 100 g^{-1} of cheese.

23.5 Chemical Composition

As for other traditional cheeses manufactured from raw milk, the chemical composition of Fiore Sardo cheese produced by different farms can be different. In Table 23.2, we present some recent data on the chemical composition of Fiore Sardo cheese as reported by selected authors (Mangia et al. 2008; Di Cagno et al. 2003).

Generally, during Fiore Sardo ripening, a steady increase in the total solids, ashes, proteins, fats, lactic acid, and sodium chloride takes place, and at the end of ripening, these values usually approach those reported in Table 23.2.

At the end of the ripening, Fiore Sardo cheese generally shows a low water content (26.5%), while fats and proteins are both around 29%. For these characteristics, Fiore Sardo cheese can be defined as a semi-fat cheese with a high protein content. Lactose, which is quickly metabolized by the lactic microflora during the first 24 hours, is only present as trace amounts in the ripened cheese.

23.6 Proteolytic Activity and Free Amino Acid Content

During Fiore Sardo ripening, the water-soluble nitrogen (WSN) generally increases with respect to total nitrogen (TN) due to the proteolytic activity of the cheese microbiota and to the rennet proteases. Mangia et al. (2008) observed a significant increase in WSN and nonprotein nitrogen (NPN) during traditional Fiore Sardo ripening, while the ripening coefficient (WSN/TN*100) reached a mean value of 24% at 7 months (Table 23.3). In experimental Fiore Sardo cheese, manufactured with the addition of a selected starter culture, we found a higher ripening coefficient (approximately 27%) that can be a result of the proteolytic activity of mesophilic lactobacilli present in the starter culture (Mangia et al. 2008). In particular, *Lb. plantarum* that preferentially metabolizes β-casein (Khalid and Marth 1990) can be responsible for this "accelerated" cheese ripening. The high levels of NPN, which characterize the ripened Fiore

TABLE 23.3

Evolution of Nitrogenous Fractions (WSN, NPN, TN) in Fiore Sardo Cheese during Ripening

	Nitrogenous Fraction			Ripening Indices (%)	
	TN	**WSN**	**NPN**	**NPN/TN**	**WSN/TN**
Curd	2.68 ± 0.05	0.31 ± 0.01	0.15 ± 0.06	6	11
30 days	4.48 ± 0.15	1.05 ± 0.06	0.63 ± 0.05	14	23
210 days	5.18 ± 0.15	1.23 ± 0.05	0.95 ± 0.17	18	24

Note: TN, total nitrogen; WSN, water soluble nitrogen; NPN, nonprotein nitrogen.

Sardo cheese, can be indicative of an intense activity of the lactic microflora (Furtado and Partridge 1988). More generally, the increase in the NPN/TN*100 and WSN/TN*100 ratios during cheese ripening is indicative of the development of a variety of nitrogenous compounds deriving from proteolysis. This process first leads to the formation of polypeptides and subsequently to small- and medium-size peptides and eventually to free amino acids (FAAs; Tavaria et al. 2003).

At the end of the ripening, the total FAA content of Fiore Sardo cheese varies considerably among authors, being about 1000 mg/100 g of cheese for Mangia et al. (2008) and about 100 mg/kg of cheese for Di Cagno et al. (2003). On the contrary, there is a general agreement among authors about the more abundant FAAs in the ripened cheese. Leucine, valine, isoleucine, and phenylalanine are the more abundant amino acids after 210 days of ripening (Mangia et al. 2008; Di Cagno et al. 2003) accounting together for the 61% of total FAAs in cheese (Table 23.4). Also proline is present at important levels, and its content increases progressively during the ripening as well as the amount of several other FAAs (e.g., tyrosine, cysteine, treonine, and serine; Table 23.4). The increased content of proline during Fiore Sardo ripening can be due to a substantial hydrolysis of the β-casein (a proline-rich polypeptide) during the maturation process. Of particular interest is the evolution of glutamic acid and lysine during Fiore Sardo maturation. These amino acids reach their highest content after 90 days of ripening, while subsequently their content decreases remarkably (Table 23.4). An explanation for this behavior could be associated with the high numbers of viable LAB recorded up to the 90 days, which reinforces the likely role of these microorganisms in flavor generation via FAA conversion into volatile compounds (Tavaria et al. 2003). In general, amino acids released from peptides may be transformed by other enzymes, provided they are present in cheese, to acids, aldehydes, and amines, which will contribute to flavor development (Khalid and Marth 1990).

TABLE 23.4

Evolution of FAAs (mg 100 g^{-1}) in Fiore Sardo Cheese during Ripening (Mean Values ± S.D.)

	Ripening Time			
	1 day	**5 days**	**30 days**	**210 days**
Aspartic acid	0.73 ± 0.3	2.10 ± 0.5	3.50 ± 0.21	1.63 ± 0.4
Glutamic acid	4.28 ± 0.2	6.14 ± 0.31	35.74 ± 1.38	28.71 ± 7.3
Asparagine	0.35 ± 0.1	1.00 ± 0.2	3.94 ± 1.75	15.95 ± 6.4
Serine	0.90 ± 0.2	0.94 ± 0.3	2.29 ± 0.24	16.81 ± 2.1
Glutamine	0.88 ± 0.4	4.62 ± 0.2	9.89 ± 2.89	7.76 ± 2.3
Histidine	1.35 ± 0.4	1.47 ± 0.3	1.12 ± 0.36	4.54 ± 0.9
Glycine	0.34 ± 0.1	0.58 ± 0.1	4.19 ± 0.09	10.93 ± 1.7
Threonine	0.62 ± 0.0	1.49 ± 0.6	5.02 ± 0.07	14.82 ± 1.6
Alanine	2.26 ± 0.2	5.58 ± 0.4	16.09 ± 0.89	48.83 ± 3.5
Arginine	N.D.	1.34 ± 0.1	2.24 ± 0.06	N.D.
Tyrosine	1.80 ± 0.1	2.25 ± 0.5	8.53 ± 1.72	27.34 ± 1.4
Cystine	N.D.	N.D.	2.33 ± 0.26	9.31 ± 0.5
Valine	2.70 ± 0.1	9.64 ± 2.5	54.54 ± 1.33	128.29 ± 11.6
Methionine	2.29 ± 0.1	4.20 ± 0.9	16.60 ± 0.41	74.07 ± 7.8
Tryptophan	0.12 ± 0.0	0.31 ± 0.0	1.53 ± 0.49	7.46 ± 2.4
Phenylalanine	1.59 ± 0.4	5.78 ± 1.7	43.29 ± 1.26	106.85 ± 14.7
Isoleucine	0.91 ± 0.2	3.12 ± 0.7	18.53 ± 0.27	122.54 ± 11.1
Ornithine	0.91 ± 0.1	3.14 ± 2.2	21.25 ± 0.82	9.86 ± 3.0
Leucine	3.01 ± 0.1	11.42 ± 3.0	83.52 ± 5.36	269.27 ± 36.8
Lysine	1.99 ± 0.2	6.25 ± 2.7	42.04 ± 1.89	30.09 ± 1.9
Proline	3.70 ± 1.2	5.21 ± 1.7	16.85 ± 1.10	92.53 ± 7.7
∑ FAA	30.70 ± 1.6	76.56 ± 18.2	393.00 ± 8.08	1027.58 ± 113.8

23.7 Lipolytic Activity and Free Fatty Acid Content

The hydrolysis of fats due to lipolytic activity plays a significant role in Fiore Sardo ripening. This process appears very intense during the entire maturation and leads to a significant buildup of free fatty acids (FFAs) in the end product (Table 23.5).

The concentration of FFAs in a cheese partly derives from the fat content of the milk of origin (in our case 5%–6%). In this regard, it should be mentioned that the amount of FFAs detected in Fiore Sardo cheese at the end of the ripening is much higher with respect to the relative amounts detected in other Sardinian ewe's milk PDO cheeses (Madrau et al. 2006; Mangia et al. 2011).

FFAs are important components of the flavor of many cheese types. In particular, the short-chain fatty acids can impart a desirable piquant taste typical of several Italian cheeses such as Fiore Sardo (Larrayoz et al. 1999).

In general, the high amount of FFAs in Fiore Sardo cheese originates from the combined action of the lipases present in milk (constitutive lipases) and those of microbial origin. Moreover, a significant role in this sense is also played by the lipases present in the paste rennet used in the cheese-making (Deiana et al. 1980). In this regard, it is well known that the paste rennet used for the Fiore Sardo cheese-making presents a high lipase activity (Larrayoz et al. 1999). As shown in Table 23.5, palmitic, oleic, myristic, and butyric acids are the most representative FFAs in Fiore Sardo cheese at 210 days of ripening (Mangia et al. 2008). During cheese maturation, a significant buildup of short FFAs such as butyric, capric, and caproic acid can be observed (Table 23.5). The high content of butyric acid in the cheese is likely due to the specific activity of the lipases present in the lamb rennet (Virto et al. 2003). The presence of this FFA in significant amounts is important for cheese flavor because butyric acid imparts a desirable piquant taste to Fiore Sardo (Larrayoz et al. 1999). Interestingly, butyric acid is revealed as a potent anticancer agent in various cancer cell lines (Archer et al. 1998; Yanagi et al. 1993; Belobrajdic and McIntosh 2000).

Moreover, substantial amounts of linoleic acid (C18:2) can be found at the end of the ripening (approximately 145 mg/100 g of cheese). Despite the fact that the release of linoleic acid can be mainly attributed to the constitutive milk lipases and to the rennet lipases, this fatty acid can also originate from the lipolytic activity of the mesophilic lactic microflora. Specific lipases from *Lb. casei*, a LAB species frequently isolated from Fiore Sardo cheese (Mangia et al. 2008; Pisano et al. 2006; Comunian et al. 2010), could contribute to the accumulation of this long chain fatty acid (Yu 1986).

TABLE 23.5

Evolution of FFAs (mg/100 g^{-1}) in Fiore Sardo Cheese during Ripening (Mean Values ± S.D.)

	Ripening Time			
	1 day	**5 days**	**30 days**	**210 days**
Butyric acid C4	9.94 ± 1.9	26.18 ± 1.6	104.27 ± 4.7	254.97 ± 49.4
Caproic acid C6	3.23 ± 0.1	8.83 ± 0.8	41.46 ± 3.2	124.35 ± 14.1
Caprylic acid C8	0.50 ± 0.2	3.32 ± 0.7	21.30 ± 0.9	79.80 ± 5.3
Capric acid C10	4.63 ± 0.2	15.05 ± 0.7	71.11 ± 3.2	228.74 ± 15.1
Lauric acid C12	7.36 ± 1.2	12.52 ± 0.5	40.84 ± 1.9	123.42 ± 10.6
Myristic acid C14:0	7.52 ± 2.2	17.17 ± 0.4	90.25 ± 5.4	318.88 ± 30.9
Palmitic acid C16:0	17.42 ± 3.2	47.16 ± 1.4	247.22 ± 14.7	760.90 ± 46.6
Stearic acid C18:0	15.64 ± 1.1	26.09 ± 0.9	74.90 ± 7.5	243.31 ± 19.5
Oleic acid C18:1	20.64 ± 3.5	51.21 ± 5.1	147.00 ± 16.0	490.16 ± 55.3
Linoleic acid C18:2	5.72 ± 0.3	11.06 ± 1.7	35.44 ± 6.2	135.87 ± 20.7
Linolenic acid C18:3	N.D.	7.69 ± 0.1	13.79 ± 0.9	26.40 ± 4.6
Σ FFA	92.58 ± 6.33	226.26 ± 8.5	887.57 ± 52.3	2786.80 ± 209.3

REFERENCES

Archer S, Meng S, Wu J, Johnson J, Tang Raymond, Hodin R. 1998. Butyrate inhibits colon carcinoma cell growth through two distinct pathways. *Surgery* 124:248–253.

Belobrajdic DP, McIntosh GH. 2000. Dietary butyrate inhibits NMU-induced mammary cancer in rats. *Nutr Cancer* 36:217–223.

Bottazzi V, Arrizza S, Ledda A, 1978. Impiego di colture di fermenti lattici nella produzione del Fiore Sardo. *Sci Tec Latt Cas* 29:160–168.

Comunian R, Paba A, Daga ES, Dupre I, Scintu MF. 2010. Traditional and innovative production methods of Fiore Sardo cheese: a comparison of microflora with a PCR-culture technique. *Int J Dairy Technol* 2:224–233.

Cossu G. 1787. Discorso georgico riguardante le pecore sarde. Stamperia Reale, Cagliari, 54–55.

Deiana P, Farris GA, Fatichenti F, Carini S, Lodi R, Todesco R. 1980. Impiego di caglio di agnello e capretto in polvere nella fabbricazione di formaggio Fiore Sardo: aspetti microbiologici e tecnologici. *Il Latte* 5:191–200.

Di Cagno R, Banks J, Sheehan L, Fox PF, Brechany EY, Corsetti A, Gobbetti M. 2003. Comparison of the microbiological, compositional, biochemical, volatile profile and sensory characteristics of three Italian PDO ewes' milk cheeses. *Int Dairy J* 13:961–972.

EC Regulation 1236/1996. Regolamento (CE) n. 1263/96 della Commissione del 1 luglio 1996 relativo alla registrazione delle indicazioni geografiche e delle denominazioni di origine nel quadro della procedura di cui all'articolo 17 del regolamento (CEE) n. 2081/92. GUCE, L163, 19–21.

EU Council Directive 92/46/EC. 1992. Official Journal L 268, 14/09/1992, 1–32.

Furtado MM, Partridge JA. 1988. Characterization of nitrogen fractions during ripening of a soft cheese made from ultrafiltration retentates. *J Dairy Sci* 71:2877–2884.

Gemelli F. 1776. Rifiorimento della Sardegna. In: *Giammichele Briolo*, Ed. Torino 515–522.

Giraffa G, Carminati D, Neviani E. 1997. Enterococci isolated from dairy products: a review of risk and potential technological use. *J Food Prot* 60:732–738.

Khalid NM, Marth EH. 1990. Proteolytic activity by strains of *Lactobacillus plantarum* and *Lactobacillus casei*. *J Dairy Sci* 73:3068–3076.

Larrayoz P, Martınez MT, Barron LJR, Torre P, Barcina Y. 1999. The evolution of free fatty acids during the ripening of Idiazabal cheese: Influence of rennet type. *Eur Food Res Tech* 210:9–12.

Ledda A, Scintu MF, Pirisi A, Sanna S, Mannu L. 1994. Caratterizzazione tecnologica di ceppi di lattococchi ed enterococchi per la produzione di formaggio pecorino Fiore Sardo. *Sci Tec Latt Cas* 45:443–456.

Manca dell'Arca A. 1780. Delle pecore. In: *Illisso Edizioni. Agricoltura di Sardegna*. 2nd ed. Nuoro, pp. 311–319.

Madrau MA, Mangia NP, Murgia MA, Sanna MG, Garau G, Leccis L, Caredda M, Deiana P. 2006. Employment of autochthonous microflora in Pecorino Sardo cheese manufacturing and evolution of physicochemical parameters during ripening. *Int Dairy J* 16:876–885.

Mangia NP, Murgia MA, Garau G, Deiana P. 2011. Microbiological and physicochemical properties of Pecorino Romano cheese produced with a selected starter culture. *J Agric Sci Technol* 13:585–600.

Mangia NP, Murgia MA, Garau G, Sanna MG, Deiana P. 2008. Influence of selected lab cultures on the evolution of free amino acids, free fatty acids and Fiore Sardo cheese microflora during the ripening. *Food Microbiol* 25:366–377.

Mannu L, Comunian R, Scintu ME. 2000. Mesophilic lactobacilli in Fiore Sardo cheese: PCR-identification and evolution during cheese ripening. *Int Dairy J* 10:383–389.

Milesi MM, McSweeney PLH, Hynes ER. 2008. Viability and contribution to proteolysis of an adjunct culture of *Lactobacillus plantarum* in two model cheese systems: Cheddar cheese-type and soft-cheese type. *J Appl Microbiol* 105:884–892.

Pisano MB, Fadda ME, Deplano M, Corda A, Casula M, Cosentino S. 2007. Characterization of Fiore Sardo cheese manufactured with the addition of autochthonous cultures. *J Dairy Res* 74:255–261.

Pisano MB, Fadda ME, Deplano M, Corda A, Cosentino S. 2006. Microbiological and chemical characterization of Fiore Sardo, a traditional Sardinian cheese made from ewe's milk. *Int J Dairy Technol* 59:171–179.

Poveda JM, Cabezas L, McSweeney PLH. 2004. Free amino acid content of Manchego cheese manufactured with different starter cultures and change throughout ripening. *Food Chem* 84:213–218.

Tavaria FK, Franco I, Carballo FJ, Malcata FX. 2003. Amino acid and soluble nitrogen evolution throughout ripening of Serra da Estrela cheese. *Int Dairy J* 13:537–545.

Virto M, Chavarri F, Bustamante MA, Barron LJR, Aramburu M, Vicente MS, Perez-Elortondo FJ, Albisu M, de Renobales M. 2003. Lamb rennet paste in ovine cheese manufacture. Lipolysis and flavour. *Int Dairy J* 13:391–399.

Vodret A, Campus RL, Deiana P, Catzeddu P, Soletta M, Cicu I. 1996. Fiore Sardo. In: *Consiglio Nazionale delle Ricerche* Ed. I Prodotti Caseari Del Mezzogiorno. Roma, pp. 459–462.

Yanagi S, Yamashita M, Imai S. 1993. Sodium butyrate inhibits the enhancing effect of high fat diet on mammary tumorigenesis. *Oncology* 50:201–204.

Yu JH. 1986. Studies on the extracellular and intracellular lipase of Lactobacillus casei. I. On the patterns of free fatty acids liberated from milk reacted with the lipases. *Korean J Dairy Sci* 8:167–176.

24

Zamorano Cheese

D. Fernández, R. Arenas, R. E. Ferrazza, M. E. Tornadijo, and Jose Maria Fresno Baro

CONTENTS

24.1 Ewe's Milk and Cheese Production

Around 34% of the world's ewe-milk production, some 2.8 million tons, is concentrated in Europe. This production is mainly centered on countries around the Mediterranean, especially Greece, Romania, and Italy, which together account for 72% of all ewe's milk produced in Europe, followed by Spain, France, and Portugal (Food and Agriculture Organization of the United Nations [FAO] 2010).

Ewe's milk production in Spain has increased steadily since 1990, reaching a total of 489,800 MT in 2009 (Ministerio de Medio Ambiente y Medio Rural y Marino [MARM] 2010a; FAO 2010). This production represents 6% of all milk produced by the various species (cow, goat, and sheep).

By Autonomous Community, Castilla y León and Castilla-La Mancha account for the majority of ewe's milk produced in Spain, with 64.9% and 26.3%, respectively (MARM 2010b). However, although ewe's milk production in other communities, especially the Basque Country, Navarre, Extremadura, and Madrid, is limited in overall terms, these regions nevertheless have a long-standing reputation for producing high-quality ewe-milk cheese.

Ewe's milk production in Castilla y León reached a total of 317,951 tons in 2009, which is 26.2% of all milk produced in this region (MARM 2010b) and 5% of overall nationwide milk production.

Ewe's milk production in the Autonomous Community of Castilla y León is concentrated around Zamora (28%), Valladolid (26.1%), León (13.6%), Palencia (12.5%), and Salamanca (9.8%).

As is the case nationally, essentially all the ewe's milk produced in Castilla y León (99.94%) is used to make cheese, around 80% of which is manufactured using mixtures of milk from different species (cow, goat, and sheep). Indeed, nationally, only 13% of all cheeses with a Protected Designation of Origin (PDO), and 1% of those from Castilla y León, are manufactured exclusively or partially using ewe's milk (MARM 2010c). The following cheeses with a PDO are manufactured exclusively using ewe's milk: Manchego, Idiazábal, La Serena, Torta del Casar, Roncal, and Zamorano.

Zamorano cheese production is restricted to the province of Zamora, and the quality of this cheese was recognized by the award of a Designation of Origin in May 1993 (Boletín Oficial del Estado [BOE] 1993).

TABLE 24.1

Production and Marketing of Zamorano Cheese PDO

	2001	2002	2003	2004	2005	2006	2007	2008
Ewe's milk used in Zamorano cheese production (L)	1,533,306	2,277,579	NA	NA	2,435,384	2,375,576	2,540,217	2,397,987
Zamorano cheese produced (kg)	296,177	304,263	NA	NA	478,786	474,284	507,560	477,814
Zamorano cheese sold nationally (kg)	251,602	275,046	257,000	284,000	371,180	337,157	338,611	314,214
Zamorano cheese exported to EU (kg)	7675	10,729	11,000	14,000	9583	8978	15,533	16,126
Zamorano cheese exported to other countries (kg)	16,355	16,628	13,000	15,000	20,968	29,048	24,650	4408

Source: Ministerio de Medio Ambiente y Medio Rural y Marino (MARM). Gobierno de España. 2010. Cifras y datos de las Denominaciones de Origen Protegidas (D.O.P.) e Indicaciones Geográficas Protegidas (I.G.P.) de Productos Agroalimentarios. Años 2001–2008. Available from: http://www.mapa.es/es/alimentacion/pags/Denominacion/htm/cifrasydatos.htm. Accessed Dec. 15, 2010.

Note: NA = not available.

The production of Zamorano cheese PDO in 2008 reached 477,814 kg, some 4.15% of all Spanish ewe's cheese with a PDO (MARM 2010c). Although this value is still relatively low, especially when compared to Manchego and, to a lesser extent, Idiazábal cheese, which accounted for 76.7% and 10.6%, respectively, the growth in production experienced in the last few years (see Table 24.1), increasing by 62% between 2001 and 2008 despite the fact that the number of dairy sheep has remained essentially unchanged and the number of dairy farms has decreased by 65%, is highly promising (MARM 2010d). The only aspect of this industry that has remained practically constant over the past few years is the number of Zamorano cheese PDO producers (ten; five industrial and five traditional). The majority of the cheese produced is sold in Spain, with only between 6% and 10% being exported, mainly to the rest of the European Union (EU) and the United States.

24.2 Cheese-Making Technology of Zamorano Cheese

Zamorano cheese is currently produced both industrially and traditionally. Traditional producers typically use milk produced on their own farms, and the amount of milk used for each batch does not generally exceed 1000 L. In contrast, the milk for industrial producers is collected from various farms, and the volume used for each batch is much higher (typically 25,000 L). Despite this, the manufacturing process followed is very similar in both cases. Thus, the milk used must come from the Churra (Figure 24.1) and Castellana breeds (Figure 24.2) and can optionally be pasteurized (72°C–75°C/15–20 s). Traditional producers tend not to pasteurize their milk. The milk is warmed to 30°C–32°C, and calcium chloride (0.2 g/L) and nitrates or lysozyme are added. A lyophilized starter culture (1%–1.5%) is added irrespective of whether the milk has been pasteurized or not. The most commonly used starter culture is made of a mixture of *Lactococcus lactis* subsp. *lactis* and *L. lactis* subsp. *cremoris*, and its main role is to acidify the milk. However, some industries use a starter culture containing the aforementioned *Lactococcus* spp. together with *Streptococcus salivarius* subsp. *thermophilus* in order to improve the texture of the cheese.

The mixture of milk and starter culture is then left to stand for around 30 minutes. After this time, 30 ml of liquid lamb rennet (strength 1:10,000; 75% chymosin) is added per 100 L of milk (30°C–32°C); coagulation occurs within 30–45 minutes. The resulting coagulum is then cut with knives for 10–20 minutes

FIGURE 24.1 Churra sheep.

FIGURE 24.2 Castellana sheep.

until the lumps are the size of a grain of rice, and then the curds are stirred for a further 30 minutes while slowly increasing the temperature to 36°C–38°C at a rate of 1°C every 5 minutes. Once the curds have acquired the optimal consistency, the mass is pre-pressed while in the vat to remove the whey before it is introduced into the molds. These molds were traditionally made from esparto grass, although nowadays, cylindrical plastic molds are used for the sake of cleanliness and hygiene. Modern molds have a pattern on their inner surface to simulate that produced by esparto grass. Industrially, the molds are filled automatically in large cylinders, which perform the pre-pressing and cut the curds into suitably sized blocks, whereas the molds for traditional cheeses are filled by hand. Likewise, the "despizque" (Figure 24.3), whereby the curd grains are broken up to improve oxygenation and matting, is also performed by hand. Before pressing, the molded curds are transferred into different molds containing a cloth in order to promote whey removal from the mass. This use of cloths in the molds tends to be limited to small cheese makers, although some industrial manufacturers also use this technique. However, the majority of industrial manufacturers are replacing them with microperforated molds. Pressing of the mass is performed with horizontal pneumatic presses in two stages. In the first stage, the cloth-wrapped curds are pressed for 20–30 minutes at a pressure of 28 psi, whereas in the second stage, the cloths are removed and the curds turned in the mold and pressed at 50 psi (for approximately 4–5 hours) until they reach a pH of 5.4–5.5.

Traditional cheese makers remove the curds from the molds by hand, whereas their industrial counterparts perform this mechanically. The curds, now with their characteristic cylindrical shape, are salted in brine (17°Bé–18°Bé; pH 5.4 and temperature 8°C–10°C) for between 18 and 24 hours. After removal of the salted curds from the brine, they are then treated differently depending on whether the manufacturer follows an industrial or a traditional process. Thus, in traditional cheese making, the curds are stored at 8°C–10°C for 12–24 hours to allow the rind to dry, whereas industrial cheese makers pass them through a bath to apply a plastic, antifungal coating before storing them at 8°C–10°C for drying. Finally, the cheeses are stored in ripening rooms at a temperature of approximately 10°C–12°C and a relative humidity (RH) of 85%, where they remain until they are ready to be sold. Zamorano cheese PDO must be ripened for a minimum of 100 days, after which ripening is extended by several more weeks depending

FIGURE 24.3 Break-up by hand of curd grains or "despizque."

on the size of the cheese. Thus, cheeses weighing 1.3–1.5 kg ripen in approximately five months, whereas larger cheeses (3.0–3.2 kg) need between six and ten months, depending on the type of manufacture (industrial or traditional).

Traditional cheeses develop a rind upon which molds grow during ripening (Figure 24.4). These molds are removed by brushing before the cheeses are sold.

Once ripening is complete, Zamorano cheese is cylindrical in shape (diameter: 18–22 cm; height: 8–12 cm), with a hard, dark-gray or straw-yellow colored rind, and the patterns of the mold (flower) on one of its flatter faces and the esparto grass bands ("pleitas") around its sides. The cheese itself is hard and compact, with an ivory or yellowish color, and usually presents small mechanical eyes spread unevenly throughout. It has an intense smell and taste, leaving a persistent buttery sensation on the palate. It generally weighs between 1.3 and 3.2 kg. All cheeses with a PDO contain a casein plaque (Figure 24.5) and a back label from the Regulatory Council, whose ornamental motif is a Romanic-style rosette (Figure 24.6).

FIGURE 24.4 Traditional Zamorano cheese at the end of ripening.

FIGURE 24.5 Typical casein label embedded in Zamorano cheese PDO.

FIGURE 24.6   Specific label of Regulatory Council of Zamorano cheese.

24.3 Composition of Milk Used in Zamorano Cheese Manufacture

The regulations of the Regulatory Council for Zamorano cheese PDO specify the chemical and physico-chemical parameters that the ewe's milk used to make this cheese must meet. These include fats ≥ 7%, protein ≥ 5%, lactose ≥ 4%, maximum titratable acidity 23°D, and total solids (TS) ≥ 17.5%.

Our research group has analyzed the chemical and physicochemical composition of samples of ewe's milk from various traditional and industrial cheese makers belonging to the Zamorano cheese PDO and found that the requirements established by the Regulatory Council were met in all cases (see Table 24.2). We also analyzed the ash content and the main (Ca, P, Na, K, and Mg) and trace (Fe and Cu) mineral elements.

The TS, fats, protein, and lactose contents were lower than those reported by Pardo et al. (1996) for ewe's milk from the Manchega breed, González et al. (1997) for ewe's milk from the Merina breed, and Micari et al. (2002) for ewe's milk from the Sarda breed, but higher than those reported by Ibáñez et al. (1993) for ewe's milk from the Latxa breed.

TABLE 24.2

Average Values ± Standard Deviation of Compositional and Physicochemical Parameters and Main and Trace Mineral Elements of Milk and Whey

	Milk	Whey
Total solids[a]	17.38 ± 0.97	8.40 ± 0.63
Fat[a]	6.85 ± 0.64	1.00 ± 0.48
Protein[a]	5.17 ± 0.22	1.65 ± 0.23
Lactose[a]	4.46 ± 0.15	4.74 ± 0.39
Ash[a]	1.13 ± 0.02	0.77 ± 0.03
Chloride[a]	0.13 ± 0.01	0.15 ± 0.01
pH	6.62 ± 0.13	6.31 ± 0.05
Titratable acidity[b]	19.13 ± 0.42	13.00 ± 0.83
Phosphorus (g/kg)	0.95 ± 0.09	0.42 ± 0.04
Calcium (g/kg)	1.85 ± 0.14	0.51 ± 0.02
Sodium (g/kg)	0.68 ± 0.04	0.71 ± 0.03
Potassium (g/kg)	1.14 ± 0.06	1.21 ± 0.12
Magnesium (g/kg)	0.18 ± 0.01	0.12 ± 0.01
Iron (mg/kg)	0.62 ± 0.07	0.16 ± 0.02
Copper (mg/kg)	0.37 ± 0.05	0.20 ± 0.06

[a] Expressed as g/100 g of milk or whey.
[b] Expressed as degrees Dornic.

The ash content was high, and the pH and titratable acidity were similar to those noted by other authors for ewe's milk (Park et al. 2007).

Our values for TS, fats, and protein were slightly lower than those described by Rodríguez-Nogales et al. (2007) and Revilla et al. (2009b). Indeed, the latter authors noted that the chemical composition was affected by the breed from which the milk was obtained, with milk from the Castellana breed having lower TS, fats, and protein contents than that from the Assaf or Churra breeds. Likewise, the same authors found that ewe's milk with a low somatic cell count had a higher fat content, whereas milk with a high somatic cell count had higher protein content due to an increase in the amount of extracellular serum compounds, which have a negative influence on cheese yield, in the milk.

The fats/protein ratio, which is a key parameter in cheese making, averaged 1.32 and was similar to that reported by Pellegrini et al. (1994) and Raynal-Ljutovac et al. (2008).

The main mineral content in the milk used to manufacture Zamorano cheese was similar to that reported by other authors for ewe's milk (Martín-Hernández et al. 1992; Rincón et al. 1994) but lower than that reported by Park et al. (2007) and Raynal-Ljutovac et al. (2008). The results obtained upon analyzing the trace elements in milk (Fe and Cu) were similar to those reported by other authors (Moreno-Rojas et al. 1993; Rincón et al. 1994).

The overall whey composition obtained during the manufacture of Zamorano cheese PDO varies depending on the manufacturing technology used. Average whey composition values are listed in Table 24.2. As a result of the action of the rennet during clotting and the intense whey draining process to which the curds are submitted, the TS content in whey was half that of the milk. Likewise, the fat and protein contents were 15% and 32%, respectively, of the values for the starting milk. The whey obtained during traditional cheese making had higher TS content than that obtained during the industrial process (8.88 versus 7.8 g/100 g, respectively) due to the lesser degree of standardization of the manufacturing technologies. These differences in TS content mainly occurred due to the loss of fats and proteins. The whey obtained during traditional cheese making had 50% more fat and 22% more protein than that collected during the industrial process, and this translated, in general terms, to a lower cheese yield for the traditional process. Nevertheless, these differences in cheese yield were not that high, as the milk obtained from the traditional cheese makers tended to have higher fat and protein contents than that obtained from their industrial counterparts.

The mineral content in whey showed two different patterns, with Na and K in one group and Ca, P, and Mg in the other. Thus, the chemical elements in the former group are completely dissolved in the milk, and therefore, their content in whey will always be similar to that in this milk (Renner et al. 1989). In contrast, a significant proportion of the elements in the second group are present as colloids in the milk; therefore, they tend to bind to the curds during whey draining. The Ca, P, and Mg contents in the whey obtained during the manufacture of Zamorano cheese were 28%, 45%, and 67%, respectively, of the values in the starting milk. These results are in accordance with those presented by Martín-Hernández and Juárez (1989) and Moreno-Rojas et al. (1995). The higher Fe and Cu contents in whey were related to their stronger or weaker bonding to the casein micelles.

24.4 Microbiological Characteristics of Zamorano Cheese

The number of studies concerning the microbiological characteristics of Zamorano cheese PDO is limited. Fernández-García et al. (2004) performed a study of the main microorganism counts in the Zamorano cheese manufactured at three dairy sites (two traditional and one industrial) at different times of the year (winter, summer, spring, and autumn). These authors observed significant differences in the microbial counts depending on the manufacturer, with higher counts being found in the industrially manufactured cheeses. The season during which the cheeses were manufactured also had a very significant influence on the lactobacilli and micrococci counts. In a similar study, Etayo et al. (2006) found no *Salmonella*, *Listeria monocytogenes*, or sulfite-reducing clostridia in milk used to manufacture Zamorano cheese, although they did detect high coliform (4.9 log cfu g^{-1}) and staphylococci (3.2 log cfu g^{-1}) counts, thus indicating inappropriate hygiene conditions during milking and subsequent collection and storage of the milk. However, these authors also noted that the microbiological counts decreased during ripening, with coliforms, sulfite-reducing clostridia, and staphylococci essentially disappearing, which thus means that, when sold, the Zamorano cheese PDO complied with all the microbiological criteria established by the EU for cheeses manufactured from raw milk.

Ferrazza et al. (2004) studied the influence of the accelerated ripening of Zamorano cheese (by varying the temperature) on the microbial flora and found that an increase in temperature during ripening had a notable effect on lactobacilli, which presented higher counts throughout ripening than in cheeses matured in the traditional manner. The most significant effect was observed after the 60th day of ripening, when similar lactobacilli counts to those expected for a cheese manufactured from raw milk were obtained. In contrast, no *Enterobacteriaceae*, *Micrococcaceae*, or *enterococci* were detected after the first or second month of ripening.

There is currently great public demand for safe foodstuffs with a long shelf life and that have been processed as little as possible. In this respect, biopreservation is a possible alternative that is being evaluated with great interest by the food industry.

Cheeses manufactured from raw milk are an excellent source of new strains of lactic acid bacteria, which are known to inhibit undesirable microbiota by synthesizing bacteriocins (Cogan et al. 1997; Cleveland et al. 2001). Bravo et al. (2009) detected the production of bacteriocins (nisin and lacticin 481) by lactococci strains isolated from Zamorano cheese PDO manufactured with raw ewe's milk. Indeed, some of the strains studied had the rare ability to produce two bacteriocins. These authors concluded that the lactococci strains isolated from Zamorano cheese could be used industrially to develop new native starter cultures for the manufacture of dairy products.

24.5 Changes in Compositional and Physicochemical Parameters during Ripening of Zamorano Cheese

Table 24.3 shows the evolution of the TS, fat, lactose, sodium chloride, and ash contents during the ripening of Zamorano cheese PDO.

The TS content increased significantly during the ripening process, reaching a final value of 68.89±1.51 g TS/100 g cheese. These results were similar to those obtained by Ortigosa et al. (2006) for Roncal-type cheese and Mangia et al. (2008) for Fiore Sardo cheese, but higher than those for Manchego (Serrano et al. 1997) or Portuguese Serpa cheese (Bettencourt et al. 1998).

The protein and fat content, expressed as a percentage of TS, did not differ notably during ripening, with average final values of 34.42 ± 0.62 and 60.53 ± 1.78 g/100 g TS, respectively. These values are higher than the minimum values stipulated by the Zamorano cheese PDO Regulatory Council's regulations and similar to those found for other ewe's milk cheeses (González-Viñas et al. 2001a).

The change in NaCl content of Zamorano cheese reflected the process used during its salting. A significant increase in salt content, reaching mean final values of 3.25 ± 0.82 g/100 g TS, occurred from the first day of ripening as a result of the immersion of the cheeses in brine. These values were lower than those reported previously for Idiazábal cheese (Ibáñez et al. 1995), ewe's milk cheeses from Extremadura (Plaza-Carabantes et al. 1999), or Castellano cheese (Román-Blanco et al. 1999), but similar to those reported by Cabezas et al. (2007) for Manchego cheese.

The total ash content increased throughout the ripening process, reaching a mean final value of 6.82 ± 0.73 g/100 g TS. These results were similar to those reported by González-Viñas et al. (2001b) for mature ewe's milk cheese and slightly lower than those reported for Idiazábal (Ibáñez et al. 1995) and Castellano cheeses (Román-Blanco et al. 1999).

Finally, the rapid disappearance of lactose from Zamorano cheese within the first week of ripening should be noted. This trend appears to be related to the effects that various factors, such as the addition of starter cultures to the raw or pasteurized milk, the intense whey draining to which the curds are submitted, and the cheese pressing and salting processes, have on the degree of lactose fermentation or its elimination in the whey. This behavior differs from that noted previously by Román-Blanco et al. (1999) for Castellano cheese and Mangia et al. (2008) for Fiore Sardo cheese, where small amounts of lactose could still be detected after ripening for more than three months.

TABLE 24.3

Changes in Compositional and Physicochemical Parameters (Average Values ± Standard Deviations) during Ripening of Zamorano Cheese PDO

	Ripening Time (days)							
	1	7	15	30	60	120	180	240
Total Solids[a]	57.54 ± 2.64	59.39 ± 1.98	59.87 ± 1.99	60.83 ± 2.26	62.02 ± 1.78	64.35 ± 2.59	66.19 ± 2.68	68.89 ± 1.51
Protein[b]	37.56 ± 0.75	37.09 ± 1.07	36.76 ± 1.31	36.33 ± 1.09	35.49 ± 1.53	36.54 ± 1.86	34.77 ± 0.45	34.42 ± 0.62
Fat[b]	58.13 ± 3.03	57.81 ± 2.61	59.31 ± 3.19	59.91 ± 3.12	61.15 ± 2.98	60.48 ± 1.61	60.49 ± 1.95	60.53 ± 1.78
Lactose[b]	0.64 ± 0.08	0.06 ± 0.01	ND	ND	ND	ND	ND	ND
Ash[b]	3.84 ± 0.27	4.71 ± 0.61	5.23 ± 0.46	5.89 ± 0.58	6.48 ± 0.83	6.84 ± 0.78	6.89 ± 0.83	6.82 ± 0.73
Chloride[b]	0.24 ± 0.09	1.31 ± 0.29	2.04 ± 0.44	2.35 ± 055	2.77 ± 0.65	3.12 ± 0.84	3.43 ± 0.89	3.25 ± 0.82
pH	5.71 ± 0.02	5.53 ± 0.13	5.49 ± 0.16	5.49 ± 0.12	5.45 ± 0.11	5.41 ± 0.06	5.49 ± 0.07	5.89 ± 0.12
Titratable acidity[c]	0.92 ± 0.52	1.36 ± 0.13	1.39 ± 0.12	1.49 ± 0.15	1.64 ± 0.12	1.71 ± 0.21	1.83 ± 0.26	1.82 ± 0.24
a_w	0.995 ± 0.004	0.976 ± 0.002	0.974 ± 0.003	0.965 ± 0.006	0.955 ± 0.007	0.944 ± 0.008	0.941 ± 0.007	0.925 ± 0.005

Note: ND = not detected.

[a] Expressed as g/100 g cheese.

[b] Expressed as g/100 g of TS.

[c] Expressed as g of lactic acid/100 g of TS.

The pH values decreased during the first 15 days of cheese ripening but subsequently remained constant until its completion, when a significant increase to reach a final mean value of 5.90 ± 0.21 was noted. This value is higher than that reported by other authors for ewe's milk cheeses with similar characteristics (González-Viñas et al. 2001a; Ballesteros et al. 2006; Mangia et al. 2008).

The titratable acidity content increased steadily throughout the ripening process to reach a mean value of 1.82 ± 0.24 g lactic acid/100 g TS, twice the initial value. The change in acidity content in Zamorano cheese PDO is related to the influence of certain parameters such as moisture content, salt concentration, and water activity on the growth and activity of the microbial flora present in the cheese.

Finally, the water activity values decreased throughout the ripening process, although with two notable inflexion points in this trend. The first of these (first week of ripening) was a result of the salting process to which the cheese was submitted and the other (final months of ripening) to the important degree of proteolysis, which increases the number of ionic groups capable of binding free water, that the cheese underwent. The mean a_w value at the end of the ripening process was 0.925 ± 0.005. These values were similar to those reported for other mature ewe's milk cheeses (Fontecha et al. 1994; González-Viñas et al. 2001a), but lower than those for Idiazábal (Ibáñez et al. 1995) and Castellano cheeses (Román-Blanco et al. 1999).

The pH, moisture, fat (expressed as percentage of TS), and lactose contents (%) we obtained were higher than those reported by Revilla et al. (2007) for Zamorano cheese, although the protein content was very similar. These differences can be explained by the technique used by these authors to manufacture the cheese at a pilot plant level, especially the use of a greater proportion of starter culture than would normally be used during the industrial production of Zamorano cheese and a higher salting temperature (18°C). Likewise, the temperature (18°C) and RH (70%) of the ripening chamber were very different from those commonly used by industrial and traditional cheese manufacturers during the production of Zamorano cheese PDO (10°C–12°C and 80%–85%, respectively).

In contrast, Revilla et al. (2009b) noted that the presence of high levels of somatic cells in the milk used to make the Zamorano cheese had a significant effect on its physicochemical composition. Thus, the TS and fat contents were lower, whereas the pH and fat acidity index were much higher than in cheeses produced from milk containing low levels of somatic cells. Furthermore, from a sensory point of view, the cheeses produced from milk containing high somatic cell levels were much more pungent and granular and less creamy, and were less well accepted by the consumers.

24.6 Proteolysis during Ripening of Zamorano Cheese

Proteolysis is perhaps the most important and complex biochemical process that takes place during the ripening of Zamorano cheese PDO. The peptides, free amino acids, and catabolites of the latter have a major influence on the flavor and textural characteristics of this type of uncooked, pressed-paste cheese.

The main changes in the soluble nitrogenous compounds [total soluble nitrogen (TSN), nonprotein nitrogen (NPN), ammonia nitrogen (NH_3–N), amino acid nitrogen (NH_2–N), polypeptide nitrogen included after polypeptide nitrogen (ppN), and peptide nitrogen (pN)] during the ripening of Zamorano cheese are shown in Figure 24.7.

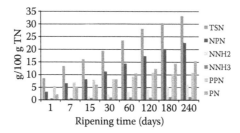

FIGURE 24.7 Changes in soluble nitrogenous compounds during ripening of Zamorano cheese (TSN: total soluble nitrogen; NPN: nonprotein nitrogen; NNH2: amino acid nitrogen; NNH3: ammonia nitrogen; PPN: polypeptide nitrogen; PN: peptide nitrogen).

The degree of proteolysis of Zamorano cheese, expressed as % TSN/TN, was notably high throughout ripening, reaching a final value of 33.30 g TSN/100 g TN. These values were higher than those reported previously for other varieties of ewe's milk cheese such as Idiazábal (Ibáñez et al. 1995), Roncal (Irigoyen et al. 2001), Fiore Sardo and Pecorino-Romano (Di Cagno et al. 2003), and Manchego (Ballesteros et al. 2006), but lower than those for Los Pedroches (Sanjuan et al. 2002) and Serra da Estrela (Tavaria et al. 2003) cheeses. However, it should be noted that the latter two were produced using vegetable rennet, which is more proteolytic than its animal counterpart.

The % NPN/TN, expressed in terms of % TN, increased significantly throughout the ripening process, reaching a final mean value of 22.45 g NPN/100 g TN. This represented 67.4% of the TSN, thereby indicating a notable proteolytic activity of the microbiota present in the cheese. These high % NPN values were mainly a result of the pN content, expressed as % TN, which accounted for 68% of the NPN at the end of the ripening process, and the NH_2–N, expressed as % TN, which represented 27% of the NPN. The remaining 5% of NPN was made up of NH_3–N. The aminopeptidase activity resulting from the natural or added microbiota present in the Zamorano cheese PDO was more important than the peptidase activity. Thus, the % NH_2–N/TN increased almost 10-fold during ripening, whereas the % pN/TN only increased sevenfold. The % NH_2–N/TN at the end of the ripening of Zamorano cheese (6.07) was very similar to that reported by Cabezas et al. (2007) for Manchego cheese.

A typical electropherogram showing the changes in the caseins and their degradation products during the ripening of Zamorano cheese PDO is shown in Figure 24.8. It can be seen from this figure that α_s-casein undergoes the greatest degree of hydrolysis during ripening, with its content decreasing by almost 50% with respect to the initial value. This drop in α_s-casein content resulted in an almost 2.4-fold increase in its degradation products, as represented by the pre-α_s-casein fraction. These findings suggest that the clotting enzyme is mainly responsible for primary proteolysis in Zamorano cheese.

The alkaline protease activity in this type of cheese was not particularly high, as can be seen from the limited degradation of β_s-casein, which did not exceed 20% of the initial content. These results are to be expected for uncooked, pressed-paste cheeses such as Zamorano cheese, where the pH (5.5), water activity (approximately 0.910), and salt/moisture concentration (approximately 4–8%) do not favor the optimal activity of this enzyme.

Revilla et al. (2007) found no hydrolysis of α_s- or β_s-casein in Zamorano cheese ripened for three months. This lack of primary proteolysis in the cheese was related to the great variability between the different batches of cheese studied.

The changes in free amino acid content, expressed as mg/100 g TS, during the ripening of Zamorano cheese are shown in Table 24.4. The free amino acid content is a good indicator of the degree of proteolysis in the cheese and its sensory qualities as these components contribute to the flavor, either alone or as a result of their catabolites. The free amino acid content in Zamorano cheese increased almost 20-fold

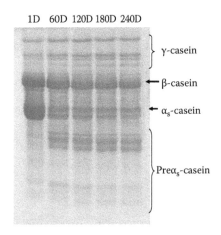

FIGURE 24.8 Typical electrophoretogram of casein fractions during ripening of Zamorano cheese (1D: 1 day; 60D: 60 days; 120D: 120 days; 180D: 180 days; and 240D: 240 days).

TABLE 24.4

Changes in Concentration of Free Amino Acids during Ripening of Zamorano Cheese PDO[a]

Amino Acids	Ripening Time (days)							
	1	**7**	**15**	**30**	**60**	**120**	**180**	**240**
Asp	9.45	11.39	14.61	21.38	29.40	46.95	73.78	98.86
Glu	23.56	32.08	47.23	60.33	122.10	243.66	356.96	503.82
Asn	7.79	17.81	24.08	48.67	99.31	175.67	204.23	282.33
Ser	8.18	12.96	17.34	23.35	32.85	42.80	59.13	72.71
Gln	16.49	27.66	37.38	65.78	98.46	152.97	189.40	251.44
Gly	2.43	3.47	4.48	6.53	9.93	16.55	22.01	31.74
His	4.48	5.63	7.17	8.75	12.19	16.58	25.40	36.08
Arg	0.80	6.70	7.98	9.98	14.90	22.10	33.61	45.06
Tau + GABA	18.42	37.45	58.65	104.24	184.07	209.88	263.23	272.16
Thr	5.05	6.80	9.63	15.21	26.13	42.48	56.81	73.47
Ala	9.89	14.81	19.45	30.48	49.97	82.37	104.55	144.54
Pro	11.43	15.66	20.20	24.29	39.37	59.54	75.82	97.02
Tyr	6.47	10.42	13.02	18.42	28.78	48.17	59.43	75.88
Val	8.73	17.80	22.29	47.37	93.71	138.50	175.12	202.30
Met	3.24	5.78	9.74	13.85	24.70	41.31	55.85	75.85
Cys	ND	6.14	9.12	11.45	15.18	20.26	23.25	51.65
Ile	4.39	7.12	9.98	16.18	32.33	54.64	77.93	102.55
Leu	13.80	27.78	34.04	69.16	117.57	207.16	269.96	331.52
Phe	8.19	15.86	20.95	40.35	84.26	111.60	131.28	164.91
Trp	10.33	26.48	33.77	56.65	100.95	144.56	164.90	217.52
Lys	15.99	27.88	39.38	57.00	127.09	210.76	304.37	436.34

[a] Expressed as mg/100 g of TS.

throughout the ripening process, from an initial value of 189.11 mg/100 g TS to a final value of 3567.75 mg/100 g TS. These values were higher than those reported previously for Idiazábal (Vicente et al. 2001) and Manchego cheese (Poveda et al. 2004) but similar to those reported by Tavaria et al. (2003) for Serra da Estrela cheese and Irigoyen et al. (2007) for Roncal-type cheese.

The amino acid profile of Zamorano cheese PDO is characterized by a high Glu, Lys, Leu, Asn, Gln, Trp, and Val content, which together represent 62.4% of all free amino acids. This profile was similar to that observed for other types of uncooked pressed-paste ewe's milk cheeses (Vicente et al. 2001; Tavaria et al. 2003; Poveda et al. 2004).

24.7 Lipolysis during Ripening of Zamorano Cheese

The hydrolysis of triglycerides to free fatty acids (FFAs) plays a key role in the development of flavor during cheese ripening. Lipolysis in uncooked pressed-paste cheeses ripened by bacteria is not as important as proteolysis in sensory terms. However, low levels of lipolysis influence the final aroma of the cheese, especially in cheeses made from ewe's milk due to its relatively high short- to medium-chain-length fatty acid content (C_4–C_{10}).

A study undertaken by Fernández-García et al. (2006) regarding the FFA content of Zamorano cheese showed that, as is the case with other ewe's milk cheeses such as Manchego and La Serena, the predominant FFAs were palmitic ($C_{16:0}$) and oleic ($C_{18:1}$). The degree of lipolysis in Zamorano cheese was relatively low, with values ranging between 1460 and 2864 mg/kg of total FFAs after ripening for eight months (see Table 24.5). These values were similar to those reported previously by the same authors for Manchego cheese, but much lower than those for La Serena cheese. In a similar study undertaken by Etayo et al. (2006), the FFA content of Zamorano cheese was found to be significantly lower than that

TABLE 24.5

Means ± Standard Deviation of FFA Concentrations (mg/kg) Found in Zamorano Cheeses Made from Raw Milk throughout the Year

Free Fatty Acid	Age (months)	Spring	Summer	Autumn	Winter	Age	Season	Dairy
						Significance of Effects		
Acetic (C2:0)	4	1102 ± 246[a]	1197 ± 253[a]	812 ± 301[b]	866 ± 256[b]	***	***	***
	8	1344 ± 228[a]	1248 ± 226[ab]	942 ± 168[c]	1096 ± 233[bc]		***	***
Butyric	4	110 ± 34[a]	102 ± 29[a]	59 ± 14[b]	76 ± 30[b]	***	***	***
(C4:0)	8	201 ± 65[a]	167 ± 46[b]	113 ± 26[c]	136 ± 36[c]		***	***
Caproic	4	58 ± 17[a]	49 ± 13[ab]	32 ± 6[c]	43 ± 19[b]	***	***	***
(C6:0)	8	105 ± 37[a]	65 ± 12[b]	63 ± 12[b]	71 ± 23[b]		***	***
Caprilic	4	102 ± 29[a]	88 ± 16[b]	80 ± 13[b]	87 ± 25[b]	***	***	***
(C8:0)	8	146 ± 55[a]	102 ± 14[b]	108 ± 18[b]	114 ± 34[b]		***	***
Capric	4	152 ± 65[a]	116 ± 36[b]	109 ± 33[b]	122 ± 61[b]	***	***	***
(C10:0)	8	254 ± 121[a]	152 ± 25[b]	176 ± 41[b]	178 ± 81[b]		***	***
Lauric	4	93 ± 33[a]	72 ± 17[b]	70 ± 19[b]	77 ± 33[b]	***	***	***
(C12:0)	8	152 ± 61[a]	96 ± 15[b]	110 ± 22[b]	108 ± 45[b]		***	***
Myristic	4	192 ± 61[a]	172 ± 40[a]	132 ± 33[b]	143 ± 53[b]	***	***	***
(C14:0)	8	320 ± 114[a]	238 ± 43[b]	218 ± 45[b]	218 ± 75[b]		***	***
Palmitic	4	419 ± 112[a]	392 ± 100[a]	277 ± 61[b]	309 ± 91[b]	***	***	***
(C16:0)	8	672 ± 194[a]	528 ± 131[b]	419 ± 74[c]	468 ± 125[bc]		***	***
Stearic	4	154 ± 26[a]	153 ± 32[a]	114 ± 14[b]	120 ± 21[b]	***	***	***
(C18:0)	8	218 ± 43[a]	154 ± 38[bc]	137 ± 13[c]	155 ± 28[b]		***	***
Oleic (C18:1)	4	390 ± 143[a]	404 ± 125[a]	253 ± 67[b]	271 ± 111[b]	***	***	***
	8	627 ± 259[a]	518 ± 186[b]	377 ± 78[c]	420 ± 147[c]		***	***
Linoleic	4	118 ± 32[a]	114 ± 18[a]	89 ± 13[b]	98 ± 23[b]	***	***	***
(C18:2)	8	168 ± 62[a]	131 ± 21[b]	117 ± 17[b]	129 ± 35[b]		***	***
Sum	4	1789 ± 552[a]	1663 ± 428[a]	1216 ± 272[b]	1346 ± 467[b]	***	***	***
C4:0–C18:2	8	2864 ± 1011[a]	2153 ± 530[ab]	1460 ± 345[c]	1998 ± 629[b]		***	***

Source: Fernández-García, E. et al., 2006. *Int Dairy J*, 16, 252–261, 2006. With permission.
Note: Significance of effects: *$P < 0.05$; **$P < 0.01$; and ***$P < 0.001$.
[abc] Means followed by the same superscript letter within the same row are not significantly different ($P > 0.05$).

for Manchego cheese but similar to that for Idiazábal cheese. Fernández-García et al. (2006) also found that the type of cheese manufacturer, ripening time, and season in which the cheeses were produced had a very significant influence on the degree of lipolysis of Zamorano cheese PDO. Thus, the FFA content increased with ripening time and was higher for those cheeses produced in spring and summer.

The short- (SCFA), medium- (MCFA), and long-chain fatty acid (LCFA) content with respect to total FFA content was very similar in Zamorano and Manchego cheeses but different in the values reported for La Serena cheese. Furthermore, the La Serena cheese had a higher unsaturated fatty acid/LCFA ratio than Zamorano and Manchego cheeses, whereas the MCFA content was lower. These differences between cheeses could be due to the type of feed provided for the sheep flocks.

The proportion of SCFAs, MCFAs, and LCFAs with respect to total FFAs reported by Etayo et al. (2006) for Zamorano cheese were very different than those reported by Fernández-García et al. (2006), especially as regards SCFAs and LCFAs. Indeed, the % SCFAs and LCFAs determined by Etayo et al. (2006) were 40% and 48%, respectively, whereas the values reported by Fernández-García et al. (2006) were 22.6% and 60%, respectively. These differences could be related to the natural lipase activity of the milk and the lipolytic activity of the microbiota present in the cheese during ripening.

The addition of lamb rennet paste to the milk is common practice during the production of some varieties of Spanish ewe's milk cheese such as Idiazábal but not for others such as Manchego or Zamorano.

The use of rennet pastes was very common many years ago in traditional cheese making, although currently, it is much rarer due, above all, to the lack of standardization of the rennet, its difficult preparation, and its low microbiological quality when compared with commercial liquid or powdered rennets (Etayo et al. 2006). However, the absence of rennet paste results in an important loss of the flavors for which traditional cheeses were renowned and therefore an excessive standardization of the flavors of many ewe's milk cheeses belonging to a PDO. In this respect, Etayo et al. (2006) showed in a study performed with Zamorano cheeses manufactured with rennet paste and commercial rennet that the former were graded slightly higher in terms of smell and flavor than the latter due to their higher degree of lipolysis. However, in contrast to the findings with Manchego and Idiazábal cheese, these differences were not significant. Etayo et al. (2006) concluded that larger amounts of lipase-containing rennet paste (>25 g 100 L^{-1}) should be added to enhance the pungent taste of Zamorano cheese PDO. In our opinion, one problem that may arise from the widespread use of rennet paste in the manufacture of Zamorano cheese is that, although it would be acceptable to a significant number of consumers, the vast majority prefer a much milder cheese.

24.8 Sensory Characteristics of Zamorano Cheese

Traditional cheese varieties are normally characterized on the basis of studies of their manufacturing process and their physicochemical and microbiological characteristics. However, to date, such characterization does not generally include an analysis of their sensory characteristics. This has begun to change, however, and an analysis of sensory parameters, using either tasting panels or instrumental methods, has formed a major part of numerous studies published in the past few years. Indeed, the color, texture, and flavor are key factors that are taken into account by consumers when assessing cheese quality.

Despite this, studies concerning the sensory characteristics of Zamorano cheese are very scarce in comparison with those regarding other types of uncooked pressed-paste ewe's milk cheeses such as Manchego, Idiazábal, and Roncal. The majority of such studies have concentrated on the influence of factors such as the time of year when the cheese was manufactured (Fernández-García et al. 2004), the type of animal (bovine or lamb) rennet added (Barrón et al. 2005) and lamb rennet paste (Etayo et al. 2006), and the somatic cell content of the milk and the breed from which it was obtained (Lurueña-Martínez et al. 2010) on the volatile components and sensory characteristics.

The typical flavor of each variety of cheese is determined by the equilibrium between volatile and nonvolatile chemical constituents released from the fats, proteins, and carbohydrates during ripening (Fox and Wallace 1997).

A total of 90 constituents have been identified in the volatile fraction of Zamorano cheese (Fernández-García et al. 2004). The majority of these constituents have also been detected in other varieties of cheese and include aldehydes (2-propenal, 3-methyl-1-butanal), ketones (2-butanone and 2,3-butanedione), alcohols (ethanol and propanol), and ethyl and propyl esters (see Tables 24.6 and 24.7). However, these authors detected three ketones, namely, 3-buten-2-one, 4-methyl-2-pentanone and 3-methyl-2-pentanone in Zamorano cheese, which had not been found in other varieties of PDO cheese such as Manchego, La Serena, Idiazábal, and Roncal. Likewise, Barrón et al. (2005) only detected the presence of 3-hydroxy-2-butanone in Zamorano cheese.

The overall odor and flavor intensities were higher in Zamorano cheese than those reported for Manchego and Roncal cheese, but similar to those for Idiazábal cheese (Barrón et al. 2005). The most highly scored attributes, which differentiated it from other cheese varieties, were the buttery and toasty odors. Barrón et al. (2005) associated these odors with the higher content of acetic acid, methyl ketones, and their reduction products in the Zamorano cheese.

The odor and aroma qualities of Zamorano cheese are strongly influenced by the manufacturer and the season in which the cheeses are produced, whereas the odor and aroma intensities are affected by the season and ripening time (Fernández-García et al. 2004). The highest scoring cheeses were those with higher values for attributes such as lactic and animal family and lower values for rancidity level. The toasted flavor, which was more intense in those cheeses produced in winter, and pungent flavor, which was more intense in those cheeses produced in the spring, were both evaluated positively.

TABLE 24.6

Relative Abundance* (Mean ± SD) of Carbonyl Compounds Detected in the Volatile Fraction of 8-Month-Old Zamorano Cheese Made in Different Seasons

	Spring	Summer	Autumn	Winter	HSC[†]	Significant Effects[‡]		
						Dairy	Season	Age
Linear Aldehydes								
Acetaldehyde	0.74 ± 0.28^b	0.94 ± 0.31^{ab}	1.05 ± 0.26^a	0.83 ± 0.31^{ab}	0.87 ± 0.31	**	*	***
Propanal	0.52 ± 0.31^{ab}	1.03 ± 1.18^a	0.43 ± 0.44^b	0.58 ± 0.47^{ab}	0.62 ± 0.46	***	*	***
2-Propenal	4.55 ± 6.19	3.38 ± 2.87	3.92 ± 3.61	3.16 ± 3.64	6.16 ± 6.12	***	NS	NS
n-Hexanal	1.31 ± 0.90^a	0.73 ± 0.21^b	0.58 ± 0.35^b	0.85 ± 0.54^{ab}	0.93 ± 0.51	***	**	*
n-Heptanal	0.16 ± 0.09	0.61 ± 1.20	0.04 ± 0.03	0.08 ± 0.04	0.11 ± 0.04	NS	NS	NS
n-Nonanal	0.13 ± 0.06^{ab}	0.09 ± 0.06^b	0.10 ± 0.09^{ab}	0.20 ± 0.16^a	0.18 ± 0.17	NS	*	***
n-Decanal	0.11 ± 0.06	0.13 ± 0.10	0.09 ± 0.08	0.09 ± 0.05	0.10 ± 0.05	NS	NS	NS
Branched Chain Aldehydes								
2-Methyl propanal	1.30 ± 0.47^a	0.96 ± 0.53^b	0.50 ± 0.68^b	1.17 ± 0.47^a	1.10 ± 0.38	NS	***	***
2-Methyl-1-butanal	1.42 ± 0.52^a	1.13 ± 0.45^{ab}	0.72 ± 0.49^b	1.49 ± 0.73^a	1.34 ± 0.57	NS	**	***
3-Methyl-1-butanal	4.99 ± 1.70^a	4.39 ± 1.75^{ab}	2.91 ± 1.86^b	4.71 ± 1.89^a	4.51 ± 1.35	NS	*	***

Methyl-ketones									
2-Propanone	2.50 ± 0.93^b	4.44 ± 2.64^b	7.09 ± 5.14	4.47 ± 3.44^b	4.58 ± 3.28	***	***	***	***
2-Butanone	220 ± 194^b	289 ± 368^ab	451 ± 329^a	2753 ± 288^ab	160 ± 91.8	***	*	*	NS
2-Pentanone	52.25 ± 50.79	65.00 ± 72.37	84.32 ± 98.76	72.89 ± 73.55	51.26 ± 75.62	***	***	NS	***
2-Hexanone	1.20 ± 1.18	1.05 ± 0.99	1.19 ± 1.43	1.51 ± 1.39	0.88 ± 1.33	***	***	NS	***
2-Heptanone	22.26 ± 20.16^ab	20.70 ± 20.85^ab	14.89 ± 19.29^b	33.63 ± 28.83^a	15.02 ± 17.81	***	***	**	***
2-Nonanone	1.83 ± 2.02^ab	2.13 ± 2.83^ab	0.56 ± 0.48^b	2.34 ± 1.68^a	1.55 ± 1.51	***	***	*	***
4-Methyl-2-pentanone	0.20 ± 0.26	0.09 ± 0.13	0.32 ± 0.41	0.29 ± 0.17	0.31 ± 0.26	*	*	***	*
3-Methyl-2-pentanone	0.16 ± 0.36	0.17 ± 0.15	0.08 ± 0.11	0.11 ± 0.11	0.03 ± 0.05	**	**	NS	NS
Other Ketones									
3-Buten-2-one	0.22 ± 0.17	0.79 ± 1.04	0.58 ± 0.31	0.76 ± 1.37	0.20 ± 0.20	*	*	NS	***
2,3-Butanedione	34.38 ± 8.96^b	22.96 ± 8.91^b	127 ± 126^a	36.54 ± 34.14^b	32.70 ± 34.81	***	***	***	NS
2,3-Pentanedione	0.37 ± 0.16^b	0.21 ± 0.11^c	0.54 ± 0.52^a	0.30 ± 0.18^b	0.36 ± 0.19	***	***	***	NS
3-Hydroxy-2-butanone	1.26 ± 1.07^b	0.32 ± 0.21	3.78 ± 4.60^a	0.68 ± 0.96^b	1.11 ± 1.03	**	**	***	*

Source: Fernández-García, E. et al., *Int Dairy J*, 14, 701–711, 2004. With permission.

^abc Means followed by the same letter within the same row are not significantly different ($P > 0.05$).

* Relative abundance expressed as percentage of the cyclohexanone peak.

† HSC: highest scored cheeses.

‡ NS: not significant; *($P < 0.05$), **($P < 0.01$), and ***($P < 0.001$).

TABLE 24.7

Relative Abundance* (Mean ± SD) of Alcohols Detected in the Volatile Fraction of 8-Month-Old Zamorano Cheese Made in Different Seasons

	Spring	Summer	Autumn	Winter	HSC[†]	Significant Effects[‡]		
						Dairy	Season	Age
Primary Alcohols								
Ethanol	132 ± 92.4^{ab}	109 ± 57.06^{b}	141 ± 86.20^{ab}	267 ± 364^{a}	333 ± 334	***	*	***
1-Propanol	121 ± 107^{b}	160 ± 135^{a}	82.55 ± 92.30^{c}	166 ± 167^{a}	215 ± 166	***	***	*
2-Propen-1-ol	51.20 ± 72.42^{a}	10.46 ± 12.15^{b}	45.70 ± 65.22^{ab}	50.95 ± 92.03^{a}	98.98 ± 93.27	***	*	***
1-Butanol	14.12 ± 9.54	16.69 ± 10.09	8.27 ± 6.44	17.04 ± 21.26	20.42 ± 20.98	***	NS	**
1-Pentanol	0.62 ± 0.27^{a}	0.54 ± 0.14^{a}	0.34 ± 0.11^{b}	0.56 ± 0.16^{a}	0.65 ± 0.27	**	***	**
1-Hexanol	1.31 ± 1.02^{ab}	2.68 ± 3.38^{a}	0.73 ± 0.57^{b}	1.20 ± 1.26^{b}	1.36 ± 1.25	**	**	**
Secondary Alcohols								
2-Propanol	24.57 ± 7.98	31.38 ± 10.92	19.60 ± 10.28	28.84 ± 20.36	31.91 ± 19.30	NS	NS	***
2-Butanol	1134 ± 525^{ab}	1287 ± 767^{a}	1012 ± 595^{b}	1024 ± 582^{b}	812 ± 321	***	*	NS
2-Pentanol	34.88 ± 21.97^{a}	29.07 ± 20.37^{a}	11.82 ± 11.37^{b}	33.69 ± 30.00^{a}	12.70 ± 13.02	***	***	***
2-Hexanol	0.31 ± 0.24^{a}	0.32 ± 0.29^{a}	0.11 ± 0.11^{b}	0.39 ± 0.39^{a}	0.12 ± 0.12	***	***	***
2-Heptanol	1.92 ± 1.08^{a}	2.38 ± 1.68^{a}	0.60 ± 0.43^{b}	2.45 ± 2.37^{a}	1.04 ± 0.74	***	***	*
Branched Chain Alcohols								
2-Methyl-2-propanol	0.52 ± 0.75	0.75 ± 1.15	0.78 ± 1.08	0.84 ± 1.16	1.32 ± 1.02	***	NS	***
2-Methyl-1-propanol	8.26 ± 6.50	24.88 ± 46.05	3.83 ± 1.31	3.43 ± 1.76	7.92 ± 6.92	NS	NS	NS
3-Methyl-1-butanol	20.52 ± 19.12	85.44 ± 168	8.70 ± 4.13	7.00 ± 3.73	19.88 ± 19.78	NS	NS	NS
4-Methyl-2-pentanol	0.43 ± 0.54	0.25 ± 0.22	0.19 ± 0.15	0.17 ± 0.14	0.32 ± 0.56	***	NS	NS
2-Methyl-3-butenol	0.91 ± 0.24^{a}	0.59 ± 0.26^{b}	0.66 ± 0.23^{b}	0.74 ± 0.3^{ab}	0.90 ± 0.28	**	**	***
3-Methyl-2-butenol	0.13 ± 0.13^{ab}	0.04 ± 0.07^{b}	0.10 ± 0.07^{ab}	0.16 ± 0.20^{a}	0.25 ± 0.16	***	*	***
Other Alcohols								
1-Methoxy-2-propanol	8.99 ± 15.53^{a}	1.35 ± 1.25^{ab}	0.99 ± 0.83^{b}	1.19 ± 0.84^{b}	1.84 ± 1.51	NS	*	*
2-Butoxyethanol	29.77 ± 43.77^{b}	43.27 ± 64.63^{a}	29.77 ± 44.96^{b}	17.08 ± 26.23^{c}	46.41 ± 39.22	***	***	***
Cyclohexanol	0.22 ± 0.45	0.86 ± 10.99	0.12 ± 0.15	0.48 ± 0.72	0.04 ± 0.08	**	NS	**

Source: Fernández-García, E. et al., *Int Dairy J*, 14, 701–711, 2004. With permission.

^{abc} Means followed by the same letter within the same row are not significantly different ($P > 0.05$).

* Relative abundance expressed as percentage of the cyclohexanone peak.

[†] HSC: highest scored cheeses.

[‡] NS: not significant; *($P < 0.05$), **($P < 0.01$), and ***($P < 0.001$).

Etayo et al. (2006) found no significant differences in terms of odor and flavor profiles between Zamorano cheese produced with commercial rennet and that produced with lamb rennet paste other than a nutty flavor. These results were somewhat unexpected and were attributed to differences in the lipolytic parameters (total FFA content, percentage of each FFA, and total glyceride content), and interactions between them, which could affect the perceived intensity of the individual sensory attributes.

The sensory characteristics of Zamorano cheese can also be affected by the somatic cell content of the milk used to manufacture the cheese and the breed from which it is obtained. Indeed, Lurueña-Martínez et al. (2010) observed significant differences in terms of hardness, taste intensity, and pungency of cheeses in relation to their somatic cell counts. Thus, the Zamorano cheese manufactured from milk containing higher levels of somatic cells was softer, had greater taste intensity, and was more pungent due to its higher degree of proteolysis and lipolysis.

The same authors also found significant differences in terms of color and hardness depending on the breed of sheep. Thus, Zamorano cheese manufactured with milk from Churra breed was yellower than that manufactured using milk from the Castellana and Assaf breeds. Likewise, the Warner–Bratzler Shear Force was significantly higher in the Zamorano cheese manufactured using milk from Castellana breed. Lurueña-Martínez et al. (2010) found a good correlation between the hardness and color determined both instrumentally and using a consumer panel.

Likewise, Revilla et al. (2009a) noted that it is possible to determine the texture of Zamorano cheese by near-infrared (NIR) spectroscopy, together with a fiber optic probe, in the range between 0 and 49 N. The main advantages of this technique are that it can be performed quickly and is nondestructive.

REFERENCES

Ballesteros C, Poveda PM, González-Viñas MA, Cabezas L. 2006. Microbiological, biochemical and sensory characteristics of artisanal and industrial Manchego cheeses. *Food Control* 17:249–255.

Barrón LJR, Redondo Y, Flanagan CE, Pérez-Elortondo FJ, Albisu M, Nájera AI, De Renobales M, Fernández-García E. 2005. Comparison of the volatile composition and sensory characteristics of Spanish PDO cheeses manufactured from ewes' raw milk and animal rennet. *Int Dairy J* 15:371–382.

Bettencourt CMV, Matos CAP, Batista T, Canada J, Fialho JBR. 1998. Preliminary data on the ewe breed effect on the quality of Portuguese Serpa cheese. In *Basis of the Quality of Typical Mediterranean Animal Products. European Association of Animal Production* no. 90:234–238 [JC Flamant, D Gabiña, and M Espejo Díaz, editors]. Wageningen, The Netherlands: Wageningen Pers.

Boletín Oficial del Estado (BOE). 1993. Order of 6-05-1993 of Ministry of Agriculture, Fish and Food that approves the Regulation board of Queso Zamorano with Designation of Origin and the Regulatory Council. (BOE no. 120, May 20, 1993). Available from: http://www.mapa.es/alimentacion/pags/Denominacion/Quesos/Zamorano/BOE_120_200593.pdf. Accessed Dec 15, 2010.

Bravo D, Rodríguez E, Medina M. 2009. Nisin and lacticin 481 coproduction by *Lactococcus lactis* strains isolated from raw ewes' milk. *J Dairy Sci* 92:4805–4811.

Cabezas L, Sánchez I, Poveda JM, Seseña S, Palop ML. 2007. Comparison of microflora, chemical and sensory characteristics of artisanal Manchego cheeses from two dairies. *Food Control* 18:11–17.

Cleveland J, Montville TJ, Nes IF, Chikindas ML. 2001. Bacteriocins: safe, natural antimicrobials for food preservation. *Int J Food Microbiol* 71:1–20.

Cogan TM, Barbosa M, Beuvier E, Bianchi-Salvadori B, Cocconcelli PS, Fernandes I, Gómez J, Gómez R, Kalantzopoulos G, Ledda A, Medina M, Rea MC, Rodríguez E. 1997. Characterization of the lactic acid bacteria in artisanal dairy products. *J Dairy Res* 64:409–421.

Di Cagno R, Banks J, Sheehan L, Fox PF, Brechany EY, Corsetti A, Gobbetti M. 2003. Comparison of the microbiological, compositional, biochemical, volatile profile and sensory characteristics of three Italian PDO ewes' milk cheeses. *Int Dairy J* 13:961–972.

Etayo I, Pérez Elortondo FJ, Gil PF, Albisua M, Virto M, Conde S, Rodríguez Barrón LJ, Nájera AI, Gómez-Hidalgo ME, Delgado C, Guerra A, De Renobales M. 2006. Hygienic quality, lipolysis and sensory properties of Spanish Protected Designation of Origin ewe's milk cheeses manufactured with lamb rennet paste. *Lait* 86:415–434.

Fernández-García E, Carbonell M, Calzada J, Núñez M. 2006. Seasonal variation of the free fatty acids contents of Spanish ovine milk cheeses protected by a designation of origin: a comparative study. *Int Dairy J* 16:252–261.

Fernández-García E, Carbonell M, Gaya P, Núñez M. 2004. Evolution of the volatile components of ewes' raw milk Zamorano cheese. Seasonal variation. *Int Dairy J* 14:701–711.

Ferrazza RE, Fresno JM, Ribeiro JI, Tornadijo ME, Mansur Furtado M. 2004. Changes in the microbial flora of Zamorano cheese (P.D.O.) by accelerated ripening process. *Food Res Int* 37:149–155.

Fontecha J, Peláez C, Juárez M. 1994. Biochemical characteristics of a semi-hard ewe's-milk cheese. *Z Lebensm Unters Forsch* 198:24–28.

Food and Agriculture Organisation of the United Nations (FAO). 2010. World sheep milk production. Available from: http://faostat.fao.org/site/339/default.aspx. Accessed Dec 15, 2010.

Fox PF, Wallace JM. 1997. Formation of flavour compounds in cheese. *Adv Appl Microbiol* 45:17–85.

González J, Lozano M, Mas M, Mendiola FJ, Roa I. 1997. Características de las leches y los quesos de la D.O. queso La Serena. *Alimentaria* 285:35–39.

González-Viñas MA, Poveda J, García Ruiz A, Cabezas L. 2001b. Changes in chemical, sensory and rheological characteristics of Manchego cheeses during ripening. *J Sens Stud* 16:361–371.

González-Viñas MA, Poveda JM, Cabezas I. 2001a. Sensory and chemical evaluation of Manchego cheese and other cheese varieties available in the Spanish market. *J Food Qual* 24:157–165.

Ibáñez FC, Torres MI, Ordóñez AI, Barcina Y. 1995. Effect of composition and ripening on casein breakdown in Idiazábal cheese. *Chem Mikrobiol Technol Lebensm* 171:37–44.

Ibáñez FC, Torres MI, Pérez-Elortondo FJ, Barcina Y. 1993. Physicochemical changes during ripening of Idiazábal induced by brining time. *Chem Mikrobiol Technol Lebensm* 15:79–83.

Irigoyen A, Izco JM, Ibáñez FC, Torre P. 2001. Influence of rennet milk-clotting activity on the proteolytic and sensory characteristics of an ovine cheese. *Food Chem* 72:137–144.

Irigoyen A, Ortigosa M, Juansaras I, Oneca M, Torre P. 2007. Influence of an adjunct culture of *Lactobacillus* on the free amino acids and volatile compounds in a Roncal-type ewe's-milk cheese. *Food Chem* 100:71–80.

Larueña-Martínez MA, Revilla I, Severiano-Pérez P, Vivar-Quintana AM. 2010. The influence of breed on the organoleptic characteristics of Zamorano sheep's raw milk cheese and its assessment by instrumental analysis. *Int J Dairy Technol* 63:216–223.

Mangia NP, Murgia MA, Garau G, Sanna MG, Deiana P. 2008. Influence of selected lab cultures on the evolution of free amino acids, free fatty acids and Fiore Sardo cheese microflora during the ripening. *Food Microbiol* 25:366–377.

Martín-Hernández MC, Amigo L, Martín-Álvarez PJ, Juárez M. 1992. Differentiation of milks and cheeses according to species based on the mineral content. *Z Lebensm Unters Forsch* 194:541–544.

Martín-Hernández MC, Juárez M. 1989. Retention of main and trace elements in four types of goat cheese. *J Dairy Sci* 72:1092–1097.

Micari P, Caridi A, Colacino T, Caparra P, Cufari A. 2002. Physicochemical, microbiological and coagulating properties of ewe's milk produced on the Calabrian Mount Poro plateau. *Int J Dairy Technol* 55:204–210.

Ministerio de Medio Ambiente y Medio Rural y Marino (MARM). Gobierno de España. 2010a. Anuario Estadístico Agrario. Available from: http://www.mapa.es/estadistica/pags/anuario/2009/AE_2009_14_03_01_05 .pdf. Accessed Dec 15, 2010.

Ministerio de Medio Ambiente y Medio Rural y Marino (MARM). Gobierno de España. 2010b. Anuario Estadístico Agrario. Available from: http://www.mapa.es/estadistica/pags/anuario/2009/AE_2009_14_03_01_06 .pdf. Accessed Dec 15, 2010.

Ministerio de Medio Ambiente y Medio Rural y Marino (MARM). Gobierno de España. 2010c. Datos de las Denominaciones de Origen Protegidas (D.O.P.) e Indicaciones Geográficas Protegidas (I.G.P.) de Productos Agroalimentarios. Año 2008. Available from: http://www.mapa.es/alimentacion/pags/ denominacion/documentos/Agroalimentarios2008.pdf. Accessed Dec 15, 2010.

Ministerio de Medio Ambiente y Medio Rural y Marino (MARM). Gobierno de España. 2010d. Cifras y datos de las Denominaciones de Origen Protegidas (D.O.P.) e Indicaciones Geográficas Protegidas (I.G.P.) de Productos Agroalimentarios. Años 2001–2008. Available from: http://www.mapa.es/es/alimentacion/ pags/Denominacion/htm/cifrasydatos.htm. Accessed Dec 15, 2010.

Moreno-Rojas R, Amaro-López MA, García-Gimeno RH, Zurera-Cosano G. 1995. Effects of Manchego-type cheese-making process on contents of mineral elements. *Food Chem* 53:435–439.

Moreno-Rojas R, Amaro-López MA, Zurera-Cosano G. 1993. Micronutrients in natural cow, ewe and goat milk. *Int J Food Sci Nutr* 44:37–46.

Ortigosa M, Arizcun C, Irigoyen A, Oneca M, Torre P. 2006. Effect of lactobacillus adjunct cultures on the microbiological and physicochemical characteristics of Roncal-type ewes'-milk cheese. *Food Microbiol* 23:591–598.

Pardo JE, Pérez J, Gómez R, Tardáguila J, Martínez M, Serrano CE. 1996. Calidad fisico-química del queso manchego. *Alimentaria* 278:95–100.

Park YW, Juárez M, Ramos M, Haenlein GFW. 2007. Physico-chemical characteristics of goat and sheep milk. *Small Ruminant Res* 68:88-113.

Pellegrini O, Remeuf FY, Rivemale M. 1994. Evolution des caracteristiques physicochimiques et des parametres de coagulation du lait de brebis collecte dans la región de Roquefort. *Lait* 74:425–442.

Plaza-Carabantes JP, González-Crespo J, Roa-Ojalvo I. 1999. Quesos extremeños: Composición química y características sensoriales. *Alimentaria* 299:41–45.

Poveda JM, Cabezas L, McSweeney PLH. 2004. Free amino acid content of Manchego cheese manufactured with different starter cultures and changes throughout ripening. *Food Chem* 84:213–218.

Raynal-Ljutovac K, Lagriffoul G, Paccard P, Guillet I, Chilliard Y. 2008. Composition of goat and sheep milk products: an update. *Small Ruminant Res* 79:57–72.

Renner E, Schaafsma G, Scott KJ. 1989. Micronutrients in milk. In *Micronutrients in Milk and Milk-Based Food Products*. [E Renner, editor]. London: Elsevier Applied Science Publishers, pp. 1–70.

Revilla I, González-Martín I, Hernández-Hierro JM, Vivar-Quintana A, González-Pérez C, Lurueña-Martínez MA. 2009a. Texture evaluation in cheeses by NIRS technology employing a fibre-optic probe. *J Food Eng* 92:24–28.

Revilla I, Lurueña-Martínez MA, Vivar-Quintana AM. 2009b. Influence of somatic cell count and breed on physico-chemical and sensory characteristics of hard ewes'-milk cheeses. *J Dairy Res* 76:283–289.

Revilla I, Rodríguez-Nogales JM, Vivar-Quintana AM. 2007. Proteolysis and texture of hard ewes' milk cheese during ripening as affected by somatic cell counts. *J Dairy Res* 74:127–136.

Rincón F, Moreno R, Zurera G, Amaro M. 1994. Mineral composition as a characteristic for the identification of animal origin of raw milk. *J Dairy Res* 61:151–154.

Rodríguez-Nogales JM, Vivar-Quintana AM, Revilla I. 2007. Influence of somatic cell count and breed on capillary electrophoretic protein profiles of ewes' milk: a chemometric study. *J Dairy Sci* 90:3187–3196.

Román-Blanco C, Santos-Buelga J, Moreno-García B, García-López ML. 1999. Composition and microbiology of Castellano cheese (Spanish hard cheese variety made from ewes' milk). *Milchwissenschaft* 54:255–257.

Sanjuan E, Millán R, Saavedra P, Carmona MA, Gómez R, Fernández-Salguero J. 2002. Influence of animal and vegetable rennet on the physicochemical characteristics of Los Pedroches cheese during ripening. *Food Chem* 78:281–289.

Serrano CE, García C, Medina LM, Serrano E. 1997. Estudio físico-químico del queso Manchego con Denominación de Origen. *Alimentaria* 281:81–82.

Tavaria FK, Franco I, Carballo FJ, Malcata FX. 2003. Amino acid and soluble nitrogen evolution throughout ripening of Serra da Estrela cheese. *Int Dairy J* 13:537–545.

Vicente MS, Ibáñez FC, Barcina Y, Barrón LJR. 2001. Changes in the free amino acid content during ripening of Idiazabal cheese: influence of starter and rennet type. *Food Chem* 72:309–317.

25

Hispánico Cheese

Sonia Garde, Marta Ávila, Antonia Picon, and Manuel Nuñez

CONTENTS

25.1 General Characteristics and Manufacturing Procedure

Spain is the seventh producer of milk in the European Union, contributing 6% of the total volume. Cow milk accounted for 85.2% of the year 2010 national milk production, ewe milk for 7.8%, and goat milk for 7.0% (MARM 2010). Most of ewe and goat milk, mixed with cow milk, is used to manufacture blended milk cheeses, which include milk from two or three species. Blended milk cheeses represent around 40% of the total cheese production and consumption in Spain, as shown in Table 25.1. Traditionally, proportions of milk from the different species in blended milk cheeses showed great variability, depending basically on the seasonality of production and the price of milk. Gradually, the sector is tending to standardize milk blends, with the aim of achieving higher homogeneity in cheese composition and flavor. The majority of Spanish blended milk cheeses are included in three main types: Hispánico cheese, Ibérico cheese, and de la Mesta cheese. Each of these types has its own quality regulations (Anon. 1987) that establish milk proportions and define the requirements for composition, characteristics, and manufacturing procedures that each type of cheese must meet.

Hispánico cheese stands as a representative of blended milk cheeses (Figure 25.1). It is a semihard variety, manufactured from a mixture of at least 50% cow milk and at least 30% ewe milk in many regions of Spain. The ripening period typically lasts from 1 to 6 months. Different sizes of Hispánico cheese can be found in the market, but all present a cylindrical shape (up to 12 cm in height and 24 cm in diameter), with a weight generally ranging from 2.5 to 3.5 kg, although they can reach up to 5 kg. This cheese variety should have a minimum of 55% (w/w) dry matter content and 45% (w/w) fat content in dry matter. Hispánico cheese presents a firm, compact, and white-yellowish internal paste. Eyeholes may appear, regular or irregularly distributed, with different forms but always small sized. The rind of Hispánico cheese is hard to the touch, dry, and smooth. It usually has a yellowish color, but when molds are not removed from the surface, cheeses can develop blackish- or greenish-colored rinds. At retail points, Hispánico cheeses can be sold as a whole, in wedges, or sliced.

Hispánico cheese can be manufactured from either raw or pasteurized milk. Pasteurization requirements for milk establish that it should be heated at 72°C–78°C for 15 s or at 63°C for 30 min. The processing steps in Hispánico cheesemaking are resumed in Table 25.2. Calcium chloride and annatto

TABLE 25.1

Distribution of Spanish Cheese Production in 2010

Cheese	Tons
Cow milk	124,100
Ewe milk	44,800
Goat milk	16,200
Blended milk (two or more species)	116,800
Total (except for melting cheeses)	301,900
Melting cheeses	33,800

FIGURE 25.1   Pasteurized milk Hispánico cheese aged for 50 days.

TABLE 25.2

Processing Steps in Hispánico Cheese Manufacture

1. Milk pasteurization (optional), minimum of 15 s at 72°C–78°C or 30 min at 63°C
2. Calcium chloride and annato addition to milk (optional)
3. Addition (optional for raw milk) of lactic starter cultures (mesophilic + optional thermophilic) to milk, 15–45 min at 30°C–33°C
4. Enzymatic coagulation of milk, 25–40 min at 30°C–33°C (rennet, chymosin, or other authorized coagulating enzymes)
5. Curd cutting with 1–2 cm of cutting harps for 8–10 min
6. Curd settling for 5 min
7. Curd heating to 36°C–38°C and stirring for 15–30 min
8. Curd settling for 10 min and whey drainage
9. Shaping in cylindrical molds and pressing for 6–16 h at 16°C–18°C
10. Salting in sodium chloride brine for 24–48 h at 12°C–13°C
11. Cheese ripening at 10°C–15°C at 80%–85% RH (pasteurized milk, minimum of 30 days; raw milk, minimum of 60 days)

colorant, at a maximum dose of 600 mg/kg of cheese, may be added. Addition of lactic cultures is optional, and regulations do not specify the bacterial species allowed. Lactic cultures used as starters include mesophilic lactic acid bacteria (LAB) or mixtures of mesophilic and thermophilic LAB. After inoculation, milk is held at 30°C–33°C for 15–45 min to favor acid production. Milk is coagulated by means of rennet, chymosin, or other authorized coagulants of animal, vegetal, or microbial origin, with a typical clotting time of 25–40 min at 30°C–33°C. Curd is cut into cubes with 1–2 cm of cutting harps for 8–10 min, with minimal agitation, and then, the curd is let resting for 5 min. Afterward, the temperature of the curd is gradually raised up to 36°C–38°C while gently stirring, which allows uniform heating. Once that temperature is reached, vigorous stirring starts for 15–30 min, which reduces curd grains to the size of rice grains. The curd is then allowed to settle down for 10 min, whey is drained off, and the finished curd is placed into cylindrical molds. Cheeses are pressed for 6–16 h at 16°C–18°C, gradually increasing the pressure applied from 2 to 8 kg per 1 kg of cheese. Salting of cheeses is usually carried out by immersing the cheese in sodium chloride brine in a cool room at 12°C–13°C for 24–48 h.

Aging of Hispánico cheese takes place at 10°C–15°C and a relative humidity of 80%–85%. Pasteurized milk cheeses are usually ripened for at least 30 days, but raw milk cheeses must ripen for at least 60 days according to Spanish regulations. Cheeses may be coated in mid ripening with paraffin, wax, or plastic, usually containing an antifungal, or wrapped with a film of polymeric material.

25.2 Main Changes during Manufacture and Ripening

The complex processes occurring during cheese ripening involve microbiological and biochemical changes that result in the characteristic flavor and texture of each particular variety. Microbiological changes in cheese during manufacture and ripening include the growth, death, and lysis of starter LAB and the growth of an adventitious microbiota, mostly nonstarter LAB. In Hispánico cheese, LAB counts reach about 10^9 cfu/g on day 1 and decline during ripening, at a rate depending on the lactic cultures used. Lysis of LAB in cheese can be inferred from levels of released intracellular enzymes such as aminopeptidases. Aminopeptidase activity usually increases during ripening, although it may also increase during the early stages of ripening to remain constant or decrease afterward. With respect to nonstarter LAB present in Hispánico cheese, little information is available. *Enterococcus faecalis* and *Lactobacillus paracasei* subsp. *paracasei* were the predominant species of tyrosine decarboxylase-positive LAB isolated from raw-milk Hispánico cheese (Fernández-García et al. 2000). In pasteurized-milk Hispánico cheese, nonstarter lactobacilli counts increased during ripening from less than 10^4 cfu/g on day 1 to 10^8 cfu/g on day 75 (Ávila et al. 2005).

The metabolism of lactose, lactate and citrate, lipolysis, and proteolysis are the main events during cheese manufacture and early ripening, whereas the metabolism of fatty acids and amino acids mainly occurs during mid and late ripening. Acidification, one of the primary phenomena taking place during the manufacture of Hispánico cheese, involves the fermentation of lactose to lactic acid by starter cultures, generally composed of mesophilic LAB, although the use of thermophilic LAB as adjunct cultures is a growing trend. The pH of Hispánico cheese ranges from 4.9 to 5.2 on day 1 and tends to increase with the cheese age. Dry matter content, close to 50% on day 1, increases up to 55%–60% in 60-day-old cheese.

Cheese lipolysis can be carried out by esterases and lipases originating in milk, starter and/or nonstarter microorganisms, and some coagulants. Both esterases and lipases catalyze the hydrolysis of the ester bond in milk triglycerides, yielding free fatty acids (FFAs), glycerol, and monoglycerides or diglycerides. In bacterially ripened cheeses such as Hispánico cheese, lipolysis, although not very intense, can be essential for the development of typical cheese flavor (Ávila et al. 2007a,b; Picon et al. 2010a,b). Total FFAs content increases during the ripening of Hispánico cheese from around 500 mg/kg on day 15 to around 700 mg/kg on day 60. Palmitic, oleic, and stearic acids are the most abundant FFAs, and their concentrations reflect their abundance in milk fat rather than a preferential release from triglycerides. Short-chain FFAs comprise around 5% of total FFAs, medium-chain FFAs around 15%, and long-chain FFAs around 80%—a pattern hardly varying along ripening. Butyric acid (3% of total FFAs), myristic acid (9% of total FFAs), and palmitic acid (30% of total FFAs) are the main FFAs of the respective groups.

Proteolysis is the most complex and relevant biochemical event that occurs during the ripening of most cheese varieties. The proteinases and peptidases involved in cheese proteolysis originate in milk, rennet, starter LAB, and nonstarter LAB in a wide variety of cheeses, such as Hispánico cheese. These enzymes transform caseins into small peptides and free amino acids (FAAs), which contribute to flavor and serve as aroma precursors. Because of its importance, proteolysis has been extensively studied in Hispánico cheese (Medina et al. 1992; Garde et al. 1997, 2002b, 2003, 2006; Gómez et al. 1997; Mohedano et al. 1998; Oumer et al. 1999, 2000, 2001b; Ávila et al. 2005, 2006b; Picon et al. 2010a,b). Cheese overall proteolysis as determined by the *o*-phthaldialdehyde (OPA) test, which detects free α-amino groups and, thus, is an index of primary and secondary proteolysis, increases considerably (from 3- to 14-fold) during the ripening of Hispánico cheese, depending on the LAB cultures used. A more pronounced breakdown of α_s-casein than β-casein occurs during ripening, with degradation percentages of 75%–98% and 10%–74%, respectively, in 60-day-old Hispánico cheese. Levels of pH 4.6, trichloroacetic acid–soluble N, and phosphotungstic acid–soluble N increased from values of 9%, 4%, and 1% of total N, respectively, on day 1 to values of 18%, 9%, and 4%, respectively, on day 60. Concentrations of hydrophobic and hydrophilic peptides in cheese, and, hence, their ratio depend on the balance between their formation through casein breakdown and their degradation to small-size peptides and amino acids. Levels of hydrophobic and hydrophilic peptides usually increase during Hispánico cheese ripening, but their ratio depends on the relative increases of these two types of peptides. FAAs accumulate in Hispánico cheese during ripening, reaching concentrations of total FAAs up to 12 g/kg of dry matter in 75-day-old cheese. Leu, His, Lys, and Phe are the most abundant FAAs throughout Hispánico cheese ripening, accounting for 18%, 12%, 11%, and 11% of total FAAs content, respectively.

Only the lower molecular weight compounds contribute significantly to cheese flavor. An important group of low-molecular-weight molecules are the volatile compounds. For this reason, many works have been focused on the volatile profile of Hispánico cheese (Oumer et al. 2000, 2001b; Garde et al. 2002a, 2003, 2005, 2007; Ávila et al. 2006a; Picon et al. 2010a,b). Up to 68 compounds have been identified in its volatile fraction, including aldehydes, ketones, alcohols, carboxylic acids, esters, benzene compounds, hydrocarbons, and sulfur compounds, all of them previously found in other cheese varieties. Ethanol was the most abundant compound in the volatile fraction of Hispánico cheese, and although it has a limited aromatic role in cheeses, it is the precursor of ethyl esters, which are volatile compounds relevant to cheese aroma. Among the compounds identified in the volatile fraction, ethyl butanoate, ethyl hexanoate, acetaldehyde, 3-methylbutanal, 2-methyl-1-propanol, 3-methyl-3-buten-1-ol, 3-methyl-1-butanol, 2-pentanone, 2-heptanone, 2,3-pentanedione, 2,3-butanedione, and 3-hydroxy-2-butanone have been repeatedly correlated with favorable sensory aroma characteristics of Hispánico cheese and can be considered to impact aroma compounds of Hispánico cheese.

The flavor and texture characteristics of Hispánico cheese are the result of the above-mentioned microbiological and biochemical changes occurring during manufacture and ripening. Aroma, odor, and taste intensity increase with cheese age, and this increase is generally accompanied by an improvement of flavor quality. Using a descriptive test for Hispánico cheese based on the guidelines to the odor and aroma evaluation of hard and semihard cheeses of Berodier et al. (1997), the odor and the aroma of Hispánico cheese have been defined with the descriptors "milky," "buttery," "yogurt-like," and "cheesy" from the "lactic" family, "caramel" from the "toasted" family, and "sheepy" and "meat broth" from the "animal" family (Garde et al. 2005). Scores for the buttery and yogurt-like descriptors of both odor and aroma tend to decrease during the ripening of Hispánico cheese, whereas scores for the cheesy, sheepy, and meat broth attributes tend to increase. Umami and bitter taste scores generally increase with cheese age in Hispánico cheese.

Textural parameters such as fracturability, hardness, and elasticity increase during the ripening of Hispánico cheese, reflecting a firmer texture, most probably due to the strengthening effect of moisture loss during ripening, which predominates over the weakening effect of caseinolysis. The color of Hispánico cheese evolves during ripening (Ávila et al. 2008). Thus, the L^* parameter (lightness) reaches its maximum value during the first month, but afterward, it decreases due to the concentration of cheese components caused by moisture loss. The b^* (blue to yellow) and C (color saturation) values increase with cheese age, shifting the hue angle (h) to the yellow direction.

25.3 Effect of Bacteriocin-Producing Cultures

The ripening of hard and semihard cheese varieties is a long and costly process because of capital immobilization, large refrigerated storage facilities, weight losses, and spoilage caused by undesirable fermentations. Consequently, a shortened ripening period would lead to a considerable reduction in manufacturing costs. LAB are an important source of enzymes that transform milk constituents retained in the curd into flavor compounds and aroma precursors. As most of these enzymes are located in the interior of the cell, lysis of LAB cells will favor the access of enzymes to their substrates and may thus accelerate the development of cheese flavor and, hence, cheese ripening. Lysis of LAB during early ripening may be enhanced by milk inoculation with bacteriocin-producing (BP) adjunct cultures. Bacteriocins are antimicrobial proteins produced by bacteria, generally active against bacterial species closely related to the producing organism. Their production is widespread among LAB present in milk and dairy products. The major application of bacteriocins produced by LAB has been their use as food preservatives against pathogenic and spoilage bacteria, but they have been also employed, based on their lytic effect on starter bacteria, to accelerate cheese ripening.

The addition of enterocin AS-48-producing *E. faecalis* INIA 4 (Garde et al. 1997; Oumer et al. 1999, 2001b), nisin Z– and lacticin 481–producing *Lactococcus lactis* subsp. *lactis* INIA 415 (Garde et al. 2002a,b, 2005; Ávila et al. 2005, 2006a,b, 2007a), and lacticin 481-producing *Lc. lactis* subsp. *lactis* INIA 639 (Garde et al. 2006, 2007; Ávila et al. 2007b) as adjuncts to the starter in the manufacture of Hispánico cheese had significant effects on starter viability, release of intracellular enzymes, proteolysis, lipolysis, texture, volatile compounds, and sensory characteristics of cheese.

The use of *E. faecalis* INIA 4, a nonvirulent hemolysin-negative enterocin AS-48-producing strain, as adjunct culture to a commercial mixed-strain LD-type starter in the manufacture of Hispánico cheese accelerated cell lysis, proteolysis, and flavor development but retarded acidification (Garde et al. 1997). The cheese obtained showed lower levels of hydrophobic peptides, associated with bitterness in Hispánico cheese (Gómez et al. 1997), and higher concentrations of 3-methyl-1-butanal, 2,3-butanedione, and 3-hydroxy-2-butanone and received higher scores for flavor quality than control cheese throughout ripening (Oumer et al. 2001b). Hispánico cheese made with a defined strain starter system consisting of *Lc. lactis* subsp. *lactis* strain H6, with a low sensitivity to enterocin AS-48, and *E. faecalis* INIA 4 exhibited higher aminopeptidase activity and more extensive proteolysis than control cheese, receiving higher flavor quality and intensity scores (Oumer et al. 1999).

As the dairy industry is reluctant to employ enterococci as lactic starters in cheese manufacture, *Lc. lactis* subsp. *lactis* INIA 415, a strain harboring the structural genes of bacteriocins nisin Z and lacticin 481, was used as adjunct culture in the manufacture of Hispánico cheese with a mesophilic starter, comprising *Lc. lactis* subsp. *lactis* INIA 437 and *Lc. lactis* subsp. *cremoris* INIA 450, or with the mesophilic starter and a thermophilic starter, comprising *Streptococcus thermophilus* INIA 463 and INIA 468 (Garde et al. 2002a,b). The addition of the BP strain did not retard acid production and promoted the early lysis of mesophilic and thermophilic starter bacteria, which resulted in higher aminopeptidase activities in cheeses made with the BP culture than in the respective control cheeses made without it. The release of intracellular enzymes into the cheese matrix accelerated proteolysis and enhanced the formation of 2-methylpropanal, 2-methylbutanal, 3-methylbutanal, 2-methyl-1-propanol, 3-methyl-1-butanol, 1-octanol, 2-butanone, and 2,3-butanedione. Cheese made with mesophilic and thermophilic starters plus the BP strain received the highest scores for flavor quality and flavor intensity and reached in 25 days the flavor intensity score of a 75-day-old control cheese made without the BP strain. Similar results were obtained by Ávila et al. (2005, 2006a,b) and Garde et al. (2005) when using *Lc. lactis* subsp. *lactis* INIA 415 as a BP adjunct culture, *Lc. lactis* subsp. *lactis* INIA 415-2, a spontaneous mutant not producing bacteriocins, as a mesophilic starter culture, and a commercial *S. thermophilus* culture in the manufacture of Hispánico cheese. The addition of the BP culture promoted the early lysis of thermophilic starter bacteria, with the concomitant release of intracellular aminopeptidases and the acceleration of secondary proteolysis. However, the inhibition of thermophilic starter by the BP culture retarded α_s-casein proteolysis, a fact that resulted in a firmer texture. The formation of 2-methylpropanal, 3-methylbutanal, 2-methyl-1-propanol, 3-methyl-1-butanol, 2-pentanone, 2-hexanone, 2-heptanone, and 2-nonanone was enhanced in cheese made from milk inoculated with the BP culture. This cheese received higher taste and aroma intensity scores than control cheese. When the BP culture was added to milk, a significant increase in the release of intracellular esterases was observed, which resulted in a higher rate of FFA accumulation during cheese ripening (Ávila et al. 2007a). Cheese made with the BP culture reached in 17 days the proteolysis level (OPA test) and in 28 days the flavor intensity score of a 50-day-old control cheese made without the BP strain. Experimental cheeses not only ripened at a faster rate but also exhibited a more ripe visual appearance as a result of BP addition, with higher yellow component and saturation values from day 25 onward than those of cheese made without the BP culture (Ávila et al. 2008).

The combination of lacticin 481–producing *Lc. lactis* subsp. *lactis* INIA 639, non-bacteriocin-producing *Lc. lactis* subsp. *lactis* INIA 437, and a *Lb. helveticus* culture sensitive to lacticin 481 as lactic starter for Hispánico cheese manufacture optimized the release of intracellular aminopeptidases during early cheese ripening and accelerated proteolysis (Garde et al. 2006). Hydrophobic and hydrophilic peptides and their ratio were at the lowest levels in cheese made with the three lactic cultures, which received the lowest scores for bitterness and the highest scores for umami taste and for taste quality. Milk inoculation with the lacticin 481 producer enhanced the formation of some volatile compounds such as 2-methyl-propanal, 2-methylbutanal, ethanol, 1-propanol, ethyl acetate, ethyl butanoate, and ethyl hexanoate, increased aroma intensity scores, and improved aroma quality (Garde et al. 2007). Furthermore, the use of the BP enhanced the release of intracellular esterases, causing a more rapid evolution of lipolysis, which resulted in higher concentrations of most individual FFAs without affecting the overall pattern of lipolysis during the ripening of Hispánico cheese (Ávila et al. 2007b). Cheese made with the three lactic cultures reached in 28 days the same proteolysis level (OPA test) and in 38 days the same flavor intensity score of a 50-day-old cheese made without the BP strain.

Based on the studies commented on above, it can be concluded that the use of a BP strain at an adequate dose as an adjunct to the starter culture is a feasible and noncostly procedure for the acceleration of Hispánico cheese ripening through the lysis of starter LAB and the subsequent release of intracellular enzymes.

25.4 Effect of Proteolytic Adjunct Cultures

Starter cultures used in cheese manufacturing are generally composed of different LAB strains belonging to one or more species. The selection of LAB for the formulation of cheese starters and adjuncts is usually done on the basis of their technological properties such as acidification, proteolytic and lipolytic activities, redox potential control capacity, salt tolerance, compatibility with other strains, and flavor production. Most of these characteristics depend on the bacterial enzymatic pool (proteinases, peptidases, amino acid catabolic enzymes, and esterases), as well as on their ability to lyse and release enzymes to the cheese matrix, which eases the contact enzyme–substrate. In addition to the degree of starter cell lysis, both the specificity and level of intracellular enzymes may affect cheese ripening and flavor development. Hence, the use of highly peptidolytic strains as adjunct cultures in cheese manufacturing appears as another approach for the acceleration of proteolysis and the general ripening process of Hispánico cheese.

S. thermophilus, one of the most important LAB used by the dairy industry, possesses two additional peptidases with respect to *Lc. lactis* and shows higher specific activities of PepX, PepN, and PepC (Rul and Monnet 1997). Aminopeptidase activities of different thermophilic starters comprising *S. thermophilus* strains were considerably higher than those of mesophilic starters comprising *Lc. lactis* strains (Oumer et al. 2001a). *S. thermophilus* may play an important role in the catabolism of amino acids and the production of aroma compounds because of its aminotransferase and threonine aldolase activities (Chaves et al. 2002). On top of that, glutamate dehydrogenase activity is widespread, often at higher levels, in *S. thermophilus* than in other LAB (Helinck et al. 2004). These *S. thermophilus* characteristics are advantageous in cheese manufacturing and ripening.

The use of *S. thermophilus* INIA 463 and INIA 468 as an adjunct thermophilic culture for the production of Hispánico cheese in combination with a mesophilic starter (Garde et al. 2002a,b) considerably increased the values of aminopeptidase activity and proteolysis (OPA test) and accelerated the degradation of both α_s- and β-caseins, probably due to an additive effect of lactococcal and streptococcal proteinases. Levels of hydrophobic peptides decreased while total levels of FAAs increased in cheese made with the thermophilic starter, doubling those reached in cheese made only with the mesophilic starter. Furthermore, the relative abundance of 25 out of the 46 compounds identified in the volatile fraction of Hispánico cheese was influenced by the addition of the thermophilic starter. Ethanol, ethyl butanoate, and ethyl hexanoate were at higher levels in cheese made with the thermophilic starter, because *S. thermophilus* produces high amounts of ethanol compared to other homofermentative LAB (Beshkova et al. 1998). Besides, levels of acetaldehyde, 3-methyl-2-buten-1-ol, 3-methyl-3-buten-1-ol, 2-butanone, and 2,3-butanedione were increased in cheeses made with the thermophilic starter. Consequently, the addition of the thermophilic starter enhanced both flavor quality and flavor intensity, accelerating flavor development in Hispánico cheese. It was estimated that the flavor intensity score reached in 75 days by cheese made using only the mesophilic starter would be reached in only 50 days by cheese additionally containing the thermophilic adjunct.

Strains from some *Lactobacillus* species, normal constituents of thermophilic starters for cheese varieties such as parmesan, mozzarella, and Swiss types, also present a wide-range complex of peptidolytic enzymes, which can influence flavor development (Williams and Banks 1997; Williams et al. 1998). Comparisons between *Lb. helveticus* and other LAB species have demonstrated that *Lb. helveticus* strains possess, by far, the highest aminopeptidase and dipeptidase activities (Hickey et al. 1983; Frey et al. 1986; Sasaki et al. 1995). The addition of a commercial *Lb. helveticus* LH 92 culture to milk for Hispánico cheese making, together with a mesophilic starter (Garde et al. 2006, 2007), considerably influenced cheese ripening in comparison with Hispánico cheese made using only a mesophilic starter (Garde et al. 2002a,b). Aminopeptidase activity values were around 100-fold higher in cheese made with *Lb. helveticus* LH 92 than in cheese made without it. After 50 days of ripening, proteolysis values (OPA test) of cheese were threefold higher when *Lb. helveticus* LH 92 was added than in cheese elaborated only with the mesophilic starter. In addition, levels of residual caseins declined more rapidly during the ripening of Hispánico cheese containing *Lb. helveticus* LH 92. This cheese showed up to 7.6-fold higher FAA concentrations than cheese made without *Lb. helveticus* LH 92. Relative abundances of volatile compounds increased in cheese made with *Lb. helveticus* LH 92, especially those derived from amino acid metabolism such as acetaldehyde, 2-methylpropanal, 2-methylbutanal, and 3-metylbutanal, and 2,3-butanedione and 2,3-pentanedione, whose formation from Asp (Kieronczyk et al. 2004) and Ile (Imhof et al. 1995), respectively, has been postulated. Biochemical changes resulting from the use of *Lb. helveticus* LH 92 in Hispánico cheese manufacture led to higher flavor intensity scores.

The above studies prove that the use of selected highly peptidolytic strains as adjunct cultures in Hispánico cheese manufacture is a simple and inexpensive method for the acceleration of biochemical phenomena occurring during cheese ripening, with a concomitant enhancement of flavor intensity.

25.5 Effect of High-Pressure Treatment of Cheese

High-pressure (HP) processing (100–1000 MPa) is a nonthermal technology that has attracted considerable research activity over the past 20 years. Its main advantages, as summarized by Rastogi et al. (2007), are that (1) it enables food processing at ambient temperature or even lower temperatures; (2) it enables the instant transmittance of pressure throughout the system, irrespective of size and geometry; (3) it causes microbial death while virtually eliminating heat damage and the use of chemical preservatives/additives, thereby leading to improvements in the overall quality of foods; and (4) it can be used to create ingredients with novel functional properties. Most applications of HP treatment in cheeses are related to the inactivation or reduction of pathogenic and spoilage microorganisms. Nevertheless, studies on the HP treatment of cheeses with the aim of accelerating cheese ripening have also emerged. Because HP treatment may increase cell membrane permeability (Cheftel 1992; Malone et al. 2002), the release of intracellular enzymes such as peptidases (Trujillo et al. 2000) and esterases to the medium can be favored, and therefore, cheese ripening may be accelerated. Furthermore, HP treatment may induce changes in the cheese matrix that may favor proteolysis, promoting conformational changes in the structure of caseins, which render the proteins more susceptible to the action of proteases (Kunugi 1993), improving water retention, maybe increasing the activity of some enzymes, and giving rise to higher pH values, which may favor enzyme action. On the contrary, some recent studies indicate the benefits of HP treatment to control excessive cheese ripening based on the inactivation of enzymes at certain pressure levels. HP treatment effects will depend on the pressure level, time, temperature, and number of pulses of the treatment, as well as on the food composition and, in the case of microorganisms, on the strain and the growth stage.

In relation with Hispánico cheese, HP treatments have been applied, aiming at the acceleration of its ripening process. Hispánico cheeses were made with *Lc. lactis* subsp. *lactis* INIA 415-2 and a commercial *S. thermophilus* culture. After 15 days of ripening, half of the cheeses were treated at 400 MPa for 5 min at 10°C. All cheeses, treated and untreated, continued ripening until day 50 (Ávila et al. 2006a,b, 2007a, 2008). HP treatment resulted in pH increase after treatment, but differences were reversible from day 25 onward, whereas dry matter was not affected. With respect to LAB, pressurized cheese showed lower counts of *S. thermophilus* and a decrease in activity levels of released intracellular aminopeptidases, because enzyme inactivation caused by the HP treatment apparently prevailed over the higher release of intracellular enzymes after the death of thermophilic LAB caused by HP treatment. Bacterial esterases seemed to be notably barotolerant under the tested HP conditions, and close esterase activity values were found in HP-treated and HP-untreated cheeses. As a result of the HP processing, casein degradation in cheese was accelerated, and the FAA content increased, what the authors attributed to conformational changes in the structure of caseins and peptides that made them more susceptible to the action of enzymes. Volatile compounds of Hispánico cheese were also influenced by HP treatment. Pressurized cheeses showed higher levels of hexanal, 3-hydroxy-2-pentanone, 2-hydroxy-3-pentanone, and hexane and lower levels of acetaldehyde, ethanol, 1-propanol, ethyl acetate, ethyl butanoate, ethyl hexanoate, 2-pentanone, and butanoic acid than untreated cheeses. In general, HP treatment caused negligible differences in individual FFA levels and did not affect the lipolysis pattern of Hispánico cheese because of the observed barotolerance of esterases. Most individual FFAs were at lower levels immediately after HP treatment, with a reduction in total (C4:0–C18:2) FFA concentration, but those differences were no longer found for most individual FFAs at the end of ripening.

The HP treatment of Hispánico cheese enhanced caseinolysis, what led to a softer texture, probably related with modifications in cheese microstructure. Those microstructure modifications may also influence cheese color. Thus, the lightness of HP-treated cheese was lower on day 15, and the yellow component and color saturation were higher than those of HP-untreated cheese at 50 days, meaning that the color of HP-treated Hispánico cheese on day 15 resembled that of a much older ripe control cheese. HP treatment did not affect taste quality or taste intensity but had a detrimental effect on cheese odor quality and intensity due to the lower content of volatile compounds. The odor description of HP-treated cheeses was also affected, and cheeses received higher "milky" odor descriptor scores and lower scores for "buttery," "yogurt-like," and "caramel" odor descriptors.

The combined effect of HP treatment (400 MPa for 5 min at 10°C) with the addition of a nisin Z– and lacticin 481–producing *Lc. lactis* subsp. *lactis* INIA 415 culture to milk has also been studied in Hispánico cheese, with the aim of accelerating the ripening process (Ávila et al. 2006a,b, 2007a, 2008). The combination had a synergistic effect on the reduction of mesophilic LAB counts, because the presence of the bacteriocin in cheese seemed to increase the lethality of the HP treatment and also reduced aminopeptidase activity values. It also resulted in higher levels of both hydrophobic and hydrophilic peptides but had no significant effect on the FAA content, taste quality, or taste intensity scores compared with those of HP-untreated cheese made with the BP culture. The HP treatment of the cheese elaborated with the BP culture generally limited the formation of volatile compounds, producing a slight decrease of odor quality and intensity scores of cheeses. Esterase activity increased immediately after HP treatment, but FFA levels were similar to those of HP-untreated cheese. The combination of BP culture addition and HP treatment decreased the lightness, yellowness, and color saturation of cheese up to day 25, but this effect was reversed after 50 days of ripening. The cheese texture of pressurized cheese became more elastic than in HP-untreated cheese. Flavor development was delayed in comparison with the HP-untreated cheese made with the BP strain.

In view of the results obtained to date, it can be said that the HP treatment of Hispánico cheese (400 MPa for 5 min at 10°C), by itself or combined with the addition of the BP strain *Lc. lactis* subsp. *lactis* INIA 415 to milk, does not accelerate cheese ripening and may even arrest some of the related biochemical events. The complex phenomena induced by the HP treatment of cheese require further research before the ripening process is fully mastered. Investigation of other HP treatment conditions may contribute to the acceleration of Hispánico cheese ripening.

25.6 Effect of the Use of Frozen Ewe Milk Curd

The maintenance of a desirable constant proportion of cow and ewe milk for Hispánico cheese manufacture throughout the year may become difficult because of the seasonality in ewe milk production, with a peak in spring months and a valley in summer and autumn. A number of studies have been carried out, with the aim of overcoming seasonal shortages of ovine milk and its effect on cheese production. Research to date has focused on freezing a product, curd, or cheese, molded to its final size and shape, in order to stop or to retard as much as possible, for a certain period of time, the biochemical changes occurring during manufacture and ripening. Afterward, curds or cheeses were thawed, and ripening proceeded with the concomitant biochemical changes.

A novel approach has been reported to overcome ewe milk shortage during summer and autumn, consisting of freezing curds made from spring ewe milk and, some months later, thawing and employing them for Hispánico cheese manufacture. The use of curds made from pasteurized ewe milk [scalded at different temperatures (32°C, 35°C, or 38°C), pressed for 30 min, and then frozen at −24°C for 4 months] was investigated (Picon et al. 2010a). After thawing, the curd was cut into 2-mm pieces and added to fresh cow milk curd for the manufacture of experimental Hispánico cheeses. Control cheese was made from a mixture of pasteurized cow and ewe milk in the same proportion. No significant effect of the addition of frozen ewe milk curd or scalding temperature was found for LAB counts, dry matter content, hydrophilic and hydrophobic peptides, 45 out of the 65 volatile compounds, texture, and sensory characteristics throughout a 60-day ripening period. Differences between cheeses of low magnitude and minor practical significance were found for pH value, aminopeptidase activity, proteolysis (OPA test), FAAs, FFAs, and the remaining 20 volatile compounds.

The use of pasteurized ewe milk curds pressed for different times (15, 60, or 120 min) and frozen at −24°C for 4 months was also studied (Picon et al. 2010b). Hispánico cheese was manufactured by mixing thawed ewe milk curd with fresh cow milk curd and compared with control cheese made from a mixture of pasteurized cow and ewe milk in the same proportion. No significant differences were found for LAB counts, dry matter, hydrophilic peptides, 47 out of the 68 volatile compounds, instrumental texture, or sensory flavor characteristics. On the other hand, minor differences between cheeses were found for pH value, aminopeptidase activity, proteolysis (OPA test), hydrophobic peptides, FAAs, FFAs, and the remaining 21 volatile compounds, which could not be related to the use of frozen ewe milk curd or the pressing time of ewe milk curd before freezing.

The addition of frozen ewe milk curd, previously thawed, to fresh cow milk curd in the manufacture of Hispánico cheese may serve to overcome the seasonal shortage of ewe milk without altering cheese characteristics. Furthermore, the freezing of only a portion of the curd, instead of the totality as in previous studies, is economically advantageous to the cheese manufacturing industry.

25.7 Effect of the Use of Pressurized Ewe Milk Curd

Milk pasteurization has a negative effect on the activity of milk native enzymes and, consequently, on the sensory quality of cheese (Grappin and Beuvier 1997). HP treatment is an alternative procedure to thermal processes, because it achieves a reduction of the bacterial population without the detrimental effect on enzymatic activities characteristic of pasteurization. The HP treatment of raw ewe milk curd may thus be applied instead of ewe milk pasteurization for Hispánico cheese manufacture.

At our laboratory, raw ewe milk curd was HP treated at 400 and 500 MPa for 10 min at 10°C (Alonso et al. 2011). HP treatment had a drastic effect on the microstructure of the curd (Figure 25.2). After

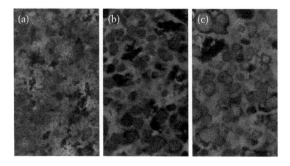

FIGURE 25.2 Confocal scanning laser micrographs showing the structure (protein matrix and fat globules) of nonpressurized curd (a) and curds subjected to HP treatment at 400 (b) or 500 (c) MPa for 10 min at 10°C.

TABLE 25.3

Effect of Different Manufacturing Strategies on Hispánico Cheese Characteristics

Strategy	Dry Matter	pH	Proteolysis	Lipolysis	Volatile Compounds	Flavor	Texture	References
Bacteriocin-producing cultures	Unaffected	Increase or unaffected	Increase	Increase	Increase	Enhancement of intensity and quality	Firmer or softer	Garde et al. 1997; Oumer et al. 1999, 2001b; Garde et al. 2002a,b, 2005, 2006, 2007; Ávila et al. 2005, 2006a,b, 2007a,b
Proteolytic adjunct cultures	Unaffected	Unaffected	Increase	Increase	Increase	Enhancement of intensity and quality	Softer	Garde et al. 2002a,b, 2006, 2007; Ávila et al. 2007a,b
High-pressure treatment	Unaffected	Reversible increase	Increase	Unaffected	Decrease	Detrimental effect on odor intensity and quality	Softer	Ávila et al. 2006a,b, 2007a, 2008
Frozen ewe milk curd	Unaffected	Minor differences	Minor differences	Minor differences	Minor differences	Unaffected	Unaffected	Picon et al. 2010a,b
Pressurized ewe milk curd	Decrease	Minor differences	Increase	Increase	Increase	Unaffected	Softer	Alonso et al. 2011

treatment, the curd was frozen and stored for 4 months at −24°C. Once thawed, it was cut into 2-mm pieces and added to fresh cow milk curd for the manufacture of experimental Hispánico cheeses. Control cheese was made from a mixture of pasteurized cow and ewe milk in the same proportion. Differences found for LAB counts, aminopeptidase activity, pH value, proteolysis (OPA method), and water-soluble peptides between experimental and control cheeses were of minor significance. Experimental cheeses had lower dry matter content, which resulted in a softer texture and higher esterase activity and total FAA concentration than control cheese. The experimental cheese made using curds pressurized at 400 MPa showed the higher concentration of total FFA and of some relevant aroma compounds such as 1-propanol, 2-propanol, 2-butanol, 2-pentanol, and ethyl hexanoate. The addition of pressurized raw ewe milk curd to fresh curd from pasteurized cow milk did not alter the sensory characteristics of Hispánico cheese and increased the yield of ripe cheese because of the higher moisture retention.

The effect of the above-discussed strategies on Hispánico cheese characteristics are summarized in Table 25.3.

ACKNOWLEDGMENT

The authors thank the Spanish Ministry of Education and Science for funding Project AGL2004-06051.

REFERENCES

Alonso, R., Picon, A., Rodríguez, B., Gaya, P., Fernández-García, E., and Nuñez, M. 2011. Microbiological, chemical and sensory characteristics of Hispánico cheese manufactured using frozen high-pressure treated curds made from raw ovine milk. *Int. Dairy J.* 21:484–492.

Anon. 1987. Normas de composición y características específicas para los quesos "Hispánico," "Ibérico" y "De La Mesta," destinados al mercado interior (B.O.E. July 17, 1987).

Ávila, M., Calzada, J., Garde, S., and Nuñez, M. 2007a. Effect of a bacteriocin-producing *Lactococcus lactis* strain and high-pressure treatment on the esterase activity and free fatty acids in Hispánico cheese. *Int. Dairy J.* 17:1415–1423.

Ávila, M., Garde, S., Calzada, J., and Nuñez, M. 2007b. Lipolysis of semihard cheese made with a lacticin 481–producing *Lactococcus lactis* strain and a *Lactobacillus helveticus* strain. *Lait* 87:575–585.

Ávila, M., Garde, S., Gaya, P., Medina, M., and Nuñez, M. 2006a. Effect of high-pressure treatment and a bacteriocin-producing lactic culture on the odor and aroma of Hispánico cheese: Correlation of volatile compounds and sensory analysis. *J. Agric. Food Chem.* 54:382–389.

Ávila, M., Garde, S., Gaya, P., Medina, M., and Nuñez, M. 2006b. Effect of high-pressure treatment and a bacteriocin-producing lactic culture on the proteolysis, texture, and taste of Hispánico cheese. *J. Dairy Sci.* 89:2882–2893.

Ávila, M., Garde, S., and Nuñez, M. 2008. Effect of a bacteriocin-producing lactic culture and high-pressure treatment on the color of Hispánico cheese. *Milchwissenschaft* 63:406–409.

Ávila, M., Garde, S., Pilar, G., Medina, M., and Nuñez, M. 2005. Influence of a bacteriocin-producing lactic culture on proteolysis and texture of Hispánico cheese. *Int. Dairy J.* 15:145–153.

Berodier, F., Lavanchy, P., Zannoni, M., Casals, J., Herrero, L., and Aramo, C. 1997. A guide to the sensory evaluation of smell, aroma and taste of hard and semihard cheeses. *Lebensm.-Wiss. Technol.* 30:653–664.

Beshkova, D., Simova, E., Frengova, G., and Simov, Z. 1998. Production of flavour compounds by yogurt starter cultures. *J. Ind. Microbiol. Biotechnol.* 20:180–186.

Chaves, A. C., Fernandez, M., Lerayer, A. L., Mierau, I., Kleerebezem, M., and Hugenholtz, J. 2002. Metabolic engineering of acetaldehyde production by *Streptococcus thermophilus*. *Appl. Environ. Microbiol.* 68:5656–5662.

Cheftel, J. 1992. Effects of high hydrostatic pressure on food constituents: An overview. In: *High Pressure and Biotechnology*, Vol. 224, edited by C. Balny, R. Hayashi, K. Heremans, and P. Masson, pp. 195–209. Colloque INSERM. London: John Libbey and Co., Ltd.

Fernández-García, E., Tomillo, J., and Nuñez, M. 2000. Formation of biogenic amines in raw milk Hispánico cheese manufactured with proteinases and different levels of starter culture. *J. Food Prot.* 63:1551–1555.

Frey, J. P., Marth, E. H., Johnson, M. E., and Olson, N. F. 1986. Peptidases and proteases of lactobacilli associated with cheese. *Milchwissenschaft* 41:622–624.

Garde, S., Ávila, M., Fernández-García, E., Medina, M., and Nuñez, M. 2007. Volatile compounds and aroma of Hispánico cheese manufactured using lacticin 481–producing *Lactococcus lactis* subsp. *lactis* INIA 639 as adjunct culture. *Int. Dairy J.* 17:717–726.

Garde, S., Ávila, M., Medina, M., and Nuñez, M. 2005. Influence of a bacteriocin-producing lactic culture on the volatile compounds, odour and aroma of Hispánico cheese. *Int. Dairy J.* 15:1034–1043.

Garde, S., Ávila, M., Pilar, G., Medina, M., and Nuñez, M. 2006. Proteolysis of Hispánico cheese manufactured using lacticin 481–producing *Lactococcus lactis* ssp. *lactis* INIA 639. *J. Dairy Sci.* 89:840–849.

Garde, S., Carbonell, M., Fernández-García, E., Medina, M., and Nuñez, M. 2002a. Volatile compounds in Hispánico cheese manufactured using a mesophilic starter, a thermophilic starter and bacteriocin-producing *Lactococcus lactis* subsp. *lactis* INIA 415. *J. Agric. Food Chem.* 50:6752–6757.

Garde, S., Gaya, P., Fernández-García, E., Medina, M., and Nuñez, M. 2003. Proteolysis, volatile compounds and sensory evaluation in Hispánico cheese manufactured with addition of a thermophilic starter, nisin and calcium alginate–nisin microparticles. *J. Dairy Sci.* 86:3038–3047.

Garde, S., Gaya, P., Medina, M., and Nuñez, M. 1997. Acceleration of flavor formation in cheese by a bacteriocin-producing adjunct lactic culture. *Biotechnol. Lett.* 19:1011–1014.

Garde, S., Tomillo, J., Gaya, P., Medina, M., and Nuñez, M. 2002b. Proteolysis in Hispánico cheese manufactured using a mesophilic starter, a thermophilic starter and bacteriocin-producing *Lactococcus lactis* subsp. *lactis* INIA 415 adjunct culture. *J. Agric. Food Chem.* 50:3479–3485.

Gómez, M. J., Garde, S., Gaya, P., Medina, M., and Nuñez, M. 1997. Relationship between levels of hydrophobic peptides and bitterness in cheese made from pasteurized and raw milk. *J. Dairy Res.* 64: 289–297.

Grappin, R. and Beuvier, E. 1997. Possible implications of milk pasteurization on the manufacture and sensory quality of ripened cheese. *Int. Dairy J.* 7:751–761.

Helinck, S., Le Bars, D., Moreau, D., and Yvon, M. 2004. Ability of thermophilic lactic acid bacteria to produce aroma compounds from amino acids. *Appl. Environ. Microbiol.* 70:3855–3861.

Hickey, M. W., Hillier, A. J., and Jago, G. R. 1983. Peptidase activities in lactobacilli. *Aust. J. Dairy Technol.* 38:118–123.

Imhof, R., Glaettli, H., and Bosset, J. O. 1995. Volatile organic compounds produced by thermophilic and mesophilic single strain dairy starter cultures. *Lebensm.-Wiss. Technol.* 28:78–86.

Kieronczyk, A., Skeie, S., Langsrud, T., Le Bars, D., and Yvon, M. 2004. The nature of aroma compounds produced in a cheese model by glutamate dehydrogenase positive *Lactobacillus* INF15D depends on its relative aminotransferase activities towards the different amino acids. *Int. Dairy J.* 14:227–235.

Kunugi, S. 1993. Modification of biopolymer functions by high pressure. *Prog. Polym. Sci.* 18:805–838.

Malone, A. S., Shellhammer, T. H., and Courtney, P. D. 2002. Effects of high pressure on the viability, morphology, lysis, and cell wall hydrolase activity of *Lactococcus lactis* subsp. *cremoris*. *Appl. Environ. Microbiol.* 68:4357–4363.

MARM. 2010. Anuario de Estadística 2010. Ministerio de Medio Ambiente y Medio Rural y Marino. http://www.marm.es/es/estadistica/temas/anuario-de-estadistica/.

Medina, M., Gaya, P., Guillen, A. M., and Nuñez, M. 1992. Characteristics of Burgos and Hispánico cheeses manufactured with calf rennet or with recombinant chymosin. *Food Chem.* 45:85–89.

Mohedano, A. F., Fernandez, J., Gaya, P., Medina, M., and Nuñez, M. 1998. Effect of the cysteine proteinase from *Micrococcus* sp. INIA 528 on the ripening process of Hispánico cheese. *J. Dairy Res.* 65:621–630.

Oumer, A., Fernandez-Garcia, E., Serrano, C., and Nuñez, M. 2000. Flavour of Hispánico cheese manufactured with *Lactococcus lactis* subsp. *lactis* and *L. lactis* subsp. *cremoris* as starter cultures. *Milchwissenschaft* 55:325–328.

Oumer, A., Garde, S., Gaya, P., Medina, M., and Nuñez, M. 2001a. The effects of cultivating lactic starter cultures with bacteriocin-producing lactic acid bacteria. *J. Food Prot.* 64:81–86.

Oumer, A., Garde, S., Medina, M., and Nuñez, M. 1999. Defined starter system including a bacteriocin producer for the enhancement of cheese flavor. *Biotechnol. Tech.* 13:267–270.

Oumer, A., Gaya, P., Fernández-García, E., Mariaca, R., Garde, S., Medina, M., and Nuñez, M. 2001b. Proteolysis and formation of volatile compounds in cheese manufactured with a bacteriocin-producing adjunct culture. *J. Dairy Res.* 68:117–129.

Picon, A., Alonso, R., Gaya, P., Fernández-García, E., Rodríguez, B., de Paz, M., and Nuñez, M. 2010a. Microbiological, chemical, textural and sensory characteristics of Hispánico cheese manufactured using frozen ovine milk curds scalded at different temperatures. *Int. Dairy J.* 20:344–351.

Picon, A., Gaya, P., Fernández-García, E., Rivas-Cañedo, A., Ávila, M., and Nuñez, M. 2010b. Proteolysis, lipolysis, volatile compounds, texture, and flavor of Hispánico cheese made using frozen ewe milk curds pressed for different times. *J. Dairy Sci.* 93:2896–2905.

Rastogi, N. K., Raghavarao, K. S. M. S., Balasubramaniam, V. M., Niranjan, K., and Knorr, D. 2007. Opportunities and challenges in high-pressure processing of foods. *Crit. Rev. Food Sci. Nutr.* 47:69–112.

Rul, F. and Monnet, V. 1997. Presence of additional peptidases in *Streptococcus thermophilus* CNRZ 302 compared to *Lactococcus lactis*. *J. Appl. Microbiol.* 82:695–704.

Sasaki, M., Bosman, B. W., and Tan, P. S. T. 1995. Comparison of proteolytic activities in various lactobacilli. *J. Dairy Res.* 62:601–610.

Trujillo, A. J., Capellas, M., Buffa, M., Royo, C., Gervilla, R., Felipe, X., Sendra, E., Saldo, J., Ferragut, V., and Guamis, B. 2000. Applications of high-pressure treatment for cheese production. *Food Res. Int.* 33:311–316.

Williams, A. G. and Banks, J. M. 1997. Proteolytic and other hydrolytic activities in nonstarter lactic acid bacteria (NSLAB) isolated from cheddar cheese manufactured in the United Kingdom. *Int. Dairy J.* 7:763–774.

Williams, A. G., Felipe, X., and Banks, J. M. 1998. Aminopeptidase and dipeptidyl peptidase activity of *Lactobacillus* spp. and nonstarter lactic acid bacteria (NSLAB) isolated from cheddar cheese. *Int. Dairy J.* 8:255–266.

26

Asiago and Other Cheeses: Standards, Grades, and Specifications*

Y. H. Hui

CONTENTS

26.1 Introduction

Food manufacturers in the United States must produce a product that consumers like in order to sell the product nationally. To achieve this goal, they depend on developments in science, technology, and engineering in order to produce a quality product. However, they must also pay attention to two other important considerations:

---

* The information in this chapter has been modified from *Food Safety Manual* published by Science Technology System (STS) of West Sacramento, CA. Copyrighted 2012©. With permission.

1. Are their products safe for public consumption?
2. Have they assured the economic integrity of their products? This means many things. The most important one is the assurance that their products contain the proper quality and quantity of ingredients, that is, "no cheating."

To protect the consumers, U.S. government agencies have established many checks and balances to prevent "injuries" to the consumers in terms of health and economic fraud. The two major responsible agencies are

1. U.S. Food and Drug Administration (www.FDA.gov)
2. U.S. Department of Agriculture (www.USDA.gov)

The FDA establishes standards of identities for a number of cheeses for interstate commerce. This makes sure that a particular type of cheese contains the minimal quality and quantity of the ingredients used to make this cheese. The FDA also issues safety requirements and recommendations in the manufacturing of cheeses. It makes periodic inspections of establishments producing cheeses to ensure full compliance. In effect, the FDA is monitoring the safety production of cheeses and their economic integrities.

The USDA plays the same important role in the commerce of cheeses in the United States. For example, national restaurant chains that make and sell pizzas will not buy cheeses from any manufacturer if the cheeses do not bear the USDA official identification or seal of approval. The USDA has issued official requirements and processing specifications that cheese manufacturers must follow if they want to obtain the USDA seal of approval for the products. To achieve this goal, the USDA sets up voluntary services of inspection under a standard fee arrangement. The inspection focuses on the safety and economic integrity of manufacturing cheeses. When the inspection demonstrates that the manufacturer has complied with safety and economic requirements, the USDA issues its official identification or seal of approval.

In essence, the FDA and USDA are the federal agencies that make sure that the product is safe and will not pose any economic fraud. Using the public information distributed in their Web sites, this chapter discusses FDA standards of identities and USDA specifications for establishments in relation to the production of selected cheeses.

The information in this chapter is important because it facilitates commerce of cheeses and protects consumers from injuries of eating cheeses in the market. It also protects the consumers from economic fraud. For example, without FDA and USDA intervention, one may purchase a cheese that contains less than the normal amount of milk used. If so, the consumers lose money in the transaction.

All FDA standards of identities for cheeses are provided in 21 CFR 133, and USDA grade standards and dairy plant specifications are provided in 7 CFR 58. For explanation, 21 CFR 133 refers to United States Code of Federal Regulations, Part 133. The following focuses on the standards of identities issued by the FDA with selected discussion on USDA specifications and grade standards.

26.2 Asiago Fresh and Asiago Soft Cheese

Asiago fresh and asiago soft cheese is the food prepared from milk and other ingredients specified by the FDA. It contains not more than 45% of moisture, and its solids contain not less than 50% of milkfat. It is cured for not less than 60 days.

Milk, which may be pasteurized or clarified or both, and which may be warmed, is subjected to the action of harmless lactic-acid producing bacteria, present or added. Harmless artificial blue or green coloring in a quantity that neutralizes any natural yellow coloring in the curd may be added. Sufficient rennet, or other safe and suitable milk-clotting enzyme that produces equivalent curd formation, or both, with or without purified calcium chloride in a quantity not more than 0.02% (calculated as anhydrous calcium chloride) of the weight of the milk, is added to set the milk to a semisolid mass. The mass is cut, stirred, and heated to promote and regulate separation of the whey from the curd. The whey is drained off. When the curd is sufficiently firm, it is removed from the kettle or vat, further drained for a short time, packed into hoops, and pressed. The pressed curd is salted in brine and cured in a well-ventilated room. During curing, the surface of the cheese is occasionally rubbed with a vegetable oil. A harmless preparation of enzymes of animal or plant origin capable of aiding in the curing or development of flavor of asiago fresh cheese may be added during the procedure in such quantity that the weight of the solids of such preparation is not more than 0.1% of the weight of the milk used.

Milk may be bleached by the use of benzoyl peroxide or a mixture of benzoyl peroxide with potassium alum, calcium sulfate, and magnesium carbonate; but the weight of the benzoyl peroxide is not more than 0.002% of the weight of the milk bleached, and the weight of the potassium alum, calcium sulfate, and magnesium carbonate, singly or combined, is not more than six times the weight of the benzoyl peroxide used. If milk is bleached in this manner, sufficient vitamin A is added to the curd to compensate for the vitamin A or its precursors destroyed in the bleaching process, and artificial coloring is not used.

Safe and suitable antimycotic agent(s), the cumulative levels of which should not exceed current good manufacturing practice, may be added to the surface of the cheese.

Asiago medium cheese contains not more than 35% moisture; its solids contain not less than 45% of milkfat, and it is cured for not less than 6 months. Asiago old cheese contains not more than 32% moisture; its solids contain not less than 42% of milk fat, and it is cured for not less than 1 year.

26.3 Blue Cheese

Blue cheese is the food prepared by the procedure set forth by the FDA. It is characterized by the presence of bluish-green mold, *Penicillium roquefortii*, throughout the cheese. The minimum milkfat content is 50% by weight of the solids, and the maximum moisture content is 46% by weight. The dairy ingredients used may be pasteurized. Blue cheese is at least 60 days old.

One or more of the dairy ingredients specified may be homogenized, bleached, or warmed and are subjected to the action of a lactic acid–producing bacterial culture. One or more of the clotting enzymes specified are added to set the dairy ingredients to a semisolid mass. The mass is cut into smaller portions and allowed to stand for a time. The mixed curd and whey is placed in forms permitting further drainage. While the curd is being placed in forms, spores of the mold *P. roquefortii* are added. The forms are turned several times during drainage. When sufficiently drained, the shaped curd is removed from the forms and salted with dry salt or brine. Perforations are then made in the shaped curd, and it is held at a temperature of approximately 50°F at 90%–95% relative humidity, until the characteristic mold growth has developed. During storage, the surface of the cheese may be scraped to remove surface growth of undesirable microorganisms. Antimycotics may be applied to the surface of the whole cheese. One or more of the other optional ingredients specified may be added during the procedure.

The following safe and suitable ingredients may be used:

1. Milk, nonfat milk, or cream used alone or in combination.
2. Rennet and/or other clotting enzymes of animal, plant, or microbial origin.
3. Blue or green color in an amount to neutralize the natural yellow color of the curd.
4. Calcium chloride in an amount not more than 0.02% (calculated as anhydrous calcium chloride) of the weight of the dairy ingredients, used as a coagulation aid.
5. Enzymes of animal, plant, or microbial origin, used in curing or flavor development.
6. Antimycotic agents, applied to the surface of slices or cuts in consumer-sized packages or to the surface of the bulk cheese during curing.
7. Benzoyl peroxide or a mixture of benzoyl peroxide with potassium alum, calcium sulfate, and magnesium carbonate used to bleach the dairy ingredients. The weight of the benzoyl peroxide is not more than 0.002% of the weight of the milk being bleached, and the weight of the

potassium alum, calcium sulfate, and magnesium carbonate, singly or combined, is not more than six times the weight of the benzoyl peroxide used. If milk is bleached in this manner, vitamin A is added to the curd in such quantity as to compensate for the vitamin A or its precursors destroyed in the bleaching process, and artificial coloring is not used.

8. Vegetable fats or oils, which may be hydrogenated, used as a coating for the rind.

26.4 Brick Cheese

Brick cheese is the food prepared from dairy ingredients and other ingredients specified. The minimum milkfat content is 50% by weight of the solids, and the maximum moisture content is 44% by weight. If the dairy ingredients used are not pasteurized, the cheese is cured at a temperature of not less than 35°F for at least 60 days.

If pasteurized dairy ingredients are used, the phenol equivalent value of 0.25 g of brick cheese is not more than 5 μm.

One or more of the dairy ingredients specified are brought to a temperature of about 88°F and subjected to the action of a lactic acid–producing bacterial culture. One or more of the clotting enzymes specified are added to set the dairy ingredients to a semisolid mass. The mass is cut into cubes with sides approximately 3/8 in. long, and stirred and heated so that the temperature rises slowly to about 96°F. The stirring is continued until the curd is sufficiently firm. Part of the whey is then removed, and the mixture is diluted with water or salt brine to control the acidity. The curd is transferred to forms and drained. During drainage, it is pressed and turned. After drainage, the curd is salted, and the biological curing agents characteristic of brick cheese are applied to the surface. The cheese is then cured to develop the characteristics of brick cheese. One or more of the other optional ingredients specified may be added during the procedure.

The following safe and suitable ingredients may be used:

1. Milk, nonfat milk, or cream used alone or in combination.
2. Rennet and/or other clotting enzymes of animal, plant, or microbial origin.
3. Coloring.
4. Calcium chloride in an amount not more than 0.02% (calculated as anhydrous calcium chloride) of the weight of the dairy ingredients, used as a coagulation aid.
5. Enzymes of animal, plant, or microbial origin, used in curing or flavor development.
6. Antimycotic agents, the cumulative level of which should not exceed current good manufacturing practice, may be added to the surface of the cheese.

26.5 Caciocavallo Siciliano Cheese

Caciocavallo siciliano cheese is the food prepared from cow's milk, sheep's milk, goat's milk, or mixtures of two or all of these and other ingredients specified. It has a stringy texture and is made in oblong shapes. It contains not more than 40% moisture, and its solids contain not less than 42% milkfat. It is cured for not less than 90 days at a temperature of not less than 35°F.

Milk, which may be pasteurized or clarified or both, and which may be warmed, is subjected to the action of harmless lactic acid–producing bacteria, present in such milk or added thereto. Harmless artificial blue or green coloring in a quantity that neutralizes any natural yellow coloring in the curd may be added. Sufficient rennet, rennet paste, extract of rennet paste, or other safe and suitable milk-clotting enzyme that produces equivalent curd formation, singly or in any combination (with or without purified calcium chloride in a quantity not more than 0.02%, calculated as anhydrous calcium chloride, of the weight of the milk), is added to set the milk to a semisolid mass. The mass is cut, stirred, and heated so as to promote and regulate the separation of whey from curd. The whey is drained off, and the curd is removed to another vat containing hot whey, in which it is soaked for several hours. This whey is withdrawn, the curd is allowed to mat, and is cut into blocks. These are washed in hot whey until the desired elasticity is obtained. The curd is removed from the vat, drained, pressed into oblong forms, dried, salted in brine, and cured. It may be paraffined. A harmless preparation of enzymes of animal or plant origin capable of aiding in the curing or development of flavor of caciocavallo siciliano cheese may be added during the procedure, in such quantity that the weight of the solids of such preparation is not more than 0.1% of the weight of the milk used.

The word "milk" means cow's milk, goat's milk, sheep's milk, or mixtures of two or all of these. Such milk may be bleached by the use of benzoyl peroxide or a mixture of benzoyl peroxide with potassium alum, calcium sulfate, and magnesium carbonate; but the weight of the benzoyl peroxide is not more than 0.002% of the weight of the milk bleached, and the weight of the potassium alum, calcium sulfate, and magnesium carbonate, singly or combined, is not more than six times the weight of the benzoyl peroxide used. If milk is bleached in this manner, sufficient vitamin A is added to the curd to compensate for the vitamin A or its precursors destroyed in the bleaching process, and artificial coloring is not used.

Safe and suitable antimycotic agent(s), the cumulative levels of which should not exceed current good manufacturing practice, may be added to the cheese during the kneading and stretching process and/or applied to the surface of the cheese.

When caciocavallo siciliano cheese is made solely from cow's milk, the name of such cheese is "caciocavallo siciliano cheese." When made from sheep's milk or goat's milk or mixtures of these, or one or both of these with cow's milk, the name is followed by the words "made from ___," with the blank being filled in with the name or names of the milks used, in order of predominance by weight.

26.6 Cheddar Cheese

Cheddar cheese is the food prepared according to FDA standard of identities. The minimum milkfat content is 50% by weight of the solids, and the maximum moisture content is 39% by weight. If the dairy ingredients used are not pasteurized, the cheese is cured at a temperature of not less than 35°F for at least 60 days.

If pasteurized dairy ingredients are used, the phenol equivalent value of 0.25 g of cheddar cheese is not more than 3 μm.

One or more of the dairy ingredients specified may be warmed or treated with hydrogen peroxide/catalase and are subjected to the action of a lactic acid–producing bacterial culture. One or more of the clotting enzymes specified are added to set the dairy ingredients to a semisolid mass. The mass is cut, stirred, and heated with continued stirring so as to promote and regulate the separation of whey and curd. The whey is drained off, and the curd is matted into a cohesive mass. The mass is cut into slabs, which are piled and handled so as to promote the drainage of whey and the development of acidity. The slabs are then cut into pieces, which may be rinsed by sprinkling or pouring water over them, with free and continuous drainage; but the duration of such rinsing is so limited that only the whey on the surface of such pieces is removed. The curd is salted, stirred, further drained, and pressed into forms. One or more of the other optional ingredients may be added during the procedure.

The following safe and suitable ingredients may be used:

1. Milk, nonfat milk, or cream used alone or in combination.
2. Rennet and/or other clotting enzymes of animal, plant, or microbial origin.
3. Coloring.
4. Calcium chloride in an amount not more than 0.02% (calculated as anhydrous calcium chloride) of the weight of the dairy ingredients, used as a coagulation aid.
5. Enzymes of animal, plant, or microbial origin, used in curing or flavor development.
6. Antimycotic agents, applied to the surface of slices or cuts in consumer-sized packages.
7. Hydrogen peroxide, followed by a sufficient quantity of catalase preparation to eliminate the hydrogen peroxide. The weight of the hydrogen peroxide should not exceed 0.05% of the weight of the milk, and the weight of the catalase should not exceed 20 parts per million of the weight of the milk treated.

For cheddar cheese for manufacturing, the milk is not pasteurized, and curing is not required.

Low sodium cheddar cheese is the food prepared from the same ingredients and in the same manner prescribed for cheddar cheese, except that

1. It contains not more than 96 mg of sodium per pound of finished food.
2. The name of the food is "low sodium cheddar cheese." The letters in the words "low sodium" should be of the same size and style of type as the letters in the words "cheddar cheese," wherever such words appear on the label.
3. If a salt substitute is used, the label should bear the statement "___ added as a salt substitute," with the blank being filled in with the common name or names of the ingredient or ingredients used as a salt substitute.

26.7 USDA Specifications for Reduced Fat Cheddar Cheese

Reduced fat cheddar cheese should comply with all applicable FDA regulations for nutrient content claims for fat and for foods named by use of a nutrient content claim and a standardized term.

Reduced fat cheddar cheese should contain between 1/4 and 1/3 less fat than that of traditional cheddar cheese. The reduced fat cheddar cheese should contain not less than 19.2% total fat (as marketed) and not more than 22.9% total fat (as marketed). The moisture content should not exceed 49.0%. The cheese should contain not less than 1.4% but not more than 2.0% salt. No vat may have a pH value higher than 5.30 using the quinhydrone method.

All dairy ingredients should be pasteurized at a temperature of not less than 161°F for a period of not less than 15 s, or for a time and at a temperature equivalent thereto in phosphatase destruction. The shelf life of the reduced fat cheddar cheese should be 5 months after manufacture, when stored at 40°. The cheese should demonstrate satisfactory meltability characteristics, as represented by the USDA.

The original, unheated sample is equivalent to no. 1. The melted sample is assigned the value of the concentric circle, which any portion of the melted sample reaches.

Poor: meltability scale reading is no. 2 or less

Good: meltability scale reading equal to no. 3

Very good: meltability scale reading equal to no. 4 or greater

Be sure the oven and interior shelving are kept level.

The cheese should have a pleasing flavor. It may be lacking in flavor development or may possess slight characteristic cheddar cheese flavor; it may also possess very slight acid, bitter, or slight feed, but should not possess any undesirable flavors and odors.

The cheese body and texture should be firm, compact, and slightly translucent and shiny. It may have a few mechanical openings (one to three openings per plug) if not large and connecting. The cheese should be free from gas holes. The body may be very slight gummy, slight mealy or coarse, and definitely curdy.

The cheese should have a fairly uniform, bright attractive appearance. It may be slightly wavy, and it may be colored or uncolored, but if colored, it should be a medium yellow-orange. The cheese surfaces should be free of mold but may be slightly soiled or rough. The cheese blocks may be slightly lopsided.

The offeror must comply with the following requirements:

1. Cheese must be at least 30 days old when graded.
2. Cheese must be located in a cooler place and stored at temperatures between 35°F and 45°F until delivered to CCC.

The cheese should be packaged in commercially acceptable wrappers or bags; air should be permanently excluded between the wrapper and cheese surface through shrinkage, vacuumizing, mechanical means, or a combination of these methods. Closure of the bag should be by heat sealing. The bag may be slightly wrinkled but should be of such condition as to fully protect the surface of the cheese.

26.8 USDA Specifications for Shredded Cheddar Cheese

Cheddar cheese should be manufactured, shredded, and packaged in accordance with the USDA requirements.

It should be aged no less than 10 days at 38°F–42°F (3.5°C–5.5°C) prior to inspection. This aging may take place before or after the cheese is shredded. Shredded cheddar cheese should contain not less than 49.0% milkfat by weight of the solids; it should contain not more than 39.0% moisture. Shredded cheddar cheese should not have a pH exceeding 5.35 using the quinhydrone method.

Shredded cheddar cheese samples should be tempered to 45°–55°F (7°C–13°C) prior to product evaluation.

Shredded cheddar cheese should have a mild pleasing flavor. It may be lacking in flavor development or may possess characteristic cheddar cheese flavor. Shredded cheddar cheese should meet the flavor requirements of U.S. Grade A or better according to the U.S. Standards for Grades of Cheddar Cheese.

Shredded cheddar cheese may have the following body and texture characteristics to a slight degree: mealy, weak, or pasty. The cheese should be free from all foreign and extraneous materials. Shredded cheddar cheese should have a height and width up to 3/16 in. in either dimension. It should be free flowing and should not be matted. An approved anticaking agent may be added as a processing aid. If an anticaking agent is used, the amount used should be the minimum required to produce the desired effect, but should not exceed 2.0% of the weight of the shredded cheddar cheese.

Shredded cheddar cheese should not contain more than 6.0% fines. For shreds whose height and/or width is 1/16 in. or less, the fine content should be determined using a standard test sieve #14 (1.4 mm). For shreds whose height and/or width are greater than 1/16 in., the fine content should be determined using a standard test sieve #8 (2.36 mm).

Note that the manufacturer should provide shred size information. This information will determine the sieve used to measure the fine content. If shred size information is not provided, the standard test sieve #8 will be used.

Shredded cheddar cheese may be colored or uncolored; if colored, it should be a medium yellow-orange. The cheese should have a uniform bright color and an attractive sheen. No visible signs of mold should be permitted.

26.9 Colby Cheese

Colby cheese is the food prepared from milk and other ingredients specified. It contains not more than 40% of moisture, and its solids contain not less than 50% of milkfat. If the milk used is not pasteurized, the cheese so made is cured at a temperature of not less than 35°F for not less than 60 days.

Milk, which may be pasteurized or clarified or both, and which may be warmed, is subjected to the action of harmless lactic acid–producing bacteria, present in such milk or added thereto. Harmless artificial coloring may be added. Sufficient rennet, or other safe and suitable milk-clotting enzyme that produces equivalent curd formation, or both, with or without purified calcium chloride in a quantity not more than 0.02% (calculated as anhydrous calcium chloride) of the weight of the milk, is added to set the milk to a semisolid mass. The mass is cut, stirred, and heated with continued stirring so as to promote and regulate the separation of whey and curd. A part of the whey is drained off, and the curd is cooled by adding water, the stirring being continued so as to prevent the pieces of curd from matting. The curd is drained, salted, stirred, further drained, and pressed into forms. A harmless preparation of enzymes of animal or plant origin capable of aiding in the curing or development of flavor of colby cheese may be added during the procedure in such quantity that the weight of the solids of such preparation is not more than 0.1% of the weight of the milk used.

Milk should be deemed to have been pasteurized if it has been held at a temperature of not less than 143°F for a period of not less than 30 minutes, or for a time and at a temperature equivalent thereto in phosphatase destruction. Colby cheese should be deemed not to have been made from pasteurized milk if 0.25 g shows a phenol equivalent of more than 3 μm.

During the cheese-making process, the milk may be treated with hydrogen peroxide/catalase.

Colby cheese in the form of slices or cuts may have added to it a clear aqueous solution prepared by condensing or precipitating wood smoke in water. Colby cheese in the form of slices or cuts in consumer-sized packages may contain an optional mold-inhibiting ingredient consisting of sorbic acid, potassium sorbate, sodium sorbate, or any combination of two or more of these, in an amount not to exceed 0.3% by weight calculated as sorbic acid. If colby cheese has added to it a clear aqueous solution prepared by condensing or precipitating wood smoke in water, the name of the food is immediately followed by the words "with added smoke flavoring," with all words in this phrase of the same type size, style, and color without intervening written, printed, or graphic matter.

If colby cheese in sliced or cut form contains an optional mold-inhibiting ingredient, the label should bear the statement "___ added to retard mold growth" or "___ added as a preservative," with the blank being filled in with the common name or names of the mold-inhibiting ingredient or ingredients used.

26.10 USDA Grade Standards for Colby Cheese

26.10.1 Definitions

Colby cheese is cheese made by the colby process or by any other procedure that produces a finished cheese having the same organoleptic, physical, and chemical properties as the cheese produced by the colby process. The cheese is made from cow's milk with or without the addition of artificial coloring. It contains added common salt and not more than 40% moisture, its total solid content is not less than 50% milkfat, and it conforms to FDA requirements.

26.10.2 Types of Surface Protection

The following are the types of surface protection for colby cheese.

1. Rinded and paraffin-dipped

   The cheese that has formed a rind is dipped in a refined paraffin, amorphous wax, microcrystalline wax, or other suitable substance. Such coating is a continuous, unbroken, and uniform film adhering tightly to the entire surface of the cheese rind.

2. Rindless

   Wrapped. The cheese is properly enveloped in a tight-fitting wrapper or other protective covering, which is sealed with sufficient overlap or satisfactory closure. The wrapper or covering should not impart color or objectionable taste or odor to the cheese. The wrapper or covering should be of sufficiently low permeability to air so as to prevent the formation of a rind.

   Paraffin-dipped. The cheese is dipped in a refined paraffin, amorphous wax, microcrystalline wax, or other suitable substance. The paraffin should be applied so that it is continuous, unbroken, and uniformly adheres tightly to the entire surface. If a wrapper or coating is applied to the cheese prior to paraffin dipping, it should completely envelop the cheese and not impart color or objectionable taste or odor to the cheese.

26.10.3 U.S. Grades

The nomenclature of U.S. grades is as follows:

- U.S. Grade AA
- U.S. Grade A
- U.S. Grade B

26.10.4 Basis for Determination of U.S. Grade

1. The cheese should be graded no sooner than 10 days of age.
2. The rating of each quality factor should be established on the basis of characteristics present in any vat of cheese.
3. The U.S. grades of colby cheese are determined on the basis of rating the following quality factors:
   a. Flavor
   b. Body and texture
   c. Color
   d. Finish and appearance
4. The final U.S. grade should be established on the basis of the lowest rating of any one of the quality factors.

26.10.5 Specifications for U.S. Grades

U.S. Grade AA colby cheese should conform to the following requirements (see Tables 26.1 through 26.4):

1. *Flavor.* The cheese should possess a fine and highly pleasing colby cheese flavor that is free from undesirable tastes and odors, or it may be lacking in flavor development. The cheese may possess a very slight feed flavor (see Table 26.1).
2. *Body and texture.* A plug drawn from the cheese should be firm. Dependent upon the method of manufacture, a satisfactory plug may exhibit evenly distributed small mechanical openings or a close body. The cheese should not possess sweet holes, yeast holes, or other gas holes. The texture may be definitely curdy (see Table 26.2).
3. *Color.* The color should be uniform and bright. If colored, the cheese should be a medium yellow-orange (see Table 26.3).

TABLE 26.1

Classification of Flavor with Corresponding U.S. Grade for Colby Cheese

Flavor Characteristics	AA	A	B
Acid	--	VS	D
Barny	--	--	S
Bitter	--	--	S
Feed	VS	S	D
Flat	--	--	S
Fruity	--	S	S
Malty	--	--	S
Old milk	--	--	S
Onion	--	--	VS
Rancid	--	--	S
Sour	--	--	VS
Utensil	--	--	S
Weedy	--	--	S
Whey-taint	--	--	S
Yeasty	--	--	S

Note: (--) = not permitted; VS = very slight; S = slight; D = definite.

TABLE 26.2

Classification of Body and Texture with Corresponding U.S. Grade
for Colby Cheese

Body and Texture Characteristics	AA	A	B
Coarse	--	--	S
Corky	--	--	S
Crumbly	--	--	S
Curdy	D	D	D
Gassy	--	--	S
Loosely knit	--	VS	S
Mealy	--	--	S
Pasty	--	--	S
Short	--	--	S
Slitty	--	--	S
Sweet holes	--	--	S
Weak	--	--	S

Note: (--) = not permitted; VS = very slight; S = slight.

4. *Finish and appearance.*

   a. *Rinded and paraffin-dipped.* The bandage should be evenly placed over the entire surface of the cheese and be free from unnecessary overlapping and wrinkles, and not be burst or torn. The rind should be sound, firm, and smooth and should provide good protection to the cheese. The surface should be smooth and bright and should have a good coating of wax or coating that adheres firmly to all surfaces. The cheese should be free from mold under the paraffin. It should also be free from high edges, huffing, or lopsidedness, but may possess soiled surface to a very slight degree (see Table 26.4).

   b. *Rindless and wrapped.* The wrapper or covering should be practically smooth and properly sealed with adequate overlapping at the seams or sealed by any other satisfactory type of closure. The wrapper or covering should be neat and should adequately and securely envelop the cheese, but may be slightly wrinkled. Allowance should be made for slight wrinkles caused by crimping or sealing when vacuum packaging is used. The cheese should be free from mold under the wrapper or covering and should not be huffed or lopsided (see Table 26.4).

TABLE 26.3

Classification of Color with Corresponding Grades for Colby Cheese

Color Characteristics	AA	A	B
Acid-cut	--	--	S
Bleached surface (rindless)	--	--	S
Dull or faded	--	--	S
Mottled	--	--	S
Salt spots	--	--	S
Seamy	--	--	S
Unnatural	--	--	S
Wavy	--	VS	S

Note: (--) = Not permitted; VS = very slight; S = slight.

TABLE 26.4

Classification of Finish and Appearance with Corresponding U.S. Grade

Finish and Appearance Characteristics	AA	A	B
Rindless:			
Defective coating (paraffin-dipped: scaly, blistered, and checked)	--	--	S
High edges	--	S	D
Irregular press cloth (uneven, wrinkled, and improper overlapping)	--	S	D
Lopsided	--	S	D
Mold under wrapper or covering	--	--	VS
Rough surface	--	S	D
Soiled surface	--	--	S
Soiled surface (paraffin-dipped)	VS	VS	S
Surface mold	--	--	S
Wrinkled wrapper or covering (paraffin-dipped)	S	S	D
Rinded:			
Checked rind	--	--	S
Defective coating (scaly, blistered, and checked)	--	--	S
High edges	--	S	D
Irregular press cloth (uneven, wrinkled, and improper overlapping)	--	S	D
Lopsided	--	S	D
Mold under paraffin	--	--	VS
Rough surface	--	S	D
Soiled surface	VS	VS	S
Sour rind	--	--	S
Surface mold	--	VS	S
Weak rind	--	--	S

Note: (--) = not permitted; VS = very slight; S = slight; D = definite.

c. *Rindless and paraffin-dipped.* The cheese surface should be smooth and bright and should have a good coating of paraffin that adheres firmly. If a wrapper or coating is applied prior to paraffin dipping, it should completely envelop the cheese. The cheese should be free from high edges, huffing, lopsidedness, or mold. The cheese may possess soiled surface to a very slight degree. The wrapper may be wrinkled to a slight degree (see Table 26.4).

26.10.6 U.S. Grade A Colby Cheese

The cheese should conform to the following requirements (see Tables 26.1 through 26.4):

1. *Flavor.* The cheese should possess a pleasing colby cheese flavor, which is free from undesirable tastes and odors, or it may be lacking in flavor development. The cheese may possess very slight acid flavor or feed flavor to a slight degree (see Table 26.1).
2. *Body and texture.* A plug drawn from the cheese should be reasonably firm. Dependent upon the method of manufacture, a satisfactory plug may exhibit evenly distributed mechanical

openings or a close body. The plug should be free from sweet holes, yeast holes, or other gas holes. The body may be very slightly loosely knit or definitely curdy (see Table 26.2).

3. *Color.* The color should be fairly uniform and bright. If colored, the cheese should be a medium yellow-orange. The cheese may possess waviness to a very slight degree (see Table 26.3).

4. *Finish and appearance.*
   a. *Rinded and paraffin-dipped.* The bandage should be evenly placed over the entire surface of the cheese and not be burst or torn. The rind should be sound, firm, and smooth and should provide good protection for the cheese. The surface should be practically smooth and bright and should have a good coating of paraffin that adheres firmly to all surfaces. The cheese should be free from mold under the paraffin. It may possess the following characteristics: to a very slight degree—soiled surface or surface mold; to a slight degree—high edges, irregular press cloth, lopsided, or rough surface (see Table 26.4).
   b. *Rindless and wrapped.* The wrapper or covering should be practically smooth, properly sealed with adequate overlapping at the seams, or sealed by any other satisfactory type of closure. The wrapper or covering should be neat and should adequately and securely envelop the cheese, but may be slightly wrinkled. Allowance should be made for slight wrinkles caused by crimping or sealing when vacuum packaging is used. The cheese should be free from mold under the wrapper or covering and should not be huffed, but may possess to a slight degree the following characteristics: high edges, lopsided, irregular press cloth, or rough surface (see Table 26.4).
   c. *Rindless and paraffin-dipped.* The cheese surface should be bright and have a good coating of paraffin that adheres firmly. If a wrapper or coating is applied prior to paraffin dipping, it should completely envelop the cheese and have a good coating of paraffin that adheres firmly. The cheese may possess a soiled surface to a very slight degree. It should be free from mold and may possess to a slight degree the following characteristics: high edges, lopsided, irregular press cloth, or rough surface. The wrapper may be wrinkled to a slight degree (see Table 26.4).

26.10.7 U.S. Grade B Colby Cheese

The cheese should conform to the following requirements (see Tables 26.1 through 26.4):

1. *Flavor.* The cheese may possess a fairly pleasing colby cheese flavor, or it may be lacking in flavor development. The cheese may possess the following flavors: to a very slight degree—onion or sour; to a slight degree—barny, bitter, flat, fruity, malty, old milk, rancid, utensil, weedy, whey-taint, or yeasty; and to a definite degree—acid or feed (see Table 26.1).

2. *Body and texture.* A plug drawn from the cheese should be moderately firm. Dependent upon the method of manufacture, a satisfactory plug may exhibit mechanical openings or a close body. The cheese may possess the following characteristics: to a slight degree—coarse, corky, crumbly, gassy, loosely knit, mealy, pasty, short, slitty, sweet holes, or weak; to a definite degree—curdy (see Table 26.2).

3. *Color.* The cheese may possess the following characteristics to a slight degree: acid-cut, dull, faded, mottled, salt spots, seamy, unnatural, or wavy. In addition, rindless colby cheese may have a bleached surface to a slight degree (see Table 26.3).

4. *Finish and appearance.*
   a. *Rinded and paraffin-dipped.* The bandage should be placed over the entire surface of the cheese and may be uneven and wrinkled, but not be burst or torn. The rind should be reasonably sound and free from soft spots, rind rot, cracks, or openings of any kind. The surface may be rough and unattractive but should possess a fairly good coating of paraffin. The paraffin may be scaly or blistered, with very slight mold under the bandage or paraffin, but there should be no indication that mold has entered the cheese. The cheese may possess

the following characteristics: to a slight degree—checked rind, defective coating, soiled surface, sour rind, surface mold, or weak rind; to a definite degree—high edges, irregular press cloth, lopsided, or rough surface (see Table 26.4).

b. *Rindless and wrapped.* The wrapper or covering should be unbroken and should adequately and securely envelop the cheese. The following characteristics may be present: to a very slight degree—mold under the wrapper but not entering the cheese; to a slight degree—soiled surface or surface mold; and to a definite degree—high edges, irregular press cloth, lopsided, rough surface, or wrinkled wrapper or cover (see Table 26.4).

c. *Rindless and paraffin-dipped.* The wrapper or coating applied prior to paraffin dipping should adequately and securely envelop the cheese and have a coating of paraffin that adheres firmly to the cheese wrapper or should be unbroken but may be definitely wrinkled. The paraffin may be scaly or blistered, with very slight mold under the paraffin, but there should be no indication that mold has entered the cheese. The cheese may possess the following characteristics: to a slight degree—defective coating, soiled surface, or surface mold; and to a definite degree—high edges, lopsided, irregular press cloth, or rough surface (see Table 26.4).

26.10.8 Grade Not Assignable

Colby cheese should not be assigned a U.S. grade for one or more of the following reasons:

1. The cheese fails to meet or exceed the requirements for U.S. grade B.
2. The cheese is produced in a plant that is rated ineligible for USDA grading service or is not USDA-approved.

26.10.9 Explanation of Terms

With respect to types of surface protection:

1. *Paraffin.*   Refined paraffin, amorphous wax, microcrystalline wax, or any combination of such or any other suitable substance
2. *Paraffin dipped.*   Cheese that has been coated with paraffin
3. *Rind.*   A hard coating caused by the dehydration of the surface of the cheese
4. *Rinded.*   A protection developed by the formation of a rind
5. *Rindless.*   Cheese that has not formed a rind due to the impervious type of wrapper, covering, or container enclosing the cheese
6. *Wrapped.*   Cheese that has been covered with a transparent or opaque material (plastic film type or foil) next to the surface of the cheese
7. *Wrapper or covering.*   A plastic film or foil material next to the surface of the cheese, used as an enclosure or covering of the cheese

With respect to flavor:

1. *Very slight.*   Detected only upon very critical examination
2. *Slight.*   Detected only upon critical examination
3. *Definite.*   Not intense but detectable
4. *Undesirable.*   Those listed in excess of the intensity permitted or those characterizing flavors not listed
5. *Acid.*   Sharp and colby to the taste, characteristic of lactic acid
6. *Barny.*   A flavor characteristic of the odor of a poorly ventilated cow barn

7. *Bitter.* Distasteful, similar to the taste of quinine

8. *Feed.* Feed flavors (such as alfalfa, sweetclover, silage, or similar feed) in milk carried through into the cheese

9. *Flat.* Insipid, practically devoid of any characteristic colby cheese flavor

10. *Fruity.* A fermented, sweet, fruit-like flavor resembling apples

11. *Lacking in flavor development.* No undesirable and very little, if any, colby cheese flavor development

12. *Malty.* A distinctive, harsh flavor suggestive of malt

13. *Old milk.* Lacks freshness

14. *Onion.* A flavor recognized by the peculiar taste and aroma suggestive of its name; present in milk or cheese when the cows have eaten onions, garlic, or leeks

15. *Rancid.* A flavor suggestive of rancidity or butyric acid; sometimes associated with bitterness

16. *Sour.* An acid, pungent flavor resembling vinegar

17. *Utensil.* A flavor that is suggestive of improper or inadequate washing and sterilization of milking machines, utensils, or factory equipment

18. *Weedy.* A flavor present in cheese when cows have eaten weedy hay or grazed on weed-infested pasture

19. *Whey-taint.* A slightly acid flavor characteristic of fermented whey

20. *Yeasty.* A flavor indicating yeasty fermentation

With respect to body and texture:

1. *Very slight.* Detected only upon very critical examination and present only to a minute degree.

2. *Slight.* Barely identifiable and present only to a small degree.

3. *Definite.* Readily identifiable and present to a substantial degree.

4. *Coarse.* Feels rough, dry, and sandy.

5. *Corky.* Hard, tough, over-firm cheese that does not readily break down when rubbed between the thumb and fingers.

6. *Crumbly.* Tends to fall apart when rubbed between the thumb and fingers.

7. *Curdy.* Smooth but firm; when worked between the fingers, it is rubbery and not waxy or broken down.

8. *Firm.* Feels solid, not soft or weak.

9. *Gassy.* Gas holes of various sizes and may be scattered.

10. *Loosely knit.* Curd particles are not well-matted and fused together.

11. *Mealy.* Short body, does not mold well and looks and feels like corn meal when rubbed between the thumb and fingers.

12. *Mechanical openings.* Irregular shaped openings that are caused by variations in make procedure and not caused by gas fermentation.

13. *Pasty.* Is usually a weak body and when the cheese is rubbed between the thumb and fingers, it becomes sticky and smeary.

14. *Pinny.* Numerous very small gas holes.

15. *Reasonably firm.* Somewhat less firm but not to the extent of being weak.

16. *Short.* No elasticity in the cheese plug and when rubbed between the thumb and fingers, the cheese tends toward mealiness.

17. *Slitty.* Narrow, elongated slits generally associated with a cheese that is gassy or yeasty. Sometimes referred to as "fish-eyes."

18. *Sweet holes.* Spherical gas holes that are glossy in appearance and usually about the size of BB shots. These gas holes are sometimes referred to as "shot holes."

19. *Weak.* The cheese plug is soft but is not necessarily sticky like a pasty cheese and requires little pressure to crush.

With respect to color:

1. *Very slight.* Detected only upon very critical examination and present only to a minute degree

2. *Slight.* Barely identifiable and present only to a small degree

3. *Acid-cut.* A bleached or faded color that sometimes varies throughout the cheese and appears most often around mechanical openings

4. *Bleached surface.* A faded color beginning at the surface and progressing inward

5. *Dull or faded.* A color condition lacking in luster or translucency

6. *Mottled.* Irregular shaped spots or blotches in which portions are light-colored and others are of higher color; also an unevenness of color due to combining the curd from two different vats, sometimes referred to as "mixed curd"

7. *Salt spots.* Large light-colored spots or areas

8. *Seamy.* White thread-like lines that form when the curd is not properly matted or fused

9. *Unnatural.* Deep orange or reddish color

10. *Uncolored.* Absence of added coloring

11. *Wavy.* Unevenness of color that appears as layers or waves

With respect to finish and appearance:

1. *Very slight.* Detected only upon very critical examination and present to a minute degree.

2. *Slight.* Barely identifiable and present to a small degree.

3. *Definite.* Readily identifiable and present to a substantial degree.

4. *Adequately and securely enveloped.* The wrapper or covering is properly sealed and entirely encloses the cheese with sufficient adherence to the surface of the cheese to protect it from contamination or dehydration.

5. *Bandage.* Cheese cloth used to wrap cheese prior to dipping in paraffin.

6. *Bandage evenly placed.* Placement of the bandage so that it completely envelops the cheese and overlaps evenly about 1 in.

7. *Bright surface.* Clean, glossy surface.

8. *Burst or torn bandage.* A severance of the bandage usually occurring at the side seam; or when the bandage is otherwise snagged or broken.

9. *Checked rind.* Numerous small cracks or breaks in the rind that sometimes follows the outline of curd particles.

10. *Defective coating.* A brittle coating of paraffin that breaks and peels off in the form of scales or flakes; flat or raised blisters or bubbles under the surface of the paraffin; checked paraffin, including cracks, breaks, or hairline checks in the paraffin or coating of the cheese.

11. *Firm sound rind.* A rind possessing a firmness and thickness (not easily dented or damaged) consistent with the size of the cheese and that is dry, smooth, and closely knit, sufficient to protect the interior quality from external defects; free from checks, cracks, breaks, or soft spots.

12. *High edge.* A rim or ridge on the side of the cheese.

13. *Huffed.* A block of cheese that is swollen because of gas fermentation. The cheese becomes rounded or oval in shape instead of having flat surfaces.

14. *Irregular press cloth.* Press cloth improperly placed in the hoop resulting in too much press cloth on one end and insufficient on the other causing overlapping; wrinkled and loose fitting.

15. *Lopsided.* One side of the cheese is higher than the other side.

16. *Mold under bandage and paraffin.* Mold spots or areas under the paraffin.

17. *Mold under wrapper or covering.* Mold spots or areas under the wrapper or covering.

18. *Rind rot.* Soft spots on the rind that have become discolored and are decayed or decomposed.

19. *Rough surface.* Lacks smoothness.

20. *Smooth surface.* Not rough or uneven.

21. *Soft spots.* Areas soft to the touch and that are usually faded and moist.

22. *Soiled surface.* Milkstone, rust spots, or other discoloration on the surface of the cheese.

23. *Sour rind.* A fermented rind condition, usually confined to the faces of the cheese.

24. *Surface mold.* Mold on the exterior of the paraffin or wrapper.

25. Wax or paraffin that adheres firmly to the surface of the cheese. A coating with no cracks, breaks, or loose areas.

26. *Weak rind.* A thin rind that possesses little or no resistance to pressure.

26.11 Cold-Pack and Club Cheese

Cold-pack and club cheese is the food prepared by comminuting, without the aid of heat, one or more cheeses of the same or two or more varieties, except cream cheese, neufchatel cheese, cottage cheese, low-fat cottage cheese, cottage cheese dry curd, hard grating cheese, semisoft part-skim cheese, part-skim spiced cheese, and skim milk cheese for manufacturing, into a homogeneous plastic mass. One or more of the optional ingredients may be used.

All cheeses used in a cold-pack cheese are made from pasteurized milk or are held for not less than 60 days at a temperature of not less than 35°F before being comminuted.

The moisture content of a cold-pack cheese made from a single variety of cheese is not more than the maximum moisture content prescribed by the definition and standard of identity, if any there be, for the variety of cheese used. If there is no applicable definition and standard of identity, or if such standard contains no provision as to maximum moisture content, no water is used in the preparation of the cold-pack cheese.

The fat content of the solids of a cold-pack cheese made from a single variety of cheese is not less than the minimum prescribed by the definition and standard of identity, if any there be, for the variety of cheese used, but in no case is less than 47%, except that the fat content of the solids of cold-pack Swiss cheese is not less than 43%, and the fat content of the solids of cold-pack Gruyère cheese is not less than 45%.

The moisture content of a cold-pack cheese made from two or more varieties of cheese is not more than the arithmetical average of the maximum moisture contents prescribed by the definitions and standards of identity, if any there be, for the varieties of cheese used, but in no case is the moisture content more than 42%, except that the moisture content of a cold-pack cheese made from two or more of the varieties of cheddar cheese, washed curd cheese, colby cheese, and granular cheese is not more than 39%.

The fat content of the solids of a cold-pack cheese made from two or more varieties of cheese is not less than the arithmetical average of the minimum percent of fat prescribed by the definitions and standards of identity, if any there be, for the varieties of cheese used, but in no case is less than 47%, except that the fat content of the solids of a cold-pack cheese made from Swiss cheese and Gruyère cheese is not less than 45%.

The weight of each variety of cheese in a cold-pack cheese made from two varieties of cheese is not less than 25% of the total weight of both, except that the weight of blue cheese, nuworld cheese, roquefort cheese, or gorgonzola cheese is not less than 10% of the total weight of both, and the weight of limburger cheese is not less than 5% of the total weight of both. The weight of each variety of cheese in a cold-pack cheese made from three or more varieties of cheese is not less than 15% of the total weight of all, except that the weight of blue cheese, nuworld cheese, roquefort cheese, or gorgonzola cheese is not less than 5% of the total weight of all, and the weight of limburger cheese is not less than 3% of the total weight of all. These limits do not apply to the quantity of cheddar cheese, washed curd cheese, colby cheese, and granular cheese in mixtures that are designated as "American cheese." Such mixtures are considered as one variety of cheese.

Cold-pack cheese may be smoked, or the cheese or cheeses from which it is made may be smoked, before comminuting and mixing, or it may contain substances prepared by condensing or precipitating wood smoke.

The optional ingredients are as follows:

1. An acidifying agent consisting of one or any mixture of two or more of the following: a vinegar, lactic acid, citric acid, acetic acid, and phosphoric acid, in such quantity that the pH of the finished cold-pack cheese is not below 4.5. For the purposes of this section, vinegar is considered to be acetic acid.
2. Water.
3. Salt.
4. Harmless artificial coloring.
5. Spices or flavorings, other than any that singly or in combination with other ingredients simulate the flavor of a cheese of any age or variety.
6. Cold-pack cheese in consumer-sized packages may contain an optional mold-inhibiting ingredient consisting of sorbic acid, potassium sorbate, sodium sorbate, or any combination of two or more of these, in an amount not to exceed 0.3% by weight, calculated as sorbic acid or consisting of not more than 0.3% by weight of sodium propionate, calcium propionate, or a combination of sodium propionate and calcium propionate.

26.12 Cook Cheese (Koch Kaese)

Cook cheese, or koch kaese, is the food prepared by a specified procedure. The maximum moisture content is 80% by weight. The dairy ingredients used may be pasteurized.

The phenol equivalent value of 0.25 g of cook cheese is not more than 3 µg. One or more of the dairy ingredients specified may be warmed and are subjected to the action of a lactic acid–producing bacterial culture. One or more of the clotting enzymes specified are added to set the dairy ingredients to a semisolid mass. The mass is cut, stirred, and heated with continued stirring so as to separate the curd and whey. The whey is drained from the curd, and the curd is cured for 2 or 3 days. It is then heated to a temperature of not less than 180°F until the hot curd drops from a ladle with a consistency like that of honey. The hot cheese is filled into packages and cooled. One or more of the other optional ingredients specified may be added during the procedure.

The following safe and suitable ingredients may be used:

1. Nonfat milk
2. Rennet and/or other clotting enzymes of animal, plant, or microbial origin
3. Calcium chloride in an amount not more than 0.02% (calculated as anhydrous calcium chloride) of the weight of the dairy ingredients, used as a coagulation aid
4. Culture of white mold
5. Pasteurized cream
6. Caraway seed
7. Salt

26.13 Edam Cheese

Edam cheese is the food prepared by a specified procedure. The minimum milkfat content is 40% by weight of the solids, and the maximum moisture content is 45% by weight. If the dairy ingredients used are not pasteurized, the cheese is cured at a temperature of not less than 35°F for at least 60 days.

If pasteurized dairy ingredients are used, the phenol equivalent value of 0.25 g of edam cheese is not more than 3 μg.

One or more of the dairy ingredients specified may be warmed and are subjected to the action of a lactic acid–producing bacterial culture. One or more of the clotting enzymes specified are added to set the dairy ingredients to a semisolid mass. After coagulation, the mass is cut into small cube-shaped pieces with sides approximately 3/8 in. long. The mass is stirred and heated to about 90°F and handled by further stirring, heating, dilution with water or salt brine, and salting so as to promote and regulate the separation of curd and whey. When the desired curd is obtained, it is transferred to forms permitting drainage of whey. During drainage, the curd is pressed and turned. After drainage, the curd is removed from the forms and is salted and cured. One or more of the other optional ingredients specified may be added during the procedures.

The following safe and suitable ingredients may be used:

1. Milk, nonfat milk, or cream, used alone or in combination
2. Rennet and/or other clotting enzymes of animal, plant, or microbial origin
3. Coloring
4. Calcium chloride in an amount not more than 0.02% (calculated as anhydrous calcium chloride) of the weight of the dairy ingredients, used as a coagulation aid
5. Enzymes of animal, plant, or microbial origin, used in curing or flavor development
6. Antimycotic agents, the cumulative levels of which should not exceed current good manufacturing practice, may be added to the surface of the cheese

26.14 Gammelost Cheese

Gammelost cheese is the food prepared from nonfat milk. The maximum moisture content is 52% by weight.

The dairy ingredients are subjected to the action of a lactic acid–producing bacterial culture. The development of acidity is continued until the dairy ingredients coagulate to a semisolid mass. The mass is stirred and heated until a temperature of about 145°F is reached and is held at that temperature for at least 30 minutes. The whey is drained off, and the curd is removed and placed in forms and pressed. The shaped curd is placed in whey and heated for 3 or 4 hours and may again be pressed. It is then stored under conditions suitable for curing.

26.15 Gorgonzola Cheese

Gorgonzola cheese is the food prepared by a specified procedure. It is characterized by the presence of bluish-green mold, *P. roquefortii,* throughout the cheese. The minimum milkfat content is 50% by weight of the solids, and the maximum moisture content is 42% by weight. The dairy ingredients used may be pasteurized. Gorgonzola cheese is at least 90 days old.

One or more of the dairy ingredients specified may be warmed and are subjected to the action of a lactic acid–producing bacterial culture. One or more of the clotting enzymes specified are added to set the dairy ingredients to a semisolid mass. The mass is cut into smaller portions and allowed to stand for a time. The mixed curd and whey is placed into forms permitting further drainage. While being placed in forms, spores of the mold *P. roquefortii* are added. The forms are turned several times during drainage. When sufficiently drained, the shaped curd is removed from the forms and salted with dry salt or brine. Perforations are then made in the shaped curd, and it is held at a temperature of approximately 50°F at 90% to 95% relative humidity until the characteristic mold growth has developed. During storage, the surface of the cheese may be scraped to remove surface growth of undesirable microorganisms. One or more of the other optional ingredients specified may be added during the procedure.

The following safe and suitable ingredients may be used:

1. Milk, nonfat milk, or cream used alone or in combination.
2. Rennet and/or other clotting enzymes of animal, plant, or microbial origin.
3. Blue or green color in an amount to neutralize the natural yellow color of the curd.
4. Calcium chloride in an amount not more than 0.02% (calculated as anhydrous calcium chloride) of the weight of the dairy ingredients, used as a coagulation aid.
5. Enzymes of animal, plant, or microbial origin, used in curing or flavor development.
6. Antimycotic agents, the cumulative levels of which should not exceed current good manufacturing practice, may be added to the surface of the cheese.
7. Benzoyl peroxide, or a mixture of benzoyl peroxide with potassium alum, calcium sulfate, and magnesium carbonate used to bleach the dairy ingredients. The weight of the benzoyl peroxide is not more than 0.002% of the weight of the dairy ingredients being bleached, and the weight of the potassium alum, calcium sulfate, and magnesium carbonate, singly or combined, is not more than six times the weight of the benzoyl peroxide used. If the dairy ingredients are bleached in this manner, vitamin A is added to the curd in such quantity as to compensate for the vitamin A or its precursors destroyed in the bleaching process, and artificial coloring is not used.
8. Vegetable fats or oil, which may be hydrogenated, used as a coating for the rind.

27

Gouda and Other Cheeses: Standards, Grades, and Specifications*

Y. H. Hui

CONTENTS

---

* The information in this chapter has been modified from the *Food Safety Manual* published by the Science Technology System (STS) of West Sacramento, CA. Copyrighted 2011©. With permission.

27.1 Introduction

Food manufacturers in the United States must produce a product that consumers like in order to sell the product nationally. To achieve this goal, they depend on developments in science, technology, and engineering in order to produce a quality product. However, they must also pay attention to two other important considerations:

1. Are their products safe for public consumption?
2. Have they assured the economic integrity of their products? This means many things. The most important is the assurance that their products contain the proper quality and quantity of ingredients, that is, "No cheating."

To protect the consumers, U.S. government agencies have established many checks and balances to prevent "injuries" to consumers in terms of health and economic fraud. The two major responsible agencies are

1. U.S. Food and Drug Administration (FDA; www.FDA.gov)
2. U.S. Department of Agriculture (USDA; www.USDA.gov)

The FDA establishes standards of identities for a number of cheeses for interstate commerce. This ensures that a particular type of cheese contains the minimal quality and quantity of the ingredients used to make this cheese. The FDA also issues safety requirements and recommendations in the manufacturing of cheeses. It makes periodic inspections of establishments producing cheeses to ensure full compliance. In effect, the FDA is monitoring the safety production of cheeses and their economic integrities.

The USDA plays the same important role in the commerce of cheeses in the United States. For example, a national restaurant chain that makes and sells pizzas will not buy cheeses from any manufacturer if the cheeses do not bear the USDA official identification or seal of approval. The USDA has issued official requirements and processing specifications that cheese manufacturers must follow if they want to obtain the USDA seal of approval for the products. To achieve this goal, the USDA sets up voluntary services of inspection under a standard fee arrangement. The inspection focuses on the safety and economic integrity of manufacturing cheeses. When the inspection demonstrates that the manufacturer has complied with safety and economic requirements, the USDA issues its official identification or seal of approval.

In essence, the FDA and USDA are the federal agencies that ensure that the product is safe and will not pose any economic fraud. Using the public information distributed at their web sites, this chapter discusses FDA standards of identities and USDA specifications for establishments in relation to the production of selected cheeses.

The information in this chapter is important, because it facilitates the commerce of cheeses and protects consumers from injuries of eating cheeses in the market. It also protects the consumers from economic fraud. For example, without FDA and USDA intervention, one may purchase cheese that contains less than the normal amount of milk used. If so, the consumers lose money in the transaction.

All FDA cheese standards of identities are described in 21 CFR 133. All USDA requirements and standards for grade are described in 7 CFR 58. For explanation, CFR refers to the United States Code of Federal Regulations. This chapter focuses on two areas:

1. Part 133: standards of identities issued by the FDA.
2. Part 58 (selected sections): USDA specifications and grade standards.

27.2 Gouda Cheese

Gouda cheese conforms to FDA's definition and standard of identity for edam cheese (see discussion later), except that the minimum milkfat content is 46% by the weight of the solids and the maximum moisture content is 45% by weight.

27.3 Granular and Stirred Curd Cheese

Granular cheese (stirred curd cheese) is a food prepared by a specified procedure. The minimum milkfat content is 50% by the weight of the solids, and the maximum moisture content is 39% by weight. If the dairy ingredients used are not pasteurized, the cheese is cured at a temperature of not less than 35°F for at least 60 days.

If pasteurized dairy ingredients are used, the phenol equivalent value of 0.25 g of granular cheese is not more than 3 µg.

One or more of the dairy ingredients specified may be warmed, treated with hydrogen peroxide/catalase, and is subjected to the action of a lactic acid–producing bacterial culture. One or more of the clotting enzymes specified is added to set the dairy ingredients to a semisolid mass. The mass is cut, stirred, and heated with continued stirring so as to promote and regulate the separation of whey and curd. A part of the whey is drained off. The curd is then alternately stirred and drained to prevent matting and to remove whey from the curd. The curd is then salted, stirred, drained, and pressed into forms. One or more of the other optional ingredients specified may be added during the procedure.

The following safe and suitable ingredients may be used:

1. Milk, nonfat milk, or cream, used alone or in combination.
2. Rennet and/or other clotting enzymes of animal, plant, or microbial origin.
3. Coloring.
4. Calcium chloride in an amount not more than 0.02% (calculated as anhydrous calcium chloride) by the weight of the dairy ingredients, used as a coagulation aid.
5. Enzymes of animal, plant, or microbial origin, used in curing or flavor development.
6. Antimycotic agents, which may be added to the surface of the cheese and the cumulative levels of which should not exceed current good manufacturing practice.
7. Hydrogen peroxide, followed by a sufficient quantity of catalase preparation to eliminate the hydrogen peroxide. The weight of the hydrogen peroxide should not exceed 0.05% of the weight of the dairy ingredients, and the weight of the catalase should not exceed 20 parts per million of the weight of the dairy ingredients treated.

Granular cheese for manufacturing conforms to the definition and standard of identity prescribed for granular cheese, except that the dairy ingredients are not pasteurized and curing is not required.

27.4 Grated Cheese

Grated cheese is a class of food prepared by grinding, grating, shredding, or otherwise comminuting cheese of one variety or a mixture of two or more varieties. The cheese varieties that may be used are those for which there are definitions and standards of identity, except that cream cheese, neufchatel cheese, cottage cheese, creamed cottage cheese, cook cheese, and skim milk cheese for manufacturing may not be used. All cheese ingredients used are either made from pasteurized milk or held at a temperature of not less than 35°F for at least 60 days. Moisture may be removed from the cheese ingredients in the manufacture of the finished food, but no moisture is added. One or more of the optional ingredients specified may be used. The composition is as follows:

1. Each cheese ingredient used is present at a minimum level of 2% of the weight of the finished food.
2. When one variety of cheese is used, the minimum milkfat content of the food is not more than 1% lower than the minimum prescribed by the standard of identity for that cheese.

3. When two or more varieties of cheese are used, the minimum milkfat content is not more than 1% below the arithmetical average of the minimum fat content percentages prescribed by the standards of identity for the varieties of cheese used, and in no case is the milkfat content less than 31%.

The following safe and suitable ingredients may be used:

1. Antimycotics
2. Anticaking agents
3. Spices
4. Flavorings other than those which, singly or in combination with other ingredients, simulate the flavor of cheese of any age or variety

The name of the food is "grated cheese" or "grated cheeses," as appropriate. The name of the food should be accompanied by a declaration of the specific variety of cheese(s) used in the food and by a declaration indicating the presence of any added spice or flavoring.

Any cheese varietal names used in the name of the food are those specified by applicable standards of identity, except that the designation "American cheese" may be used for cheddar, washed curd, colby, or granular cheese or for any mixture of these cheeses.

The following terms may be used in place of the name of the food to describe specific types of grated cheese:

1. If only one variety of cheese is used, the name of the food is "grated ___ cheese," the name of the cheese filling the blank.
2. If only parmesan and romano cheeses are used and each is present at a level of not less than 25% by the weight of the finished food, the name of the food is "grated ___ and ___ cheese," the blanks being filled in with the names "parmesan" and "romano" in order of predominance by weight. The name "reggiano" may be used for "parmesan."
3. If a mixture of cheese varieties (not including parmesan or romano) is used and each variety is present at a level of not less than 25% of the weight of the finished food, the name of the food is "grated ___ cheese," the blank being filled in with the names of the varieties in order of predominance by weight.
4. If a mixture of cheese varieties in which one or more varieties (not including parmesan or romano) are each present at a level of not less than 25% by the weight of the finished food and one or more other varieties (which may include parmesan and romano cheese) are each present at a level of not less than 2% but in the aggregate not more than 10% of the weight of the finished food, the name of the food is "grated ___ cheese with other grated cheese" or "grated ___ cheese with other grated cheeses," as appropriate, the blank being filled in with the name or names of those cheese varieties present at levels of not less than 25% by the weight of the finished food in order of predominance, in letters not more than twice as high as the letters in the phrase "with other grated cheese(s)."

The following terms may be used in place of "grated" to describe alternative forms of cheese:

1. "Shredded," if the particles of cheese are in the form of cylinders, shreds, or strings
2. "Chipped" or "chopped," if the particles of cheese are in the form of chips.

27.5 Grated American Cheese

Grated American cheese is a food prepared by mixing, with or without the aid of heat, one or more of the optional cheese ingredients, with one or more of the optional ingredients prescribed in the next paragraph, into a uniformly blended, partially dehydrated, powdered, or granular mixture.

Grated American cheese food contains not less than 23% of milkfat. The optional cheese ingredients are cheddar cheese, washed curd cheese, colby cheese, and granular cheese. The other optional ingredients are

1. Nonfat dry milk
2. Dried whey
3. An emulsifying agent consisting of one or any mixture of two or more of the emulsifying ingredients approved by the FDA, in such quantity that the weight of the solids thereof is not more than 3% of the weight of the grated American cheese food
4. An acidifying agent consisting of one or more of the acid-reacting approved ingredients
5. Salt
6. Artificial coloring

27.6 Hard Grating Cheeses

The cheeses for which definitions and standards of identity are prescribed in this section are hard grating cheeses. They are made from milk and the other ingredients specified. They contain not more than 34% of moisture, and their solids contain not less than 32% of milkfat. Hard grating cheeses are cured for not less than 6 months.

Milk, which may be pasteurized, clarified, or both and may be warmed, is subjected to the action of harmless lactic acid–producing bacteria or other harmless flavor-producing bacteria, present in such milk or added thereto. Sufficient rennet, rennet paste, extract of rennet paste, or other safe and suitable milk-clotting enzyme that produces equivalent curd formation, singly or in any combination (with or without purified calcium chloride in a quantity not more than 0.02%, calculated as anhydrous calcium chloride, of the weight of the milk) is added to set the milk to a semisolid mass. Harmless artificial coloring may be added. The mass is cut into small particles, stirred, and heated. The curd is separated from the whey, drained, shaped into forms, pressed, salted, and cured. The rind may be colored or rubbed with vegetable oil or both. A harmless preparation of enzymes of animal or plant origin capable of aiding in the curing or development of flavor of hard grating cheese may be added during the procedure in such quantity that the weight of the solids of such preparation is not more than 0.1% of the weight of the milk used.

The word "milk" means cow, goat, or sheep milk or a mixture of two or all of these. Such milk may be adjusted by separating part of the fat or (in the case of cow milk) by adding one or more of the following: cream, skim milk, concentrated skim milk, and nonfat dry milk; the corresponding products from goat milk (in the case of goat milk); the corresponding products from sheep milk (in the case of sheep milk); and water in a quantity sufficient to reconstitute any such concentrated or dried products used.

Safe and suitable antimycotic agent(s), the cumulative levels of which should not exceed current good manufacturing practice, may be added to the surface of the cheese.

27.7 Gruyere Cheese

Gruyere cheese is a food prepared by the procedure specified. It contains small holes or eyes. It has a mild flavor, due in part to the growth of surface-curing agents. The minimum milkfat content is 45% by the weight of the solids, and the maximum moisture content is 39% by weight. The dairy ingredients used may be pasteurized. The cheese is at least 90 days old.

If pasteurized dairy ingredients are used, the phenol equivalent value of 0.25 g of gruyere cheese is not more than 3 µg.

One or more of the dairy ingredients specified may be warmed and is subjected to the action of lactic acid–producing and propionic acid–producing bacterial cultures. One or more of the clotting enzymes specified is added to set the dairy ingredients to a semisolid mass. The mass is cut into particles similar in size to wheat kernels. For about 30 min, the particles are alternately stirred and allowed to settle. The temperature is raised to about 126°F. Stirring is continued until the curd becomes firm. The curd is transferred to hoops or forms and pressed until the desired shape and firmness are obtained. The cheese is surface salted while held at a temperature of 48°F–54°F for a few days. It is soaked for 1 day in a saturated salt solution. It is then held for 3 weeks in a salting cellar and wiped every 2 days with brine cloth to ensure the growth of biological curing agents on the rind. It is then moved to a heating room and held at progressively higher temperatures, finally reaching 65°F with a relative humidity of 85%–90%, for several weeks, during which time small holes or the so-called eyes form. The cheese is then stored at a lower temperature for further curing. One or more of the other optional ingredients specified may be added during the procedure.

The following safe and suitable ingredients may be used:

1. Milk, nonfat milk, or cream, used alone or in combination
2. Rennet and/or other clotting enzymes of animal, plant, or microbial origin
3. Calcium chloride in an amount not more than 0.02% (calculated as anhydrous calcium chloride) of the weight of the dairy ingredients, used as a coagulation aid
4. Enzymes of animal, plant, or microbial origin, used in curing or flavor development
5. Antimycotic agents, applied to the surface of slices or cuts in consumer-sized packages

27.8 Hard Cheeses

Hard cheeses are made from milk and the other ingredients specified. They contain not more than 39% of moisture, and their solids contain not less than 50% of milkfat. If the milk used is not pasteurized, the cheese so made is cured at a temperature of not less than 35°F for not less than 60 days.

Milk, which may be pasteurized, clarified, or both and may be warmed, is subjected to the action of harmless lactic acid–producing bacteria, with or without other harmless flavor-producing bacteria, present in such milk or added thereto. Harmless artificial coloring may be added. Sufficient rennet, rennet paste, extract of rennet paste, or other safe and suitable milk-clotting enzyme that produces equivalent curd formation, singly or in any combination (with or without purified calcium chloride in a quantity not more than 0.02%, calculated as anhydrous calcium chloride, of the weight of the milk) is added to set the milk to a semisolid mass. The mass is cut into small particles, stirred, and heated. The curd is separated from the whey, drained, and shaped into forms, and may be pressed. The curd is salted at some stage of the manufacturing process. The shaped curd may be cured. The rind may be coated with paraffin or rubbed with vegetable oil. A harmless preparation of enzymes of animal or plant origin capable of aiding in the curing or development of flavor of hard cheese may be added during the procedure in such quantity that the weight of the solids of such preparation is not more than 0.1% of the weight of the milk used. Harmless flavor-producing microorganisms may be added, and curing may be conducted under suitable conditions for the development of biological curing agents.

The word "milk" means cow, goat, or sheep milk or a mixture of two or all of these. Such milk may be adjusted by separating part of the fat or (in the case of cow milk) by adding one or more of the following: cream, skim milk, concentrated skim milk, and nonfat dry milk; the corresponding products from goat milk (in the case of goat milk); the corresponding products from sheep milk (in the case of sheep milk); and water in a quantity sufficient to reconstitute any concentrated or dried products used.

Milk should be deemed to have been pasteurized if it has been held at a temperature of not less than 143°F for a period of not less than 30 min or for a time and at a temperature equivalent thereto in phosphatase destruction. A hard cheese should be deemed not to have been made from pasteurized milk if 0.25 g shows a phenol equivalent of more than 3 µg.

Safe and suitable antimycotic agent(s), the cumulative levels of which should not exceed current good manufacturing practice, may be added to the surface of the cheese.

27.9 Limburger Cheese

Limburger cheese is a food prepared by one of the specified procedures. The minimum milkfat content is 50% by the weight of the solids, and the maximum moisture content is 50% by weight. If the dairy ingredients used are not pasteurized, the cheese is cured at a temperature of not less than 35°F for at least 60 days.

If pasteurized dairy ingredients are used, the phenol equivalent value of 0.25 g of limburger cheese is not more than 4 μg.

One of the following procedures may be followed for producing limburger cheese.

One or more of the specified unpasteurized dairy ingredients is warmed to about 92°F and subjected to the action of a lactic acid–producing bacterial culture. One or more of the clotting enzymes specified is added to set the dairy ingredients to a semisolid mass. The mass is cut into cubes, with sides approximately 1/2 inch long. After a few minutes, the mass is stirred and heated, gradually raising the temperature to 96°F–98°F. The curd is then allowed to settle, most of the whey is drained off, and the remaining curd and whey is dipped into molds. During drainage, the curd may be pressed. It is turned at regular intervals. After drainage, the curd is cut into pieces of desired size and dry salted at intervals for 24–48 h. The cheese is then cured with frequent applications of a weak brine solution to the surface until the proper growth of surface-curing organisms is obtained. It is then wrapped and held in storage for the development of as much additional flavor as desired. One or more of the other optional ingredients specified may be added during the procedure.

One or more of the pasteurized dairy ingredients are brought to a temperature of 89°F–90°F after pasteurization and subjected to the action of a lactic acid–producing bacterial culture. The procedure is then the same for unpasteurized dairy ingredients, except that heating is to 94°F. After most of the whey is drained off, salt brine at a temperature of 66°F–70°F is added so that the pH of the curd is about 4.8. The mixed curd, whey, and brine is dipped into molds, and the remaining procedure is as described above.

The following safe and suitable ingredients may be used:

1. Milk, nonfat milk, or cream, used alone or in combination
2. Rennet and/or other clotting enzymes of animal, plant, or microbial origin
3. Coloring
4. Calcium chloride in an amount not more than 0.02% (calculated as anhydrous calcium chloride) by the weight of the dairy ingredients, used as a coagulation aid
5. Enzymes of animal, plant, or microbial origin, used in curing or flavor development

27.10 Monterey and Monterey Jack Cheese

Monterey (monterey jack) cheese is a food prepared by a specified procedure. The minimum milkfat content is 50% by the weight of the solids, and the maximum moisture content is 44% by weight. The dairy ingredients used are pasteurized. The phenol equivalent of 0.25 g of monterey cheese is not more than 3 μg.

One or more of the dairy ingredients specified is subjected to the action of a lactic acid–producing bacterial culture. One or more of the clotting enzymes specified is added to set the dairy ingredients to a semisolid mass. The mass is cut, stirred, and heated with continued stirring so as to promote and regulate the separation of whey and curd. Part of the whey is drained off, and water or salt brine may be added. The curd is drained and placed in a muslin or sheeting cloth, formed into a ball, and pressed, or the curd is placed in a cheese hoop and pressed. Later, the cloth bandage is removed, and the cheese may be covered with a suitable coating. One or more of the other optional ingredients specified may be added during the procedure.

The following safe and suitable ingredients may be used:

1. Milk, nonfat milk, or cream, used alone or in combination
2. Rennet and/or other clotting enzymes of animal, plant, or microbial origin
3. Calcium chloride in an amount not more than 0.02% (calculated as anhydrous calcium chloride) by the weight of the dairy ingredients, used as a coagulation aid
4. Enzymes of animal, plant, or microbial origin, used in curing or flavor development
5. Salt
6. Antimycotic agents, which may be added to the surface of the cheese and the cumulative levels of which should not exceed current good manufacturing practice
7. Vegetable oil, with or without rice flour sprinkled on the surface, used as a coating for the rind

High-moisture monterey jack cheese has a moisture content that is more than 44% but less than 50%.

27.11 USDA Grade Standards for Monterey (Monterey Jack) Cheese

27.11.1 Definitions

Monterey (monterey jack) cheese is a cheese made by the monterey process or by any other procedure that produces a finished cheese having the same organoleptic, physical, and chemical properties as the cheese produced by the monterey process. The cheese is made from pasteurized cow milk. It may contain added common salt and contains not more than 44% moisture, its total solids content is not less than 50% milkfat, and it conforms to FDA requirements.

27.11.2 Types of Surface Protection

The following are the types of surface protection for monterey (monterey jack) cheese:

1. *Rinded and paraffin dipped.* The *cheese that has formed a rind* is dipped in refined paraffin, amorphous wax, microcrystalline wax, or other suitable substance. Such coating is a continuous, unbroken, and uniform film adhering tightly to the entire surface of the cheese rind.
2. *Rindless*
   a. *Wrapped.* The cheese is completely enveloped in a tight-fitting wrapper or other protective covering, which is sealed with sufficient overlap or satisfactory closure. The wrapper or covering should not impart color, objectionable taste, or odor to the cheese. The wrapper or covering should be of sufficiently low permeability to air so as to prevent the formation of a rind.
   b. *Paraffin dipped.* The cheese is dipped in refined paraffin, amorphous wax, microcrystalline wax, or other suitable substance. The paraffin should be applied so that it is continuous, unbroken, and uniformly adheres tightly to the entire surface. If a wrapper or coating is applied to the cheese prior to paraffin dipping, it should completely envelop the cheese and not impart color, objectionable taste, or odor to the cheese.

If antimycotics are used, they should be used in accordance with FDA requirements.

27.11.3 U.S. Grades

The nomenclature of U.S. grades is as follows:

1. U.S. Grade AA
2. U.S. Grade A
3. U.S. Grade B

The basis for the determination of U.S. grade should be as follows:

1. The cheese should be graded no sooner than 10 days of age.
2. The rating of each quality factor should be established on the basis of characteristics present in any vat of cheese.
3. The U.S. grades of monterey (monterey jack) cheese are determined on the basis of rating the following quality factors:
   a. Flavor
   b. Body and texture
   c. Color
   d. Finish and appearance
4. The final U.S. grade should be determined on the basis of the lowest rating of any one of the quality factors.

27.11.4 Specifications for U.S. Grades

The general requirements for the U.S. grades of monterey (monterey jack) cheese are as follows.

27.11.4.1 U.S. Grade AA

U.S. Grade AA monterey (monterey jack) cheese should conform to the following requirements (Tables 27.1–27.4):

1. *Flavor.* The cheese should possess a fine and highly pleasing monterey (monterey jack) cheese flavor that is free from undesirable tastes and odors or may be lacking in flavor development. The cheese may possess a very slight acid or feed flavor (Table 27.1).
2. *Body and texture.* A plug drawn from the cheese should be reasonably firm. Depending on the method of manufacture, a satisfactory plug may exhibit evenly distributed small mechanical openings or a close body. The cheese should be free from sweet holes, yeast holes, or

TABLE 27.1

Classification of Flavor with Corresponding U.S. Grade for Monterey Jack Cheese

Flavor Characteristics	AA	A	B
Acid	VS	S	D
Barny	—	—	S
Bitter	—	VS	S
Feed	VS	S	D
Flat	—	VS	S
Fruity	—	—	S
Malty	—	—	S
Old milk	—	—	S
Onion	—	—	VS
Rancid	—	—	S
Sour	—	—	VS
Utensil	—	—	S
Weedy	—	—	S
Whey-taint	—	—	S
Yeasty	—	—	S

Note: —, Not permitted; VS, very slight; S, slight; and D, definite.

TABLE 27.2

Classification of Body and Texture with Corresponding U.S. Grade for
Monterey Jack Cheese

Body and Texture Characteristics	AA	A	B
Coarse	—	—	S
Corky	—	—	S
Crumbly	—	—	S
Curdy	D	D	D
Gassy	—	—	S
Loosely knit	—	VS	S
Mealy	—	—	S
Pasty	—	—	S
Short	—	—	S
Slitty	—	—	S
Sweet holes	—	—	S
Weak	VS	VS	S

Note: —, Not permitted; VS, very slight; S, slight; and D, definite.

other gas holes. The body may be very slightly weak, and the texture may be definitely curdy (Table 27.2).

3. *Color.* The color should be natural, uniform, and bright (Table 27.3).

4. *Finish and appearance*

   a. *Rinded and paraffin dipped.* The bandage should be evenly placed over the entire surface of the cheese and be free from unnecessary overlapping and wrinkles, and not be burst or torn. The rind should be sound, firm, and smooth and provide good protection to the cheese. The surface should be smooth and bright and have a good coating of wax or paraffin that adheres firmly to all surfaces. The cheese should be free from mold under the paraffin. The cheese should be free from high edges, huffing, or lopsidedness but may possess soiled surface to a very slight degree (Table 27.4).

   b. *Rindless and wrapped.* The wrapper or covering should be practically smooth and properly sealed with adequate overlapping at the seams or sealed by any other satisfactory type of closure. The wrapper or covering should be neat and should adequately and securely envelop the cheese, but it may be slightly wrinkled. Allowance should be made for slight wrinkles caused by crimping or sealing when vacuum packaging is used. The cheese should be free from mold under the wrapper or covering and should not be huffed or lopsided (Table 27.4).

   c. *Rindless and paraffin dipped.* The cheese surface should be smooth and bright and have a good coating of paraffin that adheres firmly. If a wrapper or coating is applied prior to

TABLE 27.3

Classification of Color with Corresponding U.S. Grade for Monterey Jack Cheese

Color Characteristics	AA	A	B
Acid cut	—	—	S
Bleached surface (rindless)	—	—	S
Dull or faded	—	—	S
Mottled	—	—	S
Salt spots	—	—	S
Unnatural	—	—	S
Wavy	—	VS	S

Note: —, Not permitted; VS, very slight; and S, slight.

TABLE 27.4

Classification of Finish and Appearance with Corresponding U.S. Grade for
Monterey Jack Cheese

Finish and Appearance Characteristics	AA	A	B
Rindless			
Defective coating (paraffin dipped—scaly, blistered, and checked)	—	—	S
High edges	—	S	D
Irregular press cloth (uneven, wrinkled, and improper overlapping)	—	S	D
Lopsided	—	S	D
Mold under wrapper or covering	—	—	VS
Rough surface	—	S	D
Soiled surface	—	—	S
Soiled surface (paraffin dipped)	VS	VS	S
Surface mold	—	—	S
Wrinkled wrapper or covering	S	S	D
Rinded			
Checked rind	—	—	S
Defective coating (scaly, blistered, and checked)	—	—	S
High edges	—	S	D
Irregular press cloth (uneven, wrinkled, and improper overlapping)	—	S	D
Lopsided	—	S	D
Mold under paraffin	—	—	VS
Rough surface	—	S	D
Soiled surface	VS	VS	S
Sour rind	—	—	S
Surface mold	—	VS	S
Weak rind	—	—	S

Note: —, Not permitted; VS, very slight; S, slight; and D, definite.

paraffin dipping, it should completely envelop the cheese. The cheese should be free from
high edges, huffing, lopsidedness, or mold. The cheese may possess soiled surface to a very
slight degree. The wrapper may be wrinkled to a slight degree (Table 27.4).

27.11.4.2 U.S. Grade A

U.S. Grade A monterey (monterey jack) cheese should conform to the following requirements (Tables
27.1–27.4):

1. *Flavor.* The cheese should possess a pleasing monterey (monterey jack) cheese flavor that is
   free from undesirable tastes and odors or may be lacking in flavor development. The cheese
   may possess a bitter or flat flavor to a very slight degree and an acid or feed flavor to a slight
   degree (Table 27.1).
2. *Body and texture.* A plug drawn from the cheese should be reasonably firm. Depending on the
   method of manufacture, a satisfactory plug may exhibit evenly distributed mechanical openings

or a close body. The plug should be free from sweet holes, yeast holes, or other gas holes. The body and texture may be very slightly weak or loosely knit and definitely curdy (Table 27.2).

3. *Color.* The color should be natural, fairly uniform, and bright. The cheese may possess waviness to a very slight degree (Table 27.3).

4. *Finish and appearance*

    a. *Rinded and paraffin dipped.* The bandage should be evenly placed over the entire surface of the cheese and not be burst or torn. The rind should be sound, firm, and smooth and provide good protection to the cheese. The surface should be practically smooth and bright and have a good coating of paraffin that adheres firmly to all surfaces. The cheese should be free from mold under the paraffin. The cheese may possess the following characteristics: soiled surface or surface mold to a very slight degree and high edges, irregular press cloth, lopsided, or rough surface to a slight degree (Table 27.4).

    b. *Rindless and wrapped.* The wrapper or covering should be practically smooth and properly sealed with adequate overlapping at the seams or sealed by any other satisfactory type of closure. The wrapper or covering should be neat and adequately and securely envelop the cheese but may be slightly wrinkled. Allowance should be made for slight wrinkles caused by crimping or sealing when vacuum packaging is used. The cheese should be free from mold under the wrapper or covering and should not be huffed but may possess, to a slight degree, high edges, irregular press cloth, lopsidedness, or rough surface (Table 27.4).

    c. *Rindless and paraffin dipped.* The cheese surface should be bright and have a good coating of paraffin that adheres firmly. If a wrapper or coating is applied prior to paraffin dipping, it should completely envelop the cheese and have a good coating of paraffin that adheres firmly. The cheese may possess soiled surface to a very slight degree but should be free from mold, and it may possess, to a slight degree, high edges, irregular press cloth, lopsidedness, rough surface, or wrinkled wrapper or covering (Table 27.4).

27.11.4.3 U.S. Grade B

U.S. Grade B monterey (monterey jack) cheese should conform to the following requirements (Tables 27.1–27.4):

1. *Flavor.* The cheese may possess a fairly pleasing monterey (monterey jack) cheese flavor, or it may be lacking in flavor development. The cheese may possess an onion or sour flavor to a very slight degree; a barny, bitter, flat, fruity, malty, old milk, rancid, utensil, weedy, whey-taint, or yeasty flavor to a slight degree; and an acid or feed flavor to a definite degree (Table 27.1).

2. *Body and texture.* A plug drawn from the cheese should be moderately firm. Depending on the method of manufacture, a satisfactory plug may exhibit mechanical openings or a close body. The cheese may possess the following characteristics: coarse, corky, crumbly, gassy, loosely knit, mealy, pasty, short, slitty, sweet holes, or weak to a slight degree and curdy to a definite degree (Table 27.2).

3. *Color.* The cheese may possess the following characteristics to a slight degree: acid-cut, dull, faded, mottled, salt spots, unnatural, or wavy. In addition, rindless monterey cheese may have a bleached surface to a slight degree (Table 27.3).

4. *Finish and appearance*

    a. *Rinded and paraffin dipped.* The bandage should be placed over the entire surface of the cheese and may be uneven and wrinkled, but not be burst or torn. The rind should be reasonably sound and free from soft spots, rind rot, cracks, or openings of any kind. The surface may be rough and unattractive but should possess a fairly good coating of paraffin. The paraffin may be scaly or blistered, with very slight mold under the bandage or paraffin, but there should be no indication that mold has entered the cheese. The cheese may possess the following characteristics: checked rind, defective coating, soiled surface, sour rind, surface mold, or weak rind to a slight degree and high edges, irregular press cloth, lopsided, or rough surface to a definite degree (Table 27.4).

b. *Rindless and wrapped.* The wrapper or covering should be unbroken and should adequately and securely envelop the cheese. The following may be present: to a very slight degree, mold under the wrapper but not entering the cheese; to a slight degree, soiled surface or surface mold; and to a definite degree, high edges, irregular press cloth, lopsided, rough surface, or wrinkled wrapper or cover (Table 27.4).

c. *Rindless and paraffin dipped.* The wrapper or coating applied prior to paraffin dipping should adequately and securely envelop the cheese, have a coating of paraffin that adheres firmly to the cheese wrapper, and be unbroken but may be definitely wrinkled. The paraffin may be scaly or blistered, with very slight mold under the paraffin, but there should be no indication that mold has entered the cheese. The cheese may possess the following characteristics: defective coating, soiled surface, or surface mold to a slight degree and high edges, irregular press cloth, lopsided, rough surface, or wrinkled wrapper or covering the following to a definite degree (Table 27.4).

27.11.5 U.S. Grade That Is Not Assignable

Monterey (monterey jack) cheese should not be assigned a U.S. grade for one or more of the following reasons:

1. The cheese fails to meet or exceed the requirements for U.S. Grade B.
2. The cheese is produced in a plant that is rated ineligible for USDA grading service or is not USDA approved.

27.11.6 Explanation of Terms

With respect to types of surface protection

1. *Paraffin*—Refined paraffin, amorphous wax, microcrystalline wax, or any combination of such or any other suitable substance
2. *Paraffin dipped*—Cheese that has been coated with paraffin
3. *Rind*—A hard coating caused by the dehydration of the surface of the cheese
4. *Rinded*—A protection developed by the formation of a rind
5. *Rindless*—Cheese that has not formed a rind due to the impervious type of wrapper, covering, or container enclosing the cheese
6. *Wrapped*—Cheese that has been covered with a transparent or opaque material (plastic film type or foil) next to the surface of the cheese
7. *Wrapper or covering*—A plastic film or foil material next to the surface of the cheese, used as an enclosure or covering of the cheese

With respect to flavor

1. *Very slight*—Detected only upon very critical examination
2. *Slight*—Detected only upon critical examination
3. *Definite*—Not intense but detectable
4. *Undesirable*—Those listed in excess of the intensity permitted or those characterizing flavors not listed

5. *Acid*—Sharp and puckery to the taste, characteristic of lactic acid
6. *Barny*—A flavor characteristic of the odor of a poorly ventilated cow barn
7. *Bitter*—Distasteful, similar to the taste of quinine
8. *Feed*—Feed flavors (such as alfalfa, sweet clover, silage, or similar feed) in milk that have carried through into the cheese
9. *Flat*—Insipid, practically devoid of any characteristic monterey (monterey jack) cheese flavor
10. *Fruity*—A fermented, sweet, fruit-like flavor resembling apples
11. *Lacking in flavor development*—No undesirable and very little, if any, monterey (monterey jack) cheese flavor development
12. *Malty*—A distinctive, harsh flavor suggestive of malt
13. *Old milk*—Lacks freshness
14. *Onion*—A flavor recognized by the peculiar taste and aroma suggestive of its name; present in milk or cheese when the cows have eaten onions, garlic, or leeks
15. *Rancid*—A flavor suggestive of rancidity or butyric acid, sometimes associated with bitterness
16. *Sour*—An acid, pungent flavor resembling vinegar
17. *Utensil*—A flavor that is suggestive of improper or inadequate washing and sterilization of milking machines, utensils, or factory equipment
18. *Weedy*—A flavor present in cheese when cows have eaten weedy hay or grazed on weed-infested pasture
19. *Whey-taint*—A slightly acid flavor characteristic of fermented whey
20. *Yeasty*—A flavor indicating yeast fermentation

With respect to body and texture

1. *Very slight*—Detected only upon very critical examination and present only to a minute degree.
2. *Slight*—Barely identifiable and present only to a small degree.
3. *Definite*—Readily identifiable and present to a substantial degree.
4. *Coarse*—Feels rough, dry, and sandy.
5. *Corky*—Hard, tough, overfirm cheese that does not readily break down when rubbed between the thumb and fingers.
6. *Crumbly*—Tends to fall apart when rubbed between the thumb and fingers.
7. *Curdy*—Smooth but firm, rubbery and not waxy or broken down when worked between the fingers.
8. *Firm*—Feels solid, not soft or weak.
9. *Gassy*—Gas holes of various sizes and may be scattered.
10. *Loosely knit*—Curd particles that are not well matted and fused together.
11. *Mealy*—Has short body, does not mold well, and looks and feels like corn meal when rubbed between the thumb and fingers.
12. *Mechanical openings*—Irregular-shaped openings that are caused by variations in make procedure and not caused by gas fermentation.
13. *Pasty*—Is usually a weak body and becomes sticky and smeary when the cheese is rubbed between the thumb and fingers.
14. *Pinny*—Numerous very small gas holes.
15. *Reasonably firm*—Somewhat less firm, but not to the extent of being weak.
16. *Short*—No elasticity in the cheese plug and the cheese tends toward mealiness when rubbed between the thumb and fingers.

17. *Slitty*—Narrow, elongated slits generally associated with a cheese that is gassy or yeasty; slits may sometimes be referred to as fish-eyes.

18. *Sweet holes*—Spherical gas holes that are glossy in appearance and usually about the size of BB shots; gas holes are sometimes referred to as shot holes.

19. *Weak*—Cheese plug is soft but is not necessarily sticky like a pasty cheese and requires little pressure to crush.

With respect to color

1. *Very slight*—Detected only upon very critical examination and present only to a minute degree
2. *Slight*—Barely identifiable and present only to a small degree
3. *Acid-cut*—A bleached or faded color that sometimes varies throughout the cheese and appears most often around mechanical openings
4. *Bleached surface*—A faded color beginning at the surface and progressing inward
5. *Dull or faded*—A color condition lacking in luster or translucency
6. *Mottled*—Irregular-shaped spots or blotches in which portions are not uniform in color; moreover, an unevenness of color due to combining the curd from two different vats, sometimes referred to as mixed curd
7. *Natural*—White to light cream in color
8. *Salt spots*—Large light-colored spots or areas
9. *Unnatural*—Any color that is not white to light cream
10. *Wavy*—An unevenness of color that appears as layers or waves

With respect to finish and appearance

1. *Very slight*—Detected only upon very critical examination and present to a minute degree.
2. *Slight*—Barely identifiable and present to a small degree.
3. *Definite*—Readily identifiable and present to a substantial degree.
4. *Adequately and securely enveloped*—Wrapper or covering is properly sealed and entirely encloses the cheese with sufficient adherence to the surface of the cheese to protect it from contamination or dehydration.
5. *Bandage*—Cheese cloth used to wrap cheese prior to dipping in paraffin.
6. *Bandage evenly placed*—Placement of the bandage so that it completely envelops the cheese and overlaps evenly about 1 inch.
7. *Bright surface*—Clean, glossy surface.
8. *Burst or torn bandage*—A severance of the bandage usually occurring at the side seam; or when the bandage is otherwise snagged or broken.
9. *Checked rind*—Numerous small cracks or breaks in the rind that sometimes follow the outline of curd particles.
10. *Defective coating*—A brittle coating of paraffin that breaks and peels off in the form of scales or flakes; flat or raised blisters or bubbles under the surface of the paraffin; checked paraffin, including cracks, breaks, or hairline checks in the paraffin or coating of the cheese.
11. *Firm sound rind*—A rind possessing firmness and thickness (not easily dented or damaged) consistent with the size of the cheese and is dry, smooth, and closely knit, sufficient to protect the interior quality from external defects; free from checks, cracks, breaks, or soft spots.
12. *High edge*—A rim or ridge on the side of the cheese.
13. *Huffed*—A block of cheese that is swollen because of gas fermentation; the cheese becomes rounded or oval in shape instead of having flat surfaces.

14. *Irregular press cloth*—Press cloth improperly placed in the hoop resulting in too much press cloth on one end and insufficient on the other causing overlapping; wrinkled and loose fitting.
15. *Lopsided*—One side of the cheese is higher than the other side.
16. *Mold under bandage and paraffin*—Mold spots or areas under the paraffin.
17. *Mold under wrapper or covering*—Mold spots or areas under the wrapper or covering.
18. *Rind rot*—Soft spots on the rind that have become discolored and are decayed or decomposed.
19. *Rough surface*—Lacks smoothness.
20. *Smooth surface*—Not rough or uneven.
21. *Soft spots*—Areas soft to the touch and that are usually faded and moist.
22. *Soiled surface*—Milkstone, rust spots, or other discoloration on the surface of the cheese.
23. *Sour rind*—A fermented rind condition, usually confined to the faces of the cheese.
24. *Surface mold*—Mold on the exterior of the paraffin or wrapper.
25. *Wax or paraffin that adheres firmly to the surface of the cheese*—A coating with no cracks, breaks, or loose areas.
26. *Weak rind*—A thin rind that possesses little or no resistance to pressure.

27.12 Mozzarella and Scamorza Cheese

Mozzarella cheese (scamorza cheese) is a food prepared from dairy ingredients and other ingredients specified. It may be molded into various shapes. The minimum milkfat content is 45% by the weight of the solids, and the moisture content is more than 52% but not more than 60% by weight. The dairy ingredients are pasteurized.

The phenol equivalent value of 0.25 g of mozzarella cheese is not more than 3 µg. One or more of the dairy ingredients specified is warmed to approximately 88°F (31.1°C) and subjected to the action of a lactic acid–producing bacterial culture. One or more of the clotting enzymes specified is added to set the dairy ingredients to a semisolid mass. The mass is cut, and it may be stirred to facilitate the separation of whey from the curd. The whey is drained, the curd may be washed with cold water, and the water is drained off. The curd may be collected in bundles for further drainage and for ripening. The curd may be iced, it may be held under refrigeration, and it may be permitted to warm to room temperature and ripen further. The curd may be cut. It is immersed in hot water or heated with steam and is kneaded and stretched until it is smooth and free of lumps. It is then cut and molded. The molded curd is firmed by immersion in cold water and drained. One or more of the other optional ingredients specified may be added during the procedure.

The following safe and suitable ingredients may be used:

1. Cow milk, nonfat milk, or cream, or the corresponding products of water buffalo origin, except that cow milk products are not combined with water buffalo products
2. Rennet and/or other clotting enzymes of animal, plant, or microbial origin
3. Vinegar
4. Coloring to mask any natural yellow color in the curd
5. Salt
6. Antimycotics, which may be added to the cheese during the kneading and stretching process and/or applied to the surface of the cheese and the cumulative levels of which should not exceed current good manufacturing practice

The name of the food is "mozzarella cheese" or, alternatively, "scamorza cheese." When the food is made with water buffalo milk, the name of the food is accompanied by the phrase "made with water buffalo milk."

27.13 Low-Moisture Mozzarella and Scamorza Cheese

Low-moisture mozzarella (scamorza) cheese is a food prepared from dairy ingredients and other ingredients specified. It may be molded into various shapes. The minimum milkfat content is 45% by the weight of the solids, and the moisture content is more than 45% but not more than 52% by weight. The dairy ingredients are pasteurized.

The phenol equivalent value of 0.25 g of low-moisture mozzarella cheese is not more than 3 μg. One or more of the dairy ingredients specified may be warmed and is subjected to the action of a lactic acid–producing bacterial culture. One or more of the clotting enzymes specified is added to set the dairy ingredients to a semisolid mass. The mass is cut, stirred, and allowed to stand. It may be reheated and again stirred. The whey is drained, and the curd may be cut and piled to promote further separation of whey. It may be washed with cold water, and the water is drained off. The curd may be collected in bundles for further drainage and for ripening. The curd may be iced, it may be held under refrigeration, and it may be permitted to warm to room temperature and ripen further. The curd may be cut. It is immersed in hot water or heated with steam and is kneaded and stretched until smooth and free of lumps. It is then cut and molded. In molding, the curd is kept sufficiently warm to cause proper sealing of the surface. The molded curd is firmed by immersion in cold water and drained. One or more of the other optional ingredients specified may be added during the procedure.

The following safe and suitable ingredients may be used:

1. Cow milk, nonfat milk, or cream, or the corresponding products of water buffalo origin, except that cow milk products are not combined with water buffalo products
2. Rennet and/or clotting enzymes of animal, plant, or microbial origin
3. Vinegar
4. Coloring to mask any natural yellow color in the curd
5. Salt
6. Calcium chloride in an amount not more than 0.02% (calculated as anhydrous calcium chloride) of the weight of the dairy ingredients, used as a coagulation aid
7. Antimycotics, which may be added to the cheese during the kneading and stretching process and/or applied to the surface of the cheese and the cumulative levels of which should not exceed current good manufacturing practices

The name of the food is "low-moisture mozzarella cheese" or, alternatively, "low-moisture scamorza cheese." When the food is made with water buffalo milk, the name of the food is accompanied by the phrase "made with water buffalo milk."

Part-skim mozzarella (scamorza) cheese conforms to the definition and standard of identity as prescribed for mozzarella cheese, except that its milkfat content, calculated based on the solids, is less than 45% but not less than 30%. Low-moisture part-skim mozzarella and scamorza cheeses conform to the definition and standard of identity and comply with the requirements for the label declaration of ingredients prescribed for low-moisture mozzarella and scamorza cheeses, except that their milkfat content, calculated based on the solids, is less than 45% but not less than 30%.

27.14 USDA Specifications for Mozzarella Cheeses

This document has been developed by the USDA to aid in the purchase of mozzarella cheeses that are intended for pizza, cooking, or table uses. It is designed to provide the minimum requirement for acceptable mozzarella cheeses according to flavor, body and texture, color, finish and appearance, and packing and packaging. These mozzarella cheeses may be evaluated in the loaf, slice, shredded, or diced form. The cheese should be graded no sooner than 5 days of age. The following types, forms, and styles of mozzarella cheeses are covered by this specification.

27.14.1 Type (Composition by Fat and Moisture)

1. *Type I—Mozzarella cheese.* Mozzarella cheese should contain more than 52% but not more than 60% moisture and not less than 45% milkfat on dry basis.
2. *Type II—Low-moisture mozzarella cheese.* Low-moisture mozzarella cheese should contain more than 45% but not more than 52% moisture and not less than 45% milkfat on dry basis.
3. *Type III—Part-skim mozzarella cheese.* Part-skim mozzarella cheese should contain more than 52% but not more than 60% moisture and less than 45% but not less than 30% milkfat on dry basis.
4. *Type IV—Low-moisture part-skim mozzarella cheese.* Low-moisture part-skim mozzarella cheese should contain more than 45% but not more than 52% moisture and less than 45% but not less than 30% milkfat on dry basis.

27.14.2 Form

1. Loaf
2. Sliced
3. Shredded
4. Diced

27.14.3 Style

1. Fresh
2. Frozen

27.14.4 Requirements

All types of mozzarella cheese with the approved specification should have been manufactured and/or processed in a plant approved by the USDA. Mozzarella cheeses covered by this specification should conform to or exceed the following characteristics.

The cheese should have a mild, pleasing, flavor but may possess the following flavors to a slight degree: acid or feed. A slight or definite rancid flavor, which is characteristic to certain traditional markets, may be acceptable when specified by the buyer.

27.14.5 Body and Texture

1. *Loaf form.* A slice or plug drawn from all types of mozzarella cheese should be flexible. It should not possess sweet holes or be gassy. The cheese should be smooth and pliable.
   a. Mozzarella or low-moisture mozzarella cheese (types I and II) may possess the following defects: to a slight degree, open, coarse, mealy, loosely knitted with or without pockets of free liquid, and pasty and to a definite degree, lacking flexibility and weak.
   b. Part-skim or low-moisture part-skim mozzarella cheese (types III and IV) may possess the following defects to a slight degree: open, lacking flexibility, mealy, and weak.
2. *Sliced form.* All types of sliced mozzarella cheese should not possess sweet holes or be gassy. The slices should be free and easily separate without breaking. The cheese should be smooth and pliable.
   a. Sliced mozzarella or low-moisture mozzarella cheese (types I and II) may possess the following defects to a slight degree: open, coarse, mealy, loosely knitted with or without pockets of free liquid, and pasty; the cheese may be weak to a definite degree.

b. Sliced part-skim or low-moisture part-skim mozzarella cheese (types III and IV) may possess the following defects to a slight degree: open, coarse, mealy, pasty, loosely knitted with or without pockets of free liquid, and weak.

3. *Shredded or diced form.* All types of shredded or diced mozzarella cheese may possess the following defects to a slight degree: coarse, mealy, pasty, and weak. Shredded or diced mozzarella cheeses should be loose and free from clumps, except those that readily break up with slight pressure.

27.14.6 Color

All types of mozzarella cheese should have a natural white to light cream, uniform bright color, and attractive sheen. It may be wavy to a very slight degree. There may be a slight variation in color due to salt penetration.

27.14.7 Finish and Appearance

The wrapper or covering should be properly sealed with adequate overlapping at the seams to prevent the entrance of air or drying of the cheese or sealed by any other satisfactory type of closure. The wrapper or covering should be neat and should adequately and securely envelop the cheese but may be definitely wrinkled. Allowances should be made for wrinkles caused by crimping or sealing when vacuum sealing is used. The cheese should be free from mold under the wrapper or covering but may be misshapen to a definite degree.

27.14.8 Packing and Packaging

The packaging should satisfactorily protect the cheese for its final use. All types of mozzarella cheese covered by this specification should be evaluated in accordance with the United States Standards for Condition of Food Containers.

27.14.9 Optional Requirements

There are certain requirements that may be requested at the option of an interested party. The optional requirements for pizza making are as follows:

1. *Meltability*—The melted cheese should be evenly distributed over the surface of the pizza and be free from blisters.
2. *Color*—The cheese should have a rich, even, natural white to light cream, and uniform bright color and an attractive sheen.
3. *Stretchability*—Insert the tip of a fork into the cheese and lift vertically at least 3 inches from the surface of the pizza. The cheese should be stringy and unbroken from the fork to the surface of the pizza. The cheese may be chewy but not gummy.
4. *Free fat*—There should be no free fat drippage when a wedge-shaped cut is removed from the pizza.

27.14.10 Terms Commonly Used

With respect to flavor

1. *Slight*—An attribute that is barely identifiable and present only to a small degree
2. *Acid*—Sharp and puckery to the taste, characteristic of lactic or acetic acid
3. *Feed*—Feed flavors (such as alfalfa, sweet clover, silage, or similar feed) in milk carried through into the cheese

4. *Rancid*—A flavor caused by the activity in the milk of the enzyme lipase. The lipase enzyme, in the form of rennet paste or commercial lipase powder preparations, may be added to milk for making mozzarella cheeses to impart a mild "piquant" flavor

With respect to body and texture

1. *Slight*—An attribute that is barely identifiable and present only to a small degree
2. *Definite*—An attribute that is readily identifiable and present to a substantial degree
3. *Slight pressure*—Only sufficient pressure to readily disintegrate the lumps
4. *Coarse*—Feels rough, dry, and sandy
5. *Gassy*—Gas holes of various sizes and may be scattered
6. *Flexible*—The plug may be bent without breaking
7. *Lacking flexibility*—The plug tends to break when bent
8. *Loosely knitted*—Knitted loosely enough to be separated with or without occasional pockets of liquid
9. *Mealy*—Has a short body, does not mold well, and looks and feels like corn meal when worked between the thumb and fingers
10. *Open*—Includes mechanical openings that are irregular in shape and holes that are caused by trapped air or steam during the make procedure
11. *Pasty*—Has a weak body and becomes sticky and smeary when the cheese is worked between the thumb and fingers
12. *Plug*—The cheese drawn from a block of cheese using a number 8 cheese trier
13. *Sweet hole*—Spherical gas holes, glossy in appearance; usually about the size of BB shots
14. *Weak*—Requires little pressure to crush, is soft, but is not necessarily sticky like pasty cheese

With respect to color

1. *Very slight*—An attribute that is detected only upon critical examination and present only to a minute degree
2. *Natural white to light cream*—Color that is obtained from milk
3. *Wavy*—Unevenness of color that appears as layers or waves

With respect to finish and appearance

1. *Definite*—An attribute that is readily identifiable and present to a large degree
2. *Misshapen*—Deformed from its characteristic shape
3. *Wrinkled packaging*—Rough or uneven

27.14.11 Other Requirements

1. Fresh mozzarella cheeses have an estimated shelf life of approximately 30 days when held at a temperature of 35°F (2°C). Freezing will prolong the shelf life of this product for as long as 12 months. Care should be taken when moving frozen cheese, because the wrappers will tear easily at freezer temperatures.
2. All types of frozen mozzarella cheese should be thawed gradually at a temperature between 40°F and 45°F (4°C and 7°C) and held for a minimum of 14 days to a maximum of 21 days to eliminate a crumbly cheese caused by freezing. After 21 days at this temperature, the product will return to its original body and texture and should be used within a few days.

3. For immediate use in institutions, fresh cheese or properly thawed mozzarella cheeses should be stored at a temperature between 40°F and 45°F (4°C and 70°C).

27.15 USDA Specifications for Loaf and Shredded Lite Mozzarella Cheese

1. Lite mozzarella cheese should be manufactured and packaged in accordance with USDA requirements.
2. Lite mozzarella cheese should be aged no less than 5 days at 38°F–42°F (3.5°C–5.5°C) prior to inspection, unless the cheese is shredded and frozen. If the shredded cheese is not frozen immediately after manufacture, the cheese should be stored at 38°F–42°F (3.5°C–5.5°C) until frozen.
3. Lite mozzarella cheese should comply with all applicable federal regulations.
4. Samples of shredded, frozen lite mozzarella cheese should be taken prior to tempering.
5. Lite mozzarella cheese should contain not more than 10.8% milkfat.
6. Lite mozzarella cheese should contain not less than 52% and not more than 60% moisture.
7. Lite mozzarella cheese should not have a pH exceeding 5.3 using the quinhydrone method.
8. Lite mozzarella cheese should contain not less than 1.2% and not more than 1.8% salt.
9. Lite mozzarella cheese samples should be tempered to 45°F–55°F (7°C–13°C) prior to product evaluation.
10. Lite mozzarella cheese should have a mild pleasing flavor and may possess a slight acid or feed flavor.
11. Lite mozzarella cheese in loaf form should possess a smooth, pliable body and should not contain sweet holes or be gassy. Lite mozzarella cheese should be free from all foreign and extraneous materials. The cheese may have the following body and texture characteristics to a slight degree: open (caused by entrapped steam), lacking flexibility, mealy, weak, sticky, and rubbery. Shredded lite mozzarella cheese should have a height and width up to 3/16 inch either dimension. It should be free flowing and should not be matted. An approved anticaking agent may be added as a processing aid. If an anticaking agent is used, the amount used should be the minimum required to produce the desired effect but should not exceed 2.0% of the weight of the shredded cheese.
12. Shredded lite mozzarella cheese should not contain more than 6.0% fines. For shreds whose height and/or width is 1/16 inch or less, the fines content should be determined using Standard Test Sieve No. 14 (1.4 mm). For shreds whose height and width are greater than 1/16 inch, the fines content should be determined using Standard Test Sieve No. 8 (2.36 mm). (Note: The manufacturer should provide shred size information. This information will determine the sieve used to measure the fines content. If shred size information is not provided, Standard Test Sieve No. 8 will be used.)
13. Lite mozzarella cheese should have a natural white to light cream, uniform bright color, and an attractive sheen. No visible signs of mold should be permitted. Cheese in loaf form may be wavy to a very slight degree and may have a slight variation in color due to salt penetration.
14. Meltability characteristics of lite mozzarella cheese should be tested in accordance with accepted official method, except that a pizza prepared with lite mozzarella cheese should be placed in an oven preheated to 450°F (232°C) and baked at that temperature for 10 min. The cheese should melt completely, should not exhibit shreds of unmelted cheese or excessive blistering, and should stretch to a minimum of 3 inches of unbroken strings. The melted cheese may be chewy but not gummy. The cheese may possess a slightly darker color than the color of cheese before cooking but should not exhibit burnt areas or excessive browning.

28

Muenster and Other Cheeses: Standards, Grades, and Specifications*

Y. H. Hui

CONTENTS

28.1 Introduction

Food manufacturers in the United States must produce a product that consumers like in order to sell the product nationally. To achieve this goal, they depend on developments in science, technology, and engineering in order to produce a quality product. However, they must also pay attention to two other important considerations:

---

* The information in this chapter has been modified from *Food Safety Manual* published by Science Technology System (STS) of West Sacramento, CA. Copyrighted 2012©. With permission.

1. Are their products safe for public consumption?
2. Have they assured the economic integrity of their products? This means many things. The most important one is the assurance that their products contain the proper quality and quantity of ingredients, that is, "No cheating."

To protect the consumers, U.S. government agencies have established many checks and balances to prevent "injuries" to the consumers in terms of health and economic fraud. The two major responsible agencies are

1. U.S. Food and Drug Administration (FDA; www.FDA.gov)
2. U.S. Department of Agriculture (USDA; www.USDA.gov)

The FDA establishes standards of identities for a number of cheeses for interstate commerce. This ensures that a particular type of cheese contains the minimal quality and quantity of the ingredients used to make this cheese. The FDA also issues safety requirements and recommendations in the manufacturing of cheeses. It makes periodic inspections of establishments producing cheeses to ensure full compliance. In effect, the FDA is monitoring the safety production of cheeses and their economic integrities.

The USDA plays the same important role in the commerce of cheeses in the United States. For example, a national restaurant chain that makes and sells pizzas will not buy cheeses from any manufacturer if the cheeses do not bear the USDA official identification or seal of approval. The USDA has issued official requirements and processing specifications that cheese manufacturers must follow if they want to obtain the USDA seal of approval for the products. To achieve this goal, the USDA sets up voluntary services of inspection under a standard fee arrangement. The inspection focuses on the safety and economic integrity of manufacturing cheeses. When the inspection demonstrates that the manufacturer has complied with safety and economic requirements, the USDA issues its official identification or seal of approval.

In essence, the FDA and USDA are the federal agencies that ensure that the product is safe and will not pose any economic fraud. Using the public information distributed at the web sites of FDA and USDA, this chapter discusses USDA specifications and FDA inspection in relation to the production of selected cheeses.

The information in this chapter is important, because it facilitates the commerce of cheeses and protects consumers from injuries of eating cheeses in the market. It also protects the consumers from economic fraud. For example, without FDA and USDA intervention, one may purchase cheese that contains less than the normal amount of milk used. If so, the consumers lose money in the transaction.

All FDA cheese standards of identities are described in 21 CFR 133. All USDA requirements and standard for grades are described in 7 CFR 58. For explanation, CFR refers to the United States Code of Federal Regulations. This chapter focuses on two areas:

1. Part 133: standards of identities issued by the FDA.
2. Part 58 (selected sections): USDA specifications and grade standards.

28.2 Muenster and Munster Cheese

Muenster (munster) cheese is a food prepared under FDA requirements. The minimum milkfat content is 50% by the weight of the solids, and the maximum moisture content is 46% by weight. The dairy ingredients used are pasteurized. The phenol equivalent of 0.25 g of muenster cheese is not more than 3 µg.

One or more of the dairy ingredients specified may be warmed and is subjected to the action of a harmless lactic acid–producing bacterial culture. One or more of the clotting enzymes specified is added to set the dairy ingredients to a semisolid mass. After coagulation, the mass is divided into small portions, stirred, and heated, with or without dilution with water or salt brine, so as to promote and regulate the separation of whey and curd. The curd is transferred to forms permitting drainage of the whey. During drainage, the curd may be pressed and turned. After drainage, the curd is removed from the forms and is salted. One or more of the other optional ingredients specified may be added during the procedure.

The following safe and suitable ingredients may be used:

1. Milk, nonfat milk, or cream, used alone or in combination
2. Rennet and/or other clotting enzymes of animal, plant, or microbial origin

Other optional ingredients are

1. Coloring
2. Calcium chloride in an amount not more than 0.02% (calculated as anhydrous calcium chloride) of the weight of the dairy ingredients, used as a coagulation aid
3. Enzymes of animal, plant, or microbial origin used in curing or flavor development
4. Antimycotic agents, which may be added to the surface of the cheese and the cumulative levels of which should not exceed current good manufacturing practice
5. Vegetable oil, used as a coating for the rind

The name of the food is "muenster cheese" or, alternatively, "munster cheese." Each of the ingredients used in the food should be declared on the label as required by law, except that

1. Enzymes of animal, plant, or microbial origin may be declared as "enzymes."
2. The dairy ingredients may be declared, in descending order of predominance, by the use of the terms "milkfat and non-fat milk" or "non-fat milk and milkfat," as appropriate.

Muenster cheese for manufacturing conforms to the definition and standard of identity for muenster cheese prescribed by the FDA, except that the dairy ingredients are not pasteurized.

28.3 USDA Specifications for Loaf, Sliced, Shredded, and Diced Muenster Cheese

Muenster cheese should be manufactured and packaged in accordance with USDA. Muenster cheese should be aged no less than 10 days at 38°F–45°F (3.5°C–7°C) prior to inspection. Muenster cheese may be aged before or after the cheese is sliced, shredded, or diced. Muenster cheese should comply with all applicable FDA regulations.

Composition requirements are as follows:

1. Muenster cheese in loaf, sliced, and diced form should contain not less than 50% milkfat by the weight of the solids.
2. Muenster cheese in shredded form should contain not less than 49% milkfat by the weight of the solids.
3. Muenster cheese should contain not more than 46% moisture.
4. Muenster cheese should not have a pH exceeding 5.4.
5. Muenster cheese should contain not less than 1.2% and not more than 2.0% salt.

28.3.1 Quality Requirements

Muenster cheese samples should be tempered to 45°F–55°F (7°C–13°C) prior to product evaluation. It should possess a mild and pleasing flavor and may possess slight feed flavors. The cheese should also be firm, possess a smooth, pliable body, and may exhibit either small mechanical openings that are evenly distributed or a close body. It may be slightly curdy and should not possess sweet holes, yeast holes, or other gas holes.

Muenster cheese in sliced form. Muenster cheese in sliced form should easily separate from one another and not exhibit a rough surface or possess torn corners.

Muenster cheese in shredded form. Shredded muenster cheese should have a height and width up to 3/16 inch in either dimension. It should be free flowing and should not be matted. An approved anticaking agent may be added to shredded cheese. If anticaking agent is used, the amount used should be the minimum required to produce the desired effect but should not exceed 2.0% of the weight of the shredded cheese.

Muenster cheese in diced form. Muenster cheese in diced form should have all dimensions of the dice relatively equal.

Shredded muenster cheese should not contain more than 6.0% fines. For shreds whose height and/or width is 1/16 inch or less, the fines content should be determined using Standard Test Sieve No. 14 (1.4 mm). For shreds whose height and/or width is greater than 1/16 inch, the fines content should be determined using Standard Test Sieve No. 8 (2.36 mm).

The manufacturer should provide shred size information. This information will determine the sieve used to measure the fines content. If the shred size information is not provided, Standard Test Sieve No. 8 will be used.

28.3.2 Color and Appearance

Muenster cheese should possess a natural, uniform, and bright color. Muenster cheese in loaf form may possess an orange surface color, typically provided on muenster cheese. If surface color–treated loaves are sliced, shredded, or diced, such slices, shreds, or dice may exhibit the surface color of the loaf. No visible signs of mold should be permitted.

28.3.3 Official Identification

Muenster cheese that is officially inspected and found to meet the requirements may be identified with the official USDA Quality Approved Inspection Shield.

28.4 Neufchatel Cheese

Neufchatel cheese is a soft uncured cheese prepared by complying with FDA requirements. The milkfat content is not less than 20% but less than 33% by the weight of the finished food, and the maximum moisture content is 65% by weight. The dairy ingredients used are pasteurized.

One or more of the dairy ingredients specified is subjected to the action of a harmless lactic acid–producing bacterial culture, with or without one or more of the clotting enzymes specified. The mixture is held until the dairy ingredients coagulate. The coagulated mass may be warmed and stirred, and it is drained. The moisture content may be adjusted with one of the optional ingredients. The curd may be pressed, chilled, worked, and heated until it becomes fluid. It may then be homogenized or otherwise mixed. One or more of the dairy ingredients specified or the other optional ingredients specified may be added during the procedure.

The following safe and suitable ingredients may be used:

1. Milk, nonfat milk, or cream
2. Rennet and/or other clotting enzymes of animal, plant, or microbial origin
3. Salt
4. Cheese whey, concentrated cheese whey, dried cheese whey, or reconstituted cheese whey prepared by the addition of water to concentrated cheese whey or dried cheese whey
5. Stabilizers, in a total amount not to exceed 0.5% of the weight of the finished food, with or without the addition of dioctyl sodium sulfosuccinate in a maximum amount of 0.5% of the weight of the stabilizer(s) used

28.5 Nuworld Cheese

Nuworld cheese is a food prepared according to FDA requirements. It is characterized by the presence of creamy-white mold, a white mutant of *Penicillium roquefortii*, throughout the cheese. The minimum milkfat content is 50% by the weight of the solids, and the maximum moisture content is 46% by weight. The dairy ingredients used may be pasteurized. Nuworld cheese is at least 60 days old.

One or more of the dairy ingredients specified may be warmed and is subjected to the action of a lactic acid–producing bacterial culture. One or more of the clotting enzymes specified is added to set the dairy ingredients to a semisolid mass. The mass is cut into smaller portions and allowed to stand for a time. The mixed curd and whey is placed into forms, permitting further drainage. While being placed in forms, spores of a white mutant of the mold *P. roquefortii* are added. The forms are turned several times during drainage. When sufficiently drained, the shaped curd is removed from the forms and salted with dry salt or brine. Perforations are then made in the shaped curd, and it is held at a temperature of approximately 50°F at 90%–95% relative humidity until the characteristic mold growth has developed. During storage, the surface of the cheese may be scraped to remove the surface growth of undesirable microorganisms. One or more of the other optional ingredients specified may be added during the procedure.

The following safe and suitable ingredients may be used:

1. Milk, nonfat milk, or cream used alone or in combination
2. Rennet and/or other clotting enzymes of animal, plant, or microbial origin
3. Blue or green color in an amount to neutralize the natural yellow color of the curd
4. Calcium chloride in an amount not more than 0.02% (calculated as anhydrous calcium chloride) of the weight of the dairy ingredients, used as a coagulation aid
5. Enzymes of animal, plant, or microbial origin, used in curing or flavor development

28.6 Parmesan and Reggiano Cheese

Parmesan (reggiano) cheese is a food prepared from milk and other ingredients specified by the FDA. It is characterized by a granular texture and a hard and brittle rind. It grates readily. It contains not more than 32% of moisture, and its solids contain not less than 32% of milkfat. It is cured for not less than 10 months.

Milk, which may be pasteurized, clarified, or both and may be warmed, is subjected to the action of harmless lactic acid–producing bacteria, present in such milk or added thereto. Sufficient rennet or other safe and suitable milk-clotting enzyme that produces equivalent curd formation or both, with or without purified calcium chloride in a quantity not more than 0.02% (calculated as anhydrous calcium chloride) of the weight of the milk, is added to set the milk to a semisolid mass. Harmless artificial coloring may be added. The mass is cut into pieces no larger than wheat kernels, heated, and stirred until the temperature reaches between 115°F and 125°F. The curd is allowed to settle and is then removed from the kettle or vat, drained for a short time, placed in hoops, and pressed. The pressed curd is removed and salted in brine or dry salted. The cheese is cured in a cool, ventilated room. The rind of the cheese may be coated or colored. A harmless preparation of enzymes of animal or plant origin capable of aiding in the curing or development of flavor of parmesan cheese may be added during the procedure, in such quantity that the weight of the solids of such preparation is not more than 0.1% of the weight of the milk used.

The word "milk" means cow milk, which may be adjusted by separating part of the fat therefrom or by adding thereto one or more of the following: cream, skim milk, concentrated skim milk, nonfat dry milk, and water in a quantity sufficient to reconstitute any concentrated skim milk or nonfat dry milk used.

Such milk may be bleached by the use of benzoyl peroxide or a mixture of benzoyl peroxide with potassium alum, calcium sulfate, and magnesium carbonate, but the weight of the benzoyl peroxide is not more than 0.002% of the weight of the milk bleached, and the weight of the potassium alum, calcium sulfate, and magnesium carbonate, singly or combined, is not more than six times the weight of the benzoyl peroxide used. If milk is bleached in this manner, sufficient vitamin A is added to the curd to compensate for the vitamin A or its precursors destroyed in the bleaching process, and artificial coloring is not used.

Safe and suitable antimycotic agent(s) may be added to the surface of the cheese, the cumulative levels of which should not exceed current good manufacturing practice.

28.7 Pasteurized Blended Cheese

Pasteurized blended cheese conforms to the definition and standard of identity required by the FDA. In mixtures of two or more cheeses, cream cheese or neufchatel cheese may be used. None of the ingredients prescribed or permitted for pasteurized process cheese is used. In case of mixtures of two or more cheeses containing cream cheese or neufchatel cheese, the moisture content is not more than the arithmetical average of the maximum moisture contents prescribed by the definitions and standards of identity for the varieties of cheeses blended, for which such limits have been prescribed. The word "process" is replaced by the word "blended" in the name prescribed for pasteurized process cheese.

28.8 Pasteurized Blended Cheese with Fruits, Vegetables, or Meat

Pasteurized blended cheese with fruits, vegetables, meat, or mixtures of these is a food that conforms to the definition and standard of identity prescribed for pasteurized blended cheese, except that

1. Its moisture content may be 1% more, and the milkfat content of its solids may be 1% less than the limits prescribed for pasteurized blended cheese.
2. It contains one or any mixture of two or more of the following: any properly prepared cooked, canned, or dried fruit; any properly prepared cooked, canned, or dried vegetable; any properly prepared cooked or canned meat.
3. The name of a pasteurized blended cheese with fruits, vegetables, or meat is the name prescribed for the applicable pasteurized blended cheese, followed by the term "with ___," the blank being filled in with the common or usual name or names of the fruits, vegetables, or meat used, in order of predominance by weight.

28.9 Pasteurized Process Cheese

Pasteurized process cheese is a food prepared by comminuting and mixing, with the aid of heat, one or more cheeses of the same or two or more varieties, except for cream cheese, neufchatel cheese, cottage cheese, low-fat cottage cheese, cottage cheese dry curd, cook cheese, hard grating cheese, semisoft part-skim cheese, part-skim spiced cheese, and skim milk cheese for manufacturing with an emulsifying agent into a homogeneous plastic mass. One or more of the optional ingredients designated may be used.

During its preparation, pasteurized process cheese is heated for not less than 30 s at a temperature of not less than 150°F. When tested for phosphatase, the phenol equivalent of 0.25 g of pasteurized process cheese is not more than 3 µg.

The moisture content of a pasteurized process cheese made from a single variety of cheese is not more than 1% greater than the maximum moisture content prescribed by the definition and standard of identity, if any, for the variety of cheese used but in no case is more than 43%, except that the moisture content of pasteurized process washed curd cheese or pasteurized process colby cheese is not more than 40%. The moisture content of pasteurized process Swiss cheese or pasteurized process gruyere cheese is not more than 44%, and the moisture content of pasteurized process limburger cheese is not more than 51%.

The fat content of the solids of pasteurized process cheese made from a single variety of cheese is not less than the minimum prescribed by the definition and standard of identity, if any, for the variety of cheese used but in no case is less than 47%, except that the fat content of the solids of pasteurized process Swiss cheese is not less than 43% and the fat content of the solids of pasteurized process gruyere cheese is not less than 45%.

The moisture content of a pasteurized process cheese made from two or more varieties of cheese is not more than 1% greater than the arithmetical average of the maximum moisture contents prescribed by the definitions and standards of identity, if any, for the varieties of cheese used but in no case is the moisture content more than 43%, except that the moisture content of a pasteurized process cheese made from two or more of the varieties cheddar cheese, washed curd cheese, colby cheese, and granular cheese is not more than 40% and the moisture content of a mixture of Swiss cheese and gruyere cheese is not more than 44%.

The fat content of the solids of pasteurized process cheese made from two or more varieties of cheese is not less than the arithmetical average of the minimum fat contents prescribed by the definitions and standards of identity, if any, for the varieties of cheese used but in no case is less than 47%, except that the fat content of the solids of a pasteurized process gruyere cheese made from a mixture of Swiss cheese and gruyere cheese is not less than 45%.

The weight of each variety of cheese in pasteurized process cheese made from two varieties of cheese is not less than 25% of the total weight of both, except that the weight of blue cheese, nuworld cheese, roquefort cheese, or gorgonzola cheese is not less than 10% of the total weight of both and the weight of limburger cheese is not less than 5% of the total weight of both. The weight of each variety of cheese in pasteurized process cheese made from three or more varieties of cheese is not less than 15% of the total weight of all, except that the weight of blue cheese, nuworld cheese, roquefort cheese, or gorgonzola cheese is not less than 5% of the total weight of all and the weight of limburger cheese is not less than 3% of the total weight of all. These limits do not apply to the quantity of cheddar cheese, washed curd cheese, colby cheese, and granular cheese in mixtures that are designated as "American cheese." Such mixtures are considered one variety of cheese.

Cheddar cheese for manufacturing, washed curd cheese for manufacturing, colby cheese for manufacturing, granular cheese for manufacturing, brick cheese for manufacturing, muenster cheese for manufacturing, and Swiss cheese for manufacturing are considered as cheddar cheese, washed curd cheese, colby cheese, granular cheese, brick cheese, muenster cheese, and Swiss cheese, respectively.

Pasteurized process cheese or the cheese or cheeses from which it is made may be smoked before comminuting and mixing, or it may contain substances prepared by condensing or precipitating wood smoke.

The emulsifying agent is one or any mixture of two or more of the following, in such quantity that the weight of the solids of such emulsifying agent is not more than 3% of the weight of the pasteurized process cheese: monosodium phosphate, disodium phosphate, dipotassium phosphate, trisodium phosphate, sodium metaphosphate (sodium hexametaphosphate), sodium acid pyrophosphate, tetrasodium pyrophosphate, sodium aluminum phosphate, sodium citrate, potassium citrate, calcium citrate, sodium tartrate, and sodium potassium tartrate.

The optional ingredients are

1. An acidifying agent consisting of one or any mixture of two or more of the following, in such quantity that the pH of the pasteurized process cheese is not below 5.3: vinegar, lactic acid, citric acid, acetic acid, and phosphoric acid.
2. Cream, anhydrous milkfat, dehydrated cream, or any combination of two or more of these, in such quantity that the weight of the fat derived therefrom is less than 5% of the weight of the pasteurized process cheese.
3. Water.
4. Salt.
5. Harmless artificial coloring.
6. Spices or flavorings, other than any that, singly or in combination with other ingredients, simulate the flavor of a cheese of any age or variety.
7. Pasteurized process cheese in the form of slices or cuts in consumer-sized packages may contain an optional mold-inhibiting ingredient consisting of not more than 0.2% by the weight of

sorbic acid, potassium sorbate, sodium sorbate, or any combination of two or more of these or consisting of not more than 0.3% by the weight of sodium propionate, calcium propionate, or a combination of sodium propionate and calcium propionate.

8. Pasteurized process cheese in the form of slices or cuts in consumer-sized packages may contain lecithin as an optional antisticking agent in an amount not to exceed 0.03% by the weight of the finished product.

9. Safe and suitable enzyme modified cheese.

The name of pasteurized process cheese for which a definition and standard of identity is prescribed as follows:

1. In case it is made from a single variety of cheese, its name is "pasteurized process ___ cheese," the blank being filled in with the name of the variety of cheese used.

2. In case it is made from two or more varieties of cheese, its name is "pasteurized process ___ and ___ cheese," "pasteurized process ___ blended with ___ cheese," or "pasteurized process blend of ___ and ___ cheese," the blanks being filled in with the names of the varieties of cheeses used, in order of predominance by weight; except that

   a. In case it is made from gruyere cheese and Swiss cheese and the weight of gruyere cheese is not less than 25% of the weight of both, it may be designated as "pasteurized process gruyere cheese."

   b. In case it is made of cheddar cheese, washed curd cheese, colby cheese, or granular cheese or any mixture of two or more of these, it may be designated as "pasteurized process American cheese," or when cheddar cheese, washed curd cheese, colby cheese, granular cheese, or any mixture of two or more of these is combined with other varieties of cheese in the cheese ingredient, any of such cheeses or such mixture may be designated as "American cheese."

The name of the food should include a declaration of any flavoring, including smoke and substances prepared by condensing or precipitating wood smoke and a declaration of any spice that characterizes the product.

Each of the ingredients used in the food should be declared on the label as required, except that cheddar cheese, washed curd cheese, colby cheese, granular cheese, or any mixture of two or more of these may be designated as "American cheese."

28.10 Pasteurized Process Cheese with Fruits, Vegetables, or Meat

Unless a definition and standard of identity specifically applicable is established by another section of this part, pasteurized process cheese with fruits, vegetables, meat, or mixtures of these is a food that conforms to the definition and standard of identity prescribed for pasteurized process cheese, except that

1. Its moisture content may be 1% more and the milkfat content of its solids may be 1% less than the limits prescribed for moisture and fat in the corresponding pasteurized process cheese.

2. It contains one or any mixture of two or more of the following: any properly prepared cooked, canned, or dried fruit; any properly prepared cooked, canned, or dried vegetable; any properly prepared cooked or canned meat.

3. The name of pasteurized process cheese with fruits, vegetables, or meat is the name prescribed for the applicable pasteurized process cheese, followed by the term "with ___," the blank being filled in with the common or usual name or names of the fruits, vegetables, or meat used, in order of predominance by weight.

28.11 Pasteurized Process Pimento Cheese

Pasteurized process pimento cheese is a food that conforms to the definition and standard of identity for pasteurized process cheese with fruits, vegetables, or meat and is subject to the requirements for the label statement of ingredients, except that

1. Its moisture content is not more than 41% and the fat content of its solids is not less than 49%.
2. The cheese ingredient is cheddar cheese, washed curd cheese, colby cheese, granular cheese, or any mixture of two or more of these in any proportion.
3. Cheddar cheese for manufacturing, washed curd cheese for manufacturing, colby cheese for manufacturing, and granular cheese for manufacturing should be considered cheddar cheese, washed curd cheese, colby cheese, and granular cheese, respectively.
4. The only fruit, vegetable, or meat ingredient is pimentos in such quantity that the weight of the solids thereof is not less than 0.2% of the weight of the finished pasteurized process pimento cheese.
5. Only optional ingredients designated by the FDA may be used.

28.12 Pasteurized Process Cheese Food

Pasteurized process cheese food is a food prepared by comminuting and mixing, with the aid of heat, one or more of approved optional cheese, dairy and/or other ingredients into a homogeneous plastic mass.

During its preparation, pasteurized process cheese food is heated for not less than 30 s, at a temperature of not less than 150°F. When tested for phosphatase, the phenol equivalent of 0.25 g of pasteurized process cheese food is not more than 3 μg.

The moisture content of a pasteurized process cheese food is not more than 44%, and the fat content is not less than 23%. The weight of the cheese ingredient prescribed constitutes not less than 51% of the weight of the finished pasteurized process cheese food.

The weight of each variety of cheese in pasteurized process cheese food made with two varieties of cheese is not less than 25% of the total weight of both, except that the weight of blue cheese, nuworld cheese, roquefort cheese, gorgonzola cheese, or limburger cheese is not less than 10% of the total weight of both. The weight of each variety of cheese in pasteurized process cheese food made with three or more varieties of cheese is not less than 15% of the total weight of all, except that the weight of blue cheese, nuworld cheese, roquefort cheese, gorgonzola cheese, or limburger cheese is not less than 5% of the total weight of all. These limits do not apply to the quantity of cheddar cheese, washed curd cheese, colby cheese, and granular cheese in mixtures that are designated as "American cheese." Such mixtures are considered one variety of cheese.

Cheddar cheese for manufacturing, washed curd cheese for manufacturing, colby cheese for manufacturing, granular cheese for manufacturing, brick cheese for manufacturing, muenster cheese for manufacturing, and Swiss cheese for manufacturing are considered as cheddar cheese, washed curd cheese, colby cheese, granular cheese, brick cheese, muenster cheese, and Swiss cheese, respectively.

Pasteurized process cheese or the cheese or cheeses from which it is made may be smoked before comminuting and mixing, or it may contain substances prepared by condensing or precipitating wood smoke.

The optional cheese ingredients are one or more cheeses of the same or two or more varieties, except for cream cheese, neufchatel cheese, cottage cheese, creamed cottage cheese, cook cheese, and skim milk cheese for manufacturing and that hard grating cheese, semisoft part-skim cheese, and part-skim spiced cheese are not used alone or in combination with each other as the cheese ingredient.

The optional dairy ingredients are cream, milk, skim milk, buttermilk, cheese whey, any of the foregoing from which part of the water has been removed, anhydrous milkfat, dehydrated cream, albumin from cheese whey, and skim milk cheese for manufacturing.

The other optional ingredients are

1. An emulsifying agent consisting of one or any mixture of two or more of the following, in such quantity that the weight of the solids of such emulsifying agent is not more than 3% of the weight of the pasteurized process cheese food: monosodium phosphate, disodium phosphate, dipotassium phosphate, trisodium phosphate, sodium metaphosphate (sodium hexametaphosphate), sodium acid pyrophosphate, tetrasodium pyrophosphate, sodium aluminum phosphate, sodium citrate, potassium citrate, calcium citrate, sodium tartrate, and sodium potassium tartrate.
2. An acidifying agent consisting of one or any mixture of two or more of the following in such quantity that the pH of the pasteurized process cheese food is not below 5.0: vinegar, lactic acid, citric acid, acetic acid, and phosphoric acid.
3. Water.
4. Salt.
5. Harmless artificial coloring.
6. Spices or flavorings other than any that, singly or in combination with other ingredients, simulate the flavor of cheese of any age or variety.
7. Pasteurized process cheese food in the form of slices or cuts in consumer-sized packages may contain an optional mold-inhibiting ingredient consisting of not more than 0.2% by the weight of sorbic acid, potassium sorbate, sodium sorbate, or any combination of two or more of these or consisting of not more than 0.3% by the weight of sodium propionate, calcium propionate, or a combination of sodium propionate and calcium propionate.
8. Pasteurized process cheese food in the form of slices or cuts in consumer-sized packages may contain lecithin as an optional antisticking agent in an amount not to exceed 0.03% by the weight of the finished product.
9. Safe and suitable enzyme-modified cheese.

The name of the food is "pasteurized process cheese food." The full name of the food should appear on the principal display panel of the label in type of uniform size, style, and color. Wherever any word or statement emphasizing the name of any ingredient appears on the label so conspicuously as to be easily seen under customary conditions of purchase, the full name of the food should immediately and conspicuously precede or follow such a word or statement in type of at least the same size as the type used in such a word or statement.

The name of the food should include a declaration of any flavoring, including smoke and substances prepared by condensing or precipitating wood smoke, which characterizes the product as specified in 21 CFR 101.22 and a declaration of any spice that characterizes the product.

Each of the ingredients used in the food should be declared on the label as required by the applicable sections of 21 CFR 101 and 130, except that cheddar cheese, washed curd cheese, Colby cheese, granular cheese, or any mixture of two or more of these may be designated as "American cheese."

For details on name and labeling, the original legal documents should be consulted since there are exceptions not presented here.

28.13 Pasteurized Process Cheese Food with Fruits, Vegetables, or Meat

Pasteurized process cheese food with fruits, vegetables, meat, or mixtures of these is a food that conforms to the definition and standard of identity and is subject to the requirements for label statement of ingredients, prescribed for pasteurized process cheese food, except that

1. Its milkfat content is not less than 22%.
2. It contains one or any mixture of two or more of the following: any properly prepared cooked, canned, or dried fruit; any properly prepared cooked, canned, or dried vegetable; any properly prepared cooked or canned meat.

3. When the added fruits, vegetables, or meat contains fat, use the method prescribed for the determination of fats specifically prescribed by the USDA.

The name of pasteurized process cheese food with fruits, vegetables, or meat is "pasteurized process cheese food with ___," the blank being filled in with the common or usual name or names of the fruits, vegetables, or meat used, in order of predominance by weight.

If the only vegetable ingredient is pimento and no meat or fruit ingredient is used, the weight of the solids of such pimentos is not less than 0.2% of the weight of the finished food. The name of this food is "pimento pasteurized process cheese food" or "pasteurized process pimento cheese food."

28.14 Pasteurized Cheese Spread

Pasteurized cheese spread is a food that conforms to the definition and standard of identity and is subject to the requirements for label statement of ingredients, except that no emulsifying agent is used.

28.15 Pasteurized Cheese Spread with Fruits, Vegetables, or Meat

Pasteurized cheese spread with fruits, vegetables, meat, or mixtures of these is a food that conforms to the definition and standard of identity and is subject to the requirements for label statement of ingredients, except that

1. It contains one or any mixture of two or more of the following: any properly prepared cooked, canned, or dried fruit; any properly prepared cooked, canned, or dried vegetable; any properly prepared cooked or canned meat.
2. When the added fruits, vegetables, or meat contains fat, use the method prescribed for the determination of fats specifically prescribed by the USDA.

The name of a pasteurized cheese spread with fruits, vegetables, or meat is "pasteurized cheese spread with ___," the blank being filled in with the name or names of the fruits, vegetables, or meat used, in order of predominance by weight.

28.16 Pasteurized Neufchatel Cheese Spread with Other Foods

Pasteurized neufchatel cheese spread with other foods is a class of foods prepared by mixing, with the aid of heat, neufchatel cheese with one or a mixture of two or more properly prepared foods (except for other cheeses), such as fresh, cooked, canned, or dried fruits or vegetables; cooked or canned meat; and relishes, pickles, or other foods suitable for blending with neufchatel cheese. It may contain one or any mixture of two or more of the optional ingredients named. The amount of the added food or foods must be sufficient to differentiate this group of cheese from the standard Neufchatel cheese. It is spreadable at 70°F.

During its preparation, the mixture is heated for not less than 30 s at a temperature of not less than 150°F. When tested for phosphatase, the phenol equivalent of 0.25 g of such food is not more than 3 μg.

1. No water other than that contained in the ingredients used is added to this food, but the moisture content is in no case more than 65%.
2. The milkfat is not less than 20% by the weight of the finished food.

The optional ingredients are

1. One or any mixture of two or more of the following: gum karaya, gum tragacanth, carob bean gum, gelatin, algin (sodium alginate), propylene glycol alginate, guar gum, sodium

carboxymethylcellulose (cellulose gum), carrageenan, oat gum, or xanthan gum. The total quantity of any such substances, including that contained in neufchatel cheese, is not more than 0.8% by the weight of the finished food. When one or more of the optional ingredients are used, dioctyl sodium sulfosuccinate may be used in a quantity not in excess of 0.5% by the weight of such ingredients.

2. Artificial coloring, unless such addition conceals damage or inferiority or makes the finished food appear better or of greater value than it is.

3. An acidifying agent consisting of one or a mixture of two or more of the following: vinegar, acetic acid, lactic acid, citric acid, and phosphoric acid.

4. A sweetening agent consisting of one or a mixture of two or more of the following: sugar, dextrose, corn syrup, corn syrup solids, glucose syrup, glucose syrup solids, maltose, malt syrup, and hydrolyzed lactose.

5. Cream, milk, skim milk, buttermilk, cheese whey, any of the foregoing from which part of the water has been removed, anhydrous milkfat, dehydrated cream, and albumin from cheese whey.

The name of the food is "pasteurized neufchatel cheese spread with ___" or "pasteurized neufchatel cheese spread and ___," the blank being filled in with the common names of the foods added, in order of predominance by weight. The full name of the food should appear on the principal display panel of the label in type of uniform size, style, and color. Wherever any word or statement emphasizing the name of any ingredient appears on the label so conspicuously as to be easily seen under customary conditions of purchase, the full name of the food should immediately and conspicuously precede or follow such a word or statement in type of at least the same size as the type used in such a word or statement. Refer to the original legal document for more details.

28.17 Pasteurized Process Cheese Spread

Pasteurized process cheese spread is a food prepared by comminuting and mixing, with the aid of heat, one or more of the optional cheese ingredients prescribed, with or without one or more of the optional dairy ingredients prescribed, with one or more of the emulsifying agents prescribed, and with or without one or more of the optional ingredients prescribed, into a homogeneous plastic mass, which is spreadable at 70°F.

During its preparation, pasteurized process cheese spread is heated for not less than 30 s at a temperature of not less than 150°F. When tested for phosphatase, the phenol equivalent of 0.25 g of pasteurized process cheese spread is not more than 3 μg.

The moisture content of pasteurized process cheese spread is more than 44% but not more than 60%, and the milkfat content is not less than 20%.

The weight of the cheese ingredient referred to in the next paragraph constitutes not less than 51% of the weight of the pasteurized process cheese spread.

The weight of each variety of cheese in pasteurized process cheese spread made with two varieties of cheese is not less than 25% of the total weight of both, except that the weight of blue cheese, nuworld cheese, roquefort cheese, gorgonzola cheese, or limburger cheese is not less than 10% of the total weight of both. The weight of each variety of cheese in pasteurized process cheese spread made with three or more varieties of cheese is not less than 15% of the total weight of all, except that the weight of blue cheese, nuworld cheese, roquefort cheese, gorgonzola cheese, or limburger cheese is not less than 5% of the total weight of all. These limits do not apply to the quantity of cheddar cheese, washed curd cheese, colby cheese, and granular cheese in mixtures that are designated as "American cheese." Such mixtures are considered one variety of cheese.

Cheddar cheese for manufacturing, washed curd cheese for manufacturing, colby cheese for manufacturing, granular cheese for manufacturing, brick cheese for manufacturing, muenster cheese for manufacturing, and Swiss cheese for manufacturing are considered cheddar cheese, washed curd cheese, colby cheese, granular cheese, brick cheese, muenster cheese, and Swiss cheese, respectively.

Pasteurized process cheese spread or the cheese or cheeses from which it is made may be smoked before comminuting and mixing, or it may contain substances prepared by condensing or precipitating wood smoke.

The optional cheese ingredients are one or more cheeses of the same or two or more varieties, except that skim milk cheese for manufacturing may not be used and cream cheese, neufchatel cheese, cottage cheese, creamed cottage cheese, cook cheese, hard grating cheese, semisoft part-skim cheese, and part-skim spiced cheese are not used, alone or in combination with each other, as the cheese ingredient.

The optional dairy ingredients are cream, milk, skim milk, buttermilk, cheese whey, any of the foregoing from which part of the water has been removed, anhydrous milkfat, dehydrated cream, albumin from cheese whey, and skim milk cheese for manufacturing.

The emulsifying agents are one or any mixture of two or more of the following, in such quantity that the weight of the solids of such emulsifying agent is not more than 3% of the weight of the pasteurized process cheese spread: monosodium phosphate, disodium phosphate, dipotassium phosphate, trisodium phosphate, sodium metaphosphate (sodium hexametaphosphate), sodium acid pyrophosphate, tetrasodium pyrophosphate, sodium aluminum phosphate, sodium citrate, potassium citrate, calcium citrate, sodium tartrate, and sodium potassium tartrate.

The other optional ingredients are

1. One or any mixture of two or more of the following: carob bean gum, gum karaya, gum tragacanth, guar gum, gelatin, sodium carboxymethylcellulose (cellulose gum), carrageenan, oat gum, algin (sodium alginate), propylene glycol alginate, or xanthan gum. The total weight of such substances is not more than 0.8% of the weight of the finished food. When one or more of the optional ingredients are used, dioctyl sodium sulfosuccinate may be used in a quantity not in excess of 0.5% by the weight of such ingredients.

2. An acidifying agent consisting of one or any mixture of two or more of the following, in such quantity that the pH of the pasteurized process cheese spread is not below 4.0—vinegar, lactic acid, citric acid, acetic acid, and phosphoric acid.

3. A sweetening agent consisting of one or any mixture of two or more of the following—sugar, dextrose, corn sugar, corn syrup, corn syrup solids, glucose syrup, glucose syrup solids, maltose, malt syrup, and hydrolyzed lactose, in a quantity necessary for seasoning.

4. Water.

5. Salt.

6. Harmless artificial coloring.

7. Spices or flavorings other than any that, singly or in combination with other ingredients, simulates the flavor of a cheese of any age or variety.

8. Pasteurized process cheese spread in consumer-sized packages may contain an optional mold-inhibiting ingredient consisting of sorbic acid, potassium sorbate, sodium sorbate, or any combination of two or more of these, in an amount not to exceed 0.2% by weight, calculated as sorbic acid or consisting of not more than 0.3% by the weight of sodium propionate, calcium propionate, or a combination of sodium propionate and calcium propionate.

9. Pasteurized process cheese spread in consumer-sized packages may contain lecithin as an optional antisticking agent in an amount not to exceed 0.03% by the weight of the finished product.

10. Safe and suitable enzyme-modified cheese.

11. Nisin preparation in an amount that results in not more than 250 parts per million nisin in the food.

The name of the food is "pasteurized process cheese spread." The full name of the food should appear on the principal display panel of the label in type of uniform size, style, and color. Wherever any word or statement emphasizing the name of any ingredient appears on the label so conspicuously as to be easily seen under customary conditions of purchase, the full name of the food should immediately and conspicuously precede or follow such a word or statement in type of at least the same size as the type used in such a word or statement.

The name of the food should include a declaration of any flavoring, including smoke and substances prepared by condensing or precipitating wood smoke, that characterizes the product and a declaration of any spice that characterizes the product.

Each of the ingredients used in the food should be declared on the label, except that cheddar cheese, washed curd cheese, colby cheese, granular cheese, or any mixture of two or more of these may be designated as "American cheese."

28.18 Pasteurized Process Cheese Spread with Fruits, Vegetables, or Meat

Pasteurized process cheese spread with fruits, vegetables, meat, or mixtures of these is a food that conforms to the definition and standard of identity and is subject to the requirements for label statement of ingredients, prescribed, except that

1. It contains one or any mixture of two or more of the following: any properly prepared cooked, canned, or dried fruit; any properly prepared cooked, canned, or dried vegetable; any properly prepared cooked or canned meat.
2. When the added fruits, vegetables, or meat contains fat, use the method prescribed for the determination of fats specifically prescribed by the USDA.

The name of pasteurized process cheese spread with fruits, vegetables, or meat is "pasteurized process cheese spread with ___," the blank being filled in with the name or names of the fruits, vegetables, or meat used, in order of predominance by weight.

28.19 Provolone Cheese

Provolone, a pasta filata-type or stretched curd-type cheese, is a food prepared according to USDA requirements. It has a stringy texture. The minimum milkfat content is 45% by the weight of the solids, and the maximum moisture content is 45% by weight. If the dairy ingredients used are not pasteurized, and the cheese is cured at a temperature of not less than 35°F for at least 60 days.

If pasteurized dairy ingredients are used, the phenol equivalent value of 0.25 g of provolone cheese is not more than 3 μg.

One or more of the dairy ingredients specified may be bleached or warmed and is subjected to the action of a lactic acid–producing bacterial culture. One or more of the clotting enzymes specified is added to set the dairy ingredients to a semisolid mass. The mass is cut, stirred, and heated so as to promote and regulate the separation of whey from the curd. The whey is drained off, and the curd is matted and cut, immersed in hot water, and kneaded and stretched until it is smooth and free from lumps. Antimycotics may be added to the curd during the kneading and stretching process. Then, it is cut and molded. During the molding, the curd is kept sufficiently warm to cause proper sealing of the surface. The molded curd is then firmed by immersion in cold water, salted in brine, and dried. It is given some additional curing. Provolone cheese may be smoked, and one or more of the other optional ingredients specified may be added during the procedure.

The following safe and suitable ingredients may be used:

1. Milk, nonfat milk, or cream, used alone or in combination.
2. Rennet and/or other clotting enzymes of animal, plant, or microbial origin.
3. Blue or green color in an amount to neutralize the natural yellow color of the curd.
4. Calcium chloride in an amount not more than 0.02% (calculated as anhydrous calcium chloride) by the weight of the dairy ingredients, used as a coagulation aid.

5. Enzymes of animal, plant, or microbial origin, used in curing or flavor development.
6. Safe and suitable antimycotic agent(s), the cumulative levels of which should not exceed current good manufacturing practice, may be added to the cheese during the kneading and stretching process and/or applied to the surface of the cheese.
7. Benzoyl peroxide or a mixture of benzoyl peroxide with potassium alum, calcium sulfate, and magnesium carbonate used to bleach the dairy ingredients. The weight of the benzoyl peroxide is not more than 0.002% of the weight of the milk being bleached, and the weight of the potassium alum, calcium sulfate, and magnesium carbonate, singly or combined, is not more than six times the weight of the benzoyl peroxide used. If milk is bleached in this manner, vitamin A is added to the curd in such quantity as to compensate for the vitamin A or its precursors destroyed in the bleaching process, and artificial coloring is not used.

The name of the food is "provolone cheese." The name of the food may include the common name of the shape of the cheese, such as "salami provolone." One of the following terms, in letters not less than 1/2 the height of the letters used in the name of the food, should accompany the name of the food wherever it appears on the principal display panel or panels:

1. "Smoked," if the food has been smoked
2. "Not smoked," if the food has not been smoked

28.20 Soft Ripened Cheeses

The cheeses for which definitions and standards of identity are prescribed by this section are soft ripened cheeses for which specifically applicable definitions and standards of identity are not prescribed by other sections of 21 CFR 133. They are made from milk and other ingredients specified. Their solids contain not less than 50% of milkfat. If the milk used is not pasteurized, the cheese so made is cured at a temperature of not less than 35°F for not less than 60 days.

Milk, which may be pasteurized, clarified, or both and may be warmed, is subjected to the action of harmless lactic acid–producing bacteria or other harmless flavor-producing bacteria, present in such milk or added thereto. Sufficient rennet, rennet paste, extract of rennet paste, or other safe and suitable milk-clotting enzyme that produces equivalent curd formation, singly or in any combination (with or without purified calcium chloride in a quantity not more than 0.02%, calculated as anhydrous calcium chloride, of the weight of the milk), is added to set the milk to a semisolid mass. Harmless artificial coloring may be added. After coagulation, the mass is treated so as to promote and regulate the separation of whey and curd. Such treatment may include one or more of the following: cutting, stirring, heating, and dilution with water or brine. The whey, or part of it, is drained off, and the curd is collected and shaped. It may be placed in forms and may be pressed. Harmless flavor-producing microorganisms may be added. It is cured under conditions suitable for the development of biological curing agents on the surface of the cheese, and the curing is conducted so that the cheese cures from the surface toward the center. Salt may be added during the procedure. A harmless preparation of enzymes of animal or plant origin capable of aiding in the curing or development of flavor of soft ripened cheeses may be added, in such quantity that the weight of the solids of such preparation is not more than 0.1% of the weight of the milk used.

The word "milk" means cow, or goat, or sheep milk or mixtures of two or all of these. Such milk may be adjusted by separating part of the fat therefrom or (in the case of cow milk) by adding one or more of the following: cream, skim milk, concentrated skim milk, nonfat dry milk; the corresponding products from goat milk (in the case of goat milk); the corresponding products from sheep milk (in the case of sheep milk); and water, in a quantity sufficient to reconstitute any such concentrated or dried products used.

Milk should be deemed to have been pasteurized if it has been held at a temperature of not less than 143°F for a period of not less than 30 min or for a time and at a temperature equivalent thereto in phosphatase destruction. The name of each soft ripened cheese for which a definition and standard of identity is prescribed by this section is "soft ripened cheese," preceded or followed by

1. The specific common or usual name of such soft ripened cheese if any such name has become generally recognized
2. An arbitrary or fanciful name that is not false or misleading in any particular if no such specific common or usual name has become generally recognized

When milk other than cow milk is used in whole or in part, the name of the cheese includes the statement "made from ___," the blank being filled in with the name or names of the milk used, in order of predominance by weight.

28.21 Romano Cheese

Romano cheese is a food prepared from cow, sheep, or goat milk or mixtures of two or all of these and other ingredients specified. It grates readily and has a granular texture and a hard and brittle rind. It contains not more than 34% of moisture, and its solids contain not less than 38% of milkfat. It is cured for not less than 5 months.

Milk, which may be pasteurized, clarified, or both and may be warmed, is subjected to the action of harmless lactic acid–producing bacteria present in such milk or added thereto. Harmless artificial blue or green coloring in a quantity that neutralizes any natural yellow coloring in the curd may be added. Rennet, rennet paste, extract of rennet paste, or other safe and suitable milk-clotting enzyme that produces equivalent curd formation, singly or in any combination (with or without purified calcium chloride in a quantity not more than 0.02%, calculated as anhydrous calcium chloride, of the weight of the milk), is added to set the milk to be a semisolid mass. The mass is cut into particles no larger than corn kernels, stirred, and heated to a temperature of about 120°F. The curd is allowed to settle to the bottom of the kettle or vat and is then removed and drained for a short time, packed in forms or hoops, and pressed. The pressed curd is salted by immersing in brine for about 24 h and is then removed from the brine, and the surface is allowed to dry. It is then alternately rubbed with salt and washed at intervals. It may be perforated with needles. It is finally dry cured. During curing, it is turned and scraped. The surface may be rubbed with vegetable oil. A harmless preparation of enzymes of animal or plant origin capable of aiding in the curing or development of flavor of romano cheese may be added during the procedure, in such quantity that the weight of the solids of such preparation is not more than 0.1% of the weight of the milk used.

The word "milk" means cow, goat, or sheep milk or mixtures of two or all of these. Such milk may be adjusted by separating part of the fat therefrom or (in the case of cow milk) by adding one or more of the following: cream, skim milk, concentrated skim milk, nonfat dry milk; the corresponding products from goat milk (in the case of goat milk); the corresponding products from sheep milk (in the case of sheep milk); and water, in a quantity sufficient to reconstitute any such concentrated or dried products used.

Such milk may be bleached by the use of benzoyl peroxide or a mixture of benzoyl peroxide with potassium alum, calcium sulfate, and magnesium carbonate, but the weight of the benzoyl peroxide is not more than 0.002% of the weight of the milk bleached, and the weight of the potassium alum, calcium, sulfate, and magnesium carbonate, singly or combined, is not more than six times the weight of the benzoyl peroxide used. If milk is bleached in this manner, sufficient vitamin A is added to the curd to compensate for the vitamin A or its precursors destroyed in the bleaching process, and artificial coloring is not used.

Safe and suitable antimycotic agent(s), the cumulative levels of which should not exceed current good manufacturing practice, may be added to the surface of the cheese.

When romano cheese is made solely from cow milk, the name of such cheese is "romano cheese made from cow milk" and may be preceded by the word "vaccino" (or "vacchino"); when made solely from sheep milk, the name is "romano cheese made from sheep milk" and may be preceded by the word "pecorino"; when made solely from goat milk, the name is "romano cheese made from goat milk" and may be preceded by the word "caprino"; and when a mixture of two or all of the milk specified in this section is used, the name of the cheese is "romano cheese made from ___," the blank being filled in with the names of the milk used, in order of predominance by weight.

28.22 Roquefort, Sheep Milk Blue-Mold, and Blue-Mold Cheese from Sheep Milk

Roquefort cheese (sheep milk blue-mold cheese or blue-mold cheese from sheep milk) is a food prepared by the procedure set forth in paragraph 2 of this section or by any other procedure that produces a finished cheese having the same physical and chemical properties. It is characterized by the presence of bluish-green mold, *P. roquefortii*, throughout the cheese. The minimum milkfat content is 50% by the weight of the solids, and the maximum moisture content is 45% by weight. The dairy ingredients used may be pasteurized. Roquefort cheese is at least 60 days old.

One or more of the dairy ingredients specified may be warmed and is subjected to the action of a lactic acid–producing bacterial culture. One or more of the clotting enzymes specified is added to set the dairy ingredients to a semisolid mass. The mass is cut into smaller portions and allowed to stand for a time. The mixed curd and whey is placed into forms permitting further drainage. While being placed in forms, spores of the mold *P. roquefortii* are added. The forms are turned several times during drainage. When sufficiently drained, the shaped curd is removed from the forms and salted with dry salt or brine. Perforations are then made in the shaped curd, and it is held at a temperature of approximately 50°F at 90%–95% relative humidity until the characteristic mold growth has developed. During storage, the surface of the cheese may be scraped to remove the surface growth of undesirable microorganisms. One or more of the other optional ingredients specified may be added during the procedure.

The following safe and suitable ingredients may be used:

1. Forms of milk, nonfat milk, or cream of sheep origin, used alone or in combination
2. Rennet and/or other clotting enzymes of animal, plant, or microbial origin
3. Enzymes of animal, plant, or microbial origin, used in curing or flavor development

The name of the food is "roquefort cheese" or, alternatively, "sheep milk blue-mold cheese" or "blue-mold cheese from sheep milk."

28.23 Samsoe Cheese

Samsoe cheese is a food prepared according to FDA requirements. It has a small amount of eye formation of approximately uniform size of about 5/16 inch (8 mm). The minimum milkfat content is 45% by the weight of the solids, and the maximum moisture content is 41% by weight, as determined by the methods described in 21 CFR 133.5. The dairy ingredients used may be pasteurized. Samsoe cheese is cured at not less than 35°F for at least 60 days.

If pasteurized dairy ingredients are used, the phenol equivalent value of 0.25 g of samsoe cheese is not more than 3 μg.

One or more of the dairy ingredients specified may be warmed and is subjected to the action of a lactic acid–producing bacterial culture. One or more of the clotting enzymes specified is added to set the dairy ingredients to a semisolid mass. After coagulation, the mass is cut into small cube-shaped pieces with sides approximately 3/8 inch (1 cm). The mass is stirred and heated to about 102°F and handled by further stirring, heating, dilution with water, and salting so as to promote and regulate the separation of curd and whey. When the desired curd is obtained, it is transferred to forms permitting drainage of the whey. During drainage, the curd is pressed. After drainage, the curd is removed from the forms and is further salted by immersing in a concentrated salt solution for about 3 days. The curd is then cured at a temperature of 60°F–70°F for 3–5 weeks to obtain the desired eye formation. Further curing is conducted at a lower temperature. One or more of the other optional ingredients specified may be added during the procedure.

The following safe and suitable ingredients may be used:

1. Milk, nonfat milk, or cream, used alone or in combination
2. Rennet and/or other clotting enzymes of animal, plant, or microbial origin
3. Coloring
4. Calcium chloride in an amount not more than 0.02% (calculated as anhydrous calcium chloride) by the weight of the dairy ingredients, used as a coagulation aid
5. Enzymes of animal, plant, or microbial origin, used in curing or flavor development
6. Antimycotic agents, applied to the surface of slices or cuts in consumer-sized packages

28.24 Sap Sago Cheese

Sap sago cheese is a food prepared according to FDA requirements. The cheese is pale green in color and has the shape of a truncated cone. The maximum moisture content is 38% by weight. Sap sago cheese is not less than 5 months old.

One or more of the dairy ingredients specified is allowed to become sour and is heated to boiling temperature, with stirring. Sufficient sour whey is added to precipitate the casein. The curd is removed, spread out in boxes and pressed, and allowed to drain and ferment while under pressure. It is ripened for not less than 5 weeks. The ripened curd is dried and ground; salt and dried clover of the species *Melilotus coerulea* are added. The mixture is shaped into truncated cones and ripened. The optional ingredient may be added during this procedure.

The following safe and suitable ingredients may be used:

1. Nonfat milk
2. Buttermilk

29

*Semi-Soft Pasteurized and Other Cheeses: Standards, Grades, and Specifications**

Y. H. Hui

CONTENTS

29.1 Introduction

Food manufacturers in the United States must produce a product that consumers like in order to sell the product nationally. To achieve this goal, they depend on developments in science, technology, and engineering in order to produce a quality product. However, they must also pay attention to two other important considerations:

1. Are their products safe for public consumption?
2. Have they assured the economic integrity of their products? This means many things. The most important one is the assurance that their products contain the proper quality and quantity of ingredients, that is, "no cheating."

---

* The information in this chapter has been modified from *Food Safety Manual* published by Science Technology System (STS) of West Sacramento, California. Copyrighted 2012©. Used with permission.

To protect the consumers, U.S. government agencies have established many checks and balances to prevent "injuries" to the consumers in terms of health and economic fraud. The two major responsible agencies are

1. U.S. Food and Drug Administration (FDA, www.FDA.gov)
2. U.S. Department of Agriculture (USDA, www.USDA.gov)

The FDA establishes standards of identities for a number of cheeses for interstate commerce. This makes sure that a particular type of cheese contains the minimal quality and quantity of the ingredients used to make this cheese. The FDA also issues safety requirements and recommendations in the manufacturing of cheeses. It makes periodic inspections of establishments producing cheeses to ensure full compliance. In effect, the FDA is monitoring the safety production of cheeses and their economic integrities.

The USDA plays the same important role in the commerce of cheeses in the United States. For example, a national restaurant chain that makes and sells pizzas will not buy cheeses from any manufacturer if the cheeses do not bear the USDA official identification or seal of approval. The USDA has issued official requirements and processing specifications that cheese manufacturers must follow if they want to obtain the USDA seal of approval for the products. To achieve this goal, the USDA sets up voluntary services of inspection under a standard fee arrangement. The inspection focuses on the safety and economic integrity of manufacturing cheeses. When the inspection demonstrates that the manufacturer has complied with safety and economic requirements, the USDA issues its official identification or seal of approval.

In essence, the FDA and USDA are the federal agencies that make sure that the product is safe and will not pose any economic fraud. Using the public information distributed at the Web sites of FDA and USDA, this chapter discusses USDA specifications and FDA inspection in relation to the production of selected cheeses.

The information in this chapter is important because it facilitates commerce of cheeses and protects consumers from injuries of eating cheeses in the market. It also protects the consumers from economic fraud. For example, without FDA and USDA intervention, one may purchase a cheese that contains less than the normal amount of milk used. If so, the consumers lose money in the transaction.

All FDA cheese standards of identities are described in 21 CFR 133; all USDA requirements and standard for grades are described in 7 CFR 58. For explanation, CFR refers to the United States Code of Federal Regulations. This chapter focuses on two areas:

1. Part 133: standards of identities issued by the FDA.
2. Part 58 (selected sections): USDA specifications and grade standards.

29.2 Semi-Soft Cheeses

The cheeses for which definitions and standards of identity are prescribed by this section are semi-soft cheeses for which specifically applicable definitions and standards of identity are not prescribed by other sections of 21 CFR 133. They are made from milk and other ingredients specified. They contain more than 39%, but not more than 50%, of moisture, and their solids contain not less than 50% of milkfat. If the milk used is not pasteurized, the cheese so made is cured at a temperature of not less than 35°F for not less than 60 days.

Milk, which may be pasteurized or clarified or both, and which may be warmed, is subjected to the action of harmless lactic acid–producing bacteria or other harmless flavor-producing bacteria, present in such milk or added thereto. Sufficient rennet, rennet paste, extract of rennet paste, or other safe and suitable milk-clotting enzyme that produces equivalent curd formation, singly or in any combination (with or without purified calcium chloride in a quantity not more than 0.02%, calculated as anhydrous calcium chloride, of the weight of the milk), are added to set the milk to a semisolid mass. Harmless artificial coloring may be added. After coagulation, the mass is treated so as to promote and regulate the separation of whey and curd. Such treatment may include one or more of the following: cutting, stirring, heating, and dilution with water or brine. The whey, or part of it, is drained off, and the curd is collected and shaped. It may be placed in forms and may be pressed. Harmless flavor-producing microorganisms may be added. It may be cured in a manner to promote the growth of biological curing agents. Salt may be added during the procedure. A harmless preparation of enzymes of animal or plant origin capable of aiding in the curing or development of flavor of semi-soft cheese may be added, in such quantity that the weight of the solids of such preparation is not more than 0.1% of the weight of the milk used.

The word "milk" means cow milk, goat milk, sheep milk, or mixtures of two or all of these. Such milk may be adjusted by separating part of the fat therefrom, or by adding one or more of the following: (in the case of cow milk) cream, skim milk, concentrated skim milk, non-fat dry milk; (in the case of goat milk) the corresponding products from goat milk; (in the case of sheep milk) the corresponding products from sheep milk; or water in a quantity sufficient to reconstitute any concentrated or dried products used.

Milk should be deemed to have been pasteurized if it has been held at a temperature of not less than 143°F for a period of not less than 30 min, or for a time and at a temperature equivalent thereto in phosphatase destruction. A semi-soft cheese should be deemed not to have been made from pasteurized milk if 0.25 g shows a phenol equivalent of more than 5 µg.

Semi-soft cheeses in the form of slices or cuts in consumer-sized packages may contain an optional mold-inhibiting ingredient consisting of sorbic acid, potassium sorbate, sodium sorbate, or any combination of two or more of these, in an amount not to exceed 0.3% by weight, calculated as sorbic acid.

The name of each semi-soft cheese for which a definition and standard of identity are prescribed by this section is "semi-soft cheese," preceded or followed by

1. The specific common or usual name of such semi-soft cheese, if any such name has become generally recognized
2. If no such specific common or usual name has become generally recognized, an arbitrary or fanciful name that is not false or misleading in any particular

When milk other than cow milk is used in whole or in part, the name of the cheese includes the statement "made from ___," with the blank being filled in with the name or names of the milk used, in order of predominance by weight.

If semi-soft cheese in sliced or cut form contains an optional mold-inhibiting ingredient as specified in paragraph 5 of this section, the label should bear the statement "___ added to retard mold growth" or "___ added as a preservative," with the blank being filled in with the common name or names of the mold-inhibiting ingredient or ingredients used.

Wherever the name of the food appears on the label so conspicuously as to be easily seen under customary conditions of purchase, the words and statements prescribed, showing the optional ingredient used, should immediately and conspicuously precede or follow such name, without intervening written, printed, or graphic matter.

29.3 Semi-Soft Part-Skim Cheeses

The cheeses for which definitions and standards of identity are prescribed by the FDA are semi-soft part-skim cheeses for which specifically applicable definitions and standards of identity are not prescribed by the FDA. They are made from partly skimmed milk and other ingredients specified. They contain not more than 50% of moisture, and their solids contain not less than 45%, but less than 50%, of milkfat. If the milk used is not pasteurized, the cheese so made is cured at a temperature of not less than 35°F for not less than 60 days.

Milk, which may be pasteurized or clarified or both, and which may be warmed, is subjected to the action of harmless lactic acid–producing bacteria or other harmless flavor-producing bacteria, present in such milk or added thereto. Sufficient rennet, rennet paste, extract of rennet paste, or other safe and suitable milk-clotting enzyme that produces equivalent curd formation, singly or in any combination (with or without purified calcium chloride in a quantity not more than 0.02%, calculated as anhydrous calcium chloride, of the weight of the milk), are added to set the milk to a semisolid mass. Harmless artificial coloring may be added. After coagulation, the mass is treated so as to promote and regulate the separation of whey and curd. Such treatment may include one or more of the following: cutting, stirring, heating, and dilution with water or brine. The whey, or part of it, is drained off, and the curd is collected and shaped. It may be placed in forms, and it may be pressed. Harmless flavor-producing microorganisms may be added. It may be cured in a manner to promote the growth of biological curing agents. Salt may be added during the procedure. A harmless preparation of enzymes of animal or plant origin capable of aiding in the curing or development of flavor of semi-soft part-skim cheese may be added in such quantity that the weight of the solids of such preparation is not more than 0.1% of the weight of the milk used.

The word "milk" means cow milk or goat milk or sheep milk or mixtures of two or all of these. Such milk may be adjusted by separating part of the fat therefrom or by adding one or more of the following: (in the case of cow milk) cream, skim milk, concentrated skim milk, or non-fat dry milk; (in the case of goat milk) the corresponding products from goat milk; (in the case of sheep milk) the corresponding products from sheep milk; or water in a quantity sufficient to reconstitute any such concentrated or dried products used.

Milk should be deemed to have been pasteurized if it has been held at a temperature of not less than 143°F for a period of not less than 30 min, or for a time and at a temperature equivalent thereto in phosphatase destruction. A semi-soft part-skim cheese should be deemed not to have been made from pasteurized milk if 0.25 g shows a phenol equivalent of more than 5 µg.

Semi-soft part-skim cheeses in the form of slices or cuts in consumer-sized packages may contain an optional mold-inhibiting ingredient consisting of sorbic acid, potassium sorbate, sodium sorbate, or any combination of two or more of these, in an amount not to exceed 0.3% by weight, calculated as sorbic acid.

The name of each semi-soft part-skim cheese for which a definition and standard of identity is prescribed by this section is "semi-soft part-skim cheese," preceded or followed by

1. The specific common or usual name of such semi-soft cheese, if any such name has become generally recognized
2. If no such specific common or usual name has become generally recognized, an arbitrary or fanciful name that is not false or misleading in any particular

When milk other than cow milk is used in whole or in part, the name of the cheese includes the statement "made from ___," with the blank being filled in with the name or names of the milk used, in order of predominance by weight.

If semi-soft part-skim cheese in sliced or cut form contains an optional mold-inhibiting ingredient as specified, the label should bear the statement "___ added to retard mold growth" or "___ added as a preservative," with the blank being filled in with the common name or names of the mold-inhibiting ingredient or ingredients used.

Wherever the name of the food appears on the label so conspicuously as to be easily seen under customary conditions of purchase, the words and statements prescribed by this section, showing the optional ingredient used, should immediately and conspicuously precede or follow such name, without intervening written, printed, or graphic matter.

29.4 Skim Milk Cheese for Manufacturing

Skim milk cheese for manufacturing is a food prepared from skim milk and other ingredients specified. It contains not more than 50% of moisture. It is coated with blue paraffin or other tightly adhering blue coating.

Skim milk or the optional dairy ingredients specified, which may be pasteurized, and which may be warmed, are subjected to the action of harmless lactic acid-producing bacteria, present in such milk or added thereto. Harmless artificial coloring may be added. Sufficient rennet, or other safe and suitable milk-clotting enzyme that produces equivalent curd formation, or both, with or without purified calcium chloride in a quantity not more than 0.02% (calculated as anhydrous calcium chloride) of the weight of separation of whey and curd. Such treatment may include one or more of the following: cutting, stirring, heating, and dilution with water or brine. The whey, or part of it, is drained off, and the curd is collected and shaped. It may be placed in forms and may be pressed. Harmless flavor-producing microorganisms may be added. It may be cured in a manner to promote the growth of biological curing agents. Salt may be added during the procedure. A harmless preparation of enzymes of animal or plant origin capable of aiding in the curing or development of flavor of semi-soft cheese may be added, in such quantity that the weight of the solids of such preparation is not more than 0.1% of the weight of the milk used.

The word "milk" means cow milk, goat milk, sheep milk, or mixtures of two or all of these. Such milk may be adjusted by separating part of the fat therefrom, or by adding one or more of the following: (in the case of cow milk) cream, skim milk, concentrated skim milk, non-fat dry milk; (in the case of goat milk) the corresponding products from goat milk; (in the case of sheep milk) the corresponding products from sheep milk; or water in a quantity sufficient to reconstitute any concentrated or dried products used.

Milk should be deemed to have been pasteurized if it has been held at a temperature of not less than 143°F for a period of not less than 30 min, or for a time and at a temperature equivalent thereto in phosphatase destruction. A semi-soft cheese should be deemed not to have been made from pasteurized milk if 0.25 g shows a phenol equivalent of more than 5 μg.

Semi-soft cheeses in the form of slices or cuts in consumer-sized packages may contain an optional mold-inhibiting ingredient consisting of sorbic acid, potassium sorbate, sodium sorbate, or any combination of two or more of these, in an amount not to exceed 0.3% by weight, calculated as sorbic acid.

The name of each semi-soft cheese for which a definition and standard of identity are prescribed by this section is "semi-soft cheese," preceded or followed by

1. The specific common or usual name of such semi-soft cheese, if any such name has become generally recognized
2. If no such specific common or usual name has become generally recognized, an arbitrary or fanciful name that is not false or misleading in any particular

When milk other than cow milk is used in whole or in part, the name of the cheese includes the statement "made from ___," with the blank being filled in with the name or names of the milk used, in order of predominance by weight.

If semi-soft cheese in sliced or cut form contains an optional mold-inhibiting ingredient as specified in paragraph 5 of this section, the label should bear the statement "___ added to retard mold growth" or "___ added as a preservative," with the blank being filled in with the common name or names of the mold-inhibiting ingredient or ingredients used.

Wherever the name of the food appears on the label so conspicuously as to be easily seen under customary conditions of purchase, the words and statements prescribed, showing the optional ingredient used, should immediately and conspicuously precede or follow such name, without intervening written, printed, or graphic matter.

29.3 Semi-Soft Part-Skim Cheeses

The cheeses for which definitions and standards of identity are prescribed by the FDA are semi-soft part-skim cheeses for which specifically applicable definitions and standards of identity are not prescribed by the FDA. They are made from partly skimmed milk and other ingredients specified. They contain not more than 50% of moisture, and their solids contain not less than 45%, but less than 50%, of milkfat. If the milk used is not pasteurized, the cheese so made is cured at a temperature of not less than 35°F for not less than 60 days.

Milk, which may be pasteurized or clarified or both, and which may be warmed, is subjected to the action of harmless lactic acid–producing bacteria or other harmless flavor-producing bacteria, present in such milk or added thereto. Sufficient rennet, rennet paste, extract of rennet paste, or other safe and suitable milk-clotting enzyme that produces equivalent curd formation, singly or in any combination (with or without purified calcium chloride in a quantity not more than 0.02%, calculated as anhydrous calcium chloride, of the weight of the milk), are added to set the milk to a semisolid mass. Harmless artificial coloring may be added. After coagulation, the mass is treated so as to promote and regulate the separation of whey and curd. Such treatment may include one or more of the following: cutting, stirring, heating, and dilution with water or brine. The whey, or part of it, is drained off, and the curd is collected and shaped. It may be placed in forms, and it may be pressed. Harmless flavor-producing microorganisms may be added. It may be cured in a manner to promote the growth of biological curing agents. Salt may be added during the procedure. A harmless preparation of enzymes of animal or plant origin capable of aiding in the curing or development of flavor of semi-soft part-skim cheese may be added in such quantity that the weight of the solids of such preparation is not more than 0.1% of the weight of the milk used.

The word "milk" means cow milk or goat milk or sheep milk or mixtures of two or all of these. Such milk may be adjusted by separating part of the fat therefrom or by adding one or more of the following: (in the case of cow milk) cream, skim milk, concentrated skim milk, or non-fat dry milk; (in the case of goat milk) the corresponding products from goat milk; (in the case of sheep milk) the corresponding products from sheep milk; or water in a quantity sufficient to reconstitute any such concentrated or dried products used.

Milk should be deemed to have been pasteurized if it has been held at a temperature of not less than 143°F for a period of not less than 30 min, or for a time and at a temperature equivalent thereto in phosphatase destruction. A semi-soft part-skim cheese should be deemed not to have been made from pasteurized milk if 0.25 g shows a phenol equivalent of more than 5 μg.

Semi-soft part-skim cheeses in the form of slices or cuts in consumer-sized packages may contain an optional mold-inhibiting ingredient consisting of sorbic acid, potassium sorbate, sodium sorbate, or any combination of two or more of these, in an amount not to exceed 0.3% by weight, calculated as sorbic acid.

The name of each semi-soft part-skim cheese for which a definition and standard of identity is prescribed by this section is "semi-soft part-skim cheese," preceded or followed by

1. The specific common or usual name of such semi-soft cheese, if any such name has become generally recognized
2. If no such specific common or usual name has become generally recognized, an arbitrary or fanciful name that is not false or misleading in any particular

When milk other than cow milk is used in whole or in part, the name of the cheese includes the statement "made from ___," with the blank being filled in with the name or names of the milk used, in order of predominance by weight.

If semi-soft part-skim cheese in sliced or cut form contains an optional mold-inhibiting ingredient as specified, the label should bear the statement "___ added to retard mold growth" or "___ added as a preservative," with the blank being filled in with the common name or names of the mold-inhibiting ingredient or ingredients used.

Wherever the name of the food appears on the label so conspicuously as to be easily seen under customary conditions of purchase, the words and statements prescribed by this section, showing the optional ingredient used, should immediately and conspicuously precede or follow such name, without intervening written, printed, or graphic matter.

29.4 Skim Milk Cheese for Manufacturing

Skim milk cheese for manufacturing is a food prepared from skim milk and other ingredients specified. It contains not more than 50% of moisture. It is coated with blue paraffin or other tightly adhering blue coating.

Skim milk or the optional dairy ingredients specified, which may be pasteurized, and which may be warmed, are subjected to the action of harmless lactic acid-producing bacteria, present in such milk or added thereto. Harmless artificial coloring may be added. Sufficient rennet, or other safe and suitable milk-clotting enzyme that produces equivalent curd formation, or both, with or without purified calcium chloride in a quantity not more than 0.02% (calculated as anhydrous calcium chloride) of the weight of the skim milk, are added to set the skim milk to a semisolid mass. The mass is cut, stirred, and heated with continued stirring, so as to promote and regulate the separation of whey and curd. The whey is drained off, and the curd is matted into a cohesive mass. Proteins from the whey may be incorporated. The mass is cut into slabs that are piled and handled so as to promote the drainage of whey and the development of acidity. The slabs are then cut into pieces, which may be rinsed by pouring or sprinkling water over them, with free and continuous drainage; but the duration of such rinsing is so limited that only the whey on the surface of such pieces is removed. The curd is salted, stirred, further drained, and pressed into forms. A harmless preparation of enzymes of animal or plant origin capable of aiding in the curing or development of flavor of skim milk cheese for manufacturing may be added during the procedure, in such quantity that the weight of the solids of such preparation is not more than 0.1% of the weight of the milk used.

The optional dairy ingredients are skim milk or concentrated skim milk or non-fat dry milk or a mixture of any two or more of these, with water in a quantity not in excess of that sufficient to reconstitute any concentrated skim milk or non-fat dry milk used.

29.5 Spiced Cheeses

Spiced cheeses are cheeses for which specifically applicable definitions and standards of identity are not prescribed by the FDA. The food is prepared by specific FDA procedure. The minimum milkfat content is 50% by weight of the solids. The food contains spices, in a minimum amount of 0.015 ounce per pound of cheese, and may contain spice oils. If the dairy ingredients are not pasteurized, the cheese is cured at a temperature of not less than 35°F for at least 60 days. The phenol equivalent of 0.25 g of spiced cheese is not more than 3 µg.

One or more of the dairy ingredients specified may be warmed and is subjected to the action of a harmless lactic acid-producing bacterial culture. One or more of the clotting enzymes specified is added to set the dairy ingredients to a semisolid mass. The mass is divided into smaller portions and handled by stirring, heating, and diluting with water or salt brine so as to promote and regulate the separation of whey and curd. The whey is drained off. The curd is removed and may be further drained. The curd is then shaped into forms and may be pressed. At some time during the procedure, spices are added so as to be evenly distributed throughout the finished cheese. One or more of the other optional ingredients specified may be added during the procedure.

The following safe and suitable ingredients may be used:

1. Milk, non-fat milk, or cream, or corresponding products of goat or sheep origin, used alone or in combination
2. Rennet and/or other clotting enzymes of animal, plant, or microbial origin
3. Coloring
4. Calcium chloride in an amount not more than 0.02% (calculated as anhydrous calcium chloride) of the weight of the dairy ingredients, used as a coagulation aid
5. Salt
6. Spice oils that do not, alone or in combination with other ingredients, simulate the flavor of cheese of any age or variety
7. Enzymes of animal, plant, or microbial origin, used in curing or flavor development
8. Antimycotic agents, applied to the surface of slices or cuts in consumer-sized packages

The name of the food is "spiced cheese." The following terms should accompany the name of the food, as appropriate:

1. The specific common or usual name of the spiced cheese, if any such name has become generally recognized
2. An arbitrary or fanciful name that is not false or misleading in any particular

Part-skim spiced cheeses conform to the definition and standard of identity and are subject to the requirements for label statement of ingredients prescribed for spiced cheeses except that their solids contain less than 50%, but not less than 20%, of milkfat.

29.6 Spiced and Flavored Standardized Cheeses

A spiced or flavored standardized cheese conforms to the applicable definitions, standard of identity, and requirements for label statement of ingredients prescribed for that specific natural cheese variety promulgated by the FDA. In addition, a spiced and/or flavored standardized cheese should contain one or more safe and suitable spices and/or flavorings, in such proportions as are reasonably required to accomplish their intended effect, provided that no combination of ingredients should be used to simulate the flavor of cheese of any age or variety.

The name of a spiced or flavored standardized cheese should include, in addition to the varietal name of the natural cheese, a declaration of any flavor and/or spice that characterizes the food.

29.7 Swiss and Emmentaler Cheese

Swiss cheese (Emmentaler cheese) is a food prepared according to FDA requirements. It has holes or eyes developed throughout the cheese. The minimum milkfat content is 43% by weight of the solids, and the maximum moisture content is 41% by weight. The dairy ingredients used may be pasteurized. Swiss cheese is at least 60 days old.

If pasteurized dairy ingredients are used, the phenol equivalent value of 0.25 g of Swiss cheese is not more than 3 μg.

One or more of the dairy ingredients specified may be bleached, warmed, or treated with hydrogen peroxide/catalase, and is subjected to the action of lactic acid–producing and propionic acid–producing bacterial cultures. One or more of the clotting enzymes specified is added to set the dairy ingredients to a semisolid mass. The mass is cut into particles similar in size to wheat kernels. For about 30 min, the particles are alternately stirred and allowed to settle. The temperature is raised to about 126°F. Stirring is continued until the curd becomes firm. The acidity of the whey at this point, calculated as lactic acid, does not exceed 0.13%. The curd is transferred to hoops or forms and pressed until the desired shape and firmness are obtained. The cheese is then salted by immersing it in a saturated salt solution for about 3 days. It is then held at a temperature of about 50°F–60°F for a period of 5–10 days, after which it is held at a temperature of about 75°F until it is approximately 30 days old, or until the so-called eyes form. Salt, or a solution of salt in water, is added to the surface of the cheese at some time during the curing process. The cheese is then stored at a lower temperature for further curing. One or more of the optional ingredients specified may be added during the procedure.

The following safe and suitable ingredients may be used:

1. Milk, non-fat milk, or cream, used alone or in combination.
2. Rennet and/or other clotting enzymes of animal, plant, or microbial origin.
3. Coloring.
4. Calcium chloride in an amount not more than 0.02% (calculated as anhydrous calcium chloride) by weight of the dairy ingredients, used as a coagulation aid.
5. Enzymes of animal, plant, or microbial origin, used in curing or flavor development.
6. Antimycotic agents may be added to the surface of the cheese, the cumulative levels of which should not exceed good manufacturing practice.
7. Benzoyl peroxide or a mixture of benzoyl peroxide with potassium alum, calcium sulfate, and magnesium carbonate used to bleach the dairy ingredients. The weight of the benzoyl peroxide is not more than 0.002% of the weight of the milk being bleached, and the weight of the

potassium alum, calcium sulfate, and magnesium carbonate, singly or combined, is not more than six times the weight of the benzoyl peroxide used. If milk is bleached in this manner, vitamin A is added to the curd in such quantity as to compensate for the vitamin A or its precursors destroyed in the bleaching process, and artificial coloring is not used.

8. Hydrogen peroxide followed by a sufficient quantity of catalase preparation to eliminate the hydrogen peroxide. The weight of the hydrogen peroxide should not exceed 0.05% of the weight of the milk, and the weight of the catalase should not exceed 20 parts per million of the weight of the milk treated.

Swiss cheese for manufacturing conforms to the definition and standard of identity prescribed for Swiss cheese, except that the holes, or eyes, have not developed throughout the entire cheese.

29.8 USDA Grade Standards for Swiss and Emmentaler Cheese

29.8.1 Definitions

The words "Swiss" and "Emmentaler" are interchangeable.

Swiss cheese is cheese made by the Swiss process or by any other procedure that produces a finished cheese having the same physical and chemical properties as cheese produced by the Swiss process. It is prepared from milk and has holes, or eyes, developed throughout the cheese by microbiological activity. It contains not more than 41% of moisture, and its solids contain not less than 43% of milkfat. It is not less than 60 days old and conforms to FDA provisions for "Cheese and Related Cheese Products."

29.8.2 Styles

1. *Rinded.* The cheese is completely covered by a rind sufficient to protect the interior of the cheese.
2. *Rindless.* The cheese is properly enclosed in a wrapper or covering that will not impart any objectionable flavor or color to the cheese. The wrapper or covering is sealed with a sufficient overlap or satisfactory closure to exclude air. The wrapper or covering is of sufficiently low permeability to water vapor and air so as to prevent the formation of a rind through contact with air during the curing and holding periods.

29.8.3 U.S. Grades

The nomenclature of the U.S. grades is as follows:

1. U.S. Grade A
2. U.S. Grade B
3. U.S. Grade C

The determination of U.S. grades of Swiss cheese should be on the basis of rating the following quality factors:

1. Flavor
2. Body
3. Eyes and texture
4. Finish and appearance
5. Color

The rating of each quality factor should be established on the basis of characteristics present in a randomly selected sample representing a vat of cheese. In the case of institutional-size cuts, samples may be selected on a lot basis.

To determine flavor and body characteristics, the grader will examine a full trier plug of cheese withdrawn at the approximate center of one of the largest flat surface areas of the sample. For some institutional-size samples, it may not be possible to obtain a full trier plug. When this occurs, a U.S. grade may be determined from a smaller portion of a plug.

To determine eyes and texture as well as color characteristics, the wheel or block should be divided approximately in half, exposing two cut surfaces, for examination. The exposed cut surfaces of institutional-size packages should be used to determine eye and texture as well as color characteristics.

The final U.S. grade should be established on the basis of the lowest rating of any one of the quality factors.

29.8.4 Specifications for U.S. Grades

29.8.4.1 U.S. Grade A Swiss Cheese

The cheese should conform to the following requirements (Tables 29.1–29.5):

1. *Flavor*: Should be a pleasing and desirable characteristic Swiss cheese flavor, consistent with the age of the cheese, and free from undesirable flavors.
2. *Body*: Should be uniform, firm, and smooth.
3. *Eyes and texture*: The cheese should be properly set and should possess well-developed round or slightly oval-shaped eyes that are relatively uniform in size and distribution. The majority of the eyes should be 3/8–13/16 inch in diameter. The cheese may possess the following eye characteristics to a very slight degree: dull, rough, and shell; and the following texture characteristics to a very slight degree: checks, picks, and streuble.
4. *Finish and appearance*

    Rinded. The rind should be sound, firm, and smooth, providing good protection to the cheese. The surface of the cheese may exhibit mold to a very slight degree. There should be no indication that the mold has penetrated into the interior of the cheese.

TABLE 29.1

Classification of Flavor for Swiss Cheese

Identification of Flavor Characteristics	U.S. Grade		
	A	B	C
Acid	–	S	D
Barny	–	–	S
Bitter	–	S	D
Feed	–	S	D
Flat	–	S	D
Fruity	–	–	S
Metallic	–	–	S
Old milk	–	–	S
Onion	–	–	S
Rancid	–	–	S
Sour	–	–	S
Utensil	–	S	D
Weedy	–	–	S
Whey-taint	–	–	S
Yeasty	–	–	S

Note: –, Not permitted; S, slight; D, definite.

TABLE 29.2

Classification of Body for Swiss Cheese

Identification of Body Characteristics	U.S. Grade		
	A	**B**	**C**
Coarse	–	–	S
Pasty	–	–	S
Short	–	–	S
Weak	–	S	D

Note: –, Not permitted; S, slight; D, definite.

Rindless. Rindless blocks of Swiss cheese should be reasonably uniform in size and well-shaped. The wrapper or covering should adequately and securely envelop the cheese, be neat, be unbroken, and fully protect the surface of the cheese, but may be slightly wrinkled. The surface of the cheese may exhibit mold to a very slight degree. There should be no indication that the mold has penetrated into the interior of the cheese.

5. *Color:* Should be natural, attractive, and uniform. The cheese should be white to light yellow in color.

TABLE 29.3

Classification of Eyes and Texture for Swiss Cheese (for Evaluations of Cut Surfaces)

Identification of Eyes and Texture Characteristics	U.S. Grade		
	A	**B**	**C**
Afterset	–	–	S
Cabbage	–	–	S
Checks	VS	S	D
Collapsed	–	–	S
Dead	–	VS	D
Dull	VS	S	D
Frog mouth	–	S	D
Gassy	–	–	S
Irregular	–	–	S
Large eyed	–	–	S
Nesty	–	VS	D
One sided	–	S	D
Overset	–	S	S
Picks	VS	S	D
Rough	VS	S	D
Shell	VS	S	D
Small eyed	–	–	S
Splits	–	–	S
Streuble	VS	S	D
Sweet holes	–	–	S
Underset	–	S	D
Uneven	–	S	D

Note: –, Not permitted; VS, very slight; S, slight; D, definite.

TABLE 29.4

Classification of Finish and Appearance for Swiss Cheese

Identification of Finish and Appearance Characteristics	U.S. Grade		
	A	**B**	**C**
Checked rind	–	–	S
Huffed	–	S	D
Mold on rind surface	VS	S	D
Mold under wrapper or covering	VS	S	D
Soft spots	–	–	S
Soiled surface (rinded)	–	S	D
Soiled surface (rindless)	–	–	VS
Uneven	–	S	D
Wet rind	–	S	D
Wet surface (rindless)	–	S	D

Note: –, Not permitted; VS, very slight; S, slight; D, definite.

29.8.4.2 U.S. Grade B Swiss Cheese

The cheese should conform to the following requirements (Tables 29.1–29.5):

1. *Flavor*: Should be a pleasing and desirable characteristic Swiss cheese flavor, consistent with the age of the cheese, and free from undesirable flavors. The cheese may possess the following flavors to a slight degree: acid, bitter, feed, flat, and utensil.
2. *Body*: Should be uniform, firm, and smooth. The cheese may possess a slight weak body.
3. *Eyes and texture*: The cheese should possess well-developed round or slightly oval-shaped eyes. The majority of the eyes should be 3/8 to 13/16 inch in diameter. The cheese may possess the following eye characteristics to a very slight degree: dead eyes and nesty; and the following to a slight degree: dull, frog mouth, one sided, overset, rough, shell, underset, and uneven. The cheese may possess the following texture characteristics to a slight degree: checks, picks, and streuble.
4. *Finish and appearance*

   Rinded. The rind should be sound, firm, and smooth, providing good protection to the cheese. The cheese may exhibit the following characteristics to a slight degree: huffed, mold, soiled, uneven, and wet rind. There should be no indication that the mold has penetrated into the interior of the cheese.

TABLE 29.5

Classification of Color for Swiss Cheese

Identification of Color Characteristics	U.S. Grade		
	A	**B**	**C**
Acid-cut	–	–	S
Bleached surface	–	S	D
Colored spots	–	–	S
Dull or faded	–	–	S
Mottled	–	–	S
Pink ring	–	–	S

Note: –, Not permitted; S, slight; D, definite.

Rindless. The wrapper or covering of rindless blocks of Swiss cheese should adequately and securely envelop the cheese, be neat, be unbroken, and fully protect the surface, but may be slightly wrinkled. The cheese may exhibit the following characteristics to a slight degree: huffed, mold, uneven, and wet surface. There should be no indication that the mold has penetrated into the interior of the cheese.

5. *Color:* The cheese should be white to light yellow. The cheese may possess to a slight degree a bleached surface.

29.8.4.3 U.S. Grade C Swiss Cheese

The cheese should conform to the following requirements (Tables 29.1–29.5):

1. *Flavor:* Should possess a characteristic Swiss cheese flavor that is consistent with the age of the cheese. The cheese may possess the following flavors to a slight degree: barny, fruity, metallic, old milk, onion, rancid, sour, weedy, whey-taint, and yeasty; and the following to a definite degree: acid, bitter, feed, flat, and utensil.
2. *Body:* Should be uniform and may possess the following characteristics to a slight degree—coarse, pasty, and short; and to a definite degree, the cheese may be weak.
3. *Eyes and texture:* The cheese may possess the following eye characteristics to a slight degree—afterset, cabbage, collapsed, irregular, large eyed, and small eyed; and the following to a definite degree—dead eyes, dull, frog mouth, nesty, one sided, overset, rough, shell, underset, and uneven. The cheese may possess the following texture characteristics to a slight degree: gassy, splits, and sweet holes; and the following to a definite degree: checks, picks, and streuble.
4. *Finish and appearance*

     Rinded. The rind should be sound, providing good protection to the cheese. The cheese may exhibit the following characteristics to a slight degree: checked rind and soft spots; and the following to a definite degree: huffed, mold, soiled, uneven, and wet rind. There should be no indication that the mold has penetrated into the interior of the cheese.

     Rindless. The wrapper or covering should adequately and securely envelop the cheese, should be unbroken, should fully protect the surface, and may be wrinkled. The cheese may exhibit a very slight soiled surface and contain soft spots to a slight degree. The cheese may possess the following characteristics to a definite degree: huffed, mold, uneven, and wet surface. There should be no indication that the mold has penetrated into the interior of the cheese.
5. *Color:* The cheese may possess the following color characteristics to a slight degree—acid-cut, colored spots, dull or faded, mottled, and pink ring; and to a definite degree—bleached surface.

29.8.4.4 U.S. Grade Not Assignable

Swiss cheese should not be assigned a U.S. grade for one or more of the following reasons:

1. Fails to meet or exceed the requirements for U.S. Grade C
2. Fails to meet composition, minimum age, or other requirements of the FDA
3. Produced in a plant found on inspection to be using unsatisfactory manufacturing practices, equipment, or facilities, or to be operating under unsanitary plant conditions

29.8.5 Explanation of Terms

With respect to style:

1. *Rinded:* Cheese that has a hard protective outer layer formed by drying the cheese surface and by the addition of salt (usually wheel shaped)

2. *Rindless*: Cheese that has been protected from rind formation and that is packaged with an impervious type of wrapper or covering enclosing the cheese (usually cube or rectangular shaped)

3. *Institutional-size packages*: Multipound, wrapped portions of cheese, generally cut from a larger piece, intended for use by restaurants, delicatessens, schools, etc.

With respect to flavor:

1. *Slight*: Detected only upon critical examination.
2. *Definite*: Not intense but detectable.
3. *Undesirable*: Identifiable flavors in excess of the intensity permitted, or those flavors not listed.
4. *Acid*: Sharp and puckery to the taste, characteristic of lactic acid.
5. *Barny*: A flavor characteristic of the odor of a cow stable.
6. *Bitter*: A distasteful flavor similar to the taste of quinine.
7. *Feed*: Feed flavors (such as alfalfa, sweet clover, silage, or similar feed) in milk carried through into the cheese.
8. *Flat*: Insipid, practically devoid of any characteristic Swiss cheese flavor.
9. *Fruity*: A sweet fruit-like flavor resembling apples; generally increasing in intensity as the cheese ages.
10. *Rancid*: A flavor suggestive of rancidity or butyric acid, sometimes associated with bitterness.
11. *Metallic*: A flavor having qualities suggestive of metal, imparting a puckery sensation.
12. *Old milk*: Lacks freshness.
13. *Onion*: This flavor is recognized by the peculiar taste and odor suggestive of its name. Present in milk or cheese when the cows have eaten onions, garlic, or leeks.
14. *Sour*: An acid, pungent flavor resembling vinegar.
15. *Utensil*: A flavor that is suggestive of improper or inadequate washing and sanitizing of milking machines, utensils, or factory equipment.
16. *Weedy*: A flavor due to the use of milk that possesses a common weedy flavor. Present in cheese when cows have eaten weedy feed or grazed on common weed-infested pastures.
17. *Whey-taint*: A slightly acid taste and odor characteristic of fermented whey, caused by too slow expulsion of whey from the curd.
18. *Yeasty*: A flavor indicating yeast fermentation.

With respect to body:

1. *Slight*: Detected only upon critical examination
2. *Definite*: Not intense but detectable
3. *Smooth*: Feels silky; not dry and coarse or rough
4. *Firm*: Feels solid, not soft or weak
5. *Coarse*: Feels rough, dry, and sandy
6. *Pasty*: Usually weak body and when the cheese is rubbed between the thumb and fingers, it becomes sticky and smeary
7. *Short*: No elasticity to the plug when rubbed between the thumb and fingers
8. *Uniform*: Not variable
9. *Weak*: Requires little pressure to crush; is soft but is not necessarily sticky like pasty cheese

With respect to eyes and texture in general:

1. *Blind*: No eye formation present
2. *Set*: The number of eyes in any given area of cheese

3. *Well-developed eyes*: Eyes perfectly developed, glossy or velvety, with smooth even walls, round or slightly oval in shape, and fairly uniform in distribution throughout the cheese

With respect to eyes and texture as it relates to cabbage, collapsed, dead, dull, frog mouth, irregular, rough, and shell:

1. *Very slight*: Characteristic exhibited in less than 5% of the eyes
2. *Slight*: Characteristic exhibited in 5% or more but less than 10% of the eyes
3. *Definite*: Characteristic exhibited in 10% or more but less than 20% of the eyes
4. *Cabbage*: Cheese having eyes so numerous within the major part of the cheese that they crowd each other, leaving only a paper-thin layer of cheese between the eyes, causing the cheese to have a cabbage appearance and very irregular eyes
5. *Collapsed*: Eyes that have not formed properly and do not appear round or slightly oval but rather flattened and appear to have collapsed
6. *Dead*: Developed eyes that have completely lost their glossy or velvety appearance
7. *Dull*: Eyes that have lost some of their bright shiny luster
8. *Frog mouth*: Eyes that have developed into a lenticular or spindle-shaped opening
9. *Irregular*: Eyes that have not formed properly and do not appear round or slightly oval and that are not accurately described by other more descriptive terms
10. *Rough*: Eyes that do not have smooth, even walls
11. *Shell*: A rough nut shell appearance on the wall surface of the eyes

With respect to eyes and texture as it relates to streuble:

1. *Very slight*: Extends no more than 1/8 inch into the body of the cheese
2. *Slight*: Extends 1/8 inch or more but less than 1/4 inch into the body of the cheese
3. *Definite*: Extends 1/4 inch or more but less than 1/2 inch into the body of the cheese
4. *Streuble:* An overabundance of small eyes just under the surface of the cheese

With respect to eyes and texture as it relates to checks, picks, and splits:

1. *Very slight*: Infrequent occurrence, not more than 1 inch from the surface
2. *Slight*: Limited occurrence, not more than 1 inch from the surface
3. *Definite*: Limited occurrence throughout the cheese
4. *Checks*: Small, short cracks within the body of the cheese
5. *Picks*: Small irregular or ragged openings within the body of the cheese
6. *Splits*: Sizable cracks, usually in parallel layers and usually clean cut, found within the body of the cheese

With respect to eyes and texture as it relates to large eyed and small eyed:

1. *Slightly large eyed*: Majority of the eyes more than 13/16 inch but less than 1 inch
2. *Slightly small eyed*: Majority of the eyes less than 3/8 inch but more than 1/8 inch
3. *Relatively uniform eye size*: The majority of the eyes fall within a 1/4 inch range

With respect to eyes and texture as it relates to gassy and sweet holes:

1. *Slight*: No more than three occurrences per any given 2 square inches
2. *Gassy*: Gas holes of various sizes that may be scattered
3. *Sweet holes*: Spherical gas holes, glossy in appearance; usually about the size of BB shot

With respect to eyes and texture as it relates to nesty:

1. *Very slight*: Occurrence limited to no more than 5% of the exposed cut area of the cheese
2. *Slight*: Occurrence more than 5% but less than 10% of the exposed cut area of the cheese
3. *Definite*: Occurrence more than 10% but less than 20% of the exposed cut area of the cheese
4. *Nesty*: An overabundance of small eyes in a localized area

With respect to eyes and texture as it relates to one sided and uneven:

1. *Slight*: Eyes evenly distributed throughout at least 90% of the total cheese area
2. *Definite*: Eyes evenly distributed throughout at least 75% but less than 90% of the total cheese area
3. *One sided*: Cheese that is reasonably developed on one side and underdeveloped on the other as to eye development
4. *Uneven*: Cheese that is reasonably developed in some areas and underdeveloped in others as to eye development

With respect to eyes and texture as it relates to afterset, overset, and underset:

1. *Very slight*: Number of eyes present exceed or fall short of the ideal by limited amount
2. *Slight*: Number of eyes present exceed or fall short of the ideal by a moderate amount
3. *Afterset*: Small eyes caused by secondary fermentation
4. *Overset*: Excessive number of eyes present
5. *Underset*: Too few eyes present

With respect to finish and appearance:

1. *Very slight*: Detected only upon very critical examination
2. *Slight*: Detected only upon critical examination
3. *Definite*: Not intense but detectable
4. *Checked rind*: Numerous small cracks or breaks in the rind
5. *Huffed*: The cheese becomes rounded or oval in shape instead of flat
6. *Mold on rind surface*: Mold spots or areas that have formed on the rind surface
7. *Mold under wrapper or covering*: Mold spots or area that have formed under the wrapper or on the cheese
8. *Soft spots*: Spots that are soft to the touch and usually faded and moist
9. *Soiled surface*: Milkstone, rust spots, grease, or other discoloration on the surface of the cheese
10. *Uneven*: One side of the cheese is higher than the other
11. *Wet rind*: A wet rind is one in which the moisture adheres to the surface of the rind and which may or may not soften the rind or cause discoloration
12. *Wet surface (rindless)*: A wet surface is one in which the moisture appears between the wrapper and the cheese surface

With respect to color:

1. *Slight*: Detectable only upon critical examination
2. *Definite*: Not intense but detectable
3. *Acid-cut*: Bleached or faded appearance that sometimes varies throughout the cheese

4. *Bleached surface*: A faded coloring beginning at the surface and extending inward a short distance

5. *Colored spots*: Brightly colored areas (pink to brick red or gray to black) of bacteria growing in readily discernible colonies randomly distributed throughout the cheese

6. *Dull or faded*: A color condition lacking in luster

7. *Mottled*: Irregular-shaped spots or blotches in which portions are light colored and others are higher colored; unevenness of color due to combining two different vats sometimes referred to as "mixed curd"

8. *Pink ring*: A color condition that usually appears pink to brownish red and occurs as a uniform band near the cheese surface and may follow eye formation

29.9 Supplement to USDA Grade Standards for Swiss and Emmentaler Cheeses

29.9.1 Alternate Method for Determination of U.S. Grades

This alternate method should be used only when requested by the applicant. With this method, the eyes and texture and color factors are rated on the basis of trier plugs rather than by slicing the cheese. A statement should appear on the grading certificate indicating that the alternate method was used as requested by the applicant.

The following quality factors should be rated when using the alternate method for determining U.S. grades:

1. Flavor
2. Body
3. Eyes and texture
4. Finish and appearance
5. Color

Flavor and body ratings should be determined by the methods prescribed by the USDA earlier. Finish and appearance ratings should be determined as prescribed by the USDA earlier.

Eyes and texture and color ratings should be determined by drawing and examining at least two full trier plugs, withdrawn at the approximate center of one of the largest flat surface areas of the sample. For some institutional-size samples, it may not be possible to obtain a full trier plug. When this occurs, a U.S. grade may be determined from a smaller portion of a plug.

The final U.S. grade should be established on the basis of the lowest rating of any one quality factor.

29.9.2 Specifications for U.S. Grades When Using the Alternate Method

29.9.2.1 U.S. Grade A Swiss Cheese

U.S. Grade A Swiss cheese should conform to the following requirements (Tables 29.1–29.5):

Eyes and texture: The cheese should be properly set and should possess well-developed round or slightly oval-shaped eyes that are relatively uniform in size and in distribution. A full plug drawn from the cheese should be free from splits and not appear gassy or large eyed; it may possess checks and picks within 1 inch from the surface and may possess a limited number of checks and picks beyond 1 inch from the surface. The majority of the eyes should be 3/8 to 13/16 inch in diameter. The cheese should have at least two but not more than eight eyes to a trier plug.

Color: Should be natural, attractive, and uniform. The cheese should be white to light yellow in color.

29.9.2.2 U.S. Grade B Swiss Cheese

The cheese should conform to the following requirements (Tables 29.1–29.5):

Eyes and texture: The cheese should possess well-developed round or slightly oval-shaped eyes. A full plug drawn from the cheese should be free from splits and not appear gassy or large eyed and may be moderately overset and have a limited amount of checks and picks. The majority of the eyes should be in the range of 3/8 to 13/16 inch in diameter. The cheese should have at least one but not more than 10 eyes to a trier plug.

Color: The cheese should be white to light yellow in color. The cheese may possess, to a slight degree, a bleached surface.

29.9.2.3 U.S. Grade C Swiss Cheese

The cheese should conform to the following requirements (Tables 29.1–29.5):

Eyes and texture: A full plug drawn from the cheese may be overset, shell, or dead eyed; may have splits, checks, picks, and gassy; and may be large-eyed to a slight degree. The cheese is not totally blind or totally gassy.

Color: The cheese may possess the following color characteristics to a slight degree—acid-cut, colored spots, dull or faded, mottled, and pink ring; and to a definite degree—bleached surface.

30

U.S. Dairy Processing Plants: Safety and Inspection*

Y. H. Hui

CONTENTS

---

* The information in this chapter has been modified from *Food Safety Manual* published by Science Technology System (STS) of West Sacramento, CA. Copyrighted 2012©. With permission.

30.1 Introduction

Food manufacturers in the United States must produce a product that consumers like in order to sell the product nationally. To achieve this goal, they depend on developments in science, technology, and engineering in order to produce a quality product. However, they must also pay attention to two other important considerations:

1. Are their products safe for public consumption?
2. Have they assured the economic integrity of their products? This means many things. The most important one is the assurance that their products contain the proper quality and quantity of ingredients, that is, "no cheating."

To protect the consumers, U.S. government agencies have established many checks and balances to prevent "injuries" to the consumers in terms of health and economic fraud. The two major responsible agencies are

1. U.S. Food and Drug Administration (FDA, www.FDA.gov)
2. U.S. Department of Agriculture (USDA, www.USDA.gov)

The FDA establishes standards of identities for a number of cheeses for interstate commerce. This makes sure that a particular type of cheese contains the minimal quality and quantity of the ingredients used to make this cheese. The FDA also issues safety requirements and recommendations in the manufacturing of cheeses. It makes periodic inspections of establishments producing cheeses to ensure full compliance. In effect, the FDA is monitoring the safety production of cheeses and their economic integrities.

The USDA plays the same important role in the commerce of cheeses in the United States. For example, a national restaurant chain that makes and sells pizzas will not buy cheeses from any manufacturer if the cheeses do not bear the USDA official identification or seal of approval. The USDA has issued official requirements and processing specifications that cheese manufacturers must follow if they want to obtain the USDA seal of approval for the products. To achieve this goal, the USDA sets up voluntary services of inspection under a standard fee arrangement. The inspection focuses on the safety and economic integrity of manufacturing cheeses. When the inspection demonstrates that the manufacturer has complied with safety and economic requirements, the USDA issues its official identification or seal of approval.

In essence, the FDA and USDA are the federal agencies that make sure that the product is safe and will not pose any economic fraud. Using the public information distributed at the Web sites of FDA and USDA, this chapter discusses USDA specifications and FDA inspection in relation to the production of selected cheeses.

The information in this chapter is important because it facilitates commerce of cheeses and protects consumers from injuries of eating cheeses in the market. It also protects the consumers from economic fraud. For example, without FDA and USDA intervention, one may purchase a cheese that contains less than the normal amount of milk used. If so, the consumers lose money in the transaction.

All FDA standards of identities for cheeses are provided in 21 CFR 131 and 133, and USDA dairy plant specifications are provided in 7 CFR 58. For explanation, 21 CFR 133 refers to United States Code of Federal Regulations, Part 133.

The basis of USDA dairy plant specifications is as follows.

30.1.1 Definitions

Bulk American cheese is American cheese that is packaged in bulk form. No single piece of cheese, whatever its shape, should weigh less than 100 lbs. American cheese includes the following varieties:

1. Cheddar cheese and cheddar cheese for manufacturing should conform to the provisions of 21 CFR 133.113 and 133.114, respectively, "Cheeses and Related Cheese Products," as issued by the FDA.
2. Washed curd cheese (soaked curd cheese) and washed curd cheese for manufacturing should conform to the provisions of 21 CFR 133.136 and 133.137, respectively, "Cheeses and Related Cheese Products," as issued by the FDA.
3. Granular cheese (stirred curd cheese) and granular cheese for manufacturing should conform to the provisions of 21 CFR 133.144 and 133.145, respectively, "Cheeses and Related Cheese Products," as issued by the FDA.
4. Colby cheese and colby cheese for manufacturing should conform to the provisions of 21 CFR 133.118 and 133.119, respectively, "Cheeses and Related Cheese Products," as issued by the FDA.

30.1.1.1 Packaging

The primary container (liner) should be new, in good condition, unbroken, fully protective of all surfaces of the cheese, and properly closed or sealed so as to protect the cheese from damage, contamination, or excessive desiccation. If the cheese is handled and stored in only a primary container after cooling, there should be a satisfactory system for cooling the cheese, retaining the desired shape, and providing reasonable protection of the cheese during transportation, storage, and handling.

The secondary container, when used, should be in good condition and should satisfactorily protect the cheese. The secondary container should be of such construction and be filled to a sufficient level so as not to cause handling, stacking, or storage problems. If antimycotics are used, they should be used in accordance with the provisions of FDA.

30.1.1.2 Degree of Curing

1. *Fresh (current)*—Cheese that is at the early stages of the curing process, usually 10 to about 90 days old
2. *Cured (aged)*—Cheese that has the more fully developed flavor and body attributes that are characteristic of the curing process, generally over 90 days old

30.1.2 U.S. Grades

30.1.2.1 Nomenclature of U.S. Grades

The nomenclature of U.S. grades is as follows:

1. U.S. Extra Grade
2. U.S. Standard Grade
3. U.S. Commercial Grade

30.1.2.2 Basis for Determination of U.S. Grades

The determination of U.S. grades of bulk American cheese should be based on the rating of the following quality factors:

1. Flavor
2. Body and texture
3. Finish and appearance (as determined by examination of at least the filling end)

The rating of each quality factor should be established on the basis of characteristics present in a randomly selected sample representing a vat of cheese. If the cheese in a container is derived from more than one vat, the container labeling should so indicate by showing both vat numbers, and the grade should be determined on the basis of the lowest grade of either vat. The cheese should be graded no sooner than 10 days after being placed into the primary container.

The final U.S. grade should be established on the basis of the lowest rating of any one of the quality factors.

30.1.2.3 Specifications for U.S. Grades

U.S. Extra Grade should conform to the requirements of Tables 30.1–30.3. Flavor should be pleasing and characteristic of the variety and type of cheese (Table 30.1). Body and texture are characterized as follows. A sample drawn from the cheese should be firm and sufficiently compact to draw a plug for examination (Table 30.2). (For detailed specifications and classification of finish and appearance characteristics, see Table 30.3.)

U.S. Standard Grade should conform to the requirements in Tables 30.1–30.3. Flavor should be pleasing but may possess certain flavor defects to a limited degree (Table 30.1). For body and texture, the cheese should be sufficiently compact to draw a plug for examination; however, it may have large and

TABLE 30.1

Classification of Flavor with Corresponding U.S. Grade

Flavor Characteristics	Fresh or Current			Cured or Aged		
	Extra	**Standard**	**Commercial**	**Extra**	**Standard**	**Commercial**
Acid	S	D	P	S	D	P
Barny	–	S	D	–	S	D
Bitter	VS	S	D	VS	S	D
Feed	S	D	P	S	D	P
Flat	–	S	D	–	S	D
Fruity	–	S	D	VS	S	D
Malty	–	S	D	–	S	D
Metallic	–	–	VS	–	–	VS
Old milk	–	S	D	–	S	D
Onion	–	VS	S	–	VS	S
Rancid	–	S	D	–	S	D
Sour	–	–	VS	–	–	VS
Sulfide	–	S	D	VS	S	D
Utensil	–	S	D	–	S	D
Weedy	–	S	D	–	S	D
Whey-taint (whey)	–	S	D	VS	S	D
Yeasty	–	S	D	–	S	D

Note: –, Not permitted; S, slight; P, pronounced; VS, very slight; D, definite.

TABLE 30.2

Classification of Body and Texture with Corresponding U.S. Grade

Body and Texture Characteristics	Fresh or Current			Cured or Aged		
	Extra	Standard	Commercial	Extra	Standard	Commercial
Coarse	S	D	P	S	D	P
Corky	–	S	P	–	S	P
Crumbly	–	D	P	D	D	P
Curdy	D	D	P	S	D	P
Gassy	–	S	D	–	S	D
Mealy	S	D	P	S	D	P
Open[a]	S	P	P	S	P	P
Pasty	–	D	P	–	D	P
Pinny	–	VS	S	–	VS	S
Short	S	D	P	S	D	P
Slitty	–	S	D	–	S	D
Sweet holes	–	D	P	S	D	P
Weak	S	D	P	S	D	P

Note: –, Not permitted; S, slight; P, pronounced; VS, very slight; D, definite.

[a] Not applicable for colby cheese.

TABLE 30.3

Classification of Finish and Appearance with Corresponding U.S. Grade (as Determined by Examination of at Least the Filling End)

Finish and Appearance Characteristics	Fresh or Current			Cured or Aged		
	Extra	Standard	Commercial	Extra	Standard	Commercial
Free whey	S	D	P	–	S	D
Mold	–	D	D	S	D	D
Rough surface[2]	S	D	P	S	D	P
Rough surface[3]	D	P	P	D	P	P
Soiled surface	S	D	P	S	D	P

Note: –, Not permitted; S, slight; P, pronounced; D, definite.

connecting mechanical openings. In addition to four sweet holes, the plug sample may have scattered yeast holes and other scattered gas holes. (For additional detailed specifications and classification of body and texture characteristics, see Table 30.2. For detailed specifications and classification of finish and appearance characteristics, see Table 30.3.)

U.S. Commercial Grade should conform to the requirements in Tables 30.1–30.3. The cheeses may possess certain flavor defects to specified degrees. (For detailed specifications and classification of flavor characteristics, see Table 30.1.) A plug drawn from the cheese may appear loosely knit with large and connecting mechanical openings. (For detailed specifications and classification of body and texture characteristics, see Table 30.2. For detailed specifications and classification of finish and appearance characteristics, see Table 30.3.)

30.1.2.4 U.S. Grade Not Assignable

Bulk American cheese should not be assigned a U.S. grade for one or more of the following reasons:

1. Fails to meet or exceed the requirements for U.S. Commercial Grade
2. Produced in a plant that is rated ineligible for USDA grading service
3. Produced in a plant that is not USDA approved

30.1.3 Explanation of Terms

The following provides explanation for some terms.

With respect to flavor:

1. *Very slight*—Detected only upon very critical examination
2. *Slight*—Detected only upon critical examination
3. *Definite*—Not intense but detectable
4. *Pronounced*—So intense as to be easily identified
5. *Undesirable*—Identifiable flavors in excess of the intensity permitted, or those flavors not listed
6. *Acid*—Sharp and puckery to the taste; characteristic of lactic acid
7. *Barny*—A flavor characteristic of the odor of a cow barn
8. *Bitter*—Distasteful, similar to the taste of quinine
9. *Feed*—Feed flavors (such as alfalfa, sweet clover, silage, or similar feed) in milk carried through into the cheese
10. *Flat*—Insipid, practically devoid of any characteristic cheese flavor
11. *Fruity*—A fermented fruit-like flavor resembling apples
12. *Malty*—A distinctive, harsh flavor suggestive of malt
13. *Metallic*—A flavor having qualities suggestive of metal, imparting a puckery sensation
14. *Old milk*—Lacks freshness
15. *Onion*—A flavor recognized by the peculiar taste and aroma suggestive of its name; present in milk or cheese when cows have eaten onions, garlic, or leeks
16. *Rancid*—A flavor suggestive of rancidity or butyric acid, sometimes associated with bitterness
17. *Sour*—An acidly pungent flavor resembling vinegar
18. *Sulfide*—A flavor of hydrogen sulfide, similar to the flavor of water with a high sulfur content
19. *Utensil*—A flavor that is suggestive of improper or inadequate washing and sterilization of milking machines, utensils, or factory equipment
20. *Weedy*—A flavor present in cheese when cows have eaten weedy hay or grazed on weed-infested pasture
21. *Whey-taint (whey)*—A slightly acid flavor and odor characteristic of fermented whey caused by too slow expulsion of whey from the curd
22. *Yeasty*—A flavor indicating yeasty fermentation

With respect to body and texture:

1. *Very slight*—An attribute that is detected only upon very critical examination and present only to a minute degree
2. *Slight*—An attribute that is barely identifiable and present only to a small degree
3. *Definite*—An attribute that is readily identifiable and present to a substantial degree
4. *Pronounced*—An attribute that is markedly identifiable and present to a large degree
5. *Curdy*—Smooth but firm; rubbery and not waxy or broken down when worked between the fingers
6. *Coarse*—Feels rough, dry, and sandy
7. *Corky*—Hard, tough, and overfirm cheese that does not readily break down when rubbed between the thumb and fingers
8. *Crumbly*—Tends to fall apart when rubbed between the thumb and fingers
9. *Gassy*—Gas holes of various sizes, which may be scattered
10. *Mealy*—Short body, does not mold well; looks and feels like corn meal when rubbed between the thumb and fingers

11. *Open*—Mechanical openings that are irregular in shape and are caused by workmanship and not by gas fermentation
12. *Pasty*—Usually weak body; when the cheese is rubbed between the thumb and fingers, it becomes sticky and smeary
13. *Pinny*—Numerous very small gas holes
14. *Short*—No elasticity in the cheese plug; when rubbed between the thumb and fingers, it tends toward mealiness
15. *Slitty*—Narrow elongated slits generally associated with a cheese that is gassy or yeasty; sometimes referred to as "fish-eyes"
16. *Sweet holes*—Spherical gas holes, glossy in appearance; usually about the size of BB shots; also known as shot holes
17. *Weak*—Requires little pressure to crush; soft but not necessarily sticky like a pasty cheese

With respect to finish and appearance:

1. *Free whey*—Whey or moisture that comes from the cheese or has not been incorporated into the curd. The free whey determination should be made on the basis of whey or moisture on the cheese or liner. The intensity is described as slight when droplets are easily detected, definite when the droplets are readily identifiable and run together, and pronounced when the droplets run together and pool.
2. *Mold*—Mold spots or areas that have formed on the surface of the cheese. The intensity is described as very slight when the total top surface area covered with mold is not greater than a dime; slight when the area covered is not greater than 10 dimes; definite when the area is more than slight but not greater than 1/4 of the top surface area; and pronounced when greater than 1/4 of the top surface area.
3. *Rough surface*—Lacks smoothness. The intensity is described as slight when the defect is easily detected, definite when readily detected, and pronounced when obvious.
4. *Soiled surface*—Discoloration on the surface of the cheese due to poor production or handling practices. The intensity is described as slight when the defect is detected upon critical examination, definite when easily detectable, and pronounced when easily identified and covers more than 1/2 of the surface.

30.2 Establishment Processing Cheese

Cheese is the fresh or matured product obtained by draining after coagulation of milk, cream, skimmed milk, partly skimmed milk, or a combination of some or all of these products and including any cheese that conforms to the requirements of the FDA. Milkfat from whey is the fat obtained from the separation of cheese whey.

30.2.1 Rooms and Compartments

A separate starter room or properly designed starter tanks and satisfactory air movement techniques should be provided for the propagation and handling of starter cultures. All necessary precautions should be taken to prevent contamination of the facility, equipment, and the air therein. A filtered air supply with a minimum average efficiency of 90% should be provided so as to obtain outward movement of air from the room to minimize contamination.

The rooms in which the cheese is manufactured should be of adequate size, and the equipment should be adequately spaced to permit movement around the equipment for proper cleaning and satisfactory working conditions. Adequate filtered air ventilation should be provided. When applicable, the mold count should be not more than 15 colonies per plate during a 15-min exposure.

A brine room, when applicable, should be a separate room constructed so it can be readily cleanable. The brine room equipment should be maintained in good repair and corrosion kept at a minimum.

When applicable, a drying room of adequate size should be provided to accommodate the maximum production of cheese during the flush period. Adequate shelving and air circulation should be provided for proper drying. Temperature and humidity control facilities should be provided, which will promote the development of a sound, dry surface of the cheese.

When applicable for rind cheese, a separate room or compartment should be provided for paraffining and boxing the cheese. The room should be of adequate size, and the temperature should be maintained near the temperature of the drying room to avoid sweating of the cheese prior to paraffining.

For rindless cheese, a suitable space should be provided for proper wrapping and boxing of the cheese. The area should be free from dust, condensation, mold, or other conditions that may contaminate the surface of the cheese or contribute to unsatisfactory packaging of the cheese.

Coolers or curing rooms where cheese is held for curing or storage should be clean and maintained at the proper uniform temperature and humidity to adequately protect the cheese and minimize the undesirable growth of mold. Proper circulation of air should be maintained at all times. The shelves should be kept clean and dry. This does not preclude the maintenance of suitable conditions for the curing of mold and surface ripened varieties.

When small packages of cheese are cut and wrapped, separate rooms should be provided for the cleaning and preparation of the bulk cheese and for the cutting and wrapping operation. The rooms should be well lighted, ventilated, and provided with filtered air. Air movement should be outward to minimize the entrance of unfiltered air into the cutting and packaging room. The waste materials and waste cheese should be disposed of in an environmentally and/or sanitary approved manner.

30.2.2 Equipment and Utensils

30.2.2.1 General Construction, Repair, and Installation

All equipment and utensils necessary to the manufacture of cheese and related products should meet standard general requirements. In addition, for certain other equipment, the following requirements should be met.

Bulk starter vats should be of stainless steel or equally corrosion-resistant metal and should be constructed according to the applicable sanitary standards. New or replacement vats should be constructed according to the applicable sanitary standards. The vats should be in good repair, equipped with tight-fitting lids, and have adequate temperature controls such as valves, indicating and/or recording thermometers.

The vats, tanks, and drain tables used for making cheese should be of metal construction with adequate jacket capacity for uniform heating. The inner liner should be minimum 16 gauge stainless steel or other equally corrosion-resistant metal, properly pitched from side to center and from rear to front for adequate drainage. The liner should be smooth, free from excessive dents or creases, and should extend over the edge of the outer jacket. The outer jacket should be constructed of stainless steel or other metal that can be kept clean and sanitary. The junction of the liner and outer jackets should be constructed so as to prevent milk or cheese from entering the inner jacket.

The vat, tank, and/or drain table should be equipped with a suitable sanitary outlet valve. Effective valves should be provided and properly maintained to control the application of heat to this equipment. If this equipment is provided with removable cloth covers, they should be clean.

The mechanical agitators should be of sanitary construction. The carriages should be of the enclosed type, and all product contact surfaces, shields, shafts, and hubs should be constructed of stainless steel or other equally corrosion-resistant metal. Metal blades, forks, or stirrers should be constructed of stainless steel and of approved material and rubber and rubber-like materials and should be free from rough or sharp edges that might scratch the equipment or remove metal particles.

The automatic curd making system should be constructed of stainless steel or of approved material. All areas should be free from cracks and rough surfaces and constructed so that they can be easily cleaned.

Curd conveying systems. The curd conveying system, conveying lines, and cyclone separator should be constructed of stainless steel or other equally corrosion-resistant metal and in such manner that it can be satisfactorily cleaned. The system should be of sufficient size to handle the volume of the curd and be provided with filtered air of the quality satisfactory for the intended use. Air compressors or vacuum pumps should not be located in the processing or packaging areas.

Automatic salter. The automatic salter should be constructed of stainless steel or other equally corrosion-resistant metal. This equipment should be constructed to equally distribute the salt throughout the curd. It should be designed to accurately weigh the amount of salt added. The automatic salter should be constructed so that it can be satisfactorily cleaned. The salting system should provide for adequate absorption of the salt in the curd. Water and steam used to moisten the curd prior to salting should be potable water or culinary steam.

Automatic curd filler. The automatic curd filler should be constructed of stainless steel or other equally corrosion-resistant metal. This equipment should be of sufficient size to handle the volume of the curd and constructed and controlled so as to accurately weigh the amount of the curd as it fills. The curd filler should be constructed so that it can be satisfactorily cleaned.

Hoop and barrel washer. The washer should be constructed so that it can be satisfactorily cleaned. It should also be equipped with temperature and pressure controls to ensure satisfactory cleaning of the hoops or barrels. It should be adequately vented to the outside.

30.2.2.2 Curd Mill and Miscellaneous Equipment

Knives, hand rakes, shovels, scoops, paddles, strainers, and miscellaneous equipment should be stainless steel or of material approved in the 3-A Sanitary Standards for Plastic and Rubber-Like Material. The product contact surfaces of the curd mill should be of stainless steel. All pieces of equipment should be so constructed that they can be kept clean and free from rough or sharp edges that might scratch the equipment or remove metal particles. The wires in the curd knives should be stainless steel, kept tight, and replaced when necessary.

30.2.2.3 Hoops, Forms, and Followers

The hoops, forms, and followers should be constructed of stainless steel, heavy tinned steel, or other approved materials. If tinned, they should be kept tinned and free from rust. All hoops, forms, and followers should be kept in good repair. Drums or other special forms used to press and store cheese should be clean and sanitary.

30.2.2.4 Press

The cheese press should be constructed of stainless steel, all joints welded, and all surfaces, seams, and openings readily cleanable. The pressure device should be the continuous type. Press cloths should be maintained in good repair and in sanitary condition. Single-service press cloths should be used only once.

30.2.2.5 Brine Tank

The brine tank should be constructed of suitable non-toxic material and should be resistant to corrosion, pitting, or flaking. The brine tank should be operated so as to assure that the brine is clean, well circulated, and of the proper strength and temperature for the variety of cheese being made.

30.2.2.6 Cheese Vacuumizing Chamber

The vacuum chamber should be satisfactorily constructed and maintained so that the product is not contaminated with rust or flaking paint. An inner liner of stainless steel or other corrosion-resistant material should be provided.

30.2.2.7 Monorail

The monorail should be constructed so as to prevent foreign material from falling on the cheese or cheese containers.

30.2.2.8 Conveyor for Moving and Draining Block or Barrel Cheese

The conveyor should be constructed so that it will not contaminate the cheese and be easily cleaned. It should be installed so that the press drippings will not cause an environmental problem.

30.2.2.9 Rindless Cheese Wrapping Equipment

The equipment used to heat-seal the wrapper applied to rindless cheese should have square interior corners, reasonably smooth interior surface, and controls that provide uniform pressure and heat equally to all surfaces. The equipment used to apply shrinkable wrapping material to rindless cheese should operate to maintain the natural intended shape of the cheese in an acceptable manner have reasonably smooth surfaces on the cheese, and tightly adhere the wrapper to the surface of the cheese.

30.2.2.10 Paraffin Tanks

The metal tank should be adequate in size, have wood rather than metal racks to support the cheese, and have heat controls and an indicating thermometer. The cheese wax should be kept clean.

30.2.2.11 Specialty Equipment and Washing Machine

All product contact areas of specialty equipment should be constructed of stainless steel or of approved material. When used, the washing machine for cheese cloths and bandages should be of commercial quality and size or of sufficient size to handle the applicable load. It should be equipped with temperature and water level controls.

30.2.3 Quality Specifications for Raw Materials

30.2.3.1 Milk

The milk should be fresh, sweet, pleasing, and desirable in flavor and should meet USDA requirements. It may be adjusted by separating part of the fat from the milk or by adding one or more of the following dairy products: cream, skim milk, concentrated skim milk, non-fat dry milk, and water in a quantity sufficient to reconstitute any concentrated or dry milk used. Such dairy products should have originated from raw milk meeting USDA requirements.

30.2.3.2 Hydrogen Peroxide

The solution should comply with the specification of the U.S. Pharmacopeia, except that it may exceed the concentration specified therein and it does not contain added preservative. Application and usage should be as specified in the "Definitions and Standards of Identity for Cheese and Cheese Products," FDA.

30.2.3.3 Catalase

The catalase preparation should be a stable, buffered solution, neutral in pH, having a potency of not less than 100 Keil units per milliliter. The source of the catalase, its application, and usage should be as specified in the "Definitions and Standards of Identity for Cheese and Cheese Products," FDA.

30.2.3.4 Cheese Cultures

Harmless microbial cultures used in the development of acid and flavor components in cheese should have pleasing and desirable taste and odor and should have the ability to actively produce the desired results in the cheese during the manufacturing process.

30.2.3.5 Calcium Chloride

Calcium chloride, when used, should meet the requirements of the Food Chemical Codex.

30.2.3.6 Color

Coloring, when used, should be annatto or any cheese or butter color that meets FDA requirements.

30.2.3.7 Rennet, Pepsin, Other Milk-Clotting Enzymes, and Flavor Enzymes

Enzyme preparations used in the manufacture of cheese should be safe and suitable.

30.2.3.8 Salt

The salt should be free flowing, white refined sodium chloride and should meet the requirements of the Food Chemical Codex.

30.2.4 Operations and Operating Procedures

30.2.4.1 Cheese from Pasteurized Milk

If the cheese is labeled as pasteurized, the milk should be pasteurized by subjecting every particle of milk to a minimum temperature of 161°F for not less than 15 s or by any other acceptable combination of temperature and time treatment approved by the administrator. High-temperature short-time (HTST) pasteurization units should be equipped with the proper controls and equipment to assure pasteurization. If the milk is held more than 2 h between the time of pasteurization and setting, it should be cooled to 45°F or lower until the time of setting.

30.2.4.2 Cheese from Unpasteurized Milk

If the cheese is labeled as "heat treated," "unpasteurized," "raw milk," or "for manufacturing," the milk may be raw or heated at temperatures below pasteurization. Cheese made from unpasteurized milk should be cured for a period of 60 days at a temperature not less than 35°F. If the milk is held more than 2 h between time of receipt or heat treatment and setting, it should be cooled to 45°F or lower until the time of setting.

30.2.4.3 Make Schedule

A uniform schedule should be established and followed as closely as possible for the various steps of setting, cutting, cooking, draining the whey, and milling the curd to promote a uniform quality of cheese.

30.2.4.4 Records

Starter and make records should be kept for at least 3 months.

30.2.4.5 Laboratory and Quality Control Tests

30.2.4.5.1 Chemical Analyses

1. *Milkfat and moisture.* One sample should be tested from each vat of the finished cheese to assure compliance with composition requirements.
2. *Test method.* Chemical analysis should be made in accordance with prescribed methods.

30.2.4.5.2 Weight or Volume Control

Representative samples of the finished product should be checked during the packaging operation to assure compliance with the stated net weight on the container of consumer-size packages.

30.2.4.6 Whey Handling

1. Adequate sanitary facilities should be provided for the handling of whey. If outside, necessary precautions should be taken to minimize flies, insects, and development of objectionable odors.
2. Whey or whey products intended for human food should at all times be handled in a sanitary manner in accordance with the procedures of this subpart as specified for handling milk and dairy products.
3. Milkfat from whey should not be more than 4 days old when shipped.

30.2.5 Packaging and Repackaging

Packaging rindless cheese or cutting and repackaging all styles of bulk cheese should be conducted under rigid sanitary conditions. The atmosphere of the packaging rooms, the equipment, and the packaging material should be practically free from mold and bacterial contamination.

When officially graded bulk cheese is to be repackaged into consumer-type packages with official grade labels or other official identification, a supervisor of packaging should be required. If the repackaging is performed in a plant other than the one in which the cheese is manufactured and the product is officially identified, the plant, equipment, facilities, and personnel should meet the same requirements as outlined in 21 CFR 133.

30.2.6 General Identification

Bulk cheese for cutting and the container for cheese for manufacturing should be legibly marked with the name of the product, code or date of manufacture, vat number, officially designated code number, or name and address of the manufacturer. Each consumer-sized container should meet the applicable regulations of the FDA.

30.2.7 USDA Official Identification Requirements

The quality requirements for cheddar cheese should be in accordance with the U.S. Standards for Grades of Cheddar Cheese. The quality requirements for colby cheese should be in accordance with the U.S. Standards for Grades of Colby Cheese. The quality requirements for monterey (monterey jack) cheese should be in accordance with the U.S. Standards for Grades of Monterey (Monterey Jack) Cheese. The quality requirements for Swiss cheese, Emmentaler cheese, should be in accordance with the U.S. Standards for Grades of Swiss Cheese, Emmentaler Cheese. The quality requirements for bulk American cheese for manufacturing should be in accordance with the U.S. Standards for Grades of Bulk American Cheese for Manufacturing.

30.3 Establishment Processing Pasteurized Process Cheese and Related Products

The USDA specifications for plant manufacturing, processing, and packaging pasteurized process cheese and related products are as follows.

30.3.1 Definitions

1. *Pasteurized process cheese and related products.* Pasteurized process cheese and related products are foods that conform to the applicable requirements of the FDA for cheeses and related cheese products.
2. *Blend setup.* The trade term for a particular group of vat lots of cheese selected to form a blend based upon their combined ability to impart the desired characteristics to a pasteurized process cheese product.
3. *Cooker batch.* The amount of cheese and added optional ingredients placed into a cooker at one time, heated to pasteurization temperature, and held for the required length of time.

30.3.2 Equipment and Utensils

The equipment and utensils used for the handling and processing of cheese products should be as specified by the USDA. In addition, for certain other equipment, the following requirements should be met.

30.3.2.1 Conveyors

Conveyors should be constructed of material that can be properly cleaned and will not rust, or otherwise contaminate the cheese, and should be maintained in good repair.

30.3.2.2 Grinders or Shredders

The grinders or shredders used in the preparation of the trimmed and cleaned cheese should be of corrosion-resistant material and of such construction as to prevent contamination of the cheese and to allow thorough cleaning of all parts and product contact surfaces.

30.3.2.3 Cookers

The cookers should be the steam jacketed or direct steam type. They should be constructed of stainless steel or other equally corrosion-resistant material. All product contact surfaces should be readily accessible for cleaning. Each cooker should be equipped with an indicating thermometer and should be equipped with a temperature recording device. The recording thermometer stem may be placed in the cooker if satisfactory time charts are obtained; if not, the stem should be placed in the hotwell or filler hopper. Steam check valves on direct steam-type cookers should be mounted flush with cooker wall, constructed of stainless steel, and designed to prevent the backup of product into the steam line, or the steam line should be constructed of stainless-steel pipes and fittings that can be readily cleaned. If direct steam is applied to the product, only culinary steam should be used.

30.3.2.4 Fillers

A strainer should be installed between the cooker and the filler. The hoppers of all filters should be covered, but the cover may have sight ports. If necessary, the hopper may have an agitator to prevent buildup on the side wall. The filler valves and head should be kept in good repair and capable of accurate measurements. Product contact surfaces should be of stainless steel or other corrosion-resistant material.

30.3.3 Quality Specifications for Raw Materials

30.3.3.1 *Cheddar, Colby, Washed or Soaked Curd, Granular, or Stirred Curd Cheese*

Cheese used in the manufacture of pasteurized process cheese products should possess pleasing and desirable taste and odor consistent with the age of the cheese; should have body and texture characteristics that will impart the desired body and texture characteristics in the finished product; and should possess finish and appearance characteristics that will permit removal of all packaging material and surface defects. The cheese should at least meet the requirements equivalent to U.S. Standard Grade for Bulk American Cheese for Manufacturing, provided that the quantity of the cheese with any one defect as listed for U.S. Standard Grade is limited to assure a satisfactory finished product.

30.3.3.2 *Swiss Cheese*

Swiss cheese used in the manufacture of pasteurized process cheese and related products should be equivalent to U.S. Grade B or better, except that the cheese may be blind or possess finish characteristics that do not impair the interior quality.

30.3.3.3 *Gruyere Cheese*

Gruyere cheese used in the manufacture of process cheese and related products should be of good wholesome quality and, except for smaller eyes and sharper flavor, should meet the same requirements as for Swiss cheese.

30.3.3.4 *Cream Cheese and Neufchatel Cheese*

These cheeses when mixed with other foods or used for spreads and dips should possess a fresh, pleasing, and desirable flavor.

30.3.3.5 *Cream, Plastic Cream, and Anhydrous Milkfat*

These food products should be pasteurized, should be sweet, should have a pleasing and desirable flavor, should be free from objectionable flavors, and should be obtained from milk that complies with USDA quality requirements.

30.3.3.6 *Non-Fat Dry Milk*

Non-fat dry milk used in cheese products should meet the requirements equivalent to U.S. Extra Grade except that the moisture content may be in excess of that specified for the particular grade.

30.3.3.7 *Whey*

Whey used in cheese products should meet the requirements equivalent to USDA Extra Grade except that the moisture requirement for dry whey may be waived.

30.3.3.8 *Flavor Ingredients*

Flavor ingredients used in process cheese and related products should be those permitted by the Food and Drug Standards of Identity and in no way deleterious to the quality or flavor of the finished product. In the case of bulky flavoring ingredients such as pimento, the particles should be, to at least a reasonable degree, uniform in size, shape, and consistency. The individual types of flavoring materials should be uniform in color and should impart the characteristic flavor desired in the finished product.

30.3.3.9 Coloring

Coloring should be annatto or any other cheese or butter color that is approved by the FDA.

30.3.3.10 Acidifying Agents

Acidifying agents, if used, should be those permitted by the FDA for the specific pasteurized process cheese product.

30.3.3.11 Salt

Salt should be free flowing, white refined sodium chloride and should meet the requirements of the Food Chemical Codex.

30.3.3.12 Emulsifying Agents

Emulsifying agents should be those permitted by the FDA for the specific pasteurized process cheese product and should be free from extraneous material.

30.3.4 Operations and Operating Procedures

30.3.4.1 Basis for Selecting Cheese for Processing

A representative sample should have been examined to determine fat and moisture contents. One sample unit from each vat of cheese should have been examined to determine the suitability of the vat for use in process cheese products in accordance with the flavor, body, and texture characteristics permitted by the USDA and to determine the characteristics that it will contribute to the finished product when blended with other cheese. The cheese included in each blend should be selected on the basis of the desirable qualities that will result in the desired finished product. Recook from equivalent blends may be used in an amount that will not adversely affect the finished product. Hot cheese from the filler may be added to the cooker in amounts that will not adversely affect the finished product.

30.3.4.2 Blending

To as great an extent as is practical, each vat of cheese should be divided and distributed throughout numerous cooker batches, with the purpose being to minimize the preponderance and consequent influence of any one vat on the characteristics of the finished product and to promote as much uniformity as is practical. In blending, also consider the final composition requirements for fat and moisture. Quantities of salt, color, emulsifier, and other allowable ingredients to be added should be calculated and predetermined for each cooker batch.

30.3.4.3 Trimming and Cleaning

The natural cheese should be cleaned free of all non-edible portions. Paraffin and bandages as well as rind surface, mold or unclean areas, or any other part that is unwholesome or unappetizing should be removed.

30.3.4.4 Cutting and Grinding

The trimmed and cleaned cheese should be cut into sections of convenient size to be handled by the grinder or shredder. The grinding and mixing of the blended lots of cheese should be done in such a manner as to ensure a homogeneous mixture throughout the batch.

30.3.4.5 Adding Optional Ingredients

As each batch is added to the cooker, the predetermined amounts of salt, emulsifiers, color, or other allowable optional ingredients should be added. However, a special blending vat may be used to mix the ground cheese and other ingredients before they enter the cooker to provide composition control.

30.3.4.6 Cooking the Batch

Each batch of cheese within the cooker, including the optional ingredients, should be thoroughly commingled and the contents pasteurized at a temperature of at least 158°F and held at that temperature for not less than 30 s or any other equally effective combination of time and temperature approved by the administrator. Care should be taken to prevent the entrance of cheese particles or ingredients after the cooker batch of cheese has reached the final heating temperature. After holding for the required period of time, the hot cheese should be emptied from the cooker as quickly as possible.

30.3.4.7 Forming Containers

Containers either lined or unlined should be assembled and stored in a sanitary manner to prevent contamination. The handling of containers by filler crews should be done with extreme care and observance of personal cleanliness. Preforming and assembling of pouch liners and containers should be kept to a minimum and the supply rotated to limit the length of time exposed to possible contamination prior to filling.

30.3.4.8 Filling Containers

Hot fluid cheese from the cookers may be held in hotwells or hoppers to assure a constant and even supply of processed cheese to the filler or slice former. Filler valves should effectively measure the desired amount of product into the pouch or container in a sanitary manner and should cut off sharply without drip or drag of cheese across the opening. An effective system should be used to maintain accurate and precise weight control. Damaged or unsatisfactory packages should be removed from production, and the cheese may be salvaged into sanitary containers and added back to cookers.

30.3.4.9 Closing and Sealing Containers

Pouches, liners, or containers having product contact surfaces should be folded or closed and sealed in a sanitary manner after filling, preferably by mechanical means, so as to assure against contamination. Each container in addition to other required labeling should be coded in such a manner as to be easily identified as to date of manufacture by lot or sublot number.

30.3.4.10 Cooling the Packaged Cheese

After the containers are filled, they should be stacked or cased and stacked in such a manner as to prevent breaking of seals due to excessive bulging and to allow immediate progressive cooling of the individual containers of cheese. As a minimum, the cheese should be cooled to a temperature of 100°F or lower within 24 h after filling. The temperature of the cheese should be reduced further before being shipped or if storage is intended.

30.3.4.11 Quality Control Tests

30.3.4.11.1 Chemical Analyses

The following chemical analyses should be performed in accordance with approved methods.

   Cheese. A representative sample of cheese used in the manufacture of pasteurized process cheese products should have been tested prior to usage to determine its moisture and fat content.

Pasteurized process cheese products. As many samples should be taken of the finished product direct from the cooker, hopper, filler, or other location as is necessary to assure compliance with composition requirements. Spot checks should be made on samples from the cooker as frequently as is necessary to indicate pasteurization by means of the phosphatase test, as well as any other tests necessary to assure good quality control.

30.3.4.11.2 Examination of Physical Characteristics

As many samples should be taken as is necessary to assure meeting the required physical characteristics of the products. Representative samples should be taken from production for examination of physical characteristics. The samples should be examined at approximately 70°F the first day of operation after the date of processing for the following characteristics: (1) finish and appearance, (2) flavor, (3) color, (4) body and texture, and (5) slicing or spreading properties.

Keeping quality. During processing or preferably from the cooled stock, select sufficient samples at random from the production run. The samples should be stored at approximately 50°F for 3 months for evaluation of physical characteristics. Additional samples may be selected and held at different temperatures or time.

30.3.4.11.3 Weight Control

During the filling operation, as many samples should be randomly selected and weighed from each production run as is necessary to assure accuracy of the net weight established for the finished products.

30.3.5 Requirements for Processed Cheese Products Bearing USDA Official Identification

Only processed cheese products manufactured and packaged in accordance with USDA requirements may be identified with official USDA Quality Approved Inspection Shield.

30.3.6 Quality Specifications for Raw Materials

Cheddar, colby, washed or soaked curd, and granular or stirred curd cheese. Cheese used in the manufacture of pasteurized process cheese products that are identified with the USDA official identification should possess pleasing and desirable taste and odor consistent with the age of the cheese; should have body and texture characteristics that will impart the desired body and texture characteristics in the finished product; and should possess finish and appearance characteristics that will permit removal of all packaging material and surface defects. The cheese should at least meet the requirements of U.S. Standard Grade for Bulk American Cheese for Manufacturing, provided that the quantity of the cheese with any one defect as listed for U.S. Standard Grade is limited, to assure compliance with the specifications of the finished product.

Swiss. Swiss cheese used in the manufacture of pasteurized process cheese and related products bearing official identification should be U.S. Grade B or better, except that the cheese may be blind or possess finish characteristics that do not impair the interior quality.

Gruyere. Gruyere cheese used in the manufacture of processed cheese and related products should be of good wholesome quality and, except for smaller eyes and sharper flavor, should meet the same requirements as for Swiss cheese.

Cream cheese and Neufchatel cheese. Mixed with other foods, or used for spreads and dips, they should possess a fresh, pleasing, and desirable flavor.

Cream, plastic cream, and anhydrous milkfat. These food products should be pasteurized, should be sweet, should have a pleasing and desirable flavor, should be free from objectionable flavors, and should be obtained from milk that complies with USDA quality requirements.

Non-fat dry milk. Non-fat dry milk used in officially identified cheese products should meet the requirements of U.S. Extra Grade except that the moisture content may be in excess of that specified for the particular grade.

Whey. Condensed or dry whey used in officially identified cheese products should meet the requirements for USDA Extra Grade except that the moisture requirement for dry whey may be waived.

Flavor ingredients. Flavor ingredients used in process cheese and related products should be those permitted by the Food and Drug Standards of Identity and in no way deleterious to the quality or flavor of the finished product. In the case of bulky flavoring ingredients such as pimento, the particles should be, to at least a reasonable degree, uniform in size, shape, and consistency. The individual types of flavoring materials should be uniform in color and should impart the characteristic flavor desired in the finished product.

Other ingredients. For coloring, acidifying agents, salt, and emulsifying agents, consult USDA requirements.

30.3.7 Quality Specifications for Finished Products

30.3.7.1 Pasteurized Process Cheese

It should conform to the provisions of the Definitions and Standards of Identity for Pasteurized Process Cheese and Related Products, FDA. The average age of the cheese in the blend should be such that the desired flavor, body, and texture will be achieved in the finished product. The quality of pasteurized process cheese should be determined on the basis of flavor, body and texture, color, and finish and appearance.

Flavor—Has pleasing and desirable mild cheese taste and odor characteristic of the variety or varieties of cheese ingredients used. If additional optional ingredients are used, they should be incorporated in accordance with good commercial practices and the flavor imparted should be pleasing and desirable. It may have a slightly cooked or very slight acid or emulsifier flavor; it is free from any undesirable tastes and odors.

Body and texture—Should have a medium-firm, smooth, and velvety body free from uncooked cheese particles. It is resilient and not tough, brittle, short, weak, or sticky. It should be free from pin holes or openings except those caused by trapped steam. The cheese should slice freely and should not stick to the knife or break when cut into approximately 1/8 inch slices. If in sliced form, the slices should separate readily.

Color—May be colored or uncolored but should be uniform throughout. If colored, it should be bright and not be dull or faded. To promote uniformity and a common reference to describe color, use the color designations as depicted by the National Cheese Institute standard color guide for cheese.

Finish and appearance—The wrapper may be slightly wrinkled but should envelop the cheese, adhere closely to the surface, and be completely sealed and not broken or soiled.

30.3.7.2 Pasteurized Process Cheese Food

It should conform to the provisions of the Definitions and Standards of Identity for Pasteurized Process Cheese Food and Related Products, FDA. The average age of the cheese in the blend should be such that the desired flavor, body, and texture will be achieved in the finished product. The quality of pasteurized process cheese food should be determined on the basis of flavor, body and texture, color, and finish and appearance.

Flavor—Has pleasing and desirable mild cheese taste and odor characteristic of the variety or varieties of cheese ingredients used. If additional optional ingredients are used, they should be incorporated in accordance with good commercial practices and the flavor imparted should be pleasing and desirable. It may have a slightly cooked or very slight acid or emulsifier flavor; it is free from any undesirable tastes and odors.

Body and texture—Should have a reasonably medium-firm smooth and velvety body and free from uncooked cheese particles. It is resilient and not tough, brittle, short, or sticky. It should be free from pin holes or openings except those caused by trapped steam. The product should slice freely with only a slight amount of sticking and should not break when cut into approximately 1/8 inch slices. If in sliced form, the slices should separate readily.

Color—May be colored or uncolored but should be uniform throughout. If colored, it should be bright and not be dull or faded. To promote uniformity and a common reference to describe color, use the color designations as depicted by the National Cheese Institute standard color guide for cheese.

Finish and appearance—The wrapper may be slightly wrinkled but should envelop the cheese, adhere closely to the surface, and be completely sealed and not broken or soiled.

30.3.7.3 Pasteurized Process Cheese Spread and Related Products

It should conform to the applicable provisions of the Definitions and Standards of Identity for Pasteurized Process Cheese Spreads, FDA. The pH of pasteurized process cheese spreads should not be below 4.0.

The quality of pasteurized process cheese spreads should be determined on the basis of flavor, body and texture, color, and finish and appearance.

Flavor—Has pleasing and desirable cheese taste and odor characteristic of the variety or varieties of cheese ingredients used. If additional optional ingredients are used, they should be incorporated in accordance with good commercial practices and the flavor imparted should be pleasing and desirable. It may have a slight cooked, acid, or emulsifier flavor; it is free from any undesirable tastes and odors.

Body and texture—Should have a smooth body free from uncooked cheese particles and when packaged should form into a homogeneous plastic mass and be free from pin holes or openings except those caused by trapped steam. Product made for slicing should slice freely when cut into approximately 1/8 inch slices with only a slight amount of sticking. Product made for spreading should be spreadable at approximately 70°F.

Color—May be colored or uncolored but should be uniform throughout. If colored, it should be bright and not be dull or faded. To promote uniformity and a common reference to describe color, the color designations as depicted by the National Cheese Institute standard color guide for cheese may be used.

Finish and appearance—Wrappers, if used, may be slightly wrinkled but should envelop the cheese, adhere closely to the surface, and be completely sealed and not broken or soiled. Other containers made of suitable materials should be completely filled, should be sealed, and should not be broken or soiled.

30.4 FDA Inspection

The following describes instructions issued by the FDA to their inspectors regarding the inspection of establishments manufacturing cheese and related cheese product. The format resembles that of a teacher lecturing a group of students or a supervisor talking to a group of inspectors. Many FDA documents use this technique, making it easy to follow and understand. It is also important to emphasize that the information is basic. The technique of inspection changes with many parameters such as urgency, size of establishments, and so on. Always visit the FDA Web site if details and latest information are needed.

Although cheese standards allow the use of unpasteurized milk, provided that the cheese is aged at not less than 35°F for 60 days (longer for some cheese), utilization of unpasteurized milk should be considered potentially hazardous. In the past, use of unpasteurized milk was justified on the belief that the curing process inactivated all pathogens. However, some pathogenic bacteria such as *Listeria* may not always be inactivated. When unpasteurized milk is used, give special emphasis to in-depth evaluation of quality control and product handling at all points from the milk producers to packaging and storage. Post-pasteurization contamination must be given close scrutiny. Therefore, direct the attention to the following areas when conducting inspections of cheese producers.

30.4.1 Pasteurized

When used to describe a dairy ingredient, it means that every particle of such ingredient should have been heated in properly designed and operated equipment to one of the temperatures specified in the table below and held continuously at or above the temperature for the specified time or other time/temperature relationship that has been demonstrated to be equivalent thereto in microbial destruction:

Temperature	Time
145°F	30 min
161°F	15 s
191°F	1 s
194°F	0.50 s
201°F	0.10 s
204°F	0.05 s
212°F	0.01 s

Note: Products with greater than 10% fat and/or added sweeteners should be pasteurized at 5°F higher.

30.4.2 Heat Treatment

Heat treatment is a process in which dairy products such as milk, cream, whey, etc., are subjected to heat at *less than* a time/temperature relationship necessary to achieve pasteurization. No standard time/temperature has been established for the process. Heat treatment is used to control undesirable bacteria that might compete with the starter cultured, or if uncontrolled, cause foodborne illnesses and/or poor-quality cheese (e.g., gassy cheese or off-flavors).

Positive phosphatase test results do not always indicate a milk pasteurization problem. The phosphatase test is not specific to bovine phosphatase, but may detect phosphatase from other sources such as vegetative cells, mold, etc., that is more heat resistant than bovine phosphatase. Deviations from the normal phosphatase reaction expected in pasteurized products sometimes occur when HTST pasteurization temperatures exceed 163°F and/or salts such as magnesium chloride are added after pasteurization.

Some of the guidance furnished below is only applicable to certain inspectional situations.

30.4.3 Raw Materials and Manufacturing and Processing

Most firms now receive milk in bulk. All milk must meet the following quality standard—less than 1 million somatic cell count (SCC) and less than 1 million standard plate count (commingled milk).

Evaluate filtration and clarification steps. These procedures affect cheese quality by removing sediment, debris, body cells from the cow's udder (somatic cells), some bacteria, etc. Using specialized tools such as blacklight, check for milkstone on the equipment. A milkstone buildup indicates improper cleaning.

30.4.4 Evaluation of the Pasteurization Process and Equipment

Conduct a phosphatase test on pasteurized milk using the method furnished with the field phosphatase test kit. This field test is applicable to milk only, *not to cheese or curd because of interfering substances.* Be sure reagents are satisfactory by conducting tests in the District's Laboratory on 0.5% raw milk in pasteurized milk and on pasteurized milk before use in the field.

HTST pasteurized products found negative for phosphatase when sampled immediately after pasteurization may yield positive results after a short period of storage without refrigeration. This phosphatase reactivation phenomenon may occur under any one of the following conditions:

1. Storage of pasteurized products at temperatures between 50°F and 93°F
2. HTST pasteurization above 163°F
3. The addition of such salts as magnesium chloride to milk or cream after pasteurization

Collect plant samples of cheese if field phosphatase test cannot be made on the milk.

30.4.5 Quality Control and Records

Evaluate the quality control program. Determine what follow-up is made on patrons whose milk is high in sediment. Ascertain the duties of the field man; determine whether his primary job is to have filtered milk delivered to the plant or to improve the sanitary conditions under which the milk is produced.

Check the firm's quality control records for drug residues in milk. Verify that every farm tanker truck of raw milk has been sampled and tested for beta-lactam drug residues prior to processing. If milk found to contain drug residues has been used, determine the ultimate disposition and follow up.

Quality control records should also be checked for abnormal milk (e.g., SCCs over 1 million per milliliter). When bacteria counts exceed 1 million per milliliter, determine the disposition of the milk.

Review cheese, make records, and determine if all critical times, temperatures, and pH readings were recorded. Vats showing slow acid development indicate a high potential for excessive staphylococcal growth and concurrent enterotoxin production. Consider sampling the cheese, whey, whey cream, whey butter, etc., for staphylococcal enterotoxin.

If cheeses are being made, which may, in lieu of pasteurization, be aged for a specific time and minimum temperature, and the plant has elected this option, or if aging is required regardless of pasteurization, check storage practices including the following:

1. The date stamping procedure and the accuracy of date stamped on cheese
2. Labeling practices to ensure that uncured cheese made from raw or unpasteurized milk is labeled to indicate what further curing or processing is necessary
3. The existence of an agreement with the storage warehouse for proper storing and handling of uncured or unaged cheeses, if the warehouse is not operated by the cheese manufacturer

Pasteurized process cheeses are required to be heated, during preparation, to a temperature not less than 150°F for not less than 30 s. If applicable, report time and temperature of this process.

30.4.6 Sample Collection

Samples should be held and shipped under refrigerated conditions. Coat or reseal the holes or surface of any cheese exposed by sampling to prevent spoilage by mold. Use the following formula for sealing compound:

Paraffin	3 oz.
Beeswax	3 oz.
White petrolatum	6 oz.

or by mixing and heating white petrolatum and paraffin (1:1).

Cheese is practically always coded with the vat (or vat series) numbers or initial and date of manufacture. Each subsample should represent one such code (vat). Record the code number on the subsample jar. Seal each subsample separately.

Collect in-plant samples, sediment pads, and other evidence necessary to document insanitary practices and the receipt and use of filthy milk.

There are usually one or more in-line filters in addition to the dump tank screen that will remove gross material. Clarifiers or modified separators are sometimes in the line. Samples of this gross filth and the entire in-line filters are excellent exhibits. Filters and clarifier sludge can be preserved for laboratory examination by shaking with 25 mL of perchloroethylene in a quart jar. Report the amount of milk that an exhibit represents. Do *not* take sediment tests after pasteurization as it could lead to criticism that your action contaminated the milk.

During inspections, collect in-line and finished product samples as necessary to document suspected or observed bacteriological problems. If inspectional evidence indicates slow acid formation, collect a 1/4 lb sample aseptically from each vat of cured cheese available (with a maximum of five vats) immediately prior to removal of the cheese from the vat for hooping and pressing. If possible, collect sample before cheese curd is salted. Staphylococci die rapidly after cheese has been pressed, and for this reason, sample only freshly made cheese.

Cottage cheese, baker's cheese, cream cheese, Neufchatel cheese, or other unripened cheese would be collected if staphylococcal or other bacterial contamination is suspected.

It is important to emphasize the following. Immediately after collection, refrigerate between 32°F and 40°F, but do not freeze. Submit sample promptly so that analysis for staphylococci can be started, within 48 h of the completion of the cheesemaking process, because the staphylococcal count may change rapidly in the period of time following sample collection. In no case should this time period for submission of the samples exceed 96 h. Submit the vat samples as separate subdivisions under one investigational sample number. Identify and handle these samples separately from any other samples collected to demonstrate plant conditions.

Part IV

Meat and Fish Products

31

Meat Fermentation

Luca Cocolin and Kalliopi Rantsiou

CONTENTS

31.1 General Introduction

Food fermentations, which have played a fundamental role in nutrition and sustainability of human beings through the centuries, are microbial processes that are often discovered by chance. Cheeses and yogurts have been discovered, for example, after transportation of fresh milk in containers made out of the stomachs of animals, where the rennet (enzyme responsible for the hydrolysis of caseins) was still present.

In general, fermentation of foods allows an extension of the shelf life of very perishable raw materials. Fermented products possess higher nutritional value with respect to the starting materials, and they are characterized by different physicochemical properties and sensory profiles. Moreover, in Europe, they are bound to strong local traditions. This last aspect has to be connected to important microbial and meat enzyme contributions, which are responsible for the production of volatile compounds such as alcohols, aldehydes, and ketones coming from proteolytic and lipolytic activities. Lastly, fermented foods are considered as safe products because during production, acidification, often combined with drying processes, takes place, thereby creating an environment that does not allow the growth of pathogenic microorganisms.

Fermented meat products have a long history and tradition. It is believed that sausages were invented by Sumerians in the modern Iraq around 3000 BC. Chinese sausage *làcháng*, which consisted of goat and lamb meat, was first mentioned in 589 BC. Homer, the poet of the ancient Greece, reported a kind of blood sausage in the *Odyssey* (book 20, verse 25), and Epicharmus (ca. 550 BC–ca. 460 BC) wrote a comedy entitled "The Sausage." Evidence suggests that sausages were already popular both among the ancient Greeks and Romans (Lücke 1974).

31.2 Manufacturing Technology

For the production of fermented sausages, fresh meat (pork and other meats) and pork fat are first subjected to grinding and then mixed with salt and natural flavorings (including sodium nitrate and sodium nitrite). Filling into natural or artificial casings follows, and finally, a fermentation and ripening step, for variable periods, is carried out (Cantoni 2007).

The production process through which fermented sausages are prepared is detailed below.

Raw materials. Usually made of two-thirds of low-fat cuts (shoulder, loin, back meat and ham trimmed, low-fat parts, and bacon) and one-third of fat cuts (lard, hog-jowl, and hard fat). These meat cuts should present a total microbial count below 10^5 colony forming units (cfu)/g, *Listeria monocytogenes* and *Salmonella* spp. have to be absent, and the pH should be between 5.5 and 6.

Meat preparation. Tendons and connective tissues are removed from the meats and subsequently cut into pieces.

Refrigeration. The raw materials have to be stored under refrigerated conditions (0°C and 2°C) in order to reduce the growth of spoilage microorganisms. Moreover, the refrigeration will facilitate the following phase of grinding.

Grinding. After storage at refrigeration temperatures, the meat is subjected to grinding. It is possible to differentiate fine-grained (1–4 mm) and coarse-grained (4–14 mm) grinding based on the type of sausage that has to be produced.

Flavoring mixture. Based on the type of sausages to be produced, the mixture that is added is different. Several ingredients may be used.

1. Salt: Influences the organoleptic characteristics of the product, but also possesses antibacterial properties. It lowers the water activity (A_w) and is able to solubilize salt-soluble proteins (mainly sarcoplasmatic and myofibrillar proteins), which are able to create a protein gel that assures the cohesion of the mixture during ripening of the product.
2. Sugars: Represented by dextrose, saccharose, and milk powder (as source of lactose). They serve as sources for the growth of lactobacilli, thereby enhancing the acidification step.
3. Spices: They influence the aroma of the product, and their mix is different for each type of sausage.
4. Additives: Mainly represented by sodium nitrate and nitrite added together with salt as the so-called curing salt. They have different roles: sodium nitrate can be reduced to nitrite through the action of reducing enzymes (nitrate reductase) possessed by *Staphylococcus* and *Kocuria* spp., whereas sodium nitrite has an important double function on the color (it allows the maintenance of an intense red color) and on the inhibition of specific spoilage and pathogenic microorganisms. As an example, nitrites are able to avoid the growth of *Clostridium botulinum* and germination of its spores during ripening.
5. Ascorbic acid: It has mainly an antioxidant activity and is thus involved in the red color formation together with nitrite.

As already indicated above, the type and quantity of flavoring mix depend on the fermented sausage to be produced. In the South of Europe, where the tradition is particularly rich in artisanal products, a vast number of sausages that differ slightly in their formulation can be differentiated. Moreover, at this stage, the addition of starter cultures can also be carried out.

Kneading. The goal of this procedure is to distribute homogeneously the ingredients and the fat in the meat portion. This phase has to be carried out at low temperature and has to be rapid. One relevant risk associated to the kneading is the rise in the temperature and the subsequent liquefaction of the low-melting fats. In these conditions, the fats will cover the meat particles with an impermeable layer, hindering water loss and thereby priming growth of spoilage microorganisms.

Filling. After 24 h at refrigeration conditions (0°C–2°C), which is important for the diffusion of salt and for color formation, the meat batter is filled into casings. The stuffing machine works under vacuum, thereby avoiding the inclusion of air bubbles that can be responsible for oxidations and growth of spoilage microorganisms. Natural or artificial casings can be used for the filling process. In both cases, certain characteristics need to be taken into consideration—good porosity, which will allow a good exchange of water from the inside to the outside of the sausage; casings have to be soft and elastic to assure the adherence of the meat batter to the casing during the whole process, and they have to allow a simple removal.

Tying. After filling, the sausage is tied. This step can be carried out manually (for traditional products) or automatically through specific machines, often associated with the filling equipment.

Fermentation. Conducted in temperature- and humidity-controlled rooms, this step lasts from 2 to 5 days. In the first 1–2 days of fermentation, the sausages are kept at 25°C–27°C and at a relative humidity (RH) of 65%–75%. In these conditions, lactic acid bacteria (LAB) responsible for the acidification process are characterized by a fast growth. Thereby, the type of LAB that has been added as starter culture dictates the temperature used. *Pediococcus* spp. usually prefer temperatures above 25°C, whereas *Lactobacillus* spp. grow better at temperatures below 25°C. A drying process follows for 2–3 days. At this stage, a decrease in the A_w is recorded, and this allows a better preservation of the product. Moreover, the lactic acid fermentation ends at this point. This process is conducted at a temperature between 16°C and 22°C and RH of 55%–65%. Ventilation can also be carried out to maintain homogeneity of the RH inside the drying rooms. At the end of the fermentation process, the pH of the sausages is usually between 4.9 and 5.3. Another important event that takes place in the last stages of the fermentation is the growth of molds on the surface of the sausages. This step is not obligatory; however, it is particularly important because the molds (more specifically *Penicillium nalgiovense* and *P. caseiculum*) enhance the water loss process by creating micro-holes on the casings, they protect the sausages from oxidation because they create a protective layer against light, and at the end of the ripening, they are able to keep the product moist, avoiding excessive drying.

Ripening. It lasts from a couple of weeks to several months, depending on the dimensions of the sausages. During ripening, a weight loss of about 18%–35% is foreseen. The conditions in which ripening is carried out are temperatures from 13°C to 15°C and RH of 75%–90%. At the end of this step, the pH of the sausages increases to 5.6–5.7 due to proteolytic activities, and the products possess specific organoleptic characteristics such as flavor and taste.

At the end of the production process, fermented sausages are characterized by a long shelf life at room temperature. This important characteristic is due to the acidification during the fermentation and the decrease in A_w during the ripening. The acidification is also essential for the consistency, the color, and the flavor. A decrease in A_w results in an increase in the ratio salt/A_w, thereby negatively influencing the development of spoilage and pathogenic microorganisms during ripening. The organoleptic profile of the fermented sausages is the result of complex set of parameters, such as the equilibrium in the spices, the acidification process, and the enzymatic reactions on the lipids and proteins (Cantoni 2007).

31.3 Microbial Ecology

Meat fermentations are complex microbial ecosystems in which bacteria, yeasts, and molds are involved. The study of their microbial ecology not only highlights the extreme interspecies heterogeneity, described by the isolation of several genera and species, but also underlines an important intraspecies biodiversity, represented by the coexistence, in the same niche, of diverse biotypes belonging to the same species. For instance, Urso et al. (2006a) have shown that during the natural fermentation of Italian sausages, a number of different *Lactobacillus sakei* types could be identified, contributing in a different way to the characteristics of the final product.

The fermentation of sausages is a microbial process that has been investigated since 1960 (Lerche and Reuter 1960; Lücke 1974; Niinivaara et al. 1964; Reuter 1972), and these studies showed that the main microorganisms responsible for the transformation are LAB (mainly *Lactobacillus* spp.) and coagulase-negative cocci (CNC) (*Staphylococcus*, and *Kocuria* spp.). These evidences have been repeatedly confirmed by several researchers during the last 20 years (Bacha et al. 2010; Comi et al. 2005; Coppola et al. 2000; Drosinos et al. 2005; Samelis et al. 1994a; Villani et al. 2007). The application of modern approaches based on molecular biology has further supported the importance of LAB and CNC as main actors during sausage fermentations (Albano et al. 2008; Aquilanti et al. 2007; Aymerich et al. 2003; Cocolin et al. 2001b; Fontana et al. 2005; Rantsiou et al. 2005c; Silvestri et al. 2007). Moreover, in some fermented sausages, especially those produced in France, Italy, and Spain, the characteristics of the final products are influenced by the activity of molds and yeasts that are developing on the surface of the product (Lücke 2000).

The type of microbiota that develops during sausage production is often closely related to the fermentation technique utilized. Sausages with a short fermentation time have more lactobacilli from the early stages of fermentation, and an "acid" flavor predominates in the products, which are commonly sold after less than 2 weeks of ripening. The intensity of the flavor depends on the pH value, but at a given pH, a high amount of acetic acid gives the product a less "pure" and more "sour" flavor (Montel et al. 1998). Longer fermentation times and greater activity of microorganisms other than LAB, such as CNC and yeasts, lead to higher levels of volatile compounds with low sensory thresholds. Lipids and peptides are precursors of most of these substances. Tissue enzymes are the main agents of initial lipolysis and proteolysis processes (Lücke 2000); however, later in ripening, bacterial enzymes play a role in the degradation of the released peptides and free fatty acids.

In naturally fermented sausages, which are produced without the addition of starter cultures, there is an evident and strong connection between the microbiota that develops during fermentation and the sensory characteristics of the final product. As a consequence, the study of the autochthonous microbial ecology is an important parameter to consider in sausage fermentation and, for this reason, is a subject of intense study.

Apart from the contribution of the raw materials to the initial contamination with the technologically important microorganisms, it should be pointed out that, in the last years, it has been repeatedly demonstrated that the processing plant is playing a crucial role in the enrichment of important biota for the production of fermented sausages. This aspect has been reported at least for plants in Greece (Gounadaki et al. 2008), France (Lebert et al. 2007; Leroy et al. 2010), and Denmark (Sørensen et al. 2008). A possible connection of high numbers of the "house-biota" with the presence of pathogenic microorganisms in the sausages, such as *L. monocytogenes*, *Salmonella* spp., and *Staphylococcus aureus*, has also been reported; however, this evidence could not always be demonstrated (Gounadaki et al. 2008).

In the 2000s (notties), scientific evidences highlighted that the use of methods that are relying on the cultivation of the microorganisms (culture-dependent techniques) does not properly profile the microbial diversity present in a specific ecosystem (Hugenholtz et al. 1998). Indeed, populations that are numerically limited or microorganisms that are stressed or in a sublethal state cannot be recovered; therefore, they are eliminated from consideration. Moreover, viable but not culturable cells that are not able to form colonies on agar plates, but possess metabolic activity, will not be picked up by culture-dependent methods. Methods that do not depend on cultivation (culture-independent techniques) have attracted the attention of many scientists in different domains of investigation, spanning from the environmental microbiology to food fermentations. Soon, this trend reached also the field of fermented sausages and a new generation of studies started to be published, exploiting approaches based on the direct extraction of DNA and/or RNA from the sausage matrix (Rantsiou and Cocolin 2006). Culture-independent methods are able to profile the microbial populations in complex microbial ecosystems without any cultivation. Once the DNA and RNA are available, they can be subjected to several types of analysis that can either be preceded by a polymerase chain reaction (PCR) step or not. The culture-independent method that has been more extensively applied to sausage fermentation is the denaturing gradient gel electrophoresis (DGGE; Figure 31.1). The method is able to differentiate DNA molecules based on their denaturation behaviors. When the method is used for microbial population profiling, the PCR is carried out with universal primers, able to prime amplification for all the microbes present in the sample. After this step, a complex mixture of DNA molecules will be obtained, which can be differentiated if separated in gels with denaturant gradients. Every single band that is visible in DGGE gels represents a component of the microbiota. The more the bands are visible, the more complex the ecosystem is. By using this method, it is possible not only to profile the microbial populations but also to follow their dynamics during time. DGGE analysis has been applied mainly to Italian fermented sausages (Cocolin et al. 2001b; Rantsiou et al. 2005c; Silvestri et al. 2007; Villani et al. 2007), but studies on the fermentation dynamics of Argentinean sausages are available as well (Fontana et al. 2005). However, this method is not quantitative. Recent technological improvements allowed the PCR to become a quantitative method and thus, direct enumeration of technologically important species during fermentation of sausages can be achieved. Such an approach was described for the first time by

Days of fermentation
3 10 20 30 45

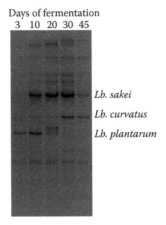

Lb. sakei

Lb. curvatus

Lb. plantarum

FIGURE 31.1 DGGE profiles of DNA extracted directly from a natural fermented sausage produced in Italy, at different days of the fermentation and ripening. The initial dominant populations of *Lb. plantarum* are accompanied by *Lb. sakei* and *Lb. curvatus* from the 10th day of fermentation. In the last two sampling points, these populations took over the fermentation process.

Martin et al. (2006), who optimized a quantitative PCR protocol for the rapid quantitative detection of *Lb. sakei* in fermented sausages.

Lastly, a promising culture-independent method, which has not yet been efficiently exploited to study the microbial diversity in fermented sausages, is fluorescence *in situ* hybridization (FISH). In this case, a set of specific probes is used to target different microorganisms directly in the sample. The probes are labeled with different fluorophores, thereby allowing a detection of several species simultaneously. Because the probes are generally designed on the ribosomal RNA, only alive cells are detected by FISH (Bottari et al. 2006). One of the most fascinating features of FISH is the possibility to localize the microorganisms directly into the food matrix. This method was first applied to dairy products (Ercolini et al. 2003). The only application of FISH to fermented and fresh sausages, in order to profile their microbial populations, has been described by Cocolin et al. (2007).

31.3.1 Lactic Acid Bacteria

LAB are mainly responsible for acidification. They are able to reduce the pH of the sausages by production of lactic acid from carbohydrates (Hammes and Knauf 1994; Hammes et al. 1990). Moreover, they influence the sensory characteristics of the fermented sausages by the production of small amounts of acetic acid, ethanol, acetoin, pyruvic acid, and carbon dioxide (Bacus 1986; Demeyer 1982), and they are able to initiate the production of aromatic substances, thanks to the proteolytic activity of muscle sarcoplasmatic proteins (Fadda et al. 1999). Lastly, a relevant function, which in the last years has been investigated more intensively, relates to the capability of certain strains on LAB to produce antimicrobial compounds, defined as bacteriocins, able to inhibit the growth of pathogenic and spoilage microorganisms (Stiles 1996). LAB are the fastest growing microbial group during the production of sausages. From the initial counts of 10^2–10^3 cfu/g, they reach values of 10^7–10^8 cfu/g in the first 3 days of fermentation (Aymerich et al. 2003; Cocolin et al. 2009; Drosinos et al. 2005; Mauriello et al. 2004), and this situation establishes both in the core and in the external layers of the sausage (Coppola et al. 2000). Their counts remain quite stable in number throughout the ripening period. Concerning the LAB ecology, it is apparent that *Lb. sakei*, *Lb. curvatus*, and *Lb. plantarum* are the best adapted species of *Lactobacillus* spp. to meat fermentations (Urso et al. 2006a). These species are described as main LAB obtained from fermented sausages produced in different countries, and usually, they are isolated together from the same product. Only in a few cases the papers report the lack of detection of one of the species mentioned. For instance, Samelis et al. (1998) and Coppola et al. (1998, 2000) did not isolate *Lb. curvatus*. Other species of *Lactobacillus* have also been identified. Samelis et al. (1994b),

Papamanoli et al. (2003), and Parente et al. (2001) described the isolation and identification of other members of LAB (i.e., *Lb. farciminis, Lb. coryniformis, Lb. casei* subsp. *pseudoplantarum, Lb. paracasei* subsp. *paracasei, Lb. buchneri, Leuconostoc carnosum, Leuc. gelidum,* and *Leuc. pseudomesenteroides*). Moreover, representatives of the genus *Weissella* have also been identified. It is interesting to point out that only in few cases wild strains of *Pediococcus* spp. were found during the fermentation process (Kaban and Kaya 2008; Parente et al. 2001; Santos et al. 1998). *Pediococcus* spp. are among the most common starter cultures used in fermented meat products in the United States, whereas in Europe, *Lactobacillus* spp. are more often used (Bacus and Brown 1981). This fact raises the issue of whether or not *Pediococcus* spp. should be used as starter cultures for the production of traditional products in Europe. As a matter of fact, it has been demonstrated that starter cultures isolated from one product have failed to lead fermentation when used to produce a different product (Marchesini et al. 1992). In the late 1990s, the advancement in molecular biology offered a number of new methodologies that could be used for a molecular identification of isolated LAB from fermented sausages. Rebecchi et al. (1998) used random amplified polymorphic DNA (RAPD) analysis to group LAB that were subsequently subjected to the 16S rRNA gene sequencing, resulting in the identification of *Lb. sakei* and *Lb. plantarum*. The RAPD-PCR approach was also used by Andrighetto et al. (2001) that identified *Lb. sakei* and *Lb. curvatus* from traditional salami produced in the Veneto region, Italy. Versatile methods, recently extensively used for strain identification, are DGGE and temperature gradient gel electrophoresis. Because the separation between strains is based on differential migrations, these techniques can be used for screening and grouping the isolates, thereby reducing the number of cultures to be identified by 16S rRNA gene sequencing. This approach was used to identify LAB during production of Italian, Greek, and Hungarian fermented sausages (Cocolin et al. 2000; Comi et al. 2005; Rantsiou et al. 2005a, 2006). It is interesting to notice that the application of PCR-DGGE and 16S rRNA gene sequencing allows the identification of a large number of strains in a quick and fast way. Lastly, the use of species-specific primers and restriction analysis of amplified DNA segments are also advantageous approaches that researchers have used to identify sausage LAB isolates (Albano et al. 2009a,b; Bonomo et al. 2008; Cocolin et al. 2009).

31.3.2 Coagulase-Negative Cocci

CNC (mainly represented by *Staphylococcus* and *Kocuria* spp.) also contribute to the final characteristics of the product due to the production of proteolytic and lipolytic enzymes responsible for the release of low-molecular-weight compounds, such as peptides, amino acids, aldehydes, amines, and free fatty acids, which influence the aromatic profile of the final product (Demeyer et al. 1986; Schleifer 1986). CNC have also a fundamental role in the development and stability of the red color through the formation of nitrosomyoglobin by nitrate reductase activity, possibly involved in the limitation of the lipid oxidation as well (Talon et al. 1999). Moreover, as described for the LAB, the CNC show a rapid increase in the first days of fermentation starting from as low as 10^3 cfu/g, reaching counts of 10^5–10^6 cfu/g after the first days of fermentation (Comi et al. 2005). Growth of CNC is also favored by the availability of oxygen; thus, cell counts inside differ from that on the surface. However, in some cases, it has been reported that fast growth of LAB, with the consequence of deep acidification of the substrate, could result in inhibitions toward CNC that exhibit a slow growth (Papamanoli et al. 2002).

Taking into consideration the CNC microbiota, the species that is always isolated, independently of the country of production, is *Staphylococcus xylosus* (Iacumin et al. 2006a). It often dominates the CNC populations, and this justifies its wide use as starter culture. Another species among *Staphylococcus* spp., which was frequently isolated from the fermentation process, is *S. saprophyticus* isolated from Greek sausages (Samelis et al. 1998; Papamanoli et al. 2002) and from Naples-type salami (Coppola et al. 2000; Mauriello et al. 2004). *S. carnosus, S. simulans, S. equorum, S. kloosii, S. sciuri, S. warneri,* and *S. lentus,* as well as *Kocuria* spp., were identified at a lesser extent. As reported for LAB identification, also in the case of CNC, an increased application of molecular methods has been observed in the last years. PCR-DGGE was exploited to identify CNC from Italian sausages (Cocolin et al. 2001a), and 16S–23S rRNA gene intergenic PCR and species-specific PCR were used as well (Blaiotta et al. 2003, 2004; Bonomo et al. 2009; Rantsiou et al. 2005b; Rossi et al. 2001).

31.3.3 Yeasts and Molds

Yeasts are present during sausage fermentations in lower numbers with respect to LAB and CNC. Cocolin et al. (2006a) showed that they can reach values of 10^6 cfu/g during production of Italian fermented sausages at low temperature, confirming results obtained previously by other authors (Metaxopoulos et al. 2001; Samelis et al. 1993). Coppola et al. (2000) found yeasts to be a predominant biota, together with LAB and CNC, in Naples-type salami. Studies carried out on different yeast starter cultures in fermented sausages have demonstrated their participation in the development of color (by removing oxygen) and flavor due to their ability to degrade peroxides, lipolytic activity, and, to a lesser extent, proteolytic activity (Lücke 2000). Moreover, yeasts, together with molds, protect the sausages from lipid oxidative reactions primed by light due to the creation of a physical barrier that does not allow its penetration. Several studies have investigated the ecology of yeasts during fermentation of sausages, and the species most often isolated is represented by *Debaryomyces hansenii*, followed by other species isolated with lower frequencies such as *Candida zeylanoides*, *Rhodotorula mucilaginosa*, and *Yarrowia lipolytica*. This evidence became available from studies considering sausages from different countries, such as Italy, Spain, and Portugal (Cocolin et al. 2006a; Encinas et al. 2000; Flores et al. 2004; Gardini et al. 2001; Saldanha-da-Gama et al. 1997). Interestingly, molecular characterization of *D. hansenii* by RAPD demonstrated a shift in its population from the beginning to the end of the maturation of the Italian sausages (Cocolin et al. 2006a). Strains present during the early stages of the fermentation were grouped in clusters that differed from those isolated in the final phases of the maturation, underlining the genetic differences between these two populations of *D. hansenii*. However, all the isolates were able to grow in the presence of 3.5% sodium chloride and at 10°C, which is an evidence that these parameters did not select the species present at the end of the maturation period.

Moreover, molds, in particular, those species belonging to the genus *Penicillium*, are involved in the development of the organoleptic profile of fermented sausages, and it is their lipolytic activity that is mainly involved in the aroma formation process. Because molds are strictly aerobic organisms, they grow only on the surface of the sausages, where they create a homogeneous white mycelium that characterizes certain productions in the south of Europe (López-Díaz et al. 2001). *P. nalgiovense* and *P. chrysogenum* are the most important species during sausage production, and they are often used as starter cultures (Leistner 1990).

31.4 Safety Aspects

Fermented sausages can be considered an excellent example of food product produced by following the philosophy of the hurdle technologies (Leistner and Gorris 1995) in which the intelligent use of combinations of different preservation factors or techniques results in mild but reliable preservation effects. In fermented sausages, from raw materials and without the use of heat treatments, the growth inhibition of pathogenic microorganisms can be guaranteed by a series of factors such as pH, A_w, organic acids, bacteriocins, sodium chloride, and nitrates/nitrites. Foodborne pathogens that can be found in the raw materials and, therefore, present in the sausages at the beginning of the fermentations, are represented by *Salmonella* spp., *L. monocytogenes*, *S. aureus*, and shiga toxin–producing *Escherichia coli* (STEC). Growth of salmonellae is inhibited by low pH, low temperature, and low A_w, whereas staphylococci are retarded in their growth by the combined action of low pH and low temperature. Listeriae, on the other hand, can grow at low temperature and low pH, and STEC can survive for a long time in these conditions that normally destroy many other foodborne pathogens (Incze 1998). An important differentiation, in terms of safety of sausages, must be made between short ripened products and long ripened and dried sausages. Short ripened sausages represent a substantial risk for human health because specific pathogenic microorganisms such as STEC, *L. monocytogenes*, and *Yersinia enterocolitica* (Lindqvist and Lindblad 2009) may survive the physicochemical conditions of the final product; therefore, heating should be considered as a suitable treatment to guarantee their absolute safety (Incze 1998). Conversely, long ripened and dried sausages possess low safety risks. During the long ripening time, the synergistic effect of different parameters allows the control of pathogenic bacteria; therefore, if they are present in high number in the raw materials, they most likely die off during ripening and drying (Heir et al. 2010;

Montet et al. 2009). Several studies investigated the presence of *L. monocytogenes* and *Salmonella* spp. during the different stages of the production. It has been defined that these foodborne pathogens are generally absent in 25 g of product (Aymerich et al. 2003; Comi et al. 2005; Rantsiou et al. 2005c). *L. monocytogenes* can contaminate the fresh sausage mix, but commonly, it is undetectable at the end of the fermentation (Samelis et al. 1998; Metaxopoulos et al. 2001; Drosinos et al. 2005).

Another aspect of safety concern in fermented sausages is represented by the production of biogenic amines (BAs). These compounds are produced by the decarboxylation of amino acids; therefore, their production is enhanced in fermented foods produced from raw materials that are particularly rich in proteins, such as sausages and cheeses. Microorganisms, responsible for BA production in fermented sausages, are represented by *Staphylococcus*, *Enterococcus*, and, at a lesser extent, *Lactobacillus* (Ansorena et al. 2002). *Enterobacteriaceae* can also be involved in their production; however, they decrease in number and activity immediately at the beginning of the fermentation process, thereby not representing a major risk for BA accumulation. Several papers have been published on the role of the different microorganisms in the production of BAs and the influence of the microbial population dynamics, starter culture adjunction, ingredients, sausage diameter, and time of ripening (Ansorena et al. 2002; Bover-Cid et al. 2009; Gardini et al. 2002; Komprda et al. 2009). Generally, it can be concluded that the control of BAs in fermented sausages can be reached by the addition of strains with negative amino acid–decarboxylase activity, although contaminations, during production, with BA producers, such as *Enterococcus* spp., seems to be inevitable. Moreover, the use of appropriate mixes of spices and the modulation of the casing diameter may have an impact in the reduction of the BA content (Komprda et al. 2009).

Lastly, an increasing concern in sausage fermentation is represented by mycotoxins. Proteinaceous foods such as ochratoxin A (OTA) have been shown to stimulate the production of mycotoxins (Larsen et al. 2001), thereby making sausages a vehicle for mycotoxin ingestion. *Aspergillus ochraceous*, *P. nordicum*, and *P. verrucosum* isolated from meat products are able to produce OTA. A recent study on the OTA contamination of artisanal and industrial dry sausages has highlighted the contamination of about 45%, in a total of 160 sausage samples, with OTA, with concentrations that were spanning from 3 to 18 μg/kg. The OTA content was reduced below the limit of quantification if sausages were brushed and washed prior to sale, highlighting that the OTA accumulation is taking place only on the surface, and it is not able to diffuse inside the sausage (Iacumin et al. 2009).

31.5 Biopreservation of Fermented Sausages

Nowadays, consumers demand more natural products, with a reduction of chemically synthesized preservatives. In this context, a new approach to food stabilization, called biopreservation, based on the antagonism displayed by one microorganism toward another, was established. According to Stiles (1996), biopreservation refers to extended storage life and enhanced safety of food using natural or controlled microbiota and (or) associated antibacterial products. Microbial interference by LAB can also arise due to the production of bacteriocins (peptides, ribosomally synthesized by bacteria, which have the capability to interfere with the growth of many foodborne spoilage and pathogenic bacteria; De Vuyst and Leroy 2007). Adding a pure culture of a viable bacteriocin-producing LAB strain, during fermented sausage production, represents an example of biopreservation. This indirect way of incorporating bacteriocins into meat products depends on the capability of the added strain to grow and produce the bacteriocin during the fermentation process. Moreover, the synthesis of a bacteriocin under laboratory conditions does not guarantee its effectiveness during real production conditions. When evaluating a bacteriocin-producing culture for sausage fermentation or biopreservation, it is important to consider that meat and meat products are complex systems; therefore, the influence of formula and fermentation technology on the performance of bacteriocin-producing strains needs to be tested (Hugas 1998). An integrated selection approach of a bacteriocinogenic *Lb. sakei* (named I151) has been described by Urso et al. (2006b,c) (Figure 31.2). After the sequencing of the bacteriocin-encoding gene, which revealed the presence of the *sakP* gene, responsible for the production of sakacin P, the strain was inoculated in a sausage fermentation, and the expression of the gene, together with the quantification of the bacteriocin, was carried out in order to assess the capability of the strain to produce the antimicrobial compound

FIGURE 31.2   Agar well diffusion assay of *Lb. sakei* I151 producing sakacin P active against *L. monocytogenes*. The halo around the well is produced by the inhibition of the bacteriocin on the pathogenic microorganism inoculated in the layer of agar.

during the transformation process. Bacteriocin gene expression and production were detected throughout the fermentation, emphasizing the potential of the strain to be used as a starter culture (Urso et al. 2006b). In a subsequent study, *Lb. sakei* I151 was used and compared with a commercial starter culture for sausage manufacturing. Sausages made with the I151 strain had an overall better microbiological quality, with low levels of enterococci and total bacterial count, compared to the sausage produced with the commercial starter, and a sensory profile that was preferred by a team of panelists (Urso et al. 2006c). Biopreservation of fermented sausages is an approach that has been repeatedly described to reach products with enhanced safety. *Lactobacillus* spp. (Liu et al. 2010; Papathomopoulou and Kotzekidou 2009) and *P. acidilactici* (Albano et al. 2007, 2009a; Nieto-Lozano et al. 2010) are the main LAB investigated *in situ* for their efficacy. Moreover, the purified bacteriocins, enterocin AS-48 (Ananou et al. 2010) and enterocins A and B (Jofré et al. 2009), in combination with high pressure, have been used to improve the safety of low-acid fermented sausages. The use of bacteriocinogenic LAB has shown a good applicability in the fermentation of sausages, although differences in reduction of the foodborne pathogens are observed.

31.6 Starter Cultures in Meat Fermentations

Starter cultures are mixtures of specific strains of microorganisms, which with their activity are able to produce relevant transformation during the sausage production process. Usually, they consist of mixtures of LAB (*Lactobacillus* and *Pediococcus*) and CNC (*Staphylococcus* and *Kocuria*); however, on the market, starter cultures of yeasts and molds are also available. Starter cultures are added to meats for different purposes: to enhance the safety aspect by reducing or eliminating pathogenic microorganisms; to improve the stability by controlling spoilage microbiota; to provide diversity at sensory level; and to provide health benefits if starters are also probiotic microorganisms (Lücke 2000). By the addition of starter cultures, fermentation of meat becomes a process that is easier to control, and with reduced fermentation time, moreover, the final organoleptic characteristics of the product are more standardized, thereby offering to the consumer a constant quality throughout the year (Hugas and Monfort 1997). On the other hand, one of the biggest criticisms addressed to the use of starter cultures is the risk related to the flattening of the sensory properties of different fermented sausages. Especially in the south of Europe, where the tradition of fermented sausages is particularly colorful and vivid, the use of starter cultures may jeopardize the unique organoleptic characteristics of these products. In order to protect and promote the unique aspects of these products, researchers have dedicated efforts to develop indigenous starter cultures for typical fermented sausages (Talon et al. 2007). These starters are combinations of strains isolated from naturally fermented sausages or from processing plants that do not use commercial starter cultures. The selection process is not a simple task and should comprise molecular identification of the isolated strains, their technological characterization, definition of the production of BAs, and the resistance to antibiotics (Ammor and Mayo 2007); finally, they have to be tested *in situ*, and their capability to take over the fermentation process should be proven by molecular typing methods (Figure 31.3).

M 1 2 3 4 5 6 7 8 9 10 11 12 13 14 15 16 17 18 M

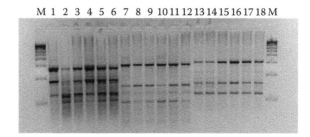

FIGURE 31.3 RAPD profiles of *Lb. curvatus* isolated from a commercial starter culture. It is obvious how the different isolates show a different pattern underlining the intraspecies variability within *Lb. curvatus* contained in the starter culture. At least three different biotypes can be observed. M, molecular weight; 1–18, *Lb. curvatus* isolates from the starter culture.

This last aspect is becoming an important step in food fermentations because it is often demonstrated that strains inoculated as starter cultures are not able to dominate; therefore, the process is guided by wild microbiota. One such example was described by Cocolin et al. (2006b) during a study to assess the performances of a commercial starter culture for sausage production, declared to contain *Lb. plantarum* and *S. carnosus*. Isolations were carried out from the lyophilized culture, and during fermentation, LAB and CNC were first identified and then subjected to molecular characterization to compare the profiles obtained by RAPD. The results obtained showed that the starter culture did not contain *S. carnosus*; *S. xylosus* was the main species isolated instead. Moreover, three *Lb. plantarum* RAPD types were detected, and only one was able to persist and dominate the fermentation process. The use of autochthonous starter cultures turned out to be a successful approach at least in the case of fermented sausages produced in Greece (Baka et al. 2011), France (Talon 2006), and Italy (Bonomo et al. 2011).

31.7 Final Remarks and Future Perspectives

Fermented meats are foodstuffs that have been investigated for the last 50 years. The transformation process as well as the technology is well described and established. The modern meat fermentation industry has been capable to translate ancient procedures in up-to-date processes in which the safety of the products is guaranteed and the quality is kept at a high level. Looking at the last half century, the main advancement that was achieved in this field was the development of starter cultures, which could be used in order to improve the safety and to standardize the quality of traditional products, as already described above. Considering the technology, it should be pointed out that no big changes in the traditional way to produce sausages have happened. Up-scaling has taken place in the industrialized countries, resulting in the necessity to process higher quantities of sausages. This traditional sector of the food fermentations has seen a particular interest from the researchers, thanks to the new possibilities given by the molecular biology field. New techniques have been produced and made available to scientists, who used them in order to confirm, or not, what was done from the 1940s with the conventional microbiological techniques. Nowadays, there are powerful methods that could analyze the microbiota without the need of its cultivation. Moreover, isolates can be identified without any biases if sequencing of the ribosomal RNA is performed. The picture of the fermented sausage microbiota that can be observed with the application of molecular methods agrees with the one produced using conventional identification. Once more, the predominance of *Lb. sakei*, *Lb. curvatus*, and *S. xylosus* is emerging. Other lactobacilli could be identified, but their numbers were found to be significantly lower than those of *Lb. sakei* and *Lb. curvatus*. *Leuconostoc* and *Weissella* spp. were identified at a lesser extent, underlining possible pitfalls in their identification by traditional methods or, more simply, their low presence in the type of sausages studied. Moreover, in the case of CNC, apart from *S. xylosus*, other species were described. *S. carnosus*, *S. simulans*, *S. condimenti*, *S. pulvereri/vitulis*, *S. equorum*, and *S. saprophyticus* were identified among the CNC isolates, and it is interesting to notice that some species were identified only when molecular methods were applied. One of the future challenges that research should face is the study of the intra-species biodiversity of the strains involved in the fermentation, in connection with their capability to produce a specific sensory profile. It is pretty amazing that fermented sausages produced in different countries, or different regions of the same country, may possess different organoleptic profiles, only partially due to the ingredients used in the manufacture. It has been demonstrated that during natural fermentation of sausages, a complex ecosystem is created, in which several biotypes of the same species can be identified. The balance of those biotypes, together with biotypes belonging to other species active in the fermentation (for instance, *Lb. sakei*, *Lb. curvatus*, and *S. xylosus*), is directly influencing the sensory profile of one specific sausage. Studies on the biodiversity of LAB (Urso et al. 2006a), CNC (Iacumin et al. 2006b), and yeasts (Cocolin et al. 2006a) have underlined that there may be a progression during fermentation of several biotypes, or, on the opposite, one or two types can take over the process and totally define the organoleptic characteristics. The study of strain interactions will be extremely important to fully understand their behavior during fermentations. In the post-genomic era, transcriptomics, proteomics, and metabolomics are becoming the experimental approaches to go for in order to better comprehend meat fermentations (Hufner et al. 2007). Moreover, interactomics, defined as the study of the interactions of different microorganisms, will have to be considered as well. The modern sequencing technology (i.e., pyrosequencing) gives a good opportunity to scientists interested in trying to investigate the physiological response of microorganisms at the transcriptional level. Of course, it should not be forgotten that meat fermentation is a complex mixture of different microorganisms; therefore, the study of single populations in laboratory conditions will not help at all in the comprehension of the phenomena taking place during the production of fermented sausages.

REFERENCES

Albano, H., Henriques, I., Correia, A., Hogg, T., and Teixeira, P. 2008. Characterization of microbial population of 'alheira' (a traditional Portuguese fermented sausage) by PCR-DGGE and traditional cultural microbiological methods. *J. Appl. Microbiol.* 105:2187–2194.

Albano, H., Oliveira, M., Aroso, R., Cubero, N., Hogg, T., and Teixeira, P. 2007. Antilisterial activity of lactic acid bacteria isolated from "alheiras" (traditional Portuguese fermented sausages): *In situ* assays. *Meat Sci.* 76:796–800.

Albano, H., Pinho, C., Leite, D., Barbosa, J., Silva, J., Carneiro, L., Magalhães, R., Hogg, T., and Teixeira, P. 2009a. Evaluation of a bacteriocin-producing strain of *Pediococcus acidilactici* as a biopreservative for "alheira," a fermented meat sausage. *Food Control* 20:764–770.

Albano, H., van Reenen, C.A., Todorov, S.D., Cruz, D., Fraga, L., Hogg, T., Dicks, L.M.T., and Teixeira, P. 2009b. Phenotypic and genetic heterogeneity of lactic acid bacteria isolated from "alheira," a traditional fermented sausage produced in Portugal. *Meat Sci.* 82:389–398.

Ammor, M.S. and Mayo, B. 2007. Selection criteria for lactic acid bacteria to be used as functional starter cultures in dry sausage production: An update. *Meat Sci.* 76:138–146.

Ananou, S., Garriga, M., Jofré, A., Aymerich, T., Gálvez, A., Maqueda, M., Martínez-Bueno, M., and Valdivia, E. 2010. Combined effect of enterocin AS-48 and high hydrostatic pressure to control food-borne pathogens inoculated in low acid fermented sausages. *Meat Sci.* 84:594–600.

Andrighetto, C., Zampese, L., and Lombardi, A. 2001. RAPD-PCR characterization of lactobacilli isolated from artisanal meat plants and traditional fermented sausages of Veneto region (Italy). *Lett. Appl. Microbiol.* 33:26–30.

Ansorena, D., Montel, M.C., Rokka, M., Talon, R., Eerola, S., Rizzo, A., Raemaekers, M., and Demeyer, D. 2002. Analysis of biogenic amines in northern and southern European sausages and role of flora in amine production. *Meat Sci.* 61:141–147.

Aquilanti, L., Santarelli, S., Silvestri, G., Osimani, A., Petruzzelli, A., and Clementi, F. 2007. The microbial ecology of a typical Italian salami during its natural fermentation. *Int. J. Food Microbiol.* 120:136–145.

Aymerich, T., Martin, B., Garriga, M., and Hugas, M. 2003. Microbial quality and direct PCR identification of lactic acid bacteria and nonpathogenic staphylococci from artisanal low-acid sausages. *Appl. Environ. Microbiol.* 69:4583–4594.

Bacha, K., Jonsson, H., and Ashenafi, M. 2010. Microbial dynamics during the fermentation of wakalim, a traditional Ethiopian fermented sausage. *J. Food Qual.* 33:370–390.

Bacus, J.N. 1986. Fermented meat and poultry products. In: *Advances in Meat and Poultry Microbiology*, edited by A.M.D. Pearson, pp. 123–164. London: Macmillan.

Bacus, J.N. and Brown, W.L. 1981. Use of microbial cultures: Meat products. *Food Technol.* 35:74–83.

Baka, A.M., Papavergou, E.J., Pragalaki, T., Bloukas, J.G., and Kotzekidou, P. 2011. Effect of selected autochthonous starter cultures on processing and quality characteristics of Greek fermented sausages. *LWT—Food Sci. Technol.* 44:54–61.

Blaiotta, G., Pennacchia, C., Parente, E., and Villani, F. 2003. Design and evaluation of specific PCR primers for rapid and reliable identification of *Staphylococcus xylosus* strains isolated from dry fermented sausages. *Syst. Appl. Microbiol.* 26:601–610.

Blaiotta, G., Pennacchia, C., Villani, F., Ricciardi, A., Tofalo, R., and Parente, E. 2004. Diversity and dynamics of communities of coagulase-negative staphylococci in traditional fermented sausages. *J. Appl. Microbiol.* 97: 271–284.

Bonomo, M.G., Ricciardi, A., and Salzano, G. 2011. Influence of autochthonous starter cultures on microbial dynamics and chemical–physical features of traditional fermented sausages of Basilicata region. *World J. Microbiol. Biotechnol.* 27:137–146.

Bonomo, M.G., Ricciardi, A., Zotta, T., Parente, E., and Salzano, G. 2008. Molecular and technological characterization of lactic acid bacteria from traditional fermented sausages of Basilicata region (Southern Italy). *Meat Sci.* 80:1238–1248.

Bonomo, M.G., Ricciardi, A., Zotta, T., Sico, M.A., and Salzano, G. 2009. Technological and safety characterization of coagulase-negative staphylococci from traditionally fermented sausages of Basilicata region (Southern Italy). *Meat Sci.* 83:15–23.

Bottari, B., Ercolini, D., Gatti, M., and Neviani, E. 2006. Application of FISH technology for microbiological analysis: Current state and prospects. *Appl. Microbiol. Biotechnol.* 73:485–494.

Bover-Cid, S., Torriani, S., Gatto, V., Tofalo, R., Suzzi, G., Belletti, N., and Gardini, F. 2009. Relationships between microbial population dynamics and putrescine and cadaverine accumulation during dry fermented sausage ripening. *J. Appl. Microbiol.* 106:1397–1407.

Cantoni, C. 2007. Fermented sausages [Il salame]. In: *The Microbiology Applied to the Food Industries [La Microbiologia Applicata alle Industrie Alimentari]*, edited by L. Cocolin and G. Comi, pp. 133–158. Rome: Aracne Editrice.

Cocolin, L., Diez, A., Urso, R., Rantsiou, K., Comi, G., Bergmaier, I., and Beimfohr, C. 2007. Optimization of conditions for profiling bacterial populations in food by culture-independent methods. *Int. J. Food Microbiol.* 120:100–109.

Cocolin, L., Dolci, P., Rantsiou, K., Urso, R., Cantoni, C., and Comi, G. 2009. Lactic acid bacteria ecology of three traditional fermented sausages produced in the north of Italy as determined by molecular methods. *Meat Sci.* 82:125–132.

Cocolin, L., Manzano, M., Aggio, D., Cantoni, C., and Comi, G. 2001a. A novel polymerase chain reaction (PCR)-denaturing gradient gel electrophoresis (DGGE) for the identification of *Micrococcaceae* strains involved in meat fermentations. Its application to naturally fermented Italian sausages. *Meat Sci.* 57:59–64.

Cocolin, L., Manzano, M., Cantoni, C., and Comi, G. 2000. Development of a rapid method for the identification of *Lactobacillus* spp. isolated from naturally fermented Italian sausage using a polymerase chain reaction—Temperature gradient gel electrophoresis. *Lett. Appl. Microbiol.* 30:126–129.

Cocolin, L., Manzano, M., Cantoni, C., and Comi, G. 2001b. Denaturing gradient gel electrophoresis analysis of the 16S rRNA gene V1 region to monitor dynamic changes in the bacterial population during fermentation of Italian sausages. *Appl. Environ. Microbiol.* 67: 5113–5121.

Cocolin, L., Urso, R., Rantsiou, K., Cantoni, C., and Comi, G. 2006a. Dynamics and characterization of yeasts during natural fermentation of Italian sausages. *FEMS Yeast Res.* 6:692–701.

Cocolin, L., Urso, R., Rantsiou, K., Cantoni, C., and Comi, G. 2006b. Multiphasic approach to study the bacterial ecology of fermented sausages inoculated with a commercial starter culture. *Appl. Environ. Microbiol.* 72:942–945.

Comi, G., Urso, R., Iacumin, L., Rantsiou, K., Cattaneo, P., Cantoni, C., and Cocolin, L. 2005. Characterization of naturally fermented sausages produced in the North East of Italy. *Meat Sci.* 69:381–392.

Coppola, R., Giagnacovo, B., Iorizzo, M., and Grazia, L. 1998. Characterization of lactobacilli involved in the ripening of soppressata molisana, a typical Southern Italy fermented sausage. *Food Microbiol.* 15:347–353.

Coppola, S., Mauriello, G., Aponte, M., Moschetti, G., and Villani, F. 2000. Microbial succession during ripening of Naples-type salami, a southern Italian fermented sausage. *Meat Sci.* 56:321–329.

De Vuyst, L. and Leroy, F. 2007. Bacteriocins from lactic acid bacteria: Production, purification, and food applications. *J. Mol. Microbiol. Biotechnol.* 13:194–199.

Demeyer, D.I. 1982. Stoichiometry of dry sausage fermentation. *Antonie van Leeuwenhoek* 48:414–416.

Demeyer, D.I., Verplaetse, A., and Gistelink, M. 1986. Fermentation of meat: An integrated process. *Belg. J. Food Chem. Biotechnol.* 41:131–140.

Drosinos, E.H., Mataragas, M., Xiraphi, N., Moschonas, G., Gaitis, F., and Metaxopoulos, J. 2005. Characterization of the microbial flora from traditional Greek fermented sausage. *Meat Sci.* 69:307–317.

Encinas, J., López-Díaz, T., García-López, M., Otero, A., and Moreno, B. 2000. Yeast populations on Spanish fermented sausages. *Meat Sci.* 54:203–208.

Ercolini, D., Hill, P.J., and Dodd, E.R. 2003. Bacterial community structure and location in Stilton cheese. *Appl. Environ. Microbiol.* 69:3540–3548.

Fadda, S., Sanz, Y., Vignolo, G., Aristoy, M.C., Oliver, G., and Toldrà, F. 1999. Hydrolysis of pork muscle sarcoplasmatic proteins by *Lactobacillus curvatus* and *Lactobacillus sake*. *Appl. Environ. Microbiol.* 65:578–584.

Flores, M., Durà, M.A., Marco, A., and Toldrà, F. 2004. Effect of *Debaryomyces* spp. on aroma formation and sensory quality of dry-fermented sausages. *Meat Sci.* 68:439–446.

Fontana, C., Cocconcelli, P.S., and Vignolo, G. 2005. Monitoring the bacterial population dynamics during fermentation of artisanal Argentinean sausages. *Int. J. Food Microbiol.* 103:131–142.

Gardini, F., Martuscelli, M., Crudele, M.A., Paparella, A., and Suzzi, G. 2002. Use of *Staphylococcus xylosus* as a starter culture in dried sausages: Effect on the biogenic amine content. *Meat Sci.* 61:275–283.

Gardini, F., Suzzi, G., Lombardi, A., Galgano, F., Crudele, M.A., Andrighetto, C., Schirone, M., and Tofalo, R. 2001. A survey of yeasts in traditional sausages of southern Italy. *FEMS Yeast Res.* 1:161–167.

Gounadaki, A.S., Skandamis, P.N., Drosinos, E.H., and Nychas, G.E. 2008. Microbial ecology of food contact surfaces and products of small-scale facilities producing traditional sausages. *Food Microbiol.* 25:313–323.

Hammes, W.P., Bantleon, A., and Min, S. 1990. Lactic acid bacteria in meat fermentation. *FEMS Microbiol. Rev.* 87:165–174.

Hammes, W.P. and Knauf, H.J. 1994. Starters in processing of meat products. *Meat Sci.* 36:155–168.

Heir, E., Holck, A.L., Omer, M.K., Alvseike, O., Høy, M., Måge, I., and Axelsson, L. 2010. Reduction of verotoxigenic *Escherichia coli* by process and recipe optimisation in dry-fermented sausages. *Int. J. Food Microbiol.* 141:195–202.

Hufner, E., Markieton, T., Chaillou, S., Crutz-Le Coq, A.M., Zagorec, M., and Hertel, C. 2007. Identification of *Lactobacillus sakei* genes induced in meat fermentation and their role in survival and growth. *Appl. Environ. Microbiol.* 73:2522–2531.

Hugas, M. 1998. Bacteriocinogenic lactic acid bacteria for the biopreservation of meat and meat products. *Meat Sci.* 49:S139–S150.

Hugas, M. and Monfort, J.M. 1997. Bacterial starter cultures for meat fermentation. *Food Chem.* 59:547–554.

Hugenholtz, P., Goebel, B.M., and Pace, N.R. 1998. Impact of culture-independent studies on the emerging phylogenetic view of bacterial diversity. *J. Bacteriol.* 180:4765–4774.

Iacumin, L., Chiesa, L., Boscolo, D., Manzano, M., Cantoni, C., Orlic, S., and Comi, G. 2009. Moulds and ochratoxin A on surfaces of artisanal and industrial dry sausages. *Food Microbiol.* 26:65–70.

Iacumin, L., Comi, G., Cantoni, C., and Cocolin, L. 2006a. Molecular and technological characterization of *Staphylococcus xylosus* isolated from Italian naturally fermented sausages by RAPD, Rep-PCR and Sau-PCR analysis. *Meat Sci.* 74:281–288.

Iacumin, L., Comi, G., Cantoni, C., and Cocolin, L. 2006b. Ecology and dynamics of coagulase-negative cocci isolated from naturally fermented Italian sausages. *Syst. Appl. Microbiol.* 29:480–486.

Incze, K. 1998. Dry fermented sausages. *Meat Sci.* 49:S169–S177.

Jofré, A., Aymerich, T., and Garriga, M. 2009. Improvement of the food safety of low acid fermented sausages by enterocins A and B and high pressure. *Food Control* 20:179–184.

Kaban, G. and Kaya, M. 2008. Identification of lactic acid bacteria and Gram-positive catalase-positive cocci isolated from naturally fermented sausage (sucuk). *J. Food Sci.* 73:M385–M388.

Komprda, T., Sládková, P., and Dohnal, V. 2009. Biogenic amine content in dry fermented sausages as influenced by a producer, spice mix, starter culture, sausage diameter and time of ripening. *Meat Sci.* 83:534–542.

Larsen, T.O., Svendsen, A., and Smedsgaard, J. 2001. Biochemical characterization of ochratoxin A-producing strains of the genus *Penicillium. Appl. Environ. Microbiol.* 67:3630–3635.

Lebert, I., Leroy, S., Giammarinaro, P., Lebert, A., Chacornac, J.P., Bover-Cid, S., Vidal-carou, M.C., and Talon, R. 2007. Diversity of microorganisms in the environment and dry fermented sausages of small traditional french processing units. *Meat Sci.* 76:112–122.

Leistner, L. 1990. Mould-fermented foods: Recent developments. *Food Biotechnol.* 4:433–441.

Leistner, L. and Gorris, L.G. 1995. Food preservation by hurdle technology. *Trends Food Sci. Technol.* 6:41–46.

Lerche, M. and Reuter, G. 1960. A contribution to the method of isolation and differentiation of aerobic "lacto-bacilli" (Genus *Lactobacillus* Beijerinck). *Arch. Hyg. Bakteriol.* 179:354–370.

Leroy, S., Giammarinaro, P., Chacornac, J.P., Lebert, I., and Talon, R. 2010. Biodiversity of indigenous sta-phylococci of naturally fermented dry sausages and manufacturing environments of small-scale proces-sing units. *Food Microbiol.* 27:294–301.

Lindqvist, R. and Lindblad, M. 2009. Inactivation of *Escherichia coli, Listeria monocytogenes* and *Yersinia enterocolitica* in fermented sausages during maturation/storage. *Int. J. Food Microbiol.* 129:59–67.

Liu, G., Griffiths, M.W., Shang, N., Chen, S., and Li, P. 2010. Applicability of bacteriocinogenic *Lactobacillus pentosus* 31-1 as a novel functional starter culture or coculture for fermented sausage manufacture. *J. Food Prot.* 73:292–298.

López-Díaz, T., Santos, J., García-López, M., and Otero, A. 2001. Surface mycoflora of a Spanish fermented meat sausage and toxigenicity of *Penicillium* isolates. *Int. J. Food Microbiol.* 68:69–74.

Lücke, F.K. 1974. Fermented sausages. In: *Microbiology of Fermented Foods,* edited by B.J.B. Wood, pp. 41–49. London: Applied Science Publishers.

Lücke, F.K. 2000. Utilization of microbes to process and preserve meat. *Meat Sci.* 56:105–115.

Marchesini, B., Bruttin, A., Romailler, N., Moreton, R.S., Stucchi, C., and Sozzi, T. 1992. Microbiological events during commercial meat fermentations. *J. Appl. Bacteriol.* 73:203–209.

Martin, B., Jofré, A., Garriga, M., Pla, M., and Aymerich, T. 2006. Rapid quantitative detection of *Lactobacillus sakei* in meat and fermented sausages by real-time PCR. *Appl. Environ. Microbiol.* 72:6040–6048.

Mauriello, G., Casaburi, A., Blaiotta, G., and Villani, F. 2004. Isolation and technological properties of coagu-lase negative staphylococci from fermented sausages of Southern Italy. *Meat Sci.* 67:149–158.

Metaxopoulos, J., Samelis, J., and Papadelli, M. 2001. Technological and microbiological evaluation of tradi-tional processes as modified for the industrial manufacturing of dry fermented sausage in Greece. *Ital. J. Food Sci.* 13:3–18.

Montel, M.C., Masson, F., and Talon, R. 1998. Bacterial role in flavour development. In: *Proceedings of the 44th International Congress on Meat Science and Technology,* Barcelona, pp. 224–233. Barcelona: IRTA/Eurocarne.

Montet, M.P., Christieans, S., Thevenot, D., Coppet, V., Ganet, S., Muller, M.L.D., Dunière, L., Miszczycha, S., and Vernozy-Rozand, C. 2009. Fate of acid-resistant and non-acid resistant shiga toxin-producing *Escherichia coli* strains in experimentally contaminated French fermented raw meat sausages. *Int. J. Food Microbiol.* 129:264–270.

Nieto-Lozano, J.C., Reguera-Useros, J.I., Peláez-Martínez, M.d.C., Sacristán-Pérez-Minayo, G., Gutiérrez-Fernández, A.J., and la Torre, A.H.D. 2010. The effect of the pediocin PA-1 produced by *Pediococcus acidilactici* against *Listeria monocytogenes* and *Clostridium perfringens* in Spanish dry-fermented sau-sages and frankfurters. *Food Control* 21:679–685.

Niinivaara, F.P., Pohja, M.S., and Komulainen, S.E. 1964. Some aspects about using bacterial pure cultures in the manufacture of fermented sausages. *Food Technol.* 18:147–153.

Papamanoli, E., Kotzekidou, P., Tzanetakis, N., and Litopoulou-Tzanetaki, E. 2002. Characterization of *Micrococcaceae* isolated from dry fermented sausages. *Food Microbiol.* 19:441–449.

Papamanoli, E., Tzanetakis, N., Lipopoulou-Tzanetaki, E., and Kotzekidou, P. 2003. Characterization of lactic acid bacteria isolated from a Greek dry-fermented sausage in respect of their technological and probiotic properties. *Meat Sci.* 65:859–867.

Papathomopoulou, K. and Kotzekidou, P. 2009. Inactivation of verocytotoxigenic *Escherichia coli* and *Listeria monocytogenes* co-cultured with *Lactobacillus sakei* in a simulated meat fermentation medium. *J. Food Saf.* 29:331–347.

Parente, E., Griego, S., and Crudele, M.A. 2001. Phenotypic diversity of lactic acid bacteria isolated from fer-mented sausages produced in Basilicata (Southern Italy). *J. Appl. Microbiol.* 90:943–952.

Rantsiou, K. and Cocolin, L. 2006. New developments in the study of the microbiota of naturally fermented sausages as determined by molecular methods: A review. *Int. J. Food Microbiol.* 108:255–267.

Rantsiou, K., Drosinos, E.H., Gialitaki, M., Metaxopoulos, I., Comi, G., and Cocolin, L. 2006. Use of molecular tools to characterize *Lactobacillus* spp. isolated from Greek traditional fermented sausages. *Int. J. Food Microbiol.* 112:215–222.

Rantsiou, K., Drosinos, E.H., Gialitaki, M., Urso, R., Krommer, J., Gasparik-Reichardt, J., Toth, S., Metaxopoulos, I., Comi, G., and Cocolin, L. 2005a. Molecular characterization of *Lactobacillus* species isolated from naturally fermented sausages produced in Greece, Hungary and Italy. *Food Microbiol.* 22:19–28.

Rantsiou, K., Iacumin, L., Cantoni, C., Comi, G., and Cocolin, L. 2005b. Ecology and characterization by molecular methods of *Staphylococcus* species isolated from fresh sausages. *Int. J. Food Microbiol.* 97:277–284.

Rantsiou, K., Urso, R., Iacumin, L., Cantoni, C., Cattaneo, P., Comi, G., and Cocolin, L. 2005c. Culture dependent and independent methods to investigate the microbial ecology of Italian fermented sausages. *Appl. Environ. Microbiol.* 71:1977–1986.

Rebecchi, A., Crivori, S., Sarra, P.G., and Cocconcelli, P.S. 1998. Physiological and molecular techniques for the study of bacterial community development in sausage fermentation. *J. Appl. Microbiol.* 84:1043–1049.

Reuter, G. 1972. Experimental ripening of dry sausages using lactobacilli and micrococci starter cultures. *Fleischwirtschaft* 52:465–468, 471–473.

Rossi, F., Tofalo, R., Torriani, S., and Suzzi, G. 2001. Identification by 16S-23S rDNA intergenic region amplification, genotypic and phenotypic clustering of *Staphylococcus xylosus* strains from dry sausages. *J. Appl. Microbiol.* 90:365–371.

Saldanha-da-Gama, A., Malfeito-Ferreira, M., and Loureiro, V. 1997. Characterization of yeasts associated with Portuguese pork-based products. *Int. J. Food Microbiol.* 37:201–207.

Samelis, J., Aggelis, G., and Metaxopoulos, J. 1993. Lipolytic and microbial changes during the natural fermentation and ripening of Greek dry sausages. *Meat Sci.* 35:371–385.

Samelis, J., Maurogenakis, F., and Metaxopoulos, J. 1994b. Characterization of lactic acid bacteria isolated from naturally fermented Greek dry salami. *Int. J. Food Microbiol.* 23:179–196.

Samelis, J., Metaxopoulos, J., Vlassi, M., and Pappa, A. 1998. Stability and safety of traditional Greek salami— A microbiological ecology study. *Int. J. Food Microbiol.* 44:69–82.

Samelis, J., Stavropoulos, S., Kakouri, A., and Metaxopoulos, J. 1994a. Quantification and characterization of microbial populations associated with naturally fermented Greek dry salami. *Food Microbiol.* 11:447–460.

Santos, E.M., Gonzalez-Fernandez, C., Jaime, I., and Rovira, J. 1998. Comparative study of lactic acid bacteria house flora isolated in different varieties of "chorizo." *Int. J. Food Microbiol.* 39:123–128.

Schleifer, K.H. 1986. Gram positive cocci. In: *Bergey's Manual of Systematic Bacteriology*, Vol. 2, edited by P.H.A. Sneath, N.H. Mair, and J.G. Holt, pp. 999–1003. Baltimore, MD: Williams & Wilkins.

Silvestri, G., Santarelli, S., Aquilanti, L., Beccaceci, A., Osimani, A., Tonucci, F., and Clementi, F. 2007. Investigation of the microbial ecology of Ciauscolo, a traditional Italian salami, by culture-dependent techniques and PCR-DGGE. *Meat Sci.* 77:413–423.

Sørensen, L.M., Jacobsen, T., Nielsen, P.V., Frisvad, J.C., and Koch, A.G. 2008. Mycobiota in the processing areas of two different meat products. *Int. J. Food Microbiol.* 124:58–64.

Stiles, M.E. 1996. Biopreservation by lactic acid bacteria. *Antonie van Leeuwenhoek* 70:331–345.

Talon, R. 2006. Saucissons secs fermiers du Massif central—Principaux résultats du projet europèen. *Viande et Produits Carnés* 25:187–188.

Talon, R., Leroy, S., and Lebert, I. 2007. Microbial ecosystems of traditional fermented meat products: The importance of indigenous starters. *Meat Sci.* 77:55–62.

Talon, R., Walter, D., Chartier, S., Barriere, C., and Montel, M.C. 1999. Effect of nitrate and incubation conditions on the production of catalase and nitrate reductase by staphylococci. *Int. J. Food Microbiol.* 52:47–56.

Urso, R., Comi, G., and Cocolin, L. 2006a. Ecology of lactic acid bacteria in Italian fermented sausages: Isolation, identification and molecular characterization. *Syst. Appl. Microbiol.* 29:671–680.

Urso, R., Rantsiou, K., Cantoni, C., Comi, G., and Cocolin, L. 2006b. Sequencing and expression analysis of the sakacin P bacteriocin produced by a *Lactobacillus sakei* strain isolated from naturally fermented sausages. *Appl. Microbiol. Biotechnol.* 71:480–485.

Urso, R., Rantsiou, K., Cantoni, C., Comi, G., and Cocolin, L. 2006c. Technological characterization of a bacteriocin-producing *Lactobacillus sakei* and its use in fermented sausage production. *Int. J. Food Microbiol.* 110:232–239.

Villani, F., Casaburi, A., Pennacchia, C., Filosa, L., Russo, F., and Ercolini, D. 2007. The microbial ecology of the *Soppressata* of Vallo di Diano, a traditional dry fermented sausage from southern Italy, and *in vitro* and *in situ* selection of autochthonous starter cultures. *Appl. Environ. Microbiol.* 73:5453–5463.

32

Dry-Cured Ham

Fidel Toldrá

CONTENTS

32.1 Introduction

The origin of dry-cured ham is lost in ancient times when man used salting as a useful preservation tool for times of scarcity. The evolution of this product has followed a traditional route, with oral transmission from generation to generation over the centuries, but very empirically, with a rather poor knowledge of the process technology (Toldrá 1992). It was just in the latest decades of the twentieth century when rapid advances in the scientific knowledge of chemistry, biochemistry, and microbiology involved in the process were reached (Toldrá 2002). This knowledge prompted successful developments in technology and a significant progress in quality standardization.

There is a wide variety of processing technologies (with important variations in the conditions for drying, ripening, smoking, etc.) as well as an important influence of the hams used as raw material (genetic type, feed, rearing system, etc.), all of these giving important variations in quality especially in sensory characteristics (Toldrá and Aristoy 2010). The main types of hams and the most important processing technologies are described in this chapter.

32.2 Types of Product

Some of the most important and well-known hams are listed in Table 32.1. The Iberian hams are produced in the southwest region of Spain with a long process that usually reaches 2–3 years and gives a unique typical flavor. Hams proceed from autochthonous heavy pigs grown in an extensive system and fattened with acorn (Toldrá et al. 1997; Estevez et al. 2007). Similarly, Corsican hams are produced for a long time (18 months) in Corsica (France) from an autochthonous heavy pig grown in an extensive system and fattened with chestnuts, although the production is restricted due to the low number of pigs (Toldrá 2004a).

Hams from certain crossbreeds of white pigs constitute the raw material for Spanish Serrano, Italian San Danielle, and French Bayonne dry-cured hams. These pigs are slaughtered at 110 kg live weight, whereas those used for Italian Parma hams are heavier (150 kg live weight). Dry-cured hams may receive different denominations in the European Union (EU) area either as Protected Designation of Origin, Protected Geographical Indication, or Traditional Speciality Guaranteed, depending on the specific region and particular regulations such as type of crossbreeds, type of feed, slaughter age, and processing technology. These hams are controlled by consortiums, such as the Parma Consortium or the Serrano Foundation, which verify the accomplishment of the specific requirements. All these hams are eaten with no further smoking or cooking. It must be mentioned that Iberian ham experiences very long processing times that usually exceed 2–3 years. Denominations of origin for Spanish Iberian hams are Huelva, Guijuelo, Dehesa de Extremadura, and Los Pedroches (Toldrá and Aristoy 2010).

Country-style ham is produced in the United States, particularly in Kentucky and Virginia. Hams are salted, dried for at least 70 days or even longer for better flavor development, and then smoked. The traditional German Westphalian ham, the German cold smoked ham (Katenschinken), and the Finnish "sauna" hams are dry-salted, left for a few weeks in vats or wooden barrels, and then smoked (Puolanne 1982). Hams are also produced in other areas such as China, where typical hams such as Ching Hua or Yunnan have great acceptance (Campbell-Platt 1995).

TABLE 32.1

Main Characteristics of Dry-Cured Hams Worldwide

Dry-Cured Ham	Country of Origin	Approximate Length of Process (Months)	Smoking
Iberian	Spain	24–36	No
Serrano	Spain	9–18	No
Bayonne	France	9–12	No
Corsican	France	24	No
Parma	Italy	12–18	No
San Danielle	Italy	9–18	No
Katenschinken	Germany	3–5	Yes
Westphalia	Germany	3–5	Yes
Country-style	USA	3–9	Yes
Sauna	Finland	2–4	Yes
Ching Hua	China	3–6	Yes
Yunnan	China	3–6	Yes

Source: Toldrá, F., *Dry-Cured Meat Products*, Food & Nutrition Press, Trumbull, CT, 2002. With permission.

32.3 Processing Technology

Pigs were traditionally reared at home and slaughtered by the end of November or early December so that hams were salted and then left for salt diffusion, just in coincidence with the coldest months. During the spring and summer, hams were ripened and dried, being ready for consumption by autumn (almost 1 year of total process). The production sites were usually located in the mountains, with cool and dry weather conditions favoring this process. The windows of the rooms were opened or closed depending on the visual and tactile assessment by an experienced operator. Of course, this type of process was transmitted from fathers to sons, but the subjective assessment originated a great variability in the final quality.

Today, most of the modern factories use computer-controlled drying chambers that allow a full control of air speed, temperature, and relative humidity. The final quality depends on the length of the process because time is needed for the enzymatic and chemical development of flavor, as will be discussed later. In general, the process is schematized in Figure 32.1 and consists of the following stages.

32.3.1 Reception

Pork legs are classified when arriving at the factory in order to facilitate their correct processing. This classification depends on each particular area, but is usually based on ham weight, pH, and fat thickness (Toldrá 2002). The fatty acid composition of the fat mainly depends on the feed (Toldrá et al. 1996) and, in a minor scale, on the crossbreed (Armero et al. 1999a), but is of extreme importance for correct flavor development. Depending on the composition in certain polyunsaturated fatty acids, hams may develop the adequate flavor or experience undesirable oxidation and develop rancid off-flavors. Fat may be controlled through the iodine index (as an indicator of unsaturation) and the acid index (as an indicator of freshness). The exudative hams, having a condition known as pale soft and exudative (PSE), have a low water-binding capacity and may reach important weight losses, substantially higher than normal hams (Arnau et al. 1995). In addition, PSE hams have a pale color and a wetted surface that facilitates the dissolution and penetration of the added salt, but, on the other hand, it originates an excessive salty taste. Other groups of hams, having high ultimate pH and known as dark firm and dry, must be rejected in order to avoid microbial contamination (Toldrá 2004b).

Although the modern meat industry uses standard pigs, hams produced from older pigs usually give better quality due to the higher amount of myoglobin (improved color) and a different enzyme profile (better flavor profile; Toldrá et al. 1996; Rosell and Toldrá 1998).

The skin is partially removed, leaving an area where salt will penetrate and water will evaporate. Hams are then registered to facilitate traceability, subjected to pressing rollers for bleeding, and left for 1 or 2 days under refrigerated storage (2°C–4°C) to reach a uniform temperature. In the case of frozen hams, they are allowed to thaw until about –4°C inside the ham.

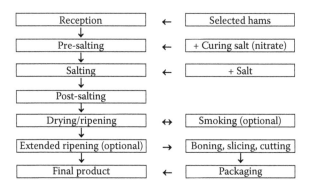

FIGURE 32.1 Process flow diagram for the processing of dry-cured ham.

32.3.2 Pre-Salting

This is a short stage where nitrate is added to the hams in the form of a curing salt (sodium chloride with 4% potassium nitrate) for a few minutes within a rotary drum (i.e., Spanish Serrano hams). The curing salt may be directly applied in the salting stage (i.e., French and country-style hams). Nitrate and/or nitrite are used as protective agents against botulism (Cassens 1995). Nitrate is reduced to nitrite by the action of nitrate reductase, a bacterial enzyme present in the natural flora (i.e., *Micrococcaceae*) of ham. The use of nitrate is very convenient for such a slow process such as dry-curing (Toldrá et al. 2009). This reduction is slow due to the low bacterial counts. Further formation of nitric oxide is achieved at slight acid pH, as found in the ham and favored by curing adjuncts, such as ascorbic or erythorbic acids, which act as reducing substances. The maximum amount allowed in the EU is 150 parts per million (ppm) potassium nitrate or 300 ppm for a combination of potassium nitrate + sodium nitrite, whereas in the United States, 156 ppm sodium nitrite (1/4 ounce per 100 pounds of meat) is allowed. In some cases, the use of nitrate and/or nitrite is banned (i.e., Italian Parma ham).

32.3.3 Salting

Salt inhibits the growth of spoilage microorganisms by reducing a_w; it also imparts a characteristic salty taste and increases the solubility of myofibrillar proteins. The main objective of the salting stage is to supply the necessary amount of salt on the outer surface of the hams. Absorbed salt is then slowly diffused through the whole piece during the post-salting stage. The amount of salt may be tightly controlled, on a weight basis, allowing enough time for its penetration into the piece (exact salt supply). Therefore, hams are weighed one by one, and the exact amount of salt per kilogram of ham is added on the lean surface. For instance, Parma hams receive 20–30 g medium-grain salt per kilogram on the lean surface and 10–20 g of wet salt per kilogram on the skin (Parolari 1996). Then, salt is hand-rubbed and left to be absorbed into the ham (14–21 days, depending on the size).

In other cases, the amount of salt is undetermined, but the time of salting is strictly controlled. Hams are entirely surrounded by rough sea salt or refined mineral salt and then placed by layers into stainless-steel bins with holes for the elimination of drippings. Salt may be rubbed onto the lean surface, and the hams are placed on shelves. This stage may last up to 13 days under refrigeration with 3%–4% weight losses. In some cases, hams are salted again. Once the salting stage is finished, the excess salt is removed by brushing and water rinsing.

32.3.4 Post-Salting or Resting

The main objective of this stage is to achieve salt equalization through the entire piece. The required time may vary between 40 and 60 days, depending on many variables such as the size of the ham, pH, amount of fat, and conditions in the chamber. The relative humidity in the chamber is progressively reduced with time, and the typical weight losses are around 4%–6%.

32.3.5 Smoking

The use of smoke is one of the oldest preservation technologies that are used for short-term processed hams such as the American country-style or the German Westphalia ham. The use of smoking is typical in areas where drying was originally more difficult (i.e., Northern countries) and gives a particular flavor to the hams. The smoke compounds also protect the hams against mold or yeast growth due to their bactericidal effects.

32.3.6 Ripening–Drying

Hams are placed into modern computer-controlled drying chambers; some may contain up to 30,000 hams per chamber. Temperature, relative humidity, and air speed must be as homogeneous as possible and are carefully controlled and registered. Each type of ham has a specific set of variables along time

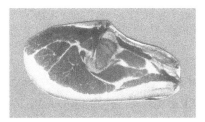

FIGURE 32.2   Details of a cross-section of a typical dry-cured ham.

of processing. For instance, Spanish hams are subjected to a progressive slow increase in temperature, whereas French hams are heated to 22°C–26°C just after the post-salting stage. In all cases, these conditions allow the action of the endogenous enzymes, as will be described later. The length of the process depends on the type of ham (pH, size, amount of intramuscular fat, etc.) and drying conditions. The final expected weight loss (around 32%–36%) is usually achieved within 6–9 months (Toldrá 2007a). Then, hams are covered with a layer of lard to avoid further dehydration and prevent any growth of molds and/ or yeasts on the outer surface. Hams' quality is controlled through a sniff test, which consists of the insertion of a small probe in a specific area of the ham prone to spoilage and then immediately smelled by an expert for detection of any off-flavor (Parolari 1996). The rapid development of commercial electronic noses and probes to get an objective assessment of flavor quality has led to their increased use for quality classification of hams (Spanier et al. 1999).

32.3.7 Extended Ripening

Hams of high quality are further ripened in cellars for several months under mild conditions in order to get a full, rich flavor development. This is the case with Iberian hams, which may have 24–30 months of total processing time.

32.3.8 Final Product

Hams may be sold either as an entire piece (usually those of higher quality) or boned. Commercial distribution of sliced ham in vacuum packages or under controlled atmosphere is increasing very fast. Boned hams are usually vacuum-packaged and distributed through retailers for final cutting into pieces or slices (see a slice in Figure 32.2). Hams are sliced by retailers or directly by consumers at home.

32.4 Microbial Evolution

The increased concentration of salt and progressive reduction in water activity constitute limiting factors for microbial growth (Toldrá et al. 2001). In fact, low bacterial counts have been found inside the hams (Toldrá and Etherington 1988). Some species of the *Kocuria* family with nitrate reductase activity are present in ham. Some microorganisms, such as *Pediococcus pentosaceus* and *Staphylococcus xylosus*, which are present in the natural flora of ham, have been studied for its enzyme activity, but no significant endoprotease activity was detected, only a minor exopeptidase activity (Molina and Toldrá 1992), although *S. xylosus* also showed an important nitrate reductase activity. Some lactic acid bacteria in their enzyme profile have also been studied because they could be used as microbial starters to accelerate the process. In fact, *Lactobacillus sakei*, *L. curvatus*, *L. casei*, and *L. plantarum* have shown good endoproteolytic and exoproteolytic activity against myofibrillar and sarcoplasmic proteins (Fadda et al. 1999a,b; Sanz et al. 1999a,b).

   Molds can grow and develop on the outer surface of the ham due to the humidity and temperature conditions in the curing chambers when no caution is taken. The most common mold is *Penicillium* (Nuñez et al. 1996a), but some yeasts, mainly *Candida zeylanoides* in early stages and *Debaryomyces hansenii*, may also grow. The isolated molds (around 75%) have shown good antimicrobial activity against *S. aureus* by inhibiting its growth (Nuñez et al. 1996b).

32.5 Physical and Chemical Changes during the Process

Main chemical changes are the result of changes in composition due to water loss and salt penetration. The diffusion of water through the ham and its evaporation when reaching the surface is a slow and difficult process. Both rates must be equilibrated to get an adequate drying, and, in this sense, it is very important to have the water sorption isotherms to predict the required time for drying. Diffusion of salt is also very slow and is affected by many variables such as temperature, size of the ham, pH, amount of moisture, and intramuscular fat. It takes around 4 months to get full salt equalization through the entire piece, although the salt profile may change a little depending on the particular moisture content in each muscle.

pH increases from initial values around 5.6–5.8 to values near 6.4 toward the end of the process. PSE hams have a similar pH evolution as normal ones (Toldrá 2007b). This evolution constitutes a narrow range where all the enzymes and chemical reactions operate (Toldrá 1998). However, even slight variations in pH might affect the action of muscle enzymes; for example, a more intense proteolysis has been reported in low-pH hams (Buscailhon et al. 1994b).

32.6 Biochemical Changes during the Process

Many biochemical changes have been reported during the processing of dry-cured ham, most of them being a consequence of enzymatic reactions (Toldrá, 1992). Some of these changes are restricted to the beginning of the process; such is the case with nucleotide breakdown reactions or the glycolysis-related enzymes and subsequent generation of lactic acid. Proteolysis and lipolysis constitute two of the most important enzymatic phenomena that are responsible for the generation of compounds with direct influence on taste and aroma.

32.6.1 Proteolysis

Proteolysis consists of the progressive degradation and breakdown of major meat proteins (sarcoplasmic and myofibrillar proteins) and the subsequent generation of peptides and free amino acids. The result is a weakening of the myofibrillar network and generation of taste compounds, but its extent depends on many factors. One of the most important is the activity of endogenous muscle enzymes, which depends on the original crossbreeds (Armero et al. 1999a,b) and the age of the pigs (Toldrá et al. 1996). Main muscle enzymes involved in these phenomena and their main properties are listed in Table 32.2. These enzymes show a great stability in long dry-curing processes such as those in hams (Toldrá et al. 1993, 1995). Other important factors are related to the processing technology. For instance, the temperature and time of ripening will determine the major or minor action of the enzymes, and the amount of added salt, which is a known inhibitor of cathepsins and other proteases, will also regulate the enzyme action (Rico et al. 1990, 1991). Proteolysis may be enhanced (Toldrá 2006a), but hams with excessive softness have been correlated to high cathepsin B activity and low salt content (Parolari et al. 1994; García-Garrido et al. 2000).

Large amounts of small peptides, in the range of 2700–4500 Da, or even below 2700 Da, are generated during the process (Aristoy and Toldrá 1995; Rodríguez-Nuñez et al. 1995), although this generation may be depressed by the level of salt that inhibits muscle peptidases (Toldrá 2006b, 2007b). Some of these peptides give characteristic tastes (Aristoy and Toldrá 1995). Recently, several tripeptides and dipeptides have been isolated and sequenced (Sentandreu et al. 2003). Furthermore, a large number of peptides resulting from the degradation of myofibrillar proteins such as titin, myosin light chains, and troponin T (Mora et al. 2009, 2010, 2011a), as well as sarcoplasmic proteins such as glycolytic enzymes (Mora et al. 2011b), have been isolated and sequenced.

TABLE 32.2

Proteolytic Muscle Enzymes and Main Properties

Enzyme	EC Number	Main Action	Main Substrate	Product	Optimum pH	Optimum Temperature (°C)	Stability	Effect of Salt	References
Cathepsin B	3.4.22.1	Endoprotease	Proteins	Polypeptides	6.0	37	Years	Inhibition	Rico et al. (1991)
Cathepsin L	3.4.22.15	Endoprotease	Proteins	Polypeptides	6.0	30	Years	Inhibition	Toldrá et al. (1993)
Cathepsin D	3.4.23.5	Endoprotease	Proteins	Polypeptides	6.8	40	Months	Inhibition	Rico et al. (1990)
Cathepsin H	3.4.22.16	Amino/Endo	Proteins	Amino acids	4.0	37	Months	Inhibition	Rico et al. (1991)
Calpain I	3.4.22.17	Endo	Proteins	Polypeptides	7.5	25	Days	Activation	Rosell and Toldrá (1996)
Calpain II	3.4.22.17	Endo	Proteins	Polypeptides	7.5	25	Days	Activation	Rosell and Toldrá (1996)
TPP I	3.4.14.9	Exoprotease	Polypeptides	Tripeptides	4.0	37	Months	Inhibition	Toldrá (2002)
TPP II	3.4.14.10	Exoprotease	Polypeptides	Tripeptides	7.0	30	Months	Inhibition	Toldrá (2002)
DPP I	3.4.14.1	Exoprotease	Polypeptides	Dipeptides	5.5	50	Months	No effect	Sentandreu and Toldrá (2000)
DPP II	3.4.14.2	Exoprotease	Polypeptides	Dipeptides	5.5	65	Months	Inhibition	Sentandreu and Toldrá (2001a)
DPP III	3.4.14.4	Exoprotease	Polypeptides	Dipeptides	8.0	45	Months	Inhibition	Sentandreu and Toldrá (1998)
DPP IV	3.4.14.5	Exoprotease	Polypeptides	Dipeptides	8.0	45	Months	Inhibition	Sentandreu and Toldrá (2001b)
Methyonyl	3.4.11.18	Aminopeptidase	Peptides	Amino acids	7.5	40	Years	Inhibition	Flores et al. (2000)
Alanyl	3.4.11.14	Aminopeptidase	Peptides	Amino acids	6.5	37	Years	Inhibition	Flores et al. (1996)
Leucyl	3.4.11.1	Aminopeptidase	Peptides	Amino acids	9.0	45	Years	No effect	Flores et al. (1997)
Pyroglutamyl	3.4.19.3	Aminopeptidase	Peptides	Amino acids	8.5	37	Weeks	Inhibition	Toldrá et al. (1992)
Arginyl	3.4.11.6	Aminopeptidase	Peptides	Amino acids	6.5	37	Months	Activation	Flores et al. (1993)

Final generation of free amino acids by endogenous muscle aminopeptidases is very important, reaching such impressive amounts as high as several hundreds of milligrams per 100 g of ham (Aristoy and Toldrá 1991; Córdoba et al. 1994; Toldrá et al. 1995, 2000).

32.6.2 Lipolysis

Lipolysis consists of the breakdown of triacylglycerols by lipases and phospholipids by phospholipases, resulting in the generation of free fatty acids. These fatty acids may contribute directly to taste and indirectly to the generation of aroma compounds through further oxidation reactions. Main lipolytic enzymes, located in muscle and adipose tissue and involved in these phenomena, are listed in Table 32.3. These enzymes show good stability through the full process (Motilva et al. 1992, 1993a,b). Although their activity also depends on pH, salt concentration, and water activity, the conditions found in the hams favor their action (Motilva and Toldrá 1993). The generation rate of free fatty acids in the muscle, especially oleic, linoleic, estearic, and palmitic acids, increases up to 10 months of processing. Most of these fatty acids proceed from phospholipid degradation. After this time, a reverse trend is observed due to further oxidative reactions (Motilva et al. 1993a; Buscailhon et al. 1994b). In the case of adipose tissue, the rate of generation, especially that of oleic, palmitic, linoleic, stearic, palmitoleic, and myristic acids, is also high up to 6 months of processing (Motilva et al. 1993b). In the same way, a decrease of 14% in the triacylglycerols is observed (Coutron-Gambotti and Gandemer 1999).

32.6.3 Oxidation

The generated monounsaturated and polyunsaturated fatty acids are susceptible to further oxidative reactions to give volatile compounds. The beginning of lipid oxidation is correlated to an adequate flavor development (Buscailhon et al. 1993). On the contrary, an excess of oxidation may lead to off-flavors. In fact, the generation of the characteristic aroma of dry-cured meat products is in agreement with the beginning of lipid oxidation. Free radical formation is catalyzed by muscle oxidative enzymes, such as peroxidases and cyclooxygenases, external light, heating, and the presence of moisture and/or metallic cations. The next step in oxidation is the formation of peroxide radicals (propagation) by reaction of free radicals with oxygen. The formed hydroxyperoxides (primary oxidation products) are flavorless but very reactive, giving secondary oxidation products that contribute to flavor (Flores et al. 1997c). The oxidation is finished when free radicals react with each other. Oxidation levels were compared among different meat products, being dry-cured ham in an intermediate position (Armenteros et al. 2009). Main products from lipid oxidation (Berdagué et al. 1991; Flores et al. 1998b) are aliphatic hydrocarbons (poor contribution to flavor), alcohols (high odor threshold), aldehydes (low odor threshold), and ketones. The last two groups are related with the aroma of dry-cured ham in French-type hams (Buscailhon et al. 1994c) and Spanish hams (Flores et al. 1997a,d). Alcohols may interact with free carboxylic fatty acids giving esters, especially when nitrate is not used, as in Parma ham where esters are generated in greater amounts and are well correlated with its aged odor (Careri et al. 1993).

32.7 Development of Sensory Characteristics

32.7.1 Color

The color of dry-cured ham mainly depends on the concentration of its natural pigment myoglobin, which depends on the type of muscle and the age of the animal (Aristoy and Toldrá 1998). Therefore, myoglobin concentration is higher in muscles with oxidative pattern and in older animals. The typical bright-red color is due to nitrosomyoglobin—a compound formed after reaction of nitric oxide with myoglobin. Those hams without added nitrate present a pinky-red color. Some surface colors on smoked hams may result from the pyrolytic decomposition of wood.

TABLE 32.3

Lipolytic Muscle and Adipose Tissue Enzymes and Main Properties

Enzyme	Main Action	Main Substrate	Optimum pH	Optimum Temperature (°C)	Stability	Effect of Salt	References
Muscle							
Acid lipase	Lipase	Triacylglycerols	5.0	37	Months	Activation	Motilva et al. (1992, 1993a)
Neutral lipase	Lipase	Triacylglycerols	7.5	45	Years	Inhibition	Motilva et al. (1992, 1993a)
Phospholipase A	Phospholipase	Phospholipids	5.0	37	Months	Activation	Toldrá (2002)
Acid esterase	Esterase	Triacylglycerols	5.0	30	Years	Inhibition	Motilva et al. (1992, 1993a)
Neutral esterase	Esterase	Triacylglycerols	7.5	20	Years	Inhibition	Motilva et al. (1992, 1993a)
Adipose Tissue							
Hormone-sensitive lipase	Lipase	Triacylglycerols	7.0	37	Months	Activation	Motilva et al. (1992, 1993b)
Monoacylglycerol lipase	Lipase	Monoacylglycerols	7.0	37	Months	Activation	Toldrá (1992)
Lipoprotein lipase	Lipase	Lipoproteins	8.5	37	Months	Inhibition	Toldrá (1992)
Acid esterase	Esterase	Triacylglycerols	5.0	60	Years	Inhibition	Motilva et al. (1992, 1993b)
Neutral esterase	Esterase	Triacylglycerols	7.5	45	Years	Inhibition	Motilva et al. (1992, 1993b)

32.7.2 Texture

The texture of dry-cured hams depends on several factors such as the extent of drying (loss of moisture), the extent of proteolysis (degree of myofibrillar protein breakdown), and the content of connective tissue. In fact, major structural proteins such as titin, nebulin, and troponin T are fully degraded, whereas myosin heavy chain and α-actinin are partly proteolyzed (Monin et al. 1997; Tabilo et al. 1999). Some small fragments (150, 85, 40, and 14.4 kDa) appear as a consequence of proteolysis (Toldrá 1998). The content in intramuscular fat also exerts a positive influence on some texture and appearance traits (Ruiz et al. 2000).

32.7.3 Flavor

32.7.3.1 Generation of Taste Compounds

Glutamic and aspartic acids impart an acid taste, whereas its sodium salts give a salty taste. Bitter taste is mainly associated with aromatic amino acids such as phenylalanine, tryptophan, and tyrosine; sweet taste is mainly associated with alanine, serine, proline, glycine, and hydroxyproline (Aristoy and Toldrá 1995). The generation of all these free amino acids is extremely important in dry-cured hams (Toldrá et al. 2000) and somehow affected by levels of salt (Martin et al. 1998). For instance, lysine and tyrosine are well correlated to an improvement in the aged taste of Parma ham (Careri et al. 1993), although in other cases, such as in French-type dry-cured ham, only a small effect on flavor development has been reported (Buscailhon et al. 1994c). An excess of proteolysis (proteolysis index higher than 29%–30%) is undesirable because it may give a bitter-like or metallic aftertaste (Careri et al. 1993; Parolari et al. 1994).

Specific tastes for dry-cured ham have been found after fractionation by gel filtration chromatography in several fractions with low molecular mass, below 2700 Da, accompanied by some nucleotides and a few compounds from protein–lipid interactions (Aristoy and Toldrá 1995). Some of these tasty peptides, mainly dipeptides and tripeptides, have been successfully purified and sequenced (Sentandreu et al. 2003). Free amino acids may also serve as a source of volatile compounds during further ripening (Flores et al. 1998b) or when the ham is heated, such as the country-style ham (McCain et al. 1968).

32.7.3.2 Generation of Aroma Compounds

Aroma development in dry-cured ham is a very complex process involving numerous reactions such as chemical or enzymatic oxidation of unsaturated fatty acids and further interactions with proteins, peptides, and free amino acids (Flores et al. 1997a,b). In fact, more than 200 volatile compounds have been reported in dry-cured hams (García et al. 1991; Ruiz et al. 1999; Sabio et al. 1998) as summarized in Table 32.4. Some volatile compounds such as 2-methyl propanal, 2-methyl butanal, and 3-methyl butanal arise from Strecker degradation

TABLE 32.4

Main Groups of Volatile Compounds Generated during the Processing of Dry-Cured Ham

Groups of Volatile Compounds	Main Origin	General Characteristic Aromas
Aliphatic hydrocarbons	Autooxidation of lipids	Alkane, crackers
Aldehydes	Oxidation of free fatty acids	Green, pungent, fatty
Branched aldehydes	Strecker degradation of amino acids	Roasted cocoa, cheesy-green
Alcohols	Oxidative decomposition of lipids	Medicinal, onion, green, alcoholic
Ketones	β-Keto acid decarboxylation or fatty acid β-oxidation	Buttery, floral, fruity
Esters	Interaction of free carboxylic acids and alcohols	Fruity
Nitrogen compounds	Maillard reaction of amino acids with carbohydrates	Meaty, nutty, toasted nuts
Sulfur compounds	Sulfur-containing amino acids	Dirty socks
Furans	Sulfur-containing amino acids with carbohydrates	Ham-like, fishy

Source: Toldrá, F., *Dry-Cured Meat Products*, Food & Nutrition Press, Trumbull, CT, 2002. With permission.

of the amino acids valine, isoleucine, and leucine, respectively. Some pyrazines, formed through Maillard reactions between sugars and free amino acids, although in low amounts, also impart some characteristic aromas such as nutty, green, and earthy. Final flavor depends on the mixture of characteristic aromas and odor thresholds for each compound, although, in general, ketones, esters, aromatic hydrocarbons, and pyrazines are correlated with the pleasant aroma of ham (Toldrá and Flores 1998). Some correlations have been found between some volatile compounds and specific characteristics of the process, for instance, the correlation of aged flavor of Parma ham with short-chain methyl-branched aldehydes, esters, and alcohols (Bolzoni et al. 1996; Flores et al. 1997c); hexanal, 3-methyl butanal, and dimethyl disulfide with short drying processes; or methyl-branched aldehydes, secondary alcohols, methyl ketones, ethyl esters, and dimethyl trisulfide with nutty, cheesy, and salty descriptors (Hinrichsen and Pedersen 1995).

The generation of volatile compounds may vary depending on the type of process. Its solid-phase extraction followed by injection into a gas chromatograph coupled to either a flame ionization detector or to a mass spectrometer allows the characterization of the aromas (Pérez-Juan et al. 2006) and even the possibility to differentiate among types of hams (Luna et al. 2006). These volatile compounds may interact more or less strongly depending on the ionic strength in the ham (Pérez-Juan et al. 2007).

32.8 Accelerated Processing of Dry-Cured Hams

Many attempts have been developed to accelerate the process, especially in country-style hams. Most of them try to accelerate the penetration and diffusion of salt into the hams such as boning and skinning of hams (Montgomery et al. 1976; Kemp et al. 1980; Marriott et al. 1983), mechanical tenderization through blade penetration prior to dry-curing (Marriott et al. 1985), tumbling in a revolving drum with baffles (Leak et al. 1984), or the direct use of nitric oxide instead of nitrate or nitrite (Marriott et al. 1992). In other cases, more intense biochemical changes occur through papain injection (Smallings et al. 1992), microbial inoculation (Marriott et al. 1987), or membrane disruption by pre-freezing and thawing of hams (Kemp et al. 1987; Motilva et al. 1994). Other recent developments are based on vacuum impregnation of salt to accelerate its penetration and diffusion through the ham (Barat et al. 2006). If combined with thawing, the reduction in time is even larger without special effects on the biochemical reactions into the hams during the processing (Flores et al. 2006a) or in their sensory quality (Flores et al. 2006b).

32.9 Salt Reduction

Salt reduction is an important area of research in latest years as a consequence of the current trend toward the reduction of sodium intake. The salt content in dry-cured ham is relatively high (4%–6%) and needs to be reduced according to such nutritional recommendations (Jiménez-Colmenero et al. 2010). Some studies for sodium reduction in dry-cured hams have been performed in recent years (Barat and Toldrá 2011). This reduction has been based on the partial replacement of NaCl with other salts such as potassium lactate (Costa-Corredor et al. 2010), potassium chloride, and also calcium and magnesium chlorides (Aliño et al. 2010; Armenteros et al. 2010; Ripollés et al. 2011). The results showed an effective reduction without noticeable sensory changes at about 40% Na$^+$ reduction.

REFERENCES

Aliño, M., Grau, R., Toldrá, F., and Barat, J.M. 2010. Physicochemical changes in dry-cured hams salted with potassium, calcium and magnesium chloride as a partial replacement for sodium chloride. *Meat Sci.* 86:331–336.

Aristoy, M.C. and Toldrá, F. 1991. Deproteinization techniques for HPLC amino acid analysis in fresh pork muscle and dry-cured ham. *J. Agric. Food Chem.* 39:1792–1795.

Aristoy, M.C. and Toldrá, F. 1995. Isolation of flavor peptides from raw pork meat and dry-cured ham. In: *Food Flavors: Generation, Analysis and Process Influence*, edited by G. Charalambous, pp. 1323–1344. Amsterdam, The Netherlands: Elsevier Science Publishers BV.

Aristoy, M.C. and Toldrá, F. 1998. Concentration of free amino acids and dipeptides in porcine skeletal muscles with different oxidative patterns. *Meat Sci.* 50:327–332.

Armenteros, M., Aristoy, M.C., Barat, J.M., and Toldrá, F. 2011. Biochemical and sensory changes in dry-cured ham salted with partial replacement of sodium by a mixture of potassium, calcium and magnesium. *Meat Sci.*, 90:361–367.

Armenteros, M., Heinonen, M., Ollilainen, V., Toldrá, F., and Estévez, M. 2009. Analysis of protein carbonyls in meat products by using the DNPH method, fluorescence spectroscopy and liquid chromatography–electrospray ionisation–mass spectrometry (LC–ESI–MS). *Meat Sci.* 83:104–112.

Armero, E., Barbosa, J.A., Toldrá, F., Baselga, M., and Pla, M. 1999a. Effect of the terminal sire and sex on pork muscle cathepsin (B, B + L and H), cysteine proteinase inhibitors and lipolytic enzyme activities. *Meat Sci.* 51:185–189.

Armero, E., Flores, M., Toldrá, F., Barbosa, J.Á., Olivet, J., Pla, M., and Baselga, M. 1999b. Effects of pig sire types and sex on carcass traits, meat quality and sensory quality of dry-cured ham. *J. Sci. Food Agric.* 79:1147–1154.

Arnau, J., Guerrero, L., Casademont, G., and Gou, P. 1995. Physical and chemical changes in different zones of normal and PSE dry cured ham during processing. *Food Chem.* 52:63–69.

Barat, J.M., Grau, R., Ibáñez, J.B., Pagán, M.J., Flores, M., Toldrá, F., and Fito, P. 2006. Accelerated processing of dry-cured ham. Part I. Viability of the use of brine thawing/salting operation. *Meat Sci.* 72:757–765.

Barat, J.M. and Toldrá, F. 2011. Reducing salts in processed meat products. In: *Processed Meats: Improving Safety, Nutrition and Quality*, edited by J.P. Kerry and J.F. Kerry. Cambridge, UK: Woodhead Publishing Ltd., in press.

Berdagué, J.L., Denoyer, C., le Quéré, J.C., and Semon, E. 1991. Volatile compounds of dry-cured ham. *J. Agric. Food Chem.* 39:1257–1261.

Bolzoni, L., Barbieri, G., and Virgili, R. 1996. Changes in volatile compounds of Parma ham during maturation. *Meat Sci.* 43:301–310.

Buscailhon, S., Berdagué, J.L., Bousset, J., Cornet, M., Gandemer, G., Touraille, C., and Monin, G. 1994c. Relations between compositional traits and sensory qualities of French dry-cured ham. *Meat Sci.* 37:229–243.

Buscailhon, S., Berdagué, J.L., and Monin, G. 1993. Time-related changes in volatile compounds of lean tissue during processing of French dry-cured ham. *J. Sci. Food Agric.* 63:69–75.

Buscailhon, S., Gandemer, G., and Monin, G. 1994b. Time-related changes in intramuscular lipids of French dry-cured ham. *Meat Sci.* 37:245–255.

Campbell-Platt, G. 1995. Fermented meats—A world perspective. In: *Fermented Meats*, edited by G. Campbell-Platt and P.E. Cook, pp. 39–52. London, UK: Blackie Academic & Professional.

Careri, M., Mangia, A., Barbieri, G., Bolzoni, L., Virgili, R., and Parolari, G. 1993. Sensory property relationships to chemical data of Italian type dry-cured ham. *J. Food Sci.* 58:968–972.

Cassens, R.G. 1995. Use of sodium nitrite in cured meats today. *Food Technol.* 49:72–81.

Córdoba, J.J., Antequera, T., García, C., Ventanas, J., López-Bote, C., and Asensio, M.A. 1994. Evolution of free amino acids and amines during ripening of Iberian cured ham. *J. Agric. Food Chem.* 42:2296–2301.

Costa-Corredor, A., Muñoz, I., Arnau, J., and Gou, P. 2010. Ion uptakes and diffusivities in pork meat brine-salted with NaCl and K-lactate. *LWT—Food Sci. Technol.* 43:1226–1233.

Coutron-Gambotti, C. and Gandemer, G. 1999. Lipolysis and oxidation in subcutaneous adipose tissue during dry-cured ham processing. *Food Chem.* 64:95–101.

Estevez, M., Morcuende, D., Ventanas, J., and Ventanas, S. 2007. Mediterranean products. In: *Handbook of Fermented Meat and Poultry*, edited by F. Toldrá, Y.H. Hui, I. Astiasarán, W.K. Nip, J.G. Sebranek, E.T.G. Silveira, L.H. Stahnke, and R. Talon, pp. 393–405. Ames, IA: Blackwell Publishing.

Fadda, S., Sanz, Y., Vignolo, G., Aristoy, M.C., Oliver, G., and Toldrá, F. 1999a. Hydrolysis of pork muscle sarcoplasmic proteins by *Lactobacillus curvatus* and *Lactobacillus sake*. *Appl. Environ. Microbiol.* 65:578–584.

Fadda, S., Sanz, Y., Vignolo, G., Aristoy, M.C., Oliver, G., and Toldrá, F. 1999b. Characterization of muscle sarcoplasmic and myofibrillar protein hydrolysis caused by *Lactobacillus plantarum*. *Appl. Environ. Microbiol.* 65:3540–3546.

Flores, M., Aristoy, M.C., Spanier, A.M., and Toldrá, F. 1997a. Non-volatile components effects on quality of Serrano dry-cured ham as related to processing time. *J. Food Sci.* 62:1235–1239.

Flores, M., Aristoy, M.C., and Toldrá, F. 1993. HPLC purification and characterization of porcine muscle aminopeptidase B. *Biochimie* 75:861–867.

Flores, M., Aristoy, M.C., and Toldrá, F. 1996. HPLC purification and characterization of soluble alanyl aminopeptidase from porcine skeletal muscle. *J. Agric. Food Chem.* 44:2578–2583.

Flores, M., Aristoy, M.C., and Toldrá, F. 1997b. Curing agents affect aminopeptidase activity from porcine skeletal muscle. *Z. Lebensm.-Unters. Forsch.* 205:343–346.

Flores, M., Barat, J.M., Aristoy, M.C., Peris, M.M., Grau, R., and Toldrá, F. 2006a. Accelerated processing of dry-cured ham. Part 2. Influence of brine thawing/salting operation on proteolysis and sensory acceptability. *Meat Sci.* 72:766–772.

Flores, M., Grimm, C.C., Toldrá, F., and Spanier, A.M. 1997c. Correlations of sensory and volatile compounds of Spanish Serrano dry-cured ham as a function of two processing times. *J. Agric. Food Chem.* 45:2178–2186.

Flores, M., Marina, M., and Toldrá, F. 2000. Purification and characterization of a soluble methionyl aminopeptidase from porcine skeletal muscle. *Meat Sci.* 56:247–254.

Flores, M., Sanz, Y., Spanier, A.M., Aristoy, M.C., and Toldrá, F. 1998a. Contribution of muscle and microbial aminopeptidases to flavor development in dry-cured meat products. In: *Food Flavor: Generation, Analysis and Process Influence*, edited by Contis et al., pp. 547–557. Amsterdam, The Netherlands: Elsevier Science BV.

Flores, M., Soler, C., Aristoy, M.C., and Toldrá, F. 2006b. Effect of brine thawing–salting for time reduction in Spanish dry-cured ham manufacturing on proteolysis and lipolysis during salting and post-salting period. *Eur. Food Res. Technol.* 222:509–515.

Flores, M., Spanier, A.M., and Toldrá, F. 1998b. Flavour analysis of dry-cured ham. In: *Flavor of Meat, Meat Products and Seafoods*, edited by F. Shahidi, pp. 320–341. London, UK: Blackie Academic & Professional.

García, C., Berdagué, J.L., Antequera, T., López-Bote, C., Córdoba, J.J., and Ventanas, J. 1991. Volatile compounds of dry cured Iberian ham. *Food Chem.* 41:23–32.

García-Garrido, J.A., Quiles, R., Tapiador, J., and Luque, M.D. 2000. Activity of cathepsin B, D, H and L in Spanish dry-cured ham of normal and defective texture. *Meat Sci.* 56:1–6.

Hinrichsen, L.L. and Pedersen, S.B. 1995. Relationship among flavor, volatile compounds, chemical changes and microflora in Italian-type dry-cured ham during processing. *J. Agric. Food Chem.* 43:2932–2940.

Jiménez-Colmenero, F., Ventanas, J., and Toldrá, F. 2010. Nutritional composition of dry-cured ham and its role in a healthy diet. *Meat Sci.* 84:585–593.

Kemp, J.D., Abidoye, D.F.O., Langlois, B.E., Franklin, J.B., and Fox, J.D. 1980. Effect of curing ingredients, skinning, and boning on yield, quality, and microflora of country hams. *J. Food Sci.* 45:174–177.

Leak, F.W., Kemp, J.D., Langlois, B.E., and Fox, J.D. 1984. Effect of tumbling and tumbling time on quality and microflora of dry-cured hams. *J. Food Sci.* 49:695–698.

Luna, G., Aparicio, R., and García-González, D.L. 2006. A tentative characterization of white dry-cured hams from Teruel (Spain) by SPME-GC. *Food Chem.* 97:621–630.

Marriott, N.G., Graham, P.P., and Claus, J.R. 1992. Accelerated dry curing of pork legs (hams): A review. *J. Muscle Foods* 3:159–168.

Marriott, N.G., Graham, P.P., Shaffer, C.K., and Phelps, S.K. 1987. Accelerated production of dry cured hams. *Meat Sci.* 19:53–64.

Marriott, N.G., Kelly, R.F., Shaffer, C.K., Graham, P.P., and Boling, J.W. 1985. Accelerated dry curing of hams. *Meat Sci.* 15:51–62.

Marriott, N.G., Tracy, J.B., Kelly, R.F., and Graham P.P. 1983. Accelerated processing of boneless hams to dry-cured state. *J. Food Prot.* 46:717–721.

Martin, J., Córdoba, J.J., Antequera, T., Timón, M.L., and Ventanas, J. 1998. Effects of salt and temperature on proteolysis during ripening of Iberian ham. *Meat Sci.* 49:145–153.

McCain, G.R., Blumer, T.N., Craig, H.B., and Steel, R.G. 1968. Free amino acids in ham muscle during successive aging periods and their relation to flavor. *J. Food Sci.* 33:142–146.

Molina, I. and Toldrá, F. 1992. Detection of proteolytic activity in microorganisms isolated from dry-cured ham. *J. Food Sci.* 57:1308–1310.

Monin, G., Marinova, P., Talmant, A., Martin, J.F., Cornet, M., Lanore, D., and Grasso, F. 1997. Chemical and structural changes in dry-cured hams (Bayonne hams) during processing and effects of the dehairing technique. *Meat Sci.* 47:29–46.

Montgomery, R.E., Kemp, J.D., and Fox, J.D. 1976. Shrinkage, palatability, and chemical characteristics of dry-cured country ham as affected by skinning procedure. *J. Food Sci.* 41:1110–1115.

Mora, L., Sentandreu, M.A., Koistinen, K.M., Fraser, P.D., Toldrá, F., and Bramley, P.M. 2009. Naturally generated small peptides derived from myofibrillar proteins in Serrano dry-cured ham. *J. Agric. Food Chem.* 57:3228–3234.

Mora, L., Sentandreu, M.A., and Toldrá, F. 2010. Identification of small troponin T peptides generated in dry-cured ham. *Food Chem.* 123:691–697.

Mora, L., Sentandreu, M.A., and Toldrá, F. 2011a. Intense degradation of myosin light chain isoforms after dry-cured ham processing. *J. Agric. Food Chem.*, in press.

Mora, L., Valero, M.L., Del Pino, M.M.S., Sentandreu, M.A., and Toldrá, F. 2011b. Small peptides released from muscle glycolytic enzymes during dry-cured ham processing. *J. Proteomics*, in press.

Motilva, M.J. and Toldrá, F. 1993. Effect of curing agents and water activity on pork muscle and adipose subcutaneous tissue lipolytic activity. *Z. Lebensm.-Unters. Forsch.* 196:228–231.

Motilva, M.J., Toldrá, F., Aristoy, M.C., and Flores, J. 1993b. Subcutaneous adipose tissue lipolysis in the processing of dry-cured ham. *J. Food Biochem.* 16:323–335.

Motilva, M.J., Toldrá, F., and Flores, J. 1992. Assay of lipase and esterase activities in fresh pork meat and dry-cured ham. *Z. Lebensm.-Unters. Forsch.* 195:446–450.

Motilva, M.J., Toldrá, F., Nadal, M.I., and Flores, J. 1994. Pre-freezing hams affects hydrolysis during dry-curing. *J. Food Sci.* 59:303–305.

Motilva, M.J., Toldrá, F., Nieto, P., and Flores, J. 1993a. Muscle lipolysis phenomena in the processing of dry-cured ham. *Food Chem.* 48:121–125.

Nuñez, F., Rodríguez, M.M., Bermúdez, M.E., Córdoba, J.J., and Asensio, M.A. 1996a. Composition and toxigenic potential of the mould population of dry-cured Iberian ham. *Int. J. Food Microbiol.* 32:185–197.

Nuñez, F., Rodríguez, M.M., Córdoba, J.J., Bermúdez, M.E., and Asensio, M.A. 1996b. Yeast population during ripening of dry-cured Iberian ham. *Int. J. Food Microbiol.* 29:271–280.

Parolari, G. 1996. Review: Achievements, needs and perspectives in dry-cured ham technology: The example of Parma ham. *Food Sci. Technol. Int.* 2:69–78.

Parolari, G., Virgili, R., and Schivazzappa, C. 1994. Relationship between cathepsin B activity and compositional parameters in dry-cured hams of normal and defective texture. *Meat Sci.* 38:117–122.

Pérez-Juan, M., Flores, M., and Toldrá, F. 2006. Generation of volatile flavour compounds as affected by the chemical composition of different dry-cured ham sections. *Eur. Food Res. Technol.* 222:658–666.

Pérez-Juan, M., Flores, M., and Toldrá, F. 2007. Effect of ionic strength of different salts on the binding of volatile compounds to sarcoplasmic protein homogenates in model systems. *Food Res. Int.* 40:687–693.

Puolanne, E. 1982. Dry cured hams—European style. Proceedings of the Reciprocal Meat Conference, Vol. 35, Blacksburg, VA, pp. 49–52.

Rico, E., Toldrá, F., and Flores, J. 1990. Activity of cathepsin D as affected by chemical and physical dry-curing parameters. *Z. Lebensm.-Unters. Forsch.* 191:20–23.

Rico, E., Toldrá, F., and Flores, J. 1991. Effect of dry-curing process parameters on pork muscle cathepsins B, H and L activities. *Z. Lebensm.-Unters. Forsch.* 193:541–544.

Ripollés, S., Campagnol, P.C.B., Armenteros, M., Aristoy, M.C., and Toldrá, F. 2011. Influence of partial replacement of NaCl for KCl, $CaCl_2$ and $MgCl_2$ in lipolysis and lipid oxidation on dry-cured ham. *Meat Sci.*, submitted.

Rodríguez-Núñez, E., Aristoy, M.C., and Toldrá, F. 1995. Peptide generation in the processing of dry-cured ham. *Food Chem.* 53:187–190.

Rosell, C.M. and Toldrá, F. 1996. Effect of curing agents on *m*-calpain activity throughout the curing process. *Z. Lebensm.-Unters. Forchs.* 203:320–325.

Rosell, C.M. and Toldrá, F. 1998. Comparison of muscle proteolytic and lipolytic enzyme levels in raw hams from Iberian and White pigs. *J. Sci. Food Agric.* 76:117–122.

Ruiz, J., Ventanas, J., Cava, R., Andrés, A., and García, C. 1999. Volatile compounds of dry-cured Iberian ham as affected by the length of the curing process. *Meat Sci.* 52:19–27.

Ruiz, J., Ventanas, J., Cava, R., Andrés, A.J., and García, C. 2000. Texture and appearance of dry cured hams as affected by fat content and fatty acid composition. *Food Res. Int.* 33:91–95.

Sabio, E., Vidal-Aragón, M.C., Bernalte, M.J., and Gata, J.L. 1998. Volatile compounds present in six types of dry-cured ham from south European countries. *Food Chem.* 61:493–503.

Sanz, Y., Fadda, S., Vignolo, G., Aristoy, M.C., Oliver, G., and Toldrá, F. 1999a. Hydrolytic action of *Lactobacillus casei* CRL 705 on pork muscle sarcoplasmic and myofibrillar proteins. *J. Agric. Food Chem.* 47:3441–3448.

Sanz, Y., Fadda, S., Vignolo, G., Aristoy, M.C., Oliver, G., and Toldrá, F. 1999b. Hydrolysis of muscle myofibrillar proteins by *Lactobacillus curvatus* and *Lactobacillus sake*. *Int. J. Food Microbiol.* 53:115–125.

Sentandreu, M.A., Stoeva, M.A., Aristoy, M.C., Laib, K., Voelter, W., and Toldrá, F. 2003. Identification of taste related peptides in Spanish Serrano dry-cured hams. *J. Food Sci.* 68:64–69.

Sentandreu, M.A. and Toldrá, F. 1998. Biochemical properties of dipeptidylpeptidase III purified from porcine skeletal muscle. *J. Agric. Food Chem.* 46:3977–3984.

Sentandreu, M.A. and Toldrá, F. 2000. Purification and biochemical properties of dipeptidylpeptidase I from porcine skeletal muscle. *J. Agric. Food Chem.* 48:5014–5022.

Sentandreu, M.A. and Toldrá, F. 2001a. Importance of dipeptidylpeptidase II in postmortem pork muscle. *Meat Sci.* 57:93–103.

Sentandreu, M.A. and Toldrá, F. 2001b. Dipeptidylpeptidase IV from porcine skeletal muscle: Purification and biochemical properties. *Food Chem.* 75:159–168.

Smallings, J.B., Kemp, J.D., Fox, J.D., and Moody, W.G. 1992. Effect of antemortem injection of papain on the tenderness and quality of dry-cured hams. *J. Anim. Sci.* 32:1107–1112.

Spanier, A.M., Flores, M., and Toldrá, F. 1999. Flavor differences due to processing in dry-cured and other ham products using conducting polymers (electronic nose). In: *Flavor Chemistry of Ethnic Foods*, edited by F. Shahihi and C.-T. Ho, pp. 169–183. New York: Kluwer Academic/Plenum Publishers.

Tabilo, G., Flores, M., Fiszman, S., and Toldrá, F. 1999. Postmortem meat quality and sex affect textural properties and protein breakdown of dry-cured ham. *Meat Sci.* 51:255–260.

Toldrá, F. 1992. The enzymology of dry-curing of meat products. In: *New Technologies for Meat and Meat Products*, edited by J.M. Smulders, F. Toldrá, J. Flores, and M. Prieto, pp. 209–231. Nijmegen, The Netherlands: Audet.

Toldrá, F. 1998. Proteolysis and lipolysis in flavour development of dry-cured meat products. *Meat Sci.* 49:S101–S110.

Toldrá, F. 2002. *Dry-Cured Meat Products*. Trumbull, CT: Food & Nutrition Press.

Toldrá, F. 2004a. Ethnic meat products: Mediterranean. In: *Encyclopedia of Meat Sciences*, edited by W. Jensen, C. Devine, and M. Dikemann, pp. 451–453. London: Elsevier Science Ltd.

Toldrá, F. 2004b. Curing: (b) Dry. In: *Encyclopedia of Meat Sciences*, edited by W. Jensen, C. Devine, and M. Dikemann, pp. 360–365. London: Elsevier Science Ltd.

Toldrá, F. 2006a. Biochemical proteolysis basis for improved processing of dry-cured meats. In: *Advanced Technologies for Meat Processing*, edited by L.M.L. Nollet and F. Toldrá, pp. 329–351. Boca Raton, FL: CRC Press.

Toldrá, F. 2006b. Biochemistry of processing meat and poultry. In: *Food Biochemistry & Food Processing*, edited by Y.H. Hui, W.K. Nip, L.M.L. Nollet, G. Paliyath, and B.K. Simpson, pp. 315–335. Ames, IA: Blackwell Publishing.

Toldrá, F. 2007a. Biochemistry of muscle and fat. In: *Handbook of Fermented Meat and Poultry*, edited by F. Toldrá, Y.H. Hui, I. Astiasarán, W.K. Nip, J.G. Sebranek, E.T.F. Silveira, L.H. Stahnke, and R. Talon, pp. 51–58. Ames, IA: Blackwell Publishing.

Toldrá, F. 2007b. Ham. In: *Handbook of Food Product Manufacturing*, edited by Y.H. Hui, R. Chandan, S. Clark, N. Cross, J. Dobbs, W.J. Hurst, L.M.L. Nollet, E. Shimoni, N. Sinha, E.B. Smith, S. Surapat, A. Titchenal, and F. Toldrá, Vol. 2, pp. 231–247. New York: John Wiley Interscience.

Toldrá, F. and Aristoy, M.C. 2010. Dry-cured ham. In: *Handbook of Meat Processing*, edited by F. Toldrá, pp. 351–362. Ames, IA: Blackwell Publishing.

Toldrá, F., Aristoy, M.C., and Flores, M. 2000. Contribution of muscle aminopeptidases to flavor development in dry-cured ham. *Food Res. Int.* 33:181–185.

Toldrá, F., Aristoy, M.C., and Flores, M. 2009. Relevance of nitrate and nitrite in dry-cured ham and its effects on aroma development. *Grasas Aceites* 60, 291–296.

Toldrá, F. and Etherington, D.J. 1988. Examination of cathepsins B, D, H and L activities in dry cured hams. *Meat Sci.* 23:1–7.

Toldrá, F. and Flores, M. 1998. The role of muscle proteases and lipases in flavor development during the processing of dry-cured ham. *CRC Crit. Rev. Food Sci. Nutr.* 38:331–352.

Toldrá, F., Flores, M., and Aristoy, M.C. 1995. Enzyme generation of free amino acids and its nutritional significance in processed pork meats. In: *Food Flavors: Generation, Analysis and Process Influence*, edited by G. Charalambous, pp. 1303–1322. Amsterdam, The Netherlands: Elsevier Science Publishers BV.

Toldrá, F., Flores, M., Aristoy, M.C., Virgili, R., and Parolari, G. 1996. Pattern of muscle proteolytic and lipolytic enzymes from light and heavy pigs. *J. Sci. Food Agric.* 71:124–128.

Toldrá, F., Flores, M., Navarro, J.L., Aristoy, M.C., and Flores, M. 1997. New developments in dry-cured ham. In: *Chemistry of Novel Foods*, edited by H. Okai, O. Mills, A.M. Spanier, and M. Tamura, pp. 259–272. Carol Stream, IL: Allured Publishing Co.

Toldrá, F., Flores, M., and Sanz, Y. 2001. Meat fermentation technology. In: *Meat Science and Applications*, edited by Y.H. Hui, W.K. Nip, R.W. Rogers, and O.A. Young, pp. 537–561. New York: Marcel Dekker Inc.

Toldrá, F., Rico, E., and Flores, J. 1992. Activities of pork muscle proteases in cured meats. *Biochimie* 74:291–296.

Toldrá, F., Rico, E., and Flores, J. 1993. Cathepsin B, D, H and L activity in the processing of dry-cured-ham. *J. Sci. Food Agric.* 62:157–161.

33

Italian Salami:
Survey of Traditional Italian Salami,
Their Manufacturing Techniques,
and Main Chemical and Microbiological Traits

Lucia Aquilanti, Cristiana Garofalo, Andrea Osimani, and Francesca Clementi

CONTENTS

33.1 Description and Classification of Salami

Salami are food products manufactured with pork, beef, or veal, added with salt, spices, and sometimes herbs and/or other ingredients. The use of additives (preservatives) is also allowed in certain cases. Meat, cut into pieces or minced, is traditionally stuffed into natural casings made from cleaned gut turned inside–out, which gives the product its characteristic cylindrical shape. Today, natural gut can be replaced by synthetic casings made of collagen, cellulose, or even plastic, especially in the case of salami manufactured in industrial plants. Salami manufacturing is a very ancient strategy for the conservation of meat by fermentation, salting, drying, and, possibly, smoking. The composition of the bacterial population carrying out the fermentation has a key role in the determination of the sensory characteristics. These microbiota mainly include lactic acid bacteria (LAB), coagulase-negative cocci (CNC), coagulase-negative staphylococci (CNS) (Rantsiou and Cocolin 2006), and, to a lesser extent, yeasts and molds (Cook, 1995). The contribution of LAB to flavor is primarily due to their acidifying activity and to the production of volatile compounds through the fermentation of carbohydrates (Urso et al. 2006), while CNC and CNS participate in color stabilization, decomposition of peroxides, proteolysis, and lipolysis (Iacumin et al. 2006). The fermentation is a key phase of the curing process of salami, since at this stage, the main physical, microbiological, and biochemical transformations take place (Villani et al. 2007). These transformations can be summarized as follows: change of the composition of the meat microflora, drop in pH, reduction of nitrates to nitrites and of nitrites to nitric oxide, formation of nitrosomyoglobin, solubilization and gelification of protein fractions, lipolysis, proteolysis, oxidation of organic compounds, and dehydration (Casaburi et al. 2007).

33.2 Art of Manufacturing Italian Salami

In Italy, the main features of pig breeding differ considerably from those of other European countries, where pigs weighing less than 100 kg are slaughtered to sell fresh meats. On the contrary, Italian swine production typically involves animals with a higher weight, which are commonly referred to as "heavy pigs." Meat destined for the manufacture of high-quality salami has to possess a high capacity for retaining juices, an appropriate fat content, and the absence of off-flavors and off-odors. The peculiarities of Italian swine meat derive from animal genetics, feeding, age, and weight at the time of slaughtering. The pigs must have specific features such as large size, no pigmented skin, and white bristles. Their body structure has to be compact, cylindrical, and with plenty of muscles; the skeleton and limbs have to be solid, the back-loin line straight or slightly curved, and the shoulders muscular and tight to the body. Eligible animals should also have wide and pulpy loins, together with a long, wide, and straight rump.

The slaughtering process, which greatly affects meat quality, consists of a series of operations, which starts when the animals are put down and finishes when the meat cuts are chilled. Improper slaughtering may cause the weight loss of the swine carcasses, discarding of skin, muscles, blood vessels, and bones, and even the deterioration of meat.

Italian salami are manufactured with different meat cuts deriving from swine carcasses deprived of blood, the digestive system, bladder, heart, breathing apparatus, spleen, and liver. Carcasses are first divided into two halves, referred to as *mezzene*, and further subjected to dissection procedures, which vary greatly depending on the geographical area and the final destination of the meat cuts.

Overall, the quality of Italian salami is influenced by multiple factors involving either the raw materials or the processing conditions. On the whole, the different procedures adopted in the various geographical areas widely contribute to the definition of the unique sensory properties of traditional Italian salami. As a general rule, salami produced in Southern Italy have more flavor than those produced in Northern Italy, which are generally characterized by a sweeter taste. However, irrespective of the area of production, the continuous effort to improve the production techniques and, at the same time, to respect local traditions is the "secret" of the high quality of traditional Italian salami, which are popular throughout the world.

33.3 History of Salami

The history of salami dates back to the Egyptian, Etruscan, and Roman eras. Remains of salami have been retrieved from the tomb of Ramses III (1166 B.C.), whereas in the archaeological site of Forcello (5th century B.C.), in the province of Mantua (Northern Italy), about 30,000 remains of animal bones from two- or three-year-old pigs have been found. The concept of charcuterie probably originated in that period.

A description of a method for the preservation of the leg ham trim can be found in *De Agricoltura* written by Cato the Censor (about 234–149 B.C.). The ancient procedure, consisting of the addition of salt and olive oil to the meat and its further drying, somehow resembles the modern manufacturing of cured ham. Thereafter, in the 1st century B.C., Marco Terenzio Varrone, in the *De re rustica*, praised the great ability of the Gauls in the preservation of swine meat and narrated that the Romans were used to importing pig quarters and cured ham from the Gallic lands in the Po valley, as is also confirmed by Polibio (206–124 B.C.) and Strabone (about 58 B.C., between 21 and 25 B.C.). The first statement about the great importance of swine meat in human nutrition dates back to Hippocrates in the early 5th century B.C., whereas Plinio the Old (23 B.C., after 1 September 79 B.C.), the author of the biggest Roman encyclopedia, wrote that this meat represents a great delight for the palate, offering almost 50 different flavors, whereas meats from all the other animals have just their own taste. The great popularity of swine meat with the Romans is also confirmed by ancient documents dealing with the manufacture of *farcimen myrtatum*, a salami flavored with myrtle berries, and *lucanica*, a smoked salami added with ground pepper, cumin, rue, parsley, bay, salt, bacon, and fennel seeds. With the successive Barbarian invasions, pigs became the most important meat source for both the villages and rural areas, being mainly destined for the manufacture of charcuterie products like ham, bacon, and salami.

In the Middle Ages, pig grazing was so important that woods were measured according to their ability to nourish pigs, rather than to their extension. The Longobards were accustomed to eating huge amounts of meat from wild pigs, which was preserved through the addition of salt, mainly extracted from the natural salt mines located in Salsomaggiore, in the proximity of Parma. Between the 12th and the 17th centuries, a great development in jobs connected with swine meat processing took place. This period saw the origin of the *norcino*, a term meaning "pork butcher from Norcia," which is a village in Central Italy where a lot of new charcuterie products began to be produced. These workers started to organize themselves into guilds or confraternities, such as the *Corporazione dei Salaroli* (Guild of Salters) in Bologna and the *Confraternita dei Facchini di San Giovanni* (Confraternity of Saint John's Servants) in Florence. With time, pig breeding and the consumption of swine meat products progressively gained importance, going from the triumphs of the Renaissance period, when there were great developments in the art of gastronomy and swine meat appeared on the most sumptuous banquet tables, up to the 19th century, when the first charcuterie shops opened.

Some documents kept in the State Archives of Genoa report that in the period between the 18th and 19th centuries, Genoese pirates carefully prepared their long-distance voyages by stowing large amounts of ham, renowned for its long life and high nutritional value.

In 1615, Pope Paul V recognized the Confraternity of Pork Butchers, which was dedicated to the Saints Benedict and Scholastica from Norcia. His successor, Pope Gregorius XV, elevated it to an Archconfraternity, which in 1677 was joined by the Pork Butchers and the Empirical Pork Physicians of the University of Norcia and Cascia (a neighboring village). From that time onward, the figure of the "graduated, blessed, and patented norcino" became well known outside the Papal State and still survives in some rural communities, especially in its native area. Part of this heritage has been transferred from homemade products to more industrialized and widespread productions, which, to a certain extent, still respect the ancient recipes.

33.4 European and Italian Regulations concerning Typical and Traditional Foods

Three EU schemes known as Protected Designation of Origin (PDO), Protected Geographical Indication (PGI) ruled by Council Regulation (EC) No. 510/2006 of 20 March 2006, and Traditional Speciality

TABLE 33.1

Italian Salami with PDO or PGI Designation

EU Designation	Italian Salami	Manufacturing Area
PDO	*Salame Brianza*	Lombardy
	Salame di Varzì	Lombardy
	Salame Piacentino	Emilia-Romagna
	Salsiccia di Calabria	Calabria
	Soppressata di Calabria	Calabria
	Salamini Italiani alla Cacciatora	Lombardy, Piedmont, Veneto, Friuli Venezia Giulia, Emilia-Romagna, Tuscany, Latium, Marche, Abruzzo, and Umbria
PGI	*Salame d'Oca di Mortara*	Lombardy
	Salame Cremona	Lombardy
	Ciauscolo	Marche
	Salame di Sant'Angelo	Sicily

Guaranteed (STG) ruled by Council Regulation (EC) No. 509/2006 of 20 March 2006 promote and protect the names of agricultural products and foodstuffs. These EU schemes encourage diverse agricultural production, protect product names from misuse and imitation, and help consumers by giving them information concerning the specific characteristics of the products. In more detail, PDO covers agricultural products and foodstuffs that are produced, processed, and prepared in a given geographical area according to well-established techniques, while PGI covers agricultural products and foodstuffs that have a less strict link with the geographical area; in this case, at least one stage of production, processing, or preparation must take place in the specific geographical area.

At present, a few Italian salami have received the PDO or PGI designation, as illustrated in Table 33.1, but a much larger number of fermented meat specialities are manufactured all over Italy, according to ancient local traditions passed down from generation to generation. Most of these salami are included in the official list of traditional Italian foods published by the Italian Ministry of Agriculture (http://www.politicheagricole.it/ProdottiQualita/ProdottiTradizionali).

33.5 Overview of Some Typical Italian Salami

In the following paragraphs, the manufacturing techniques of some traditional Italian salami and their main chemical and microbiological traits are described. This survey is divided into three sections that focus on the salami manufactured in Northern, Central, and Southern Italy, respectively.

33.5.1 Northern Italy

33.5.1.1 Salame di Varzì PDO

History. Varzì is a small town on the river Stàffora, which takes its name from the surrounding valley, in the province of Pavia. The town is famous both in Italy and abroad thanks to the PDO salami called *Salame di Varzì* whose production dates back to the Longobard invasions, when this barbarian population is hypothesized to have introduced the recipe and manufacturing of this product in Lombardy. Indeed, documents going back to the 12th century indicate that the marquises Malaspina, a noble Lombard family, used to offer this delicious fermented meat speciality to their guests (see Figure 33.1).

Raw materials. Salame di Varzì PDO is currently manufactured in a wide geographical area, including 15 municipalities located in the Oltrepò Pavese Mountain Community, in the province of Pavia, with heavy pigs (weight of at least 150 kg) bred, slaughtered, and bled in the Piedmont, Lombardy, or Emilia-Romagna regions. Lean cuts of pork, with the exception of those obtained from the head, are used to prepare the meat paste (shoulder, ham, and *lonza* and *coppa*, two lean meat cuts obtained from the loin and the neck, respectively); these cuts have to be preliminarily deprived of fat and nerve tissues,

FIGURE 33.1 *Salame di Varzì* PDO.

and added with fat obtained from the jowl, the shoulder butt, the belly, the loin, and the ham. Freezing of meats is not allowed before mincing. Marine salt, sodium nitrate, black pepper in whole grains, and filtered garlic infusion in red wine are added to the meat.

Manufacturing process. The meat is mechanically minced and extruded through holes with a diameter of 12 mm. Once salted and added with the remaining ingredients (and additives), the meat paste is stuffed into natural casings, derived from different portions of the pig gut, namely, *pelato suino, budello gentile* (obtained from the rectum), or *doppio pelato suino cucito*, also known as *cucito doppio*. Salami with a diameter of less than 5 cm can also be manufactured with *budello torto* or *aggetta*, obtained from the bovine small intestine. The product is then manually pierced and tied at the ends with a string. Drying and ripening are carried out in aerated rooms; the duration of these steps varies, depending on the salami diameter, from a minimum of 45 days to over 180 days. If *budello gentile* and *cucito doppio* are used, ripening lasts at least 100 days, irrespective of the salami diameter, due to the thickness of these natural casings.

Appearance and taste. Salame di Varzì PDO has a cylindrical shape and a soft consistency; the slice has a dark red color, which is marbled with white streaks of fat. The flavor is sweet and delicate; the characteristic aroma is strongly dependent on the duration of the ripening.

Distinctive feature. Salame di Varzì PDO is produced with coarsely minced meat added with 30%–33% fat. The size and weight of the salami may be variable, going up to 2 kg. Depending on the size, this salami is referred to as *Salame di Varzì–Filzetta* (weight: 0.5–0.7 kg; ripening duration: minimum of 45 days), *Salame di Varzì–Filzettone* (weight: 0.5–1 kg; ripening duration: minimum of 60 days), *Salame di Varzì–Sottocrespone* (weight: 1–2 kg; ripening duration: minimum of 120 days), and *Salame di Varzì–Cucito a budello doppio* (weight: 1 to over 2 kg; ripening duration: minimum of 180 days). *Salame di Varzì* PDO owes its particular sensory characteristics to the climatic conditions of the Staffora valley and to the abundant addition of salt.

33.5.1.2 Salame Brianza PDO

History. Salame Brianza PDO is traditionally produced in a well-defined territory in the Lombardy region known as Brianza, including the provinces of Lecco, Como, and Milan. The first news on the origins of this Italian salami dates back to the middle of the 16th century, but more detailed and reliable information was provided by Riberti's writings in the 19th century and by those of Faelli in the 20th century, which testify the presence of *Salame Brianza* at the Universal Exposition in Paris. Starting from these ancient origins, *Salame Brianza* has continued to be produced until today as an artisan product and since the 1970s, when some artisan producers became industrial manufacturers, on a larger scale (see Figure 33.2).

Raw materials. Salame Brianza PDO is currently manufactured with the meat from heavy pigs, weighing at least 150 kg, bred in Lombardy, Emilia-Romagna, and Piedmont. The meat paste is prepared with the following meat cuts: shoulder deprived of bones and nerve tissue, meat trimmings from ham, throat and bacon deprived of soft fat, salt, and pepper in whole grains or powder. The use of frozen meat is allowed, as well as the addition of wine, sugars (saccharose, dextrose, fructose, and/or lactose), starter cultures, garlic, and preservative agents (sodium or potassium nitrate, sodium or potassium nitrite, ascorbic acid, and its sodium salt).

FIGURE 33.2 *Salame Brianza* PDO.

Manufacturing process. Shoulders, weighing no less than 5 kg, and the other meat cuts are mechanically minced and extruded through holes of varying size depending on the weight of the salami: 4–4.5 and 7–8 mm for the salami weighing less than and more than 300 g, respectively. The ingredients may be mixed at atmospheric pressure or under vacuum. The meat paste is introduced into natural or synthetic casings and tied with a string or a net. Drying is carried out at temperatures ranging from 15°C to 25°C (heat drying) or from 3°C to 7°C (cold drying) and allows a rapid dehydration of the salami within the first days of ripening. This is carried out in well-aerated rooms at temperatures ranging from 9°C to 13°C. The duration of drying and ripening varies greatly depending on the diameter of the salami. According to the PDO production regulations, the end product has to be characterized by the following physicochemical characteristics: total protein, minimum of 23%; collagen/protein ratio, maximum of 0.10; water/protein ratio, maximum of 2.00; fat/protein ratio, maximum of 1.5; pH > 5.3. As far as the microbiological traits are concerned, the load of mesophilic aerobes should reach at least 1×10^7 CFU/g, with a dominance of LAB, CNC, and CNS.

Appearance and taste. *Salame Brianza* PDO has a cylindrical shape and a compact nonelastic consistency. The slice is homogeneous, with a uniform ruby red color. The aroma is delicate and characteristic; the taste is very sweet and never acidulous. Due to its geographical position in the hills and the typical climate of the area, Brianza is particularly suitable for the drying and ripening of *Salame Brianza* PDO, thus contributing to conferring specific sensory properties to this product.

33.5.1.3 Salame Piacentino PDO

History. The production area of *Salame Piacentino* PDO includes the whole territory of the province of Piacenza, in Emilia-Romagna. Its origins are deeply rooted in the past. In fact, in the Piacenza area, pig breeding dates back to the 1st millennium B.C., as documented by some archaeological findings. Starting from this ancient tradition, the manufacture of *Salame Piacentino* continued until the 18th century, when this salami became popular in the courts of France and Spain; Cardinal Giulio Alberoni even used this salami for diplomatic purposes: notwithstanding his numerous commitments, he made sure that *Salame Piacentino* was regularly supplied to Queen Elisabetta Farnese, who greatly appreciated this meat specialty. At the beginning of the 20th century, local salami manufacturing started to become more oriented toward industrial production, although both artisan and industrial production still conform closely to the traditional procedures followed by expert pork butchers (*massalein* in the local dialect) and handed down from father to son (see Figure 33.3).

Raw materials. *Salame Piacentino* PDO is salami obtained from heavy pigs bred in Lombardy and Emilia-Romagna. For the lean meats, different cuts are selected, excluding those from the head, whereas for the fat meats, used in a proportion of 10%–30% of the lean meat, backfat and bacon deprived of soft fat are used. The ingredients added to the meat paste are salt, whole or ground grains of black or white pepper, sugars, garlic, and wine. Sodium L-ascorbate and potassium nitrate are the allowed additives, the latter in a maximum dose of 150 mg/kg, which corresponds to half the legal limit.

Manufacturing process. Meat is first cut into small pieces, then mechanically minced and extruded through large holes with a diameter greater than 10 mm. The meat paste is then added with the other ingredients and stuffed into pig gut, tied with a string, pierced, and dried at a temperature between 15°C

FIGURE 33.3   *Salame Piacentino* PDO.

and 25°C and relative humidity (RH) in the range 40%–90%. Ripening is performed at temperatures ranging from 12°C to 19°C and RH of 70%–90% and lasts no less than 45 days. During ripening, aeration and exposure to light are allowed. According to the PDO production regulations, the end product has to be characterized by the following physicochemical characteristics: humidity, 27%–50%; total protein, 23.5–33.5%; fat, 16%–35%; salt content, 3%–5%; ashes, 4%–6.5%; collagen, 0.5%–4%; and pH 5.4–6.5.

Appearance and taste. Salame Piacentino PDO has a cylindrical shape with a variable weight, which is between 400 g and 1 kg. The slice has a dark red color, with small spots of white-rosy fat. The flavor is sweet and delicate; the characteristic aroma is strongly dependent on the duration of the ripening.

33.5.1.4 Salame Cremona PGI

History. The production area of *Salame Cremona* PGI includes the following regions: Lombardy, Piedmont, Veneto, and Emilia-Romagna. The main historical references, which clearly testify the origins of *Salame Cremona* and its link to the territory, date back to the year 1231 and are conserved in the State Archive of Cremona. These documents confirm the existence of pork and the pork meat trade between the territory of Cremona and the neighboring states. The importance of this salami also emerges from other documents dating back to the Renaissance period. Since then, its production area, which at the time of the Longobards corresponded to the current territory of the province of Cremona, has extended further, covering the whole basin of the Po valley and the foothills of the Alps and Apennines by around the middle of the 19th century (see Figure 33.4).

Raw materials. The raw material for the manufacture of *Salame Cremona* PGI comes from pigs born, bred, and slaughtered in the regions of Friuli Venezia Giulia, Veneto, Lombardy, Piedmont, Emilia-Romagna, Umbria, Tuscany, Marche, Abruzzo, Latium, and Molise. The meat cuts destined for the production of this salami are obtained from both the striated muscle and the fat fractions of swine carcasses. Salt, whole or ground pepper, other spices, and crushed garlic or garlic paste are the ingredients currently added to the minced meat. Other ingredients allowed are white or red wine and sugars (saccharose, dextrose, fructose, and/or lactose). The use of starter cultures and preservatives (potassium and/or sodium nitrite, ascorbic acid, and its sodium salt) is also permitted.

FIGURE 33.4   *Salame Cremona* PGI.

Manufacturing process. Lean meats and fat are carefully trimmed, mechanically minced, and extruded through 6-mm holes. The meat cuts are salted before mincing, whereas the remaining ingredients are added to the meat paste. All the ingredients are mixed thoroughly under vacuum or at atmospheric pressure. The mixture is stuffed into swine, bovine, horse, or sheep gut. Tying with strings is carried out either manually or mechanically. The storing of the salami in refrigerated rooms at temperatures between 2°C and 10°C for a maximum of one day (cold drying) is allowed. Drying may alternatively be performed at temperatures between 15°C and 25°C (heat drying). The salami is ripened in chambers with adequate air circulation, at temperatures ranging from 11°C to 16°C for at least five weeks. The duration of ripening varies greatly depending on the diameter of the salami. According to the PGI production regulations, the end product has to weigh at least 500 g and to be characterized by the following physicochemical characteristics: total protein, minimum of 20.0%; collagen/protein ratio, maximum of 0.10; water/protein ratio, maximum of 2.00; fat/protein ratio, maximum of 2.0; pH ≥ 5.20. As far as the microbiological traits are concerned, the load of mesophilic aerobes should reach at least 1×10^7 CFU/g, with a dominance of LAB, CNC, and CNS.

Appearance and taste. Salame Cremona PGI is a very aromatic, spicy, and fragrant salami; it is characterized by a deep red color that fades gradually close to the white adipose tissue. The shape is cylindrical and slightly uneven; the consistency is soft, and the slices are compact and homogeneous.

Distinctive feature. Salame Cremona PGI is currently produced with modern techniques that fully respect the local traditions. The sensory properties of this salami are deeply influenced by the narrow district bordering the river Po, which is characterized by a very humid climate.

33.5.1.5 Salame Bergamasco

History. Salame bergamasco, known in the local dialect as *salàm de la bergamasca,* is exclusively produced in the province of Bergamo, in Lombardy. It originated in the rural peasant society of this area, but due to the lack of reference documents, it is difficult to establish when it was first produced. Although the year 1900 marks the transition from household to artisan production and even to industrial manufacture, in the valleys close to Bergamo household pig keeping still persists. It is strongly linked to tradition and represents a supplementary form of income and food supply for the local families. Following this tradition, the pigs are slaughtered once a year, usually in December, and, on the same day, the last products left over from the previous year are consumed (see Figure 33.5).

Raw materials. The raw meat used for the manufacture of *Salame bergamasco* in either artisan or industrial plants comes from local farms located in the lowland areas. Pig breeding is commonly carried out near dairies, since the pigs are generally fed with whey, in addition to chestnuts and acorns, thus contributing to making the meat particularly tasty. Red wine, refined marine salt, fresh garlic, black pepper, other spices, and sugars (dextrose, fructose, or sucrose) are added to the meat; the use of microbial starters is allowed, as well as the use of additives within the legal limits, except for thickeners and moistening agents. The use of milk and its derivatives is forbidden.

FIGURE 33.5 *Salame bergamasco.*

Manufacturing process. After a preliminary selection, meats are minced and added with salt, pepper, spices, garlic marinated in red wine, and additives, the latter being previously mixed together (if used). The meat paste is then stuffed into swine gut and hand-tied with string. The salami is dried for 5–10 days at temperatures gradually decreasing, from about 24°C to 13°C, at appropriate RH values; the ripening lasts at least 40 days at temperatures comprised between 8°C and 15°C with RH values of between 80% and 90%. At the end of the ripening process, the product is dry and is covered in a light gray-green patina, whose color becomes progressively more intense during the ripening process.

Appearance and taste. Salame bergamasco is covered with a unifom patina of molds, which is characterized by a white-gray to green color. The paste is compact, with the fat being well blended with the meat; the slice has a red color, which is neither too dark nor tending to orange. The texture is mushy and becomes more compact if the ripening process is prolonged. The flavor is delicate, with a salty-sweet taste; the aroma has a slight garlic scent, with a hint of wine.

Distinctive feature. Manufacturers believe that the specific climate is one of the most important factors that link this salami to the place of origin: although no scientific studies have provided evidence of a relationship between the peculiar traits of *Salame bergamasco* and the local geography or climate, manufacturers claim that this salami maintains its characteristics, even up to one year, only if stored in the production area; by contrast, its shelf life tends to reduce if it is sold outside the province.

33.5.1.6 Salame Mantovano

History. Salame mantovano is traditionally produced in the province of Mantua, where pigs are mainly bred for salami manufacture. Swine meat was already eaten in the Mantua area during the Etruscan era, as indicated by the findings from the Forcello archaeological site. This practice continued throughout the centuries up to the Renaissance, at the court of Isabella d'Este Gonzaga, the Marchioness of Mantua (1474–1539), where salami, manufactured with tongue or cooked sausages, were included in the daily diet (see Figure 33.6).

The *masin* or *masalin* was a very popular figure, known as *perfecto maestro de tal mestero*, which literally means "perfect master of salami manufacturing." This art is still the symbol of prosperity and wealth in the province of Mantua, and *Salame mantovano* is one of the most representative salami of this territory due to its unique features.

Raw materials. Lean and fat meats from mature pigs bred in the sole province of Mantua are used, in addition to salt, pepper, fresh peeled and crushed garlic, cloves, nutmeg, wine, or *grappa*, an Italian brandy produced with grape pomace. Potassium nitrate can be used. Each salami is produced with meats from a single animal. Frozen meat and starter cultures are not used.

Manufacturing process. Meats are deprived of nerve, connective, and soft fat tissues. After coarsely mincing through 8- to 12-mm plates, the meat paste is added with salt and other ingredients and stuffed into pig gut. The salami is tied manually and dried for 2–7 days. Ripening is performed in a natural, cool environment, with appropriate RH conditions, in order to allow the typical microbial population to develop. The state of the casing is continuously monitored, and the salami is periodically brushed.

Appearance and taste. Salame mantovano has a cylindrical shape and is typically covered by an external patina of white-gray molds. The weight varies between 0.5 and 3 kg. The meat paste has a strawberry-red

FIGURE 33.6 *Salame mantovano.*

color and a compact but soft texture, dotted with small white or pink spots due to the fat grains; the addition of pepper and garlic gives the salami its characteristic scent and unique taste.

Distinctive feature. The recipe for this salami varies greatly depending on the geographical area.

33.5.1.7 Salame Milano

History. Salame Milano, also known as *Crespone*, is one of the most popular Italian salami. It was originally manufactured in the two towns of Codogno and San Colombano al Lambro but is now widely produced all over the Lombardy region. In the territory of Milan, the term *salame* appeared for the first time in a document dated 1475, which chronicles the wedding of Costanzo I Sforza (1447–1483) and Camilla di Aragona. At that time, the terms *salziozone* or *cervellato* were also used to refer to this salami, but these names disappeared at the end of the 18th century (see Figure 33.7).

Raw materials. Salame Milano is produced with a blend of beef and pork meats, which are finely minced into rice-sized grains, salt, pepper, and garlic.

Manufacturing process. Lean and fat meats are minced through 3- to 3.5-mm plates. The paste is then spiced and stuffed into pig gut (*crespone*) or synthetic casings (the latter being preferred for large-scale production). The salami are dried at temperatures varying between 15°C and 25°C or 3°C and 7°C, and ripened at temperatures in the range 9°C–13°C for 3 to 9 weeks, according to the salami diameter (3 weeks for the salami with a diameter < 50 mm, 4 weeks for a diameter of 50–70 mm, 6 weeks for a diameter of 71–90 mm, and 9 weeks for a diameter of 91–110 mm).

Appearance and taste. The color is always bright red, almost ruby; the meat is firm and nonelastic with a typical rice-grain texture. The length varies from 20 to 60 cm, the diameter from 6 to 11 cm, and the weight from 2 to 3 kg. This salami has a distinctive aroma and a sweet, delicate flavor.

Distinctive feature. The appearance and taste of *Salame Milano* are very similar to those of the typical salami manufactured in Hungary, despite differences in the manufacturing process. The typical texture of *Salame Milano* is due to the use of a special mincer, called *finimondo*, which literally means "end of the world." This piece of equipment consists of two series of circular steel knives, mounted on two axes, which rotate in opposite directions, thus facilitating the mincing of fat and meat into fine grains and their uniform distribution in the paste.

33.5.1.8 Soprèssa Vicentina PDO

History. The production area for *Soprèssa Vicentina* PDO covers the whole province of Vicenza, in the Veneto region; this territory is delimited by the Small Dolomites and the Asiago tableland to the North

FIGURE 33.7 *Salame Milano.*

FIGURE 33.8 *Soprèssa Vicentina* PDO.

and by the Berici foothills to the South. The close link between *Soprèssa Vicentina* and this territory and the central role of this product in local traditional gastronomy is reflected in the great number of recipes that include this salami. The price list published by the Vicenza Chamber of Commerce in 1862 distinguished between *soprèssa* and *salame*, thus suggesting that it was already acknowledged in its own right. Since the middle of the last century, several village festivals have been dedicated to this salami. Nowadays, *Soprèssa Vicentina* continues to be produced according to the local traditions handed down from father to son (see Figure 33.8).

Raw materials. Soprèssa Vicentina PDO is produced with meat from heavy pigs (minimum weight: 130 kg) born and bred in the province of Vicenza. The finest pork cuts are used, namely, ham, shoulder, *coppa*, loin, bacon, and fat from the throat. These are added with salt, ground black pepper, finely ground spices (cinnamon, cloves, nutmeg, rosemary, and sometimes garlic), sugars, and potassium nitrate under the legal limits.

Manufacturing process. The meat cuts are cooled at a temperature ranging from 0°C to 3°C for at least 24 hours. They are deprived of bones, soft fat, and nerve tissue and further minced with 6- to 7-mm plates. The meat paste is cooled at 3°C–6°C and added with the other ingredients, which have previously been mixed together. The use of autochthonous selected starter cultures is allowed. The meat paste is accurately kneaded in order to amalgamate the lean and fat components, and then stuffed into natural casings with a diameter of at least 8 cm. The salami is exclusively tied with a string and dried for about 12 hours at a temperature ranging from 20°C to 24°C and further dried for 4–5 days at temperatures progressively decreasing from 22°C–24°C to 12°C–14°C. The ripening period has a variable duration, depending on the weight of the salami, which is comprised between 1 and 8 kg. According to the PDO production regulations, the end product has to be characterized by the following physicochemical characteristics: humidity, <55%; total protein, >15%; fat, 30%–43%; ashes, 3.5%–5%; pH 5.4 – 6.5.

As far as the microbiological traits are concerned, the bacterial community has to be dominated by LAB, CNC, and CNS.

Appearance and taste. Soprèssa Vicentina PDO has a cylindrical shape, and the external surface is covered in a whitish microbial patina, which appears naturally during ripening. The slice has a rosy color, tending to red, with a compact but soft consistency and a texture with medium-sized grains. The flavor is spicy, sometimes with a fragrance of herbs or garlic. The taste is delicate, slightly sweet, and peppery.

33.5.1.9 Microbiological Traits of Northern Italy Salami

The bacterial dynamics that come into play during the ripening of *Filzetta* (*Salame di Varzì* PDO) manufactured without the addition of starter cultures in an artisan plant located in Langhirano, in the province of Parma, were investigated by Conter et al. (2005). *Enterobacteriaceae*, pseudomonads, enterococci, yeasts, molds, LAB, as well as coagulase positive cocci (CPC) and CNC were enumerated, and 90 presumptive LAB isolates were subjected to a phenotype-based identification using the API 50 CH system (Biomerieux, France). The results obtained showed that lactobacilli were the dominating micro-flora during the ripening period, with the facultative hetero-fermentative species *Lactobacillus sakei* as the prevailing taxon. In the early stages of ripening, *Lactobacillus fermentum* was also found with a high frequency, while *Lactobacillus brevis* was only rarely isolated. Among coccal-shaped LAB, leuco-nostocs ascribed to *Leuconostoc mesenteroides* occurred with the highest frequency.

More recently, the chemical, microbiological, biochemical, and aromatic traits of *Salame di Varzì* PDO, *Salame Brianza* PDO, and *Salame Piacentino* PDO have been comparatively evaluated by Di Cagno et al. (2008). Significant differences in the gross composition, especially moisture, fat, total pro-tein, and nitrate concentration, were seen among the three types of salami, as a result of the numerous differences in both the ingredients and technological parameters used for their manufacture. By contrast, a generally high homogeneity was seen in the viable counts of the major microbial groups, with the exception of *Brochothrix thermosphacta*, enterococci, and molds. The molecular identification of LAB and CNC allowed the dominant species to be determined, namely, *Lactobacillus curvatus*, *L. sakei*, and *Staphylococcus xylosus*. The species *Pediococcus pentosaceus* and *Lactobacillus coryneformis* also occurred in *Salame Brianza* and *Salame Piacentino*, respectively, although with low frequencies. As far as the proteolysis is concerned, comparable sodium dodecyl sulfate–polyacrylamide gel electropho-resis (SDS-PAGE) profiles of the sarcoplasmic protein fractions were seen for the three meat products, whereas *Salame di Varzì* was characterized by myofibrillar protein band profiles, which were mark-edly different from *Salame Brianza* and *Salame Piacentino*. Significant differences also emerged in the total concentration of free amino acids (FAAs), which was highest in *Salame di Varzì*, followed by *Salame Brianza* and then *Salame Piacentino*. The solid phase microextraction/gas chromatography–mass spectroscopy analyses allowed 52 volatile compounds grouped in six chemical classes to be identi-fied, namely, alcohols, aldehydes, ketones, acids, esters, and terpenes. Once again, significant differences were seen in the levels of some volatile components found in the three salami. By contrast, none of the free fatty acids (FFAs) distinguished the three salami, although differences were seen in both the amount of saturated fatty acids (FA), highest in *Salame Brianza*, and of polyunsaturated FA, highest in *Salame di Varzì* and *Salame Piacentino*.

Three isolates collected from *Salame Milano* and preliminarily identified as *S. xylosus* (two cultures) and *Staphylococcus epidermidis* were used as reference strains by Blaiotta et al. (2003a) in a study aimed at comparing the reliability of polymerase chain reaction–denaturing gradient gel electrophoresis (PCR-DGGE) and internal spacer region (ISR)–PCR analysis of the 16S rDNA V3 region and the 16S–23S internal spacer, respectively, for the identification of CNC and CNS of meat origin. To our knowledge, no other published data are available about the species composition of the microbial community of this salami.

The LAB ecology of *Salame mantovano*, *Salame Cremona* PGI, and *Salame bergamasco* was inves-tigated by Cocolin et al. (2009) with culture-dependent and -independent methods. Eleven plants located in the provinces of Cremona (five plants), Bergamo (two plants), and Mantua (four plants), where starter cultures had never been used, were selected for the collection of the salami samples. In all three prod-ucts, the microbial viable counts revealed the dominance of LAB from the start of the ripening period, with minor contributions of CNC, yeasts, and molds, the latter being mainly found in *Salame man-tovano*. As far as the hygiene of the three salami is concerned, total coliforms and *Escherichia coli* always showed counts below 10^2–10^3 colony forming units (CFU)/g, whereas the load of *Staphylococcus aureus* was always below 50 CFU/g. From the isolation campaign, based on the selection of 15 LAB colonies at each sampling point (at 5, 30, and 60 days of ripening), two species were found to be common to the three salami, namely, *L. sakei* and *L. curvatus*, representing more than 90% of the total number of isolates. The species *Lactobacillus plantarum*, *Lactobacillus paraplantarum*, *Leuconostoc citreum*, and *Weissella hellenica* were also isolated with clearly lower frequencies. The cluster analysis of the ran-domly amplified polymorphic DNA (RAPD) patterns obtained with primer D8635 from the 293 *L. sakei* and 177 *L. curvatus* isolates revealed a high plant, rather than product, intraspecies biodiversity. The salami samples collected at 5, 30, and 60 days of ripening were also subjected to PCR-DGGE analysis, for a better understanding of their bacterial diversity. The DGGE profiles from the analysis of the DNA extracted directly from the salami samples confirmed the dominance of the species *L. sakei* and *L. curvatus*. Other species were sporadically identified, namely, *Lactobacillus algidus* plus *Le. mesenteroides/pseudomesenteroides* and *L. paraplantarum* in *Salame mantovano* at 5 and 60 days of ripening, respectively, and *Staphylococcus saprophyticus* in *Salame Cremona* at 5 days of ripening.

As far as *Soprèssa Vicentina* PDO is concerned, the bacterial ecology was investigated by Andrighetto et al. (2001). Fifty-three lactobacilli isolated from salami samples manufactured without the addition of starter cultures and collected at different ripening times (1 week, 2 and 6 months) were subjected to RAPD analysis with primers M13 and D8635. Cluster analysis of the profiles from these isolates and five reference strains allowed the unambiguous identification of the natural cultures to be obtained. Most of the isolates (38 out of 53) belonged to the species *L. sakei*, followed by *L. curvatus* (8 isolates). These findings clearly confirmed the dominance of these two lactobacilli in salami from Northern Italy, as reported in a number of studies (Cocolin et al. 2001; Comi et al. 2005; Rantsiou et al. 2005; Urso et al. 2006). The isolation of cultures ascribed to *L. sakei* from the cold rooms used for the ripening of *Soprèssa Vicentina* PDO prompted the authors to hypothesize a high resistance and adaptability of this species to this particular environment. In the same study, a high intraspecific variability was also seen, which encouraged the authors to recommend the use of the molecular method they had applied in order to identify the strains intended for use as starter cultures and to further monitor them during the manufacturing process.

33.5.2 Central Italy

33.5.2.1 *Ciauscolo PGI*

History. *Ciauscolo* PGI is a typical salami of the Marche region; it is manufactured in the mountainous hinterland of the province of Macerata and in several municipalities located in the provinces of Ancona, Ascoli Piceno, and Fermo. The manufacture of this salami originates from the rural traditions of this Italian region, where all the products derived from the slaughtering of pigs constituted a precious source of proteins, essential to face the long winter and the hard work in the fields. According to etymologists, *Ciauscolo*, or its synonyms *Ciavuscolo* and *Ciabuscolo*, derives from the Latin term *cibusculum*, which means tiny meal or tiny food, which was consumed by the farmers during their work in the fields, between breakfast and lunch or lunch and dinner. This salami is mentioned for the first time in a notarial deed drawn up in the territory of Visso and dating back to the middle of the 18th century. A further historical trace indicating the manufacture of this product is the document known as *Prezzi dei generi*, dating back to 1851 and conserved in the notarial archive of the municipality of Camerino, where *Ciauscolo* is listed among other quoted food products (see Figure 33.9).

FIGURE 33.9 *Ciauscolo* PGI.

Raw materials. Ciauscolo PGI is manufactured with excellent meat cuts from Italian heavy pig carcasses, such as bacon, shoulder, and trimmings from the production of other traditional meat specialties, namely, ham and *Lonza*. The addition of sugars (lactose, dextrose, fructose, and saccharose) and preservatives (L-ascorbic acid, sodium ascorbate, and potassium nitrate) is also allowed, whereas milk-based powders, dairy derivatives, and coloring agents are expressly forbidden. Salt, ground black pepper, wine, and crushed garlic are added to the meat paste.

Manufacturing process. The meat cuts are accurately selected and deprived of the connective tissue and fat in excess. The meat destined for the manufacture of this salami has to be used no sooner than two days and no later than ten days after the slaughtering in order to reach the tissue maturation (*frollatura*), which allows a meat with the right consistency to be processed. The meat is subjected to two or three mincing processes through plates, whose holes are characterized by a progressively smaller diameter, as small as 2–3 mm in the last process. It is possible to keep the meat paste refrigerated for no more than 24 hours before stuffing. The latter operation is performed using natural casings, from swine or bovine intestine, which has to be previously desalted and flavored by immersion in tepid water containing vinegar or wine; stuffing must be carried out very carefully in order to avoid the formation of air bubbles. The salami is tied at the two ends with a string made of hemp, pierced manually, hung upside down, and dried for about 4–7 days at room temperature or under controlled conditions. At the end of this step, smoking is traditionally performed by saturating the ripening chamber with smoke produced by braziers fed with sweet wood sawdust. During ripening, the salami is kept for at least 15 days at temperatures ranging from 8°C to 18°C and RH of 60%–85%. According to the PGI production regulations, the end product has to be characterized by the following physicochemical characteristics: total protein, minimum of 14%; fat, 30%–45%; water/protein ratio, maximum of 3.10; fat/protein ratio, maximum of 3.10; pH ≥ 4.8. As far as the microbiological traits are concerned, the load of mesophilic aerobes and LAB should reach at least 1×10^7 CFU/g.

Appearance and taste. This salami has a cylindrical shape with a diameter ranging from 4.5 to 10 cm, length of 15–45 cm, and weight of 0.4–2.5 kg. The consistency is soft, the slice homogeneous, with a uniform rosy color, a delicate, aromatic and typical odor, and a sapid, delicate, never acidulous flavor.

Distinctive feature. Ciauscolo PGI is characterized by its distinctive soft and homogeneously rosy paste and by the ease with which it can be spread. All these particular features are due to both the high fat content (from 30% to 45%) and the characteristic production technique, which makes this salami unique in the national context.

33.5.2.2 Microbiological Traits of Central Italy Salami

The microbial ecology of ready-for-sale *Ciauscolo* salami, collected from 22 plants located in the Marche region within the original production area, was preliminarily investigated by Silvestri et al. (2007) with traditional microbiological and PCR-DGGE analyses. Detailed information about raw materials, ingredients, and manufacturing parameters applied in the 22 plants had previously been collected (Beccaceci et al. 2006). A high variability in the viable counts of LAB, CNC, and yeasts was highlighted among the products analyzed, thus reflecting the great heterogeneity in the manufacturing techniques and the environmental conditions characterizing the different geographical areas where production plants were located. Notwithstanding such a heterogeneity, a dominance of LAB over CNC and yeasts was generally seen, with LAB viable counts of at least 1×10^7 CFU/g in most of the salami samples. In all except three samples, total coliforms and *E. coli* reached viable counts lower than 10 log CFU/g, thus suggesting the right application of good manufacturing practices. As far as toxinogenic microorganisms are concerned, *S. aureus* was never detected. For the bacterial species distribution, the two psychrotropic species *L. sakei* and *L. curvatus* were the most frequently found, possibly due to the low temperatures used during the salami drying and ripening process. Among yeasts, *Debaryomices hansenii* was the most frequently detected, followed by *Candida psychrophila* and *Saccharomyces barnettii*, which were only occasionally found. Cluster analysis of the DGGE profiles showed that neither the bacterial nor the fungal ecology was influenced by plant location, production scale, or production technology, probably due to the standardization of the manufacturing processes at both artisan and industrial levels. Finally, from calculation of diversity indices, a lower heterogeneity was seen among yeasts and molds with respect to bacteria, thus confirming the secondary role of the first two microbial groups in the ripening process of this salami.

A production of *Ciauscolo*, representative of the 22 samples investigated, in terms of both process technology and microbial ecology, was later selected and subjected to a polyphasic analytical approach aimed at assessing the microbial dynamics throughout the manufacturing process (Aquilanti et al. 2007; Santarelli et al. 2007). As far as the main physicochemical parameters are concerned, a steep drop in pH was seen between the 3rd and the 10th day of curing, and a slow but progressive decrease in water activity was highlighted. As expected, high loads of LAB were found from the beginning of the ripening period, with LAB viable counts equal to or higher than 1×10^7 CFU/g, as required by PGI production regulations. From the molecular identification of the isolates collected at different sampling points during the manufacturing process, the prevalence of *L. curvatus, L. plantarum*, and *S. xylosus* was found among bacteria, whereas *D. hansenii* was the prevalent species among the yeasts. The species *Rhodotorula mucilaginosa* and *Trichosporon brassicae* were also recovered with lower frequencies from the meat paste. The profiling of the microbial community by PCR-DGGE analysis of the DNA extracted from either the salami or the bulk cells grown on the serial-dilution agar plates confirmed the occurrence of the species *L. curvatus, L. plantarum*, and *D. hansenii* from the 10th and even the 3rd day of ripening up to the end of the monitoring period (45 days). The PCR-DGGE approach also revealed the key role of *Lactococcus lactis* subsp. *lactis* in the very early stages of fermentation, as well as the occurrence of *Saccharomyces cerevisiae* in the meat paste and the 1-day-ripened salami. Two further CNC species, namely, *S. saprophyticus* and *Staphylococcus equorum*, plus *L. sakei* were detected with the PCR-DGGE approach from the 3rd to the 45th day of ripening. Although *L. plantarum* is believed to be less competitive than *L. curvatus* in salami dried and ripened at low temperatures, such as those used in *Ciauscolo* manufacture, its stable prevalence in the sample analyzed prompted the authors to suggest its inclusion in the starter cultures for the manufacture of this salami, together with *S. xylosus* and *D. hansenii*.

In a very recent study by Petruzzelli et al. (2010), 52 out of 62 *Ciauscolo* salami collected from 15 plants located in the Marche region were positive for *Listeria* spp., with 28 samples positive for *Listeria monocytogenes*. Of the latter, 17 were not acceptable for sale and consequent human consumption, according to EC regulation no. 2073/2005 and following modifications. Although *Ciauscolo* was hypothesized to be at higher risk of *Li. monocytogenes* contamination compared with other salami, due to its short ripening time, its generally high water activity, and the rare use of additives and starter cultures, the statistical analysis of data allowed such a supposition to be rejected, since no significant differences were seen between *Ciauscolo* PGI and the other salami analyzed.

33.5.3 Southern Italy

33.5.3.1 Soppressata Molisana

Soppressata molisana is produced all over the Molise region, although four towns are specifically renowned for its production, namely, Castel del Giudice, Macchiagodena, Montenero di Bisaccia, and Rionero Sannitico. Its manufacture is carried out from autumn to spring (see Figure 33.10).

History. The first documents that attest the manufacture of this salami date back to the early 19th century and mainly consist of the letters sent by families, together with parcels of homemade food, to soldiers on military service. *Soppressata* was also considered a precious salami to be given as a present to important people or benefactors.

Raw materials. Soppressata molisana is produced with meats from heavy pigs, bred according to local traditions. The finest lean meat cuts such as loin and ham are used, together with fat, the latter in a proportion not exceeding 3%–5% of the lean meats. The other ingredients are marine salt, black and white pepper (finely ground or in whole grains), and olive oil or lard. The use of additives (nitrates) is allowed.

Manufacturing process. Refrigerated meat is chopped or minced with a manual or electric mincer. Fat from the ham or the loin is also manually chopped or minced. The meat paste is added with marine salt (2.3%–2.7%), black pepper in whole grains (0.1%–0.2%), and nitrates (100–150 mg/kg). It is maintained at 2°C–4°C overnight, then stuffed into pig gut, and tied with a fine string at both ends. The salami are

FIGURE 33.10 *Soppressata molisana.*

lined up between two boards and pressed under a weight made of cement or stone for about 24 hours. Thereafter, *Soppressata molisana* is hung up and stored in aerated ripening chambers, at 12°C and about 80% RH, for about 4 weeks. After ripening, *Soppressata molisana* is vacuum-packed in plastic film or stored in olive oil or pork fat.

Appearance and taste. The slice has a brilliant red color and a compact consistency. This salami has a typical strong flavor and taste, both originating from the peculiar metabolites that are released during the fermentation, which is carried out by the autochthonous LAB population.

Distinctive feature. The main feature of this salami is the typical thin parallelepiped-like shape and the high microbiological stability, which is due to the fermentation process and the intense dehydration during ripening.

33.5.3.2 Salsiccia Lucana

The production area for this sausage includes the municipalities located in the territory of the Alto Agri Mountain Community, in the province of Potenza (see Figure 33.11).

Raw materials. Salsiccia lucana is produced from pigs born in Italy and bred in the territory of the Alto Agri Mountain Community.

Manufacturing process. After slaughtering, the pig carcasses are refrigerated for about 24 hours and then dissected. The selected meat cuts are manually chopped or minced, and the meat paste is added with marine salt, wild fennel, and sweet and/or hot pepper powder. All the ingredients are kneaded together, and the meat paste is stuffed in pig gut. The salami is hung to dry for about 20–40 days, and then kept in lard or olive oil, or is vacuum-packed and stored in cool chambers.

Appearance and taste. This salami has a variable length, depending on the production area, and a diameter of about 3 cm. The color varies from dark red, if hot pepper is added, to purple-red with white streaks due to fat, if hot pepper is not used. The consistency is soft, and the taste is aromatic, sometimes spicy.

FIGURE 33.11 *Salsiccia lucana.*

FIGURE 33.12 *Soppressata lucana.*

33.5.3.3 Soppressata Lucana

Soppressata lucana, referred to as *Soperzata* in the local dialect, is produced all over the Basilicata region, specifically in the villages of Lagonegro, Lauria, Picerno, and Tricarico (see Figure 33.12).

Raw materials. Soppressata lucana is produced with loin and ham from local black breed pigs, fat chopped into small cubes, marine salt, and black pepper in whole grains.

Manufacturing process. Meat is manually chopped until a homogeneous meat paste is obtained. This is then added with fat in small cubes, marine salt, and black pepper, and stuffed into pig gut. After tying, the salami is pressed for about 24 hours, ripened for 4–5 months, and slightly smoked. During ripening, the salami can be kept in earthenware pots or glass jars under olive oil or pork fat.

Appearance and taste. The slice, containing small white fat cubes, has a brilliant red color; its consistency is soft, and the flavor is fragrant and slightly spicy; the taste is delicate and sweet.

33.5.3.4 Soppressata del Vallo di Diano

This is a smoked salami produced in the province of Salerno, in the Cilento district (see Figure 33.13).

Raw materials. For the production of *Soppressata del Vallo di Diano*, pork, extra virgin olive oil from Cilento, marine salt, wine, and dried spices are used.

Manufacturing process. The pork is minced and added with marine salt and the remaining ingredients. The meat paste is placed in large baskets made of natural fiber called *pani*, air-dried, and then stuffed into pig gut. The salami is smoked with beech and oak wood and ripened for at least 40 days.

Appearance and taste. The end product has a diameter of 5–8 cm and a length of 10–15 cm. It is stored in glass jars filled with oil.

33.5.3.5 Salame Napoli

History. Salame Napoli is produced all over the Campania region, and especially in the city of Naples, from which this salami takes its name. In the rural areas of the region, it was commonly used as a precious merchandise, to be given in exchange for high professional services (e.g., by physicians, lawyers, etc.) or

FIGURE 33.13 *Soppressata del Vallo di Diano.*

FIGURE 33.14   *Salame Napoli.*

to be consumed during festivities or ceremonies. This explains the great care in all the manufacturing procedures, which have been handed down unchanged from generation to generation (see Figure 33.14).

Raw materials. Meats used for the manufacture of this salami are obtained from pigs bred in Campania or in other regions of Southern and, sometimes, Central Italy. Only the following meat cuts, deprived of soft fat and connective tissue, are used: shoulder, ham, *coppa*, and loin. Backfat and bacon are only used in a proportion no higher than 25% of lean meats.

Manufacturing process. Meat and fat are minced using plates with medium-sized holes. The meat paste is added with marine salt and pepper and then stuffed into pig gut. During the drying process, the salami is slightly smoked using braziers fed with oak, beech, chestnut, and alder wood. The drying and ripening are performed in well-aerated rooms. The latter process has a variable duration, depending on the salami size, but in all cases is never less than 30 days.

Appearance and taste. Salame Napoli has a long, cylindrical shape, an external surface with an intense red color and visible white fat, a compact and nonelastic consistency, a typical smoked odor, and a sweet and smoked or spicy taste.

Distinctive feature. The particular sensory traits of *Salame Napoli* are due to both the raw materials used and the specific technology applied.

33.5.3.6 Soppressata di Ricigliano

History. Soppressata di Ricigliano is produced in the province of Salerno, in Campania, and specifically in the municipalities of Ricigliano and San Gregorio Magno. This product is deeply rooted in local tradition, as documented by the manufacturing technology, which has remained almost unchanged for at least 25 years (see Figure 33.15).

Raw materials. This salami is manufactured with pork ham, loin, bacon (the latter added to the lean meat cuts in 1:10 ratio), marine salt, and pepper in whole grains.

Manufacturing process. Lean meats and fat are minced, and the meat paste is salted, peppered, and stuffed into pig gut. The salami is pressed for about 24 hours and pierced to eliminate the air trapped in the meat paste; it is then dried, smoked with braziers fed with wood at 20°C for 4–5 days, and ripened

FIGURE 33.15   *Soppressata di Ricigliano.*

at 14°C–16°C for at least 35 days. At the end of ripening, it is placed in ceramic containers and covered with olive oil or pork fat.

Appearance and taste. *Soppressata di Ricigliano* has a rough shape, which is similar to a cylinder or a parallelepiped, with longitudinal grooves caused by the string, which is used to tie up the salami. The external surface of the salami is uneven, with an intense red-brown color. The slice has a coarse-grained texture, with a dark brick-red color slightly marbled with white fat.

33.5.3.7 Soppressata di Gioi

History. *Soppressata di Gioi* was originally only produced during the winter in the municipality of Gioi, a village located in the Cilento hinterland at 600 m above sea level. This salami is currently also produced in other municipalities in the Cilento National Park: Cardile, Salento, Stio, Gorga, Orria, and Piano Vetrale. The typical manufacturing technique for this salami was probably brought to the Cilento area by shepherds moving with their flocks from nearby Abruzzo (see Figure 33.16).

Soppressata di Gioi is already mentioned in the *Compendio di agricoltura pratica* (Practical agriculture compendium) edited in 1835, where we can read that among different *Soppressata* salami, those produced in the Cilento area, and especially in Gioi, are greatly appreciated, thus underlying the ancient tradition and the unique production technology for this salami.

Raw materials. *Soppressata di Gioi* is produced with only high-quality pork cuts, namely, loin, ham, and shoulder. Other ingredients are marine salt, pepper and more rarely hot pepper, and wild fennel.

Manufacturing process. Meat is finely chopped by hand, added with marine salt, pepper and sometimes hot pepper, and wild fennel. The meat paste is accurately kneaded to amalgamate all the ingredients, left to rest for about 10 hours, and stuffed into casings from the pig large intestine. A narrow piece of lard, as long as the salami, is inserted within the casing before stuffing. *Soppressata di Gioi* is ripened for about 40 days in rooms with little aeration, and then smoked using braziers, in order to bring out the characteristic intense flavor of pork and spices.

Appearance and taste. The end product has a diameter of 8–10 cm and a dark red color with a white core. It is stored in olive oil or, more rarely, in pork fat.

Distinctive feature. *Soppressata di Gioi* is the only salami manufactured in Campania with a fat portion, referred to as *lardello*, inserted directly within the meat paste. Besides its decorative function, this fat contributes to keeping the meat paste humid during the smoking and ripening processes.

33.5.3.8 Salame Sant'Angelo PGI

History. This salami is produced in the municipal area of Sant'Angelo di Brolo, in the province of Messina. It has a long history, since it was first manufactured in the 11th century, when the Normans introduced new habits into Sicily, such as the eating of swine meat, in contrast with those imposed by the Muslim religion of the Arab conquerors, which, as is well known, forbids the consumption of swine meat products. Over the years, Sant'Angelo di Brolo has become the jealous guardian of this unique Sicilian tradition, which has been passed down from generation to generation until the present day. This is confirmed by numerous historical documents, such as a municipal resolution dating back to the middle

FIGURE 33.16 *Soppressata di Gioi.*

FIGURE 33.17   *Salame Sant'Angelo* PGI.

of the 19th century, which introduced a tax on a list of local products, including this salami. In 1851, a soft of certification, called *Rivelo*, was also introduced, which obliged the salami manufacturers to publish details about the amount of salami produced and where these were ripened. A few years later, a resolution by the same municipality introduced regulations for the production and commercialization of *Salame Sant'Angelo* (see Figure 33.17).

Raw materials. Salame Sant'Angelo PGI is manufactured with the most valuable cuts, including the shoulder, the ham, the belly, and the neck. Ground marine salt, black pepper, and preservative agents (potassium nitrate) are also used. The use of frozen meats is forbidden, whereas their maintaining at temperatures ranging from 0°C to 4°C for no later than six days after the slaughtering is allowed.

Manufacturing process. Selected meats are manually chopped to obtain a coarse-grained meat paste, which is salted, peppered, and stuffed into pig gut. Lean and fat meats can be alternatively extruded through holes with a diameter of 12 mm. The salami is tied with a string at the ends, dried, and ripened for a variable period, depending on its size, in well-aerated rooms, which should allow the product to absorb the typical perfumes of the local flora. According to the PGI production regulations, the end product has to be characterized by the following physicochemical and microbiological traits: pH 5.1–6.2; water activity (a_w), 0.81–0.96; load of mesophilic aerobes, maximum of 50×10^7 CFU/g.

Appearance and taste. Salame Sant'Angelo PGI has a cylindrical shape and an uneven surface. The slice has a ruby-red color and a tender and compact consistency, with well-distributed white fat granules. The flavor is slightly spicy, whereas the aroma is fragrant, sweet, and characteristic.

Distinctive feature. Depending on the size and the portion of the pig gut used for stuffing, *Salame Sant'Angelo* PGI is referred to as *Sottocularino* (weight: 0.2–0.7 kg; ripening duration: minimum of 30 days), *Fellata* (weight: 0.3–0.6 kg; ripening duration: minimum of 30 days), *Cularino* (weight: 0.7–1.5 kg; ripening duration: minimum of 50 days), and *Sacco* (weight: 1–3.5 kg; ripening duration: minimum of 60 days). The sensory properties of this salami are due to the unique traditional production method, as well as to the microclimate of the production area.

33.5.3.9 Salsiccia Sarda

History. Salsiccia sarda, known in the local dialect as *Saltizza* or *Sartizza*, is a dry-cured sausage produced all over Sardinia, thus representing the typical salami of this territory. Its origin dates back to ancient times, when this fermented meat product was mainly destined for home consumption or barter, and only rarely commercialized (see Figure 33.18).

Raw materials. Salsiccia sarda is produced with lean meats and fat (10%–15%) from pigs bred in Sardinia. Salt, pepper, wild fennel or aniseed, and even parsley, garlic, nutmeg, cinnamon, vinegar, or turbid white wine undergoing malolactic fermentation can be used, depending on the geographical area and the local tradition.

FIGURE 33.18 *Salsiccia sarda.*

Manufacturing process. The selected meat cuts are manually chopped, and the meat paste is added with the various ingredients. This is then stuffed into natural casings, derived from swine or bovine intestine. These have to be repeatedly washed with salted water, vinegar, or *Vernaccia* (a typical Italian white wine). The salami is tied, folded in the typical horseshoe shape, left to drip for 2–3 hours in a cool and aerated room, and dried for 4–5 hours. During this step, the product can also be naturally smoked using braziers fed with aromatic myrtle or lentisk wood. The product is ripened at 10°C–14°C for a variable period, ranging from 8–15 days to 1 month, depending on the desired consistency.

Appearance and taste. This dry-cured sausage is characterized by a perfect balance between fat and lean meats. At the end of the ripening period, the external surface has a nut brown-ochre color, with a white-gray microbial patina, which is not uniformly distributed. The slice has a brilliant or brick-red color, depending on the ingredients used, a coarse-grained texture, and a homogeneous distribution of fat and lean meat.

Distinctive note. This dry-cured sausage has a typical horseshoe shape. Its unique features are due to the typical local recipes as well as to the climate specific to each production area.

33.5.3.10 Microbiological Traits of Southern Italy Salami

The biochemical and technological properties of CNC and CNS isolated during the ripening of *Soppressata molisana* were assessed by Coppola et al. (1997). Among the 138 isolates, 80 were ascribed to the genus *Staphylococcus*, with a predominance of *S. xylosus* (60 isolates), whereas 58 were assigned to the genus *Micrococcus*, with a predominance of *Micrococcus kristinae* (38 isolates). Other species were also identified, but with lower frequencies, namely, *Staphylococcus simulans* (six isolates), *S. equorum* (six isolates), *Staphylococcus kloosii* (one isolate), *Micrococcus roseus* (19 isolates), and *Micrococcus varians* (one isolate), respectively. As far as the technological traits of the isolates are concerned, most of them were able to grow in the presence of a high salt concentration (15% NaCl) and were capable of reducing nitrate at 18°C, a temperature close to that adopted for the ripening of *Soppressata molisana*. When the lipolytic activity was tested, a diverse response of the isolates was seen in the different substrates used, whereas a widespread proteolytic activity was shown by CNC.

One year later, 183 cultures of lactobacilli isolated during the ripening of *Soppressata* manufactured in Molise were further investigated by Coppola et al. (1998) in order to analyze their main technological traits (salt tolerance, lipolytic activity against pork fat, and production of acetoin, slime, and H_2O_2). The purpose of this work was to select distinctive strains to be used as starter cultures in the manufacture of this salami. Among the isolates, the species *L. sakei* was predominant (125 isolates), followed by *L. plantarum* (13 isolates), *Lactobacillus paracasei* subsp. *paracasei* (11 isolates), *Lactobacillus viridescens* (10 isolates), *L. coryneformis* subsp. *torquens* (6 isolates), *Lactobacillus paralimentarius*

(6 isolates), *L. brevis* (6 isolates), *Lactobacillus graminis* (4 isolates), and *L. curvatus* (2 isolates). All the cultures were able to grow in the presence of 8% NaCl, and most of them tolerated up to 10% of this salt.

A number of authors have investigated the microbial ecology and dynamics of *Salsiccia lucana* and *Soppressata lucana*.

Parente et al. (2001a) assessed the diversity of 414 LAB isolated during the fermentation and ripening of these salami from Southern Italy on the basis of 28 phenotypic tests. Ten batches of *Salsiccia*, including 30 samples collected at different ripening times, and nine of *Soppressata*, including 27 samples, were analyzed. These were manufactured either with or without the use of starter cultures in industrial or artisan plants. About 50% of the isolates were identified as *L. sakei*, 22% as *Pediococcus* spp., 7% as *Leuconostoc carnosum*, 7% as *Leuconostoc gelidum*, 7% as *Le. pseudomesenteroides*, 6% as *L. plantarum*, and 1% as *L. curvatus*. If the sole salami produced without the addition of starter cultures are considered, the proportion of the isolates ascribed to *L. sakei* rise to 62%. Pediococci were isolated from the sole salami manufactured with starter cultures ascribed to this microbial group, and their progressive replacement by *L. sakei* during the ripening of these salami was seen.

In a further study, Parente et al. (2001b) evaluated the occurrence of biogenic amines, as well as their relative concentration, in the two above-mentioned types of salami, manufactured at either artisan or industrial plants. The salami were sampled during ripening and subjected to microbiological analyses, pH measurement, and biogenic amine determination. Overall, no correlation was found between the type of product, use of starter cultures, ripening duration, pH, microbial viable counts, and biogenic amine content. As far as the latter parameter is concerned, a great variability was seen, although a generally low concentration was found, which was ascribed by the authors to the high-quality hygiene of the raw materials, the occurrence of microorganisms with amino acid decarboxylase and monoamine oxidase activities, and the specific physicochemical conditions used for the processing and ripening of the salami studied.

In the same year, Gardini et al. (2001) investigated the evolution and biodiversity of the yeast population during the manufacturing and curing of different batches of *Salsiccia sotto sugna*, a dry-cured sausage stored in pork fat. Sausages were produced at farm level in plants located in Cancellara, Genzano, Viaggianello, and Rotonda (Basilicata). The yeasts were identified at species level by evaluating phenotype-based traits, including growth at increasing NaCl concentrations and at different temperatures, as well as proteolytic, lipolytic, and urease activity; the API ID 32C system (Biomerieux, France) was also used. Interestingly, each batch was characterized by a specific yeast population, although the species *D. hansenii* and its anamorph counterpart *Candida famata* dominated all the yeast communities, the latter with the two varieties *C. famata* var. *famata*, and *C. famata* var. *flareri*. *Yarrowia lipolytica* and *R. mucilaginosa* were also isolated from three out of four batches, together with minority species (e.g., *Candida citrea*, *Candida diversa*, *Candida maltosa*, etc.), which were more rarely found. The isolates ascribed to *Y. lipolytica* were further subjected to RAPD typing with primers M13 and RF2 and to assessment of the lipolytic activity in pork fat agar. A high intraspecific genetic heterogeneity was seen among yeasts, as well as a widespread lipolytic activity.

The composition of CNS and CNC communities of six batches of *Salsiccia sotto sugna* was investigated by Rossi et al. (2001) with traditional microbiological (viable counts) and RAPD analyses. The results showed that high loads of these microorganisms were found throughout the ripening process (60 days), with *S. xylosus* as the dominant species.

Ninety-six isolates from *Salsiccia lucana* and *Soppressata lucana* and other salami from Southern Italy, namely, *Salame Napoli*, *Soppressata di Gioi*, and *Soppressata di Ricigliano*, were used by Blaiotta et al. (2003a) as reference cultures for the validation of a combined molecular approach aimed at identifying CNC and CNS cultures. The isolates were definitively assigned to *S. xylosus* (53 isolates), *S. saprophyticus* (8 isolates), *Staphylococcus intermedius* (2 isolates), *Staphylococcus haemolyticus* (1 isolate), *S. epidermidis* (6 isolates), *Staphylococcus caseolyticus* (1 isolate), *Staphylococcus carnosus* (1 isolate), *Staphylococcus capitis* (2 isolates), *S. aureus* (2 isolates), *Micrococcus luteus* (2 isolates), and *Kokuria varians* (1 isolate).

In the same year, Blaiotta et al. (2003b) published a species-specific PCR assay for the reliable identification of the species *S. xylosus*. The assay was based on the use of primers targeting the *xyl*B and *hsp*60

genes, which code xylulokinase and 60 kDA heat-shock proteins, respectively. Further species-specific PCR assays, relying on the amplification of the *sod*A gene, were developed soon afterward for the unambiguous identification of *S. equorum* (Blaiotta et al. 2004a), *S. carnosus*, and *S. simulans* (Blaiotta et al. 2005). In both these studies, wild isolates collected from *Salsiccia lucana*, *Soppressata lucana*, *Salame Napoli*, *Soppressata di Gioi*, and *Soppressata di Ricigliano* were used as reference strains.

A combined culture-dependent approach, based on both phenotype- and genome-based analyses (16S rRNA gene sequencing, species-specific PCR assays, ISR-PCR, and PCR-DGGE), was used to identify 471 cultures isolated from 36 samples of *Salsiccia* and *Soppressata* collected during curing (Blaiotta et al. 2004b). In both these salami, the CNS population was largely dominated by *S. xylosus*, followed by *Staphylococcus pulvereri/vitulus*, *S. equorum*, and *S. saprophyticus*. Other *Staphylococcus* species, namely, *S. succinus*, *S. pasteuri*, *S. epidermidis*, *S. warneri*, and *Macrococcus caseolyticus*, were also found to a lesser extent. Statistical data processing showed a great variability in the composition and evolution of the CNS community, depending on the type of product, plant, and duration of the ripening. A succession of species, and even biotypes, was also highlighted in some salami samples. *S. xylosus* was isolated throughout the ripening process from almost all the salami, whereas *S. saprophyticus* was recovered with lower frequencies; the occurrence of this latter species during ripening progressively decreased until its complete replacement by *S. xylosus*. Members of the species *S. epidermidis*, *S. pasteuri*, *S. succinus*, *S. pulvereri*, and *S. vitilus* were almost exclusively isolated from the artisan salami, and their load rapidly decreased during ripening.

More recently, Bonomo et al. (2008) identified 49 LAB cultures isolated from *Soppresssata lucana* using the two molecular techniques RAPD and amplified ribosomal DNA restriction analysis. All the isolates also underwent phenotype-based tests aimed at assessing: (1) growth and acid production in the presence of different temperatures and NaCl and nitrite concentrations; (2) the hydrolysis of sarcoplasmic and myofibrillar proteins; (3) the release of antimicrobial substances, peptides, and FAA; and (4) the nitrate reductase activity. *L. sakei* was the dominant species, followed by *P. pentosaceus*, *Le. carnosum*, *L. plantarum*, *L. brevis*, and *Le. pseudomesenteroides*. Under the *in vitro* conditions used, the assessment of the acidifying ability allowed a good distinction among different species and even strains to be achieved. As far as the traits of technological interest are concerned, most of the isolates ascribed to *L. sakei* showed a high proteolytic activity toward the sarcoplasmic proteins, whereas leuconostocs were particularly efficient in the reduction of nitrates.

The technological properties of 37 CNS isolates from *Salsiccia lucana* and *Soppressata lucana* were again investigated by Bonomo et al. (2009) to better understand the role of these microorganisms in the fermentation and ripening of these two salami and to find out which strains could be good candidates for use as starter cultures. Thirty-five of these isolates had previously been identified by phenotype-based tests and molecular analyses in the course of several research studies (Blaiotta et al. 2003a,b, 2004a,b, 2005). Once again, *S. xylosus* was the dominant species, followed by *S. equorum*, *S. pulvereri/vitilus*, and *S. warneri*. All the isolates were screened for their ability to grow at different salt concentrations, temperatures, and pH values, for their proteolytic and lipolytic activities, and for those of nitrate reductase and amino acid decarboxylase. The possible presence of antimicrobial activity was also assessed. All the cultures grew well at 20°C and 30°C under the conditions used. At 10°C, most of them showed a low capacity for growth, whereas only a few isolates were able to grow well at pH values ranging from 6.0 to 5.2, which is the typical pH range found in the salami manufactured in Basilicata. At pH 6.0 and 5.6, a positive effect of NaCl and nitrites on the growth of the staphylococci was seen. Most of these microorganisms hydrolyzed the sarcoplasmic protein fraction, whereas about 30% of them were also able to hydrolyze myofibrillar proteins. *S. pulvereri* was the most proteolytic species, whereas all the isolates ascribed to *S. equorum* showed a high nitrate-reductase activity. None of the isolates were lipolytic, whereas 62% of them were able to reduce nitrates. As far as the production of biogenic amines is concerned, about 80% of the staphylococci studied decarboxylated at least one of the tested FAA, whereas a lower proportion (13%) was able to decarboxylate more than one amino acid.

The microbial and physicochemical characteristics of *Salame Napoli* (Naples-type salami) manufactured without the addition of sugars and fat were investigated by Coppola et al. (1995). The monitoring of pH, volatile acidity, moisture, proteins, and FAA content revealed rapid ripening, which was also confirmed by a fast weight loss in the salami during its prolonged smoking (until the 10th day of curing).

This evidence was in line with the results of the microbiological analyses, which highlighted a succession in the composition of the microbial community.

A few years later, the changes in the composition of the microbial population associated with the natural fermentation and curing of this salami from Southern Italy were assessed by Coppola et al. (2000) who studied five batches of Naples-type salami produced by a local artisan plant without the use of starter cultures. These were sampled immediately after stuffing and throughout the ripening process, at 2, 7, 14, 23, and 41 days of maturation. The isolation campaign and the phenotype-based identification of the isolates revealed that *L. curvatus* and *L. sakei* dominated the lactobacilli community, whereas leuconostocs prevailed among coccal-shaped LAB. As far as CNS are concerned, *S. xylosus* was the species most frequently recovered (18 cultures) followed by *S. saprophyticus*, *Staphylococcus chromogenes*, *Staphylococcus hominis*, *S. warneri*, *Staphylococcus lugdumensis*, and *S. epidermidis*. Among yeasts, *D. hansenii* prevailed, followed by *Cryptococcus albidus*, *Trichosporon terrestre*, *Trichosporon pullulans*, and *Candida incommunis*.

The presence of CNC carrying antibiotic resistances in artisan Naples-type salami was assessed by Mauriello et al. (2000) in order to determine the contributions of these nonpathogenic microorganisms to the occurrence and spread of antibiotic resistances in the food chain. Accordingly, a pool of isolates ascribed to *S. xylosus* (30 cultures), *S. capitis*, *S. saprophyticus*, *S. hominis*, *S. simulans*, *Staphylococcus cohnii*, and other *Staphylococcus* spp. (one culture for each taxon) was assayed against 25 antibiotics.

Two years later, 27 strains of *S. xylosus*, previously isolated from five samples of Naples-type salami, were assayed for their proteolytic ability on pork myofibrillar and sarcoplasmic proteins in order to evaluate their potential as autochthonous starter cultures (Mauriello et al. 2002).

Mauriello et al. (2004) assessed the biochemical characteristics of 96 strains of six different *Staphylococcus* species in order to select the most suitable as starters for fermented dry sausage production. The 96 cultures ascribed to *S. saprophyticus* (28 isolates), *S. equorum* (28 isolates), *S. xylosus* (23 isolates), *S. warneri* (6 isolates), *S. succinicus* (6 isolates), and *Staphylococcus lentus* (5 isolates) were chosen among those isolated from samples of Naples-type salami, *Soppressata di Ricigliano*, and *Soppressata di Gioi* at the end of ripening. These strains were characterized by their ability to grow at different temperatures and pH values, by their lipolytic and proteolytic activities, and by nitrate reductase. Most *S. xylosus* strains showed technological properties that would make them eligible as starter cultures for salami manufacture, whereas only two strains of *S. saprophyticus* were able to reduce nitrate, the main criterion in the starter strain selection.

Several cultures isolated from *Soppressata del Vallo di Diano* and ascribed to *S. simulans* (2 isolates) and *S. carnosus* (4 isolates) were assayed by Casaburi et al. (2005) for growth at different temperatures, NaCl concentrations, and pH values, as well as for their lipolytic and proteolytic activities and a number of other key enzymatic activities (nitrate reductase, catalase, superoxide dismutase, and decarboxylase). The impact of a laboratory-made starter culture, including two strains of *S. xylosus* and one strain of *L. plantarum*, on the physicochemical, microbiological, biochemical, and sensory properties of *Sopressata del Vallo di Diano* was also investigated by Casaburi et al. (2007). To that end, salami samples manufactured at a local plant were assayed for viable counts of mesophilic LAB, CNC, yeasts, molds, and enterobacteria, as well as for moisture, water activity, pH, color, content of FAA, total lipids, FFA, and SDS-PAGE profiles of sarcoplasmic and myofibrillar muscle proteins. The results of this study showed that the salami inoculated with the starter culture were characterized by a significantly lower pH and different sensory properties; in particular, they were chewier and easier to peel, and the intensity of both the red color and the cured flavor was less than in the control sample.

In the same year, the microbial ecology of *Soppressata del Vallo di Diano* was investigated by Villani et al. (2007) with the double aim of (1) identifying the microbial species that carry on fermentation and ripening of this salami through both culture-dependent and -independent methods and (2) isolating and selecting LAB and CNS strains suitable for use as starter cultures through *in vitro* and *in situ* technological characterization. Viable counts showed that ripened salami were characterized by high loads of both lactobacilli and staphylococci. The PCR-DGGE analysis of the variable regions V3 and V1 of the 16S rRNA gene and the direct sequencing of the DGGE bands allowed the identification of *S. xylosus*, *Staphylococcus succinicus*, and *S. equorum* among CNS and *L. curvatus* and *L. plantarum* among

LAB. In parallel, the amplification and DGGE analysis of the yeast 26S rRNA gene by DGGE revealed the dominance of *D. hansenii*. A pool of selected cultures ascribed to *S. xylosus*, *L. curvatus*, and *L. plantarum* was assayed for some traits of technological interest, namely, nitrate reduction, lipolysis, antioxidant activity, and proteolysis of the myofibrillar and sarcoplasmic protein fractions. Both *in vitro* and *in situ* assays were performed, the latter using each selected strain in a laboratory-scale manufacture of *Soppressata del Vallo di Diano*, evaluating the microbiological and chemical changes at the end of the ripening time.

More recently, three further starter cultures including strains of *L. curvatus* and *S. xylosus* with distinctive lipolytic and proteolytic activities were used for the manufacture of *Soppressata del Vallo di Diano* by Casaburi et al. (2008). Once again, the aim of this work was the evaluation of the impact of the experimental starter cultures on the main biochemical (extent of proteolysis and lypolysis, changes in FAA and FFA) and sensory (taste and aroma) properties of the end product.

The chemical, microbiological, and sensory properties of *Salame di Sant'Angelo* were evaluated by Moretti et al. (2004) as a contribution to the exploitation of this traditional Sicilian salami. A further aim of the study was to compare two alternative (traditional and industrial) ripening processes. Accordingly, two batches, each including 60 salami, were manufactured following traditional recipes. The salami from the first batch were ripened for 90 days in a traditional ripening chamber located in the Nebrodi area (Northeastern Sicily) at temperatures ranging from 6.5°C to 18.3°C and RH from 67% to 82%. The products from the second batch were ripened for 83 days under controlled conditions, namely, temperatures of 9.2°C–10.5°C and RH in the range 67%–82%. Samples of 50-day-ripened salami available on the local market were purchased and used as a control. All the salami were subjected to the determination of chemical composition (pH and water activity, nonprotein nitrogen, FAA, and volatile compounds), viable counts (LAB, total mesophilic bacteria, coliforms, enterobacteria, CNC and CNS, yeasts, and molds), and sensory attributes (color intensity, saltiness, acidity, rancidity, hardness, elasticity, cohesiveness, and overall acceptability). All the data were subjected to statistical processing for the assessment of the significant differences. On the whole, ripening under controlled conditions did not seem to negatively affect the sensory properties of *Salame di Sant'Angelo*. However, physicochemical analyses revealed that the salami ripened at fixed temperature, and RH values were characterized by lower amounts of aldehydes, higher moisture, and lower weight loss. By contrast, viable counts of LAB, CNC, and CNS were positively influenced by the industrial ripening conditions. Good quality hygiene of the products from both batches was seen.

As far as *Salsiccia sarda* is concerned, Mangia et al. (2007) carried out a study aimed at characterizing this typical Sardinian fermented meat specialty, with a special focus on the microbial hygiene indicators. The evolution of the autochthonous population was monitored throughout the manufacturing process. Both the meat paste and the sausages sampled during ripening were subjected to microbiological analysis. Some concerns emerged about the sausage hygiene standards because of the high occurrence of *Escherichia coli*, although neither *Salmonella* and *Listeria* spp. nor sulfite reducing clostridia were ever found. Overall, 58 bacterial isolates were collected and assayed for both the morphology and the main physiological and biochemical features (catalase reaction, growth in the presence of furazolidone and lisostaphin, coagulase activity, and fermentation profile). *Salsiccia sarda* was mainly dominated by *L. curvatus*, *L. plantarum*, *S. xylosus*, *Kocuria varians*, *Micrococcus* spp., *D. hansenii*, and *C. famata*. The species *L. curvatus* was predominant in the early stage of the manufacturing process, presumably being the main actor in the fermentation process together with *L. plantarum*. CNC occurred only on the first days of maturation, with *S. xylosus* as the prevailing species.

One year later, Mangia et al. (2008) investigated the microbiota involved in the fermentation and ripening of *Salsiccia sarda* manufactured with lean sheep meats. The microbiological analyses showed an initial high contamination by fecal coliforms, although *Salmonella* and *Listeria* species were never detected. Overall, 161 cultures ascribed to *L. plantarum*, *S. xylosus*, *S. lentus*, *K. varians*, *Micrococcus* spp., *Trichosporon* spp., *D. hansenii*, and *C. famata* were collected throughout the manufacturing process. As far as the microbial viable counts are concerned, yeasts progressively increased in number during ripening, whereas molds remained almost stable. Among CNC and CNS, those ascribed to the genus *Micrococcus* were dominant during the early maturation stage, being later replaced by the staphylococci, with *S. xylosus* as the dominant species. Although LAB showed a slow growth rate during the first days of ripening, after the 5th day, they were found to dominate the microbial population.

33.6 Conclusions

In this survey, the task of the authors was to give an idea of the huge heritage of history, culture, expertise, flavor, taste, and microbial biodiversity that is involved in the production of traditional Italian salami. Every Italian region offers one or more of these specialties, only a few of which have been awarded the PDO and PGI labels. For this reason, it is outside the scope of this survey to provide an exhaustive overview of all these valuable products.

Since it was necessary to select a reasonable number of products to exemplify the great Italian tradition in salami manufacturing, our choice was oriented toward those products whose microbial traits have so far been investigated by the international scientific community. This choice was basically driven by two factors: firstly, the key role played by the microbiota involved in salami fermentation and curing in the definition of the specific and unique properties of these meat-based specialties and, secondly, our scientific background as microbiologists. We hope the readers are not disappointed.

ACKNOWLEDGMENTS

The authors wish to thank Salumificio Valverde–Valverde (Pavia, Italy), Consorzio Salumi DOP Piacentini (Piacenza, Italy), Consorzio Salame Cremona (Cremona, Italy), Salumificio Cattini–Suzzara (Mantova, Italy), Sapori Mediterranei–Cirigliano (Matera, Italy), Salumificio Gioi–Gioi (Salerno, Italy), and Oro dei Nebrodi–S. Angelo di Brolo (Messina, Italy).

REFERENCES

Andrighetto C, Zampese L, Lombardi A. 2001. RAPD-PCR characterization of lactobacilli isolated from artisanal meat plants and traditional fermented sausages of Veneto region (Italy). *Lett Appl Microbiol* 33:26–30.

Aquilanti L, Santarelli S, Silvestri G, Osimani A, Petruzzelli A, Clementi F. 2007. The microbial ecology of a typical Italian salami during its natural fermentation. *Int J Food Microbiol* 120:136–45.

Beccaceci A, Silvestri G, Santarelli S, Aquilanti L, Osimani A, Petruzzelli A, Clementi F. 2006. Indagine sulla popolazione di batteri lattici del salame "Ciauscolo" mediante PCR-DGGE. *Ind Conserve* 81:297–303.

Blaiotta G, Casaburi A, Villani F. 2005. Identification and differentiation of *Staphylococcus carnosus* and *Staphylococcus simulans* by species-specific PCR assays of *sod*A genes. *Syst Appl Microbiol* 28:519–26.

Blaiotta G, Ercolini, D, Mauriello G, Salzano G, Villani F. 2004a. Rapid and reliable identification of *Staphylococcus equorum* by a species-specific PCR assay targeting the *sod*A gene. *Syst Appl Microbiol* 27:696–702.

Blaiotta G, Pennacchia C, Ercolini D, Moschetti G, Villani F. 2003a. Combining denaturing gradient gel electrophoresis of 16S rDNA V3 region and 16S–23S rDNA spacer region polymorphism analyses for the identification of staphylococci from Italian fermented sausages. *Syst Appl Microbiol* 26:423–33.

Blaiotta G, Pennacchia C, Parente E, Villani F. 2003b. Design and evaluation of specific PCR primers for rapid and reliable identification of *Staphylococcus xylosus* strains isolated from dry fermented sausages. *Syst Appl Microbiol* 26:601–9.

Blaiotta G, Pennacchia C, Villani F, Ricciardi A, Tofalo R, Parente E. 2004b. Diversity and dynamics of communities of coagulase-negative staphylococci in traditional fermented sausages. *J Appl Microbiol* 97:271–84.

Bonomo MG, Ricciardi A, Zotta T, Parente E, Salzano G. 2008. Molecular and technological characterization of lactic acid bacteria from traditional fermented sausages of Basilicata region (Southern Italy). *Meat Sci* 80:1238–48.

Bonomo MG, Ricciardi A, Zotta T, Sico MA, Salzano G. 2009. Technological and safety characterization of coagulase-negative staphylococci from traditionally fermented sausages of Basilicata region (Southern Italy). *Meat Sci* 83:15–23.

Casaburi A, Aristoy MC, Cavella S, Di Monaco R, Ercolini D, Toldrá F, Villani F. 2007. Biochemical and sensory characteristics of traditional fermented sausages of Vallo di Diano (Southern Italy) as affected by the use of starter cultures. *Meat Sci* 76:295–307.

Casaburi A, Blaiotta G, Mauriello G, Pepe O, Villani F. 2005. Technological activities of *Staphylococcus carnosus* and *Staphylococcus simulans* strains isolated from fermented sausages. *Meat Sci* 71:643–50.

Casaburi A, Di Monaco R, Cavella S, Toldrá F, Ercolini D, Villani F. 2008. Proteolytic and lipolytic starter cultures and their effect on traditional fermented sausages ripening and sensory traits. *Food Microbiol* 25:335–47.

Cocolin L, Dolci P, Rantsiou K, Urso R, Cantoni C, Comi G. 2009. Lactic acid bacteria ecology of three traditional fermented sausages produced in the north of Italy as determined by molecular methods. *Meat Sci* 82:125–32.

Cocolin L, Manzano M, Cantoni C, Comi G. 2001. Denaturing gradient gel electrophoresis analysis of the 16S rRNA gene V1 region to monitor dynamic changes in the bacterial population during fermentation of Italian sausages. *Appl Environ Microbiol* 67:5113–21.

Comi G, Urso R, Iacumin L, Rantsiou K, Cattaneo P, Cantoni C, Cocolin L. 2005. Characterisation of naturally fermented sausages produced in the north east of Italy. *Meat Sci* 69:381–92.

Conter M, Muscariello T, Zanardi E, Ghidini S, Vergara A, Campanini G, Ianieri A. 2005. Characterization of lactic acid bacteria isolated from an Italian dry fermented sausage. *Ann Fac Medic Vet Parma* XXV:167–74.

Cook PE. 1995. Fungal ripened meats and meat products. In: Campbell-Plott G, Cook PE (eds). *Fermented Meats*. Glasgow: Chapman & Hall, pp. 110–29.

Coppola R, Giagnacovo B, Iorizzo M, Grazia L. 1998. Characterization of lactobacilli involved in the ripening of soppressata molisana, a typical southern Italy fermented sausage. *Food Microbiol* 15:347–53.

Coppola R, Iorizzo M, Saotta R, Sorrentino E, Grazia L. 1997. Characterization of micrococci and staphylococci isolated from soppressata molisana, a Southern Italy fermented sausage. *Food Microbiol* 14:47–53.

Coppola R, Marconi E, Rossi F, Dellaglio F. 1995. Produzioni artigianali di salame tipo Napoli: aspetti chimici e microbiologici. *Italian J Food Sci* 7:57–62.

Coppola S, Mauriello G, Aponte M, Moschetti G, Villani F. 2000. Microbial succession during ripening of Naples-type salami, a southern Italian fermented sausage. *Meat Sci* 56:321–9.

Di Cagno R, Chaves Lopez C, Tofalo R, Gallo G, De Angelis M, Paparella A, Hammes W, Gobbetti M. 2008. Comparison of the compositional, microbiological, biochemical and volatile profile characteristics of three Italian PDO fermented sausages. *Meat Sci* 79:224–35.

Gardini F, Suzzi G, Lombardi A, Galgano F, Crudele MA, Andrighetto C, Schirone M, Tofalo R. 2001. A survey of yeasts in traditional sausages of Southern Italy. *FEMS Yeast Res* 1:161–7.

Iacumin L, Comi G, Cantoni C, Cocolin L. 2006. Ecology and dynamics of coagulase-negative cocci isolated from naturally fermented Italian sausages. *Syst Appl Microbiol* 29:480–6.

Mangia NP, Murgia MA, Garau G, Deiana P. 2007. Microbiologia e valutazione igienico sanitaria della salsiccia sarda. *Ind Aliment* XLVI:533–36.

Mangia NP, Murgia MA, Garau G, Merella R, Deiana P. 2008. Sardinian fermented sheep sausage: microbial biodiversity resource for quality improvement. *Opt Méditerranéennes* A 78:273–77.

Mauriello G, Casaburi A, Blaiotta G, Villani F. 2004. Isolation and technological properties of coagulase negative staphylococci from fermented sausages of Southern Italy. *Meat Sci* 67:149–58.

Mauriello G, Casaburi A, Villani F. 2002. Proteolytic activity of *Staphylococcus xylosus* strains on pork myofibrillar and sarcoplasmic proteins and use of selected strains in the production of Naples-type salami. *J Appl Microbiol* 92:482–90.

Mauriello G, Moschetti G, Villani F, Blaiotta G, Coppola S. 2000. Antibiotic resistance of coagulase–negative staphylococci isolated from artisanal Naples-type salami. *Int J Food Sci Nutr* 51:19–24.

Moretti VM, Madonia G, Diaferia C, Mentasti T, Paleari MA, Panseri S, Pirone G, Gandini G. 2004. Chemical and microbiological parameters and sensory attributes of a typical Sicilian salami ripened in different conditions. *Meat Sci* 66:845–54.

Parente E, Grieco S, Crudele MA. 2001a. Phenotypic diversity of lactic acid bacteria isolated from fermented sausages produced in Basilicata (Southern Italy). *J Appl Microbiol* 90:943–52.

Parente E, Martuscelli M, Gardini F, Grieco S, Crudele MA, Suzzi G. 2001b. Evolution of microbial populations and biogenic amine production in dry sausages produced in Southern Italy. *J Appl Microbiol* 90:882–91.

Petruzzelli A, Blasi G, Masini L, Calza L, Duranti A, Santarelli S, Fisichella S, Pezzotti G, Aquilanti L, Osimani A, Tonucci F. 2010. Occurrence of *Listeria monocytogenes* in salami manufactured in the Marche region (Central Italy). *J Vet Med Sci* 72:499–502.

Rantsiou K, Cocolin L. 2006. New developments in the study of the microbiota of naturally fermented sausages as determined by molecular methods: a review. *Int J Food Microbiol* 108:255–67.

Rantsiou K, Urso R, Iacumin L, Cantoni C, Cattaneo P, Comi G, Cocolin L. 2005. Culture-dependent and -independent methods to investigate the microbial ecology of Italian fermented sausages. *Appl Environ Microbiol* 71:1977–86.

Rossi F, Tofalo R, Torriani S, Suzzi G. 2001. Identification by 16S-23S rDNA intergenic region amplification, genotypic and phenotypic clustering of *Staphylococcus xylosus* strains from dry sausages. *J Appl Microbiol* 90:365–71.

Santarelli S, Aquilanti L, Silvestri G, Osimani A, Petruzzelli A, Clementi F. 2007. Evoluzione delle comunità microbiche nel corso della maturazione del Ciauscolo. *Ind Aliment* XLVI:744–8.

Silvestri G, Santarelli S, Aquilanti L, Beccaceci A, Osimani A, Tonucci F, Clementi F. 2007. Investigation of the microbial ecology of Ciauscolo, a traditional Italian salami, by culture-dependent techniques and PCR-DGGE. *Meat Sci* 77:413–23.

Urso R, Comi G, Cocolin L. 2006. Ecology of lactic acid bacteria in Italian fermented sausages: isolation, identification and molecular characterization. *Syst Appl Microbiol* 29:671–80.

Villani F, Casaburi A, Pennacchia C, Filosa L, Russo F, Ercolini D. 2007. Microbial ecology of the Soppressata of Vallo di Diano, a traditional dry fermented sausage from Southern Italy, and *in vitro* and *in situ* selection of autochthonous starter cultures. *Appl Environ Microbiol* 73:5453–63.

34

Mold-Ripened Sausages

Kálmán Incze

CONTENTS

34.1 Introduction

Following the collection of plants and fruits, food of animal origin also became available to humankind in the form of predators' leftover and later by fishing and hunting. They must have realized that meat cannot be left "as is" because in some hours or some days, depending on the actual ambient temperature, it spoils or becomes poisonous and consequently cannot be eaten. Since meat as animal protein is not only a good-tasting food but also an excellent source of essential nutrients and some vitamins, it has always been the target of pathogenic and spoilage microorganisms that break down and transform these substances for the sake of their own growth. Keeping in mind this effect and its consequences, humans felt forced to find ways to keep these hard-to-get meats available to them for a longer time.

It is probably a fairly acceptable guess that our ancestors arrived at the idea that there are preparation methods that help meats preserve their advantageous properties for a longer period of time. In addition to cooking, we may assume that drying, frying, cooling, and freezing are the most ancient ways of preservation, and yet, while cooling and freezing could take place only temporarily and not under warm climate, drying was an applicable method everywhere. In early times, meat drying took place in caves over fires and open fires; this also allowed for the advantageous effects of smoke (Incze 2007).

It seems reasonable to presume on the basis of written relicts (Rixson 2000) that our ancestors not only dried but also salted and then dried meat. Since salt possibly contained nitrate as a contaminant, some meat pieces might have been cured to some extent by chance. Cured and dried meat was known around 1000 B.C. (Rixson 2000), while fish was salted and dried in ancient Egypt.

Salted and dried meat played an important role in the Roman Empire as high-energy food for legions in battles. Soldiers received a salt portion on a regular basis (called "salarium," where the word salary possibly comes from) as an important material for preservation and making the food tasty. Also Hungarian warriors used salted and dried meat more than 1000 years ago, but this type of preservation still exists in many parts of the world [biltong in South Africa, pemmican of Native Americans in North America, or tasajo in South America, as also Lawrie (1998) mentions].

We also know that Romans already prepared "botulus" (sausage, hence the name botulism, i.e., sausage poisoning) from smaller pieces of meat, fat mixed with salt and spices and stuffed into intestines of killed animals. According to Pederson (1979), Homer mentions a dried sausage in the 9th century B.C., while Leistner (1986, 2005) suggests that sausage and salami manufacture may be considered as 250–300 years old.

Developing this way step by step the preparation of dried sausages by collecting the experience of our ancestors, this experience has been supported more and more by research activities and scientific evidences in the last century, but research activities are still continued in order to get further detailed information.

In the earlier stage, dry sausages were produced only from meat, fat, (curing) salt, and spices. Although scientists were aware of the existence of microorganisms in the 19th century, and their possible role in fermentation was known thanks to Pasteur's work, this referred to dairy and wine and no relation between microbial fermentation and dry sausage manufacture was thought of before the 1950s when Niinivaara and Niven did the excellent work in this field. (Niinivaara wrote the history of starter cultures in 1993.)

As a result of their initiating research, a determinant change came through in the dry sausage technology: with their contribution, we learned about the important role of microorganisms, and we are able to exploit their abilities in producing raw sausages, thus ensuring consistency in quality and food safety.

Interestingly enough, the conquest of useful microbes, starter cultures, as they are called, was overwhelmingly successful in sausage technology, but nevertheless was not so successful in the production of traditional raw ham (Iberian and Parma ham). Some of the reasons of this phenomenon are large intact instead of chopped muscle and very long ripening–drying time.

Mold-ripened sausages, salamis, are raw fermented products with usually longer ripening and drying time. During traditional technology, no starter cultures and/or chemical acidulants are applied, but the old technology has been changed exceedingly to the starter technology all over the world with the application of lactic acid bacteria, staphylococci, yeasts, and moulds.

Because of chemical changes, final pH is usually higher (pH > 5.5) due to the longer ripening–drying time even if the pH value after incubation with the starter culture is lower. For this reason, it is of vital importance, from the food safety point of view, that the final a_w value be sufficiently low, which is fulfilled by a long drying period.

34.2 Characterization of Mold-Ripened Sausages

Manufacturing (selection of raw material, chopping of meat and fat, blending with additives, ripening, and drying) of mold-ripened sausages is rather similar to that of sausages without mold with the basic difference that a short period of ripening–drying of mold-ripened sausages needs higher temperature and relative humidity in the ripening room in order to support the mold growth by ensuring optimum conditions.

Under wet conditions, any kind of molds can grow, but for technological, organoleptic, and safety reasons, growth of selected molds is desired that can be ensured either by a house mycoflora being developed during decades or by inoculation with mold starter.

In order to reduce the original a_w value of about 0.97 to below 0.90–0.91, traditionally manufactured mold-ripened sausages are dried for a longer period of time than acidified (by starter) sausages where

TABLE 34.1

Characterization of Mold-Ripened Sausages (Diameter: 65 mm)

Data Measured	Traditional Technology	Starter Culture in Meat Batter
Drying time[a]	10–12 weeks	5–6 weeks
pH[b]	5.6–6.2 or above	≤5.3 after incubation
		5.5–5.6 or above in final product
a_w[c]	<0.90–0.91	<0.93

[a] Data given can vary depending on the composition of raw material (e.g., beef dries faster), drying room parameters, etc. Evidently, the tenures given can be extended depending on the actual sensory properties to be attained. Data given in the table can be considered as average figures.

[b] pH of sausages with starter technology is lower than that of traditional ones, yet, because of longer drying, it is usually higher than 5.3.

[c] While an a_w value below 0.95 in combination with low pH is sufficient with starter sausage without mold for prevention of spoilage and pathogenic growth, a lower a_w value serves for a better safety, preventing lactic acid breakdown in case of mold-ripened sausages.

reduced pH (<5.3) ensures safety in combination with lowered a_w. Nevertheless, it has to be mentioned that even in the case of mold-ripened starter sausages, drying time is long enough to promote biochemical reactions responsible for raising the pH value unlike in short ripened–dried sausages without mold.

Some characteristic properties of mold-ripened sausages are shown in Table 34.1.

In certain cases, moisture/protein ratios are given for specifying dry and semidry sausages, but a_w values can be considered less equivocal (Incze 2007).

34.3 Technology of Mold-Ripened Sausages

34.3.1 Raw Material

High-quality product can be manufactured only from high-quality ingredients, that is, from hygienically processed raw material with good technological and sensoric attributes produced under good manufacturing practice (GMP). This high hygienic level is even more important in the case of dry sausage manufacture, since there is no way of pasteurization for reducing undesired microbial load in raw materials as it is in dairy industry. Structure and texture, as well as sensory value, are influenced (Incze 2000) by the following factors:

- Type of meat (pork, beef, poultry, game, young, old animal, etc.)
- Type of fat (saturated-to-unsaturated fatty acid ratio)
- Type of fat tissue (marbling, back fat)
- Meat-to-fat ratio in sausage batter (higher fat/protein ratio = softer texture)
- Particle size
- Temperature of chopping, grinding, and stuffing
- Type and concentration of seasonings
- Application of carbohydrate (type and concentration)
- Casing diameter
- Smoking
- Pattern of pH changes
- Salt concentration
- Microflora (starters or spontaneous)
- Drying parameters
- Final pH and water activity

As for the technological quality of meat, the "drier" muscle of sows—with higher intramuscular fat content—is more suitable than the lean muscle of young animals if a low-water-activity dry salami is the aim, which is achieved through long ripening. Meat with higher fat and lower moisture content allows for better control of drying; also, less shrinking of sausage diameter occurs, which gives a more attractive product.

Presence of pale, soft, exudative and dark, firm, dry meat is less of a problem because of the low incidence of such types; their presence is compensated for by comminution and mixing with normal meat. Formation of texture greatly depends on pH value: when lactic acid is produced by starter cultures, gelifying a protein solution between meat and fat particles, a firm texture and sliceability are attained more rapidly than when the sausage pH is far from the isoelectric point of muscle. This means that with a higher sausage pH (low acidity), a significant rise in salt concentration (through long drying) is the only factor responsible for gelifying the protein solution and for sliceability, which is the result of the gel structure of the protein matrix. Further details on influencing factors will be discussed in the following sections. Process flow of dry sausage manufacture is shown in Figure 34.1.

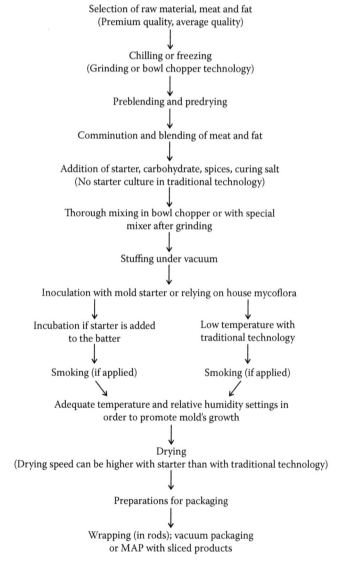

FIGURE 34.1   Process flow of dry sausage manufacture.

34.3.2 Prefreezing and Comminution Techniques

Depending on the comminution technique (chopping or grinding), prefeezing is either a must or can be omitted. If raw material is ground (e.g., in Italy), no prefreezing is needed; chilling is adequate. But if chopping (bowl chopper) is applied (e.g., in Germany, Hungary, the Netherlands, etc.), prefreezing of fat (–5°C to –7°C) and meat cannot be avoided; otherwise, smearing of fat particles during chopping and stuffing causes drying failure and loss of sensory quality and also leads to hygienic risk.

During comminution and blending by bowl chopper meat, fat and other ingredients, curing salt, seasonings, carbohydrates, and starter cultures (if applicable) are added and thoroughly mixed. Care should be taken with the order of addition: simultaneous addition of curing salt and starter culture should be avoided for the sake of protection of starter culture against initial high salt concentration. If grinding technology is applied, in addition to mixing the ingredients before grinding another mixing after grinding may be necessary.

Even if comminution is done properly, adequate stuffing is needed to avoid smearing of sausage batter; in addition to the appropriate temperature, the right choice of vacuum stuffing machine is of crucial importance because air bubbles in the sausage could cause spoilage in form of rancidity and discoloration, not to mention the visual defect of the cut surface with air holes. In order that moisture can evaporate from sausage during drying, casings with proper vapor permeability are used. Details on casings can be found in the book of Savic and Savic (2002).

After stuffing, sausages are washed, dried, and incubated (if starters are used), the surface is inoculated with mold starters, and smoking is eventually applied.

During ripening, the temperature, relative humidity, and air velocity are adjusted to promote mold growth. Major critical points during dry sausage manufacture are somewhat different whether traditional or starter culture technology is applied (Incze 2004a).

34.3.3 Fermentation

In original sense, in the case of fermentation, we think of the breakdown of added and tissue carbohydrate, but with raw sausage fermentation, it can be used in a broader sense, consisting of other metabolic activities as well. We may make a distinction between the first days' fermentation processes and the ripening that follows, but these activities of tissue and microbial enzymes are continuous with descending intensity as a_w decreases causing partial or total inactivation at a later stage. Meat fermentation principles and application are discussed in detail by Demeyer (2004) and Petäjä-Kanninen and Puolanne (2007).

34.3.4 Use of Starter Culture

In long-ripened sausages, just as in short-ripened ones, lactic starters reduce pH and contribute to aroma; apathogenic staphylococci play a role mainly in color and aroma formation. Since staphylococci require higher pH, the pH drop caused by lactobacilli may retard their growth. Preripening (keeping the sausage at 5°C for some days) can give staphylococci a chance to grow before the pH drop (Sanz et al. 1997).

Depending on whether starter cultures (lactic starters alone or in combination with staphylococci and yeasts) are used or not, initial temperature (incubation) is basically different. In the presence of starter culture, the temperature has to be adjusted to the microorganisms' need, usually below 30°C (18°C –24°C) in Europe and above 30°C (27°C –32°C) in the United States.

The presence and metabolic activity of the starter culture contribute to product safety through its inhibitory effect against undesired microbes. But if no starter culture is added to the sausage batter, another inhibiting factor—low temperature—has to be relied on. This low temperature (usually 10°C –12°C), together with common and curing salt, ensures sufficient protection against growth of pathogenic and spoilage microorganisms until adequately reduced water activity ensures inhibition also at higher temperatures.

Types and application of starter cultures are discussed in several publications; some of them are mentioned here (Leistner 1990; Knauf 1995; Cocconcelli 2007; Cocconcelli and Fontana 2010).

Earlier starter cultures were applied mainly for controlling pH changes or nitrate reduction and color stability, respectively, but nowadays, other tasks are also accomplished by microbial cultures: they increase safety by inhibiting growth of undesired, that is, pathogenic and spoilage, microorganisms. This is specially the case with starters and protective cultures that, in addition to lactic acid, produce also specific antibacterial substances, also called bacteriocins.

These protective or bioprotective cultures can be effective against bacteria such as *Listeria monocytogenes*, *Clostridium botulinum*, and *Staphylococcus aureus*.

Some more details concerning mold-ripened sausages can be found in the work of Incze (2004b, 2010).

34.3.5 Color Formation

Raw fermented sausages are usually manufactured with curing salt in earlier times and, in some cases, also today with KNO_3, but mostly with nitrite ($NaNO_2$), or a mixture is applied. Since sodium nitrite ($NaNO_2$) is poisonous also in small concentrations, for the sake of prevention of eventual poisoning, a curing salt, that is, a mixture of 99.5% NaCl and 0.5% $NaNO_2$, is used that excludes the possibility of overdosage of nitrite. (In Europe, 0.4% $NaNO_2$ content in curing salt is also applied; in the United States, much higher $NaNO_2$/NaCl ratio is applied, in which the mixture is diluted by common salt during technology.) At the end of the 1970s, with the growing concern on nitrosamine poisoning, use of nitrite and nitrate was reduced, in some countries even prohibited, but a few years later, application of curing salts is further allowed with preference of nitrite. In this change of situation, basically two facts played a decisive role: the very intensive investigations did not find the danger unequivocally as real if there was no abuse in nitrite concentration, temperature, etc., and the reducing agent was applied on the one hand, and never since has been found any other compound that could substitute the favorable effects of nitrite with emphasis on its *C. botulinum* inhibitive capability on the other.

If nitrate is used, it has to be reduced by microorganisms to nitrite, which then reacts with myoglobin to give nitroso-myoglobin. The more stable pigment is formed most likely by denaturation through increasing salt concentration during drying. Dry sausage can also be manufactured with common salt only, yet the microbiological risk is higher if the antibacterial effect of nitrite cannot be relied on. This is even more pronounced if, for health purposes, sodium content is reduced. In this case, safety requirements can be met only if other a_w-reducing substances are applied. In the absence of curing salt, color stability of sausage is reduced, but intensive smoking improves it to some extent because NO compounds from smoke deposit and diffuse inside (Incze 1987). Details on color formation are discussed thoroughly by Møller and Skibsted (2007) and Honikel (2010).

34.3.6 Smoking and Mold Growth

Most mold-ripened sausages are not smoked; this is explained by the long traditions that offer some advantages, mainly in recent decades since artificial inoculation of the surface by starter molds like *Penicillium nalgiovense*, *Penicillium chrysogenum*, and *Penicillium camemberti* (Knauf 1995) has been applied. Starter molds are sensitive to smoke constituents in general, and this retards mold growth, resulting in scarce and uneven growth. There are, nevertheless, mold-ripened sausage technologies in which after intensive smoking, surface inoculation with mold starters is applied, or spontaneous mold growth covers the surface (e.g., Hungarian salami). For even growth on the whole surface of this latter type, sporadic colonies are spread by brushing. Smoking has manifold advantages:

- Inhibition of growth of yeasts, (slime-producing) *Kocuria* strains, and undesired, toxigenic molds
- Strong antioxidative effect retarding rancidity very efficiently
- Significant positive influence on sensory characteristics

As a result, smoked products are significantly different in taste and aroma.

Mold layer on sausages has several advantages:

- Drying is more balanced, thus reducing frequency of drying failure.
- It slows down moisture loss in the final product.

- Because of its light protection, discoloration and rancidity are retarded
- Molds metabolize peroxides, thus protecting against rancidity (Spotti and Berni 2007)

The source of molds is either the air in old ripening rooms with selected mold strains (it is actually a kind of "autoselection" controlled by temperature, relative humidity, eventually by smoke, and certainly by manufacturing for decades), which deposit on the surface of sausages, or a mold starter culture that is used for artificial inoculation. Inoculation may take place in the form of dipping the sausages in starter culture suspension after stuffing or in the form of atomizing of the suspension. This spraying can be applied either before or after smoking if the product is smoked. If applied after smoking, molds sensitive to smoke may have better chance to grow.

Growth of molds, be it inoculated or spontaneous, has to be supported by suitable temperature and relative humidity parameters. This is a process with risk involved because with too low a temperature and relative humidity, mold growth is retarded, and undesired microbes might grow instead; on the other hand, with too high a temperature and relative humidity, pathogenic and/or spoilage microorganisms are able to grow inside the sausage, mainly if a_w is not low enough, and molds metabolize lactate, which allows growth and toxin production of staphylococci (Rödel et al. 1993). Consequently, strict control of these parameters in accordance with the requirement of mold applied is of vital importance.

If mold growth is well supported, an even mold cover on the surface is formed that sticks well to the product and remains intact during handling, unless abuse occurs during storage: either too high relative humidity or an impermeable packaging film.

A well-developed mold layer plays an important role during further technological steps and also during storage in retail and in households, as mentioned earlier. After the mold layer is completed, relative humidity, and in some cases also temperature, is decreased and drying is continued. Thanks to this mold layer, risk of too rapid drying causing case hardening is significantly less.

34.3.7 Ripening–Drying

Ripening is the most complicated process controlled only to some extent by technologists and completed mostly by tissue enzymes and by metabolic activity of inoculated and/or spontaneous microorganisms.

Intentional means for control are application of starter culture, type and concentration of additives (intrinsic factors), and, mainly, adjustment of temperature, relative humidity, and air velocity as a function of time (extrinsic factors), controlling in this way the drying rate (Leistner 1990). Uniform air distribution is a precondition for uniform and consistent quality. Contrary to short-fermented sausages, in which, to some extent, also tissue enzymes and definitely bacterial enzymes are active in the final product, this process can slow down or even stop after an extremely long drying period of several months, resulting in a_w values as low as 0.85.

Since there is a difference between the pH value before drying of traditional and starter sausages, the latter can be dried faster without the risk of case hardening. The well-known reason is that at a pH close to the isoelectric point of muscle protein (pH 5.3), the less strongly bound moisture can be given off easier than in traditional sausage with a typical initial pH value of 5.6–5.8. This also means that greater care has to be taken in the latter case with drying in order to avoid case hardening. This explains why there are significant differences between the drying times of traditional and starter sausages as shown in Table 34.1.

For details on theory and practice of sausage drying, one can consult with the work of Zukál and Incze (2010).

34.4 Changes during Ripening–Drying: Aroma Formation

Controlling temperature and drying rate plays a decisive role in food safety by inhibiting growth of undesired microbes, but this control is needed also for a consistent sensory quality. As a result of drying (i.e., a_w reduction), less resistant microorganisms gradually disappear (members of Enterobacteriaceae, pseudomonads, bacilli, etc.); less sensitive ones (*Kocuria*, staphylococci, enterococci, listeriae, enterohemorrhagic

Escherichia coli (EHEC), etc.) may survive for a longer period of time. In the final product of long-ripened dried sausages, spoilage microflora has practically no chance to grow and cause deterioration. The situation is somewhat similar with pathogenic microorganisms, yet their very presence is not accepted by food authorities requiring sometimes zero tolerance—a standard that is hard to support scientifically with all types of microorganisms.

In order to meet safety requirements, a complex strategy has to be followed by adding and/or supporting growth of useful microorganisms in the hygienically produced sausage and by controlling ripening–drying parameters, thus favoring growth of useful microorganisms and inhibiting growth of undesired ones, as discussed above. While doing so, not only are safety requirements met but also this has contributed to aroma formation.

Starter cultures and bacteriocin-producing microorganisms can help in inhibiting pathogenic microbes (Demeyer and Vandekerckhove 1979; Rödel et al. 1993; Aymerich et al. 1998; Hugas et al. 1998; Leroy et al. 2001), where inhibition of staphylococci and listeriae is desired mostly because these bacteria are difficult to combat as a result of their resistance to low a_w, the main controlling factor in long-dried sausages.

It should be kept in mind that bacteriocin producers are, in general, more effective under laboratory conditions on media than in a food matrix (Leroy et al. 2001).

Unlike in dried ham, tissue endoenzymes in long-dried sausages play a less important role in aroma formation, and these are active mainly in the first part of ripening (Harnie et al. 2000). In long-dried sausages, the metabolic activity of microorganisms contributes more to aroma, but other factors such as spices and lipolysis may also rather be involved in this complex process.

As mentioned already, smoking and pH changes (low- and high-acid sausages) do influence aroma characteristics, but the types and concentrations of spices used determine aroma even more (Chizzolini et al. 1998). In this respect, spices that can be evenly distributed in sausage batter and react immediately with meat and fat matrix have a much more pronounced impact on aroma characteristics than occurs in long-ripened hams, where spices are generally not applied or they are rubbed on the muscle surface only. Natural spices in sausages influence aroma as well as safety indirectly by enhancing the metabolic activity of starter cultures (Nes and Skjelkvale 1982; Hugas et al. 1998).

As for taste and aroma formation by microbial metabolism, the breakdown of proteins, peptides, lipids, and carbohydrates will be discussed next.

Depending on application of lactic starters and addition of carbohydrate, lactic acid is produced through homofermentation, reducing the pH level (Demeyer and Vandekerckhove 1979; Roca and Incze 1990). Concentration of D-lactate may be five times higher in high-acid sausages than in low-acid products (Montel MC et al. 1998). At a later stage, pH also increases in sausages with lactic starters; this is related to the formation of ammonia and some amino acids (Demeyer and Vandekerckhove 1979). Nevertheless, this occurs mainly in products in which apathogenic staphylococci are used without lactobacilli (Marchesini et al. 1992). In traditionally manufactured molded dry sausage (e.g., genuine Hungarian salami), no starter culture and very small amounts of carbohydrate are added; as a result, pH in the final product is around 6.0 (Incze 1987).

As for breakdown of muscle proteins, endoenzymes (cathepsin D) are active at the beginning of ripening (Harnie et al. 2000) and are enhanced by lower pH (Montel MC et al. 1998). Lower pH, on the other hand, does not favor endogenic aminopeptidases. Microbial aminopeptidases that break down peptides into amino acids also contribute to aroma development (Flores et al. 1998). Their activity is also influenced by a_w, pH, and temperature. Bacterial enzymes play a role in proteolysis at a later period of ripening (Verplaetse et al. 1992).

With regards to the microorganisms, the proteolytic activity of lactobacilli, micrococci, and molds has been more thoroughly investigated, and it has been found that some lactobacilli (*Lactobacillus casei* and *Lactobacillus plantarum*) can break down sarcoplasmic and myofibrillar protein (Fadda et al. 2000).

Intensity of protein metabolism is higher by micrococci (staphylococci) that are also responsible for formation of volatile and nonvolatile aroma compounds (Stahnke et al. 2000; Sunesen et al. 2000).

Nielsen and Coban (2001) found that *P. nalgiovense* has proteolytic activity, breaking down peptides in addition to muscle protein stimulated by salt. Ordóñez et al. (2000) found strong proteolytic (and lipolytic) activity with *Penicillium aurantiogriseum*.

In lipolysis, endoenzymes and exoenzymes participate, yet microbial enzymes also play a more intensive role here (Talon et al. 1992). Typical flavors and aromas are related to the hydrolytic and oxidative changes occurring in the lipid fraction during ripening. Lipid oxidation may cause off-flavor but may also contribute to the development of desirable flavor (Talon et al. 2000). Lipid oxidation is mostly associated with unsaturated fatty acids and is often autocatalytic. For this reason, risk is higher if the raw material contains a higher ratio of unsaturated fatty acid (Rödel et al. 1993; Bloukas et al. 1997) by either partially replacing fat with vegetable oil or using the fat of animals that consumed elevated levels of unsaturated fatty acid. Risk of oxidation, on the other hand, can be reduced by the antioxidative effect of smoking and by mold ripening because of the light protection and direct oxygen consumption provided by mold, as pointed out by Ordónez et al. (2000).

It is interesting to note that as a result of microbial lipolysis in mold-ripened Hungarian salami, a significant increase in the acid number of the fat fraction takes place, reaching a value of 15 without any sign of rancidity as judged by peroxide or thio-barbituric acid (TBA) number or organoleptically (Nagy et al. 1988). Such a high acid number can otherwise be detected only in strongly rancid lard with a high peroxide number, where, because of high temperature rendering no microbial activity takes place and moisture is also very low.

Because physical, chemical, biochemical, and microbiological changes in mold-fermented sausages are rather similar to those of dry sausages without mold, the reader is referred to publications on these topics (Navarro et al. 1997; Chizzolini et al. 1998; Flores et al. 1998; Incze 1998, 2000; Montes et al. 1998; Stahnke et al. 2000; Sunesen et al. 2000; Talon et al. 2000, 2004; Zanardi et al. 2000; Paramithiotis et al. 2010); mainly those features have been discussed that are characteristic to mold-ripened sausage. It has to be mentioned that sensory characteristics may significantly differ depending on the diameter for two reasons: (1) the metabolic pathway is different in the case of anaerobiosis (large-diameter salamis) compared with small-diameter sausages, in which anaerobiosis occurs to a lesser extent, and (2) a thinner sausage has much less time for aroma formation because of the significantly faster drying.

34.5 Safety of Mold-Ripened Sausages

34.5.1 Bacterial Risk

The problem of possible growth, survival, and/or toxin production of salmonellae, EHEC, toxinogenic clostridia, staphylococci, listeriae, and mycotoxic molds has to be dealt with, as well as the potential risk of biogenic amines.

Safety requirements can only be met if hygienically produced raw materials, ingredients, and additives are used during Good Manufacturing Practice (GMP), applying a well-based Hazard Analysis Critical Control Points (HACCP) system. For traditional technologies, long-existing safety records prove the right way of manufacturing (Incze 1998). Should a new manufacturing technology be started, challenge tests with pathogens must be used. (With enterohemorrhagic *Escherichia coli*, the Food Safety Inspection Service (FSIS) requires the following challenge test with raw fermented sausages: it has to be proved that a 5 logarithmic reduction is caused by the given technology.)

It is relatively easy to inhibit the growth of salmonellae and clostridia because both are sensitive to low pH and low a_w, and the latter are also inhibited by nitrite. Staphylococci are inhibited by low temperature and low pH, and although they are rather resistant to low a_w, no enterotoxin can be produced at an a_w value that is common with long-ripened sausages ($a_w < 0.90$).

Listeriae are rather resistant to the pH and a_w values of fermented sausages, but if initial count is low and GMP is applied, they are affected by low pH, low a_w, growth, or presence of starter cultures causing reduction in their number mainly if bacteriocinogenic microbes are applied (Aymerich et al. 1998; Hugas et al. 1998). A new, possibly efficient way of inactivating listeriae in foods of animal origin may be the application of bacteriophages specific for destroying *L. monocytogenes*. Soni and Nannapaneni (2010) found that LISTEX P 100 phage (qualified as Generally Recognized as Safe [GRAS]) caused significant reduction on raw salmon tissue. Further research is needed to find out whether this phage works under the conditions of dry sausage.

Enterohemorrhagic *E. coli* are also rather resistant to low pH and low water activity, but they are reduced in number if a low a_w ($\leq$0.91) condition exists for longer time at elevated ambient temperature (Nissen and Holck 1998). This phenomenon, also called metabolic exhaustion, is a reliable means that helps ensure a five-log reduction in case of long-ripened dry sausages.

Further details on spoilage and pathogenic risk and control are discussed by Skandamis and Nychas (2007) and by Labadie (2007).

34.5.2 Mycotoxic Molds

Since the discovery of aflatoxin, numerous other mycotoxins have been detected and identified on agricultural products, in foods, and in nature. Ever since we became aware of the potential risk of mycotoxins, foods with mold growing inside or outside, spontaneously or intentionally, are considered "suspicious." With this in mind, mold-ripened sausages (and hams) have been investigated thoroughly, and the isolated molds were tested for toxin production (Bullerman et al. 1969; Cigler et al. 1972; Alperden et al. 1973). It was possible to isolate mycotoxic mold from mold-fermented meat products, and mycotoxins were also found in products artificially inoculated with pure culture of toxic molds; but, in general, no mycotoxins were detected in commercially manufactured meat products. In further experiments, the effects of extrinsic and intrinsic factors on toxin production potential were investigated (Incze and Frank 1976a,b). In these latter experiments, mixed culture ("house flora") of apathogenic molds, low temperature (13°C), and low a_w value (0.94) as well as intensive smoking have been shown to be growth inhibitors of mycotoxic molds (8 aflatoxin and 11 sterigmatocystin producers were inoculated). This finding has been supported by investigation of 800 salami samples taken from retail shops: testing analytically as well as biologically for mycotoxins gave proof of the absence of mycotoxins (Incze et al. 1976). On the basis of these results, "house mycoflora" of ripening rooms can be considered harmless if traditional technologies with reliable hurdles are applied during ripening–drying; nevertheless, technologies have been changed in general for application of mold starters.

Summing up, we can conclude that fermented, dried meat products are stable and safe without refrigeration because of the low a_w value (<0.90) or because of a combination of low a_w and low pH value.

34.5.3 Biogenic Amines

These compounds are mainly formed by decarboxylation of amino acids or by amination and transamination of aldehydes and ketones (Eerola et al. 1992). Occurrence and concentration of biogenic amines (tyramine, putrescine, cadaverine, etc.) in fermented foods have been intensively investigated (Pipek et al. 1992), and it has been found that starter cultures may play a role in the extent of biogenic amine formation, but freshness of raw material can be considered a major factor in keeping low the occurrence and concentration of biogenic amines (Hernández-Jover et al. 1997; González-Fernandez et al. 2000). Consequently, dry sausages properly manufactured from fresh raw materials can also be considered safe from this point of view. For further details on biogenic amines, see Vidal-Carou et al. (2007).

34.6 Shelf Life

From a microbiological point of view, long-ripened sausages are stable because of their low water activity or because of the combination of low water activity and low pH. This refers to sausages stored under adequate conditions; otherwise, at high relative humidity, mold cover may loosen and get off, or secondary mold growth may occur.

Too low relative humidity and too high temperature during storage shorten the shelf life but only because they lower sensory quality and have nothing to do with microbiological activity, that is, with spoilage or with safety risk. While vacuum and modified atmosphere packaging (MAP) is an excellent method for extending the shelf life of sausages without mold, it is no real alternative for mold-ripened sausages for many different reasons.

The problem with both types of packaging is that moisture migrating from the core to the surface cannot evaporate, which makes it wet; thus, mold layer loosens and comes off, giving the product a bad appearance, and wild strains of yeasts and some bacteria eventually present under the mold layer start to grow, which also causes bad odor. For shorter storage of several weeks, either wrapping in cellophane can be done or perforated polypropylene can be used in order to let moisture evaporate, with awareness that weight loss occurs even if at lower speed. Evidently, if mold-ripened dry sausage is sliced after peeling of the casing and packaged under vacuum or in MAP, a much longer shelf life, lasting several months, can be guaranteed. Zanardi et al. (2002) claim that because of less residual oxygen in MAP, less oxidation effect was detected than in vacuum packaging, which is not always the case and depends very much on how precisely the packaging technology is carried out. On the other hand, it is true that MAP is favored because slices do not stick together as strongly as in vacuum packages and can be separated easier.

As mentioned above, long-ripened dry sausages are stable and safe at ambient temperature, and cold storage in a refrigerator is rather disadvantageous because phosphate crystals can appear on the product. This can occur even if no phosphate was added: muscle tissue's own phosphate compounds are sufficient for causing this phenomenon, supported by decreasing solubility of phosphates at low a_w and higher pH.

34.7 Quality Defects

Reasons of quality defects can be as follows: use of poor-quality raw materials, technological failures, and abuses during storage. Some of the more common faults and causes are discussed below.

34.7.1 Defects in Appearance

A properly manufactured mold-ripened dry sausage has a uniform, even mold layer that sticks to the surface. If mold comes off, possibly yeasts and/or bacteria grew first on the surface of sausage, thus eventually making adhesion of molds to the surface and/or wet conditions are responsible as mentioned. If pieces of mold layer come off, quite often with the cellulose casing itself, followed by greasing through, it can be explained by presence and activity of cellulase-producing molds. As for adequate prevention, use of proper mold starter in necessary concentration and activity and/or use of casings that cannot be attacked by cellulase enzyme (e.g., collagene casing) can be considered. The other reason for an uneven mold layer may be technological in nature (temperature and relative humidity unfavorable for mold growth, inefficient mold starter) or storage abuse (high relative humidity). The other type of defect is if yellow, green, or black molds grow on the surface distorting the required grayish-white mold layer.

Deformation of sausages is caused by improper drying (too rapid, mainly in the first period), resulting in case hardening that does not allow evaporation of moisture from the core. As a consequence of case hardening, spoilage and pathogenic microbes may grow because a_w reduction is slowed down significantly. Another reason for case hardening may be the wrong comminution temperature and wrong selection of the stuffing machine causing fat smearing that inhibits proper drying of the inner part of sausage.

In some cases, there are consumers who are accustomed with excessively dried, distorted, deformed sausages with dents (e.g., in France, Spain, Switzerland, China, etc.), be it with or without mold, having the impression of a very low moisture product suggesting good safety record, long shelf life, and higher nutritive value (high muscle protein content). This actual demand is met quite often by local smaller manufacturers. These sometimes intentionally pressed, oval cut surface sausages can be of excellent quality, but deformed by chance, and rancid products also occur.

34.7.2 Color Defects

Color defects can be caused by insufficient amount or absence of curing salt by mistake and by wrong way of smoking. If sausages with wet surface are smoked or are at high relative humidity in the smoking chamber, brownish-grayish discoloration occurs not only on the surface but also in a thicker layer, which is a fault that is most of the time irreversible. Color defect can occur also if air holes are numerous and extended caused by failure during stuffing or by faulty drying, causing case hardening with longer holes. In both cases, discoloration and rancidity effect of oxygen are the causative agents.

34.7.3 Odor Defects

Odor defects can be caused if surface is wet for a long period, and yeasts as well as micrococci grow, making the surface slimy and malodorous. This slimy layer inhibits good adhesion of molds, which can slip off easily. If raw material is not of good hygienic quality or ripening–drying is not done properly (too high temperature, too slow drying, inefficient starter culture, etc.), spoilage can occur with unpleasant odor and taste; health risk cannot be excluded either.

As mentioned earlier, it is very important that the initial microbial load of the raw material be low because the means are not plenty for inhibiting spoilage and pathogenic microorganisms in case of raw sausage manufacture: only low temperature and low pH (if starter is applied) can combat their undesired growth.

To this inhibitive effect, common and curing salts also contribute, and for this reason, either reduction of common salt or omission of nitrite increases health and spoilage risk.

34.7.4 Flavor Defects

Flavor defects can be caused if product is too acidic (higher carbohydrate concentration, higher temperature for a long period) or rancid. The latter can occur if raw material was not fresh, fat ingredients contain too high concentration of unsaturated fatty acids, drying and storage are done at too high temperatures and for too long, nitrite is not used, and sausages are not smoked.

An intensive smoking treatment, evidently at low temperature (<15°C) because the sausage is raw dried, has a very efficient antioxidative effect preventing rancidity, thus extending shelf life. If the sausage is manufactured with fresh meat and fat with proper redox state and the product is intensively smoked, practically no rancidity occurs in spite of slight abuse during storage. On the other hand, sausages not smoked, which is the kind of technology typical in the Mediterranean area, are not only significantly different in flavor from smoked products but also have more chance to become rancid retarded to some extent by the mold cover.

Some more details on quality defects as well as their reasons and possible prevention can be found in the literature (Incze 2010).

34.8 Conclusion

Mold-ripened sausages are commodities that have been known for thousands of years; however, somewhat detailed knowledge about them in Europe has a history of some hundred years only. Controlled fermentation is even younger, and application of known microorganisms aiming at influencing the organoleptic properties as well as the safety of this type of product for ensuring a reproducible consistency in quality has a history of 50 odd years.

Mold-ripened sausages are air-dried or smoked and dried products with a special appearance of the mold layer on the surface contributing to a balanced drying process and extended shelf life, as well as a unique sensory value highly esteemed by conscious consumers. Because of the low water activity (traditional products) or combination of low pH and low water activity (starter culture technology), raw dried mold-ripened sausages manufactured under hygienic conditions with properly controlled technology have an excellent safety record and can be considered as most valuable meat products beloved by connoisseurs.

REFERENCES

Alperden I, Mintzlaff H-J, Tauchmann F, Leistner L. 1973. Bildung von Sterigmatocystin in mikrobiologischen Nährmedien und in Rohwurst durch Aspergillus versicolor. *Fleischwirtschaft* 53: 707–710.
Aymerich T, Artigas MG, Garriga M, Monfort JM, Hugas M. 1998. Optimization of enterocin A production and application in Spanish type dry fermented sausages. *Proceedings of International Congress of Meat Science and Technology*, Barcelona, pp. 340–341.

Bacus J. 1984. *Utilization of Microorganisms in Meat Processing.* Research Studies Press Ltd., Letchworth; John Wiley & Sons Inc., p. 8.

Bloukas JG, Paneras ED, Fournitzis GC. 1997. Effect of replacing pork backfat with olive oil on processing and quality characteristics of fermented sausages. *Meat Sci* 45: 133–144.

Bullerman LB, Hartman PA, Ayres JC. 1969. Aflatoxin production in meats. II. Aged dry salamis and aged country cured hams. *Appl Microbiol* 18: 718–722.

Chizzolini R, Novelli E, Zanardi E. 1998. Oxidation in traditional Mediterranean meat products. *Proceedings of International Congress of Meat Science and Technology*, Barcelona, pp. 132–141.

Cigler A, Mintzlaff H-J, Weisleder D, Leistner L. 1972. Potential production and detoxification of penicillic acid in mould fermented sausage (salami). *Appl Microbiol* 24: 114–119.

Cocconcelli PS. 2007. Starter Cultures: Bacteria In: Toldrá F. editor. *Handbook of Fermented Meat and Poultry.* Ames, IA: Blackwell Publishing Professional, pp. 137–146.

Cocconcelli PS, Fontana C. 2010. Starter cultures for meat fermentation In: Toldrá F. editor. *Handbook of Meat Processing.* Ames, IA: Wiley-Blackwell, pp. 199–218.

Demeyer D. 2004. Meat Fermentation: Principles and Applications. In: Hui YH. editor. *Handbook of Food and Beverage Fermentation Technology.* New York: Marcel Dekker, pp. 353–368.

Demeyer DI, Vandekerckhove P. 1979. Compounds determining pH in dry sausage. *Meat Sci* 3: 161–167.

Dirinck P, Opstaele F V. 1999. GC-MS characterization of the flavour of dry sausages from different European origins. *Proceedings Euro Food Chem X.* Budapest. Functional Foods—A New Challenge for the Food Chemist, pp. 815–821.

Eerola S, Maijala R, Hill P. 1992. The influence of nitrite on the formation of biogenic amines in dry sausages. *Proceedings of International Congress of Meat Science and Technology*, Clermont-Ferrand, pp. 783–786.

Fadda SG, Vignolo GM, Sanz Y, Aristoy M-C, Oliver G, Toldrá F. 2000. Muscle proteins hydrolysis by Lactobacillus isolated from dry sausages effect of curing conditions. *Proceedings of International Congress of Meat Science and Technology*, Buenos Aires, pp. 232–233.

Flores M, Sanz Y, Toldrá F. 1998. Curing agents as regulators of muscle and microbial aminopeptidase activity in dry fermented sausages. *Proceedings of International Congress of Meat Science and Technology*, Barcelona, pp. 868–869.

González-Fernández C, Rovira J, Jaime Isabel. 2000. Influence of starter cultures on sensory properties and biogenic concentration in chorizo. *Proceedings of International Congress of Meat Science and Technology*, Buenos Aires, pp. 222–223.

Harnie E, Claeys E, Raemaekers M, Demeyer D. 2000. Proteolysis and lipolysis in an aseptic meat model system for dry sausages. *Proceedings of International Congress of Meat Science and Technology*, Buenos Aires, pp. 238–239.

Hernández-Jover T, Izquierdo-Pulido M, Veciana-Nogués MT, Mariné-Font A, Carmen M, Vidal-Caron. 1997. Effect of starter cultures on biogenic amine formation during fermented sausage production. *J Food Prot* 60: 825–830.

Honikel KO. 2010. Curing. In: Toldrá F. editor. *Handbook of Meat Processing.* Ames, IA: Wiley-Blackwell, pp. 125–141.

Hugas M, Garriga M, Pascual M, Aymerich MT, Monfort JM. 1998. Influence of technological ingredients on the inhibition of *Listeria monocytogenes* by a bacteriocinogenic starter culture in dry fermented sausages. *Proceedings of International Congress of Meat Science and Technology*, Barcelona, pp. 338–339.

Incze K. 1987. The technology and microbiology of Hungarian salami. Tradition and current status. *Fleischwirtschaft* 67: 445–447.

Incze K. 1998. Dry fermented sausages. *Meat Sci* 49: 169–177.

Incze K. 2000. Research priorities in fermented and dried meat products. *Proceedings of International Congress of Meat Science and Technology*, Buenos Aires, pp. 210–215.

Incze K. 2004a. Dry and semi-dry sausages. In: Jensen WK. editor. *Encyclopedia of Meat Sciences.* Oxford: Elsevier Academic Press, pp. 1207–1216.

Incze K. 2004b. Mold-ripened sausages. In: Hui YH. editor. *Handbook of Food and Beverage Fermentation Technology.* New York: Marcel Dekker. Inc., pp. 417–428.

Incze K. 2007. European products. In: Toldrá F. editor. *Handbook of Fermented Meat and Poultry.* Ames: Blackwell Publishing, pp. 307–318.

Incze K. 2010. Mold-ripened sausages. In: Toldrá F. editor. *Handbook of Meat Processing.* Ames, IA: Wiley-Blackwell, pp. 363–377.

Incze K, Frank HK. 1976a. Besteht eine Mykotoxingefahr bei der ungarischen Salami? I. Einfluss von Substrat, a_w-Wert and Temperatur auf die Toxinbildung bei Mischkulturen. *Fleischwirtschaft* 56: 219–225.

Incze K, Frank HK. 1976b. Besteht eine Mykotoxingefahr bei der ungarischen Salami? II. Die Bedeutung der Hülle und der Räucherung. *Fleischwirtschaft* 56: 866–868.

Incze K, Mihályi Vilma, Frank HK. 1976. Besteht eine Mykotoxingefahr bei der ungarischen Salami? III. Chemisch-analytische und biologische Untersuchungen an schnittfesten Salamiproben. *Fleischwirtschaft* 56: 1616–1618.

Knauf H. 1995. Starterkulturen für die Herstellung von Rohwurst und Rohpökelwaren: Potential, Auswahlkriterien und Beeinflussungsmöglichkeiten. *Die Fleischerei* 46: 4–14.

Labadie J. 2007. Spoilage microorganisms: risks and control. In: Toldrá F. editor. *Handbook of Fermented Meat and Poultry*. Ames: Blackwell Publishing, pp. 421–426.

Lawrie RA. 1998. *Lawrie's Meat Science*. 6th ed. Cambridge: Woodhead Publication Ltd.

Leistner L. 1986. Allgemeines über Rohwurst. *Fleischwirtschaft* 66: 290–300.

Leistner L. 1990. Fermented and intermediate moisture products. *Proceedings of International Congress of Meat Science and Technology*, Havana, pp. 842–855.

Leistner L. 2005. European Raw Fermented Sausage: Fundamental Principles of Technology and Latest Developments. *Proceedings International Symposium on Fermented Foods*. Chonbuk Natl Univ Jeonju South Korea, pp. 29–39.

Leroy F, Verluyten J, Messens W, de Wuyst L. 2001. The success of using bacteriocin-producing starter cultures for sausage fermentation is strongly strain dependent. *Proceedings of International Congress of Meat Science and Technology*, Kraków, pp. 2–3.

Marchesini Barbara, Brutin Anne, Romailler Nicole, Moreton RS, Stucchi C, Stozzi T. 1992. Microbial events during commercial meat fermentations. *J Appl Bacteriol* 73:203–209.

Møller JKS, Skibsted LH. 2007. Color. In: Toldrá F. editor. *Fermented Meat and Poultry*. Ames: Blackwell Publishing, pp. 203–216.

Montel MC, Masson F, Talon R. 1998. Bacterial role in flavour development. *Proceedings of International Congress of Meat Science and Technology*, Barcelona, pp. 224–233.

Nagy A, Mihályi V, Incze K. 1988. Reifung und Lagerung ungarischer Salami. Chemische und organoleptische Veränderungen. *Fleischwirtschaft* 68: 431–435.

Navarro JL, Nadal MI, Izquierdo L, Flores J. 1997. Lipolysis in dry cured sausages as effected by processing conditions. *Meat Sci* 45: 161–168.

Nes JF, Skjelkvale R. 1982. Effect of natural spices and oleoresins on *Lactobacillus plantarum* in the fermentation of dry sausage. *J Food Sci* 47:1618–1621, 1625.

Nielsen H-JS, Coban I. 2001. Proteolytic activity of *Penicillium nalgiovense*. *Proceedings of International Congress of Meat Science and Technology*, Kraków, pp. 6–7.

Niinivaara FP. 1993. Geschichtliche Entwicklung des Einsatzes von Starterkulturen in der Fleischwirtschaft. I. Stuttgarter Rohwurtsforum: Gewürzmüller, pp. 9–20.

Nissen H, Holck AL. 1998. Survival of *Escherichia coli* O157:H7, *Listeria monocytogenes* and Salmonella kentucky in Norwegian fermented, dry sausage. *Food Microbiol* 15: 273–279.

Ordónez JA, Bruna JM, Fernández M, Herranz B, Hoz de la L. 2000. Role of *Penicillium aurantiogriseum* on ripening: Effect of the superficial inoculation and/or the addition of an intracellular cell free extract on the microbial and physico-chemical parameters of dry fermented sausages. *Proceedings of International Congress of Meat Science and Technology*, Buenos Aires, pp. 234–235.

Paramithiotis S, Eleftherios H, Drosinos H, Sofos JN, Nychas GJE. 2010. Fermentation: microbiology and biochemistry. In: Toldrá F. editor. *Handbook of Meat Processing*. Ames, IA: Wiley-Blackwell, pp. 185–198.

Pederson CS. 1979. *Microbiology of Food Fermentation*. 2nd ed. Westport Connecticut: AVI Publishing Co.

Petäjä-Kanninen E, Puolanne E. 2007. Principles of meat fermentation. In: Toldrá F. editor. *Handbook of Fermented Meat and Poultry*. Ames, IA: Blackwell Publishing, pp. 31–36.

Pipek P, Bauer F, Seiwald G. 1992. Formation of histamine in vacuum packed fermented sausages. *Proceedings of International Congress of Meat Science and Technology*, Clermont-Ferrand, pp. 819–822.

Rixson D. 2000. *The History of Meat Trading*. Nottingham, UK: Nottingham University Press.

Roca M, Incze K. 1990. Fermented sausages. *Food Rev Int* 6: 91–118.

Rödel W, Stiebing A, Kröckel L. 1993. Ripening parameters for traditional dry sausages with a mould covering. *Fleischwirtschaft* 73: 848–853.

Sanz Y, Flores J, Toldrá F, Feria A. 1997. Effect of pre-ripening on microbial and chemical changes in dry fermented sausage. *Food Microbiol* 14: 575–582.

Savic Z, Savic I. 2002. *Sausage Casings*. Vienna: Victus Lebensmittelindustriebedarf Vertriebsgesellschaft m b H.

Skandamis O, Nychas GJE. 2007. Pathogens: risk and control. In: Toldrá F. editor. *Handbook of Fermented Meat and Poultry*. Ames: Blackwell Publishing, pp. 427–454.

Soni KA, Nannapaneni R. 2010. Bacteriophage significantly reduces *Listeria monocytogenes* on raw salmon fillet tissue. *J Food Prot* 73: 32–38.

Spotti E, Berni E. 2007. Starter cultures. In: Toldrá F. editor. *Handbook of Fermented Meat and Poultry*. Ames: Blackwell Publishing, pp. 171–176.

Stahnke LH, Holck A, Jensen A, Nilsen A, Zanardi E. 2000. Flavour compounds related to maturity of dried fermented sausage. *Proceedings of International Congress of Meat Science and Technology*, Buenos Aires, pp. 236–237.

Sunesen LO, Dorigoni V, Zanardi E, Stahnke L. 2000. Volatile compounds during ripening in Italian dried sausages. *Proceedings of International Congress of Meat Science and Technology*, Buenos Aires, pp. 298–299.

Talon R. Barriére C, Centeno D. 2000. Role of staphylococci in the oxidation of free fatty acids. *Proceedings of International Congress of Meat Science and Technology*, Buenos Aires, pp. 290–290.

Talon R, Leroy-Sétrin S, Fadda S. 2004. Dry fermented sausages. In: Hui YH. editor. *Handbook of Food and Beverage Fermentation Technology*. New York: Marcel Dekker, Inc., pp. 397–416.

Talon R, Montel MC, Cantonett M. 1992. Lipolytic activity of Micrococcaceae. *Proceedings of International Congress of Meat Science and Technology*, Clermont-Ferrand, pp. 843–846.

Verplaetse A, Demeyer D, Gerard S, Buys Elke. 1992. Endogenous and bacterial proteolysis in dry sausage fermentation. *Proceedings of International Congress of Meat Science and Technology*, Clermont-Ferrand, pp. 851–854.

Vidal-Carou MC, Veciana Noques MT, Latorre-Noratala ML, Bover-Cid S. 2007. Biogenic amines: risk and control. In: Toldrá F. editor. *Handbook of Fermented Meat and Poultry*. Ames: Blackwell Publishing, pp. 455–468.

Zanardi E, Dorigoni V, Badiani A, Chizzolini R. 2002. Lipid and colour stability of Milano-type sausages: effect of packing conditions. *Meat Sci* 61: 7–14.

Zanardi E, Eerola S, Hartikainen K, Rizzo A, Badiani A, Dorigoni V, Chizzolini R. 2000. Characterisation of Mediterranean and North Europe fermented meat products: chemistry and oxidative stability aspects. *Proceedings of International Congress of Meat Science and Technology*, Buenos Aires, pp. 300–301.

Zukál E, Incze K. 2010. Drying. In: Toldrá F. editor. *Handbook of Meat Processing*. Ames: Wiley-Blackwell, pp. 219–229.

35

Ecosystem of Greek Spontaneously Fermented Sausages

G.-J. E. Nychas, Eleftherios H. Drosinos, and S. Paramithiotis

CONTENTS

35.1 Introduction

Meat fermentation has been the epicenter of thorough research over the last decade. Availability of meat, as a raw material, was the key element for the development of fermented products with enhanced shelf life and modified organoleptic and nutritional properties. Dry fermented sausages are one of the most important meat products. They are characterized by a long shelf life and a remarkable safety record due to many factors, namely, curing salts, lactic acid produced by the lactic microbiota, nitrite produced from nitrate by members of the genus *Kocuria*, and the drying that occurs during the ripening process. In the present chapter, the available literature concerning traditional Greek fermented sausages is reviewed.

35.2 Historical Aspects

Records on meat processing in ancient Greece have been presented in a monograph on sausage casings by Savic and Savic (2002). Sausage-making appeared as a common practice in classical Greece, and thus, many references can be found in the respective literature. Among others, in the *Knights* by Aristophanes, Nicias says: "A sausage-seller! Ah by Poseidon! What a fine trade," highlighting the high appreciation for this type of food. Moreover, Agoracritus, a sausage seller, is himself a character in the same play, and young Sadocus is mentioned for being greedy of the sausages eaten at the Apaturia, a 3-day family or clan festival held annually. In the *Frogs* by Aristophanes, the landlady says: "And I should like to get a scythe, and cut that throat that swallowed all my sausages" referring to Hyperbolus, an unscrupulous politician who was ridiculed as a lamp-maker. Athenaeus, in his *Deipnosophists*, the oldest known Greek cookbook, mentions sausages repeatedly. He refers to them as a meat most saleable and places them among the delicacies of the period such as scent of Syrian myrrh, spices, and fishes. Furthermore, a kind of recipe is described that includes pettitoes of a young porker and his ears as well as snout of pig sprinkled with savory assafoetida. Finally, Aphthonetus is mentioned for his ability to make rich sausages, and Aponetus was held unrivaled for his sausages (Anonymous 1854, 1892, 1912).

TABLE 35.1

Typical Recipes Used for the Production of Greek Spontaneously Fermented Sausages

	Ingredient (%)	Drosinos et al. (2005a)	Samelis et al. (1993)	Papamanoli et al. (2002) I	Papamanoli et al. (2002) II	Rantsiou et al. (2005)	Samelis et al. (1994a)	Samelis et al. (1998)	Metaxopoulos et al. (2001) I	Metaxopoulos et al. (2001) II	Metaxopoulos et al. (2001) III	Drosinos et al. (2007)
Lean meat	Beef	20.0	37.53	38.82	31.61	20	28.15	–	33.46	–	–	31.5
	Pork	55.0	28.15	24.68	31.61	55	23.45	70.89	33.46	65.38	70.89	34.5
	Mutton	–	–	–	–	–	14.07	–	–	–	–	–
Fat additives—other ingredients	Pork back	20.0	28.15	31.71	31.61	20	28.15	22.38	28.68	28.02	22.38	28.7
	Sodium chloride	2.3	2.81	2.79	2.62	2.30	2.81	2.33	2.48	2.80	2.33	2.20
	Sodium nitrite	0.02	0.02	0.01	0.01	0.02	0.02	0.02	0.02	0.02	0.02	0.02
	Sodium nitrate	0.02	0.02	0.02	0.02	0.02	0.02	0.02	0.02	0.02	0.02	0.02
	Sodium ascorbate	0.06	0.03	0.07	0.07	0.07	0.03	0.05	0.05	0.02	0.05	0.07
	Sodium glutamate	–	–	–	–	–	–	–	0.19	–	–	–
	Sugars mix	1.5	1.87	1.15	1.15	1.50	1.87	1.39	0.95	2.24	1.39	1.20
	Spices mix	0.3	0.28	–	–	0.32	0.28	0.27	0.67	0.56	0.28	0.33
	Garlic	–	0.18	0.03	0.03	0.08	0.18	0.09	–	–	0.09	–
	White wine	–	–	0.35	0.35	0.20	–	0.18	–	–	0.18	–
	Skim milk powder	1.2	0.9	–	–	1.18	0.93	2.33	–	0.93	2.33	1.18

35.3 Microbiological and Physicochemical Characteristics

In Table 35.1, typical recipes used for the production of fermented sausages are shown. Lean meat, fat, and additives are used at 63.2%–75.0%, 20.0%–31.7%, and 4.4%–6.7%, respectively. Beef and pork lean meat are typically used at varying percentages, and the utilization of mutton lean meat has been reported in only one case (Samelis et al. 1994a). On the contrary, no variability exists regarding the type of fat that is employed; in all cases, only pork back fat was used. As far as the additives were concerned, variability is recorded mainly due to the variety of mixtures that are commercially available and secondarily due to the utilization of extra materials such as garlic and white wine.

Production of fermented sausages is a two-step process and is divided into fermentation and ripening (Table 35.2). During the fermentation step, temperature is set to equilibrate from 21°C–24°C to 15°C–20°C within 6–7 days with a simultaneous decrease in relative humidity from 93%–96% to 83%–90%. During this step, smoking may occur. Then, ripening takes place, which may last from 10 days to a minimum of 4 weeks, during which the temperature is held at 12°C–16°C and relative humidity at 75%–80%.

The physicochemical and microbiological characteristics of Greek spontaneously fermented sausages are summarized in Table 35.3. Despite the quite satisfactory number of studies that have

TABLE 35.2

Fermentation and Ripening Conditions Used for the Production of Greek Spontaneously Fermented Sausages

Fermentation	Ripening	Smoking	Reference
Within 7 days *T*: from 24.0°C to 19.0°C RH: from 95% to 88%	For 3 weeks at *T*: 14°C–16°C RH: 80%	At 2nd day for 90 min	Drosinos et al. (2005a)
Within 7 days *T*: from 21.0°C to 18.0°C RH: from 93% to 83%	For 3 weeks at *T*: 15°C RH: 75%–80%	After 3rd day overnight	Samelis et al. (1993)
Within 7 days *T*: from 24.0°C to 18.0°C RH: from 94% to 86%	For 10 days at *T*: 12°C–13°C RH: 75%–78%	Smoking during fermentation	Papamanoli et al. (2002)
Within 7 days *T*: from 24.0°C to 18.0°C RH: from 94% to 86%	For 20 days at *T*: 12°C–13°C RH: 75%–78%	Smoking during fermentation	
Within 7 days *T*: from 24.0°C to 19.0°C RH: from 95% to 88%	Minimum 4 weeks at *T*: 14°C–16°C RH: 78%–80%	At 2nd day for 90 min and at 3rd day for 30 min	Rantsiou et al. (2005)
Within 7 days *T*: from 21.0°C to 18.0°C RH: from 93% to 83%	For 3 weeks at *T*: 15°C RH: 75%–78%	After 3rd day overnight	Samelis et al. (1994a)
Within 7 days *T*: from 24.0°C to 20.0°C RH: from 94% to 90%	For 3 weeks at *T*: 15°C–16°C RH: 80%	At 2nd day at 24°C for 2–3 h	Samelis et al. (1998)
Within 6 days *T*: from 23.0°C to 18.0°C RH: from 96% to 80%	For 3 weeks at *T*: 15°C RH: 80%	After 4th day overnight	Metaxopoulos et al. (2001)
Within 6 days *T*: from 24.0°C to 15.0°C RH: from 95% to 80%	For 3 weeks at *T*: 15°C RH: 80%	From 2nd to 5th day for 15–30 min each time	
Within 7 days *T*: from 24.0°C to 20.0°C RH: from 94% to 86%	For 3 weeks at *T*: 15°C–16°C RH: 80%	After 2nd day for 2–3 h	
Within 7 days *T*: from 24.0°C to 19.0°C RH: from 95% to 88%	For 3 weeks at *T*: 14°C–16°C RH: 80%	None	Drosinos et al. (2007)

TABLE 35.3

Physicochemical and Microbiological Characteristics of Greek Spontaneously Fermented Sausages (Mean Values)

	Drosinos et al. (2005a)	Samelis et al. (1993)	Papamanoli et al. (2002) I	II	Samelis et al. (1994a)	Samelis et al. (1998)	Metaxopoulos et al. (2001) I	II	III	Drosinos et al. (2007)
pH	4.78–5.10[a]	5.5–5.8	4.9	4.8	5.2–5.6	5.0–5.2	4.9	4.7	5.1	4.83–4.85
a_w	—[b]	—	—	—	—	—	0.925	0.895	0.871	—
Moisture (%)	28.22–34.70	—	33.35	32.34	18.9–30.2	27.7–30.3	38.0	35.1	29.0	29.12–29.38
NaCl (%)	3.88–4.22	—	4.1	4.1	4.3–5.1	4.3–5.1	3.6	4.8	4.1	4.11–4.25
Nitrite (mg kg⁻¹)	0.30–0.61	—	5	4	—	<10	—	—	—	0.42–0.50
Nitrate (mg kg⁻¹)	5.36–5.80	—	47	53	—	27–43	—	—	—	5.10–5.31
TAMC[c]	6.48–8.58	7.4–7.9	8.08	8.24	7.71–8.26	8.00–8.17	8.20	7.76	8.33	8.46–8.64
LAB	7.43–8.31	7.4–7.9	7.30	7.00	7.63–8.20	7.78–8.23	8.22	8.46	8.25	8.12–8.45
Micrococcaceae	1.70–4.50	6.8–7.5	1.65	1.0	6.43–7.28	4.15–5.89	4.24	<2.0	5.63	5.20–5.23
Enterobacteriaceae	<1.0	—	1.00	1.00	—	<1.00	<1.0	<1.0	<1.0	—
Enterococci	4.50–5.61	—	2.95	2.40	—	4.69–5.17	4.35	<2.0	4.57	—
Pseudomonas	<2.0	—	—	—	—	<2.0	<2.0	<2.0	<2.0	—
Aerobic spore formers	1.80–1.85	—	—	—	—	<2.0–2.69	2.41	<2.0	2.46	—
Yeasts	<2.0–4.37	4.6–5.0	—	—	5.08–5.82	3.65–4.95	3.29	2.33	4.12	—
Sulfite-reducing clostridia	<1.0	—	—	—	—	<1.0	<1.0	<1.0	<1.0	<1.0
Pathogenic staphylococci	<2.0	—	—	—	—	<2.0–2.60	<2.0	<2.0	<2.0	<2.0
Listeria sp.	Absence	—	—	—	—	Absence	Absence	Absence	Absence	Absence
Salmonella sp.	Absence	—	—	—	—	Absence	Absence	Absence	Absence	Absence

[a] Min–max values of different batches studied.

[b] —, Not available.

[c] TAMC, total aerobic mesophilic count (log CFU g⁻¹).

been conducted so far, only a few of them can be regarded as conclusive, and thus, our ability to draw safe conclusions is rather restricted. The pH value depends upon factors affecting the development of the microbial consortium, such as formulation, temperature, and relative humidity. In the case of Greek spontaneously fermented sausages, the pH value has been reported to range from 4.3 to 5.8. In most of the studies, different batches of the same product were analyzed, leading to the conclusion that despite the spontaneous character of the fermentation, carefully controlled conditions can lead to reproducible end products, at least as far as the pH value is concerned. Water activity and moisture (%) are characteristics that depend solely upon technological parameters such as setting of the ripening chamber. Most of the studies determine moisture, which has been found to range from 18.9 to 38.0, despite the significance of water activity. The latter has been determined in only a few studies and was found to range from 0.670 to 0.925 (Arkoudelos et al. 1997). Only one study report a_w values from different batches of the same product, underlining the difficulty for accurate control of this parameter (Metaxopoulos et al. 2001). Regarding NaCl, nitrite, and nitrate concentrations, they are expected to change during fermentation and ripening. The concentration of sodium chloride is expected to increase due to the gradual dehydration of the product. On the other hand, the drastic decrease in the concentration of nitrate and nitrite is due to the subsequent reduction of nitrate to nitrite and of the latter into nitric oxide. In all cases, the values that have been determined can be linked to the initial ones supplied in the recipes (Table 35.1).

Successful spontaneous meat fermentation is driven by lactic acid bacteria (LAB). In Greek spontaneously fermented sausages, LAB population has been reported to range from 5.2 to 8.5 log colony forming units (CFU) g^{-1}, which, in most of the cases, justifies the respective pH value of the end product. Micrococcaceae, enterococci, and yeasts form a secondary microbiota with populations ranging from 1.0 to 7.5, <2.0 to 5.6, and 1.1 to 5.9 log CFU g^{-1}, respectively. Occasional presence of aerobic spore formers and pathogenic staphylococci has also been reported.

In Figure 35.1, the dynamics of key microbial populations, namely, LAB, Micrococcaceae, enterococci, and yeasts, during spontaneous fermentation of several Greek traditional sausages, is shown. LAB generally dominate meat fermentations through the reduction of the pH value, water activity, and oxygen

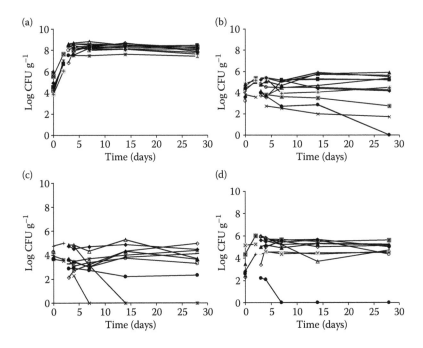

FIGURE 35.1 Dynamics of key microbial populations, namely, (a) LAB, (b) Micrococcaceae, (c) enterococci, and (d) yeasts during spontaneous fermentation of several Greek traditional sausages reported in Drosinos et al. (2007) (□, MSE1; ■, MSE2), Drosinos et al. (2005a) (X, b1; (Ж b2; +, b3), Samelis et al. (1998) (△, bA; ▲, bB; ◇, bC; ♦, bD), and Metaxopoulos et al. (2001) (○, pA; ●, pB; -, pC).

availability. In the case of Greek traditional meat fermentations, LAB dominate and reach populations up to 10^9 CFU g^{-1} after 4 days of fermentation. Members of the Micrococcaceae family are able to survive the rather demanding conditions and retain a maximum population of 10^6 CFU g^{-1}. The deviations between the populations of different batches of the same product that are observed in Figure 35.1b highlight the difficulty in maintaining the given populations in a desired level in spontaneous fermentations. On the contrary, control of the enterococcal population seems to be more feasible, and it has been reported to range between 10^4 and 10^6 CFU g^{-1} in traditional Greek meat fermentations. Finally, as far as yeasts were concerned, their population remains constant throughout fermentation and ripening and usually does not exceed 10^5 CFU g^{-1}.

35.4 Lactic Acid Bacteria

In Table 35.4, the number of LAB strains isolated from Greek spontaneously fermented sausages is shown. The spontaneous character of the fermentation along with the effect of the differences in the formulation, especially regarding the percentage and type of lean meat, as well as the production procedure is depicted. *Lactobacillus plantarum* strains dominated fermentation in only one case (Drosinos et al. 2005a) and co-dominated with *Lb. sakei* strains in another (Drosinos et al. 2007). The latter dominated fermentation in three cases (Rantsiou et al. 2006; Samelis et al. 1998; Papamanoli et al. 2003). Moreover, a co-domination of *Lb. sakei* with *Lb. curvatus* (Samelis et al. 1994b) and *Lb. paracasei* subsp. *paracasei* (Papamanoli et al. 2003) has been mentioned. Interestingly, in only one case, a quite rich *Weissella* sp. microcommunity has been revealed (Samelis et al. 1994b). However, it should be mentioned that characterization of the above-mentioned isolates was performed by culture-dependent approaches—either by phenotypic traits or by sequencing of their 16S rRNA gene.

The role of LAB in meat fermentation, apart from the formation of lactic acid and the concomitant inhibition of pathogenic and spoilage bacteria, is continuously under investigation. Proteolytic and lipolytic activities, as well as antimicrobial compound production and decarboxylation of amino acids, are among the most important and most studied technological and safety properties of LAB in fermented meat products. The initial breakdown of sarcoplasmic and myofibrillar proteins is due to muscle proteinases (Luecke 2000). Then, complete hydrolysis of oligopeptides into free amino acids is performed by endogenous peptidases along with bacterial ones (Sanz et al. 1999). The proteolytic capacity of LAB seemed to be rather rare as only 6 out of 300 strains, belonging to *Lb. sakei* species, were found to be proteolytic and only against the myofibrillar protein fraction (Drosinos et al. 2007) by a mode quite different from the one already described by Fadda et al. (1999), referring to *Lb. plantarum* CRL 681; the complete decomposition of actin, myosin, and all myofibrillar proteins ranging from 200 to 12 kDa was evident, comparing to the partial hydrolysis of only actin and myosin by *Lb. plantarum* CRL 681. Several authors have studied the lipolytic activities of LAB in pork fat. As far as the lipolytic activities of LAB strains isolated from Greek spontaneously fermented sausages were concerned, the reports of El Soda et al. (1986) and Montel et al. (1998), generally stating the weak lipolytic system of LAB, have been confirmed by Drosinos et al. (2007), Samelis et al. (1993), and Samelis and Metaxopoulos (1997).

The ability of LAB to compete effectively and dominate meat fermentation has been partly attributed to specific metabolic traits, such as the energetic catabolism of nucleosides abundant in meat and arginine (Chaillou et al. 2005; Champomier-Verges et al. 1999; Zuriga et al. 2002) and partly to the production of antimicrobial metabolites such as organic acids, hydrogen peroxide, and bacteriocins. The production of the latter has attracted special attention over the last few decades, and the characterization and purification of several such molecules have taken place.

The production of sakacin B, a bacteriocin secreted by *Lb. sakei* strain 251, has been reported by Samelis et al. (1994c). Sakacin B has been described as a hydrophobic peptide of 6.3 kDa molecular weight, secreted during late logarithmic phase, heat (100°C for 20 min) and pH (2.0–9.0) stable, as well as effective against indicator strains belonging to the genera *Lactobacillus*, *Leuconostoc*, and *Streptococcus*, but not against common pathogens such as *Listeria monocytogenes* and *Staphylococcus aureus*. Moreover, *in vitro* production of sakacin B by the producer strain in a mixed culture caused a strong biocidal effect on growing indicator cells.

TABLE 35.4

Number of Lactic Acid Bacteria Isolates and Their Relative Percentage (in Parentheses) from Greek Spontaneously Fermented Sausages

	Drosinos et al. (2005a)	Papamanoli et al. (2003)		Samelis et al. (1994b)	Drosinos et al. (2007)	Samelis et al. (1998)	Rantsiou et al. (2005)	Rantsiou et al. (2006)
		I	II					
Carnobacterium sp.[a]						24 (10.3)		
Lactobacillus alimentarius							2 (1.8)	2 (0.6)
Lb. brevis	2 (0.7)			1 (0.3)	2 (0.6)			
Lb. buchneri		10 (13.3)	13 (17.9)					
Lb. casei		1 (1.3)	1 (1.4)					
Lb. casei subsp. *pseudoplantarum*				1 (0.3)				
Lb. coryniformis subsp. *coryniformis*		1 (1.3)		1 (0.3)				
Lb. curvatus	21 (7.3)			88 (24.6)	13 (4.3)	5 (2.1)	55 (48.2)	69 (20.3)
Lb. curvatus subsp. *curvatus*		14 (18.6)	10 (13.8)					
Lb. farciminis				10 (2.8)				
Lb. paracasei subsp. *paracasei*	3 (1.0)	2 (2.6)	20 (27.6)					
Lb. paracasei subsp. *tolerans*		3 (3.9)	1 (1.4)		6 (5.3)			
Lb. paraplantarum			1 (1.4)				6 (5.3)	7 (2.0)
Lb. paraplantarum/ plantarum[a]							1 (0.9)	1 (0.3)
Lb. pentosus	17 (5.9)							
Lb. plantarum	107 (37.2)	6 (7.9)	1 (1.4)	34 (9.5)	137 (45.6)	2 (0.8)	17 (14.9)	69 (20.3)
Lb. plantarum/pentosus[a]	72 (25.0)							1 (0.3)
Lb. rhamnosus	10 (3.5)	1 (1.3)			8 (2.6)			
Lb. sakei	10 (3.5)	32 (42.5)	17 (23.4)	76 (21.3)	130 (42.9)	180 (77.4)	22 (19.2)	173 (51.0)
Lb. sakei/curvatus[a]				5 (1.4)		16 (6.9)		
Lb. salivarius	1 (0.3)							
Lb. casei/paracasei[a]							6 (5.3)	6 (1.7)
Lactococcus lactis subsp. *lactis*	14 (4.9)							
Leuconostoc sp.[a]		1 (1.3)						
Ln. mesenteroides subsp. *mesenteroides*	1 (0.3)			1 (0.3)				
Ln. pseudomesenteroides		1 (1.3)	1 (1.4)					
Pediococcus sp.[a]		2 (2.6)	1 (1.4)	1 (0.3)				
Weissella sp.[a]				60 (16.8)				
Ws. halotolerans				3 (0.8)				
Ws. hellenica				11 (3.1)				
Ws. minor				31 (8.7)				
Ws. paramesenteroides				11 (3.1)				1 (0.3)
Ws. paramesenteroides/ hellenica								1 (0.3)
Ws. viridescens		1 (1.3)		4 (1.1)				1 (0.3)
Not identified	26 (9.0)				10 (3.3)			

[a] The applied technique did not allow a more accurate identification.

The bacteriocins produced by *Lb. curvatus* strain L442 and *Leuconostoc mesenteroides* strains L124 and E131, all active against *L. monocytogenes* strains, have been extensively studied. The influence of environmental factors on the production of the latter, which was found to be identical to mesenterocin Y105, has been studied by Drosinos et al. (2005b). It has been shown that bacteriocin production was enhanced by growth at suboptimal pH value (5.5) and temperatures close to the optimum for growth (25°C). The effect of the aforementioned parameters along with the effect of sodium chloride and carbon source on the production kinetics has been mathematically described, and a model was created for the effective prediction of the kinetic parameters of growth and bacteriocin production within the pH and temperature range examined (Drosinos et al. 2006a). Furthermore, the non-thermal inactivation of *L. monocytogenes* during sausage fermentation caused by the addition of this bacteriocin in a semi-purified form at a concentration of 1280 arbitrary units per kilogram, as well as the concomitant case study for risk assessment, has been described by Drosinos et al. (2006b). It has been exhibited that addition of the bacteriocin alone was not sufficient for the effective reduction of *L. monocytogenes* numbers, and the incorporation of a starter culture, preferably bacteriocinogenic, was necessary. Detailed characterization of the bacteriocin activity revealed that it was active at pH values between 4.0 and 9.0, and it retained activity after incubation at 100°C for 1 h. Proteolytic enzymes caused inactivation after 1 h of incubation, rennin after 24 h, and lipase after 4 h. Finally, it was purified by 50% ammonium sulfate precipitation, cation exchange, and reverse-phase chromatography (Xiraphi et al. 2008). As far as the bacteriocins produced by strain L124 and curvaticin L442, the antimicrobial peptide produced by *Lb. curvatus* strain L442, were concerned, they were also studied to some extent. It has been exhibited that the optimum conditions for bacteriocin production did not coincide with those for growth, that is, optimum growth was obtained at pH 6.0–6.5 at 30°C, whereas optimum bacteriocin production was obtained at pH 5.5 and temperature of 25°C (Mataragas et al. 2003a). Moreover, the kinetic behavior in de Man Rogosa Sharpe (MRS) broth with varying the concentration of carbon, yeast extract, and nitrogen has been studied and mathematically described (Mataragas et al. 2004). The antagonistic activity against *L. monocytogenes* in sliced cooked cured pork shoulder stored under vacuum or modified atmosphere at 4 ± 2°C has also been studied (Mataragas et al. 2003b). It has been shown that addition of the bacteriocins reduced the listeriae population below the enumeration limit (10^2 CFU g^{-1}) under both vacuum and modified atmosphere packaging. Moreover, detailed characterization of curvaticin L442 revealed that it is active at pH values between 4.0 and 9.0, and it retains activity even after incubation for 5 min at 121°C with 1 atm of overpressure. Proteolytic enzymes and α-amylase resulted in inactivation, whereas the effect of lipase was not severe. Partial N-terminal sequence analysis using Edman degradation revealed 30 amino acid residues, revealing high homology with the amino acid sequence of sakacin P. Finally, the bacteriocin was purified by 50% ammonium sulfate precipitation, cation exchange, reverse-phase, and gel filtration chromatography (Xiraphi et al. 2006).

The biogenic amine content of fermented meat products is generally characterized by great variability (Suzzi and Gardini 2003) that occurs as a result of a complex equilibrium between the abiotic sausage environment and the enzymatic activities of the microbial population (Bover-Cid et al. 2001). The capability of LAB isolated from spontaneously fermented sausages to *in vitro* decarboxylate amino acids, namely, lysine, tyrosine, ornithine, and histidine, to the respective amines, has also been studied to some extent. It has been shown with the utilization of a screening medium developed by Bover-Cid and Holzapfel (1999) that only one lactic acid bacterium, namely, *Lb. sakei* strain LQC 1011, that is, 0.3% of the tested LAB strains, possessed decarboxylase activity on tyrosine (Drosinos et al. 2007). Generally, several lactobacilli strains belonging to the species *Lb. buchneri*, *Lb. alimentarius*, *Lb. plantarum*, *Lb. curvatus*, *Lb. farciminis*, *Lb. bavaricus*, *Lb. homohiochii*, *Lb. reuteri*, and *Lb. sakei* have been found amine-positive either *in vitro* or *in situ*, with tyramine being quantitatively the most important biogenic amine produced (Masson et al. 1996; Montel et al. 1999; Bover-Cid et al. 2001; Pereira et al. 2001).

35.5 Micrococcaceae

In Table 35.5, the number of the strains belonging to the Micrococcaceae family isolated from Greek spontaneously fermented sausages is shown. The spontaneous character of the fermentation is perfectly

TABLE 35.5

Number of Micrococcaceae Isolates and Their Relative Percentage (in Parentheses) from Greek Spontaneously Fermented Sausages

	Papamanoli et al. (2002)		Drosinos et al. (2005a)	Drosinos et al. (2007)
	I	II		
Staphylococcus aureus/intermedius			7 (3.2)	
S. auricularis	1 (1.7)	2 (4.8)	1 (0.5)	
S. capitis	1 (1.7)		18 (8.2)	
S. capitis capitis				5 (1.6)
S. caprae			13 (5.9)	
S. carnosus subsp. *carnosus*	15 (25.8)	5 (12.0)		
S. cohnii subsp. *cohnii*	4 (6.9)	1 (2.4)	2 (0.9)	18 (5.9)
S. cohnii subsp. *urealyticus*	4 (6.9)	2 (4.8)	3 (1.4)	
S. epidermidis	3 (5.1)	3 (7.2)	1 (0.5)	
S. equorum				3 (1.0)
S. haemolyticus		5 (12.0)	17 (7.8)	
S. hominis	3 (5.1)		4 (1.8)	
S. hyicus	1 (1.7)	2 (4.8)		
S. lentus		1 (2.4)		4 (1.3)
S. saprophyticus			68 (31.1)	151 (49.8)
S. sciuri	1 (1.7)		17 (7.8)	10 (3.3)
S. vitulinus				5 (1.6)
S. warneri	1 (1.7)		1 (0.5)	
S. xylosus	3 (5.1)	7 (16.8)	42 (19.2)	38 (12.5)
S. gallinarum				16 (5.3)
S. hominis novobiosepticus				4 (1.3)
S. saprophyticus subsp. *saprophyticus*	16 (27.5)	6 (14.4)		
S. simulans		4 (9.6)	25 (11.4)	46 (15.2)
Arthrobacter agilis	1 (1.7)	1 (2.4)		
Dermacoccus nishinomiyaensis	1 (1.7)	1 (2.4)		
Kocuria varians	3 (5.1)	2 (4.8)		

highlighted by the diversity of species that have been isolated. In the cases reported by Papamanoli et al. (2002), no unambiguous dominance can be reported; in the first case, *S. carnosus* subsp. *carnosus* and *S. saprophyticus* subsp. *saprophyticus* dominated fermentation, whereas in the second case, a consortium consisting of *S. xylosus*, *S. saprophyticus* subsp. *saprophyticus*, *S. carnosus* subsp. *carnosus*, and *S. haemolyticus* seemed to dominate. On the contrary, in the cases reported by Drosinos et al. (2005a, 2007), dominance by *S. saprophyticus* was evident. However, as in the case of LAB, it should be mentioned that characterization of the above-mentioned isolates was performed by culture-dependent approaches—mostly by phenotypic traits.

The role of Micrococcaceae in meat fermentation is, as in the case of LAB, under continuous investigation. Enhancement of color stability, rancidity prevention, spoilage reduction, and contribution to flavor development are among the properties that have been assigned to them (Nychas and Arkoudelos 1990). Additionally, lipolytic and proteolytic activities, as well as decarboxylation of amino acids, are among the most important and most studied technological and safety properties of Micrococcaceae in fermented meat products. Although lipolysis in dry fermented sausages has been attributed partly to members of the *Staphylococcus* genus and partly to tissue lipases (Hugas and Monfort 1997) with the latter being more important for aroma formation (Montel et al. 1996; Luecke 2000), in a study conducted by Drosinos et al. (2007), none of the 300 isolates was found to be lipolytic either by an agar plate assay containing triolein or by the liquid medium assay containing pork fat. On the contrary, many strains were found to exhibit proteolytic activity in myofibrillar and/or sarcoplasmic fractions. More accurately, the complete decomposition of the protein profile obtained by SDS-PAGE of the myofibrillar fraction and a multiple, strain-dependent proteolytic pattern on the sarcoplasmic fraction was reported (Drosinos et al. 2007).

Regarding their capability for biogenic amine formation, it was found to be a rather common property. Drosinos et al. (2007) reported that 185 strains (61.6% of total isolates) decarboxylated at least one of the examined amino acids, namely, lysine, tyrosine, ornithine, and histidine, to the respective amines. Interestingly enough, decarboxylation was reported to follow a pattern, and the most common patterns were the decarboxylation of both lysine and tyrosine and the decarboxylation of tyrosine by 46 and 42 strains, respectively. The staphylococci species with the highest decarboxylase activity were *S. saprophyticus* and *S. xylosus* with 75.5% and 68.4%, respectively.

35.6 Enterococci

Identification of enterococci has been rather complicated due to the high degree of heterogeneity regarding their phenotypic features. Thus, genotyping is a prerequisite for accurate identification. Rantsiou et al. (2005, 2006) reported the presence of *Enterococcus faecium/durans* and *E. faecalis* on the basis of their 16S rRNA gene partial sequence, whereas Paramithiotis et al. (2008) adopted a polyphasic approach—phenotypic characterization was combined with Pulsed-Field Gel Electrophoresis using *Sma*I for clustering of different pulsotypes, and representative strains were selected for sequencing of the 16S rRNA gene. Lack of correlation between the clusters obtained according to phenotypic traits and the pulsotypes was evident. It was reported that as many as 43.4% of the isolates would have been misidentified without the application of genotypic methods. All strains were identified as *E. faecium* according to their 16S rRNA gene partial sequence. It has been suggested that certain pulsotypes could be a persistent part of the production of environment microbiota, as they were isolated from all three batches that have been examined. In the same study, it was reported that out of 108 enterococcal strains that have been isolated, none exhibited proteolytic or lipolytic activity and none decarboxylated lysine, tyrosine, ornithine, or histidine. On the contrary, 42 strains produced antimicrobial compounds of proteinic nature with activity against *L. monocytogenes*.

35.7 Yeasts–Molds

Yeasts and molds are very often involved in the ripening process of dry fermented sausages. Their role seems to be related to the increase in ammonia and reduction of lactic acid contents (Garriga and Aymerich 2007). Moreover, they contribute to oxygen depletion and through their enzymatic activities to flavor development. In the case of Greek spontaneously fermented sausages, yeasts are generally ignored despite the fact that their population might even reach 5.8 log CFU g^{-1} (Table 35.3). Thus, scarce literature is currently available. Samelis et al. (1994a) reported a dominance of *Debaryomyces* sp. accompanied by lower levels of *Cryptococcus* sp., *Torulopsis* sp., and *Trichosporon* sp. Furthermore, Metaxopoulos et al. (1996) isolated 100 yeast strains during spontaneous sausage fermentation of five batches and characterized them on the basis of their phenotypic traits. The prevalence of *Debaryomyces hansenii* was evident, and a secondary yeast microcommunity, consisting of *D. marama*, *D. polymorphus*, *Candida famata*, *C. zeylanoides*, *C. guilliermondii*, *C. parapsilosis*, *C. kruisii*, *C. humicola*, *Cryptococcus albidus*, *Cr. skinneri*, and *Trichosporon pullulans*, has been reported. These results are in accordance with the majority of the literature available that describes the salt-tolerant, non-nitrate-reducing yeast *D. hansenii* as a prevalent species in spontaneous meat fermentations (Encinas et al. 2000; Coppola et al. 2000; Selgas et al. 2003; Wolter et al. 2000; Osei Abunyewa et al. 2000).

35.8 Safety Aspects of Greek Fermented Sausages

Survival of post-process *L. monocytogenes* contamination on sliced salami packaged under vacuum or air and stored under the temperatures associated with retail and domestic storage was assessed in a challenge study by Gounadaki et al. (2007). It was clearly indicated that the kinetics of *L. monocytogenes* inactivation were highly dependent on the interaction of factors such as storage temperature, packaging conditions, and initial level of contamination. More accurately, all survival curves of *L. monocytogenes* were characterized by an initial rapid inactivation within the first days of storage, followed by a second, slower inactivation phase or "tailing." Greater reduction of *L. monocytogenes* was observed at the high storage temperature (25°C), followed by ambient (15°C) and chill (5°C) storage conditions. Moreover, vacuum packaging resulted in a slower destruction of *L. monocytogenes* than air packaging, and this effect increased as storage temperature decreased.

Evaluation of the effectiveness of currently applied hygienic practices within high-throughput small-scale facilities producing traditional fermented and/or dry sausages was investigated by Gounadaki et al. (2008). It was shown that the microbial stability and safety of the final products largely depended upon the hygienic status of the processing environment and equipment. The sampling sites tested were highly (>4 log CFU/cm^2) contaminated by spoilage biota (*Pseudomonas*, Enterobacteriaceae, and yeasts/molds), with knives, tables, and mincing machines being the most heavily contaminated surfaces. Furthermore, *L. monocytogenes*, *Salmonella* spp., and *S. aureus* were detected in 11.7%, 26.4%, and 11.7% of the food contact surfaces, respectively. Thus, the need for strict control measures within small-scale-processing facilities has been highlighted. Regarding the safety of the final products, low risk for the consumers remains as the products studied were characterized as unable to support growth of *L. monocytogenes* (products with pH ≤ 4.4 or a_w ≤ 0.92; products with pH ≤ 5.0 and a_w ≤ 0.94). Efforts have also been made to implement food safety management systems in small-scale meat processing enterprises (Trianti et al. 2008).

REFERENCES

Anonymous. 1854. *The Deipnosophists or Banquet of the Learned of Athenaeus*, Volumes I, II, and III, translated by C.D. Yonge, 1252 pp. London: Henry G. Bohn.

Anonymous. 1892. *Aristophanes' the Knights*, edited by W.C. Green, 204 pp. London: Longmans, Green and Co.

Anonymous. 1912. *The Frogs of Aristophanes*, translated by G. Murray, 136 pp. London: George Allen & Company.

Arkoudelos, J.S., Nychas, G.-J.E., and Samaras, F. 1997. Vorkommen von Staphylokokken in griechischen Rohwursten. *Fleischwirtschaft* 77:571–577.

Bover-Cid, S. and Holzapfel, W.H. 1999. Improved screening procedure for biogenic amine production by lactic acid bacteria. *Int. J. Food Microbiol.* 53:33–41.

Bover-Cid, S., Hugas, M., Izquierdo-Pulido, M., and Vidal-Carou, M.C. 2001. Amino acid-decarboxylase activity of bacteria isolated from fermented pork sausages. *Int. J. Food Microbiol.* 66:185–189.

Chaillou, S., Champomier-Verges, M.C., Cornet, M., Crutz-Le Coq, A.-M., Dudez, A.-M., Martin, V., Beaufils, S., Darbon-Rongere, E., Bossy, R., Loux, V., and Zagorec, M. 2005. The complete genome sequence of the meat-borne lactic acid bacterium *Lactobacillus sakei* 23K. *Nat. Biotechnol.* 23:1527–1533.

Champomier-Verges, M.C., Zuniga, M., Morel-Deville, F., Perez-Martinez, G., Zagorec, M., and Ehrlich, S.D. 1999. Relationships between arginine degradation, pH and survival in *Lactobacillus sakei*. *FEMS Microbiol. Lett.* 180:297–304.

Coppola, S., Mauriello, G., Aponte, M., Moschetti, G., and Villani, F. 2000. Microbial succession during ripening of Naples-type salami, a southern Italian fermented sausage. *Meat Sci.* 56:321–329.

Drosinos, E.H., Mataragas, M., and Metaxopoulos, J. 2006a. Modeling of growth and bacteriocin production by *Leuconostoc mesenteroides* E131. *Meat Sci.* 74:690–696.

Drosinos, E.H., Mataragas, M., Nasis, P., Galiotou, M., and Metaxopoulos, J. 2005b. Growth and bacteriocin production kinetics of *Leuconostoc mesenteroides* E131. *J. Appl. Microbiol.* 99:1314–1323.

Drosinos, E.H., Mataragas, M., Veskovic-Moracanin, S., Gasparik-Reichardt, J., Hadziosmanovic, M., and Alagic, D. 2006b. Quantifying nonthermal inactivation of *Listeria monocytogenes* in European fermented sausages using bacteriocinogenic lactic acid bacteria or their bacteriocins: A case study for risk assessment. *J. Food Prot.* 69:2648–2663.

Drosinos, E.H., Mataragas, M., Xiraphi, N., Moschonas, G., Gaitis, F., and Metaxopoulos, J. 2005a. Characterization of the microbial flora from a traditional Greek fermented sausage. *Meat Sci.* 69:307–317.

Drosinos, E.H., Paramithiotis, S., Kolovos, G., Tsikouras, I., and Metaxopoulos, I. 2007. Phenotypic and technological diversity of lactic acid bacteria and staphylococci isolated from traditionally fermented sausages in Southern Greece. *Food Microbiol.* 24:260–270.

El Soda, M., Korayem, M., and Ezzat, N. 1986. The esterolytic and lipolytic activities of lactobacilli. III: Detection and characterisation of the lipase system. *Milchwissenschaft* 41:353–355.

Encinas, J.P., Lopez-Diaz, T.M., Garcia-Lopez, M.L., Otero, A., and Moreno, B. 2000. Yeast populations on Spanish fermented sausages. *Meat Sci.* 54:203–208.

Fadda, S., Sanz, Y., Vignolo, G., Aristoy, M.-C., Oliver, G., and Toldra, F. 1999. Characterization of muscle sarcoplasmic and myofibrillar protein hydrolysis caused by *Lactobacillus plantarum*. *Appl. Environ. Microbiol.* 65:3540–3546.

Garriga, M. and Aymerich, T. 2007. The microbiology of fermentation and ripening. In: *Handbook of Fermented Meat and Poultry*, edited by F. Toldra, pp. 125–136. Ames, IA: Blackwell Publishing Professional.

Gounadaki, A.S., Skandamis, P.N., Drosinos, E.H., and Nychas G.-J.E. 2007. Effect of packaging and storage temperature on the survival of *Listeria monocytogenes* inoculated postprocessing on sliced salami. *J. Food Prot.* 70:2313–2320.

Gounadaki, A.S., Skandamis, P.N., Drosinos, E.H., and Nychas, G.-J.E. 2008. Microbial ecology of food contact surfaces and products of small-scale facilities producing traditional sausages. *Food Microbiol.* 25:313–323.

Hugas, M. and Monfort, J.M.A. 1997. Bacterial starter cultures for meat fermentation. *Food Chem.* 29:547–554.

Luecke, F.-K. 2000. Utilization of microbes to process and preserve meat. *Meat Sci.* 56:105–115.

Masson, F., Talon, R., and Montel, M.C. 1996. Histamine and tyramine production by bacteria from meat products. *Int. J. Food Microbiol.* 32:199–207.

Mataragas, M., Drosinos, E.H., and Metaxopoulos, J. 2003b. Antagonistic activity of lactic acid bacteria against *Listeria monocytogenes* in sliced cooked cured pork shoulder stored under vacuum or modified atmosphere at 4 ± 2°C. *Food Microbiol.* 20:259–265.

Mataragas, M., Drosinos, E.H., Tsakalidou, E., and Metaxopoulos, J. 2004. Influence of nutrients on growth and bacteriocin production by *Leuconostoc mesenteroides* L124 and *Lactobacillus curvatus* L442. *Antonie van Leeuwenhoek* 85:191–198.

Mataragas, M., Metaxopoulos, J., Galiotou, M., and Drosinos, E.H. 2003a. Influence of pH and temperature on growth and bacteriocin production by *Leuconostoc mesenteroides* L124 and *Lactobacillus curvatus* L442. *Meat Sci.* 64:265–271.

Metaxopoulos, J., Samelis, J., and Papadelli, M. 2001. Technological and microbiological evaluation of traditional processes as modified for the industrial manufacturing of dry fermented sausage in Greece. *Int. J. Food Sci.* 13:3–18.

Metaxopoulos, J., Stavropoulos, S., Kakouri, A., and Samelis, J. 1996. Yeasts isolated from traditional Greek dry salami. *Ital. J. Food Sci.* 1:25–32.

Montel, M.C., Masson, F., and Talon, R. 1998. Bacterial role in flavor development. *Meat Sci.* 49:S111–S123.

Montel, M.C., Masson, F., and Talon, R. 1999. Comparison of biogenic amine content in traditional and industrial French dry sausages. *Sci. Aliments* 19:247–254.

Montel, M.C., Reitz, J., Talon, R., Berdague, J.-L., and Rousset-Akrim, S. 1996. Biochemical activities of Micrococcaceae and their effects on the aromatic profiles and odours of a dry sausage model. *Food Microbiol.* 13:489–499.

Nychas, G.-J.E. and Arkoudelos, J.S. 1990. Staphylococci: Their role in fermented sausages. *J. Appl. Bacteriol. Symp. Suppl.* 19:167S–188S.

Osei Abunyewa, A.A., Laing, E., Hugo, A., and Viljoen, B.C. 2000. The population change of yeasts in commercial salami. *Food Microbiol.* 17:429–438.

Papamanoli, E., Kotzekidou, P., Tzanetakis, N., and Litopoulou-Tzanetaki, E. 2002. Characterization of Micrococcaceae isolated from dry fermented sausage. *Food Microbiol.* 19:441–449.

Papamanoli, E., Tzanetakis, N., Litopoulou-Tzanetaki, E., and Kotzekidou, P. 2003. Characterization of lactic acid bacteria isolated from a Greek dry-fermented sausage in respect of their technological and probiotic properties. *Meat Sci.* 65:859–867.

Paramithiotis, S., Kagkli, D.-M., Blana, V.A., Nychas, G.-J.E., and Drosinos, E.H. 2008. Identification and characterization of *Enterococcus* spp. in Greek spontaneous sausage fermentation. *J. Food Prot.* 71:1244–1247.

Pereira, C.I., Barreto Crespo, M.T., and San Romao, M.V. 2001. Evidence for proteolytic activity and biogenic amines production in *Lactobacillus curvatus* and *L. homohiochii*. *Int. J. Food Microbiol.* 68:211–216.

Rantsiou, K., Drosinos, E.H., Gialitaki, M., Metaxopoulos, I., Comi, G., and Cocolin, L. 2006. Use of molecular tools to characterize *Lactobacillus* spp. isolated from Greek traditional fermented sausages. *Int. J. Food Microbiol.* 112:215–222.

Rantsiou, K., Drosinos, E.H., Gialitaki, M., Urso, R., Krommer, J., Gasparik-Reichardt, J., Toth, S., Metaxopoulos, I., Comi, G., and Cocolin, L. 2005. Molecular characterization of *Lactobacillus* species isolated from naturally fermented sausages produced in Greece, Hungary and Italy. *Food Microbiol.* 22:19–28.

Samelis, J. and Metaxopoulos, J. 1997. Lipolytische Aktivitat von Laktobazillen aus naturlich gereifter griechischer Rohwurst. *Fleischwirtschaft* 77:165–168.

Samelis, J., Aggelis, G., and Metaxopoulos, J. 1993. Lipolytic and microbial changes during the natural fermentation and ripening of Greek dry sausages. *Meat Sci.* 35:371–385.

Samelis, J., Maurogenakis, F., and Metaxopoulos, J. 1994b. Characterisation of lactic acid bacteria isolated from naturally fermented Greek dry salami. *Int. J. Food Microbiol.* 23:179–196.

Samelis, J., Metaxopoulos, J., Vlassi, M., and Pappa, A. 1998. Stability and safety of traditional Greek salami— A microbiological ecology study. *Int. J. Food Microbiol.* 44:69–82.

Samelis, J., Roller, S., and Metaxopoulos, J. 1994c. Sakacin B, a bacteriocin produced by *Lactobacillus sakei* isolated from Greek dry fermented sausages. *J. Appl. Bacteriol.* 76:475–486.

Samelis, J., Stavropoulos, S., Kakouri, A., and Metaxopoulos, J. 1994a. Quantification and characterization of microbial populations associated with naturally fermented Greek dry salami. *Food Microbiol.* 11:447–460.

Sanz, Y., Fadda, S., Vignolo, G., Aristoy, M.-C., Oliver, G., and Toldra, F. 1999. Hydrolytic action of *Lactobacillus casei* CRL 705 on pork muscle sarcoplasmic and myofibrillar proteins. *J. Agric. Food Chem.* 47:3441–3448.

Savic, Z. and Savic, I. 2002. *Sausage Casings*. Vienna: Victus.

Selgas, M.D., Ros, J., and Garcia, M.L. 2003. Effect of selected yeast strains on the sensory properties of dry fermented sausages. *Eur. Food Res. Technol.* 217:475–480.

Suzzi, G. and Gardini, F. 2003. Biogenic amines in dry fermented sausages: A review. *Int. J. Food Microbiol.* 88:41–54.

Trianti, I., Drosinos, E.H., and Zoiopoulos, P.E. 2008. Establishing a HACCP system in a small-scale traditional meat processing enterprise. *Int. J. Food Sci.* 20:427–432.

Wolter, H., Laing, E., and Viljoen, B.C. 2000. Isolation and identification of yeasts associated with intermediate moisture meats. *Food Technol. Biotechnol.* 38:69–75.

Xiraphi, N., Georgalaki, M., Rantsiou, K., Cocolin, L., Tsakalidou, E., and Drosinos, E.H. 2008. Purification and characterization of a bacteriocin produced by *Leuconostoc mesenteroides* E131. *Meat Sci.* 80:194–203.

Xiraphi, N., Georgalaki, M., Van Driessche, G., Devreese, B., Van Beeumen, J., Tsakalidou, E., Metaxopoulos, J., and Drosinos, E.H. 2006. Purification and characterization of curvaticin L442, a bacteriocin produced by *Lactobacillus curvatus* L442. *Antonie van Leeuwenhoek* 89:19–26.

Zuriga, M., Miralles, M., and Perez-Martinez, G. 2002. The product of *arcR*, the sixth gene of the *arc* operon of *Lactobacillus sakei*, is essential for expression of the arginine deiminase pathway. *Appl. Environ. Microbiol.* 68:6051–6058.

36

Formulations for Fermented Sausages with Health Attributes

Diana Ansorena and Iciar Astiasarán

CONTENTS

36.1 Introduction

In contrast with other types of fermented foods, fermented meat products are generally not considered among the most recommended foods in the dietary guidelines, and a negative perception of the role of red meat in health is noticed (McAfee et al. 2010). The main reasons underlying this situation are the relatively high supply of energy, animal fat, and sodium and the use of some additives as nitrites, synthetic antioxidants, and colorants in these foods.

However, fermented sausages also have some interesting components from the nutritional point of view, which should be taken into account by nutritionists when designing diets. In addition, current lifestyles very much appreciate some of the convenience properties shown by these products, such as their sensory properties, their feasibility for consumption, and their relatively long shelf life.

These considerations and the relevance of the meat industry in many countries have encouraged the development of different strategies to obtain new meat products and new formulations with healthier nutritional attributes (Jiménez-Colmenero et al. 2001, 2006; Muguerza et al. 2004a).

36.2 Nutritional Properties of Traditional Fermented Sausages

Traditional fermented sausages are a very wide group of products with different formulations, including meat from different species (i.e., beef, pork, lamb, chicken, duck, buffalo, horse, and reindeer), backfat or other types of animal fats, salt, simple sugars, and a great variety of additives, aromas, and spices. The major factors influencing composition of the finished product are the composition and ratio of raw materials and the rest of the ingredients used, processing procedures utilized, and the drying method.

Nevertheless, it can be assumed that, in general, fermented sausages show a relatively high energetic value, with significant supply of protein (whose biological value depends on the part of the animal), fat, and cholesterol and low supply of carbohydrates. The biological value of proteins depends on the quality of the meat used in the formulation (muscle or connective tissues). As the fat is from animal origin, it presents a high percentage of saturated fatty acids and also (especially in the case of pork fat) a relatively high percentage of monounsaturated fatty acids (oleic acid). The amount of carbohydrates depends on the nature of other ingredients and some additives used in the formulation. In addition, some significant amounts of minerals as iron, zinc, phosphorus, and magnesium and vitamins of group B can be supplied by these products. An essential ingredient of fermented sausages is sodium chloride, which contributes to the sensory properties and is also important for the maturation process and the shelf life of the products, as it allows lactic acid bacteria (LAB) to grow and inhibits several unwanted microorganisms.

The development of healthier fermented sausages has been raised by means of different strategies, among which the following can be found—to modify the composition of the traditional formulations in the way of decreasing the supply of energetic value, total fat, saturated fatty acids, cholesterol, and amount of sodium or increasing the supply of vitamins, minerals, antioxidants, fiber, or other compounds with demonstrated beneficial effects.

36.3 Strategies to Develop Healthier Fermented Sausages

36.3.1 Modifications in Lipid Fraction: Use of Vegetable and/or Marine Oils

Most research works on healthier dry fermented sausages dealing with the lipid fraction have been focused on decreasing the total fat content or on substituting the animal fat by other lipid or non-lipid ingredients or in combination of these strategies. By means of animal production practices or by using different processing strategies (encapsulation, pre-emulsification, addition of oils, or solid fats), successful results have been obtained, despite the fact that the manufacturing technology of these products is more difficult compared to that of the other types of meat products (Jiménez-Colmenero 2007). A list of papers dealing with different strategies used for formulating healthier dry fermented sausages is presented in Table 36.1.

Reducing the total fat content at high levels presents certain drawbacks from the sensorial point of view, mainly due to consequences on the texture and palatability of the products. Olivares et al. (2010) aimed at reducing fat in slow fermented sausages, concluding that the limit to produce high-acceptability, low-fat fermented sausages was 16% fat content in the raw mixture, that is, half the usual content of dry fermented sausages. In fact, low-fat Greek sausages (10% fat) had lower sensory scores than high-fat ones (30% fat), although methods such as vacuum packaging can contribute to improve their external appearance (Liaros et al. 2009).

Changing formulations by substitution of animal fat by other lipid sources aims at decreasing the content of saturated fatty acids and cholesterol—both compounds are linked to the development of cardio-vascular diseases. Depending on the type of product, the level of substitution that can be reached without noticing technological defects is different. Olive oil has been added to different types of fermented products, with the objective of increasing the monounsaturated fatty acid (MUFA) content—traditional Greek slow fermented sausages (Bloukas et al. 1997; Koutsopoulos et al. 2008; Muguerza et al. 2002), chorizo de Pamplona (Ansorena and Astiasarán 2004a; Muguerza et al. 2001), salami (Del Nobile et al. 2009; Severini et al. 2003), and sucuk (a Turkish fermented meat product; Kayaardi and Gök 2004).

For increasing the polyunsaturated fatty acid (PUFA) fraction, different works with linseed oil have been carried out (Ansorena and Astiasarán 2004b; García-Iñiguez de Ciriano et al. 2009; Valencia et al. 2006a), as well as with inter-esterified palm and cottonseed oil (Vural 2003), soy oil (Muguerza et al. 2003), and hazelnut oil (Ilikkan et al. 2009; Yildiz-Turp and Serdaroglu 2008). All these papers reported improved MUFA + PUFA/SAT fractions and good sensorial characteristics if adequate proportions of oils are used, which differ depending on the type of meat product.

Long-chain omega-3 PUFA [eicosapentaenoic acid and docosahexaenoic acid (DHA)] incorporation into dry fermented products has been performed by using fish oil extracts (Muguerza et al. 2004b), deodorized fish oil (Valencia et al. 2006b), or microalgae oil from *Schizochytrium* sp. (García-Iñiguez

TABLE 36.1

Healthier Fermented Meat Products

Strategy	Reference
Modification of lipid fraction	Bloukas et al. 1997
	Muguerza et al. 2001, 2002, 2003, 2004a,b
	Severini et al. 2003
	Vural 2003
	Ansorena and Astiasarán 2004a,b
	Kayaardi and Gök 2004
	Valencia et al. 2006a,b, 2007
	Jiménez-Colmenero 2007
	Pelser et al. 2007
	Koutsopoulos et al. 2008
	Yildiz-Turp and Serdaroglu 2008
	Liaros et al. 2009
	Ilikkan et al. 2009
	Del Nobile et al. 2009
	García-Iñiguez de Ciriano et al. 2009, 2010a,b
	Delgado-Pando et al. 2010
	Olivares et al. 2010
	Ospina et al. 2010
Use of natural antioxidants	Ghiretti et al. 1997
	Bozkurt 2006
	Nam et al. 2006
	Miliauskas et al. 2007
	Sebranek et al. 2005
	Karabacak and Bozkurt 2008
	Park and Kim 2009
	García-Iñiguez de Ciriano et al. 2009, 2010b
	Salem and Ibrahim 2010
	Mercadante et al. 2010
Addition of vitamins	Decker and Hultin 1992
	Schaefer et al. 1995
	Jensen et al. 1998
	Torrissen 2000
	Decker and Park 2010
Modifications in the mineral content	Gou et al. 1996
	Gimeno et al. 1998, 1999, 2001a,b
	Daengprok et al. 2002
	Wichittra et al. 2002
	Gelabert et al. 2003
	Flores et al. 2005
	Ruusunen and Puolanne 2005
	Guàrdia et al. 2006, 2008
	Selgas et al. 2009
	García-Iñiguez de Ciriano et al. 2010b
Enrichment in fiber	Mendoza et al. 2001
	García et al. 2002
	Aleson-Carbonell et al. 2003
	Fernández-López et al. 2007, 2008

(continued)

TABLE 36.1 (Continued)

Healthier Fermented Meat Products

Strategy	Reference
Probiotics	Eim et al. 2008
	Salazar et al. 2009
	Pidcock et al. 2002
	Jahreis et al. 2002
	Pennacchia et al. 2004, 2006
	Klingberg et al. 2005
	Muthukumarasamy and Holley 2006
	Ruiz-Moyano et al. 2008
	De Vuyst et al. 2008
	Yuksekdag and Aslim 2010
Endogenous bioactive compounds	Brown 1981
	Kawasaki et al. 2000
	Hipkiss and Brownson 2000
	Seppo et al. 2003
	Wahle et al. 2004
	Arihara 2006
	Park and Pariza 2007
	Fang et al. 2008
	Park et al. 2005
	Park and Kim 2009
	Decker and Park 2010
Nitrite reduction	Yilmaz and Zorba 2010
	Rojsuntornkitti et al. 2010

de Ciriano et al. 2010a; Valencia et al. 2007). As algae oil is particularly rich in DHA, a low amount of oil permits achieving significant DHA levels in the final product. In fact, when substituting 15% of pork backfat by an emulsion containing algae oil, a concentration of 1.30 g/100 g product of DHA is reached, with an interesting n–6/n–3 ratio and good stability of products stored under vacuum.

Most of these experiments used the pre-emulsification as the choice method for incorporating the different types of oils. In this sense, current research on different models of oil-in-water emulsions tries to achieve stable and nutritionally improved new sources of fat for the development of functional meat products. Mixtures of olive, linseed, and fish oils were emulsified with different protein sources, obtaining systems suitable for use as fat replacers (Delgado-Pando et al. 2010). In addition, emulsions of linseed and algae oil have been stabilized with artificial (butylated hydroxytoluene) or natural antioxidants (water extract from *Melissa officinalis*), controlling the potential oxidation process that could have been enhanced by the presence of a high content of long-chain PUFAs (García-Iñiguez de Ciriano et al. 2010b).

Besides emulsions, chemically modified vegetable oils have also been studied as pork backfat substitutes in sausage formulations (Ospina et al. 2010). Another technological approach that has been used is encapsulation of oils: encapsulated fish oil and flaxseed oil gave good results from the physical and sensorial point of view in Dutch-style fermented sausages, confirming that encapsulation is able to protect oxidative sensitive lipids (Pelser et al. 2007).

36.3.2 Use of Natural Antioxidants

The use of natural antioxidants in fermented sausages is becoming a useful strategy to obtain new meat products with greater stability against lipid oxidation and, at the same time, a new dietary source of antioxidant compounds. The need to use plant extracts to maintain the chemical stability of the unsaturated lipids is certainly not an issue because consumers readily accept products containing natural, nonsynthetic components. The industry's quest for natural plant extracts with a high antioxidant potential and low impact on taste and flavor continues (Weiss et al. 2010).

Traditional fermented sausages usually include in their formulations herbs and spices that contain antioxidant compounds that help in increasing their shelf life (Palic et al. 1993). At the same time, these compounds (phenolic, diterpenes, flavonoids, tannins, and phenolic acids) could develop antioxidant, anti-inflammatory, and anticancer activities in the organism (Dawidowicz et al. 2006; Zhang et al. 2010).

Catechin, sesamol (both phenolic antioxidants extracted from green tea leaves and sesame seeds, respectively), and phytic acid (present in seeds and cereal grains) were used in salame Milano and mortadella (Ghiretti et al. 1997). The results showed their efficacy to decrease the lipid oxidation process, although some color problems appeared in the products.

Rosemary extracts have been used in different meat products to stabilize color and to control lipid oxidation (Formanek et al. 2003; Nam et al. 2006; Sebranek et al. 2005; Yu et al. 2002). The main phenolic compounds with antioxidant activities present in rosemary plants are carnosol, carnosic acid, rosmanol, epirosmanol, isorosmanol, rosmarinic acid, rosmaridiphenol, and rosmariniquinone (Fernández-López et al. 2003; Ibanez et al. 2003). Rosemary extracts used in wiener sausages have been proved to efficiently control lipid oxidation during long-term frozen storage (Coronado et al. 2002). Rosemary powder has been used due to its antioxidant efficacy to stabilize different meat and meat products (Martínez et al. 2006a,b). Rosmarinic acid has been described also as the main antioxidant compound present in *M. officinalis* L. (medicinal plant), whose lyophilized water extract has shown capacity to stabilize an algae and linseed oil-in-water emulsion and dry fermented sausages elaborated with them (García-Iñiguez de Ciriano et al. 2010a,b).

The effects of other plant extracts as those from *Rhus coriaria* L. (sumac), *Urtica dioica*, *Hibiscus sabdariffa*, green tea, and *Thymbra spicata* oil were used in "sucuk," a Turkish dry fermented sausage (Bozkurt 2006; Karabacak and Bozkurt 2008). The results of these studies showed that most of these extracts enhance the sensory quality of the product and provide safer products, decreasing the concentration of bioamines (putrescine, histamine, and tyramine).

Sage oil extract (containing two phenolic diterpenes with antioxidant activity: carnosic acid and carnosol) added to dry fermented buffalo sausages also showed final products with acceptable lipid oxidation and biogenic amine levels as well as improved sensory quality (Salem and Ibrahim 2010). Garlic (rich in allicin, a compound with antimicrobial activity) has shown an antioxidant activity during cold storage of emulsified sausage beyond its antimicrobial activity (Park and Kim 2009).

In addition, plant extracts from by-products have been used in dry fermented sausages. This is the case of lyophilized water extracts of *Borago officinalis* leaves, which have been proved to be efficient in stabilizing dry fermented sausages enriched in omega-3 PUFAs (elaborated with linseed oil; García-Iñiguez de Ciriano et al. 2009).

The interest on the antioxidant capacity of plants has driven to some preliminary studies over the activities of some usual plants as *Geranium macrorrhizum* and *Potentilla fructicosa* (Miliauskas et al. 2004). Different extracts from these plants have been used in Dutch-style fermented sausages, although the obtained results do not seem to be especially significant (Miliauskas et al. 2007).

Natural pigments norbixin, lycopene, zeaxanthin, and beta-carotene have been used in sausage formulations showing antioxidant effects over storage under refrigeration, especially in the case of zeaxanthin and norbixin in which intermediate polarities allow them to concentrate in the membrane lipids or emulsion interface where oxidation is most prevalent (Mercadante et al. 2010).

In conclusion, all the studies showed that natural antioxidants enhance the quality of fermented sausages and show positive effects from the technological standpoint, as they contribute to stabilize enriched PUFA formulations and provide safer products. However, there are no studies yet proving further benefits in the organism as a consequence of the supply of antioxidant compounds by means of fermented meat products.

36.3.3 Addition of Vitamins

The increase in vitamins in dry fermented sausage formulations is not widely extended, although some experiments have been carried out in order to improve the vitamin composition of meat by means of animal dietary treatments. In this sense, the content of antioxidants with vitamin function can be increased in the raw material used for the meat product elaboration (Decker and Park 2010). In the case of vitamin C, this enrichment is not easy because of its instability, its tendency to promote lipid oxidation, and its interaction with meat pigments (Decker and Hultin 1992; Haak et al. 2009). Vitamin A (beta-carotene, in general) is also unstable and alters the color of the product (Torrissen 2000). In the case of vitamin E, the most efficient way to increase its presence in the products is through the incorporation of vitamin E in the diet of the animals (Schaefer et al. 1995; Jensen et al. 1998). Dietary vitamin changes in the animal nutrition can induce changes in the mineral pattern. Meats from animals supplemented with vitamin D in their diets showed higher total cytosolic concentrations of Ca, P, and Mg (Montgomery et al. 2004).

Folic acid is a soluble vitamin that helps in preventing certain diseases such as macrocytic and megaloblastic anemia, cardiovascular diseases, and several types of cancer (breast, colon, and pancreatic; Caudill 2004; Wen et al. 2008). In addition, adequate daily intake of this vitamin decreases the risk of neural tube birth defects (Caudill 2004; Gregory 2004; Wald et al. 2006). Although its stability has been proved in cooked sausages (Cáceres et al. 2008) and also in hamburgers (Galán et al. 2010), there are no works published in fermented meat products.

36.3.4 Modifications in the Mineral Content

Meat and meat products are good sources of interesting minerals such as iron, zinc, and selenium. However, meat products also constitute great suppliers of sodium in the diet, as it is needed in their formulation due to its importance on flavor, texture, shelf life, and safety. Sodium intake exceeds the nutritional recommendations in many industrialized countries; thus, the improvement of the mineral pattern of this type of product has been traditionally focused on the salt and/or sodium reduction. The use of mineral salt mixtures has been a good way to reduce the sodium content of meat products, although it is difficult for the consumers to get used to weaker perceived saltiness of low-salt meat products (Ruusunen and Puolanne 2005).

First approaches in fermented products tried substitutions of NaCl with individual components: KCl (0%–60%), potassium lactate (0%–100%), and glycine (0%–100%) in fuet (Gou et al. 1996). Subsequently, studies in which mixtures of these salts were used were carried out, concluding that the partial substitution (above 40%) of NaCl with different mixtures of KCl/glycine and K-lactate/glycine showed important flavor and textural defects that did not permit an increase in the level of substitution compared to those obtained with the individual components (Gelabert et al. 2003). Better results have been found in these products when eliminating glycine from the mixtures. A reduction of 50% of the NaCl can be done by molar substitution with KCl (50%) or with a mixture of KCl/potassium lactate (4:1, 3:2, and 2:3), without modifying acceptability or preference of the product. Substitution levels of 40%–50% by potassium lactate showed lower overall acceptance than the control batch (Guàrdia et al. 2006, 2008).

In chorizo, a typical paprika-containing Spanish fermented sausage, Gimeno et al. (1998, 1999, and 2001a) substituted NaCl by combinations of sodium, potassium, magnesium, and calcium chlorides, reaching significant reductions in the Na content and obtaining improved supplies of Mg, K, and Ca. Calcium chloride concentration, when used as a source of calcium and not merely as a sodium chloride substitute, should be well adjusted to avoid lipid oxidation inducement. Flores et al. (2005) found that the addition of this salt at a final concentration of 0.05% contributed to proper dry fermented sausage aroma, whereas larger amounts (0.5%) favored lipid oxidation and led to worse scored products in the sensory evaluation.

Besides chloride, other sources of calcium have also been assayed in order to enrich fermented meat products with this mineral. Different combinations of calcium ascorbate and NaCl were tested by Gimeno et al. (2001b), with 29.17 g/kg of calcium ascorbate and 14 g/kg of salt, resulting in a successful combination from the technological and sensorial point of view. A reduction of 45% in the sodium content and an increment of calcium supply (399.4 mg/100 g in the modified product and 130 mg/100 g in the control one) were achieved. Although products were scored as less salty, acceptability was still well evaluated.

Calcium lactate, gluconate, and citrate were added to dry fermented sausages (salchichón) to give 20% and 30% of the recommended daily allowance of calcium established by the Institute of Medicine of the

USA (1000 mg/day; Selgas et al. 2009). High-calcium products resulted in low-acceptability products due to off-flavors and high hardness. However, the lower amount of the three calcium salts supplied significant amounts of calcium without negative consequences on the sensorial acceptability.

Nhams (Thai-style fermented pork sausages) were fortified with commercial or hen eggshell calcium lactate at different concentration levels (150–450 mg/100 g). Those with a calcium level of 150 mg/100 g did not differ from the control in sensory scores of sour taste, flavor, and overall acceptability (Wichittra et al. 2002). Texture measurements showed that fortifying these products with eggshell calcium lactate significantly decreased the shear force compared to controls, whereas fortifying Nhams with commercial calcium lactate did not alter their texture (Daengprok et al. 2002).

Iodine deficiency is a major health problem in Europe and other areas (WHO 2007), and salt iodization has been proposed as the main strategy to control this deficiency. Using iodized salt (60 ppm iodine in the salt) at a usual concentration in dry fermented sausage formulation, the resulting products contain approximately 208 µg iodine/100 g product (García-Iñiguez de Ciriano et al. 2010b). As current international recommended intakes for iodine are 150 µg/day for adults and 220 µg/day for pregnant and lactating women, 50 g of this product would contain approximately 70% of the recommended daily intake for adults and 47% for pregnant and lactating women. This research work also modified the selenium content of the product by using Se yeast in the formulation (2 g/kg). Recommended Se intakes for humans (55 µg/day) are not currently achieved, and furthermore, this recommended intake does not take into account that higher levels of Se intake appear to confer additional health benefits besides its action on the activity of selenoenzymes (Rayman 2004). The glutathione peroxidase and thioredoxin reductase are the most abundant selenium-containing proteins that play key roles in redox regulation via removing and decomposing hydrogen peroxide and lipid hydroperoxides (Ursini et al. 1997). After the addition of Se yeast, total Se found in the new products was 364 µg/100 g, with selenomethionine as the main source (60%–85% of total Se), with other organic selenium compounds including Se–Cys not exceeding 10% and inorganic residue being less than 1%. Thus, the addition of Se yeast into dry fermented sausages achieved a good concentration of an excellent bioavailable Se source. No modifications from the sensorial standpoint were noticed with the use of iodized salt and Se yeast (García-Iñiguez de Ciriano et al. 2010b).

36.3.5 Enrichment in Fiber

The healthy effects of fiber on the organism are well known and scientifically demonstrated. The protective effect of dietary fiber against cardiovascular diseases, irritable colon, colon cancer, obesity, and diabetes (Cho and Dreher 2001; Rodríguez et al. 2006) has driven to consider dietary fiber as a nutrient with an adequate intake established around 20–25 g per day for adults (American Dietetic Association 2008).

The addition of fiber to meat products has been one of the most frequent strategies to develop healthier products. Concerning dry fermented sausages, different formulations partially substituting fat with inulin (Mendoza et al. 2001), wheat, oat, and fruit fibers (García et al. 2002) have been developed, obtaining, in general, lower scores for acceptability compared to control products. The higher the percentage of added fiber, the lower the acceptability scores. In this sense, Eim et al. (2008) concluded that carrot dietary fiber added to sobrassada (dry fermented sausage) at levels over 3% did not allow a proper fermentation process; in addition, textural parameters were negatively affected. By-products of citrus fruits have also been widely used in fermented sausages (Aleson-Carbonell et al. 2003; Fernández-López et al. 2007, 2008), showing not only a supply of fiber but also other beneficial effects related to the decrease of nitrite level and a decrease in the oxidation process. These effects are a consequence of the presence of biocompounds such as hesperidine in the case of orange fiber. Salazar et al. (2009), using short-chain fructooligosaccharides as potential functional ingredient in dry fermented sausages (until 6%), also found good acceptability, especially in batches with an intermediate fat level (15% of backfat).

36.3.6 Probiotics

Probiotics are living microorganisms that exert health benefits in the organism after their ingestion in certain numbers. As it happens with the rest of healthy compounds, the presence of probiotics in a food has to be related to some beneficial effects. The health-promoting effects of probiotics deal with the reduction of the risk of intestinal and bowel disturbances, atopic diseases and allergies, inhibition of undesirable and pathogenic bacteria, modulation of the immune system, and anticarcinogenic and hypocholesterolemic effects (Ouwehand et al. 2002).

Different potential probiotic bacteria have been isolated from fermented sausages (Pidcock et al. 2002; Pennacchia et al. 2004, 2006; Klingberg et al. 2005; Muthukumarasamy and Holley 2006; Ruiz-Moyano et al. 2008; Yuksekdag and Aslim 2010). However, there are a few research experiments designed to demonstrate the beneficial effects of these potential probiotic products. In 2002, the effect of a sausage with *Lactobacillus paracasei* on healthy volunteers was studied, providing some evidences about the modulation of immunity system (Jahreis et al. 2002). De Vuyst et al. (2008) published a brief paper about the use of probiotics in fermented sausages, concluding that, although fermented meat products are adequate for the carriage of probiotic bacteria, their viability may be reduced due to the high content in curing salt and the low water activity and pH. They pointed out that there are not yet enough available results to evaluate the effect of probiotic fermented meats on human health.

It can be concluded that although it is true that strains used as starter in the production of fermented sausages (most of them belonging to the group of LAB) have been associated with health effects, their role in the product has been related to technological and sensory quality. Nowadays, there are no scientific evidences on beneficial effects of probiotic fermented sausages. New approaches to the use of microorganisms, such as nanotechnologies applied to the encapsulation of microorganisms, could help in the development of new healthier fermented products.

36.3.7 Endogenous Bioactive Compounds

The possibilities of fermented meat products being considered as healthier foods are focused not only in the development of new formulations elaborated with less problematic ingredients or with the addition of other beneficial ones but also in the study of some components that are present in traditional formulations but which have not been deeply studied from nutritional and health points of view. This is the case of meat-based bioactive compounds such as histidyl dipeptides (carnosine and anserine), carnitine, and other meat protein-derived bioactive peptides, and the conjugated linoleic acid (CLA; Arihara 2006).

CLA (especially the two major isomers *cis*-9,*trans*-11 and *trans*-10,*cis*-12) has been investigated in the last two decades in relation to its effects as anticancer, antiobesity, antiadipogenic, antiatherogenic, anti-inflammatory, and antidiabetogenic (Wahle et al. 2004; Park and Pariza 2007). A number of clinical trials of CLA with effects on body composition have been reported, but effects on coronary heart disease risk factors have been inconsistent (Park 2009). In relation to cancer prevention, there are some significant effects in a number of animal cancer models; however, it is not conclusive if naturally occurring CLA has a significant health impact on prevention of cancer on humans (Decker and Park 2010).

Carnosine and anserine can be considered as antioxidants based on their ability to chelate transition metals such as copper (Brown 1981). As antioxidant activity is considered an important beneficial effect for the preventive effect on oxidative damage of the body, the presence of these histidyl dipeptides could be taken into account from a nutritional point of view. The bioavailability of carnosine after ingestion of beef has been demonstrated (Park et al. 2005). Moreover, prevention of diseases and aging has been related to these peptides (Hipkiss and Brownson 2000). However, there is no evidence of a significant correlation between the ingestion of meat (neither of fermented meat products) and the prevention or decrease in oxidative stress in the human body.

Meat products could also be a supply of angiotensin I-converting enzyme (ACE) inhibitors, which work by competitively binding to ACE, thus blocking the possibility of converting angiotensin I to angiotensin II, which induces the constriction of arteries, thereby increasing the blood pressure. The efficacy of these ACE inhibitor peptides obtained from natural food proteins as antihypertensive compounds has been scientifically demonstrated in different studies (Seppo et al. 2003; Kawasaki et al. 2000; Fang et al. 2008). Their presence in fermented meat products could be a consequence of the proteolysis during the fermentation and ripening stages, or they could be added as a functional ingredient obtained from other sources. In the last case, the addition could be made using new technologies such as encapsulation to maintain the peptides in their active forms (Decker and Park 2010). Nowadays, there are no commercially available fermented meat products showing these antihypertensive effects as a consequence of the presence of ACE inhibitor peptides.

Lastly, it could be mentioned that the production of bacteriocins (peptides with antibacterial properties) produced by LAB during fermentation can also be considered as a subject of interest from the safety point of view (Nieto-Lozano et al. 2010).

36.3.8 Nitrite Reduction

Nitrites are added to dry fermented sausages for several purposes such as inhibiting spoilage and potentially pathogenic microorganisms, for contributing to stabilize the bright red-pink color that characterizes cured meat, for contributing to the achievement of its characteristic aroma and flavor, and for retarding the development of rancidity and off-odors and flavors during storage. However, residual nitrite can favor the production of nitrosamines, which have been shown to be carcinogenic in experimental animals and as such are also potential human carcinogens (Pegg et al. 2000). Besides this, both the presence of heme iron and the endogenous formation of *N*-nitroso compounds seem to be the most likely potential factors in the contribution of processed meats to colorectal cancer (Demeyer et al. 2008). In consequence, a reduction in the use of nitrites has become a key issue for the industry, and some experiments have been successfully applied. Yilmaz and Zorba (2010) studied glucono-delta-lactone and ascorbic acid as partial substitutes of nitrite in Turkish-type fermented sausage (sucuk), achieving good results with ascorbic acid. In addition, Chinese red broken rice powder (angkak; RBR) from *Monascus purpureus* was tested as a substitute for nitrite in Thai traditional fermented pork sausage (nham; Rojsuntornkitti et al. 2010).

36.4 Conclusions

Modification of the traditional composition of fermented sausages to make them healthier is viable from the technological point of view. Sensory properties are also maintained if adequate proportions of new ingredients are introduced into the different types of products. Besides this, the specific support needed to demonstrate the improvement of physiological functions or reduced risk of certain diseases by the intake of new products requires deep and precise studies. Such demonstration of the beneficial effects will allow us to consider these new formulations as functional meat products.

REFERENCES

Aleson-Carbonell, L., Fernández-López, J., Sayas-Barbera, E., Sendra, E., and Perez-Alvarez, J. 2003. Utilization of lemon albedo in dry-cured sausages. *J. Food Sci.* 68(5):1826–1830.

American Dietetic Association. 2008. Position of the American Dietetic Association: Health Implications of Dietary Fiber. *Journal of the American Dietetic Association.* 108(10):1716–1731.

Ansorena, D. and Astiasarán, I. 2004a. Effect of storage and packaging on fatty acid composition and oxidation in dry fermented sausages made with added olive oil and antioxidants. *Meat Sci.* 67(2):237–244.

Ansorena, D. and Astiasarán, I. 2004b. The use of linseed oil improves nutritional quality of the lipid fraction of dry-fermented sausages. *Food Chem.* 87(1):69–74.

Arihara, K. 2006. Strategies for designing novel functional meat products. *Meat Sci.* 74(1):219–229.

Bloukas, J., Paneras, E., and Fournitzis, G. 1997. Effect of replacing pork backfat with olive oil on processing and quality characteristics of fermented sausages. *Meat Sci.* 45(2):133–144.

Bozkurt, H. 2006. Utilization of natural antioxidants: Green tea extract and *Thymbra spicata* oil in Turkish dry-fermented sausage. *Meat Sci.* 73(3):442–450.

Brown, C. 1981. Interactions among carnosine, anserine, ophidine and copper in biochemical adaptation. *J. Theor. Biol.* 88(2):245–256.

Cáceres, E., García, M., and Selgas, M. 2008. Effect of pre-emulsified fish oil—as source of PUFA *n*–3—on microstructure and sensory properties of mortadella, a Spanish bologna-type sausage. *Meat Sci.* 80(2):183–193.

Caudill, M. 2004. The role of folate in reducing chronic and developmental disease risk: An overview. *J. Food Sci.* 69(1):SNQ55–SNQ59.

Cho, S.S. and Dreher, M.L. 2001. Dietary fiber in health and disease. In: *Handbook of Dietary Fiber*, edited by S.S. Cho and M.L. Dreher, pp. 117–194. New York: Marcel Dekker.

Coronado, S., Trout, G., Dunshea, F., and Shah, N. 2002. Antioxidant effects of rosemary extract and whey powder on the oxidative stability of wiener sausages during 10 months frozen storage. *Meat Sci.* 62(2):217–224.

Daengprok, W., Garnjanagoonchorn, W., and Mine, Y. 2002. Fermented pork sausage fortified with commercial or hen eggshell calcium lactate. *Meat Sci.* 62(2):199–204.

Dawidowicz, A., Wianowska, D., and Baraniak, B. 2006. The antioxidant properties of alcoholic extracts from *Sambucus nigra* L. (antioxidant properties of extracts). *Food Sci. Technol.* 39(3):308–315.

De Vuyst, L., Falony, G., and Leroy, F. 2008. Probiotics in fermented sausages. *Meat Sci.* 80(1):75–78.

Decker, E. and Hultin, H. 1992. Lipid oxidation in muscle foods via redox iron. *Lipid Oxid. Food* 500:33–54.

Decker, E. and Park, Y. 2010. Healthier meat products as functional foods. *Meat Sci.* 86(1):49–55.

Del Nobile, M., Conte, A., Incoronato, A., Panza, O., Sevi, A., and Marino, R. 2009. New strategies for reducing the pork back-fat content in typical Italian salami. *Meat Sci.* 81(1):263–269.

Delgado-Pando, G., Cofrades, S., Ruiz-Capillas, C., Solas, M., and Jiménez-Colmenero, F. 2010. Healthier lipid combination oil-in-water emulsions prepared with various protein systems: An approach for development of functional meat products. *Eur. J. Lipid Sci. Technol.* 112(7):791–801.

Demeyer, D., Honikel, K., and de Smet, S. 2008. The World Cancer Research Fund report 2007: A challenge for the meat processing industry. *Meat Sci.* 80(4):953–959.

Eim, V., Simal, S., Rossello, C., and Femenia, A. 2008. Effects of addition of carrot dietary fibre on the ripening process of a dry fermented sausage (sobrassada). *Meat Sci.* 80(2):173–182.

Fang, H., Luo, M., Sheng, Y., Li, Z., Wu, Y., and Liu, C. 2008. The antihypertensive effect of peptides: A novel alternative to drugs? *Peptides* 29(6):1062–1071.

Fernández-López, J., Sendra, E., Sayas-Barbera, E., Navarro, C., and Perez-Alvarez, J. 2008. Physico-chemical and microbiological profiles of "salchichon" (Spanish dry-fermented sausage) enriched with orange fiber. *Meat Sci.* 80(2):410–417.

Fernández-López, J., Sevilla, L., Sayas-Barbera, E., Navarro, C., Marin, F., and Perez-Alvarez, J. 2003. Evaluation of the antioxidant potential of hyssop (*Hyssopus officinalis* L.) and rosemary (*Rosmarinus officinalis* L.) extracts in cooked pork meat. *J. Food Sci.* 68(2):660–664.

Fernández-López, J., Viuda-Martos, M., Sendra, E., Sayas-Barbera, E., Navarro, C., and Perez-Alvarez, J. 2007. Orange fibre as potential functional ingredient for dry-cured sausages. *Eur. Food Res. Technol.* 226(1–2):1–6.

Flores, M., Nieto, P., Ferrer, J., and Flores, J. 2005. Effect of calcium chloride on the volatile pattern and sensory acceptance of dry-fermented sausages. *Eur. Food Res. Technol.* 221(5):624–630.

Formanek, Z., Lynch, A., Galvin, K., Farkas, J., and Kerry, J. 2003. Combined effects of irradiation and the use of natural antioxidants on the shelf-life stability of overwrapped minced beef. *Meat Sci.* 63(4):433–440.

Galán, I., García, M., and Selgas, M. 2010. Effects of irradiation on hamburgers enriched with folic acid. *Meat Sci.* 84(3):437–443.

García, M., Dominguez, R., Galvez, M., Casas, C., and Selgas, M. 2002. Utilization of cereal and fruit fibres in low fat dry fermented sausages. *Meat Sci.* 60(3):227–236.

García-Iñiguez de Ciriano, M., García-Herreros, C., Larequi, E., Valencia, I., Ansorena, D., and Astiasarán, I. 2009. Use of natural antioxidants from lyophilized water extracts of *Borago officinalis* in dry fermented sausages enriched in omega-3 PUFA. *Meat Sci.* 83(2):271–277.

García-Iñiguez de Ciriano, M., Larequi, E., Rehecho, S., Calvo, M.I., Cavero, R.Y., Navarro-Blasco, Í., Astiasarán, I., and Ansorena, D. 2010b. Selenium, iodine, ω–3 PUFA and natural antioxidant from *Melissa officinalis* L.: A combination of components from healthier dry fermented sausages formulation. *Meat Sci.* 85(2):274–279.

García-Iñiguez de Ciriano, M., Rehecho, S., Calvo, M., Cavero, R., Navarro, I., and Astiasarán, I. 2010a. Effect of lyophilized water extracts of *Melissa officinalis* on the stability of algae and linseed oil-in-water emulsion to be used as a functional ingredient in meat products. *Meat Sci.* 85(2):373–377.

Gelabert, J., Gou, P., Guerrero, L., and Arnau, J. 2003. Effect of sodium chloride replacement on some characteristics of fermented sausages. *Meat Sci.* 65(2):833–839.

Ghiretti, G., Zanardi, E., Novelli, E., Campanini, G., Dazzi, G., and Madarena, G. 1997. Comparative evaluation of some antioxidants in salame Milano and mortadella production. *Meat Sci.* 47(1–2):167–176.

Gimeno, O., Astiasarán, I., and Bello, J. 1998. A mixture of potassium, magnesium, and calcium chlorides as a partial replacement of sodium chloride in dry fermented sausages. *J. Agric. Food Chem.* 46(10):4372–4375.

Gimeno, O., Astiasarán, I., and Bello, J. 1999. Influence of partial replacement of NaCl with KCl and CaCl₂ oil texture and color of dry fermented sausages. *J. Agric. Food Chem.* 47(3):873–877.

Gimeno, O., Astiasarán, I., and Bello, J. 2001a. Influence of partial replacement of NaCl with KCl and CaCl₂ on microbiological evolution of dry fermented sausages. *Food Microbiol.* 18(3):329–334.

Gimeno, O., Astiasarán, I., and Bello, J. 2001b. Calcium ascorbate as a potential partial substitute for NaCl in dry fermented sausages: Effect on colour, texture and hygienic quality at different concentrations. *Meat Sci.* 57(1):23–29.

Gou, P., Guerrero, L., Gelabert, J., and Arnau, J. 1996. Potassium chloride, potassium lactate and glycine as sodium chloride substitutes in fermented sausages and in dry-cured pork loin. *Meat Sci.* 42(1):37–48.

Gregory, J. 2004. Dietary folate in a changing environment: Bioavailability, fortification, and requirements. *J. Food Sci.* 69(1):S59–S60.

Guàrdia, M.D., Guerrero, L., Gelabert, J., Gou, P., and Arnau, J. 2006. Consumer attitude towards sodium reduction in meat products and acceptability of fermented sausages with reduced sodium content. *Meat Sci.* 73(3):484–490.

Guàrdia, M., Guerrero, L., Gelabert, J., Gou, P., and Arnau, J. 2008. Sensory characterisation and consumer acceptability of small calibre fermented sausages with 50% substitution of NaCl by mixtures of KCl and potassium lactate. *Meat Sci.* 80(4):1225–1230.

Haak, L., Raes, K., and de Smet, S. 2009. Effect of plant phenolics, tocopherol and ascorbic acid on oxidative stability of pork patties. *J. Sci. Food Agric.* 89(8):1360–1365.

Hipkiss, A. and Brownson, C. 2000. A possible new role for the anti-ageing peptide carnosine. *Cell. Mol. Life Sci.* 57(5):747–753.

Ibanez, E., Kubatova, A., Senorans, F., Cavero, S., Reglero, G., and Hawthorne, S. 2003. Subcritical water extraction of antioxidant compounds from rosemary plants. *J. Agric. Food Chem.* 51(2):375–382.

Ilikkan, H., Ercoskun, H., Vural, H., and Sahin, E. 2009. The effect of addition of hazelnut oil on some quality characteristics of Turkish fermented sausage (sucuk). *J. Muscle Foods* 20(1):117–127.

Jahreis, G., Vogelsang, H., Kiessling, G., Schubert, R., Bunte, C., and Hammes, W. 2002. Influence of probiotic sausage (*Lactobacillus paracasei*) on blood lipids and immunological parameters of healthy volunteers. *Food Res. Int.* 35(2–3):133–138.

Jensen, C., Lauridsen, C., and Bertelsen, G. 1998. Dietary vitamin E: Quality and storage a stability of pork and poultry. *Trends Food Sci. Technol.* 9(2):62–72.

Jiménez-Colmenero, F. 2007. Healthier lipid formulation approaches in meat-based functional foods. Technological options for replacement of meat fats by non-meat fats. *Trends Food Sci. Technol.* 18(11):567–578.

Jiménez-Colmenero, F., Carballo, J., and Cofrades, S. 2001. Healthier meat and meat products: Their role as functional foods. *Meat Sci.* 59(1):5–13.

Jiménez-Colmenero, F., Reig, M., and Toldra, F. 2006. New approaches for the development of functional meat products. *Adv. Technol. Meat Process.* Chapter 11, 275–308.

Karabacak, S. and Bozkurt, H. 2008. Effects of *Urtica dioica* and *Hibiscus sabdariffa* on the quality and safety of sucuk (Turkish dry-fermented sausage). *Meat Sci.* 78(3):288–296.

Kawasaki, T., Seki, E., Osajima, K., Yoshida, M., Asada, K., and Matsui, T. 2000. Antihypertensive effect of Valyl–Tyrosine, a short chain peptide derived from sardine muscle hydrolyzate, on mild hypertensive subjects. *J. Hum. Hypertens.* 14(8):519–523.

Kayaardi, S. and Gök, V. 2004. Effect of replacing beef fat with olive oil on quality characteristics of Turkish soudjouk (sucuk). *Meat Sci.* 66(1):249–257.

Klingberg, T., Axelsson, L., Naterstad, K., Elsser, D., and Budde, B. 2005. Identification of potential probiotic starter cultures for Scandinavian-type fermented sausages. *Int. J. Food Microbiol.* 105(3):419–431.

Koutsopoulos, D., Koutsimanis, G., and Bloukas, J. 2008. Effect of carrageenan level and packaging during ripening on processing and quality characteristics of low-fat fermented sausages produced with olive oil. *Meat Sci.* 79(1):188–197.

Liaros, N., Katsanidis, E., and Bloukas, J. 2009. Effect of the ripening time under vacuum and packaging film permeability on processing and quality characteristics of low-fat fermented sausages. *Meat Sci.* 83(4):589–598.

Martínez, L., Cilla, I., Beltrán, J., and Roncalés, P. 2006a. Antioxidant effect of rosemary, borage, green tea, pu-erh tea and ascorbic acid on fresh pork sausages packaged in a modified atmosphere: Influence of the presence of sodium chloride. *J. Sci. Food Agric.* 86(9):1298–1307.

Martínez, L., Cilla, I., Beltrán, J., and Roncalés, P. 2006b. Comparative effect of red yeast rice (*Monascus purpureus*), red beet root (*Beta vulgaris*) and betanin (E-162) on colour and consumer acceptability of fresh pork sausages packaged in a modified atmosphere. *J. Sci. Food Agric.* 86(4):500–508.

McAfee, A., McSorley, E., Cuskelly, G., Moss, B., Wallace, J., and Bonham, M. 2010. Red meat consumption: An overview of the risks and benefits. *Meat Sci.* 84(1):1–13.

Mendoza, E., García, M., Casas, C., and Selgas, M. 2001. Inulin as fat substitute in low fat, dry fermented sausages. *Meat Sci.* 57(4):387–393.

Mercadante, A., Capitani, C., Decker, E., and Castro, I. 2010. Effect of natural pigments on the oxidative stability of sausages stored under refrigeration. *Meat Sci.* 84(4):718–726.

Miliauskas, G., Mulder, E., Linssen, J., Houben, J., van Beek, T., and Venskutonis, P. 2007. Evaluation of antioxidative properties of *Geranium macrorrhizum* and *Potentilla fruticosa* extracts in Dutch style fermented sausages. *Meat Sci.* 77(4):703–708.

Miliauskas, G., Venskutonis, P., and van Beek, T. 2004. Screening of radical scavenging activity of some medicinal and aromatic plant extracts. *Food Chem.* 85(2):231–237.

Montgomery, J., Blanton, J., Horst, R., Galyean, M., Morrow, K., and Allen, V. 2004. Effect of supplemental vitamin D-3 concentrations of calcium phosphorus, and magnesium relative to protein in subcellular components of the *longissimus* and the distribution of calcium within *longissimus* muscle of beef steers. *J. Anim. Sci.* 82(9):2742–2749.

Muguerza, E., Ansorena, D., and Astiasarán, I. 2003. Improvement of nutritional properties of chorizo de Pamplona by replacement of pork backfat with soy oil. *Meat Sci.* 65(4):1361–1367.

Muguerza, E., Ansorena, D., and Astiasarán, I. 2004b. Functional dry fermented sausages manufactured with high levels of n-3 fatty acids: Nutritional benefits and evaluation of oxidation. *J. Sci. Food Agric.* 84(9):1061–1068.

Muguerza, E., Fista, G., Ansorena, D., Astiasarán, I., and Bloukas, J.G. 2002. Effect of fat level and partial replacement of pork backfat with olive oil on processing and quality characteristics of fermented sausages. *Meat Sci.* 61(4):397–404.

Muguerza, E., Gimeno, O., Ansorena, D., and Astiasarán, I. 2004a. New formulations for healthier dry fermented sausages: A review. *Trends Food Sci. Technol.* 15(9):452–457.

Muguerza, E., Gimeno, O., Ansorena, D., Bloukas, J.G., and Astiasarán, I. 2001. Effect of replacing pork backfat with pre-emulsified olive oil on lipid fraction and sensory quality of Chorizo de Pamplona—A traditional Spanish fermented sausage. *Meat Sci.* 59(3):251–258.

Muthukumarasamy, P. and Holley, R. 2006. Microbiological and sensory quality of dry fermented sausages containing alginate-microencapsulated *Lactobacillus reuteri*. *Int. J. Food Microbiol.* 111(2):164–169.

Nam, K., Ko, K., Min, B., Ismail, H., Lee, E., and Cordray, J. 2006. Influence of rosemary-tocopherol/packaging combination on meat quality and the survival of pathogens in restructured irradiated pork loins. *Meat Sci.* 74(2):380–387.

Nieto-Lozano, J., Reguera-Useros, J., Pelaez-Martínez, M., Gutierrez-Fernández, A., and de la Torre, A. 2010. The effect of the pediocin PA-1 produced by *Pediococcus acidilactici* against *Listeria monocytogenes* and *Clostridium perfringens* in Spanish dry-fermented sausages and frankfurters. *Food Control* 21(5):679–685.

Olivares, A., Navarro, J.L., Salvador, A., and Flores, M. 2010. Sensory acceptability of slow fermented sausages based on fat content and ripening time. *Meat Sci.* 86(2):251–257.

Ospina, J., Cruz, A., Perez-Alvarez, J., and Fernández-López, J. 2010. Development of combinations of chemically modified vegetable oils as pork backfat substitutes in sausages formulation. *Meat Sci.* 84(3):491–497.

Ouwehand, A., Salminen, S., and Isolauri, E. 2002. Probiotics: An overview of beneficial effects. *Antonie van Leeuwenhoek* 82(1–4):279–289.

Palic, A., Krizanec, D., and Dikanoviclucan, Z. 1993. The antioxidant properties of spices in dry fermented sausages. *Fleischwirtschaft* 73(6):684–687.

Park, W. and Kim, Y. 2009. Effect of garlic and onion juice addition on the lipid oxidation, total plate counts and residual nitrite contents of emulsified sausage during cold storage. *Korean J. Food Sci. Anim. Resour.* 29(5):612–618.

Park, Y. 2009. Conjugated linoleic acid (CLA): Good or bad trans fat? *J. Food Compos. Anal.* 22:S4–S12.

Park, Y. and Pariza, M. 2007. Mechanisms of body fat modulation by conjugated linoleic acid (CLA). *Food Res. Int.* 40(3):311–323.

Park, Y., Volpe, S., and Decker, E. 2005. Quantitation of carnosine in humans plasma after dietary consumption of beef. *J. Agric. Food Chem.* 53(12):4736–4739.

Pegg, R., Fisch, K., and Shahidi, F. 2000. The replacement of conventional meat curing with nitrite-free curing systems. *Fleischwirtschaft* 80(5):86–89.

Pelser, W., Linssen, J., Legger, A., and Houben, J. 2007. Lipid oxidation in *n*–3 fatty acid enriched Dutch style fermented sausages. *Meat Sci.* 75(1):1–11.

Pennacchia, C., Ercolini, D., Blaiotta, G., Pepe, O., Mauriello, G., and Villani, F. 2004. Selection of *Lactobacillus* strains from fermented sausages for their potential use as probiotics. *Meat Sci.* 67(2):309–317.

Pennacchia, C., Vaughan, E., and Villani, F. 2006. Potential probiotic *Lactobacillus* strains from fermented sausages: Further investigations on their probiotic properties. *Meat Sci.* 73(1):90–101.

Pidcock, K., Heard, G., and Henriksson, A. 2002. Application of nontraditional meat starter cultures in production of Hungarian salami. *Int. J. Food Microbiol.* 76(1–2):75–81.

Rayman, M. 2004. The use of high-selenium yeast to raise selenium status: How does it measure up? *Br. J. Nutr.* 92(4):557–573.

Rodríguez, R., Jiménez, A., Fernández-Bolanos, J., Guillen, R., and Heredia, A. 2006. Dietary fibre from vegetable products as source of functional ingredients. *Trends Food Sci. Technol.* 17(1):3–15.

Rojsuntornkitti, K., Jittrepotch, N., Kongbangkerd, T., and Kraboun, K. 2010. Substitution of nitrite by Chinese red broken rice powder in Thai traditional fermented pork sausage (Nham). *Int. Food Res. J.* 17(1):153–161.

Ruiz-Moyano, S., Martin, A., Benito, M., Nevado, F., and Cordoba, M. 2008. Screening of lactic acid bacteria and bifidobacteria for potential probiotic use in Iberian dry fermented sausages. *Meat Sci.* 80(3):715–721.

Ruusunen, M. and Puolanne, E. 2005. Reducing sodium intake from meat products. *Meat Sci.* 70(3):531–541.

Salazar, P., García, M.L., and Selgas, M.D. 2009. Short-chain fructooligosaccharides as potential functional ingredient in dry fermented sausages with different fat levels. *Int. J. Food Sci. Technol.* 44(6):1100–1107.

Salem, F.A. and Ibrahim, H. 2010. Dry fermented buffalo sausage with sage oil extract: Safety and quality. *Grasas Aceites* 61(1):76–85.

Schaefer, D., Liu, Q., Faustman, C., and Yin, M. 1995. Supranutritional administration of vitamin-E and vitamin-C improves oxidative stability of beef. *J. Nutr.* 125(6):S1792–S1798.

Sebranek, J.G., Sewalt, V.J.H., Robbins, K.L., and Houser, T.A. 2005. Comparison of a natural rosemary extract and BHA/BHT for relative antioxidant effectiveness in pork sausage. *Meat Sci.* 69:289–296.

Selgas, M., Salazar, P., and García, M. 2009. Usefulness of calcium lactate, citrate and gluconate for calcium enrichment of dry fermented sausages. *Meat Sci.* 82(4):478–480.

Seppo, L., Jauhiainen, T., Poussa, T., and Korpela, R. 2003. A fermented milk high in bioactive peptides has a blood pressure-lowering effect in hypertensive subjects. *Am. J. Clin. Nutr.* 77(2):326–330.

Severini, C., De Pilli, T., and Baiano, A. 2003. Partial substitution of pork backfat with extra-virgin olive oil in 'salami' products: Effects on chemical, physical and sensorial quality. *Meat Sci.* 64(3):323–331.

Torrissen, O.J. 2000. Chapter 11: Dietary delivery of carotenoids. In: *Antioxidants in Muscle Foods*, edited by E.A. Decker, C. Faustman, and J.C. López-Bote, pp. 289–313. Wiley-Interscience.

Ursini, F., Maiorino, M., and Roveri, A. 1997. Phospholipid hydroperoxide glutathione peroxidase (PHGPx): More than an antioxidant enzyme? *Biomed. Environ. Sci.* 10(2–3):327–332.

Valencia, I., Ansorena, D., and Astiasarán, I. 2006a. Stability of linseed oil and antioxidants containing dry fermented sausages: A study of the lipid fraction during different storage conditions. *Meat Sci.* 73(2):269–277.

Valencia, I., Ansorena, D., and Astiasarán, I. 2006b. Nutritional and sensory properties of dry fermented sausages enriched with *n*–3 PUFAs. *Meat Sci.* 72(4):727–733.

Valencia, I., Ansorena, D., and Astiasarán, I. 2007. Development of dry fermented sausages rich in docosahexaenoic acid with oil from the microalgae *Schizochytrium* sp.: Influence on nutritional properties, sensorial quality and oxidation stability. *Food Chem.* 104(3):1087–1096.

Vural, H. 2003. Effect of replacing beef fat and tail fat with interesterified plant oil on quality characteristics of Turkish semi-dry fermented sausages. *Eur. Food Res. Technol.* 217(2):100–103.

Wahle, K., Heys, S., and Rotondo, D. 2004. Conjugated linoleic acids: Are they beneficial or detrimental to health? *Prog. Lipid Res.* 43(6):553–587.

Wald, D., Morris, J., Law, M., and Wald, N. 2006. Cardiovascular disease—Folic acid, homocysteine, and cardiovascular disease: Judging causality in the face of inconclusive trial evidence. *Br. Med. J.* 333(7578):1114–1117.

Weiss, J., Gibis, M., Schuh, V., and Salminen, H. 2010. Advances in ingredient and processing systems for meat and meat products. *Meat Sci.* 86(1):196–213.

Wen, S., Chen, X., Rodger, M., White, R., Yang, Q., and Smith, G. 2008. Folic acid supplementation in early second trimester and the risk of preeclampsia. *Am. J. Obstet. Gynecol.* 198(1):45e1–45e7.

WHO. Andersson, M., de Benoist, B., Darnton-Hill, I., and Delange, F. 2007. Iodine deficiency in Europe. A continuing public health problem.

Wichittra, D., Wunwiboon, G., and Mine, Y. 2002. Fermented pork sausage fortified with commercial or hen eggshell calcium lactate. *Meat Sci.* 62(2):199–204.

Yildiz-Turp, G. and Serdaroglu, M. 2008. Effect of replacing beef fat with hazelnut oil on quality characteristics of sucuk—A Turkish fermented sausage. *Meat Sci.* 78(4):447–454.

Yilmaz, M. and Zorba, O. 2010. Response surface methodology study on the possibility of nitrite reduction by glucono-delta-lactone and ascorbic acid in Turkish-type fermented sausage (Sucuk). *J. Muscle Foods* 21(1):15–30.

Yu, L., Scanlin, L., Wilson, J., and Schmidt, G. 2002. Rosemary extracts as inhibitors of lipid oxidation and color change in cooked turkey products during refrigerated storage. *J. Food Sci.* 67(2):582–585.

Yuksekdag, Z. and Aslim, B. 2010. Assessment of potential probiotic and starter properties of *Pediococcus* spp. isolated from Turkish-type fermented sausages (sucuk). *J. Microbiol. Biotechnol.* 20(1):161–168.

Zhang, W., Xiao, S., Samaraweera, H., Lee, E.J., and Ahn, D.U. 2010. Improving functional value of meat products. *Meat Sci.* 86(1):15–31.

37

Turkish Pastirma: A Dry-Cured Beef Product

Ersel Obuz, Levent Akkaya, and Veli Gök

CONTENTS

37.1 Introduction

Pastirma, a popular dry-cured beef product made from whole muscle, can be considered an intermediate-moisture food (Gök et al. 2008). Several whole meat dried products such as ham, bacon, corned beef, Bündnerfleisch, and pastrami are produced all over the world. Pastirma is generally produced from whole muscles obtained from certain parts of cow and water buffalo carcasses (Aktaş and Gürses 2005). In order to produce pastirma, all the exterior fat and connective tissue are removed from the meat, and then the processes of curing, drying, pressing, and coating the resultant meat with cemen paste are applied (Aksu and Kaya 2002a). Although pastirma is defined as a cured and dried meat product, partial fermentation takes place during the production step, thanks to natural microbial flora (Katsaras et al. 1996; Aksu and Kaya 2001a).

37.2 Processing Technology

Pastirma is derived from the Turkish verb "bastırma," meaning a strong pressing action (Tekinşen and Doğruer 2000). Although beef is the main meat used in pastirma production (Aksu et al. 2005; Gök et al. 2008), meat from camel, water buffalo, sheep, turkey, and chicken can also be used (Doğruer 2001). Pastirma production takes place under environmental conditions in a "master–apprentice" fashion. Therefore, skill rather than technology is at the center of a successful pastirma production. This brings

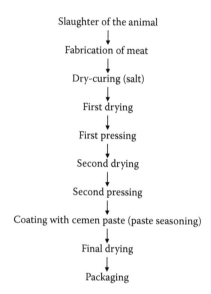

FIGURE 37.1   Pastirma production scheme. (From Gök, V. et al., *Meat Sci.* 80, 335–344, 2008. With permission.)

up economical, quality-centered, hygienic, and lack-of-standardization issues to our attention (Tekinşen and Doğruer 2000). "Pastirma summer" is the best period of time for natural pastirma production beginning in the second half of October and ending in the first half of November. The temperature difference between day and night is around 15°C; the days are sunny, and the relative humidity is the optimum during "pastirma summer" (Öztan 2005). However, a more recent trend is to produce pastirma in so-called "climatic rooms" where temperature and relative humidity can be effectively controlled. The production scheme for pastirma is shown in Figure 37.1.

37.2.1 Raw Material

Meat from 3- to 6-year-old cow and water buffalo is generally used in pastirma production (Anıl 1988). Meat from too-young animals is not preferred because the meat from these animals has too-high moisture content that causes excessive moisture loss in drying. Meat from too-old animals is also not preferred due to its coarse and tough texture. Moreover, fatty meat from too-old animals results in a low-quality pastirma and is not preferred (Gökalp et al. 1999).

The amount of stress before animal slaughter significantly changes the course of pH decline postmortem (Lewis et al. 1989). The rate and extent of this pH decline affect several meat quality attributes including water holding capacity and color stability (Briskey 1964). Meat having pH values ranging from 5.4 to 5.8 is optimum. After slaughtering the animal, the blood is effectively drained, and rigor mortis develops. The muscles are then fabricated, and they are ready to be processed for pastirma making (Öztan 2005).

37.2.2 Fabrication

Fabrication of meat is a critical step in pastirma production that affects the final quality grade of pastirma (Gürbüz 1994). Eight pieces of meat are taken from the carcass (two pieces from loin, two from rib, two from flank, and two from round). Then, they are fabricated to be used for pastirma production. To easily remove meat pieces from the carcass, rigor must be completed, and adequate cooling must be applied to the carcass (Doğruer 1992). Up to 16 different pastirmas having different names and quality attributes can be produced from a cow carcass (Anonymous 1991; Aksu et al. 2005; Ceylan 2009). For example, pastirma from loin has a rectangular shape (longer side measures

60–70 cm; shorter side measures 15–20 cm) and is about 3–4 kg. It has optimum eating qualities and is the most preferred pastirma (Gökalp et al. 1999; Tekinşen and Doğruer 2000). The next step is "şaklama," where deep cuts, up to half the thickness of the meat, are made on the meat surface with a 45°–60° angle, by hand (Anıl 1986).

37.2.3 Curing

Curing agents and curing methods have a direct effect on final pastirma quality (Gökalp et al. 1999). Salt along with nitrite, nitrate, and ascorbic acid are used as curing agents. Using salt as a sole curing agent results in a tough, dry, and bland-flavored pastirma (Pearson and Tauber 1984). The curing agents are responsible for esteemed and stable red color; nitrite is a strong antioxidant and prevents or retards microbial growth and gives a pleasant flavor (Honikel 2008). Dry curing is the traditionally used curing method in pastirma production. Salt is a major ingredient in cured meat products and has a bacteriostatic effect, that is, it inhibits the growth of undesirable microorganisms. It also contributes to salty taste and increases myofibrillar protein solubility (Toldra 2002). Salt used in pastirma production is medium size and is called "ant head" (Tekinşen and Doğruer 2000). Coarse salt cannot be adequately absorbed by meat and can negatively affect the histology of meat by causing salt burn. Too-fine salt is readily absorbed by meat and causes several quality defects such as oxidation, black color, and excessive salty taste (Gökalp et al. 1999). Curing mixture for pastirma production usually contains 94.5% NaCl, 1.5% KNO_3, 2% glucose, and 2% sucrose (Aksu et al. 2005). Salt is applied to deeply cut meat surfaces by hand. The number of cuts and the distance between them are very critical. Cuts too close to each other may result in excessive salt absorption, leading to a tough and bland product. On the other hand, too few cuts are responsible for less salt absorption and decreased product shelf life. Salted meat pieces are piled up and measure about a meter in height. Then, the piled meat pieces are turned upside down (Gökalp et al. 1999). Pieces that could not absorb enough salt are re-salted. Salted meat pieces are called "salty."

37.2.4 Drying

Dried meat products achieve shelf stability after drying and curing (sometimes salting) processes because these processes decrease water activity (Comaposada et al. 2000). After the curing process, excess salt is drained away by washing the cured meat for at least 60 s. (Özeren 1980; Gökalp et al. 1999). After the washing step, drying is carried out by hanging the meat pieces. The time frame of the drying period depends on environmental conditions. Drying lasts about 3–5 days during the pastirma summer, and it lasts about 15 days in winter months. However, recently, drying has been finished much more quickly in controlled climatic rooms where parameters such as temperature, relative humidity, and speed of air can be controlled. The adequacy of drying is subjectively judged by smashing the processed meat between two fingers and smelling it (Anıl 1986).

37.2.5 Pressing

The first pressing (cold pressing) step follows the drying process (Gürbüz 1994). This process is accomplished in a fairly large area made from cement or mosaic in small plants, which traditionally produce pastirma (Tekinşen and Doğruer 2000). Cured and dried meat pieces are piled up on each other, and a weight of 250 kg is put onto this pile (Gökalp et al. 1999). A pressure of 0.9–1.0 kg is applied per centimeter square area (Özeren 1980). The time of first pressing ranges from 6 to 16 h (Gökalp et al. 1999). Pressed meats start losing water immediately due to the effect of curing and applied pressure. The pressing process also modifies the shape of meat pieces (Öztan 2005), and deep cuts on the meat surface vanish (Tekinşen and Doğruer 2000).

37.2.6 Second Drying and Hot Pressing

Second drying follows the cold pressing and is accomplished under minimal air speed. Second drying takes about 3 days at 20°C (Aksu and Kaya 2002a). Fats on the meat surface melt and form a white layer.

Pastirma manufacturers call this process "sweating" or "bleaching" (Gökalp et al. 1999). After the second drying, second pressing or hot pressing starts. Because meat pieces have a high amount of fat, they are soft and fairly warm. Therefore, they will change their shape readily and will lose water. Fats on the surfaces migrate to the interior of the meat, modifying meat texture significantly (Tekinşen and Doğruer 2000). The meat becomes very tender during this stage (Öztan 2005).

37.2.7 Cementing (Coating with Cemen Paste)

Cementing involves coating pastirma with cemen paste, which gives pastirma its characteristic taste, color, aroma, and flavor (Anıl 1988). A typical cemen mixture has 500 g flour of *Trigonella foenum graecum* seed, 350 g smashed fresh garlic, 150 g paprika, and 1200 mL water (Anonymous 1983, 1991). Cemen significantly affects appearance, color, texture, taste, and flavor of pastirma and helps to inhibit microbial contamination and excess drying of pastirma (Işıklı and Karababa 2005). Cemen (paste seasoning) is reported to have a protective effect on some pathogens (Yetim et al. 2006) and mold growth (Kaya et al. 1996; Berkmen 1960). Cemen paste is very sticky and encloses meat much like a casing (Öztan 2005). Cementing lasts between 16 and 36 h. The thickness of the cemen paste is about 3–4 mm (Gökalp et al. 1999). Cemented meat pieces are dried for 1–2 days during summer and 6–7 days during winter.

37.3 Pastirma Quality

To evaluate pastirma quality, physical, chemical, microbiological, and sensorial analyses can be used. Pastirma should meet certain physical, chemical, and microbiological standards and should not cause harm to human health. Turkish standards for pastirma include maximum fat of 40%, salt of 8.5%, pH of 4.5–5.8, maximum moisture content of 50%, maximum cemen (cement) level of 10%, no rancidity, and zero tolerance to pathogens (TS 1071). Based on the Turkish Standards (TS 1071) (Anonymous 1983), pastirma is classified into one of three quality grades. The first grade of pastirma is of the highest quality. It has an abundant level of marbling and the least amount of intermuscular fat. It is the most tender and has a fine structure and texture. The color of its interior is bright red, and it does not have a visible crust. The thickness of its cement (paste seasoning) does not exceed 4 mm. The second grade of pastirma has less marbling but more intermuscular fat than the first-grade pastirma. Its cement is not very homogeneous. The color of its interior is dark red, and it has a visible crust. The third grade of pastirma does not have a fine structure and texture and is very dry. Its cement is excessive. It does not have a significant amount of marbling but has the most intermuscular fat. The color of its interior varies from dark red to brown (Gökalp et al. 1999).

Therefore, a high-quality pastirma

1. Should have an abundant and homogeneous amount of marbling
2. Should have a minimum amount of intermuscular fat
3. Should have a typical cement color on the interior and bright red color inside
4. Should have a fine texture, should be easily sliced, and should not be excessively tough or extremely tender

37.3.1 pH

TS 1071 (Anonymous 1983) gives the pH of pastirma as ranging from 4.5 to 5.8. However, pH values lower than 5 are very unlikely for pastirma because starter cultures are not used in pastirma production. pH values higher than 6.2 mark the beginning of pastirma spoilage. Storage period affects pH values of pastirma. Nitrogenous compounds formed by proteolysis during the extended storage period may be responsible for this increase, as suggested by Lücke (1998). Aksu and Kaya (2005) and Gök et al. (2008) reported increased pH values with storage time for pastirmas produced from fresh beef. As reported by Kaban (2009), pastirma does not undergo true lactic acid fermentation. The absence of true lactic fermentation has also been reported in other raw cured meat products such as French dry-cured hams (Buscailhon et al. 1994), Italian dry-cured ham (Virgili et al. 2007), and dry-cured lacón (Lorenzo et al. 2003), and dry-cured lacón (Marra et al. 1999). The pH, albeit less significant for pastirma than in other fermented meat products, should not be less than 5.5 for sensorial reasons (Leistner 1988).

37.3.2 Water Activity and Moisture Content

Pastirma can be considered an intermediate-moisture food and is shelf stable due to its relatively low water activity (a_w). Leistner (1988) reported that a_w of pastirma must be between 0.85 and 0.90. Meat products having a_w values less than 0.90 do not have to be refrigerated but can be stored at room temperature (Öztan 2005). Kaban (2009) reported that a_w values of pastirma changed from 0.98 (before curing) to 0.87 (final product). The new Turkish Food Codex on pastirma will be in effect shortly and suggests a moisture content of not more than 40% for pastirma. Moisture content of less than or equal to 40% should ensure that the a_w of pastirma does not exceed 0.90. The chemical analysis of the pastirma samples sold in Turkish markets revealed that the moisture content ranged from 39% to 52% (Aksu and Kaya 2001a; Soyutemiz and Özenir 1996). This is very alarming because moisture content exceeding 40% may bring food safety issues to our attention and bring about the need for refrigeration of the pastirma. In the related literature, moisture content as low as 32% (Berkmen 1940) and as high as 61.34% (Goma et al. 1978) has been reported. Setting the drying procedure for pastirma in a way that the final moisture content is less than 40% should ensure that the a_w for pastirma will fall between 0.85 and 0.90.

37.3.3 Thiobarbituric Acid Reactive Substance Content

Thiobarbituric acid reactive substance (TBARS) is one of the significant lipid oxidation markers in meat and meat products (Shahidi et al. 1987). TBARS is reported to change from 0.16 to 2.46 mg malonaldehyde/kg for pastirmas (Askar et al. 1993; Yagli and Ertas 1998; Aksu et al. 2005). TBARS is mostly affected by storage time, storage temperature, or method of packaging. Aksu et al. (2005) reported that TBARS value for pastirmas increased from 0.64 mg malonaldehyde/kg on day 0 of storage to 1.91 mg malonaldehyde/kg on day 150 of storage. Similarly, Gök et al. (2008) suggested that TBARS value for pastirmas increased with increased storage time. Storing pastirmas at 10°C resulted in higher TBARS values than storing pastirmas at 4°C. Gök et al. (2008) reported that the lowest TBARS values occurred for pastirmas packaged in modified atmosphere packaging, followed by those packaged by vacuum packaging and aerobic packaging. Modified atmosphere packaging should be the right choice of packaging to retard oxidative rancidity over the pastirma storage period.

37.3.4 Color (*L**, *a**, and *b** Values)

The attractive pink–red color of dry-cured meat products is mainly contributed by nitrite (Pearson and Gillett 1997). Instrumental redness (a^*) is a reliable indicator of dry-cured meat product quality. With storage time, a^* was reported to decrease, indicating loss of cured color of pastirmas (Gök et al. 2008; Aksu et al. 2005). Packaging method has certain implications on cured meat color. Gök et al. (2008) reported that modified atmosphere packaging preserved typical cured meat color of pastirmas much better than vacuum packaging or aerobic packaging. Storage temperature also affects a^* values. Aksu et al. (2005) concluded that storing pastirmas at 10°C resulted in higher a^* values than storing pastirmas at 4°C. L^* and b^* values of pastirmas were reported to decrease with increased storage time, which may be attributed to myoglobin (pigment) oxidation (Gök et al. 2008; Aksu et al. 2005). However, L^* (lightness) and b^* (yellowness) are not very significant as far as pastirma quality is concerned.

37.3.5 Microbiological Properties

Results of studies on the microbial flora of pastirma in the Turkish market were as follows—total viable mesophilic count and *lactobacilli* count ranged from 4 to 8 log cfu/g, *micrococci/staphylococci* from 4 to 7 log cfu/g, Enterobacteriaceae from <2 log to 4 log cfu/g, coliform bacteria from <2 log to 3 log cfu/g, enterococci from <2 log to 4 log cfu/g, *Clostridium perfringens* from <1 log cfu/g, *Pseudomonas* and yeasts from <2 log cfu/g, and molds from <2 log to 5 log cfu/g (Kılıç 2009). However, *Salmonella, C. botulinum,* and *Listeria monocytogenes* were not detected in pastirma. Özdemir et al. (1999) and Soyutemiz and Özenir (1996) reported that lactic acid bacteria, *Micrococcus,* and *Staphylococcus* were the dominant flora of pastirma. The type of packaging and storage time may have a commanding effect on the microbial stability of pastirma. Modified atmosphere packaging was, by far, the best method of packaging for pastirmas as far as overall microbiological quality is concerned (Gök et al. 2008). Kaban (2009) concluded that a significant increase in microbial counts occurred during the course of pastirma processing. On the other hand, Gök et al. (2008) and Aksu et al. (2005) reported decreased mesophilic bacteria count with increased storage time. Microbiological quality problems should be of less concern for pastirma because it is an intermediate-moisture food and has relatively low water activity. However, post-processing contamination or post-processing operations, such as slicing, may increase the microbial risk. Therefore, it is highly recommended that pastirmas be packaged appropriately (modified atmosphere packaging) and be refrigerated (at 4°C) to optimize shelf life.

37.4 New Trends in Pastirma Processing

Pastirma is generally made from certain beef cuts; however, there has been research on the use of alternative meats such as turkey, chicken, and fish in the manufacture of pastirma (Arslan et al. 1997a,b; Doğruer 2001; Gök and Uzun 2010). The use of turkey, chicken, or fish may increase the market share of pastirma because these meats are much cheaper than beef.

Using brine solution instead of drying in the processing of pastirma is another new approach in pastirma manufacture. Güner et al. (2008) investigated the effect of the use of tumbling and brine injection on pastirma quality. They concluded that brine injection of curing agents minimally affected pastirma quality. However, tumbling significantly decreased microbial count and improved the sensory properties of dry-cured pastirma.

Application of heat process in pastirma manufacture was also investigated. Ibrahim (2001) reported that heat treatment (71°C) decreased moisture content, pH, the residual nitrite, and aerobic plate count but increased total fat, ash, and protein contents. Heat treatment also positively affected sensory properties.

Effects of *Lactobacillus sakei* and *Staphylococcus xylosus* (starter cultures) on the inhibition of *Escherichia coli* O157: H7 in pastirma were investigated by Aksu et al. (2008). They reported that the processing step of drying rather than the use of starter culture was the significant factor for the inhibition of *E. coli* O157: H7. Use of starter cultures decreased nitrite and nitrate residues in pastirma (Aksu and Kaya 2001b; Aksu and Kaya 2002b). The authors further concluded that the use of starter cultures (especially *S. xylosus + L. sakei* combination) in pastirma improved chemical quality parameters such as pH, residual nitrite and nitrate, and instrumental color.

37.5 Conclusions

Pastirma, as a traditional meat product, has not gained enough recognition and market share among traditional meat products. Its relatively high price, longer production time, and non-standardized production procedures may be responsible for this problem. Standardizing production steps, especially pressing stage, using controlled artificial drying methods, and optimizing raw meat quality are among some of the important issues that need to be addressed. Moreover, a better production control system to satisfy consumer demands, but to avoid potential food safety hazards, is promptly needed. In addition, alternative ingredient technologies (use of starter cultures) and processing steps (i.e., tumbling) may be implemented to further improve the quality of pastirma.

REFERENCES

Aksu, M.I. 1999. Pastirma Uretiminde Starter Kullanim Imkanlari (Research on the possibility of starter culture use in pastirma production). PhD dissertation, Graduate Institute of Science, Atatürk University, Erzurum, Turkey.

Aksu, M.I. and Kaya, M. 2001a. Some microbiological, chemical and physical characteristics of pastirma marketed in Erzurum. *Turk. J. Vet. Anim. Sci.* 25:319–326.

Aksu, M.I. and Kaya, M. 2001b. The effect of starter culture use in pastirma production on the properties of end product. *Turk. J. Vet. Anim. Sci.* 6:847–854.

Aksu, M.I. and Kaya, M. 2002a. Production of pastirma with different curing methods and using starter culture. *Turk. J. Vet. Anim. Sci.* 26:909–916.

Aksu, M.I. and Kaya, M. 2002b. Effect of commercial starter cultures on the fatty acid composition of pastirma (Turkish dry meat product). *J. Food Sci.* 67:2342–2345.

Aksu, M.I. and Kaya, M. 2005. Effect of storage temperatures and time on shelf-life of sliced and modified atmosphere packaged pastirma, a dried meat product, produced from beef. *J. Sci. Food Agric.* 85:1305–1312.

Aksu, M.I., Kaya, M., and Ockerman, H.W. 2005. Effect of modified atmosphere packaging and temperature on the shelf life of sliced pastirma produced from frozen/thawed meat. *J. Muscle Foods* 16:192–206.

Aksu, M.I., Kaya, M., and Oz, F. 2008. Effect of *Lactobacillus sakei* and *Staphylococcus xylosus* on the inhibition of *Escherichia coli* O157:H7 in pastirma, a dry-cured meat product. *J. Food Saf.* 28:47–58.

Aktaş, N. and Gürses, A. 2005. Moisture adsorption properties and adsorption isosteric heat of dehydrated slices of Pastirma (Turkish dry meat product). *Meat Sci.* 71:571–576.

Anıl, A. 1986. Türk pastırması: Modern yapım tekniğinin geliştirilmesi vakumla paketlenerek saklanması. Türkiye Bilimsel Araştırmalar Kurumu, Veterinerlik ve hayvancılık araştırma Grubu, Proje No: VHAG: 574 (in Turkish).

Anıl, A. 1988. Türk pastırması: Modern yapım tekniğinin geliştirilmesi ve vakumla paketlenerek saklanması. *S.Ü. Vet. Fak. Derg.* 4:363–373 (in Turkish).

Anonymous. 1983. Pastırma. TS 1071. Institute of Turkish Standards. Ankara, Turkey (Türk Standardlari Enstitüsü) Necatibey Cad. 112, Bakanliklar (in Turkish).

Anonymous. 1991. Pastırma Yapım Kuralları. TS. 9268. Institute of Turkish Standards. Ankara, Turkey (Türk Standardlari Enstitüsü) Necatibey Cad. 112, Bakanliklar (in Turkish).

Arslan, A., Çelik, C., Gönülalan, Z., Ates, G., Kok, F., and Kaya, A. 1997b. Analysis of microbiological and chemical qualities of vacuumed and unvacuumed mirror carp (*Cyprinus carpino* L.) pastrami. *Turk. J. Vet. Anim. Sci.* 21:23–29.

Arslan, A., Gönülalan, Z., and Çelik, C.C. 1997a. Effect of storage time on mirror carp (*Cyprinus carpio* L.) pastrami stored in market temperature. *Turk. J. Vet. Anim. Sci.* 21:215–220.

Askar, A., El-Samahy, S.K., Shehata, H.A., Tawfik, M. 1993. Pasterma and beef bouillon. The effect of substituting KCl and K-lactate for sodium chloride. *Fleischwirtschaft* 73:289–292.

Berkmen, L. 1940. Türkiye'de Ette, Et Müstehzaratında ve Bilhassa Pastırmada Hastalık Amillerinin Mevcudiyetiyle, Dayanma Müddetleri Üzerinde Araştırmalar. T.C. Ziraat Vekaleti, Y. Zir. Ens. Çalış. 72 (in Turkish).

Berkmen, L. 1960. Studies on the endurance of pathogens in specific Turkish meat food product. *Fleischwirtschaft* 11:926–931.

Briskey, E.J. 1964. Etiological status and associated studies of pale, soft, exudative porcine musculature. *Adv. Food Res.* 13:89–178.

Buscailhon, S., Berdague, J.L., Gandemer, G., Touraille, C., and Monin, G. 1994. Effects of initial pH on compositional changes and sensory traits of French dry-cured hams. *J. Muscle Foods* 5:257–270.

Ceylan, S. 2009. Free amino acid composition of some pastirma types (sirt, bohça and şekerpare). MSc thesis, Department of Food Engineering, Natural Science Institute, Erzurum, Turkey, 60 pp.

Comaposada, J., Gou, P., and Arnau, J. 2000. The effect of sodium chloride content and temperature on pork meat isotherms. *Meat Sci.* 55:291–295.

Doğruer, Y. 1992. The investigation on the effect of varying salting periods and pressing weights on the quality of Pastirma. DPhil dissertation, Health Science Institute, Selçuk University, Konya, Turkey, 102 pp.

Doğruer, Y. 2001. Usage possibility of turkey and chicken meat in manufacture of traditional pastirma. *J. Vet. Sci.* (Veteriner Bilim Dergisi) 17:37–42.

Gök, V., Obuz, E., and Akkaya, L. 2008. Effects of packaging method and storage time on the chemical, microbiological, and sensory properties of Turkish pastirma—A dry cured beef product. *Meat Sci.* 80:335–344.

Gök, V. and Uzun, T. 2010. Effect of using different packaging techniques on the some quality characteristics of chicken pastirma. The First International Symposium on Traditional Foods from Adriatic to Caucasus, April 15–17, Tekirdağ, Turkey, pp. 235–237.

Gökalp, H.Y., Kaya, M., and Zorba, O. 1999. *Technology of Pastirma and Some Other Dried Products: Engineering of Meat Products Processing*, 3rd edition, 459 pp. Erzurum, Turkey: Atatürk Univ. Publ. No. 786.

Goma, M., Zein, G.N., Dessouki, T.B., and Bekr, A.A. 1978. Effect of pepsin treatment on some chemical indices of pastırma processed from camel meat. *Monaufeia J. of Agri. Res.* 1:125–153.

Güner, A., Gönülalan, Z., and Doğruer, Y. 2008. Effect of tumbling and multi-needle injection of curing agents on quality characteristics of pastirma. *Int. J. Food Sci. Technol.* 43:123–129.

Gürbüz, Ü. 1994. Application of various salting techniques on pastirma production and its effects on the quality. DPhil dissertation, Health Science Institute, Selçuk University, Konya, Turkey, 87 pp.

Honikel, K.O. 2008. The use and control of nitrate and nitrite for the processing of meat products. *Meat Sci.* 78:68–76.

Ibrahim, H.M.A. 2001. Acceleration of curing period of pastrami manufactured from buffalo meat: Chemical and microbiological properties. *Nahrung* 45:293–297.

Işıklı, N.D. and Karababa, E. 2005. Rheological characterization of fenugreek paste (çemen). *J. Food Eng.* 69:185–190.

Kaban, G. 2009. Changes in the composition of volatile compounds and in microbiological and physicochemical parameters during pastirma processing. *Meat Sci.* 82:17–23.

Katsaras, K., Lautenschläger, R., and Bosckova, K. 1996. Das Verhalten von Mikroflora und Starterkulturen während der Pökelung, Trocknung, und Lagerung von Pasterma. *Fleischwirtsch* 76:308–314.

Kaya, M., Aksu, M.I., and Gökalp, H.Y. 1996. Dry curing-raw meat products. In: Symposium of Meat and Meat Products, Istanbul Univ. Vet. Fac., Istanbul, Turkey, pp. 26–34.

Kılıç, B. 2009. Current trends in traditional Turkish meat products and cuisine. *LWT—Food Sci. Technol.* 42:1581–1589.

Leistner, L. 1988. Hürden–Technologie bei Fleischerzeugnissen und anderen Lebensmitteln. Lebensmittelqualität Wissenchaft und Technik, R. Stufe(Hrsg), Wissenschaftliche Arbeitstagung 25 Jahre Institut für Forschung und Entwicklung der Maizena Ges. MbH's, 323–340 in Heilbornn, 2. bis 4 März.

Lewis, Jr., P.K., Rakes, L.Y., Brown, C.J., and Noland, P.R. 1989. Effect of exercise and pre-slaughter stress on pork muscle characteristics. *Meat Sci.* 26:121–129.

Lorenzo, J.M., Prieto, B., Carballo, J., and Franco, I. 2003. Compositional and degradative changes during the manufacture of dry-cured 'lacon.' *J. Sci. Food Agric.* 83:593–601.

Lücke, F.K. 1998. *Microbiology of Fermented Foods, Fermented Sausages*, pp. 441–483. London, UK: Blackie Academic and Professional.

Marra, A.I., Salgado, A., Prieto, B., and Carballo, J. 1999. Biochemical characteristics of dry-cured lacón. *Food Chem.* 67:33–37.

Özdemir, H., Sireli, U.T., Sarimehmetoglu, B., and Inat, G. 1999. Investigation of the microbial flora of pastirma marketing in Ankara. *Turk. J. Vet. Anim. Sci.* 23:57–62.

Özeren, T. 1980. Pastırmanın olgunlaşması sırasında mikroflora ve bazı kimyasal niteliklerinde meydana gelen değişiklikler üzerine incelemeler. Uzmanlık Tezi, Ankara Üniversitesi, Veterinerlik Fakültesi, Ankara (in Turkish).

Öztan, Z. 2005. Et Bilimi ve Teknolojisi, p. 495. Ankara, Turkey: Gıda Mühendisliği Odası Yayınları (in Turkish).

Pearson, A.M. and Tauber, F.W. 1984. *Processed Meats*, 2nd edition. Westport, CT: The AVI Publishing Co., INC.

Pearson, M. and Gillett, T.A. 1997. *Processed Meats*, 3rd edition, pp. 53–77. New Delhi: CBS Publishers and Distributors.

Shahidi, F., Yun, J., Rubin, L.J., and Wood, D.F. 1987. The hexanal content as an indicator of oxidative stability and flavor acceptability in cooked ground pork. *Can. Inst. Food Sci. Technol. J.* 20:104–106.

Soyutemiz, G.E. and Özenir, A. 1996. Determination of residual nitrate and nitrite contents of dry fermented sausage, salami, sausage and pastirma consumed in Bursa. *Gıda* 21:471–476.

Tekinşen, O.C. and Doğruer, Y. 2000. *Pastirma from Every Aspects*, 124 pp. Konya, Turkey: Selcuk University Press.

Toldra, D.F. 2002. *Dry-Cured Meat Products*. Trumbull, CT: Food & Nutrition Press, Inc.

Virgili, R., Saccani, G., Gabba, L., Tanzi, E., and Soresi Bordini, C. 2007. Changes of free amino acids and biogenic amines during extended ageing of Italian dry-cured ham. *LWT—Food Sci. Technol.* 40:871–878.

Yagli, (Gur) H. and Ertas, A.H. 1998. Effect of sodium ascorbate on some quality characteristics of Turkish pastirma. *Turk. J. Agric. For.* 22:515–520.

Yetim, H., Sagdic, O., Doğan, M., and Ockerman, H.W. 2006. Sensitivity of three pathogenic bacteria to Turkish cemen paste and its ingredients. *Meat Sci.* 74:354–358.

38

Chinese Fermented Meat Products

Ming-Ju Chen, Rung-Jen Tu, Hsiang-Yun Wu, and Chung-Miao Tien

CONTENTS

38.1 Introduction

Traditional fermented meat products have a long history and are still very popular in China. Based on recorded Chinese history, as early as 770–470 BC, the first fermented meat product was described in the *Rites of the Zhou*, one of four extant collections of ritual matters from the Zhou Dynasty. Legend has it that the processing technology used to produce Chinese dry-cured ham was introduced in Europe by Marco Polo during the thirteenth/fourteenth century and had an important impact on the development of dry-cured ham processing technology outside China.

In the early period, the main areas producing fermented meats were located in eastern and southwestern China. The unique flavor of Chinese fermented meat products was not only due to the elaborate processing methods but also connected to the four distinct seasons and mountainous geographic terrain of these areas, which are particularly suitable for producing fermented meat products (Zhou and Zhao 2007). Generally, the pigs were slaughtered annually during the winter season. Next, the meat was trimmed, salted, and piled up. After removing the liquid, the meat was hung in rows on sunny days for maturation until autumn. The protein and fat in the muscle tissue are hydrolyzed by internal enzymes and microorganisms during the fermentation process. Various compounds, such as peptides, free amino acids, free fatty acids, and volatile compounds, are generated, which contribute to the unique flavor of these Chinese fermented meat products.

TABLE 38.1

Types of Chinese Fermented Meat Products and Areas of Production

Types	Area	Production
Dry-Cured Ham		
Jinhua ham	Six districts in the Quzhou region and nine districts in the Jinhua region	_Pig breed_—Jinhua black two-ends pigs _Green ham preparation_—weighted 5.5–7.5 kg from hind legs _Salting_—9%–10% of leg weight at 0°C–10°C for 25–35 days depending on the size of hams _Soaking and washing_—24 h _Sun drying_—about 7 sunny days _Ripening_—6–8 months in a ventilated chamber (temperature increasing from 15°C to 37°C gradually and RH controlling within 60%–80%)
Xuanwei ham	Xuanwei district of Yunnan province	_Pig breed_—Wujin breed _Green ham preparation_—weighted 5–7 kg from hind legs _Salting_—salted at least three times about 1.5%–2% of leg weight at 3°C–10°C and 65%–85% RH for 25–35 days depending on the size of hams
Rugao ham	Rugao district of Jiangsu province	_Pig breed_—Jiangquhai pig _Preparation of green ham_—weighted 4–7 kg from hind legs _Salting_—50 kg of green ham (a total of 8.75 kg of salt with 15 g of sodium nitrite per 50 kg of green ham) at the temperature 0°C–8°C and RH 60%–80% for around 40 days _Soaking and washing_—24 h _Sun drying_—5–7 sunny days _Ripening_—6–8 months in a ventilated chamber (temperature increasing from 15°C to 37°C gradually and RH controlling within 60%–80%) _Retrimmed and oiled_
Semi-Dried Sausage		
Chinese Cantonese sausage		_Grinding/mixing_—6–8 mm of taper hole disc for pork and 3–4 mm for back fat _Casings and stuffing_—stuffing into nature casing with diameter 37 mm _Drying and fermentation_—sun dried for 10–14 days or oven dried at 50°C for 24 h, followed by a reduction of temperature to 45°C for 48 h

Source: Du, M. and Ahn, D.U., _J. Food Sci._ 66, 827–831, 2001; Guo, J. et al., _J. Anim. Sci._ 88(E), 275, 2010; He, Z.-F. et al., _Food Sci. (China)_ 29, 190–196, 2008; Jiang, D.F. et al., _J. Yunnan Univ._ 12, 71–75, 1990; Lu, L. et al., _Sci. Technol. Food Ind. (China)_ 29, 202–203, 2008; Pan, L.H. et al., _J. Anhui Agric. Sci._ 35, 7610–7611, 2007; Wei, F.H. et al., _Meat Sci._ 81, 451–455, 2009; Zhou, G.H. and Zhao, G.M., _Meat_ Sci. 77, 114–120, 2007.

There are numerous traditional fermented meat products in China. The most famous meat products are the dry-cured hams produced over a long ripening period, during which the lipids and proteins are degraded by the natural enzymes and microorganisms in meat. Table 38.1 presents the various types of Chinese fermented meat products and the areas where they are produced. Below, the processing technology of these traditional Chinese fermented meat products and the latest studies on microorganisms and flavor in these products are outlined.

38.2 Chinese Fermented/Dry-Cured Ham

Jinhua, Xuanwei, and Rugao hams are the three most famous hams produced in China. The names of hams correspond to the districts where the products are manufactured. Typical Chinese dry-cured ham

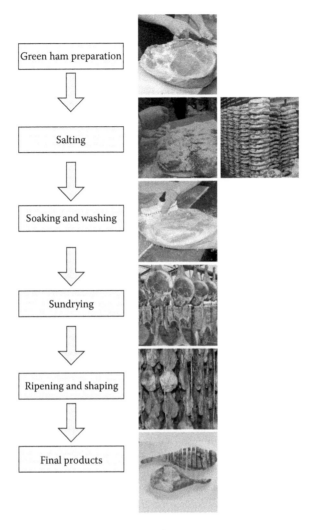

FIGURE 38.1 Flowchart of Chinese fermented ham production.

processing generally takes 8–10 months and involves green ham preparation, salting, soaking, washing, sun drying, and ripening (Figure 38.1).

38.2.1 Jinhua Ham

Jinhua ham is produced in Jinhua District, Zhejiang Province, in People's Republic of China (PRC). Jinhua ham has a long production history. According to the legend, the manufacturing of Jinhua ham can be traced back to the Tang Dynasty (AD 618–907), but the name "Jinhua ham" was formally given by the first emperor of the South Song Dynasty, Zhao Gou, approximately 800 years ago (Wu et al. 1959). Among traditional meat products, Jinhua ham is one of the most famous products used in Chinese cuisine. This ham has an attractive color (pink muscle, slightly golden brown skin with white fat), a unique flavor, and a typical bamboo-leaf shape. Since 2002, Jinhua ham production area has been recognized by General Administration of Quality Supervision Inspection and Quarantine (AQSIQ) of the PRC and is divided into two regions: Jinhua and Quzhou (Wang 2009).

38.2.1.1 Manufacturing Processes

Pig breed. In general, Jinhua black two-ends pigs bred in Lanzi (Zhejiang Province) are used for processing into Jinhua ham. The Jinhua pigs, also named Liangtouwu, are one of a number of fine breeds of pig found in the Jinhua region, and these pigs are famous for their thin skin and bone, tasty flavor, and high propagation rate. Like other slow-growing breeds, the meat of Jinhua pigs has a high fat content. Guo et al. (2010) indicated that the Jinhua breed has a higher intramuscular fat content ($p < 0.01$) and a lower drip loss when compared with Landrace pigs ($p < 0.05$). This characteristic gives the ham a softer texture and more flavor than ham from modern breeds of pig.

Green ham preparation. Jinhua pigs are slaughtered at 5–6 months old and 90–100 kg. The legs are removed from the body (the weight being 5.5–7.5 kg for each hind leg from a Jinhua pig after trimming). The trimmed green legs (the shape should be like a bamboo leaf) are then held for 12–24 h at 0°C–10°C after slaughter.

Salting. Dry-curing with salt is a key operation and depends heavily on weather. The hams are thoroughly rubbed with salt (9%–10% of leg weight) at a temperature 0°C–10°C and a relative humidity (RH) of 75%–90%. Traditionally, salting is conducted during winter season because the temperature and humidity conditions in Jinhua are perfect for salting hams. Several studies (Zhao and Zhou 2003; Zhao et al. 2004; Zhou and Zhao 2007) have reported that the most desirable temperature and humidity for salting range from 5°C to 10°C and 75% to 85% RH, respectively. If the humidity is lower than 70% RH, the ham will become dry, whereas, if it is higher than 90% RH, microorganisms may propagate and spoil the ham. The salting temperature also significantly affects the quality of hams. When the ham is salted at a temperature above 15°C, the proliferating microorganisms may ruin the ham (Zhou and Zhao 2007). However, the salt content was found to have little effect on the lipolysis during ripening (Lu et al. 2008).

The duration of the salting ranges from 25 to 35 days depending on the size of the ham, and during this time, salt is added five to seven times. The first salting requires the green hams to be covered with salt to remove the remaining blood and superabundant water from the muscle. The second salting should be conducted no later than 24 h after the first one.

Soaking, washing, and sun drying. After salting, the excess salt and waste substances on the surface of the hams are removed by soaking the ham in water for 4–8 h and by washing with brushes. After changing the initial water, the hams are soaked in water for another 16–18 h. Two hams with similar weights are then tied together and hung on a rack for sun-drying to prevent spoilage. The sun-dried process is completed when the hams start to drip liquefied fat, which usually needs about seven sunny days.

Ripening. This stage is the critical process for generating flavor compounds in the hams. The hams are hung on the strings to ripen for 6–8 months in a ventilated chamber (temperature increasing from 15°C to 30°C gradually and RH controlling within the range 60%–80%). During the ripening process, the meat proteins and fat are hydrolyzed by endogenous enzymes and microorganisms, which results in increased peptides, free fatty acids, and amino acids within the ham. These organic compounds are the major substances that create the unique flavor of the ham. In addition, the skin and muscle of ham shrink due to moisture loss, and the bones at the joints begin to stick out from the surface of the ham. Each pair of hams is then removed from the strings and retrimmed. Traditionally, the retrimming process is performed in mid-April. After reshaping the hams, they are hung once again on string until the ripening process is completed (generally in the mid-August). The salt level, moisture content, and pH of the final products are around 12.45%, 43.75%, and 6.8, respectively (He et al. 2008).

38.2.1.2 Microorganisms

The microorganisms in Jinhua ham, mainly coagulase-negative cocci, lactic acid bacteria, yeast, and mold, play a very important role in developing the ham's unique flavor during ripening. The microbial counts are low both on the surface and in the center area of the ham samples at the raw meat stage. After salting, no or very few microorganisms are found due to the high concentration of sodium chloride (He et al. 2008). Zhen (2003) studied the surface microorganisms of Jinhua ham during ripening. The microorganisms initially grew fast during the first 3-month ripening period. Mold increased from an initial value of 10^3 to reach 10^7 colony forming units (cfu)/g. The numbers of coagulase-negative cocci and lactic acid bacteria also increased from 10^4 and reached 10^7 cfu/g. However, after the 3-month ripening, the microbial counts started to drop and reached a mean value lower than 10^1 cfu/g at the end of ripening. Similar findings were found for the center of Jinhua ham. Thus, during the overall process, microorganism numbers quickly increase during the first 3 months of ripening, but then decrease rapidly so that the microorganism counts at the center of Jinhua ham are very low ($<10^1$ cfu/g) at the end of ripening. He et al. (2008) reported that the presence of mold on the surface of the ham affects the microflora in the center of the ham, and that an increase in the surface mold resulted in a decrease in the internal bacterial number during the ripening period.

Many different microorganisms have been isolated and identified at different stages during the processing of Jinhua hams (He et al. 2008; Zhen 2003). Five dominant bacterial strains, namely, *Staphylococcus xylosus*, *S. lugunensis*, *Pediococcus pentosaceus*, *P. urinaeequi*, and *Lactobacillus alimentarius*, have been observed in the Jinhua ham samples during the ripening stage. In terms of yeasts, *Candida zeylanoides*, *Debaryomyces hansenii*, *Hansenula sydowiorum*, and *Rhodotorula glautinis* are most abundant in the Jinhua ham samples. Among molds, members of the genus *Penicillium* (*Penicillium italicum*, *Pen. simplicissimum*, *Pen. aurantiogriseum*, *Pen. solitum*, *Pen. implicatum*, *Pen. viridicatum*, *Pen. fellutanum*, and *Pen. citrinum*) and members of the genus *Aspergillus* (*Aspergillus sydowi*, *A. glaucus*, *A. versicolor*, *A. sydowii*, *A. candidus*, and *A. flauipes*) are predominant at the beginning of ripening (He et al. 2008). Other molds such as *Eurotium rubrum* and *E. amstelodami* have also been isolated and identified in the Jinhua ham (Chen and Meng 2009).

38.2.1.3 Flavor

The flavor of Jinhua ham is mostly generated by its fatty acid composition and the presence of high levels of free amino acids and nucleotides (Zhao et al. 2008). The high proportion of the free compounds created by the long ripening process brings about the distinctive taste of the ham. Du and Ahn (2001) reported that the fatty acid content of Jinhua ham is about 61%–62% unsaturated fatty acids and 36%–37% saturated fatty acids. The major fatty acids present were octadecenoic acid, hexadecenoic acid, and octadecanoic acid (Table 38.2). The flavor compounds are shown in Table 38.3 and can be grouped as follows: alkanes and alkenes, aromatic hydrocarbons, alcohols, aldehydes, ketones, carboxylic acids, esters, oxygenous heterocyclic compounds, nitrogenous compounds, sulfur compounds, chloride compounds, amides, and terpenes (Huan et al. 2005).

TABLE 38.2

Intramuscular Fatty Acid Composition of Biceps Femoris Muscle from Chinese Fermented Meat Products

Fatty Acid	Chinese Fermented Meat Products (%)			
	Jinhua Ham	Xuanwei Ham	Rugao Ham	Cantonese Sausage
Saturated	36.86	39.82	35.74	35.73
C14:0	1.22	4.97	0.66	1.21
C16:0	23.25	23.70	20.96	22.86
C18:0	12.39	11.04	14.35	11.66
C20:0	–	0.14	–	–
Monounsaturated	46.41	29.70	29.22	47.06
C16:1	2.34	2.49	1.35	2.51
C18:1	44.07	27.41	27.86	44.55
C22:1	–	0.80	–	–
Polyunsaturated	14.24	28.81	30.20	14.80
C18:2	8.98	20.19	23.7	9.12
C18:3	3.90	1.20	1.02	4.23
C20:4	1.36	6.07	5.37	1.45
C22:4	–	1.35	0.63	–

Source: Du, M. and Ahn, D.U., *J. Food Sci.* 66, 827–831, 2001; Yang, H.G. et al., *Meat Sci.* 71, 670–675, 2005, doi:10.1016/j.meatsci.2005.05.019; Zhang, X.L., Research on lipoxygenase and flavor in Rugao ham, Master paper, Nanjing Agriculture University, Weigang, China, 2008.

TABLE 38.3

Volatile Compounds in Chinese Fermented Meat Products

		Chinese Fermented Meat Products			
Compounds	**Description**	**Jinhua Ham**	**Xuanwei Ham**	**Rugao Ham**	**Cantonese Sausage**
Aldehydes (%)		45.07			17.43
3-Methyl-butanal	Sour cheese	v	v	v	v
2-Methyl-butanal	Butter, caramel	v	v		v
Pentanal	Green leaves	v			v
Hexanal	Green	v	v	v	v
Heptanal	Potato, green leaves	v	v	v	v
Octanal	Orange, spicy	v	v	v	
Nonanal	Herb, citrus	v	v	v	v
Decanal	Cucumber, dry grass		v	v	
Tetradecanal			v		
Octadecanal			v		
2-Hexenal			v	v	
2-Octenal			v	v	
2,4-Nonadienal	Fatty		v		
2,4-Decadienal	Deep fried		v		
Benzene-acetaldehyde		v	v	v	
2-Methyl-propanal		v		v	v
4-Methyl-3-pentenal			v		
2-Nonenal		v	v		
2-Decenal			v		
2-Dodecenal			v		
Undecanal			v		
2-Butyl-2-octenal			v		
2-Tridecenal			v		
13-Tetradecenal			v		
17-Octadecenal			v		
2-Pentenal			v		
2-Heptenal			v		v
2,4-Heptadienal			v		
16-Octadecenal			v		
Alkanes and Alkenes (%)		13.89			1.07
Butane		v		v	
Heptane		v	v		v
Hexane		v			v
Tetradecane		v	v		
Pentadecane		v	v		
Hexadecane		v	v		v
Butylatedhydroxytoluene			v		
Octane		v	v		v
Nonane			v		
Undecane		v	v		
Naphthalene			v		
Butyl-cyclopentane			v		
Pentane			v		
Cyclododecane			v		
1,1-Diethoxy-ethane			v		v

(continued)

TABLE 38.3 (Continued)

Volatile Compounds in Chinese Fermented Meat Products

Compounds	Description	Chinese Fermented Meat Products			
		Jinhua Ham	Xuanwei Ham	Rugao Ham	Cantonese Sausage
Hexadecyl-oxirane			v		
1-Octene		v	v		v
2-Octene		v			
1-Pentene		v			
1-Nonene			v		
5-Decene			v		
1-Tetradecene			v		
3-Hexadecene			v		
8-Heptadecene			v		
5-Octadecene			v		
3-Ethyl-2-methyl-1,3-hexadiene			v		
1-Methyl-benzene					v
Alcohols (%)		13.93			30.01
Ethanol	Wine			v	v
1-Pentanol			v	v	
1-Hexanol		v	v	v	
1-Heptanol			v		
1-Octanol			v	v	v
1-Penten-3-ol			v		v
1-Octen-3-ol	Mushroom	v	v	v	
2,2-Dichloro-ethanol			v		
3-Methyl-1-butanol			v	v	
3-Ethyl-3-heptanol			v		
1-Nonanol			v		
6-Pentadecen-1-ol			v		
3-Furanmethanol			v		
2-Ethyl-1-dodecanol					v
Ketones (%)		9.00			0.10
3-Hydroxy-2-butanone	Butter	v	v	v	
2-Heptanone	Pelargonium, nettle, fruity	v	v		
4-Heptanone					
5-Ethyldihydro-2(3*H*)-furanone			v		
2-Octanone		v			
3,5-Octadien-2-one			v		
3-Nonen-2-one			v		
2-Pentadecanone		v	v		
2-Pentanone			v	v	v
2-Nonanone	Plant, herb		v		
2-Ethyl-cyclohexanone			v		
6-Teidecanone			v		
Dihydro-5-propyl-2(3*H*)-furanone			v		
3-Ethylcyclopentanone			v		
Acids (%)		18.37			
Acetic acid	Vinegar	v		v	
Propanoic acid	Sour, sweaty	v		v	

(*continued*)

TABLE 38.3 (Continued)

Volatile Compounds in Chinese Fermented Meat Products

Compounds	Description	Jinhua Ham	Xuanwei Ham	Rugao Ham	Cantonese Sausage
		Chinese Fermented Meat Products			
2-Methyl-propanoic acid		v		v	
Butanoic acid	Cheesy, sweaty	v		v	
3-Methyl-butanoic acid	Cheesy, sweaty	v		v	
2-Methyl-butanoic acid		v			
Pentanoic acid	Sweaty	v		v	
2-Methyl-2-butanoic acid		v			
3-Hydroxy-butanoic acid		v			
Hexanoic acid	Sweaty, putrid	v	v	v	
Octanoic acid		v		v	
Nonanoic acid		v			
Decanoic acid		v			
Esters		1.21			50.12
Ethyl acetate	Artificial fruity	v		v	v
1-Propen-2-ol acetate				v	
Ethyl propionate	Fruity				v
Ethyl butanoate	Fruity, candy like				v
Ethyl lactate	Sweet				v
Ethyl-3-methylbutyrate	Fruity, candy like				v
Ethyl pentanoate	Fruit				v
Ethyl hexanoate					v
Decanoicacidethylester			v		
Tetradecanoicacidethylester			v		
Diethylphthalate			v		
11-Tetradecyn-1-olacetate			v		
Dibutylphthalate			v		v
Ethyloleate			v		
Ethyl nonanoate					v
Ethyl 9-decenoate					v
Ethyl decanoate					v
Ethyl dodecanoate					v
Ethyl tetradecanoate					v
Ethyl hexadecanoate					v
Phenols					
4-Ethyl-phenol			v	v	
Other Compounds					
2-Methyl-furan		v			
2-Pentyl-furan			v		
2-Hexyl-furan			v		
Anthracene			v		

Source: Du, M. and Ahn, D.U., *J. Food Sci.* 66, 827–831, 2001; Sun, W.Z. et al., *Food Chem.* 121, 319–325, 2010; Yao, P. et al., *Food Sci. (China)* 25, 146–150, 2004; Tian, H.S., Research on flavor components of Jinhwa ham and preparation of its flavor base, PhD paper, Jiangnan University, Wuxi, China, 2005; Tjener, K. and Stahnke, L.H., Flavor, In: *Handbook of Fermented Meat and Poultry*, edited by F. Toldrá, Chapter 22, pp. 227–239, Oxford, UK: Blackwell Publishing, 2007; Zhang, X.L., Research on lipoxygenase and flavor in Rugao ham, Master paper, Nanjing Agriculture University, Weigang, China, 2008.

During the ripening process, the aldehydes, acids, lactones, pyrazines, and sulfur compounds were found to increase, whereas the various alcohols and ketone decreased (Zhou and Zhao 2007). The major volatile compounds in Jinhua ham are aldehydes (48%), followed by carboxylic acids (23%) and ketones (9%), which are derived from the breakdown and rearrangement of amino acids and fatty acids during auto-oxidation and fermentation (Tian 2005).

The presence of molds during the fermentation contributes significantly to the flavor of the ham. The fermentation of the Jinhua ham samples is characterized by a rapid increase in the number of molds on the surface. Molds on the surface of Jinhua hams may be linked to increases in the levels of glutamic acid and aspartic acid, which give the umami flavor (Chen and Meng 2009). Members of the genus *Penicillium*, especially *Pen. fellutanum* and *Pen. canescens*, possess high lipolytic activity. On the other hand, members of the genus *Aspergillus*, particularly *A. ochaceus*, have high neutral and alkaline protease activity (Chen and Meng 2009). These two fungal genera seem to play main roles in the hydrolyzation of protein and lipid in Jinhua ham samples. Two studies (Toldrá et al. 1997; Zhou and Zhao 2007) have found that proteolysis and lipolysis are the major biochemical reactions associated with flavor generation.

38.2.2 Xuanwei Ham

Chinese Xuanwei ham has been produced in Xuanwei district of Yunnan province for more than 1000 years. The total production of this ham is around 10,000 tons per year and is still growing. Xuanwei ham is characterized by a flavor that has strong intensity. This is probably because of the use of traditional techniques for the manufacture of this ham and to the fact that the pigs used to make the ham are raised under free-range conditions. Just like Jinhua ham, the Xuanwei ham production area has also been recognized by AQSIQ of the PRC since 2001.

38.2.2.1 Manufacturing Processes

Pig breed. Usually, pigs of the Wujin breed from the Xuanwei district are used for processing into Xuanwei ham. Wujin pigs are a typical local breed that can be raised outdoors or indoors using a feeding strategy based on the availability of natural resources. Wujin pigs, which grow slowly, make an ideal material for producing the high-quality Xuanwei ham because of the pigs' high body fat content and good muscle quality (Yang et al. 2008).

Green ham preparation. Wujin pigs are slaughtered at 12–14 months old and 90–100 kg. The trimmed green legs (oval shape) are held for 12–24 h at 5°C–10°C after slaughter.

Salting. During the processing of Xuanwei ham, the green legs need to be salted at least three times. At the first salting, the hams are thoroughly rubbed with salt (2 kg/100 kg green legs) and then placed on platforms and held for 1 week at 4°C. The green legs are then salted a second time (1.75 kg/100 kg hams) and held for 3 days at 5°C–10°C and 65%–85% RH. The third salting is conducted with 2% salt, and the legs are held for a further 30 days at 5°C–10°C and 65%–85% RH.

Ripening. The hams are hung on the strings to ripen for 8–12 months in a ventilated chamber (temperature ranging from 10°C to 20°C and 60%–80% RH). Traditionally, the ham ripening is started in March (Yang 2007). The salt level and water activity of the final products is around 9%–11% and 0.8%, respectively. The approximate makeup of the hams is 42% water, 30% protein, 11% lipid, and 7% ash (Jiang et al. 1990).

38.2.2.2 Microorganisms

The natural microflora of Chinese Xuanwei ham has been investigated in a number of studies (Li et al. 2003; Wang et al. 2006). Micrococci, staphylococci, yeast, and mold play key roles in developing the unique flavor of Xuanwei ham. Members of the genera *Penicillium* and *Aspergillus* were found to be the most dominant molds present in the ham samples. Eight species of *Aspergillus* were found. *A. fumigates*, which accounts for one-third of the *Aspergillus* species present, was the most abundant member of this genus in the ham samples and was followed by *A. flavus* and *A. versicolor*. Four *Penicillium* species (*Pen.*

lanate, Pen. divricate, Pen. velutin, and *Pen. monoverticillate*) were identified (Wang et al. 2006). The *Penicillium* and *Aspergillus* counts varied with the humidity and temperature.

Three main yeast genera were isolated and identified, namely, *Saccharomyces, Schizosaccharomyces,* and *Hansenula* (Wang et al. 2006). These yeasts increased quickly starting from the beginning of the salting process and reached a peak after a 3-month ripening. The yeasts present in the ham then started to decrease and become stable (10^7 cfu/g). Wang et al. (2006) concluded that there was a close relationship between yeast and ham quality because more than 50% of the total microorganisms present on matured Xuanwei ham are yeasts. In contrast, Li et al. (2003) suggested that the presence of salt-tolerant micrococci, staphylococci, and mold is highly correlated with the unique flavor of Xuanwei ham. The molds on the surface of Xuanwei hams are able to produce lipases and protease that break down muscle protein and lipids and thus generate flavor compounds. In addition, it is thought that the molds are able to form an anaerobic layer on the surface of ham that prevents lipid oxidation and inhibits the growth of certain spoilage microorganisms (Zhang 2005). In addition to molds, both micrococci and staphylococci also seem to play an important role in creating the unique flavor of Xuanwei ham because they are highly capable of hydrolyzing protein and lipid (Cai et al. 1999), which limits lipid oxidation and prevents rancidity (Talon et al. 1999). Unlike fermented sausage products, lactic acid bacteria are not the dominant microorganisms in Xuanwei ham, and this may be a result of the high salt content.

38.2.2.3 Flavor Compounds

The characteristics of the pork meat, the processing conditions (salting and ripening conditions), and the curing ingredients determine the flavor of Xuanwei ham. The fatty acid composition and the ratio of various proteases and lipases in muscle are important factors that affect the quality of this dry-cured ham (Tjener and Stahnke 2007). The fatty acid composition of biceps femoris muscle is listed in Table 38.2. Yang et al. (2005) analyzed lipolysis of the intramuscular lipids during the processing of Chinese Xuanwei ham. The results showed that triglycerides slightly increased during the processing; they changed from an initial 73.2% of the total lipid content in fresh ham to 80.2% in the final product. Furthermore, phospholipids were found to significantly decrease from an initial 25.3% to 5.8%, whereas free fatty acids significantly increased from an initial 2.3% to 11.2%. The rapid lipolysis of the phospholipids produced palmitic, linoleic, and arachidonic acids. The above findings suggested that phospholipids might be the major substrate during the lipolysis of the intramuscular lipids of Chinese Xuanwei ham.

In addition to lipolysis, other volatile compounds in Xuanwei ham have also been investigated using simultaneous distillation–extraction and gas chromatography/mass spectrometry (Yao et al. 2004). A total of 84 compounds were identified to be made up of 27 aldehydes, 22 hydrocarbons, 12 alcohols, 12 ketones, 6 esters, 1 acid, 1 phenol, and 3 other compounds. Four very potent odorants, namely, hexanal, 3-methyl-butanal, 1-octen-3-ol, and 2-methy-3-furanthiol, have been found in Iberian ham (Carrapiso et al. 2002; Carrapiso and Garcia 2004; Tjener and Stahnke 2007), and these were also identified in Xuanwei ham (Yao et al. 2004). Seven new aldehydes, namely, 2-tridecenal, 13-tetradecenal, 17-octadecenal, 2-pentenal, 2-heptenal, 2,4-heptadienal, and 16-octadecenal, were also recognized. These compounds all have very low odor thresholds and may contribute to the unique aroma of Chinese Xuanwei ham.

38.2.3 Rugao Ham

Chinese Rugao ham is produced in the Rugao district of Jiangsu province. Rugao ham has a long production history. According to the historical record, the first ham store was opened in Rugao district in 1851 (Wang 2009). The unique flavor of Rugao ham has been popularly accepted, and in 1929, there were 31 ham stores in Rugao.

38.2.3.1 Manufacturing Processes

Pig breed. Generally, Jiangquhai pigs from Rugao are used for making Rugao ham. Jiangquhai pigs are a typical local breed with thin skin, small trotters, and a low body fat, which is ideal for producing high-quality Rugao ham. The processing of Rugao ham, which is similar to that of Jinhua ham, consists of six steps: preparation of green ham, salting, soaking and washing, sun drying, and ripening (Wei et al. 2009).

The Jiangquhai pigs are slaughtered at a weight of 60–80 kg. The legs are removed from the body, and the hind legs of Jiangquhai pigs weigh 4–7 kg. The trimmed green legs (the shape should be like a lute) are thoroughly rubbed with salt (a total of 8.75 kg of salt plus 15 g of sodium nitrite per 50 kg of green ham) at a temperature of 0°C–8°C and 60%–80% RH; the legs are then stored in piles with 10–12 layers on a platform. The length of the salting process is around 40 days depending on the size of hams, during which time salt is added four times (1.5, 3.5, 3, and 0.75 kg salt/50 kg green ham in time order). The piles of ham are turned seven to eight times.

After salting, excess salt and waste substances on the surface of the hams are removed by soaking ham in water for 4–8 h and by washing/cleaning with brushes. After changing the water, the hams are soaked in water for another 16–18 h. The hams are then hung on a rack for sun drying. The sun-drying process is completed when the hams start to drip liquefied fat, which usually needs about five to seven sunny days.

After sun drying, the hams start to ferment. This stage is the critical process needed to generate the flavor compounds in the hams. The hams are hung on the strings to ripen for 5–6 months in a ventilated chamber, during which time mold covers the whole ham. During ripening, the skin and muscle of the ham shrink due to moisture loss, and the bones around the joints stick out of the surface of the ham. The hams are next retrimmed and oiled (flour/edible oil = 3:2). The hams are then rehung on strings to complete the ripening (generally in August). The salt level, moisture content, and pH of the biceps femoris of final products are about 11.60%, 40.21%, and 5.7, respectively (Pan et al. 2007).

38.2.3.2 Microorganisms

The natural microflora of Chinese Rugao ham has been investigated (Jiang et al. 2004; Pan 2007), and cocci, bacilli, yeasts, and molds are the main microorganisms found in Rugao ham. During the early ripening stage, bacteria (cocci and bacilli) and yeasts (2.9×10^2 cfu/g) are the predominant microorganisms found in the center of the ham with a ratio of 1.7:1 (Pan 2007), whereas only cocci (2.3×10^3 cfu/g) and yeasts (2×10^1 cfu/g) are observed during the later ripening stage (Jiang et al. 2004). Molds are only presently growing on the surface of the ham and eventually cover the surface completely. Many different microorganisms have also been isolated and identified from Chinese Rugao ham (Pan 2007). Nonetheless, there are three dominant staphylococci strains, namely, *S. epidermidis*, *S. auricularis*, and *S. xylosus*, together with two *Bacillus* species and one yeast species (*D. hansenula*).

38.2.3.3 Flavor Compounds

In Chinese Rugao ham, a total of 152 compounds were identified at the various different ripening stages, including 12 hydrocarbons (alkanes, alkenes, and alkynes), 14 aromatic and cyclic hydrocarbons, 27 aldehydes, 24 alcohols, 15 ketones, 10 carboxylic acid, 10 esters, 5 ether compounds, 4 oxygenous heterocyclic compounds, 11 nitrogenous compounds, 10 sulfur compounds, 4 chloride compounds, and 7 amides. However, only 79 compounds were found in the final product, of which 53, made up of 11 alcohols, 6 esters, 5 carboxylic acids, 5 nitrogenous compounds, 4 sulfur compounds, 3 aldehydes, 2 nitrogenous compounds, and 6 other compounds, were not identified in the raw meat (Zhang 2008). The aldehydes were the most important volatile compounds in Rugao ham, followed by alcohols and carboxylic acids.

Lipolysis and lipid oxidation of the intramuscular lipids during the processing of Chinese Rugao ham are two of the major pathways that form odorants (Pan et al. 2007). Four alcohols (1-hexanol, 1-octanol, 1-octen-3ol, and 2-octen1ol), five aldehydes [hexanal, heptanal, (Z)-2-heptenal, octanal, and nonanal], and three ketones (2-butanoue, 2-pentanone, and 1-octen-3one), which are directly derived from lipids, make up 12.03% of the total volatile compounds present in Chinese Rugao ham. Free amino acids and other proteolysis compounds are also major contributors to the flavor. Zhang (2008) reported the presence of three alcohols (2-methyl-1-propanol, 3-methyl-1-butanol, and 3-methyl-3-buten-1-ol), two aldehydes (2-methyl-propanal and 3-methyl-butanal), two ketones (6-methyl-5-hepten-2-one and

2-methyl-3-octanone), and four sulfur compounds (carbon disulfide, dimethyl disulfide, dimethyl trisulfide, and 3-methylthio-1-propanol) that are the result of proteolysis or amino acid breakdown; these make up 36.16% of the total volatile compounds present in Chinese Rugoa ham.

38.3 Chinese-Style Fermented Sausage

There is no similar product to Chinese-style fermented sausages found in any other country (Du and Ahn 2001). These fermented sausages are semi-dry sausages with wine and sugar added. The amount of wine and sugar in the sausage varies depending on the fermentation conditions and ingredients. Cantonese sausage is one of the most famous Chinese-style fermented sausages and has gained much popularity across the world because of its unique flavor and texture.

38.3.1 Manufacturing Process of Chinese-Style Fermented Sausages

Most Chinese-style fermented sausages are made using pork. It is important to select meat that is of good quality and has a low microbial load. Pork meat from the shoulder area is preferable. The color of the lean meat and back fat should be red and white, respectively. The ratio of back fat/lean meat in Chinese-style fermented sausages is 30/70. Other ingredients include salt (2%–3.5%), sugar (5%–20%), white wine (3%–4%, ethanol content 52%–58%, v/v), and sodium nitrite (0.02%). Addition of a high level of white wine into Chinese-style fermented sausages during manufacturing may help to provide the specific flavor to this product (Du and Ahn 2001). Traditionally, no lactic acid bacteria starter culture is added to Chinese-style fermented sausages. The high level of sugar may also contribute to flavor formation during the production of the Chinese-style sausage.

The processing of Chinese-style fermented sausages consists of five steps (Figure 38.2), namely, grinding, mixing, casing stuffing, fermentation, and drying. Grinding and mixing are important to prevent fat smearing and ensure particle distinction. Pork is ground using a 6- to 8-mm taper hole disc, and back fat is cut to a size of 3–4 mm. Then the meat, fat, and other ingredients are mixed using a paddle mixer or by hand. The mixing should be just sufficient to distribute the ingredients uniformly. After mixing, all the raw materials are stuffed into a casing. Natural casings are usually used for Chinese sausages.

Traditionally, the sausages are then sun-dried for 10–14 days depending on temperature and RH. However, in order to obtain better quality control, the sausages may be fermented and oven-dried at 50°C for 24 h, followed by holding at a reduction of temperature to 45°C for 48 h. Wu et al. (2010a) reported that the chemical compositions of Cantonese sausage are 15.18% ± 3.40% of moisture, 33.75% ± 7.26% of protein, 42.12% ± 6.64% of fat, 9.11% ± 3.73% of carbohydrate, and 6.13% ± 0.41% sodium chloride. The pH value was 6.22 ± 0.14.

38.3.2 Microorganisms

The natural microflora of Chinese-style fermented sausages has been investigated by a number of groups (Lin et al. 2009a,b; Wu et al. 2010b; Zheng et al. 2009). Overall, three *Pediococcus* species (*P. acidilactici*, *P. dextrinicus*, and *P. pentosaceus*), five lactic acid bacteria (*L. delbrueckii* subsp. *delbrueckii*, *L. delbrueckii* subsp. *bulgaricus*, *L. sake*, *L. alimentarius*, and *L. pentosus*), three *Staphylococcus* species (*S. xylosus*, *S. simulans*, and *S. sciuri*), and one *Micrococcus* species (*Micrococcus varians*) were isolated and identified. Wu et al. (2010b) analyzed the microbial conditions in eight different Cantonese sausages. The average total viable counts for lactic acid bacteria, staphylococci, and micrococci were 5.55 ± 1.48, 3.59 ± 0.68, and 3.96 ± 1.48 log cfu/g, respectively. The staphylococci and micrococci were the dominant microbial groups in Cantonese sausage, which was found to be different from the other sausages.

38.3.3 Flavor

The flavor of the Chinese-style fermented sausages is mostly generated by alcohol and lipid oxidation. The majority of volatile compounds are ethanol, ethanol derivatives, and aldehydes. Sun et al. (2010)

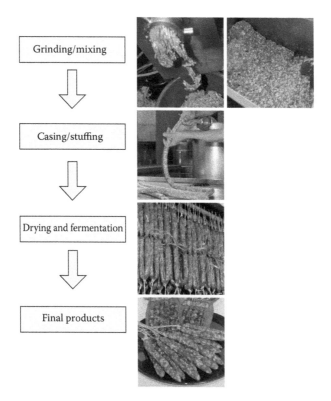

FIGURE 38.2 Flowchart of Cantonese sausage production.

identified the volatile compounds in Cantonese sausage. A total of 104 volatile compounds, mainly hydrocarbons, esters, alcohols, aldehydes, nitrogenous compounds, ketones, amines, sulfur compounds, furans, and ethers, were found. Only 22 compounds, 2 alcohols, 13 esters, 1 nitrogenous compound, 3 hydrocarbons, and 3 aldehydes were detected throughout the whole process and storage. Among the volatile compounds present in Cantonese sausage, ethanol and various ethyl esters of carboxylic acids were the most dominant compounds. The high level of ethanol and ethyl esters demonstrates that the addition of alcohol is important to Cantonese sausage flavor development (Du and Ahn 2001; Sun et al. 2010). Thus, the significant amount of wine added to the Chinese-style fermented sausages not only helps to provide its unique flavor but also aids in the prevention of spoilage during processing.

Other volatiles such as octane, heptane, pentane, heptanone, octane, and hexane are possibly formed by the degradation of unsaturated fatty acids or amino acids. Endogenous enzymes in the meat and enzymes from the microorganisms, as well as the degradation of sugar and its reaction with amino acids or lipids during fermentation, may also contribute to the formation of these volatiles (Du and Ahn 2001; Sun et al. 2010).

Du and Ahn (2001) reported that the total unsaturated fatty acid content of Cantonese sausage was about 61%–62%, whereas the total saturated fatty acid content was about 36%–37% (Table 38.2). The major fatty acids were octadecenoic acid, hexadecenoic acid, and octadecanoic acid. When the thiobarbituric acid reactive substances of Cantonese sausage are compared to those of Jinhua ham, the Cantonese sausage has a significantly higher value. A possible explanation for this is the high fat content of Cantonese sausage and the grinding steps used to produce the product.

38.4 Future

Food safety and health concepts will become a major concern affecting the future fermented meat products in China. A knowledge of and the control of each producer's typical in-house microbiota, as well as an understanding of the production processes as a whole, are critical to controlling the microbiological quality and organoleptic characteristics of such products. Such information has become essential because food safety is now a top priority. In this context, the use of starter cultures for fermented meat production is becoming increasingly necessary in order to guarantee safety and to standardize the product's properties in terms of a consistent flavor, a standard color, and a shorter ripening time. Several different studies have focused on the use of starter cultures rather than the use of natural fermentation in order to modify the processing procedures used to make Chinese fermented meat products (Wu et al. 2009). It is expected that the results of this research will be transferred to the industrial production, and this will help to optimize and standardize the processing of these products. In turn, this will help to assure the safety and quality of the Chinese fermented meat products.

REFERENCES

Cai, Y., Kumai, S., Ogawa, M., Benno, Y., and Nakase, T. 1999. Characterization and identification of *Pediococcus* species isolated from forage crops and their application for silage preparation. *Appl. Environ. Microbiol.* 65:2901–2906.

Carrapiso, A.I. and Garcia, C. 2004. Iberian ham headspace: Odorants of intermuscular fat and differences with lean. *J. Sci. Food Agric.* 84:2047–2051.

Carrapiso, A.I., Jurado, Á., Timón, M.L., and Garcia, C. 2002. Odor-active compounds of Iberian hams with different aroma characteristics. *J. Agric. Food Chem.* 50:6453–6458.

Chen, J. and Meng, Y. 2009. Study of the relationship between molds and quality of Jinhua ham during ripening. *Food Ferment. Ind. (China)* 35:138–142.

Du, M. and Ahn, D.U. 2001. Volatile substances of Chinese traditional Jinhua ham and Cantonese sausage. *J. Food Sci.* 66:827–831.

Guo, J., Shan, T.Z., Wu, T., Zhang, Y.F., and Wang, Y.Z. 2010. Comparisons of different muscles metabolic enzymes and muscle fiber types in Jinhua and Landrace pigs. *J. Anim. Sci.* 88(E):275.

He, Z.-F., Zhen, Z.-Y., Li, H.-J., Zhou, G.-H., and Zhang, J.-H. 2008. Microorganisms flora study on Jinhua ham fermentation. *Food Sci. (China)* 29:190–196.

Huan, Y.J., Zhou, G.H., Zhao, G.M., Xu, X.L., and Peng, Z. 2005. Changes in flavor compounds of dry-cured Chinese Jinhua ham during processing. *J. Meat Sci.* 71:291–299.

Jiang, D.F., Duan, R.L., Ma, P., Yang, Y.R., Chai, P., Zhang, W.P., and Qian, J.K. 1990. The studies of Xuanwei hams and functional microflora. II. Nutrient component and color–savoury-taste for Xuanwei hams. *J. Yunnan Univ.* 12:71–75.

Jiang, Y.S., Guo, B.G., Xi, J., and Dong, J. 2004. Analysis of microbial bacteria groups for Rugao ham. *Culinary Sci. J. Yangzhou Univ.* 21:10–12.

Li, P.L., Shen, Q.W., and Lu, Y.N. 2003. Analysis of main microorganisms in Xuanwei ham. *Chin. J. Microecol.* 15:262–263.

Lin, W.T., Xu, S.M., Li, J.J., Ju, B., and Wang, C.H. 2009a. Isolation, screening preliminary identification and culture of *Pediococcus* strains from the Chinese-style dry fermented sausage. *Food Sci. Technol. (China)* 34:23–27.

Lin, W.T., Xu, S.M., Li, J.J., Ju, B., and Wang, C.H. 2009b. Isolation, screening preliminary identification and culture of lactic acid bacteria strains from the Chinese-style dry fermented sausage. *Sci. Technol. Food Ind. (China)* 31:194–197.

Lu, L., Liu, X.W., and Zhang, X.Y. 2008. Study on optimization of traditional processing parameters of fermented meat. *Sci. Technol. Food Ind. (China)* 29:202–203.

Pan, L.H., Cho, K.H., Hsu, X.L., and Pon, X. 2007. Oxidation and hydrolysis analysis of esters in Rugao ham. *J. Anhui Agric. Sci.* 35:7610–7611.

Pan, M. 2007. Study on microbiology of Rugao ham. Master paper, Yangzhou University, Yangzhou, China.

Sun, W.Z., Zhao, Q.Z., Zhao, H.F., Zhao, M.M., and Yang, B. 2010. Volatile compounds of Cantonese sausage released at different stages of processing and storage. *Food Chem.* 121:319–325.

Talon, R., Walter, D., Chartier, S., Barrière, C., and Montel, M.C. 1999. Effect of nitrate and incubation conditions on the production of catalase and nitrate reductase by staphylococci. *Int. J. Food Microbiol.* 52:47–56.

Tian, H.S. 2005. Research on flavor components of Jinhwa ham and preparation of its flavor base. PhD paper, Jiangnan University, Wuxi, China.

Tjener, K. and Stahnke, L.H. 2007. Flavor. In: *Handbook of Fermented Meat and Poultry*, edited by F. Toldrá, Chapter 22, pp. 227–239. London, UK: Blackwell Publishing.

Toldrá, F., Flores, M., and Sanz, Y. 1997. Dry-cured ham flavour: Enzymatic generation and process influence. *Food Chem.* 59(4):523–530.

Wang, G.H. 2009. Application of geographical indication systems in China: Jinhua ham case study. Case study on quality products linked to geographical origin in Asia carried out for FAO.

Wang, P.P. 2009. Rugoa ham. *Meat Res.* 128:1.

Wang, X.H., Ma, P., Jiang, D.F., Peng, Q., and Yang, H.Y. 2006. The natural microflora of Xuanwei ham and the no-mouldy ham production. *J. Food Eng.* 77:103–111.

Wei, F.H., Wu, X.L., Zhou, G.H., Zhao, G.M., Li, C.B., Zhang, Y.J., Chen, L.Z., and Qi, J. 2009. Irradiated Chinese Rugao ham: Changes in volatile *N*-nitrosamine, biogenic amine and residual nitrite during ripening and post-repining. *Meat Sci.* 81:451–455.

Wu, A.F., Sun, C.Y., and Sun, G.Q. 1959. *Jinhua Ham*. Beijing, China: Light Industry Press.

Wu, Y., Cui, C., Sun, W., Yang, B., and Zhao, M. 2009. Effects of *Staphylococcus condimenti* and *Micrococcus caseolyticus* on the volatile compounds of Cantonese sausage. *J. Food Process Eng.* 32:844–854.

Wu, Y., Zhao, M., Wu, N., Sun, W., and Cao, B. 2010a. Effect of *Staphylococcus* condiment on the volatile compounds of Cantonese sausage. *Meat Sci. (China)* 51:175–183.

Wu, Y., Zhao, M., Yang, B., Sun, W., Cui, C., and Mu, L. 2010b. Microbial analysis and textural properties of Cantonese sausage. *J. Food Process Eng.* 33:2–14.

Yang, G.M., Guo, M., Zhao, Y.G., Gu, P.S., Xie, P., and Chen, J.H. 2008. The screening test on material swine of Xuanwei ham. *Acta Ecol. Anim. Domest. (China)* 31:64–67.

Yang, H.G. 2007. Production of Xuanwei ham. *Yingyoung Tuiguang (China)* (10):28–29.

Yang, H.G., Ma, C.W., Qiao, F.D., Song, Y., and Du, M. 2005. Lipolysis in intramuscular lipids at processing of traditional Xuanwei ham. *Meat Sci.* 71:670–675, doi:10.1016/j.meatsci.2005.05.019.

Yao, P., Qiao, F., Yan, H., and Ma, C. 2004. Isolation and identification of volatile compounds of Xuanwei ham. *Food Sci. (China)* 25:146–150.

Zhang, B.Y. 2005. The research of dominant microorganisms in Xuanwei ham. *Sichuan Food Ferment.* 41:38–40.

Zhang, X.L. 2008. Research on lipoxygenase and flavor in Rugao ham. Master paper, Nanjing Agriculture University, Weigang, China.

Zhao, G.M., Tian, Z., Liu, Y.X., Zhou, G.H., Xu, X.L., and Li, M.Y. 2008. Proteolysis in biceps femoris during Jinhua ham processing. *Meat Sci.* 79:39–45.

Zhao, G.M. and Zhou, G.H. 2003. The properties and traditional processing technology of 'Jinhua Huotui.' *Food Sci. Technol. (China)* 12(z1):322–326.

Zhao, G.M., Zhou, G.H., Xu, X.L., Jin, Z.M., Huan, Y.J., and Chen, M.W. 2004. Optimization of traditional processing technology of Jinhua ham and correlation analysis on main technological parameter. *Food Ferment. Indust. (China)* 30:119–123.

Zhen, Z.Y. 2003. A study of microbial flora in Jinhua ham. Doctoral paper, Southwestern University, Chongqing, China.

Zheng, X.Y., Lin, W.T., Xu, S.M., Lu, J.S., Ju, B., and Wang, C.H. 2009. Isolation, screen, identification and culture of staphylococci and micrococci strains from the Chinese-style dry fermented sausage. *China Brewing* 210:33–37.

Zhou, G.H. and Zhao, G.M. 2007. Biochemical changes during processing of traditional Jinhua ham. *Meat Sci.* 77:114–120.

39

Sucuk: Turkish Dry-Fermented Sausage

Hüseyin Bozkurt and K. Bülent Belibağlı

CONTENTS

39.1 Introduction

Sausage is one of the most important products in total meat consumption. Bratwurst, mettwurst, wienerwurst, and the many hard salamis of Italian, German, and Swiss origin are very popular in markets. Sausages termed as "fermented sausage" are categorized into two groups: dry and semi-dry sausages. Ripening periods of dry sausages (Italian salami or Italian-style salami "Milano," "Sicilian," and "Genoa," German salami, pepperoni, hard salami, and Turkish-style "Sucuk") vary between 10 and 100 days, and these types of products are not smoked. Dry sausages are ready to be sold and consumed because they contain about 35% of water.

Sucuk, a term used for Turkish fermented dry meat product, is a very popular meat product in Turkey and in countries located in Balkans, Middle East, and Caucasus. Products of similar type are also known in most Middle Eastern and European countries such as Bulgaria, Iraq, Syria, Lebanon, Armanian, and Israel (Kılıç 2009).

Sucuk consists of minced fresh meat (beef, lamb, and mutton, or mixture of them) and sheep tail fat and ingredients (nitrite and/or nitrate, potassium sorbate, and ascorbic acid), with various spices and seasonings such as cumin, allspice, cinnamon, garlic, salt, and black and red pepper. A typical formula for sucuk is as follows: meat (90 kg meat with 18% fat) is mixed with tail fat (10 kg), table salt (2 kg), sugar (sucrose 0.4 kg), clean-dry garlic (1 kg), spices (cumin, cinnamon, allspice, clove, red pepper, and black pepper), $NaNO_3$ (0.033 kg), $NaNO_2$ (0.005 kg), and vegetable oil; this mixture is known as sucuk dough. Some products can contain flavor enhancers, polyphosphates and other emulsifiers, vitamin C (or ascorbic acid), coloring matters, and various protein supplements (derived from soy milk and, possibly, from cereals and peanuts; Kolsarıcı et al. 1986; Bozkurt 2002; Kılıç 2009). After filling into natural or artificial casings, sucuk is hung up to ferment, at 22°C–23°C, by either microorganisms naturally present or added starter cultures and allowed to dry for several weeks at ambient temperature and humidity (Ensoy et al. 2010). Drying can be done at rooms with controlled temperature and humidity and stored at refrigeration temperatures later (Erkmen 1997). The process is usually considered to be finished when moisture content has decreased to 40%. The average composition of sucuks collected from retail markets in Turkey is given in Table 39.1 (Kolsarıcı et al. 1986; Bozkurt 2002; Kılıç 2009).

TABLE 39.1

Composition of Sucuk

Components	Composition		
	Minimum	**Maximum**	**Mean**
Water (%)	36.84	49.22	42.7
Fat (%)	25.13	40.15	33.23
Protein (%), $N \times 6.25$	10.85	21.13	17.31
Ash (%)	2.73	4.66	3.79
Salt (%)	2.18	3.30	2.65

Small-scale producers do not pay much attention to controlling temperature, humidity, and ripening during sucuk manufacture. However, the manufacturing technology of sucuk has been modified, and during the last three decades, heat treatment has become a part of production, although thermal processing is not a part of sucuk processing.

The rate of drying must be carefully controlled to prevent the outside of sucuk from drying out too rapidly, creating a case-hardened surface that would inhibit the drying of the center and, hence, cause spoilage. Humidity control is, therefore, important throughout the drying stages. The percentage of salt also increases proportionally as sucuk dries.

Sucuk should be eaten cooked. It is often cut into slices and cooked without additional oil, with its own fat being sufficient to fry it. At breakfast, it is cooked in a way similar to bacon or Spam. It is also fried in a pan, often with eggs. Sucuk production in Turkey is around 19,000 tons per year (Gökalp et al. 1999; Kılıç 2009).

39.2 Raw Materials

39.2.1 Meat and Fat

In sucuk production, meats (beef, lamb mutton, or a mixture of them) must be of high quality and fresh because they directly affect the quality of sucuk. Meats should complete the rigor mortis phase; if not, sucuk would be tough and may be unacceptable (Öztan 1999). The use of very old animal meats is not preferred because of excessive connective tissue, resulting in hard product texture. If very young animal meats are used, water proportion is very high; therefore, yield decreases, thus resulting in soft product texture. Excessive fat of meat and sinew tissue must also be removed to obtain high-quality sucuks (Gökalp 1986). Beef from 3- to 7-year-old animals is recommended for high-quality sucuks.

In sucuk production, dark-firm-dry (DFD) meat must not be used. DFD meat has high water-holding capacity and creates a suitable environment for bacterial growth. Pale-soft-exudative (PSE) meat is also not preferred for sucuk manufacturing because PSE meat has low water-holding capacity, and it contains less amount of pigment (myoglobin). Thus, the rate of water removal from sucuk will be too high during the drying period (Gökalp et al. 1999).

The meat that would be used in sucuk production must have preferably low microbial load and be free from pathogens. It must be taken from a healthy, nonstressed animal (at the time of slaughter) and deboned carefully; it should not include fat, fibers, and tissues. The meat's pH must not exceed 5.9 because higher-pH meats have higher water-holding capacities that can cause some problems during ripening. The meat to be used should be chilled at a maximum temperature of 2°C or kept frozen.

The fat to be used in sucuk production is generally from sheep tail and/or dorsal cattle fat. It must have a hard textural structure; otherwise, during mincing, it displays a sticky behavior that prevents a homogeneous structure. Therefore, 1–2 days before production, the fats must be stored at temperatures of –6°C to –12°C. If it is sliced before storage, grinding will be easier.

Beef and/or lamb fat are used for producing sucuks that are rich in saturated fatty acids and cholesterol. The effect of replacing beef fat with olive oil on the chemical and sensory quality characteristics and nutritional aspects of sucuk was studied (Kayaardi and Gök 2004; Kılıç 2009). It was observed that up to 60% of beef fat can be replaced by olive oil as pre-emulsified fat with soy protein isolate in the production of sucuks; this replacement of beef fat with olive oil has a positive effect on sensory quality.

Vural et al. (2004) studied that beef fat and tail fat may be substituted with inter-esterified palm and cottonseed oils in the manufacture of sucuks. Replacement of beef fat with pre-emulsified hazelnut oil up to 50% in sucuk formulation significantly affected nutritional quality without adversely affecting the ripening quality (Yıldız-Turp and Serdaroğlu 2008).

Because of the high content of fat in sucuk, poultry meat could find a use in sucuk-like products. Blending red meat with hen's meat was studied by Sarıçoban et al. (2006). This made lower fat content sucuks while preserving red meat-type flavors. The 50:50 (beef/hen meats) mixture provided the best sensory scores.

39.2.2 Additives

Salt. Salt must be added in 2%–2.5% of total mass. The maximum allowed value is 5% (Anon. 2002). Salt takes a role in extracting and solubilizing the muscle proteins to form a "sticky" film around the meat particles, creating an emulsion-type structure. It also provides flavor and controls the growth of microflora.

Nitrite/nitrate. Sodium nitrite (or nitrate) is the most important cure additive, is responsible for the typical color and flavor, and plays a role in ripening. It also provides oxidative stability to meat by preventing lipid oxidation and prevents the growth of anaerobic microorganisms such as *Clostridium botulinum.* However, if it is added excessively, there may be a risk for carcinogenic nitrosamine formation.

Sugar. Sugars in a variety of forms—sucrose, dextrose, corn syrup, and so on—are most commonly used in sucuks. Almost all sucuk/sausage products contain sugar in one form or another.

Glycogen stores are quickly depleted during the postmortem period; therefore, fresh meat contains little fermentable sugar, and addition of sugar is necessary for fermentation. Sugar accelerates fermentation and helps with taste and color formation. It is usually added as 0.2%–1.0% by mass.

Ascorbic acid/ascorbate. The primary function of ascorbic acid is to prevent oxidation and improve color stability. Ascorbic acid and ascorbate are added at maximum values of 0.03% and 0.05%, respectively.

Spices. Red pepper, black pepper, cummin, garlic, and pimento are generally used. They provide taste and odor. Unlike most other processed meats, sucuks are seasoned products. Different spices and flavorings are added in sucuks, and their levels of use are primarily dictated by producers or consumer's demands and not by regulations. Spices are aromatic vegetable substances in whole, broken, or ground form. They may be added as natural spices or spice extracts. In the latter case, they must be labeled as "flavoring." Flavorings refer to extractives that contain flavor constituents from fruits, vegetables, herbs, roots, meat, seafood, poultry, eggs, dairy products, and other food sources. Most flavorings are oil-based extracts. Because of their high flavor intensity, they can be more accurately applied in sucuk to obtain the desired flavor intensity compared to their natural counterparts (spices).

Starter culture. Microorganisms play an important role in stability and safety improvement and developing some characteristics of sucuks such as flavor, odor, texture, and color by their metabolic activities during fermentation and drying. A starter culture produces lactic acid that decreases pH. Moreover, starter culture addition competes with spoilage and pathogenic microorganisms for substrates and nutrition.

39.2.3 Casings

Generally, casings are divided into two major groups, namely, natural and artificial casings (Honikel 1989).

Natural casings are processed from various parts of the alimentary tract of cattle, hogs, or sheep. The weight of casing is about 1% of the sucuk weight. Casings are normally packed in salt. Before use, they should be thoroughly soaked and washed in lukewarm water at about 29°C–32°C. After washing, the casings should be kept wet at all times until they are filled. Some smaller manufacturers still prefer to use them (Gökalp et al. 1999).

Artificial casings are cellulose, polyvinyl chloride (PVDC), PVDC/fibrous, and collagen casings (Oliphant 1998). The advantages of *artificial casings* over natural casings lie in their uniformity of size, the absence of risk of contamination from improper preparation, and the ability in most cases to be used without preliminary washing and soaking.

39.3 Processing

39.3.1 Traditional Processing

Traditional sucuks are generally prepared by butchers. Butchers usually prepare their own formulas or, sometimes, people ask for specially formulated sucuk from the butcher. Sucuk formulation shows

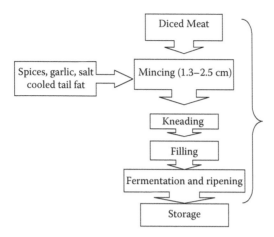

FIGURE 39.1 Production flowchart of traditional sucuk.

differences from region to region. However, general sucuk dough (Figure 39.1) is prepared from the following: lamb (generally from a 1-year-old animal) is mixed with tail fat, salt, sugar, garlic, cumin, cinnamon, allspice, clove, red and black pepper, and vegetable oil (Bozkurt and Erkmen 2002a,b). Starter culture, antimicrobials, and antioxidants are not added into traditionally produced sucuk dough. Traditional sucuks are prepared with chance inoculation.

After the mixing of ingredients, sucuk dough is stuffed into natural casings by using a simple filling apparatus connected to a meat grinder. They are then hung in the butcher shops for ripening.

Traditional sucuks are ripened and dried under climatic conditions. Small-scale producers do not pay much attention in controlling climatic conditions during sucuk manufacturing. Ripening conditions are not controlled (Bozkurt 2002). Traditional sucuks are usually prepared and ripened from September to March in Turkey. In these seasons, climatic conditions are suitable for sucuk ripening. Temperature is changed between 10°C and 15°C, and relative humidity (RH) ranges between 50% and 80%. Ripening periods and temperatures change from 6 to 20 days and from 12°C–14°C to 18°C–20°C, respectively. Color of traditional sucuk is very dark (Erkmen and Bozkurt 2004). Texture of traditional sucuk may be soft or very hard. Traditional sucuk must have a distinct flavor.

Sucuks may be sold during the ripening period, and afterwards, consumers hang the sucuks at home for drying.

39.3.2 Commercial Sucuk Processing

Commercial sucuks are produced in modern factories. Sucuk formulation again could be changed from factory to factory. A general commercial sucuk production flowchart is given in Figure 39.2. Beef/lamb is mixed with sheep tail fat, starter culture, salt, sugar, antimicrobials (nitrate, nitrite, and potassium sorbate), antioxidants (ascorbic acid), garlic, spices (cumin, cinnamon, allspice, clove, red pepper, and black pepper), and vegetable oil (Bozkurt 2002).

Refrigerated meat and frozen fat are passed through a 2.0, 2.5 cm-diameter grinder resulting in small chunks of meat. During grinding, the starter culture must be added, whereas salt, sugar, spices, and garlic would be added. This mixture is held for 20–24 hours at 0–4°C for penetration of additives to meat (Ensoy et al. 2010).

The treated mixture is ground in a 3-mm-diameter grinder; then, it is kneaded and mixed in a blender with all ingredients. Blending gives a mosaic appearance. Nowadays, a "cutter" is used for the grinding and mixing processes. Cutter can also knead the mixture. The most important factor in both mincing and blending is the temperature, and it should not exceed +4°C. The mixture is homogenized under vacuum to eliminate the O_2 before the filling.

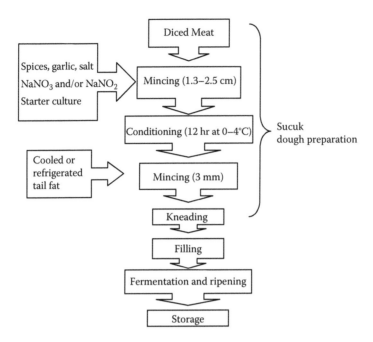

FIGURE 39.2 Production flowchart of commercial sucuk.

Previously wetted and prepared casings of the same size are then filled using a stuffing machine. Natural casings are made from small intestines, whereas artificial casings are made from plastic, cellulose, collagen, or cellophane. Before the stuffing, casings are softened by washing, so that they can expand more. Moreover, pores are opened, thus providing rapid drying of sucuk.

It is important that water must be removed from casings. Otherwise, during drying, some spots could be seen on the internal surface of casings. In addition, air space should not be formed between casings and mixture. Otherwise, color change and mold growth can occur because of the presence of O_2. This problem could be overcome by using stuffing machines with vacuum.

Stuffing temperature should not exceed +4°C. If the temperature is high, a fat layer is formed on the inside surface of the casings. After stuffing, if it is necessary, holes are made by needles in the casings to remove O_2.

Filled casings are cut into the desired length, and their edges are tied individually and then joined together (like a ring). Sucuks are hung, but not touching each other. Surfaces are washed with pressurized water. After waiting for 12–18 h at 10°C–15°C, the sucuks are then taken to the ripening room.

39.3.3 Fermentation and Ripening

Meat fermentation is a solid substrate-type fermentation with bacteria growing in microcolonies (Incze 1998; Gökalp et al. 1999). In contrast to many other fermentation substrates, meat cannot be pasteurized in sucuk production as it is done in other dry sausages.

Fermentation is the metabolic process in which carbohydrates and related compounds are oxidized with the release of energy in the absence of any external electron acceptors. The content of fermentable sugars depends on the glycogen content of the muscle. As a rule, meat with a pH above 5.9 contains too little lactate and sugar for safe fermentation; it holds water tightly and provides a better condition for growth of acid-labile bacteria. Such muscles should be sorted out and used for other purposes (Lücke 1994; Ordonez et al. 1999).

In ripening the sucuk, fermentation and drying both occur. During fermentation, carbohydrates naturally occur in meat and added afterwards to be consumed by lactic acid bacteria (LAB).

Fermentation is started naturally by lactic acid–producing bacteria. These bacteria break down carbohydrates, and therefore, quantities of L-(+)-lactic acid, L-(+)-lactates, or other organic acids increase (Öztan 1999). Desired changes at the end of ripening must occur in terms of color, consistency, and flavor.

Fermentation of sucuks with little or no catalase-positive cocci may result in production of acceptable color and a sufficient shelf life. However, these sucuks may have high residual nitrate content, high rancidity, and aroma. This type of sucuk may not be acceptable because of high lactic acid formation (Lücke 1994; Ordonez et al. 1999; Kaban and Kaya 2008).

Traditionally produced sucuks have poor color stability (Gökalp et al. 1988). Recently, commercial firms have started using starter cultures to improve the quality of sucuks. However, it has been observed that the products do not have good flavor as it has been in the traditionally produced sucuks (Özdemir et al. 1984; Gökalp et al. 1999).

pH is decreased during fermentation due to acid formation; water is also removed from the sucuk. Water removal rate depends on the ripening method. Enviromental parameters affecting ripening are RH, temperature, and air flowrate. These parameters classify the ripening methods.

Rapid method. It occurs in a room with controlled temperature, RH, and air flowrate (air-conditioned room).

1. In the first day, RH is adjusted at 95%, temperature at 22($\pm$2)°C, and air flow at 0.5–0.8 m/s
2. In the second and third days, temperature and air flowrate are kept constant; humidity is decreased to 85%
3. In the fourth day, RH is further decreased to 83% while keeping temperature and air flowrates the same
4. In the fifth and sixth days, temperature is maintained at 18($\pm$2)°C, humidity is at 80%, and air flowrate is adjusted to 0.1–0.2 m/s

Slow method. It happens in an air-conditioned room because the temperature must be controlled.

1. The temperature should never exceed 18°C
2. The humidity is decreased from 95% to 85% slowly
3. The drying happens very slowly; the ripening occurs in 15–20 days

Regular method. It is dependent on climate conditions, does not take place in air-conditioned rooms, and is especially done during the fall season.

1. Sucuk is dried by air at 12°C–21°C
2. If the air is too dry, sucuks can be moistened
3. Ripening time takes 15–30 days

39.3.4 Packaging and Storing

Sucuk is packaged mainly to prevent physical and chemical changes and microorganism-caused deterioration.

The finished product should have 40% moisture, a specialized color, cuttable texture, special taste, and odor (Anon. 2002).

If the product does not go for sale immediately, it is generally stored at 65%–70% RH, at temperatures lower than 10°C, and at an air flowrate of 0.05 m/s.

Kaya and Aksu (2005) studied the effect of modified atmosphere (50% N_2 + 50% CO_2 or 100% CO_2) and vacuum packaging on some quality characteristics of sliced "sucuk" produced using probiotic cultures (*Lactobacillus acidophilus* and *Bifidobacterium lactis*). Packaging had a significant effect on LAB, and the highest value was determined with 50% N_2 + 50% CO_2.

39.4 Types of Sucuks

39.4.1 Semi-Dry Sucuks

Semi-dry (quickly fermented) sucuks differ greatly from sucuks by their "tangy" flavor resulting from lactic acid accumulation. Semi-dry sucuks are usually sold as vacuum packed. The length of production (fermentation and drying) of these sucuks depends on ripening methods, but rarely exceeds several days. The pH of semi-dry sucuks is slightly acidic at 4.8–5.4.

Nowadays, the manufacturing technology of semi-dry sucuk has been modified, and heat treatment after the ripening period has become a part of production. Therefore, semi-dry sucuks are slightly cooked at the thermal center of 60°C for 1 h. After heat processing, the sucuks are usually air-dried for a relatively short time.

39.4.2 Dry Sucuks

Dry sucuks that are produced by slow fermentation are made from selected, mainly coarsely chopped, meat. Their water content is about 40%. Properties of dry sucuks not only depend on fermentation but are also strongly influenced by biochemical and physical changes occurring during the long drying or aging process.

The length of production and drying periods depends on some factors—physical properties of casings, sucuk formulation, methods of preparing meat, and conditions of drying—but overall processing time takes up to 90 days. The final pH of dry sausages is usually between 4.8 and 5.4.

The formulation, degree of grinding, level of fermentation, temperature of aging, and type and size of casing, as well as other factors, determine the properties of the final product.

In the preparation of dry sucuks, natural casings are preferred because they adhere closely to the sucuks as they dry.

The shelf life of dry sucuk is higher compared to that of other types, which may be especially attributed to the high salt-to-moisture ratio. The sucuks are normally stored without refrigeration.

39.5 Factors Affecting Fermentation, Ripening, and Storage of Sucuk

Parameters affecting the quality of sucuks and other dry sausages are given in Table 39.2 (adapted from Lücke 1994). They are divided into two major groups: external factors and internal factors (Lücke 1994; Oliphant 1998; Gökalp et al. 1999; Ordonez et al. 1999; Ensoy et al. 2010).

39.5.1 External Factors

External factors important for the microbial growth and chemical reactions in a food include the environmental conditions in which it is produced and stored. These are RH, temperature, and air velocity.

39.5.1.1 Temperature

Temperature greatly affects the microbial growth and chemical reactions. Chemical reaction rates (such as lipid oxidation and browning) are decreased with reduced temperature. When the foods are exposed to temperatures beyond the maximum and minimum growth temperatures of microorganisms, microbial cells die rapidly as a result of the irreversible denaturation of proteins. Temperature has an effect on the pH, texture, a_w, and water content of sucuks. Normally, sucuks are ripened at 12°C–24°C and stored at 10°C–15°C (Gökalp 1995; Gökalp et al. 1999).

39.5.1.2 Relative Humidity

Dehydration during ripening contributes to stabilizing the product by decreasing the water activity value (a_w). Water transfer from the sucuk surface to the surroundings depends on its water content and on its

TABLE 39.2

Parameters Affecting the Quality of Sausages

Parameter	Variables (Examples)	Guidelines
Raw material	Animal species (beef/poultry)	pH ≤ 5.8; good microbial quality; no antibiotics
	Age at slaughter	
	Fats/oils	Olive oil
	Types of fatty tissue (back/belly)	Tail fat
	Formulation (fat content)	
Additives	Sodium chloride	Initially a_w 0.955–0.965
	Nitrite/nitrate	Addition of 150/300 ppm
	Sugar content	0.2%–0.8%
	Sugar type (glucose, sucrose, and lactose)	0.2%–0.5% of rapidly fermentable sugar
	Lactic acid bacteria	pH reduction to ≤5.0 during fermentation
	Micrococcaceae	
	Ascorbate	
	Spices (garlic, allspice, etc.)	
Comminution	Method (grinder/cutter)	Low temperature (to avoid melting of fat)
	Degree (coarse/fine)	3 mm
Filling	Filling equipment	No air inclusions
	Casing material (natural/collagen based/cellulose based)	Permeability high for vapor and smoke, low for oxygen; shrinkable, peelable
	Casing diameter	22–25 mm diameter
Ripening	*Fermentation Climate*	
	Temperature	22°C–25°C
	Time	Until pH ≤ 5.0
	Humidity (ERH %)	85%–95% RH
	Aging/Drying Climate	
	pH	5.0–5.8
	Temperature	16–22
	Humidity (ERH %)	60%–85% RH
	Air movement	Uniform drying
	Time	Until 50% of the moisture removed
Surface treatment	Mold starter	No growth of undesired molds

Source: Lücke, F.K., *Food Res. Int.* 27, 299–307, 1994.

composition. Nevertheless, temperature and RH of the environment are important. A few reports are available on the effect of air RH on the ripening process (Ercoşkun et al. 2010).

Drying rate is important because sucuk should be sliceable without breakage and has a proper texture. After filling of sucuks, they are stored at 60% RH between 2 and 6 h at 20°C–25°C. During ripening, in the first 3 days, 20°C–25°C and 90% RH are applied and then RH is decreased to 85%. Through the end of maturity, RH must be gradually reduced to 60%. The RH of the environment changes between 70% and 75% (Gökalp et al. 1999; Ensoy et al. 2010).

39.5.1.3 Air Velocity

Air velocity affects the drying rate of sucuks. If air velocity is very high, the edges of a sucuk dry quickly and a dry layer of film cover occurs at the outside. Air circulation must be turbulent and uniform to get proper drying. In the first days of ripening, air circulation velocity must be 0.5–1 m/s and decreased to 0.1–0.2 m/s in the following days. In storage of sucuks, air circulation must be 0.05–0.1 m/s (Gökalp 1995).

39.5.2 Internal Factors

39.5.2.1 Raw Material

The raw materials for sucuk are meat and fat. Meat composition (amount of carbohydrate, lipid, water, and protein) determines the progress of the fermentation and ripening (Heperkan and Sözen 1988; Ordonez et al. 1999). Microorganisms, like all living systems, have specific carbon, nitrogen, mineral, and vitamin requirements. Meat and meat products are generally considered a rich source of nutrients. Accordingly, they can support the growth of a wide variety of microbial species. However, addition of specific nutrients to meat formulations has been shown to influence the physiological characteristics of contaminating microorganisms. For example, the addition of certain spices to sucuk stimulates bacterial lactic acid production. This stimulatory effect is due to presence of manganese and trace minerals in the spices that acting as a cofactor for lactic acid production in *Lactobacillus* and *Pediococcus*.

39.5.2.2 pH

The pH decreases with lactic acid accumulation during fermentation of sucuk. Normally, the initial pH of sucuk dough is about 6.1–6.2 and decreases rapidly during fermentation. The final pH of sucuk is largely dependent on the amount of fermentable carbohydrate in the formulation. For sucuk, pH should be in the range of 4.80–5.40 with the optimum values of 5.10–5.20. If pH is higher than 5.80, that sucuk easily spoils (Gökalp et al. 1999; Anon. 2000).

The initial pH (6.1–6.2) of sucuk dough is well within the range that permits growth of most bacteria, yeasts, and molds, and it would be expected that a wide range of microbial species could effectively compete in such an environment.

39.5.2.3 Water Activity

Microorganisms need water to survive and multiply. Water activity is the measure of water availability. Meat and meat products that have high water activity spoil faster than those of low water activity, such as sucuk. The water activity level of sucuks decreases with water loss during ripening. Adjustment of water activity of sucuks is possible by changing the amount of salt and fat added.

In general, decreasing the water activity of a meat system by addition of NaCl and other humectants or by partial dehydration shifts the microflora to more salt-tolerant genera such as *Lactobacillus*, *Streptococcus*, *Staphylococcus*, *Vibrio*, yeasts, and molds (Buchanan 1986).

During ripening of sucuks, the initial RH (90%) should be decreased after the third day (Gökalp et al. 1999).

39.5.2.4 Oxygen

Processed meat products are highly affected by oxygen content. In fermented meat products such as sucuks, the surface of the meat is highly aerobic, favoring the growth of aerobes and facultative anaerobes (Lücke 1986). However, a short distance beneath the surface, the amount of available oxygen is greatly decreased and microaerophiles, anaerobes, and facultative anaerobes would be expected to predominate. It could be anticipated that more aerobically oriented genera such as *Moroxella* or *Brochothrix* would predominate on the surface, whereas microaerophiles such as *Lactobacillus* would thrive in the interior (Buchanan 1986).

The use of a casing that reduces the oxygen content of the atmosphere surrounding the product retards the growth of aerobic or noncompetitive facultative anaerobic species (e.g., *Pseudomonas*, *Brochothrix*, and *Moroxella*), but faster growth of microaerophiles such as *Lactobacillus* and yeast.

39.5.2.5 Salt and Sugar

Acid formation during fermentation and ripening depends on the amount and kind of the sugar added. Growth of acid reduces surface microflora. Enough pH reduction can be provided with the addition of fermentable sugar as little as 0.2%. If excess sugar is added, undesired flavor and acid formation (lactic acid) occur; thus, a sour taste is obtained. Salt is used for taste and affects the chemical and microbiological changes (Alperdem 1986). The salt content of sucuk must not exceed 5% (Anon. 2002).

39.5.2.6 Initial Number of Starter Culture

The length of the apparent lag phase before measurable acid formation depends on the initial number of starter culture and their adaptation to the conditions. Addition of LAB reduces the fermentation time, provides color stability and aroma during the drying period, reduces nitrate, and inhibits pathogenic microorganisms (Turantaş and Ünlütürk 1993; Gökalp et al. 1999).

39.5.2.7 Antimicrobial Preservatives

The inhibition of the growth and activity of microorganisms is the main purpose for the use of antimicrobial preservatives. These antimicrobials deteriorate the cell wall and membrane of microorganisms and inhibit the enzymes that play important roles in the metabolism of microorganisms.

The most common preservatives used in sucuk and other meat products are sodium nitrate and sodium nitrite. These compounds are well known for their activity against clostridial species such as *Clostridium botulinum* and *C. perfringens*. Nitrate has also been shown to inhibit *Staphylococcus aureus*, although this activity is effectively limited to anaerobically cultured cells (Turantaş and Ünlütürk 1993).

Nitrates are decomposed to nitric acid, which forms nitrosomyoglobin when it reacts with the heme pigments in meats, thereby forming a stable red color (nitrosomyoglobin). Nitrates probably act as a reservoir for nitrite, and their use is being restricted. Moreover, nitrites can react with secondary and tertiary amines to form nitrosamines, which are known to be carcinogenic (Frazier and Westhoff 1988).

The effects of different nitrite doses and starter cultures (*L. plantarum* and *S. carnosus*) on sucuk samples inoculated with *Escherichia coli* 0157:H7 were studied by Öz et al. (2002). The numbers of *E. coli* 0157:H7 slowly decreased during the ripening/storage period. However, they observed that the effects of starter culture and nitrite on *E. coli* 0157:H7 counts were insignificant.

The other antimicrobial preservatives used in sucuk production are ascorbic acid and its osmotic pressure by decreasing water activity, and therefore, salt and sugar have indirect antimicrobial activity (Gökalp et al. 1999; Kaban et al. 2007).

39.6 Microbial Activity in Sucuks during Ripening and Storage Periods

Two basic microbial activities proceed during the fermentation period. They are reduction of pH by LAB and formation of nitric oxide by nitrate and nitrite reductase activity of microorganisms (Turantaş and Ünlütürk 1993; Gökalp et al. 1999; Kaban et al. 2007; Ensoy et al. 2010). The most important microbial activity occurring during the ripening and storage periods of dry sucuk belongs to LAB (genera *Lactobacillus* and *Staphylococcus*), molds, and yeasts. Microbial load in sucuks ranges from 10^5 to 10^6 colony forming units (cfu)/g before fermentation and increased up to 10^8–10^9 cfu/g during fermentation.

39.6.1 Lactic Acid Bacteria

LAB presently consist of the following four genera: *Lactobacillus*, *Leuconostoc*, *Pediococcus*, and *Streptococcus* (Erkmen and Fadıloğlu 2001; Çon et al. 2001; Kaban and Kaya 2006; Ensoy et al. 2010).

The most commonly isolated LAB from dry-fermented sucuk used in starter cultures are *L. sake*, *L. curvatus*, *L. plantarum*, *Pediococcus pentosaceus*, and *P. acidilactici* (Gürakan et al. 1995; Kaban and Kaya 2008). The nitrate and nitrite reducing bacteria present in the fermented sucuk belong to *Staphylococcus* and *Micrococcus* genera (*S. carnosus*, *S. xylosus*, and *Micrococcus varians*). A sharp increase in aerobic and lactobacilli counts and a small increase in Micrococcaceae count are observed during the fermentation period (Gökalp et al. 1999). LAB have a number of beneficial effects on the manufacturing process, quality, and shelf life of sucuks.

39.6.1.1 Lactobacillus

Lactobacilli are rod-shaped bacteria that react positively in Gram staining, are unable to form spores, and are usually not motile. They "ferment" carbohydrates into lactic acid. A subgroup of lactobacilli, known as heterofermentative strains, produces not only lactic acid but also large quantities of acetic acid, ethanol, and carbon dioxide. In contrast to "obligate" anaerobic bacteria, lactobacilli can usually tolerate air oxygen and, thus, also grow on aerobically incubated culture media. They are relatively easy to recognize because, unlike almost all other Gram-positive, aerobically growing rods, they produce no catalase. Catalase breaks down hydrogen peroxide, a product of the reaction of air oxygen with certain cell constituents. Hydrogen peroxide is also produced in large quantities from air oxygen by many *Lactobacillus* strains. The fact that they grow aerobically is due to their great tolerance to hydrogen peroxide (Erkmen and Fadıloğlu 2001).

The acids formed by sugars from LAB are a great help in suppressing undesirable bacteria in the dry sucuk. Lactobacilli are greatly responsible for carbohydrate breakdown and, thus, for the acid aroma component of dry sucuk; they also affect texture and hygienic stability (Gökalp et al. 1999).

39.6.1.2 Micrococcus and Staphylococcus

Bacteria of the genera *Staphylococcus* and *Micrococcus* are Gram-positive spherical bacteria. They have catalase and grow anaerobically; the species *S. xylosus*, *S. saprophyticus*, *S. carnosus*, and *M. varians* have been mainly found in dry sucuk (Erkmen and Fadıloğlu 2001; Kaban and Kaya 2008). The numbers of these bacteria increase during the first days of ripening period (fermentation) due to the high level of water activity and then decrease during the further ripening and storage periods (Gökalp et al. 1999). They are sensitive to the decrease in pH. Therefore, fermentation is a critical process in the manufacturing of sucuks, and pH reduction should be conducted slowly to allow the growth of *Staphylococcus* spp. If the Micrococcaceae do not grow well, reduction of nitrates and nitrites is hindered (Lücke 1994).

Bacteria of the Micrococcaceae family, naturally developed or added as starter culture, accomplish an essential role to the reduction of nitrate to nitrite by means of their nitrate reductase activity in dry-fermented sausages. This activity allows the production of nitrite, which contributes to preservation of both color and flavor (Sanz et al. 1997; Ordonez et al. 1999). The catalase produced breaks down peroxides (hydrogen peroxide) produced by lactobacilli and reduces rancidity. Micrococcaceae also has an important role in the fermentation of dry-fermented sausages. Their ability to break down fat (lipolysis) and protein (proteolysis) has great importance in the development of flavor and stabilization of color in sucuks (Lücke 1994; Bover-Cid et al. 1999; Erkmen and Fadıloğlu 2001).

39.6.2 Molds and Yeasts

A surface flora consisting of molds and yeasts grows during the drying of sucuks provided that the RH of the ripening room is not too low. Because of application of low temperatures (16°C–22°C) during drying, the mold flora consists principally of *Penicillium* species, with *Penicillium verrucosum* species being those isolated most often. *Aspergillus* species (mainly representatives of the *Aspergillus glaucus* group) are more likely to be found on products that have been stored too warm. The yeasts present are basically strains of the genus *Debaryomyces* that are adapted to high common salt concentrations (Erkmen and Fadıloğlu 2001).

The surface flora of many air-dried dry sucuks, which consists mainly of molds of the genus *Penicillium*, protects the products from harmful effects of air oxygen and light, makes drying easier, and gives them their typical aroma by breaking down fats, proteins, and lactic acid (Erkmen and Fadıloğlu 2001).

39.6.3 Starter Cultures and Their Effects

The primary genera of bacteria that have been successfully utilized as fermented sausage starter cultures are *Lactobacillus*, *Micrococcus*, *Pediococcus*, and *Staphylococcus*; that of yeasts is *Debaryomyces* and that of molds is *Penicillium* (Erkmen and Fadıloğlu 2001; Kaban and Kaya 2006, 2008). Table 39.3 lists

TABLE 39.3

Starters and Their Roles in Sausage Fermentation

Microbial Group	Species Used as Starters	Useful Metabolic Activity	Benefits to Sausage Fermentation
Lactic acid bacteria	*Lactobacillus plantarum, L. pentosus, L. sake, L. curvatus, Pediococcus pentosaceus, P. acidilactici*	Formation of lactic acid and bacteriocins	Inhibition of pathogenic and spoilage bacteria Acceleration of color formation and drying
Catalase-positive cocci	*Staphylococcus carnosus, S. xylosus, Micrococcus varians*	Nitrate reduction and oxygen consumption	Color formation and stabilization Removal of excess nitrate
		Peroxide destruction	Delay of rancidity
		Lipolysis	Aroma formation
Yeasts	*Debaryomyces hansenii*	Oxygen consumption	Delay rancidity
		Lipolysis	Aroma formation
Molds	*Penicillium nalgiovense* biotypes 2, 3, and 6	Oxygen consumption	Color stability
		Peroxide destruction	Delay of rancidity
		Lactate oxidation	Aroma formation
		Proteolysis	Aroma formation
		Lipolysis	Aroma formation

the starters and their roles in sausage fermentation. Many of the microbial species available as a starter culture in the fermentation of sausages are, in particular, the psychotrophic, salt-tolerant *Lactobacillus* species (such as *L. sake* and *L. curvatus*), *Micrococcus* species, nonpathogenic *Staphylococcus* species, *Debaryomyces hansenii*, and *Pen. nalgiovense* (Kaban and Kaya 2008).

Sucuks can be manufactured using microbial strains belonging to the genera *Lactobacillus* (*L. plantarum, L. pentosus, L. curvatus,* and *L. sake*), *Pediococcus* (*P. pentosaceus* and *P. acidilactici*), Micrococcaceae (*Staphylococcus* and *Micrococcus*), and *Debaryomyces* (*D. hansenii*) as starter cultures (Anon. 2002; Kaban and Kaya 2008). The commercial starter cultures commonly used in sucuk production are selected according to their properties, namely, fermentative, proteolytic, or lipolytic characteristics. However, they are not always suitable for sucuk production and may result in losses of some desirable sensory characteristics (Bozkurt and Erkmen 2002a).

The use of a suitable starter culture is very important in the production of safe and high-quality sucuks (Kaban and Kaya 2006). Quality (chemical and microbiological) characteristics of sucuk were improved by use of starter cultures such as *S. xylosus*; *P. pentosaceus* + *L. plantarum*; *S. carnosus*; *S. carnosus* + *L. pentosus* + *S. xylosus*; *P. pentosaceus* + *P. acidilactici* + *S. carnosus*; *P. pentosaceus*; and *S. xylosus* + *L. alimentarus*.

Starter cultures used in fermented meat products consist of pure or mixed microorganisms. The starter culture limits the fermentation time, standardizes the products, helps in the development of color, and provides the inhibition of pathogenic microorganisms that can be found naturally in foods during fermentation, preventing the formation of some biogenic amines such as histamine and tyramine, inhibiting the formation of nitrosamine from nitrates, which is added to sucuk dough as a curing component, and increasing the nutritive values of the product. Standard high quality and long shelf life are provided (Vural and Öztan 1991). Use of *L. sake* and *P. acidilactici* as starter cultures inhibit inoculated *Yersinia enterocolitica* in sucuk (Ceylan and Fung 2000). Erkkila (2001) and Kaya and Gökalp (2004a) reported that starter cultures (*L. plantarum* and *S. carnosus*) stopped the growth of *L. monocytogenes* in the sucuk. In addition, Kaya and Gökalp (2004b) observed that *L. monocytogenes* was inhibited by the use of starter culture (*P. pentosaceus, P. acidilactici, L. sakei,* or *L. plantarum*) in sucuk.

Traditional sucuks have been produced with chance inoculation of LAB (Gökalp et al. 1999). The pure cultures of *L. casei* var. *alactosus* or *P. cerevisiae* (Özer and Özalp 1968) and *L. plantarum* and *P. cerevisiae* could be used as sucuk starter cultures (Kaban and Kaya 2008).

Kaban and Kaya (2008) identified 10 different LAB species. *L. plantarum* is the dominant species in about 46% of the isolates. The other major *Lactobacillus* species isolated from sucuk were *L. pentosus*, *L. curvatus*, *L. fermentum*, *L. brevis*, *P. pentosaceus*, and *P. acidilactici*. Catalase-positive cocci mainly of the *Staphylococcus* genus were isolated from sucuk. *S. xylosus*, *S. carnosus*, and *S. equorum* are the dominant species isolated. Therefore, these species could be used as a starter culture in sucuk.

39.7 Chemical Changes in Sucuk during Ripening and Storage

39.7.1 Acid Formation (pH Reduction)

In fermented products, LAB, especially lactobacilli, have a great importance. Acids are formed (mainly lactic acid) by sugars from LAB following glycolytic pathway. The rate and extent of acid formation are critical in the manufacture of sucuks. These must be adjusted carefully to achieve both favorable sensory quality and safety from pathogens (prevent the multiplication of undesired microorganisms in fermented products).

The rate of acid production is determined by the type and amount of added carbohydrate, the species of LAB, the rate of drying, and the temperature. The initial water activity also affects the rate of acid formation. Acid formation may be too slow to inhibit *S. aureus* at a_w values below 0.96 (Turantas and Ünlütürk 1993). Generally, glucose, sucrose, and maltose are fermented rapidly, and lactose slowly, but the different sugar fermentation patterns of various LAB must be taken into account (Lücke 1994).

39.7.2 Lipid Oxidation (Malonaldehyde Formation)

One of the most important quality deteriorations of sucuk is the lipid oxidation that affects fatty acids and polyunsaturated fatty acids (Bozkurt 2002). The acceptability of a sucuk depends on the extent to which this deterioration has occurred. Lipid oxidation may have significant problems on the quality, for example, color, flavor, texture, and nutritional value changed due to the lipid oxidation in sucuks (Chizzolini et al. 1998; Bozkurt 2006a).

Malonaldehyde is formed as a result of oxidative rancidity. The amount of malonaldehyde limit is important in terms of human health because it has been linked to cancer and mutation. It is desired to occur at low levels during the storage of product and their marketing (Stu and Draper 1978). Although sensory analysis is one of the most sensitive methods available, it is not practical for routine analyses and generally lacks reproducibility.

A number of methods are available for determination of lipid oxidation in foods (Pikul et al. 1989). The 2-thiobarbituric acid test has been widely used for measuring oxidative rancidity in fat-containing foods, especially muscle-type foods (Bozkurt 2002).

39.7.3 Color Formation (Nitrosomyoglobin Conversion)

The characteristic color of sucuk is produced by the interaction between the meat pigments (myoglobin) and the added nitrite and nitrate (Figure 39.3). Nitrite is the only toxic substance used in foodstuffs and/ or allowed for consumption. It has been forbidden to be added directly to meat products in many countries, where it has been used as their sodium or potassium salts.

The color of the meat products depends on the reaction of meat pigments such as myoglobin and hemoglobin (Brewer et al. 1991; Han et al. 1994; Bozkurt and Bayram 2006) with curing components. Nitrosomyoglobin—from the terms *myoglobin* and *nitrite*—is the main component in the color formation of meat products as a stable color substance. The red cured color is formed when the muscle pigment myoglobin reacts with the nitric oxide (NO) produced from nitrite in the acidic medium (Figure 39.3). Micrococcaceae are responsible for reduction of nitrates and nitrites to nitric oxide, and this NO is used for the production of nitrosomyoglobin during the fermentation period (Vural and Öztan 1992; Talon et al. 1999).

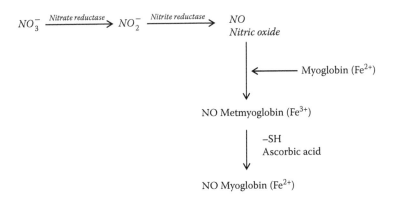

$$NO_3^- \xrightarrow{\text{Nitrate reductase}} NO_2^- \xrightarrow{\text{Nitrite reductase}} \begin{array}{l} NO \\ \textit{Nitric oxide} \end{array}$$

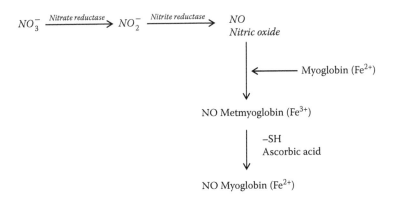

FIGURE 39.3 Formation of nitrosomyoglobin.

The color formation in meat products has a close relationship with the pH (Öztan 1999). The development of color is optimum between pH 5.4 and 5.7 (Kamarei and Karel 1982).

39.7.4 Formation of Biogenic Amines

Biogenic amines may be classified as aliphatic (monoamines, diamines, and polyamines), aromatic, or heterocyclic in their structure (Shalaby 1996). Biogenic amines could be found in meat, sausages, sucuk, milk, chocolate, cheese, fish, and some beverages (Eerola et al. 1993; Shalaby 1993; Hernandez-Jover et al. 1997a,b; Ordonez et al. 1997; Durlu-Özkaya et al. 2001; Bozkurt and Erkmen 2002a,b, 2004; Çolak and Ugur 2002; Şenöz et al. 2000; Gençalp et al. 2007, 2008). Biogenic amines can be found in meat products due to the decarboxylase activity of microorganisms during the manufacturing and storage. The formation of biogenic amines depends on the ripening and storage conditions of product, presence of amine-positive microorganisms, free amino acids required for amine formation, quality of raw materials, and hygienic conditions of processing environment (Vidal-Carou et al. 1990; Maijala et al. 1995; Shalaby 1996; Gonzalez-de-Liano et al. 1998; Ordonez et al. 1999; Durlu-Özkaya et al. 2001).

Biogenic amines are toxic substances; they can cause nausea, respiratory distress, hot flushes, sweating, heart palpitation, bright red rash, oral burning, gastric and intestinal problems, and hypertension or hypotension (Gonzales-de-Liano et al. 1998; Durlu-Özkaya et al. 2001). Histamine, putrescine, cadaverine, tyramine, tryptamine, β-phenyl ethylamine, spermine, and spermidine are the most important and common biogenic amines found in foodstuffs (Shalaby 1996). Histamine intake of 8–40, 40–100, and higher than 100 mg may cause slight, intermediate, and intensive poisoning, respectively (Parente et al. 2001). Nout (1994) pointed out that the maximum histamine content is 50–100 mg/kg for sausages. The allowable maximum level of tyramine in foods is 100–800 mg/kg, and 1080 mg/kg of tyramine is toxic for humans (Shalaby 1996). Putrescine, spermine, spermidine, and cadaverine do not have adverse health effects, but they may react with nitrite to form carcinogenic nitrosamines and also can be proposed as indicators of spoilage (Hernandez-Jover et al. 1997b; Eerola et al. 1997).

Biogenic amine formation can be observed at higher levels in sucuks stored at room temperature compared to those stored at refrigeration temperatures (Bozkurt 2002). Temperature and percentage of RH may be used to control the formation of biogenic amines in sucuk with other factors such as additives (nitrite/nitrate, ascorbic acid, α-tocopherol, *Urtica dioica*, sumac extracts, and *Thymbra spicata* oils), use of high-quality meat (low microbial load), use of frozen meat, shortening the ripening time, and use of amine-negative starter cultures (Aksu and Kaya 2004; Bozkurt 2006a,b, 2007; Gençalp et al. 2007, 2008).

It was observed that *Salvia officinalis* L. and *Rosmarinus officinalis* L., sesame, *T. spicata* oils, green tea, and sumac extracts have very strong antioxidant activity. These additives reduce histamine, tyramine, and putrescine formation significantly in sucuk (Bozkurt 2006a,b, 2007; Karabacak and Bozkurt 2008).

39.8 Organoleptic Properties of Sucuks

Organoleptic characteristics of sucuks could be divided into three groups: appearance, cooking, and eating properties.

39.8.1 Appearance Properties

Casings should not tear easily and should have a characteristic taste, odor, and good appearance. Desired taste in sucuk is lactic acid taste because sucuk is a fermented product.

When sucuk is cut, the cross-sectional surface should have a pink color (Bozkurt and Bayram 2006). Application of force with a finger should give a resistance but should not collapse. It should not be too hard or too soft. When it is cut with a sharp knife, it should not be brittle or should not stick on the knife.

There should be no color differences between the center and the outer surface. The color of meat and fat should not mix with each other.

Sucuk should have a smooth cross-sectional surface without air bubbles. Fat, water, and salt ratios should not exceed 30%, 40%, and 5%, respectively. Protein content should be at least 22%. High-quality sucuks should have a pH of 5.2–5.4 (Gökalp et al. 1999).

39.8.2 Cooking Properties

Sucuks should not lose more than 10% of their weight by frying or grilling. Losses exceeding 20% are not acceptable. Sucuks should not shrink or distort excessively after being cooked.

Deformation of sucuks by cooking is also an important criterion of quality. This may be in the form of excessive shrinkage. When natural casings are used, the skin may contract over the length of the sucuk, giving a dumbbell appearance. It is not always easy to determine whether this is caused by expansion of the filling or contraction of the casing, but the latter is probably the more likely cause especially because this rarely occurs with synthetic casings.

39.8.3 Eating Properties

Sucuk should be solid, firm, and succulent but not rubbery. Sucuk casings should be easily peelable. Flavor is highly subjective and difficult to quantify.

39.9 Quality of Sucuk

There are four basic quality criteria for sucuks: physical, chemical, microbiological, and organoleptic (Anon. 2002). These are given in Table 39.4.

Color and appearance of sucuks can be detected according to their physical structures, texture, and appearance. Sucuks having uniform dimensions should be packaged under vacuum. Packing material must be transparent and have a pleasant appearance. Sucuk should have its special taste and flavor. Fat layer must not be present on the surface. Outer color of sucuk must be red-brown; inner color must be pink-red. When it is pressed with a finger, it must be resistant and not exceedingly hard or soft. When it is cut with a knife, it must not stick on the knife (Turgut 1986; Anon. 2002).

Bright red is an appropriate color of sucuk that might occur when nitrites are reduced to nitric oxide by the starter culture. In sucuk, the stability of the color is assisted by the residual nitrite in the product. In recent years, nitrite levels are being kept as low as possible because of formation of biogenic amines or nitrosamines from nitrite (Özçelik 1982).

Several studies focused on the effects of additives on the quality and safety of sucuk during the ripening and storage periods (Bozkurt and Erkmen 2007; Erkmen and Bozkurt 2004).

TABLE 39.4

Standards for Sucuk

Criteria	Limits
Chemical Criteria	
Moisture (by mass)	Max. 40%
Salt (by mass)	Max. 5%
Coloring matter	None
pH	Min. 5.4–max. 5.8
Fat (by mass)	
for first quality	Max. 30%
for second quality	Max. 40%
for third quality	Max. 50%
Protein (by mass), $N \times 6.25$	
for first quality	Min. 22%
for second quality	Min. 20%
for third quality	Min. 20%
Microbial Criteria	
Total aerobic mesophilic bacteria (cfu/g)	10^5–10^6 in two out of five samples
Escherichia coli (cfu/g)	None
Staphylococcus aureus (cfu/g)	0–100
Salmonella (cfu/g)	None in 25 g sample
Mold and yeast (cfu/g)	0–100
Clostridium perfringens (cfu/g)	10–100 in two out of five samples
Coliform bacteria (cfu/g)	Maximum 10
Organoleptic Criteria	
Flavor	Must be in its original flavor
Taste	No rancid, sour, or bitter flavors
Appearance	Must be in appropriate color and texture

It has been observed that nitrate, nitrite, potassium pyrophosphate, dipotassium hydrogen phosphate, ascorbic acid, α-tocopherol, *U. dioica*, and potassium sorbate increase redness values (Hunter *a*-values) and overall sensory scores (derived from color, flavor, and cutting scores) of sucuk as well as decrease the aerobic plate count, mold and yeast, biogenic amine formation, and lipid oxidation (thiobarbituric acid reactive substances value). It was observed that *Sal. officinalis* L. and *Ros. officinalis* L., sesame, *T. spicata* oils, green tea, and sumac extracts have very strong antioxidant activity. These additives are more effective than buthylatedhydroxytoluene for reduction of lipid oxidation (Bozkurt 2006a,b, 2007; Karabacak and Bozkurt 2008).

The maximum allowable residual nitrite in sucuk at the retail point must not be higher than 50 ppm according to the Turkish Food Codex (Anon. 2000). However, in European countries, residual nitrite limit in sausage should have a maximum of 15 ppm and should not be higher than 30 ppm in sausages in the Codex Alimentarius. Turkish Food Codex states that the pH value for fermented sucuks should have a maximum of 5.4 (Anon. 2000). For biogenic amine contents, only histamine and tyramine limit are declared in the literature and are stated as follows: histamine contents should be in the range of 50–100 mg/kg, and the allowable maximum level of tyramine is 100–800 mg/kg in sausages (Nout 1994; Shalaby 1996).

The typical aroma of sucuk not only is attributed to volatile substances but also comes from a large number of volatile and nonvolatile compounds present in the product. The volatile compounds contribute to aroma of dry-fermented sausages.

Kaban (2010) analyzed the volatile compounds of sucuk by the use of gas chromatography/mass spectrometry with solid phase microextraction. About 120 compounds were isolated, and 92 of these were tentatively identified and estimated as follows: 5 acids, 7 esters, 10 aliphatic hydrocarbons, 7 alcohols, 5 aldehydes, 11 sulfur compounds, 2 ketones, 7 aromatic hydrocarbons, 27 terpenes, 2 nitrogen compounds, and 3 phenols. Higher level of terpenes from spices, which had strong aromatic flavor, could play an important role in the overall aroma of sucuk. Acids (especially acetic acid), sulfur compounds (diallyldisulfide, 1, propene 3-thiobis, and disulfide methyl 2-phenyl), and aldehydes (*p*-cumic aldehyde) were also found in sucuk.

39.10 Defects

General defects detected in sucuks are appearance, color, aroma, and flavor (Turgut 1986; Oğan 1996).

39.10.1 Appearance Defects

Shape deformation, microbial colonies, saltiness, oil diffusion, unstabilized color, deformation of casings, and points of brown, yellow, and black color on the surface of sucuks are the appearance defects. These defects occur during the manufacturing and storage of sucuk. Shape and color deformation occur in the manufacturing of sucuk because of the improper drying conditions such as one-way air drying, rapid drying, very hot air temperature use, and very dry air usage. Very humid storage conditions and inadequate fillings cause cavity in sucuk. Microbial colonies appear at different colors on sucuk. Most of these colonies are produced by mold and yeast. These occurred because of the unhygienic manufacturing and storage conditions. They may oxidize lipids (*Candida lipolytica*) and cause the greening of the surface (*Penicillium* and *Aspergillus* spp.). Saltiness of the surface appears due to the diffusion of small salt crystals during the hanging period. Oil diffusion may take place because of the diffusion of high-melting-point fats through the surface of sucuk during the manufacturing and storage. Unstabilized color appearances occur due to low nitrite reduction with the microorganisms, too much drying, and undesired mold growth on the surface of sucuk (Vural and Öztan 1996; Göğüş 1986).

39.10.2 Color Defects

Insufficient and unstabilized color formation, core color defects, and dark and rancid color formation are the types of color defects. Insufficient and unstabilized color formation take place due to the addition of low amount of nitrite, low amount of myoglobin present in the meat (because of young animal meat), high fat content, high pH reduction during the fermentation, and low inoculum levels of the starter culture used. Core color defects may be due to microbial origin as the presence of homofermentative LAB and proteolytic bacteria that produce H_2O_2 and H_2S, respectively. These compounds bind to meat pigments to produce a green color in the inner part of the sucuk. Dark color formation occurs because of the high amount of myoglobin in meat. Rancid color formation takes place due to the high ripening and storage temperatures. Lipids are oxidized at high temperatures and desired colors of sucuk are formed (Turgut 1986; Oğan 1996).

39.10.3 Flavor and Taste Defects

Souring, bitterness, and rancidity are considered as flavor and taste defects. Heterofermentative bacteria produce acetic acid, tartaric acid, and lactic acid. These acids affect the flavor and taste of sucuk. Having too much sugar and starter culture and high fermentation temperatures (above 25°C) may cause sour defects. Production of NH_3 and H_2S from the microorganisms adversely affects the flavor and taste of sucuk. These defects may result from unhygienic manufacturing and storage conditions. Bitter taste defects are due to excess sodium nitrate usage and the presence of Mg in salt. Rancidity is formed due to the oxidation of unsaturated fatty acids. Unsaturated fatty acids naturally are low in meat, but they are produced by means of microbial actions. Therefore, sucuks are susceptible to rancidity.

REFERENCES

Aksu, M.İ. and Kaya, M. 2004. Effect of usage *Urtica dioica* L. on microbiological quality of Turkish sucuk (Turkish style dry-fermented sausages). *Food Control* 15:591–595.

Alperdem, İ. 1986. Et ürünleri teknolojisinde katkı maddeleri. In: Et mamülleri üretimi ve muhafazası. İstanbul Ticaret Odası, Yayın No. 1987:7–31.

Anon. 2000. Turkish Food Codex, Türk Gıda Kodeksi Et Ürünleri Tebliği, Tebliğ No. 2000/4. Ankara: Agriculture and Rural Affairs Minister.

Anon. 2002. TS-1070, *Turkish Style Fermented Sausage*. Ankara: Turkish Standard Institute.

Bover-Cid, S., Hugas, M., Izquierdo-Pulido, M., and Vidal-Carou, M.C. 1999. Reduction of biogenic amine formation using a negative amino acid-decarboxylase starter culture for fuet sausage fermentation. *J. Food Prot.* 63:237–243.

Bozkurt, H. 2002. Effects of some storage conditions and additives on quality and stability of sucuk (Turkish dry-fermented sausage). PhD Thesis, University of Gaziantep, Gaziantep, Turkey.

Bozkurt, H. 2006a. Utilization of natural antioxidants: Green tea extract and *Thymbra spicata* oil in Turkish dry-fermented sausage. *Meat Sci.* 73:442–450.

Bozkurt, H. 2006b. Investigation of the effect of sumac extract and BHT addition on the quality of sucuk (Turkish dry-fermented sausage). *J. Sci. Food Agric.* 86:849–856.

Bozkurt, H. 2007. Comparison of the effects of sesame and *Thymbra spicata* oil during the manufacturing of Turkish dry-fermented sausage. *Food Control* 18:149–156.

Bozkurt, H. and Bayram, M. 2006. Colour and textural attributes of sucuk during ripening. *Meat Sci.* 73:344–350.

Bozkurt, H. and Erkmen, O. 2002a. Effects of starter cultures and additives on the quality of Turkish style sausage (sucuk). *Meat Sci.* 61:149–156.

Bozkurt, H. and Erkmen, O. 2002b. Formations of biogenic amines in Turkish style sausage during ripening and storage periods. *J Food Qual.* 25:317–332.

Bozkurt, H. and Erkmen, O. 2004. Effects of temperature, humidity and additives on the formation of biogenic amines in sucuk during ripening and storage periods. *Food Sci. Technol. Int.* 10:21–28.

Bozkurt, H. and Erkmen, O. 2007. Effects of some commercial additives on the quality of sucuk (Turkish dry-fermented sausage). *Food Chem.* 101:1465–1473.

Brewer, M.S., McKeith, F., Martin, S.E., Dallmier, A.W., and Meyer, J. 1991. Sodium lactate effects on shelf-life, sensory, and physical characteristics of fresh pork sausage. *J. Food Sci.* 56:1176–1178.

Buchanan, R.L. 1986. Processed meats as a microbial environment, *Food Technol.* April:134–138.

Ceylan, E. and Fung, D.Y.C. 2000. Destruction of *Yersinia enterocolitica* by *Lactobacillus sake* and *Pediococcus acidilactici* during low-temperature fermentation of Turkish dry sausage (sucuk). *J. Food Sci.* 65:876–879.

Chizzolini, R., Novelli, E., and Zanardi, E. 1998. Oxidation in traditional Mediterranean meat products. *Meat Sci.* 49:87–99.

Çolak, H. and Ugur, M. 2002. The effect of different temperature and time in storage on the formation of biogenic amines in fermented sucuks. *Turk. J. Vet. Anim. Sci.* 26:779–784.

Çon, A.H., Gökalp, H.Y., and Kaya, M. 2001. Antagonistic effect on *Listeria monocytogenes* and *L. innocua* of a bacteriocin-like metabolite produced by lactic acid bacteria isolated from sucuk. *Meat Sci.* 59:437–441.

Durlu-Özkaya, F., Ayhan, K., and Vural, N. 2001. Biogenic amines produced by Enterobacteriaceae isolated from meat products. *Meat Sci.* 58:163–166.

Eerola, S., Hinkkanen, R., Lindorfs, E., and Hirvi, T. 1993. Liquid chromatographic determination of biogenic amines in dry sausage. *J. AOAC Int.* 76:575–577.

Eerola, S., Sagues, A.X.R., Lilleberg, L., and Aalto, H. 1997. Biogenic amines in dry sausages during shelf-life storage. *Z. Lebensm.-Unters. Forsch. A* 205:351–355.

Ensoy, Ü., Kolsarıcı, N., Candoğan, K., and Karslıoğlu, B. 2010. Changes in biochemical and microbiological characteristics of Turkey sucuks as affected by processing and starter culture utilization. *J. Muscle Foods* 21:142–165.

Ercoşkun, H., Tağı, Ş., and Ertaş, A.H. 2010. The effect of different fermentation intervals on the quality characteristics of heat-treated and traditional sucuks. *Meat Sci.* 85:174–181.

Erkkila, S. 2001. Bioprotective and probiotic meat starter cultures for the fermentation of dry sausages. Dissertation, Department of Food Technology, University of Helsinki, Finland.

Erkmen, O. 1997. Behavior of *Staphylococcus aureus* in refrigerated and frozen ground beef and in Turkish style sausage and broth with and without additives. *J. Food Process. Preserv.* 21:279–288.

Erkmen, O. and Bozkurt, H. 2004. Quality characteristics of retailed sucuk (Turkish dry-fermented sausage). *Food Technol. Biotechnol.* 42:63–69.

Erkmen, O. and Fadıloğlu, S. 2001. Gıda fermentasyonunda mikroorganizmaların kullanımı. Gıda, Mayıs 56–61.

Frazier, W.C. and Westhoff, D.C. 1988. *Food Microbiology.* Singapore: McGraw-Hill, Inc.

Gençalp, H., Kaban, G., and Kaya, M. 2007. Effects of starter cultures and nitrite levels on formation of biogenic amines in sucuk. *Meat Sci.* 77:424–430.

Gençalp, H., Kaban, G., and Kaya, M. 2008. Determination of biogenic amines in sucuk. *Food Control* 19:868–872.

Göğüş, A.K. 1986. Et teknolojisi. Faculty of Agriculture, University of Ankara, Publication No. 991.

Gökalp, H.Y. 1986. Turkish style fermented sausage (soudjuck) manufactured by adding different starter cultures and using different ripening temperatures. II. Ripening period, some chemical analysis, pH values, weight loss, color values and organoleptic evaluations. *Fleischwirtschaft* 66:573–575.

Gökalp, H.Y. 1995. Fermente et ürünleri, Sucuk üretim teknolojisi. Standard, Special Issue on "Geleneksel Türk Et Ürünleri" July:48–55.

Gökalp, H.Y., Kaya, M., and Zorba, Ö. 1999. Et Ürünleri İşleme Mühendisliği, 3rd edition. Erzurum, Turkey: Faculty of Agriculture, Food Engineering Department, Atatürk Üniversity.

Gökalp, H.Y., Yetim, H., Kaya, M., and Ockerman, H.W. 1988. Saprophytic and pathogenic bacteria levels in Turkish soudjucks manufactured in Erzurum, Turkey. *J. Food Prot.* 51:121–125.

Gonzales-de-Liano, D., Cuesta, P., and Rodriquez, A. 1998. Biogenic amine production by wild lactococcal and leuconostoc strains. *Lett. Appl. Microbiol.* 26:270–274.

Gürakan, G.C., Bozoğlu, T.F., and Wiess, N. 1995. Identification of *Lactobacillus* strains from Turkish-style dry fermented sausage. *Lebensm.-Wiss. Technol.* 28:139–144.

Han, D., McMillin, K.W., and Godber, J.S. 1994. Hemoglobin, myoglobin, and total pigments in beef and chicken muscles: Chromatographic determination. *J. Food Sci.* 59:1279–1282.

Heperkan, D. and Sözen, M. 1988. Fermente et ürünleri üretimi ve mikrobiyal proseslerin kaliteye etkisi. *Gıda* 13:371–378.

Hernandez-Jover, T.H., Pulido, M.I., Nogues, M.T.V., Font, A.M., and Carou, M.C.V. 1997a. Biogenic amine and polyamine contents in meat and meat products. *J. Agric. Food Chem.* 45:2098–2102.

Hernandez-Jover, T.H., Pulido, M.I., Nogues, M.T.V., Font, A.M., and Carou, M.C.V. 1997b. Effect of starter cultures on biogenic amine formation during fermented sausages production. *J. Food Prot.* 60:825–830.

Honikel, K.O. 1989. The meat aspects of water and food quality In: *Water and Food Quality*, edited by T.M. Hardman, pp. 277–304. London: Elsevier Applied Sci.

Incze, K. 1998. Dry fermented sausages. *Meat Sci.* 49:169–177.

Kaban, G. 2010. Volatile compounds of traditionally Turkish dry fermented sausage (Sucuk). *Int. J. Food Prop.* 13:525–534.

Kaban, G., Aksu, M.I., and Kaya, M. 2007. Behavior of *Staphylococcus aureus* in sucuk with nettle (*Urtica dioica* L.). *J. Food Saf.* 27:400–410.

Kaban, G. and Kaya, M. 2006. Effect of starter culture on growth of *Staphylococcus aureus* in sucuk. *Food Control* 17:797–801.

Kaban, G. and Kaya, M. 2008. Identification of lactic acid bacteria and gram-positive catalase-positive cocci isolated from naturally sausage (sucuk). *J. Food Sci.* 73:M385–M388.

Kamarei, A.R. and Karel, M. 1982. An improved method for preparation of nitric oxide mygoblobin. *J. Food Sci.* 47:683–685.

Karabacak, S. and Bozkurt, H. 2008. The effect of *Salvia officinalis* L. and *Rosmarinus officinalis* L. on the ripening of sucuk. *Fleischwirtschaft* 88:103–108.

Kaya, M. and Aksu, M.I. 2005. Effect of modified atmosphere and vacuum packaging on some quality characteristics of sliced 'sucuk' produced using probiotics culture. *J. Sci. Food Agric.* 85:2281–2288.

Kaya, M. and Gökalp, H.Y. 2004a. The effects of starter cultures and different nitrite doses on the growth of *Listeria monocytogenes* in sucuk production. *Turk. J. Vet. Anim. Sci.* 28:1121–1127.

Kaya, M. and Gökalp, H.Y. 2004b. The behavior of *Listeria monocytogenes* in sucuks produced with different lactic starter cultures. *Turk. J. Vet. Anim. Sci.* 28:1113–1120.

Kayaardi, S. and Gök, V. 2004. Effect of replacing beef fat with olive oil on quality characteristics of Turkish soudjouk (sucuk). *Meat Sci.* 66:249–257.

Kılıç, B. 2009. Current trends in traditional Turkish meat products and cuisine. *LWT—Food Sci. Technol.* 42:1581–1589.

Kolsarıcı, N., Ertaş, A.H., and Şahin, M.E. 1986. Research work on the chemical composition of sucuk sold in the markets of Ankara, Afyon and Aydin, Turkey. *Gida (J. Turk. Food)* 86:34–39.

Lücke, F.K. 1986. Microbiological processes in the manufacture of dry sausage and raw ham. *Fleischwirtschaft* 66:1505–1509.

Lücke, F.K. 1994. Fermented meat products. *Food Res. Int.* 27:299–307.

Maijala, R., Eerola, S., Lievonen, S., Hill, P., and Hirvi, T. 1995. Formation of biogenic amines during ripening of dry sausage as affected by starter culture and thawing time of raw materials. *J. Food Sci.* 60:1187–1190.

Nout, M.J.R. 1994. Fermented foods and food safety. *Food Res. Int.* 27:291–298.

Oğan, H. 1996. *Gıda insan sağlığı ilgili yasalar*, pp. 57–64. İstanbul, October 1996.

Oliphant, G.G. 1998. Meat and meat products. In: *Food Industries Manual*, edited by M.D. Ranken, R.C. Kill, and C. Baker, pp. 1–45. London: Blackie Academic & Professional.

Ordonez, A.I., Ibanez, F.C., Torre, P., and Barcina, Y. 1997. Formation of biogenic amines in Idiazabal Ewe's milk cheese: Effect of ripening, pasteurization, and starter. *J. Food Prot.* 60:1371–1375.

Ordonez, J.A., Hierro, E.M., Bruna, J.M., and Hoz, L. 1999. Changes in the components of dry-fermented sausages during ripening. *Crit. Rev. Food Sci. Nutr.* 39:329–367.

Öz, F., Kaya, M., and Aksu, M.I. 2002. The effect of different nitrite doses and starter culture on the growth of *Escherichia coli* O157:H7 in sucuk (Turkish style dry sausage) processing. *Turk. J. Vet. Anim. Sci.* 26:651–657.

Özçelik, S. 1982. Bazı gıdalarda nitrit ve nitrozaminlerin oluşumu ve sağlığa zararlı etkileri. *Gıda* 7:183–188.

Özdemir, M., Batı, B., and Gökalp, H.Y. 1984. Nitrate, nitrite, and *N*-nitrosoamine contents of Turkish soud-jucks. *Fleischwirtschaft* 64:1476–1477.

Özer, I. and Özalp E. 1968. Food Hygiene Technology Association of Turkey, Bulletin No. 3. Ankara, Turkey.

Öztan, A. 1999. Et bilimi ve teknolojisi, 3rd edition. Hacettepe Üniversitesi Mühendislik Fakültesi Yayınları, Publication No. 19, Ankara.

Parente, E., Martuscelli, M., Gardini, F., Grieco, F., Crudele, M.A., and Suzzi, G. 2001. Evolution of microbial populations and biogenic amine production in dry sausages produced in Southern Italy. *J. Appl. Microbiol.* 90:882–891.

Pikul, J., Leszcynski, D.E., and Kummerow, F.A. 1989. Evaluation of three modified TBA methods for measuring lipid oxidation in chicken meat. *J. Agric. Food Chem.* 37:1309–1313.

Sanz, Y., Villa, R., Toldra, F., Nieto, P., and Flores, J. 1997. Effect of nitrate and nitrite curing salts on microbial changes and sensory quality of rapid ripened sausages. *Int. J. Food Microbiol.* 37:225–229.

Sarıçoban, C., Karakaya, M., and Caner, C. 2006. Properties of Turkish-style sucuk made with different combinations of beef and hen meat. *J. Muscle Foods* 17:1–8.

Şenöz, B., Isikli, N., and Coksoyler, N. 2000. Biogenic amines in Turkish sausages (sucuks). *J. Food Sci.* 65:764–767.

Shalaby, A.R. 1993. Survey on biogenic amines in Egyptian foods: Sausage. *J. Sci. Food Agric.* 62:291–293.

Shalaby, A.R. 1996. Significance of biogenic amines to food safety and human health. *Food Res. Int.* 29:675–690.

Stu, G.M. and Draper, H.H. 1978. A survey of the malonaldehyde contents of retail meats and fish. *J. Food Sci.* 43:1147–1149.

Talon, R., Walters, D., Chartier, S., Barriere, C., and Montel, M.C. 1999. Effect of nitrate and incubation conditions on the production of catalase and nitrate reductase by Stapphylococci. *Int. J. Food Microbiol.* 52:47–56.

Turantaş, F. and Ünlütürk, A. 1993. Sucukta nitrit, sarımsak ve starter kullanımının *Staphylococcus aureus*' un gelişmesi üzerine etkisi. *Turk. J. Eng. Environ. Sci.* 17:275–280.

Turgut, H. 1986. Et ürünleri teknolojisinde kaliteye etki eden faktörler. In: Et mamülleri üretimi ve muhafazası. İstanbul Ticaret Odası, Publication No. 1987-3:7–31.

Vidal-Carou, M.C.V., Pulido, M.L.I., Morro, M.C.M., and Font, M. 1990. Histamine and tyramine in meat products: Relationship with meat spoilage. *Food Chem.* 37:239–249.

Vural, H. and Öztan, A. 1991. Et ürünlerinde nitrosamin oluşumunun laktik asit bakterileri kullanımıyla önlenmesi. *Gıda* 16:237–240.

Vural, H. and Öztan, A. 1992. Fermente et ürünlerinde nitrosomyoglobin oluşumu ve etkileyen faktörler. *Gıda* 17:191–196.

Vural, H. and Öztan, A. 1996. Et ve ürünleri kalite kontrol laboratuvarı uygulama kılavuzu. Food Engineering Department, Hacettepe University, Publication No. 36.

Vural, H., Javidipour, I., and Ozbas, O.O. 2004. Effects of interesterified vegetable oils and sugarbeet fiber on the quality of frankfurters. *Meat Sci.* 67:65–72.

Yıldız-Turp, G. and Serdaroğlu, M. 2008. Effect of replacing beef fat with hazelnut oil on quality characteristics of sucuk—A Turkish fermented sausage. *Meat Sci.* 78:447–454.

40

Safe Practices for Sausage Production in the United States

Y. H. Hui

CONTENTS

40.1 History of Sausages

The remaining information in this chapter has been modified from the document "Safe Practices for Sausage Production: Distance Learning Course Manual, 1999," available at the Web sites of

1. Food Safety Inspection Service of the United States Department of Agriculture (USDA, FSIS, www.FSIS.USDA.gov)
2. Food and Drug Administration (FDA, www.FDA.gov)
3. Association of Food and Drug Officials (AFDO, www.AFDO.org)

The process of preserving meats by stuffing salted, chopped meats flavored with spices into animal casings dates back thousands of years, to the ancient Greeks and Romans, and earlier. The word "sausage" is derived from the Latin word "salsus," which means salted or preserved by salting. Sausages and sausage products have since evolved into a wide variety of flavors, textures, and shapes resulting from variations in ingredients and manufacturing processes.

In the United States, expansion in the meat packing industry during the Civil War, along with development of refrigeration for use in railroad cars and slaughtering facilities, provided an incentive for meat processors to create sausage products that could utilize cheaper, perishable cuts of meat, along with scrap trimmings and offal products. In addition, persons from various nationalities and ethnic groups immigrated to the United States, bringing with them traditional recipes and manufacturing skills for creating a wide range of sausage types.

In the early 1900s, a series of scandals, including the publication of Upton Sinclair's *The Jungle*, exposed undesirable practices in the meat processing industry, including the practice of adding variety meats and offal products to a sausage product without identifying the ingredients to the consumer. The federal government responded in 1906 with the Federal Meat Inspection Act.

Today, the sausage manufacturing industry must adhere to government standards for ingredients and processes. In addition, accurate labeling requirements ensure that the consumer is informed of the ingredients of a sausage product. The objective of these standards is to help ensure that sausage products maintain a consistent quality and are safe to consume.

This chapter discusses the manufacture of fermented sausages and sausage products with an emphasis on the safety issues associated with the production of sausage products. Throughout, we cover a range of issues related to sausage production, including

1. The sausage production process
2. Safety and sanitation issues at each stage of the process
3. Pathogens of concern to sausage makers

The objectives of this chapter include

1. Describe the stages of the sausage production process, along with the production, safety, and sanitation issues as appropriate to each stage of the process
2. Identify the range of pathogens of concern to sausage makers

Many of the principles that apply to large food processors also apply to retail operations. This chapter reviews the sausage production process, identifies critical food safety areas within the process, and reviews the procedures that should be followed to ensure that only safe products are produced.

Increasingly stringent food safety standards are driving food industries internationally, nationally, and at the state level to adopt a hazard analysis critical control point (HACCP) system or plan. HACCP is now used to increase product safety. HACCP is also good business: it results in decreased liability potential for the establishment and increased consistency and consumer satisfaction in the products produced.

HACCP plans are specifically required for retail facilities that use smoking, curing, and acidifying, use food additives, use alternative cooking time/temperature combinations, or use reduced oxygen packaging. The FSIS technical staff has access to recognized meat safety standards that have been developed based on years of scientific study and practical experience.

Details about HACCP plans are explored in several chapters in this book.

40.2 Examples of Sausage Products

40.2.1 Fresh Pork Sausage

Several categories of sausage products are typically consumed in the United States, each with specific production processes and storage requirements. These categories include fresh sausages such as fresh pork sausages, cooked sausages such as frankfurters, and dry sausages such as meat sticks.

Fresh pork sausages are produced from selected cuts of fresh and sometimes frozen pork, pork trimmings, and water, along with seasonings. Because fresh sausages do not contain curing agents, and are neither cooked nor smoked, they require refrigeration. These types of sausages must be thoroughly cooked before serving.

40.2.2 Frankfurter

Frankfurters are examples of cooked and smoked sausages. They are produced from fresh meat that is fully cooked and have flavors that are imparted through the addition of curing ingredients and via various cooking and smoking processes. Although they are fully cooked, they are not shelf-stable and must be refrigerated until the time of consumption.

40.2.3 Meat Stick

Meat snack sticks are produced using controlled, bacterially induced fermentation to preserve the meat and impart a special flavor, along with a long drying period to cure and preserve the meat. These processes produce a shelf-stable product.

In this chapter, the production process and production requirements for the last category of sausages are discussed.

40.3 Producing Sausages

In this section, the main steps that are utilized to produce standard dry sausage products are enumerated and explained.

40.3.1 Basic Procedures

40.3.1.1 Grinding Meat Ingredients

The main processes used to produce sausages are as follows:

1. Grinding meat ingredients

2. Adding non-meat ingredients
3. Blending
4. Stuffing
5. Packaging

The first step in sausage production is grinding the ingredients. The grinding stage reduces the meat ingredients into small, uniformly sized particles. Ground meat is the primary ingredient in a sausage formulation. The characteristics of the meat ingredients used to create the sausage define the type of sausage—the overall taste, texture, and aroma, along with the protein and fat content.

A variety of raw meat ingredients are utilized in the sausage production process. Each ingredient contributes a specific property to the final sausage formulation.

Meats must be clean, sound, and wholesome. These products should be inspected when arriving at the facility, and just prior to use, to ensure that they were not contaminated during transit or handling.

The specific meats used in a sausage formulation must be correctly identified by type and quantity.

Prior to grinding, the meat is held in cold storage. Often processors prefer to chill the meat to below 30°F to minimize the potential for fat smearing. The grinder blades must be sharp and matched with the grinding plate to ensure an efficient grind without generating extra heat during the grinding process.

Grinding processes will vary according to the manufacturer and the nature of the product. Some sausage products use coarsely ground meats; others use more finely ground meat ingredients. Some manufacturers grind the lean and fat trimmings separately, grinding the lean trimmings to a finer consistency than the fat meats.

40.3.1.2 Adding Non-Meat Ingredients

There are many non-meat ingredients that are essential to the sausage making process. These non-meat ingredients stabilize the mixture and add specific characteristics and flavors to the final product. Ingredients used in sausage include water, salt, and antioxidants, along with traditional spices, seasonings, and flavorings.

Paprika is a spice that is considered as both a flavoring and a coloring agent because of its strong red color. For this reason, paprika or oleoresin of paprika is traditionally expected in sausages, such as Italian sausage and chorizo.

The amount of non-meat ingredients, such as spices, is determined by the overall weight of the product mixture. Because the amounts of these ingredients must be carefully controlled, and measuring very small amounts of numerous specific ingredients within a manufacturing environment is often not practical, many manufacturers use a commercially pre-measured and packaged mix of these ingredients.

40.3.1.3 Meat Ingredients' Binding Ability

Skeletal beef muscle tissue from bull, cow, and shank meat have the highest binding capabilities. Intermediate- or medium-binding meats include head meat, cheek meat, and lean pork trimmings. Low-binding meats contain higher levels of fat and are typically non-skeletal meats, such as jowls, fat, briskets, hearts, and tongue trimmings.

Some meat by-products have minimal binding ability, so the use of these products in sausage is generally avoided. These products include tripe, pork stomachs, lips, snouts, and skin.

Non-meat proteins, such as cereal flours and non-fat dry milk must also be uniformly distributed; however, these ingredients usually are added last because they absorb water that is required for protein emulsification.

40.3.1.4 Emulsification

Some manufacturers run the mixture through an emulsifier after the blending stage to further reduce the size of the meat particles to achieve a very fine texture. In the emulsification process, fat, protein, salt, and water are mixed and combined into a semi-fluid emulsion. The meat muscle protein, called "myosin," is solubilized or released from the muscle fibers by contact with salt. The solubilized protein and water combine and surround the fat globules and hold the fat particles in a disbursed suspension within the mixture, along with spices and seasonings.

Careful control of the amount of each ingredient and the grinding process is essential. The manufacturer must select a mix of raw meat materials with the appropriate emulsifying and binding characteristics. Emulsifying properties are dependent on the hydrophobic (water repulsing) and hydrophilic (water attracting) characteristics of the specific protein. Myosin protein has the best emulsifying properties (possessing both hydrophobic and hydrophilic properties); whereas proteins such as collagen have little or no emulsifying properties and tend to break down at higher temperatures during the cooking. Manufacturers typically limit the use of low-binding materials to less than 15% of the product formulation to ensure finished product quality.

If too much fat is added, or overchopping of the ingredients exposes too much fat, there may not be sufficient protein to encapsulate the fat. This creates a condition called "shorting out," "greasing out," or "fat capping." In this case, fat droplets migrate to the product surface and form small pockets of fat called "fat caps." The emulsion can also break down when the mixture is chopped too long, causing too short protein fibers, or the product temperature is increased, causing some of the fats to partially render. In this condition, the protein is unable to hold liquid fat in suspension. In addition, emulsion breakdown can occur in products with a high acid content, or when heating coagulates the protein before it has a chance to surround the fat and "set" the emulsion.

40.3.1.5 Blending

Manufacturers carefully control the blending of the meat and non-meat ingredients to create the desired characteristics for a specific sausage formulation.

The meat and non-meat ingredients are placed in a mixer and thoroughly blended.

The manufacturer must monitor and control the blending process because excessive mixing can cause the salts in the formulation to break down excessive amounts of protein, or the friction created by the blending process can increase the product temperature and cause fats to partially render. Excessive handling also cuts protein fibers too short. All of these problems could result in product quality defects.

The blending process must also obtain a uniform distribution of any non-meat ingredients within the product formulation. For example, flavorings, salts, and other ingredients must be consistently mixed throughout a sausage formulation.

40.3.1.6 Regenerated Collagen Sausage Casings

Regenerated collagen sausage casings are made from collagen extracted from cattle hides and hog skins in a process called regeneration. The extracted collagen is dissolved, then hardened, washed, swelled with acid, and finally formed into the tubular casing shape in an extrusion process. This final shape is then fixed in an alkali bath.

These types of lower strength casings are typically used for smaller diameter products.

40.3.1.7 Synthetic and Cellulose Casings

Synthetic or artificial casings are made from special papers impregnated with cellulose, saran casings made from synthetic plastics, and hydro-cellulose casings made from regenerated cellulose. Cellulose casings are created from dissolved fibers extracted from cotton seeds or paper pulp. Each of these types of casings is available in a wide range of sizes and characteristics and is easy to handle; however, these types of casings are not edible and must be removed from the sausage prior to consumption. Artificial casings provide high strength and are available with excellent permeability to moisture and smoke, or a semi-permeable casing for use in producing water-cooked products such as braunschweiger.

40.3.1.8 Stuffing

After the blending is complete, the blended ingredients may be bulk packaged, or they may be extruded into a casing. This process is called stuffing.

Sausages are stuffed into casings using standard equipment. However, manufacturers of these products utilize a wider range of casing materials.

Natural casings are derived from the stomach and intestines of hogs, the intestines, bung, and bladder of cattle, and the intestines of sheep. They are edible, and allow smoke and moisture to permeate the sausage during processing.

Hog middles are produced from the large intestine. The small intestine is utilized to create small hog casings. In addition, hog bladders are also used as casings. Sheep casings are generally made from the small intestine.

Beef rounds are casings that are produced from the small intestine. Beef middles are larger casings created from the large intestine and are used to produce sausages such as bologna, cervelat, and salami.

Sometimes manufacturers use casings that are not derived from the same type of meat as the main meat ingredients of the sausage. For example, a manufacturer can use a pork casing on a beef-and-lamb combination sausage.

40.3.1.9 Stuffing Process

The stuffing process can be accomplished in a number of ways. Natural casings are typically flushed with water, and the mixture is injected into the casing at a pressure that is sufficient to fill the casing without leaving any air pockets and without tearing the casing. The stuffing process is also sometimes conducted at lower temperatures (<35°F–38°F) to minimize fat smearing on the casing.

Smaller volume or specialty producers may stuff the formulation into the casing by hand or from a screw feed. These small operations may also bypass choppers, mixers, and stuffers and stuff the output of the grinder directly into the casing. Larger manufacturers may use air or water piston-type automatic stuffers.

The stuffed casings are then separated into uniform segments of equal length in a process called linking. These segments form the single sausage portions. The linking process is typically accomplished by twisting the casing.

40.3.1.10 Cooking

Manufacturers cook sausage products to enhance the flavor and color, to produce the desired final product, and to inhibit the bacteria responsible for spoilage. In order to produce a safe product, cooking must also destroy parasites and pathogenic bacteria. Sausages can be cooked through immersion in a heated water bath, within a smokehouse environment, or within ovens.

The cooking process is carefully controlled to ensure that the product reaches a specific internal temperature for a defined period of time. Thermocouples are used to monitor the temperature during the cooking process.

40.3.1.11 Showering/Chilling

After the smoking and cooking phase, the sausage product undergoes a cold water shower. Showering maintains the product humidity and stops the cooking process by reducing the product temperature as quickly as possible. Showering also helps to prevent shrinkage and wrinkling of the product casing.

The cooling process is also carefully monitored and controlled. The temperature of the cooked product must be lowered to a specific temperature within a desired time. The FDA Food Code specifies that a product must be lowered from 140°F to 70°F within 2 h, and then from 70°F to 41°F or less within an additional 4 h. To meet the FSIS stabilization requirements of no more than 1-log growth of *Clostridium perfringens*, the uncured product can be cooled from 130°F to 80°F with 1.5 h and 80°F to 40°F in 5 h.

40.3.1.12 Peeling

Peelers are used by the sausage manufacturer to remove non-edible casings from sausage products during the later stages of the production process.

40.3.1.13 Packaging

The fresh sausage product is sometimes packaged for sale to the customer. The product may be wrapped in a gas impermeable plastic and placed into refrigerated storage or display. The specific packaging will vary according to the needs of the end user; however, the processor must follow hygienic standards when packaging any sausage product to avoid contaminating the product. Often retail fresh sausage is tray packed.

40.3.2 Dry Sausages

Producers of dry sausages and semi-dry sausages utilize controlled, bacterially induced fermentation to preserve the meat and impart flavor. The most common examples of dry sausages are salami and pepperoni. The dry category also includes shelf-stable non-fermented products such as beef jerky. Again, there are many variations of process steps and ingredients, resulting in the vast array of products available to the consumer. We are not going to deal with all possible products; rather, we concentrate on the most common examples of these types of products. In addition to the standard production stages identified for fresh, cooked, and smoked sausages, dry sausages have additional production steps, including

1. Blending special curing ingredients
2. Drying process

Let us briefly review each of these additional special production steps used to produce dry and semi-dry sausages.

40.3.2.1 Blending Special Curing Ingredients

The meats used in dry sausages are typically ground or chopped at low meat temperatures (20°F–25°F) to maintain the well-defined fat and lean particles that are desired in this type of sausage. The ground meats and spices are then mixed with curing ingredients, such as nitrates, nitrites, antioxidants, and bacterial starter cultures. Salts have traditionally been used to help preserve sausages. Eventually, producers learned that nitrates and nitrites in the salts were essential to the curing process. Manufacturers of dry and semi-dry sausages use a curing agent consisting of salt and nitrates and/or nitrites. Approved antioxidants may also be added to protect flavor and prevent rancidity and are limited to 0.003% individually or 0.006% in combination with other antioxidants. These ingredients must be uniformly distributed throughout the mixture to achieve the maximum microbiological stability.

Bacterial fermentation is then used to produce the lower pH (4.7–5.4) that results in the tangy flavor associated with this type of sausage. In earlier days of sausage making, the bacteria growth was uncontrolled, resulting from bacterial contamination of the meat or production equipment, producing unreliable results. Traditionally, sausage makers held the salted meat at low temperatures for a week or more to allow plenty of time for the lactic acid bacteria coming from the environment to reproduce in the meat mixture. Unfortunately, this traditional method was not always reliable and subject to several errors such as cutting the time short, adding too much or too little salt, or growth of the wrong lactic acid bacteria. Therefore, modern producers of dry sausage use a commercial lactic bacteria starter culture and simple sugars, such as dextrose or corn sugar, which promote lactic acid bacterial growth by serving as food to fuel the bacteria during fermentation. The bacteria starters are harmless and are limited to 0.5% in both dry and semi-dry sausage formulations. Commercial bacteria starter cultures typically consist of a blend of microorganisms such as *Pediococcus*, *Micrococcus*, and *Lactobacillus*, using specific species such as *Pediococcus cerevisiae*, *P. acidilactici*, *Micrococcus aurantiacus*, and *Lactobacillus plantarum*. The bacterial fermentation lowers the sausage pH by producing lactic acid. This lower pH also causes the proteins to give up water, resulting in a drying effect that creates an environment that is unfavorable to spoilage organisms, which helps to preserve the product. However, mold may grow and could become a problem.

Some small producers may still allow the mixture to age in a refrigerator for several days to encourage the fermentation process, even though they are using starter cultures. However, most modern processors stuff the mixture directly into casings, which then undergoes a fermentation process at about 70°F–110°F (depending on the starter type). This fermentation process is designed to allow the bacteria to continue to incubate. The fermentation occurs during a 1- to 3-day process that takes place in a "greening room" that provides a carefully controlled environment designed to obtain specific fermentation results. Temperatures are typically maintained at approximately 75°F and relative humidity at 80%. Semi-dry sausages are usually fermented for shorter periods at slightly higher temperatures.

40.3.2.2 Cooking/Smoking

What happens after fermentation depends on the type of product being produced. Semi-dry sausages like summer sausage are almost always smoked and cooked before drying. Dry sausages like pepperoni are rarely smoked and may or may not be cooked. Today, some establishments choose to heat-treat these dry sausages as a critical step designed to eliminate *Escherichia coli*. A moist heat process may be used for some products. This process utilizes a sealed oven or steam injection to raise the heat and relative humidity to meet a specific temperature/time requirement sufficient to eliminate pathogens (e.g., 130°F minimum internal temperature for 121 min, or 141°F minimum internal temperature for 10 min). The product is then sent on to the drying stage of the process.

40.3.3 Drying Process

The drying process is a critical step in ensuring product safety. FSIS requires that these products undergo a carefully controlled and monitored air drying process that cures the product by removing moisture from the product. Pork-containing products must be treated to destroy trichinae. Sausages not containing pork have no such requirement.

Manufacturers are required to control the ratio of moisture to protein in the final product. The moisture/protein ratio (MPR) is controlled by varying the amount of added water based on the overall product formulation and primarily by the drying procedure. In some products, the MPR can affect the final microbiological stability of the product. In other products, the MPR is important to ensure elements of the overall product quality, such as the texture. The minimum requirement for all products produced in FSIS-inspected facilities is that they must meet the FSIS policy standards for MPR (see Table 40.1). However, these prescribed treatments have been proven to be insufficiently lethal for some bacterial pathogens. Thus, most of the industry has volunteered to implement a more rigorous treatment.

The drying process consists of placing the product in a drying room under a relative humidity of 55%–65%, in a process that can last from 10 days to as long as 120 days, depending on the product diameter, size, and type. The drying process is designed to produce a final product with approximately 30%–40% moisture and an MPR generally of 1.9:1 or less to ensure proper drying and a safe product. Facilities must keep accurate records of the temperature and the number of days in the drying room for each product manufacturing run to help ensure product safety and consistency. The drying environment is controlled to ensure that the drying rate is slightly higher than the rate required to remove moisture from the sausage surface as it migrates from the sausage center. Drying too quickly will produce a product with a hard and dry casing. Drying too slowly results in excessive mold and yeast growth and excessive bacterial slime on the product surface.

The controlled drying process is designed to reduce moisture to the point where the final product has a specific MPR. This process will vary depending on the specific facility operation and the specifications desired for the final product. For example, pepperoni is required to maintain an MPR of 1.6:1, indicating a requirement of 1.6 parts moisture to one part protein. Genoa salami should have an MPR of 2.3:1, and all other dry sausages an MPR of 1.9:1. Inspectors may sample the products to determine compliance with the specified MPR.

TABLE 40.1

Moisture/Protein Ratio (MPR) for Dry and Semi-Dry Sausages

Sausage Product	Maximum Moisture to One Part Protein
Jerky, Pemmican	0.75:1
Pepperoni	1.6:1
Dry sausage (e.g., hard salami)	1.9:1
Summer sausage, shelf-stable	3.1:1 + pH 5.0
Dry salami	1.9:1
Genoa salami	2.3:1
Sicilian salami	2.3:1
Italian salami	1.9:1
Thuringer (semi-dry sausage)	3.7:1
Ukrainian sausage	2.0:1
Farmer Summer sausage	1.9:1

Semi-dry sausages are prepared in a similar manner but undergo a shorter drying period, producing a product with a moisture level of about 50%. These products are often fermented and finished by cooking in a smokehouse, at first at a temperature of approximately 100°F and a relative humidity of 80%. The temperature is later increased to approximately 140°F–155°F to ensure that microbiological activity is halted. Because the moisture level of the final product is about 50%, semi-dry sausages must be refrigerated to prevent spoilage. Examples of semi-dry sausages include summer sausage, cervelat, chorizo, Lebanon bologna, and thuringer.

Table 40.3 identifies the FSIS MPR policy standards for some dry and semi-dry sausage products. Although these are the FSIS standards for dry and semi-dry sausages, they can provide a basis for state and local decisions about whether specific processes are adequately designed to ensure product safety.

To ensure that the fermentation and drying processes are sufficient to reduce or eliminate any pathogens present in the product, the procedures must be validated to demonstrate that they achieve a specific reduction in organisms (e.g., a 5-log reduction in *E. coli* 0157:H7).

40.4 Sausage Production Equipment

In this section, we take a brief look at some of the standard equipment and processes that are used to produce sausages, along with the safety issues that should be addressed for each item. There are many different manufactures of equipment, and many variations of these basic equipment types can be seen.

40.4.1 Grinders

Grinders are used to chop the meat ingredients into small pieces. Meat is fed from a hopper, passed along a cylinder with an auger or worm, to a perforated plate where it is sliced away by a series of revolving blades. The friction produces a rise in temperature, which potentially could result in bacterial growth if not controlled.

40.4.2 Blenders

Blenders use screw-like agitators to mix the solid ground meat mixture with any additional non-meat ingredients to create a consistent distribution of all ingredients within the formulation.

40.4.3 Bowl Choppers

Bowl choppers (silent cutters) utilize a series of knives that chop and mix the product formulation. The friction of the knives passing through the meat will raise the temperature of the meat mixture.

40.4.4 Stuffers

Stuffers are used to extrude the sausage formulation into the casing. The ground mixture is fed from a hopper into a reservoir and forced through a nozzle into the casing.

Because grinders, blenders, choppers, and stuffers come into contact with the product formulation, they must be cleaned daily, cleaned every time that a new formulation is processed, and kept free of excess product during the process. This will ensure that any contamination is not passed on to the product formulation.

40.5 Packaging Equipment

The final sausage product is often packaged prior to sale. Packaging typically consists of wrapping the product in plastic film. The packaging protects the product from cross-contamination and helps to retard spoilage caused by contact with the air.

40.6 Ingredients in Sausages

In this section, we go over the types of ingredients that are used to produce standard dry sausages.

40.6.1 Ensuring Wholesomeness

All meat and meat by-products used in sausage formulations must be clean, wholesome, and properly labeled. Receiving of raw ingredients is an important step in ensuring food safety. The manufacturing facility should inspect all incoming meat to ensure that it is not contaminated. Even previously inspected meat should be re-inspected to ensure that it has not become contaminated during transit.

40.6.2 Properly Identifying Meat Ingredients

FSIS regulations require that the manufacturer properly identify the types and amounts of meat and meat by-products present in the final product. The actual ingredients must match the ingredients listed on the label.

However, once the meat products are chopped or ground, they lose their visual identity, making it difficult to identify specific ingredients in a formulation. Sausage manufacturers are therefore required to carefully identify and track ingredients throughout the production process and ensure that unlisted meat items are not accidentally or purposely substituted during the production process.

40.6.3 "Rework"

Sometimes a product that has partially or fully completed the production cycle is not salable but still wholesome and can be used for food. For example, the casing of some sausages may split during the cooking or smoking cycle. Manufacturers are allowed to reuse these edible but un-salable products by removing the casing and adding the contents to the grinder to include in another run of the same product. Manufacturers are not allowed to use this "rework" in a product with a different list of ingredients. Rework has little binding ability because the proteins are coagulated, so that the amounts added are self-limiting, in that it has a detrimental effect on product quality.

40.6.4 Non-Meat Ingredients

A number of non-meat ingredients are essential to the sausage making process. These non-meat ingredients stabilize the mixture and add specific characteristics and flavors. These typically include extenders and binders, water, salt, nitrite, nitrate, ascorbates, erythorbates, sugars, antioxidants, phosphates, mold inhibitors, and extenders, along with traditional spices, seasonings, and flavorings. In this section, we look at the effect and use of each of these ingredients.

40.6.4.1 Binders and Extenders

Binders and extenders have a number of uses in a sausage formulation. Manufacturers use extenders such as dry milk powder, cereal flours, and soy protein as a lower cost method to increase the overall yield of the formulation, to improve binding qualities and slicing characteristics, and to add specific flavor characteristics. A sausage formulation can include up to 3.5% of these substances.

40.6.4.2 Water

While water is a naturally occurring component of meat, manufacturers also add water to the formula in specific amounts to improve the consistency of the mixture and to substitute for fats. FSIS regulations permit manufacturers of fresh sausages to add water up to 3% of the total product weight.

Cooked sausage manufacturers are allowed to vary the amount of added water according to the amount of fat. The maximum fat content is limited to 30%, and the amount of fat and water combined is limited to 40%, so the manufacturer can increase water to substitute for reduced fat. Typically, the amount of naturally occurring water is determined by computing four times the protein content. Any moisture above that amount is considered added water.

40.6.4.3 Salt

Salt is an essential ingredient of any sausage formulation. It is used to preserve the product, enhance the flavor, and solubilize the meat proteins in order to improve the binding properties of the formulation.

Since the advent of refrigeration, the preservative properties are the least important use of salt, though dry sausages still use salt for preservation. A salt concentration of around 17% is necessary for preservation to be effective.

The most important use of salt in a sausage product is its ability to solubilize proteins. This enhances the product texture and improves water and fat binding.

Because sodium chloride (NaCl) salt has been linked to hypertension, other non-sodium salts such as potassium and calcium chlorides are sometimes substituted for a portion of the sodium chloride.

40.6.4.4 Curing Agents

Curing agents such as nitrite and nitrate have traditionally been used in sausage formulations, originally as a contaminant present in salts, and later added intentionally in the form of saltpeter. Nitrites provide bacteriostatic and antioxidant properties and improve the taste and color of the sausage. They help to control the growth of bacteria in vacuum-packed products such as frankfurters and luncheon meats and prevent the outgrowth of bacteria such as the lethal *Clostridium botulinum* bacterium that causes botulism. Nitrites also inhibit the oxidation of fats in meats, reducing the development of oxidative rancidity. They produce the desired reactions much faster and are much more commonly used than nitrates. The use of nitrate by large processors is rare because the process of converting the nitrate into nitrite within the product is much slower and less reliable than addition of nitrite directly.

Because nitrites and nitrates can be toxic to humans, the use of these ingredients in sausage formulations is carefully controlled. They are sometimes referred to as "restricted ingredients." Supplies of sodium nitrite and potassium nitrite and mixtures containing them must be kept securely under the care of a responsible employee of the establishment. The specific nitrite content of such supplies must be known and

TABLE 40.2

Maximum Levels for Nitrate and Nitrite Permitted in Chopped Meat

Substance	Amount
Sodium or potassium nitrate	2¾ oz. per 100 lb. chopped meat
Sodium or potassium nitrite	1/4 oz. per 100 lb. chopped meat

clearly marked accordingly. The maximum level of these additives that is acceptable is spelled out in the FSIS regulations, as indicated in Table 40.2, showing the maximum levels for nitrate and nitrite permitted in chopped meat. The amount of nitrite added to product must be regulated at the formulation step based on the total amount of meat and meat by-products. Nitrites dissipate quickly in the finished product, and the parts per million in the finished product does not necessarily reflect the amount that was used in formulation. This makes sampling the finished product for nitrite an impractical control measure.

40.6.4.5 Cure Accelerators

Cure accelerators such as ascorbates and erythorbates are used to speed the curing process. They also stabilize the color of the final product (Table 40.3).

40.6.4.6 Sugars

Sugars are used in sausage formulations to reduce the flavor intensity of the salt and flavorings and to provide a food source to enable microbial fermentation. Sugars used in sausage products include sucrose and dextrose.

40.6.4.7 Antioxidants

Antioxidants are approved for use in fresh sausages to retard oxidative rancidity and protect flavor. Approved antioxidants include butylated hydroxytoluene, butylated hydroxyanisole, propyl gallate, tertiary butylhydroquinone, and tocopherols. These compounds are added to the spice mixtures based on the actual percentage of fat in the fresh product formulations (typically 0.01% separately and 0.02% in combination) or the total meat block weight for dry sausage formulations (typically 0.003%).

40.6.4.8 Phosphate

Phosphates are used to improve the water-binding capacity of the meat, solubilize proteins, act as antioxidants, and stabilize the flavor and color of the product. Their maximum benefit to the processor is to

TABLE 40.3

Maximum Levels for Cure Accelerators Permitted in Meat

Ingredient	Maximum Amount
Ascorbic acid	3/4 oz. per 100 lb. meat
Erythorbic acid	3/4 oz. per 100 lb. meat
Sodium erythorbate	7/8 oz. per 100 lb. meat
Citric acid	May replace up to 50% of above listed ingredients
Sodium citrate	May replace up to 50% of above listed ingredients
Sodium acid pyrophosphate	Alone or in combination with others may not exceed 8 oz. (0.5%)
GDL	8 oz. per 100 lb. meat

reduce purge or water that is cooked out of product. Phosphates also help to increase the shelf life of a product. The maximum amount of phosphates approved for sausage products is limited to 0.5% of the finished product weight. If used, they must be food grade.

40.6.4.9 Mold Inhibitors

Mold is a commonly encountered problem in the production of dry sausages. The common technique for inhibiting the growth of mold is to dip the sausage in a mold inhibitor solution, typically 2.5% solution of potassium sorbate or a 3.5% solution of propylparaben.

40.6.4.10 Glucono-Delta-Lactone

Glucono-delta-lactone (GDL) is a cure accelerator that produces an acid tangy flavor similar to flavor resulting from natural fermentation. It is allowed at 8 oz. per 100 lb. of meat.

40.6.4.11 Spices, Seasonings, and Flavorings

Spices, seasonings, and flavorings are used to add flavor to the sausage and also affect the consistency of the ground mixture. The wide range of available spices, seasonings, and flavorings is a primary reason for the variety available in sausages.

Spices are defined as any aromatic vegetable substance that is intended to function as contributing flavoring in food instead of contributing to the nutritional substance of the food. The active aromatic or pungent properties of spices that contribute the most to the flavoring effect are mostly present in the volatile oils, resins, or oleoresins of the spice. These properties are present in the whole spice or in extracts of the active components. The use of spice extracts has some advantages over using whole spices, including providing more control over the intensity of the flavor, less opportunity for microbial contamination, easier storage, and a less conspicuous visual appearance compared with spice particles.

Spice extracts must be labeled as *Flavorings* in the product ingredients list. Flavorings are substances that are extracted from a food (such as fruits, herbs, roots, meats, and seafoods) that are also intended to contribute flavoring instead of nutritional substance.

Seasonings is another general term that refers to any substances that are used to impart flavor to the food product. Some examples of common spices and seasonings include allspice, pepper, cardamom, caraway, coriander, cumin, garlic, sage, mustard, nutmeg, paprika, pepper, rosemary, sage, thyme, and turmeric.

40.7 Sanitation Issues

As with any food product, proper worker hygiene, raw ingredient handling and storage procedures, and the final product handling and storage procedures are essential to control product contamination by organisms that are harmful to humans. Requirements governing the safe production of sausages have two-prong approaches:

1. Develop and implement sanitation standards
2. Develop and implement HACCP

Details on these programs are presented in several chapters. In this section, we look at some of the sanitation issues. The next section uses the production of pepperoni and salami to provide an outline of developing and implementing HACCP.

Raw ingredients are typically inspected prior to entering the sausage production facility/area. However, the manufacturer should not assume that these ingredients are free of pathogenic organisms. The manufacturer must ensure that raw ingredients are properly stored in refrigerated areas to minimize the opportunities for growth of pathogens. Workers handling these raw materials must not also handle completed sausage products because cross-contamination can occur. Equipment and facilities should be designed to prevent cross-contamination between raw and cooked products. In addition, of chapter, sausage production workers should follow the standard hygienic procedures that are required for all food production workers.

Sausage production equipment must be maintained in a clean and sanitary condition under conditions that minimize the potential for growth of pathogens. Grinding, blending, and stuffing equipment must be completely disassembled and cleaned after each use or at least daily. In addition, when changing to a batch of another species of meat, the entire grinding, blending, and stuffing assembly must be disassembled and cleaned.

Grinders of fresh sausage products should develop and implement rework, carry-over, and lot designation procedures that reflect an acceptable degree of product exposure (i.e., economic risk) in the event that a health risk is identified that results in recalling the product that is suspected of presenting a potential hazard to the public. This may include developing a rework tracking system.

Finally, the final products must be handled and stored according to acceptable standards to minimize the opportunities for cross-contamination and spoilage.

1. Safety of water—Water must be potable. Private wells or sources must be certified.
2. Condition and cleanliness of food contact surfaces—The portions of this area that should be monitored daily are cleaning and sanitizing of equipment, utensils, gloves, and outer garments that come in contact with food and the condition of gloves and outer garments.
3. Prevention of cross-contamination—The issues in the area of cross-contamination that should be monitored are employee practices and physical separation of raw and cooked products.
4. Maintenance of hand-washing, hand-sanitizing, and toilet facilities—The issues that should be monitored are the concentration of hand-sanitizing solutions and that the toilet facilities are in good repair.
5. Protection from adulterants and toxic compounds—Food must be protected from contaminants such as condensation, floor splash, glass, and toxic chemicals.
6. Employee health conditions—Employee health conditions must be monitored daily.
7. Pest control—Pests must be excluded from food handling areas.

40.8 Pepperoni and Salami Production: HACCP

All information (text and tables) in this section has been derived from public documents available at the website of USDA, www.FSIS.USDA.GOV.

The complete process of producing pepperoni and salami is governed by

1. Technical know-how
2. Legal requirements for safety

This chapter has discussed the technical know-how on how to produce sausages in the United States. Professional reference sources will be needed to manufacture specifically pepperoni and salami.

An outline of the legal requirements to produce these sausages safely is provided in four figures:

1. *Figure 40.1.* Pepperoni and salami: process flow diagram
2. *Figure 40.2.* Pepperoni and salami: product description
3. *Figure 40.3.* Pepperoni and salami: hazard analysis
4. *Figure 40.4.* Pepperoni and salami: HACCP plan

When studying the figures, please note the basic premise. It is assumed that the user has a minimal knowledge of the requirements of FSIS in meat and meat processing, especially in terms of terminology and abbreviations.

Process Flow Diagram

Process category: not heat treated, shelf stable
Product: pepperoni and salami

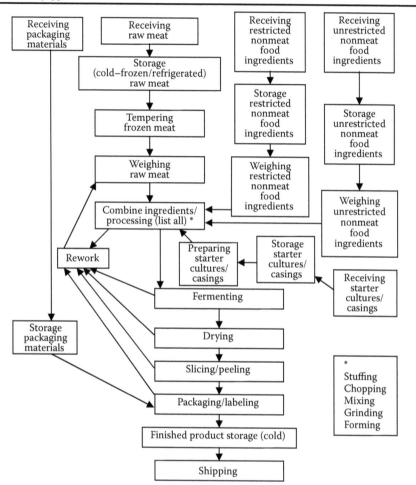

FIGURE 40.1 Pepperoni and salami: process flow diagram.

Product Description

Process category: not heat treated, shelf stable
Product: pepperoni and salami

1. Common name?	Pepperoni
	Salami
2. How is it to be used?	Consumed as purchased (ready to eat)
3. Type of package?	Bulk-packed (e.g., plastic bag, vacuum packed)
4. Length of shelf life, at what temperature?	Varies with packaging and storage temperature: may last 3 months non-refrigerated and 12 months under refrigeration
5. Where will it be sold? Consumers? Intended use?	Wholesale to distributors only
6. Labeling instructions?	Keep refrigerated
7. Is special distribution control needed?	Keep refrigerated

FIGURE 40.2 Pepperoni and salami: product description.

Hazard Analysis – Not Heat Treated, Shelf Stable – Pepperoni, Salami

Process Step	Food Safety Hazard	Reasonably Likely to Occur?	Basis	If Yes in Column 3, What Measures Could Be Applied to Prevent, Eliminate, or Reduce the Hazard to an Acceptable Level?	Critical Control Point
Receiving–Raw Meat	*Biological: Pathogens* *Salmonella* *E. coli O157:H7* *Listeria monocytogenes*	Yes	*Salmonella* and *E. coli* may be present on incoming raw product. Incoming presence of Lm may impact process control & growth.	Certification from suppliers that product has been sampled for *Salmonella* and *E. coli* O157:H7 meets FSIS performance standards. Fermentation & drying or use of post processing kill step could effectively control level.	1B
	Chemical–None				
	Physical–Foreign materials such as broken needles	No	Plant records show that there has been no incidence of foreign materials in products received into the plant.		
Receiving–Restricted and Unrestricted Nonmeat Food Ingredients; Starter Cultures/ Casings; Packaging Materials	Biological–None				
	Chemical–Not acceptable for intended use	No	Letters of guaranty are received from all suppliers of starter cultures, casings, and packaging materials.		
	Physical–Foreign materials (metal, glass, wood, etc.)	No	Plant records demonstrate that foreign material contamination has not occurred during the past several years.		

FIGURE 40.3 Pepperoni and salami: hazard analysis.

Hazard Analysis – Not Heat Treated, Shelf Stable – Pepperoni, Salami

Process Step	Food Safety Hazard	Reasonably Likely to Occur?	Basis	If Yes in Column 3, What Measures Could Be Applied to Prevent, Eliminate, or Reduce the Hazard to an Acceptable Level?	Critical Control Point
Storage–Restricted and Unrestricted Nonmeat Food Ingredients; Starter Cultures/ Casings; Packaging Materials	Biological–None Chemical–None Physical–None				
Storage (Cold–Frozen/ Refrigerated)–Raw Meat	Biological *Salmonella* *E. coli* O157:H7	Yes	*Salmonella* and *E.coli* O157:H7 are reasonably likely to grow in this product if temperature is not maintained at or below a level sufficient to preclude their growth.	Maintain product temperature at or below a level sufficient to preclude pathogen growth.	2B
	Chemical–None Physical–None				
Tempering Frozen Meat	Biological–Pathogens	Yes	Pathogenic microorganisms present are likely to grow if time/ temperature is not maintained at or below a level sufficient to preclude growth.	Control of time/temperature during thawing process. No water tempering.	
	Chemical–None Physical–None				

FIGURE 40.3 (Continued)

Hazard Analysis – Not Heat Treated, Shelf Stable – Pepperoni, Salami

Process Step	Food Safety Hazard	Reasonably Likely to Occur?	Basis	If Yes in Column 3, What Measures Could Be Applied to Prevent, Eliminate, or Reduce the Hazard to an Acceptable Level?	Critical Control Point
Weighing Raw Meat	Biological–None				
	Chemical–None				
	Physical–None				
Weighing Restricted and Unrestricted Nonmeat Food Ingredients; Preparing Starter Cultures/Casings	Biological–None				
	Chemical–None				
	Physical–None				
Combine Ingredients/ Processing (Includes one or more of the following: grinding, chopping, mixing, stuffing, forming, and slicing)	Biological–None				
	Chemical–None				
	Physical–Metal Contamination	Yes	Plant records show that during mechanical processing metal contamination is likely to occur.	Visual inspection prior to stuffing and/or metal detectors are installed prior to packaging.	3P
Rework	Biological–Pathogens	No	Rework at the end of the day is condemned.		
	Chemical–None				
	Physical–None				
Fermenting (Used for flavor development and pH reduction)	Biological–Pathogens (*Staphylococcus aureus*)	Yes	Potential growth and subsequent toxigenesis of pathogens with the failure of the fermentation process.	pH sufficient to ensure correct pH reached after fermentation.	4B
	Chemical–None				
	Physical–None				

FIGURE 40.3 (Continued)

Hazard Analysis – Not Heat Treated, Shelf Stable – Pepperoni, Salami

Process Step	Food Safety Hazard	Reasonably Likely to Occur?	Basis	If Yes in Column 3, What Measures Could Be Applied to Prevent, Eliminate, or Reduce the Hazard to an Acceptable Level?	Critical Control Point
Drying	Biological–Pathogens *Salmonella, Staphylococcus aureus, Trichina*	Yes	Potential growth and subsequent toxigenesis of pathogens with the failure of the drying process.	Room temperature can be controlled to assure the drying process is achieving the desired goal of moisture protein ratio.	5B
	Chemical–None				
	Physical–None				
Slicing/Peeling	Biological–Pathogens (*Listeria monocytogenes*)	Yes	Potential contamination from environmental sources and/or employee handling.	Sanitizer known to be effective against *Listeria monocytogenes* will be applied every 4 hours on product contact equipment.	6B
	Chemical–None				
	Physical–None				
Packaging/Labeling	Biological–None				
	Chemical–None				
	Physical–None				
Finished Product Storage (Cold)	Biological–None				
	Chemical–None				
	Physical–None				
Shipping	Biological–None				
	Chemical–None				
	Physical–None				

FIGURE 40.3 (Continued)

HACCP Plan

Process Category: Not Heat Treated, Shelf Stable

Product Example: Pepperoni and Salami

CCP# and Location	Critical Limits	Monitoring Procedures and Frequency	HACCP Records	Verification Procedures and Frequency	Corrective Actions
1B Receiving–Raw Meat	Supplier certification that product has been sampled for *Salmonella* must accompany shipment.	Receiving personnel will check each shipment for *Salmonella* certification.	Receiving log Corrective actions log	Every two months QA will request *Salmonella* data results from company for at least 2 suppliers.	Product without certification will not be accepted if a supplier fails to meet performance standards for 2 sample set. Supplier will not be used until a full sample set meets performance standards.
2B Storage (Cold–Frozen/ Refrigerated– Raw Meat/Poultry	Raw product storage areas will not exceed 40°F in refrigerated rooms or exceed 30°F in freezer rooms.	Maintenance personnel will record raw product storage area temperature every two hours.	Room temperature log Thermometer calibration log Corrective actions log	Maintenance supervisor will verify accuracy of the Room Temperature Log once per shift. QA will check all thermometers used for monitoring and verification for accuracy daily and calibrate to within 2°F accuracy as necessary. QA will observe maintenance taking & recording temperatures weekly.	QA will reject or hold product dependent on time and temperature deviation. Product disposition will be determined by effects of deviation. Process Authority will be consulted or cooling curves will be used to make a determination. QA will identify the cause of the deviation and prevent reoccurrence by adjusting maintenance schedule & repairing equipment as required.

Signature: _____ Date: _____

FIGURE 40.4 Pepperoni and salami: HACCP plan.

HACCP Plan
Process Category: Not Heat Treated, Shelf Stable
Product Example: Pepperoni and Salami

CCT# and Location	Critical Limits	Monitoring Procedures and Frequency	HACCP Records	Verification Procedures and Frequency	Corrective Actions
3P Combine Ingredients/ Processing	No metal particles to exceed 1/32 inches. All kick out product will be reworked to meet critical limit.	Maintenance personnel will check the metal detector every two hours to assure the kick out mechanism is working properly. All kick out product will be visually examined at the end of the shift or product line and results recorded.	Metal detection log Corrective actions log	Maintenance supervisor will verify metal detectors are functioning. QA will verify that the metal detectors are functioning as intended by running a seeded sample through the detector prior to start of each shift. QA will observe examination and rework of kick out product once per week. Kick out device will be tested each shift to determine it is functioning as intended.	Mechanical separation line supervisor will control and segregate affected product. Maintenance personnel will identify and eliminate the problem with the metal detector. Preventive maintenance program will be implemented. QA will run seeded sample through metal detector after repair. All potentially contaminated product will be run through metal detector, X-ray, or visually examined prior to further processing. No adulterated product will be shipped.
4B Fermenting	pH ... 5.3 within 6 hours.	QA technician will test pH of 10 sticks from each lot by probe during the fermentation process every 2 hours and at completion.	Fermentation log pH log Corrective actions log	QA supervisor will observe QA technician perform pH test once per shift. QA will check all pH meters used for monitoring and verification for accuracy daily and calibrate for accuracy daily.	QA will segregate and hold all affected product until correct pH is achieved or other appropriate disposition is determined based on the nature of the deviation, time at pH of product and food safety parameters. Starter culture will be checked for appropriate amount used, dispersion, and storage parameters. HACCP plan and process controls will be changed as required. QA will identify the cause of the deviation and prevent reoccurrence.

Signature: _____   Date: _____

FIGURE 40.4   (Continued)

HACCP Plan

Process Category: Not Heat Treated, Shelf Stable Product Example: Pepperoni and Salami

CCP# and Location	Critical Limits	Monitoring Procedures and Frequency	HACCP Records	Verification Procedures and Frequency	Corrective Actions
5B Drying	Reach established Moisture: Protein Ratio (MPR) Pepperoni 1.6:1, Salami 1.9:1	MPR checks will be done on each lot by production employee. Drying time/temperature will be monitored using room recorder charts.	Drying room recorder charts Thermometer calibration log Corrective actions Log MPR log	QA supervisor will review MPR log and drying room recorder charts once per shift and observe MPR check once per week. Maintenance supervisor will verify the accuracy of the drying room recorder once per shift. QA will check all thermometers used for monitoring and verification activities for accuracy daily and calibrate to within 2°F accuracy as necessary.	If a deviation from a critical limit occurs, the following corrective actions will be taken: 1. The cause of the deviation will be identified and eliminated. 2. The CCP will be monitored hourly after the corrective action is taken to ensure that it is under control. 3. When the cause of the deviation is identified, measures will be taken to prevent it from recurring, e.g., if the cause is equipment failure, preventive maintenance program will be reviewed and revised, if necessary. 4. QA will reject or hold product dependent on time/temperature deviation.

Signature: _____   Date: _____

FIGURE 40.4 (Continued)

HACCP Plan

Process Category: Not Heat Treated, Shelf Stable Product Example: Pepperoni and Salami

CCP# and Location	Critical Limits	Monitoring Procedures and Frequency	HACCP Records	Verification Procedures and Frequency	Corrective Actions
6B Slicing/ Peeling	Application every 2 hours of a sanitizer known to be effective against *Listeria monocytogenes* on product contact equipment.	QA will monitor the application and record the time of sanitizer application.	Sanitizer usage log Corrective actions log	QA will conduct a *Listeria* sampling program (both environmental and end product) as detailed in the FSIS issuance "*Listeria* Guidelines for Industry" to verify effectiveness of the sanitizer and its method of usage.	QA will address positive *Listeria* samples as detailed in the FSIS issuance "*Listeria* Guidelines for Industry." QA will stop slicing/peeling operations if time since last application of sanitizer exceeds 2 hours. All end product produced after the 2 hour limit is exceeded will be held until *Listeria monocytogenes* test results are final. If positive, product will be condemned and additional environmental and end product testing will be done until the source is determined. Further corrective actions will be done as detailed in FSIS regulations § 417.3.

Signature: _____ Date: _____

FIGURE 40.4 (Continued)

41

Malaysian Fermented Fish Products

Nurul Huda

CONTENTS

41.1 Introduction

In 2009, peninsular Malaysia landed around 1,064,422 tons of marine fish, which was 76.5% of the total landing of marine fish in Malaysia (Department of Fisheries 2010). The majorities of landed marine fish (63%) are consumed while fresh; the rest are frozen (1%), cured (11%), made into fish meal (20%), or disposed of by other means (5%). The category of cured fish includes dried/salted/smoked products (28%), steamed/boiled products (5%), fermented products (28%), and other products such as fish crackers and fish balls (39%). The most significant fermented fish product produced in Malaysia is *belacan*, followed by *budu* and *cincalok*. Another fermented fish product produced in Malaysia is *pekasam*. This chapter describes general processing methods of fermented fish products in Malaysia and presents some data describing their composition and microbial quality.

41.2 Malaysian Fermented Fish Products

41.2.1 Belacan

The largest share of fermented fish products from Malaysia is *belacan* or shrimp paste. *Belacan* is similar to the Indonesian fermented fish product *terasi*. *Belacan* is produced in all Malaysian states, in particular, Selangor. *Belacan* production has increased annually, from 3277 tons in 2000 to 27,305 tons in 2009 (Department of Fisheries 2010). The main raw material used to make *belacan* is small shrimp of the *Acetes* sp. or Mysid shrimp, known locally as *udang baring* or *udang geragok* (Daud 1978; Karim 1993). Physically, *belacan* is a thick, salty paste with a strong pungent shrimp odor (Figure 41.1). *Belacan* is consumed as a condiment or used as a flavoring ingredient in many Malaysian dishes (Yeoh and Merican 1978; Karim 1993). Jinap et al. (2010) found that dishes containing *belacan* had significant meat, shrimp, and salt flavors and astringency. *Belacan* is used as an ingredient in local dishes such as chili *belacan*, fried rice with *belacan*, spicy noodle soup *belacan*, stir-fried chili paste *belacan*, sour and spicy fish stew with *belacan*, a tamarind-flavored dish with *belacan*, stir-fried water convolvulus with *belacan*, and *belacan* fried chicken. Leong et al. (2009) reported that the frequency of *belacan* consumption by Malaysians was relatively regular, as the majority claimed to consume dishes containing *belacan* up to two to three times per week.

FIGURE 41.1 *Belacan* (shrimp paste).

The production of *belacan* begins with the washing of fresh small shrimp in seawater (Figure 41.2). The clean shrimp is mixed with salt in containers, such as bamboo baskets or wooden tubs, and then spread in a thin layer for drying. The semi-dried shrimp are then pounded or passed through a meat mincer (Yeoh and Merican 1978). The paste produced is then pressed in a tightly sealed container and allowed to ferment. The fermented shrimp is then sun-dried for a day and minced again. The fermentation process is repeated until the desired texture is achieved. The *belacan* is then packed in containers of different shapes and sizes. The finished product, which is normally dark in color, has a salty taste and a strong shrimp odor. Approximately 40–50 kg of *belacan* is usually obtained from 100 kg of wet shrimp (Karim 1993).

The Malaysian Food Act and Regulation states that *belacan* should not contain less than 15% salt and 24% protein and should not contain more than 40% water and 35% ash (Anon. 2009). Sharif et al. (2008) performed a proximate analysis and reported that the commercial *belacan* sample used in their studies did not comply with the protein requirement (<25%). Table 41.1 shows the chemical composition of *belacan*.

Many types of bacteria are known to play a role in the fermentation of *belacan*. Merican et al. (1980) reported the presence of *Bacillus*, *Pediococcus*, *Lactobacillus*, *Micrococcus*, *Sarcina*, *Clostridium*, *Brevibacterium*, *Flavobacterium*, and *Corynebacterium* in samples of *belacan*. Studies by Rosnizam et al. (2007) found that approximately 68% of commercial *belacan* is with $1–9.9 \times 10^4$ cfu/g of total plate count and 14% with $1–9.9 \times 10^5$ cfu/g of total plate count. *Bacillus cereus* was detected at up to

Fresh small sized shrimp
Wash with seawater
Mix with salt in a ratio 10:0.5 to 10:1
Sun dry
Pound or pass through a meat mincer
Press tightly into a sealed container
Ferment for 7 days
Sun dry
Pound or pass through a meat mincer
Press tightly into a sealed container
Ferment for 7 days
Sun dry
Pound or pass through a meat mincer
Achieve desired fine texture
Pack
Belacan

FIGURE 41.2 General procedure for *belacan* processing.

TABLE 41.1

Chemical Composition of *Belacan*

Parameter	Labeled as Brand A	Labeled as Brand B	Tee et al. (1997)
Moisture	–	–	43.1
Protein (% wb)	28.1	31.0	28.7
Fat (% wb)	1.5	0	2.1
Ash (% wb)	–	–	19.3
Carbohydrate (% wb)	1.6	5.0	6.8
Energy (kcal)	134	146	161

2.9×10^4 cfu/g, and 73% of the samples were found between 1 and 9.9×10^2 cfu/g. In addition, 2% of the samples contained 4 and 9 MPN/g of *Staphylococcus aureus*. None of the samples contained detectable *Escherichia coli*, *Salmonella*, or coliforms. High levels of arsenic (6.16 mg/kg) were found in *belacan*, which is higher than the maximum permissible concentration of arsenic in food (1.0 mg/kg; Sharif et al. 2008).

41.2.2 Budu

Budu is the second largest category of fermented fish product from Malaysia. *Budu* production is concentrated in two states in Malaysia—Kelantan and Terengganu. *Budu* production has increased annually from 173 tons in 2000 to 4309 tons in 2009 (Department of Fisheries 2010). Physically, *budu* is a dark brown liquid that is produced by fermentation of marine fish (Figure 41.3). The main raw material used to make *budu* is the anchovy (*Stolephorus* sp.), locally known as *ikan bilis*. *Budu* has a strong and distinctive aroma and is primarily used as a condiment and a flavoring ingredient in many dishes and gravies in Malaysia. It is especially popular among the people of the east coast states of West Malaysia (Yeoh and Merican 1978; Karim 1993; Rosma et al. 2009). *Budu* is also produced and is popular among people in the northern parts of Thailand, where it is produced from sardines (*Sardinella* sp.; Choorit and Prasertsan 1992).

The production of *budu* begins by washing fresh anchovies and mixing them with salt (Figure 41.4). The mixture of fish and salt is then arranged in a fermentation tank. The tank is closed, and fermentation proceeds for 3–12 months. Some manufacturers also add tamarind juice and palm sugar before fermentation (Yeoh and Merican 1978) or after fermentation (Karim 1993; Rosma et al. 2009). The dark brown supernatant liquid that results is collected and is known as unprocessed *budu*. Unprocessed *budu* is then boiled, cooled, and bottled.

FIGURE 41.3 *Budu* (fermented anchovy).

Fresh anchovies
Mix with salt at a ratio of 3:2 to 3:1
Allow to ferment for 3–12 months
Collect dark brown liquid (unprocessed *budu*)
Add tamarind juice, palm sugar, etc.
Boil
Filter
Cool
Bottle
Commercial *budu*

FIGURE 41.4 General procedure for *budu* processing.

Afiza et al. (2007) found that the quality of *budu* is dependent on the type of fish, the salt content, the duration of fermentation, and the production method. The Malaysian Food Act and Regulation states that the protein and the salt levels in *budu* must not be lower than 5% and 15%, respectively (Anon. 2009). Rosma et al. (2009) reported that the protein and the salt contents of commercial *budu* are approximately 9.92%–24.88% and 21.50%–25.70%, respectively. Previously, Beddows et al. (1979) reported that the salt content of commercial *budu* is 26.3%. The maximum volume of liquid is produced after 140 days, and proteolysis continues to occur until 200 days, when 56% of the insoluble fish protein has been hydrolyzed into soluble form. Table 41.2 shows the chemical composition of *budu*.

Various studies have identified the bacteria associated with the manufacture of *budu*. *Bacillus subtilis* and *B. licheniformis* were isolated from commercial *budu* by Choorit and Prasertsan (1992). Liasi et al. (2009) identified lactic acid bacteria isolated from *budu* as *Lactobacillus case*, *Lb. plantarum*, and *Lb. paracasei*. All three showed antimicrobial activity, inhibiting the growth of a range of gram-positive and gram-negative microorganisms. In terms of food safety, studies by Rosma et al. (2009) reported that *budu* collected from 12 producers were free from pathogenic coliforms, *E. coli*, *Vibrio parahaemolyticus*, and *V. cholerae* contamination. Carcinogenic monochloropropane-1,2-diol, a chemical contaminant resulting from the reaction of chloride with lipids in food, was below the detection level of 2 parts per billion in all samples. However, about 58% of the *budu* samples had histamine concentrations greater than the hazardous levels of 50 mg/100 g of sample, with a range of 22.21–106.40 mg/100 g of sample. Saaid et al. (2009) found that histamine was the most common biogenic amine found in *budu*, at a concentration of 187.7 mg/kg, followed by tyramine (174.7 mg/kg), tryptamine (82.7 mg/kg), putrescine (38.1 mg/kg), and spermidine (5.1 mg/kg).

41.2.3 Cincalok

Cincalok, also known as *cincaluk* or pickled shrimp, is the third most common of the fermented fish products from Malaysia (Figure 41.5). *Cincalok* production is concentrated in the state of Melaka, and its production has increased annually from 54 tons in 2000 to 215 tons in 2009 (Department of Fisheries 2010). The usual raw material used to make *cincalok* is similar to *belacan*, that is, a small shrimp of the *Acetes* sp. or Mysid shrimp, known locally as *udang baring* or *udang geragok* (Karim 1993). The product is a suspension of tiny pink shrimp in sauce that tastes salty and has a strong shrimp smell (Yeoh

TABLE 41.2

Chemical Composition of *Budu*

Parameter	Labeled as Brand A	Labeled as Brand B	Tee et al. (1997)
Moisture	–	–	57.6
Protein (% wb)	11.4	10.2	5.1
Fat (% wb)	0.6	0.7	1.4
Ash (% wb)	–	–	17.3
Carbohydrate	0.9	4.3	18.6
Energy (kcal)	53.9	64	107

FIGURE 41.5 *Cincalok* (pickled shrimp).

Fresh small-sized shrimp
Wash with seawater
Mix with salt (20%) and cooked rice (6%)
Keep in a sealed container for fermentation
for 20–30 days
Add red coloring agent or tomato sauce
(optional)
Bottle
Cincalok

FIGURE 41.6 General procedure for *cincalok* processing.

and Merican 1978; Rahman and Atan 1985). A similar product is made by the Malay people of Riau province, Indonesia.

The production of *cincalok* begins with the washing of small shrimp in seawater. The clean shrimp is then mixed with salt and cooked rice (Figure 41.6). The mixture is packed in earthenware jars or other suitable containers, covered with a piece of cloth or suitable lid, and allowed to ferment (Yeoh and Merican 1978; Karim 1993). In order to improve the color, processors may add a red coloring agent and benzoic acid at the end of the fermentation process. Some processors also blend the *cincalok* with tomato sauce. The *cincalok* is then bottled and is ready for distribution.

The Malaysian Food Act and Regulation state that the *cincalok* shall contain not less than 10% protein and 10% salt, and shall contain not more than 15% ash (Anon. 2009). Table 41.3 shows the chemical composition of *cincalok*. The major problem with *cincalok* is its high histamine concentration. Azudin and Saari (1990) reported that the histamine content of commercial *cincalok* is within the range of 28.9–84.2 mg%, which is much higher than the regulatory limit of 50 mg%, defined as the hazard action

TABLE 41.3

Chemical Composition of *Cincalok*

Parameter	Labeled as Brand A	Tee et al. (1997)
Moisture	–	74.6
Protein (% wb)	13	9.1
Fat (% wb)	2	0.8
Ash (% wb)	–	12.9
Carbohydrate	2	2.6
Energy (kcal)	74	54

level. Saaid et al. (2009) showed that tyramine is the most common biogenic amine found in *cincalok*, at concentrations of 448.8 mg/kg, followed by putrescine (30.7 mg/kg) and histamine (126.1 mg/kg). The high content of histamine is caused by poor handling, processing, and storage conditions of *cincalok*.

41.2.4 Pekasam

Pekasam is the only fermented fish product from Malaysia that uses freshwater fish as the raw material. *Pekasam* is usually consumed deep fried or prepared as a side dish that goes well with rice (Figure 41.7). The production of *pekasam* is concentrated at the northern end of the Malaysian peninsula in states such as Perlis, Kedah, and Perak. Some *pekasam* is also produced in Kelantan (Karim 1993; Anon. 2010). Popular materials for *pekasam* production include tilapia (*Oreochromis mossambica*), spotted gourami (*Trichogaster trichopterus*), catfish (*Clarias batracus*), java barb (*Puntius javanicus*), and snake head (*Channa striatus*; Ismail and Zain 1978; Hassan et al. 2001). A product similar to *pekasam* is produced in Indonesia. This product, *bekasam*, is made from the common carp (*Cyprinus caprio*).

The production of *pekasam* begins with the evisceration and washing of fresh fish (Figure 41.8). The fish is then mixed with salt and tamarind and kept in a sealed container for 2–3 days for the first fermentation (Anon. 2010). A longer fermentation process (12 months) is used to make a premium product. After the first fermentation, excessive salt is removed from the drained fish by washing with water; the fish is then mixed with roasted rice powder and brown sugar. The mixture is kept in a sealed earthenware container for 2–3 weeks of second fermentation. Larger fish take longer to ferment than smaller fish (Karim 1993). After fermentation, *pekasam* is packed and is ready for distribution and marketing.

The Malaysian Food Act and Regulation states that the salt content of *pekasam* shall not be less than 10% (Anon. 2009). Hassan (1980) reported that the moisture, protein, fat, ash, and salt contents of a commercial sample of *pekasam* fall within the range of 57.0%–73.0%, 15.0%–25.0%, 3.0%–8.0%, 6.0%–14.0%, and 10.0%–16.0%, respectively. Karim (1993) found that during the fermentation of *pekasam*, lactic acid bacteria produce acid, lowering the pH and preserving the fish. The addition of roasted rice powder supports the growth of lactic acid bacteria, as well as masking the fishy odor and promoting the characteristic color of *pekasam*. Hassan (1980) reported that the pH and lactic acid content of commercial *pekasam* is within the range of 4.5%–6.1% and 0.1%–0.4%, respectively.

FIGURE 41.7 *Pekasam.*

Fresh water fish
Eviscerate and wash
Mix with 15% salt and 0.5% *asam keping* (*Garcinia atroviridis*)
First fermentation (2–3 days)
Wash and drain
Mix with 50% water, 25% roasted rice powder,
1.5% brown sugar and tamarind
Second fermentation (2–4 weeks)
Pack
Pekasam

FIGURE 41.8 General procedure for *pekasam* processing.

Pekasam has a low histamine concentration. The histamine content of *pekasam* reported by Azudin and Saari (1990) is only 15.6 mg%, which is much lower than the regulatory limit of 50 mg% (defined as the hazard action level). However, commercial *pekasam*, especially that sold at the wet market, is often contaminated with *Listeria*. Hassan et al. (2001) found that commercial *pekasam* is contaminated with *Listeria ivanovii*, *L. innocua*, *L. seeligeri*, *L. welshimeri*, and *L. denitrificans* due to the lack of proper handling. In addition, these authors presented evidence of high cross-contamination rates in samples sold at the wet market.

41.3 Conclusion

At least four fermented fish products are produced in Malaysia, including *belacan*, *budu*, *cincalok*, and *pekasam*. Products similar to *budu* are produced in the northern region of Thailand, and products similar to *belacan*, *cincalok*, and *pekasam* are produced in some regions of Indonesia. The production of fermented fish products in Malaysia increases yearly, providing an opportunity to the processors to improve the quality of their product.

REFERENCES

Afiza, T.S., Rosma, A., Faradila, B., Wan Nadiah, W.A., and Ibrahim, C.O. 2007. Quality index of Kelantan unprocessed "Budu." In: Proceedings of the 9th Symposium of the Malaysian Society of Applied Biology, Exploring The Science of Life as a Catalyst for Technological Advancement, pp. 344–347. Penang: Universiti Sains Malaysia.

Anon. 2009. *Malaysian Food Act and Regulation*, 2009 edition. Kuala Lumpur: MDC Publisher Sdn Bhd.

Anon. 2010. *Pemprosesan produk berasaskan ikan air tawar*. Putrajaya: Bahagian Pengembangan Perikanan Jabatan Perikanan Malaysia.

Azudin, M.N. and Saari, N. 1990. Histamine content in fermented and cured fish products in Malaysia. FAO Fisheries Report No. 401 Supplement. Seventh Session of the Indo-Pacific Fishery Commission Working Party on Fish Technology and Marketing, pp. 105–111. Rome: FAO-UN.

Beddows, C.G., Ardeshir, A.G., and Daud, W.J. 1979. Biochemical changes occurring during the manufacture of Budu. *J. Sci. Food Agric.* 30:1097–1103.

Choorit, W. and Prasertsan, P. 1992. Characterization of proteases produced by newly isolated and identified proteolytic microorganisms from fermented fish (*Budu*). *World J. Microbiol. Biotechnol.* 8:284–286.

Daud, W.J.W. 1978. Processing of traditional fish products in Malaysia with special reference to the state of Terengganu. In: Proceedings of Indo-Pacific Fishery Commission, pp. 7–11. Bangkok: FAO-UN.

Department of Fisheries. 2010. Annual Fisheries Statistics. Putrajaya: Department of Fisheries Malaysia.

Hassan, Z. 1980. Pekasam—A fermented fish product (in Peninsular Malaysia). Paper presented at Seminar on Modernization of Malaysian Cottage Food Industries, Serdang, Selangor, Malaysia, 4 Feb 1980.

Hassan, Z., Purwati, E., Radu, S., Rahim, R.A., and Rusul, G. 2001. Prevalence of *Listeria* spp. and *Listeria monocytogenes* in meat and fermented fish in Malaysia. *Southeast Asian J. Trop. Med. Public Health* 32(2):402–407.

Ismail, M.S. and Zain, A. 1978. Utilization of tilapia, a trash freshwater fish. In: Proceedings of the Indo-Pacific Fishery Commission, pp. 303–309. Bangkok: FAO-UN.

Jinap, S., Ilya-Nur, A.R., Tang, S.C., Hajeb, P., Shahrim, K., and Khairunnisak, M. 2010. Sensory attributes of dishes containing shrimp paste with different concentrations of glutamate and 50-nucleotides. *Appetite* 55:238–244.

Karim, M.I.A. 1993. Fermented fish products in Malaysia. In: *Fish Fermentation Technology*, edited by C.H. Lee, K.H. Steinkraus, and P.J.A. Reilly, pp. 95–106. Tokyo: United Nations University Press.

Leong, Q.L., Ab Karim, S., Selamat, J., Mohd Adzahan, N., Karim, R., and Rosita, J. 2009. Perceptions and acceptance of 'belacan' in Malaysian dishes. *Int. Food Res. J.* 16:539–546.

Liasi, S.A., Azmi, T.I., Hassan, M.D., Shuhaimi, M., Rosfarizan, M., and Ariff, A.B. 2009. Antimicrobial activity and antibiotic sensitivity of three isolates of lactic acid bacteria from fermented fish product, Budu. *Malays. J. Microbiol.* 5(1):33–37.

Merican, Z., Lau, Y.Q., and Idrus, A.Z. 1980. Malaysian fermented foods. ASEAN Protein Project, Occasional paper no. 10.

Rahman, A.A. and Atan, M. 1985. Method for preparing 'cincalok' (fermented product from shrimp in Malaysia). *Teknologi Makanan* 4(1):40–43.

Rosma, A., Afiza, T.S., Wan Nadiah, W.A., Liong, M.T., and Gulam, R.R.A. 2009. Microbiological, histamine and 3-MCPD contents of Malaysian unprocessed 'budu': Short communication. *Int. Food Res. J.* 16:589–594.

Rosnizam, I., Nurhazwani, N., Muridah, M.N., and Selar, Z.A. 2007. Microbiological quality survey on shrimp paste (belacan) in Selangor. In: Proceedings of the 9th Symposium of the Malaysian Society of Applied Biology, Exploring the Science of Life as a Catalyst for Technological Advancement, pp. 341–343. Penang: Universiti Sains Malaysia.

Saaid, M., Saad, B., Hashim, N.H., Ali, A.S.M., and Saleh, M.I. 2009. Determination of biogenic amines in selected Malaysian food. *Food Chem.* 113:1356–1362.

Sharif, R., Ghazali, A.R., Rajab, N.F., Haron, H., and Osman, F. 2008. Toxicological evaluation of some Malaysian locally processed raw food products. *Food Chem. Toxicol.* 46:368–374.

Tee, E.S., Noor, M.I., Azudin, M.N., and Idris, K. 1997. *Nutrient Composition of Malaysian Foods*, 4th edition. Kuala Lumpur: Institute for Medical Research.

Yeoh, Q.L. and Merican, Z. 1978. Processing of non-commercial and low-cost fish in Malaysia. In: Proceedings of the Indo-Pacific Fishery Commission, pp. 572–580. Bangkok: FAO-UN.

42

Indonesian Fermented Fish Products

Nurul Huda

CONTENTS

42.1 Introduction

Indonesia consists of 17,480 islands and is surrounded with waters that provide a high production of marine fish, for example, approximately 4,701,933 tons of fish in 2008 (Departemen Kelautan dan Perikanan 2010). Approximately 41% of the total marine fish landed is subjected to different preservation and processing methods. Traditional preservation and processing methods, such as salting, boiling, smoking, and fermentation, are responsible for 54.67% of the total preserved and processed fish products. Fermented fish products, categorized as *terasi*, *peda*, and *kecap ikan*, contribute to less than 2% of the total preserved and processed fish products. In addition to *terasi*, *peda*, and *kecap ikan*, other fermented fish products are produced in Indonesia. These include *bakasang*, *bekasem*, *budu*, *cincaluk*, *jambal roti*, *naniura*, *petis*, *picungan*, *pudu*, *rusip*, and *tukai*. This chapter elaborates on the general processing methods of fermented fish products in Indonesia and presents some data related with their chemical composition and microbial characteristics.

42.2 Indonesian Fermented Fish Products

42.2.1 *Bakasang*

Bakasang is a fermented fish product that is mainly produced in the North Sulawesi province and the Moluccas Island. *Bakasang* is a dark brown liquid product with a strong fishy flavor (Figure 42.1). It is

FIGURE 42.1 *Bakasang.*

usually used as a flavoring agent for many dishes or mixed with red chilies, tomato, red onion, and garlic, which are then sautéed with coconut oil and eaten with hot porridge mixtures of rice and vegetables
called *tinutuan* (Ijong and Ohta 1995). The main raw materials for *bakasang* processing are the internal
organs (viscera and roe) of tuna and tuna-like fish obtained as a waste product from the processing of
smoked tuna. Subroto et al. (1984) reported that *bakasang* made from ground viscera showed better
quality compared to those from whole viscera. Smoked skipjack tuna (*Katsuwonus pelamis*) is a popular
product in the North Sulawesi province and the Moluccas Island. The internal organs of yellowfin tuna
(*Thunnus macropterus*) and big eye tuna (*T. obesus*) are also good raw materials for the processing of
bakasang (Poernomo 1996).

In addition to the internal organs of tuna, *bakasang* is also produced from the whole body of small
marine fish. Ijong and Ohta (1996) and Harikedua et al. (2009) reported that some small marine fish,
such as sardine (*Sardinella* sp.), anchovy (*Stolepherus* sp. or *Engraulis* sp.), and orange sunkist shrimp
(*Caridina wycki*), are also suitable for *bakasang* processing.

The traditional processing of *bakasang* starts with the collection of internal organs, including the
viscera and roe of skipjack tuna (Figure 42.2). An appropriate amount of the internal organs is washed
with seawater and drained to remove excessive seawater. The cleaned internal organs are mixed with
15%–20% salt and poured into bottles. The bottles are kept for 3–6 weeks to allow fermentation to occur
until a dark and sticky liquid is formed. Traditionally, the bottles are stored in a warm place and usually
near a source of fire. Subroto et al. (1984) stated that *bakasang* can be made from the visceral parts of
bonito by adding 10% salt and allowing the mixture to ferment for 10 days.

As a liquid food product, *bakasang* contains a high moisture content, which is approximately 66%–
69% (Ijong and Ohta 1995). Harikedua et al. (2009) reported the moisture content of *bakasang* to be

Fresh internal organ (viscera and roes) of skipjack tuna

Washed with seawater and drained to remove excessive water

Mixed with 15–20% salt

Poured into bottles

Kept at a warm temperature for 3–6 weeks

Bakasang

FIGURE 42.2 General procedure for *bakasang* processing.

TABLE 42.1

Chemical Composition of *Bakasang*

Parameter	Chasanah et al. (1994)	Ijong and Ohta (1995)	Harikedua et al. (2009)
Moisture	73.1	66.3–68.9	59.0–70.9
Protein (% wb)	12.8	14.0–17.4	–
Fat (% wb)	2.6	1.0–3.0	–
Ash (% wb)	8.8	–	–
Salt (% wb)	6.7	8.4–18.0	7.6–9.8
pH	–	5.8–6.3	6.4–6.5
TVN (mg/100 g)	–	–	345.7–351.4
Titratable acid (%)	–	–	1.2–1.9
Free amino acid nitrogen (mg N/mL)	–	–	802.2–921.6

approximately 59%–70%. The salt content will influence the amount of salt used during *bakasang* preparation. Previously, Ijong and Ohta (1995) reported that the salt content of *bakasang* was approximately 84–180 g/kg per sample; however, the report by Harikedua et al. (2009) showed that the salt content decreased to 78–98 g/kg per sample. The chemical composition of *bakasang* is presented in Table 42.1.

A quality *bakasang* product is related to a high moisture content, low viscosity, low titratable acid, low salt content, and low free amino nitrogen content (Harikedua et al. 2009). A longer fermentation process will produce *bakasang* with a higher free amino nitrogen content, which is associated with lower consumer acceptability.

In terms of microbial quality, Ijong and Ohta (1995) reported that *Staphylococcus* sp. and *Lactobacillus* sp. were the predominant bacteria isolated from *bakasang*. Other bacteria isolated in *bakasang* are *Micrococcus* sp., *Streptococcus* sp., *Bacillus* sp., and *Clostridium* sp. *Bakasang* prepared from small sardines (*Engraulis* sp.) contained additional bacteria such as *Enterobacter* sp., *Moraxella* sp., and *Pediococcus* sp. (Ijong and Ohta 1996).

42.2.2 *Bekasem*

Bekasem or *bekasam* is a fermented freshwater fish product popular among people from Central Java and the South Sumatra province. In Central Kalimantan, *bekasem* is referred to as *wadi*. Traditionally, the main raw material used for *bekasem* processing is ikan common carp (*Cyprinus carpio*) (Murtini 1992). However, some other freshwater fish, such as catfish (*Clarias batracus*), java barb (*Puntius javanicus*), tilapia (*Oreochromis mossambica*), and spotted gourami (*Trichogaster trichopterus*), are also suitable raw materials (Murtini et al. 1997). Candra (2006) prepared *bekasem* using milk fish (*Chanos chanos*); meanwhile, Vonistara et al. (2010) used nile tilapia (*O. niloticus*) as a raw material.

A specific part of *bekasem* processing is the addition of cooked rice, roasted rice, or fermented rice as a source of carbohydrate (energy) for the growth of the preferred bacteria. The traditional processing of *bekasem* starts with eviscerating and washing the freshwater fish (Figure 42.3). The eviscerated fish is then split into a butterfly shape (if a larger fish is used) and fully soaked in 15%–16% brine solution for 2–3 days. The soaked fish is drained to remove the excess brine solution and then mixed with a carbohydrate source (cooked rice, roasted rice, or fermented rice) at a ratio of 3:4. The mixed rice and fish are placed into traditional clay or plastic jars and stored for 1–2 weeks (Murtini 1992; Putro 1993).

The chemical composition of *bekasem* is shown in Table 42.2. *Bekasem* is classified as a product with an intermediate moisture content. The moisture content of commercial *bekasem* is approximately 55%–65%. As an effect of soaking in brine solution, the salt content of the product increases to 6%–17% wet base. The protein and fat content will vary depending on the raw material used. Murtini et al. (1991) reported that the protein and fat contents of *bekasem* were around 4.80%–6.91% wb and 5.0%–5.72% wb, respectively.

Eviscerated fresh water fish

Washed and split into a butterfly shape

Soaked into a 15–16% brine solution for 2–3 weeks

Drained

Mixed with a carbohydrate source

Placed in clay or plastic jars and stored for 1–2 weeks

Fermentation process

Bekasem

FIGURE 42.3 General procedure for *bekasem* processing.

TABLE 42.2

Chemical Composition of *Bekasem*

Parameter	Murtini (1992)	Putro (1993)
Moisture	66.95	55–66
Protein (% db)	45.23	41–64
Fat (% db)	17.31	11–23
Ash (% db)	17.42	13–28
Salt (% wb)	15.55	6–17
pH	–	4.46–4.98

As a result of bacterial fermentation, especially from lactic acid bacteria, *bekasem* has a lower pH—approximately 4.57–5.33 according to Murtini et al. (1991) and approximately 4.46–4.98 according to Putro (1993). Sugiyono et al. (1999) reported that *bekasem* contains the lactic acid bacteria *Lactobacillus* sp., *Pediococcus* sp., *Lactobacillus coryneformis*, and *Pediococcus damnosus*. However, Candra (2006) reported that the lactic acid bacteria isolated from commercial bekasem prepared from milk fish (*Cha. chanos*) were *Staphylococcus* sp., *Erysipelothrix* or *Lactobacillus* sp., and *Streptococcus* sp. or *Gemella*. To improve the quality of *bekasem*, the liquid of cabbage and Chinese leaf pickles can be added (Murtini et al. 1997). The results showed that the addition of the pickled liquid had significant effects on the lactic acid bacteria count, aerobic and anaerobic bacteria counts, yeast counts, pH value, total volatile acid and fat content. The *bekasem* used in this study is still accepted by panelists after 8 weeks of storage.

42.2.3 *Budu*

Budu is a fermented fish product produced mainly in the area of Pasaman, approximately 300 km from Padang, the capital city of West Sumatra. *Budu* is also produced in Sorkam, a neighboring area of Pasaman that is part of the province of North Sumatra. The product is different from the Malaysian *budu*, in terms of both the fish used and the processing steps. The *budu* from West Sumatra is normally made from larger marine fish such as Spanish mackerel (*Scomberomorus* sp.) and leatherskin (*Chorinemus* sp.), locally known as *ikan tenggiri* and *ikan talang*, respectively. The Spanish mackerel is more popular for making *budu* than the leatherskin (Figure 42.4).

There are limited reports on West Sumatra *budu* processing and quality characteristics. Huda and Rosma (2006) reported that the traditional processing of West Sumatra *budu* starts with hanging the fresh fish by its caudal fin at room temperature for 4 h (Figure 42.5). The fish is cut into a butterfly style, and all of the intestinal organs and gills are removed. The fish is then washed to ensure that it is free of blood and other intestinal residues. The fish is stacked in a *pasu*, a traditional container, and layered with coarse salt at a ratio of 1:5–1:3. The container is covered and stored at room temperature for a day, after

FIGURE 42.4   Spanish mackerel *budu.*

Fresh fish

Hanged at room temperature for 4 hrs

Eviscerated into a butterfly style

Washed and intestinal organs removed

Stacked in a *pasu* and layered with coarse salt at ratio of 1:5–1:3

Allowed to stand overnight

Rinsed

Sun-dried until a moisture content of 50% is reached

Budu

FIGURE 42.5   General procedure for *budu* processing.

which the fish is rinsed to remove the excess salt. The fish is then sun-dried for 5 days. During drying, condiments, such as white paper and garlic, may be sprinkled onto the fish to improve the quality. The product, which has a specific flavor, aroma, and texture, is then ready to be packed and marketed.

The normal shelf life of West Sumatra *budu* at room temperature is approximately 3–4 weeks. Low-temperature storage can prolong the shelf life. The approximate composition of *budu* is as follows—51% moisture content, 33% protein, 0.5% fat, and 14% ash (Huda and Rosma 2006). The higher ash content corresponds to the salt used during fermentation. Because of the strong flavor and higher salt content, *budu* cannot be consumed as a main dish in a daily meal. Normally, it is used as an ingredient during the preparation of fish curry. Some people also roast or fry the *budu* and blend it with ground chilies.

42.2.4 *Cincaluk*

Cincaluk is a traditional fermented shrimp product popular among the Malay tribe people of the Bengkalis Island. Small shrimp-like crustaceans (*Mytis* sp. or *Atya* sp.) called *udang pepai* are used for *cincaluk* production (Maamoen et al. 2003). *Cincaluk* is produced in backyard activities by local people of the Bengkalis Island. The processing of *cincaluk* begins with mixing descaled washed shrimp with tapioca flour, salt, and sugar (Figure 42.6). The tapioca flour is dissolved in water, gelatinized, and allowed to cool. The shrimp are then mixed thoroughly with salt, sugar, and gelatinized tapioca flour.

Fresh small shrimp-like crustaceans

Mixed with tapioca flour, salt and sugar at a ratio of 20:1:1:1 (Method A)
or cooked rice at a ratio of 2:1:1 (Method B)

Stored in a container for the fermentation process for 1–2 weeks (method A)
or 4 days (Method B)

Cincaluk

FIGURE 42.6   General procedure for *cincaluk* processing.

The mixture is poured into bottles that are sealed firmly for the fermentation process (Irianto and Irianto 1998). Another method by which to process cincaluk is by washing small shrimp and mixing them with cooked rice and salt. The mixture is stored in a closed container for the fermentation process. The *cincaluk* produced is poured into glass jars for marketing purposes.

Due to the fermentation process, the pH of *cincaluk* is low. Irianto and Irianto (1998) reported that the pH of *cincaluk* was approximately 4.82, with a lactic acid content of 2.34%. The chemical composition of *cincaluk* shows that the moisture, protein, fat, ash, and salt contents are 69.76%, 16.23%, 1.57%, 12.43%, and 10.11%, respectively. In terms of the microbial characteristics, it has been reported that the bacteria isolated from commercial *cincaluk* are *Lb. coryneformis*, *P. damnosus*, and *Pediococcus* sp.

42.2.5 *Jambal Roti*

Jambal roti is a fermented and dry-salted fish product that is popular in the West Java province and neighboring areas. *Jambal roti* is one of the most highly valued dried fish products in Indonesia (Figure 42.7). It is a fish with a crunchy cracker-like texture, after the frying process. Traditionally, *jambal roti* is prepared from marine catfish (*Arius thalassinus*); however, other fish species with similar characteristics are also suitable as raw materials. Sani (2001) used striped catfish (*Pangasius hypophthalmus*) as the raw material to produce a *jambal roti*-like product.

The processing of *jambal roti* may vary slightly by location, but the basic processing methods applied are similar (Burhanuddin et al. 1987; Putro 1993; Irianto and Irianto 1998). There are two types of *jambal roti*—plain *jambal roti* for larger fish and salty *jambal roti* for smaller fish. Plain *jambal roti* processing begins with the beheading, gutting, and washing of the fresh fish. An appropriate amount of salt is added into the abdominal cavity of the fish, and it is then layered in a concrete tank. Each layer is sprinkled with salt and then left for 24 h. The fish is removed from the tank, and excessive salt from the abdominal cavity is removed. The fish is sprinkled with salt and left in the tank for 48 h before washing to remove the excess salt. The fish is cut into a butterfly shape and dried for 3–5 days. During the drying process, the surface of the fish is wiped with sugar solution. The sugar solution is prepared by mixing 20% palm sugar and 10% garlic in 1 L of water (Suharna et al. 2006).

The processing of salty *jambal roti* begins with eviscerating and washing the fresh fish (Figure 42.8). The cleaned fish is then split into a butterfly shape and soaked in freshwater for 24 h. After draining, the

FIGURE 42.7   *Jambal roti.*

Fresh marine catfish

Beheaded, gutted and washed

25–45% salt added into abdominal cavity, layered in a concrete tank and sprinkled with salt

Allowed to ferment for 24 hours

Remove excess salt from the abdominal cavity, layered in a concrete tank and sprinkled with salt

Allowed to ferment for 2–3 days

Washed and split the fish into butterfly shape

Dried for 3–5 days

Jambal roti

FIGURE 42.8   General procedure for *jambal roti* processing.

fish are arranged in a basin and mixed with 30%–40% salt and then left for 24 h. After the salting process, the fish are washed. The cleaned fish are then dried for 3–5 days. Rochima (2005) reported that using 30% salt and a fermentation period of 24 h produces *jambal roti* with a higher sensory acceptability score.

Marine catfish is known as a fish with a high protein and low fat content. Damayanthi (1991) reported that the protein and fat contents of fresh marine catfish are 16.20% and 1.04% wb, respectively. The higher protein content will remain unchanged during *jambal roti* processing. The protein content of *jambal roti* is approximately 33.76%–46.56% wb. The salt content of *jambal roti* will depend on the amount of ash used. Salamah et al. (1995) reported that the salt content of *jambal roti* was approximately 4.15%–7.25%, whereas Nuraniekmah (1996) reported that the salt content was approximately 7.38%–8.53%. Table 42.3 shows the chemical composition of *jambal roti*.

The specific flavor of *jambal roti* is related to the presence of nitrogen compounds such as creatine, phenylalanine, and tyrosine. Peptides contain glutamic and aspartic, which also contribute to this specific flavor (Dewi 2005). *Staphylococcus* sp. is the dominant bacteria found in a *jambal roti*. Two isolates of *Staphylococcus* sp. were found in the product, one of which hydrolyzes fat, whereas the other does not (Santoso et al. 1995). Previously, Sarnianto et al. (1984) reported that the *Staphylococcus* sp. and *Lactobacillus* sp. contents in *jambal roti* were 0.4–0.9×10^2 and 0.8–2.9×10^2 cfu/g, respectively.

The main problem with *jambal roti* processing is insect infestation, especially infestation of *Chrysomya megacephala*, *Musca domestica*, *Phyophila*, and dermestid beetles (Putro 1993). It is estimated that the losses of product due to insect infestation reach 30%. Screening methods, such as placing the fish in closed containers during the fermentation and salting processes followed by covering the fish with a screen during drying, were found to be effective in reducing insect infestation. However, fish processors are reluctant to adopt such practices for their lack of practicality. Hidayat (2000) studied the efficacy of less harmful insecticides and found that theta-cypermethrin and pirimiphos-methyl were similarly

TABLE 42.3

Chemical Composition of *Jambal Roti*

Parameter	Salamah et al. (1995)	Salamah et al. (1997)
Moisture	42.14–43.39	38.40–44.05
Protein (% wb)	33.76–36.81	26.38–33.56
Fat (% wb)	2.76–4.36	1.78–3.29
Ash (% wb)	9.06–13.89	16.26–16.50
Salt (% wb)	4.15–7.25	–

effective in reducing insect infestation. Spraying the raw material with theta-cypermethrin at a concentration of 5 mL/m² is the most effective treatment to reduce insect infestation in *jambal roti*.

42.2.6 *Kecap Ikan*

Kecap ikan or fish sauce is a clear yellow to brown liquid with a salty taste and specific aroma (Figure 42.9). In Indonesia, *kecap ikan* is the third largest fermented fish product produced and concentrated along Java Island. The popularity of *kecap ikan* has increased yearly from 9 tons in 2002 to 748 tons in 2007. There is no specific fish species for *kecap ikan* production. Rahayu et al. (1992) reported that the raw material for *kecap ikan* production could be any marine fish, such as anchovy (*Stolephorus* sp.), herring (*Clupea* sp.), or ponyfish (*Leiognathus* sp.). Putro (1993) reported that along the Bali strait, a *kecap ikan* processor is using sardines (*Sardinella* sp.) as the raw material. Desniar et al. (2007) produced *kecap ikan* using jack (*Caranx* sp.) as the raw material. Some freshwater fish, such as java barb (*Puntius* sp.) and bony-lipped barb (*Osteochilus* sp.), are also suitable raw materials (Rahayu et al. 1992). Rachmi et al. (2008) produced *kecap ikan* using nile tilapia (*O. niloticus*) as the raw material. Suptijah (1998) reported that white shrimp (*Penaeus monodon*) were also suitable for producing a *kecap ikan*-like product.

The processing of *kecap ikan* begins by eviscerating, washing, and mincing the fresh fish (Figure 42.10). The cleaned fish are mixed with salt and arranged in the fermentation tank. The fermentation tank must be equipped with an outlet to collect the fermentation product. The fish are arranged in layers, and each layer is covered with salt. The tank is then closed, and fermentation is allowed to take place for 4–12 months. The *kecap ikan* liquid collected from the outlet is then filtered and mixed with brown sugar and spices (Putro 1993). To accelerate the fermentation process, some processors have applied enzymatic or acidic hydrolysis, which reduces the fermentation process to 3 days or 24 h, respectively (Basmal 1992). *Kecap ikan* produced using bromelein and papian showed a higher quality compared to that produced using neutrase (Suparman 1993). However, the quality of accelerated fermentation in *kecap ikan* produced using an enzyme and chemicals is lower compared to *kecap ikan* produced using spontaneous fermentation.

Because *kecap ikan* is a liquid product, the moisture content can reach above 70%. The salt content is influenced by the amount of salt added. Sarnianto et al. (1984) reported that the salt content of *kecap ikan* was 19.70%–22.35%. According to the Indonesian National Standard (Badan Standarisasi Nasional 1996), the salt content of *kecap ikan* should be in the range of 19%–26%. The Indonesian National Standard also stated that the pH value, amino nitrogen content, and total plate count (TPC) values should be 5–6, a minimum of 5%, and a maximum of 10^4 cfu/g, respectively. Table 42.4 shows the chemical composition of *kecap ikan*.

FIGURE 42.9 *Kecap ikan.*

Fresh fish

Washed, drained and minced

Mixed with 20–30% salt

Arranged in layers in a fermentation tank

Fermented for 4–12 months

Collected and filtered the sauce

Added brown sugar and spices (ratio 1:2.5)

Bottled

Kecap ikan

FIGURE 42.10   General procedure for *kecap ikan* processing.

TABLE 42.4

Chemical Composition of *Kecap Ikan*

Parameter	Poernomo et al. (1984)	Desniar et al. (2007)
Moisture	66.67–76.89	64.12–70.88
Protein (% wb)	10.17–10.51	6.11–6.40
N-amine (% wb)	–	5.37–5.84
Fat (% wb)	0.50–0.70	1.68–1.99
Ash (% wb)	21.95–23.50	16.30–21.82
Salt (% wb)	11.60–21.16	–
pH	–	5.01–5.18
TPC (cfu/g)	$8.0–8.5 \times 10^4$	$1.25–4.85 \times 10^3$
LAB (cfu/g)	–	$1.16–1.25 \times 10^3$

Halophilic bacteria are the main bacteria found in *kecap ikan*. Indiyati and Arbianto (1986) reported that *Bacillus* sp. was the dominant bacteria found in *kecap ikan*, although they also isolated *Flavobacterium* sp. Rahayu et al. (1992) reported that the mold species found in *kecap ikan* were *Cladosporium herbarum*, *Aspergillus fumigatus*, and *Penicillium nonatum*, whereas the yeast found in *kecap ikan* is from the species *Caudida claussenii*.

42.2.7 *Naniura*

Naniura is a fermented fish product popular in the Batak tribe community around the area of Toba lake in the North Sumatra provinces and in some parts of the Riau province. *Naniura* in the Batak tribe language means raw fish. *Naniura* is a product that is directly consumed without any further processing. Traditionally, the raw material used for *naniura* processing is freshwater fish such as common carp (*C. carpio*) and tilapia (*O. mossambica*). Silalahi (1994) processed *naniura* using snakehead and nile tilapia (*Ophiocephalus striatus*) as the raw materials. Irianto and Irianto (1998) stated that *naniura* processing is similar to that of bekasam; however, ground rice is used instead of cooked rice. The processing of *naniura* begins with eviscerating and washing the fresh fish (Figure 42.11). The fish is then split into a butterfly shape. The cleaned fish is fully soaked in a local lemon juice (*asam jungga*) solution. To improve the acceptability of the product, some processors mix the lemon juice with condiments including onion, ginger, turmeric, and galangal. The excessive water is removed, and the fish is mixed with ground rice. The fermentation process is allowed to take place for 1–4 days (Irianto and Irianto 1998).

Fresh water fish

Eviscerated and washed

Split into a butterfly shape

Soaked in lemon juice solution for 3–5 hours

Drained

Coated with ground rice

Fermented for 4 days

Naniura

FIGURE 42.11   General procedure for *naniura* processing.

There is limited scientific information on the *naniura* product. Silalahi (1994) processed *naniura* by soaking the fish in 2% acetic acid solution. The product showed lower pH (5.5) and higher water activity (0.8). The product can be kept at room temperature for at least 4 days and still remain acceptable for human consumption.

42.2.8 *Peda*

Peda or *pedah* is the second-largest fermented fish product produced in Indonesia, and its popularity increases yearly. In 2002, *peda* production was 6829 tons, and it increased to 16,556 tons in 2007. *Peda* production is mainly found in Java Island, and mackerel (*Rastrelliger* sp.) is traditionally used as the raw material (Figure 42.12). Some other marine fish, such as sardine (*Sardinella* sp.), scad mackerel (*Decapterus* sp.), and jack (*Caranx* sp.), have also been reported suitable for *peda* processing (Syafii 1988; Rahayu 1992; Rahayu et al. 1992; Desniar et al. 2009). However, freshwater fish are not suitable raw materials for *peda* processing (Sukarsa 1979).

There are two species of mackerel used in *peda* processing—chub mackerel (*Rastrelliger neglectus*) and indian mackerel (*R. kanagurta*). Due to their different fat contents, red *peda* is produced from chub mackerel, and the lighter-color peda is produced from Indian mackerel (Van Veen 1965). The processing methods of *peda* are similar, but each producer has his or her own process, especially in terms of the amount of salt used. The salt used must be more than 10%; otherwise, the fish will spoil (Menajang 1988). The processing of *peda* begins with eviscerating and washing the fresh fish (Figure 42.13). Syafii (1988) showed that the evisceration treatment improves the quality of *peda*. The cleaned fish is mixed with salt and then arranged in layers in the salting container. Saturated brine solution is poured into the container, and the salting process is allowed to take place for 3 days. Excessive salt is removed from the fish, which is washed with brine solution, packed in a bamboo container, and sprinkled with fine salt. A

FIGURE 42.12   *Peda.*

Fresh mackerel fish

Eviscerated and washed

Mixed with 20–50% salt and arranged in a container

Salting process for 3 days

Excessive salt was removed and fish was packed in a bamboo basket

Fermentation process for 1–2 weeks

Peda

FIGURE 42.13 General procedure for *peda* processing.

dried banana leaf is used as a separator if more than one layer of fish is arranged in the basket. The basket is kept at room temperature for 1–2 weeks while the fermentation process takes place. Putro (1993) allowed the fermentation process to continue for 2–3 months.

Syafii (1988) reported that the moisture and salt contents of commercial *peda* were 44.02%–53.12% and 18.03%–21.95%, respectively. Previously, Sarnianto et al. (1984) reported that the moisture and salt contents of commercial *peda* were 49.42%–56.78% and 10.99%–14.41%, respectively. *Peda* prepared from chub mackerel contains more fat compared with *peda* prepared from Indian mackerel. Van Veen (1965) reported that the fat content of chub mackerel *peda* was 7%–14%, whereas for Indian mackerel *peda*, it was 1.5%–7%. Using a larger amount of salt during the salting process and smaller fish will produce *peda* with a higher salt content. The maximum salt content of *peda* suggested by Van Veen (1965) is 18%. Table 42.5 shows the chemical composition of *peda*.

Sarnianto et al. (1984) reported that the number of TPC, *Lactobacillus*, and *Staphylococcus* of *peda* were 3–10 × 10⁵, 0.08–0.2 × 10², and 0–2 × 10² cfu/g, respectively. Studies by Syafii (1988) on commercial *peda* found that the number of TPC, halophilic bacteria, proteolytic bacteria, lipolytic bacteria, acid-forming bacteria, and yeast of mold were present at 4.8–4.48, 3.08–5.26, 4.11–5.50, 3.93–5.54, 3.99–4.93, and 1.00–3.00 log cfu/g, respectively. Rahayu et al. (1992) hypothesized that the bacteria responsible for *peda* fermentation were *Acinetobacter* sp., *Flavobacterium* sp., *Cytophaga* sp., *Halobacterium* sp., *Micrococcus* sp., *Staphylococcus* sp., and *Corynebacterium* sp.

Mackerel is one of the scromboid fish families related to scromboid poisoning due to the high amount of free histidine in the fish flesh. Sarnianto et al. (1984) showed that the histamine content of *peda* is the highest among Indonesian fermented fish products. The histamine content reported is approximately 107.32–133.43 mg %, which is over the limit of 50 mg % regulated in Indonesia. Therefore, utilization of fresh fish and hygienic practices during *peda* processing is necessary to produce higher quality *peda*.

TABLE 42.5

Chemical Composition of *Peda*

Parameter	Poernomo et al. (1984)	Desniar et al. (2009)
Moisture	50.35	52.71%–53.94%
Protein (% wb)	26.67	20.15%–21.54%
Fat (% wb)	6.36	1.25%–1.37%
Ash (% wb)	18.89	15.96%–16.90%
Salt (% wb)	13.72	10%–13%
pH	–	6.0
TVB (mg/100 g)	–	16.78–18.42
TMA (mg/100 g)	–	2.23–3.35
aW	0.87	0.73–0.74

42.2.9 *Petis*

Petis is a dark brown to black sticky paste produced from an extract of shrimp or fish, which is used as an appetizer or as an ingredient in some traditional foods. *Petis* is produced and is famous among people in areas of eastern Java such as Sidoarjo, Gresik, Lamongan, Tuban, and Madura (Suprapti 2001; Nisrina 2007). Various raw materials may be used to make *petis*. Leksono and Irasari (1996) used freshwater shrimp *Macrobrachium rosenbergii* as raw materials; Poernomo et al. (2004) used an extract of scad mackerel (*Decapterus* sp.). Fakhrudin (2009) used an extract of bivalve *Corbula faba*, and Pramono et al. (2009) used an extract of beef meat as raw material for *petis* production.

The production of *petis* begins with the preparation of the shrimp/fish/bivalve/meat extract (Figure 42.14). Often, the extract is the liquid waste from the processing of another product such as ebi shrimp, boiled fish, fish balls, shredded fish, meat balls, or shredded meat. The extract is mixed with squid ink as a natural coloring agent and specific spices and then boiled for 1 h (Suprapti 2001). The residue is removed, and the filtered extract is reboiled. After cooling, the extract is added to a solution of tapioca or rice flour and boiled for another hour. After cooling for 24 h, the boiling process is repeated for a better quality of *petis*. The last step is cooling, bottling, and pasteurization for 30 min.

According to the Indonesian National Standard for shrimp *petis*, the moisture content should be in the range of 20%–30%. The protein, carbohydrate, and ash contents should be at least 10%, 40% at most, and 8% at most, respectively (Badan Standarisasi Nasional 2006). Suprapti (2001) reported that the moisture, protein, fat, ash, and carbohydrate contents of marine shrimp *petis* were 39.0%, 15.0%, 0.1%, 40.0%, and 5.9%, respectively. The chemical composition of *petis* is different, depending on the raw material used. Table 42.6 shows the chemical composition of non-shrimp *petis*.

Limited information is available as to the microbial content of *petis*. The Indonesian National Standard states that the maximum TPC for shrimp *petis* is 5×10^2 cfu/g or 5×10^5 cfu/mL (Badan Standarisasi Nasional 2006). Leksono and Irasari (1996) reported that the TPC for *petis* prepared from shrimp waste was in the range of 2.4–2.6×10^4 cfu/g per sample. Studies by Nisrina (2007) of commercial *petis* marketed in Madura Island found that the range of TPC was 3500–235,500 cfu/mL, and the most probable number (MPN) of coliforms ranged from 0.115 to 1.065 MPN/mL.

Extract of shrimp/fish/bivalve

Mix with spices and squid ink

Boil for 30 minutes

Filter and remove the residue

Reboil the liquid and mix with a solution containing tapioca or rice flour

Continue the boiling process for 1 hour

Cool for 24 hours

Boil again

Cool and bottle

Pasteurize for 30 minutes

Commercial *petis*

FIGURE 42.14 General procedure for *petis* processing.

TABLE 42.6

Chemical Composition of *Petis*

Parameter	Fish (Suprapti 2001)	Bivalve (Fakhrudin 2009)	Shrimp Waste (Leksono and Irasari 1996)
Moisture	56.0	25.2	41.7–43.6
Protein (% wb)	20.0	16.3	2.4–3.1
Fat (% wb)	0.2	0.98	0.2–0.3
Ash (% wb)	23.8	8.9	9.4–11.2
Carbohydrate (% wb)	24.0	48.79	41.9–46.2

42.2.10 *Picungan*

Picungan is a unique fermented fish product that is available only in the Banten province. *Picungan* refers to the fermentation process of fish using the seeds of the picung tree (*Pangium edule*). The picung tree is a tall tree native to the mangrove swamps of Southeast Asia. The raw material for *picungan* processing is a marine fish such as mackerel (*Rastrelliger* sp.), scad (*Decapterus* sp.), anchovy (*Stolephorus* sp.), and hairtail/ribbon fish (*Trichiurus savala*). Irianto (1999) reported that the *picungan* process is similar to the bekasem process. The function of the picung seed is to provide a carbohydrate source for the growth of lactic acid bacteria due to the lower pH of the end product. Widyasari (2006) demonstrated that the antimicrobial substance contained in the picung kernel may be used as a natural preservative for fish products and seems to increase the shelf life of the fish by 6 days, when stored at ambient temperature.

The traditional processing of *picungan* begins with preparation of the picung seeds. Picung seeds contain hydrogen cyanide and are deadly if consumed without prior preparation. To release the toxin, traditionally, the skin of the seeds is divided into two pieces and submerged into flowing water (river) for 2 days or sun-dried for 2 days. The seeds need to be shredded before being mixed with the other material. The fish used as the raw material must be considered fresh fish. The evisceration and washing process is applied to small fish; it is recommended that larger fish be cut to a smaller size or filleted (Figure 42.15). The fish are mixed with the shredded picung seed and salted at a ratio of 4:2:1. The fish are then layered in traditional baskets, and each layer is covered with a banana leaf. The fermentation process is allowed to take place for 3–7 days (Irianto 1999).

The approximate composition of *picungan* is as follows—66.35% moisture content, 21.69% protein content, 3.08% fat content, and 6.79% ash content. The approximate composition will vary depending on the fish used, the amount of *picungan*, and the fermentation period. As a result of the fermentation process, the pH of *picungan* drops to 5.26 and the lactic acid content increases to around 0.36%. *Lactobacillus* sp. and *Lb. murinus* have been successfully isolated from the *picungan* product (Irianto et al. 2003).

Fresh marine fish

Eviscerated and washed

Filleted or cut (for large fish)

Mixed with shredded picung seed and salt

Layered in a traditional basket and covered with a banana leaf

Allowed to ferment for 4–7 days

Picungan

FIGURE 42.15   General procedure for *picungan* processing.

42.2.11 *Pudu*

Pudu is a fermented fish product originally from the Riau province. The raw material of *pudu* is fresh-water fish, and it is usually referred to as tilapia (*O. mossambica*). *Pudu* processing is similar to *bekasem* processing, which involves the addition of cooked rice as a carbohydrate source. Maamoen et al. (2003) stated that the processing of *pudu* begins by eviscerating and washing the fresh fish. The cleaned fresh fish is then mixed with a solution that contains 20% salt, 5% cooked rice, and an appropriate amount of *asam kandis*, a fruit of the *Garcinia parvifolia* tree. The formulation is then poured into glass jars and allowed to ferment (Figure 42.16).

There is limited scientific information on the *pudu* product. Maamoen et al. (2003) reported that the pH and salt content of *pudu* were 4.4% and 17%, respectively. There are no significant differences in terms of moisture content, protein, and fat between commercial *pudu* collected from a traditional market and *pudu* prepared in a laboratory. The moisture, protein, and fat contents are within the range of 58%–59%, 13.5%–15.7%, and 1.1%–1.2% wb, respectively.

42.2.12 *Rusip*

Rusip is a fermented fish product that originated in the Bangka–Belitung province. Physically, *rusip* is a dark brown liquid product with a strong fishy flavor (Figure 42.17). It is usually used as a flavoring agent in many dishes made by the local people of the Bangka–Belitung province. Madani et al. (2010) reported that the number of backyard processors of *rusip* in the Bangka–Belitung province is around 68. The main raw material for *rusip* processing is anchovy (*Stolephorus* sp.). Some of the processors also produce *rusip* prepared from Spanish mackerel (*Scomberomorus* sp.) and frigate mackerel (*Katsuwonus* sp.).

The traditional processing of *rusip* begins with removing the head of the fresh anchovy and washing with seawater (Figure 42.18). The cleaned, beheaded anchovy is then mixed with 20%–30% salt and kept

Fresh tilapia

Eviscerated and washed

Mixed with salt, cooked rice and *asam kandis*

Poured into glass jars

Allowed to ferment for 2 days

Pudu

FIGURE 42.16   General procedure for *pudu* processing.

FIGURE 42.17   *Rusip.*

Anchovy is beheaded and washed

Mixed with 20–30% salt

Kept in a closed container for one day

Mixed with 10% palm sugar and vinegar

Poured into bottles and kept sealed for 2–3 weeks

Rusip

FIGURE 42.18 General procedure for *rusip* processing.

TABLE 42.7

Chemical Composition of *Rusip*

Parameter	Ibrahim et al. (2009)	Madani et al. (2010)
Moisture	62.49–66.62	62.2–83.7
Protein (% wb)	16.43–16.71	10.5–14.5
Fat (% wb)	0.71–0.90	1.8–3.1
Ash (% wb)	9.23–14.41	–
Salt (% wb)	6.35–10.30	17.0–30.0
pH	4.33–4.56	5.1–6.1
TVN (mg/100 g)	–	165.0–2384.5
TMA (m/100 g)	–	11.6–94.6

in a closed container for 1 day. The container is opened, and 10% palm sugar and vinegar are added. The mixture is then poured into bottles. The mixture in the bottles is allowed to ferment for 2 weeks until a gray sticky liquid with a specific fishy aroma is produced.

Madani et al. (2010) reported that the moisture content of commercial *rusip* was 62.2%–83.7%. The moisture content of *rusip* depends on the fermentation period and the amount of salt added. Nurulita et al. (2010) reported that the moisture content of *rusip* fermented for 16 days was 50.6%, whereas Ibrahim et al. (2009) reported that a 14-day fermentation period produced *rusip* with a moisture content of 66.49%. As a result of the higher amount of salt added during *rusip* preparation, the salt content of *rusip* can reach 30% (Madani et al. 2010). Table 42.7 shows the chemical composition of *rusip*.

As a result of the fermentation process, *rusip* contains a higher amount of total lactic acid bacteria, in the range of 7.6–10.2 log cfu/g (Madani et al. 2010). Nurulita et al. (2010) reported that the total lactic acid bacteria of *rusip* were 10.5 log cfu/g. Kurniati et al. (2010) reported that the main bacteria isolated from rusip were *Leuconostoc* sp., *Streptococcus* sp., and *Lactococcus* sp. Early on in the *rusip* fermentation process, the main bacteria found are *Streptococcus* sp., while at the end of the fermentation process, the main bacteria found are *Lactococcus* sp. *Leuconostoc* sp. is more likely to be present in the middle of the fermentation process. Kusmarwati et al. (2001) reported that the *Staphylococcus aureus* count of the *rusip* product ranged from 6.0×10^2 to 1.6×10^5 cfu/g.

42.2.13 *Terasi*

Terasi is the largest fermented fish product produced in Indonesia. However, the production of *terasi* has decreased yearly, from 29.884 tons in 2002 to 19.915 tons in 2007. *Terasi* is popular as an appetizer in the daily diet of Indonesian families and is usually served along with chili, garlic acid, and salt in a mixture called *sambal terasi* (Figure 42.19). There are two types of *terasi*—*terasi udang* or shrimp paste and *terasi ikan* or fish paste. *Terasi udang* is more popular than *terasi ikan*. The raw materials for *terasi*

FIGURE 42.19   *Terasi udang.*

udang processing are small shrimp-like crustaceans (*Mytis* sp. or *Atya* sp.) called *udang rebon*, whereas the raw materials for *terasi ikan* are small anchovies (*Stolephorus* sp. or *Engraulis* sp.) (Putro 1993; Rahayu et al. 1992). In addition to being produced in a pasta form, *terasi* is also produced in a powder form. This is accomplished by applying an additional sun-drying process and fine grinding (Suparno and Murtini 1992).

The production of *terasi* is spread along Sumatra and Java Island. Although the actual processing is slightly different among producers and places, the basic methods applied for *terasi* processing are similar. Normally, shrimp-like crustaceans caught in the sea are first salted with 10% salt in the fishing boat. *Terasi udang* processing begins with washing the small shrimp-like crustaceans, draining, and sun-drying them until a half-dried product is obtained (Figure 42.20). The shrimp are then mixed with salt, pounded and sun-dried, and stored in a container at ambient temperature for 2–3 days. The shrimp are pounded again and at the same time mixed with salt. During the second pounding process, some processors also mix in a coloring agent such as rhodamine B or carthamine D to produce red *terasi* (Putro 1993). The pounded shrimp are sun-dried and stored at ambient temperature for the next 2–3 days until soft. The shrimp are then ground through a grinder until fine and formed into cubes or cylinders. The cubes or cylinders are fermented for a week or more at ambient temperature until the desired *terasi udang* aroma has developed (Yunizal 1998; Irianto and Irianto 1998). *Terasi ikan* is produced following similar procedures and more often colored with synthetic dye. The fish paste smell is more repulsive in this product than in the shrimp paste, which is the reason why this product is the less popular of the two (Putro 1993).

Shrimp-like crustaceans

Washed, drained and sun-dried until half-dried

Left overnight

Pounded and mixed with 2–3% salt

Sun-dried and stored at ambient temperature for 2–3 days

Pounded a second time and mixed with 2–3% salt

Sun-dried a second time and stored at ambient temperature for 2–3 days

Grounded and formed into cubes or cylinders

Fermented for 1–2 weeks

Terasi udang

FIGURE 42.20   General procedure for *terasi udang* processing.

TABLE 42.8

Chemical Composition of *Terasi Udang*

Parameter	Nio (1992)	Surono and Hosono (1994a)
Moisture	40.0	37.4
Protein (% wb)	30.0	25.4
Fat (% wb)	3.5	6.1
Ash (% wb)	23.0	29.2
Salt (% wb)	–	16.8
Carbohydrate (% wb)	3.5	1.9
pH	–	7.5

Terasi is also produced by mixing with other ingredients such as coconut sugar and tamarind. These ingredients are mixed during the second pounding, with the objective of accelerating the fermentation process. *Terasi* mixed with other ingredients showed a higher carbohydrate content, in the range of 20.50%–21.52% (Poernomo et al. 1984). The salt content is influenced by the amount of salt added during *terasi* preparation. Sarnianto et al. (1984) reported that the salt content of *terasi* was approximately 8.85%–17.24%. Table 42.8 shows the chemical composition of *terasi udang*.

Aerobic bacteriological analyses showed that *terasi udang* had a total viable bacterial count of 4.0×10^5 cfu/g and a halophilic count of 1.1×10^5 cfu/g. The predominant microbial flora in *terasi* were *Bacillus* sp. (65.7%), *Pseudomonas* sp. (21.4%), *Micrococcus* sp. (7.2%), *Kurthia* sp. (4.3%), and *Sporolactobacillus* sp. (1.4%) (Surono and Hosono 1994a). Moreover, the *terasi* starter was composed of *Bacillus brevis*, *B. pumilus*, *B. megaterium*, *B. coagulans*, *B. substilis*, and *Micrococcus kristinae*. *Pseudomonas* sp., *Kurthia* sp., and *Sporolactobacillus* sp. were not detectable in the *terasi*, starter and the lack of these bacteria is due to contamination during the *terasi udang* processing (Surono and Hosono 1994b). Studies by Kobayashi et al. (2003) found lactic acid bacteria in *terasi udang*, which consist of *Tetragenococcus halophilus* and *Tet. muriaticus*.

42.2.14 *Tukai*

Tukai is a fermented fish product produced in the Painan area, approximately 60 km from Padang, the capital city of the West Sumatra province. Normally, *tukai* is processed from a marine fish called seapike or barracuda (*Sphyraena* sp.), locally known as *tete* or *alu-alu* (Figure 42.21). The processing of *tukai* is closely related to that of another Indonesian fermented product, *peda*, except for the fermentation process, which is performed by burying the fish underground (Efendi 1995; Huda and Rosma 2006).

The traditional processing of *tukai* starts with removal of the scales, gills, and intestinal organs (Figure 42.22). The fish are then rinsed with water and submerged in 20% coarse salt solution for 2 h. The fish are drained to remove excess salt solution and sun-dried for 2 days. The fish are then wrapped individually in taro leaves and buried underground for 3 days. After that, they are unpacked and sun-dried for two additional days until the moisture content reaches 50%. The product, which has a specific flavor, aroma, and texture, is then ready to be packed, distributed, and marketed. Like *budu*, *tukai* cannot be consumed as a main dish. The product is normally mixed with ground chili and fried before being served.

The normal shelf life of *tukai* is about 3–4 weeks. Low-temperature storage can prolong the shelf life. The approximate composition of *tukai* is as follows—50% moisture content, 40% protein, 2% fat, and

FIGURE 42.21 *Tukai.*

Fresh barracuda

Washed and eviscerated

Dipped into 20% brine solution for 2 hrs

Drained and sun-dried for 2 days

Individual fish wrapped with taro leaves and buried underground for 3 days

Taro leaf is removed and fish is sun-dried for 2 days

Tukai

FIGURE 42.22   General procedure for *tukai* processing.

TABLE 42.9

Chemical Composition of *Tukai*

Parameter	Efendi (1995)	Huda and Rosma (2006)
Moisture	51.0	50.0
Protein (% wb)	–	40.0
Fat (% wb)	–	2.0
Ash (% wb)	–	7.0
Salt (% wb)	5.0	–
pH	6.9	–
TVN (mg/100 g)	113.2	–
VRS (meq/g)	40.9	–

7% ash. The higher ash content is correlated with the amount of salt used during salt soaking (Huda and Rosma 2006). The chemical composition of *tukai* is presented in Table 42.9.

In terms of microbial quality, Efendi (1995) reported that the main bacteria isolated in *tukai* are *Micrococcus* sp., *Staphylococcus* sp., *Pediococcus* sp., *Lactobacillus* sp., and *Pseudomonas* sp. Other bacteria isolated from the *tukai* sample include *Streptococcus* sp., *Bacillus* sp., *Corynebacterium* sp., *Clostridium* sp., and *Escherichia coli*.

42.3 Conclusions

There are many different fermented fish products available in Indonesia. The processing method of each product varies slightly among producers and production areas. The scientific information available about some products is limited and, therefore, provides an opportunity for Indonesian and international researchers to contribute to the development of fish fermentation technology in Indonesia.

REFERENCES

Badan Standarisasi Nasional. 1996. Standar Nasional Indonesia (SNI). SNI-01-4271-1996. Produk Kecap Ikan. Jakarta: Dewan Standarisasi Indonesia.

Badan Standarisasi Nasional. 2006. Standar Nasional Indonesia (SNI). SNI-01-2346-2006. Produk Petis Udang. Jakarta: Dewan Standarisasi Indonesia.

Basmal, J. 1992. Pembuatan kecap ikan. In: *Kumpulan hasil-hasil penelitian pasca panen perikanan*, edited by Suparno, S. Nasran, and E. Setiabudi, pp. 140–141. Jakarta: Balitbang Perikanan Indonesia.

Burhanuddin, S., Djamali, A., Martosewojo, S., and Hutomo, M. 1987. *Sumber daya ikan manyung di Indonesia.* Jakarta: Lembaga Oseanologi Nasional—LIPI.

Candra, J.I. 2006. *Isolasi dan karakterisasi bakteri asam laktat dari produk bekasam ikan bandeng (Chanos chanos).* Bogor: Skripsi Institut Pertanian Bogor.

Chasanah, E., Hutuely, L., and Hanafiah, T.A.R. 1994. Produk fermentasi ikan dari Maluku: Perubahan selama fermentasi dan kandungan nutrisi bekasang dari jeroan ikan cakalang (*Katsuwonus pelamis*). *J. Pasca Panen Perikanan* IV(3):8–14.

Damayanthi, E. 1991. *Pengaruh pengolahan terhadap nilai gizi protein dan lemak ikan manyung (Arius thalassinus) dan ikan bandeng (Chanos chanos).* Bogor: Program Pascasarjana Institut Pertanian Bogor.

Departemen Kelautan dan Perikanan, 2010. *Statistik perikanan tangkap.* Jakarta: Direktorat Jenderal Perikanan Tangkap Departemen Kelautan dan Perikanan Indonesia.

Dewi, S. 2005. *Identifikasi peptida gurih dan senyawa bernitrogen serta karakterisasi sensori senyawa bernitrogen yang berkontribusi terhadap rasa gurih ekstrak ikan asin jambal roti.* Bogor: Sekolah Pascasarjana Institut Pertanian Bogor.

Desniar, D., Peornomo, D., and Timoryana, V.D.B. 2007. Studi pembuatan kecap ikan selar (*Caranx leptoleptis*) dengan fermentasi spontan. In: *Prosiding Seminar Nasional Tahunan V Hasil Penelitian Perikanan dan Kelautan Tahun 2007.* Yogyakarta: Universitas Gajah Mada and Badan Riset Kelautan dan Perikanan.

Desniar, Peornomo, D. and Wijatur, W. 2009. Pengaruh konsentrasi garam pada peda ikan kembung (*Rastrelliger* sp.) dengan fermentasi spontan. *J. Pengolahan Hasil Perikanan Indones.* XII(1):73–87.

Efendi, Y. 1995. Studi pendahuluan tentang pengolahan ikan tukai. In: *Prosiding Symposium Nasional Perikanan Indonesia,* pp. 152–163. Jakarta: Ikatan Sarjana Perikanan Indonesia (ISPIKANI) and Japan International Cooperation Agency (JICA).

Fakhrudin, A. 2009. *Pemanfaatan air rebusan kupang putih (Corbula faba Hinds) untuk pengolahan petis dengan penambahan berbagai pati-patian.* Bogor: Fakultas Perikanan dan Ilmu Kelautan Institut Pertanian Bogor.

Harikedua, S.D., Wijaya, C.H., and Adawiyah, D.R. 2009. Keterkaitan mutu fisiko-kimia dengan atribut sensori: "Bekasang." Paper presented at Seminar Nasional PATPI, Jakarta, 3–4 November 2009.

Hidayat, Y.W. 2000. *Mempelajari efikasi insektisida Antiset 15 EC dan Antise 1.5 L terhadap lalat hijau Chrosomya megacepjala, lalat rumah Musca domestica dan kumbang Dermetes sp. pada ikan asin jambal roti.* Bogor: Fakultas Perikanan dan Ilmu Kelautan Institut Pertanian Bogor.

Huda, N. and Rosma, A. 2006. *Budu* and *tukai*: Endemic fermented fish products from West Sumatra. *INFOFISH Int.* 3/2006:49–51.

Ibrahim, B., Zahiruddin, W., and Sastra, W. 2009. Fermentasi rusip. In: *Prosiding Seminar Nasional Perikanan Indonesia,* edited by A. Permadi, Y.H. Sipahutar, Safuridjal, A. Basith, E. Sugriwa, A.N. Siregar, E.A. Thaib, R. Surya, and N.S. Wulandari, pp. 314–320. Jakarta: Sekolah Tinggi Perikanan.

Idiyanti, T. and Arbianto, P. 1986. Identifikasi bakteri halofilik pengurai protein pada fermentasi ikan sisa/kecap ikan. *Buletin Limbah Pangan* II(3):149–159.

Ijong, F.G. and Ohta, Y. 1995. Microflora and chemical assessment of an Indonesian traditional fermented fish sauce "Bakasang." *J. Fac. Appl. Biol. Sci. Hiroshima Univ.* 34:95–100.

Ijong, F.G. and Ohta, Y. 1996. Physicochemical and microbiological changes associated with bakasang processing—A traditional Indonesian fermented fish sauce. *J. Sci. Food Agric.* 71:69–74.

Indiyati, T. and Arbianto, P. 1986. Identifikasi bakteri halofilik pengurai protein pada fermentasi ikan sisa/kecap ikan. *Bul. Limbah Pangan* II(3):149–159.

Irianto, H.E. 1999. Picungan, produk tradisional ikan fermentasi dari daerah Banten. *Warta Penelitian Perikanan Indones.* V(1):20–25.

Irianto, H.E., Indriati, N., Amini, S., and Sugiyono. 2003. Study on the processing of picungan, a traditional fermented fish product from Banten. In: *Proceedings of the JSPS—DGHE International Workshop on Processing Technology of Fisheries Products,* edited by R. Ibrahim, pp. 139–144. Semarang: University Diponegoro.

Irianto, H.E. and Irianto, G. 1998. Traditional fermented fish products in Indonesia. In: *Fish Utilization in Asia and the Pacific. Proceedings of the APFIC Symposium,* edited by D.G. James, pp. 67–75. Bangkok: RAP Publication 1998/24.

Kobayashi, T., Kajiwara, M., Wahyuni, M., Kitakado, T., Hamada-Sato, N., Imada, C., and Watanabe, E. 2003. Isolation and characterization of halophilic lactic acid bacteria isolated from "terasi" shrimp paste: A traditional fermented seafood product in Indonesia. *J. Gen. Appl. Microbiol.* 49(5):279–286.

Kurniati, Y., Yuliana, N., and Wardana, D.K. 2010. *Isolasi dan identifikasi bakteria asam laktat pada fermentasi ikan rusip*. Bandar Lampung: Fakultas Pertanian Universitas Lampung.

Kusmarwati, A., Darmosuwito, S., and Lelana, I.Y.B. 2001. Pemeriksaan bakteri *Staphylococcus aureus* pada produk perikanan yang diolah secara tradisional di Kabupaten Bangka. *Jurnal Perikanan* 3(1):1–8.

Leksono, T. and Irasari, N. 1996. Studi pemanfaatan limbah udang untuk pembuatan petis. *Terubuk* 23(66):111–117.

Maamoen, A., Dahlia, and Lukman, S. 2003. Cencaluk, Pudu some of traditional food from Riau. In: *Proceeding of the JSPS-DGHE International Workshop on Processing Technology of Fisheries Products*, edited by R. Ibrahim, pp. 202–206. Semarang: University Diponegoro.

Madani, A., Wardani, D.K., and Susilawati. 2010. *Karakterisasi rusip dari Pulau Bangka*. Bandar Lampung: Fakultas Pertanian Universitas Lampung.

Menajang, J.I. 1988. *Aspek mikrobiologi dalam pembuatan peda ikan kembung (Rastrelliger branchysoma)*. Bogor: Fakultas Teknologi Pertanian Institut Pertanian Bogor.

Murtini, J.T. 1992. Bekasam ikan mas. In: *Kumpulan hasil-hasil penelitian pasca panen perikanan*, edited by Suparno, S. Nasran, and E. Setiabudi, pp. 135–136. Jakarta: Balitbang Perikanan Indonesia.

Murtini, J.T., Ariyani, F., Anggawati, A.M., and Nasran, S. 1991. Pengolahan bekasam ikan mas (*Cyprinus carpio*). *Jurnal Penelitian Pascapanen Perikanan* 71:11–23.

Murtini, J.T., Yuliana, E., Nurjanah, and Nasran, S. 1997. Pengaruh penambahan starter bakteri asam laktat pada pembuatan bekasam ikan sepat (*Trichogaster trichopterus*) terhadap mutu dan daya awetnya. *J. Penelitian Perikanan Indones.* 3(2):71–82.

Nio, O.K. 1992. *Daftar analisis bahan makanan*. Jakarta: Fakultas Kedokteran, Universitas Indonesia.

Nisrina, H. 2007. *Analisis mikrobiologis petis ikan produksi kabupaten Sumenep–Madura*. Surabaya: Fakultas Biologi Institut Teknologi Sepuluh, November.

Nuraniekmah, S.R. 1996. *Pengaruh suhu perendaman terhadap aktifitas enzim propteolitik dan perkembangan bakteri pada pembuatan jambal roti dari ikan manyung (Arius thalassinus)*. Bogor: Fakultas Perikanan dan Ilmu Kelautan Institut Pertanian Bogor.

Nurulita, E., Susilawati, and Yuliana, N. 2010. *Pengaruh penambahan kultur cair bakteri asam laktat pada rusip*. Bandar Lampung: Fakultas Pertanian Universitas Lampung.

Poernomo, J. 1996. Pengaruh tapioka dan garam dalam fermentasi bakteri asam laktat jeroan ikan tuna (*Thunnus* sp.). *Bul. Teknol. Hasil Perikanan* II(2):64–73.

Poernomo, A., Suryaningrum, T.D., Ariyani, F., and Putro, S. 1984. Studies on the nutritive value and microbiology of traditional fishery products. *Laporan Penelitian Teknol. Perikanan* 30:9–19.

Poernomo, D., Suseno, S.H., and Wijatmoko, A. 2004. Pemanfaatan asam cuka, jeruk nipis (*Citrus aurantifolia*) dan belimbing wuluh (*Averrhoa bilimbi*) untuk mengurangi bau amis petis ikan layang (*Decapterus* spp.). *J. Pengolahan Hasil Perikanan Indones.* 7(2):10–15.

Pramono, Y. B., Rahayu, E. Y., Suparmo, and Utami, T. 2009. Aktivitas antagonisme bakteri asam laktat hasil isolasi fermentasi petis daging tradisional. *J. Indones. Tro. Anim. Agric.* 34(1):22–27.

Putro, S. 1993. Fish fermentation technology in Indonesia. In: *Fish Fermentation Technology*, edited by C.H. Lee, K.H. Steinkraus, and P.J.A. Reilly, pp. 107–128. Tokyo: United Nations University Press.

Rachmi, A., Ekantari, N., and Budhiyanti, S.A. 2008. Penggunaan papain pada pembuatan kecap ikan dari limbah filet nila. In: *Prosiding Seminar Nasional Tahunan V Hasil Penelitian Perikanan dan Kelautan Tahun 2008*. Yogyakarta: Universitas Gajah Mada and Badan Riset Kelautan dan Perikanan. PP08.

Rahayu, S. 1992. Pengolahan ikan peda. In: *Kumpulan hasil-hasil penelitian pasca panen perikanan*, edited by Suparno, S. Nasran, and E. Setiabudi, pp. 133–134. Jakarta: Balitbang Perikanan Indonesia.

Rahayu, W.P., Ma'oen, S., Suliantari, and Fardiaz, D. 1992. *Teknologi fermentasi hasil perikanan*. Bogor: PAU Pangan dan Gizi Institut Pertanian Bogor.

Rochima, E. 2005. Pengaruh fermentasi garam terhadap karakteristik jambal roti. *Bul. Teknol. Hasil Perikanan* VIII(2):46–56.

Salamah, E., Assik, A.N., and Yuliati, I. 1997. Upaya menurunkan kandungan timbal (pb) ikan manyung (*Arius thalassinus*) dan evaluasi mutu jambal roti yang dihasilkan. *Bul. Teknol. Hasil Perikanan* IV(2):5–7.

Salamah, E., Sukarsa, D.E., and Damayanti, N.K. 1995. Pengaruh konsentrasi gula dan garam terhadap mutu jambal roti. *Bul. Teknol. Hasil Perikanan* II(2):25–30.

Sani, M. 2001. *Upaya pengolahan ikan Patin (Pagasius hypophthalmus) sebagai bahan baku ikan jambal roti*. Bogor: Skripsi Institut Pertanian Bogor.

Santoso, J., Suwandi, R., Setyaningsih, L., and Santoso. 1995. Isolasi dan Indentif!kasi bakteri darl ikan asin jambal roti (*Arius thalassinus*) serta evaluasi karakteristlknya terhadap konsentrasi garam dalam media fish broth. *Bul. Teknol. Hasil Perikanan* I (1):70–79.

Sarnianto, P., Irianto, H.E., and Putro, S. 1984. Studies on the histamine contents of fermented fishery products. *Laporan Penelitian Teknol. Perikanan* 32:35–39.

Silalahi, J.R.S. 1994. *Studi pengolahan naniura ikan gabus (Ophiocephalus striatus) dengan penambahan asam asetat berbeda.* Pekanbaru: Fakultas Perkanan Universitas Riau.

Subroto, W., Setiabudi, E., and Sjahrul, B. 1984. Preliminary study on production of bekasang. *Laporan Penelitian Teknol. Perikanan* 26:9–17.

Sugiyono, Irianto, H.E., Indriati, N., Amini, S., Rahayu, U., Sabarudin, and Suarga, E.J. 1999. Isolasi dan identifikasi bakteri asam laktat dari produk bekasam. Laporan Teknis Bagian Proyek Penelitian dan Pengembangan Perikanan Slipi. Balai Penelitian Perikanan Laut. Jakarta. (Short Report).

Suharna, C., Sya'rani, L., and Agustini, T.W. 2006. Kajian sistem manajemen mutu pada pengolahan ikan jambal roti di Pangandaran, Kabupaten Ciamis. *J. Pasir Laut* 2(1):13–25.

Sukarsa, D. 1979. Pembuatan peda dari ikan air tawar. In: *Laporan Lokakarya Teknologi Pengolahan Ikan Secara Tradisional.* pp. 94–100. Jakarta: Lembaga Penelitian Teknologi Perikanan.

Suparman, A. 1993. *Pembuatan kecap ikan dengan cara kombinasi hidrolisa enzimatis dan fermentasi.* Bogor: Fakultas Teknologi Pertanian Institut Pertanian Bogor.

Suparno and Murtini, J. 1992. Terasi bubuk. In: *Kumpulan hasil-hasil penelitian pascapanen perikanan*, edited by Suparno, S. Nasran, and E. Setiabudi, pp. 137–193. Jakarta: Baitbang Perikanan Indonesia.

Suprapti, M.L. 2001. *Teknologi tepat guna membuat petis.* Yogyakarta: Penerbit Kanisius.

Suptijah, P. 1998. Pemanfaatan ekstrak protease dalam fermentasi kecap udang. *J. Pengolahan Hasil Perikanan Indones.* V(2):19–23.

Surono, I.S. and Hosono, A. 1994a. Chemical and aerobic bacterial composition of "Terasi," a traditional fermented product from Indonesia. *J. Food Hygien. Soc. Japan* 35(3):299–304.

Surono, I.S. and Hosono, A. 1994b. Microflora and their enzyme profile in Terasi starter. *Biosci. Biotechnol. Biochem.* 58(6):1167–1169.

Syafii, A. 1988. *Mutu mikrobiologi beberapa ragam peda.* Bogor: Fakultas Pertanian Institut Pertanian Bogor.

Van Veen, A.G. 1965. Fermented and dried seafood products in Southeast Asia. In: *Fish as Food*, Volume III: Processing Part 1, edited by G. Borsgstrom, pp. 227–250. New York: Academic Press.

Vonistara, F.T., Palupi, R.D., and Muhammad, A.N. 2010. *Proses fermentasi bekasam ikan nila (Oreochromis niloticus) sebagai salah satu upaya pengawetan produk perikanan.* Bangka: Buletin FISHTECH Universitas Bangka.

Widyasari, H.E. 2006. *Teknologi pengawetan ikan kembung segar dengan menggunakan bahan alami biju picung (Pangium edule).* Bogor: Sekolah Pasca Sarjana Institut Pertanian Bogor.

Yunizal. 1998. Pengolahan terasi udang. *Warta Penelitian dan Pengembangan Pertanian* XX(1):4–6.

Part V

Probiotics and Fermented Products

43

Probiotics: An Overview*

Y. H. Hui

CONTENTS

43.1 Introduction

Unless otherwise stated, the information in this chapter has been obtained from the websites of

1. United States National Institute of Health, National Center for Complementary and Alternative Medicine (NIH, NCCAM), www.NCCAM.NIH.gov/
2. United States Food and Drug Administration (FDA), www.FDA.gov/

Experts have debated on how to define probiotics. One widely used definition, developed by the World Health Organization and the Food and Agriculture Organization of the United Nations, is that probiotics are "live microorganisms, which, when administered in adequate amounts, confer a health benefit on the host." Microorganisms are tiny living organisms—such as bacteria, viruses, and yeasts—that can be seen only under a microscope.

Probiotics are live microorganisms (in most cases, bacteria) that are similar to beneficial microorganisms found in the human gut. They are also called "friendly bacteria" or "good bacteria." In this country, the health contribution of probiotics is recognized in two sources—foods and dietary supplements.

In many parts of the world that consume fermented milk from cows, goats, sheep, buffaloes, camels, and other animals, it has been accepted by their citizens that such products offer us good health. They are believed to prevent and cure many human disorders including menstrual cramps and cancer, among others. Some probiotic foods date back to ancient times, such as fermented foods and cultured milk products. In the last 20 years, scientific development has shown that during the fermentation process, certain bacteria are formed or developed that can offer us health benefits. As such, they are called probiotic foods, including many fermented milk products. The quality and quantity of such probiotic microorganisms in such foods depend on many factors—ingredients, starter cultures, processing conditions, and other parameters such as storage, packaging, and so on. Table 43.1 shows those microorganisms that have potenial probiotic properties.

Probiotics are not the same thing as prebiotics—nondigestible food ingredients that selectively stimulate the growth and/or activity of beneficial microorganisms already in people's colons. When probiotics

---

* The information in this chapter has been modified from *Food Safety Manual* published by Science Technology System (STS) of West Sacramento, CA. Copyrighted 2012©. With permission.

TABLE 43.1

Probiotics—Microorganisms

Bifidobacteria	Lactobacilli	Other Microorganisms
Bifidobacterium adolescentis	*Lactobacillus acidophilus*	*Bacillus cereus*
Bifidobacterium animalis	*Lactobacillus casei* Shirota	*Enterococcus faecalis*
Bifidobacterium bifidum	*Lactobacillus crispatus*	*Enterococcus faecium*
Bifidobacterium breve	*Lactobacillus delbrueckii* spp. *bulgaricus*	*Escherichia coli* Nissle 1917
Bifidobacterium infantis	*Lactobacillus gallinarum*	*Lactococcus lactis*
Bifidobacterium lactis	*Lactobacillus gasseri*	*Leuconostoc mesenteroides*
Bifidobacterium longum	*Lactobacillus johnsonii*	*Pediococcus acidilactici*
	Lactobacillus plantarum	*Propionibacterium freudenreichii*
	Lactobacillus reuteri	*Saccharomyces boulardii*
	Lactobacillus rhamnosus	*Saccharomyces cerevisiae*
	Lactobacillus salivarius	*Streptococcus thermophilus*

Source: Cho, S.S., and Finocchiaro, T., Handbook of Prebiotics and Probiotics Ingredients: Health Benefits and Food Applications, CRC Press, Boca Raton, FL, 2009. With permission. Kneifel, W., and Salminen, S., Probiotics and Health Claims, Wiley, Hoboken, NJ, 2011. With permission. With kind permission from Springer Science+Business Media: Probiotics, 2011, Liong, M.T. Mozzi, F., et al., *Biotechnology of Lactic Acid Bacteria: Novel Applications.* 2010. Copyright Wiley-VCH Verlag GmbH & Co. KGaA. Reproduced with permission. Nair, G.B., and Takeda, Y., *Probiotic Foods in Health and Disease*, CRC Press, Boca Raton, FL, 2011. With permission. Salminen, S., et al., *Lactic Acid Bacteria: Microbiological and Functional Aspects*, CRC Press, Boca Raton, FL, 2004. With permission. von Wright, A., and Lahtinen, S., *Lactic Acid Bacteria: Microbiological and Functional Aspects*, CRC Press, Boca Raton, FL, 2011. With permission.

and prebiotics are mixed together, they form a synbiotic. Probiotics are available in foods and dietary supplements (e.g., capsules, tablets, and powders) and in some other forms as well. Examples of foods containing probiotics are yogurt, fermented and unfermented milk, miso, tempeh, and some juices and soy beverages. In probiotic foods and supplements, the bacteria may have been present originally or added during preparation.

Most probiotics are bacteria similar to those naturally found in people's guts, especially in those of breastfed infants (who have natural protection against many diseases). Most often, the bacteria come from two groups, *Lactobacillus* or *Bifidobacterium*. Within each group, there are different species (e.g., *Lactobacillus acidophilus* and *Bifidobacterium bifidus*), and within each species, different strains (or varieties). A few common probiotics, such as *Saccharomyces boulardii*, are yeasts, which are different from bacteria.

43.2 Foods

Apart from fermented dairy products with potential probiotic attributes, some fermented plant foods are recognized to process such beneficial microorganisms. Table 43.2 shows a list of the most popular fermented food products with potential probiotic properties. There are other foods with potential probiotic activities, such as cheese and fermented sausages.

Once we know what nature can do and has done, the next logical step is to improve and/or duplicate nature. In the last two decades, scientists worldwide have accomplished some preliminary success on both counts:

1. Foods such as fermented milk and yogurt are tempered with scientific tools so that some of such products have improved quality and quantity of beneficial bacteria.
2. New foods have been created so that they offer potential probiotic bacteria. Probiotic cheese and healthier sausage are covered in two separate chapters in this book. Table 43.3 lists some novel plant foods with probiotic activities. Some of these foods are referred to as functional

TABLE 43.2

Popular Traditional Probiotic Foods

Dairy	Yogurt	Most advertised by manufacturers to contain probiotics. The amount will depend on the inclusion of proper microorganisms.
	Buttermilk	Buttermilk is preferred by some consumers. The product has a tangy flavor. Some like adding buttermilk to smoothies with fresh fruit.
	Kefir	Kefir contains milk fermented with kefir grains. The milk can be cow, goat, or sheep milk. It is mixed with the grain and fermented for 12–24 h. The drink is very sour and is usually consumed with fruit or sweeteners.
	Indigenous fermented milk products	
Soy	Tempeh, tofu, natto, miso, sufu, soy sauce, shoyu	Tempeh is a fermented soy product. It is chewy. It is made by fermenting whole soybeans. In western countries, it is used in many vegetarian dishes. For many, it offers a high-quality protein.
Vegetables	Sauerkraut	Sauerkraut is made from fermented or pickled cabbage. Fresh cabbage is prepared and fermented in brine for various period of time, depending on the processing conditions. Most of us are familiar with sauerkraut.
	Kim Chee	A popular Korean preparation, Kim Chee is now known all over the world, available in our neighboring grocery stores. A pungent side dish or a relish, Kim Chee is cabbage fermented using a highly specialized formula.
Grains	Brewer's yeast	Brewer's yeast is a by-product of beer making. It is used in home and commercial preparations of a variety of food and beverage products.
	Miso	Miso is a Japanese condiment made by fermenting different beans or grains. In Japan, it is used in soups, sauces, and spreads. It can be white miso, black miso, and red miso. It is served in many Japanese restaurants in western countries.

Source: Cho, S.S., and Finocchiaro, T., *Handbook of Prebiotics and Probiotics Ingredients: Health Benefits and Food Applications*, CRC Press, Boca Raton, FL, 2009, With permission. Kneifel, W., and Salminen, S., *Probiotics and Health Claims*. 2011. Copyright Wiley-VCH Verlag GmbH & Co. KGaA. Reproduced with permission. Lee, Y.K., and Salminen, S., *Handbook of Probiotics and Prebiotics*, 2nd edition. 2008. Copyright Wiley-VCH Verlag GmbH & Co. KGaA. Reproduced with permission. With kind permission from Springer Science+Business Media: *Probiotics*, 2011, Liong, M.T., Nair, G.B., and Takeda, Y., *Probiotic Foods in Health and Disease*, CRC Press, Boca Raton, FL, 2011. With permission.

foods in research publications. Unfortunately, the legal status of such a name is still not yet settled in the United States.

It must be emphasized that there is a difference between "the claim of health benefits" and "health benefits." Yogurt is claimed to offer health benefits because of the presence of probiotic bacteria. This may be true. However, to prove that yogurt can prevent or cure a particular clinical disorder will require many years of scientific research in spite of some preliminary data demonstrating such potentials. The same is especially true of novel foods such as those shown in Table 43.3.

43.3 Dietary Supplements

Probiotic products taken by mouth as a dietary supplement are manufactured and regulated as foods, not drugs—products that contain vitamins, minerals, herbs or other botanicals, amino acids, enzymes, and/or other ingredients intended to supplement the diet. The FDA has special labeling requirements for dietary supplements.

TABLE 43.3

Commercial Functional Plant Foods Using Probiotic Microorganisms

Function Plant Foods	Microorganisms
Artichokes	*Lactobacillus plantarum*
	Lactobacillus paracasei
Barley flours	*Lactobacillus plantarum*
	Lactobacillus fermentum
Barley malt	*Lactobacillus reuteri*
	Lactobacillus acidophilus
	Lactobacillus rhamnosus
Maize fermented cereal gruel (Ogi)	*Lactobacillus plantarum*
Maize flour	*Lactobacillus reuteri*
	Lactobacillus acidophilus
	Lactobacillus rhamnosus
Maize pudding	*Bifidobacterium animali*
	Lactobacillus acidophilus
	Lactobacillus rhamnosus
Oat bran pudding (Yosa)	*Bifidobacterium lactis*
	Lactobacillus acidophilus
Oat cereal bar	*Bifidobacterium lactis*
Olives, table	*Lactobacillus rhamnosus*
	Lactobacillus paracasei
	Bifidobacterium bifidum
	Bifidobacterium longum
Rice pudding	*Bifidobacterium animali*
	Lactobacillus acidophilus
	Lactobacillus rhamnosus
Rice, brown rice flour	*Bifidobacterium longum*
Soy frozen dessert	*Lactobacillus acidophilus*
	Lactobacillus paracasei
	Bifidobacterium lactis
	Lactobacillus rhamnosus
	Saccharomyces boulardii
Wheat flours	*Lactobacillus plantarum*
	Lactobacillus fermentum

Source: Bagchi, D., et al., *Biotechnology in Functional Foods and Nutraceuticals*, CRC Press, Boca Raton, FL, 2010. With permission. Guo, M. *Functional Foods: Principles and Technology*, CRC Press, Boca Raton, FL, 2009. With permission. Kneifel, W., and Salminen, S., *Probiotics and Health Claims*. 2011. Copyright Wiley-VCH Verlag GmbH & Co. KGaA. Reproduced with permission. Mozzi, F., et al., *Biotechnology of Lactic Acid Bacteria: Novel Applications*. 2010. Copyright Wiley-VCH Verlag GmbH & Co. KGaA. Reproduced with permission. Smith, J., and Charter, E., *Functional Food Product Development*. 2010. Copyright Wiley-VCH Verlag GmbH & Co. KGaA. Reproduced with permission.

To be classified as a dietary supplement, a botanical must meet the definition given below. Many botanical preparations meet the definition. As defined by Congress in the Dietary Supplement Health and Education Act (http://www.fda.gov/opacom/laws/dshea.html#sec3), which became law in 1994, a dietary supplement is a product (other than tobacco) that

1. Is intended to supplement the diet
2. Contains one or more dietary ingredients (including vitamins, minerals, herbs or other botanicals, amino acids, and other substances) or their constituents
3. Is intended to be taken by mouth as a pill, capsule, tablet, or liquid
4. Is labeled on the front panel as being a dietary supplement

With all these restrictions, how do probiotics fit in with dietary supplements? Because of so much publicity and advertisement, many consumers firmly believe that probiotics are beneficial to their health. Manufacturers of dietary supplements for the public use the following approaches to introduce antibiotics into their products, assuming they implement what they describe on the labels.

1. There are many probiotic dietary supplements that contain one or multiple groups of specific beneficial bacteria specifically marketed as follows:
   a. Probiotics—benefits specific
   b. Probiotics—age specific
   c. Probiotics—sex specific
   d. Probiotics—disorder specific
   e. Probiotics—body part specific
2. Probiotics can be added to a standard dietary supplement for essential nutrients. The dietary supplement for vitamin C is a good example. The manufacturer can add beneficial bacteria to it, and it becomes a "vitamin C + probiotics" dietary supplement. Because there are more than 10 nutrients, the types of product can cover a big spectrum.
3. Probiotics can be added to a dietary supplement that contains a botanical (herbs, plant parts, and so on). Ginseng is a good example. The manufacturer can add beneficial bacteria to it, and it becomes a "ginseng + probiotics" dietary supplement. Because there are more than literally thousands of botanicals sold as dietary supplements in this country, the types of product can cover a big spectrum.
4. Probiotics can be added to a dietary supplement that contains legally approved enzymes. Lactase is a good example; it helps some ethnic groups to digest milk. The manufacturer can add beneficial bacteria to it, and it becomes a "lactase + probiotics" dietary supplement. The types of product that can be marketed will be limited to enzymes legally approved for human consumption.

All manufacturers of such products pay special attention to their marketing to avoid fraud and prosecution for making false claims. There are two standard practices.

1. They use many terms that imply certain behavioral objectives including
   Help—This product will help you to lose weight.
   Boost—This product will boost your immunity.
   Improve—This product will improve your skin color.
   Increase—This product will increase your resistance to infection.
   Decrease—This product will decrease your chance of having cancer.
   Note that none of the terms implies cure and prevention.
2. All their labels and advertisement will carry the following statement:
   These statements have not been evaluated by the Food and Drug Administration. This product is not intended to diagnose, treat, cure, or prevent any disease.

43.4 Health Benefits

Interest in probiotics in general has been growing; Americans' spending on probiotic supplements is increasing at a fast pace. There are several reasons that people are interested in probiotics for health purposes.

First, the world is full of microorganisms (including bacteria), and so are people's bodies—in and on the skin, in the gut, and in other orifices. Friendly bacteria are vital to proper development of the immune system, to protection against microorganisms that could cause disease, and to the digestion and absorption of food and nutrients. Each person's mix of bacteria varies. Interactions between a person and the microorganisms in his body, and among the microorganisms themselves, can be crucial to the person's health and well-being.

This bacterial "balancing act" can be thrown off in two major ways. The first is by antibiotics, when they kill friendly bacteria in the gut along with unfriendly bacteria. Some people use probiotics to try to offset side effects from antibiotics, like gas, cramping, or diarrhea. Similarly, some use them to ease symptoms of lactose intolerance—a condition in which the gut lacks the enzyme needed to digest significant amounts of the major sugar in milk, and which also causes gastrointestinal symptoms. (2) "Unfriendly" microorganisms such as disease-causing bacteria, yeasts, fungi, and parasites can also upset the balance. Researchers are exploring whether probiotics could halt these unfriendly agents in the first place and/or suppress their growth and activity in conditions such as

1. Infectious diarrhea
2. Irritable bowel syndrome
3. Inflammatory bowel disease (e.g., ulcerative colitis and Crohn's disease)
4. Infection with *Helicobacter pylori* (*H. pylori*), a bacterium that causes most ulcers and many types of chronic stomach inflammation
5. Tooth decay and periodontal disease
6. Vaginal infections
7. Stomach and respiratory infections that children acquire in day care
8. Skin infections

Another part of the interest in probiotics stems from the fact that there are cells in the digestive tract connected with the immune system. One theory is that if you alter the microorganisms in a person's intestinal tract (as by introducing probiotic bacteria), you can affect the immune system's defenses.

Scientific understanding of probiotics and their potential for preventing and treating health conditions is at an early stage but moving ahead. According to the NIH, some uses of probiotics for which there is some encouraging evidence from the study of specific probiotic formulations are as follows:

1. To treat diarrhea (this is the strongest area of evidence, especially for diarrhea from rotavirus)
2. To prevent and treat infections of the urinary tract or female genital tract
3. To treat irritable bowel syndrome
4. To reduce recurrence of bladder cancer
5. To shorten the duration of an intestinal infection that is caused by a bacterium called *Clostridium difficile*
6. To prevent and treat pouchitis (a condition that can follow surgery to remove the colon)
7. To prevent and manage atopic dermatitis (eczema) in children

In studies on probiotics as cures, any beneficial effect was usually low, and more research (especially in the form of large, carefully designed clinical trials) is needed in order to draw firmer conclusions. Some other areas of interest to researchers on probiotics are

1. What is going on at the molecular level with the bacteria themselves and how they may interact with the body (such as the gut and its bacteria) to prevent and treat diseases. Advances in technology and medicine are making it possible to study these areas much better than in the past.
2. Issues of quality. For example, what happens when probiotic bacteria are treated or are added to foods—is their ability to survive, grow, and have a therapeutic effect altered?
3. The best ways to administer probiotics for therapeutic purposes, as well as the best doses and schedules.
4. Probiotics' potential to help with the problem of antibiotic-resistant bacteria in the gut.

5. Whether they can prevent unfriendly bacteria from getting through the skin or mucous membranes and traveling through the body (which can happen with burns, shock, trauma, or suppressed immunity, for example).

Some live microorganisms have a long history of use as probiotics without causing illness in people. Probiotics' safety has not been thoroughly studied scientifically, however. More information is especially needed on how safe they are for young children, elderly people, and people with compromised immune systems. Probiotics' side effects, if they occur, tend to be mild and digestive (such as gas or bloating). More serious effects have been seen in some people. Probiotics might theoretically cause infections that need to be treated with antibiotics, especially in people with underlying health conditions. They could also cause unhealthy metabolic activities, too much stimulation of the immune system, or gene transfer (insertion of genetic material into a cell).

In the last decade, animal husbandry has explored the potentials of probiotics to provide health benefits to animals raised for human foods. There are now special feed or feed-related products that provide benefits to cattle and poultry, for example

1. Reduce diseases in the animals
2. Improve the quality and yield of beef, pork, and chicken

43.5 Regulatory Consideration (FDA)

At present, the FDA's position is as follows.

Probiotics may be regulated as dietary supplements, foods, or drugs under the law, depending on the product's intended use. Other factors may also affect the classification of the product, for example, whether the product contains a "dietary ingredient," whether it is represented as a conventional food or as a meal replacement, and, for probiotics used as ingredients in a conventional food, whether the ingredient is generally recognized as safe for its intended use. In addition to any requirements that apply based on the product's classification under the law, probiotics may also be subject to legal provisions concerning the prevention of communicable disease, due to potential disease-causing microorganisms that might be contained in such products. Finally, if a probiotic is a drug, it may be subject to regulation as a biological product.

43.6 Future Developments

At present, many researchers worldwide are working on probiotics from three important perspectives:

1. Health benefits of probiotics for people
2. Improving the health potentials of probiotics in known and traditional animal and plant foods especially those involving fermentation
3. Developing new foods with probiotic potentials, especially plant foods

In the next few decades, we will see results from this intense effort in research and development in the science and technology of food.

REFERENCES

Ahmad, I. and Aqil, F. 2008. *New Strategies Combating Bacterial Infection.* Hoboken, NJ: Wiley.
Bagchi, D., Lau, F.C., and Ghosh, D.K. 2010. *Biotechnology in Functional Foods and Nutraceuticals.* Boca Raton, FL: CRC Press.

Cho, S.S. and Finocchiaro, T. 2009. *Handbook of Prebiotics and Probiotics Ingredients: Health Benefits and Food Applications*. Boca Raton, FL: CRC Press.

Fuller, R. and Peridigon, G. 2003. *Gut Flora, Nutrition, Immunity and Health*. Hoboken, NJ: Wiley.

Gibson, G.R. and Roberfroid, M. 2008. *Handbook of Prebiotics*. Boca Raton, FL: CRC Press.

Goktepe, I., Juneja, V., and Ahmedna, M. 2005. *Probiotics in Food Safety and Human Health*. Boca Raton, FL: CRC Press.

Guo, M. 2009. *Functional Foods: Principles and Technology*. Boca Raton, FL: CRC Press.

Jardine, S. 2009. *Prebiotics and Probiotics*, 2nd edition. Hoboken, NJ: Wiley.

Kneifel, W. and Salminen, S. 2011. *Probiotics and Health Claims*. Hoboken, NJ: Wiley.

Lee, Y.K. and Salminen, S. 2008. *Handbook of Probiotics and Prebiotics*, 2nd edition. Hoboken, NJ: Wiley.

Liong, M.T. 2011. *Probiotics*. New York: Springer.

Mozzi, F., Raya, R.R., and Vignolo, G.M. 2010. *Biotechnology of Lactic Acid Bacteria: Novel Applications*. Hoboken, NJ: Wiley.

Mullin, G.E., Matarese, L.E., and Palmer, M. 2011. *Gastrointestinal and Liver Disease Nutrition Desk Reference*. Boca Raton, FL: CRC Press.

Nair, G.B. and Takeda, Y. 2011. *Probiotic Foods in Health and Disease*. Boca Raton, FL: CRC Press.

Salminen, S., von Wright, A., Ouwehand, A.C., and Lahtinen, S. 2004. *Lactic Acid Bacteria: Microbiological and Functional Aspects*. Boca Raton, FL: CRC Press.

Smith, J. and Charter, E. 2010. *Functional Food Product Development*. Hoboken, NJ: Wiley.

Tamang, J.P. 2009. *Himalayan Fermented Foods: Microbiology, Nutrition, and Ethnic Values*. Boca Raton, FL: CRC Press.

Tamime, A. 2006. *Probiotic Dairy Products*. Hoboken, NJ: Wiley.

Tannis, A. 2009. *Probiotic Rescue: How You Can Use Probiotics to Fight Cholesterol, Cancer, Superbugs, Digestive Complaints and More*. Hoboken, NJ: Wiley.

von Wright, A. and Lahtinen, S. 2011. *Lactic Acid Bacteria: Microbiological and Functional Aspects*. Boca Raton, FL: CRC Press.

44

Cheese as Probiotic Carrier: Technological Aspects and Benefits

Flávia Carolina Alonso Buriti, Cínthia Hoch Batista de Souza, and Susana Marta Isay Saad

CONTENTS

44.1 Introduction

Probiotics are defined as live microorganisms that exert health benefits on the host when ingested in appropriate numbers (FAO/WHO 2001). According to this definition, the viability and metabolic activity of probiotic bacteria should be maintained from production to consumption and throughout gastrointestinal transit. It is generally believed that probiotic bacteria must survive the passage through the gastrointestinal tract (GIT) and reach the distal part in sufficient numbers (10^6–10^7 CFU/g) to exert their beneficial effects (Talwalkar et al. 2004). So far, only a few studies have been conducted to define the effective dose of probiotic strains, but it is generally accepted that a daily intake of at least 10^8–10^9 viable cells, obtained through the consumption of at least 100 g of foods containing probiotic bacteria, has been suggested as a minimum to obtain health benefits (Roy 2001). Therefore, the presence of probiotic bacteria at minimum levels of 10^6–10^7 CFU/g or mL is recommended in functional foods, as well as their daily intake (Lee and Salminen 1995; Jayamanne and Adams 2006).

Lactobacillus and *Bifidobacterium* species are the most commonly used probiotics in foods for human consumption given the significant health benefits associated with ingestion of these microorganisms (Shah 2007). These bacteria have a beneficial influence on human health by improving the balance of intestinal microbiota and improving mucosal defenses against pathogens. Additional health benefits include enhanced immune response, reduction of serum cholesterol, anticarcinogenic activity, vitamin synthesis, and antibacterial activity (Boylston et al. 2004).

Although yogurt and fermented milks have received the most attention as carriers of probiotic bacteria, some cheese varieties have also been studied as vehicles of probiotic microorganisms. Cheese has been considered an important matrix to probiotic delivery because it can improve probiotic viability throughout the GIT (Ranadheera et al. 2010).

44.2 General Aspects of Cheeses

Cheese is the generic name for a group of fermented milk-based food products produced throughout the world in a great diversity of flavors, textures, and forms (Fox et al. 2000; Lucey 2008).

Cheese consists mainly of fat, protein, minerals, lactic acid, and water. Proteins form the matrix of the cheese, and fat acts as inert filler. Casein micelles contribute to the hardness, and fat and moisture contribute to the smoothness and creaminess of cheese (Fox et al. 2000; Konuklar et al. 2004).

The whole cheesemaking process contributes to the development of a distinct and more complex matrix than the departing feedstock milk. The method used to clot milk for cheesemaking influences the overall structure, characteristics, and firmness of the cheese. The two basic methods for clotting milk for cheese manufacture are by rennet or acid addition, leading to the respective terms *rennet-* or *acid-coagulated* cheeses. In general, acid-coagulated cheeses are soft, whereas rennet-set cheeses are firm (Farkye 2004).

During coagulation (which, in most cases, is enzyme driven), rennet enzymes bring about breakdown of κ-casein, which is present especially on the surface of casein micelles. After such a chemical step, physical agglomeration takes place—which gives rise to a more uniform protein mass. The altered casein is referred to as paracasein; it cannot be dissolved, nor dispersed, in milk serum. Because of this, the paracasein micelles in the milk aggregate, provided that the Ca^{2+} activity is high enough (Walstra et al. 2006a,b). The rennet-induced milk coagulum is cut and homogenized to expel moisture in a process called syneresis (Grundelius et al. 2000; Pereira et al. 2009). Curd is later drained, salted, and packaged into fresh cheese. Many cheeses need additional time to achieve their own sensory features, particularly flavor and aroma. To achieve this purpose, they are maintained in a special room, with controlled environmental conditions, for a determined time—this process is called ripening, and the final product is called ripened cheese (Everett and Auty 2008).

44.3 Probiotic Cheese Development

Cheese currently suffers from an adverse nutritional image largely due to a perceived association between saturated fatty acid, cholesterol, and the salt content of cheese and cardiovascular disease. However, cheese is also a rich source of essential nutrients, such as proteins, lipids, vitamins, and minerals that play an integral part in a healthy diet (Ash and Wilbey 2010). Cheese is a versatile food product, appealing to many palates, and is suitable for all age groups. It provides a valuable alternative to yogurt and fermented milk as a vehicle in probiotic delivery (Wilkinson et al. 2001). Consumption of cheese has increased in many countries during the last decade, providing an additional advantage for the use of cheese products in probiotic delivery (Cruz et al. 2009).

Most of the technological processes involved in the production of a probiotic product represent a threat to the viability of the strains (e.g., presence of oxygen, acidification, processing and storing temperatures, freeze-drying, and spray-drying because these factors can limit the viability of probiotic bacteria in probiotic food.

During the probiotic cheesemaking process, probiotic bacteria are exposed to various changes in environmental parameters (heat, acidity, salt, cold, etc.) and induce adaptive responses, allowing them to cope with the resulting physiological stresses (Kosin and Rakshit 2006). Therefore, probiotic bacteria should be viable and present at high numbers in the product at the time of consumption, in order to exert its beneficial effect. In order to exert the beneficial effects, probiotic microorganisms added to probiotic food must maintain their viability during the steps of the food processing operations, as well as during the whole period of storage. Moreover, probiotic microorganisms must be able to survive in the human GIT, reaching the colon in adequate amounts (Sanz 2007).

44.4 Cheese as Probiotic Carrier

Incorporation of probiotic cultures in cheeses provides potential not only to improve health status and quality of products but also to increase the range of probiotic products available for consumers. Moreover, the production of functional cheeses was proposed as a suitable and promising alternative as vehicle for probiotics, when compared to fermented milk and yogurt (Roy 2005).

Cheeses have a number of advantages over fresh fermented products, such as yogurt and fermented milk, as a delivery system for viable probiotic to GIT, as they tend to have higher pH, more solid consistency, and relatively higher fat content. It has been postulated that the embedding of probiotic bacteria in the fat–protein matrix of cheese, as well as the buffering capacity of cheeses, may assist survival of probiotic bacteria during GIT passage (Heller et al. 2003; Bergamini et al. 2005; Ong et al. 2006).

In fact, studies conducted by Stanton et al. (1998) and Gardiner et al. (1999) revealed the protective effect of cheddar cheese, compared with yogurt, as a food carrier for delivery of viable probiotic lactobacilli and enterococci to the GIT.

Several cheeses were tested as vehicles for probiotic strains of *Lactobacillus* and *Bifidobacterium*, and most of them succeeded in maintaining the viability of these microorganisms as well as achieving appropriate technological properties of the final product. These have included cheddar (McBrearty et al. 2001; Ong et al. 2007b), Gouda (Gomes et al. 1998), Crescenza (Gobbetti et al. 1997), cream cheese (Buriti et al. 2007a), prato cheese (Cichoski et al. 2008), white cheese (Kasımoğlu et al. 2004; Gursoy and Kinik 2010), Arzúa-Ulloa (Menéndez et al. 2000), whey Portuguese cheese (*requeijão*) (Madureira et al. 2005a), *petit-suisse* cheese (Cardarelli et al. 2008; Pereira et al. 2010), soft cheese (Modzelewska-Kapitula et al. 2007), Pategrás cheese (Bergamini et al. 2010), caprine cheese (Kalavrouzioti et al. 2005), Estonian Pikantne cheese (Songisepp et al. 2004), Canestrato Pugliese (Corbo et al. 2001), Swiss cheese (Montoya et al. 2009), and fresh cheeses (Vinderola et al. 2000; Suárez-Solís et al. 2002; Buriti et al. 2005a,b; Souza and Saad 2009; Fritzen-Freire et al. 2010).

Probiotic microorganisms can also be combined in a food product with prebiotic ingredients. Prebiotics are nondigestible food ingredients that beneficially affect the host by selectively stimulating the growth and/or activity of one or a limited number of bacteria in the colon, and thus improve host health (Gibson 2004). Inulin and oligofructose, nondigestible fermentable fructans, are among the most studied and well-established prebiotics (Gibson et al. 2004). Because of potential synergy between probiotics and prebiotics, foods containing a combination of these ingredients are often referred to as synbiotics (Mattila-Sandholm et al. 2002). A few studies available have shown the development of synbiotic cheeses. However, satisfactory probiotic populations were observed when synbiotic cheeses were developed (Buriti et al. 2007a; Cardarelli et al. 2008; Araújo et al. 2010; Rodrigues et al. 2011).

Although the use of probiotic cultures may contribute to improve the cheese's properties, it is important to highlight that cheeses supplemented with probiotic bacteria should retain their sensorial and related properties during the storage period (Souza et al. 2008).

44.5 Cheese Cultures with Functional and Probiotic Properties

Probiotic microorganisms include several genera of bacteria and yeasts, and among those, strains of enterococci, lactobacilli, and propionibacteria are important for manufacturing some cheeses and are well adapted to the cheese environment (Grattepanche et al. 2008). Common probiotics used in cheese production also include bifidobacteria (Roy 2005). Probiotic enterococci and lactobacilli are members of the lactic acid bacteria group (Klein et al. 1998). These microorganisms comprise Gram-positive, nonsporing, nonrespiring cocci, or rods that produce lactic acid as the major end product during the fermentation of carbohydrates (Axelsson 2004). They are also proteolytic, with fastidious amino acid growth requirements (Forsythe 2000). Lactobacilli and enterococci (low mol% G + C in DNA) belong to phylum *Firmicutes* and are phylogenetically unrelated to bifidobacteria (Axelsson 2004; Franz and Holzapfel 2004; Vasiljevic and Shah 2008; Trebichavsky et al. 2009). Bifidobacteria and propionibacteria (high mol% G + C in DNA) are included in the phylum Actinobacteria (Ventura et al. 2007a,b). Propionibacteria do not belong to the group of lactic acid bacteria but are increasingly being considered as potential probiotics (Champagne et al. 2005; Meile et al. 2008; Falentin et al. 2010).

Probiotic cultures that contain *Lactobacillus* and *Bifidobacterium* strains have an extensive safety record for use in the generally healthy population (Douglas and Sanders 2008). On the other hand, there are concerns surrounding the complete safety of use of *Enterococcus* as probiotics, particularly because this genus may contain some opportunistic pathogenic species (Boyle et al. 2006; Farnworth 2008).

The enterococci form an important part of environmental, food, and clinical microbiology. They are typical commensals of the GIT of humans (10^5–10^8 CFU/g of feces). Under normal conditions, enterococci do not show any harmful effects on the host (Bhardwaj et al. 2009). Enterococci were the focus of numerous investigations on bacteriocin production, mainly because bacteriocin production seems to be a common trait among strains associated with food systems (Izquierdo et al. 2009). *Enterococcus faecium* and *Ent. faecalis* are the most common enterococci species associated with dairy products (Franciosi et al. 2009). Like other microorganisms, depending on the strain, they are used as starters, adjunct starters, and probiotic organisms for the development of typical taste, flavor, and beneficial health claims, respectively (Bhardwaj et al. 2009). These properties allied with their ability to produce bacteriocins are, therefore, important characteristics for their application in food technology (Franciosi et al. 2009).

Various adjunct cultures used in cheese production to improve product quality and to provide probiotic benefits belong to the *Lactobacillus casei* group, which comprises the facultatively heterofermentative species *Lb. casei*, *Lb. paracasei*, and *Lb. rhamnosus* (Desai et al. 2006; Buriti and Saad 2007; Milesi et al. 2009). These species also represent a considerable part of the lactobacilli microbiota of the human intestinal mucosa (Vásquez et al. 2005). Some bacteria of this group, particularly *Lb. casei* and *Lb. paracasei*, are found naturally in raw milk and in high numbers in matured cheeses, such as cheddar and Grana cheeses, where they contribute to positive and negative aspects of cheese maturation and flavor development (Ferrero et al. 1996; Sullivan et al. 2001; Desai et al. 2006). Other lactobacilli with probiotic potential used in cheese production include *Lb. acidophilus*, *Lb. brevis*, *Lb. fermentum*, and

Lb. plantarum (Tamime et al. 2005; Madureira et al. 2005a; Lavermicocca 2006; Järvenpää et al. 2007; Kılıç et al. 2009).

Bifidobacterium spp. belong to the dominant anaerobic microbiota of the colon. *B. bifidum*, *B. longum* subsp. *infantis*, *B. longum*, and *B. animalis* subsp. *lactis* are commonly used by the food industry (Roy 2005). Contrary to starter cultures, bifidobacteria do not have a significant role either in the acidification of milk or in the formation of texture and/or flavor (Meile et al. 2008). On the other hand, *Bifidobacterium* spp. can ferment a wide variety of oligosaccharides, some of which are not digested by their host, and thus have found application in enhancing bifidobacterial numbers *in situ* (Ventura et al. 2007a). *B. animalis* subsp. *lactis* BB-12 is the most frequently used probiotic *Bifidobacterium* strain in dairy products worldwide (Meile et al. 2008).

The dairy propionibacteria are used to produce typical flavor and eyes in Swiss-type cheese and include the species *Propionibacterium freudenreichii*, *P. jenenii*, *P. thoenii*, and *P. acidipropionici*, which are not typical of the human microbiota (Champagne et al. 2005). They belong to the propionic acid bacteria group, and are characterized as Gram-positive, nonsporing, and nonmotile pleomorphic rods. Propionibacteria are basically anaerobes, although there are some aerotolerant species, which produce propionic acid, acetic acid, and CO_2 as their main fermentation products (Ouwehand 2004). Among them, *P. freudenreichii* resists against harsh physical and chemical stresses (particularly significant heat and salt stresses) during the cheesemaking process. Some dairy propionibacteria, including *P. freudenreichii*, have been demonstrated to survive after passage through the digestive stresses (Leverrier et al. 2003).

Ouwehand et al. (2002) verified that four strains of propionibacteria (*P. freudenreichii* subsp. *shermanii* JS, *P. freudenreichii* P2, *P. freudenreichii* P6, and *P. freudenreichii* subsp. *freudenreichii* P131) were able to adhere to the intestinal mucus. The authors also reported that adhesion could be increased when these microorganisms were in combination with other probiotic strains. The production of extracellular growth stimulators for bifidobacteria by *P. freudenreichii* has been reported. Mori et al. (1997) purified a bifidogenic growth factor produced by *P. freudenreichii* 7025. Experimental analyses indicated that the chemical structure of the bifidogenic growth stimulator was 2-amino-3-carboxy-1,4-naphthoquinone, which showed growth stimulatory activity for bifidobacteria at a concentration of 0.1 ng/mL. The production of vitamin B_{12}, CO_2, bacteriocins, and organic acids is also associated with probiotic effects of propionibacteria (Champagne et al. 2005).

44.6 Evaluation of Probiotic Potential of New Bacterial Strains and Their Application in Cheese Production

The demand for probiotic products has been increasing over the years, leading to a growing interest in improving the variety of probiotic cultures to attend the food industry. Although there is a reasonable number of well-characterized probiotic strains available for commercial use around the world, the isolation and characterization of new strains is still desirable. In developing countries, for example, the small dairy plants have restricted access to probiotic strains for the formulation of probiotic foods (Vinderola et al. 2008). The safety assessment of probiotic strains is usually a challenge due to lack of requisite facilities as well as technical manpower. Moreover, several traditional and indigenous fermented products with known functional or medicinal properties still did not have their natural microbiota identified properly (Ukeyima et al. 2010). In this way, efforts have been made in order to identify new strains from human, dairy, and other origins and to evaluate their functional potential and possible applications in different food products. Some research involving cheeses is discussed here.

In a study conducted by Coeuret et al. (2004), four strains from unpasteurized Camembert—two *Lb. plantarum* strains (UCMA 2997 and UCMA 3037) and two *Lb. paracasei/casei* strains (UCMA 3061 and UCMA 3063)—were found to resist low pH and bile and other potentially probiotic features. Those included resistance to lysozyme, adhesion to CACO-2 cells, antimicrobial effects against common foodborne pathogens (*Listeria monocytogenes*, *Staphylococcus aureus*, *Salmonella* spp., and *Escherichia coli*), innocuity following the ingestion of high doses by mice, and appropriate antibiotic susceptibility profiles. In addition, *Lb. plantarum* strain UCMA 3037 was incorporated into a Pont-l'Eveque cheese, a registered designation of origin soft cheese. According to the authors, the viability of *Lb. plantarum* UCMA 3037 in cheese was higher than 10^7 CFU/g throughout its shelf life, without changing the quality score of the product during the 75 days after production.

Fernández et al. (2005) investigated the potential of the strain *Lb. delbrueckii* subsp. *lactis* UO 004 as a novel probiotic and tested its ability to survive during manufacturing and ripening of Vidiago-type cheese, a washed-curd cheese made with goat milk. *Lb. delbrueckii* UO 004 strain was firstly isolated from infant feces and showed acid and bile tolerance, adherence to intestinal epithelial cells, and inhibition of the growth of certain enteropathogens, results that would support its potential use as a probiotic strain. The authors reported very high viability of *Lb. delbrueckii* UO 004 in cheeses after 28 days of ripening, 10^8–10^9 CFU/g, without negatively affecting appearance and other sensory criteria of Vidiago-type cheese.

In Brazil, the *Lb. delbrueckii* UFV H2b20 strain was first isolated in the Universidade Federal de Viçosa. It was shown to produce organic acids, hydrogen peroxide, and other inhibitors; to exhibit antagonistic activity toward a variety of pathogenic microorganisms; and to be capable of stimulating mononuclear phagocytic activity in germ-free Swiss rats, which suggested that this strain elicited a nonspecific immune response (Araújo et al. 2010). Due to these potential probiotic features, the authors incorporated the *Lb. delbrueckii* UFV H2b20 in a synbiotic cottage cheese containing inulin, reaching, successfully, levels of probiotic cells around 8.2 log CFU/g.

In addition, important endeavors were undertaken in order to characterize the probiotic potential and other technological properties of cultures isolated from artisanal (homemade) and traditional cheeses. Some studies in that area are presented in Section 44.14.

44.7 Problems for the Probiotic Bacteria Incorporation into Cheese

Numerous technological operations are used in the processing of foods. Many have an effect on how the probiotics grow and survive in the food product. It is a well-established fact that, although it is widely strain dependent, the viability of probiotic strains in cheeses is restricted by the presence of salt, as well as other factors, including oxygen and temperature (Champagne et al. 2005; Özer et al. 2008).

44.7.1 Packing

Oxygen toxicity is considered a significant factor that influences the viability of probiotic bacteria (especially bifidobacteria) in foods (Dave and Shah 1997; Champagne et al. 2005). Exposure to oxygen during processing and storage is highly detrimental to lactobacilli and bifidobacteria, which are categorized, respectively, as microaerophilic and strict anaerobe cultures (Talwalkar et al. 2004).

Many strategies have been studied to protect bacteria from the deleterious effects of oxygen toxicity, such as the use of special high-oxygen-consuming strains (Lourens-Hattingh and Viljoen 2001), the use of ascorbic acid as an oxygen scavenger in yogurts (Dave and Shah 1997), microencapsulation (Talwalkar and Kailasapathy 2004a), the use of packaging material that is less permeable to oxygen (Dave and Shah 1997), and oxidative stress adaptation (Talwalkar and Kailasapathy 2004b).

The presence of oxygen in probiotic foods affects the probiotic cultures in two ways. The first is a direct toxicity to cells. Certain probiotic cultures are very sensitive to oxygen and die in its presence, presumably due to the intracellular production of hydrogen peroxide. Unlike aerobic bacteria, which completely reduce oxygen to water, the oxygen-scavenging system in these probiotic bacteria is either reduced or completely absent. Consequently, accumulation of toxic oxygen metabolites in the cell occurs, eventually leading to its death (Talwalkar and Kailasapathy 2004a; Champagne et al. 2005; Vasiljevic and Shah 2008). The second way oxygen affects the probiotic cultures is indirectly. When oxygen is in the medium, certain cultures, particularly *Lb. delbrueckii*, excrete peroxide in the medium. A synergistic inhibition of bifidobacteria by acid and hydrogen peroxide has been demonstrated. This suggests that the probiotic strains, therefore, may be affected by the H_2O_2 produced by other cultures in the environment (Champagne et al. 2005).

Therefore, a proper selection of packaging for probiotic cheeses is necessary, in order to avoid the presence of oxygen and, consequently, its action over probiotic viability. Plastic films with low permeability to oxygen should be chosen for packing these functional products; alternatively, the practice of adopting other practices, such as the use of vacuum packaging, should be followed (Cruz et al. 2009). In fact, Daigle et al. (1999) found that *B. infantis* survived very well in cheddar cheese packed in vacuum-sealed bags kept at 4°C for 84 days and remained above 3×10^6 CFU/g cheese.

44.7.2 Temperature

Probiotics are usually anaerobic bacteria and do not survive well during temperature changes (Rouzaud 2007). After probiotic cheese production, inadequate storage conditions affect probiotic survival. Therefore, strict control of the storage temperature is required, in order to obtain high probiotic populations during storage (Cruz et al. 2009).

Ong and Shah (2008a) studied the viability of the probiotic strain *Lb. acidophilus* L10 in cheddar cheeses during ripening for 24 weeks at 4°C, 8°C, and 12°C. The probiotic bacteria population was significantly affected by the temperature employed during ripening. *Lb. acidophilus* L10 survived better at lower ripening temperatures, and the probiotic bacteria population remained above 10^7 CFU/g in cheeses ripened at 4°C and 8°C after 24 weeks. There was a 1 log reduction in the probiotic population in cheeses after 24 weeks of ripening at 12°C.

In general, dairy products containing living bacteria have to be cooled during storage. This applies in particular to products containing live probiotic bacteria. Cooling is necessary to guarantee high probiotic survival rates and enough stability of the product (Heller et al. 2003). Therefore, the temperature found in most retail stores has a negative impact on the survival of probiotic microorganisms and causes undesirable changes in texture, color, and flavor of cheeses, leading the product to be rejected by the consumer. Strict monitoring of this parameter is very important, as failure may endanger the functional status of the product and lead to the growth of contaminant and pathogenic microorganisms (Cruz et al. 2009).

44.7.3 Salting

Sodium chloride (NaCl) is widely used in the food industry as a preservative agent, in order to contribute to favorable sensory characteristics and to satisfy the human daily requirements. In the specific case of cheeses, salt acts as a flavoring and functional ingredient, as well as a preservative. Additionally, sodium chloride is important for controlling cheese ripening (Ravishankar and Juneja 2000).

However, the salt content of dairy products might jeopardize the cell viability of the probiotic cultures employed (Gomes et al. 1998). Contrary to the findings of Gomes et al. (1998), the results observed by Vinderola et al. (2002) revealed that *Bifidobacterium* strains were more sensitive to salts than *Lb. acidophilus* strains, because only some strains of *Bifidobacterium* were inhibited by 2% of NaCl among the probiotic strains used in that study.

In addition, the studies conducted by Vinderola et al. (2002) revealed that, in general, KCl was less inhibitory than NaCl. According to these results, the NaCl concentration used (0.9%) in Argentinean probiotic cheese (Vinderola et al. 2000) seems to jeopardize neither probiotic nor lactic acid starter bacteria cell viability.

It is a well-established fact that, even though it is widely strain dependent, the viability of probiotic strains in cheese is restricted by the presence of salt, as well as other factors, including oxygen and temperature. It was postulated that the populations of most probiotic strains fell dramatically when the salt level in cheese exceeded the upper limit of 4 g salt/100 g (wt/wt) (Gomes et al. 1998). Similar results were reported by Yilmaztekin et al. (2004), who monitored the viability of *Lb. acidophilus* LA-5 and *B. bifidum* Bb-02 in white-brined Turkish cheese.

Blanchette et al. (1995) reported that the addition of 1.8% salt caused a 0.5 log drop in *B. infantis* viability during fermentation for the production of cultured cottage cheese. In addition, Gomes et al. (1998) observed that high salt levels caused higher losses in *B. lactis* and *Lb. acidophilus* populations during Gouda cheese ripening. In this study, the final average salt levels ranged between 2% and 4% (wt/wt).

Kasımoğlu et al. (2004) observed populations of 10^9 and 10^{10} CFU/g of *Lb. acidophilus* on the seventh day of refrigerated storage at 4°C in cheeses produced and ripened in vacuum pack and in brine, respectively. The viable cell numbers of *Lb. acidophilus* began to decrease after 15 days of ripening, due to decrease in moisture level and increase in salt content. However, the authors observed that even though *Lb. acidophilus* populations decreased until the end of the ripening period, it did not decrease below 10^7 and 10^6 CFU/g in ripened cheese without brine and with brine, respectively.

Gomes and Malcata (1998) reported that a maximum of 3.5% (wt/wt) salt in goat cheese provided counts of *B. lactis* and *Lb. acidophilus* high enough for a probiotic effect. In contrast, Fortin et al. (2011), in a study carried out with cheddar cheese, observed that salting resulted in approximately 13% loss of cells in whey during processing and in a 3 log CFU/g higher viability loss of *B. longum* after 3 days of storage.

In order to protect the viability of probiotic bacteria against undesirable growth conditions in cheese, various alternatives, including cell incubation under sublethal conditions, cell propagation in an immobilized biofilm, and microencapsulation, have been developed. Among these techniques, cell microencapsulation has offered some advantages for improving the viability of probiotic bacteria (Krasaekoopt et al. 2003).

44.8 Microencapsulation

During processing and storage of functional foods, the fastidious probiotic bacteria may be exposed to high temperatures, low pH, high osmotic pressure, high levels of oxygen, and low levels of nutrients (Dave and Shah 1997; Talwalkar and Kailasapathy 2004a). These factors often have deleterious effects on the viability of probiotic organisms. Furthermore, the survival of probiotic bacteria is also affected by high concentrations of acid and bile that are found in the stomach and intestinal tract (Krasaekoopt et al. 2003). Microencapsulation of probiotic bacteria is the process of applying a layer of protective material to protect the sensitive probiotic organisms from their external environment (Ding and Shah 2009).

Özer et al. (2008) evaluated the viability of encapsulated *Lb. acidophilus* LA-5 and *Bifidobacterium* strain BB-12 by emulsion and extrusion techniques in Kasar cheese throughout 90 days of storage. No difference was noted between the two encapsulation techniques with regard to probiotic counts. However, while control cheese (manufactured with nonencapsulated probiotic bacteria) presented a decrease in probiotic populations, the encapsulated bacteria remained above 10^7 CFU/g in the other cheeses during storage for 90 days, besides showing metabolic activity.

On the other hand, Godward and Kailasapathy (2003), in a study conducted with cheddar cheese manufactured with the probiotic strains *Lb. acidophilus* CSCC2401, *Lb. acidophilus* 910, *B. lactis* 920, and *B. infantis* CSCC1912, observed that, after 24 weeks of storage, with the exception of *Lb. acidophilus* 910, the encapsulated cell counts decreased approximately 1–2 log cycles more when compared to free cells. *Lb. acidophilus* strains showed better survival than *Bifidobacterium* strains during the storage period studied. This study showed that free cells of probiotic bacteria survived better than encapsulated cells in the cheddar cheese matrix; hence, encapsulation does not significantly increase the survival of probiotic bacteria during cheddar cheese maturation and storage. According to the authors, the findings occurred possibly due to the generation of metabolites, such as organic acids, within the calcium alginate capsules, hence causing cell death.

Similarly, Kailasapathy and Masondole (2005) studied the survival and the effect of free and encapsulated probiotic bacteria (*Lb. acidophilus* DD 910 and *B. lactis* DD 920) when incorporated into feta cheese. Addition of encapsulated probiotic cultures increased the water-holding capacity of the cheese, due to the production of exopolysaccharides and to the polymer (alginate as encapsulant material) and filler material (Hi Maize starch) added. In addition, the authors observed a decrease of approximately 2 and 3 log cycles in free and encapsulated probiotic populations, respectively. The authors concluded that microencapsulation did not offer protection to the probiotic bacteria tested, due to the open texture of cheese, the possible disintegration of microcapsules in brine solution, and the higher salt uptake when encapsulated cultures were incorporated.

According to Kailasapathy (2002), these kinds of studies demonstrate that microencapsulation may not be necessary for increasing the viability of probiotic bacteria in some cheese varieties. This may not be the case, however, in acidic, fresh, low pH-type of cheeses, such as cottage cheese.

44.9 Addition of the Probiotic Inoculum in Probiotic Cheese

There are two options for the addition of probiotic bacteria during cheese production that can directly affect the survival rates of these microorganisms—probiotic bacteria may be added before the fermentation (together with the starter culture) or after it. Following the former option implies performing preliminary tests to know the amount of probiotic cells lost in the whey during its drainage. The ideal rate of probiotic inoculum to be added must be checked according to the process. If the second option is chosen, immediate cooling must follow (below 8°C, preferably), as metabolic activities of starters and probiotics are drastically reduced at these temperatures (Heller et al. 2003).

Bergamini et al. (2005) evaluated the viability of *Lb. acidophilus* and *Lb. paracasei* subsp. *paracasei* in a semi-hard Argentinean cheese. Two different approaches for the probiotic culture incorporation into cheese were tested—a lyophilized powder dispersed in milk and a substrate composed of milk and milk fat. According to the results obtained by the authors, direct addition of the probiotic culture was the approach with the best performance. The addition of the probiotics after preincubation in the substrate did not improve their survival during cheese ripening, and this procedure was considered as a more complex approach than the direct addition of lyophilized culture. However, the approach presented an advantage—preincubation in the substrate increased the probiotic population in the inoculum almost in 1 log cycle. This fact might contribute to diminish the costs of probiotic cultures for the dairy industry.

Fortin et al. (2011) evaluated the effect of probiotic time of inoculation on viable counts of five probiotic bacteria in the curd and whey during cheddar cheese production. The authors observed that inoculation of probiotics in milk before renneting resulted in almost half the cell losses in whey, compared with the addition just before cheddaring. Inoculation of probiotics in milk improved their subsequent stability by about 1 log over a 20-day storage period as compared with cells added during cheddaring. Similarly, Songisepp et al. (2004) concluded that addition of lactobacilli into milk should be the preferred method of incorporation, when compared with inoculation into drained curds.

Many scientific papers available revealed that a major challenge associated with the application of probiotic cultures in functional foods is the retention of viability during processing. Therefore, once a desirable culture is selected, the technological demands placed on probiotic strains are great, and new manufacturing process and formulation technologies may often be required to retain viability and, hence, functional health properties (Kosin and Rakshit 2006). Therefore, different approaches for probiotic inoculation into dairy products should be studied, in order to improve yield, to improve probiotic viability throughout the shelf life established, and to diminish the costs of probiotic cultures for the dairy industry.

44.10 Individual and Mixed Probiotic Cultures

The main choice of any probiotic microbial strain to be used in blend cultures is based on the health aspects beneficial to humans. Mixed probiotic cultures were initially developed for manufacture of fermented milks. These cultures contain blends of *Lb. acidophilus* and *Bifidobacterium* spp. (AB cultures), *Lb. acidophilus*, *Bifidobacterium* spp., and *Lb. casei* (ABC cultures), or *Lb. acidophilus*, *Bifidobacterium* spp., and *Streptococcus thermophilus* (ABT cultures) (Tamime et al. 2005).

Difficulties concerning the use of these cultures together have been related to the different temperatures and rates required for an optimal growth of each strain (Ordóñez Pereda et al. 2005; Tamime et al. 2005). Different LAB species and/or subspecies can confer distinctive characteristics upon the food product. The metabolic end products of certain strains, or combinations of strains, sometimes can render the product sensorially undesirable (McCartney 2005). For example, bifidobacteria produce acetic acid and lactic acid from lactose via a fructose-6-phosphate shunt pathway. This fermentation pathway results in 3 mol of acetic acid and 2 mol of lactic acid per 2 mol of glucose, generating a theoretical molar ratio (acetic/lactic) of 3:2 (Ong et al. 2006), which has been associated with the presence of off-flavors in probiotic cheeses. In dairy fermentations, nitrogen is limiting, and cultures initially compete for the free amino acids (FAAs) and small peptides available in milk. The ability to utilize amino acids efficiently is of primary importance in determining growth rate of microorganisms (Irlinger and Mounier 2009). Additionally, some cheese starters are typically propagated between 19°C and 28°C, which is not appropriate for extensive development of probiotics (Champagne et al. 2005). On the other hand, combination of *Bifidobacterium* with *Lb. acidophilus* or *S. thermophilus* may be advisable to assure balanced concentrations of lactic and acetic acids, avoiding the presence of undesirable flavor in the final product (Ordóñez Pereda et al. 2005).

Because controlling the ratios of the different mixed strains might be complicated, the trend is toward employing pure cultures (Champagne et al. 2005). Pure probiotic cultures have also been successfully combined for the production of different kinds of probiotic cheeses (Vinderola et al. 2000; Ong et al. 2007a; Cardarelli et al. 2008; Bergamini et al. 2010).

44.11 Fresh Acid- and Acid/Heat-Coagulated, Fresh Rennet-Coagulated, and Ripened Probiotic Cheeses

44.11.1 Fresh Acid- and Acid/Heat-Coagulated Probiotic Cheeses

Acid- and acid/heat-coagulated cheeses are typically fresh (unripened) soft cheese varieties produced through the coagulation of milk, cream or whey, or blends thereof via direct chemical acidification, culture acidification, or a combination of chemical acidification and high heat treatment (Farkye 2007). A relatively small amount of rennet is added to milk (Schulz-Collins and Senge 2004). For their production, pretreated milk undergoes slow acidification and gelation close to the isoeletric point of casein (pH 4.6). The curd is stirred, and the whey is removed (Litopoulou-Tzanetaki 2007). Because of their physical and rheological consistencies, many acid- and acid/heat-coagulated cheeses are classified as soft cheeses (Farkye 2007). Nowadays, most commercial probiotic cheeses are fresh, cottage, or soft-type cheeses, because they are easy to handle and offer a simple food matrix to deliver probiotics (Bergamini et al. 2010). Several acid- and acid/rennet-curd fresh cheeses are ready for consumption immediately after production (Fox and Cogan 2004; Litopoulou-Tzanetaki 2007). Their storage occurs at refrigeration temperatures, shelf life is rather limited, and no prolonged periods of ripening are necessary (Heller et al. 2003). These conditions make it easier for cheeses to maintain probiotic viability during shelf life (Bergamini et al. 2010). Cream cheese, cottage cheese, Quark or Tvorog, Fromage frais, and Ricotta are among the better-known acid- and acid/rennet-curd cheese types (Schulz-Collins and Senge 2004).

Fresh acid cheeses, such as cottage and Quark, have a "creamy" structure. Due to this fact, Heller et al. (2003) indicated two possibilities for the incorporation of probiotic culture in cottage cheese—together with the starter culture or together with cream and salt. Nonetheless, for addition together with the starter culture, the authors stated some problems—the considerable loss of probiotic culture during drainage of whey and the rather high scalding temperature of up to 55°C—as these conditions may reduce considerably the number of probiotic cells in the final product. Therefore, the addition of probiotic culture together with cream and salt would help to control more accurately the population of probiotics in the final product and would avoid the adverse effects of the high scalding temperature.

The addition of probiotics together with the starter culture was successfully used in the manufacture of *petit-suisse* cheese (a Quark-based cheese), with good results regarding probiotic viability after manufacture and during storage. In a study carried out with potentially synbiotic *petit-suisse* manufactured with inulin, oligofructose, and/or honey, Cardarelli et al. (2008) obtained *Lb. acidophilus* and *B. animalis* subsp. *lactis* populations above 1×10^6 and 1×10^7 CFU/g, respectively, during the refrigerated storage at $4 \pm 1°C$. Similarly, Pereira et al. (2010) manufactured a probiotic *petit-suisse* cheese supplemented with *Lb. acidophilus* LA-5 and *B. animalis* subsp. *lactis* BL04 solely or in co-culture, and observed that the populations of LA-5 and BL04 remained at 7.0 and 8.0 log CFU/g, respectively, during storage for up to 28 days.

44.11.2 Fresh Rennet-Coagulated Probiotic Cheeses

The manufacture of cheese involves several different processes, some of which are essential for nearly all cheese varieties. In fresh enzymatic coagulated cheeses, clotting of the milk is accomplished by means of enzyme activity. There are several coagulants of animal and microbial origin. However, chymosin is the main milk-clotting enzyme used in cheesemaking. As discussed earlier, the enzymes involved remove the caseinomacropeptide "hairs" from κ-casein; the resulting paracasein micelles will then aggregate. The aggregation causes formation of a space-filling network, which encloses the milk serum and the fat globules (Farkye 2004; Walstra et al. 2006a,b). Rennet is used in cheese manufacture primarily to coagulate milk. However, residual rennet, which is kept in the curd, also plays an important role in the generation of flavor compounds in cheese. Between 3% and 6% of the rennet added to the cheese milk is retained in the curd (Heller et al. 2003).

Several authors reported that studies on the addition of probiotic cultures to fresh enzymatic coagulated probiotic cheeses, containing recognized and potentially probiotic cultures, have been published, which described suitable viable counts and a positive influence on the texture and sensorial properties of these cheeses. Having in mind that portions of around 100 g of cheese are usually consumed daily, populations of about 10^6 CFU/g lead to an ingestion of 10^8 CFU/daily portion.

Suárez-Solís et al. (2002) manufactured fresh cheese supplemented with *B. bifidum* and *Lb. casei* and obtained cheese with very good sensory quality with viable populations of 1×10^7 CFU/g during 15 days of storage. Buriti et al. (2005a) tested the supplementation of Minas fresh cheese with *Lb. paracasei* subsp. *paracasei* LBC 82. The cheeses studied by the authors presented populations above 1×10^6 CFU/g during cheese production, and population increased during the whole storage, reaching 10^8 CFU/g after 21 days. In the same way, Buriti et al. (2005b) studied the addition of *Lb. acidophilus* LA-5 solely and in co-culture with a mesophilic type O lactic culture during Minas fresh cheese production and observed levels above 1×10^6 CFU/g for probiotic bacteria during the whole storage period. In both studies, *Lb. paracasei* and *Lb. acidophilus* did not alter the texture and sensorial characteristics of Minas fresh cheeses.

Masuda et al. (2005) evaluated the viability of three strains isolated from human intestine with high tolerance for acid and bile—*Lb. acidophilus* JCN11047, *Lb. acidophilus* 1132ᵀ, and *Lb. gasseri* JCM11657—in fresh cheeses stored at 7°C for 4 weeks. Negligible reductions in the viability were observed for all strains throughout the shelf life of cheeses, which remained above 8×10^7 CFU/g up to the end of storage. Fritzen-Freire et al. (2010) observed *B. animalis* BB-12 counts of 7.74 and 7.73 log CFU/g in Minas fresh cheese on days 1 and 28, respectively.

Marcatti et al. (2009) studied the manufacture of probiotic Minas fresh cheese developed from buffalo milk. Cheeses were supplemented with probiotic *Lb. acidophilus* and analyzed for up to 28 days of storage. At the end of shelf life, the authors reported *Lb. acidophilus* counts always above 10^6 CFU/g.

Buriti et al. (2007a) tested the addition of *Lb. paracasei* subsp. *paracasei* LBC 82 in co-culture with *S. thermophilus* to potentially probiotic and synbiotic fresh cream cheeses (without and with inulin, respectively). Viable counts of *Lb. paracasei* remained above 1×10^7 CFU/g during the entire storage period, 21 days, for both cheeses. In another study, Buriti et al. (2007b) observed the viability of *Lb. acidophilus* LA-5 and *B. animalis* BB-12 added to Minas fresh cheese. Both probiotic cultures were present in high levels throughout storage, above 1×10^6 CFU/g, and resulted in cheeses with texture comparable to the traditional ones and with favorable sensorial features.

Souza et al. (2008) and Souza and Saad (2009) studied the manufacture of Minas fresh cheese supplemented with the probiotic strains of *Lb. acidophilus* LA-5 solely or in co-culture with *S. thermophilus*. Cheeses manufactured with LA-5 solely presented populations above 1×10^6 CFU/g, reaching 1×10^7 CFU/g on the 14th day of storage. Moreover, the addition of LA-5 strain resulted in good acceptance of Minas fresh cheeses, improving the sensory performance of these products during storage (Souza et al. 2008).

However, even though cheese is likely to be one of the best carriers for probiotics, the addition of high numbers of viable and metabolically active cells can affect product quality, especially sensory properties (Grattepanche et al. 2008). An example is the observation reported by Modzelewska-Kapitula et al. (2007), who studied the addition of the potentially probiotic culture of *Lb. plantarum* 14 to a soft cheese. Although suitable populations for a probiotic food, between 10^6 and 10^7 CFU/g, were observed in the study, cheeses containing this strain presented slightly lower scores in the sensory analysis.

44.11.3 Ripened Probiotic Cheeses

It is important to consider that cheese ripening generally covers a longer time than the shelf life of fermented milks, and consequently, probiotic bacteria must remain viable for a longer time. In fact, cheese ripening can take as little as a few weeks or longer than a year, depending on the cheese type (Bergamini et al. 2009a). However, several reports for ripened cheese supplemented with probiotic bacteria have been published (Heller et al. 2003).

Rogelj et al. (2002) developed a semi-hard cheese, reported that the population of a probiotic strain of *Lb. acidophilus* was higher than 10^7 CFU/g after 6 months of ripening, and concluded that semi-hard cheeses are suitable food vehicles for probiotic bacteria. Similarly, Kılıç et al. (2009) evaluated the survival of the probiotic strains *Lb. fermentum* (strains AB5-18 and AK4-120) and *Lb. plantarum* (strains AB16-65 and AC18-82) incorporated to a semi-hard cheese (Turkish Beyaz cheese). Cheeses were ripened at 4°C for 120 days and presented probiotic populations of 7.42×10^7 CFU/g at the end of the storage period.

Montoya et al. (2009) reported high probiotic populations in Swiss cheese ripened anaerobically for 0, 7, and 10 days at 37°C. Following ripening, counts of the probiotic bacteria (*B. breve* R0070, *B. infantis* R0033, *B. longum* R0175, and *Pediococcus acidilactici* R1001) added to cheese increased to 9–10 log CFU/g, with no significant difference in the viability of the four probiotic bacteria.

Perotti et al. (2009) manufactured a probiotic Pategrás cheese, supplemented with *Lb. acidophilus*, *Lb. paracasei* subsp. *paracasei*, and *B. animalis* subsp. *lactis*. Pategrás cheese revealed satisfactory populations for a probiotic food, because counts of the three strains assayed were higher than 6.5 log CFU/g in 3-day-old cheeses and increased at 60 days of ripening (8.00 log CFU/g for *Lb. acidophilus*, 7.46 log CFU/g for *B. animalis* subsp. *lactis*, and 9.11 log CFU/g for *Lb. paracasei* subsp. *paracasei*).

Beyond the cheeses mentioned above, it has been shown that cheddar cheese is a suitable carrier of probiotic bacteria. Ong et al. (2007c) produced cheddar cheeses supplemented with probiotic cultures of *Lb. acidophilus* 4962, *Lb. casei* 279, *B. longum* 1941, *Lb. acidophilus* LAFTI L10, *Lb. paracasei* LAFTI L26, and *B. lactis* LAFTI B94. The authors observed populations above 7.5 log CFU/g at the end of a ripening period of 6 months at 4°C. According to the findings, the authors concluded that cheddar cheese could be an effective vehicle for delivery of probiotic microorganisms.

Similarly, McBrearty et al. (2001) studied the incorporation of two commercially available bifidobacteria on cheddar cheese. The authors observed that *B. longum* BB536 and *B. animalis* (formerly *B. lactis*) BB-12 survived in high populations after 6 months of ripening.

In another study, carried out by Phillips et al. (2006), six batches of cheddar cheese were produced containing different combinations of commercially available probiotic cultures—*Bifidobacterium* spp., *Lb. acidophilus*, *Lb. casei*, *Lb. paracasei*, and *Lb. rhamnosus*. The results indicated cheddar cheese as a good vehicle for probiotic delivery, although counts of *Lb. acidophilus* strains needed to be improved.

44.12 Cheese with Beneficial Properties Produced with Microorganisms from Kefir

Kefir is a consortium of microorganisms that is mainly used in the production of the low-alcohol, traditional Russian drink kefir, where milk constitutes the initial fermenting substrate. Lactic acid bacteria and yeasts coexist in a symbiotic association and are responsible for acid–alcoholic fermentation (Kourkoutas et al. 2006b). The main bacterial species already identified in kefir are *Lb. brevis*, *Lb. helveticus*, *Lb. delbrueckii* subsp. *delbrueckii*, *Lb. delbrueckii* subsp. *bulgaricus*, *Lb. kefir*, *Lb. acidophilus*, *Ent. faecalis*, *Ent. faecium*, *S. thermophilus*, *Lactococcus lactis* subsp. *cremoris*, and *Leuconostoc mesenteroides*. The dominant yeasts in kefir are members of the genera *Zygosaccharomyces*, *Candida*, *Saccharomyces*, *Pichia*, and *Kluyveromyces* (Goncu and Alpkent 2005; Sarkar 2007). Consumption of kefir has been linked to a variety of health benefits. Besides, lactic acid bacteria that exist in kefir grains have attracted a lot of attention, due to their ability to inhibit the development of spoilage and pathogenic microorganisms, either by the production of lactic acid or of antimicrobial agents (Kourkoutas et al. 2006b), such as bacteriocins.

Kefir has been successfully used as a starter culture in the production of cheese. White pickled cheese was reported to exhibit improved quality characteristics, in terms of appearance, structure, and odor, when kefir was used as a starter culture (Goncu and Alpkent 2005). Kourkoutas et al. (2007) presented an economic alternative for freeze-drying kefir starter culture in an industrial scale. The use of freeze-dried kefir cultures in the production of feta-type cheese resulted in products with improved quality and extended shelf life, which received scores comparable to a commercial feta cheese by the panel in the sensory evaluation (Kourkoutas et al. 2006b).

44.13 Probiotic "Cheese-Like" Soy Foods

Tofu, also known as soybean curd, is a soft, "cheese-like" unfermented food product made by curdling fresh hot aqueous extract of whole soybean with a coagulant. It is produced traditionally by curdling fresh hot aqueous extract of soybean with either salt ($CaCl_2$ or $CaSO_4$) or an acid (glucono-δ-lactone). The coagulant produces a soy protein gel, which traps water, soy lipids, and other constituents in the matrix forming curds. Although the word "tofu" is Japanese, the food seems to have originated in ancient China, where the Mandarin term is "doufu." Tofu is a popular food product in east and southeastern Asia countries and is gaining an increasing popularity around the world (Endres 2001; Oboh 2006; Obatolu 2008). Another soy food with physical appearance, texture, composition, and the basic processing technology quite similar to the milk cheese is known as "sufu," "furu," or fermented soybean curd, a traditional Chinese product obtained through fungal or bacterial fermentation of aqueous extracts of whole soybean (Ahmad et al. 2008).

Cheese-like soybean foods, such as tofu and sufu, are rich in protein, unsaturated fatty acids, lecithin, and isoflavones; contain no cholesterol or lactose; and may be consumed by those suffering from lactose intolerance (Lin et al. 2004). Soy products have been associated with prevention of chronic diseases, such as menopausal disorder, cancer, atherosclerosis, and osteoporosis. Sucrose, raffinose, and stachyose are sugars present in soybean products, which can be used by most of the strains that belong to the genus *Bifidobacterium*. Metabolism of such sugars by probiotic bacteria in food products may reduce flatulence in the human body (Liu et al. 2006). Fermentation of soy products with probiotic microorganisms that possess high α-galactosidase activity has been found to minimize the content of flatulence-causing soybean oligosaccharides (Liong et al. 2009).

Liu et al. (2006) developed a potential probiotic soy food containing *Lb. rhamnosus* 6013 based on Chinese sufu. The authors observed that after 6 h of fermentation, *Lb. rhamnosus* 6013 was capable of growing in aqueous extract of soybean up to as high as 10^8–10^9 CFU/mL, besides metabolizing stachyose completely. During storage for 30 days at 10°C, viable counts of *Lb. rhamnosus* reduced slightly, around 10^7 CFU/g. In that study, addition of the potential probiotic strain had no negative effect on the sensory properties of the soy cheese.

Addition of potential probiotic strains *Lb. bulgaricus* FTCC 0411 and *Lb. fermentum* FTD 13 in tofu was evaluated by Ng et al. (2008). Viability of strains was higher than 10^6 CFU/g during the whole storage. Hydrolysis of oligosaccharides and consumption of reducing sugars by those strains were also observed in the potential probiotic tofu.

In another study, Liong et al. (2009) incorporated *Lb. acidophilus* FTCC 0291, a strain with high α-galactosidase-specific activity, into "soy cream cheese" for a storage study of 20 days at 4°C and 25°C. According to the authors, *Lb. acidophilus* FTCC 0291 in soy cream cheese at both storage temperatures maintained viability above 10^7 CFU/g during the entire shelf life. Moreover, *Lb. acidophilus* FTCC 0291 was capable of utilizing the existing reducing sugars in aqueous extract of soybean and concurrently hydrolyzing the oligosaccharides into simple sugars for growth.

44.14 Artisanal (Homemade) Cheeses with Probiotic Potential

The interest in the probiotic potential of traditional fermented products has grown during the last few years (Vizoso Pinto et al. 2006; Klayraung et al. 2008; Mathara et al. 2008). Moreover, lactic acid bacteria strains isolated from traditionally made cheeses may constitute a reservoir of potential applications in biotechnology (Radulović et al. 2010). Ukeyima et al. (2010) reported that various indigenous fermented foods containing potential probiotic bacteria have been part of local diets in Africa, due to their functional properties. Among them, kindirmo, nono, and warankasi are common fermented dairy products in Nigeria. Warankasi is known among indigenous African consumers as cheese, and kindirmo and nono are considered as equivalent to yogurt. Warankasi is a dairy-based product that is fermented by the artisans, using an overnight portion of warankasi containing mainly *Lactococcus*, *Streptococcus*, and *Lactobacillus* strains of lactic acid bacteria, which ferment the heated milk within a period of 8–10 h.

In the study carried out by Corsetti et al. (2008), strains belonging to the *Lb. casei* group isolated from Pecorino d'Abruzzo cheese were investigated, in order to access their functional potential, particularly focusing on antigenotoxic properties. Pecorino d'Abruzzo is described by the authors as artisanal ripened cheese produced using raw ewe milk in the Abruzzo region, central Italy. In that study, the strains evaluated showed activity against two potent DNA-reactive agents, 4-nitroquinoline-1-oxide and *N*-methyl-*N'*-nitro-*N*-nitrosoguanidine. Despite the fact that these strains showed susceptibility to simulated gastric fluid (pH 2.0) with pepsin, their viability was recovered after the strains were transferred to the intestinal fluid (pH 7.4) containing bile and pancreatin. Based on these results, the authors suggested that these indigenous microorganisms might reach the gut as viable cells and prevent DNA damage.

Topisirovic et al. (2006) described a very long tradition for regional cheese production in the Balkan region. According to the authors, these cheeses are produced at specific ecological localities, such as high mountains (over 1200 m above the sea level), mountain plateaus, river valleys, islands, Adriatic coast, etc. Studying the microbiota from these artisanal cheeses, the authors found that autochthonous lactococci and lactobacilli were able to produce bacteriocins. One of them, *Lb. paracasei* subsp. *paracasei* BGSJ2-8 (isolated from the semi-hard homemade white cheese produced in Sjenica, Pester Plateau, Serbia), produced a heat-stable bacteriocin with activity in the pH range from 2 to 11. In another study conducted in Serbia, Radulović et al. (2010) verified that three *Lb. paracasei* strains isolated from traditionally made white-brined cheese presented properties very close to *Lb. rhamnosus* GG, concerning antimicrobial activity against foodborne pathogens, ability to survive to simulated gastrointestinal conditions, sensitivity to antibiotics, and autoaggregation.

Saavedra et al. (2003) evaluated the probiotic potential of wild *Ent. faecium* strains isolated from Tafí cheese. According to the authors, Tafí is a homemade cheese made for centuries in the Tafí Valley (2300 m high), Tucumán, Northwest Argentina. In that study, among 11 strains selected based on growth in oxgall assay, nine strains presented bile salt hydrolase activity and were able to reduce cholesterol. Two strains, *Ent. faecium* CRL 988 and CRL 1385, showed inhibitory activity against *Lis. innocua* 7 and *Lis. monocytogenes* Scott A. Virulence factors were not detected in the strains selected, suggesting that they might be used as nontraditional starter cultures in this kind of cheese.

44.15 Proteolysis in Probiotic Cheeses

Proteolysis is one of the major biochemical events in flavor development occurring during the storage of most cheese varieties and contributes to the cheese's taste, through the production of peptides and FAAs, and texture (Martínez-Cuesta et al. 2001; Sousa et al. 2001). Proteolysis is caused by enzymes present in milk (e.g., plasmin), rennet (e.g., pepsin and chymosin), and microbial enzymes released by starter cultures (Ong and Shah 2008b). Hydrolysis of caseins (α_{s1}-, α_{s2}-, β-, and κ-casein) by such enzymes leads to the formation of large- and intermediate-size peptides (Ong et al. 2007c). In general, lactic acid bacteria possess a complex proteolytic enzymatic system, due to their requirements of amino acids (Settanni and Moschetti 2010). Proteolytic enzymes from the starter bacteria, non-starter lactic acid bacteria, and probiotic adjuncts may further hydrolyze the large- and intermediate-size peptides into smaller peptides and FAAs. Proteolytic enzymes produced by certain probiotic adjuncts were also found to degrade bitter peptides (Ong et al. 2007b).

The starter cultures commonly used in cheese manufacture include mesophilic *Lactococcus* and *Leuconostoc* spp. and thermophilic *Lactobacillus* spp. The use of *S. thermophilus* as starter for cheese production has also been described (Sousa et al. 2001; Souza and Saad 2009). The use of more proteolytic strains of *S. thermophilus, Lb. delbrueckii* subsp. *bulgaricus, Lb. acidophilus,* and *Bifidobacterium* spp. could enhance the growth and viability of probiotic bacteria in products over the storage period (Shihata and Shah 2000). Starter culture also contributes to cheese aroma and flavor development during storage through carbohydrate metabolism, proteolysis, and, to a lesser extent, lipolysis (Candioti et al. 2002).

Ong et al. (2007c) observed that the addition of probiotic cultures influenced the proteolytic pattern of cheddar cheese in different ways—*Lb. casei* 279 and *Lb. paracasei* LAFTI L26 showed higher hydrolysis of casein, and higher concentrations of FAAs were found in all probiotic cheeses. Even though *Bifidobacterium* spp. were found to be weakly proteolytic, cheeses supplemented with those strains presented the highest concentration of FAAs.

Bergamini et al. (2006) observed that the addition of *Lb. acidophilus* and *Lb. paracasei* subsp. *paracasei* in semi-hard cheese influenced the proteolysis pattern of cheeses in different ways. According to the authors, these results occurred probably due to different proteolytic systems of probiotic bacteria and their activity in the food matrix.

Milesi et al. (2009) evaluated the contribution to cheese proteolysis and sensory profile of four potentially probiotic non-starter lactobacilli in soft and semi-hard cheeses. Potentially probiotic cultures of *Lb. plantarum* I91 and *Lb. casei* I90 caused a moderate increase in secondary proteolysis, did not overacidify, and changed sensory profiles of cheeses favorably. On the other hand, *Lb. rhamnosus* I73 and I75 exhibited a stronger increased cheese secondary proteolysis, but they did not improve cheese overall quality and caused post-acidification in some cases.

In many cheese varieties, secondary cultures are added intentionally and/or encouraged to grow by controlling environmental conditions. These cultures have a diverse range of functions, depending on the organisms. The addition of probiotic cultures can affect the proteolytic patterns of cheese in different ways, as discussed above.

Proteolysis contributes to cheese taste through the production of peptides and FAAs. The sapid flavor compounds generally partition into the soluble fraction on extraction of cheese with water. Large peptides do not contribute directly to cheese flavor but are important for the development of the correct texture. On the other hand, large peptides can be hydrolyzed by proteinases into shorter peptides that may be sapid. The peptides responsible for bitterness of cheese may be hydrolyzed into smaller peptides and non-bitter amino acids through the activity of enzymes from lactic acid bacteria, a behavior that was observed in some strains of *Lactobacillus* (Sousa et al. 2001).

44.16 Texture of Probiotic Cheeses

Texture is intrinsically related to the arrangement of various chemical components within distinct micro- and macrostructure levels—for example, protein network or fat fraction; it is the external expression of such structures that eventually determines the uniqueness and distinctive character of a cheese product (Pereira et al. 2009). Probiotic addition to the cheese matrix may result in changes in texture profile. However, these changes do not necessarily have a negative effect on cheese features. It is essential that the added probiotic bacteria incorporated into these cheeses do not adversely affect the flavor or texture attributes of the cheese (Boylston et al. 2004).

Souza and Saad (2009) observed a significant increase in hardness during the storage period and in the derivative texture parameters—chewiness and gumminess ($p < 0.05$)—in probiotic Minas fresh cheese manufactured with *Lb. acidophilus* LA-5 solely and in co-culture with *S. thermophilus*. In the same way, Buriti et al. (2005b) reported that Minas fresh cheese supplemented with *Lb. acidophilus* and lactic acid presented an increase in hardness, as did cheeses prepared only with lactic acid.

Menéndez et al. (2000) observed differences between Arzúa-Ulloa cheeses produced with starter cultures containing *Lc. lactis* subsp. *lactis* var. *diacetylactis* and those manufactured with the addition of five different strains of *Lactobacillus* (*Lb. casei* subsp. casei, *Lb. plantarum*, *Lb. casei* subsp. *pseudoplantarum* [two strains], and *Lb. casei*). The authors reported that control cheeses showed higher firmness when compared to cheeses containing *Lactobacillus*.

Buriti et al. (2007b) analyzed the texture profile of Minas fresh cheeses manufactured with the addition of a probiotic ABT culture (containing *Lb. acidophilus*, *B. animalis*, and *S. thermophilus*). The product revealed a greater stability with regard to the different texture parameters evaluated (hardness, cohesiveness, adhesiveness, springiness, chewiness, and gumminess) during refrigerated storage for up to 21 days, when compared to Minas fresh cheeses produced according to the traditional dairy technology, employing the type O lactic culture (*Lc. lactis* subsp. *lactis* + *Lc. lactis* subsp. *cremoris*).

In some scientific papers, texture profiles of probiotic cheeses were sensory evaluated by consumers. This kind of evaluation is very important because sensory methods provide a prompt measurement of the human perception (Ross 2009).

Kalavrouzioti et al. (2005) evaluated the texture profile of hard cheeses (Kefalotyri-like) manufactured from goat milk with yogurt cultures as starters and with their partial replacement by the probiotic cultures *Lb. rhamnosus* LC 705 and/or *Lb. paracasei* subsp. *paracasei* DC 412. Consumers did not notice any texture defect in any cheese sample evaluated, and all cheeses tested received favorable grades for all features analyzed.

Gomes and Malcata (1998) developed a probiotic cheese manufactured from goat milk and supplemented with *B. lactis* and *Lb. acidophilus*. Panelists were requested to evaluate cheese consistency (2 = very brittle, up to 6 = very smooth) and firmness (1 = very soft, up to 7 = very firm). The results of the study showed that *B. lactis* and *Lb. acidophilus* could be used for the successful production of a goat cheese with good flavor and texture features. Similarly, Kasımoğlu et al. (2004) evaluated body and texture of probiotic Turkish white cheese manufactured with *Lb. acidophilus*. According to the authors, *Lb. acidophilus* had a positive effect regarding sensory evaluation, also when texture profile was included.

44.17 Enumeration of Starter and Probiotic Bacteria in Cheeses

Selective media and/or incubation conditions were employed in order to selectively stimulate isolation of the bacterial group of interest. Such approaches exploited the growth characteristics of certain organisms, including their nutritional requirements; preferred incubation conditions (temperature, pH, and redox potential); and susceptibility to, or prerequisite for, certain compounds (such as antibiotics, vitamins, and blood) (McCartney 2005). Traditionally, lactic acid bacteria are counted in De Man Rogosa Sharpe (MRS) agar, but this medium is not selective for enumeration of probiotic microorganisms when non-probiotic starter cultures are present (Lima et al. 2009). In general, "selective" or differential media are used when monitoring specific probiotic or starter cultures (McCartney 2005).

To enumerate probiotic bacteria in a mixed population with starter lactic acid bacteria in cheeses, selective media should allow the growth of the organisms of interest and inhibit other microorganisms encountered. Differentiation between species strongly relies on differences in colonial morphology, but it is not always a stable phenotypic trait to identify and quantify probiotic organisms in a product (Van de Casteele et al. 2006). In some cases, differential media that will allow easy identification of probiotic colonies in the presence of other colonies can be used, provided that the probiotic bacteria are in sufficient numbers (Darukaradhya et al. 2006). Consequently, isolation media with truly selective properties are preferred for selective enumeration on a routine basis (Van de Casteele et al. 2006), but studies in this field have shown variable results, depending on the starter and probiotic strain (or commercial supplier) evaluated.

Several media have been suggested for the enumeration of *Lb. acidophilus*, including bile medium; Rogosa agar; MRS medium containing maltose, raffinose, or melibiose in the place of dextrose; cellobiose–esculin agar; and agar medium based on X-Glu. Nalidixic acid–neomycin sulfate–lithium chloride–paromomycin sulfate agar was developed to enumerate pure cultures of *Bifidobacterium* spp. However, it may not be suitable for selective enumeration of this microorganism in the presence of *Lb. acidophilus* and of some starter cultures. *Propionibacterium* could be enumerated on sodium lactate agar, under anaerobic incubation at 30°C for 7–9 days (Cogan et al. 2007).

Darukaradhya et al. (2006) evaluated 12 media for selective or differential enumeration of *Lb. acidophilus*, *Bifidobacterium* spp., *Lb. casei* group bacteria, and starter lactic acid bacteria from cheddar cheese. The authors tested pure cultures (*Lc. lactis* subsp. *cremoris* plus *Lc. lactis* subsp. *lactis*—LL50C; *Lb. acidophilus*—LAFTI L10, CSCC 2400, and CSCC 2422; *B. animalis* subsp. *lactis*—LAFTI B94; *B. breve*—CSCC 1900; *B. bifidum*—CSCC 1903 and CSCC 1909; *B. infantis*—CSCC 1912; *B. longum*—CSCC 5188; *Lb. casei* subsp. *casei*—CSCC 2603; *Lb. rhamnosus*—CSCC 2625; and *Lb. paracasei* CSCC 5437 and LAFTI L26), cultures from a commercial cheddar cheese containing *Lb. acidophilus*, and cultures from an experimental cheddar cheese manufactured with *Lb. acidophilus* LAFTI L10 and *B. lactis* LAFTI B94. According to the authors, some reported selective media (e.g., MRS–salicin for *Lb. acidophilus*, and MRS neomycin, paromomycin, nalidixic acid, and lithium chloride—NPNL—agar or MRS Ox-Bile for bifidobacteria) failed to inhibit *Lb. casei* group and starter bacteria. In that study, reinforced clostridial agar with bromocresol green and clindamycin was the most effective to select and recover *Lb. acidophilus*, and reinforced clostridial agar with bromocresol green and vancomycin was found suitable for *Lb. casei* group. Although reinforced clostridial agar with aniline blue and dicloxacillin allowed the growth of the starter LL50C, its colonies could be well differentiated from those of bifidobacteria, due the presence of the aniline blue dye.

Van de Casteele et al. (2006) observed that MRS with clindamycin (0.5 ppm) was the most effective medium to recover commercial *Lb. acidophilus* strains La-145, and LAFTI L10, in the presence of *Lb. delbrueckii* subsp. *bulgaricus* plus *S. thermophilus* (three commercial yogurt cultures), and *Lc. lactis* subsp. *lactis* plus *Lc. lactis* subsp. *cremoris*, *Lc. lactis* subsp. *lactis* biovar. *diacetylactis*, and *Leuconostoc* sp. (one commercial cheese culture). For commercial *Lb. rhamnosus* (strains LBA and LBC 80) and *Lb. paracasei* (LBC 81, LBC 82, and LAFTI L26), MRS AC (with acetic acid enough to reach pH 5.2) was effective only when in the presence of cheese culture. On the other hand, according to the authors, LC medium [basic medium with bromocresol green, 1% L-(D)-ribose and HCl enough to reach pH 5.1] shall be adopted when yogurt cultures are present. The culture media MRS LP (with 0.05% L-cystein–HCl, 0.3% LiCl, and 0.9% sodium propionate) was the most effective to enumerate commercial *B. animalis* subsp. *lactis* strains BL and LAFTI B-94, in the presence of either cheese or yogurt cultures. For another strain of *B. animalis* subsp. *lactis* (B-420), NPLN agar (basic medium with liver extract solution plus 100 mg/L neomycin sulfate, 200 mg/L paromomycin sulfate, and 15 mg/L nalidixic acid) shall be used when cheese culture is present, according to the same authors.

44.18 Screening for Probiotic and Prebiotic Potential of Cheeses through *In Vitro* Assays

In vitro tests are critical to assess the safety of probiotic microorganisms and useful to acquire knowledge of strains and the mechanism of their probiotic effect. However, the currently available tests are not fully adequate to predict the functionality of probiotic microorganisms in the human body. It was also noted that *in vitro* data available for particular strains are not sufficient for describing them as probiotic, once probiotics for human use will require substantiation of efficacy with human trials. *In vitro* tests, consequently, require validation with *in vivo* performance (FAO/WHO 2002).

On the other hand, the study of the tolerance to *in vitro* gastrointestinal conditions conducted with probiotic bacteria incorporated in the final product seems to be helpful in the selection of an adequate food matrix that contributes for probiotic survival in the GIT (Schillinger et al. 2005; Buriti et al. 2010). Therefore, it is suitable to assess probiotic potential of cheeses. Appropriate target-specific *in vitro* tests that correlate with *in vivo* results are frequently recommended. In this manner, the main currently used *in vitro* tests are resistance to gastric acidity, bile acid resistance, adherence to mucus and/or human epithelial cells and cell lines, antimicrobial activity against potentially pathogenic bacteria, ability to reduce pathogen adhesion to surfaces, and bile salt hydrolase activity (FAO/WHO 2002).

Madureira et al. (2005b) evaluated the survival of *B. animalis* (BLC-1, BB-12, and Bo), *Lb. acidophilus* (LAC-1 and Ki), *Lb. paracasei* subsp. *paracasei* (LCS-1), and *Lb. brevis* (LMG 6906) incorporated into requeijão. Requeijão is a Portuguese whey cheese obtained through heat processing, which denatures whey proteins. The authors exposed the product to *in vitro* simulated gastrointestinal conditions at 37°C for 60 or 120 min at pH 2.5–3.0, containing pepsin, and the following 120 min in the presence of bile salts. In the study, *B. animalis* BB-12 and Bo and *Lb. brevis* LMG 6906 exhibited the highest viability after the *in vitro* test. *Lb. acidophilus* LAC-1 and Ki resisted during exposure to acidic conditions but showed variable results in the presence of bile salts. According to the authors, *Lb. paracasei* subsp. *paracasei* LCS-1 and *B. animalis* BLC-1 were the most vulnerable to the *in vitro* gastrointestinal conditions tested.

The effect of the matrix, a semi-soft Gouda cheese, on the survival of *Lb. acidophilus* NCFM and *Lb. rhamnosus* HN001 through models simulating the human GIT (oral cavity, stomach, small intestine, and colon) was evaluated by Mäkeläinen et al. (2009). The authors reported that the probiotic microorganisms survived well in cheese submitted to the simulated upper GIT model. Lactobacilli populations increased during the colonic fermentation simulations of the probiotic cheese when compared to the non-probiotic cheese used as a control. Furthermore, the cheese matrix beneficially affected cyclooxygenases 1 and 2 gene expression of colonocytes in a cell culture model, up-regulating cyclooxygenase 1 (associated with maintaining the gastrointestinal integrity) and down-regulating cyclooxygenase 2 (induced by inflammatory stimuli and overexpressed in the early stages of colon cancer), which is considered to be beneficial for preserving healthy intestinal function.

The survival of the potential probiotic strain *Lb. delbrueckii* UFV H2b20 incorporated into a synbiotic cottage cheese supplemented with inulin, during exposure to simulated gastrointestinal conditions, was also evaluated by Araújo et al. (2010). Exposure to pH 7.0 or 3.5 for 4 h did not change bacterial populations, which remained around 8 log CFU/g during this period. However, exposure to pH 2.5 resulted in a reduction of 1 log cycle in bacterial viability for the 5-day-old cheeses and of 4-log cycles for the 15-day-old cheeses. The authors also observed a 2-log cycle decrease in the survival of *Lb. delbrueckii* UFV H2b20, after the cheese exposure to 0.5% or 1% bile salts for 4 h.

In relation to prebiotics, they are classified as specific colonic nutrients. Such classification requires a scientific demonstration that the nutrient resists against gastric acidity, is not hydrolyzed by mammalian enzymes, is not absorbed in the upper GIT, is fermented by the intestinal microbiota, and selectively stimulates the growth and/or activity of intestinal bacteria potentially associated with health and well-being (Roberfroid 2008). *In vitro* models may offer an interesting alternative to *in vivo* protocols for studying the prebiotic effect of ingredients and food products because they are generally inexpensive to operate, easy to set up with no ethical needs, and provide a dynamic overview of gut microbial activity and composition over several weeks (Le Blay et al. 2010). Such models have already proven to be useful for evaluating the prebiotic potential of *petit-suisse* cheeses (Cardarelli et al. 2007).

In their study, Cardarelli et al. (2007) measured the prebiotic effect of five *petit-suisse* cheese trials, combining potential prebiotics (inulin, oligofructose, and honey) and probiotics (*Lb. acidophilus* LA-5 and *B. lactis* BL04) through *in vitro* tests, using sterile, stirred, batch culture fermentations with human fecal slurry. The authors verified that the combination of the prebiotics, associated with probiotic bacteria added during cheesemaking, resulted in fast fermentation and almost no growth of detrimental bacteria (as clostridia and *E. coli*), showing it to be selective for beneficial bacteria (lactobacilli and bifidobacteria) and, therefore, the most promising *petit-suisse* cheese trial as a potential functional food.

44.19 Health Properties of Probiotic and Prebiotic Cheeses Supported by *In Vivo* Assays

According to the FAO/WHO Working Group on Evaluation of Probiotics in Food (FAO/WHO 2002), animal studies can be useful to provide substantiation of *in vitro* effects and determination of probiotic mechanisms. In addition, the FAO/WHO (2002) encourages the use of animal models prior to human trials.

With reference to animal models employed to corroborate the benefits obtained by the consumption of probiotic cheeses, Medici et al. (2004) studied the immunomodulating capacity of probiotic fresh cheese containing the probiotic adjuncts *B. bifidum* A12 (7.1 log CFU/g), *Lb. acidophilus* A9 (6.9 log CFU/g), and *Lb. paracasei* A13 (6.8 log CFU/g). The authors observed that BALB/c mice fed with probiotic fresh cheese, during a 5-day period, presented a significant increase in the phagocytic activity of peritoneal macrophages, in the number of IgA$^+$-producing cells, and in the CD4$^+$/CD8$^+$ ratio in the small intestine. In addition, *B. bifidum* and *Lb. paracasei* were identified mainly in the Peyer's patches (small intestine), whereas *Lb. acidophilus* was mainly located in the large intestine. The authors reported that 8 days following the removal of probiotic fresh cheese from the mice diet, the parameters evaluated returned to values similar to those of control mice.

Klobukowski et al. (2009) studied the influence of the consumption of white cheeses containing a potentially probiotic *Lb. plantarum* strain (10^7 CFU/g), supplemented or not with the prebiotic inulin (HPX, Orafti, at 2.5%), on calcium bioavailability in rats. The authors concluded that, even though without a statistical difference ($p > 0.05$) when compared to the control group (diet absent both of the probiotic and the prebiotic ingredient), the consumption of cheese containing *Lb. plantarum* tended to increase the apparent calcium retention (mineral intake minus fecal and urinary excretion) in rats. On the other hand, the consumption of white cheeses containing *Lb. plantarum* plus inulin resulted in the highest increase of both calcium apparent absorption (mineral intake minus fecal excretion) and calcium apparent retention.

44.20 Clinical Studies with Probiotic Cheese

The FAO/WHO (2002) pointed out important recommendations with respect to clinical trials for testing probiotic foods. Some of these recommendations are listed below:

1. Placebo would comprise the food carrier devoid of the test probiotic in double-blind, randomized, placebo-controlled (DBPC) human studies.
2. When a claim is made for a probiotic altering a disease state, the claim should be made based on sound scientific evidence in human subjects.
3. Human trials should be repeated by more than one research center for confirmation of results.
4. No adverse effects related to probiotic administration should be experienced when food is considered. Adverse effects should be monitored and incidents reported.
5. Information accumulated to show that a strain(s) is a probiotic, including clinical trial evidence, should be published in peer-reviewed scientific or medical journals.
6. Publication of negative results is encouraged, as these contribute to the totality of the evidence to support probiotic efficacy.

Important clinical trials involving the consumption of probiotic cheeses and their effects on human health are commented on this section.

Probiotics have been studied as a treatment to promote oral health. Once in the oral cavity, probiotics can create a biofilm, acting as a protective lining for oral tissues against oral diseases, and competing with cariogenic bacteria and periodontal pathogen growth (Flichy-Fernández et al. 2010). Regarding this subject, Ahola et al. (2002) investigated the relationship between the daily consumption of 5 × 15 g Edam probiotic cheese containing *Lb. rhamnosus* GG during 3 weeks and the decrease of caries-associated salivary microbial counts in young adults, through a DBPC study. The authors observed that during the post-treatment period, there was a significantly higher reduction ($p = 0.05$) in the microbial counts in the intervention group, compared to the control group. In that study, there was a trend indicating that probiotic intervention with *Lb. rhamnosus* GG might reduce the risk of the highest level of *S. mutans* and salivary yeasts.

In another study focusing on oral health, Hatakka et al. (2007) evaluated the association between the consumption of 50 g per day Emmental-type probiotic cheese containing *Lb. rhamnosus* GG, *Lb. rhamnosus* LC705, and *P. freudenreichii* subsp. *shermanii* JS during 16 weeks and the decreased prevalence of oral *Candida* in the elderly, through a DBPC study. The authors observed a reduction in the prevalence and risk of high salivary yeast counts by 32% and 75%, respectively, and the risk of hyposalivation by 56% in the group that consumed the probiotic cheese.

The role of probiotic foods in the prevention of *Helicobacter pylori* has been the target of some investigations (Wang et al. 2004; Sachdeva and Nagpal 2009; Lionetti et al. 2010). Boonyaritichaikij et al. (2009) studied the effects of a 12-month intervention with a probiotic cheese containing *Lb. gasseri* OLL2716 (LG21) on *H. pylori* eradication or prevention in *H. pylori*-positive and -negative preschool children, respectively. In the study, 29% of children who were *H. pylori*-positive at the beginning of the study became negative after the 12-month intervention with the probiotic cheese. For those who were initially *H. pylori*-negative, 4.1% and 8.1% of the children within the groups that received probiotic and control cheese, respectively, became *H. pylori*-positive after 12 months. As these observations are still unclear to prove efficacy of the probiotic cheese containing *Lb. gasseri* against *H. pylori* in preschoolers, the authors recommended additional studies for getting results that are more reliable.

The immunostimulatory potential is among the most important properties in clinical studies with probiotics, and nowadays, great interest is being given toward studies with the elderly population (Schiffrin et al. 2010; Tiihonen et al. 2010). The implications of the consumption of a commercial probiotic Gouda cheese containing 10^9 CFU/daily portion of *Lb. rhamnosus* HN001 and *Lb. acidophilus* NCFM for 4 weeks on age-related changes in the immune system of elderly volunteers (≥70 years old) were assessed by Ibrahim et al. (2010). In that study, the consumption of probiotic cheeses improved significantly ($p < 0.05$) some parameters associated with innate immunity, such as NK cell ability to kill target tumor cells and the phagocytosis activity of granulocytes, in comparison with the control group.

44.21 Production of Protective Compounds by Probiotic Bacteria and Implications on Cheese Quality and Biopreservation

Biopreservation with certain lactic acid cultures in food manufacture is an attractive tool to reduce the use of food preservatives and to monitor microbial hazards (Kourkoutas et al. 2006b). Organic acids, such as lactic, acetic, and propionic acids, produced by lactic acid bacteria, interfere with the proton motive force and the active transport mechanisms of the cytoplasmic membrane of important pathogen and spoilage bacteria (Forsythe 2000). When a large proportion of the acid is in the undissociated form, at a pH value that is below the pK_a value of the organic acids, the inhibitory effect is more pronounced. The pK_a values of formic acid, acetic acid, lactic acid, propionic acid, benzoic acid, and phenyllactic acid are 3.75, 4.76, 3.86, 4.87, 4.20, and 3.46, respectively (Tharmaraj and Shah 2009a,b). Lactic acid bacteria are slightly affected by these constituents, once most of them are able to grow or survive at a pH lower than 3.8 (Forsythe 2000). On the other hand, most of the pathogen and spoilage microorganisms are not able to grow at low pH values, and they can be prevented or controlled by the use of protective cultures producing organic acids. Other antagonistic metabolites produced by lactic acid bacteria include bacteriocins, peptides, and/or low-weight non-proteinaceous compounds, such as fatty acids and H_2O_2 (Irlinger and Mounier 2009). Cultures of bifidobacteria, enterococci, lactobacilli, and propionibacteria can produce organic acids in varying quantities, and they have been proposed as suitable for use as biopreservatives (Kourkoutas et al. 2006b; Tharmaraj and Shah 2009a).

Buriti et al. (2007a) observed a biopreservative effect against *E. coli*, *Staphylococcus* spp., and DNAse-positive *Staphylococcus* by the presence of *Lb. paracasei* subsp. *paracasei* LBC-82 in co-culture with *S. thermophilus* TA-040 in fresh cream cheeses stored under refrigeration at 4°C for 21 days. In that study, inhibition by bacteriocin or H_2O_2 production was discarded, and the authors concluded that inhibition was probably due to acid production.

Tharmaraj and Shah (2009a) observed that probiotic cultures consisting of *Lb. acidophilus* LA-5, *B. animalis* BB-12, *Lb. paracasei* subsp. *paracasei* (LC01 and LCS1), *Lb. casei* Shirota, and *Lb. rhamnosus* (LC 705, LBA, GG, and LR1524) were able to inhibit *Bacillus cereus* by 1–2 log cycles in contaminated French onion dip maintained at 4°C. In a complementary research, the same authors (Tharmaraj and Shah 2009b) did not observe a noticeable biopreservation effect of the same probiotic bacteria used in the other study on yeast and mold growth in contaminated French onion dip during a 6-week period at a storage temperature of 4°C. On the other hand, adding metabolites produced in the early stationary phase by the probiotic bacteria (5% wt/wt) showed an effective control against *Aspergillus niger*,

Fusarium spp., and *Candida albicans* during a 10-week shelf life period at 4°C. Those metabolites were mainly composed of acetic, lactic, benzoic, butyric, and phenyllactic acids. Particularly, metabolites of *Lb. acidophilus* LAC1, *B. animalis* BB-12, and *P. freudenreichii* subsp. *shermanii* P inhibited the production of visible colonies of fungi on the cheese surface throughout the 10-week storage period.

Grattepanche et al. (2008) reported the promising industrial application of commercial protective cultures to prevent spoilage by yeasts and molds in cheese production. Among them, two cultures are supplied by Danisco—HOLDBAC YM-B and YM-C. HOLDBAC YM-B is a combination of *P. freudenreichii* subsp. *shermanii* with *Lb. rhamnosus*, and HOLBAC YM-C contains *P. freudenreichii* subsp. *shermanii* and *Lb. paracasei*. Both cultures mentioned are recommended for application in fresh cheeses (Danisco 2005).

44.22 Probiotic Cheeses with Antioxidative Properties

Human nutrition is clearly associated with oxidative metabolism, which, besides production of energy, is involved in a number of vital functions of the host (Songisepp et al. 2005). A wide variety of reactive oxygen species are continuously produced in the human body and in food (Kullisaar et al. 2003). Under physiological conditions, the reactive species (including peroxyl radicals, nitric oxide radical, and superoxide anion) outline a crucial role in primary immune defense of the human body through phagocytic cells against harmful microorganisms. On the other hand, a prolonged excess of reactive species is highly damaging for the host biomolecules and cells, resulting in an unbalanced functional antioxidative network of the organism and leading to a substantial escalation of the pathological inflammation (Songisepp et al. 2005; Mikelsaar and Zilmer 2009). Damage caused by reactive oxygen species plays a substantial role in the pathogenesis of cancer, cardiovascular diseases, allergies, and atherosclerosis.

The antioxidative activity expressed by some microbial strains used as food components and probiotics may have a substantial impact on human welfare (Kullisaar et al. 2003). Bacteria producing antioxidative factors have been considered to play an important role in ameliorating the aging process, cardiovascular diseases, diabetes, as well as ulcers of the GIT and infection of the urogenital tract (Kaushik et al. 2009).

An antioxidant defense system consisting of enzymes, such as superoxide dismutase, glutathione peroxidase and catalase, and non-enzymatic compounds (e.g., glutathione), might prevent oxidative damage of lipoproteins in the plasma (Zhang et al. 2010). A *Lb. fermentum* strain, deposited in the Deutsche Sammlung von Mikroorganismen und Zellkulturen (DSM 14241, assigned as ME-3), was largely studied concerning its antimicrobial and antioxidative activity (Kullisaar et al. 2003; Mikelsaar et al. 2008; Mikelsaar and Zilmer 2009). This strain, formerly denoted as *Lb. fermentum* E-3, was found to express manganese superoxide dismutase (0.859 ± 0.309 U/mg of protein), scavenge hydroxyl radicals (~75%), and contain reduced glutathione (9.95 ± 3.30 µg/mL) (Kullisaar et al. 2002). The *Lb. fermentum* ME-3 strain was shown to be very suitable for the production of probiotic "Pikantne" cheese (an Estonian open-texture, semi-soft Tilsit-type cheese) and of a cheese spread manufactured with vegetable oils. Viability of *Lb. fermentum* ME-3 was higher than 10^7 CFU/g during the shelf life of both cheeses, 66 days and 28 days for "Pikantne" and cheese spread, respectively (Songisepp et al. 2004; Järvenpää et al. 2007). Particularly for Pikantne cheese, there was a tendency to increase total antioxidative activity during storage—on the 66th day of storage, total antioxidative activity of probiotic cheese and of pure culture of *Lb. fermentum* ME-3 was nearly the same (Songisepp et al. 2004). For cheese spread containing *Lb. fermentum* ME-3 and vegetable oils, no clear evidence of fatty acid oxidation was found by the authors throughout 28 days of storage (Järvenpää et al. 2007).

44.23 Probiotic Cheeses as a Source of Bioactive Peptides

Bioactive peptides were defined as "peptides with hormone- or drug-like activity that eventually modulate physiological function through binding interactions to specific receptors on target cells leading to induction of physiological responses" (Ong and Shah 2008b). Caseins and whey proteins can be important sources of biologically active peptides. These peptides are in an inactive state inside the protein molecule and are released during enzymatic hydrolysis with digestive enzymes or through the metabolism of proteolytic microorganisms. Bioactive peptides usually contain 3–20 amino acid residues per molecule. Biologically active peptides have been found to have specific activities, such as antihypertensive, antioxidative, antimicrobial, immunomodulatory, opioid, or mineral-binding activities (Korhonen and Pihlanto-Leppälä 2006).

Among various bioactive peptides, the antihypertensive peptides or angiotensin-converting enzyme inhibitors (ACE-I) are the most widely studied. ACE inhibition leads to a decreased level of the vaso-constricting peptide, angiotensin II, and a corresponding increased level of the vasodilatory peptide, bradykinin, therefore yielding an overall reduction in blood pressure (Donkor et al. 2007; Papadimitriou et al. 2007). ACE inhibition may also modulate immune defense and nervous system activity (Ong et al. 2007a).

Fermentation of dairy products with highly proteolytic strains of lactic acid bacteria seems to be the most effective way to increase the number of bioactive peptides. Selection of microorganisms with such capability for use in functional fermented products is gaining importance (Ramchandran and Shah 2008).

Ryhänen et al. (2001) evaluated the ACE-inhibitory activity of a probiotic "Festivo" cheese containing starter cultures of *Lactococcus* sp., *Leuconostoc* sp., *Propionibacterium* sp., and *Lactobacillus* sp., in combination with *Lb. acidophilus* and *Bifidobacterium* sp. The authors found ACE-inhibitory peptides in samples collected after 13 weeks of ripening, which corresponded to the α_{s1}-casein N-terminal peptides f(1–9), f(1–7), and f(1–6).

Ong et al. (2007a) investigated the release of ACE-inhibitory peptides in cheddar cheeses produced with *Lc. lactis* subsp. *lactis* and *Lc. lactis* subsp. *cremoris* as starters, and with the probiotic cultures *Lb. casei* 279 or *Lb. casei* LAFTI L26 during ripening. In that study, the concentrations of ACE needed to inhibit 50% of ACE activity (IC_{50}) was the lowest after 24 weeks of ripening in cheeses with *Lb. casei* 279 and *Lb. casei* LAFTI L26 (0.23 and 0.25 mg/mL, respectively), compared to 36 weeks for cheeses without any probiotic (0.28 mg/mL). Additionally, the authors observed that various ACE-inhibitory peptides corresponded to the α_{s1}-casein N-terminal peptides [f(1–6), f(1–7), f(1–9), f(24–32), and f(102–110)] and β-casein N-terminal peptides [f(47–52) and f(193–209)].

Various probiotic strains (*B. longum* 1941, *B. animalis* subsp. *lactis* LAFTI B94, *Lb. casei* 279, *Lb. casei* LAFTI L26, *Lb. acidophilus* 4962, and *Lb. acidophilus* LAFTI L10) were compared by Ong and Shah (2008b), concerning the release of ACE-inhibitory peptides in cheddar cheeses during ripening at 4°C and 8°C for 24 weeks. After 24 weeks, at both storage temperatures, *Lb. casei* 279, *Lb. casei* LAFTI L26, or *Lb. acidophilus* LAFTI L10 resulted in significantly higher ($p < 0.05$) ACE-inhibitory activity in cheddar cheese, compared to those without any probiotic adjunct. Even though without any significant difference ($p > 0.05$), the authors observed that the average IC_{50} value for *Lb. acidophilus* L10 cheeses was lower than that of the other probiotic cheeses. The improvement in ACE-inhibitory activity by *Lb. acidophilus* LAFTI L10 in cheddar cheese was also reported in another study (Ong and Shah 2008a).

Production of ACE-inhibitory peptides was also reported for probiotic cheese-like soy products. Liong et al. (2009) found significantly increased ACE-inhibitory activity of probiotic soy cream cheeses containing *Lb. acidophilus* FTCC 0291, compared to the control product, without any probiotic addition.

44.24 Usefulness of Experimental Designs and Statistical Models for Development of Probiotic and Synbiotic Cheeses

Product formulation development is a difficult and challenging task that includes performing various tests, often on a trial-and-error basis, to evaluate the performance of the prototypes in the product development process. These tests may be very expensive and time consuming (Omidbakhsh et al. 2010). Due to consumer's needs and trends, food product development forces food companies to respond rapidly to marketplace changes. The application of scientifically sound experimental designs enables companies to decrease the cost and time between concept and marketplace. A large number of experimental designs, such as response surface methodology (RSM) and modeling of mixtures, and statistical procedures, such as multivariate analysis techniques, are available for the optimization of food product development (Arteaga et al. 1994; Castro et al. 2000). These experimental designs and statistical trials have proven to be useful in studies of probiotic and synbiotic cheeses (Gomes and Malcata 1998; Cardarelli 2006; Bergamini et al. 2009a; Milesi et al. 2009; Özer et al. 2009; Rodrigues et al. 2011).

The RSM is a set of techniques based on the use of factorial planning and least squares fit applied with great success for the modeling of various industrial processes, through the construction of empirical models, usually linear or quadratic, which describe the behavior of the system under study based on experimental assays. The models define quantitative relations between variables and responses, covering the entire experimental range tested, and including the interactions when they are present (Castro et al. 2000).

In order to optimize the manufacturing process of a Portuguese semi-hard goat cheese by increasing the probiotic population (*Lb. acidophilus* and *B. lactis*) and improving its biochemical and sensory properties, Gomes and Malcata (1998) varied the relative concentrations of the probiotic inoculum, the salt concentration in dry matter, and the addition of a milk hydrolysate, using a 2^3 full factorial design. The duration of the ripening period was also varied by a 2^1 design nested in the 2^3 design. According to the authors, probiotic bacterial growth contributed significantly to ripening with the formation of low-molecular-mass peptides and amino acids, even though affecting lipolysis in minor extension, and without being related to the production of lactic and acetic acids. In that study, statistical analyses using RSM indicated that the manufacture of goat cheese could be optimized through the addition of 3×10^7 CFU of *B. lactis* and 7×10^6 CFU of *Lb. acidophilus* per milliliter of milk, 0.30% (vol/wt) milk hydrolysate, 3.50% (wt/wt) salt, and a ripening period of 70 days.

Using the simplex centroid design, Cardarelli (2006) studied the effects of the factors inulin, fructooligosaccharides (FOS), and honey for the development of a synbiotic *petit-suisse* cheese containing *Lb. acidophilus* Lac4 and *B. animalis* subsp. *lactis* Bl04. In that study, the proportion of honey tended to reduce the viability of *B. lactis*, whereas *Lb. acidophilus* was less influenced by this ingredient. The RSM with the desirability function was applied to optimize the probiotic viability, besides firmness, fructan content, and product cost. According to the author, the global desirability of 99.55% could be obtained through the addition of 25% FOS, 70% inulin, and 5% honey to the *petit-suisse* cheese.

Principal component analysis (PCA) is a multivariate analysis technique that reduces an original number of variables to a smaller number of new, uncorrelated variables (principal components), based on their linear correlations (Botelho et al. 2010). Cluster analysis is another statistical multivariate technique able to group samples according to their similarities and differences. In cluster analysis, groups can be formed by maximizing the differences between samples that belong to different groups and minimizing the differences between samples that belong to the same group (Capitani et al. 2009).

Bergamini et al. (2009a) applied PCA and hierarchical cluster analysis for the data of peptide profiles and FAAs of 60-day-old Pategrás cheeses. The authors intended to individually assess the influence of six probiotic cultures (three *Lb. acidophilus* strains—A1, A2, and A3; *Lb. casei* C2; *Lb. paracasei* C1; and *Lb. rhamnosus* C3). Moreover, they investigated how the way of probiotic inoculation (with direct incorporation of the freeze-dried culture into milk or by a two-step fermentation procedure) influenced the proteolysis profile, in comparison with control cheeses (without any probiotics). The multivariate analysis techniques showed that peptide profiles of most probiotic cheeses were different from those of the control cheeses. The effect of *Lb. acidophilus* strains on cheese peptide profiles was shown to be the strongest, and *Lb. acidophilus* A3 was the only strain that showed a distinguishable effect on peptidolysis when a two-step fermentation was carried out. Control cheeses and probiotic cheeses with *Lb. paracasei* and *Lb. rhamnosus* were characterized by the lowest concentration of FAAs. According to these results, the authors suggested that *Lb. acidophilus* might be suitable to influence secondary proteolysis of Pategrás cheeses, contributing to ripening and flavor formation.

Rodrigues et al. (2011) performed a PCA to assess the relationship between the microbiological and the chemical parameters of curdled milk matrices in different storage periods, up to 60 days. The matrices contained different probiotic cultures (*Lb. acidophilus* L10, LA-5, and Ki; *Lb. casei* 01; *Lb. paracasei* L26; *B. animalis* BB-12; *B. lactis* B94; or *B. lactis* Bo), with or without a FOS/inulin mix (synbiotic), and the control matrices did not contain probiotic or FOS/inulin. Probiotic and synbiotic matrices on day 0 and synbiotic matrix with *Lb. acidophilus* on day 15 did not correlate with any specific parameter, except with higher pH. Probiotic samples with *Lb. acidophilus* between 15 and 60 days were correlated with lower probiotic populations. All 60-day probiotic and synbiotic matrices and the 45-day synbiotic matrices with *B. lactis* B94 or *Lb. acidophilus* LA-5 were characterized by higher values of water-soluble nitrogen, total free fatty acids, and conjugated linoleic acid. Probiotic and synbiotic matrices on days 15 or 30 were correlated mainly with viable counts of probiotic bacteria, and 45-day samples were apparently better correlated with non-protein nitrogen. According to the authors, the distribution obtained was consistent with the influence of the FOS/inulin mix, justifying the differences obtained concerning proteolysis and lipolysis among the samples, mainly those containing *Lb. acidophilus* LA-5.

44.25 Factors Influencing Sensory Properties of Probiotic and Synbiotic Cheeses

Probiotics are not supposed to change the sensory features of the final cheese (Souza et al. 2008; Settanni and Moschetti 2010). Any changes in flavor that occur in a cheese may be undesirable and discourage the consumer from future purchase of that product (Montoya et al. 2009).

Proteolysis is among the most important events that contribute to cheese microbial ecology and mainly to sensory features (Settanni and Moschetti 2010). The amino acids produced during proteolysis contribute directly to the basic taste of the cheese and indirectly to cheese flavor because they are precursors for the other catabolic reactions, giving rise to volatile aroma compounds. Amino acid catabolism (deamination, decarboxylation, and transamination) and side-chain changes may yield keto acids, ammonia, amines, aldehydes, acids, and alcohols, which are essential contributors to cheese taste and aroma (Kieronczyk et al. 2001; Settanni and Moschetti 2010).

Addition of lactobacilli adjunct cultures in cheeses has been associated with an increased proteolysis and intensification of flavor. The selection of probiotic microorganisms for use as adjunct cultures, in order to improve proteolysis, increase casein hydrolysis, and enhance flavor, seems to be very useful for the manufacture of ripened cheeses (Ong et al. 2007b; Ong and Shah 2008b). In studies with semi-hard Pategrás cheeses (Bergamini et al. 2006, 2009a,b), the authors verified that *Lb. acidophilus* strains would be advantageous for increasing the ripening rate or promoting flavor enhancement, through a significantly higher peptidolytic activity and an increased production of FAAs. Milesi et al. (2009) found that the potential probiotic candidates *Lb. casei* I90 and *Lb. plantarum* I91 were suitable as adjunct cultures either in a semi-hard cheese model (Pategrás cheese) or in a soft cheese model (Cremoso cheese). According to the authors, these strains increased the secondary proteolysis moderately, did not over-acidify, and changed sensory profiles of cheeses likewise to control cheeses.

For ripened cheeses, lipolysis also plays an important role in the development of sensory characteristics. Nonetheless, the addition of probiotic cultures does not seem to affect the free fatty acid profile of cheese, likely due to a higher lipolytic activity of starters and some non-starter lactic acid bacteria, when compared to probiotic cultures (Grattepanche et al. 2008).

Glycolysis was reported to influence the flavor of cheddar cheeses. The conversion of lactose to lactate is normally mediated by the starter culture during curd preparation or the early stages of cheese ripening. During ripening, residual lactose is primarily metabolized to lactate, which may be oxidized to acetate. Acetate may also be produced by starter bacteria or probiotic adjuncts, such as *Lactobacillus* and *Bifidobacterium* from lactose or citrate or amino acid, and is usually present at high concentrations in cheddar cheeses (Fox et al. 2000; Ong et al. 2007b).

In fact, most cheeses containing probiotic lactobacilli and bifidobacteria have high acetate content, due to heterofermentation. Bifidobacteria produce mainly acetic and lactic acid, in a molar ratio of 2:3, from lactose fermentation via the fructose-6-phosphate shunt pathway. Some lactobacilli are also able to produce acetic acid, but to a lesser extent, compared to bifidobacteria. Acetic acid contributes to the typical flavor of different cheeses, although high concentrations may cause off-flavors (Grattepanche et al. 2008; Ong et al. 2007b). Ong et al. (2007b) verified that *Bifidobacterium* sp., *Lb. casei* 279, and *Lb. paracasei* L26 significantly increased the concentration of acetic acid in cheddar cheeses. Nonetheless, vinegary scores did not significantly influence the cheese acceptability in that study.

TABLE 44.1

Studies Showing Positive Results Related to Sensory Properties of Probiotic and Synbiotic Cheeses

Cheese Kind	Probiotic Cultures or Potential Probiotic Bacteria/ Prebiotic Ingredients	Starter Cultures/Other Adjunct Cultures/ Alternative Acidification	Ripening/ Storage Conditions	Research Article
Argentinean fresh cheese	*Lactobacillus acidophilus* A3, *Bifidobacterium bifidum* A1, and *Lactobacillus paracasei* A13	*Lactococcus lactis* A6 and *Streptococcus thermophilus* A4	5°C, 60 days	Vinderola et al. (2009)
Cheddar cheese	*Bifidobacterium animalis* subsp. *lactis* BB-12 or *Bifidobacterium longum* BB536	*Lactococcus lactis* subsp. *cremoris* strains 223 and 227	8°C, 7 months	Mc Brearty et al. (2001)
Cheddar cheese	Spray-dried spontaneous rifampicin-resistant *Lactobacillus paracasei* NFBC 338	*Lactococcus lactis* subsp. *cremoris* strains 227 and 303	8°C, 3 months	Gardiner et al. (2002)
Cheddar cheese	*Bifidobacterium longum* 1941, *Bifidobacterium animalis* B94, *Lactobacillus casei* L26, *Lactobacillus acidophilus* 4962, and *Lactobacillus acidophilus* L10 or B94 + L26 + L10	*Lactococcus lactis* subsp. *lactis* and *Lactococcus lactis* subsp. *cremoris*	4°C, 9 months	Ong et al. (2007b)
Cheddar cheese	*Bifidobacterium longum* 1941, *Bifidobacterium animalis* B94, *Lactobacillus casei* 279, *Lactobacillus casei* L26, *Lactobacillus acidophilus* 4962, or *Lactobacillus acidophilus* L10	*Lactococcus lactis* subsp. *lactis* and *Lactococcus lactis* subsp. *cremoris*	4°C or 8°C, 24 weeks	Ong and Shah (2009)
Cremoso cheese	*Lactobacillus casei* I90, *Lactobacillus plantarum* I91, *Lactobacillus rhamnosus* I73, or *Lactobacillus rhamnosus* I77	*Streptococcus thermophilus*	5°C, 45 days	Milesi et al. (2009)
Kasar cheese	Free or microencapsulated *Lactobacillus acidophilus* LA-5 and *Bifidobacterium animalis* subsp. *lactis* BB-12 by emulsion or by extrusion techniques	*Lactococcus lactis* subsp. *lactis* and *Lactococcus lactis* subsp. *cremoris*	10°C, 90 days	Özer et al. (2008)
Minas fresh cheese	ABT-4 mixed culture (*Lactobacillus acidophilus* LA-5 + *Bifidobacterium animalis* subsp. *lactis* BB-12 and *Streptococcus thermophilus*)	Direct acidification with lactic acid	5°C–7°C, 7 days	Buriti et al. (2007b)
Minas fresh cheese	*Lactobacillus acidophilus* LA-5	Direct acidification with lactic acid	5°C, 7 days	Buriti et al. (2005b)
Minas fresh cheese	*Lactobacillus acidophilus* LA-5	*Streptococcus thermophilus* TA040 or direct acidification with lactic acid	4°C–5°C, 14 days	Souza et al. (2008)
Minas fresh cheese	*Lactobacillus paracasei* subsp. *paracasei* LBC-82	Direct acidification with lactic acid	4°C–5°C, 7 days	Buriti et al. (2005a)
New low-fat "Festivo" cheese	*Lactobacillus acidophilus* and *Bifidobacterium* sp.	*Lactococcus* sp., *Leuconostoc* sp., *Propionibacterium* sp., and *Lactobacillus* sp.	10°C–18°C, 13 weeks	Ryhänen et al. (2001)

(continued)

TABLE 44.1 (Continued)

Studies Showing Positive Results Related to Sensory Properties of Probiotic and Synbiotic Cheeses

Cheese Kind	Probiotic Cultures or Potential Probiotic Bacteria/ Prebiotic Ingredients	Starter Cultures/Other Adjunct Cultures/ Alternative Acidification	Ripening/ Storage Conditions	Research Article
New-type ewe's cheese	*Lactobacillus casei* ATCC 393 (free or immobilized on apple or pear pieces)	No addition	4°C–6°C, 30 days	Kourkoutas et al. (2006a)
Pategrás cheese	*Lactobacillus casei* I90, *Lactobacillus plantarum* I91, *Lactobacillus rhamnosus* I73, or *Lactobacillus rhamnosus* I77	*Streptococcus thermophilus*	12°C, 80% relative humidity, 60 days	Milesi et al. (2009)
Pategrás cheese	*Lactobacillus acidophilus, Lactobacillus paracasei, Bifidobacterium lactis,* or the three strains in a mixed culture	*Streptococcus thermophilus*	12°C, 80% relative humidity, 2 months	Bergamini et al. (2009b)
Pategrás cheese	*Lactobacillus acidophilus, Lactobacillus paracasei* subsp. *paracasei,* and *Bifidobacterium animalis* subsp. *lactis*	*Streptococcus thermophilus*	12°C, 80% relative humidity, 60 days	Perotti et al. (2009)
Pecorino Sardo cheese	*Lactobacillus casei* subsp. *casei* and *Lactobacillus helveticus*	*Lactococcus lactis* subsp. *lactis* and *Streptococcus thermophilus*	10°C–12°C, 85%–90% relative humidity, 210 days	Madrau et al. (2006)
Petit-suisse cheese	*Lactobacillus acidophilus* LA-5 and *Bifidobacterium animalis* subsp. *lactis* Bl04	*Streptococcus thermophilus* TA040	4 ± 1°C, 14 days	Pereira et al. (2010)
Synbiotic cottage cheese	*Lactobacillus delbrueckii* subsp. *bulgaricus* UFV H2b20 and inulin	Mesophilic mixed culture RA 073	5°C, 15 days	Araújo et al. (2010)
Synbiotic fresh cream cheese	*Lactobacillus paracasei* subsp. *paracasei* LBC-82 and inulin	*Streptococcus thermophilus* TA040	4 ± 1°C, 7 days	Buriti et al. (2008)
Synbiotic *petit-suisse* cheese	*Lactobacillus acidophilus* Lac4, *Bifidobacterium animalis* subsp *lactis* Bl04, fructooligosaccharides, and inulin	*Streptococcus thermophilus* TA040	4 ± 1°C, 28 days	Cardarelli et al. (2008)
Turkish Beyaz cheese	*Lactobacillus fermentum* (AB5-18 and AK4-120) and *Lactobacillus plantarum* (AB16-65 and AC18-82)	No addition or *Lactococcus lactis* subsp. *cremoris* and *Lactococcus lactis* subsp. *lactis*	4°C, 120 days	Kılıç et al. (2009)
Turkish white cheese	*Lactobacillus acidophilus* 593N	*Lactococcus lactis* subsp. *lactis* and *Lactococcus lactis* subsp. *cremoris* R 707 culture	4°C, 90 days	Kasımoğlu et al. (2004)
White-brined Turkish cheese	Free or microencapsulated *Lactobacillus acidophilus* LA-5 and *Bifidobacterium animalis* subsp. *lactis* BB-12 by emulsion or by extrusion techniques	*Lactococcus lactis* subsp. *lactis* and *Lactococcus lactis* subsp. *cremoris*	4°C, 90 days	Özer et al. (2009)

Several lactic acid bacteria strains also can contribute to improve the texture, viscosity, mouthfeel, taste perception, and stability of fermented products, including cheeses, through the synthesis of exopolysaccharides. Some probiotic lactobacilli and bifidobacteria are able to produce exopolysaccharides and secrete them into milk (Ruas-Madiedo et al. 2002; Won et al. 2008). Kailasapathy and Masondole (2005) reported an increased exopolysaccharide production by *Lb. acidophilus* DD 910 and *B. lactis* DD 920 (free or microencapsulated with an alginate–starch polymer) when incorporated into feta cheese.

In the development of synbiotic cheeses, prebiotics also can be applied to produce an improved sensory quality. The use of inulin and nondigestible oligosaccharides as fiber ingredients often leads to improved taste and texture. FOS are successfully combined with fruit preparations for application in dairy products, such as fresh cheeses, allowing for an improved mouthfeel. The synergistic taste effects of FOS, when combined with high-potency sweeteners, are also known. Inulin, on the other hand, presents specific gelling characteristics and allows for the development of low-fat foods. In cream cheeses and processed cheeses, inulin allows for the replacement of significant amounts of fat and/or stabilization of the emulsion, while providing a short and spreadable texture. In low-fat fresh cheeses, the addition of 2%–3% inulin imparts a better balanced round flavor and a creamier mouthfeel (Franck 2008). Buriti et al. (2008) observed that *Lb. paracasei* subsp. *paracasei* LBC 82 interfered negatively in the flavor of fresh cream cheeses. When inulin was incorporated into fresh cream cheeses containing that microorganism, not only flavor was improved, but also texture was enhanced. Cardarelli et al. (2008) reported that the combination of FOS and inulin (50:50, 10 g/100 g) resulted in the best sensory features of synbiotic *petit-suisse* cheeses containing *Lb. acidophilus* Lac4 and *B. animalis* subsp. *lactis* BL04.

Some studies with probiotic and synbiotic cheeses showing general positive results concerning sensory characteristics are listed in Table 44.1.

44.26 Future Perspectives

The incorporation of probiotic cultures into cheeses provides potential, not only to improve health status and quality of products but also to increase the range of probiotic products available for consumers. The use of cheese as a probiotic food carrier presents potential advantages and is a valuable alternative for the dairy industry. Before producing probiotic cheese containing a specific probiotic strain or a combination of strains, the manufacturer ought to verify the behavior and the performance of the culture in the cheese environment and the compatibility between the probiotic strains employed and between the probiotic and the starter cultures. Nevertheless, preliminary *in vitro* tests regarding probiotic potential of cheeses containing probiotic cultures are extremely important and should be followed by animal studies and also by double-blind, placebo-controlled clinical human studies. Such tests must be completed with laboratorial analysis testing parameters that are decisive for the marketing of the product, such as organic acid profile, and typical aroma compounds of the product. Additionally, the use of sensorial techniques certainly helps in determining the important attributes of probiotic cheeses and how these attributes could be improved.

REFERENCES

Ahmad, N., Li, L., Yang, X.Q., Ning, Z.X., and Randhawa, M.A. 2008. Improvements in the flavor of soy cheese. *Food Technol. Biotechnol.* 46(3):252–261.

Ahola, A.J., Yli-Knuuttila, H., Suomalainen, T., Poussa, T., Ahlström, A., Meurman, J.H., and Korpela, R. 2002. Short-term consumption of probiotic-containing cheese and its effect on dental caries risk factors. *Arch. Oral Biol.* 47:799–804.

Araújo, E.A., Carvalho, A.F., Leandro, E.S., Furtado, M.M., and de Moraes, C.A. 2010. Development of a symbiotic cottage cheese added with *Lactobacillus delbrueckii* UFV H2b20 and inulin. *J. Funct. Foods* 2:85–89.

Arteaga, G.E., Li-Chan, E., Vazquez-Arteaga, M.C., and Nakai, S. 1994. Systematic experimental designs for product formula optimization. *Trends Food Sci. Technol.* 5:243–254.

Ash, A. and Wilbey, A. 2010. The nutritional significance of cheese in the UK diet. *Int. J. Dairy Technol.* 63(3):305–319.

Axelsson, L. 2004. Lactic acid bacteria: Classification and physiology. In: *Lactic Acid Bacteria: Microbiological and Functional Aspects*, 3rd edition, edited by S. Salminen, A. Wright, and A. Ouwehand, pp. 1–66. New York: Marcel Dekker.

Bergamini, C., Hynes, E., Meinardi, C., Suárez, V., Quiberoni, A., and Zalazar, C. 2010. Pategrás cheese as a suitable carrier for six probiotic cultures. *J. Dairy Res.* 77(3):265–272.

Bergamini, C.V., Hynes, E.R., Candioti, M.C., and Zalazar, C.A. 2009a. Multivariate analysis of proteolysis patterns differentiated the impact of six strains of probiotic bacteria on a semi-hard cheese. *J. Dairy Sci.* 92:2455–2467.

Bergamini, C.V., Hynes, E.R., Palma, S.B., Sabbag, N.G., and Zalazar, C.A. 2009b. Proteolytic activity of three probiotic strains in semi-hard cheese as single and mixed cultures: *Lactobacillus acidophilus, Lactobacillus paracasei* and *Bifidobacterium lactis. Int. Dairy J.* 19:467–475.

Bergamini, C.V., Hynes, E.R., Quiberoni, A., Suárez, V.B., and Zalazar, C.A. 2005. Probiotic bacteria as adjunct starters: Influence of the addition methodology on their survival in a semi-hard Argentinean cheese. *Food Res. Int.* 38:597–604.

Bergamini, C.V., Hynes, E.R., and Zalazar, C.A. 2006. Influence of probiotic bacteria on the proteolysis profile of a semi-hard cheese. *Int. Dairy J.* 16:856–866.

Bhardwaj, A., Kapila, S., Mani, J., and Malik, R.K. 2009. Comparison of susceptibility to opsonic killing by *in vitro* human immune response of *Enterococcus* strains isolated from dairy products, clinical samples and probiotic preparation. *Int. J. Food Microbiol.* 128:513–515.

Blanchette, L., Roy, D., and Gauthier, S.F. 1995. Production of cultured Cottage cheese dressing by bifidobacteria. *J. Dairy Sci.* 78:1421–1429.

Boonyaritichaikij, S., Kuwabara, K., Nagano, J., Kobayashi, K., and Koga, Y. 2009. Long-term administration of probiotics to asymptomatic pre-school children for either the eradication or the prevention of *Helicobacter pylori* infection. *Helicobacter* 14:202–207.

Botelho, P.B., Fioratti, C.O., Abdalla, D.S.P., Bertolami, M.C., and Castro, I.A. 2010. Classification of individuals with dyslipidaemia controlled by statins according to plasma biomarkers of oxidative stress using cluster analysis. *Br. J. Nutr.* 103:256–265.

Boyle, R.J., Robins-Browne, R.M., and Tang, M.L.K. 2006. Probiotic use in clinical practice: What are the risks. *Am. J. Clin. Nutr.* 83:1256–1264.

Boylston, T.D., Vinderola, C.G., Ghoddusi, H.B., and Reinheimer, J.A. 2004. Incorporation of bifidobacteria into cheeses: Challenges and rewards. *Int. Dairy J.* 14:375–387.

Buriti, F.C.A., Cardarelli, H.R., and Saad, S.M.I. 2007a. Biopreservation by *Lactobacillus paracasei* in coculture with *Streptococcus thermophilus* in potentially probiotic and synbiotic fresh cream cheeses. *J. Food Prot.* 70(1) 228–235.

Buriti, F.C.A., Cardarelli, H.R., and Saad, S.M.I. 2008. Influence of *Lactobacillus paracasei* and inulin on instrumental texture and sensory evaluation of fresh cream cheese. *Braz. J. Pharm. Sci.* 44(1):75–84.

Buriti, F.C.A., Castro, I.A., and Saad, S.M.I. 2010. Viability of *Lactobacillus acidophilus* in synbiotic guava mousses and its survival under *in vitro* simulated gastrointestinal conditions. *Int. J. Food Microbiol.* 137:121–129.

Buriti, F.C.A., Okazaki, T.Y., Alegro, J.H.A., and Saad, S.M.I. 2007b. Effect of a probiotic mixed culture on texture profile and sensory performance of Minas fresh cheese in comparison with the traditional products. *Arch. Latinoam Nutr.* 57(2):179–185.

Buriti, F.C.A., Rocha, J.S., and Saad, S.M.I. 2005b. Incorporation of *Lactobacillus acidophilus* in Minas fresh cheese and its implications for textural and sensorial properties during storage. *Int. Dairy J.* 15:1279–1288.

Buriti, F.C.A., Rocha, J.S., Assis, E.G., and Saad, S.M.I. 2005a. Probiotic potential of Minas fresh cheese prepared with the addition of *Lactobacillus paracasei. LWT—Food Sci. Technol.* 38:173–180.

Buriti, F.C.A. and Saad, S.M.I. 2007. Bacteria of *Lactobacillus casei* group: Characterization, viability as probiotic in food products and their importance for human health. *Arch. Latinoam Nutr.* 57:373–380.

Candioti, M.C., Hynes, E., Quiberoni, A., Palma, S.B., Sabbag, N., and Zalazar, C.A. 2002. Reggianito Argentino cheese: Influence of *Lactobacillus helveticus* strains isolated from natural whey cultures on cheese making and ripening processes. *Int. Dairy J.* 12:923–931.

Capitani, C.D., Carvalho, A.C.L., Rivelli, D.P., Barros, S.B.M., and Castro, I.A. 2009. Evaluation of natural and synthetic compounds according to their antioxidant activity using a multivariate approach. *Eur. J. Lipid Sci. Technol.* 111:1090–1099.

Cardarelli, H.R. 2006. Desenvolvimento de queijo *petit-suisse* simbiótico. PhD Thesis, Faculdade de Ciências Farmacêuticas, Universidade de São Paulo, São Paulo, Brazil.

Cardarelli, H.R., Buriti, F.C.A., Castro, I.A., and Saad, S.M.I. 2008. Inulin and oligofructose improve sensory quality and increase the probiotic viable count in potentially synbiotic *petit-suisse* cheese. *LWT—Food Sci. Technol.* 41:1037–1046.

Cardarelli, H.R., Saad, S.M.I., Gibson, G.R., and Vulevic, J. 2007. Functional *petit-suisse* cheese: Measure of the prebiotic effect. *Anaerobe* 13:200–207.

Castro, I.A., Tirapegui, J., and Silva, R.S.S.F. 2000. Protein mixtures and their nutritional properties optimized by response surface methodology. *Nutr. Res.* 20(9):1341–1353.

Champagne, C.P., Gardner, N.J., and Roy, D. 2005. Challenges in the addition of probiotic cultures to foods. *Crit. Rev. Food Sci. Nutr.* 45:61–84.

Cichoski, A.J., Cunico, C., Di Luccio, M., Zitkoski, J.L., and Carvalho, R.T. 2008. Efeito da adição de probióticos sobre as características de queijo prato com reduzido teor de gordura fabricado com fibras e lactato de potássio. *Ciência Tecnol. Aliment.* 28(1):214–219.

Coeuret, V., Gueguen, M., and Vernoux, J.P. 2004. *In vitro* screening of potential probiotic activities of selected lactobacilli isolated from unpasteurized milk products for incorporation into soft cheese. *J. Dairy Res.* 71:451–460.

Cogan, T.M., Beresford, T.P., Steele, J., Broadbent, J., Shah, N.P., and Ustunol, Z. 2007. Invited review: Advances in starter cultures and cultured foods. *J. Dairy Sci.* 90:4005–4021.

Corbo, M.R., Albenzio, M., De Angelis, M., Sevi, A., and Gobbetti, M. 2001. Microbiological and biochemical properties of Canestrato Pugliese hard cheese supplemented with bifidobacteria. *J. Dairy Sci.* 84:551–561.

Corsetti, A., Caldini, G., Mastrangelo, M., Trotta, F., Valmorri, S., and Cenci, G. 2008. Raw milk traditional Italian ewe cheeses as a source of *Lactobacillus casei* strains with acid-bile resistance and antigenotoxic properties. *Int. J. Food Microbiol.* 125:330–335.

Cruz, A.G., Buriti, F.C.A., Souza, C.H.B., Faria, J.A.F., and Saad, S.M.I. 2009. Probiotic cheese: Health benefits, technological and stability aspects. *Trends Food Sci. Technol.* 20:344–354.

Daigle, A., Roy, D., Bélanger, G., and Vuillemard, J.C. 1999. Production of probiotic cheese (Cheddar-like cheese) using enriched cream fermented by *Bifidobacterium infantis*. *J. Dairy Sci.* 82:1081–1091.

Danisco. 2005. HOLDBAC™ YM Protective Cultures—A natural way to control yeast and mould growth in fresh fermented dairy products. Technical Memorandum, TM 2074-1e.

Darukaradhya, J., Phillips, M., and Kailasapathy, K. 2006. Selective enumeration of *Lactobacillus acidophilus*, *Bifidobacterium* spp., starter lactic acid bacteria and non-starter lactic acid bacteria from Cheddar cheese. *Int. Dairy J.* 16:439–445.

Dave, R.I. and Shah, N.P. 1997. Viability of yoghurt and probiotic bacteria in yoghurts made from commercial starter cultures. *Int. Dairy J.* 7:31–41.

Desai, A.R., Shah, N.P., and Powell, I.B. 2006. Discrimination of dairy industry isolates of the *Lactobacillus casei* group. *J. Dairy Sci.* 89:3345–3351.

Ding, W.K. and Shah, N.P. 2009. An improved method of microencapsulation of probiotic bacteria for their stability in acidic and bile conditions during storage. *J. Dairy Sci.* 74:M53–M61.

Donkor, O.N., Henriksson, A., Singh, T.K., Vasiljevic, T., and Shah, N.P. 2007. ACE-inhibitory activity of probiotic yoghurt. *Int. Dairy J.* 17:1321–1331.

Douglas, L.C. and Sanders, M.E. 2008. Probiotics and prebiotics in dietetics practice. *J. Am. Diet. Assoc.* 108:510–521.

Endres, J.G., editor. 2001. *Soy Protein Products: Characteristics, Nutritional Aspects and Utilization*, 53 pp. Champaign, IL: AOCS.

⋮y, M.A.E. 2008. Cheese structure and current methods of analysis. *Int. Dairy J.*

.M., Jan, G., Loux, V., Thierry, A., Parayre, S., Maillard, M.B., Dhebécourt, J., Cousin, ⋮uier, P., Couloux, A., Barbe, V., Vacherie, B., Wincker, P., Gibrat, F., Gaillardin, C., ⋮0. The complete genome of *Propionibacterium freudenreichii* CIRM-BIA1ᵀ, a hardy with food and probiotic applications. *PLoS ONE* 5(7):e11748.

⋮se technology. *Int. J. Dairy Technol.* 57:91–98.

Farkye, N.Y. 2007. Acid and acid/heat-coagulated cheeses. In: *Cheese Problems Solved*, edited by P.L.H. McSweeney, pp. 343–361. Cambridge: Woodhead.

Farnworth, E.R. 2008. Evidence to support health claims for probiotics. *J. Nutr.* 1250S–1254S.

Fernández, M.F., Delgado, T., Boris, S., Rodríguez, A., and Barbés, C. 2005. A washed-curd goat's cheese as a vehicle for delivery of a potential probiotic bacterium: *Lactobacillus delbrueckii* subsp. *lactis* UO 004. *J. Food Prot.* 68:2665–2671.

Ferrero, M., Cesena, C., Morelli, L., Scolari, G., and Vescovo, M. 1996. Molecular characterization of *Lactobacillus casei* strains. *FEMS Microbiol. Lett.* 140(2–3):215–219.

Flichy-Fernández, A.J., Alegre-Domingo, T., Peñarrocha-Oltra, D., and Peñarrocha-Diago, M. 2010. Probiotic treatment in the oral cavity: An update. *Med. Oral Patol. Oral Cir. Bucal* 15(5):677–680.

Food and Agriculture Organization of United Nations; World Health Organization. FAO/WHO. 2001. Evaluation of health and nutritional properties of probiotics in food including powder milk with live lactic acid bacteria. Report of a joint FAO/WHO expert consultation. Córdoba, Argentina, October 1–4, 2001. Available from: http://www.who.int/foodsafety/publications/fs_management/en/probiotics.pdf. Accessed July 18, 2010.

Food and Agriculture Organization of United Nations; World Health Organization. FAO/WHO. 2002. Guidelines for the evaluation of probiotics in food. Report of a joint FAO/WHO working group on drafting guidelines for the evaluation of probiotics in food. London, ON, Canada, April 30 and May 1, 2002. Available from: http://www.who.int/foodsafety/fs_management/en/probiotic_guidelines.pdf. Accessed September 12, 2010.

Forsythe, J.S. 2000. *The Microbiology of Safe Foods*, 412 pp. Oxford: Blackwell.

Fortin, M.H., Champagne, C.P., St-Gelais, D., Britten, M., Fustier, P., and Lacroix, M. 2011. Effect of time of inoculation, starter addition, oxygen level and salting on the viability of probiotic cultures during Cheddar cheese production. *Int. Dairy J.* 21:75–82.

Fox, P.F. and Cogan, T.M. 2004. Factors that affect the quality of cheese. In: *Cheese: Chemistry, Physics and Microbiology*, Vol. 1—General Aspects, 3rd edition, edited by P.F. Fox, P.L.H. McSweeney, T.M. Cogan, and T.P. Guinee, pp. 584–608. London: Elsevier.

Fox, P.F., Guinee, T.P., Cogan, T.M., and McSweeney, P.L.H. 2000. *Fundamentals of Cheese Science*, pp. 236–281. Gaithersburg, MD: Aspen.

Franciosi, E., Settanni, L., Cavazza, A., and Poznanski, E. 2009. Presence of enterococci in raw cow's milk and "Puzzone Di Moena" cheese. *J. Food Process. Preserv.* 33:204–217.

Franck, A. 2008. Food applications of prebiotics. In: *Handbook of Prebiotics*, edited by G.R. Gibson and M.B. Roberfroid, pp. 437–448. Boca Raton, FL: CRC Press.

Franz, C.M.A.P. and Holzapfel, W.H. 2004. The genus *Enterococcus*: Biotechnological and safety issues. In: *Lactic Acid Bacteria: Microbiological and Functional Aspects*, 3rd edition, edited by S. Salminen, A. Wright, and A. Ouwehand, pp. 199–247. New York: Marcel Dekker.

Fritzen-Freire, C.B., Müller, C.M.O., Laurindo, J.B., and Prudêncio, E.S. 2010. The influence of *Bifidobacterium* BB-12 and lactic acid incorporation on the properties of Minas Frescal cheese. *J. Food Eng.* 96(4): 621–627.

Gardiner, G.E., Bouchier, P., O'Sullivan, E., Kelly, J., Collins, J.K., Fitzgerald, G., Ross, R.P., and Stanton, C. 2002. A spray-dried culture for probiotic Cheddar cheese manufacture. *Int. Dairy J.* 12:749–756.

Gardiner, G.E., Ross, R.P., Wallace, J.M., Scanlan, F.P., Jagers, P.P., Fitzgerald, G.F., Collins, J.K., and Stanton, C. 1999. Influence of a probiotic adjunct culture of *Enterococcus faecium* on the quality of cheddar cheese. *J. Agric. Food Chem.* 47:4907–4916.

Gibson, G.R. 2004. Fibre and effects on probiotics (the prebiotic concept). *Clin. Nutr. Suppl.* 1:25–31.

Gibson, G.R., Probert, H.M., Van Loo, J., Rastall, R.A., and Roberfroid, M.B. 2004. Dietary modulation of the human colonic microbiota: Updating the concept of prebiotics. *Nutr. Res. Rev.* 17:259–275.

Gobbetti, M., Corsetti, A., Smacchi, E., Zocchetti, A., and De Angelis, M. 1997. Production of Crescenza cheese by incorporation of bifidobacteria. *J. Dairy Sci.* 81:37–47.

Godward, G. and Kailasapathy, K. 2003. Viability and survival of free and encapsulated probiotic bacteria in Cheddar cheese. *Milchwissenschaft* 58:624–627.

Gomes, A.M.P., Vieira, M.M., and Malcata, F.X. 1998. Survival of probiotic microbial strains in a cheese matrix during ripening: Simulation of rates of salt diffusion and microorganism survival. *J. Food Er* 36:281–301.

Gomes, A.M.P. and Malcata, F.X. 1998. Development of probiotic cheese manufacture from go? Response surface analysis via technological manipulation. *J. Dairy Sci.* 81:1492–1507.

Goncu, A. and Alpkent, Z. 2005. Sensory and chemical properties of white pickled cheese produced using kefir, yoghurt or a commercial cheese culture as a starter. *Int. Dairy J.* 15:771–776.

Grattepanche, F., Miescher-Schwenninger, S., Meile, L., and Lacroix, C. 2008. Recent developments in cheese cultures with protective and probiotic functionalities. *Dairy Sci. Technol.* 88:421–444.

Grundelius, A.U., Lodaite, K., Östergren, K., Paulsson, M., and Dejmek, P. 2000. Syneresis of submerged single curd grains and curd rheology. *Int. Dairy J.* 10:489–496.

Gursoy, O. and Kinik, O. 2010. Incorporation of adjunct cultures of *Enterococcus faecium, Lactobacillus paracasei* subsp. *paracasei* and *Bifidobacterium bifidum* into white cheese. *J. Food Agric. Environ.* 8:107–112.

Hatakka, K., Ahola, A.J., Yli-Knuuttila, H., Richardson, M., Poussa, T., Meurman, J.H., and Korpela, R. 2007. Probiotics reduce the prevalence of oral *Candida* in the elderly—A randomized controlled trial. *J. Dent. Res.* 86:125–130.

Heller, K.J., Bockelmann, W., Schrezenmeir, J., and DeVrese, M. 2003. Cheese and its potential as a probiotic food. In: *Handbook of Fermented Functional Foods*, edited by E.R. Farnworth, pp. 203–225. Boca Raton, FL: CRC Press.

Ibrahim, F., Ruvio, S., Granlund, L., Salminen, S., Viitanen, M., and Ouwehand, A.C. 2010. Probiotics and immunosenescence: Cheese as a carrier. *FEMS Immunol. Med. Microbiol.* 59:53–59.

Irlinger, F. and Mounier, J. 2009. Microbial interactions in cheese: Implications for cheese quality and safety. *Curr. Opin. Biotechnol.* 20:142–148.

Izquierdo, E., Marchioni, E., Aoude-Werner, D., Hasselmann, C., and Ennahar, S. 2009. Smearing of soft cheese with *Enterococcus faecium* WHE 81, a multi-bacteriocin producer, against *Listeria monocytogenes*. *Food Microbiol.* 26:16–20.

Järvenpää, S., Tahvonen, R.L., Ouwehand, C., Sandell, M., Järvenpää, E., and Salminen, S. 2007. A probiotic, *Lactobacillus fermentum* ME-3, has antioxidative capacity in soft cheese spreads with different fats. *J. Dairy Sci.* 90:3171–3177.

Jayamanne, V.S. and Adams, M.R. 2006. Determination of survival, identity and stress resistance of probiotic bifidobacteria in bioyoghurts. *Lett. Appl. Microbiol.* 42:189–194.

Kailasapathy, K. 2002. Microencapsulation of probiotic bacteria: Technology and potential applications. *Curr. Issues Intest. Microbiol.* 3:39–48.

Kailasapathy, K. and Masondole, L. 2005. Survival of free and microencapsulated *Lactobacillus acidophilus* and *Bifidobacterium lactis* and their effect on texture of Feta cheese. *Aust. J. Dairy Technol.* 60(3):252–258.

Kalavrouzioti, I., Hatzikamari, M., Litopoulou-Tzanetaki, E., and Tzanetakis, N. 2005. Production of hard cheese from caprine milk by the use of two types of probiotic cultures as adjuncts. *Int. J. Dairy Technol.* 58:30–38.

Kasımoğlu, A., Göncüoğlu, M., and Akgün, S. 2004. Probiotic white cheese with *Lactobacillus acidophilus*. *Int. Dairy J.* 14:1067–1073.

Kaushik, J.K., Kumar, A., Duray, A.K., Mohanty, A.K., Grover, S., and Batish, V.K. 2009. Functional and probiotic attributes of an indigenous isolate of *Lactobacillus plantarum*. *PLoS ONE* 4(12):e8099.

Kieronczyk, A., Skeie, S., Olsen, K., and Langsrud, T. 2001. Metabolism of amino acids by resting cells of non-starter lactobacilli in relation to flavour development in cheese. *Int. Dairy J.* 11:217–224.

Kılıç, G.B., Kuleaşan, H., Eralp, İ., and Karahan, A.G. 2009. Manufacture of Turkish Beyaz cheese added with probiotic strains. *LWT—Food Sci. Technol.* 42:1003–1008.

Klayraung, S., Viernstein, H., Sirithunyalug, J., and Okonogi, S. 2008. Probiotic properties of lactobacilli isolated from Thai traditional food. *Sci. Pharm.* 76:485–503.

Klein, G., Pack, A., Bonaparte, C., and Reute, G. 1998. Taxonomy and physiology of probiotic lactic acid bacteria. *Int. J. Food Microbiol.* 41:103–125.

Klobukowski, J., Modzelewska-Kapitula, M., and Kornacki, K. 2009. Calcium bioavailability from diets based on white cheese containing probiotics or synbiotics in short-time study in rats. *Pak. J. Nutr.* 7(8):933–936.

Konuklar, G., Inglett, G.E., Carriere, C.J., and Felker, F.C. 2004. Use of a β-glucan hydrocolloidal suspension in the manufacture of low-fat Cheddar cheese: Manufacture, composition, yield and microstructure. *Int. J. Food Sci. Technol.* 39:109–119.

Korhonen, H. and Pihlanto-Leppälä, A. 2006. Milk derived bioactive peptides: Formation and prospects for health promotion. In: *Handbook of Functional Dairy Products*, edited by C. Shortt and J. O'Brien, 293 pp. Boca Raton, FL: CRC Press.

Kosin, B. and Rakshit, S.K. 2006. Microbial and processing criteria for production of probiotics: A review. *Food Technol. Biotechnol.* 44(3):371–379.

Kourkoutas, Y., Bosnea, L., Taboukos, S., Baras, C., Lambrou, D., and Kanellaki, M. 2006a. Probiotic cheese production using *Lactobacillus casei* cells immobilized on fruit pieces. *J. Dairy Sci.* 89:1439–1451.

Kourkoutas, Y., Kandylis, P., Panas, P., Dooley, J.S.G., Nigam, P., and Koutinas, A.A. 2006b. Dried kefir coculture as starter in Feta-type cheese production. *Appl. Environ. Microbiol.* 72(9):6124–6135.

Kourkoutas, Y., Sipsas, V., Papavasiliou, G., and Koutinas, A.A. 2007. An economic evaluation of freeze-dried kefir starter culture production using whey. *J. Dairy Sci.* 90:2175–2180.

Krasaekoopt, W., Bhandari, B., and Deeth, H. 2003. Evaluation of encapsulation techniques of probiotics for yoghurt. *Int. Dairy J.* 13:3–13.

Kullisaar, T., Songisepp, E., Mikelsaar, M., Zilmer, K., Vihalemm, T., and Zilmer, M. 2003. Antioxidative probiotic fermented goats' milk decreases oxidative stress-mediated atherogenicity in human subjects. *Br. J. Nutr.* 90:449–456.

Kullisaar, T., Zilmer, M., Mikelsaar, M., Vihalemm, T., Annuk, H., Kairane, C., and Kilk, A. 2002. Two antioxidative lactobacilli strains as promising probiotics. *Int. J. Food Microbiol.* 72:215–224.

Lavermicocca, P. 2006. Highlights on new food research. *Dig. Liver Dis.* 38(Suppl. 2):S295–S299.

Le Blay, G., Chassard, C., Baltzer, S., and Lacroix, C. 2010. Set up of a new in vitro model to study dietary fructans fermentation in formula-fed babies. *Br. J. Nutr.* 103:403–411.

Lee, Y.K. and Salminen, S. 1995. The coming age of probiotics. *Trends Food Sci. Technol.* 6:241–245.

Leverrier, P., Dimova, D., Pichereau, V., Auffray, Y., Boyaval, P., and Jan, G. 2003. Susceptibility and adaptive response to bile salts in *Propionibacterium freudenreichii*: Physiological and proteomic analysis. *Appl. Environ. Microbiol.* 69:3809–3818.

Lima, K.G.C., Kruger, M.F., Behrens, J., Destro, M.T., Landgraf, M., and Franco, B.D.G.M. 2009. Evaluation of culture media for enumeration of *Lactobacillus acidophilus*, *Lactobacillus casei* and *Bifidobacterium animalis* in the presence of *Lactobacillus delbrueckii* subsp. *bulgaricus* and *Streptococcus thermophilus*. *LWT—Food Sci. Technol.* 42:491–495.

Lin, F.M., Chiu, C.H., and Pan, T.M. 2004. Fermentation of a milk–soymilk and *Lycium chinense* Miller mixture using a new isolate of *Lactobacillus paracasei* subsp. *paracasei* NTU101 and *Bifidobacterium longum*. *J. Ind. Microbiol. Biotechnol.* 31:559–564.

Lionetti, E., Indrio, F., Pavone, L., Borrelli, G., Cavallo, L., and Francavilla, R. 2010. Role of probiotic in pediatric patients with *Helicobacter pylori* infection: A comprehensive review of literature. *Helicobacter* 15:79–87.

Liong, M.T., Easa, A.M., Lim, P.T., and Kang, J.Y. 2009. Survival, growth characteristics and bioactive potential of *Lactobacillus acidophilus* in a soy-based cream cheese. *J. Sci. Food Agric.* 89:1382–1391.

Litopoulou-Tzanetaki, E. 2007. Soft-ripened and fresh cheeses: Feta, Quark, Halloumi and related varieties. In: *Improving the Flavor of Cheese*, edited by B.C. Weimer, pp. 474–493. Cambridge: Woodhead.

Liu, D.M., Li, L., Yang, X.Q., Liang, S.Z., and Wang, J.S. 2006. Survivability of *Lactobacillus rhamnosus* during the preparation of soy cheese. *Food Technol. Biotechnol.* 44(3):417–422.

Lourens-Hattingh, A. and Viljoen, B.C. 2001. Yogurt as probiotic carrier food. *Int. Dairy J.* 11:1–17.

Lucey, J.A. 2008. Some perspectives on the use of cheese as a food ingredient. *Dairy Sci. Technol.* 88:573–594.

Madrau, M.A., Mangia, N.P., Mugia, M.A., Sanna, M.G., Garau, G., Leccis, L., Caredda, M., and Deiana, P. 2006. Employment of autochthonous microflora in Pecorino Sardo cheese manufacturing and evolution of physicochemical parameters during ripening. *Int. Dairy J.* 16:876–885.

Madureira, A.R., Gião, M.S., Pintado, M.E., Gomes, A.M.P., Freitas, C., and Malcata, F.X. 2005a. Incorporation and survival of probiotic bacteria in whey cheese matrices. *J. Food Sci.* 70(3):M161–M165.

Madureira, A.R., Pereira, C.I., Truszkowskab, K., Gomes, A.M., Pintado, M.E., and Malcata, F.X. 2005b. Survival of probiotic bacteria in a whey cheese vector submitted to environmental conditions prevailing in the gastrointestinal tract. *Int. Dairy J.* 15:921–927.

Mäkeläinen, H., Forssten, S., Olli, K., Granlund, L., Rautonen, N., and Ouwehand, A.C. 2009. Probiotic lactobacilli in a semi-soft cheese survive in the simulated human gastrointestinal tract. *Int. Dairy J.* 19:675–683.

Marcatti, B., Habitante, A.M.Q.B., Sobral, P.J.A., and Favaro-Trindade, C.S. 2009. Minas-type fresh cheese developed from buffalo milk with addition of *L. acidophilus*. *Sci. Agric.* 66(4):481–485.

Martínez-Cuesta, M.C., Palencia, P.F., Requena, T., and Peláez, C. 2001. Enzymatic ability of *Lactobacillus casei* subsp. casei IFPL731 for flavour development in cheese. *Int. Dairy J.* 11:577–585.

Masuda, T., Yamanari, R., and Itoh, T. 2005. The trial for production of fresh cheese incorporated probiotic *Lactobacillus acidophilus* group lactic acid bacteria. *Milchwissenschaft* 60(2):167–171.

Mathara, J.M., Schillinger, U., Guigas, C., Franz, C., Kutima, P.M., Mbugua, S.K., Shin, H.K., and Holzapfel, W.H. 2008. Functional characteristics of *Lactobacillus* spp. from traditional Maasai fermented milk products in Kenya. *Int. J. Food Microbiol.* 126:57–64.

Mattila-Sandholm, T., Myllärinen, P., Crittenden, R., Mogensen, G., Fondén, R., and Saarela, M. 2002. Technological challenges for future probiotic foods. *Int. Dairy J.* 12:173–182.

McBrearty, S., Ross, R.P., Fitzgerald, G.F., Collins, J.K., Wallace, J.M., and Stanton, C. 2001. Influence of two commercially available bifidobacteria cultures on Cheddar cheese quality. *Int. Dairy J.* 11:599–610.

McCartney, A.L. 2005. Enumeration and identification of mixed probiotic and lactic acid bacteria starter cultures. In: *Probiotic Dairy Products*, edited by A. Tamime, pp. 98–119. Oxford: Blackwell.

Medici, M., Vinderola, C.G., and Perdigón, G. 2004. Gut mucosal immunomodulation by probiotic fresh cheese. *Int. Dairy J.* 14:611–618.

Meile, L., Le Blay, G., and Thierry, A. 2008. Safety assessment of dairy microorganisms: *Propionibacterium* and *Bifidobacterium*. *Int. J. Food Microbiol.* 126:316–320.

Menéndez, S., Centeno, J.A., Godínez, R., and Rodríguez-Otero, J.L. 2000. Effects of *Lactobacillus* strain on the ripening and organoleptic characteristics of Arzúa-Ulloa cheese. *Int. J. Food Microbiol.* 59:37–46.

Mikelsaar, M., Hütt, P., Kullisaar, T., Zilmer, K., and Zilmer, M. 2008. Double benefit claims for antimicrobial and antioxidative probiotic. *Microb. Ecol. Health Dis.* 20:184–188.

Mikelsaar, M. and Zilmer, M. 2009. *Lactobacillus fermentum* ME-3—An antimicrobial and antioxidative probiotic. *Microb. Ecol. Health Dis.* 21:1–7.

Milesi, M.M., Vinderola, G., Sabbag, N., Meinardi, C.A., and Hynes, E. 2009. Influence on cheese proteolysis and sensory characteristics of non-starter lactobacilli strains with probiotic potential. *Food Res. Int.* 42:1186–1196.

Modzelewska-Kapitula, M., Klebukowska, L., and Kornacki, K. 2007. Influence of inulin and potentially probiotic *Lactobacillus plantarum* strain on microbiological quality and sensory properties of soft cheese. *Pol. J. Food Nutr. Sci.* 57(2):143–146.

Montoya, D., Boylston, T.D., and Mendonca, A. 2009. Preliminary screening of bifidobacteria spp. and *Pediococcus acidilactici* in a Swiss cheese curd slurry model system: Impact on microbial viability and flavor characteristics. *Int. Dairy J.* 19:605–611.

Mori, H., Sato, Y., Taketomo, N., Kamiyama, T., Yoshiyama, Y., Meguro, S., Sato, H., and Kaneko, T. 1997. Isolation and structural identification of bifidogenic growth stimulator produced by *Propionibacterium freudenreichii*. *J. Dairy Sci.* 80:1959–1964.

Ng, K.H., Lye, H.S., Easa, A.M., and Liong, M.T. 2008. Growth characteristics and bioactivity of probiotics in tofu-based medium during storage. *Ann. Microbiol.* 58(3):477–487.

Obatolu, V.A. 2008. Effect of different coagulants on yield and quality of tofu from soymilk. *Eur. Food Res. Technol.* 226:467–472.

Oboh, G. 2006. Coagulants modulate the hypocholesterolemic effect of tofu (coagulated soymilk). *Afr. J. Biotechnol.* 5(3):290–294.

Omidbakhsh, N., Dueyer, T.A., Elkamel, A., and Reilly, P.M. 2010. Systematic statistical-based approach for product design: Application to disinfectant formulations. *Ind. Eng. Chem. Res.* 49:204–209.

Ong, L., Henriksson, A., and Shah, N.P. 2006. Development of probiotic Cheddar cheese containing *Lactobacillus acidophilus*, *Lb. casei*, *Lb. paracasei* and *Bifidobacterium* spp. and the influence of these bacteria on proteolytic patterns and production of organic acid. *Int. Dairy J.* 16:446–456.

Ong, L., Henriksson, A., and Shah, N.P. 2007a. Angiotensin converting enzyme-inhibitory activity in Cheddar cheeses made with the addition of probiotic *Lactobacillus casei* sp. *Lait* 87:149–165.

Ong, L., Henriksson, A., and Shah, N.P. 2007b. Chemical analysis and sensory evaluation of Cheddar cheese produced with *Lactobacillus acidophilus*, *Lb. casei*, *Lb. paracasei* or *Bifidobacterium* sp. *Int. Dairy J.* 17:937–945.

Ong, L., Henriksson, A., and Shah, N.P. 2007c. Proteolytic pattern and organic acid profiles of probiotic Cheddar cheese as influenced by probiotic strains of *Lactobacillus acidophilus*, *Lb. paracasei*, *Lb. casei* or *Bifidobacterium* sp. *Int. Dairy J.* 17:67–78.

Ong, L. and Shah, N.P. 2008a. Influence of probiotic *Lactobacillus acidophilus* and *L. helveticus* on proteolysis, organic acid profiles, and ACE-inhibitory activity of cheddar cheeses ripened at 4, 8, and 12°C. *J. Food Sci.* 73(3):M111–M120.

Ong, L. and Shah, N.P. 2008b. Release and identification of angiotensin-converting enzyme-inhibitory peptides as influenced by ripening temperatures and probiotic adjuncts in Cheddar cheeses. *LWT—Food Sci. Technol.* 41:1555–1566.

Ong, L. and Shah, N.P. 2009. Probiotic cheddar cheese: Influence of ripening temperatures on proteolysis and sensory characteristics of cheddar cheeses. *J. Food Sci.* 74(5):S182–S191.

Ordóñez Pereda, J.A., Cambero Rodríquez, M.I., Fernádez Álvarez, L., García Sanz, M.L., de Fernando Minguillón, G.D.G., Hoz Perales, L., and Selgas Cortecero, M.D. 2005. Tecnologia de Alimentos, Volume 2—Alimentos de origem animal, pp. 67–83. Porto Alegre: Artmed.

Ouwehand, A.C. 2004. The probiotic potential of propionibacteria. In: *Lactic Acid Bacteria: Microbiological and Functional Aspects*, 3rd edition, edited by S. Salminen, A. Wright, and A. Ouwehand, pp. 159–174. New York: Marcel Dekker.

Ouwehand, A.C., Suomalainen, T., Tolkko, S., and Salminen, S. 2002. *In vitro* adhesion of propionic acid bacteria to human intestinal mucus. *Lait* 82:123–130.

Özer, B., Kırmacı, H.A., Şenel, E., Atamer, M., and Hayaloğlu, A. 2009. Improving the viability of *Bifidobacterium bifidum* BB-12 and *Lactobacillus acidophilus* LA-5 in white-brined cheese by microencapsulation. *Int. Dairy J.* 19:22–29.

Özer, B., Uzun, Y.S., and Kırmacı, H.A. 2008. Effect of microencapsulation on viability of *Lactobacillus acidophilus* LA-5 and *Bifidobacterium bifidum* BB-12 during Kasar cheese ripening. *Int. J. Dairy Technol.* 61(3):237–244.

Papadimitriou, C.G., Vafopoulou-Mastrojiannaki, A., Silva, S.V., Gomes, A.M., Malcata, F.X., and Alichanidis, E. 2007. Identification of peptides in traditional and probiotic sheep milk yoghurt with angiotensin I-converting enzyme (ACE)-inhibitory activity. *Food Chem.* 105:647–656.

Pereira, C.I., Gomes, A.M.P., and Malcata, X. 2009. Microstructure of cheese: Processing, technological and microbiological considerations. *Trends Food Sci. Technol.* 20:213–219.

Pereira, L.C., Souza, C.H.B., Behrens, J.H., and Saad, S.M.I. 2010. *Lactobacillus acidophilus* and *Bifidobacterium* sp. in co-culture improve sensory acceptance of potentially probiotic *petit-suisse* cheese. *Acta Aliment.* 39(3):265–276.

Perotti, M.C., Mercanti, D.J., Bernal, S.M., and Zalazar, C.A. 2009. Characterization of the free fatty acids profile of Pategra's cheese during ripening. *Int. J. Dairy Technol.* 62(3):331–338.

Phillips, M., Kailasapathy, K., and Tran, L. 2006. Viability of commercial probiotic cultures (*L. acidophilus*, *Bifidobacterium* sp., *L. casei*, *L. paracasei* and *L. rhamnosus*) in Cheddar cheese. *Int. J. Food Microbiol.* 108:276–280.

Radulović, Z., Petrović, T., Nedović, V., Dimitrijević, S., Mirković, N., Petrušić, M., and Paunović, D. 2010. Characterization of autochthonous *Lactobacillus paracasei* strains on potential probiotic ability. *Mljekarstvo* 60(2):86–93.

Ramchandran, L. and Shah, N.P. 2008. Proteolytic profiles and angiotensin-I converting enzyme and α-glucosidase inhibitory activities of selected lactic acid bacteria. *J. Food Sci.* 73(2):M75–M81.

Ranadheera, R.D.C.S., Baines, S.K., and Adams, M.C. 2010. Importance of food in probiotic efficacy. *Food Res. Int.* 43:1–7.

Ravishankar, S. and Juneja, V.K. 2000. Sodium chloride. In: *Natural Food Antimicrobial Systems*, edited by A.S. Naidu, pp. 705–724. Boca Raton, FL: CRC Press.

Roberfroid, M.B. 2008. Prebiotics: Concept, definition, criteria, methodologies, and products. In: *Handbook of Prebiotics*, edited by G.R. Gibson and M.B. Roberfroid, pp. 39–68. Boca Raton, FL: CRC Press.

Rodrigues, D., Rocha-Santos, T.A.P., Pereira, C.I., Gomes, A.M., Malcata, F.X., and Freitas, A.C. 2011. The potential effect of FOS and inulin upon probiotic bacterium performance in curdled milk matrices. *LWT—Food Sci. Technol.* 44:100–108.

Rogelj, I., Bogovič Matijašić, B., Majhenič, A.C., and Stojković, S. 2002. The survival and persistence of *Lactobacillus acidophilus* LF221 in different ecosystems. *Int. J. Food Microbiol.* 76:83–91.

Ross, C.F. 2009. Sensory science at the human–machine interface. *Trends Food Sci. Technol.* 20:63–72.

Rouzaud, G.C.M. 2007. Probiotics, prebiotics, and synbiotics: Functional ingredients for microbial management strategies. In: *Functional Food Carbohydrates*, edited by C.G. Biliaderis and M.S. Izydorczyk, pp. 479–509. Boca Raton, FL: CRC Press.

Roy, D. 2001. Media for the isolation and enumeration of bifidobacteria in dairy products. *Int. J. Food Microbiol.* 69:167–182.

Roy, D. 2005. Technological aspects related to the use of bifidobacteria in dairy products. *Lait* 85:39–56.

Ruas-Madiedo, P., Hugenholtz, J., and Zoon, P. 2002. An overview of the functionality of exopolysaccharides produced by lactic acid bacteria. *Int. Dairy J.* 12:163–171.

Ryhänen, E.L., Pihlanto-Leppälä, A., and Pahkala, E. 2001. A new type of ripened, low-fat cheese with bioactive properties. *Int. Dairy J.* 11:441–447.

Saavedra, L., Taranto, M.P., Sesma, F., and Font de Valdez, G. 2003. Homemade traditional cheeses for the isolation of probiotic *Enterococcus faecium* strains. *Int. J. Food Microbiol.* 88(2–3):241–245.

Sachdeva, A. and Nagpal, J. 2009. Effect of fermented milk-based probiotic preparations on *Helicobacter pylori* eradication: A systematic review and meta-analysis of randomized-controlled trials. *Eur. J. Gastroenterol. Hepatol.* 21(1):45–53.

Sanz, Y. 2007. Ecological and functional implications of the acid-adaptation ability of *Bifidobacterium*: A way of selecting improved probiotic strains. *Int. Dairy J.* 17:1284–1289.

Sarkar, S. 2007. Potential of kefir as a dietetic beverage—A review. *Br. Food J.* 109(4):280–290.

Schiffrin, E.J., Morley, J.E., Donnet-Hughes, A., and Guigoz, Y. 2010. The inflammatory status of the elderly: The intestinal contribution. *Mutat. Res. Fund. Mol. Mech. Mutagen.* 690:50–56.

Schillinger, U., Guigas, C., and Holzapfel, W.H. 2005. In vitro adherence and other properties of lactobacilli used in probiotic yoghurt-like products. *Int. Dairy J.* 15:1289–1297.

Schulz-Collins, D. and Senge, B. 2004. Acid- and acid/rennet-curd cheeses. Part A: Quark, Cream Cheese and related varieties. In: *Cheese: Chemistry, Physics and Microbiology*, Vol. 2—Major Cheese Groups, 3rd edition, edited by P.F. Fox, P.L.H. McSweeney, T.M. Cogan, and T.P. Guinee, pp. 301–328. London: Elsevier.

Settanni, L. and Moschetti, G. 2010. Non-starter lactic acid bacteria used to improve cheese quality and provide health benefits. *Food Microbiol.* 27:691–697.

Shah, N.P. 2007. Functional cultures and health benefits. *Int. Dairy J.* 17:1262–1277.

Shihata, A. and Shah, N.P. 2000. Proteolytic profiles of yogurt and probiotic bacteria. *Int. Dairy J.* 10:401–408.

Songisepp, E., Kals, J., Kullisaar, T., Mändar, R., Hütt, P., Zilmer, M., and Mikelsaar, M. 2005. Evolution of the functional efficacy of an antioxidative probiotic in healthy volunteers. *Nutr. J.* 4:22. dx.doi.org/10.1186/1475-2891-4-22.

Songisepp, E., Kullisaar, T., Hütt, P., Elias, P., Brilene, T., Zilmer, M., and Mikelsaar, M. 2004. A new probiotic cheese with antioxidative and antimicrobial activity. *J. Dairy Sci.* 87:2017–2023.

Sousa, M.J., Ardö, Y., and McSweeney, P.L.H. 2001. Advances in the study of proteolysis during cheese ripening. *Int. Dairy J.* 11:327–345.

Souza, C.H.B., Buriti, F.C.A., Beherens, J.H., and Saad, S.M.I. 2008. Sensory evaluation of probiotic Minas fresh cheese with *Lactobacillus acidophilus* added solely or in co-culture with a thermophilic starter culture. *Int. J. Food Sci. Technol.* 43:871–877.

Souza, C.H.B. and Saad, S.M.I. 2009. Viability of *Lactobacillus acidophilus* LA-5 added solely or in co-culture with a yoghurt starter culture and implications on physico-chemical and related properties of Minas fresh cheese during storage. *LWT—Food Sci. Technol.* 42(2):633–640.

Stanton, C., Gardiner, G., Lynch, P.B., Collins, J.K., Fitzgerald, G., and Ross, R.P. 1998. Probiotic cheese. *Int. Dairy J.* 8:491–546.

Suárez-Solís, V., Cardoso, F., Villavicencio, M.N., Fernández, M., and Fragoso, L. 2002. Queso fresco probiótico. *Aliment.: Rev. Tecnol. Hig. Aliment.* 333:83–86.

Sullivan, Å., Palmgren, A.C., and Nord, C.E. 2001. Effect of *Lactobacillus paracasei* on intestinal colonisation of lactobacilli, bifidobacteria and *Clostridium difficile* in elderly persons. *Anaerobe* 7:67–70.

Talwalkar, A. and Kailasapathy, K. 2004a. A review of oxygen toxicity in probiotic yoghurts: Influence on the survival of probiotic bacteria and protective techniques. *Comp. Rev. Food Sci. Food Saf.* 3:117–124.

Talwalkar, A. and Kailasapathy, K. 2004b. Oxidative stress adaptation of probiotic bacteria. *Milchwissenschaft* 59:140–143.

Talwalkar, A., Miller, C.W., Kailasapathy, K., and Nguyen, M.H. 2004. Effect of packaging materials and dissolved oxygen on the survival of probiotic bacteria in yoghurt. *Int. J. Food Sci. Technol.* 39(6):605–611.

Tamime, A.Y., Saarela, M., Søndegaard, A.K., Mistry, V.V., and Shah, N.P. 2005. Production and maintenance of viability of probiotic micro-organisms in dairy products. In: *Probiotic Dairy Products*, edited by A. Tamime, pp. 39–62. Oxford: Blackwell.

Tharmaraj, N. and Shah, N.P. 2009a. Antimicrobial effects of probiotics against selected pathogenic and spoilage bacteria in cheese-based dips. *Int. Food Res. J.* 16:261–276.

Tharmaraj, N. and Shah, N.P. 2009b. Antimicrobial effects of probiotic bacteria against selected species of yeasts and moulds in cheese-based dips. *Int. J. Food Sci. Technol.* 44:1916–1926.

Tiihonen, K., Owehand, A.C., and Rautonen, N. 2010. Human intestinal microbiota and healthy ageing. *Ageing Res. Rev.* 9:107–116.

Topisirovic, L., Kojic, M., Fira, D., Golic, N., Strahinic, I., and Lozo, J. 2006. Potential of lactic acid bacteria isolated from specific natural niches in food production and preservation. *Int. J. Food Microbiol.* 112:230–235.

Trebichavsky, I., Rada, V., Splichalova, A., and Splichal, I. 2009. Cross-talk of human gut with bifidobacteria. *Nutr. Rev.* 67(2):77–82.

Ukeyima, M.T., Enujiuha, V.N., and Sanni, T.A. 2010. Current applications of probiotic foods in Africa. *Afr. J. Biotechnol.* 9(4):394–401.

Van de Casteele, S., Vanheuverzwijn, T., Ruyssen, T., Van Assche, P., Swings, J., and Huys, G. 2006. Evaluation of culture media for selective enumeration of probiotic strains of lactobacilli and bifidobacteria in combination with yoghurt or cheese starters. *Int. Dairy J.* 16:1470–1476.

Vasiljevic, T. and Shah, N.P. 2008. Probiotics—From Metchnikoff to bioactives. *Int. Dairy J.* 18:714–728.

Vásquez, A., Molin, G., Pettersson, B., Antonsson, M., and Ahrne, S. 2005. DNA-based classification and sequence heterogeneities in the 16S rRNA genes of *Lactobacillus casei/paracasei* and related species. *Syst. Appl. Microbiol.* 28(5):430–441.

Ventura, M., Canchaya, C., Fitzgerald, G.F., Gupta, R.S., and van Sinderen, D. 2007a. Genomics as a means to understand bacterial phylogeny and ecological adaptation: The case of bifidobacteria. *Antonie van Leeuwenhoek* 91:351–372.

Ventura, M., Canchaya, C., Tauch, A., Chandra, G., Fitzgerald, G.F., Chater, K.F., and van Sinderen, D. 2007b. Genomics of *Actinobacteria*: Tracing the evolutionary history of an ancient phylum. *Microbiol. Mol. Biol. Rev.* 71(3):495–548.

Vinderola, C.G., Capellini, B., Villarreal, F., Suárez, V., Quiberoni, A., and Reinheimer, J. 2008. Usefulness of a set of simple *in vitro* tests for the screening and identification of probiotic candidate strains for dairy use. *LWT—Food Sci. Technol.* 41:1678–1688.

Vinderola, C.G., Costa, G.A., Regenhardt, S., and Reinheimer, J.A. 2002. Influence of compounds associated with fermented dairy products on the growth of lactic acid starter and probiotic bacteria. *Int. Dairy J.* 12:579–589.

Vinderola, C.G., Prosello, W., Ghiberto, D., and Reinheimer, J.A. 2000. Viability of probiotic (*Bifidobacterium*, *Lactobacillus acidophilus* and *Lactobacillus casei*) and nonprobiotic microflora in Argentinian fresco cheese. *J. Dairy Sci.* 83:1905–1911.

Vinderola, G., Proselo, W., Molinari, F., Ghiberto, D., and Reinheimer, J. 2009. Growth of *Lactobacillus paracasei* A13 in Argentinian probiotic cheese and its impact on the characteristics of the product. *Int. J. Food Microbiol.* 135:171–174.

Vizoso Pinto, M.G., Franz, C.M.A.P., Schillinger, U., and Holzapfel, W.H. 2006. *Lactobacillus* spp. with *in vitro* probiotic properties from human faeces and traditional fermented products. *Int. J. Food Microbiol.* 109:205–214.

Walstra, P., Wouters, J.T.M., and Geurts, T.J. 2006a. Cheese manufacture. In: *Dairy Science and Technology*, 2nd edition, pp. 583–638. Boca Raton, FL: CRC Press.

Walstra, P., Wouters, J.T.M., and Geurts, T.J. 2006b. Principles of cheese making. In: *Dairy Science and Technology*, 2nd edition, pp. 577–582. Boca Raton, FL: CRC Press.

Wang, K.Y., Li, S.N., Liu, C.S., Perng, D.S., Su, Y.C., Wu, D.C., Jan, C.M., Lai, C.H., Wang, T.N., and Wang, W.M. 2004. Effects of ingesting *Lactobacillus*- and *Bifidobacterium*-containing yogurt in subjects with colonized *Helicobacter pylori*. *Am. J. Clin. Nutr.* 80:737–741.

Wilkinson, M.G., Meehan, H., Stanton, C., and Cowan, C. 2001. Marketing cheese with a nutrient message. *Bulletin IDF* 363:39–45.

Won, J.S., Kim, W.J., Lee, K.G., Kim, C.W., and Noh, W.S. 2008. Fermentation characteristics of exopolysaccharide-producing lactic acid bacteria from sourdough and assessment of the isolates for industrial potential. *J. Microbiol. Biotechnol.* 18(7):1266–1273.

Yilmaztekin, M., Özer, B.H., and Atasoy, F. 2004. Survival of *Lactobacillus acidophilus* LA-5 and *Bifidobacterium bifidum* BB-02 in white-brined cheese. *Int. J. Food Sci. Nutr.* 55:53–60.

Zhang, Y., Du, R., Wang, L., and Zhang, H. 2010. The antioxidative effects of probiotic *Lactobacillus casei* Zhang on the hyperlipidemic rats. *Eur. Food Res. Technol.* 231:151–158.

Index

Page numbers followed by *f* or *t* refer to figures and tables, respectively.